Studies in Neuroscience, Psychology and Behavioral Economics

This interdisciplinary book series addresses research findings and methods in the intersection of the fields of neuroscience, psychology and behavioral economics. At the core of the series is the biological basis of human decision making, including aspects of an economic, cognitive, affective and motivational nature. Contributions from fundamental and applied sciences are equally represented and cover a broad range of scientific work, reaching from mathematical modeling to scholarly introductions into new research disciplines and related methodologies and technologies (e.g. genetics, psychophysiology, brain imaging, computational neuroscience, animal modeling). The book series presents empirical and methodological scientific innovations bridging the gap from brain to behavior. It includes research works in neuroeconomics and neuro-information systems (e.g. human-machine interaction) as well as social, cognitive and affective neuroscience, which aim at enhancing the understanding of human decision making.

Nikolai Axmacher
Editor

Intracranial EEG

A Guide for Cognitive Neuroscientists

Springer

Editor
Nikolai Axmacher
Ruhr University Bochum
Department of Neuropsychology
Institute of Cognitive Neuroscience
Faculty of Psychology
Bochum, Germany

ISSN 2196-6605 ISSN 2196-6613 (electronic)
Studies in Neuroscience, Psychology and Behavioral Economics
ISBN 978-3-031-20909-3 ISBN 978-3-031-20910-9 (eBook)
https://doi.org/10.1007/978-3-031-20910-9

This Springer imprint is published by the registered company Springer Nature Switzerland AG
The registered company address is: Gewerbestrasse 11, 6330 Cham, Switzerland

Foreword

Intracranial EEG (iEEG) is a unique tool that needs to be cherished beyond any other in Neuroscience. Here is why: brain "areas" do not operate in isolated silos, and "single" neurons do not exist. Any cognitive function or behavior is anchored in an interplay across regions of the brain, each encompassing millions and millions of neurons. We need to move to a system-level approach to understand large brains, such as the human brain. No other tool has the same power as the iEEG to address important system-wide questions through simultaneous recordings across many regions.

The iEEG method shines where the neuroimaging methods failed. The imaging studies in the last several decades gave us a beautiful yet static and frozen view of the brain that did not reveal much about the dynamics of operations across regions on a fast temporal scale when humans are engaged in a task. The subsecond temporal resolution of iEEG is a blessing.

Also, neuroimaging methods have relied heavily on the subtractive method (i.e., comparing activity between two conditions), which reveals areas of the brain in colors and the rest of the brain in black as if the dark ones are entirely silent and sleeping during the task—nothing could be farther from truth. Comparing the activity of a given site with its own baseline is another advantage of the iEEG that should not be forgotten.

Now lately, the pendulum in research has swung to the other end of the spectrum with complex computational matrices showing checkerboard patterns with little regard to the very specific anatomical architecture of the brain—almost as if the brain works in silico. The anatomical precision of the iEEG recordings makes it a method of choice for revealing the neuroanatomy of the brain's functional architecture.

Another important advantage of the iEEG is that it has a very high signal-to-noise ratio. This allows us to reach statistical significance at the single subject, or even single trial, level, allowing us to decipher the mode of a brain operation in real time in naturalistic settings. Reliance on group analysis, combined with smoothing and averaging of the signal across subjects, always will blur functional organizations in the native individual brain space and may give us an erroneous view of the brain at the individual subject level. The high signal-to-noise ratio of iEEG takes care of this potentially big problem. The iEEG signal always carries the individual anatomical source information; hence, the "i" in iEEG could also refer to the I—as in "Me"—signal.

I often get asked why iEEG and why not single-unit recordings. I answer the question by stating the obvious: iEEG is far superior to single-cell recordings in capturing the oscillatory dynamics of activity locally and across brain regions. As such, the signals captured with iEEG electrodes reveal the profile of engagement of a population of neurons at the mesoscale level rather than the behavior of a few neurons at a micro-scale level. Yes, the recorded signal only denotes the engagement of the forest underneath or around the electrode. It does not reveal the precise code of computation within a tree of neurons or across the branches of adjacent trees. But no one argues that the human mind works with single neurons detached from their local populations. No one defends the view that a given region works independently from other nodes of a distributed system. Yes, I admit that the iEEG recordings capture signals from a population of neurons, which is a coarse method. In all honesty, however, there are no single neurons in the brains as there are no single phrenological sites. Millions of neurons operate in functional columns, which work closely together and simultaneously so, with other functional units that are further away. Recording the chatter in the dendritic "forest" is the best we have gotten to capture the mode of local engagement in a widespread net of regions working together. Once we understand the oscillatory dynamics of interactions, we can dive into the operation of local circuitries and neuronal ensembles and then into single neurons. The blindness of the iEEG signal to the micrometer space may be a blessing in disguise! Neighboring populations of the brain are known to be connected with regions that perform similar computational roles. Thus, the mesoscopic level of information may have more practical utility than the microscopic level at the individual neuron level.

Lastly, from yet another practical point of view, the most significant benefit of iEEG is that implanted electrodes can deliver electrical pulses that affect the activity of a population of neurons and their interconnected nodes. The electricity propagates within a specific anatomically organized and hardwired circuit. Thus, the technique confers the ability to stimulate the human brain at specific recording sites and specific functional systems to test the causal relevance of a given brain area or a system for a particular function. It also allows one to probe the effect of perturbation of a given node of a functional system on the individual's subjective state. This sets human iEEG apart from all other electrophysiological recordings. Human subjects can report what is going on in their "mind" when we stimulate a given brain region. No human mind will be altered by changing the activity of a single neuron. Additionally, by delivering repeated single pulses in a given site and recording simultaneously in other brain regions, one can connect the immediate effect of perturbing a node to the rest of the network to which it belongs.

Overall, I remain excited about the future of iEEG and how much it will teach us about our brains. I look forward to learning new things from my colleagues when I read their chapters and I am positive that my excitement, and yours, will be much higher when we finish reading this excellent book that Nikolai has elegantly put together.

Remain excited about the iEEG and do not settle!

Dr. Josef Parvizi

Introduction: What Can You Expect in This Book?

As the name suggests, cognitive neuroscience has two aims: understanding the mind and understanding the brain. Again not surprisingly, cognitive neuroscience is built upon the premise that these two aims are inextricably intertwined: That a profound understanding of cognitive functions is only possible when their emergence from basic biological mechanisms is elucidated and that the major function of the brain is to enable cognition. The methods that are used to investigate the neural basis of cognition can be distinguished based on the precision with which they address each of these two aims. On the one end of this continuum are studies in simple organisms like C. elegans or Drosophila that allow for a very detailed analysis of neural mechanisms at subcellular resolution, but investigate only relatively rudimentary and basic cognitive processes. On the other extreme are non-invasive studies in humans that provide access to rich and complex behavior and cognitive functions, but are substantially limited in their temporal and spatial resolution (invasive recordings in rodents, primates, and other animals are somewhere in-between).

Intracranial EEG (iEEG) recordings provide the best of both worlds: They allow studying neural processes in awake and behaving human beings at millisecond temporal resolution, millimeter spatial resolution, and at a high signal-to-noise ratio, giving access to both the neocortex and to small brain structures like the hippocampus, the amygdala, or thalamic nuclei that are located deep inside the brain. In some situations—which, of course, need to be clearly limited by medical and safety considerations—iEEG electrodes cannot only be used to record data during cognitive paradigms, but also to deliver electrical stimulation, which enables a causal understanding of how cognitive functions emerge from neural processing.

Are iEEG recordings the "gold standard of cognitive" neuroscience? Well, they clearly have their challenges and disadvantages as well. First of all, such recordings are only possible in patient populations—most importantly, presurgical epilepsy patients—who are implanted with electrodes for clinical reasons. Data acquisition is thus more challenging and time-consuming than recording data in healthy participants. Moreover, iEEG experiments typically have to be shorter than recordings in young and enduring students, and the cognitive performance of patients is more variable. And once the data have been acquired, their analysis poses novel challenges—for example, they contain other types of artifacts than those which are known from EEG and MEG data, they are sampled from highly variable patterns of electrodes, and they require special considerations for referencing. Last but not least, the statistical analysis of these heterogenous data is more challenging than analysis of EEG or MEG data with standardized recording schemes.

When I started working with iEEG almost 20 years ago, I was in the lucky situation to be in a clinical and scientific environment with extensive experience of dealing with these data. Since then, I collected data from hundreds of patients and started substantial collaborations on this method with labs around the world—which is critical, because there are only very few clinical sites that provide access to large enough data for comprehensive studies in isolation. More importantly, data analysis approaches are highly heterogenous across different labs, reflecting diverse and complementary research traditions and scientific approaches. In my opinion, this heterogeneity needs to be embraced, because it reduces the risk of a limited approach to understanding brain and cognition—while there are clear obstacles and pitfalls that need to be avoided, there is more than one correct way to analyze the data.

However, this situation is difficult for newcomers who just start doing iEEG research, and also for experienced researchers, it is hard to keep track of all methodological developments in the field. This volume is supposed to serve both the beginner and the expert. It is meant to provide guidelines for basic steps, to help navigate the complexity of analysis decisions to take, and to inspire a large community to apply the most sophisticated and state-of-the-art approaches.

The book is divided into four larger sections. The first section covers some general and basic aspects of iEEG studies, including their clinical background—i.e., the patient populations (Chap. 1), their cognitive status (Chap. 2), and the impact of epilepsy on cognitive performance (Chap. 3). Most iEEG researchers do not have a background in both medicine and (neuro-)psychology, but some knowledge in both disciplines is necessary to successfully plan and conduct patient studies in a hospital setting and to interpret the findings and their limitations. Indeed, recording data in a clinical environment involves various practical challenges that need to be anticipated and solved. Since these practical issues may differ substantially between hospitals and since they can be addressed in various different ways, they are discussed in two complementary chapters (Chaps. 4 and 5). Together, Chaps. 1–5 demonstrate the substantial heterogeneity of presurgical epilepsy patients and emphasize the factors that need to be considered when interpreting data from iEEG studies; all of this information should be described in the papers that arise from these studies. These chapters are followed by two chapters on ethical questions of iEEG studies (Chaps. 6 and 7), which address many important issues—starting with the special status and possible role conflicts of the "clinician-researcher" to far-ranging philosophical questions on the impact of novel recording and stimulation approaches on personal identity, agency, and privacy. While Chap. 6 addresses more general issues, Chap. 7 specifically focuses on ethical questions related to microwires and Utah arrays, whose implantation may provide no direct clinical benefit for the patient and thus requires specific considerations, including complex question of informed consent in patients with possibly reduced decision-making abilities. A subsequent chapter addresses the overall contribution of iEEG research as compared to other methods in the cognitive neurosciences, a critical issue that needs to be considered when planning the recording modality that is optimally suited to address a specific research question (Chap. 8). Once a decision is made in favor of iEEG, a logical follow-up question concerns the required sample size. The next chapter thus discusses the amount of data that is needed for an iEEG study and critically compares the different rationale, benefits, and limitations of single-case studies, studies in small groups of patients, and large-scale investigations (Chap. 9). Large samples also allow for analyses of inter-individual differences as well as (cross-sectional) developmental studies, a relatively novel and important approach that is discussed in a subsequent chapter (Chap. 10). The concluding chapter of the first section discusses whether cognitive neuroscience research in iEEG patients has a clinical benefit (Chap. 11). While this is often implicitly assumed, it is less commonly directly demonstrated, and the chapter describes several examples where iEEG cognitive neuroscience research has fundamentally changed clinical reasoning and surgical planning in epilepsy patients.

The second section addresses the physiological basis and the functional role of iEEG signals, with a particular focus on oscillations. The section starts with chapters that discuss how iEEG signals compare to other data modalities including non-invasive methods (EEG, MEG, and fMRI) and to recordings via microelectrodes. Ideally, this comparison requires simultaneous data acquisition, and the first two chapters address the promises and challenges of simultaneous recordings of iEEG and scalp EEG data (Chap. 12) and of iEEG and MEG data (Chap. 13). In addition, some studies have now conducted eye-tracking during iEEG recordings, which may not only help to identify eye movement-related artifacts in iEEG data but also provide exciting novel insights into the neural mechanisms underlying active visual sampling of our environments (Chap. 14). Such multimodal data are not only challenging to acquire, but also their analysis raises novel questions. Chapter 15 outlines dedicated analytical frameworks for the combination of multimodal data, be they recorded simultaneously or consecutively. While iEEG allows measuring the activity of relatively small neural assemblies across a wide range of frequencies, it does not have the spatial resolution and single-cell recording capability that microelectrodes provide. This situation is described in Chap. 16 and illustrated with various recent examples. The next chapter provides an in-depth discussion of the spatial resolution of iEEG as compared to microelectrode recordings—an important topic for the interpretation of data from both modalities (Chap. 17). Relatedly, Chap. 18 describes the relationship between spikes and field potentials recorded via novel subdural hybrid electrodes. The following chapters then turn to the functional relevance of iEEG oscillations. Chapter 19 proposes a taxonomy on the relationship of specific oscillatory features to fundamental brain computations and cognitive functions. In addition to task-related activity, iEEG data can also be collected during resting states, sleep, and anesthesia; the opportunities and challenges of these recordings are addressed in Chap. 20. Subsequently, Chap. 21 discusses the cognitive function of activity during these apparent "resting" periods and reviews recent studies in this emerging field of research. The detection of narrow-band oscillations and their delineation from aperiodic activity can be challenging, and various algorithms have been suggested for this purpose, which are discussed in Chaps. 22 and 23. While the first of these chapters describes the technical background of various algorithms, which can be applied to oscillations at any frequency, the second specifically addresses the case of low-frequency oscillations during navigation. Apart from the distinction of narrow-band and aperiodic activity, the delineation of physiological and pathological brain processes can be difficult. This is particularly the case for high-frequency oscillations (ripples), and Chap. 24 is devoted to this topic. The section concludes with two chapters on more advanced topics, namely the differentiation of oscillations that reflect bottom-up vs. top-down processing of information (Chap. 25) and the relationship of iEEG oscillations to gene expression patterns and in vitro electrophysiological data that can be assessed ex-vivo, i.e., in resected brain specimens (Chap. 26).

Section III covers the complex and far-ranging issues related to data analysis in more detail. As mentioned earlier, iEEG data analysis is particularly challenging because of the heterogeneity of research approaches; nevertheless, while the exact procedures depend on the scientific question of individual studies, some general guidelines can be provided. Notably, the section focuses on questions that are specific for iEEG rather than those that need to be considered for electrophysiological data analysis in general; for example, topics such as baseline correction, analysis of event-related potentials, or time-frequency transformation that are equally relevant for EEG and MEG data are not considered here, and the reader is referred to other treatises of these topics. The section starts with an overview of procedures for electrode localization, which can be challenging because of different recording modalities of pre- and postimplantation images, brain shift, and other factors (Chap. 27). This is followed by a thorough and detailed description of the implications of different referencing schemes and their impact on data quality and the interpretation of results (Chap. 28). One of the major challenges when conducting group analyses of iEEG data as compared to EEG or MEG data are the heterogeneous implantation schemes. Several options have been proposed for how to deal with this heterogeneity, which are discussed in Chap. 29. More specific questions concern advanced approaches that have been recently applied in an increasing number of studies: The detection and analysis of "traveling waves" (Chap. 30) and the application of "frequency tagging" for the investigation of various perceptual and cognitive functions (Chap. 31). The next series of chapters discuss the wide field of bivariate or connectivity analyses. While connectivity can to a certain extent also be analyzed in EEG and MEG data, the reduced spatial resolution of these methods makes it more difficult to disentangle multiple sources whose interactions may then be analyzed. IEEG data appear ideally suited for these approaches; on the other hand, the inherently limited sampling of brain activity due to the sparse and variable implantation schemes poses important challenges again. Chapter 32 provides a comprehensive overview of various methods for connectivity analyses, discusses their respective benefits, and provides guidelines to avoid pitfalls. Patterns of functional connectivity can be conceptualized as networks, and Chap. 33 discusses the analysis and cognitive function of large-scale brain networks that can be observed during both resting periods and cognitive paradigms. Following connectivity (bivariate) analysis, the recent years have seen a rapid development in the application of multivariate analyses, which are particularly important for the identification of stimulus-specific activity patterns. These important approaches are described in Chap. 34. The next chapters concern various aspects of statistical analyses, which may be relevant for univariate, bivariate, and multivariate data: Chap. 35 provides practical guidelines and recommendations on how to conduct non-parametric surrogate analyses, and Chap. 36 describes the application of multi-level models—two approaches that are ideally suited for iEEG data and have been increasingly applied recently. Chapter 37 addresses the important question of circular analyses and how to avoid them, a crucial topic in the context of the replication crisis of all cognitive sciences, including iEEG research. Relatedly, Chap. 38 provides recommendations for data sharing and open science. While publication of research data in open-access repositories has become common practice in various research fields, it is still more rare in the field of iEEG, where concerns for data privacy are particularly important given that the data are acquired from vulnerable (patient) populations. However, open science procedures that take these concerns into account have been increasingly applied in the recent past, providing access to these rare data to an increasing number of scientists.

The final section addresses more advanced aspects related to brain stimulation, microelectrode recordings, and the application of machine learning and artificial intelligence methods for the analysis of iEEG data. Chapter 39 provides a comprehensive overview of the opportunities and challenges that electrical stimulation studies provide for cognitive neuroscientists, touching on a wide range of questions and covering a broad literature. Systematically applying stimulation in patients with widespread implantation schemes allows one to define causal networks, a topic that is addressed in Chap. 40. This is followed by Chap. 41 on closed-loop stimulation approaches that are increasingly applied in various neurological and psychiatric conditions and, for cognitive neuroscientists, offer unique insights into the neural mechanisms underlying cognitive functions. The next series of chapters addresses microelectrode recordings, starting with important practical aspects that need to be considered during the implantation of electrodes (Chap. 42). The following Chap. 43 then provides an overview of analysis pipelines for these data, including the identification of action potentials (spike detection) and their assignment to single cells (spike sorting). The next two chapters address more advanced analyses of microelectrode data: Chapter 44 focuses on the relationship of spikes to the power and phase of local field potentials, a topic of high importance for understanding mechanisms of neural coding as well as interactions between network-level and single-cell activities. Chapter 45 describes how interactions between different brain regions can be analyzed at the level of single units and local field potentials, providing a highly detailed and mechanistic account to the analysis of functional connectivity and information transfer in the brain. These chapters all refer to classical micro-macroelectrodes ("Behnke-Fried approach"), which have a substantially lower yield than state-of-the-art devices in animal neuroscience and do not provide laminar information, but are still widely used because of the substantial burden for clinical approval of novel electrode designs. However, there are now important developments of novel electrodes, which is critical since innovation in neuroscience is strongly dependent on novel technologies. These developments are described in Chap. 46, which discusses the opportunities and challenges of laminar microelectrodes.

The next chapters address novel analysis pipelines based on machine learning and artificial intelligence. These methods have recently revolutionized cognitive neuroscience: They provide direct access to the formation, task-related employment, and transformation of cognitive representations—arguably the most important constructs in cognitive psychology that could hitherto only be inferred indirectly. Chapter 47 introduces the application of AI methods in iEEG research and capitalizes on two main fields, the use of AI models as models of cognitive function and their employment for clinical purposes (e.g., brain-computer interfaces). Related to the first direction, Chap. 48 describes how machine learning and AI methods can serve to identify stimulus-driven activity, discusses differences between recording modalities, and compares these models to more basic approaches. Chapter 49 discusses a more specific application of machine learning approaches, namely their use to identify active (i.e., task-related) electrodes during various cognitive functions. Chapter 50 follows up on the multivariate analyses introduced in the preceding section and focuses on the employment of different kinds of deep neural networks as models of the neural representations that are recruited during perception, memory, and higher cognitive functions. Chapter 51 addresses a more clinical direction of AI, namely the development of brain-computer interfaces based on electrode arrays, in particular the "Utah array". The final two chapters are related to chronically implanted invasive electrodes that can be used for both recording and stimulation. Chapter 52 discusses how these electrodes may serve in the identification of novel biomarkers of neurological and psychiatric diseases, while Chap. 53 describes their usage for investigating cognitive functions during ambulatory behavior (i.e., goal-directed spatial navigation).

This broad overview shows several recurring themes. First, it becomes obvious that iEEG research is not a field in its infancy anymore, but has seen ample developments and improved standards in a number of important areas. This starts with ethical procedures and the relationship between clinical and basic research, touches the intricacies of preprocessing pipelines, and ends with the multiple facets of data analyses that are now heavily based on time-frequency decompositions and oscillations in addition to classical time-domain analysis pipelines (even though it has become clear that many seemingly oscillatory effects actually reflect changes in aperiodic 1/f signals). Various ways to cope with group analyses despite the heterogeneity of implantation schemes have been proposed, and pitfalls and circular analyses have been identified.

A second development is a clear tendency toward network analyses. Large-scale brain networks during both resting states and tasks are relevant for numerous cognitive functions and play an important role in understanding neuropsychiatric disorders. These networks can be defined by their connectivity structure using various undirected or directed connectivity measures, some of which cannot be well applied to data from non-invasive recordings. In addition, iEEG offers the unique opportunity to apply focal stimulation and define causal connectivity and networks via the ensuing responses to electric stimulation. Alternatively, networks can be determined by patterns of univariate responses that are inputted to multivariate analyses (e.g., pattern classification or representational similarity analysis). In the future, it will become important to integrate these two approaches that go back to dynamic systems theory and decoding, respectively, and understand their relationship.

Finally, there are some exciting technological developments. These include novel electrode designs for combined recordings across multiple levels of brain organization as well as closed-loop brain stimulation devices that enable recordings during natural everyday behaviors including locomotion. These developments may ultimately provide an integrated understanding of brain functions from the cell to small assemblies and large-scale networks, give access to causal mechanisms, and allow for ecologically valid and naturalistic experimental paradigms of embodied and situated cognition. In addition, they hold the promise of new therapeutic interventions in a wide-ranging spectrum of diseases such as epilepsy, Parkinson's disease, depression, or Alzheimer's disease.

This introduction cannot conclude without a series of thanks and acknowledgments. First of all, I would like to thank the authors for contributing to this book and providing such deep and comprehensive discussions on a large area of topics. This is all the more remarkable since writing book chapters is often regarded as cumbersome and less rewarding than publishing in high-profile journals. I started writing invitations in the midst of the pandemic, and it was an overwhelming experience to receive so much positive feedback and later to engage in numerous exciting discussions about the content of the chapters. I would also like to thank Leontina Di Cecco and Jayarani Premkumar for their continuing support during the extended preparation process and actually for their benevolent acceptance to extend this process repeatedly. It was a wonderful and enriching experience to edit this book, and I hope it may be similarly enriching and enjoyable for the reader!

Dr. Nikolai Axmacher

Contents

Contributors

Emmanuel J. Barbeau Centre de Recherche Cerveau et Cognition, CNRS, UMR 5549, Toulouse, France;
Faculté de Santé, Université Paul Sabatier, Toulouse, France

André M. Bastos Department of Psychology, Vanderbilt University, Nashville, TN, USA

Christian G. Bénar Aix-Marseille University, INSERM, INS, Marseille, France

Julia Berezutskaya Donders Institute for Brain, Cognition & Behaviour, Radboud University, Nijmegen, The Netherlands

Christian G. Bien Medical School OWL, Department of Epileptology (Krankenhaus Mara), Bielefeld University, Bielefeld, Germany

Tobias Bockhorst Department of Neurophysiology and Pathophysiology, University Medical Center Hamburg-Eppendorf, Hamburg, Germany

Peter Brunner Department of Neurosurgery, Washington University School of Medicine, St. Louis, MO, USA;
National Center for Adaptive Neurotechnologies, Albany, NY, USA;
Department of Neurology, Albany Medical College, Albany, NY, USA

Bassir Caravan Department of Neurosurgery, Cedars-Sinai Medical Center, Los Angeles, CA, USA

Edward F. Chang Weill Institute for Neurosciences, University of California, San Francisco, CA, USA

Qi Chen School of Psychology, South China Normal University, Guangzhou, China

Jan Cimbalnek International Clinical Research Center, St. Anne's University Hospital, Brno, Czech Republic

Jonathan Curot Centre de Recherche Cerveau et Cognition, CNRS, UMR 5549, Toulouse, France;

Département de Neurologie, Centre Hospitalier Universitaire de Toulouse, Toulouse, France;
Faculté de Santé, Université Paul Sabatier, Toulouse, France

Sarang S. Dalal Center of Functionally Integrative Neuroscience, Aarhus University, Aarhus C, Denmark

Anup Das Department of Biomedical Engineering, Columbia University, New York, NY, USA

Jonathan Daume Department of Neurosurgery, Cedars-Sinai Medical Center, Los Angeles, CA, USA

Nai Ding Key Laboratory for Biomedical Engineering of Ministry of Education, College of Biomedical Engineering and Instrument Sciences, Zhejiang University, Hangzhou, China;
Research Center for Advanced Artificial Intelligence Theory, Zhejiang Lab, Hangzhou, China

Thomas Donoghue Columbia University, New York, NY, USA

Anne-Sophie Dubarry Aix-Marseille University, CNRS, LNC, Marseille, France

Arne Ekstrom Department of Psychology, University of Arizona, Tucson, USA

Andreas K. Engel Department of Neurophysiology and Pathophysiology, University Medical Center Hamburg-Eppendorf, Hamburg, Germany

Daniel Pacheco Estefan Institute of Cognitive Neuroscience, Department of Neuropsychology, Ruhr University Bochum, Bochum, Germany

Youssef Ezzyat Department of Psychology and Program in Neuroscience and Behavior, Wesleyan University, Middletown, CT, United States

Ashley Feinsinger Department of Medicine, David Geffen School of Medicine, University of California, Los Angeles, CA, USA

Birgit Frauscher Montreal Neurological Institute and Hospital, McGill University, Montreal, QC, Canada

Pascal Fries Ernst Strüngmann Institute (ESI) gGmbH for Neuroscience in Cooperation With Max Planck Society, Frankfurt, Germany;
Donders Institute for Brain, Cognition, and Behaviour, Radboud University, Nijmegen, The Netherlands

Edgar Galindo-Leon Department of Neurophysiology and Pathophysiology, University Medical Center Hamburg-Eppendorf, Hamburg, Germany

Anne-Lise Giraud Department of Basic Neurosciences, Faculty of Medicine, University of Geneva, Geneva, Switzerland;
Institut de l'Audition - Institut Pasteur, UMR Inserm AO06, 63, rue de Charenton, Paris, France

Jean Gotman Montreal Neurological Institute and Hospital, McGill University, Montreal, QC, Canada

Philip Grewe Medical School OWL, Clinical Neuropsychology and Epilepsy Research, Bielefeld University, Bielefeld, Germany;
Medical School OWL, Department of Epileptology (Krankenhaus Mara), Bielefeld University, Bielefeld, Germany

Benjamin J. Griffiths Department of Psychology, Ludwig-Maximilians-Universität München, Munich, Germany

Mila Halgren Department of Brain & Cognitive Sciences, McGovern Institute for Brain Research, Massachusetts Institute of Technology, Cambridge, MA, USA

Biao Han School of Psychology, South China Normal University, Guangzhou, China

Simon Hanslmayr School of Psychology, Centre for Human Brain Health, University of Birmingham, Birmingham, UK;
School of Neuroscience and Psychology, Centre for Cognitive Neuroimaging, University of Glasgow, Glasgow, UK

Randolph F. Helfrich Center for Neurology, Hertie Institute for Clinical Brain Research, University of Tübingen, Tübingen, Germany

Simon Henin Department of Neurology, NYU Langone Health, New York, NY, USA;
Comprehensive Epilepsy Center, NYU Langone Health, New York, NY, USA

Dora Hermes Department of Physiology and Biomedical Engineering, Mayo Clinic, Rochester, MN, USA

Nora Alicia Herweg Department of Neuropsychology, Institute of Cognitive Neuroscience, Faculty of Psychology, Ruhr University, Bochum, Germany

Yuhao Huang Department of Neurosurgery, Stanford University Medical Center, Stanford, CA, USA

Lukas Imbach Klinik für Neurochirurgie, Universitätsspital Zürich, Zürich, Switzerland;
Swiss Epilepsy Center, Klinik Lengg, Zürich, Schweiz

Cory S. Inman Department of Psychology, University of Utah, Salt Lake City, UT, USA;
Neuroscience Program, University of Utah, Salt Lake City, UT, USA

Joshua Jacobs Department of Biomedical Engineering, Columbia University, New York, NY, USA;
Department of Neurological Surgery, Columbia University, New York, NY, USA

Karim Jerbi Cognitive & Computational Neuroscience Lab, University of Montreal, Quebec, Canada

Elizabeth L. Johnson Northwestern University, Chicago, IL, US

Jacques Jonas Université de Lorraine, CHRU-Nancy, Service de Neurologie, Nancy, France;
Université de Lorraine, CNRS, CRAN, Nancy, France

Corey Keller Department of Psychiatry and Behavioral Sciences, Stanford University Medical Center, Stanford, CA, USA;
Veterans Affairs Palo Alto Healthcare System, and the Sierra Pacific Mental Illness, Research, Education, and Clinical Center (MIRECC), Palo Alto, CA, USA

Christian Klaes Faculty of Medicine, Department of Neurotechnology, Ruhr University Bochum, Bochum, Germany

Robert T. Knight University of California, Berkeley, CA, US

Laurent Koessler Université de Lorraine, CNRS, CRAN, Nancy, France

Genevieve Konopka Department of Neuroscience, UT Southwestern Medical Center, Dallas, USA;
Peter O'Donnell Jr. Brain Institute, UT Southwestern Medical Center, Dallas, USA

Michal T. Kucewicz BioTechMed Center, Faculty of Electronics, Telecommunications and Informatics, Multimedia Systems Department, Gdansk University of Technology, Gdansk, Poland;
Department of Physiology & Biomedical Engineering, Mayo Clinic, Rochester, MN, USA;
Department of Neurology, Mayo Clinic, Rochester, MN, USA

Aaron Kucyi Department of Psychological and Brain Sciences, Drexel University, Philadelphia, PA, US

Lukas Kunz Department of Biomedical Engineering, Columbia University, New York, NY, USA

Jean-Philippe Lachaux Lyon Neuroscience Research CenterEduwell Team, INSERM UMRS 1028, CNRS UMR 5292, Université Claude Bernard Lyon 1, Université de Lyon, Lyon, France

Bradley C. Lega Peter O'Donnell Jr. Brain Institute, UT Southwestern Medical Center, Dallas, USA;
Department of Neurosurgery, UT Southwestern Medical Center, Dallas, USA

Janna D. Lendner Center for Neurology, Hertie Institute for Clinical Brain Research, University of Tübingen, Tübingen, Germany;
Dept. of Anesthesiology and Intensive Care Medicine, University of Tübingen, Tübingen, Germany

Mingli Liang Department of Psychiatry, Department of Neurosurgery, Yale University, New Haven, USA

Robin Lienkämper University of Pittsburgh, Pittsburgh, USA

Jing Liu Department of Applied Social Sciences, The Hong Kong Polytechnic University, Hong Kong SAR, China

Pengcheng Lv CAS Key Laboratory of Mental Health, Institute of Psychology, Beijing, China

Rafael Malach Department of Brain Sciences, Weizmann Institute of Science, Rehovot, Israel

Adam N. Mamelak Department of Neurosurgery, Cedars-Sinai Medical Center, Los Angeles, CA, USA

Jeremy R. Manning Dartmouth College, Hanover, NH, USA

Sabrina L. Maoz Department of Bioengineering, University of California, Los Angeles, California, USA

Pierre Mégevand Department of Basic Neurosciences, Faculty of Medicine, University of Geneva, Geneva, Switzerland;
Department of Clinical Neurosciences, Faculty of Medicine, University of Geneva, Geneva, Switzerland;
Division of Neurology, Geneva University Hospitals, Geneva, Switzerland

Lucia Melloni Department of Neurology, NYU Langone Health, New York, NY, USA;
Comprehensive Epilepsy Center, NYU Langone Health, New York, NY, USA;
Neural Circuits, Consciousness and Cognition ResearchGroup, Max Planck for Empirical Aesthetics, Frankfurt am Main, Germany

Sebastian Michelmann Department of Psychology, Princeton University, Princeton, NJ, USA;
Princeton Neuroscience Institute, Princeton University, Princeton, NJ, USA

Juri Minxha Department of Neurosurgery, Cedars-Sinai Medical Center, Los Angeles, CA, USA;
Division of Biology and Biological Engineering, California Institute of Technology, Los Angeles, CA, USA;
Center for Theoretical Neuroscience, College of Physicians and Surgeons, Columbia University, New York, NY, USA

Haley Moore Department of Neuroscience, UT Southwestern Medical Center, Dallas, USA;
Peter O'Donnell Jr. Brain Institute, UT Southwestern Medical Center, Dallas, USA;
Department of Neurosurgery, UT Southwestern Medical Center, Dallas, USA

John C. Mosher Department of Neurology, McGovern Medical School, Texas Institute for Restorative Neurotechnologies (TIRN), University of Texas Health Science Center, Houston, Texas, USA

Yitzhak Norman Department of Neurological Surgery, University of California, San Francisco, CA, USA

Carina Oehrn Department of Neurological Surgery, University of California, San Francisco, USA

Robert Oostenveld Donders Institute for Brain, Cognition, and Behaviour, Radboud University, Nijmegen, the Netherlands;
NatMEG, Karolinska Institutet, Stockholm, Sweden

George M. Parish School of Psychology, Centre for Human Brain Health, University of Birmingham, Birmingham, UK

Nader Pouratian Department of Neurological Surgery, UT Southwestern Medical Center, Dallas, TX, USA

Timothée Proix Department of Basic Neurosciences, Faculty of Medicine, University of Geneva, Geneva, Switzerland

Salman E. Qasim Department of Psychiatry, Icahn School of Medicine at Mt., New York, NY, USA

Elias M. B. Rau Ruhr University Bochum, Bochum, Germany

Supratim Ray Centre for Neuroscience, Indian Institute of Science, Bangalore, India

Bruno Rossion Université de Lorraine, CNRS, CRAN, Nancy, France;
Université de Lorraine, CHRU-Nancy, Service de Neurologie, Nancy, France

Ueli Rutishauser Department of Neurosurgery, Cedars-Sinai Medical Center, Los Angeles, CA, USA;
Computation and Neural Systems, California Institute of Technology, Pasadena, CA, USA

Krishnakant Saboo Department of Physiology & Biomedical Engineering, Mayo Clinic, Rochester, MN, USA;
Department of Electrical and Computer Engineering, University of Illinois, Urbana-Champaign, IL, USA

Sepideh Sadaghiani Psychology Department, University of Illinois at Urbana-Champaign, Urbana, IL, US

Sushmita Sadhukha Psychological and Brain Sciences, Dartmouth College, Hanover, NH, USA

Anne-Lise Saive Institut Paul Bocuse Research Center, Lyon, France

Johannes Sarnthein Klinik für Neurochirurgie, Universitätsspital Zürich, Zürich, Switzerland

Caspar M. Schwiedrzik Neural Circuits and Cognition Lab, European Neuroscience Institute Göttingen – A Joint Initiative of the University Medical Center Göttingen and the Max Planck Society, Göttingen, Germany;

Perception and Plasticity Group, German Primate Center – Leibniz Institute for Primate Research, Göttingen, Germany

Kristin K. Sellers Department of Neurological Surgery, University of California, San Francisco, CA, USA

Lu Shen School of Psychology, South China Normal University, Guangzhou, China

Ethan A. Solomon Stanford University, Stanford, CA, USA

Matthias Stangl Department of Psychiatry and Biobehavioral Sciences, Jane and Terry Semel Institute for Neuroscience and Human Behavior, University of California, Los Angeles, California, USA

Tobias Staudigl Department of Psychology, Ludwig-Maximilians-Universität München, Munich, Germany

Arjen Stolk Psychological and Brain Sciences, Dartmouth College, Hanover, NH, USA;
Donders Institute for Brain, Cognition, and Behaviour, Radboud University, Nijmegen, the Netherlands

Nanthia Suthana Department of Bioengineering, University of California, Los Angeles, California, USA;
Department of Psychiatry and Biobehavioral Sciences, Jane and Terry Semel Institute for Neuroscience and Human Behavior, University of California, Los Angeles, California, USA;
Department of Neurosurgery, David Geffen School of Medicine, University of California, Los Angeles, California, USA

Angelique Sao-Mai S. Tay Department of Neurosurgery, Cedars-Sinai Medical Center, Los Angeles, United States

Uros Topalovic Department of Electrical and Computer Engineering, University of California, Los Angeles, California, USA

Kanjana Unnwongse Ruhr-Epileptology, Department of Neurology, University Hospital Knappschaftskrankenhaus Bochum, Bochum, Germany

Luc Valton Centre de Recherche Cerveau et Cognition, CNRS, UMR 5549, Toulouse, France;
Département de Neurologie, Centre Hospitalier Universitaire de Toulouse, Toulouse, France

Marcel van Gerven Donders Institute for Brain, Cognition & Behaviour, Radboud University, Nijmegen, The Netherlands

Liang Wang CAS Key Laboratory of Mental Health, Institute of Psychology, Beijing, China

Shuang Wang Department of Neurology, Epilepsy Center, The Second Affiliated Hospital, Zhejiang University School of Medicine, Hangzhou, China

Andrew J. Watrous Baylor College of Medicine, Houston, TX, USA

Tim Wehner Ruhr-Epileptology, Department of Neurology, University Hospital Knappschaftskrankenhaus Bochum, Bochum, Germany

Jörg Wellmer Ruhr-Epileptology, Department of Neurology, University Hospital Knappschaftskrankenhaus Bochum, Bochum, Germany

John H. Wittig Jr. Surgical Neurology Branch, National Institute of Neurological Disorders and Stroke, Bethesda, MD, USA

Gregory A. Worrell BioTechMed Center, Faculty of Electronics, Telecommunications and Informatics, Multimedia Systems Department, Gdansk University of Technology, Gdansk, Poland;
Department of Physiology & Biomedical Engineering, Mayo Clinic, Rochester, MN, USA;
Department of Neurology, Mayo Clinic, Rochester, MN, USA

Weizhen Xie Surgical Neurology Branch, National Institute of Neurological Disorders and Stroke, Bethesda, MD, USA

Yihan Xiong Department of Psychology, Vanderbilt University, Nashville, TN, USA

Gui Xue State Key Laboratory of Cognitive Neuroscience and Learning, Beijing Normal University, Beijing, China;
IDG/McGovern Institute for Brain Research, Beijing Normal University, Beijing, China

Linglin Yang Department of Psychiatry, The Second Affiliated Hospital, Zhejiang University School of Medicine, Hangzhou, China

Mar Yebra Department of Neurosurgery, Cedars-Sinai Medical Center, Los Angeles, CA, USA

Michael J. Young Division of Neurocritical Care, Department of Neurology, Center for Neurotechnology and Neurorecovery, Massachusetts General Hospital, Harvard Medical School, Boston, USA

Erfan Zabeh Department of Biomedical Engineering, Columbia University, New York, NY, USA;
Mortimer B. Zuckerman Mind Brain Behavior Institute, Columbia University, New York, NY, USA

Kareem A. Zaghloul Surgical Neurology Branch, National Institute of Neurological Disorders and Stroke, Bethesda, MD, USA

Hui Zhang Department of Neuropsychology, Institute of Cognitive Neuroscience, Ruhr University Bochum, Bochum, Germany

Part I
Clinical Background, General Questions and Practical Considerations

Chapter 1
How Are Patients Selected for Intracranial EEG Recordings?

Tim Wehner, Kanjana Unnwongse, and Jörg Wellmer

Abstract Intracranial EEG recordings have been employed for decades in the presurgical workup of people with pharmacoresistant epilepsy. The global aim of this procedure is to work out if individuals can benefit from epilepsy surgery concerning seizure control and to tailor surgical intervention to minimize the risk for neurological and cognitive sequelae. On the individual patient level, intracranial EEG is recommended if incongruent or insufficient data from non-invasive examinations require further search for the seizure onset (explorative approach), if an assumed seizure onset/the epileptogenicity of a suspected lesion needs to be proven (confirmative approach), or if the spatial relationship between seizure onset zone and functionally indispensable ("eloquent") cortex has to be elucidated (functional mapping). Typical clinical scenarios for invasive EEG recordings are therefore the absence of a clear epileptogenic lesion, seizure onset potentially remote from an assumed epileptogenic lesion, potentially multifocal epilepsy, and potential overlap of epileptogenic and eloquent cortex. Depending on the given scenario, different invasive work-up method can be applied: subdural grid and strip electrodes via a craniotomy; intracerebral implantation of depth electrodes, either in the sense of few, confirmative electrodes or following the classical "French-Italian" stereo-EEG concept; variations of both. While the subdural approach allows systematic sampling and electrocortical mapping of cortical surface areas (gyral crowns), in particular the frontoparietal convexity, neocortical temporal areas, and mesial frontal and parietal areas, the depth electrodes method permits the evaluation of deep cortical areas, including temporo-mesial structures, as well as bihemispheric target areas.

1.1 Prologue

For everybody who is not working in the field of epileptology, the easiest approach to understand the rationale and the principles of intracranial electroencephalography

T. Wehner (✉) · K. Unnwongse · J. Wellmer
Ruhr-Epileptology, Department of Neurology, University Hospital Knappschaftskrankenhaus Bochum, In der Schornau 23-25, 44892 Bochum, Germany
e-mail: tim.wehner@kk-bochum.de

N. Axmacher (ed.), *Intracranial EEG*, Studies in Neuroscience, Psychology and Behavioral Economics, https://doi.org/10.1007/978-3-031-20910-9_1

(EEG) recordings is the "fuse box model" of the brain and focal epilepsies. We use this model to explain to our patients with pharmacoresistant focal epilepsy the rationale for electrical recordings directly from their brain:

> Imagine your brain is the central fuse box of your house—then epileptic seizures are like short circuits in this fuse box. If you repeatedly recognize an identical electrical dysfunction in your house (electricity first breaks down in the kitchen, then in the basement and next in the garage), you might find the explanation in the layout or "anatomy" of your central fuse box. The kitchen is protected by fuse #14, basement by fuse #15, garage by fuse #16. This makes it likely that only fuse #14 is defect and that the remaining dysfunction is just the result of a spread to neighboring fuses. After our non-invasive examination including scalp EEG delivered only an approximation where in your "fuse box" brain the dysfunction responsible for your seizures might be located, we suggest further EEG recordings directly from your brain to make sure which "fuse" exactly is responsible for your seizures. This will help us identifying a target for possible surgical intervention.

In fact, this model can explain all steps of intracranial presurgical work-up to patients and other laypersons, from candidate selection, via planning and placement of electrodes to eventual surgical recommendations and surgery.

In the following, we do no longer refer to this model but move to the epileptological literature. The goal of this chapter is to provide cognitive neuro-scientists with a brief didactic overview of the rationale for intracranial EEG recordings. We will discuss the diagnostic modalities used in the presurgical epilepsy workup, the different recording techniques of intracranial EEG, and the rationale for their use. Illustrative cases are provided. It is, however, beyond the scope of this chapter to discuss technical details and limits of each diagnostic modality.

1.2 Who is a Candidate for a Presurgical Epilepsy Work-Up?

In the vast majority of people with epilepsy, the disease is defined by recurrent unprovoked epileptic seizures [9]. Up to 70% of people with epilepsy can be rendered seizure-free by taking anti-seizure medications. These patients are not considered candidates for a neurosurgical intervention unless there is a primary neurosurgical indication such as a brain tumor or a cavernoma. In these cases, seizure freedom may be a welcomed by-effect of the operation.

In about 30% of epilepsy patients, however, seizure control is not possible by medication alone (pharmacoresistant epilepsy). Among these, a majority suffer from "focal epilepsy". For a simplified approach in this chapter, this term refers to epilepsies with an identifiable circumscribed seizure generating area and a perspective of seizure freedom after its removal.

The goal of a presurgical epilepsy workup is to define the brain area that is responsible for the generation of seizures, and to estimate the risks of its resection, destruction or disconnection. This "seizure generating area" is called the epileptogenic zone.[1] There is currently no imaging modality available to directly display the epileptogenic zone. However, if epilepsy surgery results in seizure freedom, one can infer that the epileptogenic zone was removed, destroyed or disconnected. In practical terms, the epileptogenic zone can be approximated using five different zones that in turn can be estimated by diagnostic tests [18]:

- The *symptomatogenic zone* is the brain area that generates ictal symptoms. This information is gathered from the clinical history and the observation of seizures during video-EEG-monitoring (Table 1.1).
- The *irritative zone* is the cortical area that generates interictal epileptiform discharges.
- The *seizure onset zone* is the brain area that generates the initial ictal discharges. Interictal epileptiform discharges and ictal patterns are recorded by EEG (and potentially magnetoencephalography, MEG), and interpreted via visual analysis and increasingly also using complex mathematical algorithms (source analysis, [5, 7, 10, 30]). In selected cases, ictal single-photon emission computed tomography (SPECT) is helpful to identify the area of seizure onset and early propagation if this is not possible using EEG [22].
- The *epileptogenic lesion* is a structural brain abnormality, usually identified by magnet resonance imaging (MRI), which is a (potential) cause of the epilepsy. In patients who ultimately underwent epilepsy surgery, the most common epileptogenic lesions are hippocampal sclerosis, long-term epilepsy associated tumors such as ganglioglioma and dysembryoplastic neuroepithelial tumors, malformations of cortical development including focal cortical dysplasias, cavernomatous angioma, gliotic residua of a stroke or a traumatic brain injury, and encephalitis [4]. MRI postprocessing may be used to detect subtle focal cortical dysplasias [11, 24, 25], however pharmacoresistant focal epilepsy is amenable to surgical treatment even in the absence of a structural lesion on MRI.
- The *functional deficit zone* is a brain area that is functionally compromised in the interictal state. Brain function is assessed during the clinical neurological examination, as well as detailed neuropsychological testing (see also Chap. 2). In some patients, fluoro-desoxy-glucose positron emission tomography (FDG-PET) is used to detect cerebral areas with reduced glucose uptake.
- In addition, eloquent cortical areas and white matter tracts may need to be identified if the putative epileptogenic zone is closely related to areas that are functionally indispensable. These can be approximated using functional MRI, diffusion tensor imaging (DTI), and electrocortical stimulation (see also Chap. 39) intra- and extraoperatively.

[1] For the purpose of this chapter, we use the term "epileptogenic zone" as defined by Rosenow and Lüders [18], since we find the concept of the various zones helpful from a didactic point of view.

Table 1.1 Ictal symptoms and signs with localizing or lateralizing value, adapted from [21]

Clinical feature	Description	Typical localization; Lateralization
Sensory domain		
Somatosensory aura	Unilateral, localized tingling, numbness, electrical sensation, heat, pain, sense of movement or desire to move	Primary sensory cortex, secondary somatosensory areas; contralateral (if unilateral)
Simple visual aura	Flashing or flickering lights, spots, patterns, scotoma, amaurosis	Primary visual cortex; contralateral
Complex visual aura	Visual distortions, changes in dimension	Visual association cortex; contralateral if unilateral
Auditory aura	Buzzing or drumming sounds, single tones, melodies	Primary and secondary auditory cortex; contralateral if unilateral
Olfactory aura	Burning, rotten, unnatural unpleasant smells	Amygdala, insula, frontobasal; non-lateralizing
Gustatory aura	Metallic or rubbery, unnatural unpleasant taste	Parietal operculum, basal temporal cortex; non-lateralizing
Abdominal aura	Nausea, emptiness, tightness, rising sensation in abdomen and/or chest	Insula; non-lateralizing
Vertiginous aura		Temporoparietal junction; non-lateralizing
Psychic domain		
Fear		Amygdala, hippocampus, mesial frontal cortex; non-lateralizing
Distortions of familiarity	Déjà vu: sensation of being familiar with a new situation; Jamais vu: sensation that a familiar context appears new	Mesial temporal, temporal association cortex, rhinal cortex; non-lateralizing
Multisensory hallucinations	Revocation of complex memories	Mesiobasal limbic cortex, neocortical temporal, temporo-parieto-occipital junction; non-lateralizing
Autonomic domain		
Autonomic alterations	Hot flashes, hypersecretion, piloerection, sweating, vomiting, tachycardia	Mesial temporal, insula, anterior cingulum, orbitofrontal; non-lateralizing

(continued)

Table 1.1 (continued)

Clinical feature	Description	Typical localization; Lateralization
Motor domain		
Focal clonic activity	Unilateral, localized, rhythmic repetitive movements	Primary and secondary motor cortex; contralateral
Focal tonic activity	Unilateral, sustained muscle contraction	Supplementary motor area, secondary motor cortex, anterior cingulum; contralateral
Versive head/eyes	Sustained, forced conjugate ocular, cephalic, and/or truncal rotation or lateral deviation	Secondary motor cortex, frontal eye fields; contralateral
Unilateral dystonic arm posturing	Sustained contraction of agonist and antagonist muscles, resulting in pronation and extension of the forearm, flexion of the wrist, extension of the digits	Basal ganglia activation; contralateral
Figure of four sign	Asymmetric contraction of upper limbs resembling the number “4” with arm extension contralateral to focus, ipsilateral arm flexed at elbow, occurring during initial phase of bilateral tonic–clonic seizure	Supplementary motor area, precentral areas; contralateral
Automatisms with preserved responsiveness	Coordinated, repetitive, motor activity without impaired awareness	Temporal lobe; non-dominant
Postictal paresis	Postictal weakness involving arm, face, or leg	Exhaustion of primary or secondary motor cortex; contralateral
Asymmetric termination of the clonic phase	Unilateral persistence of clonic jerks at the end of bilateral tonic–clonic seizure	Exhaustion of hemisphere of seizure onset; ipsilateral
Language domain		
Ictal/postictal aphasia	Inability to speak or comprehend with preserved consciousness	Anterior and posterior language areas; dominant
Ictal speech	Ability to speak out of situational context with impaired awareness	Temporal; non-dominant

Symptoms experienced by the patient at the beginning of a seizure (epileptic aura) and clinical signs observed during a seizure provide potential localizing and lateralizing information with regards to the hemisphere or lobe of seizure origin. Common signs and symptoms with localizing or lateralizing value are listed in Table 1.1 [21].

The first step of the presurgical workup is commonly called "non-invasive workup" and comprises a standardized set of diagnostics: a detailed history of seizure semiology, clinical neurological examination, scalp video-EEG recording of all habitual seizure types, brain MRI using a dedicated protocol with a high yield for epileptogenic lesions [3, 27], neuropsychological testing, and functional MRI to determine the language dominant hemisphere. In some patients, further imaging techniques such as FDG-PET, ictal SPECT, and MEG are used if the results of the standard workup are uninformative or contradictory. The diagnostic accuracy and yield of long-term scalp video-EEG [13], interictal and ictal source imaging [14, 19] and MRI [17] in the presurgical epilepsy evaluation have been systematically examined. However, for individual treatment decisions, the combined assessment of all available examinations supersedes the contribution of an individual technique. The results of the non-invasive workup are integrated to formulate a hypothesis about the epileptogenic zone and its extent, and an assessment of the potential risks associated with surgical treatment. Typically, this is done in a multidisciplinary conference with input from epileptologists, neuropsychologists, neuroradiologists, and neurosurgeons.

In principle, one of the following scenarios applies (Fig. 1.1):

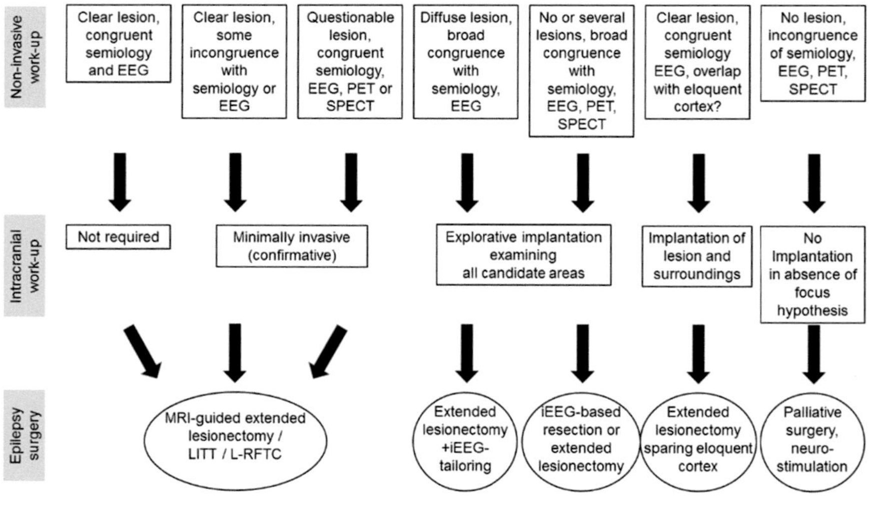

Fig. 1.1 Decision making in epilepsy surgery, adapted from [26]. Ecog—Electrocorticography, EEG—Electroencephalography ± source analysis, iEEG—intracranial EEG, LITT—Laser interstitial thermotherapy, L-RFTC—lesion-based radiofrequency thermocoagulation, MRI—Magnetic resonance imaging, PET—positron emission tomography, SPECT—single photon emission computed tomography

A. The area of seizure origin can be identified with a high degree of certainty based on seizure semiology and ictal and interictal EEG findings; MRI reveals a matching epileptogenic lesion; and the risk of its surgical treatment is low. In this scenario, surgical treatment can be recommended without the need for further investigations. A typical example is a patient with a single seizure semiology suggestive of mesial temporal sclerosis, corresponding ictal and interical EEG findings, ipsilateral hippocampal sclerosis on MRI, and a matching profile on neuropsychological examination.
B. The non-invasively obtained data result in a hypothesis about the suspected area of seizure onset, however no immediate surgical strategy can be formulated. In this case, further intracranial EEG recordings may confirm or reject the hypothesis. Typical examples are patients without clear epileptogenic lesion, with extensive or several epileptogenic lesions and/or discordant findings in the non-invasive diagnostic modalities. In representative European epilepsy centers, intracranial EEG recordings are used in approximately 30% of cases who ultimately undergo epilepsy surgery [2].
C. The putative seizure onset zone is closely related to eloquent cortical areas and/or white matter tracts. Therefore, detailed mapping of the patient's individual functional anatomy is needed to define the boundaries of a (possible) resection.
D. The suspected area is large and/or involves both cerebral hemispheres. In this situation, no surgical treatment options other than palliative measures (subpial transsections, corpus callosotomy, neurostimulation, [8]) are available (for neurostimulation, see also Chaps. 52 and 53). A typical example is a patient with multifocal epilepsy in the setting of multiple bihemispheric nodular heterotopias.

1.3 Different Methodological Approaches to Intracranial EEG

Intracranial EEG recordings have been used since the 1950s using two different approaches, their development thus predates the era of modern neuroimaging [6].

The North American school utilized subdural grid- and strip electrodes, in some cases combined with few depth electrodes. This approach allows the systematic evaluation of relatively large cortical areas, in particular the frontal and temporal convexity, the mesial frontal and parietal surface, and the temporal pole. Amygdala and hippocampus can be sampled from strip electrodes placed underneath the basal temporal surface. The implantation of grid and strip electrodes requires a craniotomy, and recordings are largely limited to structures in one cerebral hemisphere.

The French-Italian approach employs multiple stereotactically implanted depth electrodes to study the area of seizure onset and propagation. This method, also called stereo-EEG, allows sampling from non-contiguous and bihemispheric areas, as well as areas remote from the cerebral surface, such as the amygdala, hippocampus, and insula. The implantation of depth electrodes is done via multiple small skull drills, utilizing either a stereotactic frame, a robot, or a frameless approach. In order to choose safe implantation trajectories, cerebral blood vessels and sulcal boundaries are avoided.

It is important to stress that both methods of intracranial EEG suffer from sampling bias: In either approach, only a relatively small amount of cerebral cortex can be sampled. The information obtained with intracranial recordings therefore largely depends on the strength and precision of the pre-implantation hypothesis. In other words, an ill-devised intracranial EEG investigation with random sampling of cortical areas ("fishing expedition") is unlikely to result in the formulation of an individually tailored surgical treatment plan.

It is further important to note that both types of intracranial recording require a neurosurgical operation under general anesthesia for the placement of the electrodes, and are therefore associated with the general risks of neurosurgery (bleeding, infection, perioperative infarction, venous thromboembolism). According to a recent meta-analysis, the most common risks associated with subdural grid electrodes are neurologic infections (2.3%), superficial infections (3%), intracranial hemorrhage (4%), and elevated intracranial pressure (2.4%), with up to 3.5% of patients requiring additional surgical procedure(s) to manage these adverse events [1]. For SEEG implantations, the most common risks are hemorrhage (1%) and infections (0.8%, [15]). Because of the lower morbidity associated with SEEG compared to subdural grid recordings, the former is increasingly used in many epilepsy surgery programs around the world [2, 20].

For both types of implantations, the neurosurgical risks increase with the number of electrodes and electrode contacts implanted. One therefore has to balance the desire for most comprehensive EEG sampling with the need to limit the risks associated with the recording.

From the conceptional point of view, intracranial presurgical work-up that is applied for delineation of the seizure onset zone can follow a confirmative or explorative strategy, or a combination of both (Fig. 1.1, [26]).

In a *confirmative* scenario, there is a strong hypothesis about the seizure onset zone, typically based on the presence of a (suspected) epileptogenic lesion. In this case, the goal of the intracranial recording may be to prove that ictal discharges are generated from the epileptogenic lesion or its immediate vicinity, and that these predate the onset of ictal EEG changes seen on simultaneous scalp recordings as well as the initial ictal symptoms and signs. In this setting, a limited number of depth electrodes may be used. Typical clinical scenarios are:

(1) a presentation of temporal lobe epilepsy in the setting of unilateral hippocampal sclerosis, but equivocal ictal EEG patterns, and
(2) suspected focal cortical dysplasia and non-specific seizure semiology and/or ictal EEG findings (illustrative case 1, Fig. 1.2).

In an *explorative* scenario, there is often no clear epileptogenic lesion. Seizure semiology and interictal and ictal EEG findings suggest a hypothesis with regards to the epileptogenic zone, however a variety of brain areas need to be considered (or excluded) as seizure onset areas (illustrative case 2, Fig. 1.3). Typical clinical scenarios are patients with nonlesional MRI or more than one epileptogenic lesion,

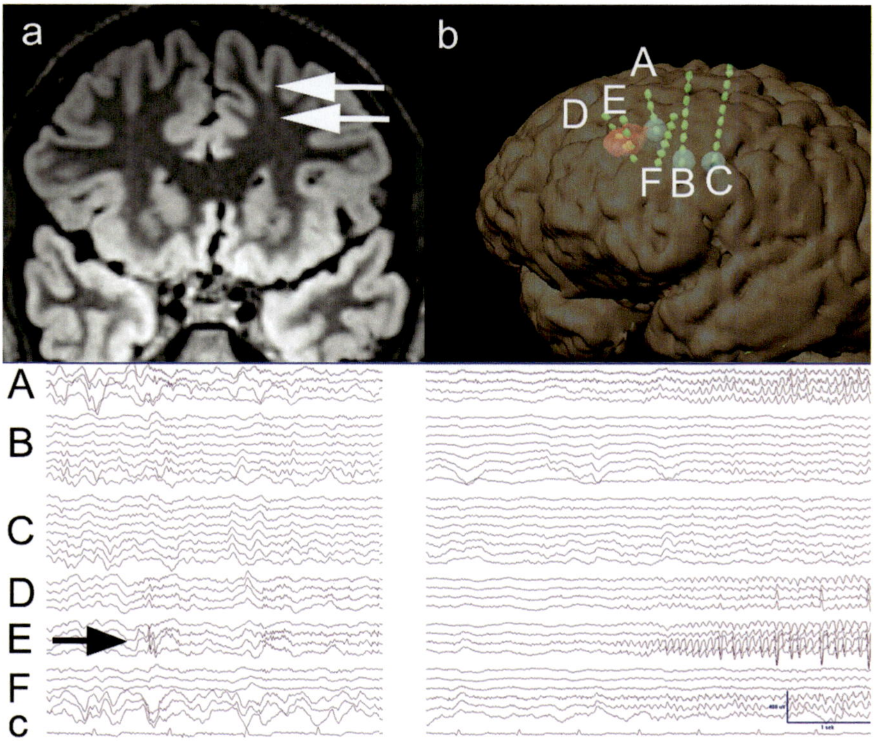

Fig. 1.2 **a** Coronal fluid attenuated inversion recovery MRI showing a small area of blurring of the grey-white matter junction in the left frontal sulcus. **b** Implantation scheme of stereotactically placed depth electrodes targeting the MRI abnormality (electrodes D + E), solutions of source localization algorithms based on interictal EEG and MEG recordings (electrodes A-C) and one interspaced gyrus (F). **c** Ictal onset on intracranial EEG, contacts E 3 + 4, and propagation to contacts F 4–6, D 1–4, and A 1–4. Tracing labels on the left refer to electrode labels in fig. B. Referential montage, the uppermost line for each tracing refers to the electrode contact furthest from the surface. Single-channel EKG at the bottom. Ten second gap between the left and the right EEG page

non-specific ictal symptoms and signs at the seizure onset, and/or equivocal surface EEG findings, or patients in whom non-invasive evaluation suggests the possibility of bilateral temporal lobe epilepsy.

The decision for a particular *implantation strategy* (grids ± few depth electrodes vs stereotactically implanted depth electrodes) takes into account individual patient circumstances and the expertise of the epilepsy surgery team. A traditional indication for an exploration with subdural grid electrodes is the need to perform extraoperative functional mapping using electrocortical stimulation, in particular if the suspected epileptogenic zone is close to eloquent cortex (illustrative case 3, Fig. 1.4). This can alternatively be done intraoperatively, with or without an "awake" approach during which the patient is woken up after the craniotomy has been performed.

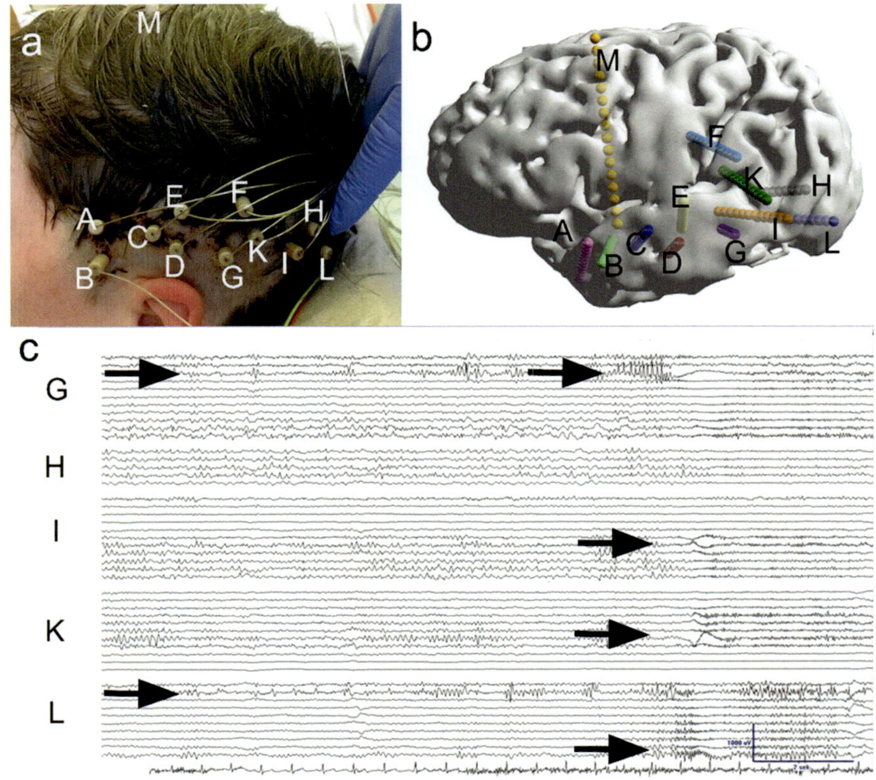

Fig. 1.3 **a** Explorative SEEG-implantation scheme used in patient 2, photograph of electrode bolts in situ. **b** Electrode position reconstruction to surface-rendered MRI. Electrode labels and corresponding anatomical targets: A—entorhinal cortex, B—amygdala, C—anterior hippocampus, D—parahippocampal gyrus, E—posterior hippocampus, F—posterior cingulate, G—anterior lingual gyrus, H—temporo-occipital junction, I—temporo-occipital base, K—temporoparietal junction, L—posterior lingual gyrus, M—posterior insula. **c** Ictal onset on intracranial EEG as explained in the text. Tracing labels on the left refer to electrode labels in Fig. 1.3 A + B. Bipolar montage of adjacent electrode contacts, the uppermost line for each tracing refers to the channel linking electrode contacts 1–2 (furthest from the cortical surface). Only selected electrodes shown. Single-channel ECG at the bottom

As outlined above, the SEEG approach is increasingly becoming the preferred approach in many epilepsy surgery programs.

The decision for a specific *implantation plan* considers all the data obtained during the non-invasive workup to define target areas for implantation. Planning is facilitated if data points can be integrated and displayed in a multimodal imaging platform, in which all datasets are coregistered to a common volumetric MPRAGE (T1-weighted) MRI sequence [16]. Although implantation schemes are tailored to

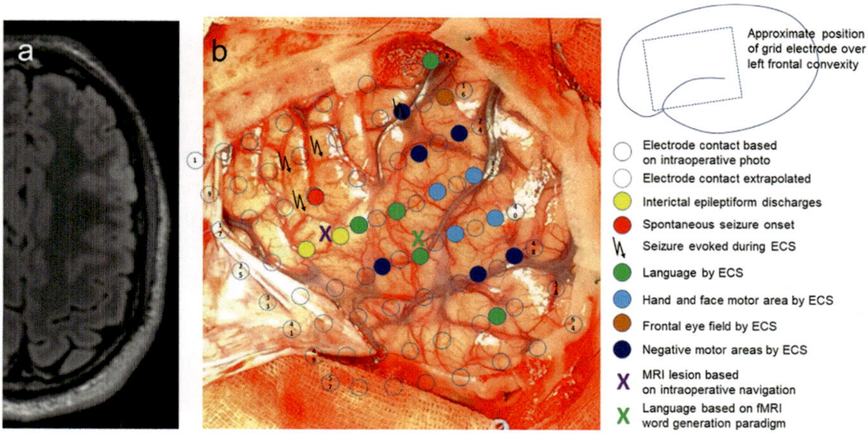

Fig. 1.4 **a** Axial FLAIR MRI demonstrates a hyperintense lesion in the left middle frontal gyrus and sulcus. **b** Summary of the intracranial EEG recording and extraoperative electrocortical stimulation (ECS), electrode contact position based on intraoperative photograph

the individual clinical scenario, there are common targets for implantation, in particular the mesial temporal structures. In our center, we investigate these with a standardized approach using five depth electrodes inserted through the temporal bone, targeting the entorhinal cortex, amygdala, anterior and posterior hippocampus, and parahippocampal gyrus.

Both methods of intracranial EEG recording have shaped our understanding of seizure initiation and propagation within the brain, in particular since they have been used in parallel with the development of modern neuroimaging techniques. As a result of this process, the relationship of the seizure onset zone and different kinds of epileptogenic lesions has been defined. For example, in small focal cortical dysplasias, the complete removal or destruction of the abnormality seen on MRI typically results in seizure freedom, suggesting a complete overlap between the epileptogenic lesion and the epileptogenic zone [23, 29]. This is also the case in some—but not all—patients with unilateral hippocampal sclerosis. In particular the SEEG method has uncovered scenarios in which the epileptogenic zone extends from the mesial temporal lobe to neocortical temporal, frontal, parietal or insular areas ("temporal plus epilepsy", [12]). For other types of MRI abnormality, the epileptogenic zone may be a subset of the visible lesion, for example in patients with extensive polymicrogyria or a large gliotic scar. Lastly, in patients with epilepsy and a brain tumor or cavernous angioma, the MRI abnormality is not necessarily the epileptogenic lesion.

In addition, intracranial EEG recordings and electrocortical stimulation provide an opportunity to study the function of the brain, for example by running neuropsychological paradigms and/or implanting combined macro- and microwire electrodes (further discussed in Chaps. 42–44).

1.4 Conclusion

Intracranial EEG is—with few exceptions—used in patients who suffer from pharmacoresistant focal epilepsy, in whom a non-invasive presurgical work up is not sufficiently able to generate a clear surgical hypothesis or in whom eloquent functions need to be delineated from the epileptological "what to remove" area. The decision for the type and number of implanted electrodes and implantation targets is based on the results of the preceding multimodal non-invasive work-up.

1.5 Illustrative Case Scenarios

Case scenario 1: A 17-year-old man presented with seizures from sleep since childhood. He was amnestic for the events, woke up in the morning with a headache and tongue bite. His parents never saw a seizure from the onset, they usually woke up from a tonic vocalization and then described bilateral stiffening with subsequent clonic movements of all extremities. On scalp video-EEG, he was found to have repetitive spikes in the left fronto-central area. Several seizures were recorded from sleep; the patient abruptly turns his head towards the right and slightly raises his arms. He loudly utters two incomprehensible words. No further symptoms, the total duration is less than ten seconds. On immediate postictal testing, he answers questions and performs tasks appropriately. In most seizures, there is no clear scalp EEG change, in some there is a subtle rhythmic delta activity over the left frontal area. MRI was initially interpreted as normal, however on further review a small area of blurring of the cortical-subcortical interface was found in the left superior frontal sulcus (Fig. 1.2a). These findings raised the suspicion of a small focal cortical dysplasia, but a vascular tract could not be excluded. The patient therefore underwent further FDG-PET imaging which revealed a circumscribed area of reduced glucose uptake coinciding with the MRI finding, and MEG as part of a research study. Neuropsychological testing revealed a profile in the high average range without localizing domains of relative underfunctioning.

Comment: In this patient, the seizure semiology is consistent with frontal lobe epilepsy, but does not provide further localizing or lateralizing information. Interictal and ictal scalp EEG findings broadly coincide with the subtle imaging findings that are, however, not evidential for a focal cortical dysplasia. The patient therefore underwent a "confirmatory" implantation of six stereotactically placed depth electrodes targeting the imaging abnormality and adjacent areas suggested by interictal EEG and MEG source localization algorithms (Fig. 1.2b). On intracranial EEG, interictal epileptiform activity was largely confined to the electrodes implanted around the subtle imaging abnormality. The patient's habitual seizures arose from the same contacts with subsequent spread to the neighboring depth electrodes (Fig. 1.2c). The findings thus confirmed epileptogenicity and seizure origin in the imaging abnormality. The patient subsequently underwent stereotactically guided radio-frequency thermocoagulation targeted at the imaging abnormality. This minimally invasive lesioning approach offers an alternative to an open resection in selected patients with very circumscribed target volumes for destruction [28]. He has been seizure free for 4.5 years since this procedure, >2 years off anti-seizure medication.

Case scenario 2. A 47-year-old woman had focal epilepsy since age 17 years. She describes a sensation of numbness in her left arm for a few seconds without propagation to other parts of her body; subsequently, she may lose awareness. Her husband reports behavioral arrest, staring and oral automatisms in some seizures. There is frequent enuresis during seizures. Seizures last up to 5 min and occur on average once per week.

Non-invasive video-EEG based source analysis revealed epileptiform discharges in the left anterior temporal area. Habitual seizures were recorded with behavioral arrest, oral automatisms, left hand fumbling, then arm tonic posturing, head version to the right, and subsequent tonic–clonic movements of the right extremities. The first ictal changes on scalp EEG were seen over the left temporoparietal area, ictal source localization suggested a seizure onset in the left temporal basal area. MRI using a dedicated protocol to detect epileptogenic lesions was repeatedly normal. FDG-PET demonstrated subtle areas of reduced glucose uptake in the temporo-occipital junction on the left. Her neuropsychological examination showed impaired simple semantic and complex phonematic fluidity and verbal memory.

Comment: In this patient without an epileptogenic lesion on MRI, the seizure semiology suggests left hemispheric origin, the initial numbness in the left arm points to the left insula as a possible symptomatogenic area. The ictal and interictal EEG findings including source localization imply the left temporal lobe and temporo-parietal junction, the PET the left temporo-occipital junction. The patient was thus further investigated using an "explorative" SEEG approach with depth electrodes implanted into the left entorhinal cortex, amygdala, anterior and posterior hippocampus, parahippocampal gyrus, posterior parahippocampal gyrus, posterior cingulate gyrus, posterior insula and the temporo-occipital-parietal junction (Fig. 1.3a, b).

On intracranial video-EEG, interictal epileptiform activity was recorded from the anterior hippocampus, the anterior temporal neocortex, the posterior cingulate, the posterior middle temporal gyrus, and the lingual gyrus (electrodes A, E, F, G and L, Fig. 1.3a, b). Four habitual seizures were recorded. The ictal EEG changes evolved from repetitive spiking in the lingual gyrus (electrode contacts G3 and L2, left arrows in Fig. 1.3c) with spread towards other contacts on these electrodes and to neighboring temporo-occipital areas (electrodes I and K, right arrows in Fig. 1.3c).

Case scenario 3: A 24-year-old left-handed man presented with focal seizures since age 16 years. His seizures start with a "woozy" feeling that he localizes to the head, then he feels his head rotating to the right in a forced manner, his body follows his head. This may result in a full 360° body turn. At this point, the seizure may terminate without any further symptoms, or he may lose consciousness. His family describes a bilateral tonic clonic seizure, sometimes associated with a right tongue bite. During scalp video-EEG, rare spikes were seen in the left frontocentral area. One habitual seizure was recorded;, in addition, five seizures were recorded during which he rose from sleep, with some incomprehensible speech and doubling of the heart rate. In all seizures, scalp EEG changes in both frontocentral regions preceded the behavioral change by 5–30 s. During the ictal phase, movement artefacts mar the EEG. MRIs since age 18 years had suggested a focal cortical dysplasia in the left middle frontal gyrus. At the time of the scalp video-EEG evaluation, the MRI features had somewhat changed to suggest a neoplastic process (Fig. 1.4a). Neuropsychological examination revealed test results in the average and above-average range without any localized functional deficit. Functional MRI using a word generation paradigm suggested left hemispheric language dominance.

Comment: Seizure semiology with initial versive rotation of the head and body to the right indicate early ictal activation of the left frontal eye field (located in the posterior part of the superior frontal sulcus), the interictal and ictal EEG and the location of the MRI lesion are corresponding. The changes in the MRI suggest a neoplastic process; hence, there is an indication for a resection of the MRI lesion, independent from the epileptological perspective. The rationale for intracranial recording in this case is the need for detailed extraoperative cortical mapping due to the proximity of the lesion to the precentral gyrus and the anterior language area in the pars triangularis of the inferior frontal gyrus. Therefore, a subdural grid electrode (8×8 contacts) is implanted covering the lesion, the central area, and the inferior frontal area. Results of the intracranial recording are summarized in Fig. 1.4b. The following aspects illustrate the principles discussed above:

(1) The spontaneous electrographic seizure onset (red dot) is adjacent to the anatomical lesion, however electrocortical stimulation elicits seizures from a more extensive area (flashes). Interictal epileptiform activity (yellow dots) is recorded from separate neighboring contacts.
(2) The cortical area of the first ictal symptoms (involuntary eye deviation to the right, brown dot) is located more than 3 cm posterior from the ictal onset area.
(3) Electrocortical stimulation revealed language functions (green dots) immediately adjacent to the contacts showing interictal epileptiform activity, extending from the area suggested by fMRI activity by ~2 cm.

The patient subsequently underwent an awake resection of the lesion in the middle frontal gyrus and the adjacent superior frontal gyrus. Histopathology revealed an angiocentric glioma. Postoperatively, he has been seizure-free for 23 months.

Acknowledgements We acknowledge the contributions of Johannes Jungilligens (neuropsychology), Yaroslav Parpaley (stereotactic neurosurgery, Fig. 1.2b) and Dorothea Miller (resective neurosurgery).

References

1. Arya R, Mangano FT, Horn PS et al (2013) Adverse events related to extraoperative invasive EEG monitoring with subdural grid electrodes: a systematic review and meta-analysis.

Epilepsia 54:828–839
2. Baud MO, Perneger T, Rácz A et al (2018) European trends in epilepsy surgery. Neurology 91:e96–e106
3. Bernasconi A, Cendes F, Theodore WH et al (2019) Recommendations for the use of structural magnetic resonance imaging in the care of patients with epilepsy: a consensus report from the international league against epilepsy neuroimaging task force. Epilepsia 60:1054–1068
4. Blümcke I, Spreafico R, Haaker G et al (2017) Histopathological findings in brain tissue obtained during epilepsy surgery. New Engl J Med 377:1648–1656
5. Brodbeck V, Spinelli L, Lascano AM et al (2011) Electroencephalographic source imaging: a prospective study of 152 operated epileptic patients. Brain 134:2887–2897
6. Chauvel P, Gonzalez-Martinez J, Bulacio J (2019) Presurgical intracranial investigations in epilepsy surgery. Handb Clin Neurol 161:45–71
7. Duez L, Tankisi H, Hansen PO et al (2019) Electromagnetic source imaging in presurgical workup of patients with epilepsy: a prospective study. Neurology 92:e576-586
8. Fauser S, Zentner J (2012) Critical review of palliative surgical techniques for intractable epilepsy. Adv Tech Stand Neurosurg 39:165–194
9. Fisher RS, Acevedo C, Arzimanoglou A et al (2014) ILAE official report: a practical clinical definition of epilepsy. Epilepsia 55:475–482
10. Foged MT, Martens T, Pinborg LH et al (2020) Diagnostic added value of electrical source imaging in presurgical evaluation of patients with epilepsy: a prospective study. Clin Neurophysiol 131:324–329
11. Hong SJ, Kim H, Schrader D et al (2014) Automated detection of cortical dysplasia type II in MRI-negative epilepsy. Neurology 83:48–55
12. Kahane P, Barba C, Rheims S et al (2015) The concept of temporal “plus” epilepsy. Rev Neurol (Paris) 171:267–272
13. Kobulashvili T, Kuchukhidze G, Brigo F et al (2018) Diagnostic and prognostic value of non-invasive long-term video-electroencephalographic monitoring in epilepsy surgery: a systematic review and meta-analysis from the E-PILEPSY consortium. Epilepsia 59:2272–2283
14. Mouthaan BE, Rados M, Boon P et al (2019) Diagnostic accuracy of interictal source imaging in presurgical epilepsy evaluation: a systematic review from the E-PILEPSY consortium. Clin Neurophysiol 130:845–855
15. Mullin JP, Shriver M, Alomar S et al (2016) Is SEEG safe? A systematic review and meta-analysis of stereo-electroencephalography-related complications. Epilepsia 57:386–401
16. Nowell M, Rodionov R, Zombori G et al (2015) Utility of 3D multimodality imaging in the implantation of intracranial electrodes in epilepsy. Epilepsia 56:403–413
17. Rados M, Mouthaan B, Barsi P et al (2021) Diagnostic value of MRI in the presurgical evaluation of patients with epilepsy: influence of field strength and sequence selection: a systematic review and meta-analysis from the E-PILEPSY Consortium. Epileptic Disord. https://doi.org/10.1684/epd.2021.1399
18. Rosenow F, Lüders H (2001) Presurgical evaluation of epilepsy. Brain 124:1683–1700
19. Sharma P, Seeck M, Beniczky S (2019) Accuracy of interictal and ictal electric and magnetic source imaging: a systematic review and meta-analysis. Front Neurol **10**:1250
20. Tandon N, Tong BA, Friedman ER et al (2019) Analysis of morbidity and outcomes associated with use of subdural grids vs stereoelectroencephalography in patients with intractable epilepsy. JAMA Neurol 76:672–681
21. Unnwongse K, Wehner T, Foldvary-Schaefer N (2010) Selecting patients for epilepsy surgery. Curr Neurol Neurosci Rep 10:299–307
22. von Oertzen TJ, Mormann F, Urbach H et al (2011) Prospective use of subtraction ictal SPECT coregistered to MRI (SISCOM) in presurgical evaluation of epilepsy. Epilepsia 52:2239–2248
23. Wagner J, Urbach H, Niehusmann P et al (2011) Focal cortical dysplasia type IIb: completeness of cortical, not subcortical, resection is necessary for seizure freedom. Epilepsia 52:1418–1424
24. Wagner J, Weber B, Urbach H et al (2011) Morphometric MRI analysis improves detection of focal cortical dysplasia type II. Brain 134:2844–2854

25. Wang ZI, Jones SE, Jaisani Z et al (2015) Voxel-based morphometric magnetic resonance imaging (MRI) post-processing in MRI-negative epilepsy. Ann Neurol 77:1060–1075
26. Wellmer J, von der Groeben F, Klarmann U et al (2012) Risks and benefits of invasive epilepsy surgery workup with implanted subdural and depth electrodes. Epilepsia 53:1322–1332
27. Wellmer J, Quesada CM, Rothe L et al (2013) Proposal for a magnetic resonance imaging protocol for the detection of epileptogenic lesions at early outpatient stages. Epilepsia 54:1977–1987
28. Wellmer J, Voges J, Parpaley Y (2016) Lesion guided radiofrequency thermocoagulation (L-RFTC) for hypothalamic hamartomas, nodular heterotopias and cortical dysplasias: review and perspective. Seizure 41:206–210
29. Wellmer J (2018) Lesion focused radiofrequency thermocoagulation of bottom-of-sulcus focal cortical dysplasia type IIb: conceptional considerations with regard to the epileptogenic zone. Epilepsy Res 142:143–148
30. Ye S, Yang L, Lu Y et al (2021) Contribution of ictal source imaging for localizing seizure onset zone in patients with focal epilepsy. Neurology 96:e366–e375

Chapter 2
Which is the Cognitive Status of Patients with Epilepsy Undergoing Intracranial Presurgical Studies, and How is This Affected by Antiepileptic Drugs?

Philip Grewe and Christian G. Bien

Abstract Patients with drug-resistant focal epilepsy undergoing presurgical assessment may be cognitively impaired. Quality and intensity of these impairments differ inter- and even intraindividually. They may depend on patient variables (epileptogenic lesion, electric ictal and interictal activity, age and epilepsy duration) and iatrogenic interventions (anti-seizure medication and electrode implantation). This chapter will both summarize the most frequent cognitive impairments found in presurgical patients with epilepsy and illustrate factors influencing their cognitive status. A special focus will be laid on how antiepileptic drugs alter patients' cognitive functioning. Finally, evidence on how these drugs may alter the signal of the intracranial EEG recording is reviewed. At the same time, we will discuss some practical implications based on this chapter's clinical neuropsychological perspective for recording and interpreting experimental data in patients with epilepsy undergoing intracranial EEG assessment.

2.1 Introduction

Cognitive impairments are among the most frequent symptoms accompanying chronic focal epilepsy. However, the spectrum of cognitive impairment may severely vary between patients, with some patients experiencing global cognitive impairment and some patients not suffering from any cognitive impairment at all. Such different cognitive phenotypes could be demonstrated applying a cluster-analytical approach on a comprehensive battery of different neuropsychological measures in a rather large cohort of 185 presurgical patients with temporal lobe epilepsy (TLE)—the

P. Grewe (✉)
Medical School OWL, Clinical Neuropsychology and Epilepsy Research, Bielefeld University, Bielefeld, Germany
e-mail: philip.grewe@uni-bielefeld.de

P. Grewe · C. G. Bien
Medical School OWL, Department of Epileptology (Krankenhaus Mara), Bielefeld University, Bielefeld, Germany
e-mail: christian.bien@mara.de

N. Axmacher (ed.), *Intracranial EEG*, Studies in Neuroscience, Psychology and Behavioral Economics, https://doi.org/10.1007/978-3-031-20910-9_2

Table 2.1 Factors with a potential influence on the cognitive status of patients with epilepsy

Epileptic activity	Medication	Brain lesion	Clinical-demographical	Psychosocial and psychiatric
Epileptic seizures (frequency, type, propagation)	Type of antiepileptic drug	Localization	Age at epilepsy onset	Cognitive reserve capacity/IQ
Interictal epileptoform discharges	Number of antiepileptic drugs	Lateralization	Duration of disease	Psychiatric comorbidities
	Dosage	Etiology	Language lateralization	
	Titration time	Prior brain surgery		
	Blood levels			

patient group most frequently presenting in presurgical settings. Whereas 21% of patients suffered from a global cognitive impairment comprising several cognitive domains, 49% of patients had selective cognitive impairments limited to single cognitive domains only; the remaining 30% of patients, though, belonged to a cluster characterized by no cognitive impairment [1]. What is more, the cognitive status of patients with epilepsy (PWE) may not always be persistent, but may fluctuate intraindividually due to potentially reversible factors such as the seizure status or antiepileptic medication. Given these heterogeneous inter- and intraindividual cognitive profiles in presurgical PWE, this chapter will both review cognitive domains potentially affected in presurgical PWE and the factors known to affect the cognitive status in individual patients most relevant to the presurgical setting (Table 2.1).

2.2 Cognitive Domains Frequently Affected in Presurgical Patients with Epilepsy

2.2.1 Memory

Among different cognitive domains, memory processes are prevalently investigated in iEEG research [2, 3]. At the same time, memory functions are most frequently affected in PWE which is mainly based on the temporal lobe's prominent role for memory processing [4, 5] and the high prevalence of TLE within the group of focal epilepsies. This especially applies to PWE undergoing presurgical evaluation due to high proportions of patients with TLE in this cohort in general [6, 7] and in the subgroup of patients undergoing invasive EEG recordings, in particular [8, 9]. When estimating the prevalence of memory disturbances in patients with TLE, numbers between 25–68% have been reported [10–12]. Variance between the numbers reported in different studies are likely to result from differences in neuropsychological measures (i.e., choice of memory measure), definitions of a "neuropsychological impairment" (e.g., scores <1 SD vs. <2 SD below average), and in sampling biases due to monocentric studies (e.g., cohorts from surgically active specialized epilepsy centres vs. cohorts from less-specialized centres). In spite of these different rates due to methodological differences between studies, complaints about memory troubles are the most frequently reported impairment among PWE [13] which underlines the preponderance of disturbed memory functions in this patient group.

As "memory" is a rather broad theoretical construct covering different mnemonic processes and functions [5, 14], one may ask which kind of memory disturbances are most prominent in PWE. For patients with TLE, *anterograde episodic memory functions* are predominantly affected. This is reflected by reduced performance in the acquisition of novel information (e.g., word-lists, design patterns, prose text, item-item-associations) and, consequently, reduced performance is most typically found for test parameters of consolidation and free recall.

While the majority of studies on memory and epilepsy stems from patients with TLE, there is an increasing number of studies investigating mnemonic functions in patients with frontal lobe epilepsy (FLE), the second-most frequent type of epilepsy presenting in presurgical cohorts of PWE. Beside reports of reduced working memory in FLE [15], there is accumulating evidence that anterograde, long-term memory, too, may be disturbed in FLE [16]. Due to the relatively low prevalence of FLE—as compared to TLE—prevalence rates of long-term memory disturbances are hard to estimate. A recent analysis among our own presurgical cohort found that 15–30% of patients with FLE presented with disturbed anterograde, verbal memory functions [12].

In excess to earlier studies focusing on anterograde memory functions, more recent studies suggest that *retrograde memory* might also be affected in PWE [17]. Most prominently, retrograde memory disturbances may manifest as reduced recall of episodic autobiographical memories, whereas recall of personal semantics is mostly unaffected or at least to a far lesser degree then personal episodes [18, 19]. Measures of episodic-autobiographical memory usually are not included in standard presurgical neuropsychological test batteries [20], making it hard to estimate the prevalence of disturbances in this function among presurgical PWE. Still, it may be supposed that episodic-autobiographical memory may be affected in a substantial number of presurgical PWE. This assumption is based on recent studies demonstrating reduced episodic-autobiographical memory recall not only in patients with TLE but with FLE as well [21]. In line with this result, our own and others' clinical experience [21] show that PWE often complain about reduced episodic-autobiographical recall, underlining the hypothesis that this type of memory disturbance may be present in a higher number of PWE than expected so far.

In addition to the aforementioned anterograde and retrograde memory disturbances, a specific deficit in long-term memory consolidation—coined "*accelerated long-term forgetting*" [22]—may be found among PWE. In this type of memory dysfunction, patients perform regular on standard clinical tests of memory functions, which usually probe memory recall with delays between 30 min and a couple of hours. Reduced recall performance can, however, be found in these patients when testing them again after longer memory delays (e.g., days or weeks). Accelerated long-term forgetting has most frequently been reported for patients with TLE and hippocampal lesions although it may occasionally also be present in TLE without structural hippocampal lesions or even extratemporal lobe epilepsy [23]. Currently, it is under debate whether such alterations in long-term consolidation as seen in accelerated long-term forgetting may eventually result in the aforementioned retrograde memory disturbances or whether these are distinct phenomena co-occurring in some PWE [24].

2.2.2 Language/Naming

Profound "classical" aphasic syndromes (such as Broca's or Wernicke's aphasia) are only rarely seen in presurgical PWE. Still, two domains of language functions are of critical interest among these patients. First, a number of studies found naming abilities to be impaired in PWE. Reduced naming abilities—usually detected by tests of confrontational naming—affect patients with TLE in particular, given the involvement of the anterior temporal lobe in object naming. Some studies propose that using measures of category-specific naming (i.e., famous faces or animals) will reveal more subtle naming deficits in TLE [25, 26]. In accordance with studies demonstrating a close link between reduced object-naming and semantic memory failures [27], these studies suggest that in addition to episodic memory, semantic memory networks may be disturbed in TLE—even though deficits may be more subtle.

A second vulnerable subfunction of language in PWE is verbal fluency. Patients with both TLE and FLE may present with impaired verbal fluency although impairments in patients with FLE may be more frequent and more pronounced than those seen in TLE [28]. It should be noted that verbal fluency to a high degree depends on executive functions, such as divergent thinking, or cognitive flexibility and therefore is occasionally taken as a neuropsychological marker of executive functions rather than language. Indeed, some researchers demonstrated distinct neurocognitive mechanisms underlying verbal fluency deficits in TLE versus FLE [29]. This suggests that verbal fluency, and semantic fluency in particular, can be used both as a marker of semantic processing and executive functions. Notwithstanding this distinction, measures of verbal fluency are a critical neuropsychological measure in PWE as they significantly contribute to the patients' subjective estimation of their own cognitive functioning [30]. As a final note, if iEEG-studies aim to investigate specific language-related aspects, it is advisory to consider adverse cognitive side effects of the antiepileptic medication as some substances (i.e., Topiramate and Zonisamide) may specifically affect this domain [31] (see below, section "antiepileptic drugs").

2.2.3 Executive Functions

Given the frontal lobes' prominent role for executive functions [32], impairments in executive functions are frequently observed in patients with FLE. Several studies investigating cognitive profiles of patients with FLE reported reduced performance in numerous executive domains such as planning, response inhibition, motor coordination, verbal fluency, cognitive estimation (i.e., generating a reasonable answer when a specific solution is not available [33]), and cognitive flexibility [34]. In addition to impairments of executive functions in patients with FLE, these impairments may also be observed in patients with TLE. However, it still appears that executive functions are more severely affected in patients with FLE as compared to patients with TLE. Although impairments in specific executive functions (i.e., task switching) were detected in a comparable proportion of patients with FLE and TLE (FLE: 27% vs. TLE: 29%), the impairment found in the FLE group was significantly more pronounced [12]. Likewise, according to a recent review [28], several studies reported significantly lower scores for patients with FLE as compared to TLE on numerous measures of executive functions (i.e., verbal and nonverbal fluency, planning, working memory, interference, task-switching, and problem solving). A certain amount of executive dysfunctions in TLE is assumed to be caused by nociferous effects of epileptic discharges propagating from temporal to interconnected frontal sites (see also Chap. 3). Reports of postoperative improvements in executive functions in TLE patients who became seizure free after temporal lobe resections support this hypothesis [35]. In addition, executive dysfunctions in TLE may reflect additional structural abnormalities, which go beyond the temporal lobe (see below, section "epileptogenic lesion"). For patients with TLE suffering from global cognitive impairment (including executive dysfunctions), structural MRI studies revealed more widespread, extratemporal grey and white matter abnormalities as compared to patients with memory impairments only or patients with minimal cognitive impairment [36]. Irrespective of the type of epilepsy, impaired executive functions may stem from adverse cognitive side effects caused by antiepileptic drugs (AEDs), given that executive functions are among the functions most frequently affected by AEDs (see below). Please note that findings about verbal fluency in PWE are discussed in detail in the language section (see above).

2.2.4 Attention

Reduced processing and psychomotor speed is a common findings among PWE [37]. As prerequisite for numerous cognitive functions, processing speed mediates the association between IQ and other cognitive functions [37] and thus significantly contributes to subjective memory complaints in PWE [38]. As reaction time and motor speed will affect performance in many experimental iEEG paradigms, measures of attention will bear valuable information for patient selection and for interpreting the results in iEEG research. In our own presurgical cohort of patients with TLE and FLE, we found about 10% of patients presenting with impaired psychomotor speed measured by a line tracing task [12]. Presumably, a large part of these reductions in psychomotor and processing speed is caused by the antiepileptic medication as many AEDs are known to affect this cognitive domain (see below). Nevertheless, it should be noted that reduced psychomotor speed was even found in drug-naïve newly diagnosed PWE [39] suggesting an additional role of disease-related variables in excess to AEDs underlying reduced attentional dysfunctions in PWE. Further attentional subfunctions (e.g., divided or selected attention) are studied less systematically in presurgical PWE.

2.2.5 Intelligence

The level of general intelligence may be of particular importance for patient selection in the setting of experimental iEEG research. Looking at the two most common patient groups undergoing presurgical assessment, somewhat lowered IQ scores were reported both for patients with TLE [40, 41] and FLE [15, 34] with no group differences when directly comparing scores between TLE and FLE [42]. Although one also has to underline that average to above-average IQ scores are not exceptional among PWE, the distribution of IQ scores in PWE is somewhat skewed to the lower end of the scale [43]. IQ scores in PWE occasionally may underestimate a patient's actual general intellectual level due to negative impact of related cognitive processes (e.g., working memory, processing speed) on IQ. This especially applies to patients with FLE or patients with AED-related cognitive adverse effects (see below), in which diminished processing speed and executive dysfunctions may negatively influence IQ scores [44]. Judging a patient's suitability for participation in an iEEG experiment based on her/his intelligence, it is recommended to not only inspect global IQ scores, but also to incorporate additional pieces of information such as the educational history, or scores of attentional and executive functions.

2.3 Variables Affecting the Cognitive Status in Presurgical Patients with Epilepsy

2.3.1 Patient Variables

2.3.1.1 Epileptogenic Lesion

Characteristics of the suspected epileptogenic lesion may most significantly influence the cognitive status in presurgical PWE. Among these lesion-related factors, the *localization* and *extent* may critically mediate the association between an epileptic lesion and cognitive disturbances. Type and severity of cognitive dysfunctions in principle follow the relationship between cognitive functions and cortical brain lesions as can be expected from functional neuroanatomy. As one example, patients with epilepsy caused by frontal lobe lesions frequently exhibit executive dysfunctions, whereas executive dysfunctions are impaired less frequently and to a lesser degree in patients with epilepsy caused by temporal lobe lesions [see above, section "executive functions"; 12]. Analogously, TLE following medial temporal lobe lesions, frequently causes mnemonic disturbances with anterograde memory functions being most frequently affected (see above, section "memory"), while memory is significantly less impaired in TLE following neocortical, lateral lesions not affecting the medial temporal lobe structures [45].

When evaluating the effects of an epileptogenic lesion's localization on a patient's cognitive performance, it is recommended to additionally take into account information on the lesion's *lateralization.* A concept closely connected to the lateralization of an epileptogenic lesion, patients' individual *language lateralization* will critically mediate the association between cognitive functions and the lateralization of a structural lesion. This is of particular relevance for the iEEG researcher because significantly increased rates of atypical language representations (i.e., bilateral or right-hemispherical) can be found among presurgical samples of PWE [46]. Thus, in presurgical settings, a classification of lesions and electrode implantation sites into "language-dominant versus non-dominant" might be more adequate than into "right versus left". We hence recommend to additionally taking into account measures of the patient's individual language lateralization when estimating potential negative effects of an epileptic lesion on a specific cognitive task in the context of iEEG-research. Language lateralization is typically measured by functional magnetic resonance imaging (fMRI), magnetoencephalography (MEG), Wada test, or functional transcranial Doppler sonography (fTCD) in the clinical context [20, 47]. During the Wada test, basic language functions are tested while temporarily deactivating the ipsilateral hemisphere with an anesthetic injected into the left versus the right internal carotid artery [48]. The aim of this technique is to determine the language capacity of each hemisphere independently of the contralateral, temporarily deactivated hemisphere. Determining language lateralization with fTCD is based on event-related changes in blood flow velocities of the left versus right middle cerebral arteries during basic language tasks such as verbal fluency [49].

The influence of an epileptogenic lesion's lateralization and localization on a patient's cognitive performance is best exemplified for memory disturbances following unilateral mesial temporal lobe lesions. Following a classical view, anterograde memory disturbances in TLE follow the principle of *material-specificity* assuming a hemispheric specialization for mnemonic processing in the temporal lobes with respect to different stimulus material [50]. This proposal assumes that verbal memory would be diminished in left (or language-dominant, respectively) TLE, whereas nonverbal memory would be affected in right TLE. Material-specific patterns of memory disturbances have yielded a high specificity for left versus right TLE, although clear cases of material-specific memory disturbances were not very frequent [51]. This low prevalence of clear material-specific memory disturbances may—among other factors—be explained by different cognitive strategies employed by patients during memory testing (e.g., verbalizing of visual stimuli or visualization of verbal stimuli) [52]. In addition, cognitive disturbances in patients with TLE may not only be caused by the epileptogenic lesion itself and disturbances in non-memory functions may also negatively influence memory functions, which complicate clear attribution of memory disturbances to an epileptogenic lesion [53]. Overall, while the specificity of verbal memory disturbances for *left* TLE has been frequently replicated across studies, specific impairment of nonverbal memory in *right* TLE could not consistently be demonstrated [54]. Refining the "classical" material-specific view of verbal versus non-verbal memory disturbances in left- versus right-sided TLE, more recent research suggests not only a role of test material (i.e., verbal vs. non-verbal), but also warrant consideration of task-specific mnemonic (sub-)processes, such as associative item binding [55–57], or the distinction between learning, recall, and recognition [58] when assessing memory functions in patients with TLE [59].

Considering the impact of the laterality of an epileptic lesion on presurgical performance, there is accumulating evidence that—at least for TLE with unilateral hippocampal sclerosis—left versus right epileptogenic lesions may differentially affect brain network alterations and cognition. Several studies have reported left-sided TLE—as compared to right-sided TLE—to be associated with more pronounced and widespread reductions in connectivity of functional networks [60–62], which may be associated with reduced performance in several cognitive domains [63, 64].

In contrast to the vast number of studies on the impact of the laterality of epileptogenic lesions in patients with TLE, results on these effects in extratemporal lobe lesions remain sparse. The few studies available have yielded mixed findings [12, 28, 65] presumably due to more heterogeneous patient cohorts and neuropsychological-methodological inconsistencies across studies on extratemporal lobe epilepsy.

Finally, the type of the epileptic lesions may be another lesion-related factor relevant to the cognitive status in presurgical PWE. Earlier studies suggested that the assumed type of epileptic lesion would not have a vast effect on the cognitive status [66, 67]. Yet, more recent studies suggest that hippocampal sclerosis may more severely affect memory and further cognitive domains than other types of lesions [68] even when these lesions affect the medial temporal lobe such as medial temporal tumors [69]. In a more recent study, Phuong et al. [45] reported analogous findings demonstrating that in patients with hippocampal sclerosis, memory and executive functions were more strongly affected than in patients with epilepsy due to medial temporal lobe foci other than hippocampal sclerosis. Interestingly, these results were even obtained after controlling for several potentially confounding factors (e.g., age, disease duration, lateralization of epileptic focus, seizure frequency, or antiepileptic medication). These results are in accordance with findings of more widespread structural–functional network alterations in TLE with hippocampal sclerosis as compared to medial TLE with other lesions [70].

As one indication for intracranial recordings in PWE may be the detection of seizure foci in non-lesional patients [71], a substantial proportion of presurgical iEEG patients will present without lesions detectable on MRI. At the same time, cognitive performance in MRI-negative patients has been found to be significantly higher as compared to patients with a structural MRI lesion as reported for presurgical verbal memory in patients with TLE [72, 73]. In contrast to this, a more recent study could not replicate these findings, reporting equally diminished memory performance in patients with TLE with and without structural MRI lesions [19]; however, this study did not solely include presurgical PWE, restricting the comparability between these and the aforementioned results. Thus far, the presence or absence of a structural alteration should be considered when deciding about the inclusion of a given presurgical iEEG patient into a cognitive study, as relatively good cognitive performance may be expected in PWE without structural lesions.

Occasionally, patients undergoing iEEG recordings have already undergone resective brain surgery at an earlier stage of their disease treatment. Prior resections may most significantly add to cognitive impairments [74] caused by the factors outlined in this chapter and may trigger cerebral reorganization. These patients, thus, may be less suitable for participation in cognitive iEEG research unless the research questions specifically aims to investigate cognitive processes after resection of particular brain structures or to address neurophysiological correlates of compensational or reorganizational mechanisms.

2.3.1.2 EEG Activity

The impact of epileptic EEG activity on cognitive functions is most certainly one of the most critical factors, which has to be considered by researchers conducting experiments with iEEG patients. For a detailed overview on this issue, the reader is referred to Chap. 3 in this volume.

2.3.1.3 Age-Related Factors

The cognitive status of presurgical PWE may also be affected by the age at onset of epilepsy and the duration of epilepsy. Although it is challenging to disentangle the effects of the age at epilepsy onset and the duration of epilepsy, an earlier disease onset is generally accepted as a risk factor for more pronounced cognitive impairment. Also, the literature suggests negative effects of longer epilepsy duration on cognitive performance demonstrating a weak but significant association between ongoing seizures and cognitive impairment—especially in patients with frequent and ongoing bilateral tonic–clonic seizures and status epilepticus [75]. Particularly, in presurgical cohorts, patients with TLE with an early age at onset (i.e., <6 years) were found to suffer from more global memory impairment as compared to those with a later onset (i.e., >9 years) of seizures [76]. Likewise, in presurgical patients with TLE, early onset of epilepsy (i.e., age <15 years) was associated with more global impairments affecting both memory and total IQ, whereas later onset was associated with more selective memory impairment only [77]. Although effects of age at onset in patients with extratemporal epilepsy have been less frequently investigated than in patients with TLE, studies on patients with FLE suggest that age at onset and the duration of epilepsy may also affect cognition in this patient group [78].

Undoubtedly, it is also critical to consider patients' chronological age at time of testing as progressing age is negatively associated with cognitive performance in patients with TLE [79, 80]. Results from a large cross-sectional study in patients with TLE suggest that lower memory performance can be attributed to an interference of the epilepsy on the development of memory performance but is not caused by progressive mental decline [68]. Accordingly, patients with TLE may fall under certain cut-offs for cognitive impairment earlier in life as compared to healthy individuals. Still, for older patients (i.e., age >65 years) there is some more recent evidence of potential associations between epilepsy and dementia [81]. Although this patient group of older patients is rarely seen among presurgical iEEG patients, it is still recommended to consider age and age-related variables as potential covariates when analyzing results from iEEG-experiments.

2.3.2 *Iatrogenic Effects*

2.3.2.1 Antiepileptic Drugs

As AEDs aim to reduce (hyper-)synchronic neural discharges underlying an epileptic seizure, it comes with little surprise that AED intake may occasionally affect neurocognitive functions by reducing neural excitability in general [82]. Though, negative effects of AEDs on cognitive performance are reversible in principle for a majority of patients and it is therefore critical to disentangle cognitive disturbances caused by the epilepsy itself (e.g., by a specific brain lesion or by epileptic seizures) and those additionally caused by AEDs [83]. Discussing potential negative effects of AEDs in the context of presurgical evaluation, one of the most critical factors is the type of AED (Table 2.2). In general, new-generation AEDs (e.g., Lamotrigine, Levetiracetam, Oxcarbazepine) are designed to cause fewer cognitive side effects as compared to older-generation AEDs (e.g., Phenytoin, Sodium valproate, Phenobarbital). Among the newer AEDs, however, Topiramate is known to frequently cause adverse cognitive side effects. Cognitive domains most frequently affected by Topiramate are language, executive functions, and processing speed, but memory may also be affected in some cases. Comparable adverse cognitive side effects were reported for Zonisamide even though effects may not be as strong as those seen in Topiramate [31]. Of note, adverse cognitive side effects caused by Topiramate or Zonisamide may not always be subjectively experienced by patients [84, 85]. Of particular interest for experimental testing during the presurgical monitoring, these adverse effects may persist for several days to even weeks after drug discontinuation – which at least in parts may be caused by the long half-life times of these AEDs. Across different types of AEDs, attention/psychomotor speed and executive functions appear to be the cognitive domains most vulnerable to adverse AED effects [86].

Table 2.2 Degree of severity of cognitive side effects of selected, commonly used AEDs [adapted from 87]

Severe	Moderate	Mild
Topiramate	Carbamazepine	Lamotrigine
Zonisamide	Phenytoin	Levetiracetam
Benzodiazepines	Pregabalin	Gabapentin
Phenobarbital	Sodium valproate	Tiagabin
	Oxcarbazepine	Vigabatrin

In excess to an increased risk of adverse cognitive effects caused by the specific type of AED, there are additional factors, which are associated with cognitive AED side effects. Polytherapy (as opposed to monotherapy) may critically affect the cognitive status in PWE. In detail, the number of AEDs was shown to be correlated with the severity of cognitive disturbances with most detrimental effects for polytherapy with more than two AEDs [88]. Apart from the mere effects of the number of AEDs [45], particular AED interactions, which may specifically affect cognition are yet to be investigated. In addition, AED-related cognitive side-effects typically are more pronounced with higher doses and blood serum levels.

Of particular interest for the iEEG researcher, AEDs are regularly reduced for patients undergoing invasive EEG monitoring to increase the likelihood of seizures during this diagnostic phase. This situation can pose a challenge for selecting an appropriate time for testing patients during this phase. On the one hand, temporary AED reductions may increase the frequency of interictal epileptic discharges and epileptic seizures and, presumably, their negative effects on cognitive functioning ([89]; also see Chap. 3). Adverse cognitive side effects related to the intake of specific AEDs, on the other hand, may be reduced during this phase. Still, one has to consider the half-life time of the AEDs administered to the patient and keep in mind that faster titration of AEDs increases the likelihood of adverse cognitive side effects. According to our own experience, best cognitive performance is achieved by avoiding experimental testing on days with (severe or frequent) seizures (particularly on days with electrocortical stimulations for functional mapping) and by avoiding days with abrupt AED tapering [90]. Experiments usually should not start before the second or third day after electrode implantation to circumvent negative effects on cognition caused by fatigue or pain during the immediate postoperative phase. We still underline that the ideal time for experimental testing may depend on the individual patient and her/his clinical features (such as frequency and type of seizure, type and number of AEDs, susceptibility to cognitive side effects, and type of implanted electrodes) and should be decided on with the clinical staff (see also Chap. 4).

Given the influence of AEDs on neural transmission on the one hand, and cognitive functions on the other hand, one might ask in which way AEDs may alter the iEEG signal and how effects of AEDs can be quantified in terms of neurophysiological markers. The existing literature on this issue is sparse. Studies based on surface EEG have typically found an AED-induced decrease in activities >6 Hz and an increase of slow activity <6 Hz [see review in 91]. Yet relatively few studies have studied the impact of AED on the iEEG signal. In a series of studies, Zaveri and colleagues have investigated potential changes in the iEEG signal caused by AED intake. Directly comparing epochs on- versus off-AEDs, the authors interestingly reported a significant *decrease* in cortical excitation after AED tapering; energy measures decreased about 35% at all recorded electrodes and for different AEDs [92]. In addition, these results were extended by showing *decreased* rates of epileptic spikes and iEEG power after AED tapering for delta, beta and—to a lesser degree—gamma frequency bands [93]. While these results offer interesting—though somewhat counterintuitive—initial findings, further studies on the effects of AEDs on different measures of the iEEG-signal are needed to draw final conclusions.

2.3.2.2 Intracranial Electrodes

An issue of particular interest for experimental research in iEEG patients is the question whether electrode implantation may cause additional alterations of cognitive functioning. Of note, there has been a recent study reporting that in patients with bilateral depth electrodes implanted along the longitudinal hippocampal axis, anterograde memory performance significantly decreased when measured immediately after electrode explantation prior to resective surgery—as compared to preimplantation scores [94]. Unfortunately, this study has not compared memory performance before electrode implantation with performance after electrode implantation. Furthermore, the study has been criticized for a number of methodological issues and contradicts earlier reports not showing negative long-term cognitive effects after electrode explantation [95]. Still, these results give rise to an important question about potential effects of implanted electrodes; future research should particularly examine these effects considering both factors such as the number, type, and localization of electrodes, and potential effects of the implantation scheme.

2.4 Conclusion

In spite of the cognitive disturbances in presurgical PWE reviewed here, it should be underlined that focal epilepsy is not per se associated with cognitive disturbances. In the abovementioned study of Elvermann and colleagues [1], no cognitive impairment was found in 30% and only circumscribed cognitive impairment in 49% of presurgical patients with TLE. Given this heterogeneity of different cognitive profiles among presurgical PWE and a number of factors affecting the cognitive status in these patients, we recommend that evaluations of an individual patient's suitability to participate in a certain iEEG paradigm should be based on the results of a formal clinical neuropsychological assessment and additional clinical information provided by the clinical staff. In addition, it may be helpful to provide practice trials of the iEEG paradigm and consider the patient's performance in these trials for evaluating her/his suitability to participate in that specific paradigm. This will be particularly relevant for paradigms, which address cognitive domains not typically included in standard clinical-neuropsychological test batteries (e.g., spatial navigation, or emotional memory). Given the high relevance of memory functions for patients undergoing iEEG recordings, the neuropsychological assessment should cover different aspects of memory functions. In addition, AED-related cognitive side effects should be estimated by evaluating scores of executive and attentional functions (e.g., verbal fluency, processing speed, task-switching). Screening instruments designed to have a high sensitivity in detecting these AED-related cognitive side effects may be an economic option [96]. Discussing individual patients with the clinical staff may answer whether a patient may be suitable for experimental testing at all, help finding the ideal time of testing during iEEG-video monitoring, or estimate the potential impact of a patient's given cognitive disturbance on a certain experimental paradigm. Such careful selection of patients at the time of experimental testing will certainly address some of the major concerns about iEEG research in PWE [as discussed in 2] and ultimately improve the quality of iEEG studies.

References

1. Elverman KH, Resch ZJ, Quasney EE, Sabsevitz DS, Binder JR, Swanson SJ (2019) Temporal lobe epilepsy is associated with distinct cognitive phenotypes. Epilepsy Behav 96:61–68
2. Lachaux J-P, Axmacher N, Mormann F, Halgren E, Crone NE (2012) High-frequency neural activity and human cognition: past, present and possible future of intracranial EEG research. Prog Neurobiol 98(3):279–301
3. Mukamel R, Fried I (2012) Human intracranial recordings and cognitive neuroscience. Annu Rev Psychol 63:511–537
4. Squire LR, Zola-Morgan S (1991) The medial temporal lobe memory system. Science 253(5026):1380–1386
5. Nadel L, Hardt O (2011) Update on memory systems and processes. Neuropsychopharmacology 36(1):251–273
6. Cloppenborg T, May TW, Blümcke I, Grewe P, Hopf LJ, Kalbhenn T, Pfäfflin M, Polster T, Schulz R, Woermann FG (2016) Trends in epilepsy surgery: stable surgical numbers despite increasing presurgical volumes. J Neurol Neurosurg Psychiatry 87(12):1322–1329
7. Bien C, Raabe A, Schramm J, Becker A, Urbach H, Elger C (2013) Trends in presurgical evaluation and surgical treatment of epilepsy at one centre from 1988–2009. J Neurol Neurosurg Psychiatry 84(1):54–61
8. Swarup O, Waxmann A, Chu J, Vogrin S, Lai A, Laing J, Barker J, Seiderer L, Ignatiadis S, Plummer C (2021) Long-term mood, quality of life, and seizure freedom in intracranial EEG epilepsy surgery. Epilepsy Behav 123:108241
9. Kalbhenn T, Cloppenborg T, Coras R, Fauser S, Hagemann A, Omaimen H, Polster T, Yasin H, Woermann FG, Bien CG, Simon M (2021) Stereotactic depth electrode placement surgery in paediatric and adult patients with the neuromate robotic device: accuracy, complications and epileptological results. Seizure 87:81–87
10. Hoppe C, Elger CE, Helmstaedter C (2007) Long-term memory impairment in patients with focal epilepsy. Epilepsia 48(s9):26–29
11. Reyes A, Kaestner E, Ferguson L, Jones JE, Seidenberg M, Barr WB, Busch RM, Hermann BP, McDonald CR (2020) Cognitive phenotypes in temporal lobe epilepsy utilizing data-and clinically driven approaches: moving toward a new taxonomy. Epilepsia 61(6):1211–1220

12. Bremm FJ, Hendriks MP, Bien CG, Grewe P (2019) Pre-and postoperative verbal memory and executive functioning in frontal versus temporal lobe epilepsy. Epilepsy Behav 101:106538
13. Hall KE, Isaac CL, Harris P (2009) Memory complaints in epilepsy: an accurate reflection of memory impairment or an indicator of poor adjustment? A review of the literature. Clin Psychol Rev 29(4):354–367
14. Markowitsch HJ, Staniloiu A (2012) Amnesic disorders. Lancet 380(9851):1429–1440
15. Ljunggren S, Andersson-Roswall L, Rydenhag B, Samuelsson H, Malmgren K (2015) Cognitive outcome two years after frontal lobe resection for epilepsy–a prospective longitudinal study. Seizure 30:50–56
16. Centeno M, Thompson PJ, Koepp MJ, Helmstaedter C, Duncan JS (2010) Memory in frontal lobe epilepsy. Epilepsy Res 91(2–3):123–132
17. McAndrews MP (2012) Remote memory and temporal lobe epilepsy. In: Zeman A, Kapur N, Jones-Gotman M (eds) Epilepsy & memory. Oxford University Press, Oxford, pp 227–243
18. Butler CR, Zeman AZ (2008) Recent insights into the impairment of memory in epilepsy: transient epileptic amnesia, accelerated long-term forgetting and remote memory impairment. Brain 131(9):2243–2263
19. Rayner G, Tailby C, Jackson G, Wilson S (2019) Looking beyond lesions for causes of neuropsychological impairment in epilepsy. Neurology. https://doi.org/10.1212/WNL.0000000000006905
20. Vogt V, Äikiä M, Del Barrio A, Boon P, Borbély C, Bran E, Braun K, Carette E, Clark M, Cross J (2017) Current standards of neuropsychological assessment in epilepsy surgery centers across Europe. Epilepsia 58(3):343–355
21. Rayner G, Jackson GD, Wilson SJ (2015) Behavioral profiles in frontal lobe epilepsy: Autobiographic memory versus mood impairment. Epilepsia 56(2):225–233
22. Jansari AS, Davis K, McGibbon T, Firminger S, Kapur N (2010) When "long-term memory" no longer means "forever": analysis of accelerated long-term forgetting in a patient with temporal lobe epilepsy. Neuropsychologia 48(6):1707–1715
23. Miller LA, Mothakunnel A, Flanagan E, Nikpour A, Thayer Z (2017) Accelerated long term forgetting in patients with focal seizures: incidence rate and contributing factors. Epilepsy Behav 72:108–113
24. Lemesle B, Planton M, Pagès B, Pariente J (2017) Accelerated long-term forgetting and autobiographical memory disorders in temporal lobe epilepsy: one entity or two? Rev Neurol (Paris) 173(7–8):498–505
25. Glosser G, Salvucci A, Chiaravalloti N (2003) Naming and recognizing famous faces in temporal lobe epilepsy. Neurology 61(1):81–86
26. Drane DL, Ojemann GA, Aylward E, Ojemann JG, Johnson LC, Silbergeld DL, Miller JW, Tranel D (2008) Category-specific naming and recognition deficits in temporal lobe epilepsy surgical patients. Neuropsychologia 46(5):1242–1255
27. Bell BD, Hermann BP, Woodard AR, Jones JE, Rutecki PA, Sheth R, Dow CC, Seidenberg M (2001) Object naming and semantic knowledge in temporal lobe epilepsy. Neuropsychology 15(4):434
28. Patrikelis P, Gatzonis S, Siatouni A, Angelopoulos E, Konstantakopoulos G, Takousi M, Sakas DE, Zalonis I (2016) Preoperative neuropsychological presentation of patients with refractory frontal lobe epilepsy. Acta Neurochir (Wien) 158(6):1139–1150
29. Drane DL, Lee GP, Cech H, Huthwaite JS, Ojemann GA, Ojemann JG, Loring DW, Meador KJ (2006) Structured cueing on a semantic fluency task differentiates patients with temporal versus frontal lobe seizure onset. Epilepsy Behav 9(2):339–344
30. O'Shea MF, Saling MM, Bladin PF, Berkovic SF (1996) Does naming contribute to memory self-report in temporal lobe epilepsy? J Clin Exp Neuropsychol 18(1):98–109
31. Ojemann LM, Ojemann GA, Dodrill CB, Crawford CA, Holmes MD, Dudley DL (2001) Language disturbances as side effects of topiramate and zonisamide therapy. Epilepsy Behav 2(6):579–584
32. Stuss DT (2011) Functions of the frontal lobes: relation to executive functions. J Int Neuropsychol Soc 17(5):759–765

33. Brand M, Fujiwara E, Kalbe E, Steingass HP, Kessler J, Markowitsch HJ (2003) Cognitive estimation and affective judgments in alcoholic Korsakoff patients. J Clin Exp Neuropsychol 25(3):324–334
34. Patrikelis P, Angelakis E, Gatzonis S (2009) Neurocognitive and behavioral functioning in frontal lobe epilepsy: a review. Epilepsy Behav 14(1):19–26
35. Martin RC, Sawrie SM, Edwards R, Roth DL, Faught E, Kuzniecky RI, Morawetz RB, Gilliam FG (2000) Investigation of executive function change following anterior temporal lobectomy: selective normalization of verbal fluency. Neuropsychology 14(4):501
36. Dabbs K, Jones J, Seidenberg M, Hermann B (2009) Neuroanatomical correlates of cognitive phenotypes in temporal lobe epilepsy. Epilepsy Behav 4(15):445–451
37. McMillan TM, Mason CA, Seidenberg M, Jones J, Hermann B (2021) The impact of processing speed on cognition in temporal lobe epilepsy. Epilepsy Behav 122:108203
38. Grewe P, Nikstat A, Koch O, Koch-Stoecker S, Bien CG (2016) Subjective memory complaints in patients with epilepsy: the role of depression, psychological distress, and attentional functions. Epilepsy Res 127:78–86
39. Taylor J, Kolamunnage-Dona R, Marson AG, Smith PE, Aldenkamp AP, Baker GA, Group SS (2010) Patients with epilepsy: cognitively compromised before the start of antiepileptic drug treatment? Epilepsia 51(1):48–56
40. Oyegbile TO, Dow C, Jones J, Bell BD, Rutecki P, Sheth R, Seidenberg M, Hermann BP (2004) The nature and course of neuropsychological morbidity in chronic temporal lobe epilepsy. Neurology 62(10):1736–1742
41. Jokeit H, Ebner A (1999) Long term effects of refractory temporal lobe epilepsy on cognitive abilities: a cross sectional study. J Neurol Neurosurg Psychiatry 67(1):44–50
42. Exner C, Boucsein K, Lange C, Winter H, Weniger G, Steinhoff BJ, Irle E (2002) Neuropsychological performance in frontal lobe epilepsy. Seizure 11(1):20–32
43. Jones-Gotman M, Smith ML, Risse GL, Westerveld M, Swanson SJ, Giovagnoli AR, Lee T, Mader-Joaquim MJ, Piazzini A (2010) The contribution of neuropsychology to diagnostic assessment in epilepsy. Epilepsy Behav 18(1–2):3–12
44. Helmstaedter C, Witt J-A, Hoppe C (2019) Evaluating the mediating role of executive functions for antiepileptic drugs' effects on IQ in children and adolescents with epilepsy. Epilepsy Behav 96:98–103
45. Phuong TH, Houot M, Méré M, Denos M, Samson S, Dupont S (2021) Cognitive impairment in temporal lobe epilepsy: contributions of lesion, localization and lateralization. J Neurol 268(4):1443–1452
46. Springer JA, Binder JR, Hammeke TA, Swanson SJ, Frost JA, Bellgowan PS, Brewer CC, Perry HM, Morris GL, Mueller WM (1999) Language dominance in neurologically normal and epilepsy subjects: a functional MRI study. Brain 122(11):2033–2046
47. Schmid E, Thomschewski A, Taylor A, Zimmermann G, Kirschner M, Kobulashvili T, Brigo F, Rados M, Helmstaedter C, Braun K (2018) Diagnostic accuracy of functional magnetic resonance imaging, Wada test, magnetoencephalography, and functional transcranial Doppler sonography for memory and language outcome after epilepsy surgery: a systematic review. Epilepsia 59(12):2305–2317
48. Wada J, Rasmussen T (1960) Intracarotid injection of sodium amytal for the lateralization of cerebral speech dominance: experimental and clinical observations. J Neurosurg 17(2):266–282
49. Knecht S, Deppe M, Ebner A, Henningsen H, Huber T, Jokeit H, Ringelstein EB (1998) Noninvasive determination of language lateralization by functional transcranial Doppler sonography: a comparison with the Wada test. Stroke 29(1):82–86
50. Milner B (1970) Memory and the medial temporal regions of the brain. In: Pribram K, Broadbent D (eds) Biology of memory, vol 23. Academic Press, New York, pp 29–50
51. Castro LH, Silva LC, Adda CC, Banaskiwitz NH, Xavier AB, Jorge CL, Valerio RM, Nitrini R (2013) Low prevalence but high specificity of material-specific memory impairment in epilepsy associated with hippocampal sclerosis. Epilepsia 54(10):1735–1742
52. Baxendale S (2008) The impact of epilepsy surgery on cognition and behavior. Epilepsy Behav 12(4):592–599

53. Baxendale S, Thompson P (2010) Beyond localization: the role of traditional neuropsychological tests in an age of imaging. Epilepsia 51(11):2225–2230
54. Barr WB, Chelune GJ, Hermann BP, Loring DW, Perrine K, Strauss E, Trenerry MR, Westerveld M (1997) The use of figural reproduction tests as measures of nonverbal memory in epilepsy surgery candidates. J Int Neuropsychol Soc 3:435–443
55. Glikmann-Johnston Y, Saling MM, Chen J, Cooper KA, Beare RJ, Reutens DC (2008) Structural and functional correlates of unilateral mesial temporal lobe spatial memory impairment. Brain 131:3006–3018
56. Grewe P, Neu D, Aengenendt J, Woermann FG, Mertens M, Bien CG, Kissler J (2020) Rhinal and hippocampal contributions to spontaneous inter-item binding and verbal memory recall: evidence from temporal lobe epilepsy. Cortex 124:204–216
57. Lillywhite LM, Saling MM, Briellmann RS, Weintrob DL, Pell GS, Jackson GD (2007) Differential contributions of the hippocampus and rhinal cortices to verbal memory in epilepsy. Epilepsy Behav 10(4):553–559
58. Smith ML, Bigel M, Miller LA (2011) Visual paired-associate learning: in search of material-specific effects in adult patients who have undergone temporal lobectomy. Epilepsy Behav 20(2):326–330
59. Saling MM (2009) Verbal memory in mesial temporal lobe epilepsy: beyond material specificity. Brain 132(3):570–582
60. Besson P, Dinkelacker V, Valabregue R, Thivard L, Leclerc X, Baulac M, Sammler D, Colliot O, Lehéricy S, Samson S (2014) Structural connectivity differences in left and right temporal lobe epilepsy. Neuroimage 100:135–144
61. de Campos BM, Coan AC, Lin Yasuda C, Casseb RF, Cendes F (2016) Large-scale brain networks are distinctly affected in right and left mesial temporal lobe epilepsy. Hum Brain Mapp 37(9):3137–3152
62. Roger E, Pichat C, Torlay L, David O, Renard F, Banjac S, Attyé A, Minotti L, Lamalle L, Kahane P (2020) Hubs disruption in mesial temporal lobe epilepsy. A resting-state fMRI study on a language-and-memory network. Hum Brain Mapp 41(3):779–796
63. Ives-Deliperi V, Butler J (2021) Mechanisms of cognitive impairment in temporal lobe epilepsy: a systematic review of resting-state functional connectivity studies. Epilepsy Behav 115:107686
64. Dinkelacker V, Dupont S, Samson S (2016) The new approach to classification of focal epilepsies: epileptic discharge and disconnectivity in relation to cognition. Epilepsy Behav 64:322–328
65. Traianou A, Patrikelis P, Kosmidis MH, Kimiskidis VK, Gatzonis S (2019) The neuropsychological profile of parietal and occipital lobe epilepsy. Epilepsy Behav 94:137–143
66. Upton D, Thompson PJ (1997) Neuropsychological test performance in frontal-lobe epilepsy: the influence of aetiology, seizure type, seizure frequency and duration of disorder. Seizure 6(6):443–447
67. Jokeit H, Schacher M (2004) Neuropsychological aspects of type of epilepsy and etiological factors in adults. Epilepsy Behav 5(Suppl 1):S14-20
68. Helmstaedter C, Elger CE (2009) Chronic temporal lobe epilepsy: a neurodevelopmental or progressively dementing disease? Brain 132(Pt 10):2822–2830
69. York MK, Rettig GM, Grossman RG, Hamilton WJ, Armstrong DD, Levin HS, Mizrahi EM (2003) Seizure control and cognitive outcome after temporal lobectomy: a comparison of classic Ammon's horn sclerosis, atypical mesial temporal sclerosis, and tumoral pathologies. Epilepsia 44(3):387–398
70. Wei W, Zhang Z, Xu Q, Yang F, Sun K, Lu G (2016) More severe extratemporal damages in mesial temporal lobe epilepsy with hippocampal sclerosis than that with other lesions: a multimodality MRI study. Medicine 95(10)
71. Jobst BC, Bartolomei F, Diehl B, Frauscher B, Kahane P, Minotti L, Sharan A, Tardy N, Worrell G, Gotman J (2020) Intracranial EEG in the 21st century. Epilepsy Curr 20(4):180–188
72. Berger J, Plotkin M, Demin K, Holtkamp M, Bengner T (2018) The relationship between structural MRI, FDG-PET, and memory in temporal lobe epilepsy: preliminary results. Epilepsy Behav 80:61–67

73. Helmstaedter C, Petzold I, Bien CG (2011) The cognitive consequence of resecting nonlesional tissues in epilepsy surgery—results from MRI-and histopathology-negative patients with temporal lobe epilepsy. Epilepsia 52(8):1402–1408
74. Sherman EMS, Wiebe S, Fay-McClymont TB, Tellez-Zenteno J, Metcalfe A, Hernandez-Ronquillo L, Hader WJ, Jetté N (2011) Neuropsychological outcomes after epilepsy surgery: systematic review and pooled estimates. Epilepsia 52(5):857–869
75. Dodrill CB (2004) Neuropsychological effects of seizures. Epilepsy Behav 5(Suppl 1):S21-24
76. Lespinet V, Bresson C, N'Kaoua B, Rougier A, Claverie B (2002) Effect of age of onset of temporal lobe epilepsy on the severity and the nature of preoperative memory deficits. Neuropsychologia 40(9):1591–1600
77. Kaaden S, Helmstaedter C (2009) Age at onset of epilepsy as a determinant of intellectual impairment in temporal lobe epilepsy. Epilepsy Behav 15(2):213–217
78. Verche E, San Luis C, Hernández S (2018) Neuropsychology of frontal lobe epilepsy in children and adults: systematic review and meta-analysis. Epilepsy Behav 88:15–20
79. Samson S, Moncomble C, Méré M, Vasseur R, Dupont S (2020) Getting older with chronic temporal lobe epilepsy: what memory profile? Rev Neurol (Paris) 176(6):439–443
80. Seidenberg M, Pulsipher DT, Hermann B (2007) Cognitive progression in epilepsy. Neuropsychol Rev 17(4):445–454
81. Sen A, Capelli V, Husain M (2018) Cognition and dementia in older patients with epilepsy. Brain 141(6):1592–1608
82. Drane DL (2015) Neuropsychological evaluation of the epilepsy surgical candidate. In: Barr WB, Morrison C (eds) Handbook on the neuropsychology of epilepsy. Springer, New York, pp 87–121
83. Witt J-A, Elger C, Helmstaedter C (2013) Impaired verbal fluency under topiramate–evidence for synergistic negative effects of epilepsy, topiramate, and polytherapy. Eur J Neurol 20(1):130–137
84. Lee S, Sziklas V, Andermann F, Farnham S, Risse G, Gustafson M, Gates J, Penovich P, Al-Asmi A, Dubeau F (2003) The effects of adjunctive topiramate on cognitive function in patients with epilepsy. Epilepsia 44(3):339–347
85. Meschede C, Witt JA, Brömling S, Moskau-Hartmann S, Rademacher M, Surges R, von Wrede R, Helmstaedter C (2020) Changes in cognition after introduction or withdrawal of zonisamide versus topiramate in epilepsy patients: a retrospective study using Bayes statistics. Epilepsia 61(7):1481–1490
86. Witt J-A, Helmstaedter C (2013) Monitoring the cognitive effects of antiepileptic pharmacotherapy—approaching the individual patient. Epilepsy Behav 26(3):450–456
87. Leeman BA, Meador KJ (2015) Cognitive effects of epilepsy therapies. In: St. Louis EK, Ficker DM, O'Brien TJ (eds) Epilepsy and the interictal state: co-morbidities and quality of life. Wiley, Chichester, UK, pp 74–87
88. Witt J-A, Elger CE, Helmstaedter C (2015) Adverse cognitive effects of antiepileptic pharmacotherapy: each additional drug matters. Eur Neuropsychopharmacol 25(11):1954–1959
89. Kleen JK, Scott RC, Holmes GL, Roberts DW, Rundle MM, Testorf M, Lenck-Santini P-P, Jobst BC (2013) Hippocampal interictal epileptiform activity disrupts cognition in humans. Neurology 81(1):18–24
90. Höller Y, Höhn C, Schwimmbeck F, Plancher G, Trinka E (2020) Effects of antiepileptic drug tapering on episodic memory as measured by virtual reality tests. Front Neurol 11
91. Höller Y, Helmstaedter C, Lehnertz K (2018) Quantitative pharmaco-electroencephalography in antiepileptic drug research. CNS Drugs 32(9):839–848
92. Zaveri HP, Pincus SM, Goncharova II, Novotny EJ, Duckrow RB, Spencer DD, Spencer SS (2009) A decrease in EEG energy accompanies anti-epileptic drug taper during intracranial monitoring. Epilepsy Res 86(2–3):153–162
93. Zaveri HP, Pincus SM, Goncharova II, Novotny EJ, Duckrow RB, Spencer DD, Blumenfeld H, Spencer SS (2010) Background intracranial EEG spectral changes with anti-epileptic drug taper. Clin Neurophysiol 121(3):311–317

94. Helmstaedter C, Gielen GH, Witt JA (2018) The immediate and short-term effects of bilateral intrahippocampal depth electrodes on verbal memory. Epilepsia 00:1–7
95. Meador KJ, Halpern CH, Hermann BP (2018) Cognitive safety of intracranial electrodes for epilepsy. Epilepsia 59(6):1132–1137
96. Lutz M, Helmstaedter C (2005) EpiTrack: tracking cognitive side effects of medication on attention and executive functions in patients with epilepsy. Epilepsy Behav 7(4):708–714

Chapter 3
(How) Does Epileptic Activity Influence Cognitive Functioning?

Linglin Yang and Shuang Wang

Abstract In patients with epilepsy, comorbid cognitive impairments are common and strongly related to both ictal and interictal epileptic activities. More specifically, the exact domains of cognitive impairment depend on the location and timing of epileptic activities, while the severity of cognitive impairment is related to the incidence, spatial extent and electrophysiological characteristics (including amplitude, duration and rhythmicity) of epileptic activities. Recent studies revealed that epileptic activities impart the deleterious influence on cognitive function through disturbance of interwoven cognitive networks. This chapter starts with a short review of cognitive networks and their hierarchical organization, highlighting the fundamental role of partial synchrony and the mediating role of subcortical regions. Subsequently, we discuss transient cognitive impairments through abrupt disruptions of network functions, as well as persistent cognitive decline via chronic remodeling of neural circuits. We emphasize the particular value of intracranial electroencephalography and brain stimulation for understanding the crucial roles of cortical-subcortical neural circuits in cognitive dysfunction.

3.1 Introduction

Although some epilepsy patients self-report normal cognitive function, detailed evaluations of various cognitive domains reveals that cognitive dysfunction is a common symptom among them, which seriously affects their psychosocial functioning and quality of life. Apart from the etiology of epilepsy, several factors including disease course, adverse effects of antiseizure medication (ASM), possible surgical resection,

---

L. Yang (✉)
Department of Psychiatry, The Second Affiliated Hospital, Zhejiang University School of Medicine, Hangzhou, China
e-mail: yang_linglin@163.com

S. Wang
Department of Neurology, Epilepsy Center, The Second Affiliated Hospital, Zhejiang University School of Medicine, Hangzhou, China

N. Axmacher (ed.), *Intracranial EEG*, Studies in Neuroscience, Psychology and Behavioral Economics, https://doi.org/10.1007/978-3-031-20910-9_3

and epileptic activities impart deleterious effects on cognitive function. Various functional domains can be involved, including attention, perception, memory, language, information processing speed and response times to stimuli [1, 2].

Obviously, more frequent and extensive epileptic activities lead to more severe cognitive decline [3–8]. Of note, various types of epileptic activities appear to exert different effects on cognitive functioning. The slowing effect of epileptic activity on reaction time was most prominent during classical spike discharges and weakest during sharp theta activity [5]. However, the effect of polyspike on cognition need further exploration. Accumulating electrophysiological evidence suggests that epileptic activities with larger amplitude and longer duration are more disruptive and correlate with more serious cognitive impairments [3, 9–11]. Focal interictal epileptiform discharges (IEDs), even infrequent, may result in transient cognitive impairment in epilepsy patients [11–15]. However, several earlier studies observed no obvious relationship between focal IEDs and cognitive impairment [4, 16]. A compelling explanation of this discrepancy is that the exact influence of epileptic activities on cognition depends on the specific location and time they occurred [12, 13, 15, 17–20]. That is, epileptic activities located in eloquent regions induce prominent and regionally-specific cognitive impairment. For example, occipital epileptic activities disturb visual perception [18]. Epileptic activities in the left middle temporal gyrus impair memory encoding during a delayed verbal free-recall memory task [11]. Surprisingly, hippocampal epileptic activities disrupt memory maintenance and retrieval rather than encoding during working memory tasks, which might be due to "buffer" functions of primary sensory cortex and prefrontal cortex during working memory encoding (described briefly in the next section) [13, 21]. Furthermore, epileptic activities with white matter propagation were associated with poorer cognitive performance [11], probably due to the involvement of a broader cognitive network [22, 23]. Another study reported that IEDs during sleep, especially non-rapid eye movement periods, were associated with visuospatial and memory impairments, while IEDs during waking state impaired attention [14]. Interestingly, the cognitive performance deteriorates with the onset of epileptic activities and recovered at the end of epileptic activities [24, 25]. These studies provide strong evidence that short-term cognitive deficits are related to transient disruption of network functions by epileptic activities [12, 13, 15, 18–20].

In addition to these transient effects, sustained cognitive deficits due to epileptic activities have been reported, especially in the developing brain [1, 26–28]. Thus, a hypothesis has been proposed that epileptic activities may result in long-term cognitive deficits through chronic remodeling of cognition-related neural circuits. Due to multiple confounding factors such as ASM administration, education, and comorbidity with other diseases, there are still insufficient data to clearly understand the long-term effects of epileptic activities on cognition [29]. Nevertheless, it has been widely documented that epileptic activities can progressively remodel neural networks including cognitive networks [30–34].

Up to now, the question remains as to how epileptic activities influence cognitive functioning. This chapter briefly reviews the interactions among cognition-related networks in the healthy brain, and then explores the deleterious influence of epileptic activities on cognitive function through transient disruptions of network functioning and chronic remodeling of neuronal circuits, respectively. By reviewing the present work, we hope to begin a journey toward the intervention of neural stimulation in patients with cognitive decline.

3.2 A Quick Look at Cognitive Networks

In the past decades, emerging neuroimaging and electrophysiological evidence demonstrated the existence and functional role of various cognitive subnetworks [35–40]. The most prominent of these networks is the default mode network (DMN) comprising the medial prefrontal cortex (mPFC), posterior cingulate cortex, precuneus and temporal-parietal junction [35, 41]. The DMN is known as an 'intrinsic' or 'task-negative' network that reflects internally oriented cognition and is deactivated during tasks driven by external stimuli [35, 38]. Conversely, there are various 'task-positive' networks. The central executive network (CEN), involving the dorsolateral prefrontal cortex and the posterior parietal cortex, is externally oriented and responsible for executive functions such as the allocation of attention, working memory, inhibitory control and decision-making [37, 38, 42]. The salience network (SN), involving the fronto-insular cortex and the dorsal anterior cingulate cortex, contributes to switching between CEN and DMN depending on task demands and subjective salience [36–38]. In addition to these three networks, the dorsal and ventral attention networks as well as the limbic system exert crucial effects on cognitive functioning [39, 40].

Intriguingly, variable spatiotemporal patterns of partial synchrony can be observed among these interleaved cognitive subnetworks during distinct cognitive tasks [43–45]. For example, the medial temporal lobe structures including the hippocampus play a vital role in memory processes via interacting with distributed areas in the neocortex: Sensory memories are initially formed in the primary sensory cortex, and then re-encoded into long-term memories by the hippocampus, entorhinal cortex, and mPFC for further consolidation, storage and retrieval [21]. Theta-band coherence in an anatomically widespread network involving frontal, temporal and parietal lobes is particularly increased when episodic memory is engaged [46]. Using subject-specific brain network models including 76 key nodes, a recent in silico experiment explored how regional brain stimulation can produce dynamically integrated patterns among nine cognitive subnetworks [44]. All cognitive subnetworks produced a so called "chimera states" which is characterized as a partially synchronized state of some of the network nodes. The chimera state contributes to the dynamic and flexible organization of large-scale networks, supporting diverse cognitive demands. In fact, individual cognitive subnetworks may be more or less specific for a given cognitive process: While auditory and visual sensory systems are mainly responsible for perceiving external stimuli, the dorsal and ventral attention networks are conjointly active with other cognitive subnetworks [44, 47], because the allocation of attention is relevant for almost all cognitive processes.

Indeed, an intricate balance between the dynamic segregation and integration of cognitive processing cannot be divorced from a robust hierarchical network architecture [44, 48, 49]. This is especially true for subcortical structures such as the thalamus [50, 51]. The thalamus is divided into multiple distinct nuclei including the anterior nucleus, the medial dorsal nucleus, the reticular nucleus, the pulvinar and the geniculate body. It is reciprocally and topographically connected to distributed cortical regions, basal ganglia, brainstem and cerebellum, participating in various cognitive processes [52]. Therefore, the thalamus has been traditionally conceptualized as a high-fidelity information relay [53]. Recent evidence has highlighted the critical function of the thalamus in filtering goal-relevant information and coordinating cognitive networks [52, 54, 55]. With abundant diverse projections to distributed cortical regions, subcortical structures play vital roles in mediating the transition between functionally separated and integrated cognitive networks [49].

In summary, optimal cognitive functioning relies on fine-tuned interactions of spatiotemporally interdependent brain networks [44, 48, 49]. Conversely, cognitive dysfunction has been considered to result from a disturbance of these networks. In the following sections, we will describe accumulating evidence that epilepsy-related alterations in functional as well as structural network connectivity may induce cognitive deficits. Moreover, these abnormal connectivity patterns are dynamic and appear to be related to the recurrence of epileptic activities.

3.3 Transient Disruption of Network Functions

It has been reported that epileptic activities may exert transient deleterious effects on various cognitive functions, including attention, perception, memory, language, and the overall speed of information processing. Since the severity and exact domain of cognitive impairment are closely related to the location and timing of epileptic activities, it is reasonable to consider that these activities directly disturb cognitive functions of eloquent brain regions [12, 13, 15, 18–20]. Moreover, when epileptic activities involve subcortical structures such as the thalamus, abnormalities in cortical-subcortical circuits are bound to result in severe cognitive impairment.

Various studies have shown that cognitive impairments fluctuate with the dynamics of epileptic activities. Cognitive performance begins to decline in the seconds preceding the onset of epileptic activities, and gradually improves towards the end of epileptic activities [5, 56]. A study combining EEG with cognitive tests (FePsy^{R}) has found that generalized IEDs result in apparent cognitive slowing (by approximately 35%) [4]. A study using functional magnetic resonance imaging (fMRI) showed that blood oxygen level dependent (BOLD) signals in the frontal and parietal cortex increase before the occurrence of generalized slow-wave discharges (SWDs) and decrease after the end of SWDs [56], suggesting that epileptic activities cause transient disruptions of cognitive networks. Moreover, a longer duration and larger amplitude of SWDs predicted poorer cognitive performance, suggesting that broader cognitive networks were disturbed [10]. Combining fMRI with analysis of magnetoencephalography (MEG) recordings, Ibrahim et al. found that cognitive impairments and large-scale network alterations preceded and followed IEDs in a group of 26 children with focal epilepsy [25]. Among the cognitive networks, the DMN is most prominently affected by epileptic activities [25, 57, 58]. More specifically, the DMN connectivity is weaker during IEDs compared to IED-free epochs [57, 58]. Another fMRI study observed IED-related attenuation of connectivity between the attentional networks and basal ganglia [59]. According to the 'network inhibition hypothesis', the suppression of DMN, attention networks and other cognitive subnetworks is interpreted as disconnection of functional networks induced by epileptic activities [25, 58]. In patients with focal epilepsy, the connectivity of widespread networks remote from the epileptogenic zone was significantly reduced, indicating global effects of the disease [60].

Besides the 'network inhibition hypothesis' of neocortical deactivation, pathological IEDs may compete with physiological neural oscillations and disturb cognitive processes [61, 62]. In epilepsy patients undergoing intracranial EEG evaluation, hippocampal IEDs during an associative memory task reduce the rate of physiological ripples and lead to marked memory deficits. More specifically, the likelihood of remembering an item was reduced by 6–23% if hippocampal IEDs occurred during memory encoding, and by 15–33% during memory retrieval [62]. These findings are consistent with those found in the rodent kindling model of temporal lobe epilepsy (TLE) [61]. In parallel with increasing rates of IEDs, this study observed lower memory performance and reduced physiological hippocampal ripples during the kindling phase. During the 'recovery' phase, cognitive performance and physiological ripple rate improved gradually. Additionally, hippocampal IEDs generated a cortical 'down' state and subsequently induced mPFC spindles during wakefulness and rapid-eye movement sleep. The pathological hippocampal-mPFC coupling surpassed the physiological ripple-spindle coupling which is an important physiological mechanism of memory consolidation.

The transient effects of epileptic activities on network function and cognition have been widely discussed. Recurrent network disruption induced by epileptic activities may accumulate and lead to chronic remodeling of neuronal circuits [30, 63], which we will discuss in the following section.

3.4 Chronic Remodeling of Neuronal Circuits

During the early stage of epileptogenesis, remodeling of brain connectivity begins with the onset of IEDs before epileptic seizures develop [33]. Brain network remodeling progresses gradually over time, accompanied by aggravating cognitive impairment [30, 31, 32, 34]. Galovic et al. have found pronounced progression of network abnormalities within the first 5 years of epilepsy [64]. It seems that cognitive deterioration is closely associated with the frequency and severity of epileptic activities [63, 65]. More specifically, the cumulative frequency of generalized seizures across lifetime is a strong predictor of cognitive decline [63]. Although cognitive deficits progress with ongoing epileptic activities, their progression can also be stopped and even reversed after successful medical or surgical treatments [65]. The degree and speed of cognitive recovery depends on the recruitment of reserve capacities in eloquent brain regions [65, 66]. Together, these studies indicate that epileptic activities exert critical effects on chronic remodeling of neuronal circuits.

Although regional atrophy of the ipsilateral hippocampus and thalamus has been well documented in TLE with mesial temporal sclerosis [67–69], enhanced ipsilateral hippocampal-thalamic connections were found at both functional and anatomical levels [70, 71], in line with the critical role of the thalamus in seizure propagation. Interestingly, however, the increased connectivity between the hippocampus and the thalamus was associated with poor cognitive performance, in particular regarding executive functioning. This suggests an abnormal and dysfunctional reorganization of hippocampal-thalamic connectivity. In the course of chronic remodeling, the cingulate midline network, comprising the cingulate cortex, bilateral thalamus and precuneus, became more synchronized as well [31]. In addition, an increased connectivity between the hippocampus and other subcortical regions, including the insula, basal ganglia, brainstem and cerebellum, has been observed [70, 72]. As described in the first section 'A Quick Look at Cognitive Networks', the hippocampus, thalamus, cingulate cortex, insula and basal ganglia are all key nodes of different cognitive subnetworks, and the abnormal synchronization among them is likely to disturb the delicate balance of various cognitive subnetworks during different tasks, resulting in cognitive decline [44]. Of note, it is hypothesized that reductions of cortico-subcortical connectivity may constitute an adaptive mechanism preventing seizure propagation. In addition, brain structures with more frequent epileptic activities seem to show more extended reductions of functional connectivity with other regions [73], further supporting the defense mechanism hypothesis. However, the decreased hippocampal-cortical and thalamocortical connectivity in mesial TLE lead to disruption of functional networks and might account for impairments in attention and memory [74–77]. One of the most often impaired cognitive subnetwork in mesial TLE is the DMN, which manifests as decreased BOLD signals in resting-state fMRI studies [32, 70, 78]. Considering the laterality of brain networks, several neuroimaging studies have revealed distinct reorganization patterns of neural circuits in the left versus right TLE [64, 73, 77, 79]. Patients with left TLE show more marked anatomical and functional alterations compared to those with right TLE [64, 73, 79], in accordance with differences in cognitive performance between the two groups [77]. Of interest, the different memory phenotypes are strongly related to the distinct reorganization patterns of brain networks in left versus right mesial TLE [77]: Verbal memory performance was positively correlated with functional connectivity in the anterior hippocampal network in left mesial TLE, while it was found to be negatively correlated with functional connectivity in the posterior hippocampal network in right mesial TLE. A recent fMRI study revealed two different network states in mesial TLE, one that was dominated by the anterior hippocampal network (segregated state) and another one that was dominated by the posterior hippocampal network (integrated state) [80]. In this study, the authors found that memory deficits were related to a less segregated and more integrated state, and the modular alterations of dynamic functional connectivity in the hippocampal networks could help identify the memory phenotypes.

Similarly, both regional and global patterns of reorganization of connectivity have been reported in focal neocortical epilepsy [81, 82]. A resting-state functional connectivity study based on MEG found decreased connectivity in both cortical and subcortical regions, including the anterior thalamus, basal ganglia, temporal-parietal junction and perisylvian cortex [82]. Subsequent analysis of mean imaginary coherence revealed decreased global mean connectivity in the delta, theta, alpha and beta frequency bands [83]. Creating functional connectivity maps that were seeded at the epileptogenetic regions in 23 patients with frontal lobe epilepsy, Luo et al. found patient-specific connectivity patterns characterized via enhanced connectivity in the neighborhood of the seeds and decreased connectivity between the seeds and remote networks [84]. Surprisingly, the reorganization pattern of functional connectivity does not occur instantaneously with the onset of IEDs, implying that the remolding of neural circuits in frontal lobe epilepsy is a gradual and chronic process [84]. Therefore, the reorganization pattern of functional connectivity may depend on structural changes. In frontal lobe epilepsy, accelerated cortical thinning was found in a widespread network involving both frontal and extra-frontal regions [64, 85]. These findings are supported by another diffusion tensor imaging study in frontal lobe epilepsy which showed increased characteristic path length and decreased network strength, global efficiency, and nodal efficiency [86]. Furthermore, reduced task activation independent of epileptic activities has been observed in epilepsy patients [59], reflecting that the chronic remodeling diminished the task-dependent synchrony of cognitive networks [46]. A recent stereotactic EEG study revealed that the absence of theta-band synchronous network might result in failure of information transfer among brain regions and memory deficits in the epilepsy patients [46].

Chronic brain network remodeling related to epileptic activities is also seen in other neuropsychiatric disorders, such as attention deficit and hyperactivity disorder (ADHD) and Alzheimer's disease [87–89]. Compared with ADHD patients without epileptic activities on scalp EEG, those with epileptic activities presented more severe executive dysfunction [87, 88]. In early stages of Alzheimer's disease, epileptic activities cooccur with cognitive decline [89, 90], and the frequency of epileptic activities is strongly associated with accelerated cognitive decline in Alzheimer's patients [91]. Notably, administration of low doses of disease-modifying ASMs over 3 weeks can improve synaptic function and cognitive performance through suppressing epileptic activities in transgenic mouse model of Alzheimer's disease [89, 92–94], and similar effects may occur in humans as well [95] (Fig. 3.1).

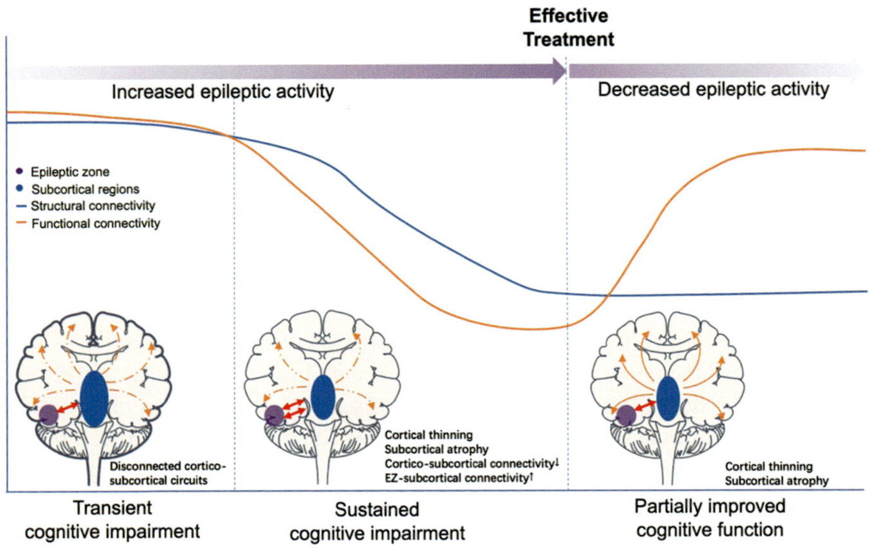

Fig. 3.1 Schematic diagram of the impact of epileptic activity on cognitive function via a disturbance of cognitive networks. Epileptic activities could exert transient deleterious effects on various cognitive functions by interrupting cortico-subcortical circuits. Recurrent disruption of networks due to epileptic activities may accumulate and lead to chronic remodeling of cognitive networks, which is characterized as enhanced connectivity between the epileptic zone and atrophic subcortical regions, thinning of distributed cortical areas and decreased cortico-subcortical connections. These reorganization patterns of cognitive networks result in sustained cognitive impairment. With the decrease of epileptic activities after effective treatments, functional connectivity of cognitive network gradually recovers in parallel with partially improved cognitive function. The red bold bidirectional arrows represent enhanced connectivity between the epileptic zone and subcortical regions, the orange dotted arrows represent interrupted cortico-subcortical circuits, and the orange solid arrows represent the normal cortico-subcortical circuits

3.5 Directions of Future Practice and Research

Detrimental effects of epileptic activities on cognitive function are closely related to functional and structural alterations of cognitive networks. On one hand, recurrent epileptic activities may directly lead to dysfunction in key nodes of neural circuits; on the other hand, adaptive mechanisms which limit the propagation of epileptic activities might contribute to long-term reorganization [60]. Suppressing epileptic activities through disease-modifying ASM and/or epilepsy surgery may attenuate these detrimental effects and improve cognitive performance [25, 65]. Even though brain reserve capacities are gradually released after successful treatments, cognitive dysfunction at the epileptic zone and its neighboring regions often do not recover completely [65]. In the future, coordinating cognitive networks via interventions that target their key nodes is a promising treatment for cognitive impairments.

To date, neuromodulation methods including transcranial magnetic stimulation, transcranial electrical stimulation, vagal nerve stimulation and intracranial (deep brain) stimulation have been investigated for improving cognitive functions such as memory, language, and perception [96–101]. Compared to non-invasive stimulation paradigms, intracranial electrical stimulation appears to exert more reliable effects on the improvement of cognitive function [97]. Direct electrical stimulation of the amygdala, the anterior nucleus of the thalamus, or of the hypothalamus/fornix could enhance memory, and that effect might be independent of seizure control [96, 102, 103]. It is hypothesized that neuromodulation initiates a molecular cascade of synaptic plasticity via regulating synchronized neural oscillations between regions and altering neurotransmitter concentrations [96, 104, 105]. Considering the high complexity of brain network, this novel therapeutic approach perturbes the neurophysiological state of the brain and faces great challenges, i.e. how to specifically and precisely improve specific cognitive functioning without compromising other functions. A better understanding of cognitive networks may help identify optimal targets and treatment parameters for neuromodulation. Given that fMRI lacks optimal temporal resolution, intracranial EEG would be optimally suited to detect transient and dynamic alterations in cognitive networks (see also Chaps. 39 and 41 on deep brain stimulation, and Chaps. 33 and 40 on the identification and analysis of cognitive networks).

References

1. Binnie CD (2003) Cognitive impairment during epileptiform discharges: is it ever justifiable to treat the EEG? Lancet Neurol 2:725–730
2. Sánchez Fernández I, Loddenkemper T, Galanopoulou AS et al (2015) Should epileptiform discharges be treated? Epilepsia 56:1492–1504
3. Siebelink BM, Bakker DJ, Binnie CD et al (1988) Psychological effects of subclinical epileptiform EEG discharges in children. II. General intelligence tests. Epilepsy Res 2:117–121
4. Aldenkamp AP, Beitler J, Arends J et al (2005) Acute effects of subclinical epileptiform EEG discharges on cognitive activation. Funct Neurol 20:23–28
5. Krestel HE, Nirkko A, von Allmen A et al (2011) Spike-triggered reaction-time EEG as a possible assessment tool for driving ability. Epilepsia 52:e126-129
6. Nicolai J, Ebus S, Biemans DP et al (2012) The cognitive effects of interictal epileptiform EEG discharges and short nonconvulsive epileptic seizures. Epilepsia 53:1051–1059
7. Ebus SC, DM IJ, den Boer JT et al (2015) Changes in the frequency of benign focal spikes accompany changes in central information processing speed: a prospective 2-year follow-up study. Epilepsy Behav 43:8–15
8. Antwi P, Atac E, Ryu JH et al (2019) Driving status of patients with generalized spike–wave on EEG but no clinical seizures. Epilepsy Behav 92:5–13
9. Berman R, Negishi M, Vestal M et al (2010) Simultaneous EEG, fMRI, and behavior in typical childhood absence seizures. Epilepsia 51:2011–2022
10. Guo JN, Kim R, Chen Y et al (2016) Impaired consciousness in patients with absence seizures investigated by functional MRI, EEG, and behavioural measures: a cross-sectional study. Lancet Neurol 15:1336–1345
11. Quon RJ, Camp EJ, Meisenhelter S et al (2021) Features of intracranial interictal epileptiform discharges associated with memory encoding. Epilepsia
12. Aarts JH, Binnie CD, Smit AM et al (1984) Selective cognitive impairment during focal and generalized epileptiform EEG activity. Brain 107(Pt 1):293–308

13. Kleen JK, Scott RC, Holmes GL et al (2013) Hippocampal interictal epileptiform activity disrupts cognition in humans. Neurology 81:18–24
14. Liu XY, Shi T, Yin WN et al (2016) Interictal epileptiform discharges were associated with poorer cognitive performance in adult epileptic patients. Epilepsy Res 128:1–5
15. Ung H, Cazares C, Nanivadekar A et al (2017) Interictal epileptiform activity outside the seizure onset zone impacts cognition. Brain 140:2157–2168
16. Fonseca LC, Tedrus GM, Pacheco EM (2007) Epileptiform EEG discharges in benign childhood epilepsy with centrotemporal spikes: reactivity and transitory cognitive impairment. Epilepsy Behav 11:65–70
17. Binnie CD, Kasteleijn-Nolst Trenité DG, Smit AM et al (1987) Interactions of epileptiform EEG discharges and cognition. Epilepsy Res 1:239–245
18. Shewmon DA, Erwin RJ (1989) Transient impairment of visual perception induced by single interictal occipital spikes. J Clin Exp Neuropsychol 11:675–691
19. Krauss GL, Summerfield M, Brandt J et al (1997) Mesial temporal spikes interfere with working memory. Neurology 49:975
20. Horak PC, Meisenhelter S, Song Y et al (2017) Interictal epileptiform discharges impair word recall in multiple brain areas. Epilepsia 58:373–380
21. Sekeres MJ, Winocur G, Moscovitch M (2018) The hippocampus and related neocortical structures in memory transformation. Neurosci Lett 680:39–53
22. Hawasli AH, Kim D, Ledbetter NM et al (2016) Influence of white and gray matter connections on endogenous human cortical oscillations. Front Hum Neurosci 10:330
23. Solomon EA, Kragel JE, Gross R et al (2018) Medial temporal lobe functional connectivity predicts stimulation-induced theta power. Nat Commun 9:4437
24. Pressler RM, Robinson RO, Wilson GA et al (2005) Treatment of interictal epileptiform discharges can improve behavior in children with behavioral problems and epilepsy. J Pediatr 146:112–117
25. Ibrahim GM, Cassel D, Morgan BR et al (2014) Resilience of developing brain networks to interictal epileptiform discharges is associated with cognitive outcome. Brain 137:2690–2702
26. Piazzini A, Turner K, Vignoli A et al (2008) Frontal cognitive dysfunction in juvenile myoclonic epilepsy. Epilepsia 49:657–662
27. Conant LL, Wilfong A, Inglese C et al (2010) Dysfunction of executive and related processes in childhood absence epilepsy. Epilepsy Behav 18:414–423
28. Filippini M, Boni A, Giannotta M et al (2013) Neuropsychological development in children belonging to BECTS spectrum: long-term effect of epileptiform activity. Epilepsy Behav 28:504–511
29. Jiang H, Ling Z, Zhang Y et al (2015) Altered fecal microbiota composition in patients with major depressive disorder. Brain Behav Immun 48:186–194
30. Jokeit H, Ebner A (1999) Long term effects of refractory temporal lobe epilepsy on cognitive abilities: a cross sectional study. J Neurol Neurosurg Psychiatry 67:44–50
31. Morgan VL, Abou-Khalil B, Rogers BP (2015) Evolution of functional connectivity of brain networks and their dynamic interaction in temporal lobe epilepsy. Brain Connect 5:35–44
32. Morgan VL, Conrad BN, Abou-Khalil B et al (2015) Increasing structural atrophy and functional isolation of the temporal lobe with duration of disease in temporal lobe epilepsy. Epilepsy Res 110:171–178
33. Chauvière L (2020) Potential causes of cognitive alterations in temporal lobe epilepsy. Behav Brain Res 378:112310
34. Kloc ML, Daglian JM, Holmes GL et al (2021) Recurrent febrile seizures alter intrahippocampal temporal coordination but do not cause spatial learning impairments. Epilepsia
35. Miller KJ, Weaver KE, Ojemann JG (2009) Direct electrophysiological measurement of human default network areas. Proc Natl Acad Sci U S A 106:12174–12177
36. Goulden N, Khusnulina A, Davis NJ et al (2014) The salience network is responsible for switching between the default mode network and the central executive network: replication from DCM. Neuroimage 99:180–190

37. Seeley WW, Menon V, Schatzberg AF et al (2007) Dissociable intrinsic connectivity networks for salience processing and executive control. J Neurosci 27:2349–2356
38. Menon V (2011) Large-scale brain networks and psychopathology: a unifying triple network model. Trends Cogn Sci 15:483–506
39. Catani M, Dell'acqua F, Thiebaut de Schotten M (2013) A revised limbic system model for memory, emotion and behaviour. Neurosci Biobehav Rev 37:1724–1737
40. Vossel S, Geng JJ, Fink GR (2014) Dorsal and ventral attention systems: distinct neural circuits but collaborative roles. Neuroscientist 20:150–159
41. Buckner RL, DiNicola LM (2019) The brain's default network: updated anatomy, physiology and evolving insights. Nat Rev Neurosci 20:593–608
42. Koechlin E, Summerfield C (2007) An information theoretical approach to prefrontal executive function. Trends Cogn Sci 11:229–235
43. Shine JM, Bissett PG, Bell PT et al (2016) The dynamics of functional brain networks: integrated network states during cognitive task performance. Neuron 92:544–554
44. Bansal K, Garcia JO, Tompson SH et al (2019) Cognitive chimera states in human brain networks. Sci Adv 5:eaau8535
45. Engel AK, Gerloff C, Hilgetag CC et al (2013) Intrinsic coupling modes: multiscale interactions in ongoing brain activity. Neuron 80:867–886
46. Young JJ, Rudebeck PH, Marcuse LV et al (2018) Theta band network supporting human episodic memory is not activated in the seizure onset zone. Neuroimage 183:565–573
47. Majerus S, Péters F, Bouffier M et al (2018) The dorsal attention network reflects both encoding load and top-down control during working memory. J Cogn Neurosci 30:144–159
48. Medaglia JD, Lynall ME, Bassett DS (2015) Cognitive network neuroscience. J Cogn Neurosci 27:1471–1491
49. Shine JM, Aburn MJ, Breakspear M et al (2018) The modulation of neural gain facilitates a transition between functional segregation and integration in the brain. Elife 7
50. Koziol LF, Budding DE (2009) Subcortical structures and cognition, vol 1. Springer, New York
51. Stoodley CJ, Schmahmann JD (2009) Functional topography in the human cerebellum: a meta-analysis of neuroimaging studies. Neuroimage 44:489–501
52. Halassa MM, Kastner S (2017) Thalamic functions in distributed cognitive control. Nat Neurosci 20:1669–1679
53. Mitchell AS, Sherman SM, Sommer MA et al (2014) Advances in understanding mechanisms of thalamic relays in cognition and behavior. J Neurosci 34:15340–15346
54. Nakajima M, Halassa MM (2017) Thalamic control of functional cortical connectivity. Curr Opin Neurobiol 44:127–131
55. Schmitt LI, Wimmer RD, Nakajima M et al (2017) Thalamic amplification of cortical connectivity sustains attentional control. Nature 545:219–223
56. Bai X, Vestal M, Berman R et al (2010) Dynamic time course of typical childhood absence seizures: EEG, behavior, and functional magnetic resonance imaging. J Neurosci 30:5884–5893
57. Yang T, Luo C, Li Q et al (2013) Altered resting-state connectivity during interictal generalized spike-wave discharges in drug-naïve childhood absence epilepsy. Hum Brain Mapp 34:1761–1767
58. Zhang Z, Liao W, Wang Z et al (2014) Epileptic discharges specifically affect intrinsic connectivity networks during absence seizures. J Neurol Sci 336:138–145
59. Shamshiri EA, Tierney TM, Centeno M et al (2017) Interictal activity is an important contributor to abnormal intrinsic network connectivity in paediatric focal epilepsy. Hum Brain Mapp 38:221–236
60. Englot DJ, Konrad PE, Morgan VL (2016) Regional and global connectivity disturbances in focal epilepsy, related neurocognitive sequelae, and potential mechanistic underpinnings. Epilepsia 57:1546–1557
61. Gelinas JN, Khodagholy D, Thesen T et al (2016) Interictal epileptiform discharges induce hippocampal-cortical coupling in temporal lobe epilepsy. Nat Med 22:641–648

62. Henin S, Shankar A, Borges H et al (2021) Spatiotemporal dynamics between interictal epileptiform discharges and ripples during associative memory processing. Brain 144:1590–1602
63. Dodrill CB (2002) Progressive cognitive decline in adolescents and adults with epilepsy. Prog Brain Res 135:399–407
64. Galovic M, van Dooren VQH, Postma TS et al (2019) Progressive cortical thinning in patients with focal epilepsy. JAMA Neurol 76:1230–1239
65. Helmstaedter C, Kurthen M, Lux S et al (2003) Chronic epilepsy and cognition: a longitudinal study in temporal lobe epilepsy. Ann Neurol 54:425–432
66. Grunwald T, Lehnertz K, Helmstaedter C et al (1998) Limbic ERPs predict verbal memory after left-sided hippocampectomy. NeuroReport 9:3375–3378
67. Yang L, Li H, Zhu L et al (2017) Localized shape abnormalities in the thalamus and pallidum are associated with secondarily generalized seizures in mesial temporal lobe epilepsy. Epilepsy Behav 70:259–264
68. Ji C, Zhu L, Chen C et al (2018) Volumetric changes in hippocampal subregions and memory performance in mesial temporal lobe epilepsy with hippocampal sclerosis. Neurosci Bull 34:389–396
69. Chen C, Li H, Ding F et al (2019) Alterations in the hippocampal-thalamic pathway underlying secondarily generalized tonic-clonic seizures in mesial temporal lobe epilepsy: a diffusion tensor imaging study. Epilepsia 60:121–130
70. Haneef Z, Lenartowicz A, Yeh HJ et al (2014) Functional connectivity of hippocampal networks in temporal lobe epilepsy. Epilepsia 55:137–145
71. Dinkelacker V, Valabregue R, Thivard L et al (2015) Hippocampal-thalamic wiring in medial temporal lobe epilepsy: enhanced connectivity per hippocampal voxel. Epilepsia 56:1217–1226
72. Maccotta L, He BJ, Snyder AZ et al (2013) Impaired and facilitated functional networks in temporal lobe epilepsy. Neuroimage Clin 2:862–872
73. Doucet G, Osipowicz K, Sharan A et al (2013) Extratemporal functional connectivity impairments at rest are related to memory performance in mesial temporal epilepsy. Hum Brain Mapp 34:2202–2216
74. Chen XM, Huang DH, Chen ZR et al (2015) Temporal lobe epilepsy: decreased thalamic resting-state functional connectivity and their relationships with alertness performance. Epilepsy Behav 44:47–54
75. He X, Doucet GE, Sperling M et al (2015) Reduced thalamocortical functional connectivity in temporal lobe epilepsy. Epilepsia 56:1571–1579
76. Voets NL, Menke RA, Jbabdi S et al (2015) Thalamo-cortical disruption contributes to short-term memory deficits in patients with medial temporal lobe damage. Cereb Cortex 25:4584–4595
77. Li H, Ji C, Zhu L et al (2017) Reorganization of anterior and posterior hippocampal networks associated with memory performance in mesial temporal lobe epilepsy. Clin Neurophysiol 128:830–838
78. Haneef Z, Lenartowicz A, Yeh HJ et al (2012) Effect of lateralized temporal lobe epilepsy on the default mode network. Epilepsy Behav 25:350–357
79. Coan AC, Appenzeller S, Bonilha L et al (2009) Seizure frequency and lateralization affect progression of atrophy in temporal lobe epilepsy. Neurology 73:834–842
80. Li H, Ding F, Chen C et al (2021) Dynamic functional connectivity in modular organization of the hippocampal network marks memory phenotypes in temporal lobe epilepsy. Hum Brain Mapp
81. Pedersen M, Curwood EK, Vaughan DN et al (2016) Abnormal brain areas common to the focal epilepsies: multivariate pattern analysis of fMRI. Brain Connect 6:208–215
82. Englot DJ, Hinkley LB, Kort NS et al (2015) Global and regional functional connectivity maps of neural oscillations in focal epilepsy. Brain 138:2249–2262
83. Torres-Platas SG, Nagy C, Wakid M et al (2016) Glial fibrillary acidic protein is differentially expressed across cortical and subcortical regions in healthy brains and downregulated in the thalamus and caudate nucleus of depressed suicides. Mol Psychiatry 21:509–515

84. Luo C, An D, Yao D et al (2014) Patient-specific connectivity pattern of epileptic network in frontal lobe epilepsy. Neuroimage Clin 4:668–675
85. Widjaja E, Mahmoodabadi SZ, Snead OC 3rd et al (2011) Widespread cortical thinning in children with frontal lobe epilepsy. Epilepsia 52:1685–1691
86. Widjaja E, Zamyadi M, Raybaud C et al (2015) Disrupted global and regional structural networks and subnetworks in children with localization-related epilepsy. AJNR Am J Neuroradiol 36:1362–1368
87. Holtmann M, Matei A, Hellmann U et al (2006) Rolandic spikes increase impulsivity in ADHD—a neuropsychological pilot study. Brain Dev 28:633–640
88. Benbadis SR, Ewen JB, Schreiber JM et al (2015) Variations in EEG discharges predict ADHD severity within individual Smith-Lemli-Opitz patients. Neurology 84:436
89. Vossel KA, Tartaglia MC, Nygaard HB et al (2017) Epileptic activity in Alzheimer's disease: causes and clinical relevance. Lancet Neurol 16:311–322
90. Vossel KA, Beagle AJ, Rabinovici GD et al (2013) Seizures and epileptiform activity in the early stages of Alzheimer disease. JAMA Neurol 70:1158–1166
91. Horvath AA, Papp A, Zsuffa J et al (2021) Subclinical epileptiform activity accelerates the progression of Alzheimer's disease: a long-term EEG study. Clin Neurophysiol 132:1982–1989
92. Sanchez PE, Zhu L, Verret L et al (2012) Levetiracetam suppresses neuronal network dysfunction and reverses synaptic and cognitive deficits in an Alzheimer's disease model. Proc Natl Acad Sci U S A 109:E2895-2903
93. Shi JQ, Wang BR, Tian YY et al (2013) Antiepileptics topiramate and levetiracetam alleviate behavioral deficits and reduce neuropathology in APPswe/PS1dE9 transgenic mice. CNS Neurosci Ther 19:871–881
94. Zhang MY, Zheng CY, Zou MM et al (2014) Lamotrigine attenuates deficits in synaptic plasticity and accumulation of amyloid plaques in APP/PS1 transgenic mice. Neurobiol Aging 35:2713–2725
95. Bakker A, Krauss GL, Albert MS et al (2012) Reduction of hippocampal hyperactivity improves cognition in amnestic mild cognitive impairment. Neuron 74:467–474
96. Inman CS, Manns JR, Bijanki KR et al (2018) Direct electrical stimulation of the amygdala enhances declarative memory in humans. Proc Natl Acad Sci 115:98–103
97. Meisenhelter S, Jobst BC (2018) Neurostimulation for memory enhancement in epilepsy. Curr Neurol Neurosci Rep 18:30
98. Winawer J, Parvizi J (2016) Linking electrical stimulation of human primary visual cortex, size of affected cortical area, neuronal responses, and subjective experience. Neuron 92:1213–1219
99. Chang EF, Kurteff G, Wilson SM (2018) Selective interference with syntactic encoding during sentence production by direct electrocortical stimulation of the inferior frontal gyrus. J Cogn Neurosci 30:411–420
100. Hill AT, Fitzgerald PB, Hoy KE (2016) Effects of anodal transcranial direct current stimulation on working memory: a systematic review and meta-analysis of findings from healthy and neuropsychiatric populations. Brain Stimul 9:197–208
101. Bonnì S, Veniero D, Mastropasqua C et al (2015) TMS evidence for a selective role of the precuneus in source memory retrieval. Behav Brain Res 282:70–75
102. Tröster AI, Meador KJ, Irwin CP et al (2017) Memory and mood outcomes after anterior thalamic stimulation for refractory partial epilepsy. Seizure 45:133–141
103. Laxton AW, Tang-Wai DF, McAndrews MP et al (2010) A phase I trial of deep brain stimulation of memory circuits in Alzheimer's disease. Ann Neurol 68:521–534
104. Akirav I, Richter-Levin G (2002) Mechanisms of amygdala modulation of hippocampal plasticity. J Neurosci 22:9912–9921
105. Stagg CJ, Best JG, Stephenson MC et al (2009) Polarity-sensitive modulation of cortical neurotransmitters by transcranial stimulation. J Neurosci 29:5202–5206

Chapter 4
Which Practical Issues Should I Consider When Planning and Conducting an iEEG Study?

Elias M. B. Rau and Robin Lienkämper

Abstract Intracranial EEG recordings enable scientists to investigate unique research questions, directly addressing the intricate interplay of neuronal activity and complex behaviors in humans. While its strength lies in measuring direct brain activity with a spatial and temporal resolution that is superior to more commonly used neuroscientific methods, its practical implementation comes with a number of unique challenges, which are particularly important to consider for novices in the field. Specifically, the large inter-subject variability and characteristics of iEEG samples raise significant considerations regarding the study design and the configuration of the behavioral task. Further, the situation of patients and their motivation to participate strikingly differs from well-defined samples of healthy subjects and requires you to behave and prepare accordingly. Additionally, specific technical aspects have to be considered concerning the general recording setup, the synchronization of neural and behavioral events, and on-site quality control checks. For each of these challenges, we propose practical solutions that may help to improve the quality of future research for all involved parties, from patients and researchers to clinical staff.

4.1 Study Design

In human behavioral and cognitive neuroscience studies, researchers usually try to recruit either a homogenous group of subjects (for example healthy, right-handed people of a certain age) or a selection of subjects that represents all age groups, genders and possible other factors. Contrariwise, in iEEG studies researchers typically cannot actively recruit subjects based on specific criteria, but can merely select among those patients who agreed to participate in scientific studies. This results

E. M. B. Rau (✉)
Ruhr University Bochum, Bochum, Germany
e-mail: elias.rau@ruhr-uni-bochum.de

R. Lienkämper
University of Pittsburgh, Pittsburgh, USA
e-mail: robin.lienkaemper@gmail.com

N. Axmacher (ed.), *Intracranial EEG*, Studies in Neuroscience, Psychology and Behavioral Economics, https://doi.org/10.1007/978-3-031-20910-9_4

in relatively small samples of patients that differ from well-defined control groups of healthy students in both their cognitive status and many demographic variables such as age, the level of education, and socioeconomic status (see also Chap. 2). As the decision to undergo invasive iEEG monitoring has to be purely based on medical judgements and not on research interests, the experimenters' control over patient and sample characteristics is very limited. Thus, iEEG samples show a high variability in many aspects which have to be considered when planning studies, approaching patients, and collecting data.

Since the number of available patients is the most limiting factor in data acquisition, waiting for "ideal" subjects to form a homogenous group is often neither recommended nor feasible. Instead, collecting data from as many suited patients as possible provides the opportunity to be mindful of other sample characteristics later on. Thus, if a given patient is both behaviorally capable of performing a task, and has electrodes implanted in regions of interest, we recommend that you consider inviting him/her for participation in your study.

The unforeseeable characteristics of a final iEEG sample complicate preparation steps when designing the experiment or comparing performance to healthy control groups. In general, pilot studies are an essential part of experimental research and are crucial for developing suitable behavioral paradigms, which afford the engagement of the cognitive processes of interest. In the non-clinical setting, such precursor experimental versions are mostly run on separate, smaller cohorts of young and healthy subjects prior to main data acquisition. Although pilot studies are invaluable for designing a suitable behavioral paradigm for iEEG as well, the scarce availability of patients and circumstances in clinical routines limit opportunities for pilot testing with a patient cohort directly. Nevertheless, especially for novel behavioral paradigms, we recommend that you test the paradigm with healthy control samples beforehand to improve your prediction on expected effect-sizes, validate the suitability of experimental conditions, adapt task difficulty of different experimental versions and identify potential technical pitfalls that may disrupt neural data acquisition. Keep in mind though, that for the same experimental version, a clinical sample will likely perform worse compared to a healthy control group. Vice versa, in a patient-optimized task, healthy controls will often perform at ceiling. Thus, transferability of performance levels between a healthy and young control group to a sample of iEEG patients is limited. To address this issue, you could test the same behavioral paradigm on a healthy age-matched sample. This will improve your estimation of the suitability of the paradigm for samples differing from student populations. An even closer comparison group to invasive patients consists in a group of other epilepsy patients who undergo non-invasive scalp-EEG monitoring, possibly even in the same clinic. This form of clinical monitoring is performed more frequently and requires fewer resources for the clinical staff, which increases the number of possible participants to be included in preliminary versions of your task. Crucially, this group of patients will share many of the sample characteristics of the final sample, and will thus provide you with better estimates of performance levels for data acquisition in implanted patients.

While diagnosis and localization of epileptic seizures strongly capitalize on inferences from EEG recordings, invasive monitoring is less common and mainly conducted in larger clinics or university hospitals with the required medical and technical expertise. Therefore, patients in need of invasive monitoring may come from a widespread area and possibly even from foreign countries. Be aware that opportunities may arise to test patients who do not speak your own language, or who only have limited skills in this language. If the patient and their implantation scheme is otherwise suitable for your experiment and may be willing to participate, it may be recommendable to invite an interpreter to the experimental session or prepare a translated version of the instructions beforehand.

Many behavioral tasks used in research nowadays contain elements like using arrow-keys or a joystick to navigate in a virtual environment. Other paradigms may contain equipment like touchscreens or even virtual reality designs presented via immersive goggles. Combined with the variation in iEEG sample demographics, differences in previous experience with such devices may introduce additional variance to the data. While navigating a 2D-world on a laptop screen using a joystick may be trivial for someone who grew up or is otherwise experienced with computers and video games, it may be a difficult challenge for someone with less experience in these areas. In your study design, it is therefore important to keep the task simple (wherever possible) and to make sure that patients have sufficient time to familiarize themselves with the paradigm and the equipment.

Besides previous experience with the input devices, physical constraints to interact with task elements should be taken into account, too. During invasive monitoring, the patient is physically connected to the recording system, which results in restricted freedom of movement. An exception to this constraint are novel approaches with wireless recordings (see Chap. 53), which are, as of this writing, far from standard practice. In addition to the large head bandages covering the head and ears, possibly complicating the wearing of glasses, medical devices for monitoring other vital parameters may also be attached to the patient. Given these circumstances, it is vital to ensure that the patient is easily able to interact with any devices that may be required for a given experiment.

Finally, similar to the variability in patient characteristics, opportunities to record neural data often cannot be easily scheduled in advance for researchers who are not directly involved in the everyday clinical routines. Since the decision to implant patients depends on a multitude of clinical considerations and available resources, surgeries may be planned or cancelled within only a few days of notice. Following successful implantation of electrodes, the duration of on-site monitoring may differ substantially between patients and can range from a few days to weeks, depending on the frequency and diagnostic value of observed seizures. These aspects are also worth considering when designing your experimental study, where single-session recordings per patient are more practically feasible, since you cannot be sure to see each patient multiple times or even on several consecutive days. The best way to address these various aspects is to establish reliable communication channels with the clinical staff. Ideally, this will allow you to plan recording sessions in advance. In other cases, it may be necessary to set up a recording on short notice.

4.2 Task Design

Considering the large variability of age and cognitive capacities of iEEG patients, you should also expect a substantial variance in their task performance. It is, for example, widely known that memory, reaction time, and motor capabilities decrease significantly with age—and they may be affected by epilepsy and antiepileptic drugs (see Chap. 2). However, given the scarcity of epilepsy patients undergoing invasive EEG monitoring, each individual patient is valuable to the research project. It is therefore important to design the experimental paradigm to be as flexible as possible. This will allow you to meet the patients' cognitive capacities and help achieve the maximum quality and quantity of recorded data. For example, we recommend implementing a practice session of the task with a sufficient number of trials for familiarization or a free exploration phase. Additionally, implement sufficient opportunity for breaks in between experimental trials and blocks. You may even allow button presses to pause the experiment whenever necessary, and save the respective information in the log file of your experiment to remember such events in the data files at a later point in time. If the paradigm allows it, you may prefer self-paced sequences of trials over fixed time limits to ensure that inter-trial intervals are long enough and allow for longer breaks in between trials. Your study design should also consider the fact that patients may abort the recording session at any point due to epileptic seizures, or because they are feeling unwell or exhausted. You should therefore make sure that behavioral data is stored on disk even if the experiment is left unfinished, and that you may continue the experiment on another day if possible. Generally, aiming for an experimental paradigm that runs no longer than one hour helps ensure that most patients complete the study. If your paradigm needs to be longer, try to split it up into several sessions and leave sufficient breaks in between. Paradigms that consist of several conditions performed in a sequence should however make sure that the risk of aborted sessions is spread evenly among conditions, for example by presenting the conditions in a randomized order.

Another important practical consideration when designing an experimental paradigm is task difficulty. A task that the patient perceives as too easy can quickly lead to boredom and decrease the subject's attention and interest in the experiment, which can negatively affect the results. On the other hand, patients can just as quickly become frustrated with tasks that they perceive as too difficult. In many cases, the paradigm may directly confront the patient with the cognitive limitations imposed on them by their neurological condition, which may cause psychological discomfort and stress. Keep in mind that patients are selected for your study because they have electrodes implanted in a target brain region whose function may be impaired by an epileptogenic lesion. Depending on the severity of accompanying neuropsychological deficits, you may have to expect a substantial variability in performance. Thus, having different versions of the task that vary in difficulty and/or duration can be helpful to adapt to the patient's condition. To select the right version of the task, you may consult results from neuropsychological testing, ask the medical staff for their impression of the patient's cognitive capacities, and/or conduct some test trials.

These recommendations will require a very flexible paradigm, but also comparability between slightly different versions of the task. An alternative way to keep a task engaging that may otherwise be perceived as too easy is to introduce entertaining elements into the paradigm (gamification). Additionally, some paradigms can be adapted to limit the feedback that patients receive about their task performance, which can greatly reduce the frustration for patients who perform worse than they expect.

4.3 Patient Handling

Empathy with the patient is of paramount importance in all medical professions, and this is no different for a scientist interacting with patients. During preparation of the recording setup, start connecting with the patient on a personal yet professional level. Describe your role as a researcher, your own motivation for doing the experiment, and offer to explain the iEEG setup. Stress the contribution of the patient's participation in the study in advancing our knowledge of mind and brain. Be aware that raising interest in the general research topic and stressing the importance of the scientific contribution is your main tool to engage the patient to participate. This is different from subjects in non-clinical settings, who often receive monetary reimbursement for their participation. Due to ethical reasons, similar reimbursements cannot be offered to patients.

However, for fundamental research, you also need to explicitly clarify that the experimental procedure, the patient's performance, and their participation is not related to the epilepsy disorder, will typically not advance the understanding and treatment of epilepsy, and most importantly, that participation in the study will by no means influence the quality of their ongoing treatment in any way. This emphasizes that your experimental procedure is completely separate from any medical procedures. Assure that while the experiment is going on, medical surveillance of the patient is never interrupted and that you know how to react and whom to contact in case of an epileptic event. Discussing beforehand how to react in case of such events clarifies behavioral guidelines and responsibilities for all involved parties.

An important aspect of patient handling in iEEG studies is to understand the vulnerable situation of the patients. At the time they participate in your study, some patients may very well have one of the worst weeks of their life. Imagine lying in a hospital bed, just days after you had a brain surgery. You still have a hole in your skull and cables coming out of your brain. Somewhere between lunch and your next change of bandages, a group of scientists comes into your room and wants you to play a memory-game on a laptop they brought with them. Understanding the patient's point of view helps to understand how to keep their motivation to participate in your study as high as possible. This also means you have to maintain a professional attitude and demonstrate some self-confidence with the situation. Minor technical difficulties, as they often occur with scientific equipment, can make you seem unprofessional in the patient's eyes. This can make patients uncomfortable or even afraid of the procedure ahead. Make sure all the equipment is tested before entering the patient's room, and that at least one person in the room is experienced in running the recording setup. While recordings can also be stressful and demanding for the operating researcher, your confidence will increase with each patient you see and as you get familiar with the clinical setting.

Additionally, patients will be much more engaged in your experiment if they understand its relevance. Briefly explain the research questions in simple terms and take time to answer any questions the patients may have. Also, emphasize what a rare opportunity it is for scientists to be able to record human iEEG, and how important it is for the advancement of your research question. While participation in a behavioral experiment can be a positive experience for the patient, it can also unveil subjective or objective cognitive deficits. Emphasize your understanding of the vulnerable situation the patients find themselves in to ease possible anxieties concerning their behavioral performance and stress that you are not expecting extraordinary performance. On the contrary, also a reduced behavioral performance combined with neural recordings may be important for scientific advancement.

4.4 Technical Setup

Besides considerations of task design and experimental procedure, correctly operating specialized iEEG hardware is a crucial factor for the success of your study. The implanted electrodes are physically connected to amplifiers that redirect the stream of information to a neural processing unit capable of interpreting the recorded signal. The recorded signal is then stored on a PC running the data acquisition software.

Research-dedicated hardware is not only advisable due to its accessibility to the researcher, but it may also outperform standard clinical systems in terms of e.g., sampling rates or superior signal-to-noise ratio. While most clinical decisions are based on recordings of macro-electrodes sampled at ~ 1 kHz, research setups may allow much higher sampling rates that are necessary for single-unit recordings of micro-wires and micro-arrays. With such high sampling rates and channel numbers, you need to ensure sufficient free disk space for storage. Recording a 60-min session of 10 electrodes sampled at 30 kHz will result in 1.08 billion data points! The size of the data file ultimately depends on the precision of data points of the recorded signal. Using neural signal oscillators or other simulated signals, you can record dummy data for the duration of an experimental run to see how large the files get.

In case of a separate research-dedicated system, the amount of implanted channels may exceed the maximum number of channels you can record. Clinical staff can help you decide for the best selection, since they may have already identified contacts that show a poor signal-to-noise ratio or frequent epileptiform activity.

All devices combined in a rack may require a considerable amount of space in the patient's monitoring room. Figure out the best way to position the equipment, the patient, and yourself to ensure that you can easily instruct and guide the patient during the course of the experiment.

4.5 Event Synchronization

Relating specific experimental events such as the onset of a stimulus or behavioral responses to recorded iEEG samples requires temporally aligning neural and behavioral events. This is usually achieved by sending temporal markers or triggers from the experimental paradigm to the recording device (i.e., the amplifier). This way, a specific sample of the neural data stream is marked with a behavioral event from the experiment. To fully exploit the high temporal resolution of intracranial EEG, it is crucial to mark such events in a temporally precise manner. At high sampling rates, transmission delays from the experimental system to the recording system can substantially bias your data. Thus, especially for time-sensitive paradigms or online data-analysis as in BCI applications (see Chap. 51), estimating the temporal delay of trigger signals—and even more important, the variance of these delays—is a crucial aspect for data quality that should be taken care of prior to the start of data acquisition. Relying on internal clocks of the running operating system may not be sufficient, since they can vary slightly between devices.

There are multiple ways to transmit triggers between the paradigm and recording hardware, and the best option depends on the specifics of your recording system. Serial or parallel ports are among the most standard approaches for sending and receiving triggers in EEG recordings. However, new-generation computers usually do not have serial or even parallel port connections. While USB-to-Serial or USB-to-Parallel adapters work well for time-insensitive applications, they may introduce jittered transmission delays that are difficult to compensate for after data acquisition. Instead, docking stations, customized hardware based on microcontrollers (e.g. Arduino), or vendor-specific hardware can be solutions to this issue.

For the presentation of complex stimuli such as sophisticated 3D virtual environments or combined audio and visual presentations, the computation time needed for rendering the stimuli may also cause a delay. When sending a trigger before presenting some complex stimulus, the time required to present the stimulus on screen is not considered. Using photodiodes to mark experimental events in your recording is the gold-standard approach and accounts for both transmission and rendering delays. Photodiodes are light-sensitive transmission devices which conduct current upon light-induced stimulation. The diodes are physically taped to e.g. the lower left corner of the presentation screen and connected to the recording device. To mark an experimental event, for example the presentation of a picture, simultaneously display a small white square on the screen-position of the diode. This will cause a light-triggered discharge of the diode and send a marker to the neural recording device. This way, the time that is needed for rendering is taken into account and the event is marked at the exact point in time the stimulus appeared.

Irrespective of your final approach for marking events in the recording, test the setup and ensure its functionality before starting each recording session. To do so, send some triggers from a separate dummy script to the recording device and verify that the events are properly recognized. Usually, such events are marked as ticks in a separate I/O channel in the recording software. To open up the possibility of realigning neural recordings and experimental events afterwards, you could send triggers at the very start and end of your experiment in predefined delays, for example 10 triggers with 100 ms delay. The temporal distance between the events in the recording file will help you estimate inherent variation in transmission and allow for realignment of data streams after the recording session is done.

4.6 Data Quality

Since data acquisition in iEEG is time-consuming and participant recruitment can be a major hurdle, researchers have to be particularly meticulous about their data quality. Importantly, ensuring good data quality prior to recording is always superior to offline data correction. Many issues concerning data quality can cost you an entire recording session if they go unnoticed. Therefore, you should not blindly trust in cables being set up correctly and recording settings being set the way they should.

Firstly, you should orient yourself in the channel layout. If the recording uses the same iEEG setup that was used for the medical recording, this step has generally been taken care of by the medical staff. If, however, a separate device is used for the scientific recording, the cables of the iEEG electrodes have to be plugged into this system first. To ensure that the channel order (and with it any information about the implantation site) is correct, it is always advisable to check some channels that are easy to discriminate from the others, and make sure they show the signal you expect. For example, some electrodes may be set up to record medical data unrelated to neural activity (for example ECG). Other electrodes may be broken and are therefore easily identifiable. Making sure that these channels are in the place where you expect them to be ensures that the channel numbering matches the implantation plan.

In a second step, you should visually inspect some channels that you expect to record neural data. Many problems with the recording setup can be identified, or at least recognized, with a short look at the data itself. The most basic system failure to be seen in this step is channels that do not show any believable neural data at all, but merely jump between oversteering and understeering. This is typically a sign that some cables in the system are not properly connected, but may also indicate issues with the reference settings or grounding. Sometimes, channels can also swing back and forth between a state showing over- and understeering and a second state with recognizable iEEG waveforms. This usually indicates a faulty connection between the electrode and the system, most commonly a loose cable connector or a broken cable.

Once you have made sure the channel layout is fine and that all relevant neural data is being recorded, a more in-depth look at the data can help identify possible further issues and subsequently improve the data quality. Visual inspection of iEEG data cannot always give you all the information needed to judge the data quality. The most prominent source of noise in iEEG recordings is line noise, caused by the electrical devices and infrastructure in the near surroundings. This noise depends on the power grid frequency of your recording site, which is generally 60 Hz in the

Americas and 50 Hz in the rest of the world. Line noise cannot always be identified visually, since it is hard to identify a certain frequency within a signal by eye. Many researchers therefore prefer to "listen" to the signal by ear, which makes it much easier to identify specific frequencies in the signal. Options for this exist for most modern recording systems. While this requires some experience, it will allow you to identify line noise on channels quickly and to look for its source. Turning off lights or unplugging devices like electrical patient beds and phone chargers can significantly improve signal quality. Other artifacts that may be picked up by iEEG electrodes could be related to muscle activity during head- or jaw movements, cable movements, or even blood flow, if the electrode is located in proximity to a blood vessel.

4.7 Conclusions

The main goal of this chapter was to provide practical guidelines on iEEG data acquisition, to discuss challenges and pitfalls and to offer solutions for commonly arising problems. We addressed several challenges concerning variability in iEEG samples and how to consider them in the development of a suitable study design and the final configuration of the behavioral task. Moreover, we raised awareness of the patient's situation and how the researcher's approach can be optimized to meet the requirements and affordances of all involved parties. Additionally, we outlined ways to ensure good data quality at the day of recording, and finally stressed the relevance of event synchronization during continuous data acquisition. Despite extensive preparation, there is and always will be room for error or unexpected events. However, good preparation, robust routines and a professional attitude are ways to adapt to the complex situation of iEEG data acquisition and ultimately gain unique insights into the neural basis of human cognition.

Chapter 5
What Are the Practical Considerations for Building a Successful Intracranial EEG and Direct Brain Stimulation Research Program?

Cory S. Inman and Peter Brunner

Abstract Intracranial EEG is performed at a wide range of clinical centers across the world and presents a unique opportunity to directly observe and modulate human brain activity at the speed of cognition. Capitalizing on this opportunity requires specific attention to the practicalities and challenges involved in establishing an effective and efficient intracranial research program. This entails identifying relevant scientific and clinical problems, building interdisciplinary, multi-institution teams that function within the clinical environment—a daunting task even for the most experienced scientists. Finally, we discuss the unique challenges to career advancement for those training in iEEG and brain stimulation. By providing a comprehensive discussion, along with practical recommendations, this chapter hopes to reduce the barriers of entry and maximize success for new investigators entering the field of intracranial research.

5.1 Introduction

Intracranial electroencephalography (iEEG) and direct brain stimulation has been used for the localization of epileptogenic regions in patients with drug-resistant epilepsy for nearly a century [1, 2]. Despite this long history of using electrodes with direct contact to the human brain to observe and provoke changes in brain activity, iEEG and direct brain stimulation have only recently gained traction over

C. S. Inman (✉)
Department of Psychology, University of Utah, Salt Lake City, UT, USA
e-mail: cory.inman@psych.utah.edu

Neuroscience Program, University of Utah, Salt Lake City, UT, USA

P. Brunner
Department of Neurosurgery, Washington University School of Medicine, St. Louis, MO, USA

National Center for Adaptive Neurotechnologies, Albany, NY, USA

Department of Neurology, Albany Medical College, Albany, NY, USA

N. Axmacher (ed.), *Intracranial EEG*, Studies in Neuroscience, Psychology and Behavioral Economics, https://doi.org/10.1007/978-3-031-20910-9_5

the past 30 years as an opportunity to study the neural correlates of human cognition and behavior on a broad scale across many laboratories and institutions across the world [3]. Insights from iEEG are now frequently used to complement longstanding evidence from lesion studies [4–9] and accumulating evidence from functional imaging modalities in healthy individuals like functional MRI and scalp EEG as to the functional role of regions throughout the human brain during cognition. In fact, relative to other brain mapping techniques, iEEG offers the unique ability to directly capture and manipulate neurophysiological processes with precise spatial localization on the order of millimeters and at the true temporal resolution of human cognition on the order of milliseconds. Proliferation of iEEG research over the past three decades has occurred due to increasing interdisciplinary recognition of iEEG data's neuroscientific and translational value, growing relationships between basic and clinical neuroscientists, and technical developments that simplify the collection of iEEG data during cognitive tasks [10]. In this chapter, we will discuss the opportunities iEEG and direct brain stimulation research offer to complementary cognitive neuroscience approaches and the necessary practical and career considerations for establishing a successful iEEG research program.

Pursuing human iEEG research requires scientists and clinicians to establish cooperative collaborations that aim to ask the strongest scientific questions in the context of the clinical team providing patients the best clinical care possible. In some cases, like clinician-scientists in neurology and neurosurgery, the scientist and the clinician are one and the same and may have received medical and PhD level training (i.e., M.D.-Ph.D.; see Chap. 6 for relevant ethical issues of this setting). Increasingly, systems and cognitive neuroscientists have been given the opportunity to pursue basic neuroscience questions with iEEG populations through collaborations with clinicians. Beyond the clinician-scientists relationship, there are many other essential team members involved in the operation of a successful iEEG research program. These essential team members include both permanent and transient collaborations with clinical staff whose primary responsibility is patient care, including neurosurgeons, neurologists, nurses, EEG technicians, and even patient family members or caregivers (see Fig. 5.1 for typical iEEG team interactions). Without some agreement and understanding between these essential team members, iEEG research can be inefficient or difficult to complete at all. Help from transient team members should be discussed and requested by clinicians involved in the patient's care to ensure that no research pursuits interfere with the clinical mission and to show support for the research program. Our first recommendation when working to establish an iEEG research program is to get to know all key personnel that might be involved in the patient's care and supporting the research endeavor. We've found it particularly helpful to give them the opportunity to learn about the research, from the basic steps of the experiment to the significance of the questions being pursued. How well various parts of the iEEG team work together is a major determinant for how successful the iEEG research program can be.

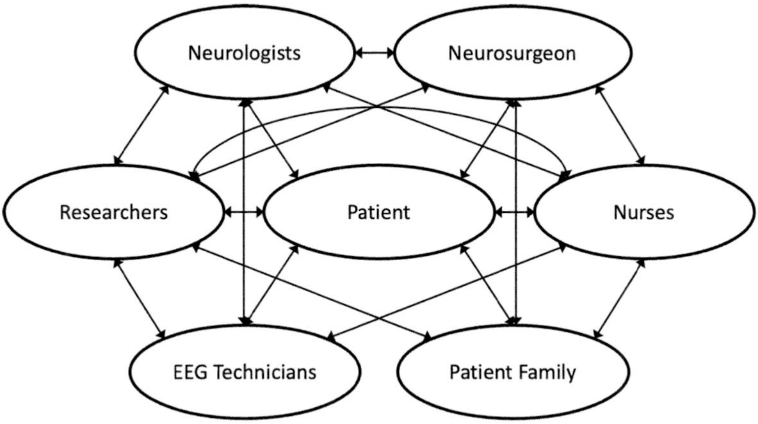

Fig. 5.1 A typical iEEG Research Team and their interactions. Every node of this network of stakeholders in the iEEG research process is essential to the success of an intracranial EEG research program. Failing to properly engage any part of the team will at least slow any iEEG research progress and at most make a research program unsustainable. Careful and purposeful communication and interactions among each of the stakeholders can help to establish, sustain, and grow a new iEEG research program

5.2 Preparing Experiments for the Clinical Environment

Intracranial recording time is highly coveted and often shared across multiple investigators and projects. To value this precious time, experiments must be well-prepared before engaging with the research subjects. Such preparations entail carefully designing the experiment to control for potential confounds, along with validation and verification to ensure that experiments yield the intended behavioral results. While validation and verification of experiments typically are conducted within the controlled laboratory setting using highly functioning neurotypical subjects, subject populations who represent the range of diversity in sex, age, and cognitive abilities that can commonly be expected within the clinical population, should be included early within the design and verification process. Soliciting qualitative feedback by these validation and verification subjects during debriefings can be used to identify issues that might otherwise only become apparent in statistical behavioral analyses. Additional consideration should be given to clinical considerations, such as limiting experimental sessions to 30 min or less, allowing for short breaks to accommodate ongoing clinical care, and capturing potential confounding variables such as distractions that might occur from the hospital public announcement system and phone calls or text messages that the patient might be receiving throughout the experimental session. Overall, experimental conditions in the clinical setting require well-justified research questions with thoughtful, careful designs that account for the unique environment in which testing will take place. Guidelines and best practices for designing and implementing iEEG research and direct brain stimulation studies are published in other chapters in this edited volume (e.g., see Chaps. 4, 8, and 39) and in recent review articles [3, 11–14].

5.3 Interactions with Clinicians and Other Stakeholders

The opportunity to do research with iEEG patients is given by neurosurgeons and neurologists caring for the patients. Without their support, iEEG research is not possible. This reliance on clinicians stems from the clinical context in which the research is being performed and the ethical obligations of researchers to not interfere with the clinical mission. The strength of support from the patient's neurologists and neurosurgeons directly contributes to the ease and success of a research program. Strong physician support makes all other research-related interactions with those involved in the patient's care easier, from the patient's family to nurses and EEG technicians. When aiming to start an iEEG research program, researchers should engage all of the physicians that might be involved in the patient's care with in-depth conversations about the goals of the research, how the physicians can be involved, the risks of the research, and how clinicians will be given credit for their participation in research through authorship on resulting publications. Having these discussions early and often will help keep clinicians engaged and interested in supporting the research endeavor.

Once support for a research program has been established with the iEEG clinical team, the most essential step to maintaining a good relationship is being respectful of the clinicians' time. Given the extremely busy and varied schedules clinicians follow on a day-to-day basis, requests for a clinician's time should be well-justified and minimized to ensure continued support. For instance, if a clinician's time is needed to do safety checks prior to a stimulation study, the procedure for performing the safety checks should be made as time efficient as possible. Communication of the research-related requests should be made well ahead of time to help the clinician fit the request into their schedule. It's important to keep in mind that every minute of time requested for clinicians to help with research is an additional minute added to their workload before they can go home for the day. Respect for all clinicians' time will go a long way towards maintaining a successful iEEG research program.

Finally, given that iEEG patients are a scarce and rare resource at many surgical epilepsy centers, researchers will need to work well with other research teams. This is best achieved through regular communication between each of the research teams and the clinical team. Ideally, independent research groups are organized under a larger iEEG research team with regular meetings prior to patient testing to establish testing priorities and order between all the independent research groups. It is often the case that many win–win situations can be found when regular communication is practiced between research groups, even across multiple institutions. We've found it particularly helpful for all research staff to be trained to administer all the tasks a patient may be asked to perform. This minimizes the number of people needing to interact with the patient and their family, as well as the number of people that need to be tracking the ongoing testing progress across projects. When this is not possible, given the complexities of certain tasks (i.e., stimulation tasks vs. recording-only tasks), then there should be regular communication between the research groups to minimize conflicts and patient confusion. Regular communication can be aided by project management software like Slack, Asana, or Microsoft Teams. The success of an iEEG research program can be determined by how well the independent research groups work together to maximize clinician support and testing efficiency while minimizing the need for clinicians' time.

5.4 Interactions with Patients and Their Families

The duration of inpatient monitoring primarily depends on reaching sufficient clinical evidence in identifying the seizure onset zone and can range from several days to multiple weeks. Duration of implantation and physical and mental condition is impossible to predict. Establishing a good working and social relationship with the patients are key to ensuring the willingness and motivation of patients to participate in research. It is also critical for the patient to feel like they're making a positive contribution to scientific knowledge. In this regard, having a small team of one to three researchers who, in pairs of two, attend to the patient for the entire duration of the inpatient monitoring tends to be more enjoyable for the patient than working with multiple teams that change from day to day which may overwhelm the patient socially and emotionally. It can be particularly helpful to get to know the patient beyond just their clinical situation. Asking the patient and their family about themselves goes a long way to establishing a good rapport throughout the patient's monitoring stay. It is possible to get to know patients and families quite well through their research experience. These short-term relationships can be a great source of satisfaction as an iEEG researcher and help maintain a healthy and caring relationship with patients and their families. Honestly one of the most incredible experiences as an iEEG researcher is bearing witness to the strength of patients and families going through the invasive experience of iEEG monitoring.

Consulting the neuropsychologist and epileptologist who often have followed a patient for years before the inpatient monitoring can be used to understand patient abilities and needs as well as the social and family dynamics that need to be considered when approaching the patient with the opportunity to participate in the research protocols (see also Chap. 2 on the neuropsychological background of epilepsy patients). Clearly explaining the scope of the research and delineating it from the ongoing clinical care is paramount to establishing and maintaining trust between patients and their families and the research team. This also entails informing patients that the research will not directly benefit nor will affect their clinical care. When clinical personnel are involved, they need to state when they visit in their capacity as researchers or clinicians to minimize any potential influence they may have (see Chap. 6). These additional steps of communication are even more important when engaging in direct brain stimulation studies. Gaining and maintaining a patient's trust during direct brain stimulation experiments requires a clear description of the potential risks associated with stimulation. Immediate explanations of any sensations or behavioral results also aids in maintaining the patient as a collaborator throughout the research process. Many patients are excited to share their experience and to have the opportunity to potentially contribute to therapies that can help future patients in their situation. Overall, patients and their families are essential parts of the research team and endeavor, and they should be treated as collaborators in any projects they participate in.

5.5 Interactions with the Nursing Staff

Patients implanted with intracranial electrodes are typically monitored within an epilepsy unit staffed by nurses dedicated to the clinical care of such patients. This ensures that nurses can rapidly respond to a clinical seizure to prevent the patient from injury. Any research experiment conducted in this patient population needs to be able to rapidly vacate the bedside within seconds to allow the nursing staff to attend to a potential generalized seizure. Consulting with the attending epileptologist about the typical manifestation, including viewing a previously recorded seizure, can be helpful in increasing the sensitivity of the research team to seizures. Especially for generalized frontal-lobe seizures, the research team needs to be prepared to be asked by the nursing staff to help subdue a violently convulsing patient. Nursing staff should be made aware of stimulation studies or any studies that might increase the chance for provoking a seizure. This may involve having anti-epileptic medications like Ativan (Lorazepam) at the ready or checking in with EEG technicians that are actively monitoring the patient's EEG data throughout the stimulation testing. Further, the daily changing nursing staff needs to be consistently kept in the loop about the scope and duration of any research protocols, and important clinical care needs to be completed before research experiments can be conducted. This allows researchers to minimize testing interruptions and nursing staff to ensure the patient has received any scheduled care or medications. Many nursing staff appreciated being involved in helping with research when clear steps to minimize any extra duties have been made by researchers. Nurses are often the most frequent caregivers for patients; respecting their expertise and time will help make the patient and researcher experience as productive as possible.

5.6 Interactions with EEG Technicians

The technical complexity of monitoring epilepsy patients requires specifically trained EEG technicians to apply EEG electrodes, connect EEG and iEEG electrodes to the amplifier, and monitor and maintain good recording quality throughout the monitoring stay. While EEG monitoring is a relatively simple and highly standardized procedure that is routinely performed by most EEG technicians on a daily basis, iEEG monitoring is highly complex as it requires connecting and managing an order of magnitude more electrodes than EEG monitoring and is only performed once or twice per month. This complexity is further increased by clinical research where additional electrodes (e.g., micro-electrodes; see Chaps. 42–46) are connected, or additional dedicated signal acquisition systems for the acquiring from these electrodes are inserted into the clinical signal path. Establishing a successful intracranial monitoring research project thus requires collaboration and a good working relationship between the EEG technicians and the research team on the technical aspects of the research. For example, connecting the dedicated research system needs to be done in the presence of the EEG technicians, and reseating of connectors or EEG electrodes throughout the monitoring should be done by the EEG technician in a timely manner to ensure stable signal quality across all research protocols. Of particular importance is that the research is slowly increased in complexity to allow for identifying and isolating issues at an early stage, rather than being faced with noise issues that cannot be isolated. This situation can create pushback by the clinical team and prevent the use of the research amplifier or stimulator system. It is often helpful to give EEG technicians the opportunity to learn about the aims of the research being done with their patients. This can be done through organizing a talk to review the current iEEG research projects, or when EEG technicians have allotted time for research, they can be brought into a more permanent role on the research team that utilizes their expertise (i.e., reviewing EEG data for interictal discharges or non-convulsive seizures or performing signal processing analyses). Like clinicians and nurses, establishing and maintaining a good rapport with EEG technicians and their supervisors by minimizing the requests on their time helps with sustaining a successful iEEG research program.

5.7 Practical Considerations for Direct Brain Stimulation Studies

Direct brain stimulation studies have become an increasingly important source of evidence for understanding causal influences brain regions have upon one another and the related cognitive behaviors (see Chaps. 39, 41, and 52). Human brain mapping studies using direct brain stimulation in awake patients began in the 1930s with the pioneering work of Foerster and Penfield [1, 2, 15–17]. Since then, our ability to deliver intracranial stimulation and to use it as a brain mapping tool outside of the operating room has improved, but there is still much to learn about how direct electrical stimulation to the human brain changes neurophysiology. Recent work has demonstrated novel capabilities to deliver stimulation responsively to epilepsy-related and memory-related brain states [18–20]. This capability to perform closed-loop stimulation in response to neurophysiological signals and states gives researchers the opportunity to deduce the causal neural dynamics of specific and widespread networks of brain regions. Whether open-loop or closed-loop, all stimulation experiments require additional consideration by the research team [21–31]. Every step of the research process is more complex when performing a stimulation study, from gaining IRB and ethics approval (see Chap. 6) to purchasing approved stimulation equipment to ensuring patient safety during testing and working with the clinical team.

Gaining IRB or ethical approval for a stimulation study starts with having a reasonable justification for performing the study that outweighs the risk. The maximum risk threshold is generally set by the common stimulation mapping practices used for clinical seizure mapping during the patient's monitoring. Research risks, based on stimulation parameters, are often kept much lower than this clinical threshold. Risks can also be minimized by involving the clinical team in discussions of the stimulation parameters (location, amplitude, frequency, duration, etc.) relative to the ongoing information coming in from the patient's monitoring. Neurologists can also be involved in the process of checking for after-discharges caused by stimulation during a pre-testing safety check where the stimulation parameters are manually controlled (i.e., stimulation delivery is not programmatically automated). These sorts of manual safety checks should be done for all parameter sets prior to testing to minimize the risk of a stimulation-induced seizure or kindling of a novel epileptic focus from repeated induction of after-discharges. Performing stimulation experiments that are safe and well-tolerated by research participants requires additional considerations relative to recording-only studies. The best way to maintain the trust of participants and clinicians is for researchers performing stimulation studies to constantly seek to minimize the risks of their experiments.

5.8 Career Pathways to Starting a Successful iEEG Research Program

Given the interdisciplinary nature of iEEG research, establishing a successful and impactful iEEG research program critically depends on three factors: (1) becoming an open-minded investigator who is well-trained in a range of disciplines; (2) building a network of open-minded and supportive collaborators who complement your skills; and (3) finding an institution that provides the necessary clinical and scientific resources and environment to engage in highly intensive basic or translational research. Of course, there are also various idiosyncrasies related to each job applicant's training background (M.D.-Ph.D., neuroscientist, psychologist, physicist, data scientist, etc.) and where they are applying for a position (R1 University, United States, Europe, etc.) that we cannot thoroughly cover here. Our primary goal with this section is to outline some suggestions for gaining training and finding a position that overcomes these hurdles, given the interdisciplinary nature of iEEG research. We suggest establishing a mentorship relationship with current iEEG investigators to get a sense of how to navigate the idiosyncrasies of your situation.

With the growth of more independent iEEG research labs, gaining the necessary hands-on training to collect and analyze data is becoming more feasible, but relative to other lab techniques, iEEG research can be difficult to break into. The easiest path to gaining the necessary expertise in iEEG research is to join an established lab in graduate school or during a postdoctoral fellowship. While these training experiences may give you the requisite experience to develop your own experiments, collect and analyze data, they do not guarantee well-rounded training across all of these aspects of a project. Well-rounded training must be sought from the beginning by the trainee. Fortunately, already collected, large iEEG and stimulation datasets have now begun to be shared via data repositories (iEEG.org, the international data portal, openneuro.org, University of Pennsylvania Computational Memory Lab; see https://github.com/openlists/ElectrophysiologyData#human-intracranial-data for a comprehensive list). These previously collected iEEG datasets can give anyone the opportunity to analyze precious iEEG data but gaining experience in iEEG experimental design and collecting iEEG data is harder to gain. There are also opportunities to gain iEEG research experience from the medical training track during M.D.-Ph.D. training, medical school, residency, or fellowship. Besides doing Ph.D. training in an iEEG lab, other medical research experiences are often time-limited, so it's often best to work with existing datasets to gain analysis experience and to shadow task development and patient testing. For those that are coming from an academic background, it may be possible to develop an outside collaboration with an existing iEEG lab at your institution. Clinician-scientists often seek support from behavioral or cognitive neuroscientists to establish a research program, given the clinician's split clinical and research responsibilities. If you are not sure if your institution has iEEG patients or research resources, reach out to the local epileptologists and functional neurosurgeons to gauge their research interests and possibilities. In general, a well-rounded iEEG researcher seeking independence will have training and experience in experimental design, patient and clinician interaction, experimental testing, signal processing, advanced statistics, and article and grant writing.

Along the way to gaining this diverse training, those interested in an iEEG research career should establish a broad network of clinicians and established iEEG researchers across institutions. Growing your network will involve the typical interactions at conferences (Society for Neuroscience, Cognitive Neuroscience Society, Human Brain Mapping, World Society of Stereotactic and Functional Neurosurgery, American Epilepsy Society, etc.) and reaching out to researchers and trainees to discuss their prior publications. Growing your iEEG network can also be aided by reaching out to neurologists and neurosurgeons in your local academic hospitals to discover what iEEG research might be occurring locally. Other avenues to developing a multi-institution iEEG research network might include working with established researchers at your institution on a multi-institution iEEG research or direct brain stimulation grant or offering your specific expertise to an ongoing project. Presenting results generated from analyses of open iEEG datasets at various conferences can also be an excellent way to get noticed by iEEG research laboratories seeking help. Finally, asking collaborators and researchers in your discipline that are at other institutions to try to make connections with the clinicians involved with the surgical epilepsy service at institutions they are familiar with can also help you determine where potential opportunities might become available. Connections established to anyone else currently engaged in iEEG research during training have the potential to open future doors along the pathway to establishing a career and independent laboratory in iEEG research.

The final and perhaps most difficult step of establishing an independent laboratory in iEEG research is finding an institution that provides the necessary clinical and scientific resources and environment to engage in highly intensive basic or translational research. There are likely hundreds of academic medical centers around the world that do not currently perform research with their intracranial EEG patients. In some cases, this is due to a lack of support from clinical administration given revenue models or physicians interest. In many cases, clinicians are interested in hosting iEEG research but simply do not have the time to establish a research program given their clinical responsibilities. There are also situations where some intracranial EEG research is taking place at an institution, but the opportunities to test patients are not yet saturated. All of these clinical situations have the potential to develop a successful iEEG and direct brain stimulation research program with the right approach. If there is interest in starting a program from the clinician's side, there might be a postdoctoral or faculty job posting in the medical school seeking researchers with iEEG, experimental design, or signal processing experience.

Most iEEG job opportunities are posted and hired through medical schools, but there are other academic departments that might support an iEEG research program (psychology, neuroscience, biomedical engineering, biology, computer science, etc.). Successfully applying to non-medical academic departments with an iEEG research program requires a certain amount of creativity in connecting your program with interests of existing faculty. Clear support from the clinicians that would give you access to work with iEEG patients is also essential for successfully landing a iEEG faculty position. Showing a clear connection between current faculty research programs in a department and a novel iEEG program is essential to gaining a search committee's interest in your application. For example, this could be done by describing how existing behavioral or cognitive neuroscience questions currently being examined with non-invasive techniques can benefit from complementary invasive studies of similar behaviors. Making this case for the search committee can also be aided by offering to help connect the current faculty in the department with the iEEG research opportunities as the program becomes established. Of course, to make these sorts of offers or connections, a potential applicant needs to establish a strong relationship and support from current researchers and clinicians involved in the patient's care that are aware of the current possibilities for adding a new iEEG research lab. Gaining clinician and researcher support will be aided by having already established a strong network of multi-institution iEEG researchers. Given the relatively small number of places to train to perform iEEG research, it is likely that someone in this broad network has a personal or collaborative connection with clinicians at iEEG monitoring centers, even those with no current iEEG research program. Once a connection is established, and you have gained clinical support for your proposed research program, a letter of support from the clinicians involved in the iEEG patient's care attached to your application can go a long way in showing a search committee the potential to grow an iEEG research program at their institution. This support letter would include details about the unique opportunities offered by iEEG research, the current availability of research time with iEEG patients, possible collaborations with current researchers, and detailed support for your specific iEEG research program. Although it is challenging to meet all these criteria to gain a tenure or researcher track faculty position, it is certainly possible with the proper training, connections, and support.

5.9 Conclusion

Intracranial EEG research offers the rare opportunity to directly record and modulate human brain activity at the speed of cognition. With rare opportunities come unique responsibilities to develop a well-justified, unique research program, gain and maintain support of clinicians caring for iEEG patients, and cultivate a rapport and trust in a limited amount of time with patients and their families. This chapter aimed to provide some insight into the practical considerations for those wanting to establish a successful, independent line of research in iEEG and direct brain stimulation. While all of these recommendations may not apply to all situations, we hope that these suggestions can help most applicants as they pursue a career in iEEG research. The iEEG research landscape will surely change over the coming years since this chapter is published with new technologies becoming available to improve long-term seizure monitoring and record electrophysiological data from the human brain. Those that can foresee and adapt to these changes in technology and data analytics will be most successful at building and sustaining a successful iEEG research program. With these rapidly developing technological and data analysis approaches, intracranial EEG and direct brain stimulation research is currently poised to transform our understanding of human brain function in the coming decades.

References

1. Foerster O, Altenburger H (1935) Elektrobiologische Vorgänge an der menschlichen Hirnrinde. Deutsche Zeitschrift f Nervenheilkunde 135:277–288. https://doi.org/10.1007/BF01732786
2. Reif PS, Strzelczyk A, Rosenow F (2016) The history of invasive EEG evaluation in epilepsy patients. Seizure 41:191–195. https://doi.org/10.1016/j.seizure.2016.04.006
3. Parvizi J, Kastner S (2018) Promises and limitations of human intracranial electroencephalography. Nat Neurosci 21:474–483. https://doi.org/10.1038/s41593-018-0108-2
4. Adolphs R (2016) Human lesion studies in the 21st century. Neuron 90:1151–1153. https://doi.org/10.1016/j.neuron.2016.05.014
5. Adolphs R, Cahill L, Schul R, Babinsky R (1997) Impaired declarative memory for emotional material following bilateral amygdala damage in humans. Learn Mem 4:291–300. https://doi.org/10.1101/lm.4.3.291

6. Adolphs R, Tranel D, Hamann S, Young AW, Calder AJ, Phelps EA et al (1999) Recognition of facial emotion in nine individuals with bilateral amygdala damage. Neuropsychologia 37:1111–1117. https://doi.org/10.1016/S0028-3932(99)00039-1
7. Anderson AK, Phelps EA (2001) Lesions of the human amygdala impair enhanced perception of emotionally salient events. Nature 411:305–309. https://doi.org/10.1038/35077083
8. Feinstein JS (2013) Lesion studies of human emotion and feeling. Curr Opin Neurobiol 23:304–309. https://doi.org/10.1016/j.conb.2012.12.007
9. Scoville WB, Milner B (1957) Loss of recent memory after bilateral hippocampal lesions. J Neurol Neurosurg Psychiatry 20:11–21
10. Brunner P, Bianchi L, Guger C, Cincotti F, Schalk G (2011) Current trends in hardware and software for brain–computer interfaces (BCIs). J Neural Eng 8:025001. https://doi.org/10.1088/1741-2560/8/2/025001
11. Mercier MR, Dubarry A-S, Tadel F, Avanzini P, Axmacher N, Cellier D et al (2022) Advances in human intracranial electroencephalography research, guidelines and good practices. NeuroImage 119438. https://doi.org/10.1016/j.neuroimage.2022.119438
12. Mobbs D, Wise T, Suthana N, Guzmán N, Kriegeskorte N, Leibo JZ (2021) Promises and challenges of human computational ethology. Neuron. https://doi.org/10.1016/j.neuron.2021.05.021
13. Suthana N, Aghajan ZM, Mankin EA, Lin A (2018) Reporting guidelines and issues to consider for using intracranial brain stimulation in studies of human declarative memory. Front Neurosci 12. https://doi.org/10.3389/fnins.2018.00905
14. Suthana N, Fried I (2014) Deep brain stimulation for enhancement of learning and memory. NeuroImage 85:Part 3:996–1002. https://doi.org/10.1016/j.neuroimage.2013.07.066
15. Foerster O, Penfield W (1930) The structural basis of traumatic epilepsy and results of radical operation. Brain 53:99–119. https://doi.org/10.1093/brain/53.2.99
16. Penfield W, Jasper H (1954) Epilepsy and the functional anatomy of the human brain. Little, Brown & Co., Oxford
17. Penfield W, Rasmussen T (1950) The cerebral cortex of man; a clinical study of localization of function. Macmillan, Oxford
18. Ezzyat Y, Wanda PA, Levy DF, Kadel A, Aka A, Pedisich I et al (2018) Closed-loop stimulation of temporal cortex rescues functional networks and improves memory. Nat Commun 9:365. https://doi.org/10.1038/s41467-017-02753-0
19. Heck CN, King-Stephens D, Massey AD, Nair DR, Jobst BC, Barkley GL et al (2014) Two-year seizure reduction in adults with medically intractable partial onset epilepsy treated with responsive neurostimulation: final results of the RNS System Pivotal trial. Epilepsia 55:432–441. https://doi.org/10.1111/epi.12534
20. Morrell MJ, Halpern C (2016) Responsive direct brain stimulation for epilepsy. Neurosurg Clin N Am 27:111–121. https://doi.org/10.1016/j.nec.2015.08.012
21. Bijanki KR, Manns JR, Inman CS, Choi KS, Harati S, Pedersen NP et al (2019) Cingulum stimulation enhances positive affect and anxiolysis to facilitate awake craniotomy. J Clin Invest 129:1152–1166. https://doi.org/10.1172/JCI120110
22. Ezzyat Y, Kragel JE, Burke JF, Levy DF, Lyalenko A, Wanda P et al (2017) Direct brain stimulation modulates encoding states and memory performance in humans. Curr Biol 27:1251–1258. https://doi.org/10.1016/j.cub.2017.03.028
23. Goldstein HE, Smith EH, Gross RE, Jobst BC, Lega BC, Sperling MR et al (2019) Risk of seizures induced by intracranial research stimulation: analysis of 770 stimulation sessions. J Neural Eng 16:066039. https://doi.org/10.1088/1741-2552/ab4365
24. Goyal A, Miller J, Watrous AJ, Lee SA, Coffey T, Sperling MR et al (2018) Electrical stimulation in hippocampus and entorhinal cortex impairs spatial and temporal memory. J Neurosci 38:4471–4481. https://doi.org/10.1523/JNEUROSCI.3049-17.2018
25. Hanslmayr S, Axmacher N, Inman CS (2019) Modulating human memory via entrainment of brain oscillations. Trends Neurosci 42:485–499. https://doi.org/10.1016/j.tins.2019.04.004
26. Inman CS, Bijanki KR, Bass DI, Gross RE, Hamann S, Willie JT (2018) Human amygdala stimulation effects on emotion physiology and emotional experience. Neuropsychologia. https://doi.org/10.1016/j.neuropsychologia.2018.03.019

27. Inman CS, Manns JR, Bijanki KR, Bass DI, Hamann S, Drane DL et al (2018) Direct electrical stimulation of the amygdala enhances declarative memory in humans. PNAS 115:98–103. https://doi.org/10.1073/pnas.1714058114
28. Jacobs J, Miller J, Lee SA, Coffey T, Watrous AJ, Sperling MR et al (2016) Direct electrical stimulation of the human entorhinal region and hippocampus impairs memory. Neuron 92:983–990. https://doi.org/10.1016/j.neuron.2016.10.062
29. Riva-Posse P, Inman CS, Choi KS, Crowell AL, Gross RE, Hamann S et al (2019) Autonomic arousal elicited by subcallosal cingulate stimulation is explained by white matter connectivity. Brain Stimul Basic Transl Clin Res Neuromodul 12:743–751. https://doi.org/10.1016/j.brs.2019.01.015
30. Sendi MSE, Inman CS, Bijanki KR, Blanpain L, Park JK, Hamann S et al (2021) Identifying the neurophysiological effects of memory-enhancing amygdala stimulation using interpretable machine learning. Brain Stimul 14:1511–1519. https://doi.org/10.1016/j.brs.2021.09.009
31. Solomon EA, Kragel JE, Gross R, Lega B, Sperling MR, Worrell G et al (2018) Medial temporal lobe functional connectivity predicts stimulation-induced theta power. Nat Commun 9:4437. https://doi.org/10.1038/s41467-018-06876-w

Chapter 6
What Ethical Issues Need to Be Considered When Doing Research with Patients Undergoing Invasive Electrode Implantation?

Ashley Feinsinger and Nader Pouratian

Abstract Directly recording and modulating human brain activity through invasive techniques takes advantage of unique neurosurgical opportunities provided by patients who have a clinical need for invasive electrode placement. These patients, including those who undergo deep brain stimulator (DBS) implantation or have wires implanted in the epilepsy monitoring unit (EMU), may consent to research to be conducted during these interventions. This research may involve extra recordings, additional behavioral tasks, temporary electrode placement, and modification of implantable wires for the purposes of gaining generalizable knowledge about the brain. The unique context of this research, including the overlap of clinical care and research and the populations of patients eligible to participate, generates a multitude of ethical issues. This chapter provides an overview of these issues and the special contextual features of iEEG research that engender them. Areas of focus will include risk/benefit assessments, informed consent, the dual-role of the physician-researcher, fair access, neurodiversity, and identity. Attention will be centered on non-therapeutic and basic science research conducted in the intraoperative setting with DBS patients and the extra-operative setting with EMU patients. This chapter is meant to inspire deliberation about research design, recruitment, and consent, and foster an appreciation for the unique ethical challenges in invasive iEEG research with neurosurgical patients.

6.1 Introduction

Intraoperative research with DBS patients is conducted during a patient's DBS implantation procedure. Crucially, these patients are seeking treatment for diseases

A. Feinsinger (✉)
Department of Medicine, David Geffen School of Medicine, University of California, 885 Tiverton Dr. Geffen Hall, Los Angeles, CA, USA
e-mail: AFeinsinger@mednet.ucla.edu

N. Pouratian
Department of Neurological Surgery, UT Southwestern Medical Center, 5303 Harry Hines Blvd, Dallas, TX, USA

N. Axmacher (ed.), *Intracranial EEG*, Studies in Neuroscience, Psychology and Behavioral Economics, https://doi.org/10.1007/978-3-031-20910-9_6

like Parkinson's, epilepsy, and dystonia, and often have exhausted other treatment options. Sometimes, the clinical activities of the surgery are paused, and patients are asked to complete additional tasks while extra recordings are taken either through the depth electrodes themselves or through a temporarily placed cortical strip (electrocorticography, or ECoG). The experiments can be conducted either in the middle of the implantation procedure or immediately prior to closure. Somewhat differently, extra-operative research is conducted during an epilepsy patient's stay in the EMU. These patients, who have wires implanted to for epilepsy diagnostics, stay in the hospital while clinicians attempt to record and localize seizure activity. During this stay, researchers may take advantage of the implanted wires to conduct iEEG recordings for research purposes, and sometimes, they may use modified wires in the implantation itself to enable these recordings during the stay [4, 26]. Importantly, both of these research activities are subsumed in what is otherwise a clinical setting and is often led by researchers who are also the patients' clinician. Crucially, all *participants* in these studies are concurrently *patients*, who in many cases come to pursue a neurosurgical intervention when other therapeutic options have been exhausted.

This duality—in space, investigator role, and participant role—may complicate traditional ethical questions about risks, benefits, and consent, but also require expanding the locus of ethical attention and developing innovative practices (Table 6.1). Similarities and differences between research contexts and patient populations in DBS research and EMU research will be noted below.

6.2 Risks and Benefits

Federal common rule requires that risks of research are "reasonable in relation to anticipated benefits, if any, to subjects, and the importance of the knowledge that may reasonably be expected to result" (45 CFR Part 46). Intraoperative and extra-operative iEEG research is made possible and judged permissible in part because a large portion of the surgical risks of intracranial recordings are already assumed in the clinical interventions (e.g., implantation of the DBS device or monitoring wires). This renders the additional surgical risks of intracranial research marginally low, contributing to the rationale that the knowledge to be gained through these recordings is reasonably balanced against the minor increase in individual risk. In fact, conducting research that takes advantage of already occurring clinical activities is ethically encouraged for just this reason [8]. But, further ethical questions persist in iEEG research, not just about (i) what risks and benefits exist, but also (ii) the importance of those risks and benefits, and (iii) how to communicate with patients about those risks and benefits. In particular, when research is non-therapeutic or guided by basic scientific questions about the brain, questions endure about how to balance individual risks against the predicted value of the expected knowledge. Since the risks of iEEG research include risks of harm to the brain, the sequelae of such harms may be consequential, making the identification and weighing of risks and

Table 6.1 Ethical considerations for iEEG research with DBS and EMU patients. The first column lists broad study features deserving ethical consideration. Rows list fine-grained sub-categories for study specific consideration. For example, when considering the ethical importance of **the participant population** for a specific study, researchers may need to consider not only the diagnoses of the patients and potential vulnerabilities for consent, but also current and potential marginalization of that population and how research may intersect with senses of identity and diverse values

Categories of ethical importance	Sub-categories to consider			
Risks:	Kinds (clinical, social)	Magnitudes	Structural and financial context	Importance to participant population
Benefits:	Kinds (therapeutic, non-therapeutic, social)	Magnitudes	Distance from translation	Relationship to participants' disease
Participant population:	Diagnoses (kinds, severity, rarity)	Potential vulnerabilities (capacity, misconceptions)	Social positionality, marginalization	Identity, diverse values
Team constitution:	Presence of clinician-investigators	Presence of non-clinicians	Involvement of technology companies	Involvement of other health professionals
Care and research overlap:	Timing (intraoperative, extra-operative, duration)	Space (in OR, hospital, clinic, home)	Tasks (similarity to clinical tasks, intimacy)	Dual-roles
Participant involvement:	Kinds (intensity, duration)	Effects of participation	Participant motivations	Opportunities for engagement
Down-stream effects:	Barriers to fair access	Potential for exploitation	Effects on neurodiversity	Effects on agency, sense of self

benefits of critical importance. This section focuses not on the empirical estimations of those risks and benefits, but on motivating the extant challenges that exist and which researchers ought to address in their own studies.

6.2.1 *What Matters When Assessing Risks?*

The risks of intraoperative research include increased risk of surgical harm, such as stroke, bleeding, and infection, as well as increased discomfort from prolonged intraoperative time. While those risks may be relatively small compared to the risks assumed for the DBS surgery itself [4, 25], it can be difficult to quantify the added risks of the research intervention. How much time, for example, correlates with how much increased surgical risk? It may also be challenging to communicate these risks to patients in comprehensible ways which distinctly separate clinical risks from research risks. Risks of discomfort and stress, for example, may be difficult for patient-participants to imagine or predict, particularly before they have experienced the intraoperative environment. While research does not extend operating time for EMU patients, risks may still include discomfort and stress related to carrying out tasks while implanted. Furthermore, since EMU stays can extend for periods up to days and weeks, patient-participants may experience these risks differently over this time and change how concerned they are about them.

There may be other research related effects on patient-participants. Some repetitive tasks may be boring, some cognitive and movement tasks may be frustrating, and tasks involving emotions or affective states may have a more profound effects. Research involving emotion regulation, or inducing or suppressing distress, for example, may itself be distressing. Patient-participants may feel differently when study tasks are more intimate or relate more closely to their sense of self. As tasks become more closely connected to patient-participants' desires, behaviors, and self-conceptions, it may become more important to assess what is it like to participate in the research. When does participation itself impact patients positively or negatively, and when should some of these effects, if there are any, be communicated *as risks*? While there is emerging data on how patients experience using DBS devices for treatment, for example [11], there is less data on how intraoperative iEEG *research* affects patients, or how EMU patients experience research in the hospital. Recent work has begun to explore these experiences, finding for example, that (i) participants in intraoperative research recall very little about the risks and were relatively unconcerned about them at the time of consent, but (ii) have varied experiences of participation, ranging from neutral and benign to uncomfortable and painful. Researchers might have obligations to explore experiences of participation with their patients and discuss this potential in study specific ways [22, 29].

6.2.2 What Matters When Assessing Benefits?

While all research intends to generate generalizable knowledge, a significant portion of iEEG research, and intracranial recording in general, is basic science research. It is governed by a scientific question, aimed at providing a greater understanding of human brain function. The benefits of this research come in the form of knowledge, and past studies have elucidated, for example, the network basis of motor decision-making as well as the role of cortical and subcortical structures in language [6, 21, 30]. However, neither the extended time in the OR nor the extra tasks performed provide patients with added clinical benefits. This has a few implications. First, there is no potential clinical benefit to offset individual risks. Second, the knowledge researchers might expect to gain may be disconnected from the patients' condition, and so patients, or others with their diagnosis, may not experience the downstream translational benefits.

While there are no direct or near-term clinical benefits to participants, there may be other sources of value in participation. The topic of non-clinical, or social, benefits has gathered some attention in research pertaining to longer term BCI studies [17, 18], including the value gained from participating and taking part in the study itself. For example, some participants cite a sense of social worth and feeling part of the research team as sources as benefits. What kinds of value might exist for patients in intraoperative and extra-operative studies? Do patients welcome participating in research? Is it a burden? It is an open question whether these kinds of benefits ought to factor into a formal risk/benefit assessment, and whether value gained from participation is an "extraneous benefit" [8] or one more directly gained from the research itself. But, if social benefits do exist for DBS and EMU patients, researchers may have obligations to maximize them and respond to them in patient interactions and research design. For example, if patients feel part of the research team, or would like to, researchers might take steps to honor this, including expressing gratitude, sharing ultimate research findings, or exploring opportunities for patient-participant engagement, etc.

Furthermore, communicating to patients about benefits, when those benefits might be unknown, in the form of specialized knowledge, and disconnected from the patient's own condition, may require special care. It may not be sufficient, for example, to disclose that research recordings will not benefit participants. There may be much more discussion required to explore how patients interpret this claim and understand potential translational benefits in the future. We return to this question in discussing consent in Sect. 6.3.

6.2.3 What Is a Reasonable Risk/Benefit Ratio?

There is no formula for risk/benefit assessment. Furthermore, comparing the clinical risks to individual patients to the benefit of advancing society's understanding requires balancing two things which are perhaps incommensurate. This may lead to vague or inconsistent judgments [23]. What should be considered in these assessments?

Determining whether the balance of risks and benefits is reasonable requires not just an empirical judgment of what those risks and benefits are, but also a value judgment of their relative worth. While the scientific community may have privileged expertise about what risks exist and what benefits are likely, the value judgments of investigators may not reflect those of society writ large or the patient-participants who volunteer. How important is gaining a particular piece of knowledge about the brain? How much risk is worth how much knowledge? Is knowledge that is more closely tied with individual participants' diseases more valuable to pursue with these patient-populations? Does the distance between the expected knowledge and eventual translational opportunity matter? Generally, lower and less likely risks may permit more "circumspect benefits" [8], and iEEG research with DBS and EMU patients often has low surgical risks. But the risk/benefit ratio must still be favorable and some have proposed that this assessment require a citizen's understanding of social values [8].

While these judgments are a challenge for any non-therapeutic study, they may carry particular import in iEEG research which can only be conducted with DBS and EMU patients. These patients need an invasive clinical intervention for a neurological or psychiatric diagnosis and may belong to various disability populations systematically marginalized in other ways. Judgments about a study's risks/benefit ratio should not be made in isolation from these contextual features, and researchers should be especially attuned to the possibility of biased judgments which may downplay the risks of harm to potentially vulnerable groups. There is a danger that broader attitudes about neurological and psychiatric disorders weight these judgments, raising the threshold for tolerable risks and lowering it for required benefits.

It is important to address the questions raised here in addition to questions about individual consent. While overall risk/benefit assessments impact the permissibility of seeking consent, they are largely meant to be determined pre-consent—before any patient is recruited. Moreover, it is not sufficient to rely on consent alone to protect DBS and EMU patient-participants from unreasonable harm, since it is possible for patients to consent to unjustified risks. This potential may be exacerbated in settings where patients may be exhausting their treatment options, or where care and research overlap. As the next section will discuss, the context of DBS and EMU research may make patients particularly vulnerable with respect to informed consent, and careful risk/benefit assessments can maximize protection against undue harm.

6.3 Consent

Informed consent, as a way to promote respect for persons and minimize harm, is a pillar of ethical research with humans and requires that the decision to participate be both informed and voluntary [27]. Research with DBS and EMU patients raises several questions along both dimensions, many of which arise from the co-occurrence of care and research. There is currently limited data on participant experiences of consent and what best practices ought to involve. But emerging studies on intracranial research document heterogeneity in consent practices [9, 19] and lack of participant recall of informed consent [29]. These studies highlight the complexity of consent interactions for this research and the need for intentional and study specific design.

6.3.1 What Do Patients Understand?

Consenting patients for intraoperative research may require additional practices to assess whether patients have the capacity to understand the risks and the benefits sufficiently to make informed decisions. Discussions of informed consent often center the notion of capacity [1]. Whether participants can give informed consent to research participation depends in part on whether they have the capacity to understand the risks and the benefits, appreciate those risks and benefits as they apply to their own situation, and rationally explain their decision to participate.

Some of the diagnoses which make patients eligible for DBS treatment (such as Parkinson's disease) may be associated with cognitive impairments, particularly in the executive domain where judgements may be affected [13]. Patients who have the capacity to understand the risks and benefits of their surgery, which is tied to their health, symptoms, and current prognosis, may not have the same capacity to understand the details of a basic science study or the additional risks involved [5, 10]. IEEG recordings, for example, may have more abstract goals, not rooted in the individual's experiences. Researchers may need to consider how to explain the study's goals and risks in ways that patients can understand, but also separately assess whether the patient *is able* to understand.

At a minimum, patients need to understand the extent to which iEEG research will not benefit them. This is in addition to understanding the purpose of the study, or that the research is basic science research, for example. More specifically, it requires understanding that the research is not directly targeted at improving their own health or, sometimes, the health of individuals with their same disease. The timing of the research in intraoperative studies and, in many instances, the similarities between research and clinical tasks, may make this fact difficult to appreciate both during initial consent and again before the initiation of research in the OR.

This raises a broader question of whether patients understand the separation of care and research, by a team which may include their own clinician. This concern may be exacerbated when clinicians approach their own patients to join research studies. These considerations make salient the possibilities of a therapeutic misconception—patients may have the false belief that research is directed at their benefit or that they will benefit from research activities [1]. The intraoperative setting may be particularly open to these pressures since care and research take place during the same operating event. Additionally, patients may not adequately understand the different clinical and research tasks in the OR as they are doing them, which raises questions about the ability to withdraw, discussed below in Sect. 6.3.2

Additionally, the temporal connection between research and care may inspire the belief that the research is more related to the patient's disease or clinical intervention that it actually is, or at least that it stands to benefit other patients with similar conditions in the future. While this is not a therapeutic misconception, it may be misguided, affect patients' motivations for enrollment, and could be mitigated in discussions between patient-participants and the research team. When research is further disconnected from patient-participants' own disorder, more explicit discussion of the distance may be warranted in consent.

Lastly, concerns about upcoming surgery may eclipse patient-participants' ability to appreciate or adequately consider the risks and benefits of research, even if they understand what those are [22, 29]. Researchers ought to consider the timing of consent for the surgery and consent for research, ensuring there is adequate time and space to discuss the research itself, and its implications for patients who might enroll.

6.3.2 When Is Participation Voluntary?

Promoting the voluntary enrollments of patients in iEEG research requires multiple considerations. Patients may feel pressure to enroll in research when that research is led by their own clinician or carried out in the same institution as their care. This may stem from worries that their clinical care will be compromised if they decline, worries that their care will be better if they join, or a general worry that receiving DBS surgery or EMU monitoring creates the expectation that they ought to help with research. Patient-participants may need in-depth conversations about these studies to ensure that they feel comfortable declining and that their motivations for joining do not including misunderstanding or undue pressures.

A crucial component to voluntary participation is the right to withdraw from research without penalty [27]. But it may be difficult for patients to withdraw from a study mid-surgery, for aforementioned reasons. For example, they may not be able to easily distinguish the research portions of the DBS surgery from the clinical portions, especially if tasks are similar, and so they may not know which tasks they can decline to do without clinical consequence. They may be exhausted, in pain, or uncomfortable, and would like to revisit their decision to join, but worry about disappointing the research team once research is initiated. Any concerns about the influence of dual-role clinician-investigators during consent outside of the OR may be exacerbated inside the OR. Researchers may need to provide explicit check-ins with patients during research and explicit opportunities to withdraw, without waiting for the patient to signal that they would like to stop or reconsider. Similar considerations exist for EMU patients, who may need additional check-ins throughout their stay to ensure that they still understand their right to withdraw and feel comfortable continuing.

Since research in the OR and EMU is initiated when patients are awake, this provides an opportunity for re-consent. Patient-participants may acquire more information about what it's like to do certain tasks, how much they care about the research, and their overall experience in brain surgery, which in turn may impact their decision to participate in research. Seeking an explicit verbal reconsent can respect patient autonomy and the ability to make continual informed decisions. In the OR, this moment in time may also let patient-participants express any concerns with the research team before initiating recordings, ask any new or persisting questions, and in general, mirror initial consent in a new context [10].

Importantly, patients may be more vulnerable mid surgery, particularly if research is pursued before clinical activities are complete. While, in principle, patients can change their minds about participation and have the right to withdraw at any time from research, making patients feel comfortable doing so may deserve careful attention in the intraoperative setting.

6.3.3 What Is the Impact of Clinical and Research Overlap?

The overlapping of clinical and research activities necessitates considerations beyond effects on patient understanding and influences on consent. It also highlights the potential for conflict of interest, as physicians are generally understood to have different obligations to patients than researchers do to participants. Appreciation of this point has led some researchers to advance the position that clinician-researchers ought to be "clinicians first". And that the design of research questions and methodologies should be governed by principles of good clinical practice, aimed at maintaining the integrity of clinical care and space [10].

Furthermore, appreciation that all members of the research team are entering into a clinical space may help promote patient welfare. While iEEG research with DBS and EMU patients has a high prevalence of clinician-researchers, it may also involve researchers unrelated to the clinical team. These researchers are nonetheless in a clinical space, and may interact with patient-participants in carrying out the study. Non-clinical members of the team may consider training in how to interact with patients when in this dual space, especially when conducting consent [5, 10].

Lastly, conducting iEEG research in these dual contexts is a relatively rare opportunity, and researchers may feel pressure to take advantage of this opportunity when it arises. There may be an inherent tension between pursing these opportunities and the rewards of doing so in the form of grants, and putting patients' clinical welfare first. Appreciating these tensions may suggest research practices which make use of checks and balances, encourage collaboration, and foster multi-team member conferences to decide when a patient is a potential candidate for intraoperative or EMU research.

6.3.4 What Consent Practices Should Be Considered?

In response to these considerations, various approaches to consent for intraoperative research with DBS patients have been proposed. What these various practices have in common is a commitment to maximizing patient-participant understanding while minimizing undue pressure and influence.

While some, including the American Medical Association, have cautioned against "dual-role consent"—against clinician-researchers seeking consent from their own patients—others have called for more nuanced views, arguing that the closer the research is to usual care, the more preferable dual consent might be [20]. Others have also proposed a hybrid model [12], on which the surgeon or other physician discusses the risks and benefits of the research, while another member of the research team discusses other aspects of enrollment. This model attempts to maximize patient understanding as clinicians may be in the best position to explain risks in the context of the upcoming surgery, but also creates space between the clinician and the patient.

Some studies may benefit from the inclusion of a third person, who is neither on the clinical team nor the research team, whose job it is to advocate for the patient-participant's interests and values. This may not be feasible at all institutions, and questions remain for how best to implement an objective perspective into consent. Some have suggested using standardized videos to supplement in depth conversations, which could provide an external voice to reinforce the right to withdraw and the separation of clinical care and research [10].

Suggestions for improving consent which make use of multiple in-depth conversations may also encourage participants to discuss research with other physicians not involved in the research itself, include their families, and devote more time to understanding the separation of care and research and the study purpose and risks. Seeking "teach-back" of important study details may increase retention of study goals and risks, but work remains on determining what relevant facts should be "taught back" [29]. These varied consent practices appreciate that patient-participants may want to discuss the study with their clinician who will be in charge of their care and with whom they may have robust trusting relationships, but also, foster communication opportunities with health professionals outside of the clinical team.

6.4 Justice

Currently, all intracranial neurophysiological research requires the participation of patients with neurological or psychiatric diseases. Individuals who do not have a condition the treatment of which DBS or EMU stays would or could ameliorate, cannot participate. This raises concerns about the fair distributions of benefits and burdens of iEEG research as well as concerns about exploitation.

6.4.1 What Are the Pressures on Fair Access?

Justice in research is often interpreted as requiring at minimum, the fair distribution of the burdens and benefits of research [27]. The burdens of this research are carried by individual patients. In the case of clinical research, participants may receive a share of the benefits in the form of therapeutic outcomes, but for basic human neuroscience research, the direct benefits are in the form of general knowledge. How, when, and if that knowledge will translate into clinical benefit may be unclear, vaguely understood, and in the distant future. How do we ensure that the patient populations who incur the risks of this research also have fair access to those clinical benefits when they are available?

When iEEG research with DBS and EMU patients involves potentially vulnerable and systemically marginalized groups, rapidly advancing cutting-edge technology, and increasing involvement of for-profit industries, concern is warranted that the fruits of this research may not be accessible to the groups who made it possible. This is in part a concern about cost, the current health care infrastructure in places like The United States, and the structural barriers many disabled individuals face in affording and accessing the resulting devices and interventions that this research may yield.

While partnering with technology companies may be crucial for scientific progress, it may also center the goals of profitability, which may likewise lead to end products and procedures that disability populations cannot easily access. Relatedly, it may encourage the end products to migrate away from therapeutic goals for particular disease groups, towards enhancement technology or general direct to consumer technology. While this technology may be likewise unaffordable, perhaps more problematic is that it may be completely unconcerned with improving the lives of disabled populations altogether.

6.4.2 What Are the Worries About Exploitation?

A central concern in any non-therapeutic study is protecting patients against exploitation—from being taken unfair advantage of or having their vulnerabilities used for another's benefit [31]. Since iEEG research with invasive electrodes must make use of existing clinical situations, a persistent question is when and under which conditions making use of these situations is fair. This chapter cannot answer this question, but considering this research through the lens of fairness may promote more varied reflection and inspire different practices than reflection on consent and risk/benefit assessments alone.

Two kinds of exploitation concerns deserve attention. The first involves the possibility that the distribution of research benefits excessively favors one party (e.g., researchers) over others (e.g., patient-participants). This lens centers which parties in the research benefit from that process, in what ways, and how much [28]. Researchers, for example, may benefit from enrolling patients in the form of future grants, publications, and promotions. Patient-participants, on the other hand, gain no such benefits and do not experience research related clinical outcomes. A second consideration is about process. It involves the worry that one party could take unfair advantage of an existing vulnerability in their interactions with another party. Patients approached to participate in intraoperative research, for example, may be particularly vulnerable to confusing care with research or feeling pressured to please their clinician [5]. A researcher might, in attempt to boost enrollment numbers, take advantage of these vulnerabilities during consent interactions. In these cases, the worry is not just that research may produce an unfair disparity in benefits between patients and researchers, but rather, that it may make unfair use of an existing disparity [16].

It is important to consider both of these questions in iEEG research design, since while good consent procedures can protect against exploitation to some extent, it's possible for patients to consent to exploitative interactions. Careful risk/benefit assessments could also protect against exploitation, but unexamined researcher perspectives may be biased. The context of intraoperative and extra-operative iEEG research highlights that these participants are different from mere volunteers and other research participants. They are participants whose involvement is made possible by the fact that they are in an operating room having brain surgery or in the hospital waiting for seizure onset. They may thus require more protections than other research participants in efforts to make sure that their situation is not taken unfair advantage of.

6.5 Future Social Considerations

As the field of invasive functional neurosurgery expands and technological capabilities improve, the kinds of diseases that invasive electrodes can treat will expand. For example, while OCD is currently the only psychiatric condition for which DBS is approved by means of a humanitarian device exemption by the FDA, studies are currently exploring the value of DBS for other disorders, such as treatment resistant depression [15, 24]. These advancements bring with them new opportunities to study the human brain, in terms of number of surgical opportunities, kinds of brain activity, and patient populations involved. As more questions become possible to ask and answer with iEEG research, different ethical complexities come to the fore, including the broader social implications and translations of this research. Furthermore, given that iEEG research may translate into advances in neurotechnology, it may inherit some of the concerns neuroscientists and ethicists have raised about the effects of these technologies on individuals and society [2, 14]. This section explores just a few of these considerations, including neurodiversity, identity, and agency.

6.5.1 How Might Potential Uses of iEEG Research Impact Neurodiversity?

Since a significant portion of iEEG recording data is for basic science research, the translational and societal uses of the eventual advances in technology may be unknown. Potential concerns have been raised, for example, about the effects of identifying biomarkers for chronic pain, including the marginalization of patients' subjective experiences (which the International Association for the Study of Pain identifies as crucial to chronic pain diagnoses), and use by employers and insurers to decline employment or insurance coverage for individuals who are particularly pain-sensitive or who exhibit a predisposition to chronic pain [7]. But Davis et al. also recognize the potential for this data to be used beneficially, to "advise patients of their personal risks and benefits of a surgical procedure, including the risks of chronic post-surgical pain, and to protect physicians from liability if such pain develops" [7].

IEEG data may have the potential to decrease stigma associated with various neurological and psychiatric diseases and disabilities, legitimizing patient claims and easing the burden of testimony. This may help marginalized populations—including women and patients of color whose reports of pain and other symptoms are often systemically disbelieved and dismissed [3]—access necessary healthcare. However, this data may also lead to an increase in stigma associated with other neurodiverse behaviors. Data from iEEG research may combine with societal biases and contribute to the view that those studied behaviors are pathological, that persons with those behaviors have something wrong with their brain function, and that patients should pursue treatment. It may encourage physicians to order imaging tests and insurance companies to require them.

While the immediate goals of iEEG research may be narrow in scope and defined in abstraction from the broader clinical and social questions raised above, the results of these studies intersect with existing social biases and social problems. Since this research involves—and will continue to involve—populations that are already marginalized, researchers ought to consider the responsible stewardship of the data and findings that might impact those populations. One of the central worries about human experimentation with vulnerable groups is that more widespread harm can come to those groups as a result of the experimentation, including further marginalization and medicalization. Researchers should consider what effects pursing certain questions about brain function will have on neurodiversity, including whether they might further exacerbate existing inequalities, and how best to take into account the social positionality of many patients who will use resulting treatments.

6.5.2 How Might Resulting Technology Affect Identity and Agency?

It has been suggested that neurotechnologies, including DBS, should be understood in terms of how they move users closer and further from their "authentic selves" [11]. This lens centers how using neurotechnology may affect potential patients' self-conception, identity, and agency—how they move around the world, and what it feels like to do so.

As Goering et al. describe, "we often think of the brain as a very important center of our values, desires, behaviors, and self-conceptions. We understandably are possessive of the neural activity that makes us who we are" [11]. When patients pursue neurotechnology therapies that change their experience of the world, such as DBS for depression or Parkinson's, this may impact how they interpret those experiences, whether they identify with them or see them as foreign, and whether they see themselves as the author of resulting actions. Patient narratives, for example, suggest that while many patients feel like themselves when using DBS devices, some may feel disconnected and unsure of the authorship of their actions [11].

Using neurotechnology may not only impact one's sense of personal identity and agency, but also aspects of one's social identity. Neurotechnology may be used, for example, to modify brain activity which accompanies various social identities (e.g., autism, deafness, blindness), and iEEG research may aim to understand brain function which accompanies those identities. Considering what patients with various identities value, including what functional abilities would be beneficial, can inform which questions researchers ought to pursue and what meaningful technological end points would be. What blind individuals might value most in assistive neurotechnology, for example, may not be what sighted researchers most value. Without patient engagement, research and resulting technology may advance ableist assumptions about various identities.

IEEG research is in some ways decoupled from these issues, as patients do not experience the outcomes of research in the same way they do with therapy. But, even basic research may contribute to the development of technologies which directly intersect with our desires, self-conceptions, and social identities. Narratives from patients can provide insight into what it's like to participate in iEEG research and to use the resulting technologies, and may prove crucial for translating iEEG research into therapies which respect diverse values, identities, and neuro-abilities.

6.6 Conclusions

The surgical opportunities provided by DBS and EMU patients enable very specific kinds of research, whose questions and tools are dependent on what the clinical context makes available. But equally important is the fact that the relationship between clinical care and iEEG research is somewhat circular. IEEG studies are not only influenced by current clinical opportunities and available technologies, but in turn, influence which clinical opportunities and technologies will exist in the future. Decisions of iEEG researchers play a crucial role in which parts of the brain are explored, and thus, which disorders will be understood and which treatments will be made realities. This chapter has explored the ethical implications of both directions of influence, with the goals of motivating further discussions among researchers, clinicians, and patient-participants.

References

1. Appelbaum P, Roth L, Lidz C (1982) The therapeutic misconception: informed consent in psychiatric research. Int J Law Psychiatry 5(3–4):319–329. https://doi.org/10.1016/0160-2527(82)90026-7
2. Cabrera Y, Evans E, Hamilton RH (2014) Ethics of the electrified mind: defining issues and perspectives on the principled use of brain stimulation in medical research and clinical care. Brain Topogr 27(1):33–45. https://doi.org/10.1007/s10548-013-0296-8
3. Carel H, Kidd IJ (2014) Epistemic injustice in healthcare: a philosophical analysis. Med Health Care Philos 17(4):529–540. https://doi.org/10.1007/s11019-014-9560-2
4. Carlson AA, Rutishauser U, Mamelak AN (2018) Safety and utility of hybrid depth electrodes for seizure localization and single-unit neuronal recording. Stereotact Funct Neurosurg 96(5):311–319. https://doi.org/10.1159/000493548

5. Chiong W, Leonard M, Chang E (2018) Neurosurgical patients as human research subjects: ethical considerations in intracranial electrophysiology research. Neurosurgery 83(1):29–37. https://doi.org/10.1093/neuros/nyx361
6. Dastolfo-Hromack C, Bush A, Chrabaszcz A et al (2022) Articulatory gain predicts motor cortex and subthalamic nucleus activity during speech. Cereb Cortex 32(7):1337–1349. https://doi.org/10.1093/cercor/bhab251
7. Davis KD, Flor H, Greely HT et al (2017) Brain imaging tests for chronic pain: medical, legal and ethical issues and recommendations. Nat Rev Neurol 13(10):624–638. https://doi.org/10.1038/nrneurol.2017.122
8. Emanuel EJ, Wendler D, Grady C (2000) What makes clinical research ethical? JAMA 283(20):2701–2711. https://doi.org/10.1001/jama.283.20.2701
9. Feinsinger A, Pham M, Pouratian N (2021) The value of heterogeneity in practices to promote ethical research. AJOB Neurosci 12(1):80–82. https://doi.org/10.1080/21507740.2020.1866116
10. Feinsinger A, Pouratian N, Ebadi H et al (2022) Ethical commitments, principles, and practices guiding intracranial neuroscientific research in humans. Neuron 110(2):188–194. https://doi.org/10.1016/j.neuron.2021.11.011
11. Goering S, Brown T, Klein E (2021) Neurotechnology ethics and relational agency. Philos Compass 16(4):e12734. https://doi.org/10.1111/phc3.12734
12. Grady C (2019) A hybrid approach to obtaining research consent. Am J Bioeth 19(4):28–30. https://doi.org/10.1080/15265161.2019.1574493
13. Halhouli O, Zhang Q, Aldridge GM (2022) Caring for patients with cognitive dysfunction, fluctuations and dementia caused by Parkinson's disease. Prog Brain Res 269(1):407–434. https://doi.org/10.1016/bs.pbr.2022.01.018
14. Hendriks S, Grady C, Ramos K et al (2019) Ethical challenges of risk, informed consent, and posttrial responsibilities in human research with neural devices: a review. JAMA Neurol 76(12):1506–1514. https://doi.org/10.1001/jamaneurol.2019.3523
15. Holtzheimer PE, Husain MM, Lisanby SH et al (2017) Subcallosal cingulate deep brain stimulation for treatment-resistant depression: a multisite, randomised, sham-controlled trial. Lancet Psychiatry 4(11):839–849. https://doi.org/10.1016/S2215-0366(17)30371-1
16. Jansen LA, Wall S (2013) Rethinking exploitation: a process-centered account. Kennedy Inst Ethics J 23(4):381–410. https://doi.org/10.1353/ken.2013.0015
17. Kögel J, Jox RJ and Friedrich O (2020) What is it like to use a BCI?—insights from an interview study with brain-computer interface users. BMC Med Ethics 21(2). https://doi.org/10.1186/s12910-019-0442-2
18. Kögel J, Schmid JR, Jox RJ et al (2019) Using brain-computer interfaces: a scoping review of studies employing social research methods. BMC Med Ethics 20(1):18. https://doi.org/10.1186/s12910-019-0354-1
19. Mergenthaler JV, Chiong W, Dohan D et al (2021) A qualitative analysis of ethical perspectives on recruitment and consent for human intracranial electrophysiology studies. AJOB Neurosci. https://doi.org/10.1080/21507740.2020.1866098
20. Morain SR, Joffe S, Largent EA (2019) When is it ethical for physician-investigators to seek consent from their own patients? Am J Bioeth 19(4):11–18. https://doi.org/10.1080/15265161.2019.1572811
21. Mosher C, Mamelak AN, Malekmohammadi M et al (2021) Distinct roles of dorsal and ventral subthalamic neurons in action selection and cancellation. Neuron 109(5):869-881.e6. https://doi.org/10.1016/j.neuron.2020.12.025
22. Peabody Smith A, Taiclet L, Ebadi H, Levy L, Weber M, Caruso EM, Pouratian N, Feinsinger A (2022) "They were already inside my head to begin with": Trust, Translational Misconception, and Intraoperative Brain Research. AJOB Empirical Bioethics. https://doi.org/10.1080/23294515.2022.2123869
23. Rid A, Wendler D (2011) A framework for risk-benefit evaluations in biomedical research. Kennedy Inst Ethics J 21(2):141–179. https://doi.org/10.1353/ken.2011.0007

24. Sheth SA, Bijanki KR, Metzger B et al (2021) Deep brain stimulation for depression informed by intracranial recordings. Biol Psychiatry S0006–3223(21):01747–01749. https://doi.org/10.1016/j.biopsych.2021.11.007
25. Sisterson ND, Carlson AA, Rutishauser U et al (2021) Electrocorticography during deep brain stimulation surgery: safety experience from 4 centers within the National Institute of Neurological Disorders and Stroke Research Opportunities in human consortium. Neurosurgery 88(5):E420–E426. https://doi.org/10.1093/neuros/nyaa592
26. Suthana NA, Parikshak NN, Ekstrom AD et al (2015) Specific responses of human hippocampal neurons are associated with better memory. Proc Natl Acad Sci USA 112(33):10503–10508. https://doi.org/10.1073/pnas.1423036112
27. The Commission (1978) The Belmont report: ethical principles and guidelines for the protection of human subjects of research
28. Wertheimer A (2010) Rethinking the ethics of clinical research: widening the lens. Oxford University Press. https://doi.org/10.1093/acprof:oso/9780199743513.001.0001
29. Wexler A, Choi RJ, Ramayya AG, Sharma N, McShane BJ, Buch LY, Donley-Fletcher MP, Gold JI, Baltuch GH, Goering S, Klein E (2022) Ethical issues in intraoperative neuroscience research: assessing subjects' recall of informed consent and motivations for participation. AJOB Empirical Bioeth 13(1):57–66. https://doi.org/10.1080/23294515.2021.1941415
30. Zavala B, Jang A, Trotta M et al (2018) Cognitive control involves theta power within trials and beta power across trials in the prefrontal-subthalamic network. Brain 141(12):3361–3376. https://doi.org/10.1093/brain/awy266
31. Zwolinski M, Wertheimer A (2017) In: Zalta EN (ed) Exploitation. The Stanford Encyclopedia of Philosophy

Chapter 7
Which Ethical Issues Need to Be Considered Related to Microwires or Utah Arrays?

Michael J. Young

Abstract Intracortical microwires and multichannel microelectrode arrays including the Utah array and its derivatives have provided formative groundwork for extraordinary insights into brain dynamics, and carry immense promise to alleviate suffering and restore function to persons with severe neurological disorders through novel neural interfaces and neuroprosthetics (see also Chap. 51). As basic and clinical research involving these novel materials and devices continues apace, with several pivotal devices at the cusp of clinical translation, proactive consideration of their ethical and philosophical dimensions is crucial. This chapter explains and critically evaluates key ethical concerns surrounding safety, signal integrity, neural data privacy, impact on personal identity, and post-trial obligations related to these devices. It also explores how rich neural data and modulatory capabilities provided by these technologies could uniquely advance philosophical inquiry into the reducibility of conscious experience, and clarify notions of causality, constitution, and identity in theories of mind.

7.1 Risks and Safety

Ethical analysis of intracortical microwires and microelectrode arrays such as the Utah array should begin with consideration of the central principles of biomedical ethics, including beneficience (i.e., maximizing benefits and wellbeing), non-maleficience (i.e., avoiding harm), justice (i.e., ensuring fair distribution and equity), and autonomy (i.e., respecting individuals' preferences and dignity). Knowledge of and transparency surrounding the risks and safety of these devices and the procedures necessary to implant them are necessary to ensure that translational research proceeds in accordance with the principles of beneficience and non-maleficience, and to ensure that consent processes are optimally informed [1–4]. Ethical issues related to justice may arise in the context of clinical research involving vulnerable patient

M. J. Young (✉)
Division of Neurocritical Care, Department of Neurology, Center for Neurotechnology and Neurorecovery, Massachusetts General Hospital, Harvard Medical School, Boston, USA
e-mail: michael.young@mgh.harvard.edu

N. Axmacher (ed.), *Intracranial EEG*, Studies in Neuroscience, Psychology and Behavioral Economics, https://doi.org/10.1007/978-3-031-20910-9_7

populations, who may be more prone to undue inducement or to taking on risks due to therapeutic misconception or false hope (see also Chap. 6 on general ethical considerations of research involving intracranial EEG and deep brain stimulation) [5].

Potential risks include those associated with neurosurgery, infection, inflammation, electrode migration, fracture of leads, skin erosion, bleeding, local disruption of the blood–brain barrier, local cytotoxicity and gliosis [2, 6–8]. Research efforts are underway to improve materials used to minimize these potential risks [9–12]. It is additionally important to consider atypical risks following implantation, including potential changes in personality, identity, mood, and cognition, especially with respect to devices that stimulate in addition to sensing and recording brain activity (see Chaps. 5, 39, 41 and 52 on the promises and challenges of deep brain stimulation) [13].

Investigators must consider these risks and proactively counsel prospective study participants in order to ensure that consent is optimally informed and that benefits outweigh risks. In situations where risks unambiguously outweigh potential benefits, research cannot proceed ethically [14–16]. As investigators themselves might harbor inherent conflicts of interest in making such determinations, institutional review boards (IRBs) have important roles to play in evaluating benefits and risks of proposed interventions and determining whether the balance is sufficiently acceptable for research to ethically proceed [17].

The system through which implantable neurotechnologies are regulated revolves primarily around the driving aims of ensuring device safety and efficacy while advancing public health [18]. While these aims are vital and the regulatory system is well-tuned to meet them, current regulations do not address a distinctive array of ethical, social and philosophical challenges that fall outside the strict purview of these aims as they are traditionally conceived. These include obstacles in ensuring adequate informed consent in contexts of neurologic disturbance that may affect patient capacity [13, 19, 20]; protection, management and sharing of uniquely sensitive and increasingly rich neural information (currently not fundamentally protected despite being of potentially equal or greater importance than genetic information which receives special protection under the Genetic Information Nondiscrimination Act [21–23] (as we will examine later in this chapter); uncertainties surrounding the phenomenological impact of implanted neural devices on users' sense of self [24], personal identity [25–27], and agency [28]; concerns surrounding potential applications for enhancement [29–31]; brain-specific risk factors (especially in the developing brain [32] for devices geared toward pediatric patients) [13, 33, 34]; and, for persons in clinical trials, the scope of sponsors and/or investigators' post-trial responsibilities toward those who opt for long-term use [35]. It is encumbant on investigators to partner with neuroethicists in proactively identifying these downstream risks and developing safeguards to mitigate and optimally counsel prospective users of these technologies about them.

7.2 Signal Integrity and Integrity in Signal Translation

Additional ethical issues arise in considering the integrity of recorded signals and the longevity of sensing apparatus. These issues become especially relevant in the context of neural interfaces to restore or augment communication. The importance of ensuring that decoded communication is reliable rises in tandem with the significance of what is being communicated and the degree to which it may impact weighty decisions. For example, the importance of ensuring the reliability of a BCI-mediated communication concerning goals of care is far greater as compared to the importance of ensuring the reliability of a BCI-mediated communication concerning one's lunch preferences. In the case of detecting command signals and translating those signals into actions, uncertainties may exist about how to determine whether the action executed accurately reflects the intention of the agent; conundrums surrounding moral responsibility and culpability for BCI-mediated action may consequently arise [28, 36]. To ensure that participants and family members are informed, investigators should counsel prospective participants on the nature and duration of training periods necessary to optimize performance. Moreover, investigators must rigorously study the integrity and longevity of recorded signals over time, and understand whether the quality of signals may gradually decline over months or years following implantation [3]. Further research is needed to understand participants' perspectives on the reliability and trustworthiness of decoded communications and commands. Furthermore, investigations of how philosophical theories of action and language may inform normative interpretations and attitudes toward BCI-mediated communication and action [25, 30]. Such approaches and the information they might yield could guide an empirically grounded ethical framework for safeguarding clinical translation and optimizing informed consent paradigms for microwires and microelectrode arrays.

7.3 Neural Data Privacy

As development and deployment of microelectrode arrays in clinical contexts advances, increasingly rich neural data will be captured and recorded at massive scales never before feasible, both within and across individuals. Profound ethical, social and legal questions accordingly emerge surrounding how to handle and protect such neural data, which may reveal uniquely sensitive information about individuals, given the centrality of the brain to human identity. Related ethical and legal questions arise concerning neural data ownership, and the ethics of monetizing such data [37]. Management and sharing of uniquely sensitive and increasingly rich neural information is currently not fundamentally protected despite being of potentially equal or greater importance than genetic information which receives special protection under GINA [21–23]. Efforts are underway to craft and codify policies to protect neural data, define rights to mental privacy, and guide responsible neural data handling and sharing [38–41]. Absent such protections, the scientific imperatives of data sharing and open science may come to conflict with the ethical imperatives of protecting individual privacy and mitigating potential harms that could accrue through advertant or inadvertant data access. This ethical imperative presents both a challenge and opportunity to investigators; proactive consideration of these issues is needed to guide development of a platform for neural data sharing that is sensitive to privacy issues while conducive to neuroscientific collaboration and progress.

Analogous challenges have indeed emerged in other research contexts involving the capture of sensitive biomedical data, leading to the development of enhanced privacy-enhancing methods to safeguard data while allowing it to be responsibly shared for specific means through vetted channels and federated platforms [42–49]. The importance of developing privacy-enhancing methods and protections is magnified in the setting of neural data, which can hypothetically reveal even more about an individual than ordinary medical data [21]. Indeed, while the topic of data sharing has been explored and dissected in a variety of research and clinical contexts to date, distinctive and under-explored challenges are posed in the context of neural data [50, 51]. The accelerating development and deployment of microelectrode arrays and microwires challenge researchers, neuroengineers and regulators to closely evaluate existing oversight and governance systems to determine how to balance the ideal of data transparency, which motivates open sharing of information in ways that are accessible and understandable to stakeholders, with countervailing considerations surrounding maintaining scientific rigor and protecting the privacy of participants sensitive neural data [52, 53]. It is recognized that maximizing data sharing and trials transparency can serve to advance science, spur further neurotechnology and neurotherapeutic development, and keep patients and the public informed, however these efforts may be met with concerns that neural data can reveal uniquely sensitive information and is potentially re-identifiable [54, 55]. Accordingly, the ideal boundaries, extent and level of generality of data-sharing are normatively constrained by principled interests to protect the privacy of research participants and patients, and preserve the rigor of neuroscientific inquiry if studies are ongoing. Neural data repository platforms have been suggested as a strategy for facilitating responsible and secure data sharing, however their resilience in terms of data protection, corporate status, long-term viability are highly variable, and systems of governance have yet to be systematically defined [56]. A tailored approach to neural data sharing may be needed to specially handle potentially re-identifiable neural information. Input from key stakeholders, captured through an empirical and embedded neuroethics approach, will be essential to inform this evolving conceptual and ethical landscape, with the aim of identifying and codifying an actionable approach to optimally balancing transparency and privacy in this evolving era of large-scale and high-dimensional neural data [57].

7.4 Personal Identity and Agency

When considering the ethical implications of microelectrode arrays or microwires in the context of brain-machine interfaces or neuroprosthetic devices, ethicists and philosophers have explored the distinctive impact that these devices may have on users' sense of self, agency and personal identity [38, 58–60]. Living with an integrated microelectrodes or microwires in the brain may carry important existential and phenomenological significance, especially when such devices exist in "closed-loop" systems (see also Chap. 41) [61]. The sense of embodiment that users might experience with relation to such devices requires careful ethical and philosophical study. Ethicists Sara Goering, Eran Klein and colleagues have described a kind of 'relational agency' that may emerge when actions and communication become mediated through technologies and others [35]. Users may experience a sense of alienation, alterations in agency, or changes in perceived authenticity when behaviors or speech are accomplished via novel neural interfaces [62–64]. The lived experiences of users of these technologies are important to understand and rigorously study not only because of their ethical gravity, but also because these very phenomenological features and experiences may reflexively influence the performance of the devices under investigation. Neuroengineers and investigators ought to consider how to incorporate these concerns upstream in the research and development process. Training, counseling and neuroethics-informed design techniques may help to proactively address and mitigate these issues [65]. On a more speculative plane, neuroethicists and legal scholars have explored the hypothetical implications of BCI-mediated behaviors and communication for users' culpability and responsibility; how should responsibility be adjudicated when an BCI-mediated action results in harm? How might such interfaces affect the fundamental rights of users or of those with whom users may interact? As legal scholars, ethicists and philosophers consider these future-oriented issues, neuroengineers and investigators should consider these potential downstream implications, especially as functional capacities enabled by microwires and microelectrode arrays increase [66–70].

7.5 Post-study Obligations

Active consideration of ethical and legal post-study responsibilities are crucial [35, 71, 72]. Who is responsible for monitoring, removing or maintaining microelectrode arrays or microwires in human participants once a research study has completed? How should investigators handle user-driven deviations from study protocols? If a participant is benefiting from an implanted neurotechnology, and perhaps might even identify with it as part of their self identity [72], may they opt to continue using it after a study ends? If so, who ought to cover the costs and carry the responsibilities of device maintenance and monitoring after study closure? A multi-stakeholder process is needed to clarify (1) the nature and scope of investigators' responsibilities to provide long-term technical and clinical support after clinical trial closure; (2) funding and budgetary processes to ensure adequate resource provisions in the event that a subject opts for device use beyond the grant funding period (e.g., via a study-end escrow); (3) plans for dissemination of and training in specialized techniques pertaining to device management and explantation, including training of additional surgeons or other personnel who could safely remove implanted devices if necessary or simply if requested; and (4) clarification of how to handle user-driven deviations from study protocols.

7.6 Neurophilosophical Implications

Microwires and microelectrode arrays promise to enable unprecedented recordings of vast data from the human brain, potentially revealing remarkable insights into its multiform functions, mechanisms and pathologies [73–80]. These data could inform not only neuroscientific inquiry to be leveraged for potential therapeutic tools, but may also importantly inform approaches to age-old philosophical questions surrounding consciousness, volition and free will, and potentially clarify notions of causality, constitution, and identity in theories of mind [81]. Such breakthroughs promise to bring previous aspirations of melding philosophy and neuroscience closer to fruition [82–84]. As high-dimensional neural data become increasingly accessible for researchers, growing opportunities for partnership with philosophers to understand how these data might be used to test philosophical theories or answer longstanding philosophical questions should be recognized and strengthened. For example, how might recent abilities to detect covert consciousness following severe brain injury by measuring neural signals inform competing theories surrounding the phylogeny, ontogeny and ontology of consciousness [4, 85–90]? Recognizing these opportunities, an emerging discipline of "neurophilosophy" has taken shape; concerted efforts to refine this nascent field's methodologies are underway [91–98].

7.7 Conclusions

To promote the successful and ethical development of neurotechnologies employing microwires and microelectrode arrays, neuroscientists, clinicians and engineers are working with neuroethicists with the common goal of proactively identifying and rigorously assessing the trajectories and potential limits of these novel technologies. The BRAIN Initiative and its funding agencies have been visionary in emphasizing that as research programs advance frontiers of neurotechnology, neuroethical considerations should be addressed. In 2018, the NIH BRAIN Initiative Neuroethics Working Group disseminated a set of Neuroethics Guiding Principles that encourage research teams to undertake robust safety assessments; to consider issues "related to capacity, autonomy and agency"; to "[p]rotect the privacy and confidentiality of neural data"; to "[a]ttend to possible malign uses of neuroscience tools and neurotechnologies"; to emphasize caution in clinical translation; to "[i]dentify and address specific concerns of the public about the brain;" to "encourage public education and dialogue;" and to "[b]ehave justly and share the benefits of neuroscience research and resulting technologies" [99, 100]. Guided by this imperative, neroengineering and translational research teams should identify opportunities to embed personnel with neuroethics expertise throughout the development lifecycle to elucidate and to be a sounding board for ethical implications and to help develop safeguards for responsible translation.

Analyses of novel ethical issues raised by implantable neurotechnologies are particularly informed when the perspectives of those whose views perhaps matter most are explicitly included, namely, those of patients, family members, surrogates [37, 101–103], in addition to those of clinicians, engineers, neuroscientists, and other researchers. In parallel with development of implantable neurotechnologies, research teams have the opportunity to improve the impact of their device development by capturing perspectives and concerns of key stakeholders during the earliest planning stages of research and development.

References

1. Ryu SI, Shenoy KV (2009) Human cortical prostheses: lost in translation? Neurosurg Focus 27:E5
2. Ferguson M, Sharma D, Ross D et al (2019) A critical review of microelectrode arrays and strategies for improving neural interfaces. Adv Healthcare Mater 8:1900558
3. Sponheim C, Papadourakis V, Collinger JL et al (2021) Longevity and reliability of chronic unit recordings using the Utah, intracortical multi-electrode arrays. J Neural Eng 18:066044
4. Young MJ, Bodien YG, Edlow BL (2022) Ethical Considerations in clinical trials for disorders of consciousness. Brain Sci 12:211
5. Largent EA, Peterson A, Lynch HF (2021) Fda drug approval and the ethics of Desperation. JAMA Intern Med 181:1555–1556
6. Delbeke J, Haesler S, Prodanov D (2020) Failure modes of implanted neural interfaces. In: Neural interface engineering. Springer, pp 123–172
7. Wang M, Mi G, Shi D et al (2018) Nanotechnology and nanomaterials for improving neural interfaces. Adv Func Mater 28:1700905
8. Tathireddy P, Solzbacher F, Hitchcock R et al (2011) Implantable microsystems. In: Springer handbook of medical technology. Springer, pp 801–819
9. Wu N, Wan S, Su S et al (2021) Electrode materials for brain–machine interface: a review. InfoMat 3:1174–1194
10. Liu S, Zhao Y, Hao W et al (2020) Micro-and nanotechnology for neural electrode-tissue interfaces. Biosens Bioelectron 170:112645
11. Wellman SM, Eles JR, Ludwig KA et al (2018) A materials roadmap to functional neural interface design. Adv Func Mater 28:1701269
12. Sharafkhani N, Kouzani AZ, Adams SD et al (2022) Neural tissue-microelectrode interaction: brain micromotion, electrical impedance, and flexible microelectrode insertion. J Neurosci Methods 365:109388
13. Hendriks S, Grady C, Ramos KM et al (2019) Ethical challenges of risk, informed consent, and posttrial responsibilities in human research with neural devices: a review. JAMA Neurol 76:1506–1514

14. Emanuel EJ, Wendler D, Grady C (2000) What makes clinical research ethical? JAMA 283:2701–2711
15. Association WM (2013) World Medical Association Declaration of Helsinki: ethical principles for medical research involving human subjects. JAMA 310:2191–2194
16. Weijer C (2000) The ethical analysis of risk. J Law Med Ethics 28:344–361
17. Rid A, Wendler D (2011) A framework for risk-benefit evaluations in biomedical research. Kennedy Inst Ethics J 21:141–179
18. Fda (2018) FDA mission. U.S. Food and Drug Administration
19. Vaishnav NH, Chiong W (2018) Informed consent for the human research subject with a neurologic disorder. In: Seminars in neurology. Thieme Medical Publishers, pp 539–547
20. Giordano J (2015) Conditions for consent to the use of neurotechnology: a preparatory neuroethical approach to risk assessment and reduction. AJOB Neurosci 6:12–14
21. Ienca M, Haselager P, Emanuel EJ (2018) Brain leaks and consumer neurotechnology. Nat Biotechnol 36:805–810
22. Hochberg L, Cochrane T (2013) Implanted neural interfaces. Neuroethics Pract 235
23. Kostiuk SA (2012) After GINA, NINA-neuroscience-based discrimination in the workplace. Vand L Rev 65:933
24. Goering S, Yuste R (2016) On the necessity of ethical guidelines for novel neurotechnologies. Cell 167:882–885
25. Young MJ (2014) Ethics and ontology in deep brain stimulation. AJOB Neurosci 5:34–35
26. Sample M, Aunos M, Blain-Moraes S et al (2019) Brain–computer interfaces and personhood: interdisciplinary deliberations on neural technology. J Neural Eng 16:063001
27. Farah MJ, Wolpe PR (2004) Monitoring and manipulating brain function: new neuroscience technologies and their ethical implications. Hastings Cent Rep 34:35–45
28. Young MJ (2020) Brain-computer interfaces and the philosophy of action. Taylor & Francis, pp 4–6
29. Emanuel P, Walper S, Dieuliis D et al (2019) Cyborg soldier 2050: human/machine fusion and the implications for the future of the DOD. CCDC CBC APG United States
30. Young MJ (2015) Bioenhancements and the telos of medicine. Med Health Care Philos 18:515–522
31. Moreno JD (2003) Neuroethics: an agenda for neuroscience and society. Nat Rev Neurosci 4:149–153
32. Peña C, Bowsher K, Samuels-Reid J (2004) FDA-approved neurologic devices intended for use in infants, children, and adolescents. Neurology 63:1163–1167
33. Obidin N, Tasnim F, Dagdeviren C (2020) The future of neuroimplantable devices: a materials science and regulatory perspective. Adv Mater 32:1901482
34. Chatterjee A, Farah MJ (2012) Neuroethics in practice. Oxford University Press
35. Goering S, Klein E (2020) Fostering neuroethics integration with neuroscience in the BRAIN initiative: comments on the NIH neuroethics roadmap. AJOB Neurosci 11:184–188
36. Steinert S, Bublitz C, Jox R et al (2019) Doing things with thoughts: brain-computer interfaces and disembodied agency. Philos Technol 32:457–482
37. Naufel S, Klein E (2020) Brain–computer interface (BCI) researcher perspectives on neural data ownership and privacy. J Neural Eng 17:016039
38. Goering S, Klein E, Specker Sullivan L et al (2021) Recommendations for responsible development and application of neurotechnologies. Neuroethics 14:365–386
39. Wajnerman Paz A (2021) Is your neural data part of your mind? Exploring the conceptual basis of mental privacy. Minds Mach 1–21
40. Baselga-Garriga C, Rodriguez P, Yuste R (2022) Neuro rights: a human rights solution to ethical issues of neurotechnologies. In: Protecting the mind. Springer, pp 157–161
41. Yuste R, Genser J, Herrmann S (2021) It's time for neuro-rights. Horiz J Int Relat Sustain Dev 18:154–165
42. Scheibner J, Raisaro JL, Troncoso-Pastoriza JR et al (2021) Revolutionizing medical data sharing using advanced privacy-enhancing technologies: technical, legal, and ethical synthesis. J Med Internet Res 23:e25120

43. White T, Blok E, Calhoun VD (2022) Data sharing and privacy issues in neuroimaging research: opportunities, obstacles, challenges, and monsters under the bed. Hum Brain Mapp 43:278–291
44. Jwa AS, Poldrack RA (2022) The spectrum of data sharing policies in neuroimaging data repositories. Hum Brain Mapp 43:2707–2721
45. Thoral PJ, Peppink JM, Driessen RH et al (2021) Sharing ICU patient data responsibly under the society of critical care Medicine/European Society of intensive care medicine joint data science collaboration: the Amsterdam University medical centers database (AmsterdamUMCdb) example. Crit Care Med 49:e563
46. Liu JC, Goetz J, Sen S et al (2021) Learning from others without sacrificing privacy: simulation comparing centralized and federated machine learning on mobile health data. JMIR Mhealth Uhealth 9:e23728
47. Thapa C, Camtepe S (2021) Precision health data: requirements, challenges and existing techniques for data security and privacy. Comput Biol Med 129:104130
48. Warnat-Herresthal S, Schultze H, Shastry KL et al (2021) Swarm learning for decentralized and confidential clinical machine learning. Nature 594:265–270
49. Mccradden MD, Anderson JA, Stephenson EA et al (2022) A research ethics framework for the clinical translation of healthcare machine learning. Am J Bioeth 1–15
50. Minielly N, Hrincu V, Illes J (2020) Privacy challenges to the democratization of brain data. Iscience 23:101134
51. Zuk P, Sanchez CE, Kostick K et al (2020) Researcher perspectives on data sharing in deep brain stimulation. Front Hum Neurosci 544
52. Rainey S, Martin S, Christen A et al (2020) Brain recording, mind-reading, and neurotechnology: ethical issues from consumer devices to brain-based speech decoding. Sci Eng Ethics 26:2295–2311
53. Higgins N, Gardner J, Carter A (2022) Recognizing a plurality of industry perspectives in the responsible innovation of neurotechnologies. AJOB Neurosci 13:70–72
54. Christen M, Domingo-Ferrer J, Draganski B et al (2016) On the compatibility of big data driven research and informed consent: the example of the human brain project. In: The ethics of biomedical big data. Springer, pp 199–218
55. Ienca M, Ignatiadis K (2020) Artificial intelligence in clinical neuroscience: methodological and ethical challenges. AJOB Neurosci 11:77–87
56. Markiewicz CJ, Gorgolewski KJ, Feingold F et al (2021) The OpenNeuro resource for sharing of neuroscience data. Elife 10:e71774
57. Murillo DG, Zhao Y, Rogovin OS et al (2022) NDI: a platform-independent data interface and database for neuroscience physiology and imaging experiments. Eneuro 9
58. Goering S, Klein E, Dougherty DD et al (2017) Staying in the loop: relational agency and identity in next-generation DBS for psychiatry. AJOB Neurosci 8:59–70
59. Yuste R, Goering S, Bi G et al (2017) Four ethical priorities for neurotechnologies and AI. Nature 551:159–163
60. Goering S, Brown T, Klein E (2021) Neurotechnology ethics and relational agency. Philos Compass 16:e12734
61. Klein E, Goering S, Gagne J et al (2016) Brain-computer interface-based control of closed-loop brain stimulation: attitudes and ethical considerations. Brain-Comput Interfaces 3:140–148
62. Schönau A, Dasgupta I, Brown T et al (2021) Mapping the dimensions of agency. AJOB Neurosci 12:172–186
63. Zawadzki P (2021) Mapping the dimensions of agency: the narrative as unifying mechanism. AJOB Neurosci 12:191–193
64. Münch N, Wagner N-F, Paul NW (2021) Mapping the other side of agency. AJOB Neurosci 12:198–200
65. Young MJ, Bernat JL (2022) Emerging subspecialties in neurology: neuroethics: an emerging career path in neurology. Neurology 98:505–508
66. Liv N (2021) Neurolaw: brain-computer interfaces. U St Thomas JL Pub Pol'y 15:328

67. Shen FX (2016) Law and neuroscience 2.0. Ariz St LJ 48:1043
68. Chandler JA (2018) Neurolaw and neuroethics. Camb Q Healthc Ethics 27:590–598
69. García LV, Winickoff DE (2022) Brain-computer interfaces and the governance system: upstream approaches
70. Jones OD, Schall JD, Shen FX (2015) Law and neuroscience. Wolters Kluwer
71. Sankary LR, Zelinsky M, Machado A et al (2021) Exit from brain device research: a modified grounded theory study of researcher obligations and participant experiences. AJOB Neurosci 1–12
72. Drew L (2020) Like taking away a part of myself'—life after a neural implant trial. Nat Med 26:1154–1156
73. Obaid A, Hanna M-E, Wu Y-W et al (2020) Massively parallel microwire arrays integrated with CMOS chips for neural recording. Sci Adv 6:eaay2789
74. Steinmetz NA, Aydin C, Lebedeva A et al. (2021) Neuropixels 2.0: a miniaturized high-density probe for stable, long-term brain recordings. Science 372:eabf4588
75. Zhang M, Tang Z, Liu X et al (2020) Electronic neural interfaces. Nat Electron 3:191–200
76. Luan L, Robinson JT, Aazhang B et al (2020) Recent advances in electrical neural interface engineering: minimal invasiveness, longevity, and scalability. Neuron 108:302–321
77. Simeral JD, Hosman T, Saab J et al (2021) Home use of a percutaneous wireless intracortical brain-computer interface by individuals with tetraplegia. IEEE Trans Biomed Eng 68:2313–2325
78. Nurmikko A (2020) Challenges for large-scale cortical interfaces. Neuron 108:259–269
79. Bettinger CJ, Ecker M, Kozai TDY et al (2020) Recent advances in neural interfaces—materials chemistry to clinical translation. MRS Bull 45:655–668
80. Lee SH, Thunemann M, Lee K et al (2022) Scalable thousand channel penetrating microneedle arrays on flex for multimodal and large area coverage brainmachine interfaces. Adv Funct Mater 2112045
81. Racine E, Illes J (2009) Emergentism at the crossroads of philosophy, neurotechnology, and the enhancement debate. In: The Oxford handbook of philosophy and neuroscience
82. Churchland PS (1989) Neurophilosophy: toward a unified science of the mind-brain. MIT Press
83. Bennett MR, Hacker PMS (2022) Philosophical foundations of neuroscience. Wiley
84. Tse P (2013) The neural basis of free will: criterial causation. MIT Press
85. Young MJ, Bodien YG, Giacino JT et al (2021) The neuroethics of disorders of consciousness: a brief history of evolving ideas. Brain 144:3291–3310
86. Young MJ, Edlow BL (2021) Emerging consciousness at a clinical crossroads. AJOB Neurosci 12:148–150
87. Peterson A, Young MJ, Fins JJ (2022) Ethics and the 2018 practice guideline on disorders of consciousness: a framework for responsible implementation. Neurology 98:712–718
88. Young MJ (2017) "Consciousness" as a vague predicate. AJOB Neurosci 8:157–159
89. Young MJ (2022) Neuroethics in the era of teleneurology. In: Seminars in neurology. Thieme Medical Publishers, Inc.
90. Solms M, Friston K (2018) How and why consciousness arises: some considerations from physics and physiology. J Conscious Stud 25:202–238
91. Klar P (2021) What is neurophilosophy: do we need a non-reductive form? Synthese 199:2701–2725
92. Northoff G (2022) Non-reductive neurophilosophy–what is it and how it can contribute to philosophy. J NeuroPhilosophy 1
93. Churchland PS (2022) What is neurophilosophy and how did neurophilosophy get started? J NeuroPhilosophy 1
94. Gouveia SS (2022) Non-reductive neurophilosophy. In: Philosophy and neuroscience. Springer, pp 181–231
95. Bickle J (2019) Lessons for experimental philosophy from the rise and "fall" of neurophilosophy. Philos Psychol 32:1–22

96. Segun ST (2019) Neurophilosophy and the problem of consciousness: an equiphenomenal perspective. In: New conversations on the problems of identity, consciousness and mind. Springer, pp 33–65
97. Carrozzo C (2019) Scientific practice and the moral task of neurophilosophy. AJOB Neurosci 10:115–117
98. Churchland PS (2002) Brain-wise: studies in neurophilosophy. MIT Press
99. Ramos KM, Grady C, Greely HT et al (2019) The NIH BRAIN initiative: integrating neuroethics and neuroscience. Neuron 101:394–398
100. Greely HT, Grady C, Ramos KM et al (2018) Neuroethics guiding principles for the NIH BRAIN initiative. J Neurosci 38:10586–10588
101. Versalovic E, Diamond M, Klein E (2020) "Re-identifying yourself": a qualitative study of veteran views on implantable BCI for mobility and communication in ALS. Disabil Rehabil Assistive Technol 1–8
102. Gilbert F (2019) An instrument to capture the phenomenology of implantable brain device use. Neuroethics 1–8
103. Gilbert F, Cook M, O'brien T et al (2019) Embodiment and estrangement: results from a first-in-human "intelligent BCI" trial. Sci Eng Ethics 25:83–96

Chapter 8
What Is the Contribution of iEEG as Compared to Other Methods to Cognitive Neuroscience?

Jing Liu and Gui Xue

Abstract Intracranial electroencephalography (iEEG) enables us to record and modulate neuronal responses from the level of macroscopic networks to small assemblies and single cells in both cortical and subcortical regions of the human brain with high spatial and temporal precision, offering significant methodological advantages over other non-invasive imaging and stimulating technologies. Leveraging these technical strengths of iEEG, in combination with sophisticated multivariate analytical approaches, researchers have obtained unprecedented insights into many long-standing problems in cognitive neuroscience. This chapter aims to illustrate these contributions, focusing on human memory. In particular, we describe how iEEG could advance our understanding of (1) the dynamic and transformative nature of short-term and long-term memory representations; (2) the role of hippocampal high-frequency neural activities, especially ripple activities, in memory formation, consolidation, and retrieval; (3) the information coding scheme of single-neuronal activity in the hippocampus and other brain regions; and (4) the common and different neural mechanisms between humans, primates and rodents. Moreover, we briefly discuss how iEEG studies can contribute to developing the state-of-the-art brain-computer interface and closed-loop brain stimulation. We conclude by summarizing the strengths and limitations of the iEEG method and providing practical guidance on how to choose between iEEG and other methods.

J. Liu
Department of Applied Social Sciences, The Hong Kong Polytechnic University, Hong Kong SAR, China

G. Xue (✉)
State Key Laboratory of Cognitive Neuroscience and Learning, Beijing Normal University, Beijing, China
e-mail: gxue@bnu.edu.cn

IDG/McGovern Institute for Brain Research, Beijing Normal University, Beijing, China

N. Axmacher (ed.), *Intracranial EEG*, Studies in Neuroscience, Psychology and Behavioral Economics, https://doi.org/10.1007/978-3-031-20910-9_8

8.1 Introduction

It is generally accepted that human brain functions, such as perception, attention, decision, learning, and memory, are supported by dynamic and interactive neuronal activities. At the microscopic level, neuronal activities can be well characterized by two main types of electrical activity, i.e., neuronal spikes and postsynaptic potentials. Neuronal spikes are action potentials that travel along the axons of neurons in an all-or-none form, with a very short duration of approximately one millisecond. Multiple spikes, which result in spike trains with highly intricate temporal patterns, serve as the code of brain information [1, 2]. Because of the brief timing and opposite electrical currents flow in the intracellular and extracellular spaces, the neuronal spikes can only be recorded with a sensor near enough to the neuron. In contrast to the spikes, the postsynaptic potentials (PSPs) are changes of the potentials in the membrane of the neurons [3], which have a longer duration ranging from tens- to hundreds- of milliseconds and can be summated in extracellular space across large populations of synchronized neurons. The resultant electrophysiological signals could propagate and thus be recorded both within and outside the skull.

For ethical reasons, most human studies rely on non-invasive brain imaging techniques, including electroencephalogram (EEG), magnetoencephalography (MEG), functional near-infrared spectroscopy (fNIRS), functional magnetic resonance imaging (fMRI), and positron emission tomography (PET). These methods provide either an indirect (e.g., fNIRS, PET, or fMRI) or a coarse (e.g., EEG and MEG) measure of neuronal responses. Intracranial EEG (iEEG) recordings, which are mainly applied in drug-resistant epilepsy patients with electrodes implanted for clinical purposes, provide a rare opportunity to directly record neuronal responses inside the human brain (see also Chap. 1 for a detailed description of presurgical epilepsy patients). IEEG can record both local field potentials (LFPs) and neuronal spikes with macroelectrodes and microelectrodes, respectively. Specifically, there are two types of macroelectrode recordings, i.e., electrocorticography (ECoG) and stereo-electroencephalography (sEEG). Whereas ECoG recordings place electrodes on the surface of the brain, sEEG uses depth electrodes perpetrating the brain to target deep brain structures. Similarly, using microelectrode arrays or microwires on the tips of the depth electrodes, iEEG can be used to record single-unit activities from the surface and deep structures of the brain (see also Chap. 17 for a detailed description of iEEG signal characteristics, and Chaps. 42–46 about microelectrode recordings). Compared with non-invasive brain imaging methods, iEEG offers several critical methodological advantages to better uncover the neuronal mechanisms of human cognition. In this chapter, we will summarize these methodological advantages and discuss how they could be leveraged to address long-standing problems in cognitive neuroscience, with a focus on learning and memory. We will conclude by providing practical guidance on how to choose between iEEG and other non-invasive methods.

8.2 The Methodological Advantages of iEEG

8.2.1 High Spatial–Temporal Resolution

FMRI and PET are well-known for their high spatial resolution (1–3 mm) and whole-brain coverage. However, they indirectly measure neuronal responses (e.g., blood flow or metabolism), with a temporal resolution at the level of seconds. The temporal resolution of EEG/MEG rivals that of iEEG, but their spatial resolution and the signal-to-noise ratio (SNR) are much lower due to the higher impedance of the electrodes, larger distances between the electrodes and the sources (i.e., neurons), and volume conduction. The irreversible problem of untangling the sources that contribute to the observed EEG/MEG signals further impedes research from precisely localizing the origins of signals, especially those from deep brain areas.

The iEEG signal has both high spatial and temporal resolutions. The grid and/or strip electrodes placed subdurally onto the cortex are ~2 mm in diameter and cover ~4 mm^2 of cortical surface. Meanwhile, each depth electrode has about 6–20 contacts, with a diameter of ~1 mm and a length of ~2 mm for each contact. Given the diameters of the electrodes and their adjacency to neurons, it is assumed that iEEG could record local field potentials from neural populations within a few millimeters [4, 5]. Microwire electrodes used for single-unit recordings are about 40 μm in diameter [6], with a spatial resolution higher than 1 mm. In addition, with a sampling rate ranging from ~1000–3000 Hz for macroelectrodes and ~30 kHz for microelectrodes, iEEG is able to observe rapid changes of local neural activities at a millisecond to submillisecond level (Table 8.1).

In summary, due to the high spatial and temporal resolution, iEEG is one of the best methods to characterize the dynamic changes of neural processes and representations across time and brain regions.

8.2.2 High Signal-to-Noise Ratio

Since iEEG can directly record neuronal responses around their origin, it can yield a greater SNR. In contrast, EEG and MEG record brain signals at the scalp, which is far away from the origins. In addition, iEEG recordings are less sensitive to environmental noise and artifacts. For example, fMRI and MEG are both sensitive to magnetic field changes in the recording environments. A slight head movement could result in motion artifacts and introduce challenges in an accurate spatial registration for fMRI and MEG. For both EEG and MEG, body movements, eye blinks, or muscle tension could introduce external noise into the recorded signals. In contrast, the iEEG is less sensitive to movement or magnetic field changes. As a result, its signal-to-noise ratio (SNR) is about 20–100 times higher than scalp EEG [7]. The high SNR of iEEG helps distinguish task-related neural activities more meticulously.

Table 8.1 Summary of the methodological advantages of iEEG as compared to common non-invasive recording methods

Recording features	iEEG	MEG	EEG	fMRI
Neural electrophysiological activities	Yes	Yes	Yes	No
Temporal resolution	<1 ms	<1 ms	<1 ms	~100 ms–3 s
Spatial resolution	<1 cm	2–10 cm (2–3 mm at the source level)	~10 cm (7–10 mm at the source level)	~1 mm
High-frequency oscillations	Up to ~150 Hz	<100 Hz	<70 Hz	No (but the BOLD activities reflect high-frequency activity)
Spike activities	Yes (only for microelectrode recordings)	No	No	No
Signal-to-noise ratio	High	Relatively low	Relatively low	High
Deep brain activities	Yes	No	No	Yes
Direct brain stimulation	Yes	No	No	No

8.2.3 High-Frequency Activity

High-frequency activity (50–150 Hz) is typically generated by local neuronal populations [8], which show synchronized activity at high frequencies and low amplitudes with high temporal dynamics. The fMRI blood-oxygen-level-dependent (BOLD) signal has been found to be correlated with high-frequency LFP power (40–80 Hz) [9]. Given the local properties of high-frequency activity and the fact that the skull acts as a low-pass filter, these high-frequency activities cannot be easily detected by non-invasive EEG recordings. Although MEG records neural activities up to about 100 Hz [10], its signal-to-noise ratio for high-frequency activities is lower than iEEG in general. This makes iEEG optimally suited to unveil high-frequency activities during both online (wake) and offline (sleep) periods.

8.2.4 Direct Electrical Stimulation of the Human Brain

In addition to recording neural activities from the brain, iEEG electrodes can also be used to deliver precise electric pulses to specific brain areas. As compared with transcranial stimulation systems, such as tDCS/tACS and TMS, direct brain stimulation via iEEG electrodes enjoys greater specificity. For example, it could target the medial temporal lobe (MTL), a region that is critical for declarative memory [11], or the orbitofrontal cortex that is involved in mood states [12]. In cognitive studies, different stimulation parameters have been applied. In general, the electrical pulses have a width of 100–300 μs, a current amplitude of 0.5–6 mA, and stimulation frequency can be either in a low-frequency range (e.g., 2–5 Hz) or at high frequencies (e.g., 50, 60, or 130 Hz) [13].

8.3 Characterizing the Dynamic and Transformative Nature of Neural Representations

Information processing in the brain is highly dynamic, showing fast and prominent changes in neural processes and representations in various brain regions. Intracranial EEG recordings can directly measure brain activities from a wide range of frequencies from 0.5 up to 150 Hz with a high temporal resolution, which is ideal for characterizing dynamic information processing at a fine spatio-temporal scale. Using this approach, many studies have uncovered the dynamic and transformative nature of neural representations, particularly during short-term and long-term memory, which have profoundly changed our perspective of human memory.

For example, the classic persistent activity model postulates that neural patterns formed during stimulus encoding persist during the short-term maintenance period [14, 15]. With high temporal resolution, a recent iEEG study showed that the neural patterns of visual items experienced rapid changes from visual formats to abstract semantic formats, and the latter was more robustly maintained during short-term memory [16] (Fig. 8.1a, b). This study further revealed that the neural representations of memory items were dynamically reactivated during the maintenance period and coupled to the phase of hippocampal low-frequency activities.

Regarding maintenance of multiple items in short-term memory, the theta-gamma coding scheme posits that individual items are represented by the synchronized activation of different subsets of neural assemblies at the gamma frequency range and the phase of theta oscillations. Specifically, hippocampal theta oscillations act as a glue that tightly links items to different theta phases in support of short-term memory maintenance [21]. Using EEG/MEG, studies have found evidence for theta-gamma coupling during the short-term memory maintenance period [22–24]. On the other hand, fMRI studies have shown that short-term memory information is stored in distributed brain regions, including the sensory cortex and higher-order brain regions, e.g., the frontal and parietal lobes [25, 26]. However, the dynamic interactions among these regions and the unique role of the hippocampus in theta-gamma coding cannot be addressed using non-invasive recordings. Using iEEG recordings, studies showed that the theta-gamma coupling is not only observed in the hippocampus [27, 28], but also in multiple neocortical sites [19] during multi-item short-term memory

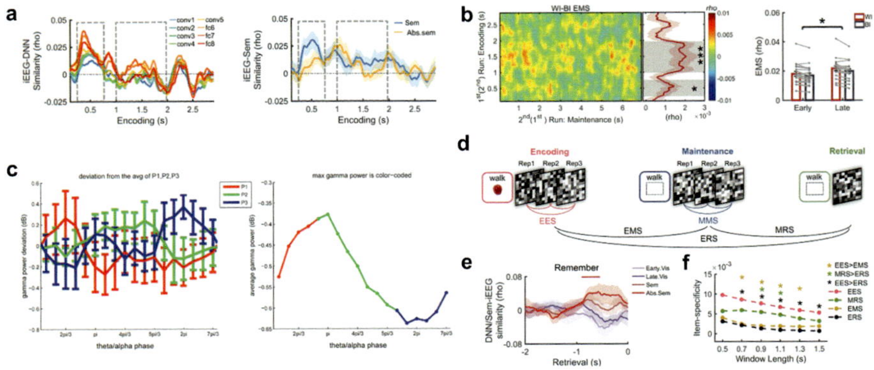

Fig. 8.1 Dynamic and transformative memory representations during encoding, maintenance, and retrieval. **a** Higher-order visual representations and abstract semantic representations during early and late encoding clusters, respectively. Left: visual representational formats observed in the iEEG data during encoding of visual objects identified using an eight-layer visual deep neural network model (AlexNet, see [17]). Right: Semantic and abstract semantic formats during encoding of visual objects measured using a Chinese word embedding model (see [18]). Dashed frames indicate encoding time clusters whose neural activities were reinstated during the short-term maintenance period. **b** Abstract semantic representations from the late encoding cluster were more strongly reactivated in short-term memory. **c** Items learned in a sequential order were locked to different phases of low-frequency oscillations in the neocortex. P1/P2/P3 indicate the positions (order) of items in a three-item sequence. **d** Schematic depiction of analysis approach to examine continuous representational transformation of memory items across task stages. **e** Abstract semantic representations of visual items are reinstated during successful retrieval. **f** Continuous transformation from encoding to maintenance and retrieval. Cross-stage transformation was examined by comparing within-stage versus between-stage item-specific pattern similarity. EES: encoding-encoding similarity; EMS: encoding-maintenance similarity; MRS: maintenance-retrieval similarity; ERS: encoding-retrieval similarity; WI: within-item similarity; BI: between-item similarity. *: $p < 0.05$. (**a**, **b**) adapted from [16]. **c** adapted from [19]. **d**–**f** adapted from [20]

(Fig. 8.1c). It further shows that the interactions between hippocampal and cortical regions increase with working memory load [29] (see also Chap. 16). Leveraging its high spatiotemporal resolution, several iEEG studies have revealed directional interactions between MTL and cortical regions at various processing stages [30–33]. Specifically, hippocampal replay preceded replay in the sensory cortex during the maintenance period, whereas a reversed pattern was observed during the encoding period [33].

Intracranial EEG studies could also shed light on the subprocesses of memory formations at a higher spatial resolution. For example, ERP analysis of EEG/MEG signals revealed that a late positive complex (LPC) component starting from ~500 ms post-stimulus onset, and an FN400 component with a peak around 400 ms, are both associated with memory retrieval [34, 35]. Using iEEG, Fernández and colleagues found that the memory-related ERPs included at least two different subprocesses that were executed sequentially [36]. The intracranial ERPs diverged between subsequently remembered and forgotten items from about 300 ms in the anterior parahippocampal cortex, followed by the reversed polarity in the hippocampus from about 500 ms post-stimulus onset, suggesting a dynamic interaction between MTL subregions.

Intracranial EEG studies could provide a detailed picture of the transformative nature of long-term memory representations. Guided by Tulving's mental time traveling perspective of episodic memory [37], extant studies have provided evidence that retrieval involves the reinstatement of neural activation during encoding, which enables humans to vividly re-experience the past [38–41]. Nevertheless, there is now increasing neural evidence to suggest that memory is rather a constructive process [42, 43] that involves a substantial transformation of neural representations. For example, an fMRI study found that item-specific encoding-retrieval similarity was not significant and substantially lower than the neural pattern similarities within the encoding and retrieval stages when considered separately [44], suggesting the neural representations were transformed from encoding to retrieval. A recent iEEG study took advantage of its high temporal resolution to systematically delineate the dynamic transformation of neural representations from encoding to maintenance and retrieval period [20] (Fig. 8.1d–f). It showed that neural representations of memory items were highly dynamic during encoding, evolving rapidly from visual to abstract semantic representations. Interestingly, greater encoding dynamicity predicted better subsequent long-term memory performance. After encoding, these neural representations experienced continuous transformation during maintenance and retrieval, and only abstract semantic representations were reactivated in successful long-term memory retrieval.

Recent iEEG studies have provided a more detailed description of the temporal dynamics of the hippocampus and cortical region during memory retrieval, which emphasized the role of the hippocampus in driving cortical activities during retrieval. Specifically, high-frequency activities (gamma band activity in the range of 40–50 Hz) in the hippocampus preceded reductions of oscillatory power in the alpha/beta band in the anterior temporal lobe during memory retrieval [45]. During long-term memory recognition, the reinstatement of item-context associations occurred within the first second of retrieval in the hippocampus, followed by reinstatement of item-specific information in the lateral temporal lobe from ~1 to ~3 s of retrieval [46]. The same study also revealed that phase synchronization between the hippocampus and lateral temporal lobe preceded the item-specific reinstatement in the lateral temporal lobe.

8.4 The Role of Hippocampal Ripple Activities in Memory

Ripple oscillations were first discovered in the rodent hippocampus with a frequency range between ~140–200 Hz [47, 48]. They are short-lived fast oscillatory patterns that are generated by pyramidal cell ensembles in the CA1 region of the hippocampus [49]. Animal studies have implicated the role of hippocampal sharp-wave ripples (SWRs), i.e., co-occurring ripples and sharp waves, in memory consolidation [50–52]. Specifically, rodent studies have observed that the sequential firing of hippocampal place cells during spatial navigation was replayed during post-learning wakeful rest [50, 51], as well as during sleep [53], in particular, slow-wave sleep stages [54]. Strikingly, these replay events were often accompanied by hippocampal SWRs. Selective disruption of hippocampal SWRs via electrical stimulation during sleep could impair memory formation [55, 56]. Using a similar method to interrupt SWRs during the encoding stage, a later study also found impairment of memory formation and memory-based decision making, suggesting an important role of hippocampal SWRs in memory encoding, consolidation, and retrieval [57].

In light of these inspiring findings from animal studies, it is natural to ask whether there are ripples in the human hippocampus and, if so, whether hippocampal ripples only support spatial memory as extensively examined in animal studies, or also other types of declarative memory in humans. Due to their high frequency and deep origin, it is a challenge to confidently isolate hippocampal ripples by non-invasive imaging methods (see recent efforts by Liu and colleagues [58]). Using iEEG recordings, several studies have investigated the functional role of hippocampal ripples in humans (see also Chap. 21). First, iEEG studies have observed ripples in the human hippocampus (Fig. 8.2a, b), although at a lower frequency band (80–140 Hz) than found in the rodent hippocampus [59, 60]. Hippocampal ripples were found during an attention-demanding cognitive task, autobiographic memory recall, wakeful rest, and sleep (Fig. 8.2c), and the duration, rate, and peak frequencies were surprisingly consistent across states [61, 62]. Notably, the ripple rate during non-rapid eye movement (NREM) sleep was higher than during other task states and sleep stages. Ripple rate was also higher in CA1 than CA2 and CA3 (Fig. 8.2d). The coincidence of individual ripples between every two hippocampal channels decreased sharply with increasing distance [62, 63], suggesting a local nature of the hippocampal ripples.

Second, iEEG studies have shed light on the functional role of ripples in memory formation and consolidation. For example, several iEEG studies have associated ripple activities during sleep with post-sleep performance. For example, the number of ripples that occurred in the rhinal cortex (but not the hippocampus) predicted memory performance after sleep [60]. Using representational similarity analysis, an iEEG study further revealed that the replay of memory representations was time-locked to hippocampal ripples during NREM sleep [64] (Fig. 8.2e) and predicted subsequent long-term memory. These results suggest that hippocampal ripples contribute to memory consolidation via memory replay. In addition to memory consolidation, emerging human iEEG studies have shown that ripples during wakeful state predict successful encoding [66] and memory retrieval [62, 67, 68]. During encoding, hippocampal ripple rates were significantly higher for novel stimuli than that for familiar stimuli, and higher ripple rates predicted better subsequent memory performance [66, 69]. During memory retrieval, hippocampal ripples modulated the high-frequency activities in the neocortex, which supported memory reinstatement [66, 68, 70]. Together, these studies highlight an important role of ripples during both wakefulness and sleep in memory [71].

Third, several iEEG studies have started to uncover the underlying mechanisms of hippocampal ripples in triggering cortical reinstatement. For example, an iEEG study found that ripples were locked to the phase of delta oscillation (0.5–4 Hz) during sleep

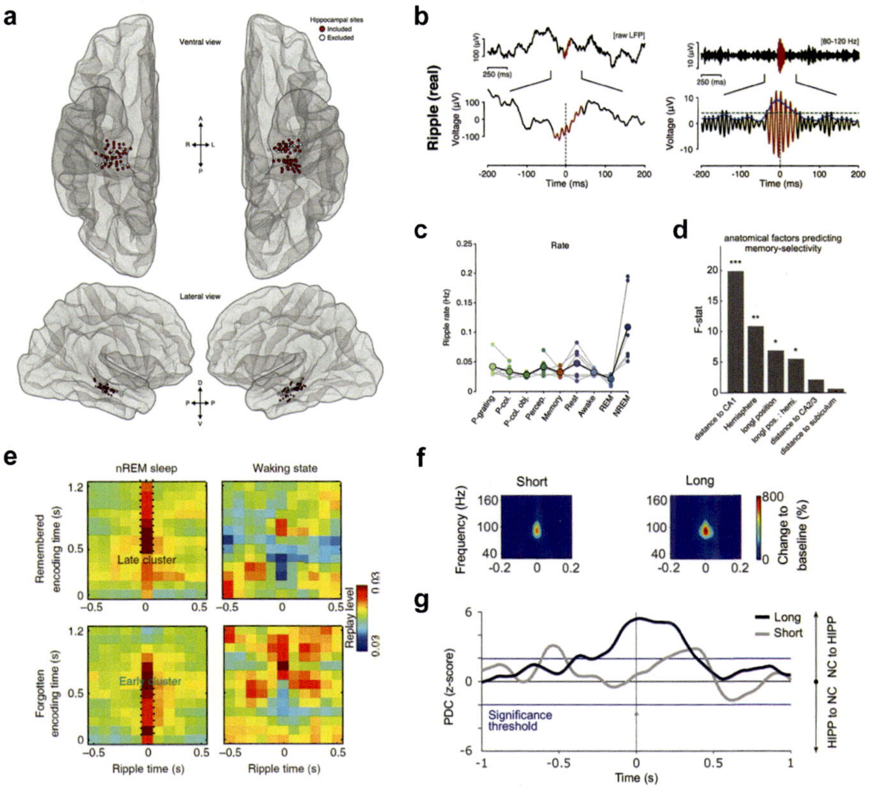

Fig. 8.2 Hippocampal ripples and memory reinstatement. **a** Depiction of hippocampal electrodes in epilepsy patients. **b** Example of ripple activities recorded from hippocampal electrodes. **c** Ripple rates across awake tasks and sleep stages. **d** Effect of anatomical locations on ripple rates in memory-selective tasks. **e** Memory replay during NREM sleep is precisely locked to ripples. **f** Ripple activities with short (mean ± SEM: duration = 0.046 ± 0.001 s) and long durations (mean ± SEM: duration = 0.059 ± 0.002 s). **g** Directional interactions from neocortex to hippocampus during the occurrence of ripples, especially long ripples. PDC: partial directed coherence in the 12–16 Hz spindle range. NC: neocortex; HIPP: hippocampus; Positive PDC value indicates information flowing from NC to HIPP, and vice versa for negative PDC values. **a**–**c** Adapted from [61]; **d** adapted from [62]; **e** adapted from [64]; **f**, **g** adapted from [65]

within the hippocampus [60], resembling the coupling between hippocampal ripples and neocortical slow oscillations (SO). By simultaneously recording MTL ripples using iEEG and slow-wave oscillations using scalp EEG, one study showed that MTL ripples and spindle activities were both time-locked to the upstate of neocortical SOs [59]. The neocortical SO-spindle coupling also coordinated MTL ripples during NREM sleep. In turn, MTL ripples mediate the information flow from the MTL to the neocortex [72]. Consistently, another iEEG study found that neocritical spindles occurred earlier than hippocampal spindles, followed by the hippocampal SWRs, especially for long but not short ripples [65] (Fig. 8.2f, g). These results suggest a top-down control from the neocortex to the hippocampus via sleep spindles, which potentially initiates the memory replay in the hippocampus. The hippocampal ripples then further drive the neocortical reinstatement of memory content.

8.5 The Information Coding Scheme of Human Single-Neuronal Activities

Using microelectrodes, iEEG recordings are capable of capturing signals from single neurons, providing direct observations of information processing at the cellular level. Nevertheless, due to technical and ethical challenges, it is extremely difficult to record single neuronal activity in the human brain. Several studies have seized this rare opportunity to uncover the information coding scheme of single-neuron activities in humans. In particular, a series of studies have examined the role of MTL neurons in signaling memory content, temporal context, and novelty, as well as in supporting memory formation and maintenance.

A classical finding from human single-unit studies is the discovery of the "concept cells", a subset of MTL neurons that are selectively firing to specific persons or items [73, 74]. For example, a subset of neurons showed an increased firing rate to all photos of Jennifer Aniston with different views or backgrounds while showing very weak or no firings to photos of other famous actresses, landmarks, animals, or objects. More interestingly, these neurons not only fired to the visual images but also the written or auditory names of the actress, suggesting an invariant, amodal conceptual representation [75] (Fig. 8.3a). These cells were initially named "Jennifer Aniston" cells and are now termed "concept cells" [74]. The proportion of modality-invariant neurons increases from the parahippocampal and perirhinal cortex to the entorhinal cortex and peaks in the hippocampus [75], which resembles the increased visual invariance along the ventral visual pathway in non-human primates [76].

Are these concept cells also involved in short-term maintenance when the stimuli are no longer presented on the screen? According to animal models, one potential neural mechanism of short-term memory maintenance is the persistent activation of neurons that are initially reactivated during the perception stage (see above; [14, 15]). This hypothesis was tested by two independent human single-unit studies, which both showed that concept cells in the MTL, which fired during the encoding phase, showed continuous firing during the maintenance period [78, 79]. Although a similar persistent activity was found in the medial frontal cortex, it was associated with the working memory load instead of memory content. It should be noted that persistent activities are generally obtained by averaging spikes across time and trials. When examining individual trials, recent studies have found that the neuronal spikes in support of working memory occurred in a sparse, transient, and dynamic manner [80, 81].

In addition to the concept cells, human single-unit studies have also observed cells that signal visual categories, novelty, familiarity, confidence, or temporal context of

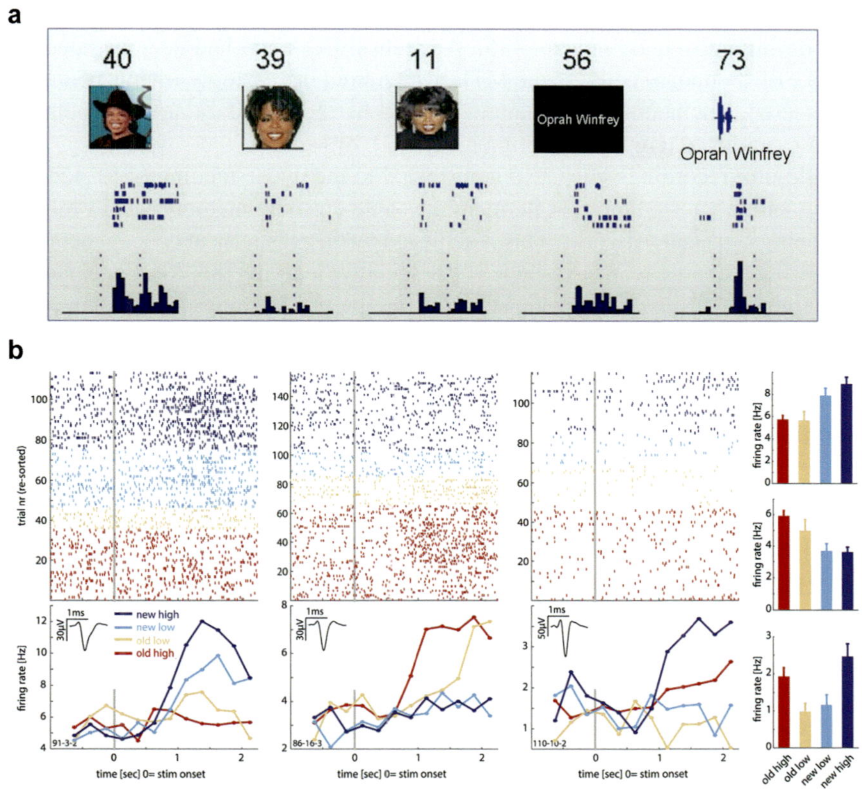

Fig. 8.3 Examples of memory-selective cells observed by human single-unit recordings. **a** Concept cells, which invariantly and selectively fire to specific concepts, for example, the picture or the written name of the television host Oprah Winfrey. **b** Memory-selective cells during the recognition task. First column: novelty cells; second column: familiarity cells; third column: cellular signal confidence level; fourth column: average firing rate of these three types of cells during a recognition task. **a** Adapted from [75]; **b** adapted from [77]

memory items in the MTL and other brain regions [77, 82, 83]. For example, by separating the neuronal responses according to subjects' behavioral responses in a recognition task, one study has found two types of neurons, with one showing greater firing to novel than familiar stimuli, and the other showing the opposing pattern [83] (Fig. 8.3b). In addition, neurons in the posterior parietal cortex could signal the subjective confidence level of memory decisions regardless of familiarity [77]. Another study found that in a memory-based decision task, neurons in the medial frontal cortex could signal the abstract task goals according to the task demands [84] (see also Chap. 45). Umbach and colleagues discovered "time cells" in the human hippocampus and entorhinal cortex, which signal the temporal positions of items within encoding/retrieval tasks and predict the temporal clustering of items during memory retrieval [85]. Another study also identified time cells in the hippocampus, which track the successive temporal information during sequential learning and replay during post-encoding gaps [86]. A recent iEEG study had demonstrated the existence of "boundary cells" in the MTL, which fired increasingly when a boundary was detected. The neural state changes elicited by the boundary at the population level also predicted temporal order memory [87].

Single-unit recordings have also contributed to the understanding of associative learning and memory. Different items in the same episodic memory would rely on overlapping neuronal networks. This is achieved by the plasticity of concept neurons, whose firing properties can be reshaped by associative learning [88, 89]. For example, a single neuron firing to a specific item during the pre-learning period showed an expanded preferred response to the associated items that co-occur with the preferred items after a few trials of learning [88]. Once the associative memory is formed, effective memory cues would trigger a predictive firing of MTL neurons before the re-appearance of items [89].

8.6 The Common and Different Neural Mechanisms Between Humans, Primates, and Rodents

Our understanding of the neuronal mechanisms of cognition has been largely advanced by animal studies. A critical question is how the findings from animal studies could inform the neural mechanisms of human cognition, given the apparent difference in brain structure and functions between animals and humans. This question is challenging to address due to different neural imaging techniques, e.g., cellular recording in animals and non-invasive imaging in humans. Human intracranial EEG associated with single-unit recordings bridges the gap, providing unprecedented insights into the shared and unique neural mechanisms between humans, non-human primates, and rodents.

Various studies have revealed that humans and rodents share similar encoding schemes during spatial navigation, memory encoding, and consolidation. For example, a well-known finding in rodents is the grid cell in the entorhinal cortex [90]. Similarly, several iEEG studies also revealed grid-like representations in the human entorhinal cortex [91–93], and this hexadirectional modulation was associated with spatial memory performance [92]. Another well-documented neural code of spatial/temporal information consists in phase precession [94]. Phase precession refers to the observation that the firing of place cells occurs with progressively earlier phase on each successive theta (~5–10 Hz) cycle, so that the spatial or temporal information can be connected into an ordered sequence within theta cycles (see also Chap. 44). This phenomenon was well observed in hippocampal place cells [94] and entorhinal grid cells in rodents [95]. Consistently, a recent iEEG study observed phase precession in the human hippocampus and entorhinal cortex when participants were performing spatial navigation tasks [96]. In addition, as described above, both rodents and human studies have implicated hippocampal SWRs in memory consolidation [52, 65, 97–100], providing another example of similar coding schemes in rodents and humans.

Despite these similarities, researchers have also revealed important differences between humans and non-human animals. For example, a human iEEG study showed that hippocampal theta oscillations during navigation and memory processing occur at a lower frequency (1–4 Hz) as compared with rodents (4–10 Hz) [101, 102]. This slower theta frequency might enable a higher working memory capacity, according to the theta-gamma coding scheme [21]. Similarly, the frequency of human SWRs is also lower than in rodents [59, 60]. More importantly, it is reasonable to argue that the modality-invariance of concept cells found in the human hippocampus [73, 103] might not be possible for rodents. Nevertheless, more studies are definitely required to test this hypothesis. Finally, it usually takes much longer time to train animals than humans to perform certain cognitive tasks, and this overtraining may introduce additional cognitive and neural processes. For example, one rodent study has revealed two types of temporal order representation in the hippocampal–entorhinal system: in addition to the temporal flow that is formed automatically with one-shot learning, there are also representations of stable event sequences which are increased in well-practiced structured events [104]. Future studies are definitely required to address these issues.

8.7 iEEG Based Brain-Computer Interface and Closed-Loop Brain Stimulation

With their superb SNRs and high temporal resolution, iEEG recordings could promote the development of brain-computer interfaces (BCI), a next-generation of rehabilitative technologies to restore impaired cognitive functions (see also Chaps. 51–53). For example, in a recent iEEG study, Chartier and colleagues recorded neural activities from the human sensorimotor cortex while participants were speaking. They identified the neural code that gives rise to the complex articulatory kinematic trajectories underlying fluent speech production [105]. These results were then used to develop neural decoders that could synthesize the audible speech from the cortical activity [106]. In another instance, ECoG has been applied to measure signals from the motor cortex, which have then been used to control prosthetic hands and arms. Compared to scalp EEG, BCI using ECoG could achieve more accurate operation with fewer training trials [107–110].

As both a recording and modulation technology, iEEG could also help to develop state-of-the-art closed-loop brain stimulation systems, which could establish a causal relationship between neural activities and behavior and also improve behavioral outcomes (see also Chap. 41). Almost 70 years ago, Wilder Penfield mapped brain functions by directly stimulating different brain areas during brain surgery [111,

112]. Although an increasing number of studies have shown that non-invasive stimulation, including transcranial current or magnetic stimulation, can modulate various brain functions, including perception [113], attention [114], decision making [115], cognitive control [116], working memory and long-term memories [117–120], the exact mechanisms are less clear due to low spatial specificity. For example, one study directly tested the effect of tACS via using iEEG recordings but failed to find any effects of applied low-frequency tACS on brain rhythms [121]. In addition, transcranial stimulations so far cannot directly target deep brain structures.

To circumvent these issues, researchers have applied direct electrical brain stimulation (EBS) to epilepsy patients. Studies on memories have found inconsistent results, with some studies showing memory enhancement [122–124], whereas others showing memory impairment [11, 125–127]. The mixed results might be explained by the interaction effects of ongoing brain states and stimulation [128]. For example, Ezzyat and colleagues showed that stimulations during a state of low encoding efficacy could promote memory and otherwise could impair memory [129]. To address this issue, they developed a closed-loop stimulation scheme, in which the stimulation was triggered by a neural state that indicates subsequent memory forgetting. Using this novel approach, they revealed a robust enhancement effect of hippocampal stimulation on long-term memory [130] (Fig. 8.4) (see also Chap. 41).

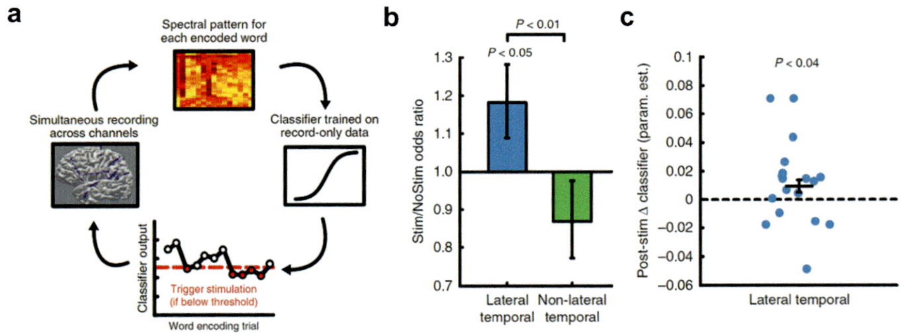

Fig. 8.4 Closed-loop brain stimulation using iEEG to modulate memory performance. **a** Illustration of closed-loop stimulation. First, a classifier is trained using an independent data set. Second, spectral decomposition analysis is performed during online recordings. Third, the pre-trained classifier is applied to the spectrally decomposed data. Fourth, an electric pulse is triggered if the decoding evidence reaches the threshold. **b** Number of successfully recalled items in the stimulation (Stim) vs. no stimulation (NoStim) condition. The stimulation was applied either to the lateral temporal lobe or other brain regions (i.e., non-lateral temporal). **c** Lateral temporal lobe stimulation increased classifier evidence during memory encoding. This figure is adapted from [130]

8.8 Summary and a Practical Guidance

So far, we have provided an overall description of the methodological advantages of iEEG and illustrated its implications in addressing fundamental questions in cognitive neuroscience. Compared with non-invasive brain imaging methods, the ability of iEEG to direct record single-neuron spikes and LFPs with greater spatial–temporal resolution and SNRs are unprecedently rare and precious. Nevertheless, it is important to notice the limitations of iEEG when planning your studies. First of all, due to ethical issues, iEEG recordings are only performed in patients—most often, drug-resistant epilepsy patients for clinical purposes—which makes iEEG less accessible than other non-invasive neuroimaging methods. Second, the electrodes are usually implanted for 2–3 weeks, during which extensive physical examinations are conducted, leaving limited time windows for conducting experiments (see Chap. 4). It usually takes much longer to collect data from the required number of participants for the above reasons. Third, the electrodes are implanted to localize the seizure onset zone, thus are often located within or near the epileptic loci. As a result, it is unlikely to get a full brain coverage of electrodes. Moreover, the signals could be smeared by the epileptic spikes, particularly when electrodes are located near the epileptic loci. Thus, iEEG is often less optimal for examining interactions across wide brain networks. Finally, implantation schemes are heterogeneous across patients, severely limiting the ability to compare subgroups (see also Chap. 29).

Given these apparent strengths and limitations, it is vital to decide when to choose iEEG recordings over other non-invasive imaging methods. Non-invasive imaging methods are always recommended if they are sufficient to address your research questions. For example, if you are interested in temporal dynamic changes of neocortical brain regions, EEG and MEG can be good options. If you are interested in localizing the brain regions where information is processed or stored, fMRI could be a good choice. Nevertheless, if you are interested in brain oscillations of deep brain regions, such as the medial temporal lobe or thalamus, iEEG would be a good choice. Moreover, epilepsy patients often show cognitive impairments, and fatigue is a common symptom in epilepsy [131] (see also Chap. 2 for a detailed description of the cognitive status of epilepsy patients). Therefore, it is important to avoid tasks that are either cognitively challenging, may introduce intense arousal and emotional responses, or take an excessively long time (e.g., >2 h). Finally, the iEEG study should be designed and conducted with potential benefits to the patients whenever possible. For example, when developing mind-controlled robotic arms for paralyzed patients, intracranial EEG recordings with high SNRs and spatiotemporal resolution might be better than other non-invasive imaging methods. Similarly, deep brain stimulation protocols can be conducted to suppress seizures or improve cognitive functions (see also Chaps. 5, 39, and 41). Still, studies that could potentially localize the brain regions involved in essential brain functions, such as motor, language, or memory, could be conducted to aid surgical planning. Together, iEEG studies are the method of choice when researchers need to record neural activities at high spatial–temporal resolutions from deep brain structures, and should be conducted with careful consideration to patients.

References

1. Zeldenrust F, Wadman WJ, Englitz B (2018) Neural coding with bursts—current state and future perspectives. Front Comput Neurosci 12:48. https://doi.org/10.3389/fncom.2018.00048
2. Azarfar A, Calcini N, Huang C et al (2018) Neural coding: a single neuron's perspective. Neurosci Biobehav Rev 94:238–247. https://doi.org/10.1016/j.neubiorev.2018.09.007
3. Purves D, Augustine GJ, Fitzpatrick D et al (2001) Excitatory and inhibitory postsynaptic potentials. Neuroscience, 2nd edn
4. Engel AK, Moll CKE, Fried I, Ojemann GA (2005) Invasive recordings from the human brain: clinical insights and beyond. Nat Rev Neurosci 6:35–47. https://doi.org/10.1038/nrn1585
5. Mukamel R, Fried I (2012) Human intracranial recordings and cognitive neuroscience. Annu Rev Psychol 63:511–537. https://doi.org/10.1146/annurev-psych-120709-145401
6. Kreiman G (2007) Single unit approaches to human vision and memory. Curr Opin Neurobiol 17:471–475. https://doi.org/10.1016/j.conb.2007.07.005
7. Ball T, Kern M, Mutschler I et al (2009) Signal quality of simultaneously recorded invasive and non-invasive EEG. Neuroimage 46:708–716. https://doi.org/10.1016/j.neuroimage.2009.02.028
8. Lachaux J-P, Axmacher N, Mormann F et al (2012) High-frequency neural activity and human cognition: past, present and possible future of intracranial EEG research. Prog Neurobiol 98:279–301. https://doi.org/10.1016/j.pneurobio.2012.06.008
9. Schölvinck ML, Maier A, Ye FQ et al (2010) Neural basis of global resting-state fMRI activity. Proc Natl Acad Sci USA 107:10238–10243. https://doi.org/10.1073/pnas.0913110107
10. Hansen P, Kringelbach M, Salmelin R (2010) MEG: an introduction to methods. Oxford University Press, New York
11. Jacobs J, Miller J, Lee SA et al (2016) Direct electrical stimulation of the human entorhinal region and hippocampus impairs memory. Neuron 92:983–990. https://doi.org/10.1016/j.neuron.2016.10.062
12. Rao VR, Sellers KK, Wallace DL et al (2018) Direct electrical stimulation of lateral orbitofrontal cortex acutely improves mood in individuals with symptoms of depression. Curr Biol 28:3893-3902.e4. https://doi.org/10.1016/j.cub.2018.10.026
13. Suthana N, Aghajan ZM, Mankin EA, Lin A (2018) Reporting guidelines and issues to consider for using intracranial brain stimulation in studies of human declarative memory. Front Neurosci 905
14. Fuster JM, Alexander GE (1971) Neuron activity related to short-term memory. Science 173:652–654. https://doi.org/10.1126/science.173.3997.652
15. Curtis CE, D'Esposito M (2003) Persistent activity in the prefrontal cortex during working memory. Trends Cogn Sci 7:415–423. https://doi.org/10.1016/S1364-6613(03)00197-9
16. Liu J, Zhang H, Yu T et al (2020) Stable maintenance of multiple representational formats in human visual short-term memory. PNAS 117:32329–32339. https://doi.org/10.1073/pnas.2006752117
17. Krizhevsky A, Sutskever I, Hinton GE (2012) ImageNet classification with deep convolutional neural networks. In: Advances in neural information processing systems. Curran Associates, Inc
18. Song Y, Shi S, Li J, Zhang H (2018) Directional skip-gram: explicitly distinguishing left and right context for word embeddings. In: Proceedings of the 2018 conference of the North American chapter of the association for computational linguistics: human language technologies, vol 2 (Short Papers). Association for Computational Linguistics, New Orleans, Louisiana, pp 175–180
19. Bahramisharif A, Jensen O, Jacobs J, Lisman J (2018) Serial representation of items during working memory maintenance at letter-selective cortical sites. PLoS Biol 16:e2003805. https://doi.org/10.1371/journal.pbio.2003805
20. Liu J, Zhang H, Yu T et al (2021) Transformative neural representations support long-term episodic memory. Sci Adv 7:eabg9715. https://doi.org/10.1126/sciadv.abg9715

21. Lisman JE, Jensen O (2013) The theta-gamma neural code. Neuron 77:1002–1016. https://doi.org/10.1016/j.neuron.2013.03.007
22. Sauseng P, Klimesch W, Heise KF et al (2009) Brain oscillatory substrates of visual short-term memory capacity. Curr Biol 19:1846–1852. https://doi.org/10.1016/j.cub.2009.08.062
23. Fuentemilla L, Penny WD, Cashdollar N et al (2010) Theta-coupled periodic replay in working memory. Curr Biol 20:606–612. https://doi.org/10.1016/j.cub.2010.01.057
24. Kamiński J, Brzezicka A, Wróbel A (2011) Short-term memory capacity (7±2) predicted by theta to gamma cycle length ratio. Neurobiol Learn Mem 95:19–23. https://doi.org/10.1016/j.nlm.2010.10.001
25. Bettencourt KC, Xu Y (2016) Decoding the content of visual short-term memory under distraction in occipital and parietal areas. Nat Neurosci 19:150–157. https://doi.org/10.1038/nn.4174
26. Christophel TB, Klink PC, Spitzer B et al (2017) The distributed nature of working memory. Trends Cogn Sci 21:111–124. https://doi.org/10.1016/j.tics.2016.12.007
27. Axmacher N, Henseler MM, Jensen O et al (2010) Cross-frequency coupling supports multi-item working memory in the human hippocampus. Proc Natl Acad Sci 107:3228–3233. https://doi.org/10.1073/pnas.0911531107
28. Leszczyński M, Fell J, Axmacher N (2015) Rhythmic working memory activation in the human hippocampus. Cell Rep 13:1272–1282. https://doi.org/10.1016/j.celrep.2015.09.081
29. Boran E, Fedele T, Klaver P et al (2019) Persistent hippocampal neural firing and hippocampal-cortical coupling predict verbal working memory load. Sci Adv 5:eaav3687. https://doi.org/10.1126/sciadv.aav3687
30. Axmacher N, Schmitz DP, Wagner T et al (2008) Interactions between medial temporal lobe, prefrontal cortex, and inferior temporal regions during visual working memory: a combined intracranial EEG and functional magnetic resonance imaging study. J Neurosci 28:7304–7312. https://doi.org/10.1523/JNEUROSCI.1778-08.2008
31. Chaieb L, Leszczynski M, Axmacher N et al (2015) Theta-gamma phase-phase coupling during working memory maintenance in the human hippocampus. Cogn Neurosci 6:149–157. https://doi.org/10.1080/17588928.2015.1058254
32. Johnson EL, Adams JN, Solbakk A-K et al (2018) Dynamic frontotemporal systems process space and time in working memory. PLoS Biol 16:e2004274. https://doi.org/10.1371/journal.pbio.2004274
33. Dimakopoulos V, Stieglitz L, Mégevand P, Sarnthein J (2021) Information flows from hippocampus to cortex during replay of verbal working memory items. BioRxiv. https://doi.org/10.1101/2021.03.11.434989
34. Woodruff CC, Hayama HR, Rugg MD (2006) Electrophysiological dissociation of the neural correlates of recollection and familiarity. Brain Res 1100:125–135. https://doi.org/10.1016/j.brainres.2006.05.019
35. Voss JL, Federmeier KD (2011) FN400 potentials are functionally identical to N400 potentials and reflect semantic processing during recognition testing. Psychophysiology 48:532–546. https://doi.org/10.1111/j.1469-8986.2010.01085.x
36. Fernández G, Effern A, Grunwald T et al (1999) Real-time tracking of memory formation in the human rhinal cortex and hippocampus. Science 285:1582–1585. https://doi.org/10.1126/science.285.5433.1582
37. Tulving E (2002) Episodic memory: from mind to brain. Annu Rev Psychol 53:1–25. https://doi.org/10.1146/annurev.psych.53.100901.135114
38. Staresina BP, Henson RNA, Kriegeskorte N, Alink A (2012) Episodic reinstatement in the medial temporal lobe. J Neurosci 32:18150–18156. https://doi.org/10.1523/JNEUROSCI.4156-12.2012
39. Staresina BP, Michelmann S, Bonnefond M et al (2016) Hippocampal pattern completion is linked to gamma power increases and alpha power decreases during recollection. eLife 5:e17397. https://doi.org/10.7554/eLife.17397
40. Ritchey M, Wing EA, LaBar KS, Cabeza R (2013) Neural similarity between encoding and retrieval is related to memory via hippocampal interactions. Cereb Cortex 23:2818–2828. https://doi.org/10.1093/cercor/bhs258

41. Yaffe RB, Kerr MSD, Damera S et al (2014) Reinstatement of distributed cortical oscillations occurs with precise spatiotemporal dynamics during successful memory retrieval. PNAS 111:18727–18732. https://doi.org/10.1073/pnas.1417017112
42. Bartlett SFC, Bartlett FC, Bartlett FC (1932) Remembering: a study in experimental and social psychology. Cambridge University Press
43. Barry DN, Maguire EA (2019) Remote memory and the hippocampus: a constructive critique. Trends Cogn Sci 23:128–142. https://doi.org/10.1016/j.tics.2018.11.005
44. Xiao X, Dong Q, Gao J et al (2017) Transformed neural pattern reinstatement during episodic memory retrieval. J Neurosci 37:2986–2998. https://doi.org/10.1523/JNEUROSCI.2324-16.2017
45. Griffiths BJ, Parish G, Roux F et al (2019) Directional coupling of slow and fast hippocampal gamma with neocortical alpha/beta oscillations in human episodic memory. PNAS 116:21834–21842. https://doi.org/10.1073/pnas.1914180116
46. Pacheco Estefan D, Sánchez-Fibla M, Duff A et al (2019) Coordinated representational reinstatement in the human hippocampus and lateral temporal cortex during episodic memory retrieval. Nat Commun 10:2255. https://doi.org/10.1038/s41467-019-09569-0
47. O'Keefe J, Nadel L (1978) The hippocampus as a cognitive map. Clarendon Press; Oxford University Press, Oxford, New York
48. Buzsáki G, Horváth Z, Urioste R et al (1992) High-frequency network oscillation in the hippocampus. Science 256:1025–1027. https://doi.org/10.1126/science.1589772
49. Csicsvari J, Hirase H, Mamiya A, Buzsáki G (2000) Ensemble patterns of hippocampal CA3-CA1 neurons during sharp wave-associated population events. Neuron 28:585–594. https://doi.org/10.1016/S0896-6273(00)00135-5
50. Wilson MA, McNaughton BL (1994) Reactivation of hippocampal ensemble memories during sleep. Science 265:676–679. https://doi.org/10.1126/science.8036517
51. Diba K, Buzsáki G (2007) Forward and reverse hippocampal place-cell sequences during ripples. Nat Neurosci 10:1241–1242. https://doi.org/10.1038/nn1961
52. Fernández-Ruiz A, Oliva A, de Oliveira EF et al (2019) Long-duration hippocampal sharp wave ripples improve memory. Science. https://doi.org/10.1126/science.aax0758
53. Nádasdy Z, Hirase H, Czurkó A et al (1999) Replay and time compression of recurring spike sequences in the hippocampus. J Neurosci 19:9497–9507. https://doi.org/10.1523/JNEUROSCI.19-21-09497.1999
54. Lee AK, Wilson MA (2002) Memory of sequential experience in the hippocampus during slow wave sleep. Neuron 36:1183–1194. https://doi.org/10.1016/S0896-6273(02)01096-6
55. Girardeau G, Benchenane K, Wiener SI et al (2009) Selective suppression of hippocampal ripples impairs spatial memory. Nat Neurosci 12:1222–1223. https://doi.org/10.1038/nn.2384
56. Ego-Stengel V, Wilson MA (2010) Disruption of ripple-associated hippocampal activity during rest impairs spatial learning in the rat. Hippocampus 20:1–10. https://doi.org/10.1002/hipo.20707
57. Jadhav SP, Kemere C, German PW, Frank LM (2012) Awake hippocampal sharp-wave ripples support spatial memory. Science. https://doi.org/10.1126/science.1217230
58. Liu Y, Mattar MG, Behrens TEJ et al (2021) Experience replay is associated with efficient nonlocal learning. Science 372:eabf1357. https://doi.org/10.1126/science.abf1357
59. Clemens Z, Molle M, Eross L et al (2007) Temporal coupling of parahippocampal ripples, sleep spindles and slow oscillations in humans. Brain 130:2868–2878. https://doi.org/10.1093/brain/awm146
60. Axmacher N, Elger CE, Fell J (2008) Ripples in the medial temporal lobe are relevant for human memory consolidation. Brain 131:1806–1817. https://doi.org/10.1093/brain/awn103
61. Chen YY, Aponik-Gremillion L, Bartoli E et al (2021) Stability of ripple events during task engagement in human hippocampus. Cell Rep 35:109304. https://doi.org/10.1016/j.celrep.2021.109304
62. Norman Y, Raccah O, Liu S et al (2021) Hippocampal ripples and their coordinated dialogue with the default mode network during recent and remote recollection. Neuron 109:2767-2780.e5. https://doi.org/10.1016/j.neuron.2021.06.020

63. Jiang X, Gonzalez-Martinez J, Halgren E (2019) Coordination of human hippocampal sharp-wave ripples during NREM sleep with cortical theta bursts, spindles, downstates, and upstates. J Neurosci 39:8744–8761. https://doi.org/10.1523/JNEUROSCI.2857-18.2019
64. Zhang H, Fell J, Axmacher N (2018) Electrophysiological mechanisms of human memory consolidation. Nat Commun 9:4103. https://doi.org/10.1038/s41467-018-06553-y
65. Ngo H-V, Fell J, Staresina B (2020) Sleep spindles mediate hippocampal-neocortical coupling during long-duration ripples. eLife 9:e57011. https://doi.org/10.7554/eLife.57011
66. Norman Y, Yeagle EM, Khuvis S et al (2019) Hippocampal sharp-wave ripples linked to visual episodic recollection in humans. Science 365:eaax1030. https://doi.org/10.1126/science.aax1030
67. Kucewicz MT, Cimbalnik J, Matsumoto JY et al (2014) High frequency oscillations are associated with cognitive processing in human recognition memory. Brain 137:2231–2244. https://doi.org/10.1093/brain/awu149
68. Vaz AP, Inati SK, Brunel N, Zaghloul KA (2019) Coupled ripple oscillations between the medial temporal lobe and neocortex retrieve human memory. Science 363:975–978. https://doi.org/10.1126/science.aau8956
69. Henin S, Shankar A, Borges H et al (2021) Spatiotemporal dynamics between interictal epileptiform discharges and ripples during associative memory processing. Brain 144:1590–1602. https://doi.org/10.1093/brain/awab044
70. Sakon JJ, Kahana MJ (2021) Hippocampal ripples signal contextually-mediated episodic recall
71. Joo HR, Frank LM (2018) The hippocampal sharp wave–ripple in memory retrieval for immediate use and consolidation. Nat Rev Neurosci 19:744–757. https://doi.org/10.1038/s41583-018-0077-1
72. Helfrich RF, Lendner JD, Mander BA et al (2019) Bidirectional prefrontal-hippocampal dynamics organize information transfer during sleep in humans. Nat Commun 10:3572. https://doi.org/10.1038/s41467-019-11444-x
73. Quian Quiroga R, Reddy L, Kreiman G et al (2005) Invariant visual representation by single neurons in the human brain. Nature 435:1102–1107. https://doi.org/10.1038/nature03687
74. Quian Quiroga R (2012) Concept cells: the building blocks of declarative memory functions. Nat Rev Neurosci 13:587–597. https://doi.org/10.1038/nrn3251
75. Quian Quiroga R, Kraskov A, Koch C, Fried I (2009) Explicit encoding of multimodal percepts by single neurons in the human brain. Curr Biol 19:1308–1313. https://doi.org/10.1016/j.cub.2009.06.060
76. Serre T, Oliva A, Poggio T (2007) A feedforward architecture accounts for rapid categorization. Proc Natl Acad Sci 104:6424–6429. https://doi.org/10.1073/pnas.0700622104
77. Rutishauser U, Aflalo T, Rosario ER et al (2018) Single-neuron representation of memory strength and recognition confidence in left human posterior parietal cortex. Neuron 97:209-220.e3. https://doi.org/10.1016/j.neuron.2017.11.029
78. Kamiński J, Sullivan S, Chung JM et al (2017) Persistently active neurons in human medial frontal and medial temporal lobe support working memory. Nat Neurosci 20:590–601. https://doi.org/10.1038/nn.4509
79. Kornblith S, Quian Quiroga R, Koch C et al (2017) Persistent single-neuron activity during working memory in the human medial temporal lobe. Curr Biol 27:1026–1032. https://doi.org/10.1016/j.cub.2017.02.013
80. Lundqvist M, Herman P, Miller EK (2018) Working memory: delay activity, yes! Persistent activity? Maybe not. J Neurosci 38:7013–7019. https://doi.org/10.1523/JNEUROSCI.2485-17.2018
81. Miller EK, Lundqvist M, Bastos AM (2018) Working memory 2.0. Neuron 100:463–475. https://doi.org/10.1016/j.neuron.2018.09.023
82. Rutishauser U, Mamelak AN, Schuman EM (2006) Single-trial learning of novel stimuli by individual neurons of the human hippocampus-amygdala complex. Neuron 49:805–813. https://doi.org/10.1016/j.neuron.2006.02.015

83. Rutishauser U, Ye S, Koroma M et al (2015) Representation of retrieval confidence by single neurons in the human medial temporal lobe. Nat Neurosci 18:1041–1050. https://doi.org/10.1038/nn.4041
84. Minxha J, Adolphs R, Fusi S et al (2020) Flexible recruitment of memory-based choice representations by the human medial frontal cortex. Science 368:eaba3313. https://doi.org/10.1126/science.aba3313
85. Umbach G, Kantak P, Jacobs J et al (2020) Time cells in the human hippocampus and entorhinal cortex support episodic memory. PNAS 117:28463–28474. https://doi.org/10.1073/pnas.2013250117
86. Reddy L, Self MW, Zoefel B et al (2021) Theta-phase dependent neuronal coding during sequence learning in human single neurons. Nat Commun 12:4839. https://doi.org/10.1038/s41467-021-25150-0
87. Zheng J, Schjetnan AGP, Yebra M et al (2022) Neurons detect cognitive boundaries to structure episodic memories in humans. Nat Neurosci 25:358–368. https://doi.org/10.1038/s41593-022-01020-w
88. Ison MJ, Quian Quiroga R, Fried I (2015) Rapid encoding of new memories by individual neurons in the human brain. Neuron 87:220–230. https://doi.org/10.1016/j.neuron.2015.06.016
89. Reddy L, Poncet M, Self MW et al (2015) Learning of anticipatory responses in single neurons of the human medial temporal lobe. Nat Commun 6:8556. https://doi.org/10.1038/ncomms9556
90. Hafting T, Fyhn M, Molden S et al (2005) Microstructure of a spatial map in the entorhinal cortex. Nature 436:801–806. https://doi.org/10.1038/nature03721
91. Jacobs J, Weidemann CT, Miller JF et al (2013) Direct recordings of grid-like neuronal activity in human spatial navigation. Nat Neurosci 16:1188–1190. https://doi.org/10.1038/nn.3466
92. Maidenbaum S, Miller J, Stein JM, Jacobs J (2018) Grid-like hexadirectional modulation of human entorhinal theta oscillations. PNAS 115:10798–10803. https://doi.org/10.1073/pnas.1805007115
93. Chen D, Kunz L, Wang W et al (2018) Hexadirectional modulation of theta power in human entorhinal cortex during spatial navigation. Curr Biol 28:3310–3315.e4. https://doi.org/10.1016/j.cub.2018.08.029
94. O'Keefe J, Recce ML (1993) Phase relationship between hippocampal place units and the EEG theta rhythm. Hippocampus 3:317–330. https://doi.org/10.1002/hipo.450030307
95. Reifenstein ET, Kempter R, Schreiber S et al (2012) Grid cells in rat entorhinal cortex encode physical space with independent firing fields and phase precession at the single-trial level. PNAS 109:6301–6306. https://doi.org/10.1073/pnas.1109599109
96. Qasim SE, Fried I, Jacobs J (2021) Phase precession in the human hippocampus and entorhinal cortex. Cell 184:3242-3255.e10. https://doi.org/10.1016/j.cell.2021.04.017
97. Buzsáki G, Logothetis N, Singer W (2013) Scaling brain size, keeping timing: evolutionary preservation of brain rhythms. Neuron 80:751–764. https://doi.org/10.1016/j.neuron.2013.10.002
98. Buzsáki G (2015) Hippocampal sharp wave-ripple: a cognitive biomarker for episodic memory and planning. Hippocampus 25:1073–1188. https://doi.org/10.1002/hipo.22488
99. Khodagholy D, Gelinas JN, Buzsáki G (2017) Learning-enhanced coupling between ripple oscillations in association cortices and hippocampus. Science 358:369–372. https://doi.org/10.1126/science.aan6203
100. Skelin I, Zhang H, Zheng J et al (2021) Coupling between slow waves and sharp-wave ripples engages distributed neural activity during sleep in humans. PNAS 118. https://doi.org/10.1073/pnas.2012075118
101. Watrous AJ, Lee DJ, Izadi A et al (2013) A comparative study of human and rat hippocampal low frequency oscillations during spatial navigation. Hippocampus 23:656–661. https://doi.org/10.1002/hipo.22124
102. Jacobs J (2014) Hippocampal theta oscillations are slower in humans than in rodents: implications for models of spatial navigation and memory. Philos Trans R Soc Lond B Biol Sci 369:20130304. https://doi.org/10.1098/rstb.2013.0304

103. Quian Quiroga R, Mukamel R, Isham EA et al (2008) Human single-neuron responses at the threshold of conscious recognition. Proc Natl Acad Sci 105:3599–3604. https://doi.org/10.1073/pnas.0707043105
104. Tsao A, Sugar J, Lu L et al (2018) Integrating time from experience in the lateral entorhinal cortex. Nature 561:57–62. https://doi.org/10.1038/s41586-018-0459-6
105. Chartier J, Anumanchipalli GK, Johnson K, Chang EF (2018) Encoding of articulatory kinematic trajectories in human speech sensorimotor cortex. Neuron 98:1042–1054.e4. https://doi.org/10.1016/j.neuron.2018.04.031
106. Anumanchipalli GK, Chartier J, Chang EF (2019) Speech synthesis from neural decoding of spoken sentences. Nature 568:493–498. https://doi.org/10.1038/s41586-019-1119-1
107. Leuthardt EC, Schalk G, Wolpaw JR et al (2004) A brain-computer interface using electrocorticographic signals in humans. J Neural Eng 1:63–71. https://doi.org/10.1088/1741-2560/1/2/001
108. Yanagisawa T, Hirata M, Saitoh Y et al (2011) Real-time control of a prosthetic hand using human electrocorticography signals. J Neurosurg 114:1715–1722. https://doi.org/10.3171/2011.1.JNS101421
109. Anderson NR, Blakely T, Schalk G et al (2012) Electrocorticographic (ECoG) correlates of human arm movements. Exp Brain Res 223:1–10. https://doi.org/10.1007/s00221-012-3226-1
110. Romanelli P, Piangerelli M, Ratel D et al (2018) A novel neural prosthesis providing long-term electrocorticography recording and cortical stimulation for epilepsy and brain-computer interface. J Neurosurg 130:1166–1179. https://doi.org/10.3171/2017.10.JNS17400
111. Penfield W, Jasper H (1954) Epilepsy and the functional anatomy of the human brain. Little, Brown & Co., Oxford
112. Penfield W, Roberts L (2014) Speech and brain mechanisms. Princeton University Press
113. Romanska A, Rezlescu C, Susilo T et al (2015) High-frequency transcranial random noise stimulation enhances perception of facial identity. Cereb Cortex 25:4334–4340. https://doi.org/10.1093/cercor/bhv016
114. Li S, Cai Y, Liu J et al (2017) Dissociated roles of the parietal and frontal cortices in the scope and control of attention during visual working memory. Neuroimage 149:210–219. https://doi.org/10.1016/j.neuroimage.2017.01.061
115. Xue G, Juan C-H, Chang C-F et al (2012) Lateral prefrontal cortex contributes to maladaptive decisions. PNAS 109:4401–4406. https://doi.org/10.1073/pnas.1111927109
116. Cai Y, Li S, Liu J et al (2016) The role of the frontal and parietal cortex in proactive and reactive inhibitory control: a transcranial direct current stimulation study. J Cogn Neurosci 28:177–186. https://doi.org/10.1162/jocn_a_00888
117. Marshall L, Mölle M, Hallschmid M, Born J (2004) Transcranial direct current stimulation during sleep improves declarative memory. J Neurosci 24:9985–9992. https://doi.org/10.1523/JNEUROSCI.2725-04.2004
118. Fregni F, Boggio PS, Nitsche M et al (2005) Anodal transcranial direct current stimulation of prefrontal cortex enhances working memory. Exp Brain Res 166:23–30. https://doi.org/10.1007/s00221-005-2334-6
119. Brunoni AR, Vanderhasselt M-A (2014) Working memory improvement with non-invasive brain stimulation of the dorsolateral prefrontal cortex: a systematic review and meta-analysis. Brain Cogn 86:1–9. https://doi.org/10.1016/j.bandc.2014.01.008
120. Lu Y, Wang C, Chen C, Xue G (2015) Spatiotemporal neural pattern similarity supports episodic memory. Curr Biol 25:780–785. https://doi.org/10.1016/j.cub.2015.01.055
121. Lafon B, Henin S, Huang Y et al (2017) Low frequency transcranial electrical stimulation does not entrain sleep rhythms measured by human intracranial recordings. Nat Commun 8:1199. https://doi.org/10.1038/s41467-017-01045-x
122. Suthana N, Haneef Z, Stern J et al (2012) Memory enhancement and deep-brain stimulation of the entorhinal area. N Engl J Med 366:502–510. https://doi.org/10.1056/NEJMoa1107212
123. Inman CS, Manns JR, Bijanki KR et al (2018) Direct electrical stimulation of the amygdala enhances declarative memory in humans. PNAS 115:98–103. https://doi.org/10.1073/pnas.1714058114

124. Kucewicz MT, Berry BM, Miller LR et al (2018) Evidence for verbal memory enhancement with electrical brain stimulation in the lateral temporal cortex. Brain 141:971–978. https://doi.org/10.1093/brain/awx373
125. Lacruz ME, Valentín A, Seoane JJG et al (2010) Single pulse electrical stimulation of the hippocampus is sufficient to impair human episodic memory. Neuroscience 170:623–632. https://doi.org/10.1016/j.neuroscience.2010.06.042
126. Merkow MB, Burke JF, Ramayya AG et al (2017) Stimulation of the human medial temporal lobe between learning and recall selectively enhances forgetting. Brain Stimul 10:645–650. https://doi.org/10.1016/j.brs.2016.12.011
127. Goyal A, Miller J, Watrous AJ et al (2018) Electrical stimulation in hippocampus and entorhinal cortex impairs spatial and temporal memory. J Neurosci 38:4471–4481. https://doi.org/10.1523/JNEUROSCI.3049-17.2018
128. Silvanto J, Muggleton N, Walsh V (2008) State-dependency in brain stimulation studies of perception and cognition. Trends Cogn Sci 12:447–454. https://doi.org/10.1016/j.tics.2008.09.004
129. Ezzyat Y, Kragel JE, Burke JF et al (2017) Direct brain stimulation modulates encoding states and memory performance in humans. Curr Biol 27:1251–1258. https://doi.org/10.1016/j.cub.2017.03.028
130. Ezzyat Y, Wanda PA, Levy DF et al (2018) Closed-loop stimulation of temporal cortex rescues functional networks and improves memory. Nat Commun 9:365. https://doi.org/10.1038/s41467-017-02753-0
131. Kwon O-Y, Ahn HS, Kim HJ (2017) Fatigue in epilepsy: a systematic review and meta-analysis. Seizure 45:151–159. https://doi.org/10.1016/j.seizure.2016.11.006

Chapter 9
How Many Data Do I Need for an iEEG Study? Treasure Maps and the Status of Variability

Jean-Philippe Lachaux

Abstract While most research projects are under strict time constraints, the acquisition of intracranial EEG data is often a long process, with few patients recorded every year with variable electrode implantations. The question of how many patients should be recorded, and how repeatedly in the same cognitive situation is therefore of utmost importance. While the implicit goal of most human cognitive neuroscience studies is to identify phenomena and mechanisms which are universal to our species (ultimate reproducibility), there can be no magic number of patients or trials that would secure that objective. Rather, we will argue that iEEG studies should be considered as a kind of "treasure maps", providing the maximum number of indications of where and how to observe the reported phenomenon. Within that framework, we propose that the exceptional quality of iEEG data, both at the anatomical and functional level, leads to a change in view point regarding variability—across trials and patients—which should be explained rather than discarded through group-level statistics. We will show how that approach applies to two complementary trends in iEEG research: the constitution of large-scale international databases and in-depth analyses of small groups of patients across multiple tasks including naturalistic conditions.

The struggle is long, the struggle is hard, the struggle is beautiful (Killing Joke, 1990).

One notorious difficulty of iEEG research is that data collection is painfully slow, in most cases. The rate of data acquisition is governed by factors that researchers cannot control, such as the volume of iEEG implantations performed by the associated epilepsy unit, the ability and willingness of patients to participate in research experiments, the number of other experiments they generously give their time to and, of course, the clinical schedule (see also Chap. 4). In addition, there are portions of the brain that are seldom recorded, because of their little clinical relevance or the difficulty to insert electrodes safely or because of specific strategies of the clinical team (such as a preference for ECoG or SEEG). To give a concrete example, patients

J.-P. Lachaux (✉)
Lyon Neuroscience Research CenterEduwell Team, INSERM UMRS 1028, CNRS UMR 5292, Université Claude Bernard Lyon 1, Université de Lyon, 69000 Lyon, France
e-mail: jp.lachaux@inserm.fr

N. Axmacher (ed.), *Intracranial EEG*, Studies in Neuroscience, Psychology and Behavioral Economics, https://doi.org/10.1007/978-3-031-20910-9_9

in Lyon or Grenoble are often implanted in the insula but almost never in the superior parietal lobe. For all those reasons, it can take years to record five or ten patients in some cortical areas, and even longer in specific pairs of regions for bivariate analysis. This is clearly incompatible with the timeline of most research projects (Ph.D. thesis, post-docs, grants, etc.).

Consequently, any researcher new to iEEG wants to know the minimal amount of data that constitutes a study, with the intuition that the usual standards of non-invasive neuroimaging—more than ten participants in most fMRI and M/EEG studies (e.g., [1, 2])—are unrealistic in this case. And the question concerns not only the number of subjects to include, but also the amount of data to record in each of them, since recording time is often limited. A well-conducted iEEG project should reach conclusions with the minimal number of patients, each recorded for a minimal amount of time. But how minimal?

9.1 Any Project Starts with a First Patient

As any iEEG project starts with recordings from a first patient, we will first discuss the minimal amount of data that is needed to reach sound conclusions in that first patient (and later, the minimal number of patients to include in a study). The answer obviously depends on the aim of the project: a few minutes of continuous recording might be sufficient to characterize functional connectivity at rest [3], but it might take a great number of trials and conditions to fit the numerous parameters of a complex neurobehavioral computational model. There can't be any universal and simple answer to our initial question, which depends ultimately on our understanding of what it really means to provide a convincing answer to a scientific question with iEEG. Although it might seem a bit trivial to experienced neuroscientists, going back to such basics is the most efficient way to decide whether a set of observations is "conclusive enough", because that decision cannot be derived from statistics alone.

We will consider the general objective of finding the neural substrate of a cognitive process or mental state with iEEG. That might not encompass all the science that can be done with intracranial recordings, but it is general enough to guide our steps throughout that chapter. Within that framework, most iEEG studies arguably try to demonstrate a relationship between "something that the patient does" and "something that her brain does", to put it simply (see also Chap. 48 for a stimulus-based perspective). "What the brain does" refers to a change at the neural level (we'll call it 'N'): it is the sudden appearance (or amplification, or reduction ...) of a neural pattern of activity that can be detected in iEEG signals. It might correspond to an Event-Related Potential, or transient neural oscillations, local broadband energy increases or decreases, long-distance synchronization between oscillations, causal influences between brain areas etc. ... with no restriction. 'What the patient does' is equally general: it refers to a change at the patient's level in a very general sense (let's tag it 'P'). P might be the experience of fear, the recollection of a life event from episodic memory, the recognition of a face, a button press ... it is usually induced by a cognitive paradigm and inferred from overt behavior: the patient is led to experience or 'do' P. Think of a patient giving a correct response during a delayed-match-to-sample paradigm. We can infer from that response that she "successfully maintained the items in working memory during the time interval between the sample and the probe": this is P, which is induced by the experimental protocol, and a researcher might search for a neural pattern N associated with that maintenance process P. In other instances, P might occur spontaneously and overtly, and be accessible only through verbal reports ("I was just thinking of my sister when she was a baby"). Let's keep that in mind when we discuss the specific case of electro-cortical stimulations studies, and naturalistic neuroscience [4, 5].

As shown in Fig. 9.1, a relationship between P and N can sometimes be inferred from single trials, because of the exquisite signal-to-noise ratio of iEEG. In that example, the onset of a face stimulus is followed by a strong energy increase between 50 and 150 Hz in the fusiform gyrus. That increase (N) seems to be related to face processing in the visual cortex (P). But how many times must that pattern repeat to be truly convincing of that N/P relationship? Our little statistical analysis reveals that six events are enough to show that the increase is significant across the trials, but is it sufficient to demonstrate the N/P relationship? We will see now that the answer is not just a matter of statistics, but ultimately … of common sense.

In daily life, it often takes very little evidence to convince ourselves that two phenomena are related. Imagine yourself in a hotel room, trying to figure out which switch turns on the bathroom light. Eventually, you will press one particular button and notice that the light goes on: do you need to repeat the procedure to be truly convinced that it is the right switch? Of course not. When you press that button, you exert a constraint on the electrical circuit (analogous to inducing P through experimental manipulation), which triggers the sudden illumination (analogous to a sudden change in brain activity, N). If things are so simple in the hotel room, why does it take several repetitions to establish convincingly a N/P relationship in iEEG research? Why does it take several trials to be convinced that a transient energy increase recorded in the fusiform gyrus is related to the processing of a face?

Well, a first difference is that activity in the fusiform gyrus never goes from zero to one in the radical fashion of the bathroom light. In other words, the spontaneous variations of the iEEG signal before P (during what is called "the baseline") complicate the detection of N. We would encounter the same difficulty in the bathroom if it was already illuminated by a couple of light bulbs with random time-varying intensities. The effect of your button press would be more difficult to detect and you would probably press the button several times to be convinced that you really changed anything. As for that bathroom light, iEEG signals often record from multiple neural sources with heterogeneous functional characteristics, that might not have the same relationship with P, which is equivalent to the situation just described. And of course, the detection would be even more difficult if you could only observe the global light of the bathroom from the outside, which explains why EEG and MEG require more repetitions than iEEG, because they record much broader brain areas. Yet, the influence of multiple neural populations on local iEEG signals, and their uncontrolled activity, makes it more difficult to detect N in single-trials.

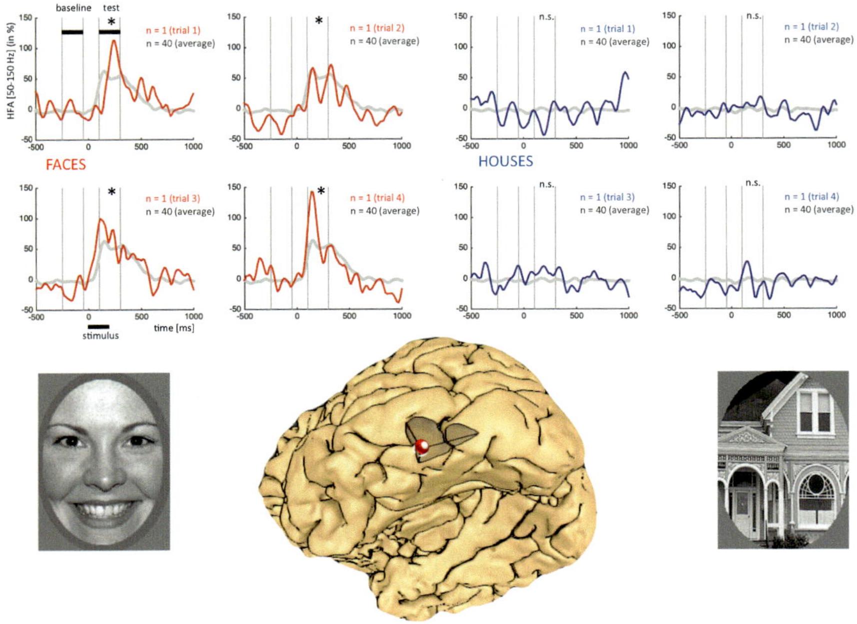

Fig. 9.1 Individual responses to face and house stimuli in the visual cortex. Each graph shows the neural activity recorded in the same SEEG contact site, in the fusiform gyrus/lateral-occipital sulcus (High-Frequency Activity, HFA, between 50 and 150 Hz, computed as in [20]). The patient watched several categories of images flashed for 200 ms during a visual oddball task (black horizontal line). The instruction was to press a button after each 'fruit' stimulus. The four red (resp. blue) curves correspond to four different individual 'face' (resp. 'house') stimuli, while the gray curve is the average response to 50 stimuli of the same category. For each of the four face stimuli, HFA increased significantly above pre-stimulus level ('*' sign: $p < 0.001$ for each, Kruskal–Wallis comparison between a test window [100: 300 ms] and the baseline [−250: −50 ms]). A Wilcoxon test was also performed to compare the average HFA during the baseline and the test window when considering only trials 1 and 2 (two trials: $p = 0.5$, n.s), or all four trials ($p = 0.125$, n.s.). When considering more trials, the significance criterion was reached for $n > 6$ ($p = 0.03125$ for six trials, $p = 0.0078$ for eight trials, $p = 0.0019$ for ten trials)

In short, it takes more trials to identify a N/P relationship when the amplitude of N is small relative to the ongoing fluctuations before P (as quantified by a z-score). Practically, it means that the number of repetitions of P that an experiment should include (the number of trials per condition, typically) cannot be estimated unless iEEG signals are measured from the region of interest, because brain activity is less stable in some regions than in others. For instance, neural activity has greater ongoing fluctuations in areas supporting spontaneous cognition or non-specific sensory and motor processing (think of the activity of the primary auditory cortex in a noisy hospital room). It follows that the number of trials required to demonstrate the N/P relationship depends on the targeted region. As a rule of thumb, it also takes more repetitions to compare a neural pattern between two experimental conditions (i.e., to show that high-gamma activity increases with memory load in the inferior frontal gyrus), because the difference that must be detected might be small. The amount of spontaneous neural fluctuations in the cortical regions of interest can be estimated from preliminary recordings, when available, and occasionally, from previous studies of the same region: for instance, figures from our study [6] show that attention-induced energy increase in the high-gamma band can be detected in the dorso-lateral prefrontal cortex from just a few trials.

One additional problem that must be taken into consideration when designing an iEEG protocol is that recording sessions might end abruptly, for multiple reasons: the fatigue of the patient, a seizure or a clinical procedure… it follows that data collection should be organized temporally in such a way that even partial recordings can be used. It would be a bad idea, for instance, to have critical experimental conditions by the end of the protocol. When possible, data should be collected equally for all conditions as time unfolds. And iEEG experiments should also be as short as possible, for the same reason (see Chaps. 4 and 5 for detailed descriptions of the practical challenges of iEEG recordings).

9.2 Confounding Factors

Until now, simple statistical considerations regarding the baseline have brought preliminary insights about our initial question ('how many data …') but additional factors must now be taken into account, which are much more difficult to address mathematically. Going back to the hotel room, let's imagine now that you came with a group of restless children busy playing with all the light switches. In the midst of such enthusiastic frenzy, how could you be sure that *you* pressed the bathroom switch? You would probably repeat the process in hope of isolating the effect of your own action. Yet, the probability that someone else pressed the right button twice in a row—exactly when you pressed your switch—increases with the number of children in the room. With iEEG, it means that even if you could detect N reliably in every trial, as in Fig. 9.1, several repetitions might be needed to establish that N is not due to causes other than P (confounds). Of course, any well-designed experiment minimizes the number of confounds and makes sure that several P's are not induced systematically at the same time. It would be suboptimal, for instance, to compare the brain response to colored faces and black-and-white landscapes because neural processes specific to faces and colors would always occur at the same time. But there are numerous other types of confounds over which experimenters have little control, such as spontaneous overt or covert actions, which can occur at the same time as P several times, unnoticed. Since such causes are impossible to list exhaustively, there can be no mathematical way to assess precisely the number of repetitions needed to demonstrate an N/P relationship. If iEEG and M/EEG studies publish such conclusions from a finite number of trials, it is only because reviewers are ready to believe that other confounds are very unlikely. A famous example is the relationship that was established between gamma-band induced responses in scalp EEG recordings (N) and the perception of visual objects (P) [7]. That relationship was widely accepted within visual neuroscience until Yuval-Greenberg et al. [8] showed that miniature saccades generate a transient artefactual gamma activity in EEG signals. Since then, the detection of gamma oscillations during object perception—however repeated and systematic—is no longer considered as a definite proof that the two phenomena are related, at least by some researchers. Unless a study provides an exhaustive list of all potential causes of their neural phenomenon of interest (the number of children in the room and how often they press the buttons), it is up to readers to appreciate whether the evidence is "convincing enough" using their common sense and not just mathematics. This limitation is primarily due to the fact that iEEG research is essentially about correlations and co-occurrences, rather than about causal effects, just as neuroimaging in general. The problem is not unique to our field, but certainly amplified by the unique complexity of the human brain and its sensitivity to such a wide variety of effects. The conclusions of any iEEG study should therefore be taken as temporary—not final—explanations, "*jusqu'à preuve du contraire*" (until proven otherwise).

9.3 When One Is Enough

One type of iEEG studies illustrates perfectly the importance of common sense: Electrical cortical stimulation studies. Stimulation studies report the behavioral, cognitive and experiential effects of electrical cortical stimulations (ECS) [9] delivered through iEEG electrodes (see also Chaps. 5, 39, 41, and 52). ECS studies have a long and rich history that dates back 80 years, and the pioneering mapping of the sensorimotor homunculus by Penfield and Boldrey [10]. Today, most clinicians still consider ECS to be the gold standard when it comes to map eloquent brain areas. Yet, conclusions are based on a few single events—if not one—because repeated ECS can induce after-discharges. For instance, Mazzola et al. [11] used ECS to build a very insightful functional map of the insula, in a study that mentions no statistical method nor the number of stimulations used, because it is not relevant in that field of research. What the authors reported was the proportion of sites triggering a specific manifestation (e.g., a gustatory sensation) in each portion of the insula, which is a way to convince others that those effects were somehow 'typical' because they could be replicated in the same area in several patients. Yet, it is up to readers to judge whether those observations were "reproducible enough" to be truly convincing. In another study, Nencha et al. [12] showed that ECS in the dorsal anterior insula can induce an ecstatic feeling. This time, the effect was demonstrated with only three stimulations of the same site in one patient only. Yet, the evidence was considered sufficiently strong to be published, with no statistical analysis, obviously (but the authors didn't claim that the effect was typical). Sometimes, important conclusions about the function of a given brain region are even derived from a single stimulation in a single patient: during awake surgery, neurosurgeons frequently use ECS to identify eloquent brain regions which must be spared by the resection [13]. For instance, if the ECS induces a speech-arrest, the region is assumed to be critical for language production. Such important decisions are based on single events and no statistics, which illustrates the "convincing" power of single events.

What makes single-trial data so convincing in ECS studies? One important aspect is timing: the fact that the change in behavior (or the experiential manifestation) is observed immediately after the stimulation. But timing cannot explain everything: if the patient blinked or coughed after the ECS, no one would conclude that it is necessarily due to the stimulation. It is also important that the behavioral or experiential manifestation be clear and rare. In the Nencha et al. study [12], the probability that the patient experienced spontaneously a state of deep clarity right after the stimulation was ridiculously small. How small? Intuitively small, based on our own experience of those states, but with no proper mathematical way to quantify that probability. To a lesser degree, the same holds for speech arrests or visual phosphenes: because such phenomena rarely occur spontaneously, their most reasonable cause is the electrical stimulation. The conclusion is an educated guess made through an observation; it is based on logical inference. It is the most logical explanation for that observation based on common sense, but certainly not a definite proof of a general principle. The followers of Penfield use the same logic we would use in our hotel room, when there is only one light that functions normally: the light goes on when you press the button, and that single event is sufficient to convince you that the two phenomena are related. It is perfectly acceptable.

With ECS, clinicians exert an influence on the neural activity and expect a change in behavior or cognition. During cognitive paradigms, experimenters exert an influence on the behavior and cognition of the participant and expect that some component of the iEEG signals will change. The two approaches are symmetric and comparable. The reason why the effect of ECS can be revealed with one single event is because (a) the change in behavior and cognition is unlikely to occur by chance, and (b) its timing is tightly linked to the ECS. It follows that during cognitive paradigms, N/P relationships can be demonstrated with very few trials—possibly one—if there is a tight relationship between changes in P and N, and if P is the 'most likely' cause of the change in N.

9.4 The Longer, the Better

So far, we emphasized the correspondence between the onsets of the stimulation and the manifestation, in ECS studies: the later occurs as soon as the former is delivered. But in most cases, it also ceases immediately when the stimulation stops, which makes the relationship even more convincing. During cognitive paradigms, a similar correspondence occurs when P and N are sustained with similar dynamics, which is often observed in associative cortical regions [14]. Figure 9.2 illustrates such a case: during a visual search task, high-frequency activity in the inferior frontal sulcus (N) is sustained above baseline level for the full duration of the search process (P). Such temporal alignment makes the relationship extremely convincing, even with few trials and especially as it holds for variable durations of P.

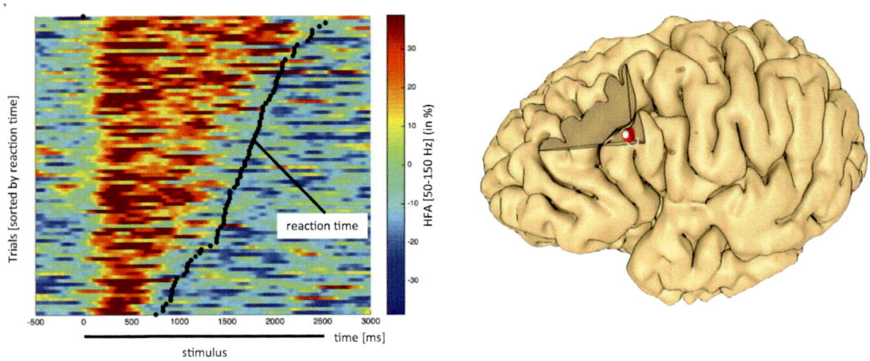

Fig. 9.2 Neural activity in the inferior frontal sulcus during a visual search task. Color codes High-Frequency Activity [50–150 Hz] expressed in percentage variation from the average HFA in that prefrontal site during the entire experiment. Trials (y-axis) are sorted according to reaction time (black dots), from the stimulus onset (t = 0 ms). This type of visualization illustrates very clearly the relationship between the sustained HFA increase and the search process, in every trial

If the bathroom light went off as soon as you released the button, you would quickly conclude that you found the right switch, even with hyperactive children in the room. In other words, a tight correlation between two sustained phenomena brings stronger support for a relationship between the two, than if they are short and transient (as in Fig. 9.1). But again, it is difficult to evaluate properly how evidence increased, because it depends on the number of independent observations of the N/P relationship which can be made from a single instance. That number depends on parameters such as the inertia of the mechanism causing the observed pattern N: in Fig. 9.2, iEEG signals were sampled at 512 Hz, but they did not provide 512 independent observations per second that "high-frequency activity is still above baseline as long as the visual search goes on". A proper statistical estimation would include factors such as the probability that "N is still above baseline level at time t + 1 knowing that it was above baseline level at time t", among others. For now, let's simply conclude that a sustained temporal relationship between a neural pattern and the cognitive or behavioral phenomenon of interest is "more convincing" than a more transient one, and requires less repetitions. This is particularly interesting for naturalistic studies, as we will see now.

9.5 Naturalistic Studies: So Hard to Replicate

In naturalistic studies, N/P relationships are studied in situations which resemble everyday life. The behavior or cognitive activity under study is elicited during the casual interaction of the patient with her environment, including people, rather than by a rigorously controlled paradigm. iEEG is the technique of choice for that purpose, because high-quality data can be recorded even when subjects are moving, which is not possible with M/EEG and fMRI. Figure 9.3a shows a sudden activity of the fusiform face area (FFA) each time a patient made eye contact with people around her, but not when she looked at inert objects in the room. Is it sufficient to conclude that in a real-life setting, the FFA reacts to faces as it does when stimuli are flashed experimentally on a screen? Using the criteria of ECS studies, the answer might be positive, because the reaction of the FFA is clear and rare—it does not happen when the patient is looking at other objects—and it also occurs with a reproducible delay after the eye-contact. Yet, our statement would not be supported by any formal statistical analysis, as in the insula stimulation studies.

Figure 9.3b shows an informal verbal exchange during which the experimenter challenges the patient to memorize a list of four words in a foreign language (in Suomi). The activity of the dorsal anterior insula (dAI) rose to reach a plateau until the patient repeated the words that he just heard. Although the behavior was observed only once, most observers would agree that the dAI was involved in that improvised memory task, because its activity (N) was sustained as long as information was maintained in working memory (P). That temporal pattern might be due to other

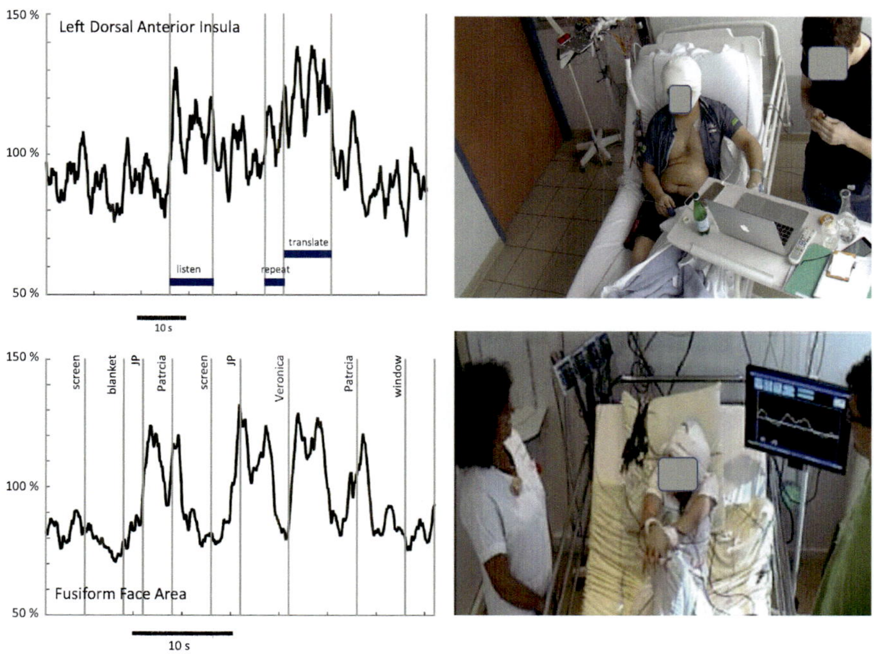

Fig. 9.3 iEEG responses during naturalistic interactions. Top panels: HFA [50–150 Hz] in the left dorsal anterior insula while the patient **a** listened to three words in a foreign language with their translation (Suomi), **b** repeated the words and **c** repeated their translation. HFA is expressed in % of the average value across the entire recording. Bottom panels: HFA in the right fusiform gyrus while the patient turned her gaze towards several objects or people in the room (JP, Veronica, Patricia: experimenters). HFA increased selectively when she turned to faces. Very few repetitions can convince observers of a relationship between a specific behavior and a focal neural activity

causes, but the possibility that working memory involves sustained dAI activation is the most plausible explanation for the tight temporal correspondence between the two phenomena. Of course, the experimenter might have challenged the patient several times with new words to accumulate more evidence for that relationship, but there are many other types of naturalistic P which are more difficult to reproduce: some emotions for instance (think of a patient receiving a good-news phone call). Neural correlates of such behavioral or cognitive events can be investigated, following the guiding principles of ECS studies ("clear, rare and coincident").

A parallel can be drawn with the seminal study of Gelbard-Sagiv et al. [15], who showed that the spontaneous recollection of a recent memory coincides with the sudden activation of individual neurons with a specificity for that memory in the medial temporal lobe ('concept cells'). During some of his talks, the senior author (I.F.) showed a striking video illustrating that effect: a silent cell responsive to the 'Simpsons concept' suddenly fired a series of action potentials a fraction of a second before the patient recalled 'he just saw a Simpsons cartoon'. I.F. would then comment with a mischievous smile: "that's when you need to hire a statistician" [to demonstrate that the cell's response is involved in memory recollection]. Again, a single occurrence can be convincing enough if the response is clear and highly selective, with the expected timing. This example is particularly compelling because the spontaneous recollection of specific episodic memories usually occurs in a random, unique and naturalistic way.

In that study, Gelbard-Sagiv et al. [15] reported several additional instances of recollection-related neural activations, which is always preferable. When some effect is observed in a region that is often recorded in the iEEG unit, it is always better to collect data for more than one patient in similar naturalistic conditions, as we will discuss later. But if the cortical region is rarely implanted, or if the natural situation is difficult to reproduce, publishing such observations, with sufficient details regarding the electrode position and the observed phenomenon (with raw data showing the correspondence between N and P) can encourage other researchers to reproduce the observation when they have a chance to record from the same brain region.

Finally, we shall mention a specific type of naturalistic study, in which patients can visualize their brain activity in real-time, thanks to a dedicated set-up [16]. During such sessions, the patients are encouraged to search for correlations between the ongoing activity in specific iEEG sites—N, for instance High-gamma activity [50–150 Hz]—and aspects of their subjective experience (what they are aware of doing or feeling: first-person data, P). This is equivalent to asking a patient if she "felt anything special", while her brain was stimulated with ECS. The real-time display allows to test N/P correspondences rapidly, using a stop-go-repeat paradigm: the patient starts and stops doing P (go/stop), then repeats the procedure. If the neural activity of interest N waxes and wanes, a mechanistic relationship between N and P is the most plausible explanation.

9.6 How Many Patients Constitute an iEEG Study?

Once an effect has been demonstrated in a first patient, the next question is whether that effect is specific to that person, or a general feature of the population she was drawn from (in most cases: the human species). The answer requires recordings from more patients, but how many? It depends essentially on what researchers try to demonstrate. Addressing that issue in an eponym article ("how many subjects constitute a study?"), Friston et al. [17] distinguished between two different objectives: showing that a trait is typical of a population (let's call it a 'type I' objective) and providing a more refined characterization with quantitative values for such traits (type II). The authors illustrated that distinction with a comparison between two objectives: "showing that farmers in Wales typically own tractors" (i.e., showing a typical trait, type I) versus "showing that farmers in Wales own on average more than 0.86 tractor(s)" (type II). With iEEG, a similar distinction could be made between "showing that the presentation of a face stimulus generates a response peak in the right fusiform gyrus later than 100 ms" versus "showing that the peak latency of the neural response to face stimuli in the fusiform gyrus is 170 ms on average". Intuitively (and mathematically), it takes smaller samples to achieve objectives of type I because they require categorical data which are less sensitive to noise (the answer is 'yes' or 'no', the feature is 'present' or 'absent'). Still, one might suspect that the 'mere' demonstration of typicality also requires samples much larger than the number of patients who can be recorded in a reasonable time frame. After all, if asking twenty random British citizens was sufficient to conclude with a high degree of certainty that 'typically, people in the UK still approve the Brexit', polling institutes would go bankrupt. Friston et al. [17] provided an estimate for the minimum number of subjects needed to establish that a trait is typical of a given population, if all subjects within that sample share that specific trait. The number is surprisingly low, but depends on the operational definition of "typicality": showing that 6 out of 6 subjects possess the specific feature is enough to demonstrate with 95% confidence that it is present in at least 60% of the population. In other words, if less than 6 individuals out of 10 have that feature in the global population, it is very unlikely that six random individuals would all have that feature. But it is impossible with a sample of six to conclude that the feature is more present in the population (more "typical"), than 60%. In other words, it is relatively easy to demonstrate that a feature is present in at least X% of the population when X is low, because it would be extremely unlikely otherwise that all individuals in your sample had that feature. For instance, it takes a group of twenty individuals with the same feature to show that X is higher than 85% (a more stringent definition of typicality) ([17], Fig. 9.2).

It follows that iEEG studies should aim for objectives of type I (to demonstrate typicality), but that the number of subjects which constitute a study depends on the level of typicality one wishes to demonstrate. According to Friston et al., one hundred patients would not be sufficient to show that the inferior frontal gyrus is involved in verbal working memory in 99% of the population. Interestingly, most iEEG studies attempt to demonstrate that a NP relationship is typical with no explicit defining criterion, which means that the actual objective is simply to convey a feeling that the relationship is «quite» typical of the human brain. It would be simpler, and clearer, if all iEEG studies tried to demonstrate "absolute typicality" (i.e., the NP relationship is present in all human beings). But that goal is mathematically unachievable and leads one to believe that iEEG projects should always include as many patients as possible. This is of course an unsatisfactory conclusion, which means that statistics alone don't provide a good answer to our initial question: how many patients constitute an iEEG study?

Let's consider electrophysiological studies conducted in non-human primates (NHP). Most NHP studies report data from two or three monkeys and not more (i.e., [18]), for obvious ethical and practical reasons. Yet reviewers and readers widely accept the conclusion that the effects are "typical", in the sense that they are not specific to those two or three individuals. That belief is not supported by mathematics, but rather by "common sense". For the reasons mentioned above, the objective of NHP studies cannot be to demonstrate effects which are general to that species, and the added value of studying one versus two monkeys is simply to show that the results are not due to some weird hidden characteristic of a specific primate. It is more psychological than mathematical and yet, NHP studies manage to convince other scientists that further recordings in the same area of the brain and the same condition would yield similar observations.

There is no reason why the same logic should not apply to iEEG studies, which rely on data very similar to some NHP recordings (local field potentials, [19]). If two monkeys constitute a NHP study, then two humans should constitute an iEEG study. The main difference is that iEEG patients suffer from a major neurological disorder with possible brain reorganization. This motivates even further the replication of the findings in a second patient, but not more. There is no mathematical reason to set the bar higher for iEEG, just because iEEG recordings are more common in humans than in monkeys.

NHP studies have long abandoned the ambition to demonstrate the perfect typicality of their findings, which is less apparent in M/EEG and fMRI but still true. But since electrophysiological and neuroimaging studies never provide a consistent, operational definition of that term, that's not a big loss after all. For instance, the study by Ossandon et al. [20] showed that high-gamma activity is reduced for the entire duration of a visual search throughout the Default-Mode Network. The claim was illustrated in several patients for each region of the DMN, but there was no attempt to quantify the probability of fellow researchers to observe a similar decrease when recording from DMN regions in the same conditions. The best we could do was to provide evidence that the phenomenon of interest was not due to a specific feature of a particular brain.

But then, why include more than two subjects in an iEEG study? We will argue that it helps other iEEG researchers to replicate the findings in other patients. Let's illustrate that point with a second metaphor: the treasure map. As everybody knows, a treasure map is a document which provides explicit cartographic information on where to find a treasure. And we expect from a good treasure map to provide numerous unambiguous details about the treasure location; for instance, that "it is buried at the bottom of a small hill, on its eastern side next to a few pine trees, on the mid-west coast of the island". In a sense, most iEEG studies are analogous to treasure maps. They provide explicit details on where a specific effect (N = the treasure) should be found. The iEEG treasure map (= the article) is here to help future neuro-adventurers to find that treasure: if they record ('dig') in that specific location, they will most likely find it. This metaphor helps understand why including more patients in a study will make the map more precise, considering that several patients are very rarely implanted in the exact same locations. Within a particular region of the cortex, reporting from more than one patient specifies further where the effect was found (and where it was not found): what were the characteristics shared by all the sites where the effect was observed? The answer might be: "they were all in the depth of the left inferior frontal sulcus, immediately above the pars opercularis of Broca (and sites more lateral didn't show the effect)". Doing so, the authors provide all the information that seems relevant to help other researchers make the same observation. And of course, the amount of information they can provide increases with the number of patients they present in their study. That includes not only detailed anatomical information, but also various factors such as behavioral measures (reaction times, hit rates ...) or the strategy used ...

That strategy somehow assumes that the effect should be present in every patient (perfect typicality), which makes sense: if we had reported, in our DMN study [20], that in some individuals, the activity of the DMN was not suppressed during a visual search, with no further explanation, it would have been very difficult to draw any insightful conclusion regarding the role of the DMN. When the effect of interest is not observed in some patients, some type of explanation is expected. The "treasure map approach" lists factors which might explain why the effect was observed only in some sites and some patients, such as: "in patient 2, where the effect was not observed, the recording site in the inferior frontal sulcus was in fact more superficial than in the other patients". The ideal study should list all distinctive properties of the sites (and patients) where the effect was present. It requires a very detailed understanding of the cortical region(s) of interest, at the single-patient level, not only in terms of anatomy, but also connectivity and functionality, ideally. For instance, the distinctive property of all sites of interest might be that "when stimulated with cortical electrical potentials, a response was observed in the primary auditory cortex" or that "in a visual localizer task, they were selectively activated by face stimuli, but not by house stimuli". And if the effect was found in all iEEG sites in a given region, in all participants, having more patients provides more information about the extent of the region where that effect should be found, given that electrode coverage of the brain is sparse and variable across patients.

9.7 Multiple-Case Studies: The More, the Merrier

None of the arguments so far impose that iEEG studies should include more than two patients: it would just add more information to the treasure map. Yet, this conclusion does not apply to studies which report more than one N/P relationship. Let's consider an attempt to show that face stimuli elicit a strong response in a specific portion of the fusiform gyrus, and in the lateral occipital cortex. Following the guidelines of NHP research, it would be good to report effects in more than one patient for each region. But since implantations vary across patients, it might take years to record from two participants with electrodes in both regions. This is not necessary, however, because the study can simply report the effect for two patients in the fusiform gyrus, and two other patients in the lateral occipital cortex. In that case, the study requires more than two patients, because it combines two independent studies in one. There is no particular need to record all regions of interest in the same patients, unless one is interested in particular relationships between those regions (such as differences in peak latencies, patterns of propagation, or long-distance coupling mechanisms ...).

To summarize, the basis of any iEEG study should be a strong and convincing observation in one individual, as for primates' studies. It must be followed by a replication in at least one subject, because the brain of an epileptic patient can have some specificities which might affect the observation. The study might of course include more patients, but never enough to demonstrate that an effect is general to the human species. Evidence will always be weak and preliminary. If the conditions of observation are sufficiently well described, with enough detailed anatomical information, other researchers can replicate the finding and provide more information regarding its exact cortical origin and factors which amplify or diminish the effect (such as strategies, patient's characteristics, ...): the 'treasure map' should be as precise as possible for future treasure hunters. If the study fails to identify distinctive features of the responsive sites and if the conclusion is that a fraction of the cortical recordings shows the pattern of interest within a well-defined region, it is up to future studies to perform new recordings of that detailed area in search for explanations of that variability. After all, there is no deadline for understanding the human brain.

The approach we described clearly differs from the common practice of non-invasive neuroimaging. In M/EEG and fMRI, inter-individual variability is usually not explained, but modeled as a random variable (random-effect analysis) (e.g., [21]). Also combining data from multiple participants facilitates the detection of effects that are not significant in single subjects. Using the same approach for iEEG is questionable: the signal-to-noise ratio is so good that the effect of interest should be detected if the electrode is correctly positioned, as in NHP studies. Nevertheless, random-effect analysis of iEEG might become popular in the near future, with the emergence of large multi-centric iEEG datasets (see below). Data will be collected in the same focal brain regions and the same paradigms in populations comparable in size to typical M/EEG and fMRI samples. All collected data could then be attributed to a single meta-subject, in the same way as many EEG studies pool together all recordings from Cz or Fz across all participants. This strategy will greatly facilitate data analysis and probably reveal effects which are absent in some patients, but it will also fail to provide any explanation for such absence. Another potential pitfall is that sites might be pooled together based on automatic labelling only (as provided by popular tools such as MAPER or Freesurfer, [22]). The parcellation induces a loss of anatomical information, which might limit data interpretation, especially if the parcellation does not take into account critical sources of response variability (such as the depth of iEEG sites within a sulcus). A detailed anatomical and functional analysis at the single-patient level will always provide a much deeper understanding of the structure–function relationship, and make full use of the spatial precision of iEEG.

9.8 Towards Large Multicentric iEEG Databases

With the advent of open-science and the emergence of technical solutions for fluid and secured data-transfer, new platforms have been created to facilitate the joint analysis of multicentric iEEG data. For instance, the Human Intracranial Platform (https://www.humanbrainproject.eu/en/medicine/human-intracerebral-eeg-platform/), funded by the Human Brain Project, will encourage researchers to submit experimental protocols to a review board, so that iEEG data can be collected in several clinical centers simultaneously. The data will then be centralized and made available to other researchers, in the spirit of open-science pioneered by the team of Michael Kahana for iEEG (https://memory.psych.upenn.edu/Data#2021). The field will no longer be a niche for a few lucky research teams, overwhelmed by the richness of their data. Open-data policies will provide more opportunities for scientists outside our field to find golden nuggets within shared iEEG datasets. iEEG studies including hundreds of patients will become more and more common, which will inevitably bring important breakthroughs. Considering our recommendation to take anatomy very seriously at the single-patient level, data analysis might become impossibly tedious, unless new software will be developed to facilitate detailed anatomical investigation in large groups of patients. One example is HiBoP, a visualization software developed within the Human Brain Project to navigate conveniently between group-level and patient-level functional data visualization, in 4D (dynamic visualization onto 3D brain reconstructions) (https://github.com/hbp-HiBoP/HiBoP) (Fig. 9.4). HiBoP helps iEEG researchers to identify regions of interest at the group level, search for reproducible effects across patients and distinctive anatomical features of the responsive sites. Hopefully, HiBoP and similar software will allow the publication of extremely precise treasure maps from large samples of patients. But that should not condemn studies based on a couple of patients, as there will be no new mathematical justification for raising the bar, in terms of sample size, beyond the criteria of NHP research.

9.9 A Final Word on Case-Reports

One might conclude from this chapter that any iEEG study should include at least two patients, which would exclude de facto single-case studies. On the contrary, publication of rare and interesting phenomena in regions seldomly accessed are

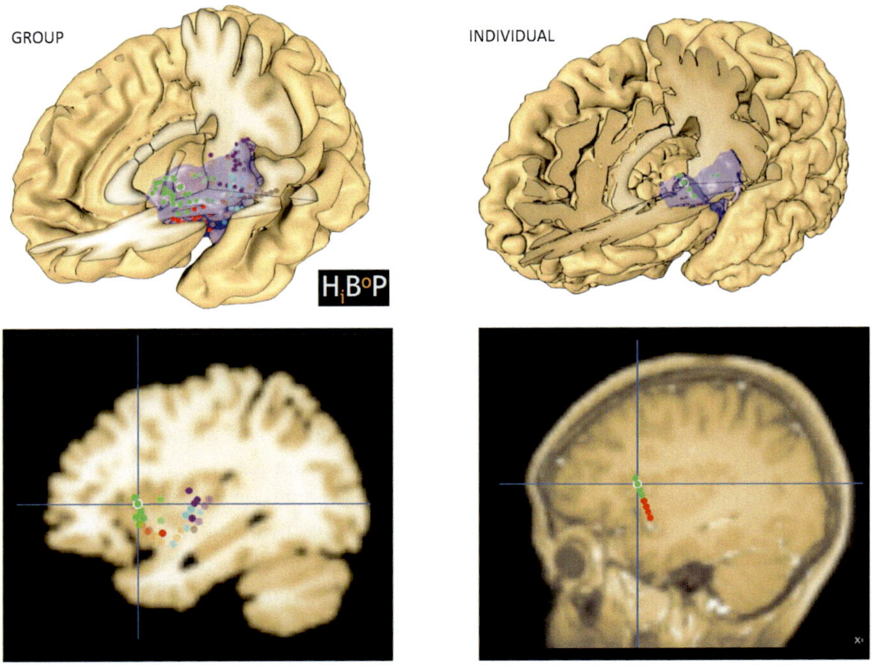

Fig. 9.4 Group versus Single-patient level visualization. The left panel displays insula iEEG sites in of twenty patients, realigned onto a 3D reconstruction of the left insula on the MNI template (color codes for subregions of the insula). The right panel displays iEEG insula sites of one patient onto her own 3D brain. The HiBoP software was designed to quickly shift between group-level and individual-level anatomy for high-precision anatomical labelling of all iEEG sites of interest (such as the green site Y'8)

invaluable, because they encourage other research teams throughout the world to reproduce those observations, combine data from several patients and reach stronger conclusions. But that implies that the initial case-report provides a very precise 'treasure map', with the most detailed information about the location of the effect. It should also list any atypicality of the patient, anatomy-, pathology- and performance-wise.

For the same reason, a multi-centric iEEG platform could become a depository of rare naturalistic observations made in single subjects. It will require new ways to share anonymized iEEG-video datasets—a technical challenge when the behavior of interest involves verbal exchanges and oro-facial expressions—but solutions will soon exist. And within a few years, researchers throughout the world will have access to large datasets of iEEG and video recordings from the same brain regions in closely-related naturalistic situations.

9.10 Conclusion

To conclude, most of what we know about the brain has come from animal studies, that is, detailed investigations of a few individuals, with high-quality recordings showing clear effects in single-trials. IEEG studies are the closest equivalent in humans, and could follow similar guidelines regarding the amount of data needed to reach sound conclusions. The specific case of Electro-Cortical Stimulation studies show that important findings can even be made from single events, for particularly clear and rare effects. This should encourage a seminal type of research which takes very seriously brain-function relationships at the finest level: a clear change of activity in specific cortical regions of one individual, in relation to one particular event, such as the observation that led to the discovery of mirror neurons in Parma [23]. With the advent of large multicentric iEEG data collection, it would be a great loss if such 'old-school, naked-eye' approach, inspired by classic electrophysiology studies (e.g., [24]) gave way entirely to efficient machine-learning algorithms which process blindly entire databases in seconds to generate global statistical reports about effects which cannot be seen in any individual patient.

Acknowledgements JPL received funding from the European Union's Horizon 2020 Research and Innovation Program under Grant Agreement No. **785907** (HBP SGA2).

References

1. Dong HW, Mills C, Knight RT, Kam JWY (2021) Detection of mind wandering using EEG: Within and across individuals. PLoS One 16:e0251490. https://doi.org/10.1371/journal.pone.0251490
2. Groot JM, Boayue NM, Csifcsák G et al (2021) Probing the neural signature of mind wandering with simultaneous fMRI-EEG and pupillometry. Neuroimage 224:117412. https://doi.org/10.1016/j.neuroimage.2020.117412
3. Cuesta P, Ochoa-Urrea M, Funke M, et al (2022) Gamma band functional connectivity reduction in patients with amnestic mild cognitive impairment and epileptiform activity. Brain Commun 4:fcac012. https://doi.org/10.1093/braincomms/fcac012
4. Jack AE, Roepstorff AE (2003) Trusting the subject?, vol 1. Imprint Academic
5. Genon S, Reid A, Langner R et al (2018) How to characterize the function of a brain region. Trends Cogn Sci 22:350–364. https://doi.org/10.1016/j.tics.2018.01.010
6. Ossandon T, Vidal JR, Ciumas C et al (2012) Efficient "pop-out" visual search elicits sustained broadband gamma activity in the dorsal attention network. J Neurosci 32:3414–3421. https://doi.org/10.1523/JNEUROSCI.6048-11.2012
7. Tallon-Baudry C, Bertrand O (1999) Oscillatory gamma activity in humans and its role in object representation. Trends Cogn Sci. https://doi.org/10.1016/S1364-6613(99)01299-1
8. Yuval-Greenberg S, Tomer O, Keren AS et al (2008) Transient Induced gamma-band response in EEG as a manifestation of miniature saccades. Neuron 58:429–441. https://doi.org/10.1016/j.neuron.2008.03.027
9. Kahane P, Hoffmann D, Minotti L, Berthoz A (2003) Reappraisal of the human vestibular cortex by cortical electrical stimulation study. Ann Neurol 54:615–624. https://doi.org/10.1002/ana.10726

10. Penfield W, Boldrey E (1937) Somatic motor and sensory representation in the cerebral cortex of man as studied by electrical stimulation. Brain 60:389–443. https://doi.org/10.1093/brain/60.4.389
11. Mazzola L, Mauguière F, Isnard J (2019) Functional mapping of the human insula: data from electrical stimulations. Rev Neurol (Paris) 175:150–156. https://doi.org/10.1016/j.neurol.2018.12.003
12. Nencha U, Spinelli L, Vulliemoz S et al (2022) Insular stimulation produces mental clarity and bliss. Ann Neurol 91:289–292. https://doi.org/10.1002/ana.26282
13. Szelényi A, Bello L, Duffau H et al (2010) Intraoperative electrical stimulation in awake craniotomy: methodological aspects of current practice. Neurosurg Focus 28:E7. https://doi.org/10.3171/2009.12.FOCUS09237
14. Haller M, Case J, Crone NE et al (2018) Persistent neuronal activity in human prefrontal cortex links perception and action. Nat Hum Behav 2:80–91. https://doi.org/10.1038/s41562-017-0267-2
15. Gelbard-Sagiv H, Mukamel R, Harel M et al (2008) Internally generated reactivation of single neurons in human hippocampus during free recall. Science 322:96–101. https://doi.org/10.1126/science.1164685
16. Lachaux J-P, Jerbi K, Bertrand O et al (2007) A blueprint for real-time functional mapping via human intracranial recordings. PLoS One 2:e1094. https://doi.org/10.1371/journal.pone.0001094
17. Friston KJ, Holmes AP, Worsley KJ (1999) How many subjects constitute a study? Neuroimage 10:1–5. https://doi.org/10.1006/nimg.1999.0439
18. Rohenkohl G, Bosman CA, Fries P (2018) Gamma synchronization between V1 and V4 improves behavioral performance. Neuron 100:953-963.e3. https://doi.org/10.1016/j.neuron.2018.09.019
19. Pesaran B, Pezaris JS, Sahani M et al (2002) Temporal structure in neuronal activity during working memory in macaque parietal cortex. Nat Neurosci 5:805–811. https://doi.org/10.1038/nn890
20. Ossandon T, Jerbi K, Vidal JR et al (2011) Transient suppression of broadband gamma power in the default-mode network is correlated with task complexity and subject performance. J Neurosci 31:14521–14530. https://doi.org/10.1523/JNEUROSCI.2483-11.2011
21. Fan X, Wang F, Shao H, et al (2020) The bottom-up and top-down processing of faces in the human occipitotemporal cortex. eLife 9:e48764. https://doi.org/10.7554/eLife.48764
22. Yaakub SN, Heckemann RA, Keller SS et al (2020) On brain atlas choice and automatic segmentation methods: a comparison of MAPER & FreeSurfer using three atlas databases. Sci Rep 10:2837. https://doi.org/10.1038/s41598-020-57951-6
23. di Pellegrino G, Fadiga L, Fogassi L et al (1992) Understanding motor events: a neurophysiological study. Exp Brain Res 91:176–180. https://doi.org/10.1007/BF00230027
24. Hubel DH, Wiesel TN (1962) Receptive fields, binocular interaction and functional architecture in the cat's visual cortex. J Physiol 160:106–154. https://doi.org/10.1113/jphysiol.1962.sp006837

Chapter 10
How Can iEEG Be Used to Study Inter-Individual and Developmental Differences?

Elizabeth L. Johnson and Robert T. Knight

Abstract Inter-individual differences, including but not limited to those that distinguish children from adolescents and younger from older adults, are a hallmark of human cognition. As described throughout this book, intracranial electroencephalography (iEEG) affords unprecedented access to the human brain, permitting insight into the neurophysiology of human cognition with high spatiotemporal and single-trial precision. However, iEEG is also limited due to brain coverage that is sparse within one patient and variable across patients. This limitation poses a fundamental challenge for the use of iEEG in controlled investigations of inter-individual differences. In this chapter, we address this challenge and describe best practices for studies that aim to elucidate inter-individual and developmental differences in the neurophysiological mechanisms of human cognition using iEEG. We first briefly discuss how iEEG data are typically handled by minimizing sources of inter-individual variability. We then present best practices for the use of iEEG in controlled investigations of inter-individual differences and describe recent studies that used iEEG to reveal signatures of memory which differ across patients. We propose that iEEG be considered a gold standard in studies of inter-individual and developmental differences in the neurophysiology of human cognition.

10.1 Introduction

No two brains are identical, and inter-individual differences are a defining feature of the human experience. This chapter focuses on intracranial electroencephalography (iEEG) as a tool to investigate inter-individual and developmental differences in human cognition, understanding of which has been hindered by common neuroscientific approaches. First, because noninvasive imaging methods offer either spatial

E. L. Johnson (✉)
Northwestern University, 633 N. St. Clair St., Chicago, IL 60611, US
e-mail: eljohnson@northwestern.edu

R. T. Knight
University of California, 132 Barker Hall, Berkeley, CA 94720, US
e-mail: rtknight@berkeley.edu

N. Axmacher (ed.), *Intracranial EEG*, Studies in Neuroscience, Psychology and Behavioral Economics, https://doi.org/10.1007/978-3-031-20910-9_10

or temporal precision, reliance on techniques such as functional magnetic resonance imaging (fMRI) and scalp EEG has limited our ability to delineate human brain activity with both spatial and temporal precision. Second, because noninvasive imaging techniques offer relatively low signal-to-noise ratio [1], many studies have relied on group-level averaging of brain data and treated inter-individual variability as a source of noise [2–5]. Third, because invasive recording, which offers high spatiotemporal resolution and signal quality, has been traditionally performed in non-human animals, resulting data cannot explain factors that distinguish one person from another. This is especially relevant to developmental inquiry, as the maturational trajectory of the human brain is more protracted and qualitatively distinct from that of even our closest primate relatives [6–10]. iEEG addresses these hindrances by providing insight into the neurophysiology of human cognition with high spatiotemporal resolution and signal-to-noise ratio enabling single-trial precision [11–14]. With appropriate controls, iEEG studies offer immense potential to advance our understanding of inter-individual and developmental differences in human cognition.

Figure 10.1 illustrates two datasets in which responses such as behavioral performance or measures of brain structure or function differ between two experimental conditions [2]. In one dataset, most individual data are consistent with group averages and averaging reveals an omnibus pattern. However, some participants show opposite trends or higher responses that are masked by averaging. In the other dataset, group averages do not differ between conditions, but the underlying individual data could be divided into two groups of participants showing opposite trends. Here, averaging may mask a systematic pattern of inter-individual differences which reflects meaningful variability in the brain. Indeed, inter-individual variability in behaviors ranging from simple motor actions to complex executive functions have been linked to inter-individual variability in brain structure [2] and function [15, 16]. Neuroimaging measures provide better predictive power of inter-individual differences in cognitive and clinical outcomes than behavioral measures alone [16], and they explain relationships between factors like socioeconomic status and adolescent development [5]. Comprehensive models in human neuroscience must account for the fact that neural phenotypes and cognitive behaviors vary widely across the population and change over time within individuals across the lifespan [17].

Due to its unparalleled spatiotemporal and single-trial precision, iEEG investigations add crucial mechanistic insight to models in human neuroscience [11–14, 18]. However, despite the advantages of iEEG, surgical electrode placement is driven solely by clinical needs. Electrodes sample brain regions that are common sources of epilepsy, such that some regions tend to be over-sampled and others under-sampled, resulting in a "corticocentric bias" that pervades iEEG literature [12]. Further, electrodes should not be placed to sample more of the brain than is necessary to identify a patient's seizure focus and, in some cases, to characterize regions critical to motor and language functions to ensure they are spared from surgical resection [19]. Electrode coverage is therefore sparse within one patient and variable from one patient to another [11], which renders the exact placement of electrodes a potential source of noise. Individual electrode placement poses a fundamental challenge for the use of iEEG in investigations of meaningful inter-individual variability in brain function.

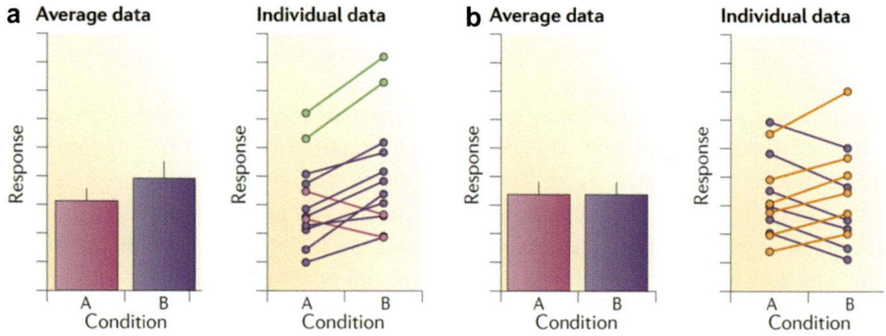

Fig. 10.1 Schematic examples of average and individual data in two experimental conditions. (**a**) The group average in condition B is larger than in condition A (left). Most individual data are consistent with the group averages (right; purple), however, some participants showed opposite trends (pink) or higher responses (green). Such inter-individual differences are masked by averaging. (**b**) Group averages do not differ between conditions A and B (left). However, the underlying individual data could be divided into two groups of participants showing opposite trends (right; orange vs. purple). Adapted from [2]

Here, we address this challenge and describe best practices for studies that aim to elucidate inter-individual and developmental differences in the neurophysiological mechanisms of human cognition using iEEG. We focus on aspects of iEEG studies that researchers can control to achieve high scientific rigor when examining systematic, generalizable patterns of inter-individual and developmental differences in the precise neurophysiology of human cognition.

10.2 Minimize Inter-Individual Variability in Study Design and Analysis

Most iEEG studies take considerable measures to minimize inter-individual variability and draw general conclusions about the neurophysiology of human cognition without considering the person to whom a brain belongs. In one common approach, patients are selected for a study based on electrode sampling of the same anatomical region-of-interest (ROI) and as few as 3-5 patients are included with results replicated per patient. This approach is akin to the standard two-sample procedure of non-human primate neurophysiology, and offers the advantage of replicability [20]. It is qualified by the high signal-to-noise ratio of intracranial data [1], which enables single-trial precision and single-subject reliability [11–14]. In another common approach, patients are selected for a study regardless of specific electrode sampling and electrodes from all patients are combined onto a population-template brain for analysis of all regions sampled. Larger sample sizes permit sampling of larger swaths of the brain [21]. These approaches are discussed in detail in Chap. 29. However, studies aiming to identify inter-individual differences cannot adopt approaches which ignore the person to whom a brain belongs. For this reason, it is important to minimize inter-individual variability not related to effects of interest across subjects during study design and analysis.

At the design stage, experiments may be designed to promote statistical testing of iEEG effects of interest prior to analysis of inter-individual differences in those effects. Specifically, researchers should pay close attention to the appropriateness of the study design to test hypotheses, and ensure that any manipulation (e.g., of experimental condition) is successful. Such research design and strict theorizing should be considered a prerequisite to rigorous statistical testing [22]. Resulting data may then be divided trial-by-trial according to the study design, whether that is a condition manipulation [23, 24], participant-defined criterion (e.g., correct versus incorrect behavioral response [25–27]), or some other task-related component (e.g., post-stimulus versus pre-stimulus epoch [24, 25]). At the analysis stage, iEEG data may be analyzed trial-by-trial at the single-subject level according to the study design. Although these steps do not directly address the issue of electrode placement, they capitalize on the high signal quality of intracranial data and isolate iEEG effects of interest per patient while minimizing other sources of noise that vary from patient to patient (e.g., hospital testing environment). Applying these steps before analyzing inter-individual differences maximizes the likelihood that iEEG measures reflect meaningful factors with unambiguous interpretation of function.

10.3 Define the Inter-Individual Factor(s) of Interest

As described above, factors reflecting the neurophysiology of cognition broadly may be tested on the single-subject level prior to analyzing inter-individual differences. These factors should be defined according to the study design [22], be they manipulations of experimental condition, participant-defined criteria, or other task-related factors. Inter-individual factors of interest, however, need not directly relate to the study design. Factors to consider include experimental task performance, demographic factors like age or sex, neuropsychological assessment data, and measures of brain structure.

Individual measures of task performance including accuracy and response time (RT) are straightforward to consider because they require no additional data collection. Here we describe two studies that used iEEG to reveal inter-individual differences in memory performance and addressed the issue of electrode placement in distinct statistical approaches. In one study, Sheehan and colleagues related individual iEEG effects to associative memory accuracy [26] (Fig. 10.2). iEEG data were analyzed for sample entropy, a measure of signal complexity posited to reflect the brain's ability to flexibly encode and process information, during the encoding of word pairs that were subsequently remembered. Individual signal complexity was found to correlate positively with associative memory accuracy across the sample of 43 participants. To address the issue of electrode placement, researchers included patients regardless of specific electrode sampling and applied spatial smoothing around each 1 × 1 cm ROI to minimize noise related to exact sampling across patients. Although this procedure attenuated the spatial resolution slightly from the mm to cm scale, it balanced the spatial precision of iEEG with the need to maintain statistical power across patients. In another study, Brzezicka and colleagues related individual iEEG effects to RT in a task that manipulated working memory load [23]. Data were analyzed for load-related changes in power in three ROIs, and theta power in the dorsolateral prefrontal cortex (PFC), but not anterior cingulate or hippocampus, was found to correlate positively with RT across electrodes from 13 patients. To address the issue of electrode placement, researchers included patients with ROI sampling and used linear mixed-effects modeling with electrodes as random samples. Although this procedure limited the spatial precision to the ROI, it minimized noise related to specific electrode sampling and increased the sample size for enhanced statistical power.

In addition to task performance, demographic factors such as patient age and sex are often obtained as part of research without additional data collection, and the information is easily de-identified [19]. Studies which aim to study inter-individual differences as they relate to development may consider age as a factor of interest, or the interaction of age and performance. Ofen, Johnson, Yin, and colleagues pioneered this approach in the first published studies of memory development using iEEG [13,

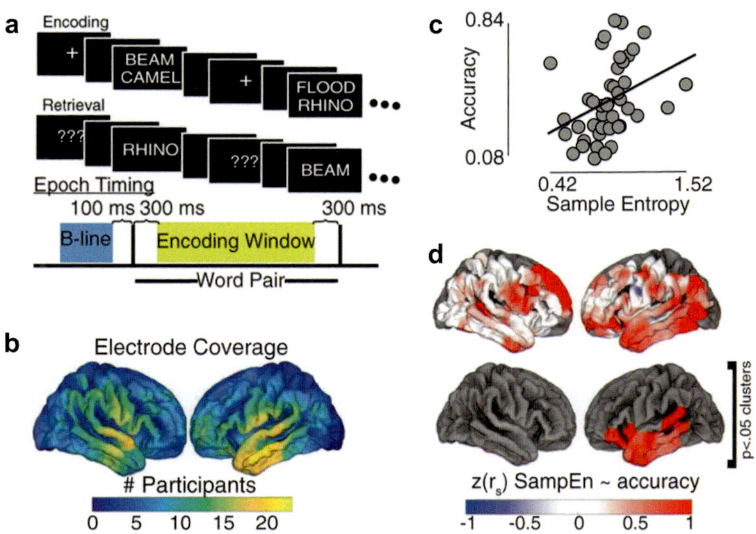

Fig. 10.2 iEEG signal complexity tracks inter-individual variability in associative memory performance. (**a**) Associative memory task in [26]. At study, participants encoded word pairs. At test, they were presented with single words and prompted to retrieve the other word in the pair. (**b**) Spatial distribution of electrode coverage color-coded by the number of participants with sampling of different regions. (**c**) Signal complexity, measured by sample entropy during the encoding window shown in (A), was positively correlated with associative memory performance across participants ($r = 0.51$, $p = 0.0007$). (**d**) Spatial distribution of correlations across all sampled regions, raw (top) and cluster-corrected for multiple comparisons at $p < 0.05$ (bottom). SampEn, sample entropy

24, 25, 27]. In these studies, researchers employed an established subsequent memory task where pediatric patients studied scenes in preparation for a recognition test [13] (Fig. 10.3A). iEEG data were analyzed per patient based on the participant-defined criterion of subsequent memory (i.e., scenes that were later remembered or forgotten at test) and then analyzed for inter-individual differences related to age and overall accuracy. The first study investigated the latency of PFC responses in 17 patients aged 6–19 years [25] (Fig. 10.3B–C). Response latency was defined as the time of peak high-frequency broadband activity, a partial proxy for multi-unit neuronal activity [28–31], and individual latency was quantified in four ROIs. The onset latency of high-frequency responses in inferior frontal gyrus was found to predict behavioral RT and explain age-related gains in recognition performance. The second study investigated alpha oscillations in the primary visual cortex of 24 patients aged 6–21 years [24] (Fig. 10.3D–E). Decreased alpha activity, which is posited to reflect increased information processing similar to signal complexity [11, 26, 32], was found to explain age-related gains in the recognition of visually complex scenes. To address the issue of electrode placement, both studies included patients with ROI sampling and used linear mixed-effects modeling with patients as random samples (see also Chap. 36 for a detailed description of this approach). Although this procedure limited the spatial precision to the ROI, it reduced noise related to specific electrode sampling across patients.

The third study, published in 2022, investigated patterns of inter-regional connectivity between medial temporal lobe (MTL) and PFC in 21 patients aged 6–21 years [27] (Fig. 10.4). Functional connectivity was assessed separately at slow and fast theta frequencies using both phase- and amplitude-based measures [33, 34]. Importantly, these analyses were performed using individually defined frequencies to capture oscillatory phenomena of interest while controlling for inter-individual differences in these phenomena. Both increased slow theta amplitude correlations [35] between MTL and inferior frontal gyrus and fast theta phase-locking values [36] between MTL and middle frontal gyrus were found to explain age-related gains in recognition performance. Patients were again included based on ROI sampling and inter-individual differences were assessed using linear mixed-effects models with patients as random samples. Finally, to identify potential underlying brain structures supporting functional connectivity effects, the researchers incorporated diffusion tractography data that had been obtained as part of the presurgical workup of 11 patients in the sample. Specifically, they tested whether distinct functional connectivity mechanisms in top-performing adolescents were more likely to reflect maturation of the same white matter tract or distinct tracts. They focused a priori on the two major white matter tracts connecting MTL and PFC, the cingulum and uncinate. Bayesian analysis provided an initial test due to limitations of the small sample [37] and suggested that age-related differences in both functional connectivity mechanisms reflected maturation of the cingulum. The high spatiotemporal precision of iEEG, combined with measures of brain structure, supported a mechanistic proposal about how brain maturation supports memory development and addressed major outstanding questions in theoretical models of memory [13].

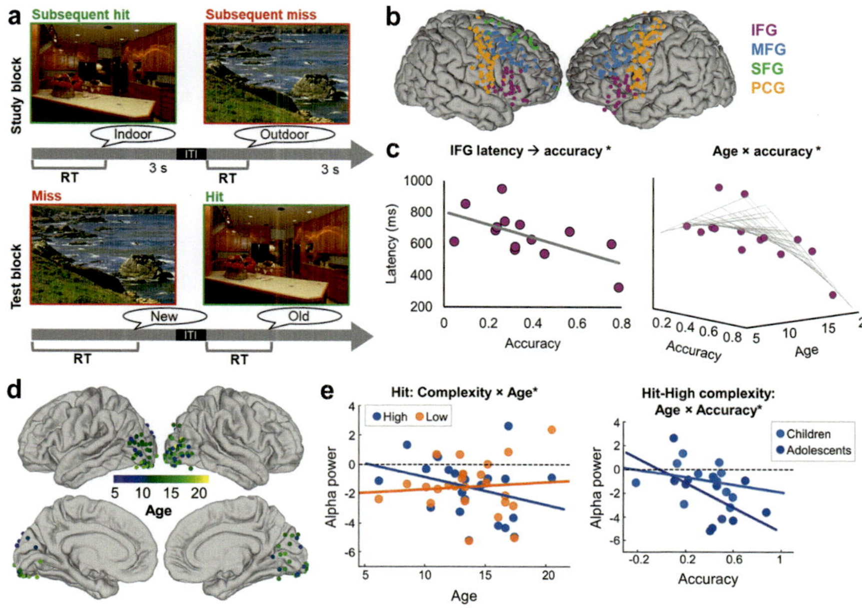

Fig. 10.3 iEEG spectral activities track age-related variability in recognition memory performance. (**a**) Recognition memory task. At study, participants encoded pictures of scenes and classified each scene as 'indoor' or 'outdoor'. At test, they were presented with studied scenes inter-mixed with new scenes and prompted to indicate whether each scene was 'old' or 'new'. (**b**) Frontal electrode coverage across participants color-coded by region of interest in [25]. IFG, inferior frontal gyrus; MFG, middle frontal gyrus; SFG, superior frontal gyrus; PCG, precentral gyrus. (**c**) The latency of peak IFG high-frequency broadband activity during encoding was negatively correlated with recognition memory performance across all participants (left; $r = -0.60$, $p = 0.0004$). The negative relationship between peak high-frequency activity latency and performance explained superior performance in adolescents (right; $p = 0.00004$). (**d**) Occipital electrode coverage across participants color-coded by participant age in [24]. (**e**) Occipital alpha power differed by age during the encoding of high- ($\geq$5 object categories) and low-complexity ($\leq$3 object categories) scenes that were subsequently recognized (left; FDR-corrected $p < 0.05$). The negative relationship between alpha power and age during the encoding of high-complexity scenes explained superior performance in adolescents (right; $p < 0.05$).

Developmental iEEG research is a burgeoning field which poses additional challenges. In studies that consider age and performance as factors, for instance, it is important to demonstrate whether patients in the sample exhibit the expected pattern of performance for their age. This may be accomplished by comparing the behavioral data from patients to a larger sample of data on the same task from non-clinical participants [11]. In the memory development studies described above [24, 25, 27], researchers related the pattern of performance by age in iEEG patient samples to larger samples of data from non-clinical participants [13]. Alternatively, researchers may present normative data from neuropsychological assessments, which may be obtained as part of routine clinical care. If patients do fall in the range of expectations, iEEG findings of inter-individual and developmental differences may generalize to

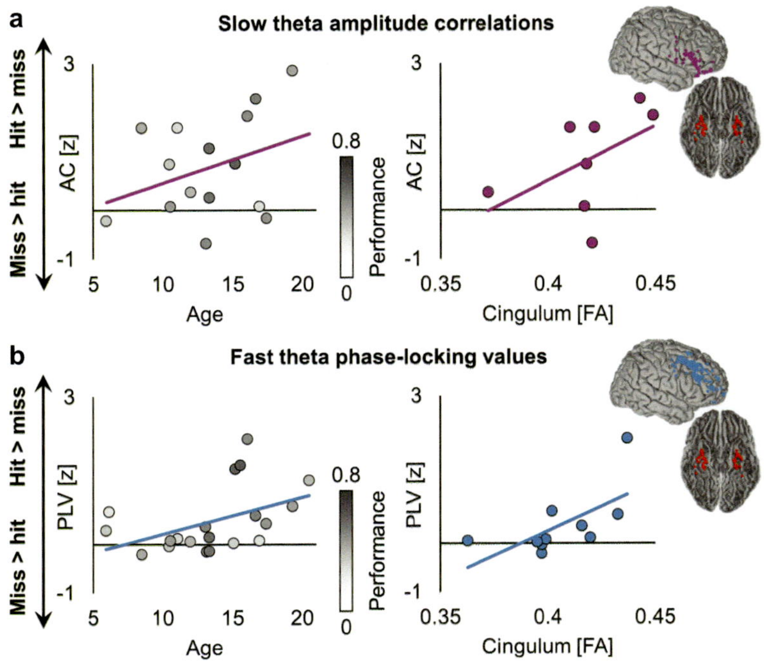

Fig. 10.4 iEEG functional connectivity tracks age-related variability in recognition memory performance and maps to brain structure. (**a**) Subsequent memory effects in slow theta amplitude correlations (AC) between MTL and inferior frontal gyrus differentiated top-performing adolescents from both low-performing adolescents and children (left; $p = 0.011$). AC subsequent memory effects correlated with individual differences the strength of the cingulum tract (right; $r = 0.50$, $BF_{10} = 1.48$). (**b**) Subsequent memory effects in fast theta phase-locking values (PLV) between MTL and middle frontal gyrus differentiated top-performing adolescents from both low-performing adolescents and children (left; $p = 0.0006$). PLV subsequent memory effects correlated with individual differences the strength of the cingulum tract (right; $r = 0.64$, $BF_{10} = 4.31$). FA, fractional anisotropy. Adapted from [27]

the population. If they do not, it is a limitation of the study sample and findings should be interpreted and acknowledged as such.

10.4 Understand (and Increase) the Sample Size

In all studies described above [23–27], analyses of inter-individual differences minimized noise related to specific electrode sampling and maintained statistical power across patients by reducing spatial precision. This illuminates a tradeoff between statistical power and spatial precision in group-level analysis of iEEG data. Because research in clinical samples is inherently constrained by the availability of patients who fit study criteria, many iEEG studies are based on few patients and examine effects in single trials, making the sample size constrained by the number of trials in an experiment as opposed to number of patients who participated. This approach capitalizes on the single-trial precision of iEEG data and, although it ignores the person to whom a brain belongs, it is relevant here as it demonstrates the reliability of the data in single subjects. The single-subject reliability of iEEG data is especially advantageous in inter-individual differences analysis because it means few patients are needed at different levels of a factor, for example, task performance for a given age. It is therefore feasible to investigate inter-individual differences in fewer participants than might be needed to achieve comparable reliability using noninvasive measures with lower signal quality [1].

Nonetheless, iEEG investigations of inter-individual differences are subject to the same rules of statistics as any other investigation and the availability of patients who fit study criteria limits the sample size, limiting statistical power [38]. For instance, samples of approximately 20 participants achieve 80% power to detect large effects and are likely to miss smaller effects (i.e., Type II error) [39]. It is likely that initial iEEG investigations of inter-individual differences [23–27] missed not only the potential to detect meaningful variability within ROIs due to spatial smoothing, but also smaller effects due to sample size constraints. This is especially relevant in developmental iEEG studies examining interactions among multiple factors. Future iEEG investigations of inter-individual and developmental differences may address both limitations by increasing sample sizes. Substantially increasing sample sizes would also permit cross-validation analysis, which is recommended over correlation to demonstrate the generalizability of findings to the population [16, 40]. As more researchers apply iEEG to examine inter-individual and developmental differences in human cognition, they may seek to increase sample sizes through multi-site collaboration and data sharing [13, 41]

10.5 Discussion

Intracranial EEG affords unprecedented access to the human brain, permitting insight into the neurophysiology of human cognition with high spatiotemporal and single-trial precision and single-subject reliability. However, because iEEG sampling is sparse within one patient and variable across patients, the technique poses a fundamental challenge in investigations of inter-individual differences. Here, we address this challenge and describe best practices for studies that aim to elucidate inter-individual and developmental differences in human cognition using iEEG. We focus on aspects of iEEG studies that researchers can control to achieve high scientific rigor when examining systematic, generalizable patterns of inter-individual and developmental differences. First, researchers should pay close attention to the appropriateness of the study design to test hypotheses and ensure that iEEG measures reflect meaningful factors with clear interpretation of function before analyzing inter-individual differences. Second, researchers should define inter-individual factors of interest based on what is feasible given sensitive, potentially identifiable patient information, and ensure that group-level analysis of inter-individual differences controls for noise in electrode sampling across patients. In developmental studies, researchers should also demonstrate whether the study sample represents the population based on non-clinical or normative data and interpret findings accordingly. Third, researchers should understand the statistical power achieved given the sample size and seek to increase the sample size when possible. With appropriate controls, we propose that iEEG be considered a gold standard in studies of inter-individual and developmental differences in the neurophysiology of human cognition.

Acknowledgements We thank K. T. Jones for support. This work was funded by grants from the National Institute of Neurological Disorders and Stroke (R00NS115918, R01NS021135, U19NS107609).

References

1. Ball T, Kern M, Mutschler I et al (2009) Signal quality of simultaneously recorded invasive and non-invasive EEG. NeuroImage 46:708–716. https://doi.org/10.1016/j.neuroimage.2009.02.028
2. Kanai R, Rees G (2011) The structural basis of inter-individual differences in human behaviour and cognition. Nat Rev Neurosci 12:231–242. https://doi.org/10.1038/nrn3000
3. Dubois J, Adolphs R (2016) Building a science of individual differences from fMRI. Trends Cogn Sci 20:425–443. https://doi.org/10.1016/j.tics.2016.03.014
4. Seghier ML, Price CJ (2018) Interpreting and utilising intersubject variability in brain function. Trends Cogn Sci xx:1–14. https://doi.org/10.1016/j.tics.2018.03.003
5. Foulkes L, Blakemore SJ (2018) Studying individual differences in human adolescent brain development. Nat Neurosci 21:1–9. https://doi.org/10.1038/s41593-018-0078-4
6. Semendeferi K, Teffer K, Buxhoeveden DP et al (2011) Spatial organization of neurons in the frontal pole sets humans apart from great apes. Cereb Cortex 21:1485–1497. https://doi.org/10.1093/cercor/bhq191
7. Semendeferi K, Armstrong E, Schleicher A et al (1998) Limbic frontal cortex in hominoids: a comparative study of area 13. Am J Phys Anthropol 106:129–155. https://doi.org/10.1002/(SICI)1096-8644(199806)106:2%3c129::AID-AJPA3%3e3.0.CO;2-L[pii]\r10.1002/(SICI)1096-8644(199806)106:2%3c129::AID-AJPA3%3e3.0.CO;2-L
8. Teffer K, Buxhoeveden DP, Stimpson CD et al (2013) Developmental changes in the spatial organization of neurons in the neocortex of humans and common chimpanzees. J Comp Neurol 521:4249–4259. https://doi.org/10.1002/cne.23412
9. Semendeferi K, Armstrong E, Schleicher A et al (2001) Prefrontal cortex in humans and apes: a comparative study of area 10. Am J Phys Antropol 114:224–241. https://doi.org/10.1002/1096-8644(200103)114:3%3c224::AID-AJPA1022%3e3.0.CO;2-I
10. Sydnor VJ, Larsen B, Bassett DS et al (2021) Neurodevelopment of the association cortices: patterns, mechanisms, and implications for psychopathology. Neuron. https://doi.org/10.1016/j.neuron.2021.06.016
11. Johnson EL, Kam JWY, Tzovara A, Knight RT (2020) Insights into human cognition from intracranial EEG: a review of audition, memory, internal cognition, and causality. J Neural Eng 17:051001. https://doi.org/10.1088/1741-2552/abb7a5
12. Parvizi J, Kastner S (2018) Human intracranial EEG: promises and limitations. Nat Neurosci. https://doi.org/10.1038/s41593-018-0108-2

13. Ofen N, Tang L, Yu Q, Johnson EL (2019) Memory and the developing brain: from description to explanation with innovation in methods. Dev Cogn Neurosci 36:100613. https://doi.org/10.1016/j.dcn.2018.12.011
14. Lachaux J-P, Axmacher N, Mormann F et al (2012) High-frequency neural activity and human cognition: past, present and possible future of intracranial EEG research. Prog Neurobiol 98:279–301. https://doi.org/10.1016/j.pneurobio.2012.06.008
15. Braver TS, Cole MW, Yarkoni T (2010) Vive les differences! Individual variation in neural mechanisms of executive control. Curr Opin Neurobiol 20:242–250. https://doi.org/10.1016/j.conb.2010.03.002
16. Gabrieli JDE, Ghosh SS, Whitfield-Gabrieli S (2015) Prediction as a humanitarian and pragmatic contribution from human cognitive neuroscience. Neuron 85:11–26. https://doi.org/10.1016/j.neuron.2014.10.047
17. Rosenberg MD, Casey BJ, Holmes AJ (2018) Prediction complements explanation in understanding the developing brain. Nat Commun 9:1–13. https://doi.org/10.1038/s41467-018-02887-9
18. Johnson EL, Knight RT (2015) Intracranial recordings and human memory. Curr Opin Neurobiol 31:18–25. https://doi.org/10.1016/j.conb.2014.07.021
19. Chiong W, Leonard MK, Chang EF (2017) Neurosurgical patients as human research subjects: ethical considerations in intracranial electrophysiology research. Neurosurgery 0:1–9. https://doi.org/10.1093/neuros/nyx361
20. Smith PL, Little DR (2018) Small is beautiful: in defense of the small-N design. Psychon Bull Rev 1–19. https://doi.org/10.3758/s13423-018-1451-8
21. Marks VS, Saboo K V, Lech M et al (2021) Independent dynamics of low, intermediate, and high frequency spectral intracranial EEG activities during human memory formation. Neuroimage 245. https://doi.org/10.1016/j.neuroimage.2021.118637
22. Fiedler K, McCaughey L, Prager J (2021) Quo Vadis, methodology? The key role of manipulation checks for validity control and quality of science. Perspect Psychol Sci 16:816–826. https://doi.org/10.1177/1745691620970602
23. Brzezicka A, Kamiński J, Reed C et al (2019) Working memory load related theta power decreases in dorsolateral prefrontal cortex predict individual differences in performance. J Cogn Neurosci 31:1290–1307. https://doi.org/10.1162/jocn
24. Yin Q, Johnson EL, Tang L et al (2020) Direct brain recordings reveal occipital cortex involvement in memory development. Neuropsychologia 148:107625. https://doi.org/10.1016/j.neuropsychologia.2020.107625
25. Johnson EL, Tang L, Yin Q, et al (2018) Direct brain recordings reveal prefrontal cortex dynamics of memory development. Sci Adv 4:eaat3702. https://doi.org/10.1126/sciadv.aat3702
26. Sheehan TC, Sreekumar V, Inati SK, Zaghloul KA (2018) Signal complexity of human intracranial EEG tracks successful associative memory formation across individuals. J Neurosci 38:180240. https://doi.org/10.1101/180240
27. Johnson EL, Yin Q, O'Hara NB et al (2022) Dissociable oscillatory theta signatures of memory formation in the developing brain. Curr Biol 32. https://doi.org/10.1016/j.cub.2022.01.053
28. Nir Y, Fisch L, Mukamel R et al (2007) Coupling between neuronal firing rate, gamma LFP, and BOLD fMRI Is related to interneuronal correlations. Curr Biol 17:1275–1285. https://doi.org/10.1016/j.cub.2007.06.066
29. Ray S, Crone NE, Niebur E et al (2008) Neural correlates of high-gamma oscillations (60–200 Hz) in macaque local field potentials and their potential implications in electrocorticography. J Neurosci 28:11526–11536. https://doi.org/10.1523/JNEUROSCI.2848-08.2008
30. Rich EL, Wallis JD (2017) Spatiotemporal dynamics of information encoding revealed in orbitofrontal high-gamma. Nat Commun 8:1139. https://doi.org/10.1038/s41467-017-01253-5
31. Leszczyński M, Barczak A, Kajikawa Y et al (2020) Dissociation of broadband high-frequency activity and neuronal firing in the neocortex. Sci Adv 6:1–13. https://doi.org/10.1126/sciadv.abb0977

32. Hanslmayr S, Staresina BP, Bowman H (2016) Oscillations and episodic memory: addressing the synchronization/desynchronization conundrum. Trends Neurosci 39:16–25. https://doi.org/10.1016/j.tins.2015.11.004
33. Mostame P, Sadaghiani S (2020) Phase- and amplitude-coupling are tied by an intrinsic spatial organization but show divergent stimulus-related changes. Neuroimage 219:117051. https://doi.org/10.1016/j.neuroimage.2020.117051
34. Siems M, Siegel M (2020) Dissociated neuronal phase- and amplitude-coupling patterns in the human brain. Neuroimage 209:116538. https://doi.org/10.1016/j.neuroimage.2020.116538
35. Adhikari A, Sigurdsson T, Topiwala MA, Gordon JA (2010) Cross-correlation of instantaneous amplitudes of field potential oscillations: A straightforward method to estimate the directionality and lag between brain areas. Journal of Neuroscience Methods 191:191–200. https://doi.org/10.1016/j.jneumeth.2010.06.019
36. Lachaux JP, Rodriguez E, Martinerie J, Varela FJ (1999) Measuring phase synchrony in brain signals. Human Brain Mapping 8:194–208
37. Rouder JN, Morey RD (2012) Default Bayes factors for model selection in regression. Multivar Behav Res 47:877–903. https://doi.org/10.1080/00273171.2012.734737
38. Button KS, Ioannidis JPA, Mokrysz C et al (2013) Power failure: why small sample size undermines the reliability of neuroscience. Nat Rev Neurosci 14:365–376. https://doi.org/10.1038/nrn3475
39. Erdfelder E, Faul F, Buchner A, Lang AG (2009) Statistical power analyses using G*Power 3.1: tests for correlation and regression analyses. Behav Res Methods 41:1149–1160. https://doi.org/10.3758/BRM.41.4.1149
40. Lebreton M, Bavard S, Daunizeau J, Palminteri S (2019) Assessing inter-individual differences with task-related functional neuroimaging. Nat Hum Behav 3:897–905. https://doi.org/10.1038/s41562-019-0681-8
41. Fair DA, Dosenbach NUF, Moore AH et al (2021) Developmental cognitive neuroscience in the era of networks and big data: strengths, weaknesses, opportunities, and threats. Annu Rev Dev Psychol 3

Chapter 11
Is IEEG-Based Cognitive Neuroscience Research Clinically Relevant? Examination of Three "Neuromemes"

Jonathan Curot, Luc Valton, and Emmanuel J. Barbeau

Abstract Much progress has been made in the field of cognitive neuroscience thanks to intracerebral EEG (iEEG) research, largely due to the possibility of directly recording brain activity with unsurpassed spatial and temporal precision while patients perform cognitive tasks. However, do these patients gain anything from the time and effort they devote to this endeavour? In this chapter, we focus on three neuromemes, the "eloquent cortex", "localisationism" and the "nociferous cortex" to provide possible answers to this question. We discuss the value of these neuromemes and show that clinical care of epilepsy and iEEG-based cognitive neuroscience are consubstantial in the sense that iEEG during epilepsy assessment provides an understanding of physiological processes of the healthy brain; but also, that cognitive iEEG research in epileptic patients has a direct impact on semiology and curative neurosurgery. Last, we highlight how recent cognitive iEEG research provides insights into interictal complaints and could improve identification of the epileptogenic zone.

11.1 Introduction

Concept neurons and the famous Jennifer Aniston cell [1–3], memories induced by direct electrical brain stimulation (EBS) as if a neurostimulator could replace a madeleine [4]... Neuroscience based on intracerebral EEG (iEEG) led to fundamental scientific discoveries that are widely popularised. Some of them became mainstream and culturally shared. Compared to scalp EEG or functional MRI, iEEG is recorded with unsurpassed spatial and temporal precision. Such an approach has been used in a large number of cognitive studies that take advantage of iEEG to investigate

J. Curot (✉) · L. Valton · E. J. Barbeau
Centre de Recherche Cerveau et Cognition, CNRS, UMR 5549, Toulouse, France
e-mail: jonathan.curot@cnrs.fr

J. Curot · L. Valton
Département de Neurologie, Centre Hospitalier Universitaire de Toulouse, Toulouse, France

J. Curot · E. J. Barbeau
Faculté de Santé, Université Paul Sabatier, Toulouse, France

N. Axmacher (ed.), *Intracranial EEG*, Studies in Neuroscience, Psychology and Behavioral Economics, https://doi.org/10.1007/978-3-031-20910-9_11

the electrophysiological correlates of cognition in humans (for recent reviews, see [5–8]). Although there is no doubt that iEEG research has significantly contributed to cognitive neuroscience, it would only be fair to ask the opposite. Has iEEG-based cognitive neuroscience contributed to the clinical workup of epileptic patients and their welfare? After all, many patients have contributed time to cognitive iEEG studies with the sincere hope it would help other patients like them in the future.

In this chapter, we will focus on three aspects that are commonplace concepts used every day by clinicians, electrophysiologists and neurosurgeons with epileptic patients in their care: the "eloquent cortex" [9], "localisationism" [10] and the "nociferous cortex" [11]. We suggest that these concepts are in fact "neuromemes", exchangeable cognitive units that spread easily from one individual to another and which can be maintained throughout time [12, 13]. Memes gather ideas, behaviours or styles, with a common goal to easily replicate within a group. Some of these memes directly concern the epilepsy pathology as analysed recently by Baxendale [14]. Neuromyths have been the focus of other authors such as Devinsky [15] who commented on three: "*epilepsy is a static disorder with minimal morbidity and mortality; epileptogenic tissue impairs only the functions of the seizure focus; and the anterior temporal lobes contain areas of non-functional, "silent" cortex.*" Hence, the medical culture, like any other culture, is rich with easily replicated and sustainable concepts and theories, i.e., "neuromemes". Although the term "neuromyth" has infiltrated the neuroscience community [16–18], "neuromeme" seems to be a more appropriate term in the context of this chapter because neuromemes are not systematically wrong are not systematically wrong or do not constantly lack rationality.

However, one characteristic of a neuromeme is that it is loosely defined. It is this malleability that makes it so useful. It can be used in many situations and can be adapted to the needs, circumstances or even the epoch. However, this represents a paradox since medicine usually relies on the efficiency of precisely defined concepts. This suggests that there is a high risk of such neuromemes causing a bias or polluting clinical practice.

In this chapter, we will show that intracranial explorations have played an important role in the genesis of neuromemes. However, we will also show that as a result of the progress made in the field, neuromemes can now be examined and refined for the benefit of patients.

11.2 IEEG Research: A Cocoon for Neuromemes

Paradoxically, while iEEG has been the historical nest for the creation or enrichment of some neuromemes, and neuromyths for that matter, it can also simultaneously play a significant role in confirming or invalidating them. First, we will review each of our selected neuromemes, which are closely related to Penfield's work either in terms of their origin or significant promotion.

The eloquent cortex

The concept of the "eloquent cortex" did not originate with Penfield, but is derived from his work. In fact, Drake was very likely the first to use the term "eloquent" after a series of lectures by Penfield in the sixties in a paper on the surgical treatment of arteriovenous malformations [9, 19, 20]. One literal definition of "eloquent" that can be found in the dictionary is "*clearly expressing or indicating something*". In the epilepsy literature, it is defined as the "*cortex related reproducibly to a given function.*" [21]. "*Cortical stimulation allows the most precise localization of eloquent cortex*" [22] or "*Fast-ripples near the resection and in distant pathologic areas could have changed the resection in 8 patients without harming functionally eloquent areas*" [23] are just two of the countless occurrences of the term "eloquent" (see [19]). In the current scientific literature, particularly in the context of preoperative functional mapping, the term "silent cortex" appeared in opposition to the parts of the cortex that were *not* "eloquent", which is consistent with previous ideas concerning brain equipotentiality [15].

However, in 1993, Fried questioned the myth of the eloquent cortex and highlighted the danger of such terminology "*which replaces the neurological reality it sets out to simplify*" [9]. In 2005, Devinsky tried to dismiss the myth of the silent cortex based on the example of temporal lobe surgery: "*Because the normal brain does not contain functionless, "silent" areas, the procedure can have negative as well as positive cognitive or behavioural consequences*" [15]. This is obviously an important issue for patients undergoing neurosurgery.

Localisationism

According to the localisationist theory, the brain is conceived as a mosaic of relatively independent functional regions, i.e., each part of the brain is associated with a particular function, in keeping with the seminal suggestions by Franz Gall in the early nineteenth century. This view appeared to be supported by later discoveries, for example, of Broca's area and the motor and sensitive homunculus proposed by Penfield.

Let us take the example of Penfield's interpretative cortex [24]. The experiential phenomena that Penfield reported were always produced by electrical stimulations applied to a large area of the temporal neocortex, which included the superior, lateral and inferomedial temporal lobes [25]. He named this area the interpretative cortex since, contrary to the pericentral and visual regions that he also studied, only stimulation here led to flashbacks of past experiences or to interpretative illusions of the present situation [25, 26]. His surgical exploration technique restricted his application of electrical stimulation to only the surface of the temporal cortex. Despite a few attempts at deep stimulation, he only exceptionally observed such responses by stimulations of the temporal uncus, and never by stimulations of the hippocampus. However, the subsequent history of intracranial EEG research clearly demonstrated that the interpretative cortex was not by any means the only region that induces experiential memory phenomena when stimulated and that the memory network extends far beyond [18, 27]. This implies that cognitive functions depend on networks of brain areas [28], the complexity of which is probably only barely understood at present [29] (see also Chaps. 33 and 40).

While no clinician will deny that complex cognitive functions are subtended by distributed networks [30, 31], daily clinical thinking remains contaminated by the localizationist approach, from which the idea of the "eloquent cortex" is derived [15]. This impacts the quality of patient care because it oversimplifies neurosurgery planning and in particular its consequences.

The nociferous cortex

This is perhaps the least identifiable of the three neuromemes we selected. In 1954, Penfield and Jasper suggested the concept of a "nociferous cortex" to describe the notion that the epileptic focus, now more appropriately defined as the epileptogenic zone, may be deleterious for widespread areas beyond the epileptogenic zone (EZ) and may disrupt normal processes [32] (see also Chap. 3). The "nociferous cortex" in epilepsy refers to the fact that local brain damage can have remote consequences, a secular concept, which could date back to the time of Brown Sequard in the middle of the nineteenth century [33, 34]. It is, for example, closely related to the concept of diaschisis described by Von Monakow [35, 36]. The distributed effects of focal lesions on brain dynamics, such as diaschisis, compensation or transneuronal degeneration, are nowadays widely supported by the study of brain connectomics and are applicable to a wide range of diseases [29, 33, 37].

In epilepsy, the "nociferous concept" was updated about 25 years ago thanks to cognitive psychology. Hermann and Seidenberg demonstrated executive disorders in temporal lobe epilepsy, whereas executive functions are thought to depend primarily on the prefrontal lobes [38]. In 2014, Coan et al. revealed decreased grey matter in extratemporal regions in patients suffering from temporal lobe epilepsy [39]. The influence of such ideas is now growing in the field of epilepsy as a result of a series of recent intracranial EEG-based neuroscience studies that demonstrate the remote influence of epileptic activities on brain functions [40–43]. Unlike the "localisationism" and "eloquent cortex" neuromemes, the meme of the nociferous cortex holds up only if networks are considered [44]. This may be the reason why it is the only one of the three selected neuromemes that appears to have some validity. It is likely that in the future, the strong interest in network neuroscience will contribute to improving the cognitive status of epileptic patients, particularly following neurosurgery.

11.3 Dispelling the Myth of the Eloquent vs the Silent Cortex

Many years after Drake (1963), Spetzler and Martin (1986) defined the eloquent cortex as "*areas that speak to readily identifiable neurological function*" and, "*if injured, result in a disabling neurologic deficit*" [19, 45].

At first glance, the etymological choice of "eloquent" to name the primary cortices that induce obvious and easily reproducible symptoms after EBS appears sound. EBS can trigger oral and written language symptoms [46–49], motor [50–52] or sensory manifestations [53], during either awake surgery [54, 55] or intracerebral explorations [56–58].

Moreover, there seems to be "silent" areas that do not respond to electrical brain stimulation: for instance, the posterior cingulate cortex where EBS *never* induces memory or other symptoms (or at least not yet reported), whereas stimulation of surrounding areas can easily trigger motor, vestibular and visual symptoms [18, 59]. A similar phenomenon can be observed in the orbitofrontal cortex, a ghost area, almost absent from the EBS literature [56]. Therefore, "silent" and "eloquent" do not appear to be inappropriate terminology.

Unfortunately, the term "eloquent" has since been extensively used and deviated from the original (literal) definition [19]. In neurosurgery, it is now a synonym of "which should be preserved" and the silent cortex is that which can be easily resected without visible or disabling deficit. In that respect, numerous articles and reviews have suggested that EBS is a good functional mapping method to determine the "eloquent cortex" before epilepsy surgery (e.g., [60, 61]). The premise is that if resecting the eloquent cortex is avoided there is a reasonable likelihood that the patient will not suffer from cognitive deficits following neurosurgery. Despite the rise of functional imaging, EBS is still considered the gold standard for mapping "functional areas" and predicting functional deficits. However, there are several conceptual difficulties with this approach.

(1) How can the limits of eloquent and silent brain areas be defined? This question is particularly relevant since a series of studies demonstrated that EBS effects are dependent on the parameters [62–66], that EBS effects may not be reproducible despite using the same parameters [18], that optimal parameters are not known [67] and that the understanding of the physiological mechanisms underlying EBS is limited [68]. Should EBS be accompanied by afterdischarges [69], or should they remain under the threshold for induction [70, 71]?
(2) The term "silent cortex" is still inappropriately used for areas where no obvious symptoms were elicited during functional mapping, while their resection may lead to deficits after surgery. For example, who is able to predict spatial and navigation deficits [72] or de novo psychiatric symptoms [73, 74] that occur after right anterior and medial temporal lobectomy guided by EBS? Do we know what type of postoperative cognitive decline to expect and how this can be adequately assessed? At present, there are too few studies regarding memory, social cognition and psychiatric consequences of right anterior temporal lobe surgery to conclude that this is a silent cortical area, if such a notion is even possible. The situation is just as unclear for other regions such as the cingulate or the prefrontal cortex.
(3) What is the value of the "eloquent cortex" when the patient is no longer eloquent? Not all patients can report their symptoms, due to either ictal amnesia or aphasia, especially in temporal lobe epilepsy [75].

(4) Are "non-eloquent" and "silent" synonymous? Does "not responding to EBS" really mean constantly and reliably silent areas? Aren't "silent" areas just "shy" areas? Patients are not always assisted to verbalise subjective experiences through specific appropriate questions, especially when the temporal lobes are explored. In case of neuropsychological disturbances, such experiences are only revealed if an appropriate cognitive task is performed at the time of stimulation [69, 76]. Coleshill et al. [71] for example, were able to demonstrate material-specific recognition memory deficits in some patients by combining unilateral electrical stimulations in the hippocampus synchronised with the presentation of verbal or visual stimuli. To date, no such routine is standardised and EBS still cannot replace the Wada test. The aim of the Wada test is to assess the hemispheric lateralization of speech or memory during preoperative evaluation of refractory epilepsy in order to avoid a risk of sequela. It consists of an intracarotid injection of an anaesthetic drug (barbiturate or propofol), which transiently inhibits the ipsilateral cerebral hemisphere in order to assess the activity of the contralateral one [77].

This also implies that theoretically, different tasks must be carried out for a single stimulation of a given region in order to test the role of this region in different functions and to determine with certainty whether the errors observed in the tests are indeed related to the EBS (and not to the preoperative cognitive state of the patient). For instance, the procedures to evaluate language are not standardised and each centre uses its own battery of tests (image naming, reading, spontaneous speech, repetition, auditory comprehension, etc.). Usually, only one of these tasks is performed for the sake of time, which strongly limits the understanding of the links between the stimulated area and its precise function in language.

(5) Cognitive functions are supported by large-scale networks. All cognitive processes are due to the emergence of a distributed, specific, transient, and synchronous neural assembly characterised by the level of synchrony of its components [30]. There is dense interregional connectivity, which makes each brain region part of an extensive network [29, 37]. These large-scale brain networks consist of nodes that share many of the same connections, sometimes reciprocally [31, 78]. There is now evidence that EBS does not only have a local effect at the EBS site but also long-range effects. Intracranial implantations are governed by hypotheses about the epileptogenic network [79, 80] and sample only a fraction of the brain. Functional and physiological processes are not considered in the surgical plan. For instance, language investigation would theoretically require bilateral implantation and should cover a large fronto-temporo-parietal surface [81], which is generally not the case, for an accurate mapping of the patient's specific language network. Indeed, about 40% of patients show a decline in image naming after left temporal lobe neurosurgery [82], whereas the preservation of regions essential for naming, for example by limiting the posterior ventral extent of the resection, could help preventing such outcome. EBS effects propagate far away from the electrode [83], with the capacity to

inhibit distant cortical regions through afferents connections [84]. In addition, networks can only be activated by some of their hubs and not by all regions involved in the network [85–87]. This is the case for memory. The rhinal cortex seems to be the "gatekeeper" of memory networks [88], the site where EBS most frequently induces memories. There are also "closed doors" to neuromodulation, such as the posterior cingulate region, where EBS has never induced memory phenomenon yet [18], even though it is a region that is known to play a role in memory. Hubs are defined as "nodes occupying a central position in the overall organization of a network", having a key role in information integration, making them "points of vulnerability that are susceptible to disconnection and dysfunction in brain disorders" [89]. However, hubs should not be conceived of as an updated and more timely formulation of the concept of "eloquent cortex". The properties of a network cannot be limited to its hubs and the use of this term should be restricted to the context of the definition above to avoid to go back to poorly defined terminology.

In conclusion, there are alternatives to EBS such as using ERPs [90, 91] or high-frequency activity induced by specific cognitive tasks [92, 93]. Some authors have suggested that these approaches could be a means of reducing the duration of the EBS procedure, which would be more efficient and at the same time more comfortable for the patient and save time for the clinician. They could also be a complement to EBS to preselect the sites to stimulate [94].

While EBS is probably one of the major sources of the misleading "eloquent" vs "silent" cortices concepts, an analysis of the limits of EBS demonstrates that the term "eloquent" should be avoided in clinical practice. New ways of thinking about the possible postoperative deficits can be imagined. iEEG-based cognitive neuroscience suggests that standard, multimodal, cognitive procedures should be performed that are tailored to each location and if possible standardised among centres, to better assess the predictive value of EBS in postoperative deficits. A collective effort to this end, guided by recent progress in the methods and concepts developed recently in the framework of iEEG-based cognitive neuroscience, would clearly benefit epileptic patients.

11.4 Eliminating the Implicit Dogma of "Localisationism"

Although clinicians may still, often implicitly or unconsciously, use localisationism to interpret brain-behaviour relationships, it is undeniable that modern neuroscience has considerably modified clinical thinking by introducing the idea that it is probable that networks are the underpinning of the cognitive or behavioural symptoms that occur during seizures.

iEEG in the areas of epileptogenic networks

One major contribution of iEEG-based cognitive science is the improvement in knowledge about seizure semiology, initiated by scalp EEG [95]. This was achieved through analysis of the sequence of symptoms that occur at the onset of seizures, that coincided temporally with the iEEG recorded in the different brain areas that were sampled [96]. Semiology is one of the keys to clinical reasoning in epilepsy and to the nosology and syndromic classification of epilepsy [97]. Semiology is the basis for building hypotheses on the epileptogenic network. It significantly influences the strategy of electrode implantation in patients who will benefit from an invasive presurgical assessment. It may also drive tailored neuromodulation treatments in the future as has recently been suggested [98, 99].

A part of the diagnostic arsenal in epileptology is video-EEG, especially during phase 2 (intracranial EEG recording), and it has simply revolutionised the understanding of seizure semiology. The primary contribution of intracranial EEG to the understanding of semiology is related to the possibility of simultaneously recording iEEG and analysing symptomatology on video during seizures and then analysing the networks underlying each symptom through the correlation of the symptoms to the distribution of the concomitant ictal discharge on EEG. The *symptomatogenic* zone ("*the area of cortex that, when activated by an epileptiform discharge, produces the ictal symptom*") should be dissociated from the *epileptogenic* zone [21] (see also Chap. 1).

Clinical symptoms have also been studied beyond a simple visual analysis of the regions concerned by the propagation of the epileptic discharge. This is particularly true of cognitive symptoms, for which a panel of spectral or network analyses have been proposed. For instance, coherence analyses (see Chap. 32) during ictal humming demonstrated the activation of a network involving the superior temporal and inferior frontal gyrus [100]. Faced with such results, any clinician who observes early ictal humming in the seizure semiology will pay specific attention to these brain areas when analysing an MRI in order to identify any structural abnormality or will want to target these two areas during the presurgical workup. Fear is another example. Fear networks identified by coherence analyses during seizures [101, 102] revealed the synchronisation of the amygdala, ventral medial PFC, cingulate and dorsomedial prefrontal cortex during fear [103]. Electrode implantations, as well as structural brain MRI to identify possible lesions, could therefore be driven by such results and target these regions in patients who are frightened while having a seizure.

Such a paradigm becomes particularly important to understand the semiology of brain areas for which EBS does not or rarely induces symptoms. By improving knowledge of the semiology of seizures, iEEG-based cognitive neuroscience can certainly play a role in the electrode implantation strategy.

iEEG in "healthy areas"

The contribution of intracranial EEG goes beyond simply understanding the sequence of symptoms that occurs during seizures. In reality, the direct benefit for a patient who agrees to participate in a cognitive protocol during his or her own intracranial exploration is not obvious and most of the time is not measurable. Nevertheless, the data acquired during each of these procedures improve clinicians' knowledge of the brain networks that underly normal cognition which incidentally influences their ability to draw anatomical-functional conclusions from the seizure semiology. The possibility to identify the "normal" networks that underly specific cognitive functions through the study of functional or effective coupling (i.e., using coherence or other measures) or cortico-cortical evoked potential (CCEP; [104, 105]; see also Chap. 40) improves the overall knowledge of clinicians. Long-term collaboration between clinicians and cognitive neuroscience researchers, and other neuroscience methods such as functional MRI constantly, if only gradually, increase the quality of interpretation of the symptoms reported by patients or observed during seizures [106].

In sum, the age-old tension between segregation and integration in the field of neuroscience also pervades the field of epilepsy. However, results clearly show that the network paradigm is a more accurate description of how the brain works and is also more nuanced (see also Chaps. 33 and 40). It opens the way to other critical questions for patients, such as how these networks reorganise in the context of the pathology or how brain plasticity can be preserved or promoted after neurosurgery. The network approach is fully embedded in the notion of the "nociferous cortex" as will be seen in the next section.

11.5 Updating and Validating the "Nociferous Cortex" Concept

The concept that a pathological area can impact remote healthy areas is not necessarily intuitive. However, it fits perfectly with the view that the brain relies on dynamic network activity. A series of recent studies contributed to renewed interest in this concept. For example, it was shown that interictal epileptic discharges can alter large-scale networks beyond the epileptogenic zone [40, 107], either by pathological coupling (e.g. spikes pathologically coupled to the spindles involved in memory consolidation [42, 43]) or by remote inhibition of healthy tissue (e.g. fast-ripples, a local phenomenon, that inhibits distant neurons outside the epileptogenic zone [108]). With regards to the nociferous cortex, combining the study of interictal epileptic activities and cognitive tasks is clinically relevant in different ways.

An impact on antiseizure medicine

The nociferous concept can help clinicians to determine how to adjust an antiseizure medicine, not only in terms of seizure persistence but also in relation to residual interictal cognitive complaints. More than 50% of patients with epilepsy complain that their memory moderately or severely limits their daily life [109]. In a recent study, we showed that approximately 75% of a large population of patients with temporal lobe epilepsy with initial *subjective* memory complaints had *objective* memory impairment after 3 weeks, regardless of whether they had a brain lesion, the type of lesion or how they responded to antiseizure medicine [110]. Such results suggest that there are other factors that explain cognitive alterations in epileptic patients, the best candidate being interictal epileptic discharges (IED). Some iEEG studies have provided convincing evidence that IEDs can disrupt cognitive processes, particularly those involved in memory [40, 111] (see Chap. 3). Beyond correlational results, it is suggested that IED disrupt memory processes because they occur at key moments (encoding and recall, [112], retrieval and maintenance, [107]), most notably because they directly disrupt the physiological interactions between neocortical (spindles) and hippocampal oscillations (ripples) necessary for memory consolidation. IEDs lead to pathological coupling with spindles [42] or they interfere with hippocampal physiological ripples during encoding and recall [41]. Such studies are almost impossible to conduct with scalp EEG or functional MRI. This clearly reinforces the importance of treating certain epileptic patients who are apparently free of seizures but who have a persistent interictal cognitive complaint probably due to IEDs [113].

A fingerprint of the EZ/non-EZ?

Recent studies have led to a renewed interest in cognitive tasks as markers of the EZ. To be succinct, the question is whether examination of the iEEG recorded during cognitive tasks can inform clinicians about the EZ or the epileptogenic network. A few studies suggest that seizure onset regions are dysfunctional at baseline and are functionally disconnected from the other functional network hubs [114, 115]. IEDs (interictal epileptic spikes, interictal high frequency oscillations) during cognitive processes [107, 116–118] or coherence analyses during cognitive paradigms [119], or even resting state analyses may reveal these fingerprints. IEDs that occur in the areas where a seizure starts do not impact cognitive scores, whereas IEDs recorded contralateral to the seizure onset zone (SOZ) or bilaterally doubled the risk of errors on a short-term memory task where letters within a sequence had to be recognized a few seconds after encoding [107]. A network synchronized in the theta-band activated during episodic memory was not found in the electrodes that belong to the seizure-onset zone [119]. The seizure onset and non-seizure onset zones may also show different electrophysical patterns during interictal cognitive tasks [118]. Increased relative entropic differences between the SOZ and non-SOZ were observed during a verbal memory task while the number of IEDs decreased during the task only in the non-SOZ, suggesting that iEEG recordings coupled with cognitive tasks might be able to unmask interictal fingerprint patterns specific to the EZ [118]. Similarly, cognitive tasks may induce a differential "reactivity" of interictal epileptic activities within or outside the EZ: for instance, an oddball task induced a significant decrease of high frequency oscillation rates, in particular ripples (see also Chap. 24), within the epileptic, but not in the non-epileptic hippocampus [116, 117]. Transient suppression of ripples in the seconds following the stimuli presentation was also only observed in the non-epileptic hippocampus.

This approach, although promising, is still in its infancy and is dominated by verbal [40, 118] or visual [120] memory tasks, attentional [116, 117] or short-term memory tasks [107]. Long-term memory tasks, especially episodic, are rare [119]. It is currently mostly relevant for EZ located in the left, dominant hemisphere, whereas only a few results are available for the right hemisphere. What is most notable is that there are still discrepancies and incongruences concerning the influence of cognitive tasks on epileptic activities according to the brain area. Results even appear to differ significantly within the same study [116, 117]. However, the study of IED during cognitive tasks could be of value to optimise presurgical assessment by distinguishing EZ and non-EZ structures, thereby completing the information derived from the examination of spontaneous seizures.

Although the concept of the nociferous cortex is somewhat outdated and rarely used, it conveys the important, yet often overlooked idea that epileptic networks may impact healthy networks and brain areas. It may help to explain some of the cognitive deficits exhibited by patients beyond simply relating them to the EZ or lesions. It might also help to provide a more nuanced and complete pattern of the electrophysiological alterations in a given patient.

11.6 Conclusion

Epilepsy and cognition are intricately intertwined in several ways. While epilepsy is related to basic dysfunctions of neural assemblies, its expression is often in the form of cognitive, affective or behavioural symptoms. In addition, epileptic patients often suffer from cognitive deficits. Last but not least, an important goal in patient management is preventing an impact on cognition by antiseizure medicine or neurosurgery. The variety of types of epilepsy also makes this specific brain disease particularly abundant in terms of cognitive symptoms and profiles (see also Chap. 2), unlike other brain diseases which usually have more stereotyped cognitive profiles.

To return to the question raised in this chapter: has iEEG-based cognitive neuroscience contributed to the clinical workup of epileptic patients and their welfare? We think this is undoubtedly so. It guides the localisation of the epileptogenic network through a thorough analysis of the symptoms and subsequently often helps to clarify the type of epilepsy. It obviously guides intracerebral electrode planning and neurosurgery. It is also an important factor in improving the understanding of the cognitive deficits that patients experience. Directly and indirectly, iEEG studies have significantly contributed to improvements in these areas. Table 11.1 provides an overview of the contribution of iEEG cognitive research to epilepsy management.

Cognitive iEEG research helps to question some neuromemes that, although widely disseminated in the field of epilepsy, appear inadequate or limiting, for example, the "eloquent cortex" and "localisationism", especially since clinicians may be unaware of their influence. Given the highly detailed cognitive paradigms currently available, these loosely defined concepts should be refined. However, this can only be done through a collective effort of clinicians and researchers across many epilepsy centres, because the task is tremendous, requires agreement and guidelines, and probably the inclusion or large groups of patients to be able to update these concepts. In contrast, the "nociferous" neuromeme appears to fit more naturally into clinical practice as we have seen, with current trends of trying to assess the impact of the epilepsy network on the global functioning of the brain. Conversely, iEEG can reveal that what appears at first sight to be nociferous can also turn out to be functional and integrated with cognitive networks, as recently described for heterotopia with similar electrophysiological responses to the healthy cortex during an attentional task [121]. In any event, identifying neuromemes in epilepsy could be highly useful as many people in the field are unaware of them and therefore do not question the validity of some of the concepts they use every day.

In general, cognitive iEEG has helped to decrease the stigmatisation of patients. For example, an often-overlooked importance of iEEG is that it is continuously recorded for several days or even weeks. Therefore, it can be correlated with the fleeting, highly subjective phenomena that sometimes occur during seizures but which are impossible to replicate using experiments in the lab. As such, iEEG is unsurpassed. For example, converging iEEG data demonstrate a pivotal involvement of the hippocampus and rhinal cortex in experiential memory phenomena such as "déjà-vu", which has helped to discard the interpretation of déjà-vu as an unconscious fantasy or being related to a past-life experience [122, 123]. Only intracranial EEG can capture the semiological diversity of such subjective phenomena such as déjà-rêvé [124], or the degree of consciousness alterations [125], while improving our understanding of their neural correlates [108].

Table 11.1 Aspects of patients' epilepsy impacted by iEEG cognitive neuroscience. CCEP: cortico-cortical evoked potential; ERP: evoked related potential; EZ: epileptogenic zone; IED: interictal epileptic discharge

Aspect	Clinical benefits	Consequence for patients	Example of IEEG methods
Semiology	More detailed understanding of symptoms, refinement of symptoms	• More accurate classification of syndromes • Improved epilepsy diagnosis • Better knowledge of daily consequences of seizures • Less stigmatisation	• Verbatim analyses (e.g., diversity of experiential phenomena) • Video analyses • Behavioural scales (e.g., graduation of consciousness) • Anatomical and functional correlations (concomitant video & iEEG)
	Identification of the networks that underly specific symptoms	• Help to focus the MRI analysis on specific areas, help to identify lesions with low visibility • Identification of targets for intracranial implantation, more relevant sample of intracranial recordings • Distinction between symptomatogenic vs epileptogenic zone • Choice of tailored neuromodulation treatments (in the future)	• iEEG visual analyses • Network analyses during symptoms
Presurgical assessment	Functional mapping	• Identification of functional/dysfunctional brain areas, not necessarily mapping the EZ • Information about the consequences of the neurosurgery • Prediction of deficit(s) • Clearer information to patients • Guiding the boundaries of the surgical resection	• EBS that induces symptom(s) • EBS during cognitive tasks, causal stimulation • High gamma activity activated by cognitive task • ERP to specific cognitive tasks • CCEPs (functional tractography)
	Epileptogenic network mapping	• More relevant EZ identification, theoretically better post-surgical outcome	• Interictal epileptic activity reactivity (e.g., ripple, fast-ripple, IEDs) during cognitive tasks • Fingerprints during resting state or cognitive tasks

(continued)

Table 11.1 (continued)

Aspect	Clinical benefits	Consequence for patients	Example of IEEG methods
Interictal cognitive complaint	Impact of IED on cognition	• Complementary tool for indication of antiseizure medicine • Curing cognitive deficits (interictal cognitive complaints) by adjusting treatment, when patients seem seizure free	• Correlation between IEDs rates and cognitive processes or performance • Correlation between IEDs rates during sleep and performance • Interferences of IEDs with cognitive processes and physiological oscillations
General knowledge about the brain	Brain activity in healthy areas during cognitive tasks	• Better identification and information of the possible postoperative deficits • Better identification and information of the consequences of epileptic activities on cognition • Questioning neuromemes	• Cognitive task proposed to patients during ECoG/SEEG/awake craniotomy
		• Extrapolation to other brain diseases	• Therapeutic neuromodulation (e.g., memory modulation in Alzheimer's disease, etc.)
		• Valuation of the patient who becomes a participant in understanding his own pathology and who contributes to neuroscience research	

References

1. Quiroga RQ, Reddy L, Kreiman G et al (2005) Invariant visual representation by single neurons in the human brain. Nature 435:1102–1107. https://doi.org/10.1038/nature03687
2. Bausch M, Niediek J, Reber TP et al (2021) Concept neurons in the human medial temporal lobe flexibly represent abstract relations between concepts. Nat Commun 12:1–12. https://doi.org/10.1038/s41467-021-26327-3
3. Reddy L, Thorpe SJ (2014) Concept cells through associative learning of high-level representations. Neuron 84:248–251. https://doi.org/10.1016/j.neuron.2014.10.004
4. Bartolomei F, Lagarde S, Médina Villalon S et al (2016) The Proust phenomenon: odor-evoked autobiographical memories triggered by direct amygdala stimulation in human. Cortex 90:173–175. https://doi.org/10.1016/j.cortex.2016.12.005
5. Jacobs J, Kahana MJ (2010) Direct brain recordings fuel advances in cognitive electrophysiology. Trends Cogn Sci 14:162–171. https://doi.org/10.1016/j.tics.2010.01.005
6. Mukamel R, Fried I (2012) Human intracranial recordings and cognitive neuroscience. Annu Rev Psychol 63:511–537. https://doi.org/10.1146/annurev-psych-120709-145401
7. Parvizi J, Kastner S (2018) Promises and limitations of human intracranial electroencephalography. Nat Neurosci 21:474–483. https://doi.org/10.1038/s41593-018-0108-2
8. Jerbi K, Combrisson E, Dalal S et al (2013) Decoding cognitive states and motor intentions from intracranial EEG: how promising is high-frequency brain activity for brain-machine interfaces? Epilepsy Behav 28:283–302. https://doi.org/10.1016/j.yebeh.2013.03.012
9. Fried I (1993) The myth of eloquent cortex, or what is non-eloquent cortex? J Neurosurg 78:1009–1010. https://doi.org/10.3171/2014.12.JNS142826
10. Duffau H (2021) The death of localizationism: The concepts of functional connectome and neuroplasticity deciphered by awake mapping, and their implications for best care of brain-damaged patients. Rev Neurol (Paris) 177:1093–1103. https://doi.org/10.1016/j.neurol.2021.07.016
11. Kleen JK, Kirsch HE (2017) The nociferous influence of interictal discharges on memory. Brain 140:2072. https://doi.org/10.1093/brain/awx143
12. Blackmore S (2000) The meme machine. Oxford Paperbacks
13. Dawkins R (1976) Memes: the new replicators. In: Press OU (ed) The Selfish Gene. Oxford, pp 1–13
14. Baxendale S (2021) Epilepsy: lessons for clinicians from popular memes on social media. Epilepsy Behav 118:107899. https://doi.org/10.1016/j.yebeh.2021.107899
15. Devinsky O (2005) The myth of silent cortex and the morbidity of epileptogenic tissue: Implications for temporal lobectomy. Epilepsy Behav 7:383–389. https://doi.org/10.1016/j.yebeh.2005.07.020
16. Dekker S, Lee NC, Howard-Jones P, Jolles J (2012) Neuromyths in education: prevalence and predictors of misconceptions among teachers. Front Psychol 3:1–8. https://doi.org/10.3389/fpsyg.2012.00429
17. Lilienfeld SO, Lynn SJ, Ruscio J, Beyerstein BL (2011) 50 great myths of popular psychology: shattering widespread misconceptions about human behavior. John Wiley & Sons
18. Curot J, Busigny T, Valton L et al (2017) Memory scrutinized through electrical brain stimulation: a review of 80 years of experiential phenomena. Neurosci Biobehav Rev 78:161–177
19. Kahn E, Lane M, Sagher O (2017) Eloquent: history of a word's adoption into the neurosurgical lexicon. J Neurosurg 127:1461–1466. https://doi.org/10.3171/2017.3.JNS17659
20. Drake CG (1979) Cerebral arteriovenous malformations: considerations for and experience with surgical treatment in 166 cases. Clin Neurosurg 26:145–208. https://doi.org/10.1093/neurosurgery/26.cn_suppl_1.145
21. Rosenow F, Luders HO (2001) Presurgical evaluation of epilepsy. Brain 1683–1700. https://doi.org/10.4103/1817-1745.40593
22. Schevon CA, Carlson C, Zaroff CM et al (2007) Pediatric language mapping: sensitivity of neurostimulation and Wada testing in epilepsy surgery. Epilepsia 48:539–545. https://doi.org/10.1111/j.1528-1167.2006.00962.x

23. van't Klooster MA, Van Klink NEC, Leijten FSS et al (2015) Residual fast ripples in the intraoperative corticogram predict epilepsy surgery outcome. Neurology 85:120–128. https://doi.org/10.1212/WNL.0000000000001727
24. Penfield W (1959) The interpretive cortex: the stream of consciousness in the human brain can be electrically reactivated. Science 129:1719–1725
25. Penfield W, Perot P (1963) The brain's record of auditory and visual experience: a final summary and discussion. Brain 86:595–696
26. Penfield W (1958) Some mechanisms of consciousness discovered during electrical stimulation of the brain. Proc Natl Acad Sci U S A 44:51–66. https://doi.org/10.1073/pnas.44.2.51
27. Bancaud J, Brunet-Bourgin F, Chauvel P, Halgren E (1994) Anatomical origin of deja vu and vivid "memories" in human temporal lobe epilepsy. Brain 117(Pt 1):71–90. https://doi.org/10.1093/brain/117.1.71
28. Barbeau E, Wendling F, Régis J et al (2005) Recollection of vivid memories after perirhinal region stimulations: synchronization in the theta range of spatially distributed brain areas. Neuropsychologia 43:1329–1337. https://doi.org/10.1016/j.neuropsychologia.2004.11.025
29. Sporns O (2014) Contributions and challenges for network models in cognitive neuroscience. Nat Neurosci 17:652–660. https://doi.org/10.1038/nn.3690
30. Varela F, Lachaux JP, Rodriguez E, Martinerie J (2001) The brainweb: phase synchronization and large-scale integration. Nat Rev Neurosci 2:229–239. https://doi.org/10.1038/35067550
31. Bressler SL, Menon V (2010) Large-scale brain networks in cognition: emerging methods and principles. Trends Cogn Sci 14:277–290. https://doi.org/10.1016/j.tics.2010.04.004
32. Penfield W, Jasper H (1954) Epilepsy and the functional anatomy of the human brain. Little, Brown & Co., Oxford, England
33. Fornito A, Zalesky A, Breakspear M (2015) The connectomics of brain disorders. Nat Rev Neurosci 16:159–172. https://doi.org/10.1038/nrn3901
34. Koehler PJ (1996) Brown-Séquard and cerebral localization as illustrated by his ideas on aphasia. J Hist Neurosci 5:26–33. https://doi.org/10.1080/09647049609525648
35. Luauté JP, Luauté J (2005) Von Monakow's diaschisis. History and future of a discovery. Ann Med Psychol (Paris) 163:329–333. https://doi.org/10.1016/j.amp.2005.03.031
36. Pearce JM (1994) Von Monakow and diaschisis. J Neurol Neurosurg Psychiatry 57:197. https://doi.org/10.1136/jnnp.57.2.197
37. Sporns O, Tononi G, Kötter R (2005) The human connectome: a structural description of the human brain. PLoS Comput Biol 1:0245–0251. https://doi.org/10.1371/journal.pcbi.0010042
38. Hermann B, Seidenberg M (1995) Executive system dysfunction in temporal lobe epilepsy: effects of nociferous cortex versus hippocampal pathology. J Clin Exp Neuropsychol 17:809–819. https://doi.org/10.1080/01688639508402430
39. Coan AC, Campos BM, Yasuda CL et al (2014) Frequent seizures are associated with a network of gray matter atrophy in temporal lobe epilepsy with or without hippocampal sclerosis. PLoS ONE 9. https://doi.org/10.1371/journal.pone.0085843
40. Ung H, Cazares C, Nanivadekar A et al (2017) Interictal epileptiform activity outside the seizure onset zone impacts cognition. Brain 140:2157–2168. https://doi.org/10.1093/brain/awx143
41. Henin S, Shankar A, Borges H et al (2021) Spatiotemporal dynamics between interictal epileptiform discharges and ripples during associative memory processing. Brain 144:1590–1602. https://doi.org/10.1093/brain/awab044
42. Gelinas JN, Khodagholy D, Thesen T et al (2016) Interictal epileptiform discharges induce hippocampal-cortical coupling in temporal lobe epilepsy. Nat Med 22:641–648. https://doi.org/10.1038/nm.4084
43. Dahal P, Ghani N, Flinker A et al (2019) Interictal epileptiform discharges shape large-scale intercortical communication. Brain 142:3502–3513. https://doi.org/10.1093/brain/awz269
44. Gilliam F (2014) "Connectionology" provides further evidence for nociferous epileptic cortex. Epilepsy Curr 14:183–185. https://doi.org/10.5698/1535-7597-14.4.183

45. Spetzler RF, Martin NA (1986) A proposed grading system for arteriovenous malformations. J Neurosurg 65:476–483. https://doi.org/10.3171/JNS/2008/108/01/0186
46. Bédos Ulvin L, Jonas J, Brissart H et al (2017) Intracerebral stimulation of left and right ventral temporal cortex during object naming. Brain Lang 175:71–76. https://doi.org/10.1016/j.bandl.2017.09.003
47. Ojemann G, a. (2003) The neurobiology of language and verbal memory: observations from awake neurosurgery. Int J Psychophysiol 48:141–146. https://doi.org/10.1016/S0167-8760(03)00051-5
48. Salanova V, Witt T, Worth R et al (2015) Long-term efficacy and safety of thalamic stimulation for drug-resistant partial epilepsy. Neurology 84:1017–1025. https://doi.org/10.1212/wnl.0000000000001334
49. Young JS, Lee AT, Chang EF (2021) A review of cortical and subcortical stimulation mapping for language. Neurosurgery 89:331–342. https://doi.org/10.1093/neuros/nyaa436
50. Signorelli F, Guyotat J, Mottolese C et al (2004) Intraoperative electrical stimulation mapping as an aid for surgery of intracranial lesions involving motor areas in children. Child's Nerv Syst 20:420–426. https://doi.org/10.1007/s00381-004-0961-z
51. Fried I, Katz A, McCarty G et al (1991) Functional organization of human supplementary studied by electrical stimulation motor cortex. J Neurosci 11:3656–3666
52. Usui N, Terada K, Baba K et al (2008) Extraoperative functional mapping of motor areas in epileptic patients by high-frequency cortical stimulation. J Neurosurg 109:605–614. https://doi.org/10.3171/JNS/2008/109/10/0605
53. Jonas J, Frismand S, Vignal JP et al (2014) Right hemispheric dominance of visual phenomena evoked by intracerebral stimulation of the human visual cortex. Hum Brain Mapp 35:3360–3371. https://doi.org/10.1002/hbm.22407
54. Duffau H (2010) Awake surgery for nonlanguage mapping. Neurosurgery 66:523–528. https://doi.org/10.1227/01.NEU.0000364996.97762.73
55. Mandonnet E, Winkler P, a, Duffau H, (2010) Direct electrical stimulation as an input gate into brain functional networks: principles, advantages and limitations. Acta Neurochir (Wien) 152:185–193. https://doi.org/10.1007/s00701-009-0469-0
56. Selimbeyoglu A, Parvizi J (2010) Electrical stimulation of the human brain: perceptual and behavioral phenomena reported in the old and new literature. Front Hum Neurosci 4:46. https://doi.org/10.3389/fnhum.2010.00046
57. Trébuchon A, Chauvel P (2016) Electrical stimulation for seizure induction and functional mapping in stereoelectroencephalography. J Clin Neurophysiol 33:511–521. https://doi.org/10.1097/WNP.0000000000000313
58. Jonas J (2018) Prédiction du devenir fonctionnel postopératoire en chirurgie de l'épilepsie grâce aux stimulations électriques corticales. Neuropsychol des épilepsies l'adulte Approch Clin Prat 109
59. Foster BL, Parvizi J (2017) Direct cortical stimulation of human posteromedial cortex. 1–7. https://doi.org/10.1212/WNL.0000000000003607
60. Aron O, Jonas J, Colnat-Coulbois S, Maillard L (2021) Language mapping using stereo electroencephalography: a review and expert opinion. Front Hum Neurosci 15:1–12. https://doi.org/10.3389/fnhum.2021.619521
61. Nandakumar N, Manzoor K, Agarwal S et al (2021) Automated eloquent cortex localization in brain tumor patients using multi-task graph neural networks. Med Image Anal 74:102203. https://doi.org/10.1016/j.media.2021.102203
62. Kahane P, Tassi L, Hoffmann SFD et al (1993) Manifestations électrocliniques induites par la stimulation électrique intracérébrale par "chocs" dans les épilepsies temporales. Neurophysiol Clin 22:305–326
63. Roux F-E, Durand JB, Djidjeli I et al (2016) Variability of intraoperative electrostimulation parameters in conscious individuals: language cortex. 126:1641–1652. https://doi.org/10.3171/2016.4.JNS152434

64. Mohan UR, Watrous AJ, Miller JF et al (2020) The effects of direct brain stimulation in humans depend on frequency, amplitude, and white-matter proximity. Brain Stimul 13:1183–1195. https://doi.org/10.1016/j.brs.2020.05.009
65. Zangaladze A, Sharan A, Evans J et al (2008) The effectiveness of low-frequency stimulation for mapping cortical function. Epilepsia 49:481–487. https://doi.org/10.1111/j.1528-1167.2007.01307.x
66. Paulk AC, Zelmann R, Crocker B et al (2022) Local and distant cortical responses to single pulse intracranial stimulation in the human brain are differentially modulated by specific stimulation parameters. Brain Stimul 15:491–508. https://doi.org/10.1016/j.brs.2022.02.017
67. Marti AS, Mirsattari SM, Steven DA et al (2022) Extraoperative electrical stimulation mapping in epilepsy presurgical evaluation: a proposal and review of the literature. Clin Neurol Neurosurg 214:107170. https://doi.org/10.1016/j.clineuro.2022.107170
68. Borchers S, Himmelbach M, Logothetis N, Karnath HO (2012) Direct electrical stimulation of human cortex—the gold standard for mapping brain functions? Nat Rev Neurosci 13:63–70. https://doi.org/10.1038/nrn3140
69. Halgren E, Wilson CL (1985) Recall deficits produced by afterdischarges in the human hippocampal formation and amygdala. Electroencephalogr Clin Neurophysiol 61:375–380
70. Suthana N, Haneef Z, Stern J et al (2012) Memory enhancement and deep-brain stimulation of the entorhinal area. N Engl J Med 366:502–510. https://doi.org/10.1056/NEJMoa1107212
71. Coleshill SG, Binnie DC, Morris RG et al (2004) Material-specific recognition memory deficits elicited by unilateral hippocampal electrical stimulation. J Neurosci 24:1612–1616. https://doi.org/10.1523/JNEUROSCI.4352-03.2004
72. Kolarik BS, Baer T, Shahlaie K et al (2018) Close but no cigar: Spatial precision deficits following medial temporal lobe lesions provide novel insight into theoretical models of navigation and memory. Hippocampus 28:31–41. https://doi.org/10.1002/hipo.22801
73. Yrondi A, Valton L, Bouilleret V et al (2020) Post-traumatic stress disorder with flashbacks of an old childhood memory triggered by right temporal lobe epilepsy surgery in adulthood. Front Psychiatry 11:1–5. https://doi.org/10.3389/fpsyt.2020.00351
74. Cleary RA, Baxendale SA, Thompson PJ, Foong J (2013) Predicting and preventing psychopathology following temporal lobe epilepsy surgery. Epilepsy Behav 26:322–334. https://doi.org/10.1016/j.yebeh.2012.09.038
75. Unterberger I, Trinka E, Ransmayr G et al (2021) Epileptic aphasia—a critical appraisal. Epilepsy Behav 121:108064. https://doi.org/10.1016/j.yebeh.2021.108064
76. Morris RG, Coleshill SG, Lacruz ME et al (2012) Hippocampal electrical stimulation and localisation of long-term episodic memory. Epilepsy Mem 358
77. Curot J, Denuelle M, Busigny T et al (2014) Bilateral Wada test: amobarbital or propofol? Seizure 23:122–128. https://doi.org/10.1016/j.seizure.2013.10.009
78. Mesulam M (2009) Defining neurocognitive networks in the BOLD new world of computed connectivity. Neuron 62:1–3. https://doi.org/10.1016/j.neuron.2009.04.001
79. Isnard J, Taussig D, Bartolomei F et al (2018) French guidelines on stereoelectroencephalography (SEEG). Neurophysiol Clin/Clin Neurophysiol 48:5–13. https://doi.org/10.1016/j.neucli.2017.11.005
80. Bartolomei F, Lagarde S, Wendling F et al (2017) Defining epileptogenic networks: contribution of SEEG and signal analysis. Epilepsia 1–17. https://doi.org/10.1111/epi.13791
81. Friederici AD, Gierhan SME (2013) The language network. Curr Opin Neurobiol 23:250–254. https://doi.org/10.1016/j.conb.2012.10.002
82. Busch RM, Love TE, Jehi LE et al (2015) Effect of invasive EEG monitoring on cognitive outcome after left temporal lobe epilepsy surgery. Neurology 85:1475–1481. https://doi.org/10.1212/WNL.0000000000002066
83. Histed MH, Bonin V, Reid RC (2009) Direct activation of sparse, distributed populations of cortical neurons by electrical microstimulation. Neuron 63:508–522. https://doi.org/10.1016/j.neuron.2009.07.016

84. Logothetis NK, Augath M, Murayama Y et al (2010) The effects of electrical microstimulation on cortical signal propagation. Nat Neurosci 13:1283–1291. https://doi.org/10.1038/nn.2631
85. Jacobs J, Lega B, Anderson C (2012) Explaining how brain stimulation can evoke memories. J Cogn Neurosci 553–563
86. Kim K, Schedlbauer A, Rollo M et al (2018) Network-based brain stimulation selectively impairs spatial retrieval. Brain Stimul 11:213–221. https://doi.org/10.1016/j.brs.2017.09.016
87. Kim K, Ekstrom AD, Tandon N (2016) A network approach for modulating memory processes via direct and indirect brain stimulation: Toward a causal approach for the neural basis of memory. Neurobiol Learn Mem 134:162–177. https://doi.org/10.1016/j.nlm.2016.04.001
88. Fernandez G, Tendolkar I (2006) The rhinal cortex: "gatekeeper" of the declarative memory system. Trends Cogn Sci 10:358–362. https://doi.org/10.1016/j.tics.2006.06.003
89. van den Heuvel MP, Sporns O (2013) Network hubs in the human brain. Trends Cogn Sci 17:683–696. https://doi.org/10.1016/j.tics.2013.09.012
90. Grunwald T, Lehnertz K, Pezer N et al (1999) Prediction of postoperative seizure control by hippocampal event- related potentials. Epilepsia 40:303–306. https://doi.org/10.1111/j.1528-1157.1999.tb00708.x
91. Elger CE, Grunwald T, Lehnertz K et al (1997) Human temporal lobe potentials in verbal learning and memory processes. Neuropsychologia 35:657–667. https://doi.org/10.1016/S0028-3932(96)00110-8
92. Kambara T, Sood S, Alqatan Z et al (2018) Presurgical language mapping using event-related high-gamma activity: the Detroit procedure. Clin Neurophysiol 129:145–154. https://doi.org/10.1016/j.clinph.2017.10.018
93. Jerbi K, Ossandón T, Hamamé CM et al (2009) Task-related gamma-band dynamics from an intracerebral perspective: review and implications for surface EEG and MEG. Hum Brain Mapp 30:1758–1771. https://doi.org/10.1002/hbm.20750
94. Cuisenier P, Testud B, Minotti L et al (2021) Relationship between direct cortical stimulation and induced high-frequency activity for language mapping during SEEG recording. 134:1251–1261. https://doi.org/10.3171/2020.2.JNS192751
95. Gastaut H, Broughton RJ (1972) Epileptic seizures; clinical and electrographic features, diagnosis and treatment. Thomas
96. Bancaud J, Talairach J (1965) La stéréo-électroencéphalographie dans l'épilepsie : informations neurophysiopathologiques apportées par l'investigation fonctionnelle stéreotaxique. Masson et Cie, Paris
97. McGonigal A, Bartolomei F, Chauvel P (2021) On seizure semiology. Epilepsia 62:2019–2035. https://doi.org/10.1111/epi.16994
98. Ren L, Yu T, Wang D et al (2020) Subthalamic nucleus stimulation modulates motor epileptic activity in humans. Ann Neurol 88:283–296. https://doi.org/10.1002/ana.25776
99. Filipescu C, Lagarde S, Lambert I et al (2019) The effect of medial pulvinar stimulation on temporal lobe seizures. Epilepsia 60:e25–e30. https://doi.org/10.1111/epi.14677
100. Bartolomei F, Wendling F, Vignal JP et al (2002) Neural networks underlying epileptic humming. Epilepsia 43:1001–1012. https://doi.org/10.1046/j.1528-1157.2002.48501.x
101. Biraben A, Taussig D, Thomas P et al (2001) Fear as the main feature of epileptic seizures. J Neurol Neurosurg Psychiatry 70:186–191. https://doi.org/10.1136/jnnp.70.2.186
102. Bartolomei F, Guye M, Wendling F et al (2003) Fear, anger and compulsive behavior during seizure: involvement of large scale fronto-temporal neural networks. Epileptic Disord 4:235–241
103. Chen S, Tan Z, Xia W et al (2021) Theta oscillations synchronize human medial prefrontal cortex and amygdala during fear learning. Sci Adv 7:1–14. https://doi.org/10.1126/sciadv.abf4198
104. Matsumoto R, Nair DR, LaPresto E et al (2007) Functional connectivity in human cortical motor system: a cortico-cortical evoked potential study. Brain 130:181–197. https://doi.org/10.1093/brain/awl257
105. Trebaul L, Deman P, Tuyisenge V et al (2018) Probabilistic functional tractography of the human cortex revisited. Neuroimage 181:414–429. https://doi.org/10.1016/j.neuroimage.2018.07.039

106. Axmacher N, Schmitz DP, Wagner T et al (2008) Interactions between medial temporal lobe, prefrontal cortex, and inferior temporal regions during visual working memory: a combined intracranial EEG and functional magnetic resonance imaging study. J Neurosci 28:7304–7312. https://doi.org/10.1523/JNEUROSCI.1778-08.2008
107. Kleen JK, Scott RC, Holmes GL et al (2013) Hippocampal interictal epileptiform activity disrupts cognition in humans. Neurology 81:18–24. https://doi.org/10.1212/wnl.0b013e318297ee50
108. Curot J, Barbeau E, Despouy E, Denuelle M, Sol JC, Lotterie JA, Peyrache A (2021) Local neuronal excitation and global inhibition during epileptic fast ripples in humans. Brain 2022 awac319. https://doi.org/10.1093/brain/awac319
109. Corcoran R, Thompson P (1993) Epilepsy and poor memory: Who complains and what do they mean? Br J Clin Psychol 32:199–208. https://doi.org/10.1111/j.2044-8260.1993.tb01044.x
110. Lemesle B, Barbeau EJ, Milongo Rigal E et al (2022) Hidden objective memory deficits behind subjective memory complaints in patients with temporal lobe epilepsy. Neurology 98:E818–E828. https://doi.org/10.1212/WNL.0000000000013212
111. Lambert I, Tramoni-negre E, Lagarde S et al (2021) Accelerated long-term forgetting in focal epilepsy: do interictal spikes during sleep matter? Epilepsia 1–7. https://doi.org/10.1111/epi.16823
112. Horak PC, Meisenhelter S, Song Y et al (2017) Interictal epileptiform discharges impair word recall in multiple brain areas. Epilepsia 58:373–380. https://doi.org/10.1111/epi.13633
113. Sánchez Fernández I, Loddenkemper T, Galanopoulou AS, Moshé SL (2015) Should epileptiform discharges be treated? Epilepsia 56:1492–1504. https://doi.org/10.1111/epi.13108
114. Warren CP, Hu S, Stead M et al (2010) Synchrony in normal and focal epileptic brain: the seizure onset zone is functionally disconnected. J Neurophysiol 104:3530–3539. https://doi.org/10.1152/jn.00368.2010
115. Burns SP, Santaniello S, Yaffe RB et al (2014) Network dynamics of the brain and influence of the epileptic seizure onset zone. Proc Natl Acad Sci U S A 111:E5321–E5330. https://doi.org/10.1073/pnas.1401752111
116. Brázdil M, Cimbálník J, Roman R et al (2015) Impact of cognitive stimulation on ripples within human epileptic and non-epileptic hippocampus. BMC Neurosci 16:1–9. https://doi.org/10.1186/s12868-015-0184-0
117. Pail M, Cimbálník J, Roman R et al (2020) High frequency oscillations in epileptic and non-epileptic human hippocampus during a cognitive task. Sci Rep 10:1–12. https://doi.org/10.1038/s41598-020-74306-3
118. Saboo KV, Balzekas I, Kremen V et al (2021) Leveraging electrophysiologic correlates of word encoding to map seizure onset zone in focal epilepsy: task-dependent changes in epileptiform activity, spectral features, and functional connectivity. Epilepsia 62:2627–2639. https://doi.org/10.1111/epi.17067
119. Young JJ, Rudebeck PH, Marcuse LV et al (2018) Theta band network supporting human episodic memory is not activated in the seizure onset zone. Neuroimage 183:565–573. https://doi.org/10.1016/j.neuroimage.2018.08.052
120. Cimbalnik J, Brinkmann B, Kremen V et al (2018) Physiological and pathological high frequency oscillations in focal epilepsy. Ann Clin Transl Neurol 5:1062–1076. https://doi.org/10.1002/acn3.618
121. Akkol S, Kucyi A, Hu W et al (2021) Intracranial electroencephalography reveals selective responses to cognitive stimuli in the periventricular heterotopias. J Neurosci 41:3870–3878. https://doi.org/10.1523/JNEUROSCI.2785-20.2021
122. Bartolomei F, Barbeau E, Gavaret M et al (2004) Cortical stimulation study of the role of rhinal cortex in deja vu and reminiscence of memories. Neurology 63:858–864. https://doi.org/10.1212/01.WNL.0000137037.56916.3F
123. Bartolomei F, Barbeau EJ, Nguyen T et al (2012) Rhinal-hippocampal interactions during déjà vu. Clin Neurophysiol 123:489–495. https://doi.org/10.1016/j.clinph.2011.08.012

124. Curot J, Valton L, Denuelle M et al (2018) Déjà-rêvé: prior dreams induced by direct electrical brain stimulation. Brain Stimul 11:875–885. https://doi.org/10.1016/j.brs.2018.02.016
125. Arthuis M, Valton L, Rgis J et al (2009) Impaired consciousness during temporal lobe seizures is related to increased long-distance corticalsubcortical synchronization. Brain 132:2091–2101. https://doi.org/10.1093/brain/awp086

Part II
Physiological Basis and Functional Role of Intracranial EEG Signals

Chapter 12
What Are the Advantages and Challenges of Simultaneous Scalp EEG and Intracranial EEG Data Recording?

Laurent Koessler

Abstract Scalp and intracranial electroencephalography are both based on the recording of field potentials, i.e., electrical potentials in the extracellular space. Thanks to recent technological developments, simultaneous recordings of EEG and iEEG provide complementary information to understand brain functions. The relationship of the cortical source activity with their scalp and intracranial EEG correlates is still not very well known. Since some cortical sources are not directly visible in scalp EEG recordings, it gives the false impression of no electrical contribution and thus that scalp electrodes are unnecessary. In this chapter, I illustrate the importance to record and precisely analyze scalp EEG in combination with intracranial EEG. First, the technical challenges imposed by combination of the two EEG methods are described. Then, historical aspects and the main findings of the first simultaneous scalp and intracranial EEG recordings since the mid-1950s are presented. Finally, applications and future perspectives in cognitive, clinical, and computational neuroscience are discussed.

Keywords Scalp EEG · Multi-scale EEG · EEG biomarkers · Biophysics · Cognition

Intracranial electroencephalography (iEEG) in humans offers closer access to neural activity generated in local brain regions than scalp EEG. Despite the growing insights obtained from intracranial human research (as recently reviewed by [1]), several limitations of iEEG as a method to understand human brain function should be mentioned. First, even though iEEG has high spatial resolution, it has limited spatial coverage. Electrode placement (both multi-contact SEEG electrodes, or subdural grids or strips) is driven solely by clinical purposes to target brain regions that are suspected to be included in a focal epileptic network or to be related to it (see Chap. 1 for the clinical background of electrode implantation). Considering the different anatomical localizations of focal epilepsy (the most frequent being temporal lobe

L. Koessler (✉)
Université de Lorraine, CNRS, CRAN, 54000 Nancy, France
e-mail: laurent.koessler@univ-lorraine.fr

N. Axmacher (ed.), *Intracranial EEG*, Studies in Neuroscience, Psychology and Behavioral Economics, https://doi.org/10.1007/978-3-031-20910-9_12

epilepsy), some brain regions may be over-sampled and some others under-sampled. Thus, a full investigation of human sensorimotor and cognitive functions is difficult to achieve with iEEG only. In contrast, scalp EEG recordings combined simultaneously with iEEG can sample more brain regions (especially those that are not sampled by iEEG) and thus offer solutions to provide additional and complementary information.

12.1 Technical Aspects and Challenges

The simultaneous recording of scalp and intracranial EEG recordings (s-iEEG) has to consider several technical aspects. First, and most importantly, the asepsis must be guaranteed during the patient's stay in the hospital. The placement of intracranial electrodes requires a (more or less extended) craniotomy in the case of subdural strip or grid electrodes or the insertion of depth electrodes via burr holes under general anesthesia. The following days and throughout the whole iEEG recording period, the head is covered with an antiseptic protection (solution/gel and sterile bandages). After the surgery, when the patient recovers, scalp electrodes must be placed following a specific protocol to respect the asepsis condition. First, the protective sterile bandage from the subject's head should be carefully removed by adding a surgical sterile field around the head, using sterile gloves and gown. Second, if the scalp electrode placement occurs several days after the surgery, a new skin disinfection with Betadine or Clorexan has to be performed, especially on the scalp regions where the electrodes will be placed. Third, the EEG net or the individual scalp electrodes used must be sterile, either by purchasing them in a sterile form or by sterilizing them (physical, radiation, ultrasonic, or chemical methods). Finally, the electrodes can be positioned on the scalp, considering a security distance (a few centimeters) from the craniotomy or the entry points of intracranial electrodes. This distance implies that the cranial positions of the scalp electrodes must often be slightly shifted from conventional positions. The 10/20 system [2, 3] cannot always be respected and new positions must be determined on the scalp surface. In this case, the new electrode positions must be recorded, ideally 3D digitized [4]. If one (or several) electrode position(s) is modified with respect to standard configurations, it is important to symmetrically displace the contralateral one(s), especially if a common average reference is computed for scalp EEG analysis. Another important reason not to place scalp electrodes next to the craniotomy is the current leakage. Since iEEG recordings requires small or large holes in the skull, physiological artefacts called breach rhythms (i.e., an increase of amplitude activity in alpha, beta, and mu rhythms) can be observed in scalp EEG recordings [5]. In a computational study, Datta et al., in 2010, demonstrated that current flow significantly changed when scalp electrodes were placed over either small (for stereoelectroencephalography, SEEG) or large (for electrocorticography, ECoG) skull defects [6].

Second, following the installation of scalp electrodes, an important issue is the synchronization of intracranial and scalp EEG signals. Ideally, scalp and intracranial electrodes must be connected to the same acquisition system (i.e., with a unique sampling clock). The use of an external trigger box to synchronize two signals coming from two different acquisition systems can be difficult to implement and risky if it is not controlled regularly over time. The sampling rate of acquisition should be at least 1 kHz in order to obtain the highest number of EEG samples for the extraction and the precise analysis of spontaneous or evoked EEG biomarkers. This high sampling rate is also required by the Nyquist-Shannon sampling theorem to record high frequency oscillations (especially in the gamma frequency range: 40–100 Hz) that have been tentatively associated with sensorimotor and cognitive processes [7, 8]. Considering the high number of scalp and intracranial electrodes to plug, a 64–128-channel recording system is recommended, or even a 256-channel system to avoid a difficult manual selection of unplugged electrodes.

The vast majority of simultaneous s-iEEG recording studies use a scalp electrode as acquisition reference (e.g., Fpz). This reference montage is particularly appropriate for simultaneous recordings because EEG amplitude is much lower, i.e., closer to zero volts, on the scalp than in the intracranial volume. This scalp reference is also convenient because it is routinely used to record evoked potentials, so that the scalp EEG signals recorded during intracranial investigation can be compared directly (i.e., without re-referencing) with those obtained before or after the pre-surgical iEEG evaluation.

12.2 Methodological Aspects and Challenges

The first simultaneous s-iEEG recordings were not performed to study cognitive brain functions but for pre-surgical evaluation of focal drug-resistant epilepsies (reviewed by [9]). These pathologies require long-term (several days) invasive EEG recordings to precisely localize the epileptogenic network [10–12], giving the unique opportunity in humans to combine simultaneously non-invasive (scalp) and invasive (intracranial) EEG recordings. Since the first intracranial EEG recordings performed in an epileptic patient in 1948 by Reginal Bickford, clinicians observed an important discordance between the abundance of discharges in iEEG and the absence of visible discharges in previous scalp EEG recordings of the same patients [13]. The current methodology of invasive techniques has inherent limitations, thus rendering multiscale comparisons particularly challenging. Indeed, invasive recordings with subdural grids or strips of electrodes provide extensive cortical coverage but are prone to sampling limitations for deep brain sources and sulci (see Chap. 39). In contrast, depth electrodes (SEEG) provide information of some selected deep brain regions but only provide irregular and incomplete coverage of the cortical surface.

The first paper with simultaneous s-iEEG recordings was published in 1958 by Abraham and Ajmone-Marsan [14]. In this early study, the authors investigated both clinical and biophysical questions. They especially dealt with the issue of *visibility*, i.e., the ability to observe by visual expertise the discharges from cortical sources in scalp EEG signals. This issue should not be confused with the ability to record discharges that are not observable by visual expertise, i.e., the issue of *contribution*. These two questions are important for cognitive neuroscience because whatever the

EEG patterns (epileptic or not), it is important to understand which EEG signals are visible and detectable with scalp or intracranial EEG recordings. In this pioneering study, the authors demonstrated that a high number of discharges at the cortical level were not visible in scalp EEG signals [14]. On the contrary, discharges that were visible in scalp EEG signals were in direct correlation with activities recorded in the intracranial volume (Fig. 12.1), although not in all cases. This last observation is spatial coverage limitation of the brain in iEEG, so that only a small subset of electrical brain sources can be recorded. This initial investigation also demonstrated that scalp EEG recordings can give additional information thanks to its global and uniform coverage of the head. Concerning the biophysical link between the two scales of EEG recordings, Abraham and Ajmone-Marson suggested that the extent of cortical activation, rather than the morphology or the duration of the discharges, determines the amplitude (and thus, the visibility) of the scalp EEG correlates. This suggested role of the scalp as "a spatial EEG average of electrical activity" was also mentioned at an early stage in animal studies [15, 16].

While these first observations provided very interesting insights about how electric fields propagate in the head volume, and showed the importance to record both signals simultaneously, several decades passed until new studies appeared at the end

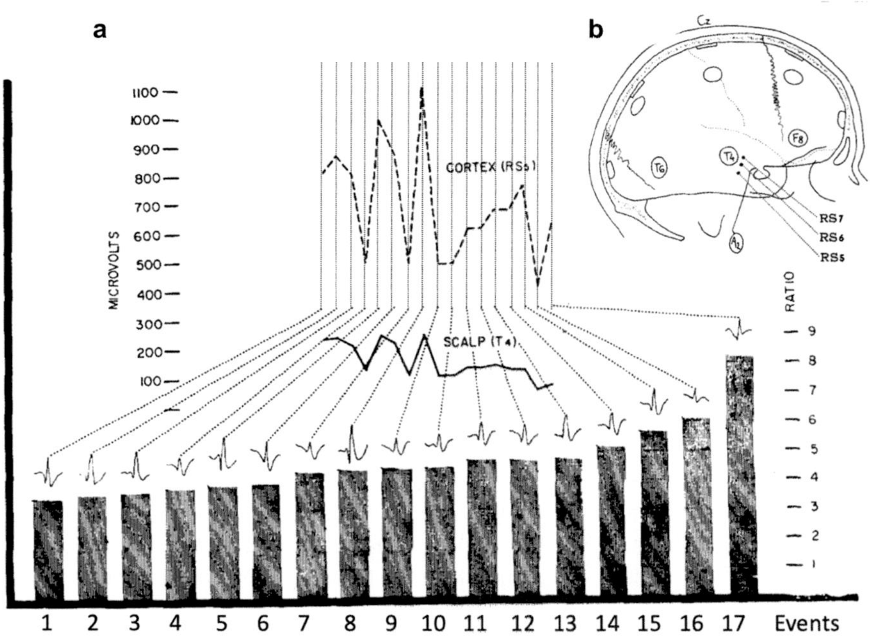

Fig. 12.1 (adapted from [14]): **a** Intracranial EEG amplitude (dashed line; RS_5 electrode) of seventeen cortical discharges and their scalp EEG correlates (continuous line; scalp T4 electrode). In this example, there is a direct relationship between the two recordings. When intracranial EEG amplitudes are high (low), scalp EEG amplitude increases (decreases). **b** Schematic view of the scalp and cortical (RS5-7) electrode positions

of the 1990s. During the second phase of investigation, two crucial questions were investigated. First, which extent of cortical activation is required to induce visible scalp EEG signals? And second, what is the amplitude, or signal-to-noise ratio (SNR), relationship between scalp and intracranial EEG recordings?

For a long time, researchers and clinicians considered that several square centimeters (from 6 to 30) of neural source activity were required to generate detectable scalp EEG activity. These values were obtained from instrumental, computational, and human in-vivo studies. Using realistic head and source models, Cosandier and colleagues, in 2008, modeled the electrical fields at the level of scalp and SEEG electrodes using different source configurations [17]. Their results showed that a cortical source with an area from 7 to 30 cm^2 can produce scalp EEG patterns with peak amplitude encompassing 1.5–2.8 times the background activity, respectively. This large cortical extent was also found using simultaneous EEG-ECoG recordings [18]. In this latter study, cortical sources with less than 6 cm^2 did not produce visible scalp potentials. Between 6 and 10 cm^2 of synchronous cerebral activity, only rare scalp-recordable EEG discharges were generated, whereas sources having more than 10 cm^2 commonly resulted in recognizable scalp potentials. Finally, this latter study concluded that prominent scalp discharges were related to areas of 30 cm^2 of activated cortex. These estimations differed importantly from Cooper et al., 1965, who found a required cortical surface of 6–8 cm^2 to obtain a similar signal decrease [19]. In the same vein, more recently, other authors demonstrated with human in-vivo studies that very focal brain sources can produce EEG signals at the level of the scalp surface. For example, high frequency oscillations from very focal sources (e.g., somatosensory cortex) were detectable in scalp EEG recordings [20–22]. Zelmann and colleagues, in 2014, found that similar scalp topographies of high frequency oscillations (HFO) corresponded to distinct cortical distributions on 1 cm intracranial grid electrodes, and that averaging similar scalp HFO resulted in focal intracranial maps [21]. This finding shows that HFO of small cortical extents are spatially undersampled with standard 10/20 scalp configurations, and also with grid electrodes with their relatively large inter-electrode distances. It also demonstrates that intracranial EEG recordings with grids do not necessarily record everything that contributes to the potentials of close scalp electrodes. Similar results were found using evoked potentials (N20 combined with HFO) in the somatosensory cortex [22]. Despite the small activated cortical surface (functional brain area 3b), the authors reported a significant correlation between HFO in intracranial and scalp EEG recordings ($r = 0.65$).

One possibility is that these focal brain sources were detectable in scalp EEG due to their superficial positions. To address this issue, the scalp contribution of both focal and deep brain sources such as the hippocampus have been investigated [23–25]. These latter studies demonstrated that the deep sources also significantly contribute to scalp EEG (Fig. 12.2). This scalp EEG contribution of deep mesial temporal lobe sources is also interesting to consider because, in addition to the depth and the small extent, these "curved" sources are very often considered as closed electrical field structures [26, 27] that thus presumably never contribute to scalp EEG. The scalp EEG contribution of hippocampal sources concerns basal temporal scalp electrodes (FT/P9 et FT/P10 electrodes). Interestingly, the scalp amplitude topographies of mesial and lateral temporal lobe sources can be distinguished using a hierarchical clustering method. The mesial sources generate only a low and focal amplitude negativity over the basal temporal electrodes (Fig. 12.2) whereas the neocortical sources generate a higher and more widespread negativity at temporal electrodes combined with a widespread positivity in the vertex (with a predominance in contralateral fronto-centro-parietal electrodes).

This spatial contribution of deep mesial temporal lobe sources has induced a recent change in the international 10/20 system for scalp EEG electrodes placement with the addition of four new electrodes in basal temporal regions [3].

It is important to mention that the contributions of these deep brain sources could only be revealed by an averaging process of a high number of events (not yet precisely determined but certainly around a hundred) to detect scalp EEG evoked potentials. Even when there is a genuine electrical contribution to the scalp EEG signals, the high amplitude background activity of superficial brain sources can mask this activity (e.g., average SNR of non-averaged EEG signals from mesial temporal sources of −2.1 dB compared to the background activity in [23]). Extracting these EEG signals from their background activity is thus a real challenge. At this level, signal analysis methods

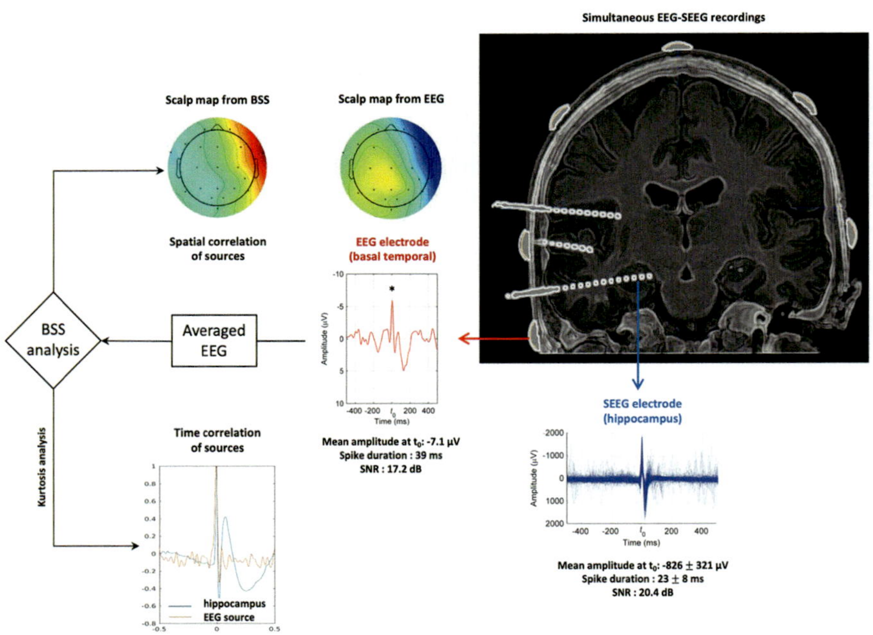

Fig. 12.2 (Adapted from [23]): Deep mesial temporal lobe contribution to scalp EEG demonstrated via simultaneous EEG-SEEG recordings. Discharges in the hippocampus (in blue) can generate (after averaging) a significant EEG signal (in red) at basal temporal scalp EEG electrodes (see scalp map from EEG). Blind source separation (BSS) method can extract a brain source with (i) a very similar scalp map as the one obtained from real data (spatial correlation) and (ii) a very similar peak of amplitude at t_0 (temporal correlation)

such as blind source separation methods and artificial intelligence (AI; e.g., machine learning or deep learning) could provide meaningful information. Fahimi Hnazaee and colleagues, in 2020 confirmed the electrical contribution of deep brain sources in scalp EEG using independent component analysis (ICA) of simultaneous s-iEEG recordings (both SEEG and ECoG) during resting state periods [28]. These authors tested whether activity from deep sources spread to the cortical surface and scalp, finding a weak but significant correlation and peaks at zero-lag between the depth electrodes, several subdural contacts and scalp electrodes. This ICA methodology of signal analysis is interesting because the analysis is carried out "blindly" and does not require tagging the intracranial EEG signals in the time domain. Another blind method, zero-crossing analysis, has also been used to detect scalp signatures of intracranial discharges. Pyrzowski and colleagues, in 2021, showed that the zero-crossing of scalp EEG patterns can be used as a reliable single-trial scalp EEG signal of intracranial discharges, with some of them originating from deep brain structures [25]. Detection based on these zero-crossing patterns can achieve high temporal precision and be generalized between patients i.e., with the same performance to automatically detect deep sources. It could allow discriminating successfully scalp EEG patterns of epileptic patients and healthy subjects without needing to access the intracranial recordings even in the absence of visible scalp discharges.

In summary, focal, deep or "curved" brain sources can produce measurable scalp EEG signals. This contribution is often invisible by a human (expert) observation but nevertheless significant after data analysis. The criteria of visibility on the scalp also (and certainly mainly) depend on the background activity of the neighboring brain regions, especially the most lateral cortices. In addition to cortical surface, the degree of synchronization between activities of the neuronal populations within a cortical surface is also a crucial determinant of the recorded EEG patterns [29]. It is also important to notice that the minimally required size of activated cortical surface does not consider that higher neural densities can produce the same amplitude within a smaller brain surface. Thus, the activated cortical surface cannot be the only or main factor that can influence the electrical contribution of brain sources to scalp EEG.

Following the influence of cortical extent on scalp EEG signals, the second important issue concerns the amplitude (or SNR) ratio between intracranial and scalp EEG signals. From a methodological point of view, amplitude ratio is better estimated with simultaneous EEG-SEEG recordings instead of EEG-ECoG. Indeed, the large craniotomy of ECoG induces an increase of electrical conductivity [30] and a breach rhythm [5] in the scalp EEG signals that can bias the ratio estimation. In addition, the presence of a non-conducting substrate like the subdural grids on the cortical surface induces modifications in the visibility of sources. Using computational modeling, Von Ellenrieder and colleagues, in 2014, have shown an important attenuation of the scalp EEG signals for generators located underneath the grid, and an amplification for generators located underneath the craniotomy [31]. These effects were spatially localized and, according to the numerical simulations, did not cancel out each other. Quantitatively, grids of 32 and 64 cm^2 can induce attenuations of 2–3 and up to 8 times, respectively. In the same way, cortical sources of 4–8 cm^2 produce scalp EEG potentials of the same maximum amplitude as generators of 10–20 cm^2 located under the center of a subdural grid. Finally, simultaneous EEG-ECoG studies can overestimate the minimum cortical extent necessary to produce visible scalp EEG activity. In simultaneous EEG-SEEG recordings, this effect is much less pronounced due to smaller intracranial electrodes (needle) and skull holes.

In the first simultaneous EEG-SEEG study described above [14], the authors mentioned that the ratio was highly variable and varied between 58:1 and 2:1 in the same patient. In the next studies, ratios from 10:1 to 2:1 were proposed by some authors [32–34] who compared the average EEG amplitudes recorded during routine scalp EEG with those recorded during electrocorticography. A recent study investigated this question using simultaneous s-iEEG recordings during complex visual stimulation, with localized face-selective brain sources in the occipital cortex of a single case [35]. To precisely quantify the amplitude and SNR in s-iEEG, the authors performed quantitative analyses of evoked EEG signals acquired during periodic stimulation ('frequency-tagging') [36–38] (see also Chap. 31). Thanks to this deterministic method, they observed scalp evoked EEG amplitudes on lateral occipital electrodes (PO8 and O2) between 7.7 and 9.9 times smaller than the corresponding intracerebral signals in the inferior occipital gyrus (Fig. 12.3). Interestingly, the SNR at the visual stimulation frequency (6 Hz in this study) on the scalp was relatively less attenuated (i.e., between 2 and 3.5 times) than the absolute amplitude compared to intracerebral recordings. This was due to the amplitude being comparatively more attenuated from intracerebral to scalp EEG recordings than the mean background (i.e., noise) amplitude in frequency bins around the 6 Hz stimulation frequency (ratio of intracerebral to scalp noise = 3.4). A main factor contributing to this decrease of the ratio between amplitudes and SNR was that, rather than being calculated over a pre-stimulus baseline as in standard evoked potentials studies, the noise (i.e., the electrophysiological background) was calculated within a small range frequency around the signal of interest (6 Hz, theta band). Therefore, scalp EEG noise was exempt from alpha activity, environmental noise, physiological artifacts (eyes and muscles activity) which greatly contaminate scalp EEG signals. Finally, the cortical extent to which a scalp EEG electrode is sensitive to is larger than that of an SEEG electrode. Noise in scalp EEG signals might be smaller compared to SEEG signals due to averaging of uncorrelated noise over a larger surface.

Several biophysical parameters can affect the amplitude ratio. Indeed, the potential field, V_e, recorded between a brain source (considered as a current dipole) and an EEG electrode, e, can be estimated as follows [39]:

$$V_e = \frac{\left|\overrightarrow{M}\right| \cos\theta}{4\pi\sigma d^2}$$

where $\left|\overrightarrow{M}\right|$ is the current dipole moment (i.e., the intensity of the brain source), θ the angle between the dipole axis and the vector r to the EEG electrode e, σ the conductivities of the volume conduction (i.e., the head tissues) and d the distance from the dipole (i.e., the brain source) to a measurement point (i.e., the EEG electrode, e).

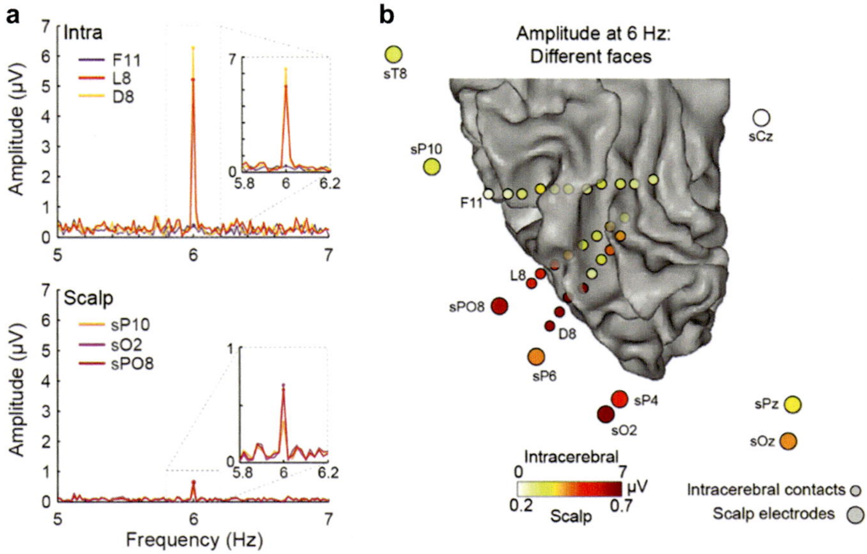

Fig. 12.3 (from [35]): Intracerebral and scalp EEG responses to visual face presentation in the frequency domain. Face stimuli were presented during fast periodic visual stimulation (sequences of 70 s using a sinusoidal contrast modulation at a rate of 6 Hz, i.e., one face stimulus every 166.66 ms). **a** Amplitude spectrum measured during fast periodic visual stimulation (FPVS) at 6 Hz with a fast frequency transform (resolution: 0.02 Hz) at the three most external intracerebral contacts of electrodes F, L, and D (top) and at three right occipito-temporal scalp electrodes (bottom). The two graphs are displayed with the same amplitude scale to compare the amplitude difference between intracerebral and scalp EEG recordings of the same visual response. Amplitude and SNR ratios of around 8:1 and 3:1, respectively, were found between scalp and intracerebral EEG signals. **b** Ventral view of the posterior half of the right hemisphere (white matter surface, the gray matter is shown as a dotted gray outline) with intracerebral contacts (small circles) and selected surrounding scalp electrodes (large circles). Circles are colored as a function of the 6 Hz-amplitude responses. Note the difference in the color scale used for intracerebral and scalp data

In a same subject, scalp and intracranial electrodes will record EEG signals with different amplitudes due to their respective distances d and angles θ with respect to the brain source. In a cohort with different subjects, the ratio is even more challenging to determine because the head tissue conductivities (σ) vary. For example, a recent computational study demonstrated that ignoring the cerebrospinal fluid in a head model leads to an overestimation of EEG SNR values [40]. Finally, whatever the study context, a global amplitude ratio is genuinely difficult to estimate in humans in-vivo because it depends on the combination of several variables (depth, position, extent, intensity, orientation) from one considered brain source. In several studies [17, 40], it has been found that the SNR of EEG signals is maximal for brain sources in the sulcal valleys or at the gyri crows where the orientations are radial to the scalp surface and minimal for brain sources with tangential orientation (e.g., those in the sulcal banks). These studies also reported, for a constant intensity and extent of a brain source placed in head model, a weak modulation of the EEG SNR values according to the source depths and angles and especially a slow SNR decrease when the depth increased.

A precise ratio estimation (for both amplitude and SNR) in human in-vivo will require a large simultaneous s-iEEG database with a high spatial resolution sampling and a lot of different brain source configurations. Cohort or meta-analysis studies with the same methodology could be conducted for this purpose.

Finally, the simultaneous combination of intracranial and scalp EEG recordings is the only "ground-truth" that allows one to investigate the influence of each of these configurations and that can provide guidelines and future directions to detect new scalp EEG biomarkers. The addition of scalp EEG is not superfluous but can help to see or detect brain sources that are not sampled by intracranial EEG recordings. Another challenge has also to be considered in cognitive neuroscience because sparse brain networks with different activated (or even co-activated) brain sources may contribute to a cognitive task. Thus, the precise comparison of intracranial and scalp EEG recordings cannot rely on a focal and unique brain source (i.e., a single equivalent current dipole model). To tackle this issue, the different brain sources need to be separated in order to estimate their respective contributions in scalp EEG recordings. This introduces a new section of this chapter regarding the key findings and advantages of adding scalp EEG electrodes during intracranial EEG investigation of cognitive processes.

12.3 Applications in Cognitive Neurosciences and Advantages

Two different types of findings can be obtained thanks to the addition of scalp EEG electrodes.

First, the identification of intracranial or scalp EEG patterns using one of them as a reference. In other words, the aim is to find new electrophysiological signals at different scales in order to better interpret them and develop specific methods to extract and analyze them. This correlation study is not achievable without the use of simultaneous s-iEEG recordings.

Second, the anatomical localization of brain sources that produced common scalp event-related potentials (ERPs), i.e., localization of brain sources using intracranial EEG recordings. This information is crucial because it can improve our knowledge about the neural basis of scalp ERPs to understand the spatiotemporal dynamics of cognition (fundamental research goal) and help to non-invasively identify brain regions that would produce abnormal ERPs (clinical goal).

Concerning the first point (i.e., the identification of new EEG signals), there are highly interesting results in sleep studies. Indeed, the sleep EEG biomarkers (i.e., change in occipital alpha waves, sleep spindles, K-complexes, delta waves, etc.) are exclusively described and taught using scalp EEG recordings. Staging sleep cycles using iEEG recordings alone is really complex, if not impossible, due to the limited spatial coverage and the signal difference in amplitude and SNR of the iEEG recordings. However, when scalp electrodes are combined with iEEG recordings, they provide crucial information for staging sleep. Thanks to this combination, hippocampal EEG activity has been described during rapid eye movement (REM) and non-REM sleep (N2 and N3 stages) [41]. Using spectral analysis of SEEG recordings in six epileptic patients, it was demonstrated that the absolute power of hippocampal activity was different during N2 and N3 stage. The power of SEEG signals in the hippocampus decreased between the first and the second sleep cycle, suggesting a homeostatic process. In term of frequency, the hippocampal sleep spindles showed a frequency closer to fast than slow cortical spindles. This simultaneous EEG-SEEG study thus pointed out for the first time the heterogeneity of human hippocampal non-REM sleep by distinguishing hippocampal N2 from N3 sleep.

In another study with a similar methodology of frequency analysis, gamma oscillations during N3 stage (slow wave sleep) were investigated in human neocortex [42]. Interestingly, intracerebral gamma activities (both in low and high frequency bands) recorded in different brain areas had amplitude peaks that were aligned with specific phases of slow waves recorded in scalp EEG. Two types of patterns were observed according to the phase of the slow phase: broadband bursts of activity in a frequency band of 30–120 Hz that correlated with the positive peaks of EEG slow waves, and other bursts in a narrow high gamma band (around 70 Hz) only that correlated with the negative troughs of EEG slow waves and especially those localized in the temporal neocortex (Fig. 12.4). These original data provide the first human evidence that reliable gamma band activity can be observed at a macroscopic scale (i.e., scalp EEG) during sleep and that these brain activity might be associated with synchronizations of neural activity during slow oscillations. This result contributes to our understanding of brain function, and especially on off-line processing of human cortical networks. It is important to note that these intracerebral findings, providing a better understanding of how the brain works during sleep, could not be possible without the use of simultaneous scalp EEG recordings.

In neurofeedback training by self-regulation, scalp EEG records electric potentials in the vertex at a very slow frequency (<0.5 Hz). These slow potentials could be related to cognitive tasks but also to peripheral physiological artifacts such as a galvanic skin response. In a simultaneous s-iEEG study, Fumuro and colleagues, in 2018, showed that the coherence (i.e., the estimation of the relation in magnitude and phase between scalp and subdural EEG signals) of this slow EEG pattern negatively correlated with the distance between subdural and scalp electrodes [43]. A significant negative correlation was noted between the linear subdural-scalp electrode distance and the coherence value ($r = -0.91$). The authors also demonstrated that the scalp slow electric potentials recorded during the neurofeedback can derive from the cortices of high lateral convexity of the brain and not from artifacts.

Finally, a highly interesting human dataset of simultaneous scalp and intracerebral (with both depth electrodes and micro-wires) EEG recordings has been reported during a verbal working memory task [44] (see also Chap. 16). In this study, subjects performed a modified Sternberg task in which the encoding of memory items, maintenance, and recall were carried out separately over time. Thanks to this dataset,

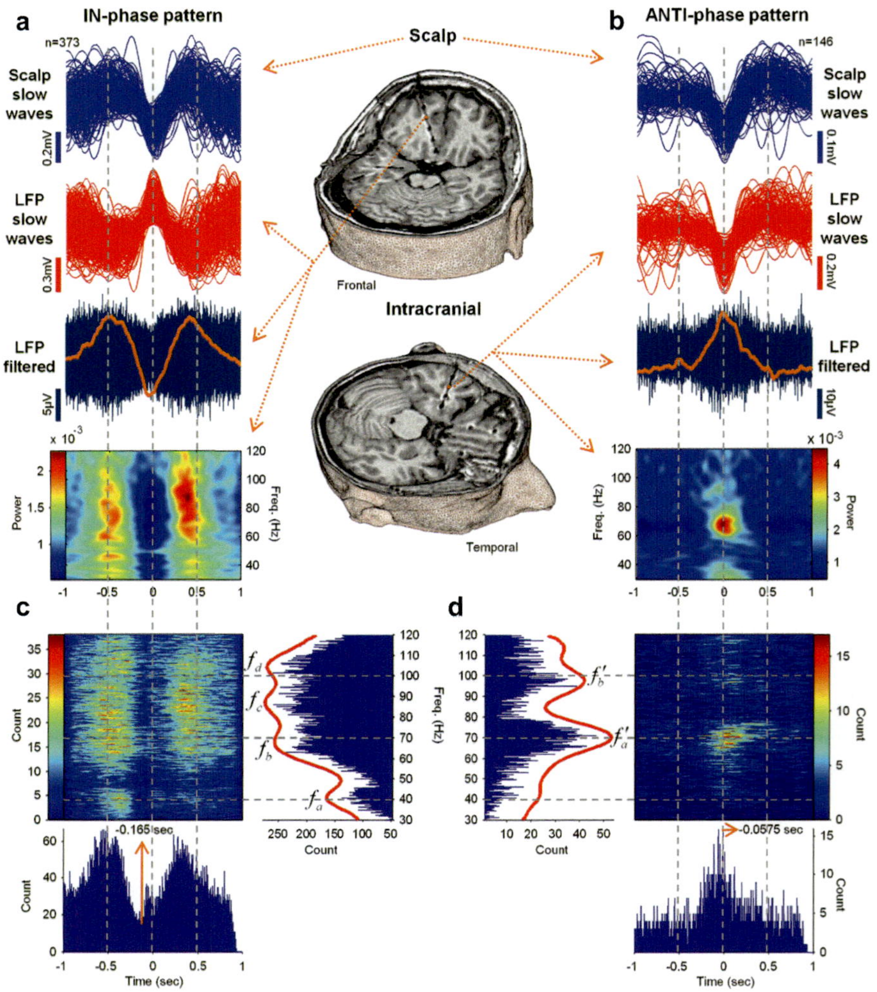

Fig. 12.4 (from [42]): Relationship between intracerebral gamma band EEG activities and slow waves scalp EEG during N3 sleep stage. Two different patterns are found and are locked to different phases of slow waves: broadband gamma activities that correlate with negative peaks of slow waves at the level of the scalp EEG (left column, IN-phase) and narrowband gamma activities (around 70 Hz) that correlate with positives peaks (ANTI-phase)

several studies could be performed, such as the analysis and the validation of the biophysical relationship between single unit recordings and local fields potentials (intracranial or scalp EEG) [45–47] or a multi-scale study (from neurons to neuronal populations and brain areas) of the electrical contribution (pattern in the time or frequency domains) of different brain sources involved in the memory processes.

Concerning the second point (i.e., source localization of functional scalp EEG components), non-invasive electrical source imaging has been around for decades, providing informative results in cognitive neuroscience [48–50]. Therefore, the usefulness of simultaneous s-iEEG recordings could be debated. However, the results of non-invasive electrical source imaging are limited due to (i) methodological reasons, such as the inaccuracies of the biophysical models (e.g., the variabilities of in-vivo head tissue conductivities; [51]) or the non-uniqueness of the solution to the inverse problem [52] and (ii) physiological reasons especially when different and sparse co-activated brain regions produce scalp ERPs. An illustration of the limits of source localization algorithms is the attempt to relate the face-evoked N170 component measured on the scalp [53] to face-selective occipito-temporal regions defined in neuroimaging studies. The vast majority of non-invasive electrical source localization studies (e.g., [54–58]) report that the N170 originates primarily from the fusiform gyrus (FG) (see also the review by [59]), which is the first and most consistent brain region associated with face-selectivity in neuroimaging studies [60–62] and also, in intracranial recordings [63, 64]. Despite the occipital projection of the N170 on the scalp surface [53, 65], a more posterior localization in the occipital cortex has rarely been considered as a potential source of the N170. In a recent study, Jacques and colleagues, in 2019, performed a correlation analysis of the N170 recorded simultaneously in s-iEEG [66]. Scalp EEG recordings identified the N170 in typical sites over the right occipito-temporal scalp regions (especially PO8 and O2 electrodes) (Fig. 12.5). Inside the brain, the N170 was observed in intracerebral electrodes localized in the right ventral and lateral sections of the occipito-temporal and posterior temporal cortex. The latency and amplitude of the scalp N170 were significantly and strongly correlated at the single-trial level with the N170 recorded in the lateral inferior occipital gyrus (IOG), i.e., close to the scalp lateral occipital surface. Most importantly, no correlation either in latency or in amplitude was found with N170 activity in the fusiform gyrus (electrode F). This study provides evidence that the IOG is a major contributor of the face-selective N170 recorded on the scalp due to its lateral position and orientation pointing to the occipital temporal scalp region. The non-significant electrical contribution of the fusiform gyrus to the scalp EEG could be due to the electrode being positioned in non-optimal localizations, but also due to the cortical geometry (down-top orientation) of that structure, projecting to the distant scalp electrodes on the vertex, with a very important amplitude attenuation.

A second example is the study of Zhang and colleagues, in 2018, who investigated how the medial temporal lobe is involved in memory retrieval with (non-simultaneous) ECoG and EEG recordings in a same cohort [67]. Three interesting points can be noticed in this study. First, despite being recorded at different times, EEG and ECoG recordings yielded concordant results about the successive cognitive processes (pre-attention, encoding, retrieving, decision making). Second, the authors found that the top five components from the canonical correlation analysis were highly correlated between intracranial and scalp EEG recordings (correlations from 0.4 to 0.87). Third and most importantly, some of these components originated from deep mesial temporal structures. This finding confirms the biophysical finding in [23] that scalp EEG recordings can detect signals from small and deep cortical structures.

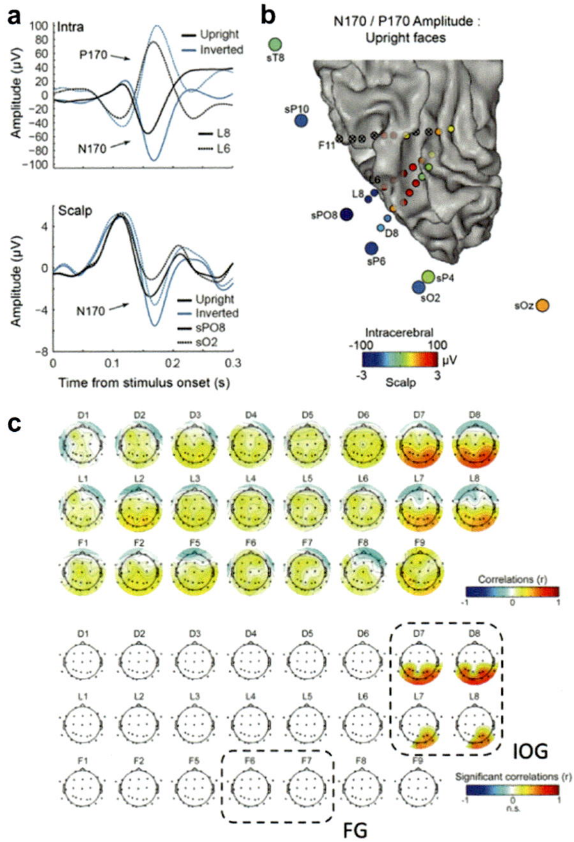

Fig. 12.5 (adapted from [66]): The inferior occipital gyrus is the main contributor to the scalp N170 component. Faces were presented in a discontinuous way (two sequences of 30 upright and inverted photographs in front view, unfamiliar faces with neutral expression). Upright and inverted faces were presented in a random order with a fixation point for 100 ms, followed between 200 and 400 ms later by a face stimulus for 300 ms. **a** Averaged event-related potentials (ERPs) to upright and inverted faces at two most external intracerebral contacts of electrodes L and D (inferior occipital gyrus) (top) and at two occipito-temporal scalp electrodes (bottom). ERPs at these two intracerebral contacts present an opposite polarity such that an N170 is measured in L8 and a P170 is measured in L6. ERPs at the scalp level are presented above. Note the scale difference between intracerebral and scalp responses but the similarity in N170 latency. **b** Same ventral view as in Fig. 12.3b. The closest scalp electrodes to intracerebral contacts D8 and L8 were sPO8, sO2, sP10 and sP6. Circles are colored as a function of the mean ERP amplitudes (N170/P170 peaks ±10 ms) for upright faces. Note the difference in the color scale used for scalp and intracerebral data. **c** Pearson correlations between intracerebral and scalp N170 amplitudes to upright faces. At the top, scalp topographical maps of the unthresholded Pearson correlation coefficients between the peak amplitude in the N170 time window measured at each intracerebral and scalp electrode. At the bottom, topographical maps showing only significant correlation coefficients ($p < 0.05$, corrected for multiple comparisons). Note that the only significant correlations were in inferior occipital gyrus (IOG) (D7-D8 and L7-L8 SEEG contacts) whereas no significant correlation was found in the fusiform gyrus (FG) (F6-F7 contacts)

Since the electric fields propagate from deep cortical structures to the scalp surface, it offers perspectives to investigate the reciprocity principle. Simply stated, the electrical path from a brain source to a recording scalp electrode is equivalent to the electrical path from a stimulating scalp electrode to the location of the brain source [68, 69]. The reciprocity principle should allow to leverage the information carried by scalp EEG recordings to guide the parameters of non-invasive transcranial stimulations. Via the addition of scalp electrodes during iEEG recordings and the application of a weak current in these electrodes (1 mA), it has been recently demonstrated in humans in-vivo that strong electrical fields (0.14 V/m in average and with a maximum value up to 0.49 V/m) can be induced in deep cortical structures like amygdala, hippocampus, or cingulate gyrus [70]. These promising results of non-invasive brain stimulations rely on the simultaneous combination of scalp and intracranial electrodes in the same subjects. In the future, transcranial electrical stimulations based on reciprocity principle should benefit from the growing development of simultaneous scalp and intracerebral EEG recordings in epilepsy units.

12.4 Conclusions

The addition of scalp EEG during presurgical intracranial EEG recordings has immediate advantages. Indeed, EEG recordings will overcome some limitations of iEEG, particularly its spatial under-sampling. First, it allows combining the focal or regional view of intracranial investigations with the global view of scalp-level investigations. This combination is particularly important to avoid misinterpretation of intracranial EEG recordings that could be due to a lack of visibility or detection of electrophysiological biomarkers. During presurgical evaluations, a precise functional mapping is required to avoid the resection of functionally relevant brain areas and thus reduce negative side effects such as cognitive impairments [71–73].

Adding scalp EEG electrodes also permits linking iEEG signals with signals from the scalp in the same individual subject. Considering that the non-invasive scalp recordings can be performed several times (before and after the intracranial recordings), they provide a robust and common standard for interpreting the intracerebral EEG signals. Indeed, if investigators can demonstrate that scalp ERPs remain stable before, during and after the invasive EEG recordings, it reinforces the validity of the cognitive biomarkers found in iEEG signals (intra-subject variability estimation). Adding scalp EEG electrodes to iEEG also permits to directly correlate the investigations performed in patients (the only population in which intracranial recordings are possible) with those performed in healthy subjects in whom intracranial recordings are not possible (inter-subject variability estimation). In the scientific literature, cognitive EEG signals are very well-described, but intracranial signals need to be more documented, with larger cohorts. Simultaneous scalp EEG correlates offer a link with the existing extensive literature about scalp ERPs and valuable information to improve the fundamental impact of cognitive intracranial EEG signals.

Looking further ahead, the addition of scalp EEG could allow to disclose new EEG signals thanks to signal analysis methods. This point is very important for clinical investigation [74] but also for cognitive neuroscience. With a precise and well-defined relationship between brain sources and EEG recording electrodes, biophysical models (especially volume conduction models) and automatic methods for scalp EEG signal extraction (blind and AI-based methods) and localization (inverse problem methods) could be validated. This approach then offers a solution to identify brain sources and to even target them by non-invasive transcranial electrical stimulations. Further down the line, these new findings coming from simultaneous s-iEEG recordings could lead to new ways to choose the number and the positioning of intracranial electrodes, or even more to avoid the placement in some brain regions where scalp EEG can record all the required information for the pre-surgical evaluation.

References

1. Johnson EL, Kam JWY, Tzovara A, Knight RT (2020) Insights into human cognition from intracranial EEG: a review of audition, memory, internal cognition, and causality. J Neural Eng 17(5):051001
2. Jasper HH (1958) The ten-twenty electrode system of the international federation. Electroencephalogr Clin Neurophysiol 10:371–375
3. Seeck M, Koessler L, Bast T, Leijten F, Michel C, Baumgartner C, He B, Beniczky S (2017) The standardized EEG electrode array of the IFCN. Clin Neurophysiol 128(10):2070–2077
4. Koessler L, Maillard L, Benhadid A, Vignal JP, Braun M, Vespignani H (2007) Spatial localization of EEG electrodes. Neurophysiol Clin 37(2):97–102
5. Brigo F, Cicero R, Fiaschi A, Bongiovanni LG (2011) The breach rhythm. Clin Neurophysiol 122(11):2116–2120
6. Datta A, Bikson M, Fregni F (2010) Transcranial direct current stimulation in patients with skull defects and skull plates: High-resolution computational FEM study of factors altering cortical current flow. Neuroimage 52:1268–1278
7. Kucewicz MT, Cimbalnik J, Matsumoto JY, Brinkmann BH, Bower MR, Vasoli V, Sulc V, Meyer F, Marsh WR, Stead SM, Worrell GA (2014) High frequency oscillations are associated with cognitive processing in human recognition memory. Brain 137(Pt 8):2231–2244
8. Lachaux JP, Axmacher N, Mormann F, Halgren E, Crone NE (2012) High-frequency neural activity and human cognition: past, present and possible future of intracranial EEG research. Prog Neurobiol 98(3):279–301
9. Ramantani G, Maillard L, Koessler L (2016) Correlation of invasive EEG and scalp EEG. Seizure 41:196–200
10. Maillard L, Koessler L, Colnat-Coulbois S, Vignal JP, Louis-Dorr V, Marie PY, Vespignani H (2009) Combined SEEG and source localisation study of temporal lobe schizencephaly and polymicrogyria. Clin Neurophysiol 120(9):1628–1636
11. Reif PS, Strzelczyk A, Rosenow F (2016) The history of invasive EEG evaluation in epilepsy patients. Seizure 41:191–195
12. Chauvel P, Gonzalez-Martinez J, Bulacio J (2019) Presurgical intracranial investigations in epilepsy surgery. Handb Clin Neurol 161:45–71
13. Klass DW, Bickford RG (1992) Reflections on the birth and early development of EEG at the Mayo Clinic. J Clin Neurophysiol 9:2–20
14. Abraham K, Marsan CA (1958) Patterns of cortical discharges and their relation to routine scalp electroencephalography. Electroencephalogr Clin Neurophysiol 10:447–461

15. Geisler CD, Gerstein GL (1961) The surface EEG in relation to its sources. Electro-encephalogr Clin Neurophysiol. 13:927–934
16. Delucchi MR, Garoutte B, Aird RB (1962) The scalp as an electroencephalographic averager. Electroencephalogr Clin Neurophysiol 14:191–196
17. Cosandier-Rimele´ D, Merlet I, Badier JM, Chauvel P, Wendling F (2008). The neuronal sources of EEG: modeling of simultaneous scalp and intracerebral recordings in epilepsy. NeuroImage 42:135–146
18. Tao JX, Ray A, Hawes-Ebersole S, Ebersole JS (2005) Intracranial EEG substrates of scalp EEG interictal spikes. Epilepsia 46(5):669–676
19. Cooper R, Winter AL, Crow HJ, Walter WG (1965) Comparison of subcortical, cortical and scalp activity using chronically indwelling electrodes in man. Electro-encephalogr Clin Neurophysiol 18:217–228
20. Von Ellenrieder N, Beltrachini L, Perucca P, Gotman J (2014) Size of cortical generators of epileptic interictal events and visibility on scalp EEG. Neuroimage 94:47–54
21. Zelmann R, Lina JM, Schulze-Bonhage A, Gotman J, Jacobs J (2014) Scalp EEG is not a blur: it can see high frequency oscillations although their generators are small. Brain Topogr 27(5):683–704
22. Burnos S, Fedele T, Schmid O, Krayenbühl N, Sarnthein J (2015) Detectability of the somatosensory evoked high frequency oscillation (HFO) co-recorded by scalp EEG and ECoG under propofol. Neuroimage Clin 14(10):318–325
23. Koessler L, Cecchin T, Colnat-Coulbois S, Vignal JP, Jonas J, Vespignani H, Ramantani G, Maillard LG (2015) Catching the invisible: mesial temporal source contribution to simultaneous EEG and SEEG recordings. Brain Topogr 28(1):5–20
24. Kokkinos V, Zaher N, Antony A, Bagić A, Mark Richardson R, Urban A (2019) The intracranial correlate of the 14&6/sec positive spikes normal scalp EEG variant. Clin Neurophysiol 130(9):1570–1580
25. Pyrzowski J, Le Douget JE, Fouad A, Siemiński M, Jędrzejczak J, Le Van QM (2021) Zero-crossing patterns reveal subtle epileptiform discharges in the scalp EEG. Sci Rep 11(1):4128
26. Jayakar P, Duchowny M, Resnick TJ, Alvarez LA (1991) Localization of seizure foci: pitfalls and caveats. J Clin Neurophysiol 8(4):414–431
27. Ebersole JS (2000) Sublobar localization of temporal neocortical epileptogenic foci by source modeling. Adv Neurol 84:353–363
28. Fahimi Hnazaee M, Wittevrongel B, Khachatryan E, Libert A, Carrette E, Dauwe I, Meurs A, Boon P, Van Roost D, Van Hulle MM (2020) Localization of deep brain activity with scalp and subdural EEG. Neuroimage 223:117344
29. Cosandier-Rimelé D, Merlet I, Bartolomei F, Badier J-M, Wendling F (2010) Computational modeling of epileptic activity: from cortical sources to EEG signals. J Clin Neurophysiol 27:465–470
30. Law SK (1993) Thickness and resistivity variations over the upper surface of the human skull. Brain Topogr 6:99–109
31. Von Ellenrieder N, Beltrachini L, Muravchik CH, Gotman J (2014) Extent of cortical generators visible on the scalp: effect of a subdural grid. Neuroimage 101:787–795
32. Penfield W, Jasper It (1954) Epilepsy and the functional anatomy of the human brain. Boston, Little Brown, p 896
33. Brazier M (1949) A study of the electrical fields at the surface of the head. II Internat. EEG Congress, Paris, pp 38–52
34. Sem-Jacobsen CW, Bickford RG, Petersen MC, Dodge HW (1953) Depth distribution of normal electroencephalographic rhythms. Proc Meet Mayo Clinic 28:156–161
35. Jacques C, Jonas J, Maillard L, Colnat-Coulbois S, Rossion B, Koessler L (2020) Fast periodic visual stimulation to highlight the relationship between human intracerebral recordings and scalp electroencephalography. Hum Brain Mapp 41(9):2373–2388
36. Adrian E, Matthews B (1934) The Berger rhythm: potential changes from the occipital lobes in man. Brain 57:355–385

37. Regan D (1966) Some characteristics of average steady-state and transient responses evoked by modulated light. Electroencephalogr Clin Neurophysiol 20:238–248
38. Norcia AM, Appelbaum LG, Ales JM, Cottereau BR, Rossion B (2015) The steady-state visual evoked potential in vision research: a review. J Vis 15(6):41–46
39. Schimpf PH, Ramon C, Haueisen J (2002) Dipole models for the EEG and MEG. IEEE Trans Biomed Eng 49(5):409–418
40. Piastra MC, Nüßing A, Vorwerk J, Clerc M, Engwer C, Wolters CH (2021) A comprehensive study on electroencephalography and magnetoencephalography sensitivity to cortical and subcortical sources. Hum Brain Mapp 42(4):978–992
41. Carpentier N, Cecchin T, Koessler L, Louis-Dorr V, Jonas J, Vignal JP, Carpentier M, Szurhaj W, Bourgin P, Maillard L (2017) Stereo-electroencephalography identifies N2 sleep and spindles in human hippocampus. Clin Neurophysiol 128(9):1696–1706
42. Valderrama M, Crépon B, Botella-Soler V, Martinerie J, Hasboun D, Alvarado-Rojas C, Baulac M, Adam C, Navarro V, Le Van QM (2012) Human gamma oscillations during slow wave sleep. PLoS ONE 7(4):e33477
43. Fumuro T, Matsuhashi M, Matsumoto R, Usami K, Shimotake A, Kunieda T, Kikuchi T, Yoshida K, Takahashi R, Miyamoto S, Ikeda A (2018) Do scalp-recorded slow potentials during neuro-feedback training reflect the cortical activity? Clin Neurophysiol 129(9):1884–1890
44. Boran E, Fedele T, Steiner A, Hilfiker P, Stieglitz L, Grunwald T, Sarnthein J (2020) Dataset of human medial temporal lobe neurons, scalp and intracranial EEG during a verbal working memory task. Sci Data 7(1):30
45. Buzsáki G, Anastassiou CA, Koch C (2012) The origin of extracellular fields and currents—EEG, ECoG, LFP and spikes. Nat Rev Neurosci 13(6):407–420
46. Bimbi M, Festante F, Coudé G, Vanderwert RE, Fox NA, Ferrari PF (2018) Simultaneous scalp recorded EEG and local field potentials from monkey ventral premotor cortex during action observation and execution reveals the contribution of mirror and motor neurons to the mu-rhythm. Neuroimage 175:22–31
47. Boran E, Fedele T, Klaver P, Hilfiker P, Stieglitz L, Grunwald T, Sarnthein J (2019) Persistent hippocampal neural firing and hippocampal-cortical coupling predict verbal working memory load. Sci Adv 5(3):eaav3687
48. Michel CM, Thut G, Morand S, Khateb A, Pegna AJ, Grave de Peralta R, Gonzalez S, Seeck M, Landis T (2001) Electric source imaging of human brain functions. Brain Res Brain Res Rev 36(2–3):108–118
49. Nunez PL, Nunez MD, Srinivasan R (2019). Multi-scale neural sources of EEG: genuine, equivalent, and representative. A tutorial review. Brain Topogr 32(2):193–214
50. Kreidenhuber R, De Tiège X, Rampp S (2019) Presurgical functional cortical mapping using electromagnetic source imaging. Front Neurol 10:628
51. Koessler L, Colnat-Coulbois S, Cecchin T, Hofmanis J, Dmochowski JP, Norcia AM, Maillard LG (2017) In-vivo measurements of human brain tissue conductivity using focal electrical current injection through intracerebral multicontact electrodes. Hum Brain Mapp 38(2):974–986
52. Dassios G, Fokas AS (2013) The definite non-uniqueness results for deterministic EEG and MEG data. Inverse Prob 29(6):065012
53. Bentin S, McCarthy G, Perez E, Puce A, Allison T (1996) Electro-physiological studies of face perception in humans. J Cogn Neurosci 8:551–565
54. Halgren E, Raij T, Marinkovic K, Jousmaki V, Hari R (2000) Cognitive response profile of the human fusiform face area as determined by MEG. Cereb Cortex 10:69–81
55. Rossion B, Joyce CA, Cottrell GW, Tarr MJ (2003) Early lateralization and orientation tuning for face, word, and object processing in the visual cortex. Neuroimage 20:1609–1624
56. Tanskanen T, Nasanen R, Montez T, Paallysaho J, Hari R (2005) Face recognition and cortical responses show similar sensitivity to noise spatial frequency. Cereb Cortex 15:526–534
57. Deffke I, Sander T, Heidenreich J, Sommer W, Curio G, Trahms L, Lueschow A (2007) MEG/EEG sources of the 170-ms response to faces are co-localized in the fusiform gyrus. Neuroimage 35:1495–1501

58. Gao Z, Goldstein A, Harpaz Y, Hansel M, Zion-Golumbic E, Bentin S (2013) A magnetoencephalographic study of face processing: M170, gamma-band oscillations and source localization. Hum Brain Mapp 34:1783–1795
59. Yovel G (2016) Neural and cognitive face-selective markers: an integrative review. Neuropsychologia 83:5–13
60. Kanwisher N, McDermott J, Chun MM (1997) The fusiform face area: a module in human extrastriate cortex specialized for face perception. J Neurosci Off J Soc Neurosci 17:4302–4311
61. Puce A, Allison T, Gore JC, McCarthy G (1995) Face-sensitive regions in human extrastriate cortex studied by functional MRI. J Neurophysiol 74:1192–1199
62. Sergent J, Ohta S, Macdonald B (1992) Functional neuroanatomy of face and object processing—a positron emission tomography study. Brain 115:15–36
63. Allison T, Ginter H, McCarthy G, Nobre AC, Puce A, Luby M, Spencer DD (1994) Face recognition in human extrastriate cortex. J Neurophysiol 71:821–825
64. Jonas J, Jacques C, Liu-shuang J, Brissart H, Colnat-Coulbois S, Maillard L (2016) A face-selective ventral occipito-temporal map of the human brain with intracerebral potentials. Proc Natl Acad Sci USA 113:E4088–E4097
65. Rossion B, Jacques C (2011) The N170: understanding the time-course of face perception in the human brain. In: Kappenman ES, Luck SJ (eds) The Oxford handbook of ERP components. Oxford University Press, New York, NY, pp 115–142
66. Jacques C, Jonas J, Maillard L, Colnat-Coulbois S, Rossion B, Koessler L (2019) The inferior occipital gyrus is a major cortical source of the face-evoked N170: evidence from simultaneous scalp and intracerebral human recordings. Hum Brain Mapp 40(5):1403–1418
67. Zhang Q, van Vugt M, Borst JP, Anderson JR (2018) Mapping working memory retrieval in space and in time: a combined electroencephalography and electrocorticography approach. Neuroimage 74:472–484
68. Rush S, Driscoll DA (1969) EEG electrode sensitivity—an application of reciprocity. IEEE Trans Biomed Eng 16:15–22
69. Dmochowski JP, Koessler L, Norcia AM, Bikson M, Parra LC (2017) Optimal use of EEG recordings to target active brain areas with transcranial electrical stimulation. Neuroimage 157:69–80
70. Louviot S, Tyvaert L, Maillard LG, Colnat-Coulbois S, Dmochowski J, Koessler L (2022) Transcranial electrical stimulation generates electric fields in deep human brain structures. Brain Stimul 15(1):1–12
71. Zumsteg D, Wieser HG (2000) Presurgical evaluation: current role of invasive EEG. Epilepsia 41(Suppl 3):S55-60
72. Taussig D, Montavont A, Isnard J (2015) Invasive EEG explorations. Neurophysiol Clin 45(1):113–119
73. Isnard J, Taussig D, Bartolomei F et al (2018) French guidelines on stereoelectroencephalography (SEEG). Neurophysiol Clin 48(1):5–13
74. Casale MJ, Marcuse LV, Young JJ, Jette N, Panov FE, Bender HA, Saad AE, Ghotra RS, Ghatan S, Singh A, Yoo JY, Fields MC (2020) The sensitivity of scalp EEG at detecting seizures—a simultaneous scalp and stereo EEG study. J Clin Neurophysiol 39(1):78–84

Chapter 13
What Are the Promises and Challenges of Simultaneous MEG and Intracranial Recordings?

Anne-Sophie Dubarry, John C. Mosher, Sarang S. Dalal, and Christian G. Bénar

Abstract Intracranial electroencephalography (iEEG) invasively measures brain activity from neurosurgical patients with higher fidelity and spatial precision than noninvasive electroencephalography (EEG) or magnetoencephalography (MEG) alone. For planning neurosurgical resection, iEEG more robustly detects lower amplitude signals that may distinguish pathological from healthy brain tissue. On the other hand, iEEG can only sample the immediate brain regions implanted for clinical reasons, while MEG synoptically measures the entire brain, albeit with lower fidelity. Relative to scalp EEG, signals recorded by MEG are less distorted by the poorly conducting skull, craniotomies, and neurosurgical hardware. By combining iEEG with simultaneous MEG recordings, we supplement the limited spatial sampling of iEEG with the superior source localization ability of MEG, yielding a combined interesting technique at two different measurement scales that can cross-validate findings from either. Setting up such simultaneous MEG-iEEG measurements involves specific considerations, and we review patient selection, patient preparation, and equipment. We then review published studies related to cognition, with emphasis on the sensitivity of MEG to source depth as well as functional connectivity between iEEG and MEG. We end with future directions opened by the unique possibility to record brain signals at different scales simultaneously.

All authors contributed equally.

A.-S. Dubarry
Aix Marseille Univ, CNRS, LNC, Marseille, France

Aix Marseille Univ, CNRS, LPL, Marseille, France

J. C. Mosher
Department of Neurology, McGovern Medical School, Texas Institute for Restorative Neurotechnologies (TIRN), University of Texas Health Science Center, Houston, Texas, USA

S. S. Dalal (✉)
Center of Functionally Integrative Neuroscience, Aarhus University, Aarhus C, Denmark
e-mail: sarang@cfin.au.dk

C. G. Bénar
Aix-Marseille Univ, INSERM, INS, Marseille, France

N. Axmacher (ed.), *Intracranial EEG*, Studies in Neuroscience, Psychology and Behavioral Economics, https://doi.org/10.1007/978-3-031-20910-9_13

13.1 Introduction and Motivation

Magnetoencephalography (MEG) is a powerful neurophysiological tool for the non-invasive investigation of brain activity at millisecond temporal precision. Clinically, however, electrodes may need to be placed intracranially for higher specificity to assist planning of neurosurgical resection, such as in the presurgical evaluation of epilepsy using electrocorticography (ECoG) and stereo-EEG (sEEG). In other cases, electrodes are chronically implanted in subcortical structures for the purpose of deep brain stimulation (DBS); these same electrodes can be alternatively used to measure local field potentials. DBS is performed most commonly in the basal ganglia for movement disorders such as Parkinson's disease, though increasingly, other applications of DBS are being investigated, such as the thalamus for essential tremor and certain forms of epilepsy.

These invasive procedures, done on purely clinical grounds, provide a unique opportunity to compare the global scale of MEG with the local scale of a ground truth recorded directly above the brain and below the dura (ECoG) or directly within the brain tissues (sEEG, DBS). The simultaneous acquisition of MEG and intracranial EEG (iEEG) represents a significant burden in terms of patient management, acquisition of signals, and data processing. Yet the recording signals of the exact same activity at different scales—local and global—provides key advantages, such as in epilepsy, where interictal discharges are spontaneous and show large variations from one event to another. Similarly, in complex cognitive protocols investigating aspects of memory or different mental strategies, the brain response cannot be assumed to be similar across repetitions of the paradigm. In either case, recording the two modalities at distinctly separate times makes it infeasible to align and average the signals in post-processing. More generally, the simultaneous recording of MEG and iEEG ensures that the brain is in the same state (vigilance, attention, etc.) at both scales.

In methodological terms, simultaneous recordings bring possibilities that are out of reach with separate recordings. In order to understand the links between depth and surface, one can quantify which intracranially observed signals are likely to be measurable extracranially by MEG. This can be detected by identifying signals with no lag between the iEEG and MEG [1, 2], implying propagation via volume conduction rather than neural connectivity (which would generally involve a lag or phase delay). Another approach is to measure trial-to-trial correlation of amplitude in the time domain (e.g., the inter-trial correlation, ITCOR introduced in [3]) or in the time–frequency domain [4]. Apart from these validation-oriented techniques, simultaneous measurements allow the computation of connectivity index between signals seen intracranially with high spatial specificity and MEG signals [5], providing a powerful means of multi-scale network analysis.

Therefore, in order to compare the exact same brain sources in the same brain states, and to take full advantage of trial-to-trial fluctuations, the invasive and non-invasive signals need to be recorded simultaneously. In this chapter, we review the technical challenges, methods, and findings of simultaneous MEG and intracranial recordings, as well as discuss future venues, with emphasis on cognitive research.

13.2 Setting up a Simultaneous Recording

In the last few years, several publications have provided detailed guidelines for the collection of MEG data [6, 7]. Clinical context involves additional challenges and simultaneous recordings require further considerations. Although simultaneous recordings of MEG and iEEG share some aspects of the context and recording procedure with clinical MEG routine [8] and iEEG procedure [9], some characteristics, which we detail here, are very specific to simultaneous recordings.

The present section describes the specific steps required to perform a simultaneous MEG-iEEG recording, including MEG site preparation, patient selection, patient preparation, and we review some of the experimental protocols found in the simultaneous MEG-iEEG literature.

13.2.1 General Considerations

One of the first considerations when deciding to perform a simultaneous recording is the accessibility of the MEG lab, particularly for inpatients in the epilepsy monitoring unit. The proximity of the MEG system facilitates the setup and the preparation of the recording session. Minimal emergency preparation is required for an MEG facility to host a simultaneous MEG-iEEG recording. Depending on the patient safety risk areas, various potential emergency situations may arise and call for best patient safety practices, including specific equipment (e.g. oxygen, suction) and documented procedures (e.g. rescue procedure, crash charts) reviewed by the MEG Medical Director.

In this context, iEEG signals can either be collected from the built-in EEG amplifier of the MEG system [3] or an external one [e.g. 10]. Both configurations have advantages and limitations. The built-in EEG system may have a limited number of channels (e.g. 64, 128) and the amplitude range may be adapted to record microvolts, whereas the iEEG signals range in the order of millivolts, requiring adjustments to the EEG gain. The built-in EEG system, usually designed for measuring scalp EEG, may need to be certified and approved by the MEG Medical Director and the Institutional Review Board for research using invasive electrodes. Nevertheless, this built-in configuration offers two main advantages: (1) reducing potential noise introduced by an external device (containing metallic parts) within the MEG environment, and (2) reducing post-recording processing, since the EEG data are digitized in the identical MEG environment. Thus a single dataset encapsulates both the MEG and iEEG signals in a single time-aligned dataset.

In other instances, however, an external EEG amplifier may be desirable. In the monitoring unit, for example, the acquisition system is typically certified for invasive clinical applications, which simplifies the procedure for obtaining institutional approval for research. The external unit may also have better gain, noise shielding, jack connectors, and channel count than that offered by the MEG system. In some instances, the analog or digital output of the EEG system can be integrated directly into the MEG system's acquisition. More often, however, the two separate acquisition systems record to two separate files, and the data need to be time-synchronised, typically by providing a periodic timing pulse into a channel of each system. As reviewed in [10], in post processing, the data need to be resampled and aligned to the same time base and filter settings, then merged into a common dataset to ease the simultaneous review of both modalities. As discussed below, the external system must also be arranged to introduce minimal additional noise into the MEG recording.

Planning the stimulation procedure involves preparing the adapted equipment in the MEG facility (e.g. headphones, speakers, relevant display, electrical stimulator). The presentation of the sequence of stimuli must be controlled by a procedure or delivered by a research software package which ensures sub millisecond precision and allows the recording of response time. The MEG facility must validate all timing with calibration measurements for the specific stimulus and computer configuration [11].

Simultaneous recordings involve placing a large amount of recording apparatus (electrodes, wires, connectors), within close reach of the MEG sensors. The presence of metal inside the MEG magnetic shielded room and nearby the sensors may thus result in higher noise levels. SEEG electrodes are usually made of platinum, while ECoG grids may be made of either platinum or nonmagnetic stainless steel. As such, the electrodes themselves, cables, and surgical hardware are typically MRI-compatible, but may nevertheless result in added MEG noise and artifacts arising from electrically conductive materials moving together with the patient's head. An example of this is shown in Fig. 1b of [3], where slow oscillatory artifacts arise from the connector placed on the patient's shoulder due to its displacement from breathing. Another source of noise comes from the amplification system that may not be perfectly isolated, and may be transmitted through the wires to the vicinity of the MEG. An example of this effect is shown in Fig. 1d of [3]. Thus, connectors need to be immobilized and removed from the patient body, and the EEG amplifier may need to be isolated from the environment (e.g., powered by battery rather than mains) [10].

13.2.2 Patient Selection

Once the hardware and MEG facility feasibility (as described in previous section) have been established, another constraint in the design of any research study is the likely selection bias of a relatively small number of patients due to their limited availability. Deep brain stimulation studies typically involve the placement of only one or two electrodes with only one pair of contacts activated, and the DBS device may or may not be capable of recording on the other remaining contacts. Thus only a small region of the brain may be available for individual or group studies, depending on the application of DBS used in the patient cohort (e.g., Parkinson's disease, essential tremor, epilepsy, depression, obsessive compulsive disorder, pain).

In contrast, ECoG and sEEG clinical studies generally span wider regions of the brain. ECoG is often performed using a silastic electrode grid placed on the cortical surface and/or sEEG depth electrodes to common targets like the hippocampus and amygdala; when used in epilepsy monitoring, ECoG surface grids can involve more than 100 electrodes spanning entire lobes. sEEG studies typically use between eight and twenty stereotactic depth probes, with five to twenty electrode contacts along each probe [9] such that simultaneous recordings of roughly 200 iEEG channels are common. Unlike DBS patients, however, these are almost always patients with epilepsy that are monitored exclusively in a hospital monitoring unit, with sensitive surgical sites, post-surgery irritability, and reduced medication level to facilitate capturing epileptiform activity. As discussed in the next section, the apparatus of sEEG channels protruding from the head makes it further difficult to find patients that will literally fit inside the rigid MEG helmet. On the other hand, the large craniotomy needed for the implantation of ECoG grids may leave the patient too vulnerable to place inside the rigid MEG helmet.

Another selection bias consideration is the medication level of the patient. The clinical examination of the iEEG patient usually involves the halting/reduction of anti-seizure medication, in order to record as many typical seizures (i.e. with typical semiology) as possible during the monitoring unit stay. Once the patient management team is satisfied that an adequate exam has been completed, the patient is returned to their medications, and explantation surgery is scheduled. Research examinations in the MEG are therefore possible in this limited time window, but with an uncertain titration of medication levels in the patient, who may also be tired and subject to brain atypical activities resulting from a series of prior seizures.

The above constraints result in a patient cohort with smaller heads (typically leading to inclusion of more females than males) or a younger pediatric cohort, with various levels of fatigue or alertness on changing medication levels, which can complicate general findings in a group study. Nonetheless, as we review below, these patients provide valuable cross-validation of studies at local and global scales of measurement, from focused invasive studies to synoptic non-invasive measurements, and the possibility to bridge the non-invasive measurements to the much broader class of non-invasive measurements of control research subjects. Put more simply, control studies with electrodes implanted in normal brains are not ethically possible, so these patient studies are invaluable, despite their unavoidable selection bias.

13.2.3 Patient Preparation at the Time of the MEG Examination

Patients with a DBS implanted may be examined in an outpatient setting, such that the procedure for preparing the patient is nearly identical to the procedures outlined in [8]. If the DBS generator is embedded in the skull, then special care may be necessary regarding any degaussing (i.e. procedure to remove minor magnetic contaminants) of the generator, and the Medical Director for the MEG site must be consulted. If the generator is considered MRI-safe, or if the generator is in the chest, then no other special handling of the patient may be necessary outside of routine preparation for a MEG exam.

Patients with either embedded depth (sEEG) electrodes or a craniotomy with cortical (ECoG) electrodes are being monitored as inpatients in a monitoring unit and therefore require special considerations and handling on the day of the MEG exam. As discussed above in Patient Selection, the opportunity to perform the MEG exam typically comes at the end of the monitoring stay, after adequate clinical data have been gathered to perform the diagnosis and/or the treatment strategy for the patient (see also Chaps. 4 and 5 on practical issues with recordings in presurgical epilepsy patients). Typically, medications are being resumed, and surgical explantation has been scheduled. We lay out the basic steps that should be considered, with an overall goal to ensure patient safety and comfort by performing many of the tasks bedside in the monitoring unit, before transport to the MEG.

First, the patient should be reconsented to ensure their continuing agreement with any research to be conducted (see also Chaps. 6 and 7 on Ethical issues). Surgical nursing is then consulted to remove the surgical head dressing, which may have been in place for many days or weeks. With the original headdress removed, the leads of the electrodes can be gathered into a bundle that drapes away from the surgical site and down the neck. The connectors should be arranged such that they land in the nape of the neck and/or outside the MEG helmet. The head can then be rewrapped in a light surgical wrapping, as shown in the figure (Fig. 13.1a, b).

At this point it is imperative to ensure that the patient and their apparatus will fit in the MEG, by using a replica helmet (provided by the MEG manufacturer) at the bedside, as shown in Fig. 13.1c. Adjustments can be made to the arrangement of the wire bundle and connectors, or it may be determined that the patient cannot fit in the MEG.

Assuming the patient fits the helmet, the MEG technician can begin the bedside process of replacing any heart or muscle (ECG or EMG) electrodes that may not be MEG-compatible, then disconnecting the patient from the monitoring room system and connecting the electrodes into the jackbox compatible with the MEG. Since the built-in EEG system of the MEG may have fewer EEG channels than the separate monitoring unit EEG system, selection of a subset of relevant electrodes has to be coordinated with the researcher, and a careful record of which contacts are retained for the MEG recording must be maintained.

Hospital transport then moves the patient to the MEG suite, with a monitoring unit nurse in attendance, where the usual patient preparation continues [8], albeit under masked (sterile) conditions. Unlike DBS outpatients, however, degaussing may not be approved by the Medical Director, since the introduction of dozens of electrodes into the brain may present unknown coupling considerations. The patient is then placed either supine or upright in the MEG, taking into account the protocols to be run, and taking into account any pressure points of the helmet with the patient's surgical sites.

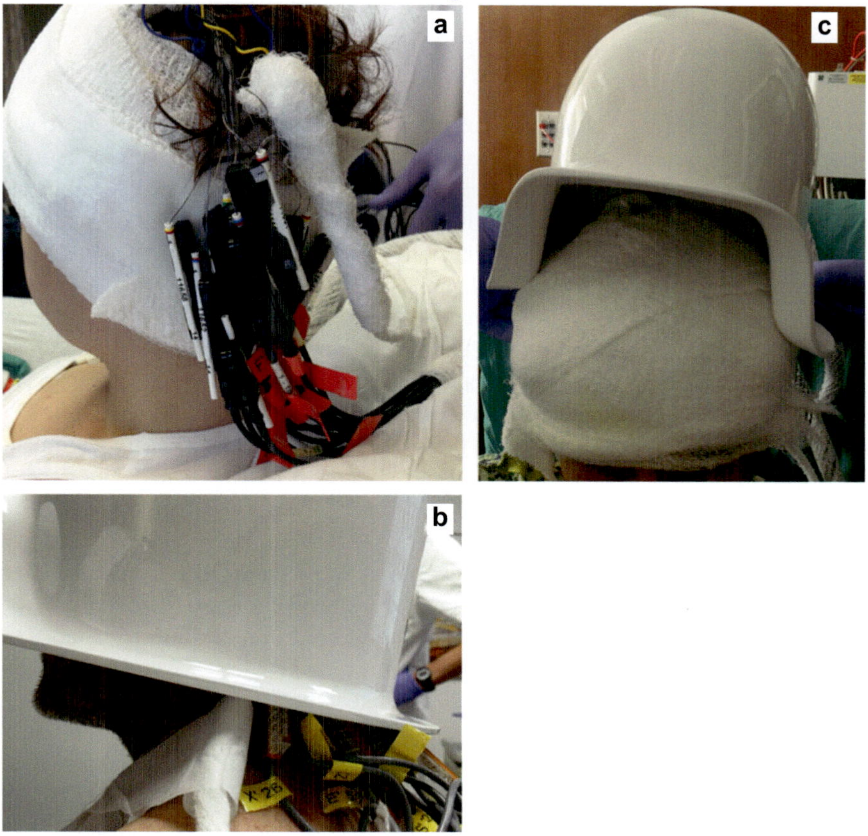

Fig. 13.1 **a** The original head dressing is removed and replaced with a light dressing, with the leads and connectors bundled together and draped away from the implantation site. **b** The nape of the neck provides a suitable place to arrange the connectors, such that they nest inside the helmet or extend just below the helmet, depending on the length of the specific brand of electrodes. **c** An exact replica of the MEG helmet is highly useful bedside to ensure that sEEG apparatus and headdress have not made it impossible for the patient to fit, as shown here

Immediately after the MEG recording is completed, the patient is directly returned to the monitoring unit for resumed care by the unit staff. Bedside, the MEG jacks and electrodes are removed, and the surgical nurse prepares a final headdress prior to explantation surgery. Depending on the timing before the explantation surgery, reconnection to the recording system in the monitoring unit may not be required.

13.2.4 Protocols with the iEEG Patient in the MEG

In this book, a wide range of possible experimental protocols are discussed, and with the general consideration laid out above, many of them could be run simultaneously with MEG. In this section, we restrict ourselves to citing a few of the published instances that have been used in simultaneous MEG-iEEG, including spontaneous recordings and evoked responses for the study of cognitive functions.

The recording of spontaneous simultaneous MEG-iEEG activity has mainly been focused on the analysis of interictal spikes in epilepsy [2, e.g. 12–14]. Because of the relatively short time of the MEG session, particularly for an sEEG patient, capturing a spontaneous seizure simultaneously with the MEG and sEEG system has not been reported to our knowledge. However, one unique study analyzed a reflexive musicogenic seizure, triggered during the recording by the playing of vocal music [15].

In contrast, studies that measured MEG simultaneously with subcortical LFPs from electrodes implanted for the purpose of DBS often investigate motor movements or associated pathology, which may be expected since DBS is most common in patients with movement disorders such as Parkinson's disease, essential tremor, and dystonia. Similar methods have also been used to examine subcortical-cortical connectivity in resting state networks. For a review, see [16].

As opposed to passive LFP recordings from deep brain electrodes, several more studies have implemented protocols measuring MEG during DBS stimulation, introducing the additional challenge of recording and analyzing the signals despite the presence of high amplitude artefact evoked by the DBS [17–19].

Less commonly, but of particular interest in this book, protocols have been designed to capture the neural activity elicited by various experimental conditions underlying specific cognitive processes, typically using protocols that were originally inspired by earlier experiments from either MEG alone or intracranial EEG alone. We will discuss these studies in greater detail in Sect. 13.3.4.

13.3 What Do Simultaneous Recordings Reveal?

13.3.1 Methodological Approaches

Classical MEG and intracranial EEG (iEEG) analyses techniques, such as evoked fields/potentials and time–frequency analysis, can be performed at the level of individual sensors. However, to explore the overall spatial information provided by the multi-sensor recordings, source reconstruction techniques are required, e.g., spatial filtering [20] or independent component analysis followed by source localization [21].

Several goals can be pursued with simultaneous recordings. One goal is to assess whether the activity measured with iEEG can be retrieved from MEG signals, either for epileptic discharges [2], evoked fields [3] or time–frequency modulations [1]. When performing correlation analysis, this boils down to finding zero-lag correlation in order to assess whether the same activity is retrieved on MEG in sEEG. Such correlation can be measured across time [2, 22] or across trials [3]. In seminal work, Dalal and colleagues [22] computed and presented the topographic maps of correlation of each sensor data with sEEG electrode in the hippocampus, showing a large-field topography that is compatible with a deep origin of the signals. Another goal is to measure delayed connectivity between an iEEG sensor and MEG signals, in order to retrieve large-scale networks and benefit from the local view of sEEG and large-scale view of MEG. This was performed by [5] using directed phase-lag index, with a seed point in the hippocampus.

13.3.2 Precision of Localization

The first tests of source localization took advantage of measuring fields generated by small currents injected in intracerebral electrodes, thus creating artificial dipoles within the head volume. The great advantage of this technique is to generate a known and well-characterized source, both in terms of location and extent, albeit with a higher amplitude and lesser spatial complexity relative to natural brain activity. Cohen and colleagues [23] injected currents in intracranial EEG electrodes while simultaneously measuring scalp EEG and MEG. No major differences in localization error were found, in contradiction with the hypothesis that MEG would yield better source localization performance. As noted in [24], the study was initially criticized on methodological grounds [25, 26]. The spatial sampling was low with only 16 channels each for MEG and EEG, which inadvertently prevented the MEG source localization from reaching its full potential; as the skull blurs MEG signals to a far lesser degree than EEG signals, MEG scalp topographies contain more nuances at higher spatial frequencies that can increase the performance of source reconstruction when sensor coverage is optimal. The exclusively radial nature of the implanted sources (due to the electrode implantation scheme) presented a strong bias favoring EEG, since MEG is much less sensitive to radial sources [27].

In [28, 29], the call was for a careful consideration of the absolute accuracies of either modality under conditions that are fair to both modalities. As noted in [30] and repeated in [29], EEG and MEG provide complementary data, and the use of both modalities can contribute to overall improved accuracy, as confirmed over a large number of theoretical cases [24].

A similar experimental study was conducted by Leahy and colleagues [31] using a human skull phantom implanted with 32 current dipoles and 64 scalp electrodes. A CT scan was used to determine ground truth, and MEG measurements were made. The results yielded a smaller error for MEG (3 mm versus 7–8 mm for EEG) which was attributed to the difficulty of accurately modelling the complexity of the human skull in EEG.

More recently, in a resting state study on patients with epilepsy, [20] measured the distance between (1) the sources found at the peak of the ICA components computed from MEG signals, and (2) the sEEG contact showing maximal correlation with component, and report a mean distance of 20 ± 12.25 mm. Two additional studies used the same localization technique on ICA components putatively corresponding to deep mesial activity, both on epileptic spikes [2] and event related responses [21]. In both cases, the confidence interval of one or two dipole scans included the mesial regions.

13.3.3 Epileptic Discharges

The first report of epileptic discharges in simultaneous MEG-sEEG from [32], compared the epileptic spikes detected on MEG recordings with the ones detected on sEEG signals in terms of detectability, amplitude and localization. The capacity of interictal MEG to detect and localize the epileptogenic zone was found to be comparable with that of sEEG when targeting convexity foci. However, the epileptiform discharges required a higher amplitude and a wider distribution to be detected and localized with only MEG signals.

Shortly after this first report, [33] provided a parametric description of MEG spikes detected thanks to sEEG, aiming at increasing the objectivity of MEG epileptiform events selection.

With a similar strategy, [12] marked the epileptic spikes on sEEG signals, and localized the sources from averaged MEG data (locked on the sEEG events). The resulting early activity was located in a plausible region, not sampled by sEEG because of physical constraints (a very posterior region where orthogonal electrodes cannot be implanted), confirming clinical hypotheses on this patient.

To investigate the visibility of high gamma oscillations on MEG, [34] performed a time–frequency analysis on the MEG signals locked to epileptic spikes that were detected from sEEG signals. The high gamma oscillations which they observed on the MEG signals were temporally aligned to the ones observed within the same frequency band on the iEEG signals [34 Supp. Fig. 2]. The oscillations formed well isolated islands in the time–frequency plane and thus do not correspond to filtered spikes, i.e. “false ripples” [35]. In a more recent study, [36] have used a beamformer analysis (i.e. a spatial filter applied to the sensor data) in order to detect and localize epileptic ripples (80–120 Hz oscillations) from MEG data. The ripples detected in MEG were validated using sEEG as a gold standard.

Finally, [2] have shown that deep epileptic discharges originating from deep mesial sources can be detected by the MEG sensors. In a first step, independent component analysis was computed on epileptic spikes measured on deep sEEG electrodes (within amygdala and hippocampus). This approach enabled separating focal deep activity from large scale (propagated) networks, whereas the analysis of the MEG signals alone showed only the propagated activity. In a second step, they have shown that in a large proportion of patients the ICA ran on the whole dataset can also extract activity from deep sources, without the prior information arising from sEEG.

13.3.4 Cognitive Potentials and Oscillations

13.3.4.1 Cortical Measurements

The first "trimodal" EEG-MEG-sEEG recording reported in a case study by [3] showed that evoked activity in response to visual presentation of a checkerboard is detected on the three recording modalities, both on average (evoked potentials/fields) and at the single trial level. The simultaneous recording enabled tracking the correlation between depth and surface fluctuations. A source analysis confirmed the consistency between the MEEG sources and the sEEG potentials. In addition, time–frequency analysis could retrieve early beta/gamma band activity (likely evoked) and alpha desynchronization. Induced gamma activations were more scarce, possibly because of the small extent of the sources activated by the experimental task.

With MEG, [5] found that decreases in theta power during spatial encoding predict greater accuracy during subsequent recall. An epilepsy patient with electrodes implanted in temporal regions allowed further investigation of this effect. By simultaneously measuring MEG and intracranial EEG, they further discovered that these theta oscillations in right anterior hippocampus and left inferior frontal phase-led the left temporal cortex (see Fig. 13.2).

These studies demonstrate different ways in which simultaneous MEG with intracranial EEG can be leveraged to validate MEG-only findings, identify novel connections that may not have been evident with either method alone, and further identify regions of interest for subsequent analyses with MEG alone.

López-Madrona et al. [21] recorded 6 patients with simultaneous MEG-sEEG during a memory task. Patients were instructed to recognize images that they previously memorized ('old/new' paradigm). A blind source separation technique (Second Order Blind Identification, SOBI) revealed MEG components sensitive to the protocol (i.e. showing evoked activity). These components showed consistent topographies across patients and were confirmed in control subjects, presenting a "large" pattern (i.e. topographies with distant positive and negative poles), suggesting the activation of a source originating from a deep origin. The source localization and the correlation analysis of the simultaneous sEEG signals revealed highest correlation with contacts located in the hippocampus and rhinal cortex (Fig. 13.3), confirming the previous findings.

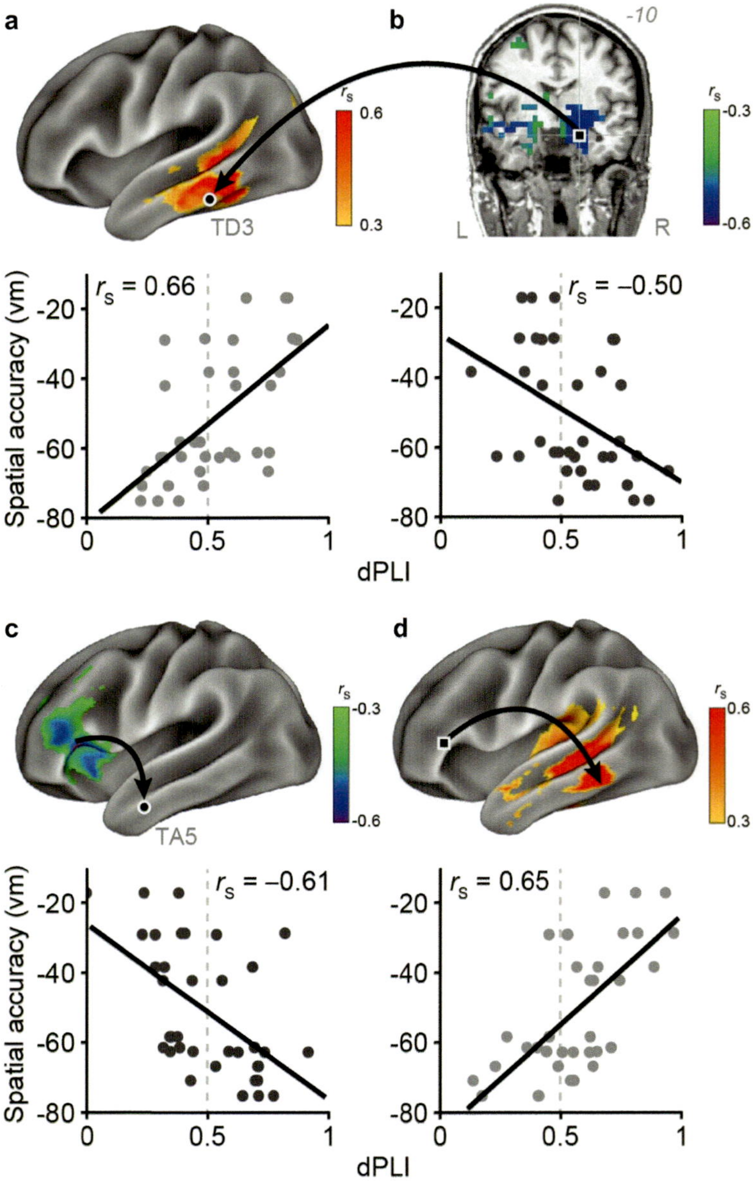

Fig. 13.2 Simultaneous MEG-iEEG analysis from a single patient to investigate connectivity between MEG sources and left temporal locations sampled directly with intracranial electrodes during a spatial encoding task, and to validate independent MEG findings. **a** *Directed* phase lag index analysis (see Chap. 32) with MEG beamformer results revealed that task performance was superior when the right hippocampus phase-led the left inferior temporal gyrus. **b** A corresponding analysis between intracranial EEG from the left temporal cortex and the right hippocampus from MEG yielded similar findings, validating the MEG results. **c** Task performance was also better when left inferior frontal gyrus, reconstructed with MEG, phase-led the anterior temporal lobe, as sampled by intracranial EEG. **d** Subsequently using the left inferior frontal gyrus as a seed for the MEG-based analyses showed that it also exhibited connectivity with the left inferior temporal gyrus, the same area found in (**a**) that exhibited connectivity with the right hippocampus. Reproduced with permission from [5]

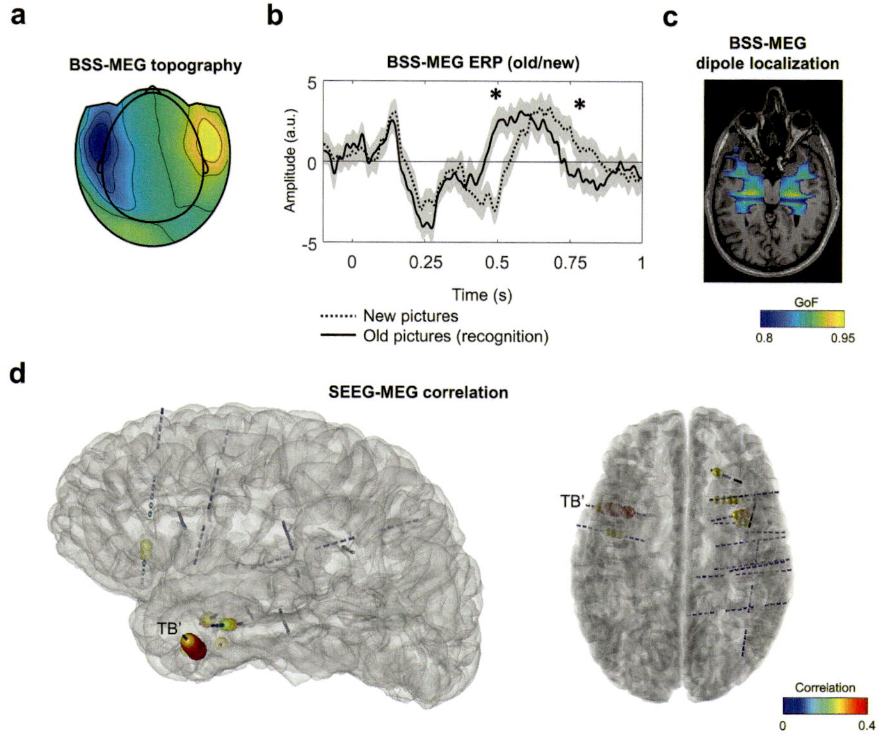

Fig. 13.3 Relation between MEG signals and simultaneously recorded intracerebral recordings in one patient [21]. **a** Topography of the deep component obtained with blind source separation (BSS) on MEG signals (BSS-MEG). This topography corresponds to the contribution of the BSS-MEG component to each MEG sensor; its broad distribution is indicative of a deep source. **b** Event-related potential (ERP) on BSS-MEG. Solid and dotted traces are the averaged ERP for old (recognition) and new pictures, respectively. Stars indicate statistically significant differences in amplitude between old and new trials. There is a clear memory-related effect, consistent with what is expected from previous intracerebral studies. **c** Source localization of the BSS-MEG topography with two symmetric dipoles, which confirms the deep origin of the signals (GoF: Goodness of Fit of the two dipoles). **d** Distribution of the correlation of BSS-MEG signals with intracerebral EEG. The highest correlation is visible on the TB (temporo-basal) electrode that targets deep mesial structures, with the peak of correlation located in the rhinal cortex. (Figure courtesy of Victor López-Madrona)

13.3.4.2 MEG-LFP of Basal Ganglia/STN via DBS Electrodes

DBS electrodes are most commonly implanted in the basal ganglia of patients with movement disorders such as Parkinson's disease. Naturally, the effects of DBS itself on cortical activity and cognition have been investigated in several studies [e.g. 37, 38] and reviews [16, 19]. Before permanent connection to the stimulator, however, DBS electrodes can be alternatively used to measure local field potentials from the implanted structure; a few groups have managed to combine such measurements with MEG.

Most such simultaneous measurements have investigated pathological oscillations or aspects of motor control. However, evidence is mounting that the basal ganglia and other DBS targets such as the thalamus are indeed involved in several other brain functions, including various aspects of cognition.

The first cognitive study employing simultaneous MEG-STN measurements came out recently. Patai et al. [39] employed an expanded judgement task in which participants needed to accumulate multiple observations of a cue with two possibilities, before deciding which of the two possibilities was the accurate choice. The study primarily aimed to investigate whether the STN was involved with "global conflict"—when a cue conflicts with several preceding ones—but did not find evidence to support that role. They found that beta oscillations in both the STN and frontal cortex instead encoded "local conflict"—i.e., when the presented cue differed from the immediately preceding one—but the beta activity in the STN persisted until the next cue, while the cortical activity subsided more quickly. They also specifically found alpha and beta band connectivity between the right dorsal premotor cortex and right STN for these conflicts. Although they could not reliably determine directionality of this relationship, the cortical activity peaked earlier and is in line with other studies which suggest that the cortical activity drives STN activity.

13.3.4.3 MEG-LFP of Thalamus via DBS Electrodes

In a pioneering study, [40] combined MEG with thalamic iEEG measurements to investigate corticothalamic circuits mediating visual perception. They found that the phase of low-frequency oscillations in the mediodorsal thalamus (7–9 Hz) predicts whether threshold-level visual stimuli were perceived. Leveraging MEG, they furthermore discovered that prefrontal cortex activity precedes these thalamic responses, as assessed by directed connectivity measures, suggesting that corticothalamic interactions ultimately mediate perceptual performance. They also found some evidence that visual cortex activity follows the thalamic responses, though did not have adequate occipital MEG coverage in enough patients to make stronger conclusions. As direct investigations of the human thalamus are rare, this provides important insights into the role of corticothalamic interactions into perceptual cognition, and an impetus for further investigations integrating both thalamic measurements with MEG.

13.4 Discussion and Future Avenues

Most effort so far in simultaneous MEG and intracranial recordings has been directed towards epileptic activity, where the spontaneous aspect of the discharges requires simultaneity in order to ensure capturing the exact same brain activity at the two levels. In cognition, the added value of simultaneous recordings may not be so obvious at a first glance. Many studies have performed cognitive protocols in intracerebral EEG alone, and may serve as a basis for assessing MEG results [e.g. 41–44]. Still, simultaneous recordings have distinct qualities that may justify the (significant) added difficulty during acquisition. Firstly, as for epilepsy, they ensure the exact same patient state (vigilance, arousal, level of medication, etc.), which can be important for subtle activity or in protocols where repetition of the same stimuli may result in different brain responses. Secondly, simultaneous recordings allow performing correlation between surface and depth at a single-trial level, which gives a stronger confirmation (in contrast to average across trials) that MEG and iEEG are indeed recording the same brain source. Finally, simultaneous recordings may allow in the future to build a "meta-modality" that combines the local view from iEEG and the global view from MEG, thus improving our knowledge of brain function across spatial scales. Of course, simultaneous recordings can only be performed in patients, thus presenting pathological activity intermingled with physiological one, and potentially reorganized brain networks. Hence the importance of combining the results from multiple patients with varying epileptic sources [45] and confirming the simultaneous MEG results with activity measured in control subjects with MEG only [21].

MEG technology is now rapidly evolving, with next-generation MEG systems employing optically pumped magnetometers (OPM) that operate without liquid helium and allow closer positioning on the head [46]. As OPMs can be placed on the head in any desired configuration rather than a rigid unisize helmet, this will relieve some of the challenges with obtaining measurements simultaneously with iEEG. Indeed, the first simultaneous measurements of OPM-MEG and iEEG for detection of interictal spikes were recently presented [47]. The impending proliferation of OPM-MEG systems will surely provide more such opportunities, with the aims of characterizing their sensitivity and source localization accuracy as well as providing further insights into functional connectivity mediating cognitive processes.

References

1. Dalal SS, Baillet S, Adam C, Ducorps A, Schwartz D, Jerbi K, Bertrand O, Garnero L, Martinerie J, Lachaux J-P (2009) Simultaneous MEG and intracranial EEG recordings during attentive reading. Neuroimage 45:1289–1304. https://doi.org/10.1016/j.neuroimage.2009.01.017
2. Pizzo F, Roehri N, Medina Villalon S, Trébuchon A, Chen S, Lagarde S, Carron R, Gavaret M, Giusiano B, McGonigal A, Bartolomei F, Badier JM, Bénar CG (2019) Deep brain activities can be detected with magnetoencephalography. Nat Commun 10:971. https://doi.org/10.1038/s41467-019-08665-5
3. Dubarry A-S, Badier J-M, Trébuchon-Da Fonseca A, Gavaret M, Carron R, Bartolomei F, Liégeois-Chauvel C, Régis J, Chauvel P, Alario F-X, Bénar C-G (2014) Simultaneous recording of MEG, EEG and intracerebral EEG during visual stimulation: From feasibility to single-trial analysis. Neuroimage 99:548–558. https://doi.org/10.1016/j.neuroimage.2014.05.055
4. Barborica A, Mindruta I, Sheybani L, Spinelli L, Oane I, Pistol C, Donos C, López-Madrona VJ, Vulliemoz S, Bénar C-G (2021) Extracting seizure onset from surface EEG with independent

component analysis: insights from simultaneous scalp and intracerebral EEG. NeuroImage Clin 32:102838. https://doi.org/10.1016/j.nicl.2021.102838
5. Crespo-García M, Zeiller M, Leupold C, Kreiselmeyer G, Rampp S, Hamer HM, Dalal SS (2016) Slow-theta power decreases during item-place encoding predict spatial accuracy of subsequent context recall. Neuroimage 142:533–543. https://doi.org/10.1016/j.neuroimage.2016.08.021
6. Gross J, Baillet S, Barnes GR, Henson RN, Hillebrand A, Jensen O, Jerbi K, Litvak V, Maess B, Oostenveld R, Parkkonen L, Taylor JR, van Wassenhove V, Wibral M, Schoffelen J-M (2013) Good practice for conducting and reporting MEG research. Neuroimage 65:349–363. https://doi.org/10.1016/j.neuroimage.2012.10.001
7. Hari R, Baillet S, Barnes G, Burgess R, Forss N, Gross J, Hämäläinen M, Jensen O, Kakigi R, Mauguière F, Nakasato N, Puce A, Romani G-L, Schnitzler A, Taulu S (2018) IFCN-endorsed practical guidelines for clinical magnetoencephalography (MEG). Clin Neurophysiol 129:1720–1747. https://doi.org/10.1016/j.clinph.2018.03.042
8. Mosher JC, Funke ME (2020) Towards best practices in clinical magnetoencephalography: patient preparation and data acquisition. J Clin Neurophysiol 37:498–507. https://doi.org/10.1097/WNP.0000000000000542
9. Mercier MR, Dubarry A-S, Tadel F, Avanzini P, Axmacher N, Cellier D, Vecchio MD, Hamilton LS, Hermes D, Kahana MJ, Knight RT, Llorens A, Megevand P, Melloni L, Miller KJ, Piai V, Puce A, Ramsey NF, Schwiedrzik CM, Smith SE, Stolk A, Swann NC, Vansteensel MJ, Voytek B, Wang L, Lachaux J-P, Oostenveld R (2022) Advances in human intracranial electroencephalography research, guidelines and good practices. Neuroimage 119438. https://doi.org/10.1016/j.neuroimage.2022.119438
10. Badier JM, Dubarry AS, Gavaret M, Chen S, Trébuchon AS, Marquis P, Régis J, Bartolomei F, Bénar CG, Carron R (2017) Technical solutions for simultaneous MEG and SEEG recordings: towards routine clinical use. Physiol Meas 38:N118–N127. https://doi.org/10.1088/1361-6579/aa7655
11. Bridges D, Pitiot A, MacAskill MR, Peirce JW (2020) The timing mega-study: comparing a range of experiment generators, both lab-based and online. PeerJ 8:e9414. https://doi.org/10.7717/peerj.9414
12. Gavaret M, Dubarry A, Carron R, Bartolomei F (2016) Simultaneous SEEG-MEG-EEG recordings Overcome the SEEG limited spatial sampling. Epilepsy Res 128:68–72. https://doi.org/10.1016/j.eplepsyres.2016.10.013
13. Kakisaka Y, Kubota Y, Wang ZI, Piao Z, Mosher JC, Gonzalez-martinez J, Jin K, Alexopoulos AV, Burgess RC (2012) Utility of simultaneous depth and MEG recording may provide complementary information regarding the epileptogenic region. 14:1–6
14. Oishi M, Otsubo H, Kameyama S, Morota N, Masuda H, Kitayama M, Tanaka R (2002) Epileptic spikes: magnetoencephalography versus simultaneous electrocorticography. Epilepsia 43:1390–1395. https://doi.org/10.1046/j.1528-1157.2002.10702.x
15. Wang ZI, Jin K, Kakisaka Y, Burgess RC, Gonzalez-Martinez JA, Wang S, Ito S, Mosher JC, Hantus S, Alexopoulos AV (2012) Interconnections in superior temporal cortex revealed by musicogenic seizure propagation. J Neurol 259:2251–2254. https://doi.org/10.1007/s00415-012-6556-9
16. Litvak V, Florin E, Tamás G, Groppa S, Muthuraman M (2021) EEG and MEG primers for tracking DBS network effects. Neuroimage 224:117447. https://doi.org/10.1016/j.neuroimage.2020.117447
17. Gopalakrishnan R, Burgess RC, Malone DA, Lempka SF, Gale JT, Floden DP, Baker KB, Machado AG (2018) Deep brain stimulation of the ventral striatal area for poststroke pain syndrome: a magnetoencephalography study. J Neurophysiol 119:2118–2128. https://doi.org/10.1152/jn.00830.2017
18. Oswal A, Jha A, Neal S, Reid A, Bradbury D, Limousin P, Foltynie T, Zrinzo L, Hariz M, Brown P, Litvak V (2015) Analysis of simultaneous MEG and intracranial LFP recordings during deep brain stimulation: a protocol and experimental validation. J Neurosci Methods 29–46. https://doi.org/10.1016/j.jneumeth.2015.11.029

19. Harmsen IE, Rowland NC, Wennberg RA, Lozano AM (2018) Characterizing the effects of deep brain stimulation with magnetoencephalography: a review. Brain Stimul 11:481–491. https://doi.org/10.1016/j.brs.2017.12.016
20. Velmurugan J, Badier J-M, Pizzo F, Medina Villalon S, Papageorgakis C, López-Madrona V, Jegou A, Carron R, Bartolomei F, Bénar C (2022) Virtual MEG sensors based on beamformer and independent component analysis can reconstruct epileptic activity as measured on simultaneous intracerebral recordings (Submitted)
21. López-Madrona VJ, Medina Villalon S, Badier J, Trébuchon A, Jayabal V, Bartolomei F, Carron R, Barborica A, Vulliémoz S, Xavier AF, Bénar CG (2022) Magnetoencephalography can reveal deep brain network activities linked to memory processes. Human Brain Mapp hbm.25987. https://doi.org/10.1002/hbm.25987
22. Dalal SS, Jerbi K, Bertrand O, Adam C, Ducorps A, Schwartz D, Martinerie J, Lachaux J-P (2013) Simultaneous MEG-intracranial EEG: new insights into the ability of MEG to capture oscillatory modulations in the neocortex and the hippocampus. In: Epilepsy, cognition, and neuropsychiatry (Epilepsy, brain, and mind, Part 2). https://doi.org/10.1016/j.yebeh.2013.03.012.
23. Cohen D, Cuffin BN, Yunokuchi K, Maniewski R, Purcell C, Cosgrove GR, Ives J, Kennedy JG, Schomer DL (1990) MEG versus EEG localization test using implanted sources in the human brain. Ann Neurol 28:811–817. https://doi.org/10.1002/ana.410280613
24. Mosher JC, Spencer ME, Leahy RM, Lewis PS (1993) Error bounds for EEG and MEG dipole source localization. Electroencephalogr Clin Neurophysiol 86:303–321. https://doi.org/10.1016/0013-4694(93)90043-U
25. Hari R, Hämäläinen M, Ilmoniemi R, Lounasmaa O (1991) MEG versus EEG localization test. Ann Neurol 30:222–222. https://doi.org/10.1002/ana.410300220
26. Williamson SJ (1991) MEG versus EEG localization test. Ann Neurol 30:222–222. https://doi.org/10.1002/ana.410300220
27. Hillebrand A, Barnes GR (2002) A quantitative assessment of the sensitivity of whole-head MEG to activity in the adult human cortex. Neuroimage 16:638–650. https://doi.org/10.1006/nimg.2002.1102
28. (1992) Assessment: magnetoencephalography (MEG). Neurology 42:1–4. https://doi.org/10.1212/WNL.42.1.1
29. Anoglanakls G, Badler JM, Barrett G, Ern S, Fenlci R, Fenwick P, Grandorl F, Harl R, Ilmonleml R, Maugulre F, Lehmann D, Perrln F, Peters M (1992) A consensus statement on relative merits of EEG and MEG. Editor Electroenceph Clin Neurophysiol 82:317–319
30. David D, Cuffin BN (1983) Demonstration of useful differences between magnetoencephalogram and electroencephalogram. Electroencephalogr Clin Neurophysiol 56:38–51. https://doi.org/10.1016/0013-4694(83)90005-6
31. Leahy RM, Mosher JC, Spencer ME, Huang MX, Lewine JD (1998) A study of dipole localization accuracy for MEG and EEG using a human skull phantom. Electroencephalogr Clin Neurophysiol 15
32. Santiuste M, Nowak R, Russi A, Tarancon T, Oliver B, Ayats E, Scheler G, Graetz G (2008) Simultaneous magnetoencephalography and intracranial EEG registration: technical and clinical aspects. J Clin Neurophysiol 25:331–339. https://doi.org/10.1097/WNP.0b013e31818e7913
33. Nowak R, Santiuste M, Russi A (2009) Toward a definition of MEG spike: parametric description of spikes recorded simultaneously by MEG and depth electrodes. Seizure 18:652–655. https://doi.org/10.1016/j.seizure.2009.07.002
34. Rampp S, Kaltenhäuser M, Weigel D, Buchfelder M, Blümcke I, Dörfler A, Stefan H (2010) MEG correlates of epileptic high gamma oscillations in invasive EEG: MEG correlates of epileptic oscillations. Epilepsia 51:1638–1642. https://doi.org/10.1111/j.1528-1167.2010.02579.x
35. Bénar CG, Chauvière L, Bartolomei F, Wendling F (2010) Pitfalls of high-pass filtering for detecting epileptic oscillations: a technical note on "false" ripples. Clin Neurophysiol 121:301–310. https://doi.org/10.1016/j.clinph.2009.10.019

36. Migliorelli C, Alonso JF, Romero S, Nowak R, Russi A, Mañanas MA (2017) Automated detection of epileptic ripples in MEG using beamformer-based virtual sensors. J Neural Eng 14:046013. https://doi.org/10.1088/1741-2552/aa684c
37. Bahners BH, Florin E, Rohrhuber J, Krause H, Hirschmann J, van de Vijver R, Schnitzler A, Butz M (2020) Deep brain stimulation does not modulate auditory-motor integration of speech in Parkinson's disease. Front Neurol 11:655. https://doi.org/10.3389/fneur.2020.00655
38. Hyder R, Højlund A, Jensen M, Johnsen EL, Østergaard K, Shtyrov Y (2021) STN-DBS affects language processing differentially in Parkinson's disease: multiple-case MEG study. Acta Neurol Scand 144:132–141. https://doi.org/10.1111/ane.13423
39. Patai ZE, Foltynie T, Limousin P, Akram H, Zrinzo L, Bogacz R, Litvak V (2020) Conflict detection in a sequential decision task is associated with increased cortico-subthalamic coherence and prolonged subthalamic oscillatory response in the beta band. Neuroscience
40. Griffiths BJ, Zaehle T, Repplinger S, Schmitt FC, Voges J, Hanslmayr S, Staudigl T (2022) Rhythmic interactions between the mediodorsal thalamus and prefrontal cortex precede human visual perception. Neuroscience
41. Barbeau EJ, Taylor MJ, Regis J, Marquis P, Chauvel P, Liégeois-Chauvel C (2008) Spatio temporal dynamics of face recognition. Cerebral cortex (New York, NY : 1991) 18:997–1009. https://doi.org/10.1093/cercor/bhm140
42. Lachaux J-P, George N, Tallon-Baudry C, Martinerie J, Hugueville L, Minotti L, Kahane P, Renault B (2005) The many faces of the gamma band response to complex visual stimuli. Neuroimage 25:491–501. https://doi.org/10.1016/j.neuroimage.2004.11.052
43. Dubarry AS, Llorens A, Trébuchon A, Carron R, Liégeois-Chauvel C, Bénar CG, Alario FX (2017) Estimating parallel processing in a language task using single-trial intracerebral electroencephalography. Psychol Sci 28:414–426. https://doi.org/10.1177/0956797616681296
44. Jerbi K, Ossandón T, Hamamé CM, Senova S, Dalal SS, Jung J, Minotti L, Bertrand O, Berthoz A, Kahane P, Lachaux J-P (2009) Task-related gamma-band dynamics from an intracerebral perspective: review and implications for surface EEG and MEG. Hum Brain Mapp 30:1758–1771. https://doi.org/10.1002/hbm.20750
45. Dubarry A-S, Liégeois-Chauvel C, Trébuchon A, Bénar C, Alario F-X (2022) An open-source toolbox for Multi-patient Intracranial EEG Analysis (MIA). Neuroimage 257:119251. https://doi.org/10.1016/j.neuroimage.2022.119251
46. Brookes MJ, Leggett J, Rea M, Hill RM, Holmes N, Boto E, Bowtell R (2022) Magnetoencephalography with optically pumped magnetometers (OPM-MEG): the next generation of functional neuroimaging. Trends Neurosci 45:621–634. https://doi.org/10.1016/j.tins.2022.05.008
47. Badier J, Schwartz D, Bonini F, Bénar C, Kansari K, Daligault S, Bartolomei F, Jung J, Carron R, Mitryukovskiy S, Fourcault W, Le Prado M, Palacios-Laloy A, Labyt E (2022) Evidence of interictal spikes recording by Helium OPM validated by simultaneous intracerebral EEG. In: 10th European Conference on Clinical Neuroimaging

Chapter 14
Why and How Should I Track Eye-Movements During iEEG Recordings?

Benjamin J. Griffiths and Tobias Staudigl

Abstract We are visual animals. How we perceive, understand and interact with the world is intimately tied to our visual sense. Yet, the value of monitoring ocular activity in neuroscientific experiments is often overlooked. In this chapter, we set out to highlight how a whole host of ocular phenomena relate to brain function and human cognition, with a special focus on intracranial electroencephalogram (iEEG) recordings. We begin by describing key ocular events, such as saccades and fixations, before discussing the extensive impact these ocular events have on common neural phenomenon and measurable behaviour. Lastly, we provide practical recommendations for combining eye tracking and intracranial EEG in neuroscientific research.

14.1 Anatomy and Activity of the Human Eye

Our eyes serve to receive incoming light, transform said light into tangible information, and project this information towards the central nervous system. The process begins with light hitting the pupil, which is then projected onto a small area of the retina called the fovea. The fovea is capable of extracting a highly-detailed image of whatever the eye is fixated upon (though the size of this image is very small due to the small size of the fovea [~1.5 mm; 0.25 mm at its most sensitive area]). Complementing the highly-detailed foveal image is a coarser surrounding image generated by the parafovea (a region of the retina that surrounds the fovea). Together, the information extracted by the fovea and parafovea is projected to the central nervous system via the optic nerve. From here, the signal is passed to the occipital cortex via the thalamus, where an image is constructed and then passed across a wide range of brain regions for additional action (e.g., interpretation, response).

Given the limited diameter of the fovea, ocular movements are essential for building a comprehensive image of our immediate environments. Such exploration is principally achieved through rapid, ballistic eye movements known as saccades.

B. J. Griffiths · T. Staudigl (✉)
Department of Psychology, Ludwig-Maximilians-Universität München, Munich, Germany
e-mail: Tobias.Staudigl@psy.lmu.de

N. Axmacher (ed.), *Intracranial EEG*, Studies in Neuroscience, Psychology and Behavioral Economics, https://doi.org/10.1007/978-3-031-20910-9_14

Table 14.1 Description of common ocular phenomena [1, 2]

Name	Description
Saccade	A rapid ballistic eye movement that orients the fovea towards the target
Fixation	A prolonged moment in which the fovea is focused upon a stationary target
Smooth pursuit	A smooth movement which allows the fovea to follow a moving target
Vestibular ocular reflex	A movement of the eye that counteracts the movement of the environment incurred by events such as head rotation
Vergence	Opposing movements of the left and right eyes as a stimulus approaches / moves away
Drift	A smooth, slow movement incurred during periods of fixation
Tremor	A small, oscillating movement (~90 Hz) incurred during periods of fixation
Microsaccade	A small, rapid jump in eye position incurred during periods of fixation (which may serve to compensate for the drifts described above)

Saccades are exceptionally brief (~20-50 ms) and involve a ~2–5-degree rotation of the eye from one target to another. Once the eye has made its saccade, it will fixate upon the visual target, allowing key visual details to be extracted by the fovea. The duration of these fixations will vary according to a whole host of factors, including attention, stimulus complexity and current goals, but typically last between 150 and 300 ms. Aside from saccades and fixations, the eye can execute other movements, such as smooth pursuits, and fixational eye movements, like tremor, and drift, and microsaccades (see Table 14.1 for details). For the rest of this chapter however, we will principally focus on the influence of saccades and fixations on electrophysiology and cognition.

14.2 The Neural Correlates of Ocular Activity

There are very few aspects of iEEG activity that do not, in some way, shape or form, correlate with oculomotor activity. Whether a researcher is interested in evoked responses, changes in spectral power, phase-locking, rhythmic fluctuations in neural excitation, inter-area coherence or single unit activity, it is essential to account for oculomotor activity induced by our innate tendency to visually explore the local environment. In this section, we will describe how electrophysiological activity changes when a saccade is executed, when this saccade stops and a fixation begins, and when the preparation for the next saccade begins.

The onset of a saccade is accompanied by a rapid spike in electrophysiological activity that is attributable to neural activity associated with the saccade [e.g., 3–5], as well as muscular activity relating to the physical movement of the eyeball [6, 7] (see "*The electro-muscular correlates of oculomotor activity*" below). The neural correlates of saccadic activity arise ~50 ms after the saccade onset [3, 8], and are readily observable in measures of local field potential, broadband power, inter-trial coherence, and single unit firing rate [e.g., 3–5, 8–10]. The magnitude of these effects scale with the size of the saccade [8, 9] and appear to be lateralised, with the ipsilateral hemisphere showing a greater saccadic spike potential than the contralateral hemisphere [3]. Importantly, these effects are not only observed in early visual regions; they extend across the entire visual hierarchy, from V1 [e.g., 4] to the hippocampus [5, 9, 10]. Functionally speaking, this spike in neural activity is thought to originate from a corollary discharge (a copy of motor output signals sent across the brain) and serves to reset the ongoing phase of oscillatory signals within the visual system so that incoming information falls upon the most excitable phase for stimulus processing [11, 12]. Moreover, by resetting phase across the entire visual hierarchy (which inherently synchronises neural activity across this network), properties of the viewed stimulus can be readily transferred between regions [13, 14] (this is discussed further in Sect. 14.3.2).

At the offset of the saccade, a second ubiquitous electrophysiological phenomenon occurs: the lambda potential [15]. Much like the neural correlates of saccades, this lambda potential arises around ~50 ms after saccade offset/fixation onset, and can be observed across the visual hierarchy using a number of different electrophysiological metrics including amplitude, inter-trial phase coherence, and single unit firing rate [12, 16–19]. This response is thought to be related to the influx of visual information [8, 17]—a theory supported by the absence of a lambda potential when saccades are made in the dark [15], and the fact that the lambda potential appears to be highly similar to the visually-evoked P100 ERP component [20].

In addition to onset and offset potentials, there is evidence to suggest that a number of regions engage in preparatory processes that precede saccades. These pre-saccadic effects present as fluctuations in amplitude, spectral power and phase clustering over the visual cortex [8, 21–23] as well as in frontal regions [24, 25]. Typically, this is thought to be related to motor planning, shifts in attention and a suppression of visual processing prior to this upcoming saccade in order to ensure perceptual stability across the saccade [26, 27]. This suppression, in conjunction with the post-saccadic excitation described above, has been interpreted as a cyclic modulation of excitability across the saccade-fixation cycle [11].

Based on the evidence above, it is clear that the brain is highly sensitive to eye movements, but the neural-ocular link is also observable in the absence of major eye movements. For example, microsaccades correlate with very similar neural responses to the saccades described above, albeit with a smaller magnitude [e.g., 28]. Moreover, pupil dilation that arises when fixating upon emotionally salient stimuli has been linked to fluctuations in amplitude and alpha/beta power [29]. In other words, even in absence of major oculomotor activity, the eyes continue to exert a noticeable influence over electrophysiological activity.

In sum, there is an extensive link between ocular and electrophysiological activity that can be observed across the brain, using many different signal processing techniques, and are sustained (in one way or another) across the fixation-saccade cycle. Given that patterns of ocular activity cannot be fully suppressed (that is, even the most well-trained participant will occasionally make a saccade), it is wise to assume that ocular activity will influence iEEG recordings under almost any circumstance. As such, one needs to consider how to account for this, and how it may interact with other variables of interest, such as behaviour.

14.2.1 A Note on the Electro-Muscular Correlates of Oculomotor Activity

Unlike scalp EEG and MEG recordings, iEEG recordings are thought to be largely unaffected by muscular artefacts induced by eye movements, owing to the local, intracerebral nature of iEEG recordings. However, this is not always the case. Saccade-related muscular artefacts can be observed in intracranial recordings taken from numerous brain regions [7]. These effects are most prominent in the temporal pole, which neighbours extraocular muscles (namely, rectus lateralis [30]). While some of these effects have been attributed to the choice of re-referencing (with a scalp-based reference producing the greatest artefact), even bipolar-referenced recordings used in conjunction with independent components analysis cannot fully correct for such artefacts [7]. Consequently, researchers should be aware of saccade-related artefacts in intracranial recordings when interpreting their results and, wherever possible, use direct measurements of ocular activity to address the possibility of such a confound.

14.3 The Functional Roles of Eye Movements for Brain Function and Behaviour

As alluded to above, the neural correlates of ocular activity are not epiphenomena; rather, they exert a profound influence over brain function and behaviour. In this section, we give two examples of how eye movements influence brain function and behaviour, and then highlight several key resources relating these eye movements to various psychological constructs.

14.3.1 Eye Movements Map Visual Space

Eye movements can be viewed as a window to cognitive processes [31] and accounting for them has been shown to reveal valuable insights into fundamental cognitive processes. One particularly impressive example pertains to the neural correlates of spatial navigation across different species, and the implications different exploration strategies in these species might have for the coding of space. Seminal research in rodents discovered place cells, head directions cells and grid cells and identified them as the major pillars underlying navigation and memory [32–34]. The quest to discover cells in primates that show analogous spatially tuned firing patterns has long been difficult (but see [35, 36]). Instead, what has been consistently found in primates are cells that code for visual space. That is, neurons in the monkey hippocampus fire with respect to where the animal is looking, rather than where the animal is located in space [e.g., 37]. A variety of gaze dependent firing patterns have been found in neurons located in the primate medial temporal lobe, some of which closely resemble their rodent counterparts (e.g., visual grid cells, saccade direction cells and visual border cells), but also cells that code for multiple aspects of visual and other spaces [38–40]. Considering the primacy of vision among primates' senses and the extent to which primates use vision to explore their environment, it seems clear that cells code essential features of visual space to build a cognitive map. In humans, single neurons coding for visual space have not yet been identified, but brain activity exhibiting grid-like coding of visual space has been shown with fMRI, MEG and iEEG [41–43].

14.3.2 Eye Movements Align Brain Rhythms

While exploring the visual environment, we rapidly move our eyes, shifting the fovea centralis from one location to another. During saccadic eye movements, neural activity is suppressed and perception is reduced, while during fixation periods in between saccades, neural activity is enhanced [44]. Saccades typically occur at a frequency of 3–4 Hz during visual exploration (with slight variations depending on the task). Each saccade interrupts the flow of visual input, making our perception of the visual world rhythmic rather than continuous. Our brains, however, compensate for the perceptual gaps during saccades and create the illusion of a continuous stream of visual input [e.g., 45, 46]. The quasi-periodicity of saccades thus produces temporal windows during which novel visual input arrives in the brain and excites both visual and non-visual areas. The brain itself also relies on temporal windows during which neural processing is favourable, located between periods when processing is impeded or even impossible which can range from a few milliseconds (e.g., refractory periods in neurons) to several hundreds of milliseconds (e.g., down-states during slow wave sleep). Brain oscillations are a prominent example of these rhythmic temporal windows, or duty cycles, defined by the phase of the oscillation. Much of the functionality ascribed to brain oscillations lies in the very fact that they provide a window of opportunity for (ensembles of) neurons to fire coherently and, thus, enable coordinated processing and communication of information [47]. Therefore, an optimised strategy to process and communicate incoming visual information would be to align one temporal window to the other, such that new input arrives during the duty cycle. At least two steps are necessary to achieve this: (1) The system needs to be able to predict upcoming temporal windows, and (2), the system needs to adjust one (or both) of the temporal windows according to the prediction.

A temporal prediction of when windows for novel visual input will occur can be made on the basis of efference copies. Efference copies are copies of the motor commands that the brain sent to produce the movement (also called corollary discharge). Whenever the brain sends a motor command towards the periphery to execute a movement, a copy of this command is kept within the system. The information about the imminent movement can then be used to prepare the brain and potentially modulate the state of a network such that it is optimised for when the predicted movement happens. Consequently, efference copies usually target brain circuits interpreting sensory inputs and are particularly important for distinguishing self-generated movements from changes in the external world. In the case of saccades, this is particularly striking: whenever we move our eyes, the image on our retinas moves in the opposite direction. However, the brain predicts and compensates for the movement in the physical world to provide a stable internal representation [27]. The neural foundations of eye movement related efference copies have principally been investigated in non-human primates [e.g., 48]. Cortical areas, in particular the frontal eye fields, send a motor command to the superior colliculus which is conveyed downstream to lower motor areas. An efference copy is sent from the superior colliculus to the mediodorsal thalamus, which holds information about the onset and direction of the upcoming saccade (see Fig. 14.1a). This information is provided to the frontal eye fields via thalamocortical neurons. Providing essential information about impending (motor) behaviour has been suggested to be one of the fundamental roles of thalamocortical loops [49] and could be a fundamental mechanism to predict an upcoming temporal window during which new input arrives.

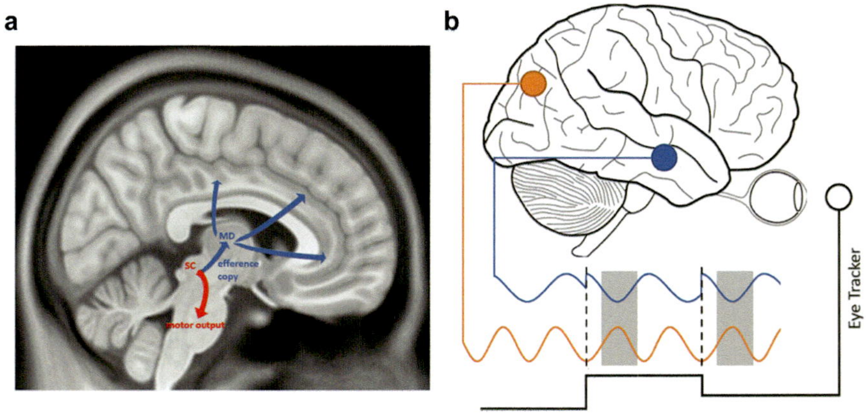

Fig. 14.1 Saccade-triggered phase alignment of brain rhythms. **a** Superior colliculus (SC) neurons send a motor command which is branched off to the mediodorsal thalamus (MD), signalling an upcoming saccade (efference copy). **b** Sketch of the proposed model. Simultaneous recording of eye movements (black lines) and iEEG from multiple brain areas (blue, orange). To prepare for processing and communication of the incoming stimulus, the phases of brain areas are aligned to the saccade (but it is also plausible to suggest that saccades align to brain rhythms). Shaded areas depict duty cycles of the respective brain rhythms. Dashed lines depict saccade onsets. Specific phase angles and brain areas are chosen for clarity purposes and not based on real data

With the system primed for an upcoming eye movement, neural activity now needs to be modulated to synchronise peak excitability with the eye movement. A potential mechanism to adjust windows of excitability according to the prediction based on an efference copy is the alignment of low frequency oscillatory phase. Such low-frequency phase-alignment has been demonstrated to co-occur with saccades in humans as well as non-human primates [9, 10, 17, 50] and linked to prioritised cognitive processing [5, 51]. Phase alignment is not necessarily limited to a single ensemble of neurons, but might, instead, be a feasible tool to organise functional networks across many brain areas [52]. In order to make this work, efference copies would need to be distributed to and induce phase alignment in multiple brain regions (which is feasible considering the widespread projections of the thalamocortical system). A coordinated alignment of low-frequency phase on a global level could set the stage for efficient neural communication [13, 47]. Recent findings combining intracranial recordings and eye tracking support this idea by showing enhanced connectivity related to saccade onset [51] (see Fig. 14.1b).

14.3.3 The Role of Eye Movements in Various Psychological Constructs

There are numerous other examples of exciting research lines investigating the relationship between cognitive processes and oculomotor behaviour. Since a comprehensive review of all these findings would be sufficient to form it's very own book, we list suggestions for the interested reader on a selection of topics: memory [e.g., 53, 54], navigation [e.g., 55, 56], attention [57, 58], perception [59], reading [60, 61] and 'active sensing' [62, 63].

14.4 Accounting for Ocular Activity in Cognitive Neuroscience Research

The previous two sections demonstrate that eye movements are intimately linked to a wide range of electrophysiological phenomena, and quite possibly mediate the interaction between this neural activity and cognition [e.g., 54]. Consequently, it is essential to account for ocular activity when exploring the links between electrophysiological activity, cognition and behaviour. In this section, we discuss how best to account for ocular activity when studying brain-behaviour correlations, and their relative strengths and weaknesses.

The first approach, which is perhaps the default for those who are unaware of the ubiquity of ocular influence over the brain and cognition, is to simply 'prevent' eye movements. This could include designing a paradigm that uses simple stimuli, presented at the centre of the screen and instructing participants to avoid making eye movements during the task. We place 'prevent' in quotes, because this 'prevent eye movements' approach does not really prevent eye movements at all. The eyes are never completely still. While these design choices may reduce the number of major saccades made in the experiment, some of these larger saccades will remain, and smaller movements (i.e., microsaccades, tremors) will continue to be prevalent. Impeding all eye movements is, on the other hand, not practical in most experimental settings anyway. If one were successful in preventing their participants' eye movements completely (including miniature ones), they would soon realise the drawback of this unnatural condition: visual perception fades when all eye movements are suppressed [64]. One can experience a mild version of this in the so-called Troxler's effect: when fixating upon the centre of a display, peripheral stimuli will start to fade after only a short period of time. The effect vanishes immediately when the eyes are moved. Neurally, this can be explained by adaptation processes during fixation, which are counteracted by micro-saccades [64]. Moreover, such an approach lacks ecological validity—in daily life, eye movements are not suppressed, and if they do indeed serve to facilitate cognition [54, 65], suppression of eye movements may lead to peculiarities in cognitive processes that undermine the experimental question at hand. Ultimately, this approach is not fit-for-purpose, and experimenters who take such an approach must accept the fact that eye movements may be acting as a latent variable that mediates the experimenter's observed link between the brain and cognition.

The second approach, which involves a deep dive into ocular activity, is to embrace "free-viewing" paradigms and employ simultaneous eye tracking-iEEG recordings. By allowing participants to fully explore the stimuli presented to them, the neural data and behaviour recorded will be most representative of what occurs in daily life, substantially boosting the ecological validity of the data. Of course, such an approach also requires careful analysis of data, focusing not only on what happens when a stimulus is presented, but also what happens when saccades and other eye movements occur (details on how to handle such analyses can be found in the next section). However, these additional analyses may be off-putting to researchers who are not focused principally on the influence of eye movements on the brain and/or behaviour, and they may prefer the third approach.

The third approach sits somewhere between the first and second approaches, in that it acknowledges the importance of eye movements in cognitive electrophysiology, but does not wish to focus a study on them. Like the second approach, simultaneous eye tracking and intracranial recordings would be taken, but in a paradigm that serves to minimise eye movements (e.g., by presenting simple stimuli at the centre of the screen). The nature of eye movements could then be calculated on a trial-by-trial level (e.g., number of saccades, length of fixations) and included as covariates in the main analyses (see the section '*"eye movements-as-covariates" analysis*' below). This approach may be best suited to those researchers who simply wish to factor out the contributions of eye movements from their experimental questions.

14.5 A "How-To" of Eye Tracking and Intracranial EEG

The sections above have highlighted the influence that ocular activity exerts over neural activity, but a question remains: what practical steps can researchers take to measure the interactions between the eyes, brain, and behaviour? Here, we provide an overview on how to conduct simultaneous iEEG-eye tracking experiments, from the practicalities of recording in the clinic to descriptions of the myriad of analytical possibilities. Given the breadth and depth of eye tracking as a research topic, this section should not be viewed as a comprehensive guide to eye tracking, but rather an introduction to the key issues in relation to intracranial recordings. For the interested reader, we have provided a number of references to other detailed guides at relevant moments which go deeper into various aspects of the methodology.

14.5.1 Recording Ocular Activity in the Clinic

It would be fair to say that most eye trackers have not been designed with iEEG patients in mind, and many lab-based eye tracking solutions have their drawbacks when brought to the patient's bedside. In this section, we aim to highlight the relative strengths and weaknesses of different eye tracking solutions when used in the clinic (see also Chaps. 4 and 5 for general considerations of recordings in a hospital environment).

Eye tracking systems can be divided into two broad categories: desk-mounted systems and wearable systems. Desk-mounted systems tend to be large units that are placed on the desk in front of a participant, either as large towers sitting directly between participants and the computer screen (e.g., Tower Mount EyeLink 1000 Plus) or smaller, remote devices that sit beneath/beside the computer screen (e.g., Tobii Pro Spectrum, Eyelink Portable Duo, TrackPixx 3). These desk-mounted systems tend to provide a more accurate and higher-sampled recording (up to 2000 Hz) of ocular activity than wearable systems, but they do have their drawbacks. Due to their weight and a lack of dedicated work surfaces available to the patient in the clinic, tower systems are not a practical option for combined iEEG-eye tracking recording. The remote systems, in contrast, can be used in the clinic settings, but care is required. For example, remote systems can be set up on the patient bedside table if this does not interfere with the patient responding to a computer task. However, the remote system is still heavy, meaning some tables will struggle to hold the weight of the system. Moreover, the system may become a hazard should the patient experience a seizure. As with other research equipment brought to the hospital ward, one needs to be careful not to block medical staff access to the patient. One solution for this particular issue is to use downscaled versions of some manufacturers' high-end products (e.g., Tobii Pro Fusion), that can be attached directly to a monitor or a laptop and are, consequently, easier to set up and more mobile. Usually, these systems come with a reduced sampling rate relative to their high-end siblings. Beyond size, it is also important to consider how tolerant the eye tracker would be to head movements. Research with iEEG patients will in most cases be conducted while the patient is in their hospital bed, and fixing their head position using a chin-/forehead rest or a bite bar will be very hard to accomplish and potentially unpleasant for the patient. A visualisation of how iEEG and eye tracking can be combined at the patient's bedside is provided by Wang and colleagues [66]. Researchers aiming to track eye movements intraoperatively (e.g., when Parkinson's patients are awake during placement of deep brain stimulation lead) will face additional challenges. The setup and access to the patient needs to be arranged in close consultation with surgeons and other medical staff, and the time available for experimental procedures during surgeries is usually very little, restricting the range of suitable tasks. However, in contrast to patients outside the operating room, head movements are less problematic since patients' heads are restricted in a head frame (though this may obscure the patient's field of view). Nonetheless, should the researcher manage to navigate these issues, remote systems can provide a robust and reliable measure of ocular activity in the clinic.

Wearable eye trackers provide a light-weight alternative to desk-mounted options. Typically, wearable eye trackers train one camera (or more) on each eye, and point an additional camera outwards to capture the participant's frame of view (e.g., Tobii Pro Glasses 3; Pupil Core, Eyelink II). As these eye trackers are worn by the participant, researchers needn't worry about additional desk space to hold the system, which may allow for more flexibility in the tasks which participants complete. However, these eye trackers also have their drawbacks. For example, the maximum sample rate tends to be smaller than desk-mounted alternatives (often <120 Hz), meaning analysis of saccades can become difficult. Furthermore, depending on where intracranial electrodes have been implanted and/or where bandages have been placed, the patient may not be able to wear these eye trackers.

Ultimately, the question of which type of eye tracker to use boils down to the experimental question at hand. If researchers are interested in fixation-related analyses (which do not require high sampling rates), or have limited space, wearable eye trackers may be more appropriate, but if researchers are focused on saccadic activity, desk-mounted options with high sampling rates are more suitable.

Regardless of which system is used however, researchers need to consider how they will synchronise the eye tracking recordings with the iEEG recordings as this is paramount to exploring the interactions between ocular activity, neural activity and behaviour. Fortunately, many systems will have the capability to receive a trigger much like how iEEG recording systems do. This means researchers may be able to send a single trigger from the experimental computer to both the iEEG and eye tracking systems, allowing the two data streams to be aligned during subsequent analysis. If the eye tracking system is not built to receive such triggers, however, researchers will need to synchronise the clocks of the experimental, iEEG-recording and eye tracker-recording computers so that timestamps are consistent across all hardware elements of the experiment (as clocks have a tendency to drift over time [67]). One popular synchronisation solution is 'Lab Streaming Layer' (LSL; https://github.com/sccn/labstreaminglayer), which can synchronise clocks up to every five seconds. While LSL may take a little additional work to set up in comparison to a simple trigger cable, it may be more effective in situations where recordings occur across multiple sites and researchers wish for a single, common means of synchronisation rather than bespoke solutions for each site.

If eye tracking equipment is unavailable, researchers may still benefit from the use of electrooculography (EOG). Depending on the position of the patient's bandages, researchers can place horizontal EOG pairs (one left of the left eye, and one right of the right eye) and vertical EOG pairs (one above the eye and one below the same eye). EOG recordings can be used to identify when a saccade or fixation occurred, as well as amplitude and direction of these ocular phenomena [68]. Moreover, EOG recordings are often directly fed into the iEEG recording system, avoiding the issues relating to synchronisation described above. However, much like the wearable eye trackers, the placement of EOG electrodes may be difficult for some patients depending on bandage placement. Nonetheless, EOG provides a feasible alternative for measuring ocular activity when eye tracking equipment is not available.

14.5.2 Detecting Ocular Events

The detection of ocular events is key to any analysis focusing on the intersection of ocular activity, neural activity and behaviour. While large events such as saccades and fixations can be readily identified in the raw eye tracker trace, manually labelling every event can be a time-consuming approach that is difficult to replicate by others. Consequently, algorithms tend to be more popular when it comes to event detection. Here, we provide a brief overview of the two most common forms of event detection algorithm, and then describe how to handle common artefacts that may impair the performance of these algorithms.

Almost all event detection algorithms involve translating the gaze co-ordinates (i.e., the x- and y-positions of where the eye is focused on the computer screen) into key ocular events (e.g., saccades, fixations). Dispersion algorithms will set out to identify key ocular events based upon the range of the x- and y-gaze positions over a predefined period of time (that is: $\text{dispersion} = [\max(\mathbf{x}) - \min(\mathbf{x})] + [\max(\mathbf{y}) - \min(\mathbf{y})]$, where $\mathbf{x}$ and $\mathbf{y}$ are vectors containing x- and y-gaze positions over the predefined time window). In the most basic of dispersion algorithms, if the dispersion measure falls below a predetermined threshold (either based on existing literature, or through visual inspection of the participant data), then that time period is considered to be a 'fixation' and the remainder of time periods are inferred to be 'saccades'. One of the most common examples of a dispersion algorithm is the 'identification by dispersion threshold' algorithm (I-DT [69]). While popular and widely-available, dispersion algorithms do have their drawbacks. For example, they are often distorted by ocular drifts, and tend not to function well when the sample rate of the eye tracker exceeds 50 Hz.

An alternative is to use velocity-based algorithms [e.g., 57]. Many of these algorithms use a derivative of gaze position (i.e., velocity = change in visual angle/change in time) to identify saccades based upon a pre-defined velocity threshold. When the data exceeds this threshold, a saccade is marked, and anything that falls below the threshold is assumed to be a fixation (though some algorithms also include a minimum fixation threshold, where velocity must fall below the threshold for a set period of time). In their most basic form, velocity-based algorithms are simpler than dispersion algorithms, but they are often built upon to help minimise distortion by noise. For example, implausibly fast saccades may be discarded, or a two-pass procedure may be implemented in which approximate saccade windows are identified in the first pass, and then the onsets/offsets are concretely defined in the second pass. However, even with these modifications, defining events based solely on velocity can be unreliable. This has led to some incorporating acceleration into the algorithm (where acceleration = change in velocity/change in time; [e.g., 70]), which can help distinguish slow saccades from moments involving fast smooth pursuits. Both velocity and velocity-plus-acceleration algorithms tend to perform better when the eye tracking data has a high sample rate, as high sample rates allow for a more robust estimation of velocity/acceleration.

Of course, no event detection algorithm is perfect, and it is important to consider how to handle potential artefacts that distort the algorithm output. One of the most common artefacts comes from blinks; the moving of the eyelid can lead to a rapid change in the eye tracker-predicted gaze position, which in turn can lead to algorithms detecting an illusionary saccade. Some researchers tackle this issue by asking participants to refrain from blinking during the key moments of a task, and then discarding trials in which participants did blink, while others tackle this issue by ignoring any events that are detected just before (e.g., 100 ms) or just after a blink. Stochastic noise can also be problematic for event detection, where momentary jumps in gaze position are erroneously defined as saccades. If these jumps are relatively small, smoothing or filtering the data will often attenuate these issues. However, in cases where these jumps are large, smoothing/filtering may make the situation worse as it may make these jumps appear more saccade-like to the automated algorithms. Visual inspection is perhaps the best approach to tackle these larger jumps, where artefactual trials are discarded in much the same way that artefactual trials are discarded when contaminated by an iEEG artefact.

In sum, many event detection algorithms are readily available for eye tracking research, each of which have their relative strengths and weaknesses. Dispersion-based algorithms search for fixations based on limited change in gaze position, and are better suited for low sample-rate data, while velocity-based algorithms detect saccades based upon changes in velocity (and also acceleration). If the reader is interested in exploring these approaches further, we can recommend Holmqvist & Andersson (2017) for a detailed review of many different forms of algorithm [1], and Nyström & Holmqvist (2010) for a description of how to build a bespoke, velocity-based, event detection algorithm [70].

14.5.3 Saccade-/Fixation-Locked Analysis

For researchers looking to directly explore the impact of saccades and fixations on a particular brain-behaviour correlation, it can be highly informative to conduct a saccade- or fixation-locked analysis in which the intracranial data is re-epoched around these saccadic or fixation events. In order to do so, however, researchers should consider three factors: (1) what ocular event to lock the iEEG data to, (2) how to address trial imbalances, and (3) how to baseline correct the data. We will consider each issue in turn.

The decision about which ocular event to lock to should be driven by the question at hand. If researchers are focused on how the brain/behaviour responds to stimulus properties, fixation-locked analyses are more appropriate as this form to time-locking is more closely locked to the 'onset' of the visual stimulus, whereas saccade-locked analyses are more effective for questions relating to the preparation, execution, and response to eye movements themselves. In instances where researchers are unsure about which method to use, they could elect to do both (though of course this will effectively double the number of epochs that will need to be analysed, as well as double the number of statistical comparisons that will be made later down the line).

After deciding which ocular event to lock to, researchers need to consider balancing trials between conditions. As described earlier, the size and direction of saccades [e.g., 8, 9] and duration of fixations [e.g., 11] can have a large impact on neural activity. If there are differences in saccadic or fixation-related activity between conditions, this has the potential to drive neural differences between conditions. So, if researchers are not interested in the effects of differences in saccadic/fixation-related activity, these parameters should be matched as best as possible between conditions. For example, researchers could, for every trial of one condition, find the most similar trial of the second condition (e.g., in terms of saccade distance, fixation duration), and use these two matched conditions for subsequent analysis.

After balancing trials, the choice of baseline needs to be considered for a range of iEEG analyses (e.g., event-related potentials or changes in amplitude/power). Typically, baselines can either be common across all saccades/fixations of a trial/recording, or individual to each saccade/fixation event. Common baselines typically take data from before the beginning of the trial and correct epochs based on this signal. Such an approach is useful if the researcher is interested in exploring how neural responses change over the course of a trial (e.g., increase in memory load, change in attentional resources). However, for long trials, or continuous recordings (e.g., movie-viewing), drifts in activity between the pre-trial baseline and later saccades can become problematic. Individual baselines are less likely to be swayed by such drift, as the baseline signal comes directly before the event of interest. It also helps control for factors such as shifts in attentional resources if this is not a variable of interest. However, individual baselines also have their drawbacks. For example, these baselines are easily swayed by differences in preceding saccadic/fixation-related activity (e.g., saccade size, fixation duration), which means greater effort is required when it comes to matching trials between conditions. This issue can also become exacerbated if the two conditions have inherent differences in saccadic-/fixation-related activity; in such instances, a common baseline may be more appropriate. Ultimately however, the decision sits with the researcher, and they should select the method they feel is most appropriate for the question at hand.

After completing these steps, researchers can proceed with their iEEG analyses as if the data were locked to a stimulus trigger. However, it is always worth remembering that for saccade-/fixation-locked recordings, the data immediately preceding the event may correspond to some other ocular activity (e.g., the offset of a fixation, saccade execution), and methods that involve temporal smoothing (e.g., wavelet-based time–frequency decomposition approaches) may lead to this preceding activity obscuring the effects of interest. For the interested reader, a deep dive into the various decisions involved in saccade-/fixation-locked analyses has been presented by Nikolaev and colleagues [8].

Most of the information provided above pertains to the analyses of local field potentials. Only a few studies have related human single cell recordings to eye movement behaviour [e.g., 50, 51, 71]. As a general approach, firing rates of putative single units can be analysed during time windows related to saccades or fixations. For example, a single cell could be classified as being visually selective if its firing rate was modulated by the category of fixated image in a defined time window after fixation onset. Obviously, a plethora of single unit analyses and approaches could be used in a similar way, the interested reader is referred to the publications cited above for inspiration (see also Chap. 45).

14.5.4 Encoding Models

There has been a rapid expansion in the use of multivariate methods to analyse neuroimaging data in the last decade. One approach that is particularly relevant to studying the link between eye movements and the brain is the *forward encoding model* (FEM).

A FEM aims to model neural activity at any given electrode/sample point as a function of the properties of a stimulus, be that visual stimuli [e.g., 72, 73], auditory stimuli [e.g., 74], or ocular activity [e.g., 75]. This is achieved by breaking the stimulus down into its constituent parts. In the case of ocular activity, breaking this down into constituent parts can be done in a myriad of ways. For example, one could simply use eye position on the x- and y-axes as two predictors in the encoding model, or one could push the boat out and model eye position as a series of circular–Gaussian kernels covering all possible ocular angles as has previously been done for virtual head direction [76]. Ultimately, the decision on how to break down the patterns of ocular activity depends on the research question at hand.

Whatever the approach used, the next step of the FEM analysis involves building a general linear model to estimate activity at a given intracranial electrode based upon the consistent parts of a stimulus (each of which will act as a separate predictor variable in the linear model). The linear model is constructed such that the outcome variable (y) reflects a vector of iEEG activity from a single iEEG channel, where every row of the y reflects a separate sample point of this iEEG activity. The predictor matrix (X) contains all the predictor variables which make up the stimulus, with every column of X representing a different stimulus feature. The beta weights (β) are then estimated using the conventional equation: $y = \beta X$.

To validate the performance of the FEM, the estimated beta weights are then applied to a second, hold-out dataset. By multiplying the beta weights with the predictor matrix of the hold-out dataset, a predicted time-series of iEEG activity is produced. This predicted time-series can then be correlated with the observed iEEG time-series of the hold-out dataset. If there is a significant correlation (or one that significantly deviates from a surrogate distribution), one can conclude that the stimulus (in this case, ocular activity) can predict neural activity at a given iEEG electrode. Beyond predicting iEEG activity, the beta weights of the FEM can be inspected to identify which parts of the stimulus best predict iEEG activity (though see [77], for a cautionary note on the interpretation of weights in multivariate models), or the weights can be inverted to allow researchers to predict stimulus properties based on iEEG activity.

FEMs are a powerful technique for probing the link between ocular and electrophysiological activity. For example, they allow for the simultaneous modelling of multiple stimulus features, which many univariate analyses cannot easily do. However, this also has drawbacks. The relative complexity of the FEM makes the analysis computationally expensive, sometimes prohibitively so. Moreover, it is sometimes unclear what parts of the stimulus feature it is best to model. This can lead to researchers having to build and compare several models, which incurs issues with multiple comparisons. Nonetheless, so long as the researchers have a clear research question in mind, and have the computational means to run the analyses, FEMs could prove to be a fruitful venture when exploring the link between ocular and electrophysiological activity.

FEMs are a complex topic in their own right, and the interested reader may wish for more detail than what is provided here. Fortunately, there are a number of great primers on the topic out there which can be explored in more detail [e.g., 78] (see also Chap. 47).

14.5.5 *"Eye Movements-As-Covariates" Analysis*

If the research question does not focus on the role of ocular activity, but the experimenters wish to control for the potential mediating effect of such activity on the brain, behaviour and cognition, the experimenters can consider ocular activity as a covariate. There are many ways to analyse brain-behaviour relationships while controlling for covariates, and here we provide two suggestions, one simple and one more sophisticated.

Perhaps the most straightforward solution is the median split, whereby the dataset is split into two based on a covariate, and analysis is conducted on the two groups separately: One could divide participants into two groups based on, say, the mean number of saccades they made per trial, and run the brain-behaviour correlation separately for the two groups. If the correlation remains significant in both groups, and there is no significant difference in the correlation coefficient between the groups, one could conclude that saccade count does not exert a substantial influence over the brain-behaviour correlation. If, however, the correlation disappears for one group, or a significant difference in the correlation coefficient is observable between the two groups, then one could conclude that saccade count mediates the brain-behaviour correlation. While this approach is easy to implement, it does have a number of drawbacks. First, by splitting the data in two, the sample size for each test is halved, substantially limiting statistical power. Second, one needs to test whether there is a difference between the two groups (i.e., a between-samples comparison) which may be difficult to robustly implement given the inconsistent placement of intracranial electrodes across participants. Third, one can only address one covariate at a time. If, for example, one wished to also explore the impact of saccade count and fixation duration on the brain-behaviour correlation, a second median split analysis using fixation duration as the criterion for dividing the groups needs to be conducted. Critically, this prevents one from simultaneously accounting for two covariates (i.e., number of saccades, fixation duration), meaning both median split analyses may still contain influence from the other covariate.

An alternative approach which does not impact statistical power, and can handle multiple covariates simultaneously, involves linear modelling. Here, a linear model would be constructed in which iEEG activity (the outcome variable) is modelled as the weighted sum of behaviour and the saccade count (the predictor variables) on a trial-by-trial level. The resulting regressor (i.e., beta) weight for behaviour can then be viewed as the link between behaviour and iEEG activity after regressing out the influence of saccade count. These beta weights can then be statistically appraised in one of two ways. The first option involves running a linear model separately for each participant, pooling the resulting beta weights and then testing whether they consistently differ from zero in a one-sample t-test. The alternative is to create a single linear mixed-effects model, which includes iEEG activity, behaviour and saccade count for every trial of every participant as well as additional random effects to account for influences of individual participants (see also Chap. 36). The linear mixed model can then be statistically appraised in a myriad of ways, including likelihood ratio tests or permutation testing (for more details, see [79]). Unlike median split analyses, linear modelling approaches preserve statistical power, which may aid in the detection of smaller effects. Moreover, multiple covariates can be inserted into the same model, allowing for the simultaneous control of several potentially-mediating factors. However, the analyses are more computationally intensive, which may cause difficulty for analytical approaches that already require large amounts of computational resources.

14.5.6 *Eye Movement Artefacts in Intracranial EEG*

As discussed throughout this chapter, treating eye movements as nothing more than artefacts is a highly problematic practice. That said, some parts of the ocular-induced change in electrophysiological signal (namely, the electromuscular component) should be suppressed prior to analysis [7, 30] as these signals do not reflect true brain activity. For example, intracranial electrodes that are proximal to the extraocular muscles are likely to pick up electrical signals generated by these muscles during saccadic activity. The magnitude of these effects depends on the referencing montage used, with bipolar montages perhaps being the most effective in suppressing the artefact [30]. However, even bipolar referencing is not sufficient to fully remove the artefact. Independent components analysis has also been proposed as a means to suppress these artefacts, and can do a good job in suppressing the electromuscular artefact [7], although this too does not totally remove the artefact. In conjunction, bipolar referencing and independent components analysis acts as perhaps the most effective means to remove electromuscular artefacts from intracranial data, but even after applying these analytical steps, researchers would be wise to view eye movement artefacts as "suppressed" rather than fully subtracted from the recordings.

14.6 Conclusion

With this chapter, we hope to convince the reader that recording eye movements and iEEG simultaneously will benefit their research. We argue that eye movements are more than just artefacts, and ignoring oculomotor behaviour in iEEG research will impede our understanding of brain processes underlying cognition. Eye tracking in clinical settings is challenging, but by pointing out several options on how to combine and analyse eye tracking and iEEG recordings, we hope to mitigate these concerns and inspire new avenues of research.

References

1. Holmqvist K, Andersson R (2017) Eye tracking: a comprehensive guide to methods, paradigms and measures, 2nd edn. Lund Eye-Tracking Research Institute, Lund, Sweden
2. Leigh R, Zee D (2015) The neurology of eye movements. Oxford University Press, United Kingdom
3. Carl C, Açık A, König P et al (2012) The saccadic spike artifact in MEG. Neuroimage 59:1657–1667. https://doi.org/10.1016/j.neuroimage.2011.09.020
4. Ito J, Maldonado P, Singer W, Grun S (2011) Saccade-related modulations of neuronal excitability support synchrony of visually elicited spikes. Cereb Cortex 21:2482–2497. https://doi.org/10.1093/cercor/bhr020
5. Jutras MJ, Fries P, Buffalo EA (2013) Oscillatory activity in the monkey hippocampus during visual exploration and memory formation. Proc Natl Acad Sci 110:13144–13149. https://doi.org/10.1073/pnas.1302351110

6. Hipp JF, Siegel M (2013) Dissociating neuronal gamma-band activity from cranial and ocular muscle activity in EEG. Front Hum Neurosci. https://doi.org/10.3389/fnhum.2013.00338
7. Kovach CK, Tsuchiya N, Kawasaki H et al (2011) Manifestation of ocular-muscle EMG contamination in human intracranial recordings. Neuroimage 54:213–233. https://doi.org/10.1016/j.neuroimage.2010.08.002
8. Nikolaev AR, Meghanathan RN, van Leeuwen C (2016) Combining EEG and eye movement recording in free viewing: Pitfalls and possibilities. Brain Cogn 107:55–83. https://doi.org/10.1016/j.bandc.2016.06.004
9. Doucet G, Gulli RA, Corrigan BW et al (2020) Modulation of local field potentials and neuronal activity in primate hippocampus during saccades. Hippocampus 30:192–209. https://doi.org/10.1002/hipo.23140
10. Hoffman KL, Dragan MC, Leonard TK et al (2013) Saccades during visual exploration align hippocampal 3–8 Hz rhythms in human and non-human primates. Front Syst Neurosci. https://doi.org/10.3389/fnsys.2013.00043
11. Barczak A, Haegens S, Ross DA et al (2019) Dynamic modulation of cortical excitability during visual active sensing. Cell Rep 27:3447-3459.e3. https://doi.org/10.1016/j.celrep.2019.05.072
12. Rajkai C, Lakatos P, Chen C-M et al (2008) Transient cortical excitation at the onset of visual fixation. Cereb Cortex 18:200–209. https://doi.org/10.1093/cercor/bhm046
13. Fries P (2015) Rhythms for cognition: Communication through coherence. Neuron 88:220–235. https://doi.org/10.1016/j.neuron.2015.09.034
14. Lakatos P, Gross J, Thut G (2019) A new unifying account of the roles of neuronal entrainment. Curr Biol 29:R890–R905. https://doi.org/10.1016/j.cub.2019.07.075
15. Evans CC (1953) Spontaneous excitation of the visual cortex and association areas—Lambda waves. Electroencephalogr Clin Neurophysiol 5:69–74. https://doi.org/10.1016/0013-4694(53)90054-6
16. Bartlett AM, Ovaysikia S, Logothetis NK, Hoffman KL (2011) Saccades during object viewing modulate oscillatory phase in the superior temporal sulcus. J Neurosci 31:18423–18432. https://doi.org/10.1523/JNEUROSCI.4102-11.2011
17. Katz CN, Patel K, Talakoub O et al (2020) Differential generation of saccade, fixation, and image-onset event-related potentials in the human mesial temporal lobe. Cereb Cortex 30:5502–5516. https://doi.org/10.1093/cercor/bhaa132
18. Leszczynski M, Chaieb L, Staudigl T et al (2021) Neural activity in the human anterior thalamus during natural vision. Sci Rep 11:17480. https://doi.org/10.1038/s41598-021-96588-x
19. Ries AJ, Slayback D, Touryan J (2018) The fixation-related lambda response: effects of saccade magnitude, spatial frequency, and ocular artifact removal. Int J Psychophysiol 134:1–8. https://doi.org/10.1016/j.ijpsycho.2018.09.004
20. Kazai K, Yagi A (2003) Comparison between the lambda response of eye-fixation-related potentials and the P100 component of pattern-reversal visual evoked potentials. Cogn Affect Behav Neurosci 3:46–56. https://doi.org/10.3758/CABN.3.1.46
21. Parks NA, Corballis PM (2008) Electrophysiological correlates of presaccadic remapping in humans. Psychophysiology 45:776–783. https://doi.org/10.1111/j.1469-8986.2008.00669.x
22. Richards JE (2003) Cortical sources of event-related potentials in the prosaccade and antisaccade task. Psychophysiology 40:878–894. https://doi.org/10.1111/1469-8986.00106
23. Staudigl T, Hartl E, Noachtar S, et al (2017) Saccades are phase-locked to alpha oscillations in the occipital nd medial temporal lobe during successful memory encoding. PLOS Biol 15:e2003404. https://doi.org/10.1371/journal.pbio.2003404
24. Gutteling TP, van Ettinger-Veenstra HM, Kenemans JL, Neggers SFW (2010) Lateralized frontal eye field activity precedes occipital activity shortly before saccades: evidence for cortico-cortical feedback as a mechanism underlying covert attention shifts. J Cogn Neurosci 22:1931–1943. https://doi.org/10.1162/jocn.2009.21342
25. Richards JE (2013) Cortical sources of ERP in prosaccade and antisaccade eye movements using realistic source models. Front Syst Neurosci. https://doi.org/10.3389/fnsys.2013.00027

26. Bremmer F, Kubischik M, Hoffmann K-P, Krekelberg B (2009) Neural dynamics of saccadic suppression. J Neurosci 29:12374–12383. https://doi.org/10.1523/JNEUROSCI.2908-09.2009
27. Sommer MA, Wurtz RH (2008) Visual perception and corollary discharge. Perception 37:408–418. https://doi.org/10.1068/p5873
28. Martinez-Conde S, Otero-Millan J, Macknik SL (2013) The impact of microsaccades on vision: towards a unified theory of saccadic function. Nat Rev Neurosci 14:83–96. https://doi.org/10.1038/nrn3405
29. Dahl MJ, Mather M, Sander MC, Werkle-Bergner M (2020) Noradrenergic responsiveness supports selective attention across the adult lifespan. J Neurosci 40:4372–4390. https://doi.org/10.1523/JNEUROSCI.0398-19.2020
30. Jerbi K, Freyermuth S, Dalal S et al (2009) Saccade related gamma-band activity in intracerebral EEG: dissociating neural from ocular muscle activity. Brain Topogr 22:18–23. https://doi.org/10.1007/s10548-009-0078-5
31. König P, Wilming N, Kietzmann TC et al (2016) Eye movements as a window to cognitive processes. J Eye Mov Res 9. https://doi.org/10.16910/jemr.9.5.3
32. Hafting T, Fyhn M, Molden S et al (2005) Microstructure of a spatial map in the entorhinal cortex. Nature 436:801–806. https://doi.org/10.1038/nature03721
33. O'Keefe J, Dostrovsky J (1971) The hippocampus as a spatial map. Preliminary evidence from unit activity in the freely-moving rat. Brain Res 34:171–175. https://doi.org/10.1016/0006-8993(71)90358-1
34. Taube S (1995) Head direction cells recorded in the anterior thalamic nuclei of freely moving rats. J Neurosci 17
35. Jacobs J, Kahana MJ, Ekstrom AD et al (2010) A sense of direction in human entorhinal cortex. Proc Natl Acad Sci 107:6487–6492. https://doi.org/10.1073/pnas.0911213107
36. Jacobs J, Weidemann CT, Miller JF et al (2013) Direct recordings of grid-like neuronal activity in human spatial navigation. Nat Neurosci 16:1188–1190. https://doi.org/10.1038/nn.3466
37. Rolls ET, O'Mara SM (1995) View-responsive neurons in the primate hippocampal complex. Hippocampus 5:409–424. https://doi.org/10.1002/hipo.450050504
38. Killian NJ, Jutras MJ, Buffalo EA (2012) A map of visual space in the primate entorhinal cortex. Nature 491:761–764. https://doi.org/10.1038/nature11587
39. Killian NJ, Potter SM, Buffalo EA (2015) Saccade direction encoding in the primate entorhinal cortex during visual exploration. Proc Natl Acad Sci 112:15743–15748. https://doi.org/10.1073/pnas.1417059112
40. Wirth S, Baraduc P, Planté A et al (2017) Gaze-informed, task-situated representation of space in primate hippocampus during virtual navigation. PLOS Biol 15:e2001045. https://doi.org/10.1371/journal.pbio.2001045
41. Julian JB, Keinath AT, Frazzetta G, Epstein RA (2018) Human entorhinal cortex represents visual space using a boundary-anchored grid. Nat Neurosci 21:191–194. https://doi.org/10.1038/s41593-017-0049-1
42. Nau M, Navarro Schröder T, Bellmund JLS, Doeller CF (2018) Hexadirectional coding of visual space in human entorhinal cortex. Nat Neurosci 21:188–190. https://doi.org/10.1038/s41593-017-0050-8
43. Staudigl T, Leszczynski M, Jacobs J et al (2018) Hexadirectional modulation of high-frequency electrophysiological activity in the human anterior medial temporal lobe maps visual space. Curr Biol 28:3325-3329.e4. https://doi.org/10.1016/j.cub.2018.09.035
44. Ross J, Morrone MC, Goldberg ME, Burr DC (2001) Changes in visual perception at the time of saccades. Trends Neurosci 24:113–121. https://doi.org/10.1016/S0166-2236(00)01685-4
45. VanRullen R, Koch C (2003) Is perception discrete or continuous? Trends Cogn Sci 7:207–213. https://doi.org/10.1016/S1364-6613(03)00095-0
46. Wurtz RH (2008) Neuronal mechanisms of visual stability. Vision Res 48:2070–2089. https://doi.org/10.1016/j.visres.2008.03.021
47. Buzsáki G (2010) Neural syntax: cell assemblies, synapsembles, and readers. Neuron 68:362–385. https://doi.org/10.1016/j.neuron.2010.09.023

48. Sommer MA, Wurtz RH (2002) A pathway in primate brain for internal monitoring of movements. Science 296:1480–1482. https://doi.org/10.1126/science.1069590
49. Sherman SM (2016) Thalamus plays a central role in ongoing cortical functioning. Nat Neurosci 19:533–541. https://doi.org/10.1038/nn.4269
50. Andrillon T, Nir Y, Cirelli C et al (2015) Single-neuron activity and eye movements during human REM sleep and awake vision. Nat Commun 6:7884. https://doi.org/10.1038/ncomms8884
51. Staudigl T, Minxha J, Mamelak AN et al (2022) Saccade-related neural communication in the human medial temporal lobe is modulated by the social relevance of stimuli. Sci Adv
52. Voloh B, Womelsdorf T (2016) A role of phase-resetting in coordinating large scale neural networks during attention and goal-directed behavior. Front Syst Neurosci. https://doi.org/10.3389/fnsys.2016.00018
53. Kragel JE, Voss JL (2022) Looking for the neural basis of memory. Trends Cogn Sci 26:53–65. https://doi.org/10.1016/j.tics.2021.10.010
54. Meister MLR, Buffalo EA (2016) Getting directions from the hippocampus: the neural connection between looking and memory. Neurobiol Learn Mem 134:135–144. https://doi.org/10.1016/j.nlm.2015.12.004
55. Nau M, Julian JB, Doeller CF (2018) How the brain's navigation system shapes our visual experience. Trends Cogn Sci 22:810–825. https://doi.org/10.1016/j.tics.2018.06.008
56. Rolls ET, Wirth S (2018) Spatial representations in the primate hippocampus, and their functions in memory and navigation. Prog Neurobiol 171:90–113. https://doi.org/10.1016/j.pneurobio.2018.09.004
57. Engbert R, Kliegl R (2003) Microsaccades uncover the orientation of covert attention. Vision Res 43:1035–1045. https://doi.org/10.1016/S0042-6989(03)00084-1
58. Popov T, Miller GA, Rockstroh B et al (2021) Alpha oscillations link action to cognition: an oculomotor account of the brain's dominant rhythm. bioRxiv. https://doi.org/10.1101/2021.09.24.461634
59. Martinez-Conde S, Macknik SL, Hubel DH (2004) The role of fixational eye movements in visual perception. Nat Rev Neurosci 5:229–240. https://doi.org/10.1038/nrn1348
60. Jensen O, Pan Y, Frisson S, Wang L (2021) An oscillatory pipelining mechanism supporting previewing during visual exploration and reading. Trends Cogn Sci 25:1033–1044. https://doi.org/10.1016/j.tics.2021.08.008
61. Rayner K (1998) Eye movements in reading and information processing: 20 years of research. Psychol Bull 124:372–422
62. Ahissar E, Arieli A (2001) Figuring space by time. Neuron 32:185–201. https://doi.org/10.1016/S0896-6273(01)00466-4
63. Leszczynski M, Schroeder CE (2019) The role of neuronal oscillations in visual active sensing. Front Integr Neurosci 13:32. https://doi.org/10.3389/fnint.2019.00032
64. Martinez-Conde S, Macknik SL, Troncoso XG, Dyar TA (2006) Microsaccades counteract visual fading during fixation. Neuron 49:297–305. https://doi.org/10.1016/j.neuron.2005.11.033
65. Liversedge SP, Findlay JM (2000) Saccadic eye movements and cognition. Trends Cogn Sci 4:6–14. https://doi.org/10.1016/S1364-6613(99)01418-7
66. Wang S, Berlyne N, Mamelak AN, Rutishauser U (2019) Simultaneous eye tracking and single-neuron recordings in human epilepsy patients. JoVE J Vis Exp
67. Artoni F, Barsotti A, Guanziroli E et al (2018) Effective synchronization of EEG and EMG for mobile brain/body imaging in clinical settings. Front Hum Neurosci 11:652. https://doi.org/10.3389/fnhum.2017.00652
68. Jia Y, Tyler CW (2019) Measurement of saccadic eye movements by electrooculography for simultaneous EEG recording. Behav Res Methods 51:2139–2151. https://doi.org/10.3758/s13428-019-01280-8
69. Salvucci DD, Goldberg JH (2000) Identifying fixations and saccades in eye-tracking protocols. In: Proceedings of the symposium on Eye tracking research & applications—ETRA '00. ACM Press, Palm Beach Gardens, Florida, United States, pp 71–78

70. Nyström M, Holmqvist K (2010) An adaptive algorithm for fixation, saccade, and glissade detection in eyetracking data. Behav Res Methods 42:188–204. https://doi.org/10.3758/BRM.42.1.188
71. Minxha J, Mosher C, Morrow JK et al (2017) Fixations gate species-specific responses to free viewing of faces in the human and macaque amygdala. Cell Rep 18:878–891. https://doi.org/10.1016/j.celrep.2016.12.083
72. Desai M, Holder J, Villarreal C et al (2021) Generalizable EEG encoding models with naturalistic audiovisual stimuli. J Neurosci 41:8946–8962. https://doi.org/10.1523/JNEUROSCI.2891-20.2021
73. Nishimoto S, Vu AT, Naselaris T et al (2011) reconstructing visual experiences from brain activity evoked by natural movies. Curr Biol 21:1641–1646. https://doi.org/10.1016/j.cub.2011.08.031
74. Hoefle S, Engel A, Basilio R et al (2018) Identifying musical pieces from fMRI data using encoding and decoding models. Sci Rep 8:2266. https://doi.org/10.1038/s41598-018-20732-3
75. Fernandes HL, Stevenson IH, Phillips AN et al (2014) Saliency and saccade encoding in the frontal eye field during natural scene search. Cereb Cortex 24:3232–3245. https://doi.org/10.1093/cercor/bht179
76. Nau M, Navarro Schröder T, Frey M, Doeller CF (2020) Behavior-dependent directional tuning in the human visual-navigation network. Nat Commun 11:1–13. https://doi.org/10.1038/s41467-020-17000-2
77. Haufe S, Meinecke F, Görgen K et al (2014) On the interpretation of weight vectors of linear models in multivariate neuroimaging. Neuroimage 87:96–110. https://doi.org/10.1016/j.neuroimage.2013.10.067
78. van Gerven MAJ (2017) A primer on encoding models in sensory neuroscience. J Math Psychol 76:172–183. https://doi.org/10.1016/j.jmp.2016.06.009
79. Singmann H, Kellen D (2019) An introduction to mixed models for experimental psychology. In: Spieler D, Schumacher E (eds) New methods in cognitive psychology, 1st edn. Routledge, pp 4–31

Chapter 15
How Can I Combine Data from fMRI, EEG, and Intracranial EEG?

Biao Han, Lu Shen, and Qi Chen

Abstract Functional Magnetic Resonance Imaging (fMRI) has become a widely used method for noninvasive mapping of cognitive functions in humans, and allows for the functional characterization of specific brain regions or large-scale neural networks. However, the temporal resolution of fMRI is limited due to the delayed hemodynamic response and the relatively poor signal-to-noise ratio. Therefore, the fine-grained temporal dynamics, the critical frequency bands, and the neural network connectivity within and across different frequency bands underlying specific cognitive functions cannot be well delineated by fMRI. Electroencephalography (EEG), on the other hand, has excellent temporal resolution but poor spatial resolution. Intracranial electroencephalography (iEEG) recordings in patients with drug-resistant epilepsy provide neurophysiological signals with superior temporal resolution, together with high anatomical precision. In the chapter, we are going to discuss how to combine data from fMRI, EEG and iEEG to best reveal the neural mechanisms underlying cognitive functions.

15.1 Introduction

As a brain signal recording technique with superior temporal and spatial resolutions, intracranial EEG has become an important tool for studying various cognitive processes in the human brain. Notably, the simultaneous employment of iEEG with existing noninvasive brain signal recording techniques (e.g., EEG and fMRI) in the same study has provided profound insights on the neural dynamics underlying a variety of cognitive processes, such as attention [1, 2], perception [1, 3], and memory [4, 5]. Theoretically, combining iEEG data with results from previous fMRI and EEG studies is straightforward since they complete each other in terms of spatial and temporal resolutions [6]. However, from a technical perspective, the rationale and the analysis workflow on how to bridge data from iEEG and fMRI/EEG experiments remain unclear. In this chapter, we describe the reasons for combining

B. Han · L. Shen · Q. Chen (✉)
School of Psychology, South China Normal University, Guangzhou 510631, China
e-mail: qi.chen27@gmail.com

N. Axmacher (ed.), *Intracranial EEG*, Studies in Neuroscience, Psychology and Behavioral Economics, https://doi.org/10.1007/978-3-031-20910-9_15

iEEG data with EEG or fMRI data, and the various approaches adopted in existing literature to combine these data.

Even though iEEG provides superior resolutions to EEG/fMRI in both the spatial and temporal domains, there are still many advantages to combine them. First, the non-invasive recordings (EEG and fMRI) provide broader coverage of brain regions. Since techniques like iEEG are most often used to identify the epileptogenic zone in patients with refractory epilepsy (see Chap. 1), this clinical rather than research motivation defines the implantation strategy. Specifically speaking, the number and the anatomical coverage of the implanted electrodes are based on diagnostic purposes regarding each patient's specific medical conditions [6]. On the contrary, many studies using non-invasive techniques such as functional MRI and EEG are designed solely for research purposes, and thus whole brain coverage is achievable. Therefore, to circumvent the limited brain coverage of iEEG, brain regions associated with a specific cognitive process can be first identified via fMRI studies, and then iEEG patients with implantations in the fMRI-localized brain regions will be recruited (see also Chap. 29). Second, since non-invasive recording methods can be used in both healthy subjects and patients, a comprehensive data pool can be acquired. In contrast, due to the invasive and clinically oriented properties of iEEG recordings, the number of patients participating in iEEG experiments is typically rather small, and accordingly the iEEG data pool is limited. By combing non-invasive recordings and iEEG, we can first detect relatively small effects using the large amount of data in the non-invasive methods, and then further confirm and characterize these effects taking advantage of the spatial and temporal resolution of iEEG. Last but not least, the combined use of iEEG and EEG/fMRI make it possible to explore the relationship between different data dimensions. For example, by combining the blood oxygen signal provided by fMRI with the intracranial signal of iEEG, it is possible to explore the relationship between the hemodynamic response and the neural oscillations during different cognitive processes.

In the subsequent sections, we will review studies that have effectively integrated iEEG data with either previous or concurrent fMRI/EEG data. We aim at providing a conceptual guidance on the data analysis pipelines for combining iEEG data with either fMRI or EEG data. Specifically, for the combination of iEEG and fMRI, we focus on the advantages of fMRI in providing anatomical precisions and whole brain coverage, together with the high temporal resolution of iEEG. For the combination of iEEG and EEG, we focus on the advantages of EEG in terms of the extensive amount of previous experiments and whole-brain coverage, together with the high spatial resolution of iEEG.

15.2 How Can I Combine Data from fMRI and Intracranial EEG?

Since the early 1990s, functional magnetic resonance imaging (fMRI) has been widely adopted in brain mapping research because it does not involve injections, surgery, the ingestion of substances, or exposure to ionizing radiation. Based on its non-invasiveness, fMRI is applicable to a relatively wide range of test subjects, including animals, healthy humans, and patients, among others. More importantly, it can provide whole brain scans with high spatial resolution. Therefore, the existing fMRI results provide the anatomical basis for iEEG studies. For example, fMRI studies can provide information on task-relevant brain areas (e.g., visual areas, auditory areas, motor areas, and high-order areas etc.) and whole-brain templates (e.g., Desikan-Killiany altas, Destrieux atlas), which help iEEG studies to purposefully identify brain areas of interest and pick out relevant electrodes [7–9]. Moreover, as fMRI records blood oxygen signals and iEEG records electrophysiological signals, the combination of the two techniques promotes an understanding of the common nature of neural signals underlying specific cognitive functions through multidimensional data analysis. Based on the existing literature, the methods of combining fMRI and iEEG can be categorized into three groups according to the complexity of data analysis: (1) direct combination of univariate signals, (2) combination of multivariate signals in decoding analyses, and (3) combination of multivariate signals in connectivity analyses (Fig. 15.1).

15.2.1 Combining Univariate Signal in fMRI and iEEG

Like all hemodynamics-based techniques, fMRI measures an indirect signal, i.e., the blood oxygen level dependent (BOLD) signal, whose spatial and temporal specificity is limited by physical and biological constraints. In general, high spatial resolution of 3 to 4 mm can be easily achieved with typical fMRI, while spatial resolution of 0.5 mm or less is possible only at higher magnetic fields (e.g., 7 T) [10]. Furthermore, the typical hemodynamic response function (HRF) of the BOLD response has a width of around 3 s, with a peak occurring approximately 5 to 6 s after the onset of a brief neural stimulus, a rate much slower than the underlying neural process, and thus the temporal information is severely blurred [10]. On the contrary, iEEG allows for a direct measurement of electrophysiological signals in the brain at millisecond resolution and minimal latency through implanted electrodes. Moreover, iEEG has the advantage of high spatial resolution. For example, typical stereotactic EEG electrodes have a contact length of ~2 mm and a diameter of ~1 mm, capturing signals from ~500,000 cells around each electrode [6, 11] (see also Chap. 17). The high temporal and spatial resolution of iEEG makes it a highly valuable tool for research. However, since the iEEG implantation schemes in patients are designed for diagnosis purposes only, unlike fMRI, these implants tend to be clustered in specific

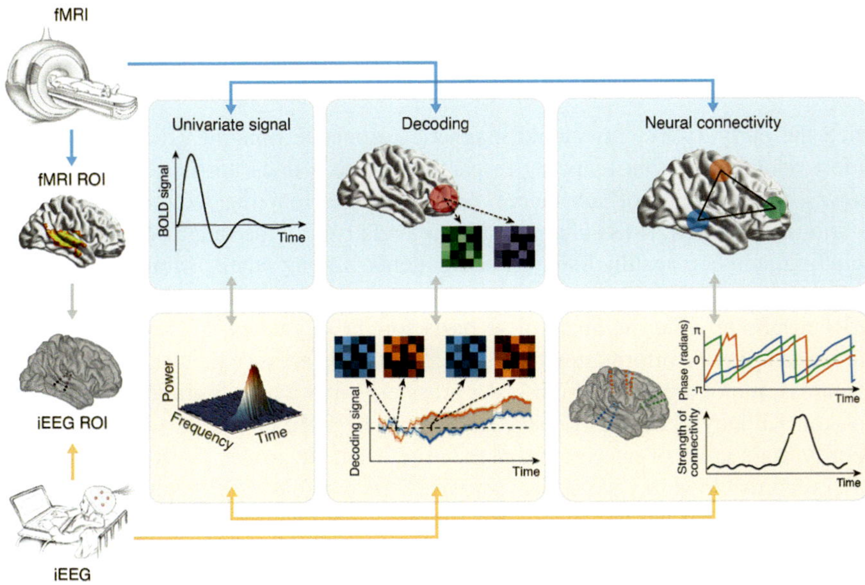

Fig. 15.1 Combining fMRI and iEEG data. The fMRI whole brain data provide information on task-relevant brain areas, i.e., regions of interest (ROIs), that may be used to select electrodes of interest in iEEG studies. For the univariate signal analysis, the amplitude of BOLD responses in fMRI can be linked to the power of neural oscillations in different frequency bands in iEEG, within the same anatomical region. For multivariate decoding analyses, neural representations of different categories may be separated within a brain region via fMRI, and temporal dynamics of the decoding performance can be further investigated via temporal decoding and temporal generalization analyses in iEEG studies. For the neural connectivity analysis, fMRI data may provide information on functionally connected brain regions, and iEEG data further show the carrier frequency and temporal dynamics of connectivity between these brain regions

brain areas and it is not possible to achieve full coverage of the whole brain. Therefore, by combining fMRI and iEEG, researchers can take advantage of both imaging techniques to achieve a more fine-grained understanding of the neural mechanisms underlying a specific cognitive task.

For example, in existing fMRI studies, it has been well documented that the default mode network (DMN) exhibits antagonistic activity with the dorsal attention network (DAN) and the saliency network (SN) during a variety of externally oriented cognitive tasks (e.g., attention and conflict tasks) [12–14] (see also Chap. 33). Using the continuous performance task sessions (GradCPT) paradigm, an iEEG study investigated the neural relevance and fine-grained temporal dynamics during attentional fluctuations [15]. To know exactly where the electrodes are located in these networks, researchers assigned each electrode to membership with the DMN, DAN, SN, or none of these networks, based on alignment with the Yeo 7 atlas (Fig. 15.2a). Later, consistent with the results of previous fMRI studies, analysis based on these three networks showed distinct profiles of task-evoked activity and provided direct electrophysiological evidence that antagonistic inter-network dynamics relate to attentional performance.

Moreover, concurrent fMRI and iEEG studies have been performed to associate the BOLD signal with neural oscillations in different frequency bands in the same cognitive task. A link between an increased BOLD signal and increased gamma-band power has been found in the visual [16, 17], auditory [1, 18], temporal [19], and sensorimotor cortex [20]. For example, Conner et al. [21] combined fMRI and iEEG in a visually cued object naming task or a verb generation task, and found a positive correlation between the BOLD signal and the power of neural activity at 60–120 Hz. Nevertheless, the direct BOLD-high gamma correspondence was not observed in all brain regions and cognitive tasks. For example, in a navigation task, Ekstrom et al. [22] did not find a coupling between BOLD and gamma power in parahippocampus; instead, a significant positive correlation was found between the BOLD signal and theta power. Unlike the high-frequency gamma activity, correspondence between the BOLD signal and low-frequency band oscillations is less consistent, depending on the specific task and the brain regions involved. For example, similar to the above-mentioned positive correlation between BOLD signal and theta power during virtual navigation in the parahippocampus, a positive correlation between BOLD responses and theta power was also found in the amygdala during a fearful face task [23]. By contrast, BOLD signal increase were found to correspond to alpha- and beta-band power decreases in human motor cortex during a simple finger movement task [20] and to beta-band power decreases in the frontal cortex [21].

15.2.2 Decoding with fMRI and iEEG

Traditional univariate analysis methods treat information in each dimension (e.g., each voxel or electrode) as independent data, which does not reflect the fact that information may be contained in the brain by non-linear combinations of activity in different dimensions. Decoding, or multivariate pattern analysis, extracts information contained in patterns of activity across multiple dimensions, such that patterns of activity between each dimension can provide information that is not available in univariate analysis. In decoding, fMRI shows great advantages along the spatial dimension while iEEG shows advantages along both the spatial, temporal and frequency dimensions. The combination of the two techniques provides us with rich multi-dimensional decoding information.

For example, the ability to recognize faces is one of the most intriguing aspects of human vision and has received a substantial amount of attention. A number of brain regions have been identified that are said to be involved in face recognition, the best known being the fusiform face area (FFA) [24, 28]. Previous fMRI studies have demonstrated that FFA activity contains information about individual faces invariant across facial expression (Fig. 15.2b, left) [24, 29–31]. However, face processing was

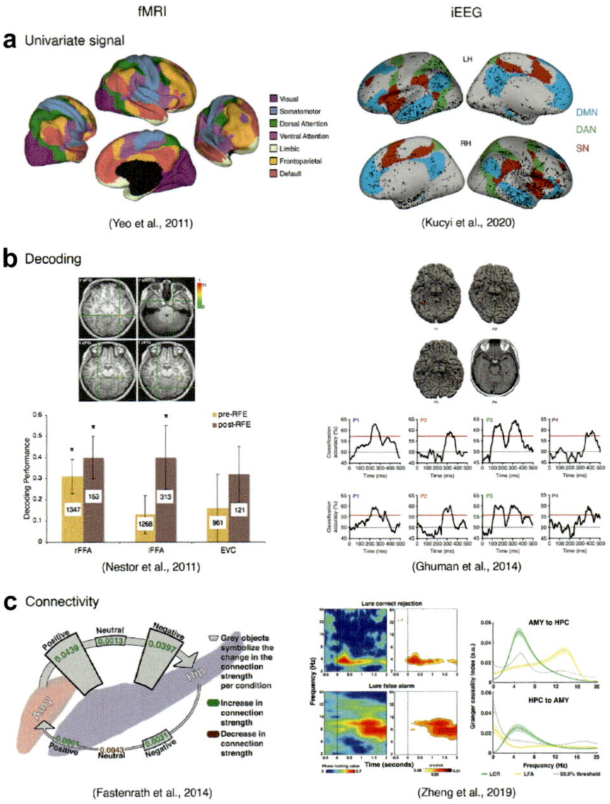

Fig. 15.2 Examples of combining fMRI and iEEG data. **a** Univariate signals. Based on fMRI data, it has been well documented that the default mode network (DMN) exhibits antagonistic activity with the dorsal attention network (DAN) and salience network (SN, also referred to as the ventral attention network) during both resting state and a variety of externally oriented cognitive tasks. Taking advantage of a brain atlas derived from previous fMRI studies (left) [9], Kucyi et al. [15] adopted iEEG and assigned implanted electrodes to the DMN, DAN, and SN, respectively, to investigate the fine-grained neural dynamics within and between these networks (right). **b** Decoding analysis. Left: fMRI data show that activity in the fusiform face area (FFA) contains information about individual faces, invariant across facial expressions (rFFA/lFFA: right/left fusiform face area, EVC: early visual cortex) [24]. Right: Based on fMRI data, Ghuman et al. [25] adopted iEEG and further tested the temporal dynamics of face individuation and categorization. The results showed that FFA is involved in multiple face processing stages (red lines indicate 95% threshold, corrected for multiple comparisons). **c** Connectivity analysis. Left: fMRI data show increases in effective connectivity from the amygdala to the hippocampus during processing emotionally arousing information [26]. Right: Based on fMRI data, Zheng et al. [27] adopted iEEG and investigated the temporal progress of the amygdala-hippocampus connectivity. The iEEG results showed that early theta-band connectivity is associated with successful processing of emotional stimuli, while later alpha-band connectivity is associated with discrimination errors (AMY: amygdala, HIP: hippocampus, LCR: lure correct rejection, LFA: lure false alarm). Figures are modified from [9, 15, 24–27]

thought to occur through a set of partially distinct stages [32]. These stages are difficult to disentangle via fMRI because of its relatively low temporal resolution, and thus this method has provided little insights on the relative timing of FFA responses to different aspects of face-related information. An iEEG study combined two experiments to tackle this question [25]. Participants were instructed to press a button when an image was repeated in Experiment 1, an approach which was adopted to examine the temporal dynamics of face sensitivity and specificity in FFA. Participants were then instructed to report whether the face was male or female in Experiment 2, which was employed to examine the temporal dynamics of face individuation and categorization invariant to facial expression. Multivariate pattern classification was then used to decode the temporal dynamics of expression-invariant face information processing using electrodes placed directly in the FFA. Results provide strong evidence that FFA is involved in multiple face processing stages: early FFA activities (50–75 ms) contained information related to face detection, while intermediate activities (200–500 ms) were related to expression-independent individuation using facial features and their configuration (Fig. 15.2b, right).

15.2.3 Neural Connectivity Within fMRI and iEEG

Efficient task performance requires neural interactions among a distributed network of areas. Such neural communications involve routing, coordinating, and integrating neural information between brain areas (see also Chap. 19). FMRI, endowed with high spatial resolution and whole-brain coverage, was adopted to localize interacting brain regions and test the direction of information flow [33–35]. The low temporal resolution of fMRI, however, deters it from detecting the fine-grained temporal and frequency dynamics of the neural communication—a problem easily tackled with iEEG. IEEG can be adopted to investigate the inter-regional interactions via the coherence of the neural oscillations, and it offers a detailed delineation of the temporal dynamics of inter-regional connectivity across different frequency bands due to its high temporal resolution [36, 37] (see Chap. 32). Therefore, based on the accurate and complete anatomical architecture defined by fMRI, one could better explore the temporal dynamics of neural connectivity via iEEG.

During emotional memory, for example, several fMRI studies highlighted an increase in the effective connectivity from amygdala to hippocampus during processing of emotionally arousing information (Fig. 15.2c, left) [26, 38]. The temporal progression of amygdala-hippocampus connectivity has remained unknown, however. A recent iEEG study tackled this question [27]. Subjects were instructed to judge whether or not a lure image was the same as an emotionally negative image presented before. The authors calculated the phase locking value (PLV) and the Granger causality index across time and frequency in the lure correct rejection (LCR) and the lure false alarm (LFA) conditions. This analysis showed that successful processing of emotional stimuli was associated with early bi-directional interactions between amygdala and hippocampus in the theta frequency band, while discrimination errors were associated with later uni-directional influence from amygdala to hippocampus in the alpha band (Fig. 15.2c, right).

Moreover, iEEG recordings allow for a detection of electrophysiological components, such as the theta frequency band, as a carrier of neural interactions [39, 40]. Inter-regional coherence of neural oscillations is thought to reflect a fundamental mechanism of neural communications [41, 42] (see also Chap. 25). With ROIs being defined via fMRI, iEEG could take a further step and unveil the carrier frequency underlying neural interactions within and between ROIs. For example, Hacker et al. [43] measured the spatial correspondence between iEEG power correlations and fMRI BOLD correlation patterns in resting human subjects. The results showed that although correspondence of fMRI correlations and iEEG power correlations was common throughout the brain, correspondence with theta (4–8 Hz) power correlations was most pronounced for the default mode network (DMN) and frontoparietal control network (FPC), while correspondence to alpha (8–12 Hz) power correlations was highest for the sensorimotor network (SMN) and dorsal attention network (DAN), suggesting specific carrier frequencies in different neural networks. Similarly, Kucyi et al. [15] assessed high frequency band (HFB) power in different large-scale brain networks during a continuous performance task and compared the latency of responses in these networks. They found that HFB responses peak fastest in the DAN, at intermediate speed in the SN, and slowest in the DMN, demonstrating the behavioral relevance of distributed activity patterns in these networks.

15.3 How Can I Combine Data from EEG and Intracranial EEG?

Scalp EEG recordings are known for their high temporal resolution but low spatial resolution. Due to its non-invasive nature, simplicity of equipment and whole brain electrode coverage, scalp EEG recordings have become one of the most widely used methods in cognitive neuroscience. Since the 1920s, EEG has provided many robust event-related potential (ERP) components, such as N75, P100 and N145 after visual stimuli [44–46] and a consistent P100 response after simple auditory stimuli [47, 48]. These stable ERP components can act as indexes for data validation. Besides, the international 10–20 system of electrode placement of EEG makes it possible to associate cortical regions with specific electrodes, by combining a sphere fitted with the spatial position of the EEG electrodes with MRI or CT images of the brain [49, 50]. For the combination of EEG and iEEG, we also categorize the data analysis methods into three groups based on their complexity: (1) ERPs averaged over simple time dimensions [51, 52]; (2) neural oscillations in the time and frequency domains [53, 54]; and (3) connectivity analyses exploring the relationship between time–frequency data from different brain areas [55, 56] (Fig. 15.3).

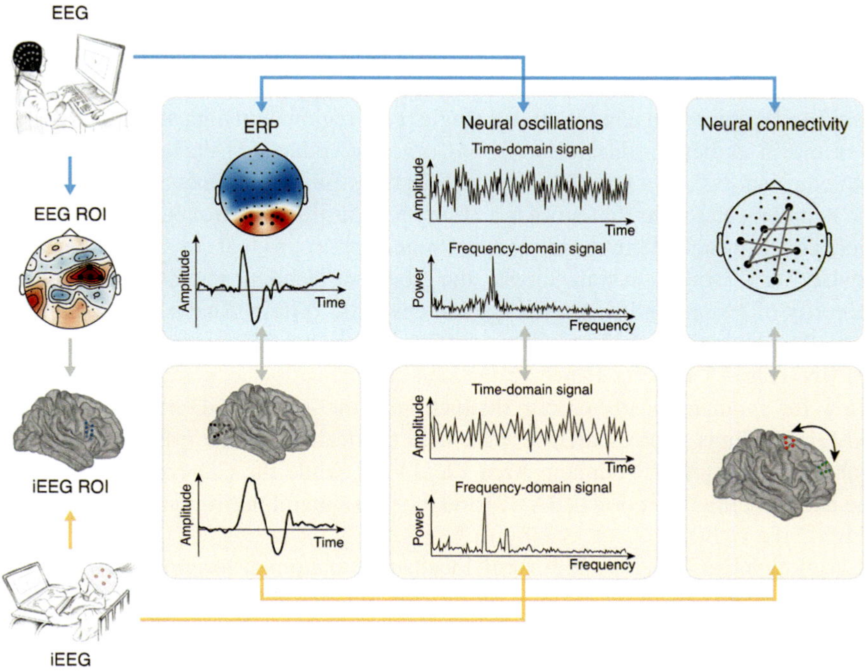

Fig. 15.3 Combining EEG and iEEG data. IEEG data provide precise anatomical localization not only for ERP components, but also for the power, frequency, and phase of neural oscillations. For connectivity analyses, the connectivity between scalp EEG electrodes can be used as a positional reference for connectivity analysis of iEEG

15.3.1 ERP Analysis Based on EEG and iEEG

Electrophysiological activity in the brain is generated by neurons in the cerebral cortex and subcortical structures [57–59]. For EEG signals, the ERP is a sum neural signal on the surface of the brain, which is assumed to be generated by synchronous activity of postsynaptic potentials in millions of neurons via volume conduction [60]. For iEEG signals, the ERP is the sum neural signal of the local field potential (LFP), which is due to both postsynaptic potentials and multi-unit spiking activity [61]. Scalp EEG recordings allow for a simultaneous assessment of ERPs across various regions throughout the brain. Compared to scalp EEG, the iEEG has a higher spatial resolution, and also allows for the recording of activity in deep brain regions that are not accessible by scalp EEG. Therefore, when the ERP results from scalp EEG are combined with the anatomically precise information in the iEEG, researchers are able to assign the differences in the ERP amplitude between different experimental conditions to specific anatomical regions [62–65]. In addition, scalp ERP components can also be accurately localized in specific brain regions taking advantage of the iEEG technique [66].

ERPs can be separated based on their peak latencies, such as P1, N2, P3, and other components, which are thought to reflect different stages of perceptual and cognitive processing. Using iEEG, researchers have explored differences in ERP amplitudes between experimental conditions in a single brain region with high anatomical precision, e.g. in studies on consciousness [64], time perception [65], decision making [63] and emotion [62]. In a joint EEG and iEEG study, Bekinschtein et al. [67] explored the neural mechanisms that process auditory regularities. In the experiment, subjects were either presented groups of identical sounds or groups that contained one "local deviant" (LD) sound. In some blocks, the groups of identical sounds constituted the minority of groups and were thus "global deviants" (GDs). Analysis of scalp EEG data showed that both LD and GD conditions induced an ERP mismatch negativity (MMN) (Fig. 15.4a, left). However, the lack of localization accuracy in EEG did not allow for an anatomically precise distinction of the LD and GD effects. Adopting iEEG recordings, the researchers were able to localize the LD effect to the superior temporal gyrus (Brodmann Area 22, BA22), while the GD effect was seen in the anterior cingulate cortex (BA 32) and the dorsolateral prefrontal cortex (BA 44) (Fig. 15.4a, right).

Although EEG can, to some extent, localize the anatomical origin of ERP components via source reconstruction, it is difficult to determine the source of the ERP components due to the spatial inaccuracy of scalp EEG. Using iEEG, researchers can localize the ERP components found in the scalp EEG to precise anatomical regions. For example, Halgren et al. [66] adopted iEEG in visual and auditory oddball paradigms, and localized the N2-P3 components in three brain regions: auditory-specific regions of the temporal lobe, areas in the parietal lobe associated with attentional orienting, and multimodal association cortex associated with event encoding.

15.3.2 Neural Oscillations in EEG and iEEG

In addition to ERPs, firing patterns in the brain generate large-scale synchronized neuronal activity, resulting in neural oscillations of different frequencies [71–74]. These oscillations can be recorded via scalp EEG or iEEG, and analyzed by converting the time domain signal to the frequency domain using Fourier or Hilbert transforms, or by exploring the temporal dynamics of neural oscillations using time–frequency analysis [53, 54]. When combining the neural oscillation results from EEG and iEEG, iEEG analysis can be based on time–frequency results obtained from scalp EEG, including power, frequency, and phase effects, and be used to further explore differences between experimental conditions in more anatomically refined regions [68, 69, 75–82].

For example, iEEG can help to allocate the neural oscillation effects observed in EEG to precise brain regions. For oscillatory frequency, taking advantage of the bistable Ternus paradigm, Shen et al. [68] found that changes in the frequency of alpha oscillations in posterior EEG electrodes predicted the outcome of bistable perception

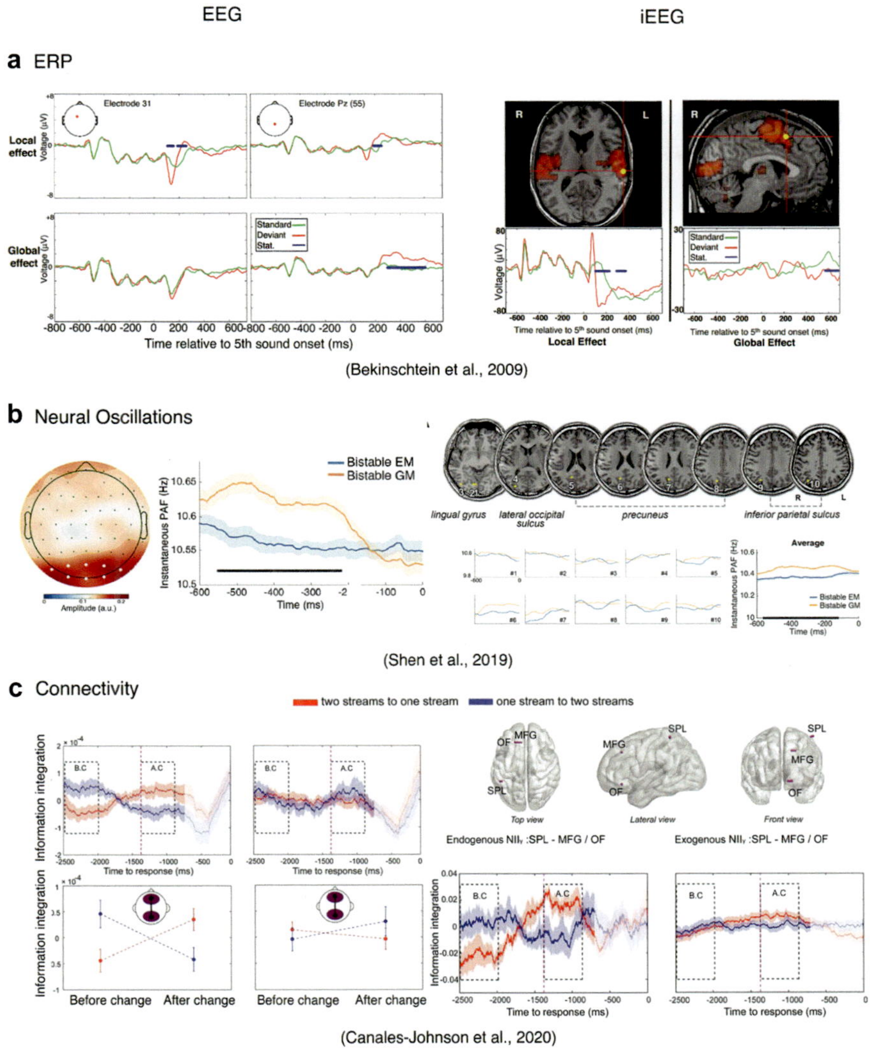

Fig. 15.4 Examples of combining EEG and iEEG data. **a** Left: Bekinschtein et al. [67] first used scalp EEG to demonstrate the local deviant and global deviant effect. Red lines indicate ERPs in the deviant condition and green lines indicate ERPs in the baseline condition. Blue lines indicate significant differences between deviant and baseline condition. Right: Using iEEG, they replicated the ERP results and found that the sources of the local deviant and global deviant effects are different. The yellow disk shows the position of the electrodes. **b** Neural oscillation analysis. Left: Shen et al. [68] first used scalp EEG to show that the frequency of alpha oscillations in posterior electrodes predicts the outcome of bistable perception in the bistable Ternus display task (EM: element motion, GM: group motion). Right: Using iEEG, they replicated the alpha frequency effect, and further localized it to several visual regions, including the lingual gyrus, lateral occipital sulcus, precuneus etc. **c** Connectivity analysis. Left: Canales-Johnson et al. [69] first used scalp EEG to demonstrate that neural connectivity between frontal and parietal areas reflected the two percepts during bistable perception of a sequence of tones. Using iEEG, they further confirmed the connectivity results and localized effects between superior parietal lobe, middle frontal gyrus, and orbitofrontal areas (B.C: before change, A.C: after change). Figures are modified from [64, 68–70]

(Fig. 15.4b, left). Using data from a smaller group of 4 subjects, the iEEG results confirmed the observed alpha frequency effect. Moreover, the authors found that the effect was present in a wide range of cortical regions including lingual gyrus, lateral occipital sulcus, precuneus and others (Fig. 15.4b, right). With regard to oscillatory power, previous EEG studies revealed theta-band activity during spatial navigation [83–85]. However, it has been unclear whether theta oscillations are generated in hippocampus or in other brain regions, and what the dynamics of theta oscillations are. Based on these EEG studies, Kahana et al. [75] conducted iEEG recordings during a spatial navigation paradigm with a short and a long maze. They found that the difference between the short and the long maze was directly reflected in theta power in multiple areas of the cortex, including the temporal lobe, verifying the important role of theta oscillations in spatial navigation. For oscillatory phase, in a joint study of EEG and iEEG, Thézé et al. [81] used computer-generated naturalistic audiovisual speech stimuli, including a keyword with a mismatched phoneme-sight pair, to trigger the McGurk effect (i.e., the percept of a "ba" sound due to visually presented lip movements despite an incongruent simultaneously presented "ga" sound). In this way, the authors created bistable stimuli in which conscious perception is driven by either visual or auditory cues. In the EEG part of the study, they found that the perception of the stimulus was determined by the pre-stimulus phase of theta oscillations. Based on this result, they submitted the time and frequency information to the iEEG part of the study to localize the observed effects and found the same phase results in posterior temporal and occipital cortex of the right hemisphere.

15.3.3 Combining Neural Connectivity in EEG and iEEG

Cognitive processes rely on a dynamic interactions within and between specialized neural networks [86–89]. While scalp EEG may be used to measure coherence between larger-scale neocortical areas, iEEG may reveal coherence between more fine-grained cortical and/or subcortical brain areas [56, 90–92]. Because of their comprehensive coverage, scalp EEG recordings allow mapping coherence patterns across all (accessible) areas of the neocortex. Therefore, the connectivity between scalp EEG electrodes can be used as a reference for connectivity analyses in iEEG studies. Reversely, iEEG connectivity results may be used to interpret EEG connectivity at a higher spatial resolution.

Connectivity results in EEG/MEG studies using the same experimental paradigm can help to further investigate the connectivity between specific brain areas in iEEG studies [36, 69, 93]. For instance, Canales-Johnson et al. [69] asked participants to listen to a sequence of tones (auditory bistable stimuli) that could be experienced as a single stream (perceptual integration) or as two parallel streams (perceptual differentiation) of sounds. They determined the neural information dynamics underlying integration and differentiation by measuring phase lag index (PLI) and weighted symbolic mutual information (wMI) in scalp EEG and iEEG recordings (see Chap. 32 on different connectivity measures). PLI and wMI between frontal and parietal regions in the EEG signals and between superior parietal lobe (SPL) and middle frontal gyrus (MFG)/orbitofrontal cortex (OF) in the iEEG signals showed significant connectivity. This suggests that perceptual integration and differentiation can be mapped to interactions between frontal and temporal networks, whereas no such effect was observed for neural connections within the frontal network or within the temporal network (Fig. 15.4c).

Moreover, simultaneous recording of EEG and iEEG signals is useful to explore the connectivity of neocortical signals (accessible via scalp EEG) and subcortical signals [5, 94, 95]. For instance, the anterior thalamic nucleus (ATN) is thought to play an important role in the formation of memories. Sweeney-Reed et al. [94] asked subjects to perform a memory task including an encoding stage and a recognition test, and simultaneously monitored the activity of each subject's ATN via implanted intracranial electrodes, and frontal and parietal areas via scalp EEG electrodes. The results showed significant neural connectivity between the frontal-parietal area and ATN during successful memory encoding: Neocortical-ATN theta phase synchrony and cross-frequency coupling of neocortical theta phases with ATN gamma power predicted later memory for scenes.

15.4 Discussion

Scalp EEG and fMRI recordings, as two widely used neuroimaging techniques, have contributed substantially to our understanding of the neural mechanisms underlying a variety of cognitive functions. IEEG, as a relatively new and less widely used technique, complements scalp EEG and fMRI in terms of both spatial and temporal resolution. In this chapter, we described relevant previous studies as examples to illustrate why and how to combine iEEG data with scalp EEG/fMRI data. Specifically, fMRI and iEEG data can be combined during univariate signal analysis, multivariate decoding analysis, and functional connectivity analysis. For scalp EEG and iEEG data, ERP components, properties of neural oscillations, and connectivity between scalp EEG electrodes can be accurately localized to specific brain structures via iEEG. Combining these datasets does not only effectively exploit the high temporal and spatial resolution of iEEG data, but also takes advantage of easy data acquisition and whole brain coverage in conventional EEG and fMRI recordings.

However, our understanding of the relationship between different types of brain signals acquired by the different techniques is still incomplete. Thus, the combination of multimodal datasets still has shortcomings and needs further improvements in future studies. For fMRI and iEEG, more studies will need to be conducted that simultaneously utilize fMRI and iEEG. In this way, the correspondence between fMRI activation and neural oscillations will be more adequately delineated. Moreover, since iEEG patients with refractory epilepsy have by definition brain pathologies, it will be ideal to perform fMRI recordings prior to iEEG implantation in the same group of patients [96]. In this way, task-relevant brain areas, which show the same pattern of activation as in previous fMRI studies in healthy adults, can be first localized via fMRI. Subsequently, neural oscillations in these fMRI-localized 'healthy' brain regions may be recorded in the same patients with the same behavioral task. For EEG and iEEG, although we often consider scalp EEG signals as direct recordings of brain activity, there is still a gap between the scalp EEG signals and the local field potentials directly recorded by iEEG. Simultaneous recording of scalp EEG and iEEG provide a better understanding of both, the difference between the electrophysiological signals recorded by scalp EEG vs. iEEG. In addition, iEEG allows for confirmation and improvement of source reconstructions of scalp EEG signals.

References

1. Esposito F, Singer N, Podlipsky I, Fried I, Hendler T, Goebel R (2013) Cortex-based inter-subject analysis of iEEG and fMRI data sets: application to sustained task-related BOLD and gamma responses. Neuroimage 66:457–468
2. Gruber MJ, Hsieh L-T, Staresina BP, Elger CE, Fell J, Axmacher N, Ranganath C (2018) Theta phase synchronization between the human hippocampus and prefrontal cortex increases during encoding of unexpected information: a case study. J Cogn Neurosci 30:1646–1656. https://doi.org/10.1162/jocn_a_01302
3. Ceolini E, Hjortkjær J, Wong DDE, O'Sullivan J, Raghavan VS, Herrero J, Mehta AD, Liu S-C, Mesgarani N (2020) Brain-informed speech separation (BISS) for enhancement of target speaker in multitalker speech perception. NeuroImage 223:117282. https://doi.org/10.1016/j.neuroimage.2020.117282
4. Axmacher N, Elger CE, Fell J (2009) Working memory-related hippocampal deactivation interferes with long-term memory formation. J Neurosci 29:1052–1060. https://doi.org/10.1523/JNEUROSCI.5277-08.2009
5. Sweeney-Reed CM, Zaehle T, Voges J, Schmitt FC, Buentjen L, Kopitzki K, Richardson-Klavehn A, Hinrichs H, Heinze HJ, Knight RT, Rugg MD (2016) Pre-stimulus thalamic theta power predicts human memory formation. Neuroimage 138:100–108. https://doi.org/10.1016/j.neuroimage.2016.05.042
6. Parvizi J, Kastner S (2018) Promises and limitations of human intracranial electroencephalography. Nat Neurosci 21:474–483. https://doi.org/10.1038/s41593-018-0108-2
7. Desikan RS, Ségonne F, Fischl B, Quinn BT, Dickerson BC, Blacker D, Buckner RL, Dale AM, Maguire RP, Hyman BT, Albert MS, Killiany RJ (2006) An automated labeling system for subdividing the human cerebral cortex on MRI scans into gyral based regions of interest. Neuroimage 31:968–980. https://doi.org/10.1016/j.neuroimage.2006.01.021
8. Destrieux C, Fischl B, Dale A, Halgren E (2010) Automatic parcellation of human cortical gyri and sulci using standard anatomical nomenclature. Neuroimage 53:1–15. https://doi.org/10.1016/j.neuroimage.2010.06.010
9. Yeo BT, Krienen FM, Sepulcre J, Sabuncu MR, Lashkari D, Hollinshead M, Roffman JL, Smoller JW, Zöllei L, Polimeni JR, Fisch B, Liu H, Buckner RL (2011) The organization of the human cerebral cortex estimated by intrinsic functional connectivity. J Neurophysiol 106:1125–1165. https://doi.org/10.1152/jn.00338.2011
10. Glover GH (2011) Overview of functional magnetic resonance imaging. Neurosurg Clin N Am 22:133–139. https://doi.org/10.1016/j.nec.2010.11.001
11. Miller KJ, Sorensen LB, Ojemann JG, Den Nijs M (2009) Power-law scaling in the brain surface electric potential. PLoS Comput Biol 5:e1000609
12. Dixon ML, Andrews-Hanna JR, Spreng RN, Irving ZC, Mills C, Girn M, Christoff K (2017) Interactions between the default network and dorsal attention network vary across default

subsystems, time, and cognitive states. Neuroimage 147:632–649. https://doi.org/10.1016/j.neuroimage.2016.12.073
13. Wang C, Ong JL, Patanaik A, Zhou J, Chee MWL (2016) Spontaneous eyelid closures link vigilance fluctuation with fMRI dynamic connectivity states. Proc Natl Acad Sci USA 113:9653–9658. https://doi.org/10.1073/pnas.1523980113
14. Kucyi A, Hove MJ, Esterman M, Matthew Hutchison R, Valera EM (2017) Dynamic brain network correlates of spontaneous fluctuations in attention. Cereb Cortex 27:1831–1840. https://doi.org/10.1093/cercor/bhw029
15. Kucyi A, Daitch A, Raccah O, Zhao B, Zhang C, Esterman M, Zeineh M, Halpern CH, Zhang K, Zhang J, Parvizi J (2020) Electrophysiological dynamics of antagonistic brain networks reflect attentional fluctuations. Nat Commun. https://doi.org/10.1038/s41467-019-14166-2
16. Niessing J, Ebisch B, Schmidt KE, Niessing M, Singer W, Galuske RA (2005) Hemodynamic signals correlate tightly with synchronized gamma oscillations. Science 309:948–951
17. Goense JBM, Logothetis NK (2008) Neurophysiology of the BOLD fMRI signal in awake monkeys. Curr Biol 18:631–640. https://doi.org/10.1016/j.cub.2008.03.054
18. Mukamel R, Gelbard H, Arieli A, Hasson U, Fried I, Malach R (2005) Coupling between neuronal firing, field potentials, and FMRI in human auditory cortex. Science 309:951–954. https://doi.org/10.1126/science.1110913
19. Ojemann GA, Corina DP, Corrigan N, Schoenfield-McNeill J, Poliakov A, Zamora L, Zanos S (2010) Neuronal correlates of functional magnetic resonance imaging in human temporal cortex. Brain 133:46–59. https://doi.org/10.1093/brain/awp227
20. Hermes D, Miller KJ, Vansteensel MJ, Aarnoutse EJ, Leijten FS, Ramsey NF (2012) Neurophysiologic correlates of fMRI in human motor cortex. Hum Brain Mapp 33:1689–1699. https://doi.org/10.1002/hbm.21314
21. Conner CR, Ellmore TM, Pieters TA, DiSano MA, Tandon N (2011) Variability of the relationship between electrophysiology and BOLD-fMRI across cortical regions in humans. J Neurosci 31:12855–12865
22. Ekstrom A, Suthana N, Millett D, Fried I, Bookheimer S (2009) Correlation between BOLD fMRI and theta-band local field potentials in the human hippocampal area. J Neurophysiol 101:2668–2678
23. Fedele T, Tzovara A, Steiger B, Hilfiker P, Grunwald T, Stieglitz L, Jokeit H, Sarnthein J (2020) The relation between neuronal firing, local field potentials and hemodynamic activity in the human amygdala in response to aversive dynamic visual stimuli. Neuroimage 213:116705
24. Nestor A, Plaut DC, Behrmann M (2011) Unraveling the distributed neural code of facial identity through spatiotemporal pattern analysis. Proc Natl Acad Sci USA 108:9998–10003. https://doi.org/10.1073/pnas.1102433108
25. Ghuman AS, Brunet NM, Li Y, Konecky RO, Pyles JA, Walls SA, Destefino V, Wang W, Richardson RM (2014) Dynamic encoding of face information in the human fusiform gyrus. Nat Commun. https://doi.org/10.1038/ncomms6672
26. Fastenrath M, Coynel D, Spalek K, Milnik A, Gschwind L, Roozendaal B, Papassotiropoulos A, de Quervain DJ (2014) Dynamic modulation of amygdala–hippocampal connectivity by emotional arousal. J Neurosci 34:13935–13947
27. Zheng J, Stevenson RF, Mander BA, Mnatsakanyan L, Hsu FPK, Vadera S, Knight RT, Yassa MA, Lin JJ (2019) Multiplexing of theta and alpha rhythms in the amygdala-hippocampal circuit supports pattern separation of emotional information. Neuron 102:887-898.e5. https://doi.org/10.1016/j.neuron.2019.03.025
28. Haxby J V, Hoffman EA, Gobbini MI (2000) The distributed human neural system for face perception. Trends Cogn Sci 4:223–233
29. Anzellotti S, Fairhall SL, Caramazza A (2014) Decoding representations of face identity that are tolerant to rotation. Cereb Cortex 24:1988–1995. https://doi.org/10.1093/cercor/bht046
30. Cowen AS, Chun MM, Kuhl BA (2014) Neural portraits of perception: reconstructing face images from evoked brain activity. Neuroimage 94:12–22. https://doi.org/10.1016/j.neuroimage.2014.03.018

31. Goesaert E, Op de Beeck HP (2013) Representations of facial identity information in the ventral visual stream investigated with multivoxel pattern analyses. J Neurosci 33:8549–8558. https://doi.org/10.1523/JNEUROSCI.1829-12.2013
32. Bruce V, Young A (1986) Understanding face recognition. Br J Psychol 77:305–327. https://doi.org/10.1111/j.2044-8295.1986.tb02199.x
33. Hopfinger JB, Buonocore MH, Mangun GR (2000) The neural mechanisms of top-down attentional control. Nat Neurosci 3:284–291
34. Finn AS, Sheridan MA, Kam CLH, Hinshaw S, D'Esposito M (2010) Longitudinal evidence for functional specialization of the neural circuit supporting working memory in the human brain. J Neurosci 30:11062–11067
35. Smith SM (2012) The future of FMRI connectivity. Neuroimage 62:1257–1266. https://doi.org/10.1016/j.neuroimage.2012.01.022
36. Watrous AJ, Tandon N, Conner CR, Pieters T, Ekstrom AD (2013) Frequency-specific network connectivity increases underlie accurate spatiotemporal memory retrieval. Nat Neurosci 16:349–356
37. Chen S, Tan Z, Xia W, Gomes CA, Zhang X, Zhou W, Liang S, Axmacher N, Wang L (2021) Theta oscillations synchronize human medial prefrontal cortex and amygdala during fear learning. Sci Adv 7:eabf4198
38. Nawa NE, Ando H (2019) Effective connectivity within the ventromedial prefrontal cortex-hippocampus-amygdala network during the elaboration of emotional autobiographical memories. Neuroimage 189:316–328
39. Johnson EL, Adams JN, Solbakk A-K, Endestad T, Larsson PG, Ivanovic J, Meling TR, Lin JJ, Knight RT (2018) Dynamic frontotemporal systems process space and time in working memory. PLoS Biol 16:e2004274
40. Kam JW, Lin JJ, Solbakk A-K, Endestad T, Larsson PG, Knight RT (2019) Default network and frontoparietal control network theta connectivity supports internal attention. Nat Hum Behav 3:1263–1270
41. Fries P (2005) A mechanism for cognitive dynamics: neuronal communication through neuronal coherence. Trends Cogn Sci 9:474–480. https://doi.org/10.1016/j.tics.2005.08.011
42. Schneider M, Broggini AC, Dann B, Tzanou A, Uran C, Sheshadri S, Scherberger H, Vinck M (2021) A mechanism for inter-areal coherence through communication based on connectivity and oscillatory power. Neuron 109:4050-4067.e12. https://doi.org/10.1016/j.neuron.2021.09.037
43. Hacker CD, Snyder AZ, Pahwa M, Corbetta M, Leuthardt EC (2017) Frequency-specific electrophysiologic correlates of resting state fMRI networks. Neuroimage 149:446–457
44. Corletto F, Gentilomo A, Rosadini G, Rossi G, Zattoni J (1967) Visual evoked potentials as recorded from the scalp and from the visual cortex before and after surgical removal of the occipital pole in man. Electroencephalogr Clin Neurophysiol 22:378–380
45. Aminoff MJ, Goodin DS (1994) Visual evoked potentials. J Clin Neurophysiol 11:493–499
46. Odom JV, Bach M, Barber C, Brigell M, Marmor MF, Tormene AP, Holder GE et al (2004) Visual evoked potentials standard. Doc Ophthalmol 108:115–123
47. Davis PA (1939) Effects of acoustic stimuli on the waking human brain. J Neurophysiol 2:494–499
48. Korpilahti P, Lang H (1994) Auditory ERP components and mismatch negativity in dysphasic children. Electroencephalogr Clin Neurophysiol 91:256–264
49. Jasper HH (1958) The ten-twenty electrode system of the international federation. Electroencephalogr Clin Neurophysiol 10:370–375
50. Homan RW (1988) The 10–20 electrode system and cerebral location. Am J EEG Technol 28:269–279
51. Picton TW, Lins OG, Scherg M (1995) The recording and analysis of event-related potentials. Handbook of Neuropsychol 10:3–3
52. Luck SJ (2012) Event-related potentials
53. Allen JB, Rabiner LR (1977) A unified approach to short-time Fourier analysis and synthesis. Proc IEEE 65:1558–1564

54. Boashash B (1992) Estimating and interpreting the instantaneous frequency of a signal. I. Fundamentals. Proc IEEE 80:520–538
55. Schoffelen J-M, Gross J (2009) Source connectivity analysis with MEG and EEG. Hum Brain Mapp 30:1857–1865
56. Sakkalis V (2011) Review of advanced techniques for the estimation of brain connectivity measured with EEG/MEG. Comput Biol Med 41:1110–1117
57. Freeman WJ et al (1975) Mass action in the nervous system. Citeseer
58. Mitzdorf U (1985) Current source-density method and application in cat cerebral cortex: investigation of evoked potentials and EEG phenomena. Physiol Rev 65:37–100
59. Speckmann E-J (1997) Generation of field potentials in the brain. J Clin Pharmacol 37:8S-10S
60. Brazier MA (1966) A study of the electrical fields at the surface of the head. Am J EEG Technol 6:114–128
61. Buzsáki G, Anastassiou CA, Koch C (2012) The origin of extracellular fields and currents–EEG, ECoG, LFP and spikes. Nat Rev Neurosci 13:407–420. https://doi.org/10.1038/nrn3241
62. Krolak-Salmon P, Hénaff M-A, Vighetto A, Bertrand O, Mauguière F (2004) Early amygdala reaction to fear spreading in occipital, temporal, and frontal cortex: a depth electrode ERP study in human. Neuron 42:665–676
63. Li Y, Vanni-Mercier G, Isnard J, Mauguière F, Dreher J-C (2016) The neural dynamics of reward value and risk coding in the human orbitofrontal cortex. Brain 139:1295–1309
64. Nourski KV, Steinschneider M, Rhone AE, Kawasaki H, Howard MA, Banks MI (2018) Auditory predictive coding across awareness states under anesthesia: an intracranial electrophysiology study. J Neurosci 38:8441–8452
65. Pfeuty M, Monfort V, Klein M, Krieg J, Collé S, Colnat-Coulbois S, Brissart H, Maillard L (2019) Role of the supplementary motor area during reproduction of supra-second time intervals: an intracerebral EEG study. Neuroimage 191:403–420
66. Halgren E, Baudena P, Clarke JM, Heit G, Liégeois C, Chauvel P, Musolino A (1995) Intracerebral potentials to rare target and distractor auditory and visual stimuli. I. Superior temporal plane and parietal lobe. Electroencephalogr Clin Neurophysiol 94:191–220
67. Bekinschtein TA, Dehaene S, Rohaut B, Tadel F, Cohen L, Naccache L (2009) Neural signature of the conscious processing of auditory regularities. Proc Natl Acad Sci USA 106:1672–1677. https://doi.org/10.1073/pnas.0809667106
68. Shen L, Han B, Chen L, Chen Q (2019) Perceptual inference employs intrinsic alpha frequency to resolve perceptual ambiguity. PLoS Biol 17:e3000025
69. Canales-Johnson A, Billig A, Olivares F, Gonzalez A, García M del C, Silva W, Vaucheret Paz E, Ciraolo C, Mikulan E, Ibanez A, Huepe D, Chennu S, Bekinschtein T (2020) Dissociable neural information dynamics of perceptual integration and differentiation during bistable perception. Cerebral cortex (New York, NY: 1991) 30. https://doi.org/10.1093/cercor/bhaa058
70. Strauss M, Sitt JD, King J-R, Elbaz M, Azizi L, Buiatti M, Naccache L, Van Wassenhove V, Dehaene S (2015) Bekinschtein. Proc Natl Acad Sci 112:E1353–E1362
71. Wang X-J, Rinzel J (1992) Alternating and synchronous rhythms in reciprocally inhibitory model neurons. Neural Comput 4:84–97
72. Başar E, Başar-Eroğlu C, Karakaş S, Schürmann M (2000) Brain oscillations in perception and memory. Int J Psychophysiol 35:95–124
73. Baddeley A (2003) Working memory: looking back and looking forward. Nat Rev Neurosci 4:829–839. https://doi.org/10.1038/nrn1201
74. Wang X-J (2006) Neural oscillations. Encyclopedia of cognitive science
75. Kahana MJ, Sekuler R, Caplan JB, Kirschen M, Madsen JR (1999) Human theta oscillations exhibit task dependence during virtual maze navigation. Nature 399:781–784
76. Rizzuto D, Madsen J, Bromfield E, Schulze-Bonhage A, Seelig D, Aschenbrenner-Scheibe R, Kahana M (2003) Reset of human neocortical oscillations during a working memory task. Proc Natl Acad Sci 100:7931–7936
77. Dastjerdi M, Ozker M, Foster BL, Rangarajan V, Parvizi J (2013) Numerical processing in the human parietal cortex during experimental and natural conditions. Nat Commun 4:2528. https://doi.org/10.1038/ncomms3528

78. Ng BSW, Logothetis NK, Kayser C (2013) EEG phase patterns reflect the selectivity of neural firing. Cereb Cortex 23:389–398
79. Aghajan ZM, Schuette P, Fields TA, Tran ME, Siddiqui SM, Hasulak NR, Tcheng TK, Eliashiv D, Mankin EA, Stern J et al (2017) Theta oscillations in the human medial temporal lobe during real-world ambulatory movement. Curr Biol 27:3743–3751
80. Fellner M-C, Gollwitzer S, Rampp S, Kreiselmeyr G, Bush D, Diehl B, Axmacher N, Hamer H, Hanslmayr S (2019) Spectral fingerprints or spectral tilt? Evidence for distinct oscillatory signatures of memory formation. PLoS Biol 17:e3000403
81. Thézé R, Giraud A-L, Mégevand P (2020) The phase of cortical oscillations determines the perceptual fate of visual cues in naturalistic audiovisual speech. Sci Adv 6:eabc6348
82. Lerousseau JP, Trébuchon A, Morillon B, Schön D (2021) Frequency selectivity of persistent cortical oscillatory responses to auditory rhythmic stimulation. J Neurosci 41:7991–8006. https://doi.org/10.1523/JNEUROSCI.0213-21.2021
83. Halgren E, Babb TL, Crandall PH (1978) Human hippocampal formation EEG desynchronizes during attentiveness and movement. Electroencephalogr Clin Neurophysiol 44:778–781
84. Arnolds D, Lopes da Silva F, Aitink J, Kamp A, Boeijinga P (1980) The spectral properties of hippocampal EEG related to behaviour in man. Electroencephalogr Clin Neurophysiol 324–328
85. Tesche C (1997) Non-invasive detection of ongoing neuronal population activity in normal human hippocampus. Brain Res 749:53–60
86. Varela F, Lachaux JP, Rodriguez E, Martinerie J (2001) The brainweb: phase synchronization and large-scale integration. Nat Rev Neurosci 2:229–239. https://doi.org/10.1038/35067550
87. Wendling F, Ansari-Asl K, Bartolomei F, Senhadji L, Lotfi S (2009) From EEG signals to brain connectivity: a model-based evaluation of interdependence measures. From EEG signals to brain connectivity: a model-based evaluation of interdependence measures. J Neurosci Methods 183:9–18. https://doi.org/10.1016/j.jneumeth.2009.04.021ï
88. Bressler SL, Menon V (2010) Large-scale brain networks in cognition: emerging methods and principles. Trends Cogn Sci 14:277–290. https://doi.org/10.1016/j.tics.2010.04.004
89. Bastos AM, Schoffelen JM (2016) A tutorial review of functional connectivity analysis methods and their interpretational pitfalls. Front Syst Neurosci. https://doi.org/10.3389/fnsys.2015.00175
90. Haufe S, Nikulin VV, Müller KR, Nolte G (2013) A critical assessment of connectivity measures for EEG data: a simulation study. Neuroimage 64:120–133. https://doi.org/10.1016/j.neuroimage.2012.09.036
91. Mierlo PV, Carrette E, Hallez H, Raedt R, Meurs A, Vandenberghe S, Roost DV, Boon P, Staelens S, Vonck K (2013) Ictal-onset localization through connectivity analysis of intracranial EEG signals in patients with refractory epilepsy. Epilepsia 54:1409–1418. https://doi.org/10.1111/epi.12206
92. Mahjoory K, Nikulin VV, Botrel L, Linkenkaer-Hansen K, Fato MM, Haufe S (2017) Consistency of EEG source localization and connectivity estimates. Neuroimage 152:590–601. https://doi.org/10.1016/j.neuroimage.2017.02.076
93. Brovelli A, Badier J-M, Bonini F, Bartolomei F, Coulon O, Auzias G (2017) Dynamic reconfiguration of visuomotor-related functional connectivity networks. J Neurosci 37:839–853. https://doi.org/10.1523/JNEUROSCI.1672-16.2016
94. Sweeney-Reed CM, Zaehle T, Voges J, Schmitt FC, Buentjen L, Kopitzki K, Esslinger C, Hinrichs H, Heinze HJ, Knight RT, Richardson-Klavehn A (2014) Corticothalamic phase synchrony and cross-frequency coupling predict human memory formation. eLife 3:e05352. https://doi.org/10.7554/eLife.05352
95. Sweeney-Reed CM, Zaehle T, Voges J, Schmitt FC, Buentjen L, Kopitzki K, Hinrichs H, Heinze H-J, Rugg MD, Knight RT, Richardson-Klavehn A (2015) Thalamic theta phase alignment predicts human memory formation and anterior thalamic cross-frequency coupling. eLife 4:e07578. https://doi.org/10.7554/eLife.07578
96. Parvizi J, Jacques C, Foster BL, Witthoft N, Withoft N, Rangarajan V, Weiner KS, Grill-Spector K (2012) Electrical stimulation of human fusiform face-selective regions distorts face perception. J Neurosci 32:14915–14920. https://doi.org/10.1523/JNEUROSCI.2609-12.2012

Chapter 16
What is the Relationship Between Scalp EEG, Intracranial EEG, and Microelectrode Activities?

Johannes Sarnthein and Lukas Imbach

Abstract Brain function evolves simultaneously at several scales in space and time. We here review studies with data recorded at the smallest scale of single neurons, local field potentials from micro-electrodes, intracranial EEG recorded from macro-electrodes, up to EEG recorded from the scalp. Simultaneous recordings of thalamic LFP and scalp EEG at rest show a strong coherence that reflects functional coupling in thalamocortical loops around 8 Hz. The iEEG in the subthalamic nucleus and the scalp EEG modulate their phase locking in the beta band (20–35 Hz) in relation to motor performance. Activation of the amygdala is evidenced by single neural firing, LFP, iEEG and BOLD. During the delay period of a working memory task, hippocampal neurons fire persistently while the iEEG and the parietal EEG show enhanced phase locking in the 8–12 Hz alpha band. For the same task, stimulus encoding elicits information flow in the 4–8 Hz theta band from auditory cortex to hippocampus while this information flow is reversed from hippocampus to cortex during active maintenance. This series of studies illustrates how local processing in specific brain areas is embedded in large-scale network communication that can be accessed through simultaneous recordings across several brain areas and neuronal scales.

16.1 Introduction

Mental activity is related both to brain activity from single neurons and also to large scale brain-wide networks. Capturing activity at different scales requires recording brain activity with different techniques. In this chapter, we review data recorded at the smallest scale (single neuron firing), local field potential (LFP) from micro-electrodes, intracranial EEG (iEEG or ECoG) recorded from macro-electrodes, up

J. Sarnthein (✉) · L. Imbach
Klinik für Neurochirurgie, Universitätsspital Zürich, Zürich, Switzerland
e-mail: Johannes.Sarnthein@usz.ch

L. Imbach
Swiss Epilepsy Center, Klinik Lengg, Zürich, Schweiz

N. Axmacher (ed.), *Intracranial EEG*, Studies in Neuroscience, Psychology and Behavioral Economics, https://doi.org/10.1007/978-3-031-20910-9_16

to EEG recorded from the scalp. As an additional technique, we review a dataset of Blood Oxygenation Level Dependent (BOLD) imaging.

How do these scales interact and what is their relationship? To characterize the functional interaction between recordings from different brain areas and scales, we focus on neuronal communication that is subserved by neuronal synchronization [1, 2]. This interaction evolves at different timescales and in frequencies that require temporal resolution down to the ms range. Therefore, it is advantageous to record the data from the different scales simultaneously (see also Chap. 53).

Simultaneous recordings in different brain regions in humans are possible in patients who undergo neurosurgical interventions [3]. Here, we review recordings performed in patients with Parkinson's disease (PD), neurogenic pain, or epilepsy who were implanted with depth electrodes. The data was acquired either during awake surgery or after implantation of the electrodes. During the recordings, patients were awake at rest or they performed cognitive or motor tasks.

We first review our recordings in the thalamus and the subthalamic nucleus (STN) and their interaction with scalp EEG recordings during rest and during task performance. We finally present iEEG recordings from the hippocampal cortex and its relationship to direct cortical recording (ECoG), scalp EEG and single neuron firing while participants perform a working memory task.

16.2 Cellular Origin of iEEG, LFP and Scalp EEG

When comparing intracranial recordings and scalp EEG, we must be aware that these signals are generated by neural networks of drastically different scales. The generation of each signal filters out the synchronized activity of neuronal assemblies. The number of neurons that form an assembly and generate the recorded signals varies widely depending on the size and physical properties of the recording electrode.

Regarding the recording setup, we define that macro-electrodes with surface >1 mm^2 record iEEG. Micro-electrodes with tip surface $\ll$1 mm^2 record LFP from which the action potentials of single neurons can be isolated.

When recording directly from the hippocampus or cerebral cortex, or from scalp EEG electrodes, the origin of these extracellular fields is well understood [4]. The scalp EEG is generated by post-synaptic dipole-like potentials along the dendrites of cortical pyramidal cells that are arranged in parallel. Their extracellular return current is averaged over several square centimeters of the cortex. Several 10,000 pyramidal cells must be active synchronously for an EEG signal to appear on the scalp [5, 6] (see also Chaps. 8 and 9). A similar reasoning holds for the iEEG recorded in the hippocampus, albeit a smaller number of neurons generate these post-synaptic potentials.

Anatomically different, the amygdala, thalamus and subthalamic nucleus lack the anatomical topography of pyramidal cells aligned in parallel. In the thalamus, ellipsoid dendritic arbors of thalamocortical cells generate a "closed field" in the sense that there is no aligned structure of dipoles that allows for the emergence of a summed dipole field as in the cortex. Therefore, recorded signals must stem from fewer neurons in closer proximity to the recording probe. For the LFP recorded with a micro-electrode in the thalamus, we assume a contribution of cells in a small cube of side length 0.1 mm [7], which corresponds to a sample of approximately four to five cells [8, 9]. Because multi-electrode LFP recordings in the thalamus show high correlation between distant sites [10], a small sample of cells may be representative for thalamic LFP activity. Similarly, the anatomy of neurons in the amygdala and STN show that also the neurons in these areas generate a closed local electrical field. Thus, the iEEG is generated in a small region of tissue closely around the depth electrode in the respective brain tissue.

16.3 Relationship Between Thalamic LFP and Scalp EEG

We analyzed spontaneous brain activity intraoperatively from patients suffering from severe PD [8]. The thalamic recordings were in the ventral anterior nucleus (VA) and in the ventral lateral nucleus (VLa). We first observed that the dominant peak in the power spectral density (PSD) appeared at around 8 Hz in both scalp EEG and in the thalamic LFP (Fig. 16.1a). When looking at the temporal relationship between LFP and scalp EEG, high thalamocortical coherence appeared at the dominant frequency of the PSD. The dominant frequency in PD is known to be shifted to frequencies lower than the common 8–12 Hz alpha band. When looking at the scalp topography of thalamocortical coherence, we found the highest coherences to frontal electrodes (Fig. 16.1b). The magnitude of thalamocortical coherence was comparable to the magnitude of coherence between two scalp EEG electrodes (Fig. 16.1c). This is much larger than the magnitude of coherence between subdural recordings, which amounts to only 20% at 3 cm and decreases with distance [10]. For the same distance, coherence in scalp EEG exceeded 80% (Fig. 16.1c). Thus, the spatial averaging, which is the basis of the EEG signal, may extract those components of cortical activity that are highly coherent over large distances [6]. Importantly, in contrast to coherence between different EEG electrodes, the thalamo-cortical coherence does not depend on the distance between electrodes in thalamus and on cortex.

In a similar setup, we analyzed spontaneous brain activity intraoperatively from patients suffering from severe neurogenic pain. The thalamic recordings were in the posterior part of the centrolateral nucleus (CL). Again, as for the recordings from anterior nuclei VA and VLa, the magnitude of thalamocortical theta coherence was comparable to the magnitude of coherence between EEG electrodes [6]. To further characterize the dominant frequency of the coherence spectrum, we asked the participants to open their eyes. Eye-opening reduced the power of the dominant frequency in the thalamus and at posterior electrode sites, reminiscent of the classic alpha desynchronization (alpha-blocking effect) in occipital scalp EEG [11].

In both studies [8, 12], we concurrently recorded firing of single neurons. We found firing patterns that were clearly associated with low threshold Ca^{2+} bursts that

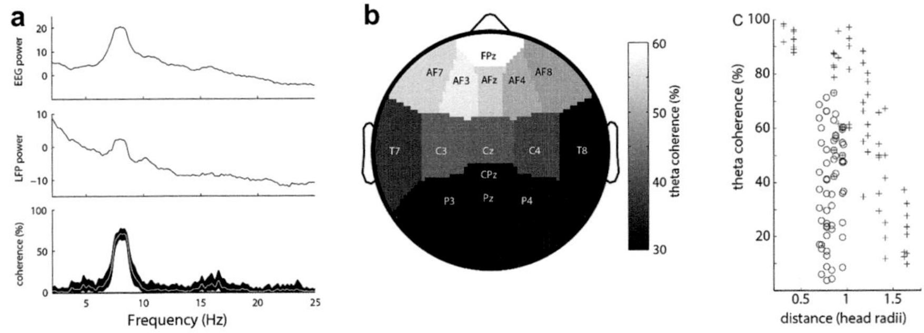

Fig. 16.1 Thalamocortical coherence from anterior thalamic nuclei. a The power spectra of scalp EEG and thalamic LFP both show a dominant frequency at 8 Hz (theta band). The thalamocortical coherence is significant only at the dominant frequency. **b** In the spatial distribution of thalamocortical coherence, the highest coherence from anterior thalamic nuclei was found with frontal electrode sites. **c** Cortico-cortical coherence (+) from frontal electrode FPz falls off with distance, which illustrates the effect of volume conduction. Thalamocortical coherence (○) is independent of distance, which speaks against volume conduction but rather for synaptic transmission. Copyright 2007 Society for Neuroscience. Modified from [8] with permission

showed the typical interburst frequency of ~4 Hz. However, there was no obvious relation of the bursts with the thalamocortical oscillations.

Our finding of high thalamocortical coherence supports the notion that the thalamus is involved in synchronizing activity between distant cortical areas. This is consistent with the anatomy of the thalamocortical network that is characterized by widely divergent thalamocortical and corticothalamic connections. In this way, very focal thalamic activity as reflected in the LFP can show high coherence with widespread cortical areas. Our data also indicates that the two electrode types, microelectrode and scalp EEG electrode, are well adapted to record manifestations of the thalamocortical network interaction at the two given sites. The comparable magnitude of thalamocortical and EEG coherence suggests that the thalamus and cortex are both relevant partners in the genesis of thalamocortical rhythmicity.

16.4 Relationship Between Subthalamic iEEG and Scalp EEG

Electrophysiological studies in Parkinson's disease (PD) aim to establish a link between motor behavior and its central electrophysiological correlates [13]. To this end, intracranial recordings from cortical and subcortical structures are used to study motor control in PD on a network level [14]. In a subgroup of patients with PD undergoing deep brain stimulation (DBS) in the subthalamic nucleus (STN), it is possible to access iEEG in the basal ganglia through the implanted deep brain electrodes post-operatively simultaneously with scalp EEG. This allows for a real time

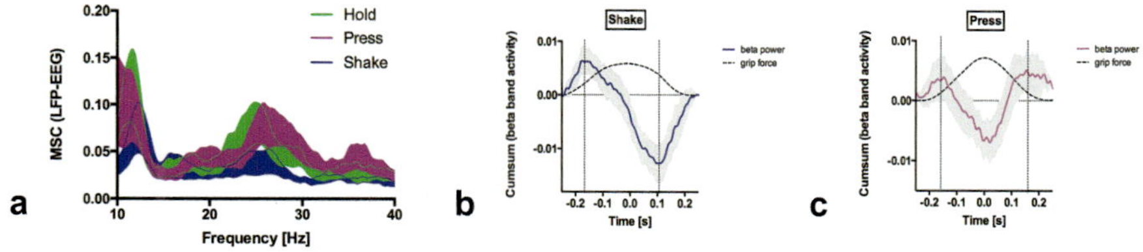

Fig. 16.2 **a** Coherence between STN and scalp electrode site (ipsilateral central EEG C3/C4). **b**, **c** The timelocking between STN beta power and motor output resembles a sine wave for habitual "shaking" (**b**) and a cosine wave for voluntary "pressing" (**c**). Modified from [15] with permission

network analysis of motor control. In addition to intracranial recordings in the resting state, this approach allows for comparative analyses of brain network activity during different behaviors. The large scale synchronicity and its modulation is measured between scalp EEG electrodes and connected subcortical structures.

In a first experiment, we asked PD patients to move a cube (volume $(6\,\text{cm})^3$, weight 300 g) with their hand. The cube recorded the time-varying force that the fingers exerted on the surface of the cube in adapting the grip force. We then recorded iEEG in the STN with simultaneous measurement of scalp EEG from PD patients who had electrodes implanted for DBS [15]. To compare the synchronization between STN and scalp EEG across different tasks, we analyzed the spectral coherence between iEEG in the STN and the ipsilateral central EEG (C3 or C4, respectively). The Magnitude Squared Coherence (MSC, Fig. 16.2a) showed a broad beta peak (20–35 Hz) that persisted during the baseline condition (hold) and for voluntary movements (press). In contrast, habitual movements (shaking) revealed a marked desynchronization between motor cortex and basal ganglia. A similarly striking difference between habitual "shaking" and voluntary "pressing" of the device appeared in the phase locking between STN beta power and the motor output, i.e. the force exerted on the cube (Fig. 16.2b, c), thus underlining the difference in STN activity for habitual and voluntary movements.

In a further experiment, we measured cortico-subcortical coupling prior to and during movements in wakefulness and REM sleep in PD patients implanted with DBS leads in the STN [16]. For voluntary self-paced movements during wakefulness, no relevant modulation of synchronicity (phase locking) in the peri-movement period 5 s before and after movement initiation was found. In contrast, during unconscious movements in REM sleep, cortico-subthalamic phase locking in the beta range was reduced after movement onset, indicating a significant decoupling of the basal ganglia from cortical neurons during sleep related movements.

In a recent study, we investigated cortico-subthalamic coupling for cued and self-paced movements on a network level [17]. Again, simultaneous recordings of iEEG in different subparts of the STN and scalp EEG were performed during different motor tasks. To differentiate the network properties of two distinct cortico-subthalamic loops for self-paced and cued motor control, the phase locking value between STN, central and frontal EEG electrodes was measured for both motor behaviors separately. Connectivity analysis revealed higher synchronicity (phase-locking value,

PLV) between the STN and central electrodes during self-paced movements and higher STN to frontal phase-locking during externally-cued movements. This finding supports the existence of functionally segregated cortico-basal ganglia networks controlling motor behavior in PD patients.

These three studies corroborate the hypothesis that during voluntary (explicit) motor control cortico-subcortical synchronicity is higher (voluntary grip force, cued tapping or wake movements), whereas more habitual (intrinsic) behavior is characterized by lower basal cortex-to-basal ganglia synchronicity and information transfer. Importantly, the conclusions on the dynamics of brain network activity rely on the large scale spatial distribution of the recording electrodes within the same functional network.

16.5 Relationship Between Neuronal Firing, Local Field Potentials and Hemodynamic Activity in the Amygdala

To examine the relation between different scales of activity within the amygdala, we recorded iEEG in epilepsy patients while they watched video sequences of dynamic fearful faces (AVERSIVE condition) interleaved with sequences of landscapes (NEUTRAL condition). Concurrently, we recorded the LFP from microelectrodes and isolated the firing of single neurons. Further, we included in our analysis the hemodynamic brain activity (BOLD) that was recorded a few months earlier while participants had watched the same video sequences [18, 19]. The comparison is valid even though BOLD was not recorded simultaneously because EEG spectra are known to be highly reproducible over several months [20].

We found a striking covariation between the neuronal firing rate (FR), the LFP filtered for high gamma power (μHGP) and the iEEG filtered for high gamma power (HGP) (Fig. 16.3). The correlations suggest that the covariation is driven by the synchronized firing within these neuronal assemblies at three different scales.

This covariation reflects common correlates of amygdalar activity across recording modalities, ranging from the neuronal firing, to high frequency microscopic and macroscopic activity, up to hemodynamic responses.

16.6 Relationship Between Scalp EEG, Hippocampal iEEG, and Single Neuron Firing

In a next set of recordings in the medial temporal lobe (MTL), we focused on signals stemming from the hippocampus, amygdala and entorhinal cortex. Multiple electrodes were implanted during the presurgical evaluation in patients with epilepsy. Simultaneous with iEEG, we recorded single neuron firing and scalp EEG [21, 22].

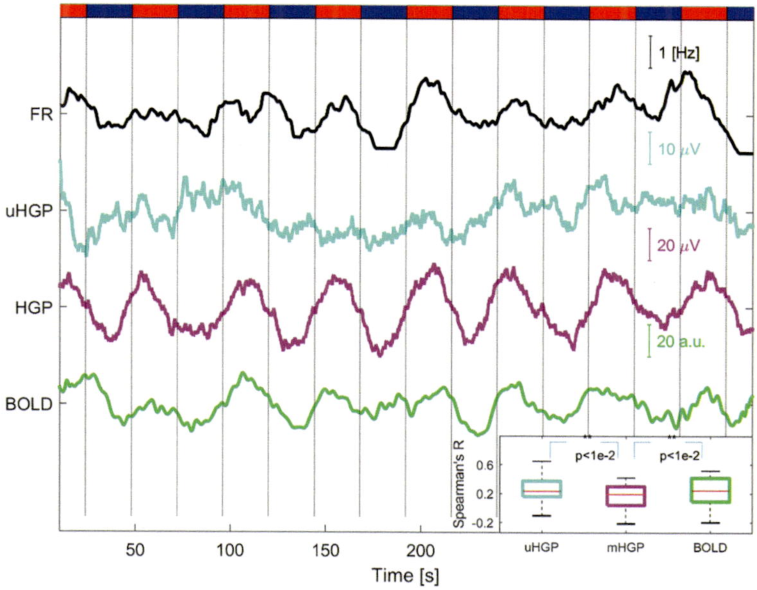

Fig. 16.3 An example of covariation between firing rate (FR), μHGP, HGP and BOLD signal in one patient along the duration of the visual stimulation (upper bar: red AVERSIVE, blue NEUTRAL). Across all neurons, the firing rate correlated positively with μHGP, HGP, and BOLD. Modified from [19] with permission

We asked the patients to participate in a task that activates their working memory (WM) because WM recruits activity in a wide network of brain regions [23]. WM describes our capacity to represent sensory input for prospective use. In our verbal WM task (Fig. 16.4), the encoding of letter items was isolated from the maintenance period in which the active rehearsal of memory items was paramount to achieve correct performance [21, 22, 24–26].

16.6.1 Phase Locking Value Between Hippocampal iEEG and Scalp EEG

We will first review how hippocampus communicates within the WM network. We investigated the PLV between hippocampal iEEG and scalp EEG and found the most salient effect in the 4–12 Hz alpha-theta range (Fig. 16.5) [21]. The PLV had a well-defined onset in the last 2 s of the maintenance period and rapidly diminished following the presentation of the probe letter. The PLV increased with workload. In our task design, this indicates that neural communication between brain regions is channeled through theta-alpha coupling within a limited time window between the onset and offset of the maintenance phase during trials with a high workload.

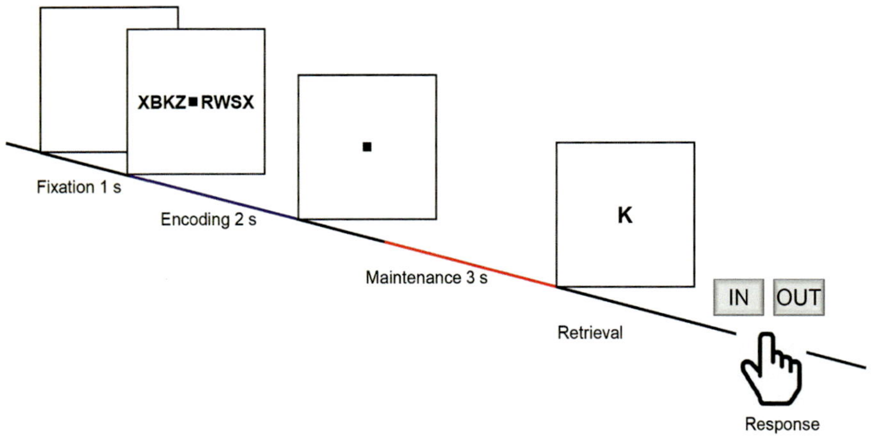

Fig. 16.4 Verbal working memory task. Sets of consonants are presented and have to be memorized. The set size (4, 6, or 8 letters) determines working memory workload. In each trial, presentation of a letter string (encoding period, [−5 −3] s) is followed by a delay (maintenance period, [−3 0] s). After the delay, a probe letter is presented. Participants indicate whether the probe was in the letter string or not

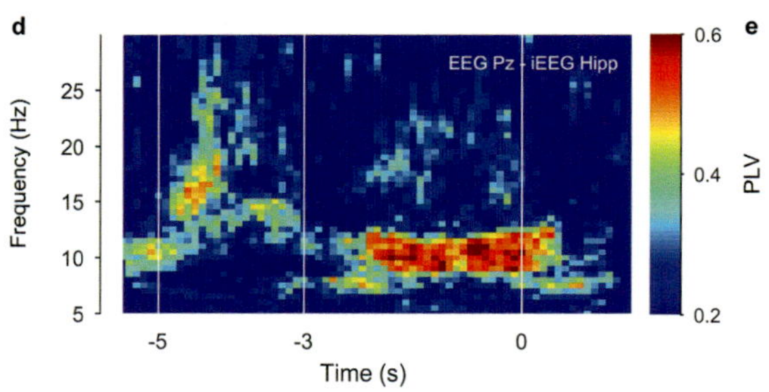

Fig. 16.5 Synchronization of hippocampal iEEG and parietal scalp EEG. The stimulus letter set was presented during the encoding period [−5 −3] s and maintained in WM during the maintenance period [−3 0] s. After presentation of the probe letter at 0 s, the participant initiated the response. Synchronization (phase locking value PLV) around 10 Hz appears during the last 2 s of the maintenance period in expectance of the probe letter. Modified from [21] with permission

Interestingly, the PLV synchronization occurred strongly between the hippocampus and occipito-parietal scalp electrodes, a common locus for alpha waves. This finding is consistent with scalp EEG findings in several healthy subjects recorded while performing the same [20, 26] or similar tasks [27]. The enhanced hippocampal coupling between iEEG and scalp EEG during maintenance suggests that hippocampal activity may be integrated into the WM network via coupling of local iEEG with long-range connectivity in the alpha-theta band. As a common finding, the PSD in the parietal alpha-band may increase during maintenance [20, 21, 26] but parietal alpha PSD may not be directly involved in WM processing as it has been proposed to be critical for protecting WM maintenance from non-relevant information [28, 29].

16.6.2 Information Flow Between iEEG from Hippocampus and Auditory Cortex

We next investigated the directionality of the information flow during encoding and maintenance in the task. To this end, we used spectral Granger causality (GC) as a measure of directed functional connectivity to determine the direction of the information flow between brain areas. As an added value compared to undirected measures of connectivity, GC quantifies the ability of a time series to predict the temporal evolution of another time series and can thus be interpreted as a measure of directional information flow (see also Chap. 32). In one study, we investigated the interaction between anterior and posterior hippocampus [24]. In another recent study, we investigated the interaction between cortex and hippocampus in a patient with iEEG recorded from a grid electrode (ECoG) over auditory cortex [25]. During encoding, the net information flow ΔGranger ($GC_{hipp \rightarrow cortex} - GC_{cortex \rightarrow hipp}$) was from auditory cortex to hippocampus in the 4–10 Hz range (Fig. 16.6a). Concurrently, encoding in the auditory cortex was accompanied by local activity in the gamma range (40–100 Hz) of the PSD. During maintenance, the net information flow ΔGranger in the 4–8 Hz range was reversed and occurred from hippocampus to auditory cortex (Fig. 16.6a, b).

To confirm this finding from three participants with ECoG, we analyzed the hippocampal iEEG and the scalp EEG of 15 patients who participated in this task [25]. We used beamforming to reconstruct the EEG sources (Fig. 16.7). Indeed, the

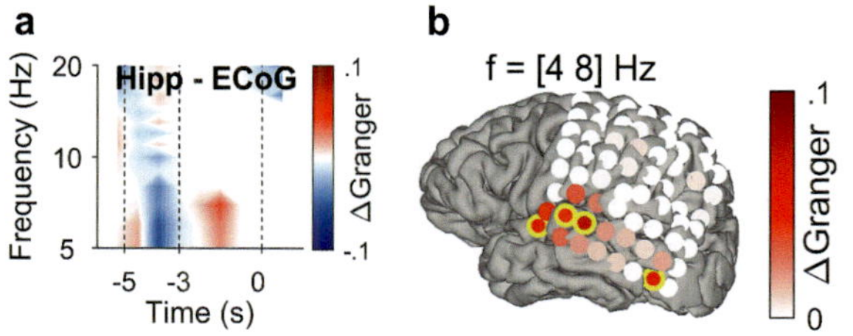

Fig. 16.6 Information flow between hippocampal iEEG and ECoG over auditory cortex. The stimulus letter set was presented during the encoding period [−5 −3] s and maintained in WM during the maintenance period [−3 0] s. After presentation of the probe letter at 0 s, the participant initiated the response. **a** The Granger time–frequency map illustrates the net information flow (ΔGranger). During encoding, net information flows from auditory cortex to hippocampus (blue). During maintenance, the information flow is reversed from hippocampus to auditory cortex (red) indicating the replay of letters in memory. **b** During maintenance, the information flow from hippocampus is most salient to electrode contacts over auditory cortex. Modified from [25] with permission

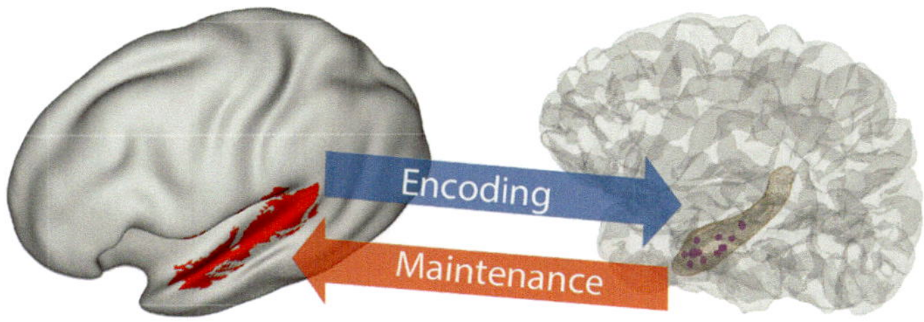

Fig. 16.7 Bidirectional information flow between cortical sites and hippocampus in the working memory network. The GC analysis suggests a surprisingly simple model of information flow during the task. During encoding, letter strings are verbalized as subvocal speech; the incoming information flows from auditory cortex to hippocampus. During maintenance, participants actively recall and rehearse the subvocal speech in the phonological loop; GC indicates an information flow from hippocampus to cortex as the physiological basis for the replay of the memory items. Modified from [25] with permission

information flow between hippocampal iEEG and EEG sources was from auditory cortex to hippocampus during encoding. During maintenance, the flow was reverse, from hippocampus to auditory cortex, indicating the active replay of memory items.

In the literature, there are several studies investigating the WM network. However, only few report directional interactions. One of these [30], reports cross-spectral directionality between intracranial recordings in frontal cortex and the medial temporal lobe in theta frequencies.

16.6.3 Neuronal Firing Patterns During the Trial

On the finest spatial scale, we isolated the firing of single neurons from the LFP recorded within the MTL. Figure 16.8 shows an example of a neuron in hippocampus that is characterized by persistent firing during the delay period after stimulus presentation. Persistent firing is considered a hallmark of WM activity [23, 31, 32].

First, during encoding, hippocampal population firing predicted whether subjects later responded correctly or incorrectly. Then we found neurons that fired in the absence of stimulation during the delay period (maintenance neurons). The firing of maintenance neurons in hippocampus predicted the number of items held in memory in each trial. Our data provide the first direct evidence that hippocampal firing predicts workload, which matches the trial difficulty as derived from the subjects' performance. The time course of activity suggests that the WM network's function goes beyond memorization towards a task-related preparation in expectance of the probe. Here, hippocampal neurons seem to participate in generic processes of working memory, possibly by establishing a contextual index in the network with distributed cortical sites involved in letter processing [23].

In their time course of activity, scalp EEG, hippocampal iEEG, PLV and hippocampal neuron firing all persistently show elevated activity in the maintenance

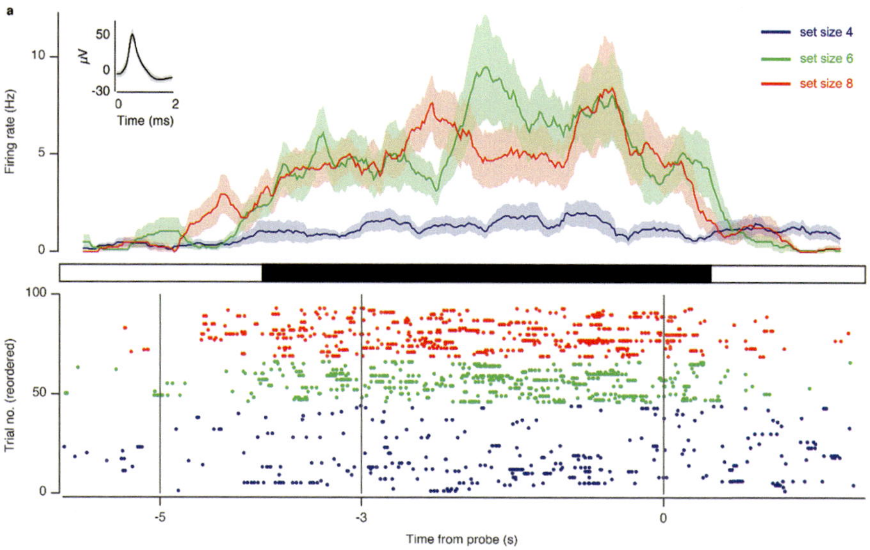

Fig. 16.8 A hippocampal neuron fires persistently during WM maintenance. The firing rate (top) of neuronal action potentials (bottom, raster plot) increases during encoding [−5 −3] s and persists during maintenance [−3 0] s until presentation of the probe letter. The firing rate increases with the workload of the trial (4, 6, or 8 letters to be memorized). Modified from [21] with permission

period after the stimulus presentation. Between single neurons and hippocampal iEEG, spike-field coherence appeared in the alpha frequency range during maintenance but not during encoding.

16.6.4 Neural Communication in Working Memory

The integration of distributed components of information across brain regions has been proposed to rely on long-range recurrent connections that support oscillatory signals [1], for example, during WM tasks [33, 34]. Regarding the frequency of oscillations, a relationship between the frequency range and the spatial extent of engaged brain networks has been proposed, with low frequencies binding large-scale networks and high frequencies coordinating smaller networks [30, 35, 36]. There are several examples in which neural oscillations play a role in coordinating functional neuronal assemblies thought to be responsible for computation and communication in large-scale brain networks [34, 37]. Thus, the interpretation that action potentials generated in the hippocampus are embedded within the WM network through coupling with iEEG and long-range recurrent connections refines our current view on neural communication within and between brain regions.

16.7 Volume Conduction

The interaction between recordings from different brain regions has to be discussed with respect to volume conduction. As an effect of volume conduction, cortico-cortical coherence in scalp EEG is highest for adjacent electrode sites and falls off with distance (Fig. 16.1c). Of note, as an unspecific effect, the coherence due to volume conduction is high over a wide range of the spectrum. Therefore, volume conduction could also lead to biased (pseudo-)connectivity in combined scalp and subcortical recordings. When looking at the synchronization between scalp EEG and iEEG, there are several aspects that speak against a large contribution of volume conduction in the studies reviewed here:

(1) The spectra of coherence and PLV show a marked frequency dependence in the range of interest (Figs. 16.1, 16.2 and 16.4).
(2) The topographical distribution of coherence and PLV is highly inhomogeneous as it varies between recording sites (Fig. 16.1, [21]).
(3) The coherence is associated with a non-zero phase lag, different from what one would expect for volume conduction [8, 11, 12, 15].
(4) The magnitude of coherence and PLV varies between subjects [8, 12, 21].
(5) The magnitude of coherence and PLV varies with task condition (Figs. 16.2 and 16.5).
(6) The direction of information flow depends on task condition (Figs. 16.6 and 16.7). Given that the relative flow may depend on the reference scheme, it is advisable to use two different reference electrodes [38].

In view of these aspects, the contribution of volume conduction to the synchronization of the signals seems to be small compared to the functional interaction that is mediated by axonal transmission.

16.8 Conclusions

It is well established that many aspects of human brain activity involve a wide network of brain areas. In each of these areas, brain activity evolves simultaneously in different spatial and temporal scales, from single neuron firing to synchronous activity of large neuronal assemblies. Thus, the most promising approaches to unravel these functions are recordings with microelectrodes, intracranial LFP, intracranial EEG, and scalp EEG. While the literature on recordings at each spatial scale is vast, the number of reports that combine simultaneous recordings is somewhat more restricted, but crucial for exploring the relevant network interaction and its modulation in space and time.

In this review, we have focused on our own studies that combine recordings at different levels and set them in the context of each other and the literature. The series of experiments presents examples on how local processing in specific brain areas is embedded in large-scale network communication that can be accessed through simultaneous recordings from several brain areas and neuronal scales.

Acknowledgements This work was funded by a grant from the Swiss National Science Foundation 204651 to JS.

References

1. Fries P (2015) Rhythms for cognition: communication through coherence. Neuron 88(1):220–235. https://doi.org/10.1016/j.neuron.2015.09.034
2. Pesaran B, Vinck M, Einevoll GT, Sirota A, Fries P, Siegel M, Truccolo W, Schroeder CE, Srinivasan R (2018) Investigating large-scale brain dynamics using field potential recordings: analysis and interpretation. Nat Neurosci 21(7):903–919. https://doi.org/10.1038/s41593-018-0171-8
3. Parvizi J, Kastner S (2018) Promises and limitations of human intracranial electroencephalography. Nat Neurosci 21(4):474–483. https://doi.org/10.1038/s41593-018-0108-2
4. Buzsaki G, Anastassiou CA, Koch C (2012) The origin of extracellular fields and currents—EEG, ECoG, LFP and spikes. Nat Rev Neurosci 13(6):407–420. https://doi.org/10.1038/nrn3241
5. Buzsaki G (2006) Rhythms of the brain. Oxford University Press
6. Nunez P, Srinivasan R (2006) Electric fields of the brain. Oxford University Press
7. Engel AK, König P, Gray CM, Singer W (1990) Stimulus-dependent neuronal oscillations in cat visual cortex: inter-columnar interaction as determined by cross-correlation analysis. Eur J Neurosci 2(7):588–606. https://doi.org/10.1111/j.1460-9568.1990.tb00449.x
8. Sarnthein J, Jeanmonod D (2007) High thalamocortical theta coherence in patients with Parkinson's disease. J Neurosci 27(1):124–131. https://doi.org/10.1523/JNEUROSCI.2411-06.2007
9. Mai JK, Assheuer J, Paxinos G (1997) Atlas of the human brain. Academic San Diego
10. Bullock TH, McClune MC, Achimowicz JZ, Iragui-Madoz VJ, Duckrow RB, Spencer SS (1995) EEG coherence has structure in the millimeter domain: subdural and hippocampal recordings from epileptic patients. Electroencephalogr Clin Neurophysiol 95(3):161–177. https://doi.org/10.1016/0013-4694(95)93347-a
11. Sarnthein J, Morel A, von Stein A, Jeanmonod D (2005) Thalamocortical theta coherence in neurological patients at rest and during a working memory task. Int J Psychophysiol 57(2):87–96. https://doi.org/10.1016/j.ijpsycho.2005.03.015
12. Sarnthein J, Jeanmonod D (2008) High thalamocortical theta coherence in patients with neurogenic pain. Neuroimage 39(4):1910–1917. https://doi.org/10.1016/j.neuroimage.2007.10.019
13. Cagnan H, Duff EP, Brown P (2015) The relative phases of basal ganglia activities dynamically shape effective connectivity in Parkinson's disease. Brain 138(Pt 6):1667–1678. https://doi.org/10.1093/brain/awv093
14. Imbach L (2018) On goals and habits: lessons from deep brain recordings in Humans. Universität Zürich
15. Imbach LL, Baumann-Vogel H, Baumann CR, Surucu O, Hermsdorfer J, Sarnthein J (2015) Adaptive grip force is modulated by subthalamic beta activity in Parkinson's disease patients. Neuroimage Clin 9:450–457. https://doi.org/10.1016/j.nicl.2015.09.010
16. Hackius M, Werth E, Surucu O, Baumann CR, Imbach LL (2016) Electrophysiological evidence for alternative motor networks in REM sleep behavior disorder. J Neurosci 36(46):11795–11800. https://doi.org/10.1523/JNEUROSCI.2546-16.2016

17. Bichsel O, Gassert R, Stieglitz L, Uhl M, Baumann-Vogel H, Waldvogel D, Baumann CR, Imbach LL (2018) Functionally separated networks for self-paced and externally-cued motor execution in Parkinson's disease: evidence from deep brain recordings in humans. Neuroimage 177:20–29. https://doi.org/10.1016/j.neuroimage.2018.05.012
18. Fedele T, Boran E, Chirkov V, Hilfiker P, Grunwald T, Stieglitz L, Jokeit H, Sarnthein J (2021) Dataset of spiking and LFP activity invasively recorded in the human amygdala during aversive dynamic stimuli. Sci Data 8(1):9. https://doi.org/10.1038/s41597-020-00790-x
19. Fedele T, Tzovara A, Steiger B, Hilfiker P, Grunwald T, Stieglitz L, Jokeit H, Sarnthein J (2020) The relation between neuronal firing, local field potentials and hemodynamic activity in the human amygdala in response to aversive dynamic visual stimuli. Neuroimage 213:116705. https://doi.org/10.1016/j.neuroimage.2020.116705
20. Näpflin M, Wildi M, Sarnthein J (2008) Test-retest reliability of EEG spectra during a working memory task. Neuroimage 43(4):687–693. https://doi.org/10.1016/j.neuroimage.2008.08.028
21. Boran E, Fedele T, Klaver P, Hilfiker P, Stieglitz L, Grunwald T, Sarnthein J (2019) Persistent hippocampal neural firing and hippocampal-cortical coupling predict verbal working memory load. Sci Adv 5(3):eaav3687. https://doi.org/10.1126/sciadv.aav3687
22. Boran E, Fedele T, Steiner A, Hilfiker P, Stieglitz L, Grunwald T, Sarnthein J (2020) Dataset of human medial temporal lobe neurons, scalp and intracranial EEG during a verbal working memory task. Sci Data 7(1):30. https://doi.org/10.1038/s41597-020-0364-3
23. Christophel TB, Klink PC, Spitzer B, Roelfsema PR, Haynes J-D (2017) The distributed nature of working memory. Trends Cogn Sci 21(2):111–124. https://doi.org/10.1016/j.tics.2016.12.007
24. Li J, Cao D, Dimakopoulos V, Shi W, Yu S, Fan L, Stieglitz L, Imbach L, Sarnthein J, Jiang T (2022) Anterior-posterior hippocampal dynamics support working memory processing. J Neurosci. https://doi.org/10.1523/JNEUROSCI.1287-21.2021
25. Dimakopoulos V, Mégevand P, Stieglitz L, Imbach L, Sarnthein J (2022) Information flows from hippocampus to auditory cortex during replay of verbal working memory items. eLife. https://doi.org/10.7554/eLife.78677
26. Michels L, Moazami-Goudarzi M, Jeanmonod D, Sarnthein J (2008) EEG alpha distinguishes between cuneal and precuneal activation in working memory. Neuroimage 40(3):1296–1310. https://doi.org/10.1016/j.neuroimage.2007.12.048
27. Tuladhar AM, ter Huurne N, Schoffelen JM, Maris E, Oostenveld R, Jensen O (2007) Parieto-occipital sources account for the increase in alpha activity with working memory load. Hum Brain Mapp 28(8):785–792. https://doi.org/10.1002/hbm.20306
28. Bonnefond M, Jensen O (2012) Alpha oscillations serve to protect working memory maintenance against anticipated distracters. Curr Biol 22(20):1969–1974. https://doi.org/10.1016/j.cub.2012.08.029
29. Iemi L, Gwilliams L, Samaha J, Auksztulewicz R, Cycowicz YM, King J-R, Nikulin VV, Thesen T, Doyle W, Devinsky O, Schroeder CE, Melloni L, Haegens S (2022) Ongoing neural oscillations influence behavior and sensory representations by suppressing neuronal excitability. Neuroimage 247:118746. https://doi.org/10.1016/j.neuroimage.2021.118746
30. Johnson EL, Adams JN, Solbakk A-K, Endestad T, Larsson PG, Ivanovic J, Meling TR, Lin JJ, Knight RT (2018) Dynamic frontotemporal systems process space and time in working memory. PLoS Biol 16(3):e2004274. https://doi.org/10.1371/journal.pbio.2004274
31. Kornblith S, Quian Quiroga R, Koch C, Fried I, Mormann F (2017) Persistent single-neuron activity during working memory in the human medial temporal lobe. Curr Biol 27(7):1026–1032. https://doi.org/10.1016/j.cub.2017.02.013
32. Kamiński J, Sullivan S, Chung JM, Ross IB, Mamelak AN, Rutishauser U (2017) Persistently active neurons in human medial frontal and medial temporal lobe support working memory. Nat Neurosci 20(4):590–601. https://doi.org/10.1038/nn.4509
33. Sarnthein J, Petsche H, Rappelsberger P, Shaw GL, von Stein A (1998) Synchronization between prefrontal and posterior association cortex during human working memory. Proc Natl Acad Sci U S A 95(12):7092–7096

34. Polania R, Nitsche MA, Korman C, Batsikadze G, Paulus W (2012) The importance of timing in segregated theta phase-coupling for cognitive performance. Curr Biol 22(14):1314–1318. https://doi.org/10.1016/j.cub.2012.05.021
35. von Stein A, Sarnthein J (2000) Different frequencies for different scales of cortical integration: from local gamma to long range alpha/theta synchronization. Int J Psychophysiol 38(3):301–313
36. Solomon EA, Kragel JE, Sperling MR, Sharan A, Worrell G, Kucewicz M, Inman CS, Lega B, Davis KA, Stein JM, Jobst BC, Zaghloul KA, Sheth SA, Rizzuto DS, Kahana MJ (2017) Widespread theta synchrony and high-frequency desynchronization underlies enhanced cognition. Nat Commun 8(1):1704. https://doi.org/10.1038/s41467-017-01763-2
37. Canolty RT, Ganguly K, Kennerley SW, Cadieu CF, Koepsell K, Wallis JD, Carmena JM (2010) Oscillatory phase coupling coordinates anatomically dispersed functional cell assemblies. Proc Natl Acad Sci U S A 107(40):17356–17361. https://doi.org/10.1073/pnas.1008306107
38. Bastos AM, Schoffelen JM (2015) A tutorial review of functional connectivity analysis methods and their interpretational pitfalls. Front Syst Neurosci 9:175. https://doi.org/10.3389/fnsys.2015.00175

Chapter 17
How Do Local Field Potentials Measured with Microelectrodes Differ from iEEG Activity?

Supratim Ray

Abstract Simultaneous recordings of local field potentials (LFPs) and iEEG from the visual cortex of macaques have revealed that the spatial spread of iEEG (i.e., the extent of cortical tissue that contributes to its activity) is highly localized (~3 mm), only about 3 times the spatial spread of LFP, even though the iEEG electrode is much larger. Further, the stimulus tuning preferences of gamma oscillations (30–70 Hz) and high-gamma power (>80 Hz) in LFP versus iEEG are surprisingly similar. In particular, high-gamma power, which is thought to reflect the average firing rate around the electrode, decreases with increasing stimulus size in both LFP and iEEG, consistent with local origins of both signals. Both signals carry information about the sensory stimulus, with maximal information in the gamma band, although iEEG has higher information and better decodability than LFP. The power of iEEG falls more rapidly than LFP between 20 and 100 Hz, leading to steeper slope of the power spectral density (PSD), but at higher frequencies the slopes are comparable. A simple model of iEEG, obtained by averaging LFPs over ~3 mm, provides an accurate description of the properties of the recorded iEEG, including its PSD slope, stimulus tuning, and information content.

Local field potential (LFP), obtained by low-pass filtering the raw signal recorded from microelectrodes and thought to mainly reflect synaptic activity of a population of neurons around the microelectrode [1], is an important signal for understanding neural circuitry and brain function [2, 3]. However, LFPs are mainly recorded in animal models, while iEEG is traditionally recorded in humans. To accurately compare the properties of LFP and iEEG, it is important to record both these signals simultaneously, so that potential differences due to differences in species, recording setup, referencing scheme and behavioral state are minimized. To this end, we recently implanted a hybrid electrode array containing both microelectrode and iEEG electrodes (9×9 Utah microelectrode array and 3×3 Adtech iEEG array;

S. Ray (✉)
Centre for Neuroscience, Indian Institute of Science, Bangalore 560012, India
e-mail: sray@iisc.ac.in

N. Axmacher (ed.), *Intracranial EEG*, Studies in Neuroscience, Psychology and Behavioral Economics, https://doi.org/10.1007/978-3-031-20910-9_17

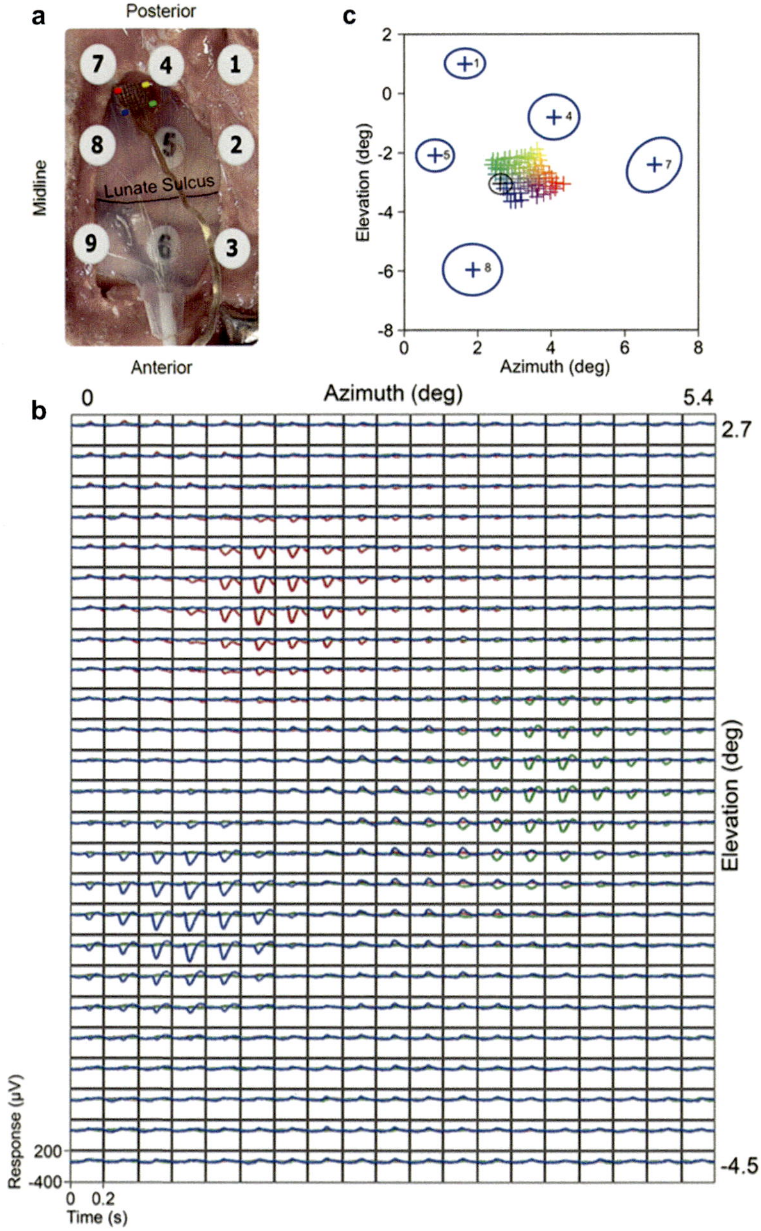

Fig. 17.1 Spatial spread of iEEG electrodes. Adapted from [4]

for details, see Fig. 17.1a) in the primary visual cortex of monkeys. In this chapter, I summarize the main findings from these studies, as reported in [4–6].

17.1 Spatial Spread

Spatial spread refers to the extent of cortical tissue around the recording electrode that contributes to its activity. For iEEG, estimation of the spatial spread has direct clinical relevance, because it determines the area of cortical tissue that needs to be resected if epileptic activity is recorded from the electrode. Unfortunately, there is very little consensus on the spread of field potentials, including LFP. For example, while some researchers have shown the LFP spread to be limited to a few hundred micrometers [4, 7–9], others have shown that it can extend up to several millimeters [10, 11]. The differences could be due to differences in recording areas and several other factors, as discussed later.

Spatial spread is most easily estimated in sensory cortices which have a well-defined topographic representation of the sensory world, such as primary visual and auditory cortices that have retinotopic and tonotopic maps, respectively. Presentation of a very small visual stimulus or a pure tone excites a small population of neurons in primary visual and auditory cortex, respectively, and shifting the location of the visual stimulus or the frequency of the tone systematically shifts the position of this activated neural population in the cortical space. For example, Fig. 17.1b shows the evoked responses recorded from iEEG electrodes when a small visual stimulus (radius of 0.3°) was presented in one of 19 × 25 grid positions on the screen, spanning [0° 5.4°] of azimuth and [−4.5° 2.7°] of elevation in visual space. The three traces at each position correspond to iEEG electrodes 1 (red), 4 (green) and 5 (blue), as shown in Fig. 17.1a. Strong evoked responses were recorded from the iEEG electrodes only when the stimulus was presented over a small and non-overlapping subset of positions.

Even before we quantitatively estimate the spatial spread, it is clear from these traces that at least in the primary visual cortex, the spatial spread of iEEG is very local. If an iEEG electrode had a spatial spread exceeding 10 mm, it is impossible that a visual stimulus would produce a strong response on one electrode but virtually no response at a neighboring electrode that is only 10 mm apart (Fig. 17.1a). In fact, the visual spreads appear to be even less than half the inter-electrode distance. Quantitative estimation (discussed below) confirms this intuition.

(A) Image of the hybrid grid, implanted in the primary visual cortex of a bonnet macaque. The microelectrode array was a 9 × 9 Utah array (Blackrock Microsystems), with inter-electrode spacing of 400 μm, length of 1 mm, active electrode made of platinum with tip-diameter of 3–5 μm. The 3 × 3 iEEG array had platinum electrodes that were 4 mm in diameter with an exposed recording area of 2.3 mm with inter-electrode spacing of 10 mm (Ad-Tech Medical Instrument). These electrodes are widely used in clinical settings. The electrodes were placed on the pial surface of the brain and recorded the electrocorticogram (ECoG; here we use iEEG and ECoG interchangeably). Four corners of the microelectrode array are color coded to provide a reference to the RF centers plotted in C. iEEG electrodes posterior to the lunate sulcus are in V1. Only iEEG electrodes 5 and 6 are visible; the remaining electrodes are under the skull and their approximate positions are indicated.

(B) Simultaneously recorded mean evoked responses for three iEEG electrodes: 1 (red), 4 (green), and 5 (blue), averaged across trials and six recording sessions at each of the 475 (19 $\times$ 25 rectangular grid) stimulus positions.

(C) Estimated RF centers of the LFP electrodes color coded based on their position on the microarray grid as shown in A. The RF size of a typical LFP electrode is shown in black. Estimated RF centers and sizes of ECoG electrodes are plotted in blue. Stable RFs were obtained from 5 iEEG electrodes (1, 4, 5, 8 and 9) and 77 LFP electrodes.

To estimate the spatial spread, a 2D Gaussian can be fitted to the average evoked response at each location, which yields the "visual spread" or receptive field (RF) for each electrode. The center and size (reported as the standard deviation (SD) of the fitted Gaussian) of the RFs of the iEEG electrodes are shown as blue ellipses in Fig. 17.1c. The RF centers of the simultaneously recorded LFPs are also shown, color coded based on their position on the microelectrode grid (as shown in Fig. 17.1a), along with the RF of the LFP recorded from one microelectrode (black ellipse). Surprisingly, even though the exposed diameter of the iEEG electrode (2.3 mm) is much larger than the tip of a microelectrode (a few micrometers), the RF sizes only differ by a factor of ~2. Specifically, the median iEEG and LFP RF sizes (1 SD of the fitted Gaussian) were 0.72° and 0.37° for this monkey, and 0.66° and 0.28° for the full dataset of 5 monkeys. These monkey iEEG RF sizes were comparable to RF sizes reported earlier in humans [12].

To convert the visual spread to cortical spread, the standard approach is to multiply the visual spread by the cortical magnification factor (MF), defined as the length of cortex that processes a unit degree of visual space. MF depends on the eccentricity of the RF center, and has been estimated by several researchers (summarized in Fig. 6 of [13]). MF can also be calculated directly from microelectrode recordings since the change in RF center can be measured for neighboring electrodes for which the inter-electrode distance is known. For example, from Fig. 17.1c, the RFs of the microelectrodes at the diagonals (marked in red and green in A) are ~2° apart in visual space (2° to 4° of azimuth). Because this array has 9 $\times$ 9 microelectrodes separated by 400 μm, the distance between the two ends of the array is $3.2 \times \sqrt{2}$ or ~4.5 mm, yielding a MF of ~2.25 mm/deg. Multiplying the MF with the visual spread yields cortical spreads of ~0.75 and ~1.73 mm for LFP and iEEG electrodes [4].

Unfortunately, the cortical spread computed this way is inflated by several factors, such as small eye movements. For example, in this task, while the visual stimuli were presented on the screen, the animals were required to maintain fixation within 2° of a small dot at the center of the screen. This fixation window is typical for these kinds of neurophysiological experiments but is much greater than the size of the stimulus and distance between neighboring stimulus positions (both 0.3°). If the animal made small eye movements (while staying within the fixation window), iEEG responses could potentially be obtained from a larger set of stimulus positions (essentially the responses will be "smeared" due to fixation jitters), leading to a larger visual spread and consequently cortical spread. This is one of many factors (discussed below) that can artifactually inflate the visual spread. Xing et al. [9] developed an elegant model to address some of these concerns, as discussed next.

Model to estimate spatial spread: To better appreciate the confounding factors, we consider an often-used analogy for describing the spatial spread: a microphone (microelectrode) in a stadium (cortical tissue). The spatial spread could be interpreted as the number of people (neurons) whose voices can be reliably picked up by the microphone. Spatial spread defined this way depends on the properties of the microphone (its size and acoustic properties), but also on the way the population responds to any event (stimulus). In particular, if the population activity is highly correlated (people singing in unison), the activity can be picked up from far away. Indeed, Lindén et al. [14] showed that the spatial spread expands when the underlying population activity is correlated, which also reconciles the differences in spatial spreads estimated by different researchers. However, while this definition of spatial spread is intuitive, there are two issues. First, spatial spread defined this way trivially depends on many factors, making its interpretation difficult. For example, recurrent connections between neurons that may differ across brain regions, stimulus size, behavioural factors such as eye movements and attentional state that changes pairwise correlations [15, 16] and RF size [17] are all expected to change the spatial spread. Second, such factors are expected to affect the spatial spread of other signals also, such as the multiunit activity (MUA). For example, since the cortical tissue is ohmic [18], the effect of correlation on spatial spread as modelled by Lindén et al. [14] should be observed at low as well as high frequencies, causing both LFP and MUA spatial spreads to become large depending on the spatial extent of correlation.

One strategy, proposed by Xing et al. [9], is to decouple the total spatial spread into two parts—one that depends on the microphone and the other that depends on the population. Only the first part is taken as the spatial spread of the signal. The total spread can be calculated by convolving the signal spread and population spread (for mathematical details, see [4, 9]). If both spread functions are assumed to be Gaussian, the resultant total spread is also a Gaussian whose variance is equal to the sum of individual variances. All the factors described above change the population spread (and hence the total spread), but not the signal spread. The signal spread now depends only on the properties of the electrode, such as its impedance and size.

Experimentally, we measure the total spread (which is the visual spread multiplied by the MF), so how can we get the signal spread, given that the population spread is unknown? The key, as proposed by Xing et al., is to compute the *difference* in the spreads of two signals (in their case, LFP and MUA). While the visual spreads of both measures will get inflated due to the population spread, they get inflated by the same amount, and therefore get cancelled when we subtract the total spreads of LFP and MUA.

We found that using this approach, the spatial spread (SD of the fitted Gaussian) of LFP reduced to ~0.5 mm, a reduction of ~30% from the earlier estimate (similar results were also obtained by Xing et al.). Because we had also collected iEEG data, we could adjust the spatial spread of iEEG by comparing with the LFP spread, which led to a ~10% reduction and an adjusted spread of ~1.5 mm (note that the spread is local even if the model is not considered). Therefore, the iEEG spread is only thrice the spread of LFP, even though the iEEG electrode is much larger. Further, if we take the diameter of spread to be twice the SD, the diameter of spread of iEEG (~3 mm) is approximately equal to the sum of LFP spread (~1 mm) and the iEEG electrode itself (2.3 mm). This is consistent with the simple idea that the spatial spread of an infinitesimally small electrode (microelectrode tip) has a diameter of ~1 mm and increasing the electrode diameter simply adds the overall spread by that amount. However, this is yet to be experimentally verified.

17.2 Stimulus Tuning Preferences of LFP and iEEG

To compare how LFP and iEEG get modulated with sensory stimulation, we presented gratings that varied in size, orientation, contrast or spatial frequency, while the monkeys fixated on a small dot at the centre of the screen as in the RF mapping task [5]. In the neural signals, we focused on two frequency bands: gamma (30–70 Hz) and high-gamma (>80 Hz). In the primary visual cortex, visual gratings are known to induce strong gamma oscillations, and both gamma power and centre frequency systematically vary when the stimulus size, orientation, spatial frequency, contrast or temporal frequency is varied [19–22]. We found that gamma tuning (i.e., how gamma power varied when the stimulus parameter was changed) was comparable between LFP and iEEG. This is in stark contrast to the gamma tuning in EEG [21], which was substantially weaker. Therefore, iEEG is more similar to LFP than EEG.

We also compared high-gamma activity, which refers to power over a broad range of frequencies above the gamma band. In iEEG, high-gamma is modulated by stimulus presentation as well as the behavioral state [23, 24]. In LFP, it is tightly correlated with the spiking activity of neurons in the vicinity of the microelectrode [25–27]. The manipulation of stimulus size has two opposing effects: larger stimulus size reduces the average firing rates of neurons due to larger surround suppression, but also activates a larger neural population. We had previously shown that the LFP high-gamma power also reduced with increasing stimulus size, following the trend observed in firing rates [27]. We might have observed an increase in iEEG high-gamma power despite a reduction in firing rate if the iEEG signal represented the average activity over a much larger cortical area than LFP. However, we found that even iEEG high-gamma power reduced with stimulus size [5], again confirming that iEEG is a local signal.

Because power in both LFP and iEEG signals varied systematically for different types of visual stimuli, they carried information about the sensory stimulus and therefore the stimulus identity could be decoded from the LFP/iEEG power. We presented a large array of natural images to the monkeys and compared the information content and decodability of LFP and iEEG. Surprisingly, iEEG had higher information and decodability than LFP [6]. Control analyses showed that higher decoding accuracy of iEEG compared with LFP was not because of differences in low-level visual features but instead because of larger spatial summation of the iEEG. Natural images tend to have more energy at low spatial frequencies, which means that image features tend to change slowly with space and consequently nearby neural assemblies process similar features. In such a situation, averaging the responses of such neural assemblies (which is more effectively done by the iEEG electrode compared to a microelectrode due to its larger size) leads to effective cancellation of random noise in the assemblies while preserving the common signal, and hence an improvement in the information content. Further, we found that low frequencies and gamma band carried more information than other frequencies, consistent with previous results in the LFP [28]. The spatial scale over which gamma rhythms remains coherent is typically several millimeters [21, 29], and therefore better captured by the iEEG electrode compared to LFP.

17.3 iEEG as an Average of LFPs

As described above, information content in iEEG far exceeded the information content in LFP even though the stimulus tuning preferences were comparable. Another feature that was found to be different was the shape of the power spectral density (PSD), which was much steeper between 20 and 100 Hz for iEEG compared to LFP [4]. We tested whether iEEG could be mimicked by simply averaging LFPs from nearby electrodes. This procedure is expected to increase the PSD slope, since lower frequencies of the LFPs are more coherent than higher frequencies, and therefore averaging the LFPs from nearby electrodes does not appreciably change the signal amplitude at low frequencies (since individual components are more or less similar) but reduces the amplitude at higher frequencies (since the signal components have different phases and therefore cancel out), leading to an increase in the slope of the PSD (for a more detailed discussion, see [4]). We found that as the LFPs were averaged over a larger set of neighboring microelectrodes, the PSD slope of the resultant averaged signal indeed increased and matched the slope of the experimentally obtained iEEG slope when ~50 microelectrodes (7×7 grid spanning ~3 mm on each side) were averaged. Similarly, the information content of the averaged signal increased and approached the information content of the iEEG signal [6].

In summary, simultaneous recordings of iEEG (obtained using electrodes with exposed diameter of 2.3 mm) and LFPs (Utah arrays with inter-electrode spacing of 400 μm) from the primary visual cortex of monkeys have revealed that the spatial spread of iEEG is about ~3 mm, only three times the LFP spread, and the iEEG signal is well approximated by simply averaging LFP signals over ~3 mm. It is likely that these results will be generalizable to humans as well. Yoshor et al. [12] measured the RFs by presenting small visual stimuli and found visual spreads that are comparable to our results. Similarly, Winawer et al. [30, 31] measured stimulus-locked and broadband components of the iEEG responses to a moving flickering bar and estimated visual spreads using a population RF (pRF) model [32]. Their iEEG spread estimated for broadband responses was comparable to our estimates. Regarding stimulus tuning, gamma oscillations recorded using iEEG from visual areas when gratings are presented show similar stimulus preferences as observed in monkeys (for example, compare [33] with [22, 34]).

Whether these results are generalizable to other brain areas remains an open question. As discussed earlier, Kajikawa and Schroeder [10] found the spatial spread of LFP to be much larger (several millimeters) in the auditory cortex. Some additional features of the estimated spatial spread can shed light on this discrepancy. For example, if the spatial spread is calculated as a function of frequency (by taking the amplitude of the response at a particular frequency instead of taking the overall magnitude of the evoked response), the spread function for LFP has a "band-pass" shape, with a substantially larger spread in the high-gamma range [7], although this band-pass effect is not seen in iEEG [4]. Further, the spread is also higher at very low frequencies in the LFP [4], likely due to the presence of the alpha rhythm that gets suppressed with visual stimulation. Since different brain areas show different oscillatory signatures (sensory-motor areas predominantly show beta rhythm), which potentially get modulated over different spatial scales, the resultant spread may depend on such factors (which cannot be accounted for in the model described above). Further studies are needed (using both microelectrode and iEEG electrodes) in other sensory areas to establish the generalizability of the results discussed here.

References

1. Buzsáki G, Anastassiou CA, Koch C (2012) The origin of extracellular fields and currents—EEG, ECoG, LFP and spikes. Nat Rev Neurosci 13:407–420. https://doi.org/10.1038/nrn3241
2. Einevoll GT, Kayser C, Logothetis NK, Panzeri S (2013) Modelling and analysis of local field potentials for studying the function of cortical circuits. Nat Rev Neurosci 14:770–785. https://doi.org/10.1038/nrn3599
3. Pesaran B, Vinck M, Einevoll GT et al (2018) Investigating large-scale brain dynamics using field potential recordings: analysis and interpretation. Nat Neurosci 21:903–919. https://doi.org/10.1038/s41593-018-0171-8
4. Dubey A, Ray S (2019) Cortical electrocorticogram (ECoG) is a local signal. J Neurosci 39:4299–4311. https://doi.org/10.1523/JNEUROSCI.2917-18.2019
5. Dubey A, Ray S (2020) Comparison of tuning properties of gamma and high-gamma power in local field potential (LFP) versus electrocorticogram (ECoG) in visual cortex. Sci Rep 10:5422. https://doi.org/10.1038/s41598-020-61961-9
6. Kanth ST, Ray S (2020) Electrocorticogram (ECoG) is highly informative in primate visual cortex. J Neurosci 40:2430–2444. https://doi.org/10.1523/JNEUROSCI.1368-19.2020
7. Dubey A, Ray S (2016) Spatial spread of local field potential is band-pass in the primary visual cortex. J Neurophysiol 116:1986–1999. https://doi.org/10.1152/jn.00443.2016
8. Katzner S, Nauhaus I, Benucci A, et al (2009) Local origin of field potentials in visual cortex. Neuron 61:35–41. https://doi.org/10.1016/j.neuron.2008.11.016
9. Xing D, Yeh CI, Shapley RM (2009) Spatial spread of the local field potential and its laminar variation in visual cortex. J Neurosci 29:11540–11549. https://doi.org/10.1523/JNEUROSCI.2573-09.2009
10. Kajikawa Y, Schroeder CE (2011) How local is the local field potential? Neuron 72:847–858. https://doi.org/10.1016/j.neuron.2011.09.029
11. Kreiman G, Hung CP, Kraskov A et al (2006) Object selectivity of local field potentials and spikes in the macaque inferior temporal cortex. Neuron 49:433–445. https://doi.org/10.1016/j.neuron.2005.12.019

12. Yoshor D, Bosking WH, Ghose GM, Maunsell JH (2007) Receptive fields in human visual cortex mapped with surface electrodes. Cereb Cortex 17:2293–2302. https://doi.org/10.1093/cercor/bhl138
13. Dow DBM, Snyder AZ, Vautin RG, Bauer R (1981) Magnification factor and receptive field size in foveal striate cortex of the monkey. Exp Brain Res 44:213–228. https://doi.org/10.1007/BF00237343
14. Lindén H, Tetzlaff T, Potjans TC et al (2011) Modeling the spatial reach of the LFP. Neuron 72:859–872. https://doi.org/10.1016/j.neuron.2011.11.006
15. Cohen MR, Maunsell JH (2009) Attention improves performance primarily by reducing interneuronal correlations. Nat Neurosci 12:1594–1600. https://doi.org/10.1038/nn.2439
16. Mitchell JF, Sundberg KA, Reynolds JH (2009) Spatial attention decorrelates intrinsic activity fluctuations in macaque area V4. Neuron 63:879–888. https://doi.org/10.1016/j.neuron.2009.09.013
17. Womelsdorf T, Anton-Erxleben K, Treue S (2008) Receptive field shift and shrinkage in macaque middle temporal area through attentional gain modulation. J Neurosci 28:8934–8944. https://doi.org/10.1523/JNEUROSCI.4030-07.2008
18. Logothetis NK, Kayser C, Oeltermann A (2007) In vivo measurement of cortical impedance spectrum in monkeys: implications for signal propagation. Neuron 55:809–823. https://doi.org/10.1016/j.neuron.2007.07.027
19. Gieselmann MA, Thiele A (2008) Comparison of spatial integration and surround suppression characteristics in spiking activity and the local field potential in macaque V1. Eur J Neurosci 28:447–459. https://doi.org/10.1111/j.1460-9568.2008.06358.x
20. Jia X, Xing D, Kohn A (2013) No consistent relationship between gamma power and peak frequency in macaque primary visual cortex. J Neurosci 33:17–25. https://doi.org/10.1523/JNEUROSCI.1687-12.2013
21. Murty DVPS, Shirhatti V, Ravishankar P, Ray S (2018) Large visual stimuli induce two distinct gamma oscillations in primate visual cortex. J Neurosci 2270-17. https://doi.org/10.1523/JNEUROSCI.2270-17.2017
22. Ray S, Maunsell JHR (2010) Differences in gamma frequencies across visual cortex restrict their possible use in computation. Neuron 67:885–896. https://doi.org/10.1016/j.neuron.2010.08.004
23. Lachaux J-P, Axmacher N, Mormann F et al (2012) High-frequency neural activity and human cognition: past, present and possible future of intracranial EEG research. Prog Neurobiol 98:279–301. https://doi.org/10.1016/j.pneurobio.2012.06.008
24. Mukamel R, Fried I (2011) Human intracranial recordings and cognitive neuroscience. Annu Rev Psychol 63:511–537. https://doi.org/10.1146/annurev-psych-120709-145401
25. Manning JR, Jacobs J, Fried I, Kahana MJ (2009) Broadband shifts in local field potential power spectra are correlated with single-neuron spiking in humans. J Neurosci 29:13613–13620. https://doi.org/10.1523/JNEUROSCI.2041-09.2009
26. Ray S, Crone NE, Niebur E, et al (2008) Neural correlates of high-gamma oscillations (60–200 Hz) in macaque local field potentials and their potential implications in electrocorticography. J Neurosci 28:11526–11536. https://doi.org/10.1523/JNEUROSCI.2848-08.2008
27. Ray S, Maunsell JHR (2011) Different origins of gamma rhythm and high-gamma activity in macaque visual cortex. PLoS Biol 9:e1000610. https://doi.org/10.1371/journal.pbio.1000610
28. Belitski A, Gretton A, Magri C et al (2008) Low-frequency local field potentials and spikes in primary visual cortex convey independent visual information. J Neurosci 28:5696–5709. https://doi.org/10.1523/jneurosci.0009-08.2008
29. Jia X, Smith MA, Kohn A (2011) Stimulus selectivity and spatial coherence of gamma components of the local field potential. J Neurosci 31:9390–9403. https://doi.org/10.1523/JNEUROSCI.0645-11.2011
30. Winawer J, Kay KN, Foster BL et al (2013) Asynchronous broadband signals are the principal source of the BOLD response in human visual cortex. Curr Biol 23:1145–1153. https://doi.org/10.1016/j.cub.2013.05.001

31. Winawer J, Parvizi J (2016) Linking electrical stimulation of human primary visual cortex, size of affected cortical area, neuronal responses, and subjective experience. Neuron 92:1213–1219. https://doi.org/10.1016/j.neuron.2016.11.008
32. Kay KN, Winawer J, Mezer A, Wandell BA (2013) Compressive spatial summation in human visual cortex. J Neurophysiol 110:481–494. https://doi.org/10.1152/jn.00105.2013
33. Bartoli E, Bosking W, Chen Y et al (2019) Functionally distinct gamma range activity revealed by stimulus tuning in human visual cortex. Curr Biol 29:3345-3358.e7. https://doi.org/10.1016/j.cub.2019.08.004
34. Shirhatti V, Ray S (2018) Long-wavelength (reddish) hues induce unusually large gamma oscillations in the primate primary visual cortex. PNAS 115:4489–4494. https://doi.org/10.1073/pnas.1717334115

Chapter 18
What Do ECoG Recordings Tell Us About Intracortical Action Potentials?

Tobias Bockhorst, Andreas K. Engel, and Edgar Galindo-Leon

Abstract Human intracranial recordings are, with few exceptions, typically confined to recording of local field potentials. Spike detection has been limited to intracortical measurements with microelectrodes or microwires. Here, we put a spotlight on an emerging class of subdural surface arrays densely packed with micro- to mesoscale electrodes. These now combine the benefits of low invasiveness, large coverage and high signal quality with the option of recording spikes in addition to spatially resolved brain waves. We discuss recent work where features of surface spiking activity have been compared to action potential recordings from intracortical electrodes, showing that substantial information about the underlying cortical columns can be retrieved form the epicortical recordings. These novel approaches have potential for large-scale network studies contributing to a deeper understanding of brain function and hold promise for refinement of clinical diagnostic or interventional procedures.

18.1 Introduction

Electric currents on the cortical surface were initially reported in rabbits and monkeys by Caton as early as 1875 [1]. In humans, the first recordings of electrical brain activity date back to 1924, when Hans Berger measured manifestations of brain waves on the scalp, setting the stage for the era of EEG [2, 3]. With the applicability of a low-cost, noninvasive method, EEG has enormously fueled insights into cognitive processes and their higher-level physiological correlates, such as synchronized neural oscillations and event-related potentials. As a drawback, however, transcranial measurement from the scalp considerably reduces spatial resolution of signal sources as well as the amplitude of the signal itself—in particular at frequencies above 100 Hz. As a remedy, approaches for more localized recording of electric fields have been developed over more recent decades, yielding the electrocorticogram (ECoG)

T. Bockhorst · A. K. Engel · E. Galindo-Leon (✉)
Department of Neurophysiology and Pathophysiology, University Medical Center Hamburg-Eppendorf, 20246 Hamburg, Germany
e-mail: e.galindo-leon@uke.de

N. Axmacher (ed.), *Intracranial EEG*, Studies in Neuroscience, Psychology and Behavioral Economics, https://doi.org/10.1007/978-3-031-20910-9_18

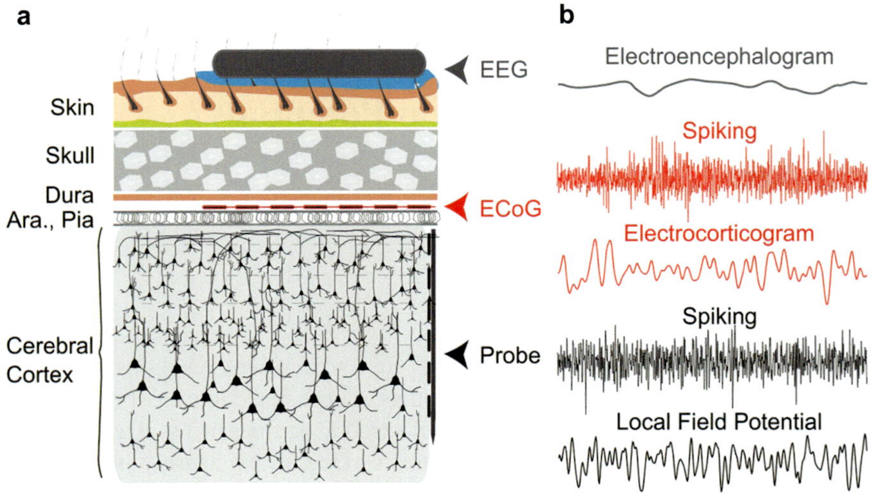

Fig. 18.1 Brain signals acquired with extracranial, epi- and intracortical recording techniques. **a** Placement of different electrode designs for extra- and intracortical recordings. EEG, electroencephalogram; ECoG, electrocorticogram; probe, intracortical extracellular probe; ara., arachnoid mater. **b** Types of signals acquired with the respective techniques. Note that contacts along the intracortical probe record the electric field at a different angle relative to the dendritic axes compared to surface electrodes

when recorded by subdural electrodes from the cortical surface. For sufficiently small electrodes the ECoG signal compares to a classical local field potential (LFP) which is usually obtained by sharp electrodes advanced into the brain (Fig. 18.1).

In primate research preparations and human patients, ECoG is usually obtained through grids of electrodes with diameters and spacing in the range of several millimeters [4–9]. Electrode size determines the number of synaptic inputs and neurons that contribute to the local raw signal—because of cell size and density as well as the specific microarchitecture of dendritic branches in the volume underneath the electrode. These microanatomical features vary between regions in the same brain [10], across species [11] and with functional organization [12], requiring careful adjustments of electrode size with respect to what sort of processes shall be resolved. Generally speaking, the relatively large electrodes common in most previous research are more prone to source cancellation by destructive interference [13] between subpopulations, which may blur topographies of both ongoing and stimulus-related activity. Moreover, the technical routine for analysis of ECoG measurements often included low-pass filtering (<100 Hz) and down-sampling for technical convenience. As a result, ECoG data available from the vast majority of past studies effectively correspond to a 'low-noise EEG'—cleaner, higher in spatial resolution, but often still spatially gross from the perspective of distinct processes within cortical regions.

Action potentials, the basic elements of neuronal information processing in every hitherto studied nervous system [14–17] are at best represented indirectly in the high-frequency component of the LFP [18, 19]. Recordings of action potentials in-vivo have so far required penetration of brain tissue with electrodes, causing stress to the neural tissue and limiting human studies to rare clinical settings, such as probing of electrode positions in deep brain stimulation [5, 20–22] or combined macro-/microelectrode recordings in epilepsy patients (see Chaps. 42, 43, 44 and 45). A possibility to overcome the major limitations of scalp recordings—spatial blur, signal attenuation and effective blindness to spiking activity—is now promised by novel types of electrode arrays for subdural ECoG.

As we will discuss in the following, such novel types of spike-sensitive ECoG arrays (spECoG) are densely populated with micro- (approx. 10 μm diameter) or mesoscale (approx. 250 μm diameter) electrodes. These are fabricated through fully customizable, waver-aided techniques. The thinness (4–10 μm) of carrier foils used in currently described designs allows folding of these arrays into fissures and sulci previously inaccessible to surface recording methods. Table 18.1 and Fig. 18.2 summarize the different designs of ECoG arrays. The spECoG approach now allows to concurrently record, from the intact cortical surface, the projection of summed synaptic inputs (i.e., graded potentials in the LFP-band) of neurons as well as their spiking outputs. While both are intrinsically related (by nonlinear transformations of post-synaptic inputs into spiking outputs) [23, 24] their informational contents are not redundant. The two signals also feature different spatial and temporal resolutions: the spike-time signal is more precise in time, yet it reflects a spatially more confined neural population [25]. As a result, analyses of spiking activity can add relevant information beyond ECoG-LFPs, when local activity needs to be resolved [26]—a prerequisite for reading out topographically organized information. This, in turn, is key to interfacing, e.g., for neuroprosthetic application, with sensory and motor cortices in the human brain, where stimulus features (e.g., retinotopy, tonotopy, somatosensory fields) and movement direction systematically vary along columnar topographies. Studies on the decoding of movement direction and grasp types revealed highest decoding performance for activity in the high gamma band, a presumed correlate of coordinated assembly spiking (Fig. 18.2c, left panel) [27, 28]. Spike information can also outperform high gamma band LFP in the decoding of more abstract cognitive processes such as attention [29]. Large-coverage recording of spatially resolved, 'clean' LFPs and action potentials from the intact cortical surface is likely to further illuminate network-level substrates of cortical processing, such as functional connectivity, criticality or travelling waves.

Table 18.1 Specifications and features of different EcoG array designs

<table>
<tr><th>Array design</th><th>Electrode diameter (spacing)</th><th>Anatomical resolution</th><th>Accessible regions</th><th>Signals</th></tr>
<tr><td>Macroscale EcoG</td><td>~5 mm (5–10 mm)</td><td>(Subsections of) gyri</td><td>Superficial gyri</td><td>Low-noise EEG</td></tr>
<tr><td>Mesoscale spECoG</td><td>250 μm (1.5 mm)</td><td>(Macro-)columns</td><td rowspan="2">Gyri and sulci; neocortex and hippocampus</td><td rowspan="2">LFP, spikes</td></tr>
<tr><td>Microscale spECoG</td><td>~5 μm (30 μm) (NeuroGrid1)</td><td>Microcolumns, cells</td></tr>
</table>

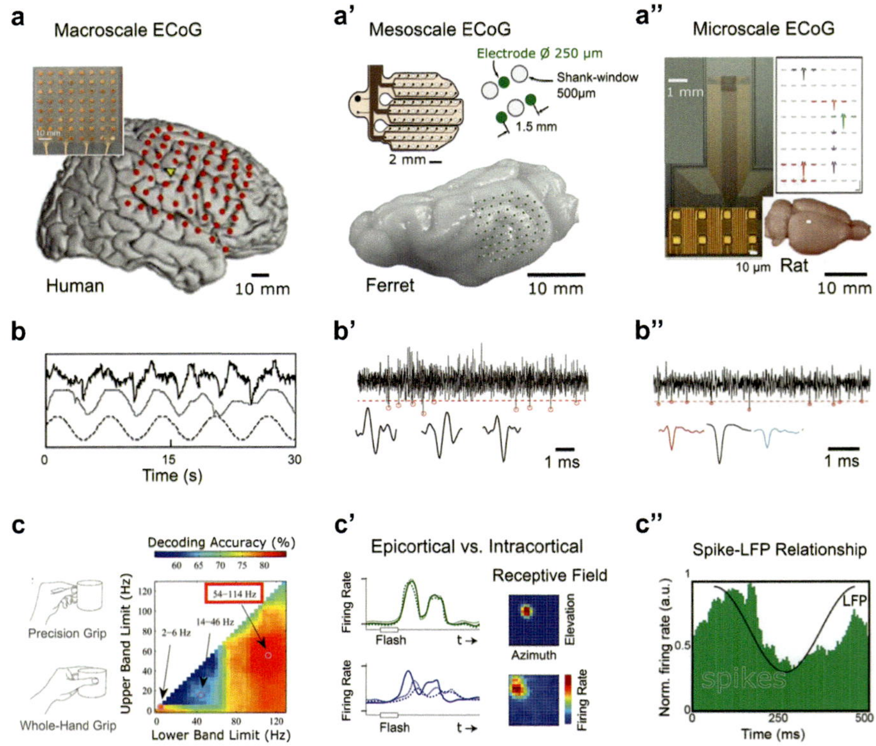

Fig. 18.2 Comparison of macro-, meso- and microscale EcoG arrays. The three columns of the figure highlight examples for macroscale (left), mesoscale (middle) and microscale (right) ECoG arrays. **a–a″** Basic design of each type and projection onto a respective brain. **a** Yellow triangle: electrode position referred to in **b**. **a′** Note that the carrier foil of the mesoscale spECoG is split into three 'fingers' to promote adherence to the cortical surface. Shank windows between the recording electrodes allow parallel measurements with penetrating intracortical probes. **a″** Optical micrograph of 256-channel NeuroGrid; inset: further magnified view of recording contacts. Note that spikes occur on a subset of contacts only (topography plot to the right), whilst almost every contact provided spiking signals in case of the mesoscale spECoG (not shown). White patch on surface of rat brain indicates the size of the first NeuroGrid implant. **b–b″** Example traces of signals recorded with the three types of EcoG arrays. **b** Upper trace: example time course of a conventional macroscale EcoG during a tracking task (electrode position indicated by yellow triangle in **a**). Dotted/dashed traces below indicate the horizontal positions of a controllable tracking cursor, and of the moving target on the screen, respectively. Note that EcoG activity rather reflects cursor- than actual target position. **b′**, **b″** Spike-band activity traces with example waveforms of detected spikes (red circles in raw trace) shown below. Red dotted line, detection threshold (corresponding to 3 times noise level). **c–c″** Representative findings. **c** Decoding accuracy for grip type in a naturalistic grasp task is best for spike-related high frequency bands (54–114 Hz) in the EcoG. **c′** Spiking responses measured with a mesoscale array are stronger, more stable (PSTH illustrations of responses to flash stimuli) and more precise in tuning (receptive field, RF, from retinotopic mapping) than concurrently obtained intracortical responses in representative single trials. **c″** Locking of spikes (green) to the averaged phase of delta-band LFP waves, recorded in a human patient under anesthesia (second instantiation of NeuroGrid). **a** photo of clinical grid from [56]. Panels **a, b** modified from [6]; **c** modified from [28]; **a′** modified from [40, 43]; **b′, c′** modified from [40]; **a″** modified from [36], except for drawing of rat brain created by E. Galindo-Leon; **b″**, **c″** modified from [36, 38]

In the following sections, we highlight recent work on epicortical recording of action potentials via such novel types of spECoG electrode arrays in animal preparations and human patients. In particular, we discuss first evidence on how the 'surface spikes' recorded through spECoGs relate to cortical action potentials.

18.2 Microscale Surface Electrodes with Single-Cell Resolution

ECoG arrays used in animal research differ mainly in electrode diameter and interelectrode spacing. These low-level features are critical in biophysical and microanatomical respects. From the technical perspective, electrode diameter is a major determinant of impedance, i.e., the frequency-dependent input resistance of the electrode. This parameter is often believed to be crucial in spike recordings, in particular for intracellular measurements, where high impedance is a sign of sufficiently sharp electrode tips that can penetrate cellular membranes. In contrast, the role of electrode impedance for spike detectability in extracellular recordings remains elusive [30–34]. A direct comparison between adjacent low- and high-impedance contacts (iridium; 177 μm^2) on the same intracortical array [35] has returned no difference in spike amplitude and detectability between the two. As the spectral composition of spikes and neural noise should be roughly similar across species, it seems plausible that this statement could hold true in general. Apart from adjusting electrode dimensions, spike registration is eased by electrode materials with mixed electronic and ionic conductivity [36] and, obviously, requires sufficient sampling frequency.

In 2015, Khodagholy and colleagues introduced the first instantiation of 'NeuroGrid', a small-coverage spECoG array composed of microscale electrodes (10×10 μm electrode area; 30 μm interelectrode distance; 256 electrodes; impedance, 10 kΩ at 1 kHz; 1 mm^2 total coverage) [36]. Its design targets recordings from single-units within cortical microcircuits (Fig. 18.2, right column). The authors report reliable, prolonged recordings of action potentials from rodent cortex (putative single-unit activity; Fig. 18.2b, right panel) and two human patients (multi-units) under minimal stress to the cortex itself. In rats, recording quality was still adequate 10 days after implantation, which corresponds to typical epochs of presurgical subdural EEG monitoring in epilepsy patients. In contrast, intraoperative recordings in patients lasted for a maximum of 30 min and the amplitude of the spikes was lower, hindering the clustering of single neurons in humans. In particular, spikes observed at the cortical surface were also differentiable into fast and slow spikes, indicative of inhibitory interneurons vs. excitatory principal cells, respectively [37], suggesting that the array can capture activity from neurons down to a depth of at least 200 μm. Previously, layer 1 neurons in particular had been difficult to monitor at such resolution by conventional methods. To investigate whether these features and limitations were specific to the neocortex, the authors also recorded activity at the alvear surface of the hippocampus, which was reached after removal of small region of the neocortex. This pilot study highlights that the novel approach eases studies into the relationship of spiking activity and LFPs (see also Chaps. 44 and 45), reporting entrainment between spikes and brain oscillations in both, neocortex and hippocampus.

Follow-up work [38] demonstrated scalability of the design under concessions to electrode density, but preservation of putative single-cell resolution. With 120 (240) tetrodes (each 4 × 10 μm × 10 μm) instead of single contact electrodes, and now spaced 2 mm apart, a cortical area of 420 (840) mm^2 was covered, sufficient to trace the propagation of local LFP activity patterns during neurosurgery. The NeuroGrid also proved more sensitive to high-frequency (gamma band) LFP than the gross conventional clinical arrays. Precise localization of sources that generate pathological activity should substantially refine tissue resection and intervention strategies in patients with neurological diseases, such as treatment-resistant focal epilepsy. The co-registration of spikes through the same electrodes provided insights into the local generation of pathological activity (Fig. 18.2c, right panel), which may help refine tissue resection and intervention strategies in diseases such as treatment-resistant focal epilepsy. Comparable data previously required intracortical recordings, which are only possible on rare occasions in the clinical setting. The study represents an important proof of principle in scaling the original 1 mm^2 NeuroGrid, originally designed for studies on cortical microcircuitry in small rodents, up to the dimensions of local networks in human cortex. Still, substantial advances in technical solutions and analysis methods for massive parallel recordings will be necessary to further scale such spECoG microarrays up to the level of networks that expand across brain areas, without further compromises in electrode density.

18.3 Mesoscale Surface Electrodes for Network-Wide Coverage

A modeling study [39] has elucidated how action potentials from layer 1 neurons can be sensed through the types of surface electrode used by Khodagholy and colleagues [36], but still the NeuroGrid studies have not addressed how the spike signals observed at the cortical surface relate to intracortical action potentials—in particular with respect to faithful representation of stimulus features and topographically mapped neural response properties. This, however, seems crucial for the assessment of the signal's relevance in processes such as perceptually driven decisions or motor control.

We have recently investigated this aspect in combined recordings of intracortical and surface spikes in anesthetized ferrets using Michigan probes (32 electrodes, 15 μm diameter, 50 μm spacing) and a spECoG array with mesoscale geometry (Fig. 18.2, middle column) (250 μm electrode diameter; 1.5 mm hexagonal spacing;

impedance at 1 kHz < 10 kΩ) [40]. The principle ECoG electrode design had previously been applied in acute and chronic LFP recordings from the cortex of monkeys and ferrets with recordings stable over months [41–43]. Epicortical spikes have been recorded using the previously described 64 electrodes instantiation [40] and a novel 96 electrodes version with denser spacing (unpublished data), both scaled to cover an extended multisensory network that spans the dorsal two thirds of the ferret cerebral hemisphere.

Using this combination of surface recordings of ongoing and stimulus-related activity with intracortical single-site recordings via the linear probe we investigated how epicortical spiking relates to action potentials of cortical cells [40]. Waveforms of epicortical spikes recorded through the mesoscale electrodes resembled those of intracortical action potentials. Epicortical responses to presentations of sensory stimuli (light flashes and/or acoustic clicks) were higher in amplitude and less variable (in amplitude, latency and duration) across trials as compared to intracortical responses. This held true even when sum activity rather than single-site measurements was considered for the intracortical data (Fig. 18.2c, middle panel). In addition to a strengthening of signal to noise ratio, the sharpness of response-tuning was found preserved (contour orientation and tonotopy) or even increased (visual receptive fields) as compared to the concurrent intracortical recordings. Hence, epicortical recordings through mesoscale spECoG electrodes can resolve topographical representations across retinotopic maps, neighboring orientation columns or tonotopical gradients. Levels of ongoing activity on the surface were below the sum of activity along the laminar probe, with rates of epicortical spiking best predicted by taking only synchronized intracortical spikes into account (see also Chap. 46 on laminar recordings).

Together, these observations render it unlikely that surface spiking merely reflects a superposition of spike trains fired by several neurons. Based on empirical and modeling data, we propose a selectivity of the mesoscale spECoG electrodes for synchronized spikes fired in concert by several neurons distributed along the depth of the underlying cortical columns. The successful recording of surface spikes with such mesoscale electrodes disproves theoretical claims that the contact area should not exceed a maximum of 100 μm^2 to maintain detectable spike amplitudes [39].

As a potential mechanism to account for both the synchrony-dependence of surface spikes and their relatively narrow waveform (as compared to dendritic calcium spikes which have been shown to be recordable at the surface [44] we proposed that sodium spikes fired by layer 5 principal cells, which can act as coincidence detector neurons, may be projected to layer 1. In fact, this mechanism is in line with previous research on sodium spikes that propagate via association fibers along layer 1 (Fig. 18.3) and play a role in attentional selection [44–48]. At the level of the LFP, a particular role of spike-synchrony for epicortical LFP in the high gamma band (60–200 Hz) was predicted by modeling [49] which revealed that an increase in ECoG high-gamma power could equally well be obtained by a large increase in firing rate or by a small increase in synchronization (see also Chap. 17).

If surface spikes trace back to intracortical action potentials fired in synchrony they may provide a particularly meaningful signal, as synchronous spiking is believed

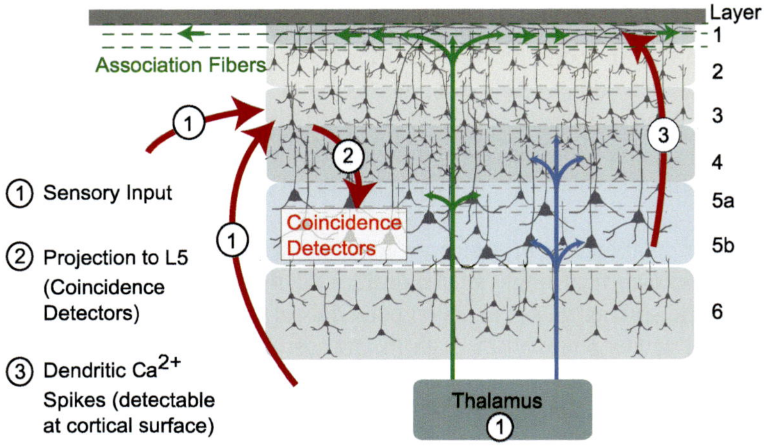

Fig. 18.3 Presumed origin of synchrony-dependent epicortical spikes. Dendritic Ca^{2+} spikes are detectable at the surface according to Suzuki and Larkum [44]. Our results suggest a similar mechanism with respect to dendritic Na^{+} spikes [48]. Sensory inputs to layers 3 and 4 are propagated to layer 5, in which some neurons act as coincident detectors and fire Na^{+} spikes. Those spikes might be projected to layer 1 in a similar way as dendritic Ca^{2+} spikes and, thus, become detectable at the cortical surface [44]

to serve the propagation of neural information, and to be key to a variety of tasks such as attentional gain modulation, feature binding or sensorimotor processing [50–54]. In line with the idea of a synchrony-based process, responses to visual stimuli have been shown to be more stable and sharpened in tuning when recorded on the surface as compared to simultaneous recordings through adjacent laminar probes [40]. Hence, chronic implants that capture synchronized spiking of neural assemblies in sensory or motor areas may improve real-time decoding of neuronal signals in neuroprosthetic device control through more robust and strong neuronal responses. From a broader perspective, mesoscale spECoG arrays may provide first insights into spiking activity of large-scale networks through minimally invasive recordings in chronically implanted, freely behaving animals or even humans.

Another proof of principle for recordings of multiunit spiking through mesoscale spECoG electrodes was reported in guinea pigs [55]. The authors describe an array of 16 electrodes (100 μm diameter, 0.5 mm spacing) scaled to cover primary auditory cortex (4 mm^2 array area). Based on comparisons of first-spike latencies between spECoG and concurrently obtained intracortical probe data (16 contacts single-shank Michigan probe), Konerding and colleagues found that surface spiking best correlated with spiking in supragranular layer IV (800–1100 μm). Moreover, frequency resolution in A1 was more resolved in the multiunit spiking data as compared to same-site LFPs (<100 Hz), again confirming the capability of mesoscale-electrode recordings to preserve topographic information.

18.4 Opportunities and Challenges

The novel spECoG arrays described here share two key features. First, they consist of miniaturized electrodes suited for spike recording mounted on flexible carrier foils that can adhere closely to gyri, fold into sulci and give access to cortices in deep fissures. Second, they are built from biocompatible materials and are stable in recording quality over clinically relevant epochs [41]. Already, spike recordings from the intact cortical surface using these arrays have been demonstrated in rats, guinea pigs, ferrets, and humans [36, 38, 40, 55]. This paves the way for a wide range of optimization, from diagnostically tailored arrays in surgery planning [38] to reading out functional brain activity at various scales of network coverage, e.g., in brain-machine interfacing [56].

Previous applications of electrocorticography were restricted to recordings of graded activity in the LFP band (<100 Hz), which may primarily reflect the concerted postsynaptic activity of inhibitory interneurons in large populations [18]. The novel spECoG approach allows the additional recording of spikes from principal cells and interneurons in spatially resolved subpopulations through the same electrodes. Piloting experiments confirmed applicability in the neurosurgical setting, where it may open novel perspectives in the treatment of epileptic patients [36, 38].

Arrays of small-size electrodes with large coverage and a high-density of contacts are essential to localize the site of microseizures in epilepsy patients. For this purpose, Chiang et al. [57] recently developed a flexible, high-resolution array based on a substrate of liquid crystal polymer (LCP) that can be implanted in humans and animals. In a follow-up study the authors observed that microseizures detected using this type of arrays may provide a more precise tool to improve pre-surgical evaluation for drug-resistant epilepsy [58]. Another remarkable achievement was recently published by Tchoe et al. [59] who developed reconfigurable thin-film, multithousand-channel recording grids using platinum nanorods (PtNRGrids) with small electrodes (30 μm). Using these novel arrays in patients the authors have monitored the patterns of seizure spread and the spatiotemporal dynamics of motor and sensory activity with high resolution [59]. In contrast to the LCP-based array by Chiang et al. [57], PtNRGrids are currently not available for chronic recording in patients. The challenges here are in the safety and durability of the thin-film. These recent studies do not report detection of spiking activity but the arrays developed may show potential to this end.

Microscale instantiations with single-cell resolution will allow deeper studies into how individual action potentials relate to LFP signals. Moreover, they may prove capable of resolving intermingled ('salt-and-pepper') subpopulations of differently tuned sensory neurons in rodent neocortex. At mesoscale resolution, coverage can be scaled up to the level of extended cortico-cortical networks, a prerequisite for the understanding of normal brain function and pathological states. Preservation of topographical maps [40, 55] and increased robustness of single-trial responses [40] promise advances in brain-machine interfacing and real time applications such as closed-loop measurement and/or stimulation regimes. In line with this perspective, macroscale LFP studies on the decoding of movement direction and grasp types revealed best predictive power for activity in the high gamma band, a presumed correlate of coordinated assembly spiking (Fig. 18.2c, left panel) [27, 28].

Naturally, the recording of spike activity through spECoG with high channel numbers also poses additional technical challenges, in particular as high sampling rates are required to cover the spike-band in the kHz range. Furthermore, any explorative analysis of the resultant data may require the development of time-efficient, unsupervised algorithms.

As discussed in the preceding section, electrodes of 250 μm in diameter allowed to observe epicortical spikes that correlated with synchronized firing of action potentials by neurons distributed along the cortical microcolumn. Arrays selective for such synchronized population activity are of particular interest to fundamental research into the propagation of neural information, sensorimotor transformation, and attentional selection, among others [50–53, 60]. Although combined recordings through spECoG and intracortical electrode arrays [40] have shed first light on the intracortical substrates of surface spikes, optogenetic interrogation of the process is clearly indicated for further clarification.

In animal models, arrays of mesoscale electrodes can be scaled up to cover multisite brain networks, and in many species, the scalability will allow a coverage equal to gross scalp EEG. This could now pave the way for novel insights into the spiking level substrate of key large-scale phenomena such as functional connectivity [61–63], neuronal avalanches [64] or traveling waves [65] by recording of LFPs and spikes through the same electrodes in an extended, yet sufficiently dense subdural grid.

References

1. Caton R (1875) Electrical currents of the brain. J Nerv Ment Dis 2:610
2. Berger H (1929) Über das Elektrenkephalogramm des Menschen. Arch Psychiatr Nervenkr 87:527–570
3. Haas LF (2003) Hans Berger (1873–1941), Richard Caton (1842–1926), and electroencephalography. J Neurol Neurosurg Psychiatry 74:9
4. Wyler AR, Ojemann GA, Lettich E, Ward AA (1984) Subdural strip electrodes for localizing epileptogenic foci. J Neurosurg 60:1195–1200
5. Engel AK, Moll CKE, Fried I, Ojemann GA (2005) Invasive recordings from the human brain: clinical insights and beyond. Nat Rev Neurosci 6:35–47
6. Schalk G, Kubánek J, Miller KJ, Anderson NR, Leuthardt EC, Ojemann JG, Limbrick D, Moran D, Gerhardt LA, Wolpaw JR (2007) Decoding two-dimensional movement trajectories using electrocorticographic signals in humans. J Neural Eng 4:264
7. Mesgarani N, Cheung C, Johnson K, Chang EF (2014) Phonetic feature encoding in human superior temporal gyrus. Science 343:1006–1010
8. Gelinas JN, Khodagholy D, Thesen T, Devinsky O, Buzsáki G (2016) Interictal epileptiform discharges induce hippocampal-cortical coupling in temporal lobe epilepsy. Nat Med 22:641–648
9. Fedele T, Schönenberger C, Curio G, Serra C, Krayenbühl N, Sarnthein J (2017) Intraoperative subdural low-noise EEG recording of the high frequency oscillation in the somatosensory evoked potential. Clin Neurophysiol 128:1851–1857

10. Collins CE, Turner EC, Sawyer EK, Reed JL, Young NA, Flaherty DK, Kaas JH (2016) Cortical cell and neuron density estimates in one chimpanzee hemisphere. Proc Natl Acad Sci USA 113:740–745
11. Herculano-Houzel S (2009) The human brain in numbers: a linearly scaled-up primate brain. Front Hum Neurosci 3:31
12. Blasdel GG, Salama G (1986) Voltage-sensitive dyes reveal a modular organization in monkey striate cortex. Nature 321:579–585
13. Lindén H, Tetzlaff T, Potjans TCC, Pettersen KHH, Grün S, Diesmann M, Einevoll GTT (2011) Modeling the spatial reach of the LFP. Neuron 72:859–872
14. Adrian ED, Zotterman Y (1926) The impulses produced by sensory nerve-endings: Part II. The response of a single end-organ. J Physiol (Lond) 61:151–171
15. Hubel DH, Wiesel TN (1998) Early exploration of the visual cortex. Neuron 20:401–412
16. Rieke F (1999) Spikes: exploring the neural code. MIT Press, Cambridge, MA
17. Grün S, Rotter S (2010) Analysis of parallel spike trains. Springer, New York
18. Teleńczuk B, Dehghani N, Quyen MLV, Cash SS, Halgren E, Hatsopoulos NG, Destexhe A (2017) Local field potentials primarily reflect inhibitory neuron activity in human and monkey cortex. Sci Rep 7:40211
19. Fedele T, Scheer HJ, Waterstraat G, Telenczuk B, Burghoff M, Curio G (2012) Towards non-invasive multi-unit spike recordings: mapping 1kHz EEG signals over human somatosensory cortex. Clin Neurophysiol 123:2370–2376
20. Fried I, Rutishauser U, Cerf M, Kreiman G (2014) Single neuron studies of the human brain: probing cognition. MIT Press, Cambridge, MA
21. Sharott A, Gulberti A, Zittel S, Jones AT, Münchau A, Köppen JA, Gerloff C, Westphal M, Buhmann C, Hamel W, Engel AK, Moll CKE (2014) Activity parameters of subthalamic nucleus neurons selectively predict motor symptom severity in Parkinson's disease. J Neurosci 34:6273–6285
22. Baaske MK, Kormann E, Holt AB, Gulberti A, McNamara C, Pötter-Nerger M, Westphal M, Gerloff C, Engel AK, Hamel W, Brown P, Moll CKE, Sharott A (2020) Parkinson's disease uncovers an underlying sensitivity of subthalamic nucleus neurons to beta-frequency cortical input in vivo. Neurobiol Dis 146:105119
23. Belitski A, Gretton A, Magri C, Murayama Y, Montemurro MA, Logothetis NK, Panzeri S (2008) Low-frequency local field potentials and spikes in primary visual cortex convey independent visual information. J Neurosci 28:5696–5709
24. Galindo-Leon EE, Liu RC (2010) Predicting stimulus-locked single unit spiking from cortical local field potentials. J Comput Neurosci 29:581–597
25. Buzsáki G, Anastassiou CA, Koch C (2012) The origin of extracellular fields and currents—EEG, EcoG, LFP and spikes. Nat Rev Neurosci 13:407–420
26. Dürschmid S, Edwards E, Reichert C, Dewar C, Hinrichs H, Heinze H-J, Kirsch HE, Dalal SS, Deouell LY, Knight RT (2016) Hierarchy of prediction errors for auditory events in human temporal and frontal cortex. Proc Natl Acad Sci USA 113:6755–6760
27. Ince NF, Gupta R, Arica S, Tewfik AH, Ashe J, Pellizzer G (2010) High accuracy decoding of movement target direction in non-human primates based on common spatial patterns of local field potentials. PLoS ONE 5:e14384
28. Pistohl T, Schulze-Bonhage A, Aertsen A, Mehring C, Ball T (2012) Decoding natural grasp types from human EcoG. Neuroimage 59:248–260
29. Tremblay S, Doucet G, Pieper F, Sachs A, Martinez-Trujillo J (2015) Single-trial decoding of visual attention from local field potentials in the primate lateral prefrontal cortex is frequency-dependent. J Neurosci 35:9038–9049
30. Cui X, Lee VA, Raphael Y, Wiler JA, Hetke JF, Anderson DJ, Martin DC (2001) Surface modification of neural recording electrodes with conducting polymer/biomolecule blends. J Biomed Mater Res 56:261–272
31. Keefer EW, Botterman BR, Romero MI, Rossi AF, Gross GW (2008) Carbon nanotube coating improves neuronal recordings. Nat Nanotechnol 3:434–439

32. Baranauskas G, Maggiolini E, Castagnola E, Ansaldo A, Mazzoni A, Angotzi GN, Vato A, Ricci D, Panzeri S, Fadiga L (2011) Carbon nanotube composite coating of neural microelectrodes preferentially improves the multiunit signal-to-noise ratio. J Neural Eng 8:066013
33. Ludwig KA, Langhals NB, Joseph MD, Richardson-Burns SM, Hendricks JL, Kipke DR (2011) Poly(3,4-ethylenedioxythiophene) (PEDOT) polymer coatings facilitate smaller neural recording electrodes. J Neural Eng 8:014001
34. Chung T, Wang JQ, Wang J, Cao B, Li Y, Pang SW (2015) Electrode modifications to lower electrode impedance and improve neural signal recording sensitivity. J Neural Eng 12:056018
35. Neto JP, Baião P, Lopes G, Frazão J, Nogueira J, Fortunato E, Barquinha P, Kampff AR (2018) Does impedance matter when recording spikes with polytrodes? Front Neurosci 12:715
36. Khodagholy D, Gelinas JN, Thesen T, Doyle W, Devinsky O, Malliaras GG, Buzsáki G (2015) NeuroGrid: recording action potentials from the surface of the brain. Nat Neurosci 18:310–315
37. Barthó P, Hirase H, Monconduit L, Zugaro M, Harris KD, Buzsáki G (2004) Characterization of neocortical principal cells and interneurons by network interactions and extracellular features. J Neurophysiol 92:600–608
38. Khodagholy D, Gelinas JN, Zhao Z, Yeh M, Long M, Greenlee JD, Doyle W, Devinsky O, Buzsáki G (2016) Organic electronics for high-resolution electrocorticography of the human brain. Sci Adv 2:e1601027
39. Hill M, Rios E, Sudhakar SK, Roossien DH, Caldwell C, Cai D, Ahmed OJ, Lempka SF, Chestek CA (2018) Quantitative simulation of extracellular single unit recording from the surface of cortex. J Neural Eng 15:056007
40. Bockhorst T, Pieper F, Engler G, Stieglitz T, Galindo-Leon E, Engel AK (2018) Synchrony surfacing: epicortical recording of correlated action potentials. Eur J Neurosci 48:3583–3596
41. Rubehn B, Bosman C, Oostenveld R, Fries P, Stieglitz T (2009) A MEMS-based flexible multichannel EcoG-electrode array. J Neural Eng 6:036003
42. Bosman CA, Schoffelen J-M, Brunet N, Oostenveld R, Bastos AM, Womelsdorf T, Rubehn B, Stieglitz T, De Weerd P, Fries P (2012) Attentional stimulus selection through selective synchronization between monkey visual areas. Neuron 75:875–888
43. Stitt I, Hollensteiner KJ, Galindo-Leon E, Pieper F, Fiedler E, Stieglitz T, Engler G, Nolte G, Engel AK (2017) Dynamic reconfiguration of cortical functional connectivity across brain states. Sci Rep 7:8797
44. Suzuki M, Larkum ME (2017) Dendritic calcium spikes are clearly detectable at the cortical surface. Nat Commun 8:276
45. Rudolph M, Destexhe A (2003) Tuning neocortical pyramidal neurons between integrators and coincidence detectors. J Comput Neurosci 14:239–251
46. Jiang X, Wang G, Lee AJ, Stornetta RL, Zhu JJ (2013) The organization of two new cortical interneuronal circuits. Nat Neurosci 16:210–218
47. Larkum M (2013) A cellular mechanism for cortical associations: an organizing principle for the cerebral cortex. Trends Neurosci 36:141–151
48. Moore JJ, Ravassard PM, Ho D, Acharya L, Kees AL, Vuong C, Mehta MR (2017) Dynamics of cortical dendritic membrane potential and spikes in freely behaving rats. Science 355:eaaj1497
49. Ray S, Crone NE, Niebur E, Franaszczuk PJ, Hsiao SS (2008) Neural correlates of high-gamma oscillations (60–200 Hz) in macaque local field potentials and their potential implications in electrocorticography. J Neurosci 28:11526–11536
50. Singer W (1999) Neuronal synchrony: a versatile code for the definition of relations? Neuron 24(49–65):111–125
51. Engel AK, Fries P, Singer W (2001) Dynamic predictions: oscillations and synchrony in top–down processing. Nat Rev Neurosci 2:704–716
52. Salinas E, Sejnowski TJ (2001) Correlated neuronal activity and the flow of neural information. Nat Rev Neurosci 2:539–550
53. Fries P (2005) A mechanism for cognitive dynamics: neuronal communication through neuronal coherence. Trends Cogn Sci 9:474–480
54. Siegel M, Donner TH, Engel AK (2012) Spectral fingerprints of large-scale neuronal interactions. Nat Rev Neurosci 13:121–134

55. Konerding WS, Froriep UP, Kral A, Baumhoff P (2018) New thin-film surface electrode array enables brain mapping with high spatial acuity in rodents. Sci Rep 8:3825
56. Schalk G, Leuthardt EC (2011) Brain-computer interfaces using electrocorticographic signals. IEEE Rev Biomed Eng 4:140–154
57. Chiang CH, Wang C, Barth K et al (2021) Flexible, high-resolution thin-film electrodes for human and animal neural research. J Neural Eng 18:045009
58. Sun J, Barth H, Qiao S, Chiang CH et al (2021) Intraoperative microseizure detection using a high-density micro-electrocorticography electrode array. Brain Commun 4:fcac122
59. Tchoe Y, Bourhis A, Cleary DR, Stedelin B, Lee J, Tonsfeldt KJ, Brown EC, Siler DA, Paulk AC, Yang JC, Oh H, Ro YG, Lee K, Russman SM, Ganji M, Galton I, Ben-Haim S, Raslan AM, Dayeh SA (2022) Human brain mapping with multithousand-channel PtNRGrids resolves spatiotemporal dynamics. Sci Transl Med 14:eabj1441
60. Engel AK, Fries P (2010) Beta-band oscillations—signalling the status quo? Curr Opin Neurobiol 20:156–165
61. Friston KJ (2011) Functional and effective connectivity: a review. Brain Connect 1:13–36
62. Engel AK, Gerloff C, Hilgetag CC, Nolte G (2013) Intrinsic coupling modes: multiscale interactions in ongoing brain activity. Neuron 80:867–886
63. Reid AT, Headley DB, Mill RD, Sanchez-Romero R, Uddin LQ, Marinazzo D, Lurie DJ, Valdés-Sosa PA, Hanson SJ, Biswal BB, Calhoun V, Poldrack RA, Cole MW (2019) Advancing functional connectivity research from association to causation. Nat Neurosci 22:1751–1760
64. Beggs JM, Plenz D (2003) Neuronal avalanches in neocortical circuits. J Neurosci 23:11167–11177
65. Han F, Caporale N, Dan Y (2008) Reverberation of recent visual experience in spontaneous cortical waves. Neuron 60:321–327

Chapter 19
What is the Functional Role of iEEG Oscillations in Neural Processing and Cognitive Functions?

Timothée Proix, Pierre Mégevand, and Anne-Lise Giraud

Abstract Oscillations of the electric field generated by neuronal populations are often observed in intracranial EEG recordings from human cortical and subcortical brain regions. The functional relevance of these oscillations for neural processing and cognitive functions remains a debated issue in modern neuroscience. In this chapter, we review evidence that iEEG oscillations constitute a key mechanism in the functional integration of neuronal activity across temporal and spatial scales. We focus on the potential role of cortical oscillations in cognitive processes, and particularly speech perception and production, which involve diverse brain regions and temporal scales in a structured hierarchy, as an ideal testbed for outlining the possible insights that iEEG oscillations offer on cognitive functions.

19.1 Introduction

Hans Berger is famous for being the first to have recorded alpha and beta rhythms using electroencephalogram [1]. Perhaps less well-known is that he was also the first to record these rhythms directly from the cortex by performing the first intracranial electroencephalogram (iEEG) of a patient undergoing surgery for a brain tumour [2]. A few other studies completed this early finding, until the opportunities to

T. Proix (✉) · P. Mégevand · A.-L. Giraud
Department of Basic Neurosciences, Faculty of Medicine, University of Geneva, Geneva, Switzerland
e-mail: Timothee.Proix@unige.ch

P. Mégevand
Department of Clinical Neurosciences, Faculty of Medicine, University of Geneva, Geneva, Switzerland

Division of Neurology, Geneva University Hospitals, Geneva, Switzerland

A.-L. Giraud
Institut de l'Audition - Institut Pasteur, UMR Inserm AO06, 63, rue de Charenton, F-75012 Paris, France

N. Axmacher (ed.), *Intracranial EEG*, Studies in Neuroscience, Psychology and Behavioral Economics, https://doi.org/10.1007/978-3-031-20910-9_19

record intracranially became more frequent in the 40s with the advent of operative explorations in conscious patients with epilepsy [3]. Nowadays, most rhythms described with non-invasive methods such as electroencephalography (EEG) and magnetoencephalography (MEG) have also been recorded directly with intracranial recordings in human. Those rhythms are often grouped into five frequency bands: Delta (0.5−4 Hz), theta (4−8 Hz), alpha (8−12 Hz), beta (12−25 Hz), and (low-) gamma (30–50 Hz). In addition, high-frequency activity (>80 Hz), a non-oscillatory signal which can be recorded with iEEG (see Chap. 23), but usually not with EEG or MEG, is now widely used as a proxy for spatially localized neuronal spiking activity, although its origin is somewhat discussed [4].

We mostly focus here on neural oscillations recorded with intracranial recordings in human, leaving aside a large part of the literature on neural oscillations, including non-human studies, alternative recording methods, and computational models. We refer the curious reader to the excellent books of György Buzsáki [5, 6] and dedicated reviews [7, 8]. Although we do report a large number of studies in the broad field of iEEG oscillations, we do not exhaustively cover all relevant publications. Also, we do not discuss here the neurophysiological underpinnings of neural oscillations, for this see for instance [7]. We only and importantly stress here that neural oscillations do not originate from a well-defined harmonic oscillator existing somewhere in the brain, but rather emerge as collective dynamics of heterogeneous populations of neurons [9]. Neuronal dynamics are by nature oscillatory in that they periodically (but often irregularly) drift out of their equilibrium for spiking and come back to it to prepare for the next spiking activity.

Neural oscillations can be recorded from a large variety of brain regions with iEEG. These oscillations can be concomitant of a specific task the subject is performing, but are also found while the subject is asleep, or during pathological neuronal activity [10]. The question whether there is a causal relationship between neural oscillations and function, or a mere correlate is intensely debated. Yet, several hypotheses on the functional role of neural oscillations and how they emerge from and influence excitability of neurons are supported by a growing amount of experimental and theoretical evidence (see below). Additional evidence might come from neurofeedback and stimulation studies that can causally influence or disturb a specific neural rhythm (see Chaps. 39, 40, and 41).

The study of neural oscillations is not trivial, and a number of methodological issues need to be considered when processing and interpreting iEEG recordings [7]. We briefly emphasize those that are relevant for this chapter. (i) Decomposition of the recordings into distinct frequency bands is often performed with predetermined frequency bands, which can be misleading if there is no real neural activity in these bands [11]. This has implications for multivariate methods that consider relationships between channels, such as phase synchronization and phase amplitude coupling [11]. (ii) A related issue is that neural oscillations are not necessarily sinusoidal, which might lead to missing important information when applying methods designed for sinusoidal signals [12]. In particular, the so-called high-gamma activity corresponds to a large frequency band (typically between 80 and 150 Hz) that does not correspond to a well-defined oscillation, and for which only the power can reliably be extracted.

Accordingly, we refer to it as broadband high-frequency activity (BHA). (iii) iEEG power spectra typically exhibit a 1/f response over a large range of frequencies, possibly due to both the low-pass properties of induced currents in the extracellular space [13], and the range of time scales involved in single neuron dynamics [14]. Whether high-frequency power represent synchronous rhythms or a change of the 1/f response is still debated and calls for caution in the interpretation [15] (see also Chaps. 22 and 23).

19.2 iEEG Oscillations Supporting Cognitive Functions

Over the last decades, human iEEG studies have shown the presence of neural oscillations in numerous brain regions and in relation to many cognitive functions, including working memory, memory formation, attention, sensory and motor activity, multi-sensory integration, sleep, and speech. Several functions have been proposed for neural oscillations in relation to these cognitive functions, including sensory selection and control, communication between neuronal populations, sequence representation, replay, and temporal prediction [7, 8]. Here, we first review the mechanisms supporting human cognitive functions, and then show the iEEG evidence for their roles in these functions.

Functional mechanisms of iEEG oscillations. Oscillations are uniquely characterized by three measures: frequency/period, phase, and amplitude. Each of those features contributes to one or several of the following five functions of iEEG oscillations. (i) Sensory selection via resonance (Fig. 19.1a). The leak and voltage-gated currents of neurons make neuronal populations respond preferentially to stimuli with specific frequencies. These resonance frequencies allow neuronal populations to select and enhance specific features of the input stimuli. (ii) Communication via coupling (Fig. 19.1b, see also Chap. 25). Coupling between (possibly distant) brain regions can take many forms, with phase coupling (phase-amplitude coupling and phase-phase coupling) being the most frequent one (see also Chap. 32). Periods of enhanced spiking activity then co-occur across brain regions, allowing for temporal windows of communication. (iii) Sequence representation via phase encoding (Fig. 19.1c). The phase advance during the on-going iEEG oscillation triggers distinct neuronal patterns representing different items. This can be exploited to encode item sequences. (iv) Replay via entrainment. Without external perturbations, iEEG oscillations persist across time, allowing previous sequences of neuronal patterns to be replayed (see also Chap. 21). (v) Temporal prediction via entrainment and phase reset. The natural frequency of the oscillation helps predicting when future stimulus or neuronal activity is expected. This prediction can be changed by simply shifting the phase of the on-going iEEG oscillations through an incoming signal.

Together, these five mechanisms allow for a large repertoire of cognitive functions, as detailed below. They are sometimes combined, and cannot always be disentangled from one another.

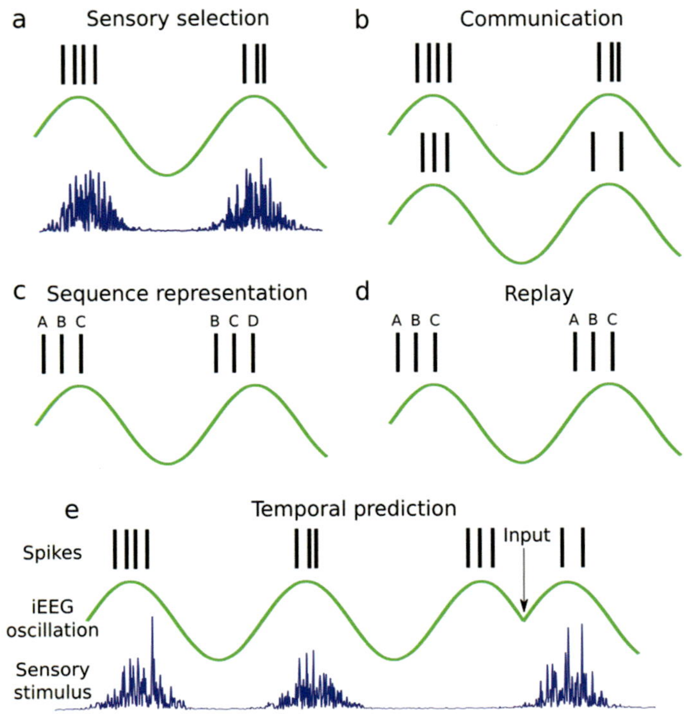

Fig. 19.1 Functional mechanisms of iEEG oscillations. (**a**) iEEG oscillations (green) facilitate neuronal spiking (black bars) for an input frequency (blue) that matches the resonant frequency of the iEEG oscillation. (**b**) Coupling between two or more iEEG oscillations allows coordinating simultaneous windows of increased or decreased excitability across brain regions. (iii) As the phase of the on-going iEEG oscillations advances, distinct neuronal patterns representing different items (A, B, C) are sequentially activated. Progressive shifting of this sequence allows for new items (D) to be represented. (**d**) Repetition of previous sequences of neuronal representation is possible thanks to the rhythmic nature of neural activity. (**e**) On-going oscillations help predicting future stimuli, and can be reset by phase-shifting through an input (arrow) signal

Working memory. The role of neural oscillations in working memory is well characterized in animal models, in particular in the mouse and rat hippocampus. In humans, theta oscillations have been described in the temporal lobe and frontal regions while subjects navigate a virtual maze [22, 23]. While it is not completely clear if maze learning only depends on working memory or also involves long-term spatial memory, further studies have confirmed that theta activity increases in frontal and temporal regions during the whole duration of working memory tasks [16, 24] (Fig. 19.2a). The phase synchronization between theta oscillations in temporal and frontal regions is argued to help co-activating and integrating information across neuronal networks, a desired feature to store items in memory (Fig. 19.1b). Gamma power also increases in working memory tasks [24, 25], allowing precise spike timing to be shared across local networks, thus supporting synaptic plasticity. The gamma rhythm is additionally phase-coupled to on-going theta rhythm in the hippocampus

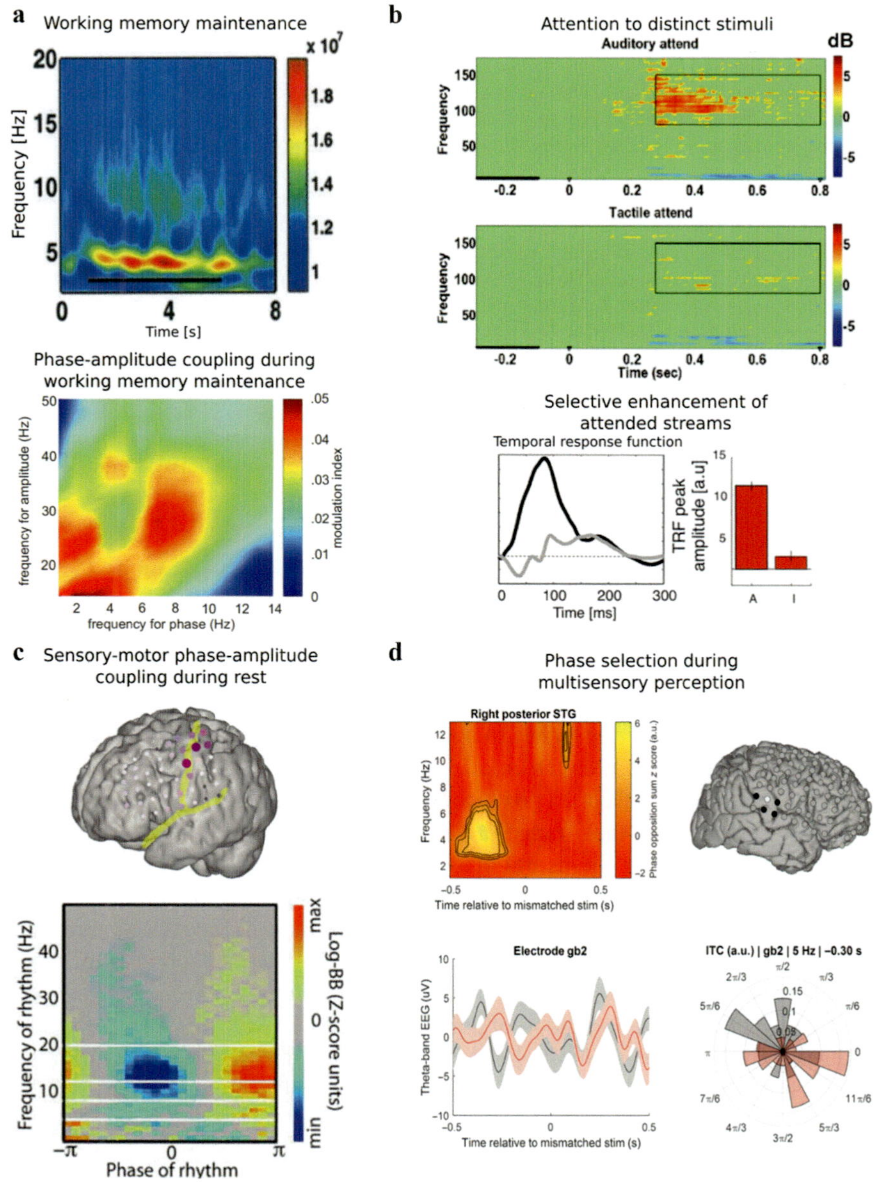

[17, 26] (Fig. 19.2a). This finding suggests that sequences of items are encoded by the phase of the on-going theta rhythm (Fig. 19.1c), while the signature of an item's content is signalled by gamma power increase. A similar mechanism is proposed for speech processing, as described below.

◂**Fig. 19.2 iEEG recordings of neural oscillations during cognitive tasks.** (a) Top: Spectrogram of iEEG recordings in the left temporal lobe during a working memory task. The black bar denotes the duration of a trial. Bottom: The gamma band power is coupled to the phase of theta iEEG oscillations in the hippocampus during a working memory task. (b) Top: Spectrogram of an ECoG electrode located in the auditory cortex, showing a BHA increase to attended auditory stimuli, but not to attended tactile stimuli. Significant effects are outlined by a rectangular box. Bottom: Time response function during a cocktail party task shows that low-frequencies are selectively enhanced for the attended stimulus. (c) During rest, broadband iEEG oscillations become phase-locked to the beta rhythm in the sensory and motor cortex. Top: Strength of the phase-locking is shown as the diameter of each iEEG electrode. Bottom: Example of phase-locking for one exemplary electrode. (d) Top left: Phase opposition sum for a mismatched audio-visual speech perception task for the electrode shown in top right. Phase opposition sum is here maximal when the theta phase angle differs as a function of perception. Bottom left: Theta phase for each perception group (grey and red) for this electrode. Bottom right: Corresponding phase distribution. The measured phase predicts the perception. Panel (a) top is adapted from [16]. Copyright 2001 Society for Neuroscience. Panel (a) bottom is adapted from [17]. Panel (b) top is adapted from [18]. Copyright 2008, with permission from Elsevier. Panel (b) bottom is adapted from [19]. Copyright 2013, with permission from Elsevier. Panel (c) is adapted from [20] Panel (d) is adapted from [21].

Attention. While functionally distinct from working memory, attention appears to share neuronal circuits with it, and both cannot always be disentangled [7]. Increases in BHA are often found in attention tasks. BHA increases in primary sensory areas, auditory or somatosensory cortex as a function of the attended stimulus (auditory vs. tactile), as well as in the prefrontal cortex [18] (Fig. 19.2b). BHA and gamma responses also increase in response to auditory deviants in the auditory association cortex [27], during motor intention tasks in the premotor cortex [28], and during visual attention tasks in occipital cortex and fusiform gyrus [29]. Critically, lower frequency oscillations in the delta-theta range drive the attention focus to a specific stream, by selectively enhancing the neural activity representing the attended stimulus in higher cortical regions hosting more abstract representations of the attended stimuli (Fig. 19.1a and Fig. 19.2b) [19, 30]. Long-range communication between prefrontal and sensory areas for selective attention is supported by phase coupling in the alpha range (Fig. 19.1b), simultaneously to a decrease of the alpha power in the sensory areas, which enables the formation of cortical ensembles [31].

Declarative memory. Similarly to working memory, the theta and gamma rhythms are thought to play an important role for memory formation. A pre-stimulus theta increase occurs in the medial temporal lobe during word learning tasks [32], and subcortical structures, e.g. dorsomedial and anterior thalamic nuclei, exhibit similar rhythms during visual learning tasks [33]. In a semantic memory task using natural language associations, strong theta rhythm can be recorded in the hippocampus [34]. The phase of the theta rhythm encodes stimulus information, thus allowing sequence encoding throughout a whole theta cycle (Fig. 19.1c) [35]. Beyond the theta rhythm, the power of gamma oscillations in the hippocampus, the left temporal lobe and the frontal cortex predicts successful encoding of verbal memory when learning a list of common words [36, 37]. Here, phase synchronization in the gamma rhythm might play a critical role in long-term memory formation by enhancing plasticity, notably between the rhinal cortex and the hippocampus[38], where it can interact with working memory [39] (Fig. 19.2c).

Sensory and motor cortical activity. Two main rhythms are classically associated with motor cortex function: The widespread mu/beta rhythm, and localized BHA activations. The large amplitude of the mu/beta rhythm had already been described long ago, both with scalp and intracranial electrode recordings [40]. The mu/beta activity, which is typically widespread across the sensory and motor regions at rest and strongly reduced during movement [3, 41, 42], plays a role in the initiation and interruption of a movement [43]. By contrast, BHA in the motor cortex is present during movement at specific cortical locations, i.e. following a somatotopic organization, and are phase-coupled to the beta rhythm [20, 44, 45] (Fig. 19.1a). The beta rhythm has a suppressive effect: By recruiting large parts of the motor cortex and phase-locking with BHA, it prevents movement-specific activity to be triggered. Conversely, its strong diminution during movement allows BHA to emerge in localized networks and trigger movement [20] (Fig. 19.2c). Dysfunction of the beta rhythm is a well-known correlate of Parkinson's disease (see below).

Other sensory areas exhibit activations of gamma oscillations and BHA when a stimulus is perceived [46–48], but no beta activity which remains specific to the motor areas, suggesting this rhythm is only needed to gate output activity, whereas inputs in sensory areas are directly triggered by stimuli.

Multisensory integration. In naturalistic environments, we simultaneously perceive events in the outside world through multiple modalities which are handled by distinct circuits and brain regions. Yet, the conscious perception of these events is merged into a single unified representation. Evidence from non-human primates indicates that crossmodal sensory influences on primary sensory cortices take place through phase reset: A visual event, say, resets the phase of ongoing oscillations in auditory cortex without modifying neuronal firing [49, 50] (Fig. 19.1e). In humans as well, crossmodal phase reset occurs in auditory cortex upon visual stimulation, and in visual cortex upon auditory stimulation [51, 52]. Phase reset, and the corresponding changes in cortical responsiveness, are argued to underlie the perceptual amplification of auditory speech by the corresponding visual cues, e.g. lip movements [53]. Buttressing this hypothesis, visual speech stimuli can cause both phase reset and power changes in auditory cortices [54].

Cortical oscillations also play a key role in selecting which modality is attended when disparate streams of stimuli arrive simultaneously to the senses [50, 55]. As noted above, the deployment of attention to one or the other stream is associated with a shift in the phase of oscillations in sensory cortex [56] (Fig. 19.1a). The phase of oscillatory activity does not only predict cortical responsiveness but also the percept itself. In the case of competing auditory and visual speech stimuli, prestimulus theta phase in the posterior superior temporal cortex determines whether the auditory or the visual component eventually drives perception [21] (Fig. 19.2d).

Sleep. During non-rapid eye movement sleep, two types of oscillations are typically recorded with scalp EEG: Slow-waves, and ripples (80–150 Hz). Both have also been found with simultaneous scalp/intracranial EEG, with activations that stays local, without propagating to other brain regions [57, 58]. The function of slow-waves and ripples is not clear but likely relates to memory consolidation, arousal and gating [58]. Ripples are found in human hippocampus during the consolidation phase of memory encoding [59–61], but also in several neocortical areas where their functions are currently not well understood [62]. Importantly, ripples are phase-locked to slow oscillations, suggesting that the latter drive the windows where replay can take place [61] (Fig. 19.1d).

19.3 Speech Perception and Production

Speech communication is a complex human behaviour that extends across multiple timescales. Accordingly, its neural representations are supported by oscillations and activities at different frequencies, including BHA, delta, theta, beta and gamma rhythms [9, 63, 64]. In the following, we explore in more details how these rhythms contribute to the neural representations of speech at various time scales.

Chunking auditory speech stimuli into syllables. Words in uttered speech are not necessarily marked by silences, but rather form a continuous stream where meaningful linguistic items need to be separated for semantic and syntactic processing. This chunking operation is performed by the theta rhythm in the primary and secondary auditory cortex [64] (Fig. 19.3a). Theta oscillations strongly correlate with speech envelope (amplitude modulation of the acoustic signal), although acoustic onset edges have also been proposed as the actual entraining events [67]. The frequency range of the theta rhythm, between 4 and 8 Hz, corresponds to the syllabic rate of all spoken languages [68] (2−8 Hz), which makes it ideally suited to chunk heard speech into distinct syllables by periodically enhancing incoming activity (Fig. 19.1a). Both rhythms must commensurate for the chunking to operate, as brilliantly demonstrated by showing that compressed speech remains intelligible if properly spaced [69, 70]. The actual nature of the theta rhythm, whether endogenous or entrained by heard speech, is nevertheless still heavily debated (see for instance [71] and the avalanche of commentaries that it triggered).

Encoding phonemes. In parallel to chunking, heard phonemes must be encoded in a neural representation. This task is carried out by the gamma rhythm, which is recorded in the auditory cortex when subjects are hearing speech [9, 63, 64, 72] (Fig. 19.2a). The amplitude of the gamma rhythm is coupled to the phase of the theta rhythm, thereby assuring that it is maximally activated during periods of interest (Fig. 19.1c, Fig. 19.3b). Simultaneously to gamma oscillations, BHA activations lay out a spatio-temporal neuronal mapping of spectral, articulatory or phonetic representations of speech utterances in the auditory and motor cortex [48, 73–75].

Binding linguistic elements. The next important step for speech processing is to combine linguistic elements together, so that complex syntactic and semantic structures can emerge. This requires binding elements in working memory over large temporal windows [61]. Slow rhythms, such as the delta rhythm, are proposed to selectively enhance items over the duration of the slow delta period (Fig. 19.1a), but additional experiments are needed to confirm this, notably using iEEG data. The

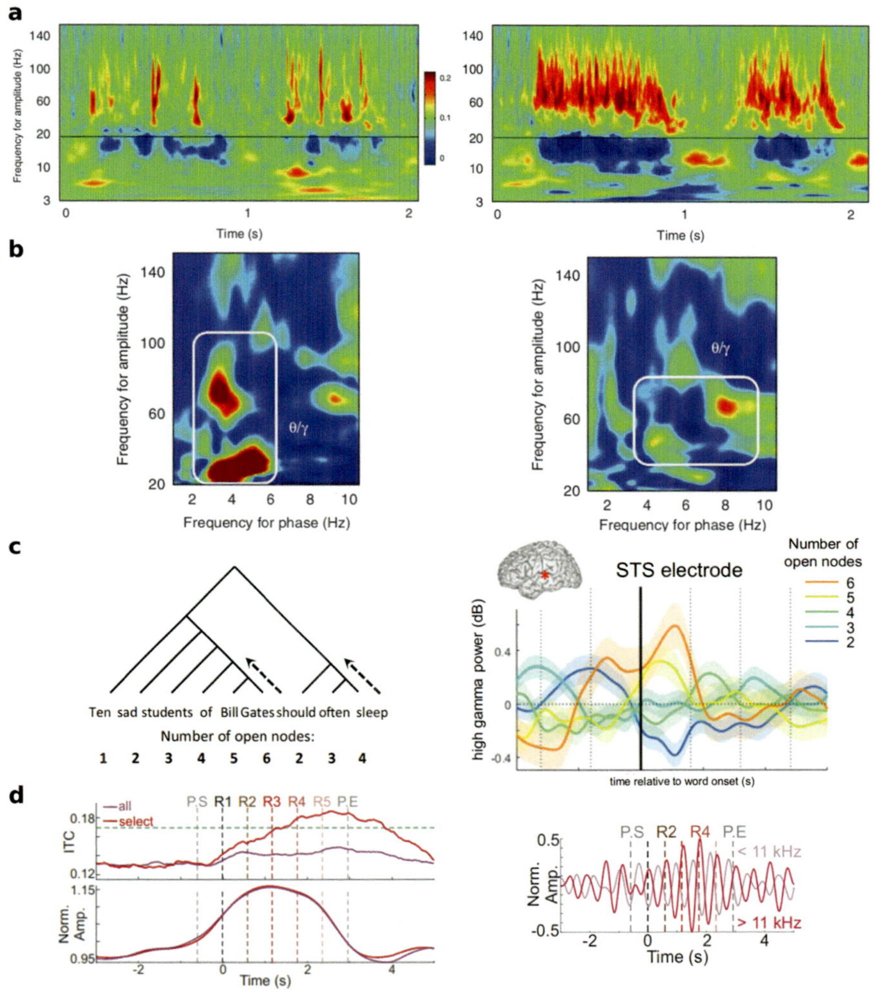

Fig. 19.3 iEEG oscillations in speech perception and production. (**a**) Spectrograms of iEEG activity responding to a sentence heard by the subject. Theta and gamma rhythms in primary (left) and association (right) auditory cortices are enhanced during speech perception. (**b**) Phase-amplitude coupling between theta and gamma bands for the same regions. (**c**) Left: The syntactic tree of a sentence heard by a subject is used to define the number of open nodes. Right: BHA increase in the superior temporal sulcus correlates with the number of open nodes at the time of word onset. (**d**) Intracranial recordings from the auditory cortex of a non-human primate during perception of repetitive stimuli. Left: Intertrial coherence (top) and amplitude (bottom) of delta oscillations increase during the repetitive patterns (R1-5), indicating a phase alignment. Right: Neuronal multi-unit activity is synchronously or anti-synchronously phase-locked to the delta rhythm for neurons tuned to frequencies < 11 kHz and > 11 kHz, respectively. Panels (a) and (b) adapted from [63]. Panel (c) adapted from [65]. Panel (d) adapted from [66]

delta rhythm is also suggested to be involved in syntactic encoding of sentences, by combining distinct syntactic elements within the same temporal windows (Fig. 19.1a) [76]. More sophisticated mechanisms, such as nested structures of progressively slower oscillations [77] or phase encoding of item positions and content [78] presumably further refine the processing of syntactic structures. Finally, BHA modulations in the inferior frontal gyrus might be a signature of compositionality of linguistic elements, as BHA power increases with working memory load and abruptly drops when a new element is bound [65] (Fig. 19.3c).

Temporal prediction. One of the key feature of speech perception is the ability of the brain to predict the next sensory input [79]. The main mechanism used by the brain to achieve this is to use oscillatory entrainment (of the theta rhythm in the auditory cortex) regulated by top-down phase resetting mechanisms (Fig. 19.1e). Although this mechanism has so far only been shown with non-invasive recordings [80, 81] or invasive recordings in non-human primate [66] (Fig. 19.3d), it will likely be found as well with invasive recordings. A good candidate for the phase resetting role is the beta rhythm, which is foreseen as the main top-down signal. Computational modelling data suggest that beta is the optimal rhythm for controlling the rate of prediction errors within a given region during speech processing [82]. The sensitivity of the theta rhythm to top-down signal would render the signal more flexible, in particular for aperiodic predictable and memorized signals [79].

Imagined speech. Interestingly, while BHA in sensory and motor cortices is the best predictor of overt or articulated speech, it becomes equally or less predictive than other frequency bands for phoneme decoding in imagined speech (i.e. without any movements), suggesting that imagined speech is presumably not primarily grounded in imagined articulatory movements [83]. In contrast, speech attempted by paralyzed persons with preserved language and cortical motor commands was well decoded using BHA recorded from the motor and sensory cortex by leveraging the articulatory representation described above [84].

Paralinguistic speech features. Several other features are relevant for the neural representation of speech, such as speaker invariant representations, emotion, prosody, etc. These features have been investigated mostly by analysing BHA, for instance with the speech prosody in the superior temporal gyrus [85], the planning of conversational answers in inferior and middle frontal gyrus[86], or even the onset of voices in rock songs [87].

19.4 Dysfunctional iEEG Oscillations

We have so far focused on the functional properties of iEEG oscillations. However, excessive neural oscillations can also be pathological, and be associated with observable symptoms. Pathological oscillations have been well characterized with iEEG for epilepsy and Parkinson's disease, where the implantation of intracranial electrodes is necessary for clinical reasons.

Epileptic seizures. Epileptic seizures constitute the prototypical example of dysfunctional oscillations that can be recorded with iEEG (Chap. 1). During a seizure, assemblies of neurons discharge hypersynchronously, creating strong oscillations that can easily be recorded with invasive and non-invasive recording techniques. This excessive synchronization disrupts the normal operative role of brain rhythms, and recruits atypical neural circuits inducing abnormal behavior. For instance, seizures that progressively recruit somatotopically arranged regions in the motor cortex via the spreading of a strong alpha/beta rhythm correspond to the apparition of motors symptoms (e.g. rhythmic muscle contractions) that appear sequentially in neighboring body parts (this particular progression is called the Jacksonian march [88]). Similar symptoms occur in the domain of language when seizures spread through the temporal lobe (mumbling of incoherent speech).

A more recently identified type of epileptic oscillatory rhythmic activity is the high frequency oscillations (HFO) between 80 and 500 Hz that appear by brief episodes[34]. Presence of HFO has been proposed to reflect pathological tissue: recent findings suggest that they reflect the presence of multiple independent pools of neurons that fire together but out-of-phase with each other [89, 90] (see also Chap. 24 for the distinction of physiological vs. pathological ripples).

Finally, thanks to recent long-term intracranial recordings (several years) obtained by chronic implantation of a neurostimulation device (see also Chaps. 52 and 53), it appeared that seizures are very often phase-locked to interictal very slow oscillations, with a period spanning weeks to months [91, 92]. These findings, in addition to known circadian rhythms of these cycles [93], have triggered new hope for the field of seizure forecasting [94].

Beyond the seizure focus itself, pathological oscillations can also invade remote cortical areas and participate in the clinical expression of seizures: During medial temporal lobe seizures, slow (delta-band) oscillations appear in frontal and parietal cortices [95, 96] and are associated with impaired consciousness [97], likely because of the large territories recruited in this synchronous regime.

Interictal epileptiform discharges (IEDs, the “spikes and waves” of epilepsy) also impair the function of the neural networks that they invade [98, 99]. Whether the pathological neuronal activity causing the IED itself suffices to alter function, or whether it is the sudden interruption of ongoing oscillations by the IED that causes the symptoms, remains an open question (see also Chap. 3). Since the occurrence of IEDs depends on the phase of ongoing oscillations in neuronal networks [100], it might be the case that the execution of a cognitive task itself influences the occurrence of IEDs [101].

Parkinson’s disease. Deep brain stimulation is often used to alleviate the symptoms of patients with Parkinson’s disease by placing electrodes in the subthalamic nucleus (STN), providing opportunities for researchers to record this brain region. Aberrant beta activity is found in the STN of 95% of patients with Parkinson’s disease [102]. This beta activity occurs in bursts, with phasic coupling across the motor network, including the globus pallidus and the thalamus [103]. As mentioned above, the beta rhythm that is enhanced in Parkinson’s disease has a suppressive effect that might be due to recruitment of local populations of neurons. This explains for instance the difficulty to initiate movements (i.e. to stop the beta rhythm). Oscillations in other bands, such as theta and alpha, might also play a role in motor but also cognitive impairments in Parkinson's disease, for which the underlying mechanisms remained to be fully characterized [104].

19.5 Conclusion

Several types of iEEG oscillations have been reported and related to many cognitive functions. Correspondingly, when these functions are impaired, neural oscillatory behaviour is abnormal, suggesting a possible causality of oscillatory phenomena and generic computational processes [105]. To further establish the functional relevance of iEEG oscillation, it will be particularly useful to be able to relate on-going iEEG oscillations with the underlying spiking activity. To this end, intracranial recording at the single neuron level simultaneously with local field potential recordings constitute an invaluable tool to further explore those mechanisms (see also Chaps. 16 and 44).

A number of oscillations found in animal or non-invasive studies have not yet been reported with iEEG in human, but are likely to be found in the future if the opportunity to record them intracranially arises. An example is in the domain of dyslexia, where scalp EEG recordings in children with dyslexia have repeatedly found a deficit in gamma power in the auditory cortex (see above for the role of gamma band in speech perception) [106]. Enhancement of this band with the use of transcranial electrical stimulation has allowed improvement of phonemic processing and reading accuracy [107], thus causally establishing the importance of the low-gamma band activity for speech processing.

To conclude, iEEG oscillations support many distinct cognitive functions using a large variety of mechanisms. Future studies will help further explore a potential role of neural oscillations in cognitive operations that are higher in the cognitive hierarchy (e.g. syntactic processing, value-based decision making etc.), for which the functional role of iEEG oscillations has so far been more difficult to establish.

References

1. Berger H (1929) Über das Elektrenkephalogramm des Menschen. Arch Für Psychiatr Nervenkrankh 87:527–570. https://doi.org/10.1007/BF01797193
2. Jung R, Berger W (1979) Hans Bergers Entdeckung des Elektrenkephalogramms und seine ersten Befunde 1924–1931. Arch Für Psychiatr Nervenkrankh 227:279–300. https://doi.org/10.1007/BF00344814
3. Jasper H, Penfield W (1949) Electrocorticograms in man: Effect of voluntary movement upon the electrical activity of the precentral gyrus. Arch Für Psychiatr Nervenkrankh 183:163–174. https://doi.org/10.1007/BF01062488
4. Leszczyński M, Barczak A, Kajikawa Y, et al (2020) Dissociation of broadband high-frequency activity and neuronal firing in the neocortex. Sci Adv 6:eabb0977. https://doi.org/10.1126/sciadv.abb0977

5. Buzsáki G (2006) Rhythms of the brain. Oxf Univ Press, Oxf, New York
6. Buzsáki G (2019) The Brain from Inside Out. Oxf Univ Press, Oxf, New York
7. Wang X-J (2010) Neurophysiological and Computational Principles of Cortical Rhythms in Cognition. Physiol Rev 90:1195–1268. https://doi.org/10.1152/physrev.00035.2008
8. Buzsaki G (2004) Neuronal oscillations in cortical networks. Science 304:1926–1929. https://doi.org/10.1126/science.1099745
9. Giraud A-L (2020) Oscillations for all. A commentary on Meyer, Sun, Martin (2020). Lang Cogn Neurosci 1–8. https://doi.org/10.1080/23273798.2020.1764990
10. Groppe DM, Bickel S, Keller CJ et al (2013) Dominant frequencies of resting human brain activity as measured by the electrocorticogram. Neuroimage 79:223–233. https://doi.org/10.1016/j.neuroimage.2013.04.044
11. Aru J, Aru J, Priesemann V et al (2015) Untangling cross-frequency coupling in neuroscience. Curr Opin Neurobiol 31:51–61. https://doi.org/10.1016/j.conb.2014.08.002
12. Cole SR, Voytek B (2017) Brain Oscillations and the Importance of Waveform Shape. Trends Cogn Sci 21:137–149. https://doi.org/10.1016/j.tics.2016.12.008
13. Bédard C, Kröger H, Destexhe A (2006) Model of low-pass filtering of local field potentials in brain tissue. Phys Rev E 73:051911. https://doi.org/10.1103/PhysRevE.73.051911
14. Hasenstaub A, Shu Y, Haider B et al (2005) Inhibitory postsynaptic potentials carry synchronized frequency information in active cortical networks. Neuron 47:423–435. https://doi.org/10.1016/j.neuron.2005.06.016
15. Miller KJ, Zanos S, Fetz EE et al (2009) Decoupling the cortical power spectrum reveals real-time representation of individual finger movements in humans. J Neurosci 29:3132–3137. https://doi.org/10.1523/JNEUROSCI.5506-08.2009
16. Raghavachari S, Kahana MJ, Rizzuto DS et al (2001) Gating of human theta oscillations by a working memory task. J Neurosci 21:3175–3183. https://doi.org/10.1523/JNEUROSCI.21-09-03175.2001
17. Axmacher N, Henseler MM, Jensen O et al (2010) Cross-frequency coupling supports multi-item working memory in the human hippocampus. Proc Natl Acad Sci 107:3228–3233. https://doi.org/10.1073/pnas.0911531107
18. Ray S, Niebur E, Hsiao SS et al (2008) High-frequency gamma activity (80–150Hz) is increased in human cortex during selective attention. Clin Neurophysiol 119:116–133. https://doi.org/10.1016/j.clinph.2007.09.136
19. Zion Golumbic EM, Ding N, Bickel S et al (2013) Mechanisms underlying selective neuronal tracking of attended speech at a "cocktail party." Neuron 77:980–991. https://doi.org/10.1016/j.neuron.2012.12.037
20. Miller KJ, Hermes D, Honey CJ et al (2012) Human motor cortical activity is selectively phase-entrained on underlying rhythms. PLoS Comput Biol 8:e1002655. https://doi.org/10.1371/journal.pcbi.1002655
21. Thézé R, Giraud A-L, Mégevand P (2020) The phase of cortical oscillations determines the perceptual fate of visual cues in naturalistic audiovisual speech. Sci Adv 6:. https://doi.org/10.1126/sciadv.abc6348
22. Kahana MJ, Sekuler R, Caplan JB et al (1999) Human theta oscillations exhibit task dependence during virtual maze navigation. Nature 399:781–784. https://doi.org/10.1038/21645
23. Caplan JB, Kahana MJ, Sekuler R et al (2000) Task dependence of human theta: The case for multiple cognitive functions. Neurocomputing 32–33:659–665. https://doi.org/10.1016/S0925-2312(00)00229-0
24. Meltzer JA, Zaveri HP, Goncharova II et al (2008) Effects of working memory load on oscillatory power in human intracranial EEG. Cereb Cortex 18:1843–1855. https://doi.org/10.1093/cercor/bhm213
25. Mainy N, Kahane P, Minotti L et al (2007) Neural correlates of consolidation in working memory. Hum Brain Mapp 28:183–193. https://doi.org/10.1002/hbm.20264
26. Dimakopoulos V, Mégevand P, Stieglitz L, et al (2022) Information flows from hippocampus to auditory cortex during replay of verbal working memory items. eLife 11:e78677. https://doi.org/10.7554/eLife.78677

27. Edwards E, Soltani M, Deouell LY et al (2005) High gamma activity in response to deviant auditory stimuli recorded directly from human cortex. J Neurophysiol 94:4269–4280. https://doi.org/10.1152/jn.00324.2005
28. Brovelli A, Lachaux J-P, Kahane P, Boussaoud D (2005) High gamma frequency oscillatory activity dissociates attention from intention in the human premotor cortex. Neuroimage 28:154–164. https://doi.org/10.1016/j.neuroimage.2005.05.045
29. Tallon-Baudry C, Bertrand O, Hénaff M-A et al (2005) Attention modulates gamma-band oscillations differently in the human lateral occipital cortex and fusiform gyrus. Cereb Cortex 15:654–662. https://doi.org/10.1093/cercor/bhh167
30. O'Sullivan J, Herrero J, Smith E et al (2019) Hierarchical encoding of attended auditory objects in multi-talker speech perception. Neuron 104:1195–1209. https://doi.org/10.1016/j.neuron.2019.09.007
31. Leszczynski M, Fell J, Jensen O, Axmacher N (2017) Alpha activity in the ventral and dorsal visual stream controls information flow during working memory. bioRxiv 180166. https://doi.org/10.1101/180166
32. Fell J, Ludowig E, Staresina BP et al (2011) Medial temporal theta/alpha power enhancement precedes successful memory encoding: evidence based on intracranial EEG. J Neurosci 31:5392–5397. https://doi.org/10.1523/JNEUROSCI.3668-10.2011
33. Sweeney-Reed CM, Zaehle T, Voges J et al (2016) Pre-stimulus thalamic theta power predicts human memory formation. Neuroimage 138:100–108. https://doi.org/10.1016/j.neuroimage.2016.05.042
34. Piai V, Anderson KL, Lin JJ et al (2016) Direct brain recordings reveal hippocampal rhythm underpinnings of language processing. Proc Natl Acad Sci 113:11366–11371. https://doi.org/10.1073/pnas.1603312113
35. Reddy L, Self MW, Zoefel B et al (2021) Theta-phase dependent neuronal coding during sequence learning in human single neurons. Nat Commun 12:4839. https://doi.org/10.1038/s41467-021-25150-0
36. Sederberg PB, Kahana MJ, Howard MW et al (2003) Theta and Gamma Oscillations during Encoding Predict Subsequent Recall. J Neurosci 23:10809–10814. https://doi.org/10.1523/JNEUROSCI.23-34-10809.2003
37. Sederberg PB, Schulze-Bonhage A, Madsen JR et al (2006) Hippocampal and neocortical gamma oscillations predict memory formation in humans. Cereb Cortex 17:1190–1196. https://doi.org/10.1093/cercor/bhl030
38. Fell J, Klaver P, Lehnertz K et al (2001) Human memory formation is accompanied by rhinal–hippocampal coupling and decoupling. Nat Neurosci 4:1259–1264. https://doi.org/10.1038/nn759
39. Axmacher N, Elger CE, Fell J (2009) Working memory-related hippocampal deactivation interferes with long-term memory formation. J Neurosci 29:1052–1060. https://doi.org/10.1523/JNEUROSCI.5277-08.2009
40. Gastaut H, Terzian H, Gastaut Y (1952) Etude d'une activité électroencéphalographique méconnue: le rythme rolandique en arceau. Mars Med 89:296–310
41. Arroyo S, Lesser RP, Gordon B et al (1993) Functional significance of the mu rhythm of human cortex: an electrophysiologic study with subdural electrodes. Electroencephalogr Clin Neurophysiol 87:76–87. https://doi.org/10.1016/0013-4694(93)90114-B
42. Aoki F, Fetz EE, Shupe L et al (1999) Increased gamma-range activity in human sensorimotor cortex during performance of visuomotor tasks. Clin Neurophysiol 110:524–537. https://doi.org/10.1016/S1388-2457(98)00064-9
43. Swann N, Tandon N, Canolty R et al (2009) Intracranial EEG reveals a time- and frequency-specific role for the right inferior frontal gyrus and primary motor cortex in stopping initiated responses. J Neurosci 29:12675–12685. https://doi.org/10.1523/JNEUROSCI.3359-09.2009
44. Miller KJ, Leuthardt EC, Schalk G et al (2007) Spectral changes in cortical surface potentials during motor movement. J Neurosci 27:2424–2432. https://doi.org/10.1523/JNEUROSCI.3886-06.2007

45. Pfurtscheller G, Graimann B, Huggins JE et al (2003) Spatiotemporal patterns of beta desynchronization and gamma synchronization in corticographic data during self-paced movement. Clin Neurophysiol 114:1226–1236. https://doi.org/10.1016/S1388-2457(03)00067-1
46. Lachaux J-P, George N, Tallon-Baudry C et al (2005) The many faces of the gamma band response to complex visual stimuli. Neuroimage 25:491–501. https://doi.org/10.1016/j.neuroimage.2004.11.052
47. Yang Q, Zhou G, Noto T et al (2022) Smell-induced gamma oscillations in human olfactory cortex are required for accurate perception of odor identity. PLOS Biol 20:e3001509. https://doi.org/10.1371/journal.pbio.3001509
48. Mesgarani N, Cheung C, Johnson K, Chang EF (2014) Phonetic feature encoding in human superior temporal gyrus. Science 343:1006–1010. https://doi.org/10.1126/science.1245994
49. Lakatos P, Chen C-M, O'Connell MN et al (2007) Neuronal oscillations and multisensory interaction in primary auditory cortex. Neuron 53:279–292. https://doi.org/10.1016/j.neuron.2006.12.011
50. Lakatos P, Karmos G, Mehta AD et al (2008) Entrainment of neuronal oscillations as a mechanism of attentional selection. Science 320:110–113. https://doi.org/10.1126/science.1154735
51. Mercier MR, Foxe JJ, Fiebelkorn IC et al (2013) Auditory-driven phase reset in visual cortex: Human electrocorticography reveals mechanisms of early multisensory integration. Neuroimage 79:19–29. https://doi.org/10.1016/j.neuroimage.2013.04.060
52. Mercier MR, Molholm S, Fiebelkorn IC et al (2015) Neuro-Oscillatory phase alignment drives speeded multisensory response times: An electro-corticographic investigation. J Neurosci 35:8546–8557. https://doi.org/10.1523/JNEUROSCI.4527-14.2015
53. Schroeder CE, Lakatos P, Kajikawa Y et al (2008) Neuronal oscillations and visual amplification of speech. Trends Cogn Sci 12:106–113. https://doi.org/10.1016/j.tics.2008.01.002
54. Mégevand P, Mercier MR, Groppe DM et al (2020) Crossmodal phase reset and evoked responses provide complementary mechanisms for the influence of visual speech in auditory cortex. J Neurosci Off J Soc Neurosci 40:8530–8542. https://doi.org/10.1523/JNEUROSCI.0555-20.2020
55. Lakatos P, O'Connell MN, Barczak A et al (2009) The leading sense: supramodal control of neurophysiological context by attention. Neuron 64:419–430. https://doi.org/10.1016/j.neuron.2009.10.014
56. Besle J, Schevon CA, Mehta AD et al (2011) Tuning of the human neocortex to the temporal dynamics of attended events. J Neurosci 31:3176–3185. https://doi.org/10.1523/JNEUROSCI.4518-10.2011
57. Nir Y, Staba RJ, Andrillon T et al (2011) Regional slow waves and spindles in human sleep. Neuron 70:153–169. https://doi.org/10.1016/j.neuron.2011.02.043
58. Andrillon T, Nir Y, Staba RJ et al (2011) Sleep spindles in humans: insights from intracranial EEG and unit recordings. J Neurosci 31:17821–17834. https://doi.org/10.1523/JNEUROSCI.2604-11.2011
59. Bragin A, Engel J, Wilson CL et al (1999) High-frequency oscillations in human brain. Hippocampus 9:137–142. https://doi.org/10.1002/(SICI)1098-1063(1999)9:2%3c137::AID-HIPO5%3e3.0.CO;2-0
60. Staba RJ, Wilson CL, Bragin A et al (2002) Quantitative analysis of high-frequency oscillations (80–500 Hz) recorded in human epileptic hippocampus and entorhinal cortex. J Neurophysiol 88:1743–1752. https://doi.org/10.1152/jn.2002.88.4.1743
61. Axmacher N, Elger CE, Fell J (2008) Ripples in the medial temporal lobe are relevant for human memory consolidation. Brain 131:1806–1817. https://doi.org/10.1093/brain/awn103
62. Alkawadri R, Gaspard N, Goncharova II et al (2014) The spatial and signal characteristics of physiologic high frequency oscillations. Epilepsia 55:1986–1995. https://doi.org/10.1111/epi.12851
63. Giraud A-L, Poeppel D (2012) Cortical oscillations and speech processing: emerging computational principles and operations. Nat Neurosci 15:511–517. https://doi.org/10.1038/nn.3063

64. Fontolan L, Morillon B, Liegeois-Chauvel C, Giraud A-L (2014) The contribution of frequency-specific activity to hierarchical information processing in the human auditory cortex. Nat Commun 5:4694. https://doi.org/10.1038/ncomms5694
65. Nelson MJ, El Karoui I, Giber K et al (2017) Neurophysiological dynamics of phrase-structure building during sentence processing. Proc Natl Acad Sci 114:E3669–E3678. https://doi.org/10.1073/pnas.1701590114
66. Barczak A, O'Connell MN, McGinnis T, et al (2018) Top-down, contextual entrainment of neuronal oscillations in the auditory thalamocortical circuit. Proc Natl Acad Sci 115:. https://doi.org/10.1073/pnas.1714684115
67. Oganian Y, Chang EF (2019) A speech envelope landmark for syllable encoding in human superior temporal gyrus. Sci Adv 5:eaay6279. https://doi.org/10.1126/sciadv.aay6279
68. Coupé C, Oh YM, Dediu D, Pellegrino F (2019) Different languages, similar encoding efficiency: Comparable information rates across the human communicative niche. Sci Adv 5:eaaw2594. https://doi.org/10.1126/sciadv.aaw2594
69. Ghitza O, Greenberg S (2009) On the possible role of brain rhythms in speech perception: intelligibility of time-compressed speech with periodic and aperiodic insertions of silence. Phonetica 66:113–126. https://doi.org/10.1159/000208934
70. Nourski KV, Reale RA, Oya H et al (2009) Temporal envelope of time-compressed speech represented in the human auditory cortex. J Neurosci 29:15564–15574. https://doi.org/10.1523/JNEUROSCI.3065-09.2009
71. Meyer L, Sun Y, Martin AE (2020) Synchronous, but not entrained: exogenous and endogenous cortical rhythms of speech and language processing. Lang Cogn Neurosci 35:1089–1099. https://doi.org/10.1080/23273798.2019.1693050
72. Crone N (1998) Functional mapping of human sensorimotor cortex with electrocorticographic spectral analysis. I. Alpha and beta event- related desynchronization. Brain 121:2271–2299. https://doi.org/10.1093/brain/121.12.2271
73. Bouchard KE, Mesgarani N, Johnson K, Chang EF (2013) Functional organization of human sensorimotor cortex for speech articulation. Nature 495:327–332. https://doi.org/10.1038/nature11911
74. Pasley BN, David SV, Mesgarani N et al (2012) Reconstructing speech from human auditory cortex. PLoS Biol 10:e1001251. https://doi.org/10.1371/journal.pbio.1001251
75. Hamilton LS, Oganian Y, Hall J, Chang EF (2021) Parallel and distributed encoding of speech across human auditory cortex. Cell 184:4626-4639.e13. https://doi.org/10.1016/j.cell.2021.07.019
76. Ding N, Melloni L, Zhang H et al (2016) Cortical tracking of hierarchical linguistic structures in connected speech. Nat Neurosci 19:10. https://doi.org/10.1038/nn.4186
77. Martin AE, Doumas LA (2019) Predicate learning in neural systems: using oscillations to discover latent structure. Curr Opin Behav Sci 29:77–83. https://doi.org/10.1016/j.cobeha.2019.04.008
78. Calmus R, Wilson B, Kikuchi Y, Petkov CI (2020) Structured sequence processing and combinatorial binding: neurobiologically and computationally informed hypotheses. Philos Trans R Soc B Biol Sci 375:20190304. https://doi.org/10.1098/rstb.2019.0304
79. Rimmele JM, Morillon B, Poeppel D, Arnal LH (2018) Proactive sensing of periodic and aperiodic auditory patterns. Trends Cogn Sci 22:870–882. https://doi.org/10.1016/j.tics.2018.08.003
80. Arnal LH, Doelling KB, Poeppel D (2015) Delta-Beta coupled oscillations underlie temporal prediction accuracy. Cereb Cortex 25:3077–3085. https://doi.org/10.1093/cercor/bhu103
81. Daume J, Wang P, Maye A et al (2021) Non-rhythmic temporal prediction involves phase resets of low-frequency delta oscillations. Neuroimage 224:117376. https://doi.org/10.1016/j.neuroimage.2020.117376
82. Hovsepyan S, Olasagasti I, Giraud A-L (2022) Rhythmic modulation of prediction errors: a possible role for the beta-range in speech processing. bioRxiv 2022.03.28.486037
83. Proix T, Delgado Saa J, Christen A et al (2022) Imagined speech can be decoded from low- and cross-frequency intracranial EEG features. Nat Commun 13:48. https://doi.org/10.1038/s41467-021-27725-3

84. Moses DA, Metzger SL, Liu JR et al (2021) Neuroprosthesis for decoding speech in a paralyzed person with anarthria. N Engl J Med 385:217–227. https://doi.org/10.1056/NEJMoa2027540
85. Tang C, Hamilton LS, Chang EF (2017) Intonational speech prosody encoding in the human auditory cortex. Science 357:797–801. https://doi.org/10.1126/science.aam8577
86. Castellucci GA, Kovach CK, Howard MA et al (2022) A speech planning network for interactive language use. Nature. https://doi.org/10.1038/s41586-021-04270-z
87. Sturm I, Blankertz B, Potes C, et al (2014) ECoG high gamma activity reveals distinct cortical representations of lyrics passages, harmonic and timbre-related changes in a rock song. Front Hum Neurosci 8:. https://doi.org/10.3389/fnhum.2014.00798
88. Jackson JH (1868) Notes on the physiology and pathology of the nervous system. Med Times Gaz ii:696
89. Jefferys JGR, de la Prida LM, Wendling F et al (2012) Mechanisms of physiological and epileptic HFO generation. Prog Neurobiol 98:250–264. https://doi.org/10.1016/j.pneurobio.2012.02.005
90. Chen Z, Maturana MI, Burkitt AN et al (2021) High-Frequency oscillations in epilepsy: what have we learned and what needs to be addressed. Neurology 96:439–448. https://doi.org/10.1212/WNL.0000000000011465
91. Baud MO, Kleen JK, Mirro EA, et al (2018) Multi-day rhythms modulate seizure risk in epilepsy. Nat Commun 9:. https://doi.org/10.1038/s41467-017-02577-y
92. Leguia MG, Andrzejak RG, Rummel C et al (2021) Seizure cycles in focal epilepsy. JAMA Neurol 78:454–463. https://doi.org/10.1001/jamaneurol.2020.5370
93. Karoly PJ, Freestone DR, Boston R et al (2016) Interictal spikes and epileptic seizures: their relationship and underlying rhythmicity. Brain 139:1066–1078. https://doi.org/10.1093/brain/aww019
94. Proix T, Truccolo W, Leguia MG et al (2021) Forecasting seizure risk in adults with focal epilepsy: a development and validation study. Lancet Neurol 20:127–135. https://doi.org/10.1016/S1474-4422(20)30396-3
95. Lieb JP, Dasheiff RM, Engel J et al (1991) Role of the frontal lobes in the propagation of mesial temporal lobe seizures. Epilepsia 32:822–837. https://doi.org/10.1111/j.1528-1157.1991.tb05539.x
96. Blumenfeld H, Rivera M, McNally KA et al (2004) Ictal neocortical slowing in temporal lobe epilepsy. Neurology 63:1015–1021. https://doi.org/10.1212/01.WNL.0000141086.91077.CD
97. Englot DJ, Yang LI, Hamid H et al (2010) Impaired consciousness in temporal lobe seizures: role of cortical slow activity. Brain 133:3764–3777. https://doi.org/10.1093/brain/awq316
98. Kleen JK, Scott RC, Holmes GL et al (2013) Hippocampal interictal epileptiform activity disrupts cognition in humans. Neurology 81:18–24. https://doi.org/10.1212/WNL.0b013e318297ee50
99. Ung H, Cazares C, Nanivadekar A et al (2017) Interictal epileptiform activity outside the seizure onset zone impacts cognition. Brain J Neurol 140:2157–2168. https://doi.org/10.1093/brain/awx143
100. Sheybani L, Mégevand P, Spinelli L et al (2021) Slow oscillations open susceptible time windows for epileptic discharges. Epilepsia 62:2357–2371. https://doi.org/10.1111/epi.17020
101. Vivekananda U, Bush D, Bisby JA et al (2019) Spatial and episodic memory tasks promote temporal lobe interictal spikes. Ann Neurol 86:304–309. https://doi.org/10.1002/ana.25519
102. Brittain J-S, Brown P (2014) Oscillations and the basal ganglia: Motor control and beyond. Neuroimage 85:637–647. https://doi.org/10.1016/j.neuroimage.2013.05.084
103. Tinkhauser G, Torrecillos F, Duclos Y et al (2018) Beta burst coupling across the motor circuit in Parkinson's disease. Neurobiol Dis 117:217–225. https://doi.org/10.1016/j.nbd.2018.06.007
104. Oswal A, Brown P, Litvak V (2013) Synchronized neural oscillations and the pathophysiology of Parkinson's disease. Curr Opin Neurol 26:662–670. https://doi.org/10.1097/WCO.0000000000000034

105. Hyafil A, Giraud A-L, Fontolan L, Gutkin B (2015) Neural Cross-Frequency coupling: connecting architectures, mechanisms, and functions. Trends Neurosci 38:725–740. https://doi.org/10.1016/j.tins.2015.09.001
106. Lehongre K, Ramus F, Villiermet N et al (2011) Altered Low-Gamma sampling in auditory cortex accounts for the three main facets of dyslexia. Neuron 72:1080–1090. https://doi.org/10.1016/j.neuron.2011.11.002
107. Marchesotti S, Nicolle J, Merlet I et al (2020) Selective enhancement of low-gamma activity by tACS improves phonemic processing and reading accuracy in dyslexia. PLOS Biol 18:e3000833. https://doi.org/10.1371/journal.pbio.3000833

Chapter 20
How Can I Run Sleep and Anesthesia Studies with Intracranial EEG?

Janna D. Lendner and Randolph F. Helfrich

Abstract The similarity of sleep and general anesthesia has fascinated scientists for a long time. At first glance, both states are characterized by similar behavioral correlates, namely decreased responsiveness, arousal and movement. Previously, non-invasive scalp electroencephalographic (EEG) recordings demonstrated highly comparable spectral signatures of both states, such as the ubiquitous presence of slow waves or delta oscillations. More recently, intracranial recordings in humans provided a more fine-grained perspective and revealed that sleep and anesthesia reflect highly distinct entities. Here, we outline how intracranial sleep and anesthesia recordings can be embedded into the clinical routine. We discuss caveats and shortcomings that need to be considered, especially in the context of epilepsy as the underlying neurological disorder. Subsequently, we provide a practical road map to obtain state-specific neural recordings and discuss technical prerequisites as well as important analytical considerations. Finally, we summarize how intracranial recordings extend our understanding about the mechanism-of-action of anesthetic drugs at the network level and to which extent these signatures overlap with physiologic sleep networks. Collectively, here we review how intracranial recordings in humans can be leveraged to gain important insights into sleep physiology and the neural correlates of (un-) consciousness.

Keywords Sleep · Anesthesia · Propofol · Unconsciousness · Intracranial EEG · Slow oscillations · Spindles · Ripples · Interictal discharges · Multitaper spectrogram

---

J. D. Lendner · R. F. Helfrich (✉)
Center for Neurology, Hertie Institute for Clinical Brain Research, University of Tübingen, Otfried-Mueller-Str. 27, 72076 Tübingen, Germany
e-mail: randolph.helfrich@gmail.com

J. D. Lendner
Dept. of Anesthesiology and Intensive Care Medicine, University of Tübingen, Hoppe-Seyler-Str. 3, 72076 Tübingen, Germany

N. Axmacher (ed.), *Intracranial EEG*, Studies in Neuroscience, Psychology and Behavioral Economics, https://doi.org/10.1007/978-3-031-20910-9_20

20.1 Introduction

"You will fall asleep now" might be the most common phrase used by anesthesiologists before administering the hypnotic drug during everyday clinical care. At first glance, sleep and anesthesia share several behavioral signatures, including decreased arousal and movement [1]. However, upon closer inspection, both states reflect distinct entities. In contrast to someone asleep, patients undergoing anesthesia remain unresponsive to painful stimuli. In addition, anesthesia impairs memory formation and can be detrimental to cognitive functioning (especially in the elderly; [2, 3]), while sleep benefits memory formation and cognition [4]. The neural correlates of unconscious brain states have fascinated scientists for decades [1, 5]. Several scalp EEG studies identified electrophysiological signatures, such as high amplitude slow waves (<1.25 Hz) and delta activity (<4 Hz) that both occur in non-rapid eye movement (NREM) sleep and under deep anesthesia [6–8].

Intracranial EEG (iEEG) offers a unique window to study cognition, sleep physiology, sleep deprivation and anesthesia on the single subject level. Patients are typically monitored for multiple days; hence, several days and nights worth of data can be obtained. Furthermore, sleep deprivation is a common intervention to trigger seizures (see Sect. 20.3.2) and provides a valuable experimental condition to test causal links between sleep and cognition. In addition, iEEG often explores deeper brain structures, such as the hippocampus, the amygdala or thalamic nuclei, which are difficult to image using non-invasive methods, but are thought to reflect key nodes of the human memory network [4, 9, 10]. The high temporal resolution of iEEG enables extraction and analysis of e.g. high-frequency band activity (~70–150 Hz; HFA; [11, 12]) or of cardinal sleep oscillations, such as sharp-wave ripples (~80–120 Hz; [9, 13–15]), which cannot be observed at the scalp level. iEEG is typically used to sample multiple nodes of the suspected epileptic network (i.e. mesial or limbic structures, including the hippocampus, cingulate and orbitofrontal cortex). To target these deeper structures, electrodes have to transverse through intact cortex (i.e. lateral temporal and frontal); thus, enabling simultaneous multisite recordings with high spatiotemporal resolution, which allows dissecting network processes in great detail. With the advent of human single neuron recordings, it is now feasible to record single unit activity (SUA), field potential, HFA, intracranial and scalp EEG all simultaneously within the same patient [16, 17] (see also Chaps. 12 and 16).

20.2 The Clinical Context for Sleep and Anesthesia Studies

20.2.1 The Peri- and Post-operative Clinical Setting

Patients undergoing invasive intracranial monitoring have typically experienced a long-lasting ordeal of seizures, failed treatments and non-invasive diagnostic tests. Once the non-invasive work-up is completed and electrode implantation has been planned to pinpoint the seizure onset zone, patients are admitted to neurosurgery for implantation of either subdural grid electrodes (ECoG), stereotactically placed depth electrodes (sEEG) or a combination of both. Once patients are out of the operating room and electrode placement has been radiologically confirmed, they are transferred to the epilepsy-monitoring unit (EMU). Depending on the duration of the procedure and the precise dosing of the anesthetics, patients might be drowsy or somnolent until all effects of the general anesthesia wear off over the course of the first few hours. Depending on the type of electrode (grid vs. depth) patients might experience different levels of discomfort. In general, depth electrodes are better tolerated with less post-operative pain, given that no craniotomy is necessary. However, depth electrodes targeting mesial temporal lobe areas typically transverse through the temporal muscles, hence, patients often report pain and discomfort while chewing or drinking. In addition, electrodes are covered in a head-wrap, often requiring a supine positioning with an elevated backrest; hence, habitual sleeping positions are not always feasible. Most patients are confined to bed rest during the entire monitoring. Therefore, patients often require one or two days (and nights) to adapt to the surroundings of the EMU during invasive monitoring, which in turn, impacts sleep quality and duration (see also Chaps. 4 and 5).

20.2.2 Factors Determining Sleep Quality in the Monitoring Unit

Peri-operative circumstances impact sleep in the first few hours, but once the immediate effects of the procedure wear off, most patients resume their habitual night-day cycle. In the context of sleep studies on the monitoring unit a few caveats apply. Depending on the medical center, some patients will be assessed clinically every few hours throughout the night to monitor their vital signs as well as their neurological state. Hence, the clinical routine might introduce sleep fragmentation and frequent arousals during the night (Fig. 20.1). Arousals during nighttime might also be triggered through alarms on the ward at night or warning sounds of intravenous infusion systems.

Another factor that influences sleep quality on the EMU is the current antiepileptic drug regime (see Sect. 20.2.3 for drug specific effects), which is typically tapered during monitoring to provoke seizures. While the patient or members of the family/staff can press a bedside button whenever the patient experiences epileptic prodromes or seizures, in some instances, especially during reduction of antiepileptic medication, patients may first present with subclinical seizure patterns. Subclinical events are noted by the EEG techs as suddenly occurring rhythmic spiking patterns in the EEG without obvious clinical correlate. In order to determine the precise clinical manifestation of a given pattern, the techs then wake up the patient and administer a series of tests to determine orientation and executive functions. Collectively, given the clinical circumstances and several contributing factors, an undisturbed night on

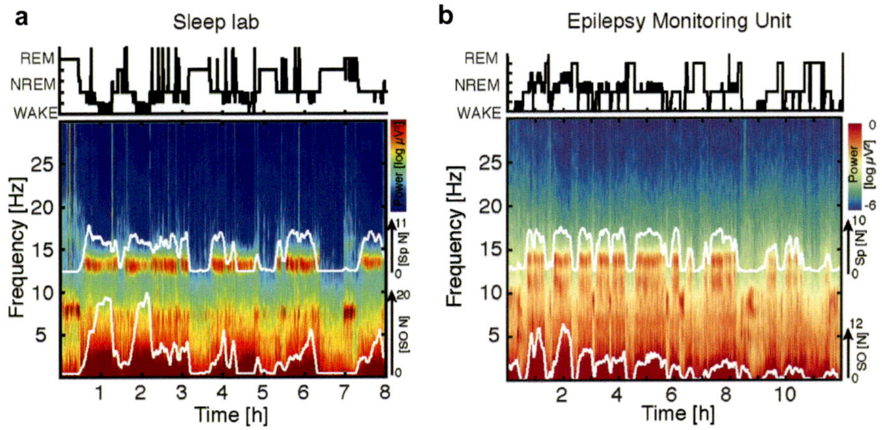

Fig. 20.1 **Sleep architecture in the sleep lab and EMU.** **a** Top: Hypnogram. Bottom: Multitaper spectral representation with number of detected slow oscillations and sleep spindles superimposed as recorded during a habitual night of sleep in a sleep lab. **b** Same conventions. Data recorded in the EMU. A comparable pattern is observed; thus, indicating the feasibility to conduct sleep studies in the EMU. Panel A is reproduced with permission from [60]. Panel B reproduced with permission from [14] under the Creative Commons Attribution (CC BY) license

the EMU is less common that in a dedicated sleep laboratory. This needs to be accounted for in studies that examine sleep physiology or sleep-dependent memory formation.

20.2.3 The Effects of Antiepileptic Drugs

Patients undergoing invasive monitoring failed multiple drug regimes and are typically being admitted while they are on a combination of different antiepileptic drugs (AEDs), which might include sodium channel blockers, GABAergic drugs or AMPA receptor antagonists. During monitoring, AEDs are typically tapered off to provoke habitual seizures. In the context of sleep studies, it is important to note that tapering off medications will increase both interictal spiking (see also Sect. 20.3.1) and the likelihood for seizures. On the other hand, AEDs themselves often impair sleep quality and its electrophysiological signatures. For instance, lamotrigine, a widely used sodium channel blocker, triggers sleep disturbances and fragmentation [18]. In contrast, GABAergic drugs, such as clobazam (a benzodiazepine), lead to daytime sleepiness and sedation [19]. AMPA receptor antagonists like perampanel are strong sedatives and are therefore taken only in the evening hours [20]. To date, the precise effects of many AEDs on sleep are not well known, but in the context of sleep studies detailed knowledge about the current medication status is helpful to interpret these findings [21–24]. This is of particular relevance, since certain anticonvulsants exhibit distinct electrophysiological signatures, such as benzodiazepines, which introduce widespread EEG beta activity (~13–30 Hz).

20.2.4 Electrode Explantation as a Window into the Neural Correlates of Anesthesia

Once intracranial electrodes are in place and patients are awake and stable, they are typically transferred from the operating room (OR) to the monitoring ward, where electrodes are first connected to the clinical and research recording setup, a process that takes between 45 min to 1.5 h. In rare instances, the setup can already be completed in the OR; thus, enabling recording electrophysiological activity during the emergence of anesthesia [25]. However, as virtually all patients undergo post-operative imaging by means of CT- or MRI-based imaging to confirm electrode placement and to rule out perioperative complications such as brain hemorrhage, electrodes would have to be disconnected for scanning and then reconnected on the EMU, making this recording strategy highly impractical in the clinical context.

The more feasible route to obtain iEEG recordings during anesthesia is recording at the end of invasive monitoring, just before electrode explantation. Again, this is only viable if electrodes are explanted under general anesthesia, which is common for grid electrodes, in patients that cannot comply with lying still (e.g. young children or patients with anxiety disorders) or when the epileptogenic tissue is removed in the same session. Nowadays, many centers remove depth electrodes in the EMU under local anesthesia, while patients are awake. If patients undergo general anesthesia for explantation, then recording equipment may be transferred to the OR and recordings continue until the wires are physically removed; hence, capturing induction and maintenance of anesthesia. A few reports are available where patients underwent light anesthesia or sedation accompanied by research testing (e.g. tasks or auditory stimuli), then recovered as part of the research protocol and then were again anesthetized for final removal of the electrodes [26]. The exact clinical set up will vary between medical centers. However, often enough it will be possible to incorporate research studies into the individual clinical context, without jeopardizing patient safety and while respecting clinical time constrains and limited OR time. In most centers, protocols can be tailored to address specific questions; however, given the scarcity of reports, there is currently no consensus or gold standard available on how to conduct these studies.

20.3 Implications of Epilepsy as the Underlying Neurological Disorder

20.3.1 Sleep Stages and Epileptic Activity

An empirical observation in numerous overnight EEG recordings is that the frequency of epileptic spikes (interictal epileptic discharges; IEDs) sharply increases during NREM sleep and that they are less common during wakefulness and REM sleep (Fig. 20.2; [21]). In addition, most nighttime seizures occur during NREM sleep. In fact, epileptic activity that occurs during REM sleep is highly informative for clinical localization of the seizure onset zone (SOZ), while NREM spikes are less specific. The precise (patho-) physiologic underpinnings of these observations remain unknown, however, it has been argued that NREM sleep reflects a hyper-synchronized brain state that facilitates propagation of synchronized volleys of epileptic activity [24]. A related clinical observation is that hippocampal IEDs are common during NREM sleep, even if the SOZ is located outside of the medial temporal lobe. It has been argued that the anatomical structure and connectivity of the hippocampus abet its susceptibility to epileptic activity [21, 27]. Uncontrolled seizure activity outside of the medial temporal lobe might 'kindle' the hippocampus [28], i.e. induce a second independent source of seizure activity, thus, rendering a focal epilepsy multi-focal and therefore, not amendable to resective surgery. In the context of sleep studies, the spatial and temporal characteristics of IEDs need to be accounted for to circumvent a systematic bias when analyzing REM and NREM sleep separately. Likewise, special caution is necessary when analyzing hippocampal activity during sleep (see also Sect. 20.4.3).

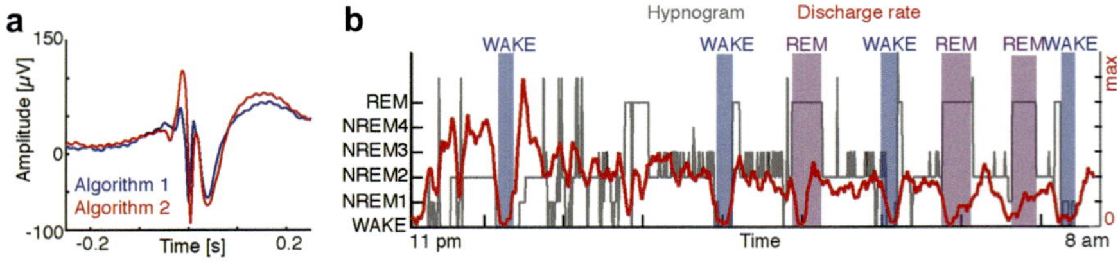

Fig. 20.2 The rate of interictal discharges increases during NREM sleep. a Inter-ictal discharges (IED) as detected by two automatic detectors [15, 27]. **b** Discharge rate (red) across the night (hypnogram in grey), highlighting more IEDs during NREM sleep as compared to wakefulness (blue) or REM sleep (purple). Panel A reproduced with permission from [14] under the Creative Commons Attribution (CC BY) license

20.3.2 Sleep Deprivation is a Powerful Trigger for Seizures

Invasive monitoring on the EMU provides a narrow time window (often between 4 and 7 days) to observe seizures and to determine the SOZ for subsequent surgical resection. In addition to tapering off the AEDs, (partial) sleep deprivation is another commonly employed tool to trigger seizures during monitoring [29, 30]. Patients are typically kept awake or are only allowed to sleep for approximately four hours, e.g. from 2 to 6 a.m. Seizures after sleep deprivation do not occur immediately, but manifest within the subsequent 24 h. The precise mechanisms are not fully understood yet, but it has been argued that sleep deprivation attenuates physiologic homeostasis for the excitatory-to-inhibitory balance; thereby, resulting in a net increase of excitation and subsequent epileptic activity [30–32]. In the context of iEEG sleep studies on e.g. overnight memory formation, sleep deprivation constitutes a valuable control condition that is already implemented in the clinical context. However, an important confound is that it also sharply increases epileptic activity, which might bias behavioral performance and electrophysiological signatures the next day.

20.3.3 The Relationship of Anesthesia and Epileptic Activity

General anesthesia induces a state of unconsciousness, often by increasing inhibition in the brain [5]. Common anesthetic agents like propofol bind to GABAergic receptors, similar to benzodiazepines, which are also used as anticonvulsants. Hence, anesthetic drugs are occasionally being used to treat a status epilepticus, i.e. a continuous epileptic seizure. This has also strong implications for intracranial EEG studies on general anesthesia. In contrast to sleep studies, where epileptic activity can be sharply increased, IEDs are typically strongly attenuated during general anesthesia; thus, potentially biasing and hampering a direct within subject comparison between both neuronal states. However, most studies focused on the neuronal correlates of the loss-of-consciousness (LOC) under anesthesia and in this scenario a strong attenuation of epileptic activity is desirable. From a clinical point-of-view, LOC from general anesthesia results from a maximum of inhibitory drive resulting in hypersynchronized medium to slow neural activity, while LOC during a seizure typically results from uncoordinated neuronal firing due to hyperexcitability.

20.4 Analysis Strategies

20.4.1 Technical Pre-requisites for Comparative Electrophysiology

In principle, data is continuously recorded during monitoring. However, in order to take full advantage of the acquired data, several prerequisites need to be met. First, it is desirable that iEEG during sleep and anesthesia are recorded using the same amplifier. Some centers run dedicated clinical and research systems that are either fully independent (parallel data streaming) or that run serially (data is streamed from the clinical to the research system). If a serial setup is employed, it might be difficult to transfer both the clinical and research system to the OR for recordings during anesthesia. In this scenario, it would be beneficial to use the clinical amplifier for both recordings. It is of critical importance to be aware of the operating room logistics where several disciplines (nurses, anesthesia techs, neurophysiologists, anesthesiologists and neurosurgeons) interact under both, time and space constrains. Continuous EEG recordings during this phase likely contain movement artifacts as well as artifacts from manipulation of wires and the head, which will require careful inspection during analysis. Wherever possible, noise should be attenuated during the recording, e.g. by shielding recording leads from surrounding noise sources or unplug unnecessary equipment in the vicinity.

To enable a direct comparison to non-invasive results and to facilitate gold-standard sleep staging, scalp EEG should be recorded simultaneously [33]. Implanted iEEG leads sometimes prohibit placement of scalp leads, but it is best practice to at least record from a few scalp locations (i.e. midline electrodes Fz, Cz and Pz along with C3/C4 to facilitate spindle detection) as well as electrooculogram (EOG) and electromyogram (EMG) electrodes to detect REM sleep.

In addition, data should be recorded at a sufficient high sampling rate (>500 Hz) to enable extraction of HFA and ripple oscillations. During recording data should be minimally processed with regard to low-pass, high-pass or band-stop filters.

20.4.2 How to Determine the Current Behavioral or Brain State?

Sleep staging from iEEG is theoretically possible, however, guidelines for sleep staging are only available for scalp EEG. Hence, it remains best practice to obtain simultaneous scalp EEG, EOG and EMG to facilitate sleep staging. The key challenge is the distinction of wakefulness and REM sleep, while NREM sleep can easily be detected given the presence of clear oscillatory key signatures, such as prominent slow waves and spindle oscillations.

With respect to anesthesia, the current gold standard to determine the loss of consciousness is based on clinical judgement by the physician. The Modified Observer's Assessment of Alertness and Sedation (MOAA/S) scale is a validated 6-point scale assessing responsiveness of patients, which has been defined by American Society of Anesthesiologists (ASA). For neuroscientific applications, sometimes a simplified categorization into awake/alert, sedated/drowsy (but arousable/responsive to predefined stimuli such as subject's name or mild prodding) and unconscious/unresponsive is employed. Lastly, anesthetic depth may be monitored with the help of special neuromonitoring devices such as the bispectral index (BIS) monitor, which was initially developed to prevent intraoperative awareness. Electrophysiological data (EEG, EMG) is measured from a few frontal sensors and then transformed into a numerical value between 0 and 100 that indicates the level of arousal (100 = wakefulness, 40–60 = sufficient anesthetic depth for surgery). However, the algorithm of this calculation is proprietary, thus, it remains unclear which EEG features are being evaluated. Furthermore, BIS has mainly been validated in propofol anesthesia and the efficacy and reliability of BIS monitoring remains controversial.

20.4.3 How to Address Epileptic Activity?

Epileptic activity is an inherent feature of iEEG data. Depending on the question, multiple approaches are conceivable. Typically, when addressing questions on sleep or cognitive physiology, it is considered best practice to exclude electrode contacts within the clinically identified SOZ and to reject any other channels that contain seizure or spiking activity [12]. However, in the context of sleep studies, these criteria might be overly conservative. As outlined above, even when the SOZ is outside of the MTL it is common to observe IEDs in hippocampal contacts. These IEDs should be rejected either based on visual inspection by a neurologist or by means of an automatic IED detector (Fig. 20.2). Several algorithms have been introduced in recent years [15, 27], but specificity and sensitivity have not fully been evaluated and detectors are not being used for clinical purposes, where the time-consuming visual inspection still constitutes the gold standard. It is common practice to only analyze nights where the patient did not experience any seizures or to discard recordings around the seizures with a large error margin of ± 2 h to avoid any pre- or postictal rhythmic slowing, which can easily be mistaken for physiologic slow waves.

20.5 Insights into Sleep and Anesthesia

20.5.1 The Human Memory Network During Sleep

Contemporary theories of memory consolidation emphasize the role of a two-step bidirectional hippocampal-neocortical dialogue, where novel information is initially encoded in hippocampal-dependent loops and overtime becomes mainly neocortex-dependent during consolidation [4, 10]. A hierarchy of sleep oscillations is thought to subserve the sleep-dependent reactivation, transfer and consolidation of mnemonic information. In this model, hippocampus-dependent information is spontaneously replayed during sleep, i.e. the very same pattern that was present during encoding is recapitulated during sleep in a time-compressed manner [34–37]. Replay is tightly linked to the expression of a hippocampal sharp-wave ripple (80–120 Hz; [37]), which in turn is nested in thalamo-cortical spindles (12–16 Hz) and neocortical slow waves (<4 Hz). Selective synchronization of these three cardinal sleep oscillations is thought to reflect an endogenous timing mechanism for the routing of information [4]. Before the advent of iEEG in humans, a major caveat of this theory was that most evidence stemmed from recordings in rodents, as non-invasive imaging of the human hippocampus did not offer a sufficiently high spatiotemporal resolution to detect ripple oscillations [9]. However, in recent years, the field of epileptology transitioned from using grid and strip electrodes on the outer surface of the MTL to employing depth electrodes that directly target the hippocampus, often in standardized bilateral implanting schemes; thus, providing the necessary resolution to examine the building blocks of systems memory consolidation in humans. Intracranial recordings from the sleeping brain have yielded important insights into sleep physiology in recent years. For instance, it has been shown that the hierarchical triple coupling is preserved in humans and that the precise SO-spindle coupling phase predicts hippocampal ripple expression [13–15, 38]. Hippocampal ripples then mediate the transfer of mnemonic information from the hippocampus to long-term neocortical storage (Fig. 20.3). Recently, intracranial recordings have been also used to also establish the presence of cortical ripples, however, their role in systems memory consolidation remains unclear [39, 40].

20.5.2 The Brain Under Anesthesia

In recent years, EEG studies under various anesthetics have revealed distinct spectral fingerprints of each drug [1, 5]. For example, the administration of GABAergic anesthetics such as propofol lead to an overall decrease of brain activity. Although this is true for most neural activity including IEDs, LOC under anesthesia is associated with a sudden increase of coherent slow and alpha oscillations (depicted in one frontal intracranial electrode Fig. 20.4a). We are currently lacking mechanistic insights into how the brain transitions from consciousness to anesthesia and how it recovers from

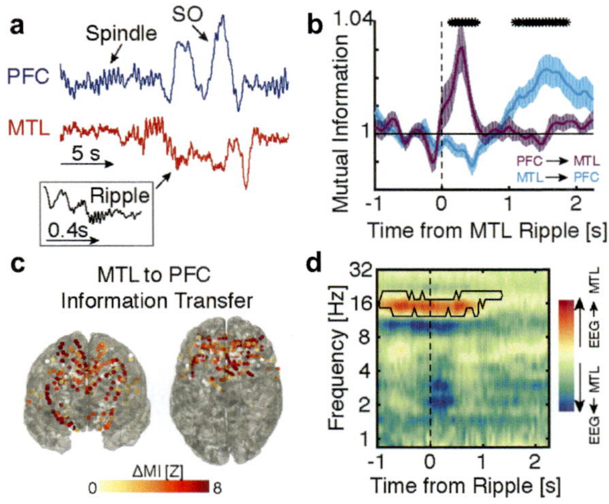

Fig. 20.3 Ripple-triggered information transfer between the hippocampus and neocortex. **a** Simultaneous recordings from frontal and hippocampal areas during NREM sleep highlight the presence of all cardinal sleep oscillations. **b** Bidirectional information exchange upon a hippocampal ripple. **c** Widespread increases in shared information between the hippocampus and neocortical intracranial contacts. **d** Spectrally-resolved information flow (transfer entropy) highlights a key role of spindle oscillations for mediating inter-areal information flow. Figure reproduced with permission from [14] under the Creative Commons Attribution (CC BY) license

the perturbation, with potential implications for coma and other states of altered arousal. Moreover, anesthesia can impact perioperative cognition beyond immediate recovery resulting in long-lasting cognitive deficits including memory impairments [3]. Although highly valuable, iEEG data during anesthesia remains scarce given the logistical challenges outlined above.

To date, iEEG has been used to illuminate the neural correlates of the loss-of-consciousness under general anesthesia. It has been demonstrated that network dynamics change dramatically upon anesthesia induction with prominent power increases in the delta- and alpha-bands [41–43]. Specifically, it has been shown that anesthesia alters spatiotemporal network configurations and alters coupling across temporal (i.e. delta-alpha or delta-gamma cross-frequency coupling; [43]) and across spatial scales (delta-, alpha- or gamma-band phase synchronization; [44–46]). It has been argued that impaired network synchronization is detrimental for information integration in large-scale cortical networks [41, 43, 46]. This consideration is in line with the observation that sensory processing in primary sensory areas remained intact under general anesthesia, while subsequent processing in secondary sensory and higher-order association areas was attenuated [47–49].

It has been proposed that neural networks operate close to criticality, i.e. at a transition point between ordered and chaotic network states, which may be optimal for information processing and transmission capacities [50–52]. Several findings indicated that anesthesia renders neural activity less critical, i.e. more predictable [53]. It has been argued that high variability close to possible network transition

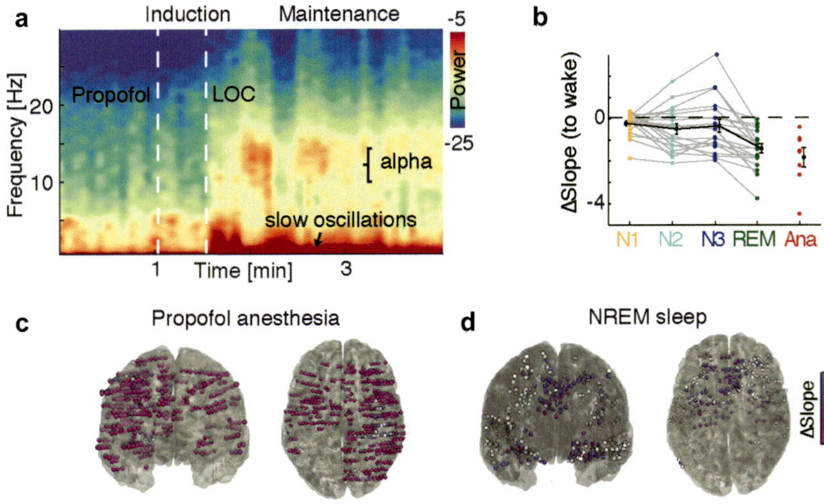

Fig. 20.4 Aperiodic activity dissociates arousal levels in sleep and anesthesia. a Multitaper spectrogram during induction of propofol anesthesia highlights the emergence of both slow waves and alpha oscillations during loss-of-consciousness. **b** Decreased arousal is associated with a reduction of the spectral exponent. States that are characterized by increased inhibition lead to a stronger reduction of the spectral slope. The steepest slope is observed during anesthesia. **c** Reduction of the slope in iEEG recordings highlights that anesthesia induces a brain wide reduction. **d** On the contrary, during sleep this reduction is confined to key nodes of the memory network, namely medial temporal and medial frontal areas. Figure reproduced with permission from [56] under the Creative Commons Attribution (CC BY) license

states is necessary to remain conscious, while anesthesia induces a shift away from the network bifurcation and thereby, promotes unconsciousness [54–57].

Collectively, this set of findings indicates that anesthesia promotes altered states of consciousness by impairing information flow and integration in cortical networks. Furthermore, the available evidence suggests that hyper-synchrony (as indicated by increased power and connectivity) heavily constrains the neural repertoire, which is necessary for consciousness. Collectively, iEEG under anesthesia provides a unique opportunity to assess the neural correlates of consciousness through the lens of pharmacologically induced unconsciousness.

20.5.3 *Comparative Electrophysiology of Sleep and Anesthesia*

In the last decade, several seminal findings were published using iEEG to understand sleep or anesthesia. However, to date only very few comparative approaches have been reported [25, 47, 56]. Hence, it remains unclear if anesthesia actually hijacks sleep pathways during induction, emergence or maintenance.

Until recently, slow oscillations have been considered a hallmark of the unconscious brain and a marker of cortical inhibition as they occur both in deep sleep and under anesthesia [1, 24]. However, this notion was in stark contrast to the presumed active role of NREM sleep in information processing [4]. Recent comparative evidence provided functional insights beyond these prominent oscillatory signatures. Computational modeling indicated increased inhibition is associated with a steepening on the electrophysiological power spectrum (reduction of the spectral exponent; [58]). Indeed, this shift towards inhibition was also observed during propofol anesthesia in rodents, monkeys and humans (Fig. 20.4; [56]). Importantly, anesthesia induced a brain-wide reduction of the spectral exponent (Fig. 20.4c). A similar exponent reduction was also observed during sleep, which had several implications [59]. First, contrary to popular belief, inhibition was maximal during REM and not NREM sleep, possibly sub-serving sleep-dependent neural homeostasis. Second, this reduction was mainly confined to the human memory network, i.e. encompassing medial temporal and medial frontal areas (Fig. 20.4d). However, it is critical to note that this line of inquiry is in early stages and we foresee that a direct, within subject comparison of sleep and anesthesia with iEEG will provide important insights into physiological as well as pathophysiological mechanism underlying e.g. post-operative cognitive decline.

20.6 Conclusions

In summary, intracranial recordings in humans can be leveraged to gain important insights into sleep physiology and the neural correlates of (un-) consciousness. We reviewed the most important technical considerations and prerequisites to successfully implement these recordings in a clinical environment. Recording sleep and anesthesia data during invasive monitoring constitutes an interdisciplinary team effort that involves multiple disciplines (neurology, neurosurgery, anesthesiology) and requires support from nurses and EEG techs. The obtained data provides a unique window in the correlates of (un-) consciousness in the human brain and therefore, constitutes an important link to recordings in non-human primates and rodents.

Acknowledgements This work was supported by the German Research Foundation (DFG Emmy Noether Program, HE 8329/2-1; R.F.H.), the Hertie Foundation (Hertie Network of Excellence in Clinical Neuroscience; R.F.H.), the Jung Foundation for Science and Research (Ernst Jung Career Advancement Award; R.F.H.) and the Medical Faculty of the University Medical Center Tübingen (JRG Plus Program; R.F.H.).

References

1. Brown EN, Lydic R, Schiff ND (2010) General anesthesia, sleep, and coma. N Engl J Med 363:2638–2650. https://doi.org/10.1056/NEJMra0808281
2. Brown EN, Purdon PL (2013) The aging brain and anesthesia. Curr Opin Anaesthesiol 26:414–419. https://doi.org/10.1097/ACO.0b013e328362d183
3. Radtke FM, Franck M, Lendner J et al (2013) Monitoring depth of anaesthesia in a randomized trial decreases the rate of postoperative delirium but not postoperative cognitive dysfunction. Br J Anaesth 110(Suppl 1):i98-105. https://doi.org/10.1093/bja/aet055
4. Klinzing JG, Niethard N, Born J (2019) Mechanisms of systems memory consolidation during sleep. Nat Neurosci 22:1598–1610
5. Brown EN, Purdon PL, Van Dort CJ (2011) General anesthesia and altered states of arousal: a systems neuroscience analysis. Annu Rev Neurosci 34:601–628. https://doi.org/10.1146/annurev-neuro-060909-153200
6. Alkire MT, Hudetz AG, Tononi G (2008) Consciousness and anesthesia. Science 322:876–880. https://doi.org/10.1126/science.1149213
7. Murphy M, Bruno M-A, Riedner BA et al (2011) Propofol anesthesia and sleep: a high-density EEG study. Sleep 34:283-291A
8. Steriade M, McCormick DA, Sejnowski TJ (1993) Thalamocortical oscillations in the sleeping and aroused brain. Science 262:679–685
9. Buzsáki G (2015) Hippocampal sharp wave-ripple: a cognitive biomarker for episodic memory and planning. Hippocampus 25:1073–1188. https://doi.org/10.1002/hipo.22488
10. Frankland PW, Bontempi B (2005) The organization of recent and remote memories. Nat Rev Neurosci 6:119–130. https://doi.org/10.1038/nrn1607
11. Leszczyński M, Barczak A, Kajikawa Y et al (2020) Dissociation of broadband high-frequency activity and neuronal firing in the neocortex. Sci Adv 6:eabb0977. https://doi.org/10.1126/sciadv.abb0977
12. Parvizi J, Kastner S (2018) Promises and limitations of human intracranial electroencephalography. Nat Neurosci 21:474–483. https://doi.org/10.1038/s41593-018-0108-2
13. Axmacher N, Elger CE, Fell J (2008) Ripples in the medial temporal lobe are relevant for human memory consolidation. Brain 131:1806–1817. https://doi.org/10.1093/brain/awn103
14. Helfrich RF, Lendner JD, Mander BA et al (2019) Bidirectional prefrontal-hippocampal dynamics organize information transfer during sleep in humans. Nat Commun 10:3572. https://doi.org/10.1038/s41467-019-11444-x
15. Staresina BP, Bergmann TO, Bonnefond M et al (2015) Hierarchical nesting of slow oscillations, spindles and ripples in the human hippocampus during sleep. Nat Neurosci 18:1679–1686. https://doi.org/10.1038/nn.4119
16. Fried I, Rutishauser U, Cerf M, Kreiman G (2014) Single neuron studies of the human brain: probing cognition. MIT Press
17. Rutishauser U (2019) Testing models of human declarative memory at the single-neuron level. Trends in cognitive sciences
18. Foldvary N, Perry M, Lee J et al (2001) The effects of lamotrigine on sleep in patients with epilepsy. Epilepsia 42:1569–1573
19. Gauthier AC, Mattson RH (2015) Clobazam: a safe, efficacious, and newly rediscovered therapeutic for epilepsy. CNS Neurosci Ther 21:543–548
20. Rocamora R, Álvarez I, Chavarría B, Principe A (2020) Perampanel effect on sleep architecture in patients with epilepsy. Seizure 76:137–142
21. Beenhakker MP, Huguenard JR (2009) Neurons that fire together also conspire together: is normal sleep circuitry hijacked to generate epilepsy? Neuron 62:612–632. https://doi.org/10.1016/j.neuron.2009.05.015
22. Derry CP, Duncan S (2013) Sleep and epilepsy. Epilepsy Behav 26:394–404
23. Méndez M, Radtke RA (2001) Interactions between sleep and epilepsy. J Clin Neurophysiol 18:106–127

24. Steriade M (2005) Sleep, epilepsy and thalamic reticular inhibitory neurons. Trends Neurosci 28:317–324
25. Lewis LD, Piantoni G, Peterfreund RA et al (2018) A transient cortical state with sleep-like sensory responses precedes emergence from general anesthesia in humans. elife 7:e33250
26. Krom AJ, Marmelshtein A, Gelbard-Sagiv H et al (2020) Anesthesia-induced loss of consciousness disrupts auditory responses beyond primary cortex. Proc Natl Acad Sci 117:11770–11780
27. Gelinas JN, Khodagholy D, Thesen T et al (2016) Interictal epileptiform discharges induce hippocampal-cortical coupling in temporal lobe epilepsy. Nat Med 22:641–648. https://doi.org/10.1038/nm.4084
28. Sato M, Racine RJ, McIntyre DC (1990) Kindling: basic mechanisms and clinical validity. Electroencephalogr Clin Neurophysiol 76:459–472
29. Badawy RA, Curatolo JM, Newton M et al (2006) Sleep deprivation increases cortical excitability in epilepsy: syndrome-specific effects. Neurology 67:1018–1022
30. Malow BA (2004) Sleep deprivation and epilepsy. Epilepsy Curr 4:193–195
31. Tononi G, Cirelli C (2014) Sleep and the price of plasticity: from synaptic and cellular homeostasis to memory consolidation and integration. Neuron 81:12–34. https://doi.org/10.1016/j.neuron.2013.12.025
32. Huber R, Mäki H, Rosanova M et al (2013) Human cortical excitability increases with time awake. Cereb Cortex 23:332–338. https://doi.org/10.1093/cercor/bhs014
33. Rechtschaffen A, Kales A (1968) A manual of standardized terminology, techniques, and scoring systems for sleep stages of human subjects
34. Antony JW, Schapiro AC (2019) Active and effective replay: systems consolidation reconsidered again. Nat Rev Neurosci. https://doi.org/10.1038/s41583-019-0191-8
35. Higgins C, Liu Y, Vidaurre D et al (2021) Replay bursts in humans coincide with activation of the default mode and parietal alpha networks. Neuron 109:882-893.e7. https://doi.org/10.1016/j.neuron.2020.12.007
36. Ólafsdóttir HF, Bush D, Barry C (2018) The role of hippocampal replay in memory and planning. Curr Biol 28:R37–R50. https://doi.org/10.1016/j.cub.2017.10.073
37. Zhang H, Fell J, Axmacher N (2018) Electrophysiological mechanisms of human memory consolidation. Nat Commun 9:4103. https://doi.org/10.1038/s41467-018-06553-y
38. Skelin I, Zhang H, Zheng J et al (2021) Coupling between slow waves and sharp-wave ripples engages distributed neural activity during sleep in humans. Proc Natl Acad Sci 118
39. Khodagholy D, Gelinas JN, Buzsáki G (2017) Learning-enhanced coupling between ripple oscillations in association cortices and hippocampus. Science 358:369–372. https://doi.org/10.1126/science.aan6203
40. Vaz AP, Inati SK, Brunel N, Zaghloul KA (2019) Coupled ripple oscillations between the medial temporal lobe and neocortex retrieve human memory. Science 363:975–978. https://doi.org/10.1126/science.aau8956
41. Breshears JD, Roland JL, Sharma M et al (2010) Stable and dynamic cortical electrophysiology of induction and emergence with propofol anesthesia. Proc Natl Acad Sci USA 107:21170–21175. https://doi.org/10.1073/pnas.1011949107
42. Lewis LD, Ching S, Weiner VS et al (2013) Local cortical dynamics of burst suppression in the anaesthetized brain. Brain 136:2727–2737. https://doi.org/10.1093/brain/awt174
43. Mukamel EA, Pirondini E, Babadi B et al (2014) A transition in brain state during propofol-induced unconsciousness. J Neurosci 34:839–845. https://doi.org/10.1523/JNEUROSCI.5813-12.2014
44. Huang Y, Wu D, Bahuri NFA et al (2018) Spectral and phase-amplitude coupling signatures in human deep brain oscillations during propofol-induced anaesthesia. Br J Anaesth 121:303–313
45. Lewis LD, Weiner VS, Mukamel EA et al (2012) Rapid fragmentation of neuronal networks at the onset of propofol-induced unconsciousness. Proc Natl Acad Sci USA 109:E3377-3386. https://doi.org/10.1073/pnas.1210907109
46. Malekmohammadi M, Price CM, Hudson AE et al (2019) Propofol-induced loss of consciousness is associated with a decrease in thalamocortical connectivity in humans. Brain 142:2288–2302

47. Banks MI, Krause BM, Endemann CM et al (2020) Cortical functional connectivity indexes arousal state during sleep and anesthesia. Neuroimage 211:116627
48. Nourski KV, Banks MI, Steinschneider M et al (2017) Electrocorticographic delineation of human auditory cortical fields based on effects of propofol anesthesia. Neuroimage 152:78–93. https://doi.org/10.1016/j.neuroimage.2017.02.061
49. Nourski KV, Steinschneider M, Rhone AE et al (2018) Auditory predictive coding across awareness states under anesthesia: an intracranial electrophysiology study. J Neurosci 0967–18. https://doi.org/10.1523/JNEUROSCI.0967-18.2018
50. Beggs JM (2008) The criticality hypothesis: how local cortical networks might optimize information processing. Philos Trans A Math Phys Eng Sci 366:329–343. https://doi.org/10.1098/rsta.2007.2092
51. Shew WL, Plenz D (2013) The functional benefits of criticality in the cortex. Neuroscientist 19:88–100. https://doi.org/10.1177/1073858412445487
52. Zimmern V (2020) Why brain criticality is clinically relevant: a scoping review. Front Neural Circuits 14
53. Toker D, Pappas I, Lendner JD et al (2022) Consciousness is supported by near-critical slow cortical electrodynamics. Proc Natl Acad Sci 119:e2024455119
54. Cocchi L, Gollo LL, Zalesky A, Breakspear M (2017) Criticality in the brain: a synthesis of neurobiology, models and cognition. Prog Neurobiol 158:132–152. https://doi.org/10.1016/j.pneurobio.2017.07.002
55. Colombo MA, Napolitani M, Boly M et al (2019) The spectral exponent of the resting EEG indexes the presence of consciousness during unresponsiveness induced by propofol, xenon, and ketamine. Neuroimage 189:631–644. https://doi.org/10.1016/j.neuroimage.2019.01.024
56. Lendner JD, Helfrich RF, Mander BA et al (2020) An electrophysiological marker of arousal level in humans. Elife 9:e55092
57. Sarasso S, Boly M, Napolitani M et al (2015) Consciousness and complexity during unresponsiveness induced by Propofol, Xenon, and Ketamine. Curr Biol 25:3099–3105. https://doi.org/10.1016/j.cub.2015.10.014
58. Gao R, Peterson EJ, Voytek B (2017) Inferring synaptic excitation/inhibition balance from field potentials. Neuroimage 158:70–78. https://doi.org/10.1016/j.neuroimage.2017.06.078
59. Helfrich RF, Lendner JD, Knight RT (2021) Aperiodic sleep networks promote memory consolidation. Trends Cogn Sci 25:648–659. https://doi.org/10.1016/j.tics.2021.04.009
60. Helfrich RF, Mander BA, Jagust WJ et al (2018) Old brains come uncoupled in sleep: slow wave-spindle synchrony, brain atrophy, and forgetting. Neuron 97:221-230.e4. https://doi.org/10.1016/j.neuron.2017.11.020

Chapter 21
What Can iEEG Inform Us About Mechanisms of Spontaneous Behavior?

Yitzhak Norman and Rafael Malach

Abstract While controlled experiments form the core of human brain research—a fundamental, yet far less studied, domain concerns the neuronal mechanism underlying freely-generated, spontaneous behavior. Intracranial recordings in conscious human patients offer an invaluable window into this question. Here we review relevant iEEG findings highlighting a universal mechanism underlying human free behavior: internally generated ultra-slow fluctuations in neuronal activity. These spontaneous firing dynamics are ubiquitous across the brain and appear at various scales—from functionally-specialized local sub-populations of neurons, to brain wide networks. Crucially, signatures of these slow fluctuations appear in every free behavior studied so far. Focusing on free recall as a paradigmatic example, we demonstrate that spontaneous activity fluctuations, manifested as slow anticipatory waves in cortico-hippocampal circuits, tend to precede spontaneous recollections by 1–2 s. Moreover, when patients attempt to constrain their spontaneous recollections to a particular category, intracranial recordings in the cortex reveal a category-specific "baseline shift", i.e., steady enhancement in the excitability of neuronal populations encoding the targeted category. Such top-down modulation can bias the free recall process by pushing the spontaneous fluctuations of the relevant neuronal subpopulations closer to the behavioral threshold. Along with evidence derived from fMRI studies, these iEEG experiments demonstrate that slow spontaneous fluctuations within the appropriate brain circuits may serve as a driving force behind the emergence of spontaneous thoughts and free human behavior.

Y. Norman
Department of Neurological Surgery, University of California, San Francisco, CA 94143, USA
e-mail: rafi.malach@weizmann.ac.il

R. Malach
Department of Brain Sciences, Weizmann Institute of Science, Rehovot 76100, Israel

N. Axmacher (ed.), *Intracranial EEG*, Studies in Neuroscience, Psychology and Behavioral Economics, https://doi.org/10.1007/978-3-031-20910-9_21

21.1 The Fundamental Importance and Characteristics of Free Behavior

Human brain research has been studied most commonly under stimulus–response paradigms. This approach, in which brain responses are measured while participants receive sensory stimulation or instructions to act has the great advantage of providing tight control and reproducibility of the experimental conditions. However, a fundamental and critically important aspect of human behavior does not lend itself to such deterministic experimental paradigms. This is the large group of internally generated behaviors that are termed free or spontaneous. It is not possible to over-estimate the importance of free behaviors in human life. The potential to act freely is an essential element for human well-being. A sub-set of free behaviors—creativity—is the fundamental generator driving human progress on all cultural and scientific fronts.

The concept of free behavior—and whether it even exists—has been a debated, and far from settled philosophical issue. For the purpose of this chapter we will define this vast field of free behaviors operationally as the group of behaviors that are not fully determined by an external stimulus or instruction. It is important to emphasize that our definition allows for a deterministic component in such behaviors (discussed below). In fact, such deterministic component is an essential part since free behavior always occurs within a particular context and is required to comply with behavioral goals or external constraints—and is therefore never completely random or chaotic. However, crucially, free behaviors will always include a significant, spontaneously generated, component that is not determined by the external instructions or conditions.

A useful strategy in searching for a likely neuronal mechanism that may drive free behaviors will be to identify what are their common and central characteristics. Identifying such common characterizations could provide important constraints on potential neuronal substrates of free behavior. Below we outline several common characteristics which are shared by all free behaviors and may assist in the identification of the neuronal mechanisms involved.

21.1.1 Heterogeneity

In light of our operational definition, the first obvious characteristic of free behaviors is their heterogeneity. In fact the range of free behaviors is so large that it is difficult to consider any domain of human behavior that does not have a spontaneous counter-part. Examples can range from improvisation theatre to creative writing, to problem solving, free associations and action painting. Thus, in terms of the brain, the neuronal mechanism driving spontaneity in all these diverse behaviors is likely to be widespread and of the same nature, regardless of the functional brain area concerned.

21.1.2 Spontaneity

Not surprising, as it is part of the defining characteristic of free behavior—a truly free behavior must have, even if only partially, an internally-generated aspect. Thus, spontaneous behaviors are not being directly determined by any informative signal from outside—not even signals that are subliminal or belong to a long association chain. An element of de-novo creation must be included.

21.1.3 Personality Constraints

Although free behaviors haves a strong element of self-generation, this does not mean that such behaviors are totally random or chaotic. In fact, it is virtually impossible to generate free behaviors that are not bounded, to some extent, by the personalities, expertise, and tendencies of the freely behaving individual: a trumpet player trained exclusively in jazz naturally improvises differently from one trained exclusively in classical music.

21.1.4 Boundary Setting

While free behavior is spontaneous, it is not devoid from the motivation and control of the individual. To the contrary—we can flexibly and rapidly define the boundaries of our free behaviors. For example, we are capable of switching, rapidly, from one free behavior say Rap, to another, e.g., Scat singing, without erroneously performing the unintended behavior. Thus, we would expect the neuronal mechanism underlying free behavior to manifest such rapid boundary setting process as well.

21.1.5 Pre-conscious Preparation

Although we have a strong sense of agency especially when behaving freely—careful psychological and physiological research has revealed the surprising observation that most, if not all, of our free behaviors are often driven by various unconscious processes. Perhaps the most common example of this fact is the "out of the blue" experience—when we freely solve a problem or recall an item without any prior premonition that this is going to occur. However, as has been beautifully described by William James, even mundane acts such as jumping out of bed on a cold morning are decided subconsciously [1]. This aspect has been studied most thoroughly in the domain of free decisions to move [2].

21.2 Spontaneous (Resting State) Fluctuations: A Plausible Neuronal Generator of Free Behaviors

Is there a potential neuronal process that satisfies these diverse characteristics of free behaviors outlined above? A broad examination of human brain research readily brings to the forefront an obvious candidate: a robust, extensively researched phenomenon termed slow neuronal activity fluctuations (also known as 'resting-state' fluctuations). These spontaneous fluctuations have been at the center of a major and rapidly expanding research field. They are characterized by ultra-slow dynamics and appear to fit all the criteria highlighted above. These fluctuations are internally generated, in the absence of external stimulation or task (hence the term "resting state"). They manifest high heterogeneity in that they have been observed in each and every cortical network studied so far. They are organized in networks that have been linked to behavioral or cognitive habits and personality traits [3–8]. Finally, they appear to operate largely below the consciousness threshold [2, 9–11] (see further evidence below).

21.3 iEEG Reveals the Neuronal Basis and Precise Dynamics of Spontaneous Fluctuations

The original observations of resting state fluctuations in humans were obtained using BOLD-fMRI imaging [12, 13]. However, this method suffers from three major limitations: First, it is based on blood flow rather than direct recordings of electrophysiological signals. In the case of the spontaneous fluctuation this raised major concerns that the phenomenon reflects non-neuronal processes (including astrocytic signals [14]). Second, the BOLD signals are notoriously sluggish—thus preventing an accurate characterization of the fluctuations' amplitude and dynamics. Finally, BOLD-fMRI at 3-T is of somewhat lower spatial resolution compared to the intracranial iEEG electrodes [15].

All these limitations can be substantially alleviated using direct intracranial recordings in patients. Thus, Nir et al. [16] took advantage of intracranial recordings that, importantly, measured homotopic regions in human auditory cortex bilaterally, as patients rested in a quiet room. Figure 21.1 depicts their finding.

First, the study demonstrated that spontaneous fluctuations can be found even in activity modulations of single neurons (Fig. 21.1b) as well as high frequency broadband (HFB) power (Fig. 21.1c). The latter signal was shown to be an excellent marker of local population firing rate [17–19]. Second, the study revealed that the spontaneous fluctuations were indeed dominated by ultra-slow frequencies (with wavelength in the seconds domain), but, importantly, the study also indicated that the fluctuations followed a power law spectrum in which the amplitude of the fluctuation was inversely related to their frequency (see Fig. 21.1d). Finally, the study demonstrated that the amplitude of the spontaneous fluctuations was about an order

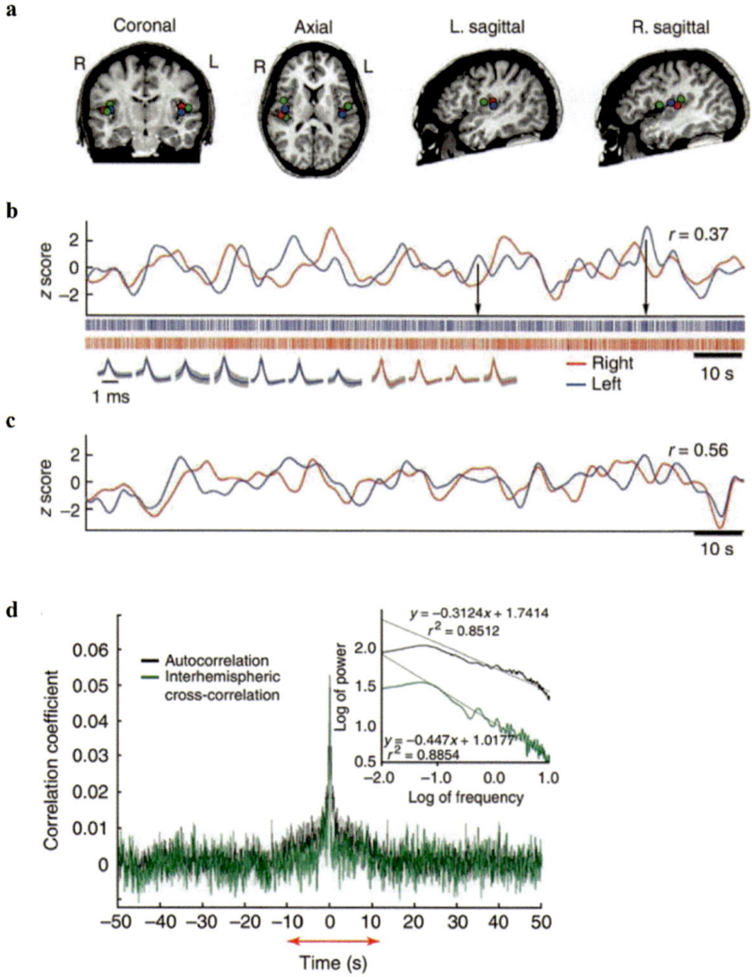

Fig. 21.1 Bi-lateral recordings of ultra-slow fluctuations of HFB and single-unit activity in human auditory cortex during quiet rest. **a** Estimated anatomical location of intracranial recording sites in auditory cortex of 3 representative patients. **b, c** Examples of slow fluctuations (<0.1 Hz) in neuronal activity showing correlation between right (red) and left (blue) hemispheres during wakeful rest. **b** Single units' activity; vertical lines show actual spike times. Black arrows indicate the relation between time courses of slow firing-rate modulations and actual spikes. Waveforms of neuronal action potentials are shown below spike trains (gray zone represents SEM across spike instances). **c** Same as in panel b, but for High frequency broadband (HFB) signals. **d** Cross- and autocorrelations of neuronal firing rates during wakeful rest (n = 8 recording sessions in two individuals). Green, cross-correlation; black, autocorrelation. Red arrows indicate very slow elevation in correlation (±10 s) corresponding to ultra-slow fluctuations <0.1 Hz. Inset shows Fourier transform of cross- and autocorrelation functions (cross-spectrum and spectrum, respectively) showing 1/f-like spectral profiles. Adapted from Ref. [16] with permission from Nature Publishing Group

of magnitude lower than the typical sensory evoked "ignitions" of activity at the same cortical sites. This work nicely illustrates the power and precision of iEEG recordings in providing a direct and detailed characterization of major neuronal processes in the human brain.

21.4 The Role of Spontaneous Fluctuations in Free Behavior: An iEEG Study of Free Visual Recall

While the slow spontaneous fluctuations appear to have features compatible with a role in free behavior, the question still remains: what direct evidence may point to such a role? A full review of this field is beyond the scope of this chapter. Here we will highlight some relevant data using a specific paradigmatic case—that of free recall, as studied, specifically, using the powerful methodology of iEEG recordings in conscious patients, both at the level of single units and local neuronal populations activity (i.e., HFB signal).

The basic experimental paradigm of free recall is rather simple and can be easily and successfully performed by the patients. Figure 21.2 illustrates the basic experimental design of a free recall task involving visual episodic memories.

In the first part of the experiment (the memory encoding stage), the patients are presented with visual content—photographs from specific categories (e.g. famous faces and places) [20–22] or short video clips depicting famous persons, places and other categories [23]. After a short distracting task (e.g., math calculation) aimed at eliminating the possibility of direct associative chains and disrupting working memory representations, the patients are then blocked from external visual stimuli and are asked to freely recall, in their mind's eye and in as much detail as possible, the materials that they had viewed earlier. Crucially, they are asked to describe verbally, in real time, any explicit content that comes to their mind. In the offline analysis,

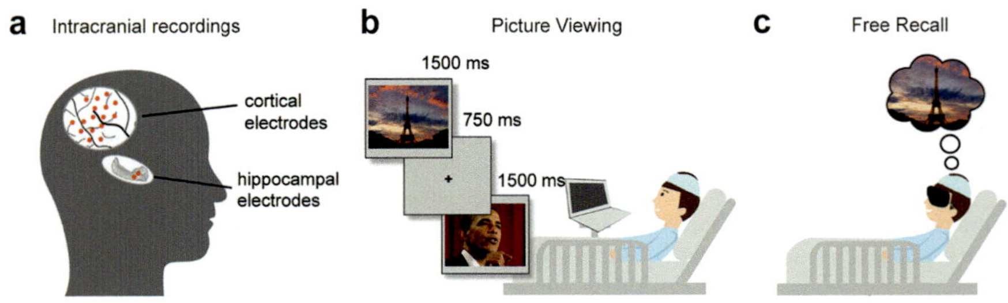

Fig. 21.2 Intracranial EEG recordings during free visual recall: experimental paradigm. **a** Intracranial recordings obtained simultaneously from the hippocampus and the cerebral cortex. **b** During the memory encoding phase, patients are presented with e.g., vivid images of famous people and places. **c** After viewing each item several times and performing a short distraction task, the patients put on blindfolds and attempt to recall as many items as possible, describing them briefly as soon as they come to mind

these verbal reports are used to determine the timing and content of the memory items that emerged spontaneously in the patients' minds.

Note, importantly, that the recall is not totally free in the sense that the patients were instructed to only recall items from the previously presented set, and in one of the experiments (to be described below) they were asked to further constrain their recall to only one category at a time, e.g., faces in one session and places in another. However, note that this type of behavior still contains a free, undetermined, component since the patients were not instructed which specific visual image to recall—allowing the "spontaneity generator" mechanism enough room to operate within this pre-determined arena.

Free visual recall is a particularly informative case since it involves two central cognitive components—memory and vision. Hence it is to be expected that the free-behavior mechanism will be manifested in both memory-related circuits in the medial temporal lobe (MTL) on the one hand, and in the visual cortex on the other. Indeed, recording single neurons in human MTL during the free visual recall of previously seen video clips, Gelbard-Sagiv et al. [23] have found re-activation of content-specific neurons in the human hippocampus (Fig. 21.3a). The study demonstrated robust activation of hippocampal neurons during the presentation of the clips. In agreement with other single-unit studies [24–26], the response profile showed a remarkable content specificity,with neurons commonly responding to only 1–2 preferred items (5–10% of stimuli presented, on average). Most important, these neurons were re-activated also when the patients freely recalled these videos later, in the absence of external stimuli, recapitulating the same content-specificity found during viewing. Thus, a neuron was reactivated during recall only when the patient recalled the specific video clip to which this neuron selectively responded earlier during the viewing session.

As for the visual cortex, a study by Khuvis et al. [20] demonstrated a similar reactivation phenomenon in face-selective neurons recorded in the inferior temporal cortex (ITC). There, the neurons were reactivated only when the patient recalled the face images, but not place images (see Fig. 21.4b; the experimental design is depicted in Fig. 21.2).

What was the mechanism that drove the patient to recall a specific picture or video clip in the absence of any external stimulation? A telling sign can be found by comparing the response of the neurons to the actual stimulus during the viewing session (Fig. 21.4, left panels), with their internally generated activation during the free recall (Fig. 21.4, right panels). Notice that in contrast to the stimulus-driven activation, the activations induced by the free recall were preceded by a slow, anticipatory buildup starting approximately 2 s before the onset of verbal recall.

The superb SNR offered by single unit recordings allows to actually observe this process at the single trial level, both in the hippocampus (Fig. 21.4a) and the visual cortex (Fig. 21.4b).

Throughout the recall period, one can observe that the neurons' activity slowly fluctuates at a low amplitude in an apparently spontaneous manner, until, just before the recall, one of the slow fluctuations becomes large enough to cross a cognitive bound or threshold—allowing the patient to become aware of the spontaneously

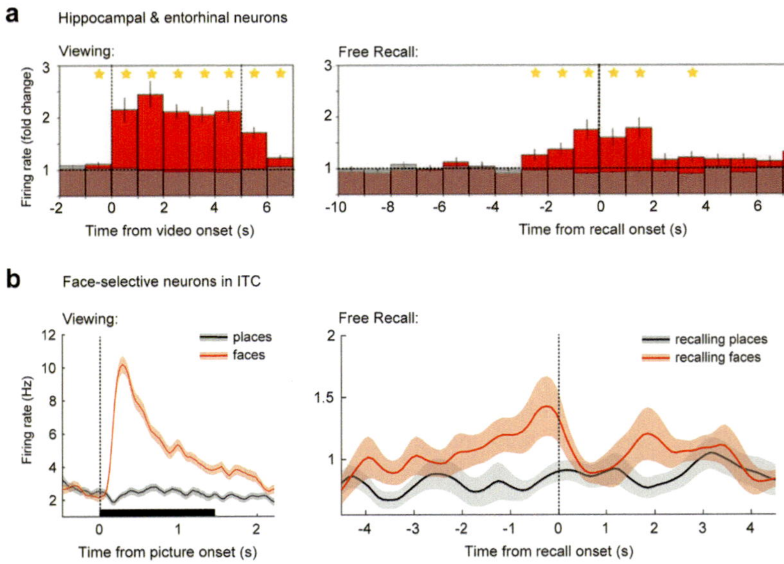

Fig. 21.3 Slow activity buildup anticipating the free recall event—averaged activity. **a** Single unit recordings of hippocampal and entorhinal neurons, showing a firing rate increase during viewing (left) and free recall (right) of video clips to which the neurons responded preferentially (compared to a 5 s blank period; gray dashed line). Firing rates were averaged across all responsive hippocampal and entorhinal neurons (n = 154). Reproduced from Ref. [23] with permission from AAAS. **b** Face-selective neurons in the visual cortex, showing a face-selective increase in firing rate during viewing (left) and free recall (right) of face images (red), but not place images (black). Reproduced from Ref. [20] with permission. Notice the sluggish buildup in firing rate anticipating the onset of verbal recall, in both structures

retrieved content, and to verbally report about it. Following the recollection, the activity declines back to the slowly fluctuating mode. Thus, taking advantage of intracranial recordings of single neuron activity, this study opens a direct window into the dynamics of slow spontaneous fluctuations and how it evolves into a conscious mental event of spontaneously recalling a specific item.

It is important to note that the slow anticipatory buildup that—we hypothesize—drives these spontaneous recall events, is also observable in fMRI [27] and is not specific to the memory domain, as similar dynamic signatures of such a process have been found in many other examples of spontaneous behavior studied so far (e.g., Refs. [2, 9, 11, 28–31]), and even creative insights [32]. Critically, a direct link between the slow spontaneous fluctuations and the slow buildup anticipating free decisions has been recently demonstrated by Broday-Dvir et al. in an fMRI study [32].

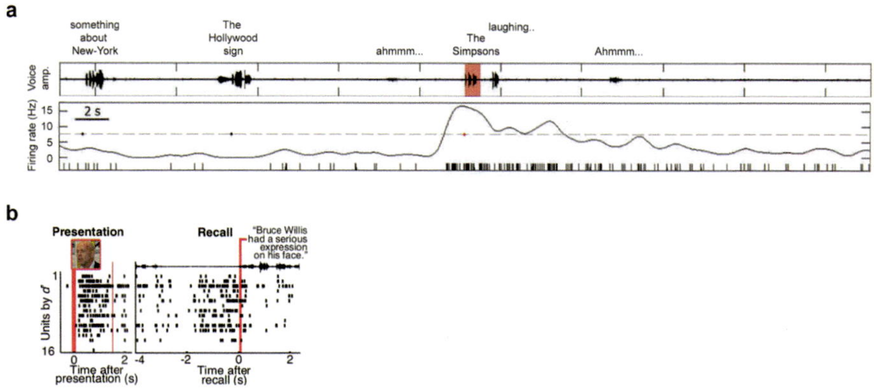

Fig. 21.4 Slow activity buildup anticipating the free recall event—single trial example. **a** A single-unit in the right entorhinal cortex, showing spontaneous activity fluctuations and then re-activation shortly before the patient recalls an episode from the TV series The Simpsons (the preferred clip of that neuron). The cell's firing rate rose significantly above baseline 1.5 s before onset of verbal report of recall (voice amplitude and the content of the verbal report are shown at top). **b** Raster plot depicting spiking activity of face-selective neurons in non-primary visual cortex of one patient. Picture viewing response is shown on the left, spontaneous recollection on the right (voice amplitude and the content of the verbal report are shown at top). Reproduced from Ref. [20] with permission

21.5 Integration of Memory and Vision: Hippocampal Ripples Anticipating Recollection

As we discussed above, the phenomenon of free visual recall is an interesting model case for research because it integrates the MTL-centered mechanisms of episodic memory with (conveniently trackable) visual content represented in high-order visual areas. What are the neuronal processes that govern such cross-system integration during free, internally-generated recall? Intracranial EEG offers a unique opportunity to explore these cross-system interactions because, as we discussed above, in a typical iEEG diagnostic session, the neurosurgeon places numerous contacts across both cortical and hippocampal sites which offer a rare experimental situation enabling simultaneous recordings from both visual and hippocampal regions during free visual recall.

Norman et al. [22] took advantage of this opportunity and probed the cortical-hippocampal interaction during visual free recall, as depicted in Fig. 21.2. Thus, after viewing various pictures of famous faces or places, patients were asked to freely recall as many images as possible, and to provide a brief visual description for each item recalled—to ensure retrieval of specific visual details and not just general semantic information.

In the electrophysiological analysis Norman et al. focused on rapid and prominent oscillations in the local field potential of the hippocampus termed hippocampal ripples. These short-lived oscillations, which have been extensively studied in animal models over the past two decades [33–35], are believed to constitute one of the most massive neuronal synchronization events in the mammalian brain. It has been estimated that hundreds of thousands of neurons (5–15% of the neurons on the hippocampal-entorhinal output pathway) participate in these events by emitting a spike within a strikingly narrow time window of about 100 ms [36–40]. Focusing on these events opens an informative window into various mnemonic signals that are neither encoded at the level of individual neurons nor at the level of an entire population, but instead, at the level of local neuronal assemblies operating together in tight temporal coordination. Moreover, these LFP markers provide a convenient and fairly accurate timestamps of informative moments in time when the hippocampus processes and communicates critical mnemonic information with other brain structures.

This last point is of particular importance for the study of spontaneous cognitive behavior, where the underlying neuronal signals are triggered spontaneously and as such their precise timing cannot be fully modeled a-priori (unlike immediate sensory responses). Thus, the value of internal timestamps obtained from the brain itself is crucial.

Importantly, hippocampal ripples can be reliably identified and detected with iEEG recordings [41–47]. Figure 21.5a, b depicts an example of a hippocampal ripple event as it appears in the raw iEEG trace from a contact located in CA1. The clear high frequency signature of such ripples is easily evident, both in the ripple-band envelope (70–180 Hz) and in the ripple-triggered spectrogram (Fig. 21.5c).

Examining ripple rate during the free recall period reveals a significant increase in ripple rate that anticipates the verbal report of recall by approximately 2 s. Figure 21.5d illustrates this finding in a single-trial example of a patient recalling an image of Bart Simpson. As can be seen, two ripples emerged in the recorded site prior to the verbal report ("Oh! I remember…"), and another one right after the patient begins to describe the picture. Importantly, just as was in the case of single neurons recorded during free recall of video clips in the Gelbard-Sagiv [23] study—the anticipatory enhancement in ripple rate was content-selective and recapitulated the picture preferences found in the same sites during the viewing session (see group level results in Fig. 21.5e). That is, specific images that generated at a given site an increase in ripple rate during the viewing session—generated a parallel, content-specific, increase in ripple rate also during the free recall event.

Notably, in agreement with the single unit studies, the findings of Norman et al. again demonstrate a slow anticipatory buildup—this time in ripple rate—beginning almost 3 s prior to the recall event. In view of the fact that the hippocampus is situated at the top of the visual processing hierarchy and is capable of integrating signals from a variety of cortical sites [48]—the increase in ripple rate prior to recall may be thought of as the result of an extensive synthesis of ongoing spontaneous activity in the cortex. In other words, considering the inherently integrative properties of the hippocampus as well as its extensive connectivity, a main factor that governs spontaneous activity in the hippocampus is likely to be the accumulation of multiple fluctuating populations within upstream cortex, culminating in hippocampus ripples (and their cognitive consequence, the recall event).

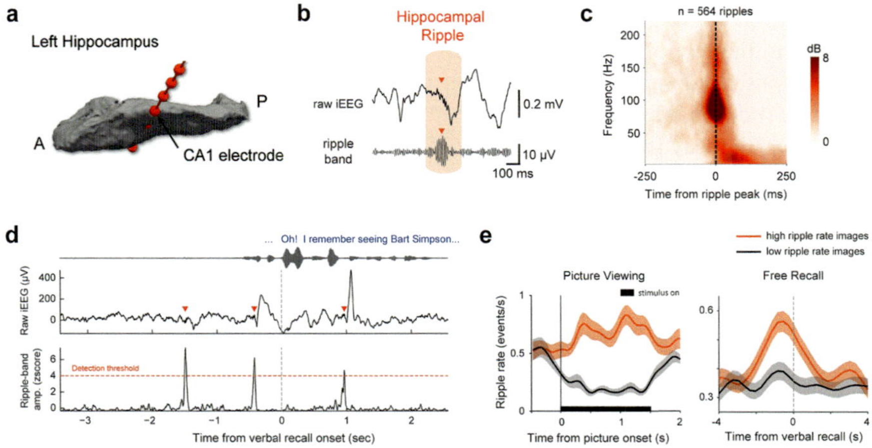

Fig. 21.5 Hippocampal ripples anticipate the free recall event. **a** Representative example of iEEG depth electrodes implanted in the hippocampus. Red circles indicate individual recording sites. Ripples could be clearly identified in recording sites located in the CA1 subfield (black arrow; panels b-d show signal from this recording site). **b** Example of an individual ripple event as it appears in the raw iEEG trace and the ripple band (70–180 Hz). **c** Ripple-triggered spectrogram averaged across all ripples detected during the experiment, demonstrating the typical spectral characteristics of human hippocampal ripples. **d** A single-trial example of spontaneously generated ripples, elicited shortly before the patient recalled an image of Bart Simpson's face. Voice amplitude and the content of the verbal report are presented at top. **e** Group-level results demonstrating the content-selective increase in ripple rate anticipating spontaneous recollections ($n = 15$ patients; shaded area represents SEM across patients in the left side, and bootstrap SE over pooled trials in the right). Left panel: during the picture viewing stage some images resulted in an increase in ripple rate (red), while others resulted in a decrease in ripple rate (black). Right panel: when patients freely recalled the images, a transient increase in ripple rate anticipated the onset of recall, driven by items that generated a higher ripple rate during viewing (compare red and black lines). Reproduced from Ref. [22] with permission from AAAS

21.6 Ripple-Mediated Cortico-Hippocampal Dialogue During Free Recall

It has been hypothesized that ripples coordinate a dialogue between the hippocampus and cortex for rapid and effective communication of mnemonic information [35, 49, 50]. In the case of free visual recall, this hypothesis predicts that when patients freely recall a vivid visual image and reexperience it in their mind, hippocampal ripples will play a central role in this cognitive process. Simultaneous iEEG recordings of ripples on the one hand and visual cortex activations on the other hand provide a unique opportunity to examine this hypothesis. Figure 21.6 compares the high order visual cortex activations when the patients first viewed the images (Fig. 21.6b) and when they freely recalled these images (Fig. 21.6c). During the viewing session, the recording sites in high order visual areas showed a strong preference for a subset of the visual images (e.g., face images in face selective sites), while remaining unresponsive to the rest of the viewed images (non-preferred images). Notably, as has been demonstrated in previous studies, the cortical HFB response elicited during the viewing condition was tightly locked to the stimulus onset—rapidly "igniting" at 170–200 ms latency [51–53].

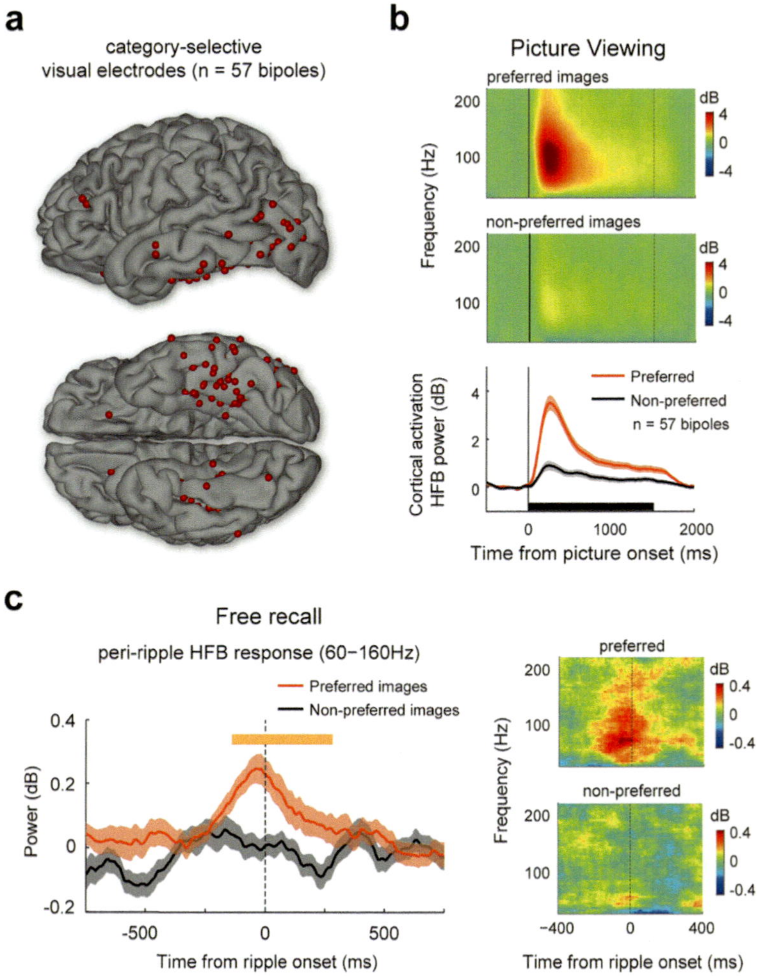

Fig. 21.6 Ripple-coupled cortical reactivation profile during free recall. **a** Anatomical location of face-selective and place-selective cortical electrodes (n = 57 bipolar pairs). **b** HFB response to preferred and non-preferred images during picture viewing. **c** During free recall, the category-selective visual sites in the cortex showed SWR-coupled reactivation, recapitulating the content selectivity found during viewing. As can be seen, activity in the cortex both precedes and follows the hippocampal ripple, consistent with a recurrent cortical-hippocampal loop. Reproduced from Ref. [22] with permission from AAAS

Time-locking the cortical HFB in those visual electrodes to the timestamps of the spontaneous hippocampal ripples elicited during the free recall, reveals a transient reactivation of visual content, with recapitulation of the content-selectivity found during viewing. Thus, when patients recalled a face, for example a picture of Barack Obama, or alternatively, a monument, such as the Eiffel Tower, cortical activity was selectively enhanced at the corresponding visual sites that represent these items. Likewise, when the non-preferred items were recalled, activity in those sites was unaffected and showed no association with the ripples. It should be noted that the activation dynamic was precisely centered on the hippocampal ripple event—supporting the hypothesis that ripple emissions participate in the emergence of a freely recalled visual percept.

21.7 Evidence for Recurrent Rather Than Unidirectional Information Flow

Examining the temporal profile of the peri-ripple cortical responses, one can discern that the cortical activation slightly preceded the hippocampal ripples by approximately 300 ms, while peaking together and persisting for additional 300 ms. This dynamic suggests that instead of participating in a unidirectional causal relationship—the ripple and cortical activations were likely engaged in recurrent, loop-like, mutual activations. In such a recurrent dialogue, spontaneously generated spiking activity in upstream cortex may play a role in seeding hippocampal representations prior to the ripple [49]. Then, by eliciting a ripple, the hippocampus can influence the cortex by coordinating reactivation of specific memory items [22, 54] as well as information related to context [44]. In such process, the cortex both precedes and follows the ripple emission event. This recurrent dynamical relationship between the ongoing cortical activity and hippocampal ripples has been confirmed and extended in a recent iEEG study examining the hippocampal cortical dialogue in human autobiographic and semantic memory recall [55] (Fig. 21.7). Similar to free visual recall, retrieval of declarative memories exhibits similar build-up and follow-up waves of cortical activation of about 400 ms duration relative to the ripple event. This recurrent relationship appears to be a wide-spread phenomenon involving a number of prominent cortical networks—dominated, in the case of autobiographical recall, by the memory-related default mode network [56]. Finally, a number of studies in rodents have demonstrated a similar recurrent dynamics between cortical activity and hippocampal ripples as shown here for the human brain [49, 57].

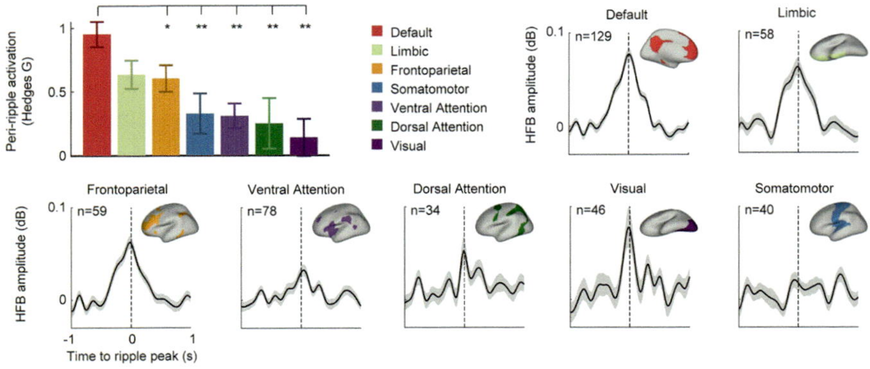

Fig. 21.7 Recurrent dynamic of ripple-coupled cortical activation during autobiographical memory recall across canonical "resting state" networks. Peri-ripple activations across seven canonical resting-state networks, based on the Yeo et al. (2011) atlas [58]. The bar plot shows the mean peri-ripple HFB response in DMN electrodes, averaged over a time window of −250 to 250 ms relative to ripple peak. Peri-ripple activity in DMN sites was significantly stronger compared to the other networks (*$p < 0.05$, **$p < 0.01$, rank-sum test, FDR adjusted). Reproduced from Ref. [55] with permission from Cell Press

21.8 Boundary Setting

As we have discussed above, despite the internally driven and spontaneous nature of free behaviors, individuals can rapidly, accurately and flexibly, define the boundary within which the free behavior is allowed to operate. Considering the example of free recall—we can easily decide that we want to recall visual images belonging to the category of famous faces, without specifying a particular individual face. And then rapidly, and effortlessly, switch to recalling images belonging only to the category of famous places. It is rare that when we attempt to recall an image from one category an image belonging to the wrong category will erroneously intrude. How does the human brain implement such well-defined and flexible boundary setting?

This question was addressed by an iEEG study [21] employing the experimental paradigm depicted in Fig. 21.2—but this time, the goal was to study the neuronal process associated with the instructions to restrict the free recall to a particular category and recall images only from one category at time (famous faces or places), in two separate sessions. Thus, in one session (recalling faces) the patients were asked to recall (and verbally report) only the faces they had seen in the earlier viewing session (which included both categories of images), whereas in the other recall session (recalling places) the patients were asked to recall only the famous places. As expected, patients' performance was high, and only a few erroneous recalls of items belonging to the wrong category were detected.

The study focused on high order visual recording sites that show category-selectivity to either faces or places. Interestingly, examining the HFB amplitude in these category-selective sites during the free recall condition, revealed a general increase in the baseline activity levels whenever participants were targeting the preferred category of the electrodes, i.e., faces in face-selective sites (red trace in Fig. 21.8a) and places in place-selective sites (green trace in Fig. 21.8b). When the opposite, non-preferred category was targeted, the same electrodes exhibited a general reduction of baseline activity (green trace and red trace in Fig. 21.8a, b, respectively). The effect was significant, and evident across the entire set of relevant electrodes (see Fig. 21.8c).

A careful analysis revealed that the baseline shift was evident not only during the moments of the actual reported recollection, but also during the memory search occurring in between recall events, and throughout the free recall period. Furthermore, power spectrum analysis revealed that in addition to the steady baseline shift—there

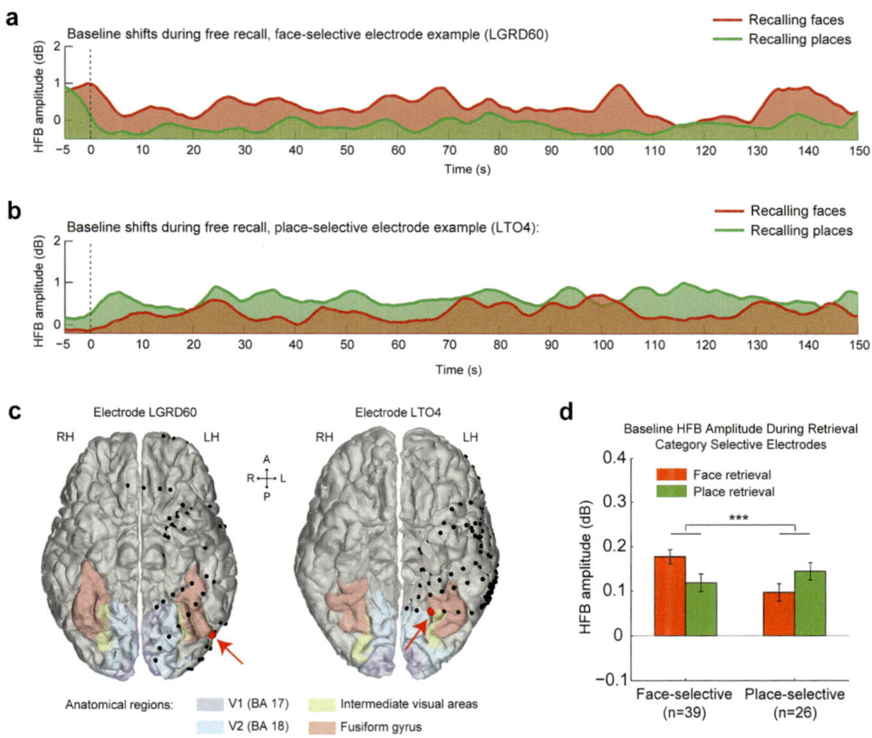

Fig. 21.8 Category-selective baseline shift during goal-directed free recall. **a**, **b** An example of a face-selective electrode (LGRD60) and a place-selective electrode (LTO4) showing a sustained baseline increase during free recall of pictures from their preferred category. HFB time-courses were normalized relative to rest and smoothed using a triangular window of 5 s for visualization purposes. **c** Anatomical locations of the two electrodes (marked in red) in relation to primary visual cortex (blue), intermediate visual areas (yellow), and the fusiform gyrus (pink). **d** Group-level results showing that the median HFB amplitude in category selective electrodes was increased in a content-selective manner throughout the free recall period (two-way mixed-effects ANOVA, interaction: $F(1,63) = 17.69$, $^{***}p < 0.0001$; data from 12 patients). Reproduced from Ref. [21]

was a significant enhancement in the amplitude of the spontaneous activity fluctuations during the targeted recall. This amplitude increase was specific to ultra-slow fluctuations (<0.25 Hz) and to the relevant high order contacts.

A plausible consequence of the baseline shift and enhancement of ongoing fluctuations is an increased probability of crossing the threshold for evoking the required memory items in downstream MTL circuits. Thanks to the baseline shift, the slow spontaneous fluctuations generated in neuronal populations representing the targeted category become more prominent relative to fluctuations generated by other populations; as a result, the cortical input that ultimately arrives at the hippocampal-entorhinal circuitry, at the top of the visual hierarchy, is inherently biased towards the required category, e.g., faces, and thus facilitate the internal activation of, e.g., face-related memories over other non-relevant items. Yet, the exact memory item will only be determined when downstream MTL representations of more specific content (e.g., Bart Simpson) are eventually activated. This ultimate memory-trace activation may take place in multiple hippocampal circuits operating in parallel at various scales—from individual concept cells [23, 59] to massive cell-assemblies orchestrated by ripples [22, 43, 54, 55, 60], to hippocampal-entorhinal pattern completion loops [59, 61]—whose mode of collective operation will require clarification in future studies.

The baseline shift mechanism constitutes an optimal candidate for boundary setting—it is volitionally controlled, rapidly and selectively activated, and thus provides a straightforward mechanism that selectively enables the emergence of free memories, thoughts and decisions within a volitionally specified boundaries.

21.9 In Summary

We would like to propose a common, universal, brain mechanism involved in the generation of free human behavior. This mechanism, consisting of ultra-slow spontaneous activity fluctuations (also termed "resting state" activity) acts by exploring and eventually moving latent neuronal representations above the cognitive thresholds of awareness and decision-making. Intracranial EEG, due to its multi-site recordings and superb spatio-temporal resolution is optimally suited to provide direct and informative characterization of this mechanism. We review here findings that demonstrate the power-law dynamics of these spontaneous fluctuations and point to the way in which they contribute to the emergence of free thoughts and spontaneous actions. The iEEG recordings highlight a central signature of these fluctuations: a slow activity buildup anticipating the onset of spontaneous mental events. This anticipatory buildup is reflected in single neuron activity, local field potentials and mass hippocampal bursts (ripples). Finally, iEEG reveals a boundary-setting mechanism, that by shifting the baseline and amplitude of the ultra-slow fluctuations enables rapid, flexible, and precise volitional control of the arena in which their free behavior operates.

References

1. James W (1890) The principles of psychology, vol 2. Classics in the history of psychology. An internet resource developed by Christopher D. Green York University, Toronto, Ontario. ISSN 1492-3173
2. Libet B, Gleason CA, Wright EW, Pearl DK (1983) Time of conscious intention to act in relation to onset of cerebral activity (readiness-potential): the unconscious initiation of a freely voluntary act. Brain 106:623–642. https://doi.org/10.1093/brain/106.3.623
3. Finn ES, Shen X, Scheinost D et al (2015) Functional connectome fingerprinting: identifying individuals using patterns of brain connectivity. Nat Neurosci 18:1664–1671. https://doi.org/10.1038/nn.4135
4. Fox MD, Raichle ME (2007) Spontaneous fluctuations in brain activity observed with functional magnetic resonance imaging. Nat Rev Neurosci 8:700–711. https://doi.org/10.1038/nrn2201
5. Hahamy A, Macdonald SN, van den Heiligenberg F et al (2017) Representation of multiple body parts in the missing-hand territory of congenital one-handers. Curr Biol 27:1350–1355. https://doi.org/10.1016/j.cub.2017.03.053
6. Harmelech T, Malach R (2013) Neurocognitive biases and the patterns of spontaneous correlations in the human cortex. Trends Cogn Sci 17:606–615. https://doi.org/10.1016/j.tics.2013.09.014
7. Seeley WW, Menon V, Schatzberg AF et al (2007) Dissociable intrinsic connectivity networks for salience processing and executive control. J Neurosci 27:2349–2356. https://doi.org/10.1523/JNEUROSCI.5587-06.2007
8. Wilf M, Strappini F, Golan T et al (2017) Spontaneously emerging patterns in human visual cortex reflect responses to naturalistic sensory stimuli. Cereb Cortex 27:750–763. https://doi.org/10.1093/cercor/bhv275
9. Fried I, Mukamel R, Kreiman G (2011) Internally generated preactivation of single neurons in human medial frontal cortex predicts volition. Neuron 69:548–562. https://doi.org/10.1016/j.neuron.2010.11.045
10. Ramot M, Grossman S, Friedman D, Malach R (2016) Covert neurofeedback without awareness shapes cortical network spontaneous connectivity. Proc Natl Acad Sci USA 113:E2413–E2420. https://doi.org/10.1073/pnas.1516857113
11. Schurger A, Sitt JD, Dehaene S (2012) An accumulator model for spontaneous neural activity prior to self-initiated movement. Proc Natl Acad Sci USA 109:E2904–E2913. https://doi.org/10.1073/pnas.1210467109
12. Biswal B, Yetkin FZ, Haughton VM, Hyde JS (1995) Functional connectivity in the motor cortex of resting human brain using echo-planar MRI. Magn Reson Med 34:537–541. https://doi.org/10.1002/mrm.1910340409
13. Nir Y, Hasson U, Levy I et al (2006) Widespread functional connectivity and fMRI fluctuations in human visual cortex in the absence of visual stimulation. Neuroimage 30:1313–1324. https://doi.org/10.1016/j.neuroimage.2005.11.018
14. Takata N, Sugiura Y, Yoshida K et al (2018) Optogenetic astrocyte activation evokes BOLD fMRI response with oxygen consumption without neuronal activity modulation. Glia 66:2013–2023. https://doi.org/10.1002/glia.23454
15. Privman E, Nir Y, Kramer U et al (2007) Enhanced category tuning revealed by intracranial electroencephalograms in high-order human visual areas. J Neurosci 27:6234–6242. https://doi.org/10.1523/JNEUROSCI.4627-06.2007
16. Nir Y, Mukamel R, Dinstein I et al (2008) Interhemispheric correlations of slow spontaneous neuronal fluctuations revealed in human sensory cortex. Nat Neurosci 11:1100–1108. https://doi.org/10.1038/nn.2177
17. Mukamel R, Gelbard H, Arieli A et al (2005) Coupling between neuronal firing, field potentials, and fMRI in human auditory cortex. Science (80-) 309:951–954. https://doi.org/10.1126/science.1110913

18. Nir Y, Fisch L, Mukamel R et al (2007) Coupling between neuronal firing rate, gamma LFP, and BOLD fMRI is related to interneuronal correlations. Curr Biol 17:1275–1285. https://doi.org/10.1016/j.cub.2007.06.066
19. Watson BO, Ding M, Buzsáki G (2018) Temporal coupling of field potentials and action potentials in the neocortex. Eur J Neurosci 48:1–16. https://doi.org/10.1111/ejn.13807
20. Khuvis S, Yeagle EM, Norman Y et al (2021) Face-selective units in human ventral temporal cortex reactivate during free recall. J Neurosci 41:JN-RM-2918-19. https://doi.org/10.1523/jneurosci.2918-19.2020
21. Norman Y, Yeagle EM, Harel M et al (2017) Neuronal baseline shifts underlying boundary setting during free recall. Nat Commun 8. https://doi.org/10.1038/s41467-017-01184-1
22. Norman Y, Yeagle EMEM, Khuvis S et al (2019) Hippocampal sharp-wave ripples linked to visual episodic recollection in humans. Science (80-) 365:eaax1030. https://doi.org/10.1126/science.aax1030
23. Gelbard-Sagiv H, Mukamel R, Harel M et al (2008) Internally generated reactivation of single neurons in human hippocampus during free recall. Science (80-) 322:96–101. https://doi.org/10.1126/science.1164685
24. Kamiński J, Sullivan S, Chung JM et al (2017) Persistently active neurons in human medial frontal and medial temporal lobe support working memory. Nat Neurosci 20:590–601
25. Quiroga RQ (2012) Concept cells: the building blocks of declarative memory functions. Nat Rev Neurosci 13:587–597. https://doi.org/10.1038/nrn3251
26. Quiroga RQ, Reddy L, Kreiman G et al (2005) Invariant visual representation by single neurons in the human brain. Nature 435:1102–1107. https://doi.org/10.1038/nature03687
27. Polyn SM, Natu VS, Cohen JD, Norman KA (2005) Category-specific cortical activity precedes retrieval during memory search. Science 310:1963–1966. https://doi.org/10.1126/science.1117645
28. American Psychological Association (2017) Clinical practice guideline for the treatment of posttraumatic stress disorder (PTSD). Washington, DC APA, Guidel Dev Panel Treat Posttraumatic Stress Disord Adults 139. https://doi.org/10.1162/jocn
29. Gelbard-sagiv H, Mudrik L, Hill MR et al (2018) Human single neuron activity precedes emergence of conscious perception. Nat Commun. https://doi.org/10.1038/s41467-018-03749-0
30. Soon CS, Brass M, Heinze HJ, Haynes JD (2008) Unconscious determinants of free decisions in the human brain. Nat Neurosci 11:543–545. https://doi.org/10.1038/nn.2112
31. Soon CS, He AH, Bode S, Haynes JD (2013) Predicting free choices for abstract intentions. Proc Natl Acad Sci USA 110:6217–6222. https://doi.org/10.1073/pnas.1212218110
32. Broday-Dvir R, Malach R (2021) Resting-state fluctuations underlie free and creative verbal behaviors in the human brain. Cereb Cortex 31:213–232. https://doi.org/10.1093/cercor/bhaa221
33. Buzsáki G (2015) Hippocampal sharp wave-ripple: a cognitive biomarker for episodic memory and planning. Hippocampus 25:1073–1188. https://doi.org/10.1002/hipo.22488
34. Jadhav SP, Kemere C, German PW, Frank LM (2012) Awake hippocampal sharp-wave ripples support spatial memory. Science (80-) 336:1454–1458. https://doi.org/10.1126/science.1217230
35. Todorova R, Zugaro M (2020) Hippocampal ripples as a mode of communication with cortical and subcortical areas. Hippocampus 30:39–49. https://doi.org/10.1002/hipo.22997
36. Chrobak JJ, Buzstiki G, Buzsáki G, Buzstiki G (1996) High-frequency oscillations in the output networks of the hippocampal-entorhinal axis of the freely behaving rat. J Neurosci 16:3056–3066. https://doi.org/10.1523/jneurosci.0640-09.2009
37. Csicsvari J, Hirase H, Mamiya A, Buzsáki G (2000) Ensemble patterns of hippocampal CA3-CA1 neurons during sharp wave-associated population events. Neuron 28:585–594. https://doi.org/10.1016/S0896-6273(00)00135-5
38. Mizuseki K, Buzsáki G (2013) Preconfigured, skewed distribution of firing rates in the hippocampus and entorhinal cortex. Cell Rep 4:1010–1021. https://doi.org/10.1016/j.celrep.2013.07.039

39. Tong APS, Vaz AP, Wittig JH Jr et al (2021) Ripples reflect a spectrum of synchronous spiking activity in human anterior temporal lobe. Elife 10:e68401
40. Le Van QM, Bragin A, Staba R et al (2008) Cell type-specific firing during ripple oscillations in the hippocampal formation of humans. J Neurosci 28:6104–6110. https://doi.org/10.1523/jneurosci.0437-08.2008
41. Axmacher N, Elger CE, Fell J (2008) Ripples in the medial temporal lobe are relevant for human memory consolidation. Brain 131:1806–1817. https://doi.org/10.1093/brain/awn103
42. Helfrich RF, Lendner JD, Mander BA et al (2019) Bidirectional prefrontal-hippocampal dynamics organize information transfer during sleep in humans. Nat Commun 10:1–16. https://doi.org/10.1038/s41467-019-11444-x
43. Henin S, Shankar A, Borges H et al (2021) Spatiotemporal dynamics between interictal epileptiform discharges and ripples during associative memory processing. Brain. https://doi.org/10.1093/brain/awab044
44. Sakon JJ, Kahana MJ (2021) Hippocampal ripples signal contextually-mediated episodic recall. PNAS 1–31. https://doi.org/10.1073/pnas.2201657119
45. Staresina BP, Bergmann TO, Bonnefond M et al (2015) Hierarchical nesting of slow oscillations, spindles and ripples in the human hippocampus during sleep. Nat Neurosci 18:1679–1686. https://doi.org/10.1038/nn.4119
46. Vaz AP, Inati SK, Brunel N, Zaghloul KA (2019) Coupled ripple oscillations between the medial temporal lobe and neocortex retrieve human memory. Science (80-) 363:975–978. https://doi.org/10.1126/science.aau8956
47. Zhang H, Fell J, Axmacher N (2018) Electrophysiological mechanisms of human memory consolidation. Nat Commun 9. https://doi.org/10.1038/s41467-018-06553-y
48. Felleman DJ, Van Essen DC (1991) Distributed hierarchical processing in the primate cerebral cortex. Cereb Cortex 1:1–47. https://doi.org/10.1093/cercor/1.1.1
49. Rothschild G, Eban E, Frank LM (2017) A cortical-hippocampal-cortical loop of information processing during memory consolidation. Nat Neurosci 20:251–259. https://doi.org/10.1038/nn.4457
50. Skelin I, Kilianski S, McNaughton BL (2018) Hippocampal coupling with cortical and subcortical structures in the context of memory consolidation. Neurobiol Learn Mem 160:0–1. https://doi.org/10.1016/j.nlm.2018.04.004
51. Davidesco I, Zion-Golumbic E, Bickel S et al (2014) Exemplar selectivity reflects perceptual similarities in the human fusiform cortex. Cereb Cortex 24:1879–1893. https://doi.org/10.1093/cercor/bht038
52. Fisch L, Privman E, Ramot M et al (2009) Neural "ignition": enhanced activation linked to perceptual awareness in human ventral stream visual cortex. Neuron 64:562–574. https://doi.org/10.1016/j.neuron.2009.11.001
53. Lachaux JP, George N, Tallon-Baudry C et al (2005) The many faces of the gamma band response to complex visual stimuli. Neuroimage 25:491–501. https://doi.org/10.1016/j.neuroimage.2004.11.052
54. Vaz AP, Wittig JH, Inati SK, Zaghloul KA (2020) Replay of cortical spiking sequences during human memory retrieval. Science (80-) 367:1131–1134. https://doi.org/10.1126/science.aaz3691
55. Norman Y, Raccah O, Liu S et al (2021) Hippocampal ripples and their coordinated dialogue with the default mode network during recent and remote recollection. Neuron 109:1–14. https://doi.org/10.1016/j.neuron.2021.06.020
56. Fox KCR, Foster BL, Kucyi A et al (2018) Intracranial electrophysiology of the human default network. Trends Cogn Sci 22:307–324. https://doi.org/10.1016/j.tics.2018.02.002
57. Abadchi JK, Nazari-Ahangarkolaee M, Gattas S et al (2020) Spatiotemporal patterns of neocortical activity around hippocampal sharp-wave ripples. Elife 9:1–26. https://doi.org/10.7554/elife.51972
58. Thomas Yeo BT, Krienen FM, Sepulcre J et al (2011) The organization of the human cerebral cortex estimated by intrinsic functional connectivity. J Neurophysiol 106:1125–1165. https://doi.org/10.1152/jn.00338.2011

59. Staresina BP, Reber TP, Niediek J et al (2019) Recollection in the human hippocampal-entorhinal cell circuitry. Nat Commun 10:1–11. https://doi.org/10.1038/s41467-019-09558-3
60. Joo HR, Frank LM (2018) The hippocampal sharp wave–ripple in memory retrieval for immediate use and consolidation. Nat Rev Neurosci 19:744–757. https://doi.org/10.1038/s41583-018-0077-1
61. Staresina BP, Michelmann S, Bonnefond M et al (2016) Hippocampal pattern completion is linked to gamma power increases and alpha power decreases during recollection. Elife 5:1–18. https://doi.org/10.7554/elife.17397

Chapter 22
How Can We Differentiate Narrow-Band Oscillations from Aperiodic Activity?

Thomas Donoghue and Andrew J. Watrous

Abstract Human intracranial recordings are composed of both periodic "narrow-band oscillations" along with aperiodic, "1/f-like", activity. While oscillations have been a consistent focus of investigation in neurophysiological data, the physiological and functional properties of aperiodic activity are less well understood, as it is often treated as "background noise". In this chapter, we provide an overview of both periodic and aperiodic activity, providing a brief historical perspective on the study of each, along with conceptual approaches and analytic assumptions researchers have made when measuring both types of activity. We then highlight recent methodological developments and available techniques for explicitly measuring (a)periodic activity, so as to evaluate which components of the neural data are changing. We propose that studies of human intracranial recordings should employ measures which can differentiate (a)periodic signals in order to determine if they have distinct generators and functional roles. Finally, we discuss putative interpretations of both periodic and aperiodic activity, as well as several unresolved issues which can be explored in future work to further model and interpret these signals.

22.1 Introduction

22.1.1 Narrowband Oscillations/Periodic Activity

Rhythmic fluctuations in electrical neural activity, or neural oscillations, have been observed since the initial use of human intracranial recordings [1, 2]. Neural oscillations are readily observable in neural field data such as the local field potential (LFP) using both macro- and micro-electrode recordings (Fig. 22.1a). Physiologically, oscillations mainly consist of synaptic activity [3] and are generated by the

T. Donoghue
Columbia University, New York, NY, USA

A. J. Watrous (✉)
Baylor College of Medicine, Houston, TX, USA
e-mail: andrew.watrous@bcm.edu

N. Axmacher (ed.), *Intracranial EEG*, Studies in Neuroscience, Psychology and Behavioral Economics, https://doi.org/10.1007/978-3-031-20910-9_22

summation of excitation and inhibition within and between large groups of neurons [4]. Neural oscillations are thought to reflect the functional organization of neural activity, possibly by aiding in information flow within the brain by flexibly aligning and misaligning periods of excitability between brain regions [5, 6]. As this chapter and others make clear, neural oscillations are a prevailing feature of human brain activity that play a key role in neural functioning [7–11].

Neural oscillations reflect "narrowband" activity with frequency-specific power which appear as peaks in the power spectrum [12, 13]. Narrowband oscillatory activity can be characterized by several features of interest, including amplitude, phase, and center frequency, and are typically assumed to exist in canonical frequency bands (e.g. delta, theta, alpha) (Fig. 22.1b). A notable aspect of neural oscillations is their variability, both within and across individuals [14, 15] and over time (Fig. 22.1b, c). This variability—in their peak frequencies, features, and spatial and temporal organization—motivates a number of methodological considerations that need to be addressed for accurately measuring neural oscillations [16]. As we discuss below, narrowband oscillatory activity coexists with, and can be dissociated from, "broadband" activity which spans the width of the measured frequency spectrum.

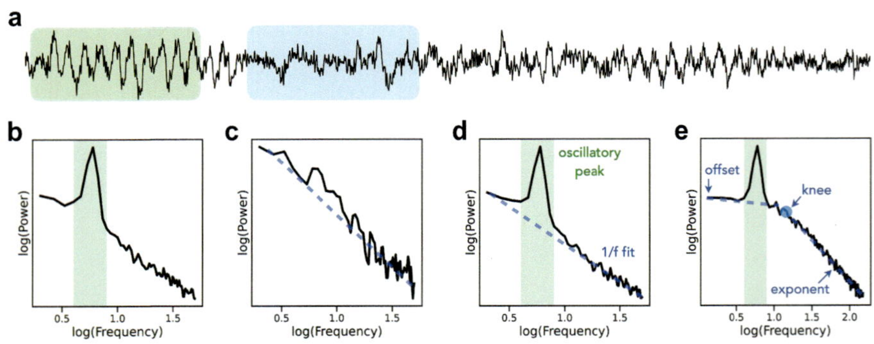

Fig. 22.1 Properties and representations of neural data. **a** A simulated neural time series, with a bursty theta oscillation combined with an aperiodic 1/f component. **b**, **c** Power spectra for different segments of the time series in A, demonstrating variability over time, including **b** for periods when the oscillation is present and **c** for periods when the oscillation is absent. **d** The power spectrum for the full time series, annotated to label signal components, including the oscillatory peak, and the fit of the aperiodic component. **e** The power spectrum for a simulated time series with annotations for the parameters of the aperiodic components. This time series was generated with a model of synaptic activity that gives rise to a 'knee' in the aperiodic component. Note that the frequency range (abscissa) is different between panels b–d and e

22.1.2 Broadband/Aperiodic Activity

In addition to neural oscillations, human intracranial recordings also show prominent aperiodic—meaning irregular, or non-periodic—activity [17]. In frequency representations, this is seen as the 1/f-like structure of neural power spectra. By 1/f, it is meant that there is an approximate power-law relation between power and frequency, reflecting exponentially decreasing power across increasing frequencies. Aperiodic neural activity, which roughly follows a power-law distribution, is sometimes described as being 'scale-free' or 'self-similar' activity [17], and will here be referred to as the 'aperiodic' component of the data. Despite the early observation of 1/f activity in electrophysiological data [18], and some early work mapping it across the cortex in human and in animal models [19–21], relatively little work has explored the properties and interpretations of aperiodic activity, as compared to, for example, the broad literature investigating neural oscillations.

In the simple case, aperiodic neural activity manifests as a linear relationship between frequency and power when plotted in log–log space. This relationship can be quantified, using a simple '1/f' function, for example, $p = a * f^x$ wherein the 'a' parameter reflects the global power of the signal, and will be referred to as the aperiodic 'offset', and the 'x' parameter reflects the steepness of the decay, and will be referred to as the aperiodic 'exponent' (Fig. 22.1d). The aperiodic exponent is analogous to the slope of the line of the log–log power-spectrum, sometimes referred to as the 'spectral slope'. This kind of power-law distributed activity is seen in many other physical systems, and as such is a feature of inquiry across areas of physics and mathematics [22]. Notably, in neural data, we refer to this aperiodic activity as '1/f-like', in that it typically does not exhibit a true 1/f structure across all frequencies because there are often 'bends' or 'knees' (Fig. 22.1e) in the aperiodic component of the data [23].

22.1.3 Overlap of Periodic & Aperiodic Components

Despite the traditional focus on analyzing neural oscillations, electrophysiological recordings are not mainly rhythmic [24] but instead contain a mixture of both oscillatory and non-oscillatory "aperiodic" activity. While most standard measures of neural oscillations assume stable background noise which can be averaged away, this simplifying assumption ignores the dynamic nature of aperiodic activity [25, 26]. Consequently, commonly employed measures can often be biased by aperiodic activity, and conflate changes in narrowband and broadband components. For example, when measuring power within a narrowband frequency range, measured changes may reflect the oscillatory power, but may also be driven by changes in the oscillatory center frequency, aperiodic offset, or aperiodic exponent (Fig. 22.2). Similarly, the available measures of 1/f activity tend to have difficulty accurately measuring aperiodic properties in the presence of concurrent periodic activity. Despite the increasingly recognized presence and variability of aperiodic neural activity, there is currently a lack of consensus for methods, interpretations, and best practices guidelines for investigations of aperiodic activity in neural field data.

Given the overlap of periodic (oscillatory) and aperiodic activity in neural recordings, this chapter focuses on how to measure, dissociate, and interpret changes in both oscillations and aperiodic activity in human intracranial recordings. Regardless of which feature is of interest, the co-occurrence of periodic and aperiodic activity is an issue for measurements that aim to measure one or both components. This can be seen in an example simulated event-related response in intracranial data (Fig. 22.3a), in which low frequency oscillatory power is higher pre-stimulus, after which there is an increase in high-frequency activity [27, 28]. This pattern of responses includes both oscillatory peaks and distributed aperiodic activity that changes with task context (Fig. 22.3b), motivating the need to apply methods that carefully adjudicate between different signal components. To address these issues, we start by considering methods which aim to disambiguate (a)periodic signals in intracranial recordings, and then discuss current findings and open questions.

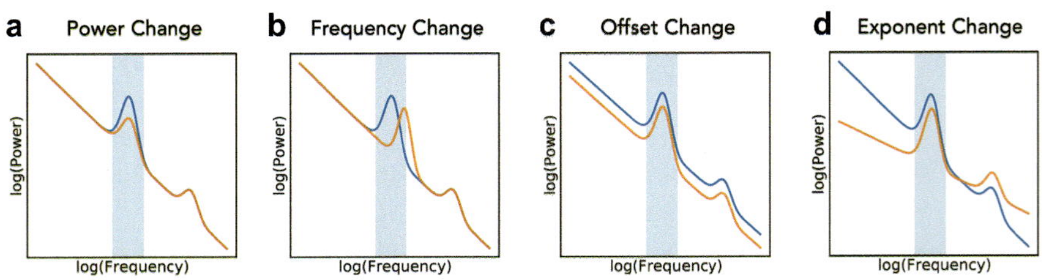

Fig. 22.2 Changes in different signal parameters may result in the same band power differences. In each panel, the power difference in the highlighted alpha range (8–12 Hz) between the two power spectra is exactly the same. This change can arise from differences in (**a**) oscillation power, (**b**) oscillation center frequency, (**c**) aperiodic offset and/or (**d**) aperiodic exponent

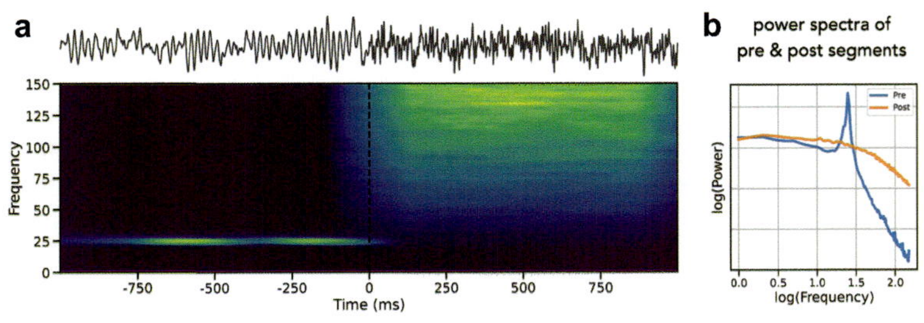

Fig. 22.3 Dynamics of spectral parameters through time. **a** An example time series (top) and corresponding spectrogram (bottom) for simulated data representing a canonical task-related response. In this example, there is a pre-stimulus oscillation (~25 Hz) that fades away post stimulus, at which point there is an increase in broadband high frequency power. **b** Power spectra, averaged across time separately for the pre- and post-stimulus windows, showing an oscillatory peak that disappears post-stimulus, as well as an overall change in the aperiodic component of the data. Note that high-frequency activity is reflected in the aperiodic exponent

22.2 Analysis Methods

22.2.1 Conventional Approaches for Analyzing Neural Time Series

The analysis of neural field data has developed a rich ecosystem of approaches, adopted and adapted from the field of digital signal processing (DSP), to measure components of interest, such as neural oscillations [29, 30]. Such analyses can be done in the time and/or frequency domain, including, for example, computing power spectra to analyze frequency ranges of interest or using time–frequency analyses to examine signal dynamics through time. Each of these analyses makes assumptions, both in the ways they operate on the data, and in how they are typically interpreted. Validating the assumptions of these methods, and interpreting the outputs accordingly, is necessary for getting appropriate results [16, 31].

Notably, the vast majority of approaches to analyze neural time-series employ Fourier-based methods. Mathematically, via the Fourier theorem, any continuous time series can be perfectly represented by a Fourier Series as a combination of sinusoidal waveforms. This mathematical convenience, and associated suite of computational tools, has led to the widespread use of frequency-domain representations and transformations. However, this powerful property that enables many of the methods researchers employ also includes some interpretational peril. This arises in the form of the Fourier fallacy; although it may be possible to analyze neural data in terms of rhythmic sine waves, it is a fallacy to a priori assume that this activity necessarily occurs as periodic phenomena in the data [32]. These issues also relate to the other similar methods, such as wavelets and Hilbert transforms, that are mathematically equivalent to Fourier analysis [33].

In practice, this means that computing a frequency domain representation of neural field data does not by itself clarify if, or to what extent, the underlying data contains rhythmic activity. Common analysis approaches, such as examining power or filtering in a predefined frequency range, can lead to measures that conflate between periodic and aperiodic activity, and/or between different features of each component [16, 26] (Fig. 22.2). Other measures that are supposed to reflect periodic activity, such as computing ratios of power between frequency bands, suffer the same limitations, and can be heavily biased by aperiodic activity, leading to erroneous interpretations [34]. Overall, these considerations indicate that the analysis of human intracranial recordings would benefit from dedicated methods that address the overlapping nature of periodic and aperiodic activity.

22.2.2 Methods for Dissociating Periodic & Aperiodic Activity

Since aperiodic activity manifests as a pattern of power across all frequencies, and periodic power exhibits as overlying peaks of frequency-specific power, initial approaches for measuring periodic and/or aperiodic activity involved fitting lines to estimate the 1/f-like component, and/or detecting peaks in the power spectrum to measure putative oscillatory activity. An estimate of the 1/f activity of the data can be obtained by fitting a line to the power spectrum in log–log space [20]. Similarly, in order to detect oscillations, peaks can be detected and measured as regions with power over and above this line, which is used as an estimate of the "background" spectrum [14, 35, 36]. Though simple line and peak fitting procedures may not generalize to all cases, early investigations successfully used these approaches to estimate both periodic and aperiodic parameters of the data.

Subsequently, variants of line and peak fitting have also been formalized into algorithms such as the 'Better OSCillation detector' (BOSC) [37, 38], the updated 'extended BOSC' (eBOSC) algorithm [39] and the 'Multiple Oscillations Detection ALgorithm' (MODAL) [40]. These methods use explicit measurements of the aperiodic component to detect frequency specific power above the 1/f activity that are then measured in a temporally resolved manner as oscillations. In doing so, these methods are able to capture large amplitude, frequently occurring oscillations with excellent temporal and spectral resolution but ignore aperiodic signal components and can fail to detect sparsely occurring oscillations. A limitation of these methods is that they typically assume that the aperiodic component is invariant/stationary, an assumption which doesn't generally hold [25, 41]. However these methods still offer a time-resolved measure of aperiodic-controlled oscillatory activity.

Previous methods based on fitting components of the power spectrum often used non-robust approaches to fit the 1/f-line, and/or were limited in their capacity to fit oscillatory peaks. To address these limitations, the spectral parameterization approach (formerly known as 'fooof') was developed to provide a more robust and generalizable method for parameterizing neural power spectra [26]. This method generalizes to an arbitrary number of oscillations, determines the number of periodic components in a data-driven manner, and measures the center frequency, aperiodic-adjusted power, and bandwidth for each. It also allows for fitting different forms of the aperiodic component and measures the aperiodic offset, exponent, and optionally, the knee. Overall, by combining elements of model selection with improved parameter fitting, spectral parameterization offers a more flexible and generalizable algorithm than previous approaches for fitting lines and peaks. A limitation is that, due to operating in the frequency domain, the time-resolution is limited to time windows of sufficient length to reliably estimate power spectra.

Another approach for investigating periodic and aperiodic activity is to use decomposition techniques which attempt to separate components of interest in the data. For example, principal component analysis (PCA) can be used to separate broadband components from periodic activity [42, 43]. In doing so, the first spectral component typically reflects broadband activity, and the next two or three high-variance components often reflect periodic activity, for example, reflecting alpha, beta and delta activity in a motor task [42]. A benefit of this spectral decomposition approach is that the obtained PCA weights can be applied to continuous measurements, such as a spectrogram, and thus allow for the analysis of time-resolved changes in the decomposed components. A disadvantage of this and other naive decompositions is that selecting and validating the top number of components is a manual process in which sparsely occurring oscillations with low variance may be discarded.

Neural oscillations display substantial variability over time and across individuals, channels, and behaviors. Aside from the aforementioned work employing new methods, most prior research has measured oscillatory activity in canonical frequency bands, which can lead to misestimations due to misalignments between the data and the analyzed frequencies [16]. Recently, two additional new techniques have been developed which employ empirically-defined frequency bands to maximize analysis sensitivity and to provide improved, time-resolved measures of amplitude, phase, and frequency in adaptively identified frequency bands. First, the 'gedBounds' method applies a data-driven procedure to identify frequency regions reflecting oscillatory activity [44]. This method finds clusters of spatially-correlated components which separate narrowband and broadband activity and has the advantage that it operates on multi-channel data. Second, the 'Oscillatory ReConstruction Algorithm' (ORCA) [15] quantifies different models of the power spectrum (i.e. frequency bands) in terms of explained variance, allowing researchers to test and compare any model(s) they wish, and enables the detection of both narrowband and broadband signal components. ORCA works well in cases in which there is substantial variability across channels, such as in intracranial recordings. A limitation of these approaches is that neither ORCA nor 'gedBounds' provide a direct measure of 1/f activity.

Another available method, designed to estimate 1/f properties of data and explicitly control for periodic components, is the irregular resampling auto-spectral analysis (IRASA) method [45]. The IRASA method leverages the scale-free property of 1/f activity, meaning it has the same structure across different scales, to separate this component of the data from periodic activity which is defined by having a characteristic scale (frequency). To do so, it uses a resampling procedure to extract frequency-specific activity, which allows for isolating the 1/f component of the data. These isolated components can then be quantified, allowing for separate measurements of the oscillatory and non-oscillatory components of the data. While IRASA is a powerful tool when its assumptions are met, a limitation of this method is that the decomposition algorithm assumes and requires true 1/f data to work optimally, and may provide biased measurements in situations where the data is not truly 1/f, such as when there is a 'knee' in the power spectrum [26, 45].

The aforementioned methods reflect many of the currently available approaches for analyzing neural data in the context of considering the combination of periodic and aperiodic activity. While there are also methods from other fields that are specific to measuring aperiodic features of the data, some of which are discussed elsewhere [46], in general these methods may not work as intended in data containing both 1/f activity and periodic components, and so may fail when applied to neural data.

22.2.3 Methodological Considerations

As many of the methods described above are quite new, systematic comparisons between methods are lacking at present but are needed in order to clarify which methods are best suited for different scientific questions. Generally, however, the above methods can be applied broadly across different types of intracranial (or scalp) electrodes. In doing so, researchers should consider the spatial scale of their analysis as well as the temporal resolution when assessing their ability to detect oscillations and/or aperiodic activity. Other factors such as hardware filtering and referencing scheme may impact which frequencies are recorded and would therefore impact oscillatory and aperiodic signal detection. Different research questions and data properties may also impose constraints on the parameters of interest, as well as the temporal and frequency resolutions needed. In sum, there may be differences between different recording methods, task designs, and analysis plans that should be considered when choosing and employing methods to separate periodic and aperiodic activity.

In particular, intracranial recordings offer better spatial resolution and also allow for analyzing the data across broader frequency ranges than is typically possible in extra-cranial recordings. Likely due to both differences in the propagation and analyzed frequency ranges, measures of the aperiodic exponent in iEEG are often steeper (in the range of 2–4 when measured across 10–100 Hz [20, 25]) than measures in M/EEG data (found to be mostly between 0.5–1.5 across the range of 3–30 Hz [26]). Relatedly, when looking at the broader frequency ranges in intracranial data, there can be a knee, whereby there are two different 1/f regimes, each with their own exponent, separated by a 'bend' at a knee frequency [23]. It may therefore be particularly important to check for knees when analyzing intracranial data across broader frequency ranges. The spatial resolution of intracranial data also allows for analyzing aperiodic and periodic components in a more localized fashion. Ultimately, more work is needed to better understand the differences observed between modalities, with intracranial data offering a particularly potent method for investigations to leverage the better temporal, spatial, and frequency resolution in order to further our understanding of periodic and aperiodic activity, and their inter-relations.

22.3 Existing Studies Separating Periodic & Aperiodic Activity

Recent empirical work has demonstrated the utility of applying methods which relate (a)periodic signal components to their behavioral and demographic correlates. For example, aperiodic activity has been shown to systematically relate to age [47, 48], a finding which recontextualizes prior reports that aging reflects a change in power within oscillation bands, by showing that the employed measures in such investigations can be biased by aperiodic activity [34, 49]. Similarly, aperiodic activity has been shown to vary within subjects across states including sleep [50] and anesthesia [51]. These studies suggest aperiodic activity may reflect between- and within-subject variation in brain state, and may therefore be useful for investigating global brain dynamics [17, 52].

More than simply being a marker of state, other studies have also demonstrated the extent to which aperiodic activity is highly dynamic within task contexts at rapid time scales, which may reflect changes in arrhythmic firing and/or bursts of excitation [41, 42, 53] For example, task related variation in aperiodic activity has been shown to relate to working memory performance [26] and cognitive processing speed [54]. Using spectral decomposition, it has been shown that in an auditory oddball task, the broadband component of the data best reflects the prediction error related response [55]. Using spectral parameterization, it has also been shown that trial-by-trial variations of aperiodic activity relate to both stimulus properties and attentional state [56]. This emerging evidence suggests that aperiodic activity may be an informative feature for investigating trial-by-trial neural activity and its relation to behavior.

Separating periodic and aperiodic components has also been important for clarifying putative functional roles of oscillatory neural activity. In aging, controlling for changes in aperiodic activity has allowed for a detailed analysis of changes in theta, alpha, and beta oscillations [47, 57]. In resting state data, controlling for aperiodic activity has enabled quantifications of oscillation occurrence [26]. In task contexts, explicitly measuring aperiodic activity has been used to clarify how theta oscillations relate to memory [58], how beta oscillations relate to motor activity [28], and how oscillatory gamma relates to visual processing [35, 59]. In cases where the focus of the investigation is primarily on oscillatory activity, these studies demonstrate that explicitly characterizing aperiodic activity can further refine our understanding of the relationship between neural oscillations and behavior.

In the context of human intracranial recordings, special consideration is due to higher frequency activity, typically referred to as the "gamma" range. This range includes gamma oscillations (~30–80 Hz), considered to reflect narrowband oscillations as well as a broader high-frequency range, sometimes called "high-gamma" or simply high-frequency activity (roughly 70–200 Hz). This higher range is generally considered to be broadband, in terms of not exhibiting any frequency specific peaks [60]. Conceptualizing, measuring, and interpreting these various 'gamma' ranges has long been a topic of debate, including differentiating between narrow- and broad-band responses [61]. Gamma oscillations, suggested to be a potential narrowband signal reflecting inter-areal communication [5], and high-frequency activity, suggested to be a general marker of neural activation likely reflecting spiking activity [27], have long been features of interest in neural data, but appropriately differentiating between the two requires careful analysis that appropriately differentiates the periodic vs. aperiodic aspects of the signal.

Empirically, animal work has demonstrated that in monkey visual cortex, narrowband, oscillatory gamma (30–80 Hz) is a distinct phenomenon that differs from broadband gamma (above 80 Hz), with each having a different relationship to spiking activity [62]. In visual tasks, it has been shown that the dominant neural response is a change in broadband activity [25, 35, 59], an effect that has also been demonstrated in motor tasks [28, 42]. Subsequent work has continued to decouple the narrowband and broadband responses in visual cortex, proposing a model for each signal, whereby the broadband activity is again found to be the dominant response, with some stimuli also inducing narrowband activity [63]. Notably, each of these investigations dissociating narrowband gamma oscillations from broadband responses employed one of the aforementioned methods for separating periodic and aperiodic activity.

22.4 Discussion

22.4.1 Interpretations of Periodic & Aperiodic Components

Thus far we have focused on methodological considerations that motivate the need to use dedicated methods to measure periodic and aperiodic components in neural data. We now turn to interpreting these components. As the physiological and theoretical underpinnings of periodic activity have been more commonly examined in prior work [3, 4], here we will briefly discuss aperiodic activity. Some investigations consider aperiodic activity in terms of the variability, and/or level of 'neural noise' in the system [48, 64]. Other functional interpretations of aperiodic neural activity focus on its scale-free properties [17]. This interpretation comes from theoretical work arguing that 1/f dynamics arise in systems that naturally evolve a self-organized critical point [22]. Based on this, criticality has been proposed and developed as a potential framework for conceptualizing neural data [65], reflected in the analysis of long-term dependencies in time series and/or critical states in dynamical systems [66]. Another distinct, though not necessarily conflicting, approach is to focus on the potential physiological underpinnings and generative models of field data that may explain the observation of ubiquitous aperiodic activity [23, 41, 42, 67]. For example, the overall power or offset of broadband activity has been related to the amount of local spiking activity [42, 67], the aperiodic exponent has been shown to relate to the balance of excitation and inhibition of underlying synaptic activity [41], and the presence and location of 'knees', which relate to the auto-correlation of the signal, have been suggested to relate to the characteristic timescale of the underlying activity [23]. Other proposals suggest that aperiodic activity could be generated by a collection of damped oscillators [68]. Overall, these functional and physiological models provide a framework for conceptualizing aperiodic neural activity, with further work needed to evaluate the most productive approaches.

22.4.2 Future Work

A key focus for future work will be to continue to expand the methodological toolkit. While there currently exists a collection of tools that have recently been developed with the goal of separating and measuring periodic and aperiodic activity in neural time series, the relationships between these different tools is currently underexplored. Methodological work including formal evaluations and comparisons between methods should be pursued, including developing best practice guides recommending which approaches to use in different scenarios. Another weakness of several available methods operating in the frequency domain is limited temporal resolution. Continued work on methods development should focus on improving the temporal resolution of these methods, in order to take full advantage of the high resolution of electrophysiological data. One way to pursue these methodological goals is by using simulated data and quantitatively comparing algorithm performance with known ground truth, as can be done with open-source tools that include simulations and algorithm implementations [69].

Beyond methodological pursuits, another focus for future work should be to refine our conceptual underpinnings and theoretical understanding of the signals under study. A particular focus should be on the aperiodic component, which has been the focus of much less work than periodic activity, but which is clearly a prominent and important facet of neural data. Another avenue for inquiry should determine how different signal components relate to each other. For example, in the context of sleep, it has been proposed that alternations between periodic and aperiodic states reflect dynamic switching between cross-areal communication (periodic activity) and local-processing (aperiodic activity) [52]. Developing and testing such theories, including how (a)periodic activity relates to physiological events at different spatial scales, such as evoked potentials and neuronal spiking, should be a focus of future theoretical and experimental work.

22.4.3 Conclusion

Electrophysiological data from intracranial recordings are complex signals that contain multiple components, including narrowband periodic and broadband, aperiodic activity. Here we have explored the evidence suggesting that periodic and aperiodic activity are ubiquitous and overlapping features of neural data, discussed the limitations of standard methods in the face of these considerations, and examined several existing methods for measuring and disentangling different signal features. Overall, when investigating periodic and/or aperiodic components in neuroelectrophysiological data, dedicated methods that account for the complexity in the data must be used.

References

1. Foerster O, Altenburger H (1935) Elektrobiologische Vorgänge an der menschlichen Hirnrinde. Deutsche Zeitschrift f Nervenheilkunde 135:277–288
2. Hanjani K, Fatehi M, Schmidt N, Aghakhani Y, Redekop GJ (2021) A history of diagnostic investigations in epilepsy surgery. Can J Neurol Sci 48:845–851
3. Buzsáki G, Anastassiou CA, Koch C (2012) The origin of extracellular fields and currents - EEG, ECoG, LFP and spikes. Nat Rev Neurosci 13:407–420
4. Wang X-J (2010) Neurophysiological and computational principles of cortical rhythms in cognition. Physiol Rev 90:1195–1268
5. Fries P (2005) A mechanism for cognitive dynamics: neuronal communication through neuronal coherence. Trends Cogn Sci 9:474–480
6. Varela FJ, Lachaux J-P, Rodriguez E, Martinerie J (2001) The brainweb: phase synchronization and large-scale integration. Nat Rev Neurosci 2:229–239
7. Buzsáki G, Draguhn A (2004) Neural oscillations in cortical networks. Science 304:1926–1929
8. Fell J, Axmacher N (2011) The role of phase synchronization in memory processes. Nat Rev Neurosci 12:105–118
9. Nyhus E, Curran T (2010) Functional role of gamma and theta oscillations in episodic memory. Neurosci Biobehav Rev 34:1023–1035
10. Voytek B, Knight RT (2015) Dynamic network communication as a unifying neural basis for cognition, development, aging, and disease. Biol Psychiat 77:1089–1097
11. Watrous AJ, Fell J, Ekstrom AD, Axmacher N (2015) More than spikes: common oscillatory mechanisms for content specific neural representations during perception and memory. Curr Opin Neurobiol 31:33–39
12. Buzsáki G, Logothetis N, Singer W (2013) Scaling brain size, keeping timing: evolutionary preservation of brain rhythms. Neuron 80:751–764
13. Lopes da Silva F (2013) EEG and MEG: relevance to neuroscience. Neuron 80:1112–1128
14. Haegens S, Cousijn H, Wallis G, Harrison PJ, Nobre AC (2014) Inter- and intra-individual variability in alpha peak frequency. Neuroimage 92:46–55
15. Watrous AJ, Buchanan RJ (2020) The oscillatory reconstruction algorithm adaptively identifies frequency bands to improve spectral decomposition in human and rodent neural recordings. J Neurophysiol 124:1914–1922
16. Donoghue T, Schaworonkow N, Voytek B (2021) Methodological considerations for studying neural oscillations. Eur J Neurosci ejn.15361
17. He BJ (2014) Scale-free brain activity: past, present, and future. Trends Cogn Sci 18:480–487
18. Motokawa K (1949) Energy of brain waves and energetics of the brain. Tohoku J Exp Med 51:119–129
19. Freeman WJ, Holmes MD, Burke BC, Vanhatalo S (2003) Spatial spectra of scalp EEG and EMG from awake humans. Clin Neurophysiol 114:1053–1068
20. Freeman WJ, Zhai J (2009) Simulated power spectral density (PSD) of background electrocorticogram (ECoG). Cogn Neurodyn 3:97–103
21. Pritchard WS (1992) The brain in fractal time: 1/F-like power spectrum scaling of the human electroencephalogram. Int J Neurosci 66:119–129
22. Bak P, Tang C, Wiesenfeld K (1987) Self-organized criticality: an explanation of the 1/f noise. Phys Rev Lett 59:381–384
23. Gao R, van den Brink RL, Pfeffer T, Voytek B (2020) Neuronal timescales are functionally dynamic and shaped by cortical microarchitecture. eLife 9:e61277
24. Bullock TH, Mcclune MC, Enright JT (2003) Are the electroencephalograms mainly rhythmic?Assessment of periodicity in wide-band time series. Neuroscience 121:233–252
25. Podvalny E, Noy N, Harel M, Bickel S, Chechik G, Schroeder CE, Mehta AD, Tsodyks M, Malach R (2015) A unifying principle underlying the extracellular field potential spectral responses in the human cortex. J Neurophysiol 114:505–519
26. Donoghue T, Haller M, Peterson EJ et al (2020) Parameterizing neural power spectra into periodic and aperiodic components. Nat Neurosci 23:1655–1665

27. Lachaux J-P, Axmacher N, Mormann F, Halgren E, Crone NE (2012) High-frequency neural activity and human cognition: past, present and possible future of intracranial EEG research. Prog Neurobiol 98:279–301
28. Miller KJ, Hermes D, Honey CJ, Hebb AO, Ramsey NF, Knight RT, Ojemann JG, Fetz EE (2012) Human motor cortical activity is selectively phase-entrained on underlying rhythms. PLoS Comput Biol 8:e1002655
29. Gross J (2014) Analytical methods and experimental approaches for electrophysiological studies of brain oscillations. J Neurosci Methods 228:57–66
30. Wacker M, Witte H (2013) Time-frequency techniques in biomedical signal analysis: a tutorial review of similarities and differences. Methods Inf Med 52:279–296
31. de Cheveigné A, Nelken I (2019) Filters: when, why, and how (not) to use them. Neuron 102:280–293
32. Jasper HH (1948) Charting the sea of brain waves. Science 108:343–347
33. Bruns A (2004) Fourier-, Hilbert- and wavelet-based signal analysis: are they really different approaches? J Neurosci Methods 137:321–332
34. Donoghue T, Dominguez J, Voytek B (2020) Electrophysiological frequency band ratio measures conflate periodic and aperiodic neural activity. eNeuro 7:ENEURO.0192–20.2020
35. Hermes D, Miller KJ, Wandell BA, Winawer J (2015) Stimulus dependence of gamma oscillations in human visual cortex. Cereb Cortex 25:2951–2959
36. Pascual-Marqui RD, Valdes-Sosa PA, Alvarez-Amador A (1988) A parametric model for multichannel EEG spectra. Int J Neurosci 40:89–99
37. Caplan JB, Madsen JR, Raghavachari S, Kahana MJ (2001) Distinct patterns of brain oscillations underlie two basic parameters of human maze learning. J Neurophysiol 86:368–380
38. Hughes AM, Whitten TA, Caplan JB, Dickson CT (2012) BOSC: a better oscillation detection method, extracts both sustained and transient rhythms from rat hippocampal recordings. Hippocampus 22:1417–1428
39. Kosciessa JQ, Grandy TH, Garrett DD, Werkle-Bergner M (2020) Single-trial characterization of neural rhythms: potential and challenges. Neuroimage 206:116331
40. Watrous AJ, Miller J, Qasim SE, Fried I, Jacobs J (2018) Phase-tuned neuronal firing encodes human contextual representations for navigational goals. eLife 7:e32554
41. Gao R, Peterson EJ, Voytek B (2017) Inferring synaptic excitation/inhibition balance from field potentials. Neuroimage 158:70–78
42. Miller KJ, Honey CJ, Hermes D, Rao RP, den Nijs M, Ojemann JG (2014) Broadband changes in the cortical surface potential track activation of functionally diverse neuronal populations. Neuroimage 85:711–720
43. Miller KJ, Zanos S, Fetz EE, den Nijs M, Ojemann JG (2009) Decoupling the cortical power spectrum reveals real-time representation of individual finger movements in humans. J Neurosci 29:3132–3137
44. Cohen MX (2021) A data-driven method to identify frequency boundaries in multichannel electrophysiology data. J Neurosci Methods 347:108949
45. Wen H, Liu Z (2016) Separating fractal and oscillatory components in the power spectrum of neurophysiological signal. Brain Topogr 29:13–26
46. Schaefer A, Brach JS, Perera S, Sejdić E (2014) A comparative analysis of spectral exponent estimation techniques for $1/f\beta$ processes with applications to the analysis of stride interval time series. J Neurosci Methods 222:118–130
47. He W, Donoghue T, Sowman PF, Seymour RA, Brock J, Crain S, Voytek B, Hillebrand A (2019) Co-Increasing neuronal noise and beta power in the developing brain. bioRxiv 1–49
48. Voytek B, Kramer MA, Case J, Lepage KQ, Tempesta ZR, Knight RT, Gazzaley A (2015) Age-related changes in 1/f neural electrophysiological noise. J Neurosci 35:13257–13265
49. Ostlund B, Donoghue T, Anaya B, Gunther KE, Karalunas SL, Voytek B, Pérez-Edgar KE (2022) Spectral parameterization for studying neurodevelopment: how and why. Dev Cogn Neurosci 54:101073
50. Lendner JD, Helfrich RF, Mander BA, Romundstad L, Lin JJ, Walker MP, Larsson PG, Knight RT (2020) An electrophysiological marker of arousal level in humans. eLife 9:e55092

51. Colombo MA, Napolitani M, Boly M et al (2019) The spectral exponent of the resting EEG indexes the presence of consciousness during unresponsiveness induced by propofol, xenon, and ketamine. Neuroimage 189:631–644
52. Helfrich RF, Lendner JD, Knight RT (2021) Aperiodic sleep networks promote memory consolidation. Trends Cogn Sci 25:648–659
53. Manning JR, Jacobs J, Fried I, Kahana MJ (2009) Broadband shifts in local field potential power spectra are correlated with single-neuron spiking in humans. J Neurosci 29:13613–13620
54. Ouyang G, Hildebrandt A, Schmitz F, Herrmann CS (2020) Decomposing alpha and 1/f brain activities reveals their differential associations with cognitive processing speed. Neuroimage 205:116304
55. Canales-Johnson A, Teixeira Borges AF, Komatsu M, Fujii N, Fahrenfort JJ, Miller KJ, Noreika V (2021) Broadband dynamics rather than frequency-specific rhythms underlie prediction error in the primate auditory cortex. J Neurosci 41:9374–9391
56. Waschke L, Donoghue T, Fiedler L, Smith S, Garrett DD, Voytek B, Obleser J (2021) Modality-specific tracking of attention and sensory statistics in the human electrophysiological spectral exponent. eLife 10:e70068
57. Cellier D, Riddle J, Petersen I, Hwang K (2021) The development of theta and alpha neural oscillations from ages 3 to 24 years. Dev Cogn Neurosci 50:100969
58. Herweg NA, Solomon EA, Kahana MJ (2020) Theta oscillations in human memory. Trends Cogn Sci 24:208–227
59. Bartoli E, Bosking W, Chen Y, Li Y, Sheth SA, Beauchamp MS, Yoshor D, Foster BL (2019) Functionally distinct gamma range activity revealed by stimulus tuning in human visual cortex. Curr Biol 29:3345-3358.e7
60. Canolty RT, Edwards E, Dalal SS, Soltani M, Nagarajan SS, Kirsch HE, Berger MS, Barbaro NM, Knight RT (2006) High gamma power is phase-locked to theta oscillations in human neocortex. Science 313:1626–1628
61. Crone NE, Korzeniewska A, Franaszczuk PJ (2011) Cortical gamma responses: Searching high and low. Int J Psychophysiol 79:9–15
62. Ray S, Maunsell JHR (2011) Different origins of gamma rhythm and high-gamma activity in macaque visual cortex. PLoS Biol 9:e1000610
63. Hermes D, Petridou N, Kay KN, Winawer J (2019) An image-computable model for the stimulus selectivity of gamma oscillations. eLife 8:e47035
64. Waschke L, Wöstmann M, Obleser J (2017) States and traits of neural irregularity in the age-varying human brain. Sci Rep. https://doi.org/10.1038/s41598-017-17766-4
65. Beggs JM (2008) The criticality hypothesis: how local cortical networks might optimize information processing. Phil Trans R Soc A 366:329–343
66. Palva JM, Zhigalov A, Hirvonen J, Korhonen O, Linkenkaer-Hansen K, Palva S (2013) Neuronal long-range temporal correlations and avalanche dynamics are correlated with behavioral scaling laws. Proc Natl Acad Sci 110:3585–3590
67. Miller KJ, Sorensen LB, Ojemann JG, den Nijs M (2009) Power-law scaling in the brain surface electric potential. PLoS Comput Biol 5:e1000609
68. Muthukumaraswamy SD, Liley DTJ (2018) 1/f electrophysiological spectra in resting and drug-induced states can be explained by the dynamics of multiple oscillatory relaxation processes. Neuroimage 179:582–595
69. Cole SR, Donoghue T, Gao R, Voytek B (2019) NeuroDSP: a package for neural digital signal processing. J. Open Source Softw. 4:1272

Chapter 23
How Can We Detect and Analyze Navigation-Related Low-Frequency Oscillations in Human Invasive Recordings?

Mingli Liang and Arne Ekstrom

Abstract Theta oscillations are a prominent semi-periodic fluctuation in the local field potential of the human hippocampus and show important links to areas of cognition like episodic memory and navigation. In this chapter, we begin by characterizing the properties of human hippocampal theta oscillations, which are more bursty and less continuous compared with rodents. Next, we introduce the Better OSCillation Detection algorithm (Whitten et al. in Neuroimage 54:860–874, 2011) for detecting oscillations based on amplitude and temporal thresholding. We compare BOSC with other oscillatory detection methods, such as those based on dual-amplitude thresholding and hidden Markov models. Additionally, we provide tutorials and a practical guide for oscillation detection for the interested reader. All codes and examples are provided freely in open-source format. Together, these provide researchers with the tools to explore novel questions about the nature of hippocampal navigation-related theta oscillations. As we demonstrate in this chapter, oscillatory detection procedures are extremely helpful for characterizing oscillatory dynamics including burst frequency and burst duration, exceeding beyond the singular dimension provided by amplitude changes measured in the power spectra.

23.1 Introduction

Low-frequency (2–12 Hz) oscillations are a prominent signature in the local field potential associated with navigation and memory. In invasive recordings they manifest in medial temporal lobe and neocortex, but they are also observed in noninvasive recordings such as scalp electroencephalogram (EEG). The bursty nature of oscillations, in contrast to the more sustained oscillations observed in animals, prompts

M. Liang
Department of Psychiatry, Department of Neurosurgery, Yale University, New Haven, USA
e-mail: mingli.liang@yale.edu

A. Ekstrom (✉)
Department of Psychology, University of Arizona, Tucson, USA
e-mail: adekstrom@arizona.edu

N. Axmacher (ed.), *Intracranial EEG*, Studies in Neuroscience, Psychology and Behavioral Economics, https://doi.org/10.1007/978-3-031-20910-9_23

the needs to detect and quantify the presence of oscillations. This chapter will first provide a background on neural oscillations as they relate to memory and navigation and include a step-by-step tutorial for detecting neural oscillations using existing algorithms.

Human navigation-related oscillations manifest in various frequencies, typically covering a range of from 1 to 12 Hz. The frequency, and prevalence of hippocampal oscillations varied across tasks and species. For example, rodents display 8–12 Hz oscillations in hippocampus associated with movement, while humans display 1–4 Hz oscillations [1, 2]. Typically, theta oscillations in rodent hippocampi are more sustained (Fig. 23.1, adapted from Vanderwolf 1969). In contrast, theta oscillations in humans are more bursty, appearing in the form of bouts [3]. Existing findings suggest that human hippocampal bouts typically last 0.57 s, with the range between 0.4 and 0.6 s, which is ~3.35 cycles for 6 Hz oscillations [4]. In contrast, 8 Hz oscillations in rodents last longer; for example, in a spatial navigation task, the average burst length measured in rats' hippocampal electrodes is ~4.3 cycles [3]. Some frequency variation in navigation-related theta may also relate to differences in sensorimotor inputs between virtual navigation and real-world locomotion. For example, patients with implanted hippocampal electrodes show more electrodes manifesting significant theta bursts in the 8–12 Hz theta band compared to virtual desktop navigation, in which oscillations tend to manifest from 1 to 4 Hz [5]. Therefore, when examining intracranial recordings from humans obtained during a spatial navigation task, it is advised to analyze a wider range of 1–12 Hz oscillatory activities.

One way to examine navigation-related responses is to analyze the time–frequency power representations. While the absolute power correlates with theta activities in the iEEG data, the absolute power can be confounded by the background noise that is independent of task-related activities. For example, successful memory encoding and retrieval involve absolute theta power decreases and absolute gamma power increases in human hippocampus [6]. However, the changes in absolute power are often confounded with the slopes of background power spectra (see Fig. 23.2 for an example of background power spectra). This is because background noise also has "power" based on the $1/f^{\alpha}$ spectrum. Absolute theta power decreases, for example, can be interpreted as relative theta power increases coupled with an increased tilt of the aperiodic background spectra. Oscillatory detection algorithms, which we will discuss in more detail here, provide a solution by standardizing the quantification of oscillations across electrodes and patients. In the next section we describe the procedure to detect oscillatory activities in intracranial recordings.

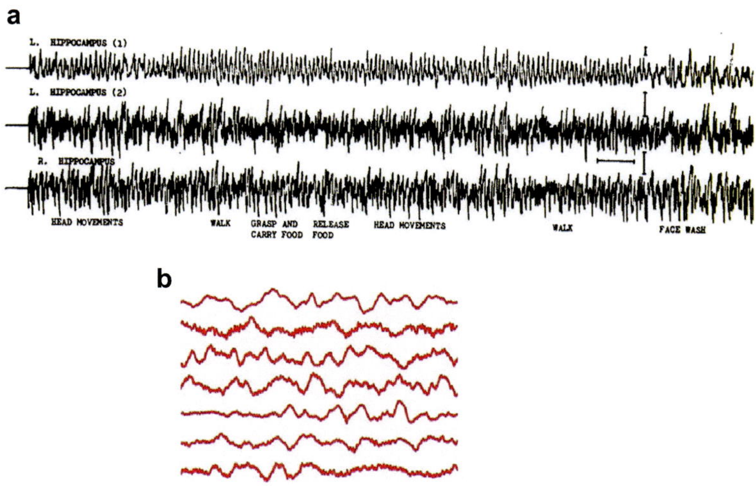

Fig. 23.1 Navigation-related hippocampal oscillations in humans and rodents manifest differently in frequency and prevalence. **a** Example of 8–12 Hz hippocampal oscillations during naturalistic behaviors in rodents. Adapted from Vanderwolf (1969) Fig. 23.4 Panel c. Note the nearly continuous nature of the oscillatory signal. Each row shows example raw traces from one electrode, and the calibrations indicate 1 s of data and 100 μV. **b** Example of navigation-related hippocampal oscillations in humans. Note the less continuous nature of human theta, even during continuous movement (similar to what is shown in rats in **a**). Adapted from Watrous et al. [3]

Fig. 23.2 Fitting the background spectra using ordinary least square regression and recursive Gaussian fitting processes. The blue solid line indicates the empirical power spectra obtained from the example recordings

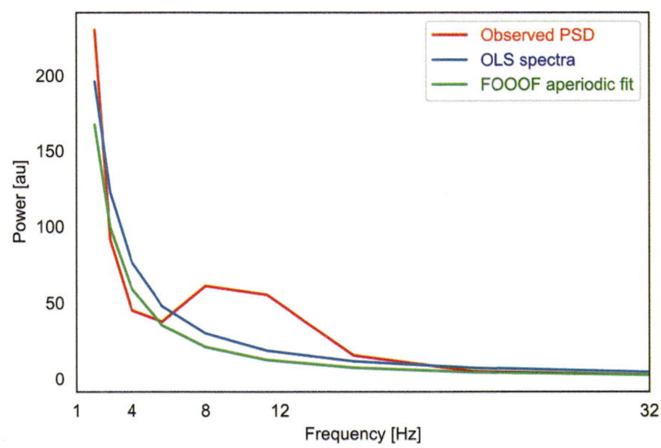

23.2 A Practical Guide for Detecting Oscillations from Human Hippocampal Recordings During a Navigation Task

In this section, we describe the procedure to detect and quantify the presence of hippocampal oscillations, using deidentified data from Liang et al. [25]. The data are available for download via https://github.com/liangmingli/chapter_scripts. The example data are recorded from an implanted electrode in the left anterior hippocampus of a patient undergoing seizure monitoring. The data were collected when the patient navigated in a T-maze in virtual reality on a laptop, and the data contains 48 repetitions of the navigation segments.

As part of the preprocessing pipeline, including filtering, line noise removal, and ictal discharges, visual inspection should also be done by the researcher. We note that researchers should properly epoch the data dependent on whether the navigation tasks are time-locked or active free exploration. In terms of free exploration, epoching is better implemented after the time–frequency decomposition on the continuous data. For an event-related design, it is better to epoch the continuous data before time–frequency power decomposition, to prevent information leak from temporal smoothing.

Amplitude thresholding is one viable method to detect and quantify navigation-related theta oscillations. They are implemented on the basis of time–frequency decomposition of iEEG data, including BOSC [7], eBOSC [8], and burst detection by neuroDSP [9]. Better Oscillation Detection algorithm detects neural oscillations based on two levels of thresholding: power threshold and duration threshold [7, 10]. The power thresholding identifies timepoints with high oscillatory power, and the duration thresholding reassures that those timepoints when clustered together are sustained based on the number of cycles.

The BOSC library can be downloaded via https://github.com/liangmingli/chapter_scripts, contributed by Whitten et al. [7]. The BOSC procedures include

(1) extracting the time–frequency power,
(2) estimating the background spectra,
(3) calculating the power threshold and duration threshold, and
(4) detecting oscillations.

When approximating the background spectra, the original BOSC algorithm uses ordinary least squares linear regression for the estimate. Other alternatives have been proposed and used, such as robust linear regression to exclude outlier values from the power spectra [8], and using recursive Gaussian processes to estimate the aperiodic spectra [11] (see also Chap. 22). After the background spectra is approximated, power threshold is estimated based on a chi square distribution with 2 degrees of freedom, and a default cutoff is 95% of the distribution. Compared to the OLS linear regression, robust linear regression and recursive Gaussian fitting produce lower power thresholds, i.e., are more sensitive to detecting navigation-related theta oscillations (Fig. 23.2). As for selecting the temporal threshold, shorter temporal thresholds enable researchers to capture the dynamics of shorter oscillatory bouts during navigation, vice versa for longer temporal thresholds (see Fig. 23.3). Three cycles or more is the preferred value for the cycle (temporal) threshold, although a shorter temporal threshold (e.g., 2 cycles) has been used by researchers as well [12].

After detecting oscillations with BOSC, the outputs are a binary sequence (i.e., timepoints with detected oscillations are noted with 1, otherwise with 0). Averaging the binary sequence yields the measure of oscillatory activities, or the percentage

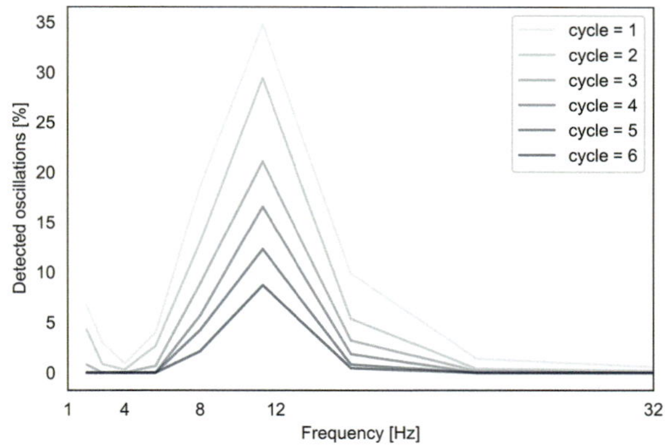

Fig. 23.3 Shorter temporal thresholds (i.e., fewer required cycles) yield an estimate of higher oscillatory activities

of time detected with oscillations at a given frequency (the $P_{episode}$ measure.) Researchers have utilized the $P_{episode}$ to investigate the relationship between oscillatory prevalence and spatial navigation [3, 4, 13]. For example, using the BOSC detection algorithm, Vass et al. showed that hippocampal theta prevalence differed between short distance and long distance traversals during navigation.

Further, oscillatory detection algorithms also enable researchers to examine the properties of bursts, such as the burst density, burst duration, and burst amplitudes. Those three aspects of burst dynamics have been implicated as relevant forms of information coding [14]. In the example dataset on trial 11, two oscillatory bursts at 8 Hz were identified: one detected around [0.10, 0.90] seconds, and the other burst around [2.72, 3.20] seconds. The first burst lasted approximately 0.79 s (6.34 cycles) while second burst lasted 0.48 s (3.83 cycles). On average, for the example trial with a four second long duration, approximately 31% of the data was detected with oscillations. To examine the amplitude of 8 Hz oscillation, the researcher can simply sample the power series during the timepoints detected with oscillations. Therefore, the BOSC detection algorithm allows researchers to more comprehensively evaluate burst dynamics including burst frequency, burst duration and burst amplitudes. For example, Aghajan et al. [4] sampled hippocampal activity from one patient with congenital blindness during real-world navigation. Compared to patients with vision, hippocampal theta oscillatory bouts were significantly longer. Therefore, characterization of the oscillatory burst dynamics enables novel insights into how neural oscillations are associated with spatial navigation.

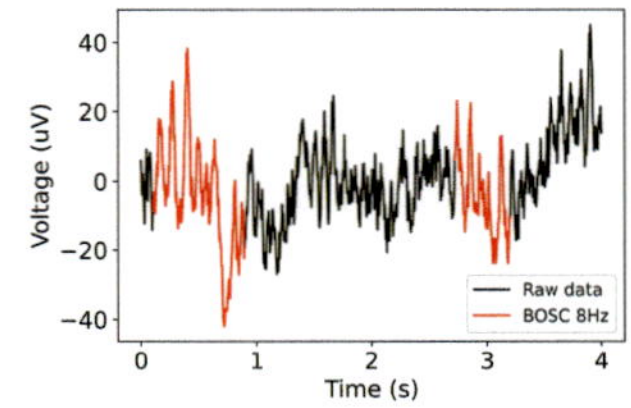

Fig. 23.4 Two oscillatory bursts at 8 Hz were detected using BOSC in the example dataset, example trial 11. Timepoints detected with 8 Hz oscillations are overlayed in red color

Table 23.1 Studies using BOSC and detecting medial temporal lobe theta oscillations

Study	Location	θ percentage	Task type
Watrous et al. [3]	Hippocampus	0–16	Navigation
Vass et al. [13]	Hippocampus	0–30	Navigation and teleportation
Aghajan et al. [4]	Medial temporal lobe	0–15	Real-world navigation
Chen et al. [15]	Entorhinal	10–20	Navigation
Kragel et al. [16]	Hippocampus	10	Recognition memory
Liang et al. [25]	Hippocampus	0–15	Navigation and teleportation

What is the typical range of detected theta oscillations in humans during spatial navigation? Here, we present several recent publications in the field of human electrophysiology and spatial navigation. During a spatial navigation task, a hippocampal electrode is expected to manifest theta bursts approximately 10–20% of the time (Table 23.1).

23.3 Convergence: Detecting Navigation-Related Oscillations Using Other Available Methods

In addition to BOSC [7, 10], as mentioned, there are other methods to cover oscillation detection. Some of these include: (1) using double amplitude thresholding in NeuroDSP [9] https://neurodsp-tools.github.io/neurodsp/index.html, and (2) Hidden Markov Model based state inferences of burst status [17, 18], via https://github.com/OHBA-analysis/Quinn2019_BurstHMM. In this section we note that we do not intend to give a formal assessment and comparison among the methods for oscillatory detection (see Chap. 22).

In contrast to BOSC using both amplitude and duration thresholds, NeuroDSP uses only power thresholds but no temporal threshold. The power series will be marked as oscillatory when it goes above the upper power threshold, and then marked as non-oscillatory when it drops below the lower power threshold. Without temporal thresholding, the detected events could include oscillatory bursts that are both short and long. Based on the example dataset, for trial 11, NeuroDSP returned two oscillatory bursts detected at the frequency range of 7–9 Hz, similar to the output reported by BOSC: one burst of 0.84 s length (6.72 cycles) and the other burst of 0.63 s length (5.0 cycles), and about 38.21% of the data were detected with 7–9 Hz oscillations (Fig. 23.5).

An alternative method is based on Hidden Markov models (HMM) and Multivariate Autoregressive models [17, 18]. Rather than thresholding with a voltage threshold, oscillatory states are inferred based on empirical observations of time–frequency data and histories of oscillatory states. The idea behind is that compared to a non-oscillatory baseline, oscillatory bursts at distinct frequencies are associated with distinct power spectra, and distinct power spectra can be detected as distinct

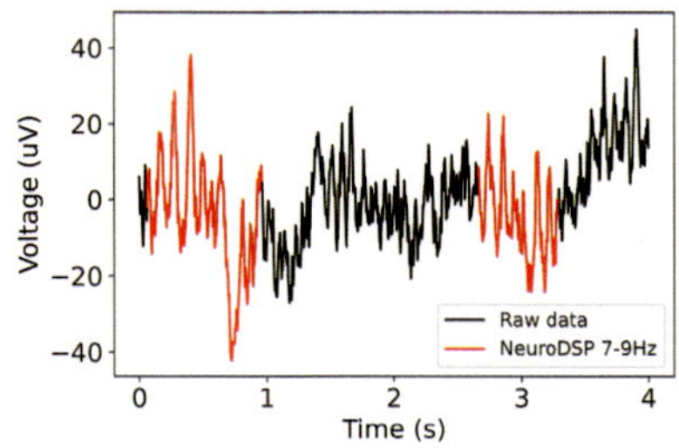

Fig. 23.5 Two oscillatory bursts within 7–9 Hz were detected using NeuroDSP dual power threshold detection. Timepoints detected with 7–9 Hz oscillations are overlayed in red color. Compared to the output reported by BOSC (Fig. 23.4), both methods discovered two oscillatory bursts with similar timestamps, and both methods detected a similar percentage of 8 Hz oscillatory activities (BOSC: 31%, NeuroDSP: 38%). We note that NeuroDSP and BOSC involved different approaches to determining the amplitude threshold(s), and therefore they could produce different detections of oscillatory onsets: note the earlier detected onsets of 8 Hz activities in the NeuroDSP outputs compared to that of BOSC

events/states. The HMM predicts the oscillatory states at each timepoint based on the history of time–frequency representations of past timepoints and the inferred states of past timepoints.

One key parameter for HMM is the possible number of states (K), which requires apriori specification. In Quinn et al. [17], the authors explored two different K values: 2 and more than 2. When K is set to 2, the possible number of states is set to 2, (i.e., oscillatory vs non-oscillatory), and the inference is operated on narrow-banded filtered signals (i.e., 7–9 Hz bandpass filtered signal for detecting 8 Hz activities). The second approach, where K is set to be more than 2, is to explore all the possible oscillatory states in the data across multiple frequency ranges. This is suited for research questions that look at multiple oscillatory frequencies in the context of spatial navigation (e.g., theta and gamma activities). One critical difference between the HMM based- and amplitude-threshold-based oscillatory detection is that the output from HMM is ambiguous and requires experimenter interpretation. For example, the HMM will output the posterior probabilities for each timepoint and for each state. But what state(s) belong to theta oscillatory occurrence? Would state 1 or state 2 be the inferred state of theta occurrence? That is the question that will need the experimenter's expertise to answer. Here, to demonstrate the HMM method for oscillatory detection, we simulated 8 Hz sinusoids added on top of *1/f* noise. HMM-based models detected the simulated 8 Hz oscillatory bursts, and after visual examination, we concluded that state 1 was associated 8 Hz oscillatory bursts and state 2 was associated with non-oscillatory timepoints (Fig. 23.6).

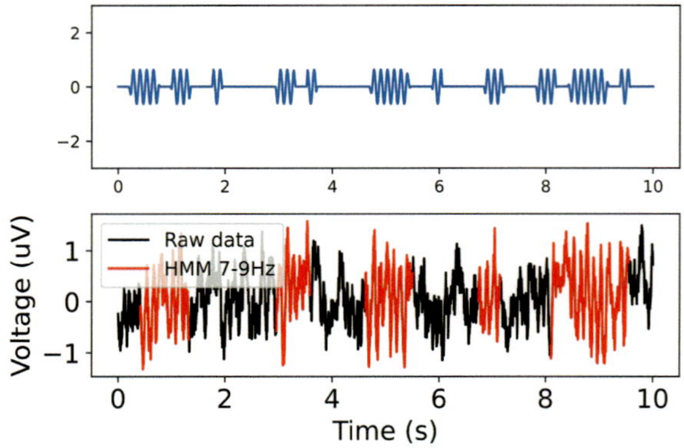

Fig. 23.6 Detecting 8 Hz simulated oscillatory bursts using Hidden Markov models [17, 18]. In the top panel, the simulated 8 Hz bursts are plotted, and the simulated bursts are added to *1/f* noise for follow-up oscillatory detection. In the bottom panel, timepoints detected with 7–9 Hz oscillations are overlayed in red color

23.4 What Do We Mean by "Oscillation?"

The mathematical definition of an oscillation refers to a continuous periodic signal that can be decomposed into frequency, amplitude, and phase. In most applications, such oscillations are approximated as sine and cosine functions using the Fourier Transform. As is fairly clear from our examples above, however, the "bursts" of signals discussed cannot be considered "oscillations" in the strict sense because they are not continuous in terms of their cycles. Consistent with this, comparing Fourier methods with methods that better account for the burstiness of these signals within the local field potential (such as BOSC) suggest that Fourier methods generally do a comparatively poor job at characterizing these signals [19].

This raises the question then why we even call such signals "oscillations" when they are clearly not, even in rodents, continuous cycles resembling sine waves. The answer to this question appears to be part nomenclature: early efforts to characterize these time-varying signals in scalp EEG and invasive recordings applied Fourier methods [20, 21] thereby assuming oscillatory signals in the stricter mathematical sense. This assumption, however, is not necessarily problematic as Fourier methods can still approximate aperiodic signals as well, and almost any function for that matter, although the numbers of coefficients can become unwieldy [22]. As mentioned, the accuracy of characterizations that assume a periodic continuous signal is weaker than methods that allow for bursts (such as BOSC or NeuroDSP). Perhaps more importantly, though, for our present consideration is that the term "oscillation" is incorrect.

As an alternative, we suggest the term "semi-periodic fluctuation" (SFP) in the local field potential. This term helps capture the idea that such signals are indeed periodic for bouts but then taper off (for reasons that have yet to be revealed). The term "fluctuation" helps to capture another issue which is that the amplitude of these bursts also varies from cycle to cycle and burst to burst. This is also in contrast to a classic sine wave in that the amplitude should be constant and therefore something that can be approximated with a single number. While we think it is likely that the term "oscillation" will persist, it is important to be aware that the signals we have discussed, particularly in humans, are not "oscillatory" in the classic Fourier sense but rather semi-periodic and varying in amplitude.

One could ask why such signals might be "bursty" in the first place, and why human signals might be more bursty than those of other species, like rats. While the reasons have yet to be uncovered, one possibility is that brief bursts of periodic fluctuations actually carry more information than long continuous bursts. This can be seen from the Fourier perspective described above; a continuous signal requires 3 parameters to describe it in Fourier space: frequency, amplitude, and phase. A semi-periodic fluctuation requires many more parameters to capture both the aperiodic nature and the fluctuating amplitude of the signal. Likely, such bursty signals are more adept at transiently coordinating disparate brain regions compared to a continuous signal, which would lead to entrainment and potentially seizures [23, 24]. While these ideas remain speculative, they provide some initial ideas about why "oscillations" may not be oscillatory at all.

23.5 Conclusions

Hippocampal low-frequency oscillations are often found when humans perform a spatial navigation task. Such low-frequency oscillations have been demonstrated to be relevant for the formation of spatial representations and the coding of multiple spatial variables. While power can provide some degree of the prevalence of navigation-related oscillations, the transient and bursty nature of human hippocampal low-frequency oscillations necessitates oscillatory detection performed on the data. Implementing oscillatory detection on the obtained intracranial recordings can quantify the amount of oscillations in the signal and allows for the extraction of parameters regarding burst frequency, burst duration, and burst amplitude. Oscillatory detection is mostly performed via amplitude thresholding, in which timepoints with sufficiently high oscillatory power are considered "oscillatory". The amplitude thresholding sometimes is accompanied by temporal thresholding to identify oscillatory bursts that last longer than a specific number of cycles. Alternative approaches to oscillatory detection also include Hidden Markov model-based approach, in which the oscillatory states are inferred rather than yielded from thresholding. Oscillatory detection can generate new analyses and insights for spatial cognition research, allow for the standardization of oscillatory prevalence across electrodes and participants, and most crucially, incorporates the transient nature of low-frequency semi-periodic fluctuations in humans versus animals.

References

1. Jacobs J (2013) Hippocampal theta oscillations are slower in humans than in rodents: implications for models of spatial navigation and memory. Philos Trans R Soc B Biol Sci 369:20130304
2. Qasim SE, Fried I, Jacobs J (2021) Phase precession in the human hippocampus and entorhinal cortex. Cell 184:3242-3255.e10
3. Watrous AJ, Lee DJ, Izadi A, Gurkoff GG, Shahlaie K, Ekstrom AD (2013) A comparative study of human and rat hippocampal low frequency oscillations during spatial navigation running. Hippocampus 1–18
4. Aghajan ZM, Schuette P, Fields TA et al (2017) Theta oscillations in the human medial temporal lobe during real-world ambulatory movement. Curr Biol 27:3743-3751.e3
5. Bohbot VD, Copara MS, Gotman J, Ekstrom AD (2017) Low-frequency theta oscillations in the human hippocampus during real-world and virtual navigation. Nat Commun 8:14415
6. Herweg NA, Solomon EA, Kahana MJ (2020) Theta oscillations in human memory. Trends Cogn Sci 24:208–227
7. Whitten TA, Hughes AM, Dickson CT, Caplan JB (2011) A better oscillation detection method robustly extracts EEG rhythms across brain state changes: the human alpha rhythm as a test case. Neuroimage 54:860–874
8. Kosciessa JQ, Grandy TH, Garrett DD, Werkle-Bergner M (2020) Single-trial characterization of neural rhythms: potential and challenges. Neuroimage 206:116331
9. Cole S, Donoghue T, Gao R, Voytek B (2019) NeuroDSP: a package for neural digital signal processing. J Open Source Softw 4:1272
10. Hughes AM, Whitten TA, Caplan JB, Dickson CT (2012) BOSC: a better oscillation detection method, extracts both sustained and transient rhythms from rat hippocampal recordings. Hippocampus 22:1417–1428
11. Donoghue T, Haller M, Peterson EJ et al (2020) Parameterizing neural power spectra into periodic and aperiodic components. Nat Neurosci 23:1655–1665
12. Stangl M, Topalovic U, Inman CS et al (2020) Boundary-anchored neural mechanisms of location-encoding for self and others. Nature. https://doi.org/10.1038/s41586-020-03073-y
13. Vass LK, Copara MS, Seyal M, Shahlaie K, Farias ST, Shen PY, Ekstrom AD (2016) Oscillations go the distance: low-frequency human hippocampal oscillations code spatial distance in the absence of sensory cues during teleportation. Neuron 89:1180–1186
14. Donoghue T, Schaworonkow N, Voytek B (2021) Methodological considerations for studying neural oscillation. https://doi.org/10.31234/osf.io/hvd67
15. Chen D, Kunz L, Lv P, Zhang H, Zhou W, Liang S, Axmacher N, Wang L (2021) Theta oscillations coordinate grid-like representations between ventromedial prefrontal and entorhinal cortex. Sci Adv 7:eabj0200
16. Kragel JE, Schuele S, VanHaerents S, Rosenow JM, Voss JL (2021) Rapid coordination of effective learning by the human hippocampus. Sci Adv
17. Quinn AJ, van Ede F, Brookes MJ, Heideman SG, Nowak M, Seedat ZA, Vidaurre D, Zich C, Nobre AC, Woolrich MW (2019) Unpacking transient event dynamics in electrophysiological power spectra. Brain Topogr 32:1020–1034
18. Smith JB, Lee AK, Jackson J (2020) The claustrum. Curr Biol 30:R1401–R1406
19. Vugt MKV, Sederberg PB, Kahana MJ, van Vugt MK, Sederberg PB, Kahana MJ (2007) Comparison of spectral analysis methods for characterizing brain oscillations. J Neurosci Methods 162:49–63
20. Brazier MAB (1968) Studies of the EEG activity of limbic structures in man. Electroencephalogr Clin Neurophysiol 25:309–318
21. Brazier MAB, Casby JU (1952) Crosscorrelation and autocorrelation studies of electroencephalographic potentials. Electroencephalogr Clin Neurophysiol 4:201–211
22. Cohen XM (2014) Analyzing neural time series data: theory and practice. MIT Press, Cambridge, USA

23. Ekstrom AD, Watrous AJ (2014) Multifaceted roles for low-frequency oscillations in bottom-up and top-down processing during navigation and memory. Neuroimage 85:667–677
24. Watrous A, Ekstrom A (2014) The spectro-contextual encoding and retrieval theory of episodic memory. Front Hum Neurosci 8
25. Liang M, Zheng J, Isham E, Ekstrom A (2021) Common and distinct roles of frontal midline theta and occipital alpha oscillations in coding temporal intervals and spatial distances. J Cogn Neurosci 33(11):2311–2327

Chapter 24
How Can I Disentangle Physiological and Pathological High-Frequency Oscillations?

Birgit Frauscher and Jean Gotman

Abstract High-frequency oscillations (HFOs) consist of ripples (80–250 Hz) and fast ripples (>250 Hz). Ripples are attributed an important role in cognition, sleep, and task-related processes and have been reported in many cortical areas of the healthy brain, with large variations in rates depending on anatomical localization. HFOs are also a promising interictal biomarker of epilepsy. HFOs were shown to be more specific to localize the epileptogenic zone than interictal epileptiform discharges. One emerging area in HFO research is the nontrivial task of separating physiological from pathological HFOs. Attempts to differentiate between both entities consist of considering their coupling with epileptic spikes, the background EEG activity, their relation to tasks, the anatomical localization of implanted electrodes, more classical neurophysiological features such as amplitude and frequency, as well as their interaction with sleep and distinct pharmacological modulation. This chapter provides an overview of the current state of evidence, discusses unresolved challenges, and finally shows how to improve the yield of HFOs for prediction of the epileptogenic zone and isolating physiological HFOs for cognitive research. Accounting for the distinct properties of physiological and pathological HFOs could be critical for the interpretation of HFO findings for clinical use or neurocognitive research.

24.1 Introduction

High-frequency oscillations (HFOs), usually subclassified as ripples (80–250 Hz) and fast ripples (250–500 Hz), have been studied in both physiological and pathological brain conditions. In physiological conditions, HFOs were attributed a significant role in memory consolidation via sharp-wave ripple complexes, sleep, and task-related processes [1–3]. In pathological conditions, HFOs were found in experimental

B. Frauscher (✉) · J. Gotman
Montreal Neurological Institute and Hospital, McGill University, 3801 University Street, Montreal, QC H3A 2B4, Canada
e-mail: birgit.frauscher@mcgill.ca

J. Gotman
e-mail: jean.gotman@mcgill.ca

N. Axmacher (ed.), *Intracranial EEG*, Studies in Neuroscience, Psychology and Behavioral Economics, https://doi.org/10.1007/978-3-031-20910-9_24

models of epilepsy to be biomarkers for the epileptic tissue [4]; fast ripples in particular were shown to be a promising predictor of seizure occurrence after traumatic brain injury [5]. These studies were performed with microelectrodes recording single cells or small cell assemblies. It was later found that HFOs could also be recorded with the much larger (contact area of several mm^2) intracerebral macroelectrodes used in the clinical investigation of epileptic patients [6]. This vastly enhanced the ability to study HFOs in humans. Indeed, HFOs recorded with macroelectrodes were shown to be more specific to localize the epileptogenic zone than interictal epileptiform discharges, the traditional biomarker of epilepsy [7]. It was also found that macroelectrode recorded HFOs, ripples in particular, were common in non-epileptic brain regions. These latter events correspond presumably to the physiological ripples studied with microelectrodes in animals [8].

24.2 Lessons Learnt for HFO Research from Human Microelectrode Studies

Studies on HFOs using microelectrode recordings in humans focused on three main lines of research: the understanding of underlying neuronal mechanisms of neurocognitive processes, the dissection of differences in the various firing patterns in physiological versus pathological HFOs to draw conclusions on underlying mechanisms, and the study of microwire correlates of HFOs using standard clinical macroelectrodes, which are necessary for a wider application in clinical research. Pioneering work using microelectrodes of 40 microns in humans showed that there are sharp wave ripple complexes in the human hippocampus which are visually comparable to those seen in experimental models [9]. Subsequently, it was shown that sharp wave ripple complexes in humans and experimental models have similar underlying mechanisms as revealed by the investigation of the relationships between ripples and unit activity [10]. Also, studies showed a complex pattern of feedforward and feedback processing between the ento- and peri-rhinal cortex during a word recognition task underlying the temporal patterns of different cell assemblies, something not possible when using macroelectrodes [11].

Surprisingly, most of this work ignored that these data were gathered from patients with drug-resistant epilepsy in whom implantations were performed for seizure focus localization. Over the past two decades, substantial work has been made to investigate the mechanisms underlying both physiological and pathological HFOs. Although HFOs may have similar spectral frequencies, both types are attributed different underlying neuronal mechanisms, with interneurons playing an important role in HFO genesis (see Fig. 24.1). In physiological sharp-wave ripples, HFOs represent extracellularly recorded, synchronous, inhibitory, postsynaptic potentials on the membranes of principal cells (Fig. 24.1a). In contrast, in epileptic ripples, the interneuronal activity and inhibitory postsynaptic potentials control the action potential firing of the active epileptic population which manifests as pathological ripples. This mechanism is dependent on intact perisomatic inhibition maintained by basket cells (Fig. 24.1b). Finally, loss of inhibition plays a role in the genesis of fast ripples. The absence of rhythmic fast inhibition results in functional clustering or asynchronous neuronal firing and the generation of fast ripples (Fig. 24.1c, d). For further details, the reader is referred to the review of Jiruska et al. [12]. Discrimination between both entities, particularly in macroelectrode recordings, is therefore not a trivial task and has remained a major challenge in HFO research [13–15].

Finally, attempts have been made to compare the ability of macro- and microelectrodes to record HFOs. As expected given the size and orientation of HFO generators [16], it was found that penetrating microcontacts of 40 microns in diameter

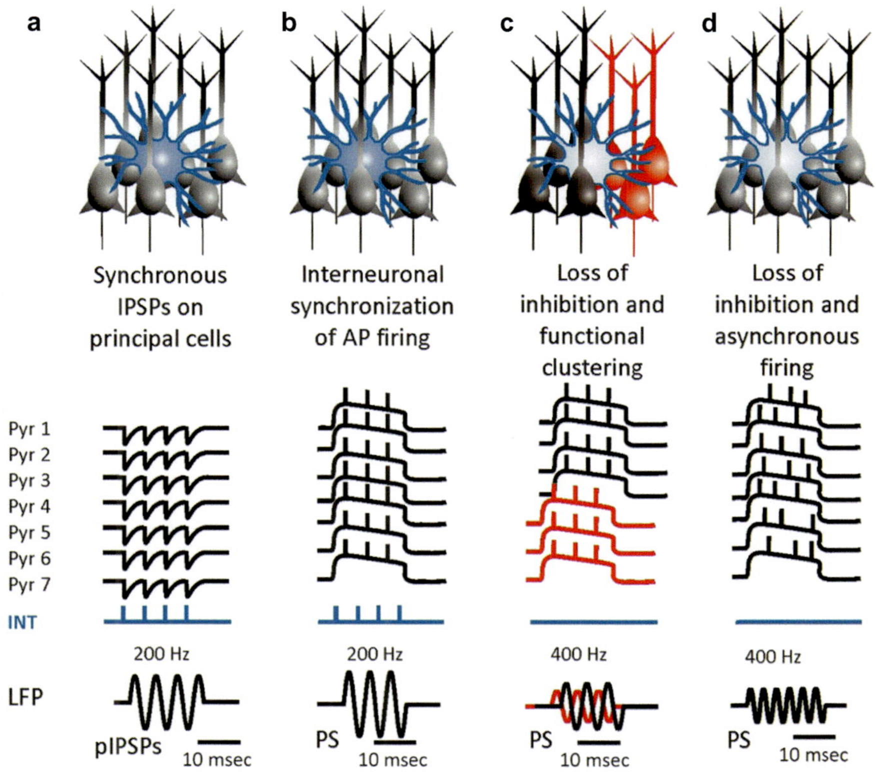

Fig. 24.1 The role of interneurons in HFOs. **(A)** HFOs may represent extracellularly recorded, synchronous, inhibitory, postsynaptic potentials on the membranes of principal cells. This mechanism underlies physiologic sharp-wave ripples. Pathologically, they may be involved in low-amplitude fast activity ictal onset but usually with frequencies lower than ripples. **(B)** In epileptic ripples, the interneuronal activity and inhibitory postsynaptic potentials control the action potential firing of the active epileptic population, which manifests as pathologic ripples. This mechanism is dependent on intact perisomatic inhibition maintained by basket cells. **(C)** Loss of inhibition may play a role in the pathogenesis of fast ripples. The absence of rhythmogenic fast inhibition may result in functional clustering or asynchronous neuronal firing and the generation of fast ripples **(D)**. *Source* Jiruska et al. [12] with permission from John Wiley and sons

detected substantially more HFOs, particularly fast ripples, than depth macroelectrodes, whereas superficial microcontacts and standard subdural electrodes did not differ in their abilities to record HFOs [17]. When considering macroelectrodes, it was found that the size of the contact area had no significant influence on the ability to record HFOs [18].

24.3 Why Is It Important to Separate Physiological from Pathological HFOs

Albeit research showed that HFOs are a promising biomarker of the epileptogenic zone when considering patient groups [7], more recent research revealed that their use as biomarker is still hampered at the individual level, with up to 30% of patients not being correctly classified using this biomarker [19, 20]. One potential explanation is the presence of physiological HFOs, which are included in the total HFO count. To identify new ways to differentiate between both entities is therefore key to improve the yield of HFOs for prediction of the epileptogenic zone on the one hand, and to isolate physiological HFOs for cognitive research on the other hand. In the following we will review various approaches that have been applied to accomplish this task focusing on studies using macroelectrodes.

24.4 Approaches to Separate Pathological from Physiologic HFOs

Distinguishing pathologic from physiologic HFOs has the potential to increase the specificity of this marker for both clinical use in epilepsy and neurocognitive research. This requires defining HFOs to be considered either as physiological or pathological. Studies attempted to separate pathological from physiological HFOs by considering the coupling with sleep spindles or epileptic spikes [21, 22], the background EEG activity [23, 24], task-induced HFOs [25–27], the anatomical location of implanted electrodes [14, 28–30], the classic features including amplitude, duration, spectral frequency, and rate [31, 32], their stereotypical appearance [15], the manner of interaction with the accompanying slow wave or phasic rapid eye movement sleep [13, 29, 33–37], and their distinct pharmacological modulation [38–41]. Whereas most of these features were able to show significant group differences reflecting that physiological and pathological HFOs are influenced by different factors, achieving sufficient performance to separate channels having predominantly one type of event or the other, or even separating individual oscillations, has remained challenging. In what follows we will discuss the most promising of these approaches in more detail: (i) building an atlas of normative HFO rates given the considerable variation in physiological HFO rates across the brain; (ii) utilizing evoked responses to separate both entities, and (iii) leveraging the different coupling to sleep physiological features to separate both entities.

24.5 Building an Atlas of HFO Normative Rates and Its Use to Improve the Yield of Identification of Epileptic Tissue

The ability of HFOs as a biomarker for epileptic tissue might be improved by correcting HFO rates according to their topographic localization. In this situation one is not trying to separate individual physiological and pathological events but one is trying to delineate epileptogenic tissue by separating channels with likely significant rates of pathological ripples from channels likely to include mostly physiological ripples. Rates of ripples vary substantially across different brain regions [14, 28, 29]. A multicenter project aiming at developing normative values of intracranial EEG (iEEG) activity [42, 43] investigated this question by carefully selecting in individual patients the small subset of iEEG channels showing normal physiologic EEG activity defined as (i) absence of interictal epileptic activity during the complete implantation period, (ii) absence of a significant slow wave anomaly, and (iii) being outside of the seizure-onset zone, and outside of lesional tissue as assessed with MRI. In a subproject of this atlas of normative iEEG activity, normative rates of HFOs (ripples and fast ripples) were assessed [30]. A total of 1,171 bipolar channels with normal physiologic activity from 71 patients were analyzed. Rates of ripples varied substantially across the different regions analyzed, with rates of up to 30/min in areas referred to as ripple-rich cortex, and other regions with rates of one or two per minute, referred as ripple-poor cortex. The mean 95th percentile of the distribution of rates of all the channels within one anatomical region was 9.6/min. The highest 95th percentile rates were recorded in the occipital cortex, the medial and basal temporal region, the transverse temporal gyrus and planum temporale, the pre- and postcentral gyri, and the medial parietal lobe (Fig. 24.2 left panel). Interestingly, all these regions correspond to cortical regions that have a higher degree of demyelination [44, 45] Indeed, myeloarchitecture was suggested to support connectivity across all bands [46]. The mean rate of fast ripples was very low with 0.038/min. Only 5% of channels had a rate of more than 0.2 fast ripples/min. This multicenter atlas is the first to provide region-specific normative values for physiologic HFOs in a common stereotactic space. It demonstrated that physiologic ripples are particularly frequent in the mesiotemporal, somatosensory and visual areas. In contrast, physiologic fast ripples are very rare, even in ripple-rich cortical areas, which makes them a better candidate for defining epileptic tissue, when present. This atlas is an open-access resource available for consultation on the web (http://mni-open-ieegatlas.research.mcgill.ca).

We give above absolute rates of number of events/min, but these are highly dependent on the sensitivity of the HFO detector. The atlas can nevertheless provide widely

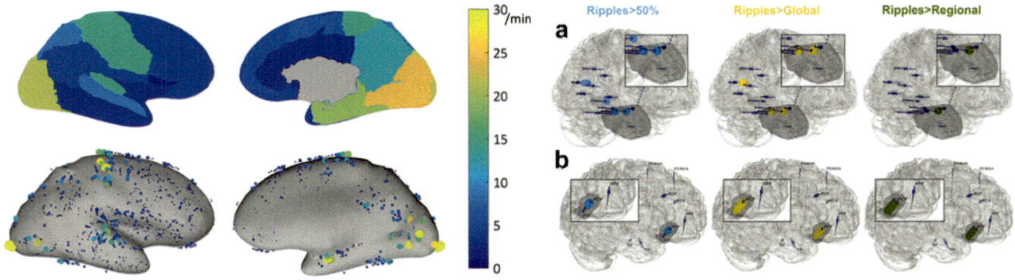

Fig. 24.2 Improving the specificity of ripples for the epileptic tissue by normalization from the atlas of physiological ripple rates. (*left*) Physiological ripple rates for bipolar channels (clinical macroelectrodes) represented on the inflated cortex. Top: 95th percentile of the physiological ripple rate per brain region. Bottom: rate of the individual channels. Each dot represents a channel, the size and color indicate its ripple rate (left: lateral view, right: medial view). (*right*) Patient examples of a case **(A)** with an epileptic focus in a region known to have high rates of physiological ripples, where normalization improves prediction, and a case **(B)** with an epileptic focus in ripple-poor cortex where the 50% and all normalization thresholds work. Both patients had Engel IA outcome. We mapped the electrodes in the patient's brain surface as dark blue cylinders and indicated which channels were above the cutoff threshold of 1 ripple/min with small dark blue spheres. The three columns indicate channels that had ripples (i) above a threshold of 50% relative to the total number of ripples in a patient as frequently used in the literature (light blue), (ii) above the global Atlas threshold defined as the 90th percentile value of the distribution of normative HFO rates of all regions combined (yellow), and (iii) above a regional Atlas (green) threshold defined as the region-specific 90th percentile values of the normative rates. The black dots indicate the resection. **(A)** This patient benefitted from regional thresholding and shifted from false negative to true positive classification. He was classified as false negative using ripple rate >1/min and as true positive using all other thresholds. *Source* Frauscher et al. [30] and Zweiphenning et al. [47] with permission from John Wiley and sons

applicable normative data in two ways: the first is that the detector is available on the web site of the atlas and the second is that users can run their own detector on the iEEGs of the atlas and obtain the normative rates that are specific to their detector.

We then evaluated whether defining abnormal HFO rates by statistical comparison to region-specific physiological HFO rates observed in the healthy brain [30] improves identification of the epileptic focus and surgical outcome prediction [47]. To do so, we performed a two-center study in 151 patients undergoing subsequent epilepsy surgery, leveraging normalization of HFOs for detection of the epileptogenic zone and prediction of surgical outcome. The main results of this work were that (1) normalization significantly improved the ability of ripples to identify the resected tissue and predict seizure freedom; (2) normalization is particularly useful in patients with a focus in ripple-rich cortex (Fig. 24.2 right panel), and that in this condition, ripple normalization was better than either considering the fast ripple rate or than using the current gold standard of seizure-onset zone localization based on ictal iEEG patterns; and (3) normalization did not improve the performance of fast ripples for focus identification or outcome prediction, presumably because physiological fast ripples are so infrequent.

24.6 Evoked Responses Are Useful to Separate Physiological from Pathological HFOs

In this case, one is trying to separate individual HFOs by examining if they occur as response to various stimuli, with the reasonable assumption that evoked HFOs are physiological. Nagasawa et al. [25] were the first to report on the possibility to distinguish physiological from epileptic HFOs by using evoked potentials. They studied 10 patients undergoing iEEG and found that visually evoked HFOs, which are presumably physiological, had significantly longer durations than epileptic HFOs. Similar results were revealed later also for somatosensory evoked or cognitive task evoked HFOs [26, 48]. Recently, it was shown that even in channels inside the epileptogenic zone, physiological HFO responses can be evoked [27], making the separation between physiological and pathological HFOs even more tricky and suggesting the need to separate HFOs not only at the channel but also the individual event level. In all patients, the authors found abundant normal physiological responses to cognitive stimuli in the epileptic sites. Regarding signal properties, the two types of HFOs were distinct, with evoked activity being of longer duration and broadband in the high-frequency range, and pathological HFOs representing short and narrow band high-frequency events (Fig. 24.3). Interestingly, evoked physiological responses were more likely to be delayed or missed when spontaneous pathological HFOs occurred. Furthermore, spontaneous pathological HFOs in the mesial temporal lobe affected the subjects' memory performance. The authors concluded that epileptic tissue is capable of generating normal physiological responses [27].

One remaining challenge here is that evoked high-frequency band responses do not have the same signal properties as spontaneously occurring and non-evoked physiological HFOs, so that it remains questionable if evoked responses will indeed be helpful to separate spontaneously occurring HFOs from epileptic HFOs with respect to clinical use in epilepsy. This issue also suggests to consider the use of one nomenclature across disciplines. Whereas for scientists in the field of epilepsy, HFOs are defined as isolated short events >80 Hz [7], several papers in neurocognitive research define high gamma activity as longer events up to 140 Hz or speak of ripples when indeed referring to high-frequency broadband activity, as demonstrated by Liu and Parvizi [27].

24.7 Separating Physiological from Pathological HFOs Using Coupling to Sleep Features

Work of various groups was able to show that there is a difference in coupling between physiological and pathological HFOs with sleep slow oscillations [13, 29, 34–37]. This distinct behavior was first reported in a study [13] investigating if epileptic activity occurs evenly across non-rapid eye movement sleep or if it is coupled to sleep slow oscillations known for their properties to orchestrate physiological brain

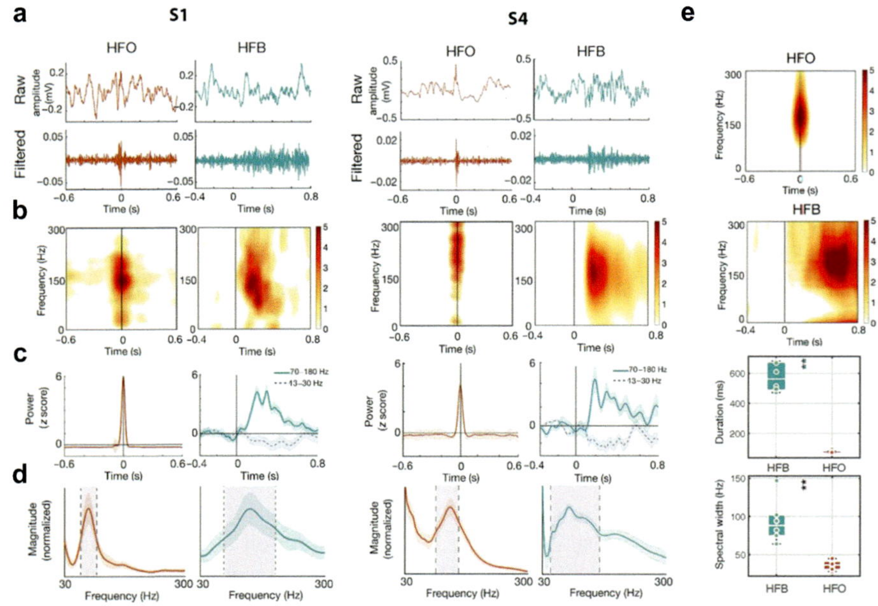

Fig. 24.3 Temporal and spectral profiles of HFO and high-frequency band (HFB) signals. **(a)** Data plots for exemplar HFO and HFB in the same sites in two representative cases (S1 and S4). For each sample event, raw data (top) and high-pass filtered data above 80 Hz (bottom) are shown. For an HFO event, time = 0 indicates the time point corresponding to the peak amplitude; for an HFB event, time = 0 indicates the stimulus onset. **(b)** Time–frequency maps averaged across all HFO and HFB events in S1 and S4. **(c)** Averaged power time course in the high-frequency band above 80 Hz for HFO and task-induced HFB in S1 and S4. Task-induced low-band deactivation is also presented. **(d)** Averaged signal spectra for HFO and HFB in S1 and S4. The spectral width is given by its full width at half maximum (dashed vertical lines). **(e)** Averaged time–frequency maps, signal duration, and spectral width for HFO and HFB activities in all subjects. $^{**}P < 0.01$. *Source* Liu and Parvizi [27] with permission from the American Association for the Advancement of Science

rhythms [49]. The authors found that the two types of HFOs are coupled to specific phases of the sleep slow wave. They selected channels in which most HFOs were likely of one type or the other. Interestingly, there was a biphasic distribution pattern with HFOs in channels with normal physiological activity occurring predominantly at the transition from the "down" to the "up" state, and HFOs in channels with epileptic activity occurring at the transition from the "up" to the "down" state [13]. Furthermore, there seems to be a difference between physiological and pathological HFOs with respect to the subtype of rapid eye movement sleep. In contrast to pathological HFOs, physiological HFOs appear predominantly during phasic as opposed to tonic REM sleep [33] and seem to increase in rate over the night during REM sleep [50]. Finally, work from the hippocampus showed that HFOs occurring at the time of sleep spindles are likely physiological in nature [51].

24.8 Conclusion and Outlook

The 2nd International Workshop on HFO in 2017 recognized that separation between physiological and pathological HFOs is important for a correct interpretation of HFO findings for clinical use or cognitive research [7]. Evidence suggests that this separation is particularly important for HFOs with frequencies in the ripple or low fast ripple band, whereas presence of fast ripples above 330 Hz was basically always linked to epilepsy [30]. In the last years, significant progress has been made to separate both entities at the channel level, whereas it remains an inaccurate and likely an impossible task to separate physiological from pathological HFOs at the event level [52]. It was shown that indeed both physiological and pathological HFOs can occur in the same channel [27], and that physiological HFOs can be highjacked by pathological HFOs [27, 53], as shown for the replacement of sleep spindles by generalized spike and wave discharges [54]. The fact however that it was recently shown that HFO normalization enables an improved delineation of epileptic tissue in areas with high rates of physiological ripples [47] indicates that a channel level separation may be sufficient for improving the yield of detecting epileptic tissue as required for HFOs to be a reliable epilepsy biomarker. Regarding neurocognitive research, several studies demonstrated that evoked physiological HFA responses are broadband high frequency events with longer duration, and that this difference in properties analyzed with supervised machine learning techniques allowed a good separation from epileptic HFOs even inside epileptic tissue [25, 27]. Future research will show if leveraging the knowledge gained from the various procedures paired with advanced techniques of artificial intelligence in large dataset will allow to further improve the performance, and ultimately result in a simpler use in epilepsy care. For cognitive research, it seems to be sufficient to advise to exclude channels with any type of epileptic activity and inside lesional tissue, as by doing so only very few pathological ripples are expected to be encountered, or alternatively to concentrate on evoked responses where separation was shown to be feasible.

Acknowledgements B.F.'s salary is supported by a salary award (Chercheur-boursier clinicien Senior) 2021–2025 of the Fonds de la Recherche du Québec - Santé. This work is supported by the Canadian Institutes of Health Research (PJT-175056 to B.F. and FDN-143208 to J.G.).

References

1. Axmacher N, Elger CE, Fell J (2008) Ripples in the medial temporal lobe are relevant for human memory consolidation. Brain 131:1806–1817
2. Buzsaki G (2015) Hippocampal sharp wave-ripple: a cognitive biomarker for episodic memory and planning. Hippocampus 25:1073–1088
3. Thomschewski A, Hincapie AS, Frauscher B (2019) Localization of the epileptogenic zone using high frequency oscillations. Front Neurol 10:94

4. Bragin A, Engel J Jr, Wilson CL, Fried I, Mathern GW (1999) Hippocampal and entorhinal cortex high-frequency oscillations (100–500 Hz) in human epileptic brain and in kainic acid-treated rates with chronic seizures. Epilepsia 40:127–137
5. Li L, Kumar U, You J et al (2021) Spatial and temporal profile of high-frequency oscillations in posttraumatic epileptogenesis. Neurobiol Dis 161:105544
6. Jirsch JD, Urrestarazu E, LeVan P, Olivier A, Dubeau F, Gotman J (2006) High-frequency oscillations during human focal seizures. Brain 129:1593–1608
7. Frauscher B, Bartolomei F, Kobayashi K et al (2017) High-frequency oscillations: the state of clinical research. Epilepsia 58:1316–1329
8. Buzsaki G, Horvath Z, Urioste R, Hetkie J, Wise K (1992) High-frequency network oscillation in the hippocampus. Science 256:1025–1027
9. Bragin A, Engel J Jr, Wilson CL, Fried I, Buzsaki G (1999) High-frequency oscillations in human brain. Hippocampus 9:137–142
10. Le Van QM, Bragin A, Staba R, Crepon B, Wilson CL, Engel J Jr (2008) Cell type-specific firing during ripple oscillations in the hippocampal formation of humans. J Neurosci 28:6104–6110
11. Halgren E, Wang C, Schomer DL et al (2006) Processing stages underlying word recognition in the anteroventral temporal lobe. Neuroimage 30:1401–1413
12. Jiruska P, Alvarado-Rojas C, Schevon CA et al (2017) Update on the mechanisms and roles of high-frequency oscillations in seizures and epileptic disorders. Epilepsia 58:1330–1339
13. Frauscher B, von Ellenrieder N, Ferrari-Marinho T, Avoli M, Dubeau F, Gotman J (2015) Brain 138:1629–1641
14. Guragain H, Cimbalnik J, Stead M et al (2018) Spatial variation in high-frequency oscillation rates and amplitudes in intracranial EEG. Neurology 90:e639–e646
15. Liu S, Gurses C, Sha Z et al (2018) Stereotyped high-frequency oscillations discriminate seizure onset zones and critical functional cortex in focal epilepsy. Brain 141:713–730
16. von Ellenrieder N, Beltrachini L, Perucca P, Gotman J (2014) Size of cortical generators of epileptic interictal events and visibility on scalp EEG. Neuroimage 94:47–54
17. Blanco JA, Stead M, Krieger A et al (2011) Data mining neocortical high-frequency oscillations in epilepsy and controls. Brain 134:2948–2959
18. Chatillon CE, Zelmann R, Hall JA, Olivier A, Dubeau F, Gotman J (2013) Influence of contact size on the detection of HFOs in human intracerebral EEG recordings. Clin Neurophysiol 124:1541–1546
19. Jacobs J, Wu JY, Perucca P et al (2018) Removing high-frequency oscillations: a prospective multicenter study on seizure outcome. Neurology 91:e1040–e1052
20. Roehri N, Pizzo F, Lagarde S et al (2018) High-frequency oscillations are not better biomarkers of epileptogenic tissue than spikes. Ann Neurol 83:84–97
21. Wang S, Wang IZ, Bulacio JC et al (2013) Ripple classification helps to localize the seizure-onset zone in neocortical epilepsy. Epilepsia 54:370–376
22. Bruder JC, Schmelzeisen C, Lachner-Piza D et al (2021) Physiological ripples associated with sleep spindles can be identified in patients with refractory epilepsy beyond mesio-temporal structures. Front Neurol 12:612293
23. Melani F, Zelmann R, Dubeau F et al (2013) Occurrence of scalp-fast oscillations among patients with different spiking rate and their role as epileptogenicity marker. Epilepsy Res 106:345–356
24. Kerber K, Dumpelmann M, Schelter B et al (2014) Differentiation of specific ripple patterns helps to identify epileptogenic areas for surgical procedures. Clin Neurophysiol 125:1339–1345
25. Nagasawa T, Juhasz C, Rothermel R et al (2012) Spontaneous and visually-driven high-frequency oscillations in the occipital cortex: intracranial recordings in epileptic patients. Hum Brain Mapp 33:569–583.88
26. Matsumoto A, Brinkmann BH, Matthew Stead S et al (2013) Pathological and physiological high-frequency oscillations in focal human epilepsy. J Neurophysiol 110:1958–1964
27. Liu S, Parvizi J (2019) Cognitive refractory state caused by spontaneous epileptic high-frequency oscillations in the human brain. Sci Transl Med 11:eaax7830

28. Dümpelmann M, Jacobs J, Schulze-Bonhage A (2015) Temporal and spatial characteristics of high frequency oscillations as a new biomarker in epilepsy. Epilepsia 56:197–206
29. von Ellenrieder N, Frauscher B, Dubeau F et al (2016) Interaction with slow waves during sleep improves discrimination of physiological and pathological high frequency oscillations (80–500 Hz). Epilepsia 57:869–878
30. Frauscher B, von Ellenrieder N, Zelmann R et al (2018) High-frequency oscillations in the normal human brain. Ann Neurol 84:374–385
31. Alkawadri R, Gaspard N, Goncharaova II et al (2014) The spatial and signal characteristics of physiological high frequency oscillations. Epilepsia 55:1986–1995
32. Malinowska U, Bergey GK, Harezlak J, Jouny CC (2015) Identification of seizure onset zone and preictal state based on characteristics of high frequency oscillations. Clin Neurophysiol 126:1505–1513
33. Frauscher B, von Ellenrieder N, Dubeau F, Gotman J (2016) EEG desynchronization during phasic REM sleep suppresses interictal epileptic activity in humans. Epilepsia 57:879–888
34. Nonoda Y, Miyakoshi M, Ojeda A et al (2016) Interictal high-frequency oscillations generated by seizure onset and eloquent areas may be differentially coupled with different slow waves. Clin Neurophysiol 127:2489–2499
35. Song I, Orosz I, Chervoneva I et al (2017) Bimodal coupling of ripples and slower oscillations during sleep in patients with focal epilepsy. Epilepsia 58:1972–1984
36. Iimura Y, Jones K, Takada L et al (2018) Strong coupling between slow oscillations and wide fast ripples in children with epileptic spasms: investigation of modulation index and occurrence rate. Epilepsia 59:544–554
37. Motoi H, Miyakoshi M, Abel TJ et al (2018) Phase-amplitude coupling between interictal high-frequency activity and slow waves in epilepsy surgery. Epilepsia 59:1954–1965
38. Ponomarenko AA, Korotkova TM, Sergeeva OA, Haas HL (2004) Multiple GABAA receptor subtypes regulate hippocampal ripple oscillations. Eur J Neurosci 20:2141–2148
39. Zijlmans M, Jacobs J, Zelmann R, Dubeau F, Gotman J (2009) High-frequency oscillations mirror disease activity in patients with epilepsy. Neurology 72:979–986
40. Toda Y, Kobayashi K, Hayashi Y, Inoue T, Oka M, Ohtsuka Y (2013) Effects of intravenous diazepam on high-frequency oscillations in EEGs with CSWS. Brain Dev 35:540–547
41. Kudlacek J, Chvojka J, Posusta A et al (2017) Lacosamide and levetiracetam have no effect on sharp-wave ripple rate. Front Neurol 8:169
42. Frauscher B, von Ellenrieder N, Zelmann R et al (2018) Atlas of the normal intracerebral electroencephalogram: neurophysiological awake activity in different cortical areas. Brain 141:1130–1144
43. Von Ellenrieder N, Gotman J, Zelmann R et al (2020) How the human brain sleeps: direct cortical recordings of normal brain activity. Ann Neurol 87:289–301
44. Glasser MF, Van Essen DC (2011) Mapping human cortical areas in vivo based on myelin content as revealed by T1- and T2-weighted MRI. J Neurosci 31:11597–11616
45. Waehnert MD, Dinse J, Schäfer A et al (2016) A subject-specific framework for in vivo myeloarchitectonic analysis using high resolution quantitative MRI. Neuroimage 125:94–107
46. Hunt BA, Tewarie PK, Mougin OE et al (2016) Relationships between cortical myeloarchitecture and electrophysiological networks. Proc Natl Acad Sci USA 113:13510–13515
47. Zweiphenning WJEM, von Ellenrieder N, Dubeau F et al (2022) Correcting for physiological ripples improves epileptic focus identification and outcome prediction. Epilepsia 63:483–496
48. Brazdil M, Cimbalnik J, Roman R et al (2015) Impact of cognitive stimulation on ripples within human epileptic and non-epileptic hippocampus. BMC Neurosci 16:47
49. Steriade M (2005) Sleep, epilepsy and thalamic reticular inhibitory neurons. Trends Neurosci 28:317–324
50. von Ellenrieder N, Dubeau F, Gotman J, Frauscher B (2017) Physiological and pathological high-frequency oscillations have distinct sleep-homeostatic properties. Neuroimage Clin 14:566–573
51. Bruder JC, Dumpelmann M, Piza DL, Mader M, Schulze-Bonhage A, Jacobs-Le VJ (2017) Physiological ripples associated with sleep spindles differ in waveform morphology from epileptic ripples. Int J Neural Syst 27:1750011

52. Alvarado-Rojas C, Huberfeld G, Baulac M et al (2015) Different mechanisms of ripple-like oscillations in the human epileptic subiculum. Ann Neurol 77:281–290
53. Foffani G, Uzcategui ZG, Gal B, Menendez de la Prida L (2007) Reduced spike-timing reliability correlates with the emergence of fast ripples in the rate epileptic hippocampus. Neuron 55:930–941
54. Gloor P (1978) Generalized epilepsy with bilateral synchronous spike and wave discharge. New findings concerning its physiological mechanisms. Electroenceph Clin Neurophysiol Suppl 34:245–249

Chapter 25
Which Rhythms Reflect Bottom-Up and Top-Down Processing?

Yihan Xiong, Pascal Fries, and André M. Bastos

Abstract Top-down processing is how the mind uses our expectations, attentional focus, and other cognitive variables to adaptively influence bottom-up sensory processing. In this Chapter we summarize and review the main sources of evidence for how different neuronal oscillations contribute to top-down and bottom-up processing. We start with a historical and methodological overview with a focus on studies that have provided rich spatio-temporal dynamics to reveal the operations that underlie cognition. We then discuss four primary sources of evidence for how dynamics in the alpha/beta (8–30 Hz) and gamma (40–100 Hz) frequency bands map onto top-down and bottom-up processing, respectively. First, we discuss task manipulations that have isolated bottom-up and top-down processing. Second, we discuss studies that have measured cortical dynamics with laminar resolution. Third, we discuss studies of inter-areal directed connectivity. Fourth, we discuss causal manipulation studies. We end with a discussion of directions for future research to elucidate how top-down versus bottom-up communication is achieved in the brain.

25.1 Introduction—Historical Overview

To understand the brain, it is imperative that we understand multiple complex and interacting spatial and temporal scales of neural activity. The relevant temporal scales of brain activity range from months to years in the case of long-term memories to less than a millisecond when considering tight spike synchrony. The relevant spatial scales are equally wide ranging. At the largest scale, multiple brain-wide networks

Y. Xiong (✉) · A. M. Bastos
Department of Psychology, Vanderbilt University, 111 21St Avenue South, 301 Wilson Hall, Nashville, TN 37240, USA
e-mail: yihan.xiong@Vanderbilt.Edu

P. Fries
Ernst Strüngmann Institute (ESI) gGmbH for Neuroscience in Cooperation With Max Planck Society, Deutschordenstr. 46, 60528 Frankfurt, Germany

Donders Institute for Brain, Cognition, and Behaviour, Radboud University, 6525 EN Nijmegen, The Netherlands

N. Axmacher (ed.), *Intracranial EEG*, Studies in Neuroscience, Psychology and Behavioral Economics, https://doi.org/10.1007/978-3-031-20910-9_25

have been discovered, and recent work pushes some of these networks beyond the brain to include the spinal cord and even the periphery [1]. At the other end of the spatial scale, dendrites and spines are defined at micrometer resolution.

Given this massive range of potentially relevant spatial and temporal scales, it is worthwhile to ask what the optimal spatial and functional scale of measurement is in neuroscience? The single neuron doctrine maintained that the spatial scale of the single neuron would be sufficient to understand the brain [2]. In this view, each neuron has a specific function, and by fine-tuning the parameters of an experiment one could decipher the function (e.g., the neuron's receptive field) of that neuron. Recording from one neuron at a time would be sufficient to understand the brain, as long as the experimenter sampled a sufficiently representative population. This is because the theory assumed that the operations of many neurons would be deducible from single-neuron properties. Many of the experiments based on this theory were indeed foundational to our present understanding of the brain. But these experiments gave us limited insight on the network properties of the brain. The emerging view is that a full understanding of brain function cannot be achieved with a single-neuron perspective alone, but requires the investigation of network properties at the appropriate scale [2, 3].

In the visual system, some of the first compelling support for a network view of the brain came from observations that two groups of neurons fired in gamma-rhythm (40–100 Hz) synchrony when oriented bars were colinear and spanned the receptive fields of both neuronal groups [4]. Other evidence suggested even tighter temporal control of synchrony at the millisecond level was important for information transmission, at least in early stages of visual processing [5].

As the non-invasive field of human neuroimaging flourished, there were multiple observations of large-scale functional networks for cognition. For example, resting-state networks were discovered [6, 7] to involve multiple areas across the brain. The dorsal attention network [8] was discovered and found to powerfully modulate neuronal responses in visual cortex. With these and other observations, the stage was set for scientists who wanted to understand both the network level (because cognition involves multiple interacting regions) as well as the single neuron level (to connect our initial foundational knowledge to the scale of network activity). These studies began to address the question of how single neurons compute information and route that information to both higher and lower order brain areas [9]. Crucially, this level of investigation requires a combination of millisecond temporal resolution, single neuron spatial resolution, with coverage of at least a few brain areas. Millisecond temporal resolution is required because that is the time scale of neural activity and communication. Single neuron spatial resolution is required to understand how spikes in one area affect others. Spatially, it requires sampling from multiple brain regions simultaneously because cognition involves interconnected networks.

In order to achieve this combination of high temporal and spatial resolution across multiple brain areas, it is currently necessary to record intracranially and directly observe neural activity in vivo. In humans, intracranial recordings are only possible in the context of performing a medical procedure, such as pre-surgical functional mapping in epilepsy. For this reason, human intracranial data is limited to recordings from locations that are suspected sites of epileptic activity, and it is difficult to acquire data with the necessary coverage to address these hypotheses. Our review will therefore be focused on animal studies, where intracranial recordings can be designed to address these scientific hypotheses. Wherever possible, we will refer to the human intracranial (and non-invasive) literature to highlight points of contact.

25.2 Methodological Summary

The main intracranial methodologies for observing these fine and fast spatiotemporal patterns are multiple microelectrodes, electrocorticography, and laminar electrodes. The multiple-microelectrode approach was a direct continuation of the methods used originally to study single-unit activity (Fig. 25.1). This method was scaled up first to a few, then a few dozen, and now a few hundred electrodes [10, 11]. This method allows for excellent single unit isolation as well as recording the local field potential which reveals local oscillations. In a modified version, this method has also been used in human intracranial recordings [12].

Electrocorticography is a method that was originally used in human patients suffering from intractable epilepsy. We and others have adapted and "reverse translated" this approach for studies in animals, where grid location, size, and position would be under direct experimental control [13, 14]. This method offers excellent temporal and spatial resolution and good coverage and can record from many cortical areas simultaneously (Fig. 25.1). Finally, multiple-contact laminar electrodes have also been used to record from all layers of cortex simultaneously [15]. Recently, this recording methodology has been expanded to allow for simultaneous recordings of between 5–6 cortical and subcortical areas [16, 17]. These recordings have also been performed in human cortex in rare cases where the recorded tissue is likely to be surgically resected [18].

These methods provide rich data across both spatial and temporal scales. They can be combined to yield even richer data [19]. They can also be complemented by other non-invasive methods such as electroencephalography and magnetoencephalography, which offer excellent temporal information but are limited in spatial scale (Fig. 25.1). On the other hand, full-brain coverage and sub-millimeter spatial scales can be achieved with functional Magnetic Resonance Imaging (fMRI) but at the sacrifice of temporal resolution. Some studies have combined EEG with fMRI to try to compensate, and where possible we will also discuss these studies.

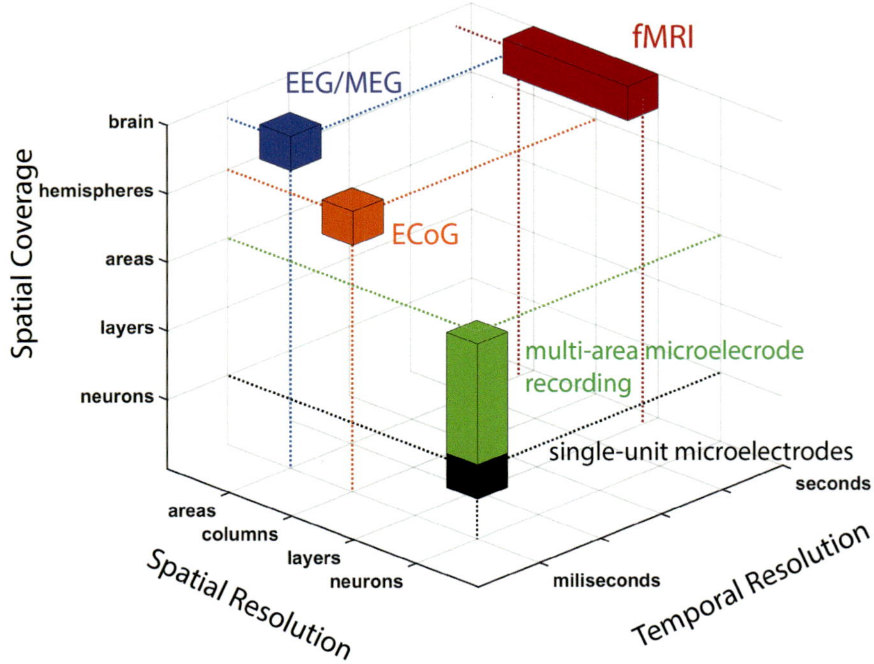

Fig. 25.1 Select current neuroimaging/neurophysiology methods and their respective capabilities. The goal is to increase resolution in the spatial and temporal domains while also increasing coverage. Note that multi-area electrophysiology recordings (e.g., multiple neuropixel probes) are approaching the optimal space in the diagram: large-scale coverage of multiple areas at a spatial resolution capable of recording single neurons at millisecond temporal resolution

25.3 Top-Down Versus Bottom-Up Processing, and Cortical Hierarchy

A high-resolution combination of spatial and temporal scales with good coverage is important for shedding light on the question of how the bottom-up and top-down streams of processing are implemented in the brain. Anatomically, the cortex appears to be constructed as a flexible hierarchy [20, 21]. The sensory periphery projects via thalamus to the first stage of processing, which in vision is area V1. Activity proceeds up the hierarchy from V1 to the temporal and parietal streams, and finally reaches the prefrontal cortex. Bottom-up processing refers to how information flows up the cortical hierarchy. It carries the sensory information about the stimuli, such as color, form, shape, and identity in vision [17, 22]. Top-down processing refers to how information flows down the hierarchy. Top-down information is the process by which our internal thoughts, goals, attention, and expectation shape and influence the bottom-up flow. Cognition arises from an interplay between these streams (Fig. 25.2).

The bottom-up and top-down counterstreams are additionally separated by laminar origin and termination patterns. The cortex is made up of six cortical layers,

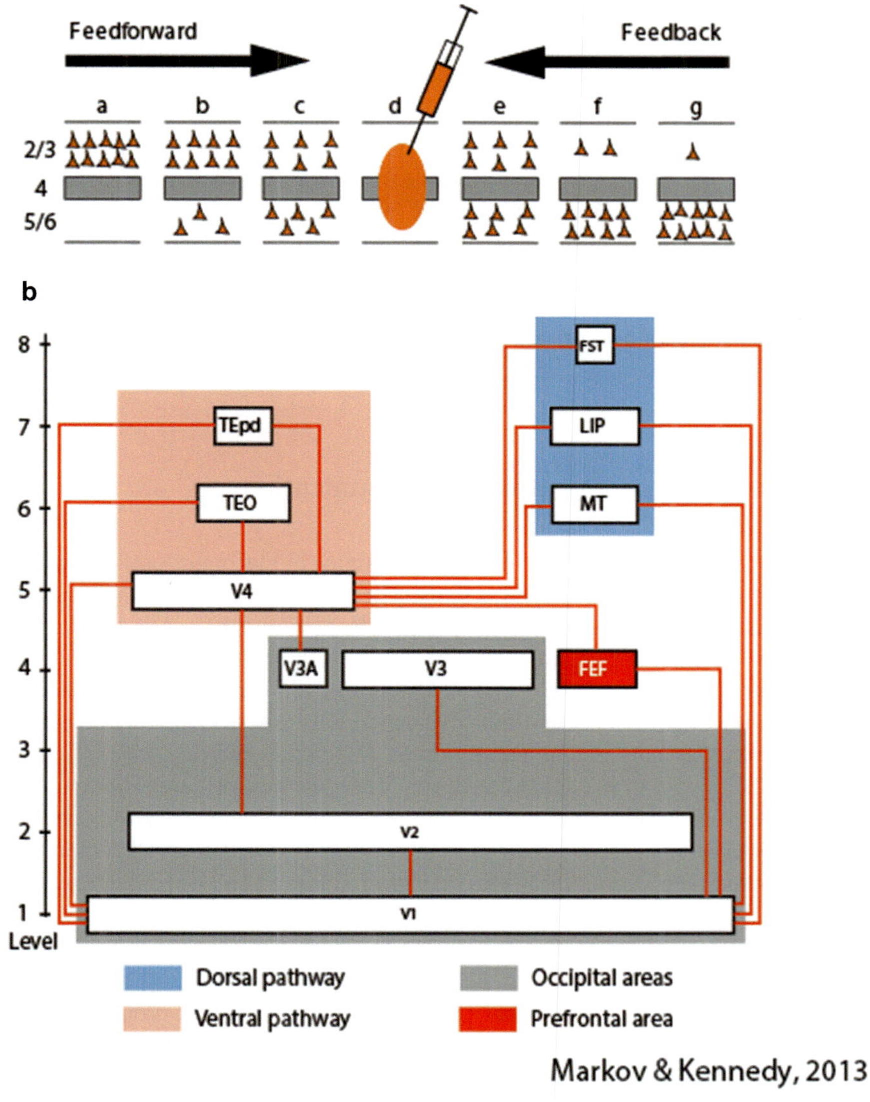

Fig. 25.2 Top-down and bottom-up counterstreams in the cortical hierarchy, with superficial layers (layer 2/3) sending feedforward information, and deep layers (layer 5/6) sending feedback information. Injections of retrograde tracers (indicated by the syringe) reveal a laminar pattern for distribution of feedforward and feedback connections. Feedforward connections arise mostly from superficial layers (patterns shown to the left of the injection site). Feedback connections arise mostly from deep layers (patterns shown to the right of the injection site). The larger the hierarchical distance between the injection site and the origin of the projection cells, the stronger these patterns become. These laminar projection patterns can be quantified and used to form a hierarchical arrangement, which is a result of taking the global model that best fits the entire dataset of multiple inter-areal laminar connection patterns. This hierarchy for the visual system of macaque monkeys is shown in the bottom subpanel. Red lines indicate anatomical connections, and the hierarchical position of each area is shown as different levels on the y-axis. Figure from Markov and Kennedy (2013), reproduced with permission

which are sometimes summarized as superficial (layers 1–3), granular (layer 4) and deep (layers 5/6). These layers largely separate the cells that project in the bottom-up (or feedforward) direction from the top-down (or feedback) direction. In primates, the majority of feedforward-projecting cells that ascend the cortical hierarchy derive from layers 2 and 3. Feedforward projections largely terminate in the middle layers, mainly layer 4, of higher areas [23]. By contrast, the majority of feedback projections arises from cells in layers 5 and 6 [21]. They largely target layers 1 and 6 of earlier cortical areas [23]. In addition to these "rules", there exist deviations from these patterns. For example, there are feedforward connections in deep layers and feedback connections in superficial layers [21]. However, across many cortico-cortical connections, the anatomical patterns explain a great deal of variance in the anatomy. Indeed, the global consistency of these patterns can define a cortical hierarchy.

25.4 Frequencies of Neuronal Communication

In parallel to the constant chatter of spiking activity in the brain, oscillations are another ubiquitous feature of neural activity. These oscillations range in frequency from less than ~1 Hz to over 100 Hz [24]. They often involve the recurrent interplay of inhibitory and excitatory neurons. In one mechanism for gamma and beta oscillations (the Pyramidal Interneuron Network Gamma, or PING model [25]), inhibitory cells shut down the activity of multiple excitatory cells. These excitatory cells tend to recover from the inhibition at a similar time. When they recover, they spike and re-activate the inhibitory cells to fire, shutting down the network and setting up the next oscillatory cycle. The inhibitory neurons are thought to control the frequency of the oscillation through the time-scale of synaptic inhibition (although other elements also contribute to the frequency of the oscillation, for a review see [26]). Modeling evidence suggests that the time constants of cortical networks robustly support these oscillations both in the beta (roughly 15–30 Hz, sometimes grouped with alpha, roughly 8–12 Hz) as well as gamma (40–100 Hz) frequency ranges [27, 28]. What these models suggest is that neuronal oscillations are an emergent property of any neural system containing recurrent inhibition and excitation. Since recurrent excitation/inhibition is a canonical cortical motif, so are the oscillations. Different areas of the cortex appear to have a natural rhythm to their activity [29, 30]. This may be due to variations in the types of cells and circuits that are present across areas [28].

These oscillations provide temporal windows in which spiking activity in a neuronal group becomes relatively more or less likely. Therefore, an oscillation entails the creation of windows of relative excitability and relative inhibition. If two or more separate neuronal groups are oscillating, then this naturally sets up a system where two sets of excitability windows can either coincide in time or not. When the rhythmic input from one neuronal group consistently arrives at the rhythmic excitability window of another, then the coherence between the rhythms likely promotes efficient neuronal communication [31]. On the other hand, if inputs and excitability windows are not aligned, then neuronal communication will be less efficient. This is the basis of the communication through coherence (CTC) hypothesis (Fig. 25.3a).

As discussed above, neuronal oscillations can occur at multiple frequencies. This allows for the interesting possibility that distinct networks of brain areas communicate at separate frequencies. It also allows for the possibility that different oscillatory frequencies can traffic information in different directions, i.e. bottom-up versus top-down [32].

In this Chapter, we will discuss the evidence for whether separate frequency oscillations carry top-down versus bottom-up information. We discuss studies that provide three primary sources of evidence. First we review studies that have relied on a cognitive operationalization of bottom-up versus top-down processing. This approach primarily uses a task contrast to reveal which neuronal dynamics are related to processes that are defined as environmental (bottom-up) versus based on internal knowledge (top-down). Second, we discuss studies that have used laminar electrodes and distinguished activity from superficial versus deep layers. The majority of bottom-up connections (also called feedforward) derive from superficial layers, and the majority of top-down connections (also called feedback) derive from deep layers (Fig. 25.2). Therefore, if certain dynamics are primarily contained within these distinct compartments, they will likely primarily affect either bottom-up or top-down processing. Third, we will discuss studies that have recorded from multiple locations in cortex simultaneously and quantified directed influences in the bottom-up versus top-down directions using analytic methods such as Granger causality. Finally, we will also discuss studies that have used causal manipulation to inform which mechanisms contribute to bottom-up versus top-down neuronal communication.

25.5 Part 1: Cognitive Operationalization

One approach to study the computations subserving top-down and bottom-up processing is by operationalizing a task which separately manipulates these factors. First, we will define these terms. By a bottom-up task factor, what we mean is a factor relating to the sensory attributes of the stimuli that can be presented to a subject. A top-down factor is one which relies on internal knowledge or rules that have been acquired through behavioral training (in the case of animal experiments) or verbal instructions (in the case of human subjects). By definition, a top-down factor cannot be deduced solely on the basis of the stimulus configuration presented to a subject at a given time.

A classic example of this is visual attention. In visual attention, "pop-out" (or "efficient") search refers to a condition in which the bottom-up features of a stimulus are sufficient to distinguish target from distractor. For example, a red target that has to be found amongst an array of green distractors is a pop-out search. Because the target differs from all distractors along a salient dimension (color), the visual system can extract the target through bottom-up mechanisms. In contrast, in a situation where the target and distractor differ through the conjunction of more than one feature

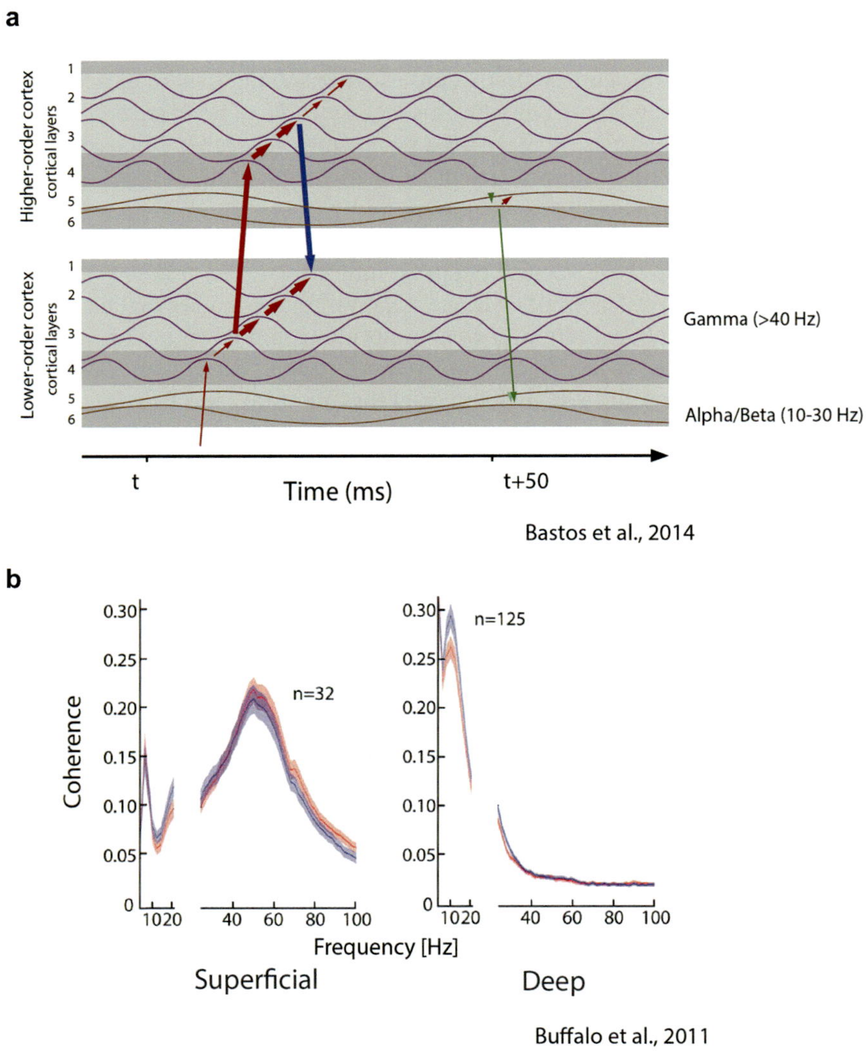

Fig. 25.3 Panel **a** shows the mechanism of Communication Through Coherence (CTC), where oscillatory coherence facilitates higher- and lower-order cortex to effectively communicate. Superficial layer gamma-frequency coherence mediates feedforward (lower to higher-order cortex, shown in red arrows) interactions and deep layer beta frequency coherence mediates feedback (higher to lower-order cortex, shown in green arrows) interactions. Activity in higher-order cortex is phase-delayed at gamma frequencies relative to activity in lower-order cortex. Activity in lower-order cortex is phase-delayed at beta frequencies relative to activity in higher-order cortex. Panel **b** shows empirical results indicating that spike-field coherence peaks in superficial layers at gamma range frequencies, while in deep layers it peaks at alpha/beta range frequencies (the red and blue lines indicate different attentional states). Red means visual spatial attention was directed into the receptive field of the recorded neurons. Blue means visual spatial attention was directed away from the receptive field of the recorded neurons. Panel **a** from Bastos et al. (2014), reproduced with permission. Panel **b** from Buffalo et al. (2011), reproduced with permission

dimension, such as color and orientation, a more deliberate search strategy is needed which entails top-down processing.

Buschman and Miller trained macaque monkeys on such a visual search task and recorded with multiple Tungsten electrodes in parietal and prefrontal cortex [33]. They recorded both spikes and local field potentials from area LIP (Lateral Intraparietal area), FEF (Frontal Eye Fields), and lateral PFC (Prefrontal Cortex). As they recorded neural activity, monkeys performed the attention task in either a bottom-up (pop-out search) or top-down (conjunction search) mode. They computed the spectral coherence between the sites in frontal and parietal cortex. Spectral coherence is a method to quantify the strength of phase alignment of oscillations in different frequency bands which is thought to reflect neuronal communication [34]. Comparing bottom-up to top-down search, they found that coherence during top-down search was stronger in the beta-band, and coherence during bottom-up search was stronger in the gamma-band. Thus, different oscillations were involved in implementing neuronal communication between frontal and parietal cortices depending on task demands.

In human intracranial work, top-down attention has also been linked to enhanced beta frequency neuronal synchronization between frontal and parietal cortex [35]. In this study, Micheli et al. recorded intracranial data from epilepsy patients as they performed a visual attention task. In the task, subjects had to report near-threshold changes to one of four targets on a computer display. During the task periods requiring sustained attention, Micheli et al. observed a larger amount of frontoparietal beta-band synchronization on trials in which subjects correctly reported a change. This supports the view that long-distance beta synchrony supports a cognitive, top-down network [36].

Another cognitive framework that is thought to map onto bottom-up versus top-down processing is predictive coding [37]. In this case, prediction refers to learning the statistical regularities of the environment. These predictions are thought to be stored as internal models that can guide perception and action, especially in noisy environments. The predictions are thought to reflect top-down processing, because they reflect internal knowledge created by experience. When predictions do not match the environment, they need to be updated. These prediction-error signals are thought to be implemented by bottom-up processing, because they reflect the features of sensory inputs that deviate from prediction.

Chao et al. were the first to test this hypothesis using intracranial data in monkeys [38]. They used an auditory oddball task to evoke simple sensory prediction errors and prediction updates. They implanted monkeys with large-scale electrocorticography (ECoG) grids to record from multiple cortical locations. These recordings spanned multiple hierarchical stages, including two stages of the auditory processing hierarchy and lateral PFC. Auditory oddballs (stimuli hypothesized to evoke "prediction errors") induced gamma frequency oscillations that emerged first in early auditory cortex, then later-stage auditory cortex, and then frontal cortex. In contrast, a prediction update signal was identified to begin later (about 400–600 ms after the occurrence of a stimulus which deviated from what was predicted) and was centered on the beta band. Granger causality analysis (a method to infer directed functional connectivity, see "Part 3: Inter-areal interactions") showed that prefrontal channels Granger-caused the temporal/auditory sites more than the other way around. This was consistent with a top-down role for beta in signaling a prediction update.

Another recent macaque study has complemented and extended this work by showing that the predictability of visual stimuli used in a working memory task changes the balance between beta and gamma oscillations [16]. Stimuli that occurred in a predictable context engaged top-down alpha and beta rhythmic networks spanning prefrontal, parietal, and visual cortex (area V4). Within area V4, alpha/beta power in deep layers was higher in response to a predictable than to an unpredictable stimulus, but only when the predictable stimulus matched the stimulus preference of the V4 neurons. Put another way, the same neurons did not show an alpha/beta modulation to predictability when the prediction concerned a non-preferred stimulus. In other words, predictions were acting at specific cortical locations to inhibit processing. In the absence of such stimulus-specific predictions, cortex was relatively uninhibited. A non-predictable stimulus evoked strong superficial-layer bottom-up gamma. Therefore, the neural implementation of prediction may work by selectively preparing, and therefore "routing" information along predictable routes via top-down beta. Unpredicted information (reflecting a prediction "error") would then naturally be re-routed in the bottom-up direction via uninhibited pathways [16].

Multiple studies in humans have also investigated the neurophysiological mechanisms for top-down prediction and bottom-up prediction error. Bauer and colleagues used temporal expectation to induce varying levels of prediction to an attended visual grating stimulus [39]. As time in the trial progressed, a change in the stimulus became more likely to occur. Given this hazard rate function, changes in the stimulus that occurred early in the trial were less predictable than changes occurring later in the trial. Attentional modulation of alpha and gamma power tracked this stimulus predictability. Gamma was negatively correlated with predictability (therefore gamma reflected unpredictable signals), and alpha was positively linked with predictability. Intracranial studies in humans have also manipulated auditory prediction using oddball tasks. A human intracranial study linked prediction errors or "surprising" aspects of an auditory stimulus to gamma oscillations in auditory cortex [40]. Other studies have also linked unpredictable auditory stimuli to gamma oscillations [41–43]. To summarize, these studies on predictable stimuli suggest gamma as a bottom-up mechanism for prediction errors and beta as a top-down mechanism for predictions or their updates.

In addition to attention and prediction, many other tasks have been used to operationalize top-down versus bottom-up processing. For example, monkeys can be trained to apply different rules (e.g. using "Color" vs. "Orientation") to the exact same bottom-up stimulus [44, 45]. They can also learn to perform concrete or abstract categorization judgements (e.g., "Dog" vs. "Cat" or abstract dot categories) or learn to associate one stimulus with another. All of these tasks engage top-down processing, because they can only be solved by applying internal knowledge to a stimulus set. Without exception, these studies have found that differences in higher-order cortex alpha/beta power and/or coherence (but not gamma) discriminate across different top-down task conditions [44–51]. This is consistent with the hypothesized role of alpha/beta in creating dynamic neuronal assemblies that facilitate top-down processing.

In addition to these studies that have focused on bottom-up versus top-down task modulations, there is a simple neurophysiological observation that links distinct neuronal oscillations to distinct forms of processing. Namely, the observation that gamma-band activity is strongly induced by a sensory stimulus is consistent with this frequency band signaling a sensory representation [31, 52]. By contrast, alpha/beta band signals are typically stronger in the pre-trial and baseline periods, which in most experiments is devoid of sensory stimulation. What is present in these pre-task, beta-rich epochs? The subject's internal chatter which is thought to be indexed by the default-mode network. In addition, this default-mode activity is highly correlated with beta band oscillations in human EEG [53]. Therefore, it makes sense that an internal status-quo is represented by beta band rhythmic activity [54].

These observations are also consistent with the positive correlation between gamma and spiking [55, 56], and with the negative correlation between alpha/beta and spiking [29, 57]. This is because task conditions that lead to large increases in spiking (and gamma) are usually associated with bottom-up sensory stimulation. In contrast, alpha/beta rhythms are frequently observed in pre-stimulation or delay-period epochs with relatively less spiking, when top-down processing is active.

25.6 Part 2: Laminar Studies

Several studies have examined neural activity in different layers of cortex within and between areas. Since the bottom-up and top-down streams are to a large extent separated between layers (Fig. 25.2), these studies should reveal whether distinct layers and oscillatory frequencies contribute to these counterstreams.

Combining the laminar and cognitive operationalization approaches, Lakatos et al. showed attention and sensory processing in distinct layers [15]. The authors recorded electrophysiological signals in area V1 and primary auditory cortex of macaques that were trained to alternate between paying attention to a stream of sensory inputs in either the visual or auditory modalities. They found that the middle layers of cortex primarily signaled the current sensory context. Interestingly, the superficial layers behaved in a more "cognitive" way. Activity there was primarily reporting the attended input. This was a landmark study that showed how important it was to consider the laminar dimension in cognition, because different layers were apparently performing distinct computations. This study paved the way for others looking further into how distinct oscillations across layers behave during cognitive tasks.

Van Kerkoerle et al. used multi-laminar probes to study the propagation of gamma and alpha band frequencies in macaque V1 [58]. The researchers phase-aligned the Local Field Potential (LFP), Current Source Density (a spatially more resolved version of the LFP), and spiking activity to the trough of the oscillation detected in layer 4 (recall that layer 4 is the input layer for feedforward connections). This trough detection procedure was performed separately for the gamma and alpha bands. They discovered a remarkable frequency dissociation: The gamma-band activity was organized as a travelling wave, with superficial and deep layers following waves in layer 4. This suggests that the gamma activity originated in layer 4 and then subsequently drove the gamma activity in more superficial and deeper layers. This functional pattern of columnar activation mirrors the intracortical feedforward canonical circuit [59]. It is also consistent with earlier observations that spiking activity in superficial cortical layers is phase delayed relative to layer 4 in squirrel monkeys [60]. Van Kerkoerle et al. next triggered their analysis to the trough of the alpha wave in layer 4. They again found a spatially coherent travelling wave pattern but with the opposite direction: layer 4 was now delayed relative to more superficial and deep layers. This suggested a feedback pattern of activity, with the alpha wave originating, or entering from other areas, in feedback recipient layers 1 and 6 and then progressing to layer 4.

These observations are consistent with other work based on within-area functional connectivity analysis (for inter-areal interactions, see Part 3). Several studies have used Granger causality [61], a time-series methodology to infer the primary direction of information flow between signals. It is beyond the scope of this review to discuss all the technical aspects of this method [62, 63]. Granger causality quantifies the amount of frequency-specific directed information flow between two signals A and B. If A is below B in the cortical hierarchy then the directed information flow from A to B is the bottom-up flow and the flow from B to A is the top-down flow. It was used to measure how alpha/beta rhythms flow between layers. In early visual cortex (V1 and V2) as well as frontal cortex, LFP signals from deeper layers (both granular layer 4 and infragranular layers 5 and 6) drive activity in more superficial layers 2 and 3 [58, 64, 65] in the alpha frequency range. Several studies have also quantified functional interactions between layers with phase-amplitude coupling analysis. This measure quantifies how the phase of a lower frequency (alpha/beta) couples to the amplitude of a higher frequency (the gamma-band). These studies examined all possible pairs of channels for providing phase and amplitude in superficial and deep layers and were therefore not biased by any possible electrode pre-selection steps. Both Spaak et al. as well as Bastos et al. found that deep alpha/beta phase coupled to gamma amplitude in superficial layers [64, 66]. This was observed in visual as well as frontal and prefrontal cortex (but see [67]).

Several additional observations from other laminar studies strengthened the hypothesis that different layers were differentially involved in generating gamma versus alpha. Maier et al. discovered separate superficial and deep laminar compartments for coherent activity [68]. Within both deep and superficial layers, LFP activity was highly coherent. But between laminar compartments, coherence levels fell quite quickly. These effects were observed both in the gamma and alpha/beta frequency ranges. They are consistent with the study of [69], which analyzed spike-field coherence between spikes and LFPs recorded either in superficial or deep layers. These authors found that superficial neurons tended to strongly synchronize to the gamma-band LFP but not very much to alpha. By contrast, neurons in deep cortical layers showed relatively little gamma-band synchronization and instead preferred to synchronize at the alpha range [69]. Finally, recordings in brain slices showed that superficial layers can resonate at gamma and deep layers at beta even after the slice has been cut at layer 4, severing the connections between layers [70].

These studies suggested separate laminar mechanisms with superficial layers resonating at gamma and deep layers resonating at alpha/beta frequency. We note that two studies have argued against this based on the observation that a local bipolar or current source density analysis will frequently eliminate nearly all alpha/beta power in deep layers while sparing alpha/beta in superficial layers [18, 71]. This observation could be explained by a source in deep layers that is highly coherent across the deep layers and which is thereby largely eliminated by the local spatial differentiation involved in the calculation of both local bipolar and of current-source-density derivations. Studies that have examined the coherence of alpha are consistent with this: alpha band synchronization is high between neighboring sites in deep layers and between nearby cortical columns. This is in contrast to coherence in the gamma-band, which is also high within superficial layers but drops off precipitously between channels in superficial and deep layers and between nearby cortical columns [72].

In addition, the invention of sub-millimeter (typically around 0.7 mm) resolution fMRI now allows non-invasive data collection in humans to corroborate the laminar mechanisms found in animal models. Notably, Scheeringa et al. combined EEG with high resolution fMRI to show that the superficial-layer BOLD signal correlates with gamma band power [73]. Also, high-resolution source modeling with MEG has shown consistently that alpha/beta sources localize to deeper locations than gamma [74]. There has also been evidence that BOLD-signal correlations reflect oscillatory synchronization sampled by MEG [75] suggesting that fMRI can provide a useful proxy for neuronal interactions. Multiple studies on working memory and prediction using layer-resolved fMRI showed that feedback information in forms of prediction [76], visual illusion [77], and working memory [78] evoked deep layer BOLD signal in V1. However, one study by Muckli et al. showed superficial layer in V1 carrying feedback information [79]. This divergence has been argued to be due to methodological differences in data collection and analysis [78]. Nevertheless, all studies concur that feedback information is signaled in non-middle layers. The literature remains to be strengthened by further investigations using neurophysiological techniques discussed in this chapter.

25.7 Part 3: Inter-Areal Interactions

Another approach to assess which oscillations are involved in bottom-up and top-down processing is to measure electrophysiological signals simultaneously from multiple brain areas and to quantify the directionality of signal flow. This requires the application of signal processing methods for directed functional connectivity, such as Granger causality.

Bastos et al. (2015) recorded electrocorticography from eight visual areas in the cortex of macaques as they performed a visual attention task that engaged both top-down processing (in this case, selective attention) as well as bottom-up processing

(processing of visual stimuli) [32]. They quantified Granger causality between the visual areas and found a common pattern. Beta frequency oscillations were stronger in the top-down direction, and theta and gamma oscillations were stronger in the bottom-up direction (Fig. 25.4a). These patterns were highly correlated with the underlying laminar anatomical patterns that define top-down versus bottom-up connectivity (Fig. 25.2). These feedforward gamma and feedback beta patterns were recently expanded to an even greater number of areas and shown to be related to the strength of anatomical projections between areas [80]. This strong anatomy-function relationship allowed us to derive a cortical hierarchy based on the functional data alone, and this functional hierarchy corresponded closely to the anatomical hierarchy (Fig. 25.4b). Selective visual attention enhanced these inter-areal gamma and beta interactions.

In a closely related study, Michalareas et al. (2016) analyzed source-projected MEG data of human subjects as they monitored a visual grating stimulus [81]. By performing Granger causality on the source-projected data they found similar functional asymmetries in the human brain as was previous detected within the macaque brain. Namely, Granger causality analysis indicated that beta oscillations were stronger in the top-down direction, and gamma was stronger in the bottom-up direction. The functional connectivity based on Granger causality from human data correlated with laminar anatomical connectivity in macaques for the cross-species homologous areas. Even more interestingly, the authors constructed a functional hierarchy amongst all source-projected areas including areas for which there is no homologue between humans and monkeys (Fig. 25.4c). This top-down direction of alpha/beta propagation has also been confirmed in an additional human MEG study which examined interactions between areas FEF and V1 [82]

The above studies have used Granger-causality (GC) based functional connectivity analysis to infer directed information flow between distinct areas. These studies have indicated that the patterns of GC correlate with layer-specific anatomy. Several studies have now also investigated the laminar specificity of GC. In a groundbreaking study, Roberts et al. (2013) recorded from retinotopically aligned locations of areas V1 and V2 with laminar probes [83]. They found that the laminar pattern of coherence and Granger causality largely matched the feedforward pattern known from anatomy. Namely, the superficial layers of V1 had strong gamma coherence and Granger causal influence onto layer 4 (the anatomical input layer) of V2. In addition, as the contrast of the presented visual image increased, the dominant frequency of inter-areal synchronization gradually shifted to faster gamma frequencies. This confirmed previous theoretical work that suggested more total excitation would lead to faster gamma [25]. However, the spatial pattern defining which layers participated in this gamma synchronization was largely conserved across contrasts and matched the anatomical feedforward pattern. Ferro and Thiele also investigated the laminar pattern of GC between visual areas V1 and V4 during an attention task. They confirmed that the bottom-up flow was dominated by gamma frequencies and the top-down flow by beta frequencies. However, in this study, the laminar pattern was not well-predicted by the anatomy and was relatively non-specific when it came to layers [84].

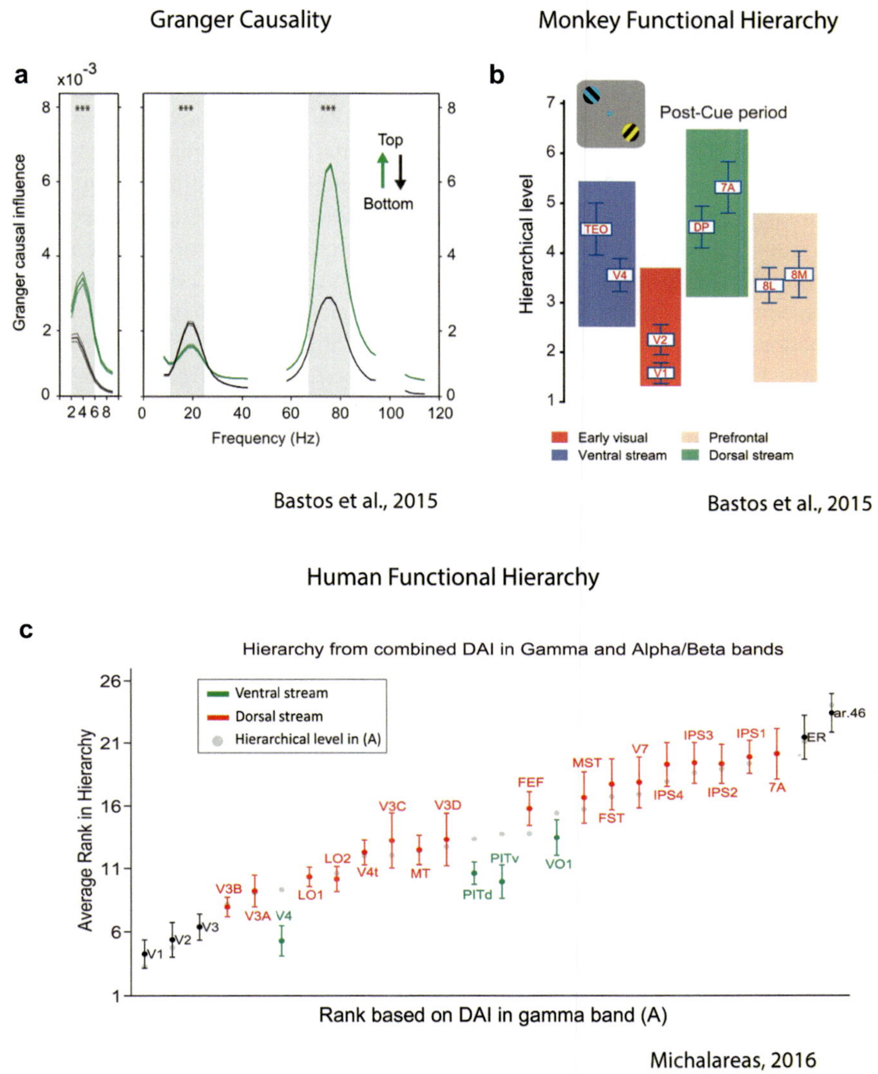

Fig. 25.4 **a** Granger causality shows functional asymmetries between bottom-up communication at gamma and theta versus top-down communication at beta frequencies. **b** The global consistency of these metrics allows us to compute a functional hierarchy in the monkey brain based on Granger causality, which is remarkably consistent with anatomy-based hierarchies (**c**). These Granger metrics can also be computed in the human brain including uniquely human areas using source-projected MEG which defines a human functional hierarchy. Panels **a**, **b** from Bastos et al. (2015), reproduced with permission, Panel **c** from Michalareas et al. (2016), reproduced with permission.

Another recent study combined the cognitive operationalization and inter-areal interaction strategies of analysis to ask which layers, frequencies, and directions of information flow were associated with visual stimuli that were either predictable or unpredictable [16]. These authors performed multi-laminar recordings in monkeys from V4, parietal areas 7A and LIP, FEF, and prefrontal cortex as monkeys performed a working memory task on either predictable or unpredictable stimuli (Fig. 25.5). They compared the patterns of GC in either predictable or unpredictable cases. In the pre-stimulus period when the visual stimulus was predictable but not yet seen, there was stronger deep-layer top-down alpha/beta GC from prefrontal cortex to the other areas. In contrast, during the visual stimulation period, there was stronger superficial-layer bottom-up gamma GC from V4 to the other areas, when the visual stimulus was unpredictable. Therefore, the bottom-up (superficial-layer) gamma flow was reflecting an unpredictable or surprising stimulus. The processing of unpredictable stimuli evoked particularly strong responses in gamma and spiking in superficial layers of V4 (Fig. 25.5). In contrast, the top-down (deep-layer) beta flow from PFC to other areas was stronger during a time period in which a predictable stimulus was about to be seen.

These patterns of functional signal flow have also been quantified in monkeys performing visual working memory tasks. These tasks require subjects to maintain information in working memory across a delay period and then report which item was

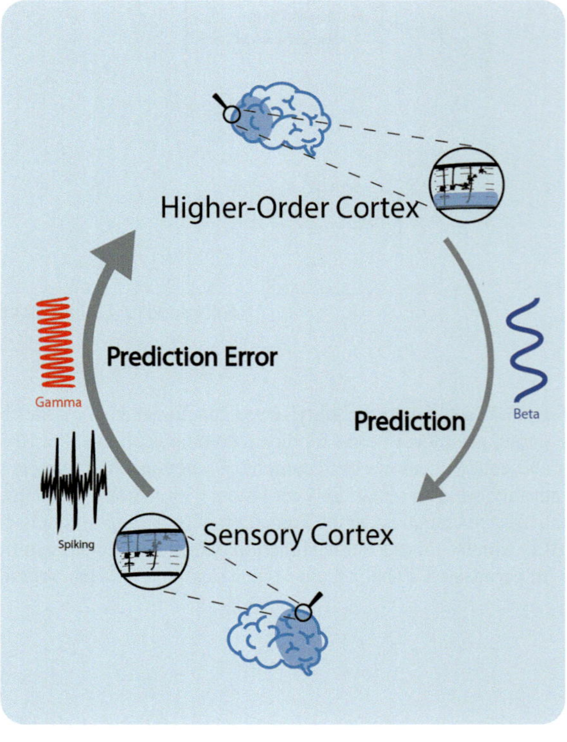

Fig. 25.5 Summary of layer-specific top-down and bottom-up information streams. Prediction error (including novel or unpredictable information) travels from sensory cortex to higher order cortex through gamma frequency and spiking in superficial layers; prediction (for example, what a subject expects to see but has not yet seen) travels from higher-order cortex through alpha/beta frequency in deep layers

previously seen. Salazar et al. (2012) recorded spiking and LFP activity in various areas of prefrontal and parietal cortex as monkeys performed the working memory task [85]. They found content-specific patterns of frontoparietal beta-frequency synchronization. In addition, they found a pattern of Granger causality in which the flow from parietal cortex to prefrontal cortex was larger than vice-versa. Both frontal and parietal cortex participate in top-down functions [8]. Thereby, these results provide additional data to suggest that at the top levels of cortical hierarchy, beta-band oscillations are critical for signaling the cognitive context [52].

Most of the available data on directed information flow has been obtained in animals. However, directed functional connectivity analysis can also be performed on human intracranial data. Indeed, Fontolan et al. recorded from two areas of the auditory cortex in human subjects [86], performed Granger causality analysis and discovered a frequency dissociation between bottom-up and top-down connectivity similar to the macaque studies reviewed above. Additional studies using intracranial data in humans are necessary to confirm and expand these observations beyond the auditory system.

25.8 Part 4: Causal Manipulation Studies

Another approach for mapping which frequencies are transmitted in the bottom-up versus top-down direction is to causally manipulate the brain. This is especially advantageous for studying top-down functions because they are notoriously difficult to experimentally control, and attempts to do so are always based on assumptions about how task modulation of top-down processing can be achieved. There are several methods capable of manipulating top-down and bottom-up communication. Some examples are optogenetics [87, 88], electrical microstimulation, transcranial magnetic stimulation (TMS) and pharmacology. A coarser approach is to lesion parts of cortex either at the top or the bottom of the hierarchy and record the electrophysiological consequences in other areas. For this chapter, we will focus on electrical stimulation and lesions in monkeys and TMS in humans, as studies with these methods have provided significant insights into top-down and bottom-up communication.

Stimulation in one area of the brain and recordings in other areas should reveal which oscillatory frequencies are transmitted between areas. The logic is that stimulating early areas of the cortical hierarchy will selectively induce bottom-up communication. Stimulating higher areas of the brain and recording in earlier areas will induce top-down communication. Van Kerkoerle et al. performed electrical microstimulation and recordings in V1 and V4 as monkeys viewed a textured stimulus [58]. Microstimulation in V1 induced gamma-power increases in V4. These effects were independent of what visual stimulus monkeys were viewing. Microstimulation in V4 induced alpha power in V1 to increase, and gamma power in V1 to *decrease*. These effects were specific to conditions when the receptive fields of the stimulated neurons fell on the background of the figure. In addition to the frequency-specificity, this suggests a role for top-down connections in modulating the background (but not foreground) of visual processing. They are also consistent with a role for top-down alpha frequencies in inhibiting neuronal activity by decreasing gamma. Overall, the results strongly support a role for gamma in bottom-up communication and alpha in top-down communication.

In humans, both bottom-up and top-down communication have been studied using a similar experimental logic. To study top-down communication, Veniero et al. used TMS to deliver a single pulse of excitation in human FEF and recorded whole-scalp EEG [89]. They examined which oscillatory components were induced by the single pulse in FEF. They observed a phase-reset of beta oscillations with a peak in magnitude occurring in occipital cortex, consistent with a top-down direction of communication. They also used dual-pulse TMS over FEF and visual cortex to test whether TMS to FEF modulates cortical excitability in visual cortex (measured as the probability of detecting a TMS-induced phosphene). They first pulsed the FEF and then applied TMS to visual cortex and asked subjects whether they perceived a phosphene. By varying the delay between the FEF and visual cortex TMS pulses they concluded that the subjects' visual phosphene detection was rhythmically modulated at a beta-rhythmic rate of ~16 Hz.

Another human study determined the probability that intracranial pulses of electrical stimulation to occipital sites would induce phosphenes [90]. The researchers titrated the levels of electrical current to the occipital areas to induce phosphenes on approximately half the trials. As they applied stimulation they also recorded from higher-level visual cortex in the Temporal-Parietal Junction (area TPJ). Trials that successfully induced phosphene perception were associated with increases in gamma-band oscillations in TPJ. These gamma oscillations were not observed in trials without phosphene perception. This supports the role of activity at gamma frequencies in human visual cortex in bottom-up communication and further suggests that gamma may be necessary for conscious processing (Fig. 25.6).

Another approach for causal studies is lesion work. Lesions to both prefrontal and primary visual cortex of macaques have been performed to study the behavioral effects on visual attention and visual processing and the neural effects in V4. Importantly, area V4 receives both bottom-up inputs from V1 as well as top-down inputs from the PFC. Schmiedt et al. trained monkeys on a visual detection task and recorded neural activity in area V4 with a chronic Utah array [91]. In the intact (non-lesioned) case, a visual grating stimulus evoked strong gamma-band power increases in V4 and alpha/beta band suppression. After a lesion to V1 that dramatically reduced the monkey's ability to perform the visual detection task, gamma (40–100 Hz) in V4 was dramatically reduced in power and delayed in time by roughly 200 ms. Alpha/beta power was no longer suppressed with a visual stimulus. Instead, alpha/beta became facilitated by a visual stimulus. Another recent study largely confirmed these results in the high-gamma band [92]. In the low-gamma band/high beta band centered at 30 Hz, Kienitz et al. also observed lesion-specific power facilitation. Therefore, in the higher-frequency gamma range above ~40 Hz, both studies suggest a massive loss of V4 gamma to V1 lesion. As the bottom-up anatomical pathway is damaged, gamma (and theta) oscillations in higher-order areas are strongly reduced.

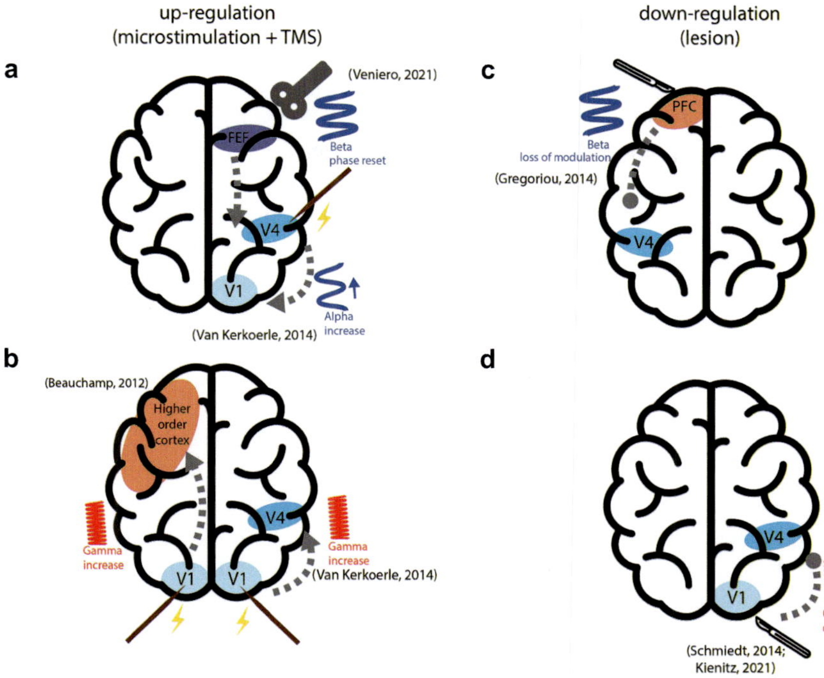

Fig. 25.6 Summary of studies involving causal manipulations of bottom-up and top-down pathways, and their effects on neuronal oscillation. **a** Microstimulation in V4 causes alpha increases in V1 (Van Kerkoerle 2014) and TMS in FEF causes beta phase-reset in visual cortex (Veniero 2021). **b** Microstimulation in V1 causes gamma enhancement in V4 (Van Kerkoerle 2014). **c** Lesion in PFC causes loss of beta attentional modulation in V4 and reduction of attentional modulation in gamma/spiking (Gregoriou 2014). D: Lesion in V1 causes loss/reduction of gamma in V4 (Schmiedt 2014; Kienitz 2021)

Gregoriou et al. performed neurophysiological recordings in macaque area V4 during a visual attention task and lesioned the PFC, a critical node of the attention network [93]. Previous studies had shown that the neurophysiological consequences of attention in V4 is to enhance firing rates and gamma-frequency power and coherence [94] while at the same time decreasing the alpha/beta power and coherence [69]. Gregoriou et al. (2014) therefore asked which of these frequency-specific attentional modulations were dependent on top-down PFC-to-V4 inputs. They recorded in both hemispheres of V4, allowing a within-subject comparison of attentional modulation in the affected versus unaffected hemispheres. PFC lesion did not have a substantial effect on overall firing rates in V4 during the sustained visual attention task to a visual grating stimulus. This indicates that PFC lesions do not alter the bottom-up excitability of V4. They did however alter the attentional modulation of both firing rates and gamma in V4. Both were strongly reduced but not eliminated. The attentional modulation on beta oscillations in V4 was eliminated.

Together, this work suggests that a V1 lesion will strongly reduce the stimulus-driven bottom-up gamma (and theta) and change the sign of alpha/beta modulation from normal suppression to facilitation. The PFC-lesion work suggests that attentional modulation of oscillatory dynamics in V4 are affected in both beta and gamma bands. Importantly, however, whereas the gamma (and spiking) attentional modulation is still weakly present in V4 after PFC lesion, the alpha/beta modulation is eliminated. Overall, the body of work supports a frequency dissociation between top-down and bottom-up neuronal communication. However, a permanent lesion will always perturb the specific pathway of interest (in these studies from V1 to V4 or from PFC to V4) but in addition cause unmasking of other inputs that may not normally contribute or induce plastic adaptation in the system. These non-specific factors can never be separated from the effects of the lesion itself. These limitations can be addressed by the use of pharmacology whereby inhibitory drugs are delivered intracortically to induce a temporary lesion. While some studies have pharmacologically inactivated PFC [95, 96] to our knowledge they have not reported on the frequency specificity of these effects on downstream structures.

What is needed in future causal studies is more precise control over the bottom-up and top-down pathways to precisely induce excitation or inhibition in cell-specific populations. Such fine control is possible with optogenetics, although few studies in non-human primates have used this technique to map out functional bottom-up versus top-down pathways and the neuronal oscillations on which they operate.

25.9 Conclusion and Future Directions

In this Chapter, we have reviewed four classes of evidence for alpha/beta and gamma oscillations as representing distinct cortical mechanisms. This evidence suggests beta as a frequency channel for top-down communication and gamma for bottom-up communication. Starting with cognitive tasks that have operationalized bottom-up versus top-down modulation, there is mounting evidence that these computations map to distinct (gamma and beta, respectively) oscillations. We next considered whether the sub-layers of cortex have distinct dynamics. Since the mapping between superficial and deep layers reflects a largely preserved segregation between bottom-up and top-down anatomical connectivity, studies suggesting different oscillatory mechanisms in these layers are suggestive of a role also in inter-areal communication. We then considered studies that have recorded neurophysiology in different areas of the hierarchy and quantified directed connectivity using Granger causality. Finally, we discussed causal manipulation studies where either the top-down or bottom-up pathways have been enhanced or eliminated. Across these broad classes of studies, there is general support for gamma and beta subserving different directions of neuronal communication and types of cognition.

Despite this body of work there are clearly many open questions for future research. A particular weakness in the current literature is that most causal manipulation studies are based on lesion studies. While important, they should be complemented by studies which enhance or eliminate top-down and bottom-up interactions on a finer scale. In addition, most studies have focused on measuring interactions using the local field potential. These studies should eventually be complemented by a more detailed picture of the underlying neuronal level and how distinct cell types support these different dynamics [97, 98] and patterns of inter-areal communication [17]. Also, we have focused in this Chapter on beta and gamma dynamics, but theta range (~2–6 Hz) dynamics have also been implicated in bottom-up processing [32, 92]. Theta and gamma dynamics have been shown to interact via phase-amplitude coupling [99], and future studies should elucidate to what extent this theta/gamma code supports bottom-up processing, rhythmic sampling [100], or other computations. Our focus here has been on a cortex-centric view. However, it is clearly the case that thalamic and other sub-cortical structures modulate beta and gamma oscillations in cortex [101–104]. Future studies should clarify and expand the underlying roles of cortex versus thalamus in controlling beta and gamma dynamics during cognition.

Another role for future research is to expand upon the computational methods used to measure signal flow. Most studies on bottom-up and top-down interactions have used Granger causality. While a useful tool, it should be complemented by other analytical methods, including computational modeling [105–107] and information theory based tools [106, 108]. This more theoretical and computational approach can help address the important question of how top-down beta and bottom-up gamma interact [109]. Finally, these studies in animal models will also be increasingly complemented and extended by richer and richer intracranial datasets in humans [110–112]. Together, these basic science studies will advance our understanding of the underlying brain mechanisms for cognition and ultimately, how they break down in a variety of cognitive disorders.

References

1. Varela F, Lachaux JP, Rodriguez E, Martinerie J (2001) The brainweb: phase synchronization and large-scale integration. Nat Rev Neurosci 2:229–239
2. Yuste R (2015) From the neuron doctrine to neural networks. Nat Rev Neurosci 16:487–497
3. Saxena S, Cunningham JP (2019) Towards the neural population doctrine. Curr Opin Neurobiol 55:103–111
4. Gray CM, König P, Engel AK, Singer W (1989) Oscillatory responses in cat visual cortex exhibit inter-columnar synchronization which reflects global stimulus properties. Nature 338:334–337
5. Usrey WM, Reppas JB, Reid RC (1998) Paired-spike interactions and synaptic efficacy of retinal inputs to the thalamus. Nature 395:384–387
6. Fox MD et al (2005) The human brain is intrinsically organized into dynamic, anticorrelated functional networks. Proc Natl Acad Sci USA 102:9673–9678

7. Raichle ME et al (2001) A default mode of brain function. Proc Natl Acad Sci 98:676
8. Hopfinger JB, Buonocore MH, Mangun GR (2000) The neural mechanisms of top-down attentional control. Nat Neurosci 3:284–291
9. Womelsdorf T et al (2007) Modulation of neuronal interactions through neuronal synchronization. Science 316:1609–1612
10. Dotson NM, Hoffman SJ, Goodell B, Gray CM (2017) A large-scale semi-chronic microdrive recording system for non-human primates. Neuron 96:769-782.e2
11. Miller EK, Wilson MA (2008) All my circuits: using multiple electrodes to understand functioning neural networks. Neuron 60:483–488
12. Quiroga RQ, Reddy L, Kreiman G, Koch C, Fried I (2005) Invariant visual representation by single neurons in the human brain. Nature 435:1102–1107
13. Bosman CA et al (2012) Attentional stimulus selection through selective synchronization between monkey visual areas. Neuron 75:875–888
14. Rubehn B, Bosman C, Oostenveld R, Fries P, Stieglitz T (2009) A MEMS-based flexible multichannel ECoG-electrode array. J Neural Eng 6:036003
15. Lakatos P, Karmos G, Mehta AD, Ulbert I, Schroeder CE (2008) Entrainment of neuronal oscillations as a mechanism of attentional selection. Science 320:110–113
16. Bastos AM, Lundqvist M, Waite AS, Kopell N, Miller EK (2020) Layer and rhythm specificity for predictive routing. Proc Natl Acad Sci USA 117:31459–31469
17. Siegle JH et al (2021) Survey of spiking in the mouse visual system reveals functional hierarchy. Nature. https://doi.org/10.1038/s41586-020-03171-x
18. Halgren M et al (2019) The generation and propagation of the human alpha rhythm. PNAS 116:23772–23782
19. Renz AF et al (2020) Opto-E-Dura: a soft, stretchable ECoG array for multimodal, multiscale neuroscience. Adv Healthcare Mater 9:2000814
20. Felleman DJ, Van Essen DC (1991) Distributed hierarchical processing in the primate cerebral cortex. Cereb Cortex 1:1–47
21. Markov NT et al (2013) The anatomy of hierarchy: feedforward and feedback pathways in macaque visual cortex. J Comp Neurol. https://doi.org/10.1002/cne.23458
22. Siegel M, Buschman TJ, Miller EK (2015) Cortical information flow during flexible sensorimotor decisions. Science 348:1352–1355
23. Shipp S (2007) Structure and function of the cerebral cortex. Science 297:253–256
24. Buzsáki G, Draguhn A (2004) Neuronal oscillations in cortical networks. Science 304:1926–1929
25. Börgers C, Kopell N (2003) Synchronization in networks of excitatory and inhibitory neurons with sparse, random connectivity. Neural Comput 15:509–538
26. Wang X-J (2010) Neurophysiological and computational principles of cortical rhythms in cognition. Physiol Rev 90:1195–1268
27. Kopell N, Ermentrout GB, Whittington MA, Traub RD (2000) Gamma rhythms and beta rhythms have different synchronization properties. Proc Natl Acad Sci USA 97:1867–1872
28. Womelsdorf T, Valiante TA, Sahin NT, Miller KJ, Tiesinga P (2014) Dynamic circuit motifs underlying rhythmic gain control, gating and integration. Nat Neurosci 17:1031–1039
29. Lundqvist M, Bastos AM, Miller EK (2020) Preservation and changes in oscillatory dynamics across the cortical hierarchy. J Cogn Neurosci 32:2024–2035
30. Rosanova M et al (2009) Natural frequencies of human corticothalamic circuits. J Neurosci 29:7679–7685
31. Fries P (2015) Rhythms for cognition: communication through coherence. Neuron 88:220–235
32. Bastos AM et al (2015) Visual areas exert feedforward and feedback influences through distinct frequency channels. Neuron 85:390–401
33. Buschman TJ, Miller EK (2007) Top-down versus bottom-up control of attention in the prefrontal and posterior parietal cortices. Science 315:1860–1862
34. Fries P (2005) A mechanism for cognitive dynamics: neuronal communication through neuronal coherence. Trends Cogn Sci 9:474–480

35. Micheli C, Kaping D, Westendorff S, Valiante TA, Womelsdorf T (2015) Inferior-frontal cortex phase synchronizes with the temporal–parietal junction prior to successful change detection. Neuroimage 119:417–431
36. Bressler SL, Richter CG (2015) Interareal oscillatory synchronization in top-down neocortical processing. Curr Opin Neurobiol 31:62–66
37. Bastos AM et al (2012) Canonical microcircuits for predictive coding. Neuron 76:695–711
38. Chao ZC, Takaura K, Wang L, Fujii N, Dehaene S (2018) Large-scale cortical networks for hierarchical prediction and prediction error in the primate brain. Neuron 100:1252-1266.e3
39. Bauer M, Stenner M-P, Friston KJ, Dolan RJ (2014) Attentional modulation of alpha/beta and gamma oscillations reflect functionally distinct processes. J Neurosci 34:16117–16125
40. Sedley W et al (2016) Neural signatures of perceptual inference. eLife 5:e11476
41. Brodski A, Paasch G-F, Helbling S, Wibral M (2015) The faces of predictive coding. J Neurosci 35:8997–9006
42. El Karoui I et al (2015) Event-related potential, time-frequency, and functional connectivity facets of local and global auditory novelty processing: an intracranial study in humans. Cereb Cortex 25:4203–4212
43. Todorovic A, van Ede F, Maris E, de Lange FP (2011) Prior expectation mediates neural adaptation to repeated sounds in the auditory cortex: an MEG study. J Neurosci 31:9118–9123
44. Antzoulatos EG, Miller EK (2016) Synchronous beta rhythms of frontoparietal networks support only behaviorally relevant representations. eLife 5
45. Buschman TJ, Denovellis EL, Diogo C, Bullock D, Miller EK (2012) Synchronous oscillatory neural ensembles for rules in the prefrontal cortex. Neuron 76:838–846
46. Antzoulatos EG, Miller EK (2014) Increases in functional connectivity between prefrontal cortex and striatum during category learning. Neuron 83:216–225
47. Brincat SL, Miller EK (2015) Frequency-specific hippocampal-prefrontal interactions during associative learning. Nat Neurosci 18:576–581
48. Brincat SL, Miller EK (2016) Prefrontal cortex networks shift from external to internal modes during learning. J Neurosci 36:9739–9754
49. Richter CG, Coppola R, Bressler SL (2018) Top-down beta oscillatory signaling conveys behavioral context in early visual cortex. Sci Rep 8:6991
50. Stanley DA, Roy JE, Aoi MC, Kopell NJ, Miller EK (2018) Low-beta oscillations turn up the gain during category judgments. Cereb Cortex 28:116–130
51. Wutz A, Loonis R, Roy JE, Donoghue JA, Miller EK (2018) Different levels of category abstraction by different dynamics in different prefrontal areas. Neuron 97:716-726.e8
52. Miller EK, Lundqvist M, Bastos AM (2018) Working memory 2.0. Neuron 100:463–475
53. Neuner I et al (2014) The default mode network and EEG regional spectral power: a simultaneous fMRI-EEG study. PLoS One 9
54. Engel AK, Fries P (2010) Beta-band oscillations—signalling the status quo? Curr Opin Neurobiol 20:156–165
55. Niessing J et al (2005) Hemodynamic signals correlate tightly with synchronized gamma oscillations. Science 309:948–951
56. Ray S, Maunsell JHR (2011) Different origins of gamma rhythm and high-gamma activity in macaque visual cortex. PLoS Biol 9:e1000610
57. Haegens S, Nácher V, Luna R, Romo R, Jensen O (2011) α-Oscillations in the monkey sensorimotor network influence discrimination performance by rhythmical inhibition of neuronal spiking. Proc Natl Acad Sci USA 108:19377–19382
58. van Kerkoerle T et al (2014) Alpha and gamma oscillations characterize feedback and feedforward processing in monkey visual cortex. Proc Natl Acad Sci USA 111:14332–14341
59. Douglas RJ, Martin KAC (2007) Mapping the matrix: the ways of neocortex. Neuron 56:226–238
60. Livingstone MS (1996) Oscillatory firing and interneuronal correlations in squirrel monkey striate cortex. J Neurophysiol 75:2467–2485
61. Bressler SL, Seth AK (2011) Wiener-Granger causality: a well established methodology. Neuroimage 58:323–329

62. Bastos AM, Schoffelen J-M (2015) A tutorial review of functional connectivity analysis methods and their interpretational pitfalls. Front Syst Neurosci 9:175
63. Dhamala M, Rangarajan G, Ding M (2008) Estimating granger causality from Fourier and wavelet transforms of time series data. Phys Rev Lett 100:018701
64. Bastos AM, Loonis R, Kornblith S, Lundqvist M, Miller EK (2018) Laminar recordings in frontal cortex suggest distinct layers for maintenance and control of working memory. PNAS 201710323. https://doi.org/10.1073/pnas.1710323115
65. Bollimunta A, Chen Y, Schroeder CE, Ding M (2008) Neuronal mechanisms of cortical alpha oscillations in awake-behaving macaques. J Neurosci 28:9976
66. Spaak E, Bonnefond M, Maier A, Leopold DA, Jensen O (2012) Layer-specific entrainment of γ-band neural activity by the α rhythm in monkey visual cortex. Curr Biol 22:2313–2318
67. Ninomiya T, Dougherty K, Godlove DC, Schall JD, Maier A (2015) Microcircuitry of agranular frontal cortex: contrasting laminar connectivity between occipital and frontal areas. J Neurophysiol 113:3242–3255
68. Maier A, Adams GK, Aura C, Leopold DA (2010) Distinct superficial and deep laminar domains of activity in the visual cortex during rest and stimulation. Front Syst Neurosci 4
69. Buffalo EA, Fries P, Landman R, Buschman TJ, Desimone R (2011) Laminar differences in gamma and alpha coherence in the ventral stream. Proc Natl Acad Sci 108:11262
70. Roopun AK et al (2006) A beta2-frequency (20–30 Hz) oscillation in nonsynaptic networks of somatosensory cortex. Proc Natl Acad Sci USA 103:15646–15650
71. Haegens S et al (2015) Laminar profile and physiology of the α rhythm in primary visual, auditory, and somatosensory regions of neocortex. J Neurosci 35:14341–14352
72. Maier A, Cox MA, Dougherty K, Moore B, Leopold DA (2014) Anisotropy of ongoing neural activity in the primate visual cortex. Eye Brain 6:113–120
73. Scheeringa R, Koopmans PJ, van Mourik T, Jensen O, Norris DG (2016) The relationship between oscillatory EEG activity and the laminar-specific BOLD signal. Proc Natl Acad Sci USA 113:6761
74. Troebinger L, López JD, Lutti A, Bestmann S, Barnes G (2014) Discrimination of cortical laminae using MEG. Neuroimage 102:885–893
75. Hipp JF, Siegel M (2015) BOLD fMRI correlation reflects frequency-specific neuronal correlation. Curr Biol 25:1368–1374
76. Aitken F et al (2020) Prior expectations evoke stimulus-specific activity in the deep layers of the primary visual cortex. PLoS Biol 18:e3001023
77. Kok P, Bains LJ, van Mourik T, Norris DG, de Lange FP (2016) Selective activation of the deep layers of the human primary visual cortex by top-down feedback. Curr Biol 26:371–376
78. Lawrence SJD et al (2018) Laminar organization of working memory signals in human visual cortex. Curr Biol 28:3435-3440.e4
79. Muckli L et al (2015) Contextual feedback to superficial layers of V1. Curr Biol 25:2690–2695
80. Vezoli J et al (2021) Brain rhythms define distinct interaction networks with differential dependence on anatomy. Neuron S089662732100725X. https://doi.org/10.1016/j.neuron.2021.09.052
81. Michalareas G et al (2016) Alpha-beta and gamma rhythms subserve feedback and feedforward influences among human visual cortical areas. Neuron 89:384–397
82. Popov T, Kastner S, Jensen O (2017) FEF-controlled alpha delay activity precedes stimulus-induced gamma-band activity in visual cortex. J Neurosci 37:4117–4127
83. Roberts MJ et al (2013) Robust gamma coherence between macaque V1 and V2 by dynamic frequency matching. Neuron 78:523–536
84. Ferro D, van Kempen J, Boyd M, Panzeri S, Thiele A (2021) Directed information exchange between cortical layers in macaque V1 and V4 and its modulation by selective attention. PNAS 118
85. Salazar RF, Dotson NM, Bressler SL, Gray CM (2012) Content-specific fronto-parietal synchronization during visual working memory. Science 338:1097–1100
86. Fontolan L, Morillon B, Liegeois C, Giraud A-L (2014) The contribution of frequency-specific activity to hierarchical information processing in the human auditory cortex. Nat Commun 5:4694

87. Gong X et al (2020) An ultra-sensitive step-function opsin for minimally invasive optogenetic stimulation in mice and macaques. Neuron 107:38-51.e8
88. Ni J et al (2016) Gamma-rhythmic gain modulation. Neuron 92:240–251
89. Veniero D et al (2021) Top-down control of visual cortex by the frontal eye fields through oscillatory realignment. Nat Commun 12:1757
90. Beauchamp MS, Sun P, Baum SH, Tolias AS, Yoshor D (2012) Electrocorticography links human temporoparietal junction to visual perception. Nat Neurosci 15:957–959
91. Schmiedt JT et al (2014) Beta oscillation dynamics in extrastriate cortex after removal of primary visual cortex. J Neurosci 34:11857–11864
92. Kienitz R et al (2021) Theta, but not gamma oscillations in area V4 depend on input from primary visual cortex. Curr Biol 31:635-642.e3
93. Gregoriou GG, Rossi AF, Ungerleider LG, Desimone R (2014) Lesions of prefrontal cortex reduce attentional modulation of neuronal responses and synchrony in V4. Nat Neurosci 17:1003–1011
94. Fries P, Reynolds JH, Rorie AE, Desimone R (2001) Modulation of oscillatory neuronal synchronization by selective visual attention. Science 291:1560–1563
95. Bichot NP, Heard MT, DeGennaro EM, Desimone R (2015) A source for feature-based attention in the prefrontal cortex. Neuron 88:832–844
96. Kar K, DiCarlo JJ (2021) Fast recurrent processing via ventrolateral prefrontal cortex is needed by the primate ventral stream for robust core visual object recognition. Neuron 109:164-176.e5
97. Boroujeni KB, Oemisch M, Hassani SA, Womelsdorf T (2020) Fast spiking interneuron activity in primate striatum tracks learning of attention cues. PNAS 117:18049–18058
98. Schneider M et al (2021) A mechanism for inter-areal coherence through communication based on connectivity and oscillatory power. Neuron S0896627321007108. https://doi.org/10.1016/j.neuron.2021.09.037
99. Canolty RT et al (2006) High gamma power is phase-locked to theta oscillations in human neocortex. Science 313:1626–1628
100. Spyropoulos G, Bosman CA, Fries P (2018) A theta rhythm in macaque visual cortex and its attentional modulation. PNAS 115:E5614–E5623
101. Bastos AM et al (2021) Neural effects of propofol-induced unconsciousness and its reversal using thalamic stimulation. eLife 10:e60824
102. Fiebelkorn IC, Pinsk MA, Kastner S (2019) The mediodorsal pulvinar coordinates the macaque fronto-parietal network during rhythmic spatial attention. Nat Commun 10:1–15
103. Saalmann YB, Pinsk MA, Wang L, Li X, Kastner S (2012) The pulvinar regulates information transmission between cortical areas based on attention demands. Science 337:753–756
104. Zhou H, Schafer RJ, Desimone R (2016) Pulvinar-cortex interactions in vision and attention. Neuron 89:209–220
105. Bastos AM et al (2015) A DCM study of spectral asymmetries in feedforward and feedback connections between visual areas V1 and V4 in the monkey. Neuroimage 108:460–475
106. Lindner M, Vicente R, Priesemann V, Wibral M (2011) TRENTOOL: a Matlab open source toolbox to analyse information flow in time series data with transfer entropy. BMC Neurosci 12:119
107. Mejias JF, Murray JD, Kennedy H, Wang X-J (2016) Feedforward and feedback frequency-dependent interactions in a large-scale laminar network of the primate cortex. Sci Adv 2:e1601335–e1601335
108. Grasso M, Albantakis L, Lang JP, Tononi G (2021) Causal reductionism and causal structures. Nat Neurosci 24:1348–1355
109. Richter CG, Thompson WH, Bosman CA, Fries P (2017) Top-down beta enhances bottom-up gamma. J Neurosci 37:6698–6711
110. Chang EF (2015) Towards large-scale, human-based, mesoscopic neurotechnologies. Neuron 86:68–78

111. Khodagholy D et al (2015) NeuroGrid: recording action potentials from the surface of the brain. Nat Neurosci 18:310–315
112. Parvizi J, Kastner S (2018) Promises and limitations of human intracranial electroencephalography. Nat Neurosci 21:474–483

Chapter 26
How Can We Study the Mechanisms of Memory-Related Oscillations Using Multimodal in Vivo and in Vitro Approaches?

Haley Moore, Genevieve Konopka, and Bradley C. Lega

Abstract The previous two decades of research in the electrophysiology of human memory have witnessed rapid expansion in the range of experiments, data availability, and technology being applied to unravel the underlying mechanisms of mnemonic cognition. A large portion of this work utilizes awake, behaving neurosurgical patients implanted with intracranial electrodes for the purposes of seizure mapping. However, this patient population also yields another opportunity for study. As these patients often undergo neurosurgical procedures to resect brain tissue, investigators with access to these patients can also ask novel questions related to gene expression, in vitro physiology, and brain morphology. We outline existing findings related to how multimethod data such as these can be used to understand potential gene expression networks linked with brain activity during mnemonic processing. We begin with an historical perspective on the first efforts to identify a genetic influence on brain activity patterns. We then introduce high-throughput sequencing technologies, followed by a discussion of how these techniques have been used to identify memory-relevant temporal lobe expression networks. We then review existing methodological treatments for culturing human tissue, and end with a description of the experimental manipulations that may help unravel key questions of human memory, including oscillatory activity.

The genetic underpinnings of human brain activity, especially memory—related brain oscillations, remain obscure. In animal models, the investigation of such genes has utilized in vitro methods such as gene editing, as these tools provide mechanistic clues to understand dissect gene networks. Neurosurgical tissue presents the opportunity to conduct analogous experiments in human tissue. The goals of this chapter

H. Moore · G. Konopka
Department of Neuroscience, UT Southwestern Medical Center, Dallas, USA

H. Moore · G. Konopka · B. C. Lega (✉)
Peter O'Donnell Jr. Brain Institute, UT Southwestern Medical Center, Dallas, USA
e-mail: Bradley.Lega@UTSouthwestern.edu

H. Moore · B. C. Lega
Department of Neurosurgery, UT Southwestern Medical Center, Dallas, USA

N. Axmacher (ed.), *Intracranial EEG*, Studies in Neuroscience, Psychology and Behavioral Economics, https://doi.org/10.1007/978-3-031-20910-9_26

are to examine the existing knowledge of the genetics of human brain oscillations, beginning with an overview of methods and results from the twentieth century and a look at how next generation sequencing technologies have revolutionized the field. We will introduce common analyses of transcriptomic data while reviewing our own studies regarding gene expression and memory-related oscillations. Finally, we will discuss the advantages of using human neurosurgical tissue in vitro to delve deeper into the molecular genetics of oscillatory activity. We will examine recent developments in long-term organotypic slice culture and propose experiments that could be conducted using this system.

26.1 The Heredity of Brain Activity—Historical Approaches

Prior to the completion of the Human Genome Project, the ability to study the genetic basis of brain oscillations (as with any trait) was limited. Genetically influenced traits were characterized using twin and family studies, in which heritability is measured by concordance. Concordance is the incidence of a particular shared trait between two people, be it twins, pairs of family members (e.g., parent-offspring, sibling-sibling), or unrelated individuals. If a trait is significantly influenced by genetic components, a twin study would theoretically find the highest rate of concordance between monozygotic (MZ) twins, then dizygotic (DZ) twins, and the lowest concordance between unrelated individuals.

Soon after the EEG was described in 1929, scientists began questioning the heritability of brain activity using twin and family studies [8, 71] A prototypical study design would compare resting state electrograms from a single brain region from twins and unrelated people. Before 1960, researchers could only use a well-trained "clinical eye" to visually estimate the similarities and differences of EEG traces recorded on paper. In 1936, the first EEG twin study concluded that resting alpha (~10 Hz) activity over the occiput or cranial vertex of MZ twins was as similar as consecutive recordings in a single person [19]. Just two years later, a family study of similar design concluded that there was no noticeable similarity in alpha activity in parent–offspring pairs [24]. Subsequent large twin studies continued to report high concordance of EEG signatures in MZ and DZ pairs. Several aspects of alpha and beta oscillations (~20 Hz), like amplitude and preferred frequency of the band, were found to be highly concordant in MZ twins, slightly less so in DZ twins, and not concordant in unrelated people [71]. Still other studies reported high MZ and low DZ concordance in multiple frequencies. A 1974 spectral analysis of 39 MZ pairs and 27 DZ pairs reported high MZ and low or absent DZ concordance in delta, theta, alpha, and beta frequency bands at the occiput [42].

1978 brought the first publication to compare sibling concordance between separate brain regions [45]. An analysis of alpha and beta power in multiple brain regions of 20 MZ pairs and 20 DZ pairs concluded that genetic influence was more pronounced in occipital and parietal leads compared to frontal and temporal leads.[46] These results were partially contradicted by a twin study in 1995 showing high heritability in the parietotemporal and anterior areas [64]. Overall, twin and family studies conducted between the introduction of the EEG and the Human Genome Project support the existence of a heritable component of brain oscillations observed via EEG [72]. However, the genetic factors that affect cognition and neurophysiological processes linked with cognition are clearly polygenic, which necessitates population level studies. In the case of complex behaviors such as memory, executive function, and related disease states such as autism, such population level genetic studies suggest that hundreds of genes affect observable phenotypes [55, 58]. Next generation sequencing technologies have drastically increased our ability to explore the genes underlying complex neurological function and disease at a population level [70].

26.2 High-Throughput Sequencing Technologies: RNA Sequencing

The advent of next generation sequencing technologies not only revolutionized the identification of genetic variants at the level of DNA to demonstrate heritability of physiological or cognitive traits but also facilitated high throughput assessment of RNA transcripts to measure gene expression [60, 75]. Such RNA sequencing (RNA-seq) approaches have been applied to brain tissue resected from surgical patients and post-mortem donors, revealing the enormous complexity of gene expression in the human brain [30]. Bulk RNA-seq results in a transcriptomic profile of an entire sample of tissue or cells as if it were a homogenous genetic unit. Tissues, particularly the brain, are composed of heterogenous cell populations with unique transcriptomic identities that cannot be easily disambiguated from bulk sequencing data. Since the human brain is comprised of dozens of cell types in any given region [66], bulk RNA-seq results can be skewed by the abundance of a given cell type or the overall transcriptional output of a cell type (e.g., neurons express greater numbers of genes than non-neurons [30]).

RNA-seq can be applied to samples of dissociated cells (single cell RNA-seq, or scRNA-seq) or isolated nuclei (single nuclei RNA-seq, or snRNA-seq) to generate transcriptomes at cellular resolution, overcoming the limitations of bulk RNA-seq [37]. In scRNA-seq, each mRNA transcript within a cell is tagged with a cell-specific barcode and a transcript-specific unique molecular identifier. Tagged mRNA transcripts are converted to cDNA, amplified, and sequenced. Subsequent data processing steps group cells into clusters based on the similarity of their transcriptomes (see Fig. 26.1). snRNA-seq, involving the isolated nuclei of cells of interests, is the preferred technique for morphologically complex cell types (e.g., neurons) [26, 38]. Although nuclei contain a fraction of a cell's total mRNA, transcriptomic data gleaned from cells versus nuclei show similar distributions of gene expression levels [26]. These single cell methods are necessary to disentangle how specific genes influence memory—related brain oscillations via the activity of glial cells or different classes of neurons.

Genome-wide association studies (GWAS) are now routinely applied to vast genomic datasets to identify associations between complex traits and many gene variants [73]. Weighted gene co-expression network analyses (WGCNA) are used to identify groups of genes, known as 'modules' in genomics parlance, with highly correlated expression levels in transcriptomic datasets [39]. The genes within these modules can be further studied using GWAS or gene ontology analyses, thus revealing which networks are enriched for trait or disease-relevant genes. Applying these methods to single cell/nuclei data allows us to associate complex traits and diseases back to specific cell types [77].

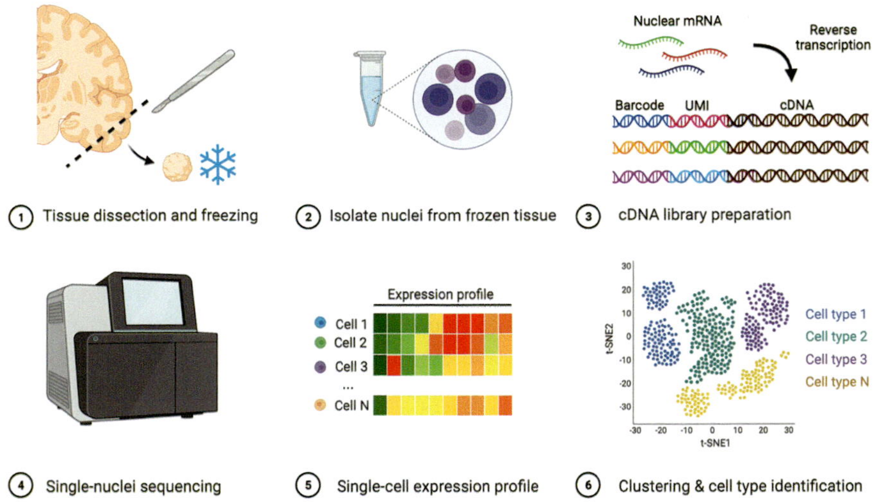

Fig. 26.1 A typical single-nuclei RNA sequencing experiment workflow. (1) Tissue is resected in the operating room. A 2–3 cm^3 piece is dissected, placed in PBS, and transported to the lab on ice. Tissue is divided into multiple tubes and flash-frozen in liquid nitrogen. (2) Tissue pieces are homogenized and nuclei are isolated using an optimized protocol (e.g., see 4). (3) Using a droplet-based microfluidics system, mRNA transcripts within individual nuclei are reverse transcribed to cDNA and tagged with transcript-specific unique molecular identifiers (UMI) and nucleus-specific barcodes. cDNA is amplified, cleaned, and prepped for sequencing. (4) Next-generation sequencing is conducted with the desired coverage. (5) Reads are aligned and mapped to individual genes and cells. (6) Clusters are generated based on similarity of gene expression profiles using dimension reduction algorithms (e.g., UMAP or tSNE) and clusters are annotated as cell types. Figure created with Biorender.com

26.3 High-Throughput Sequencing Technologies: ATAC Sequencing

Within the nucleus, DNA is wound around histones in a compact nucleoprotein structure (chromatin). Chromatin exists dynamically as dense heterochromatin (low transcription) or loose euchromatin (high transcription) depending on transcriptional demands. Gene expression occurs when chromatin structure does not sterically hinder transcription machinery. Open euchromatin, which often contains exposed transcription factor binding sites, can be accessed by transcription machinery. Chromatin structure can be characterized using the assay for transposase-accessible chromatin using sequencing (ATAC-Seq) [35]. ATAC-seq employs a hyperactive transposase to excise open regions of chromatin. After excision, these regions are tagged, sequenced, and mapped back to their cell of origin as in sc/snRNA-seq [62]. With ATAC-seq, binding sites for transcription factors can be identified within the open chromatin regions. The genes that such transcription factors might influence can be inferred by coupling the ATAC-seq data with sc/snRNA-seq. Such an approach is in contrast to chromatin immunoprecipitation-seq (ChIP-seq), where specific transcription factors are crosslinked to DNA and the bound regions purified via immunoprecipitation and then sequenced. ATAC-seq is sometimes preferred to ChIP-seq as antibodies to specific proteins and a prior knowledge of chromatin factors are not required, rendering ATAC-seq less technically intensive and more unbiased [43].

26.4 High-Throughput Sequencing Technologies: Special Considerations

After identifying transcripts of interest, the presence of the gene product should be verified (e.g., via histology or western blot). RNA transcripts and protein products do not always correlate. This is likely due to turnover rates of both transcripts and proteins as well as post-transcriptional and post-translational mechanisms, none of which are readily easily measured in a high throughput manner. For example, while parvalbumin (PVALB) positive cells (based on the presence of PVALB protein) are abundant in cortex immunohistochemistry, parvalbumin transcripts are challenging to detect in scRNA-seq datasets [5, 67]. It is also important to note that even at the single cell level, RNA-seq and ATAC-seq do not intrinsically maintain spatial information. Microdissecting individual cortical layers or hippocampus subfields, for example, then sequencing the samples separately can solve this issue [16, 34]. Adding this step increases costs and can generally only be applied when physical divisions are well-demarcated, though. Fluorescence in situ hybridization or recently developed spatial transcriptomics methods like Slide-seq can be used to determine the location of a transcript or cell type without previous knowledge of spatial information [56].

It is not possible to genetically analyze completely healthy brain tissue from a living donor who has no indication for brain biopsy. Brain tissue that is unaffected by a neurological disease is typically garnered from a deceased donor many hours or days after death, while fresh brain tissue is resected from patients diagnosed with a number of diseases, such as cancer or epilepsy. Epilepsy tissue presents less variability than infiltrating disease such as gliomas, although both types of tissue have been used in studies of human brain gene expression. The possible influence of tumors (gliomas) in other datasets may be under-explored. We have employed several control parameters to overcome the challenge of using epilepsy tissue. Epilepsy duration and seizure frequency can be included as variables in statistical models (e.g., linear mixed model) [5]. Additionally, NIH-funded studies are required to share de-identified genomic datasets in repositories that can be accessed by qualified investigators. Therefore, post-mortem "normal" transcriptome data are available from many human brain regions [34]. To confirm the validity of RNA-seq data we generated from epilepsy tissue, we have compared datasets from our epilepsy tissue to those from postmortem tissue. We found no difference between these conditions, suggesting that genetic information gleaned from these tissue conditions is comparable. This suggests that appropriate modeling can help develop generalizable insights from epilepsy patients, using pathologically normal tissue (i.e., tissue without pathological changes such as dysplasia, stroke, or other lesions). The use of fresh tissue (as compared to cadaveric tissue) offers the advantage of combining with datasets describing functional activity obtained from the same individuals, including intracranial recordings.

26.5 The Genes Underlying Memory-Associated Oscillations

We will now review our group's recent work linking genes with memory processes, providing examples of how state-of-the-art sequencing technologies can be combined with iEEG to generate and support hypotheses on the genetics of oscillations. We highlight a portion of our data and discuss the implications of identifying mnemonically relevant oscillation-related genes.

Atlases of the human brain transcriptome have facilitated comparisons of gene expression with functional brain data, including physiology and imaging [14, 74]. These atlases are derived from postmortem tissues and demonstrate robust stability of gene expression at steady state levels. Integration of gene expression with other brain activity measures or morphology in non-postmortem tissues have primarily been limited to studies that carry out ex vivo approaches such as patch-seq [13, 31]. Patch-seq integrates electrophysiology and gene expression measurements from the same individual cells. First, whole-cell patch clamp measurements (e.g., cell firing patterns in response to current injection) are taken, then the cell's cytosolic contents are aspirated and processed for RNA-seq. Unfortunately, the current patch-seq method suffers from being labor intensive and low-throughput. Building on the availability of intracranial EEG recordings during memory behavior and gene expression atlases, our group sought to investigate gene expression linked with memory-related brain oscillations.

In our initial work linking memory, oscillations, and genes, we analyzed data from patients undergoing iEEG recording for seizure mapping [10]. Focusing on surface electrodes located across the fronto-temporal-parietal cortex, we compared oscillatory activity during successful and unsuccessful recall to identify subsequent memory effects (SME) in the six standard frequency bands (delta, 2–4 Hz; theta, 4–8 Hz; alpha, 8–12 Hz; beta, 16–20 Hz; low gamma, 35–70 Hz; high gamma, 70–150 Hz). By aggregating information across hundreds of subjects, we created an estimate of memory—related oscillatory activity at these frequencies across 14 cortical and subcortical brain regions. To profile gene expression, we completed bulk RNA-seq on post-mortem samples of the same cortical areas from which we had recorded SME data. The aggregation of memory—related activity in over two hundred subjects provided data across the cortex that we could correlate with gene expression levels from these same cortical regions. Our analysis provided data to begin unraveling how expression differences can predict differences in memory—related oscillatory activity.

After comparing the gene expression and SME value correlations across brain regions, we were able to construct a list of 163 SME-associated genes. 61% of these genes were correlated with beta SME values across the cortex. Gene ontology analyses revealed that SME-associated genes were enriched for neuronal functions related to activity (e.g., synaptic transmission, ion channel activity, and neurotransmitter transport). GWAS showed that SME-related genes were highly associated with many neuropsychiatric disorders. We also found positive correlations between SME-related genes and the *WWC1* gene network. *WWC1*, encoding the protein KIBRA, is one of the few genes that has previously been associated with memory performance specifically [59]. The relationship between *WWC1* and SME-related genes supports our conclusion that the genes we identified in this study are indeed associated with memory processes. This study represented a major step forward in the efforts to uncover the relationship between memory-related brain oscillations and gene expression in the human brain.

This analysis directly complemented previous attempts to use large datasets of non—invasive data and gene expression atlases. However, we recognized that generating both gene expression and brain oscillation data from the same individuals would be a significant improvement to understand the genetic underpinnings of memory—related brain activity. Because we are unable to obtain resected surgical tissue from many brain regions from each individual, we focused this follow-up study on the temporal pole as it is frequently resected in the surgical treatment of seizures [9]. We first had patients participate in a free recall task during stereo EEG recording, then we calculated SME values for each frequency band as previously described. The same patients later underwent temporal lobectomy for epilepsy treatment. We excluded patients suffering from lesions such as cortical dysplasia, cavernomas, or low-grade tumors. None of the patients had previous laser ablation or other surgery on the temporal lobe. These data afforded us the unique opportunity to generate genetic, behavioral, and electrophysiological data from the temporal cortex of the same patients, overcoming a weakness of the previous studies.

With bulk RNA sequencing, we were able to link 300 genes to memory—related oscillations (SMEs) in individual frequency bands. Most genes were specifically correlated with a single frequency band. A high proportion of genes were correlated with 2–4 Hz 'delta' oscillations, supporting other analyses indicating that oscillations in this lower frequency range may exhibit memory—related properties analogous to those linked with 4–9 Hz 'theta' oscillations in animal models. These findings suggest that individual variability in low frequency theta/delta patterns may have an underlying basis in gene expression networks. Using GWAS, we uncovered relationships between SME modules (groups of co-expressed genes correlated with memory-related oscillations) and brain disorders. Two of these modules (named "WM4" and "WM12" in the text [9]) contained genes that were differentially expressed in several neuropsychiatric disorders including depression, schizophrenia, and autism spectrum disorders. Importantly, we did not find any enrichment for epilepsy-associated loci in the modules. This result underscores the validity of using tissue from epilepsy patients to study normal human brain processes, as well as the possible utility of identifying specific gene targets for therapeutic development using predictive models built using electrophysiological data.

We further refined our set of genes linked to memory-associated oscillations at cellular resolution by completing snRNA-seq and snATAC-seq on a subset of patients that contributed sEEG data. Excitatory neurons in cortical layers IV, V, and VI were enriched for genes in the delta-associated modules WM4 and WM12. Genes present in WM21, which is negatively associated with delta, were enriched in glial cells. This hints at a potential role for non-neuronal cells in oscillation modulation. By integrating the snATAC-seq data, we were also able to identify the transcription factor SMAD3 as a key regulator of memory-associated genes in excitatory neurons. SMAD3 is also a member of WM12, indicating a potential mechanistic link between memory-associated genes and delta oscillations.

Most recently, we used human hippocampal tissue to investigate anterior versus posterior differences in gene expression, building on cognitive models positing distinct mnemonic roles for these structures [5, 25, 53]. We identified differentially expressed genes (i.e., upregulated or downregulated) across the longitudinal axis using these human tissue samples. The anterior and posterior divisions of the human hippocampus participate in distinct neural circuits and exhibit many dichotomous functional properties [53]. For example, memory encoding and retrieval have been specifically associated with the anterior and posterior hippocampus, respectively [63]. The differentially expressed genes we identified may lead to mechanistic explanations for these specializations, potentially after further investigation in human tissue in vitro.

Taken together, our data provide several important insights to guide the next set of investigations linking mnemonic neurophysiological processes in humans with gene expression networks. At a granular level, these insights include linking *WWC1*—associated gene networks expression with memory—related oscillations, and identifying *SMAD3* gene networks as favorable targets for further experimentation. More broadly, these experiments demonstrate how investigators can use cadaveric gene expression data and cortical and hippocampal tissue from neurosurgical patients to investigate human memory from a unique perspective.

26.6 Studying the Human Brain in Vitro

In vitro investigations of human neurophysiology offer the potential to complement in vivo intracranial EEG studies to help overcome limitations inherent in the study of human subjects. The need for 'opportunistic' recordings, when subjects are in the epilepsy monitoring unit, limits the number of behavioral manipulations available. Precise anatomical distinctions, such as differences in activity across hippocampal subfields, are not reliable using in vivo recordings. Further, the use of surgical epilepsy patients entails that many variables beyond the control of the investigator, such as levels of epilepsy medicines or frequency of seizures, could potentially influence results. Most important, any experiment that poses risk of side effects or neural injury cannot be performed, including pharmacological interventions.

Tissue collected from neurosurgical cases is a valuable resource for in vitro experiments. In the acute period, tissue can be sliced for acute electrophysiology experiments, flash-frozen for gene expression, or fixed and preserved for immunocytochemistry. Acute electrophysiology experiments have helped establish key principles of neurophysiology, such as long-term potentiation and synaptic plasticity.[1] For example, studies in human and rodent hippocampus slices identified similarities in synaptic plasticity reliant on NMDA receptors, revealing a common feature of hippocampal-dependent information storage across species [7] Tissue processing and storage methods have been optimized for this purpose, with preliminary evidence showing that human hippocampal slices remain viable for up to 48 h [78]. Genetic manipulation studies (e.g., viral knockdown of a gene, optogenetics) require more time than this, typically 1–3 weeks depending on the method. The ability to define and manipulate the genetics, and therefore molecular mechanisms, of oscillatory activity in vitro is limited by the length of time brain tissue can be kept healthy and active in culture.

26.7 Organotypic Brain Slice Culture

Organotypic—A descriptive term for tissue that has been removed from its native space but continues to function as it normally would—brain slice culture is used to study the brain in vitro while maintaining in vivo-like three-dimensional architecture. The first application of mouse organotypic slice culture was in 1973 [40] and by the turn of the century the method was widely used to study normal and disordered neural processes in animal models [49]. Some have been able to maintain rodent brain slices in vitro for months at a time [44, 46]. For a review on organotypic slice culture applications in non-human animals, see [32].

While acute electrophysiology experiments using human brain slices have been commonplace for several decades, human organotypic slice culture has lagged behind due to poor viability in vitro.[51] The ability to culture human neurosurgical specimens more than a few days was only recently demonstrated. Since 2014, several groups have published methods using unique culture media formulations and different measures of viability [3, 21, 61, 66]. Electrophysiological properties have been monitored over time, with neuronal activity in cortex and hippocampal slices still appearing after four weeks in vitro [21, 61]. Some have argued that culturing cortex in unsupplemented human cerebrospinal fluid (hCSF) produces the most stable electrophysiological activity over time [61]. hCSF is a relatively easy to procure, cheap culture material compared to some defined artificial media formulations [21]. Using hCSF as a culture medium may promote tissue viability by increasing intrinsic neuronal activity, perhaps providing an ideal environment for the tissue to overcome the trauma of losing synaptic connections during dissection and slicing prior to culture [32, 79]. Long-term human organotypic slice culture may be an ideal system in which to study the genetics and molecular mechanisms of oscillations (see Fig. 26.2), particularly when animal models lack in translational relevance due to the apparent uniquely human properties of oscillations (e.g., see [33]).

26.8 Viruses and Gene Manipulation

Several molecular biology techniques have been demonstrated in human organotypic slice culture that could be useful in determining the mechanisms underlying oscillatory activity. Viral expression of reporter proteins like GFP or lacZ has been achieved in human brain tissue in less than 24 h when using adenoviral vectors [50, 66] This method could be useful in the short term to rapidly label specific cell types or to quickly introduce a short hairpin RNA (shRNA) against a gene one wishes to silence [18]. shRNA is seen as foreign genetic material by the cell's natural defenses [47]. The enzymes Dicer and the RNA-induced silencing complex (RISC) process the shRNA and "remember" its sequence; RISC will destroy that sequence any time it comes across it in the future [54]. Therefore, gene silencing can be achieved by introducing an shRNA that includes a critical portion of the gene's mRNA sequence. The mRNA will subsequently be degraded, and protein expression will decrease without needing to alter the genome as in CRISPR/Cas9 experiments. CRISPR/Cas9 requires the cell to express the RNA-guided DNA endonuclease Cas9 and a gene-specific guide RNA. The coding material for these components is delivered via viral vector. Cas9 induces a double-strand break at the point in the genome specified by the guide RNA. DNA repair via non-homologous end joining can then cause mutations that result in gene silencing [27, 28].

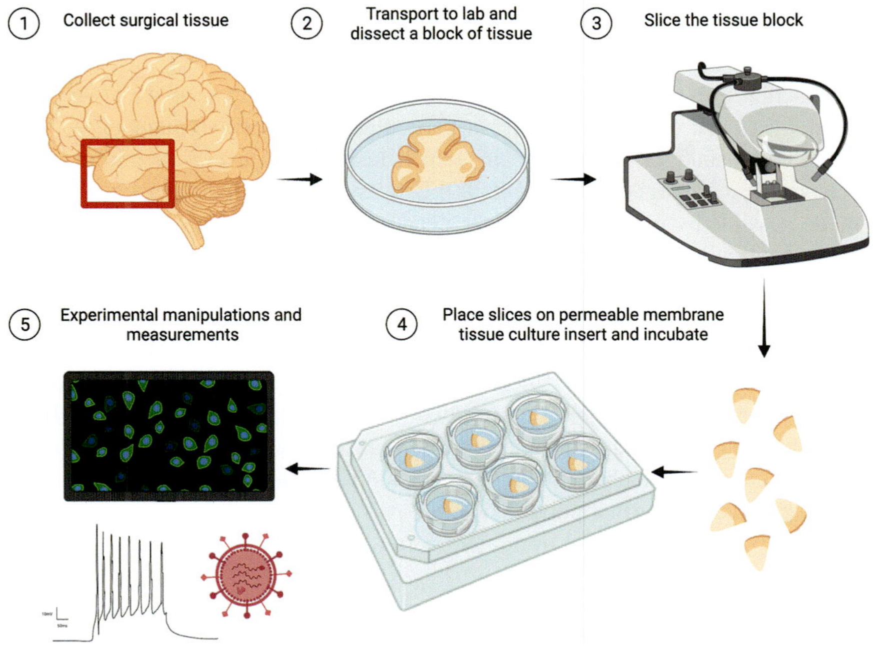

Fig. 26.2 Simplified organotypic slice culture workflow. (1) Tissue is resected in the operating room. A 2–3 cm^3 piece is dissected, placed in sterile, oxygenated transport medium, and quickly transported to the lab on ice. (2) Under sterile conditions, a ~ 1 cm^3 tissue block is dissected for slicing. (3) Tissue block is sliced at ~300–400-micron thickness. (4) Slices recover in oxygenated aCSF, then are transferred to permeable membrane tissue culture inserts in a 6-well plate with culture medium of choice (e.g., an artificial medium or human CSF). Media is refreshed or replaced at one- or two-day intervals. (5) Slices can be manipulated at any point during culture (e.g., via viral transduction), then experimental measurements can be made using many techniques like immunofluorescence, electrophysiology, or gene expression profiling. Figure created with Biorender.com

Different viral vectors have unique strengths and weaknesses that should be considered during experimental design. For example, adenoviruses act quickly but result in transient expression of the encoded material, while lentiviruses incorporate into the genome and produce stable expression after two weeks [20] Thus, a lentiviral vector would be ideal when attempting to express a protein with high-turnover rates, while an adenoviral vector would be acceptable for expression of a long-lasting protein like GFP [17]. Although the principles of manipulating genes in vitro are well-established, including in human neural stem cells and brain organoids, the methods have not yet been demonstrated in human organotypic slice culture [22].

26.9 Optogenetics

Optogenetics presents the opportunity to control the activity of a cell type of interest [41]. The cell must express a particular light-sensitive channel, such as channelrhodopsin-2 or halorhodopsin, which respectively excite or inhibit a cell in the presence of a light stimulus [48, 80] Expression of the light-sensitive channel is made possible by viral delivery of the genetic material encoding the channel and can result in cell-type specific expression using recombination systems like Cre-lox genetics [41, 69]. One group has successfully applied optogenetic manipulation of neurons within cultured human brain slices [3]. Lentiviral vectors were used to deliver the coding material for the excitatory channelrhodopsin-2 under the control of a neuron-specific promoter (meaning the channel would only be expressed in neurons). Expression was apparent after two weeks in culture, and neuronal activity was successfully induced with blue light pulses. In future experiments, the application of optogenetic techniques in human brain slice culture could assist in characterizing specific neuronal populations that participate in a range of memory and disease-relevant processes like oscillatory activity [6, 29].

26.10 Electrophysiology

Local field potentials, action potentials, and many aspects of a neuron's membrane physiology can be recorded in human brain slices using standard electrophysiology methods. Oscillatory activity analogous to what is observed with iEEG in vivo is more difficult to assess. Few groups have studied human oscillations in vitro [23, 52]. Florez et al. elicited oscillations in temporal cortex slices using electrical stimulation and a combination of glutamatergic and cholinergic agonists (kainate and carbachol, respectively), techniques that have been used extensively to study field potentials in animal models in vitro [12, 23, 68, 77]. Robust non-ictal theta and gamma oscillations, as well as ictal events, were recorded in deep and superficial cortical layers. Phase coherence in theta, but not gamma, was found between superficial and deep cortical layers. Theta-gamma cross-frequency coupling was also observed during non-ictal activity. These exciting results imply that human epilepsy tissue can be used to study the mechanistic principles of in vivo-like oscillations in vitro. Although the hippocampus is the most characterized brain region with respect to low frequency oscillations, in vitro data is absent. Only ictal-like ripples in the subiculum have been recorded in vitro [2]. Some reports suggest that ex vivo ictal activity is more prevalent in hippocampal slices compared to cortical slices [21].The use of tissue from surgical epilepsy patients may preclude our ability to study hippocampal oscillations in vitro without the interference of ictal activity.

The oscillations described above were recorded using low-impedance glass electrodes carefully placed in cortex slices submerged in a flow of warm, oxygenated artificial CSF [23]. This is a common method for recording local field potentials and can be combined with whole cell patch-clamp methods if simultaneous single neuron data is desired. [36]. Multiple recording and stimulating electrodes can be used at once, the number of which is limited by tissue size and recording rig setup. For example, Florez et al. used two recording electrodes and one stimulating electrode [23]. High-density microelectrode arrays (MEA) consist of several dozen to tens of thousands of recording electrodes packed into a small (typically $< 10\ mm^2$) chip [51, 57]. MEAs can be used to record high-resolution network activity, including local field potentials and spiking neurons, across a slice without having to place individual electrodes. MEAs have been used to study oscillatory activity in non-human animals [15]. Bursting activity in acute human cortex slices has been recorded using MEA, but oscillations were not specifically characterized [79].

There is a notable lack of research on how genes and oscillations change over time in culture, representing an important gap in knowledge that must be addressed before using this system to manipulate genes related to oscillations. It is possible to measure oscillations in vitro [23]. Few have analyzed oscillations in acute human brain slices, and no one has measured them in cultured human brain slices [2, 23]. Gene expression studies in human liver and intestine slice culture have shown myriad time-dependent changes in gene expression [11]. However, temporally influenced gene expression in brain tissue has not been studied.

26.11 Conclusion

Since 1929, our ability to study the mechanistic basis of human brain activity has increased dramatically. What began with studies of EEG heritability has most recently resulted in the identification of specific genes that are associated with memory-relevant oscillations from iEEG recordings. To understand the function of these genes in terms of human brain activity, innovative in vitro approaches must be pursued. Applying molecular techniques to human brain tissue in vitro has the potential to reveal crucial mechanisms underlying oscillatory activity and ultimately result in seminal insights into human cognition.

References

1. Abrahamsson T, Lalanne T, Watt AJ, et al (2016) In Vitro investigation of synaptic plasticity. Cold Spring Harb Protoc 2016:pdb.top087262
2. Alvarado-Rojas C, Huberfeld G, Baulac M et al (2015) Different mechanisms of ripple-like oscillations in the human epileptic subiculum. Ann Neurol 77:281–290
3. Andersson M, Avaliani N, Svensson A et al (2016) Optogenetic control of human neurons in organotypic brain cultures. Sci Rep 6:24818
4. Ayhan F, Douglas C, Lega BC et al (2021) Nuclei isolation from surgically resected human hippocampus. STAR Protoc 2:100844
5. Ayhan F, Kulkarni A, Berto S et al (2021) Resolving cellular and molecular diversity along the hippocampal anterior-to-posterior axis in humans. Neuron 109:2091-2105.e2096

6. Başar E (2013) Brain oscillations in neuropsychiatric disease. Dialogues Clin Neurosci 15:291–300
7. Beck H, Goussakov IV, Lie A et al (2000) Synaptic plasticity in the human dentate gyrus. J Neurosci 20:7080–7086
8. Berger H (1929) Über das Elektrenkephalogramm des menschen. Arch Psychiatr Nervenkr 87:527–570
9. Berto S, Fontenot MR, Seger S et al (2021) Gene-expression correlates of the oscillatory signatures supporting human episodic memory encoding. Nat Neurosci 24:554–564
10. Berto S, Wang G-Z, Germi J et al (2018) Human genomic signatures of brain oscillations during memory encoding. Cereb Cortex 28:1733–1748
11. Bigaeva E, Gore E, Simon E et al (2019) Transcriptomic characterization of culture-associated changes in murine and human precision-cut tissue slices. Arch Toxicol 93:3549–3583
12. Blatow M, Rozov A, Katona I et al (2003) A novel network of multipolar bursting interneurons generates theta frequency oscillations in neocortex. Neuron 38:805–817
13. Cadwell CR, Palasantza A, Jiang X et al (2016) Electrophysiological, transcriptomic and morphologic profiling of single neurons using Patch-seq. Nat Biotechnol 34:199–203
14. Callaway EM, Dong H-W, Ecker JR et al (2021) A multimodal cell census and atlas of the mammalian primary motor cortex. Nature 598:86–102
15. Carmeli C, Bonifazi P, Robinson H, et al (2013) Quantifying network properties in multi-electrode recordings: spatiotemporal characterization and inter-trial variation of evoked gamma oscillations in mouse somatosensory cortex in vitro. Frontiers in Computational Neuroscience 7
16. Cembrowski MS, Wang L, Sugino K et al (2016) Hipposeq: a comprehensive RNA-seq database of gene expression in hippocampal principal neurons. Elife 5:e14997
17. Csepregi R, Temesfői V, Poór M, et al (2018) Green fluorescent protein-based viability assay in a multiparametric configuration. Molecules 23
18. Davidson BL, Harper SQ (2005) Viral delivery of recombinant short hairpin RNAs. Methods Enzymol 392:145–173
19. Davis H, Davis PA (1936) Action potentials of the brain. in normal persons and in normal states of cerebral activity. Arch Neurol Psychiatry 36:1214–1224
20. Dong W, Kantor B (2021) Lentiviral vectors for delivery of gene-editing systems based on CRISPR/Cas: current state and perspectives. Viruses 13
21. Eugène E, Cluzeaud F, Cifuentes-Diaz C et al (2014) An organotypic brain slice preparation from adult patients with temporal lobe epilepsy. J Neurosci Methods 235:234–244
22. Fischer J, Heide M, Huttner WB (2019) Genetic modification of brain organoids. Frontiers in Cellular Neuroscience 13
23. Florez CM, McGinn RJ, Lukankin V et al (2015) In vitro recordings of human neocortical oscillations. Cereb Cortex 25:578–597
24. Gottlober AB (1938) The inheritance of brain potential patterns. Journal of Experimental Psychology 22
25. Goyal A, Miller J, Qasim SE et al (2020) Functionally distinct high and low theta oscillations in the human hippocampus. Nat Commun 11:2469
26. Grindberg RV, Yee-Greenbaum JL, McConnell MJ et al (2013) RNA-sequencing from single nuclei. Proc Natl Acad Sci U S A 110:19802–19807
27. Hana S, Peterson M, McLaughlin H et al (2021) Highly efficient neuronal gene knockout in vivo by CRISPR-Cas9 via neonatal intracerebroventricular injection of AAV in mice. Gene Ther 28:646–658
28. Heidenreich M, Zhang F (2016) Applications of CRISPR-Cas systems in neuroscience. Nat Rev Neurosci 17:36–44
29. Herweg NA, Solomon EA, Kahana MJ (2020) Theta oscillations in human memory. Trends Cogn Sci 24:208–227
30. Hodge RD, Bakken TE, Miller JA et al (2019) Conserved cell types with divergent features in human versus mouse cortex. Nature 573:61–68

31. Hodge RD, Miller JA, Novotny M et al (2020) Transcriptomic evidence that von Economo neurons are regionally specialized extratelencephalic-projecting excitatory neurons. Nat Commun 11:1172
32. Humpel C (2015) Organotypic brain slice cultures: A review. Neuroscience 305:86–98
33. Jacobs J (2014) Hippocampal theta oscillations are slower in humans than in rodents: implications for models of spatial navigation and memory. Philos Trans R Soc Lond B Biol Sci 369:20130304
34. Keil JM, Qalieh A, Kwan KY (2018) Brain transcriptome databases: A User's guide. J Neurosci 38:2399–2412
35. Klemm SL, Shipony Z, Greenleaf WJ (2019) Chromatin accessibility and the regulatory epigenome. Nat Rev Genet 20:207–220
36. Köhling R, Avoli M (2006) Methodological approaches to exploring epileptic disorders in the human brain in vitro. J Neurosci Methods 155:1–19
37. Kulkarni A, Anderson AG, Merullo DP et al (2019) Beyond bulk: a review of single cell transcriptomics methodologies and applications. Curr Opin Biotechnol 58:129–136
38. Lake BB, Ai R, Kaeser GE et al (2016) Neuronal subtypes and diversity revealed by single-nucleus RNA sequencing of the human brain. Science 352:1586–1590
39. Langfelder P, Horvath S (2008) WGCNA: an R package for weighted correlation network analysis. BMC Bioinformatics 9:559
40. LaVail JH, Wolf MK (1973) Postnatal development of the mouse dentate gyrus in organotypic cultures of the hippocampal formation. Am J Anat 137:47–65
41. Lee C, Lavoie A, Liu J, et al (2020) Light Up the Brain: The application of optogenetics in cell-type specific dissection of mouse brain circuits. Frontiers in Neural Circuits 14
42. Lykken DT, Tellegen A, Thorkelson K (1974) Genetic determination of EEG frequency spectra. Biol Psychol 1:245–259
43. Ma S, Zhang Y (2020) Profiling chromatin regulatory landscape: insights into the development of ChIP-seq and ATAC-seq. Molecular Biomedicine 1:9
44. Marksteiner J, Humpel C (2008) Beta-amyloid expression, release and extracellular deposition in aged rat brain slices. Mol Psychiatry 13:939–952
45. Meshkova TA, Ravich-Shcherbo IV (1978) Role of genotype in determination of individual specific features of the resting EEG. Hum Physiol 4:418–426
46. Michaelson SD, Müller TM, Bompolaki M et al (2021) Long-Lived organotypic slice culture model of the rat basolateral amygdala. Curr Protoc 1:e267
47. Moore CB, Guthrie EH, Huang MT et al (2010) Short hairpin RNA (shRNA): design, delivery, and assessment of gene knockdown. Methods Mol Biol 629:141–158
48. Nagel G, Szellas T, Huhn W et al (2003) Channelrhodopsin-2, a directly light-gated cation-selective membrane channel. Proc Natl Acad Sci 100:13940–13945
49. Noraberg J, Poulsen FR, Blaabjerg M et al (2005) Organotypic hippocampal slice cultures for studies of brain damage, neuroprotection and neurorepair. Curr Drug Targets CNS Neurol Disord 4:435–452
50. O'Connor WM, Davidson BL, Kaplitt MG et al (1997) Adenovirus vector-mediated gene transfer into human epileptogenic brain slices: prospects for gene therapy in epilepsy. Exp Neurol 148:167–178
51. Obien MEJ, Deligkaris K, Bullmann T, et al (2015) Revealing neuronal function through microelectrode array recordings. Frontiers in Neuroscience 8
52. Pennifold J (2017) Oscillatory and epileptiform activity in human and rodent cortical regions in vitro. PhD Thesis. Aston University, Birmingham.
53. Poppenk J, Evensmoen HR, Moscovitch M et al (2013) Long-axis specialization of the human hippocampus. Trends Cogn Sci 17:230–240
54. Pratt AJ, MacRae IJ (2009) The RNA-induced silencing complex: a versatile gene-silencing machine. J Biol Chem 284:17897–17901
55. Rees E, Owen MJ (2020) Translating insights from neuropsychiatric genetics and genomics for precision psychiatry. Genome Medicine 12:43

56. Rodriques SG, Stickels RR, Goeva A et al (2019) Slide-seq: A scalable technology for measuring genome-wide expression at high spatial resolution. Science 363:1463–1467
57. Ronchi S, Fiscella M, Marchetti C, et al (2019) Single-Cell electrical stimulation Using CMOS-Based High-Density Microelectrode Arrays. Frontiers in Neuroscience 13
58. Rylaarsdam L, Guemez-Gamboa A (2019) Genetic causes and modifiers of autism spectrum disorder. Frontiers in Cellular Neuroscience 13
59. Schneider A, Huentelman M, Kremerskothen J, et al (2010) KIBRA: a new gateway to learning and memory? Frontiers in Aging Neuroscience 2
60. Schuster SC (2008) Next-generation sequencing transforms today's biology. Nat Methods 5:16–18
61. Schwarz N, Hedrich UBS, Schwarz H et al (2017) Human Cerebrospinal fluid promotes long-term neuronal viability and network function in human neocortical organotypic brain slice cultures. Sci Rep 7:12249
62. Smith JP, Sheffield NC (2020) Analytical approaches for ATAC-seq data analysis. Curr Protoc Hum Genet 106:e101
63. Spaniol J, Davidson PSR, Kim ASN et al (2009) Event-related fMRI studies of episodic encoding and retrieval: Meta-analyses using activation likelihood estimation. Neuropsychologia 47:1765–1779
64. Sviderskaya NE, Korol'kova TA, (1995) Genetic features of the spatial organization of the human cerebral cortex. Neurosci Behav Physiol 25:370–377
65. Tasic B, Yao Z, Graybuck LT et al (2018) Shared and distinct transcriptomic cell types across neocortical areas. Nature 563:72–78
66. Ting JT, Kalmbach B, Chong P et al (2018) A robust ex vivo experimental platform for molecular-genetic dissection of adult human neocortical cell types and circuits. Sci Rep 8:8407
67. Tran MN, Maynard KR, Spangler A et al (2021) Single-nucleus transcriptome analysis reveals cell-type-specific molecular signatures across reward circuitry in the human brain. Neuron 109:3088-3103.e3085
68. Traub RD, Whittington MA, Colling SB et al (1996) Analysis of gamma rhythms in the rat hippocampus in vitro and in vivo. J Physiol 493(Pt 2):471–484
69. Tsien JZ (2016) Cre-Lox Neurogenetics: 20 Years of versatile applications in brain research and counting. Frontiers in Genetics 7
70. Uffelmann E, Posthuma D (2021) Emerging Methods and Resources for Biological Interrogation of Neuropsychiatric Polygenic Signal. Biol Psychiat 89:41–53
71. van Beijsterveldt CE, Boomsma DI (1994) Genetics of the human electroencephalogram (EEG) and event-related brain potentials (ERPs): a review. Hum Genet 94:319–330
72. van Beijsterveldt CEM, van Baal GCM (2002) Twin and family studies of the human electroencephalogram: a review and a meta-analysis. Biol Psychol 61:111–138
73. Visscher PM, Wray NR, Zhang Q et al (2017) 10 Years of GWAS discovery: biology, function, and translation. Am J Hum Genet 101:5–22
74. Wang GZ, Belgard TG, Mao D et al (2015) Correspondence between Resting-State Activity and Brain Gene Expression. Neuron 88:659–666
75. Wang Z, Gerstein M, Snyder M (2009) RNA-Seq: a revolutionary tool for transcriptomics. Nat Rev Genet 10:57–63
76. Watanabe K, Umićević Mirkov M, de Leeuw CA et al (2019) Genetic mapping of cell type specificity for complex traits. Nat Commun 10:3222
77. Whittington MA, Traub RD, Jefferys JG (1995) Synchronized oscillations in interneuron networks driven by metabotropic glutamate receptor activation. Nature 373:612–615
78. Wickham J, Brödjegård NG, Vighagen R et al (2018) Prolonged life of human acute hippocampal slices from temporal lobe epilepsy surgery. Sci Rep 8:4158

79. Wickham J, Corna A, Schwarz N et al (2020) Human cerebrospinal fluid induces neuronal excitability changes in resected human neocortical and hippocampal brain slices. Front Neurosci 14:283
80. Zhang F, Wang L-P, Brauner M et al (2007) Multimodal fast optical interrogation of neural circuitry. Nature 446:633–639

Part III
Data Analysis

Chapter 27
How Can I Integrate iEEG Recordings with Patients' Brain Anatomy?

Sushmita Sadhukha, Robert Oostenveld, and Arjen Stolk

Abstract Preparing intracranial electroencephalography (iEEG) datasets for analysis presents a unique set of methodological challenges that are absent in non-invasive investigative techniques. Because iEEG is primarily used in epilepsy patients with varying brain pathologies, the main challenges pertain to variability in electrode coverage and therefore the regions of the brain from which electrophysiological recordings can be obtained. In this chapter, we outline how to efficiently integrate the raw anatomical images and electrophysiological recordings during preprocessing, allowing iEEG datasets to be analyzed in an anatomically precise and consistent way.

27.1 Introduction

Intracranial electroencephalography (iEEG) has high spatiotemporal precision, making it one of the foremost investigative techniques towards gaining a rich understanding of the human brain and its cognitive processes. Notably, empirical investigations using iEEG by either placing electrodes directly on the neocortex (electrocorticography, ECoG) or in targeted brain areas (stereoelectroencephalography, sEEG), have recently shed light on the role of language areas in speech [14, 25], hippocampal circuits in learning [27], and sensorimotor rhythms in movement coordination [32]. In each case, iEEG enabled researchers to characterize neural processing and information flow with a level of spatiotemporal resolution unrivaled by non-invasive techniques such as scalp EEG, magnetoencephalography (MEG), and functional MRI. Still, iEEG's remarkable spatiotemporal precision comes with a cost; analyzing

S. Sadhukha · A. Stolk (✉)
Psychological and Brain Sciences, Dartmouth College, Hanover, NH, USA
e-mail: Arjen.Stolk@Dartmouth.edu

R. Oostenveld · A. Stolk
Donders Institute for Brain, Cognition, and Behaviour, Radboud University, Nijmegen, the Netherlands

R. Oostenveld
NatMEG, Karolinska Institutet, Stockholm, Sweden

N. Axmacher (ed.), *Intracranial EEG*, Studies in Neuroscience, Psychology and Behavioral Economics, https://doi.org/10.1007/978-3-031-20910-9_27

iEEG data is far from straightforward and presents a series of methodological challenges that are typically absent in data obtained from non-invasive methods. In this chapter, we describe these challenges and give a conceptual overview of how they are addressed during data preprocessing and analysis.

The challenges with analyzing iEEG data can be condensed to two main issues, both due to the unique circumstances under which these datasets are collected. First, there is sparse electrode coverage in each patient owing to the clinical constraints of iEEG. This technique is most commonly used to determine epileptogenic zones (EZ) in patients undergoing presurgical monitoring for intractable epilepsy. Unlike in non-invasive methods, this clinical motivation imposes case-specific considerations on the electrode implantation strategy, including how many electrodes to use and where in the brain they will be implanted to provide optimal clinical utility [40]. A consequence of sparse electrode use is that researchers cannot achieve the whole brain coverage that is characteristic of functional MRI and most other non-invasive techniques. Second, the electrode implementation schemes are not spatially consistent across patients. Because the majority of epilepsy patients suffer from limbic or frontal lobe seizures, most of the implanted electrodes tend to cover these regions of the brain [28]. However, the *type* of electrode (e.g., strip, grid, or depth electrodes) and their precise location may still vary substantially across patients in iEEG, making it challenging to aggregate data from multiple patients when performing group level analyses. Moreover, medical facilities which use iEEG also vary in the surgical tools and procedures used in their epilepsy treatment. For example, some institutes use MRI and CT scans, while others use X-rays, or different combinations thereof [7]. Thus, the anatomical images providing the neuroanatomical context in interpreting the precise location of implanted electrodes may also vary within and across datasets. Taken together, these idiosyncratic circumstances generate different sources of variation, challenging researchers to analyze iEEG datasets in an anatomically precise and consistent way.

Here we describe a workflow to efficiently integrate iEEG recordings with patients' brain anatomy at the preprocessing stage. It aims to resolve challenges of varying electrode coverage and datasets, while also providing concrete guidance and flexibility to overcome real-world obstacles in iEEG analysis, including electrode displacement and anatomical variability within and across patients. The workflow can be applied in cognitive and systems neuroscience research as well as in clinical studies of epileptogenic activity. Furthermore, each step has an implementation counterpart in an analysis protocol embedded within the open-source FieldTrip toolbox [26, 33]. This versatile MATLAB toolbox enables researchers to readily adopt reproducible preprocessing pipelines and build on a continuously growing body of advanced analysis methods developed and used by a large research community, including time–frequency, connectivity, spike-field, and statistical analysis. Starting with the anatomical images, we illustrate how the present workflow guides researchers from the multitude of raw intracranial data files to integrated observations, while also highlighting considerations for subsequent analysis.

27.2 Anatomical Images

Clinicians may acquire several anatomical images of the patient's brain before and during iEEG data collection. These images can be used to detect structural abnormalities and provide a neuroanatomical context in the clinical interpretation of iEEG recordings leading up to the surgical procedure. Most commonly, this includes a *pre-implant* structural Magnetic Resonance Imaging (MRI) scan and a *post-implant* Computed Tomography (CT) scan, but this post-implant scan can vary depending on the medical facility and other case-specific circumstances. For example, some medical institutes may collect X-ray images along one or a few directions instead of a three-dimensional CT scan, while others may acquire a second MRI scan post-implantation. Figure 27.1 below outlines the pre- and post-implant scans researchers typically encounter in iEEG analysis and how we use these raw images to interpret electrophysiological signals embedded in a neuroanatomical context. For the purpose of this chapter, we use the pre-implant MRI and post-implant CT scan as the archetypal anatomical scans; they are disproportionately common in iEEG and preprocessing them requires undertaking all the fundamental steps necessary to handle other types of scans and combinations thereof.

The MRI scans of the brain represent the soft tissue anatomy in great detail. CT scans on the other hand depict high-intensity objects such as the skull and the implanted electrodes but poorly depict neuroanatomical landmarks. During preprocessing, the pre-implant MRI and post-implant CT scans are coregistered to determine the location of the electrodes from the CT in relation to the brain's anatomy from the MRI. In some cases, clinicians will acquire post-implant MRI scans. Because these images show brain anatomy *after* electrode implantation, if the post-implant MRI is of sufficient quality for the purposes of electrode localization and cortical surface extraction, then the coregistration and the compensation for electrode displacement due to "brain shift" steps we describe further below do not apply. In a T1-weighted MRI, the electrodes are visible as dark voids in the higher-intensity brain tissue, as a consequence of magnetic susceptibility artifact. These distortions in MR signal may complicate electrode localization as well as compromise the cortical surface's structural integrity depending on the locations of the electrodes [41]. Irrespective of the type of images acquired, following the electrode localization, all scans need to be aligned to a standardized head coordinate system, such that there is a systematic way to link the electrode locations and electrophysiological recordings to the anatomical volumes across all patients.

27.2.1 Determining the Coordinate System of the Anatomical Images

MRI and CT scans are three-dimensional volumetric representations of the head; they consist of three-dimensional pixels termed "voxels", the centers of which lie

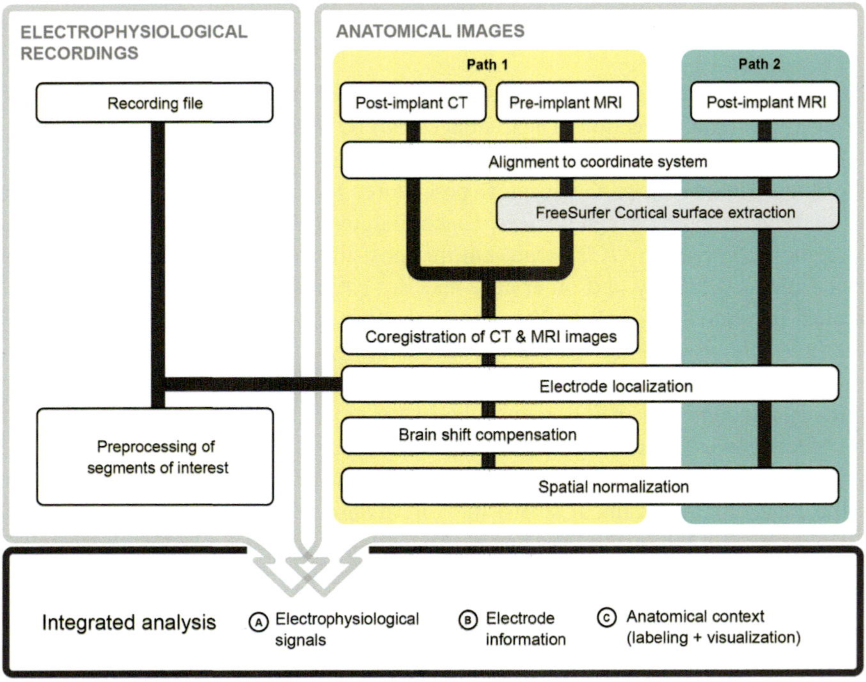

Fig. 27.1 Workflow for integrating iEEG recordings with patients' brain anatomy during data preprocessing. The initial foundation for the integration is laid during electrode localization, which offers the opportunity to directly link anatomical locations to electrode labels corresponding to the electrophysiological data as stored in the recording file (lateral connection). The integration is completed when the electrode information is combined with the preprocessed electrophysiological data (converging arrows), allowing iEEG analysis to be performed in an anatomically precise and consistent way. White boxes denote workflow steps supported by the FieldTrip toolbox

at the intersections of an equally spaced grid. In a voxel system with 256 × 256 × 256 voxels, for example, coordinate [1,1,1] is the first voxel[1] and coordinate [256,256,256] is the last voxel in the volume. Neither the voxels nor the volume need to be isometric; for instance, you can have an MRI scan with a within-slice resolution that differs from the slice spacing. It is worth noting, however, that the voxel coordinate system does not specify the position or orientation of the head within the scanned volumes, nor are the voxel indices meaningful with respect to brain anatomy (i.e., the voxels do not correspond to any specific neuroanatomical landmarks). Head and scanner coordinate systems on the other hand *do* define the real-world interpretable positions and orientations relative to a well-defined origin. In head coordinates, the origin at [0,0,0] mm is defined relative to anatomical landmarks

[1] Note that, depending on the software used, the index of the first voxel is indicated with either 1 or 0.

such as the anterior commissure, whereas in scanner coordinates the origin is a separate function of the acquisition field of view.[2]

Given that head positioning is not fixed relative to the scanner, the head coordinate system of a scan cannot typically be determined automatically from the acquired image itself. Hence, the first step in preprocessing MRI and CT scans is to determine the origin and the orientation of the coordinate system that is native to the raw anatomical image. Depending on whether we are working with the original DICOM files or with data converted to NifTI format, the direction of the cardinal axes can vary [23]. By determining the orientation of the head in the raw image, we can specify how the direction of the x, y, and z axes should be interpreted, a precondition for correct alignment and coregistration of the anatomical images in the steps described below. Although visually determining posterior-anterior and inferior-superior axes is straightforward from a sagittal slice of the brain, differentiating left from right is more challenging [15]. To this end, we recommend using FieldTrip's *ft_determine_coordsys*, which depicts an anatomical volume as three intersecting, orthogonal slices together with the cardinal axes and their x, y, and z labels. This depiction allows for rapidly determining the orientation of the left–right axis (see Box 3 in [33] for a 4-step procedure), which is necessary to demarcate the right-hemisphere landmark in the following alignment step.

27.2.2 Aligning Anatomical Images to a Standard Coordinate System

After determining the native coordinate system of the anatomical images, we align them to the ACPC coordinate system [35]. Although technically any head coordinate system could fill this role, the ACPC system is the preferred convention for the optional FreeSurfer operation we describe below. In the ACPC coordinate system, the origin is at the anterior commissure (AC), the y-axis runs along the line between the AC and the posterior commissure (PC), and the z-axis lies in the midline dividing the two cerebral hemispheres. To align the MRI to this coordinate system, we first specify the AC, the PC, an interhemispheric location along the midline at the top of the brain, and a location somewhere in the brain's right hemisphere. Contingent on the native orientation identified in the previous step, the point in the right hemisphere may have larger or smaller values for their spatial coordinates. FieldTrip's *ft_volumerealign* offers a graphical user interface (GUI) for specifying these four landmarks.

Then, we align the CT scan to the ACPC coordinate system. Because CT scans only represent high-intensity objects such as the skull and do not have any neuroanatomical landmarks (i.e., anterior and posterior commissure), we first align the CT scan to the CTF head surface coordinate system. To do this, we specify the nasion (at the root of the nose), left and right pre-auricular points (just in front of the ear canals), and

[2] For an overview and description of common head and scanner coordinate systems, see www.fieldtriptoolbox.org/faq/coordsys/.

an interhemispheric location along the midline at the top of the brain. Finally, we convert the CT's coordinate system into an approximation of the ACPC coordinate system, such that the orientations of both the MRI and CT scans are expressed in a compatible format. Note that the MRI and CT scans are not coregistered just yet. This would require estimating a detailed rigid body geometric transformation in a subsequent step. Moreover, the spatial coordinates are not consistent across patients, since that would require the additional spatial normalization of the coordinates. These steps are described further below.

27.2.3 Using FreeSurfer for Extracting Cortical Surfaces

Freesurfer is a free software package that offers several advantages for processing human brain MRI images that are beneficial for subsequent iEEG analysis and data interpretation. Notably, FreeSurfer allows extracting cortical surfaces from patients' MRI scans, enabling an anatomically realistic representation of iEEG recordings on the neocortex [8]. For instance, using cortical surfaces extracted from precisely registered MRI scans, a recent study found that two prominent and oft-considered spatially overlapping sensorimotor rhythms differed in their anatomical properties. Specifically, alpha rhythms were maximal at electrodes on the postcentral gyrus, whereas beta rhythms localized on average 12 mm anterior to both sides of the central sulcus [32]. This finding was further made possible by FreeSurfer's surface-based registration, which registers patients' brains to a template brain based on their cortical gyrification patterns, effective for generalizing across patients. As illustrated further below, cortical gyri and sulci remain difficult to accurately normalize using volume-based registration techniques due to their spatial complexity and variability across patients.

FreeSurfer is only compatible with Linux and macOS operating systems but can be used on Windows with VirtualBox. While this step is recommended, users technically would be able to do iEEG analysis without it, forgoing some of the advantages that this step offers. Furthermore, it is worth considering performing the FreeSurfer operation on both the pre- and post-implant MRI scans in case the latter is available. If the post-implant MRI yields cortical surfaces of sufficient quality for the purpose of the study, and the dark voids marking electrodes are clearly visible, then researchers may use this scan to localize and place electrodes directly in their neuroanatomical context, rendering the coregistration and brain shift compensation steps unnecessary (as seen in the reduced workflow of path 2 in Fig. 27.1).

27.2.4 Coregistering the Anatomical Images

Thus far in the workflow we describe, we have a pre-implant MRI scan aligned manually to the ACPC head coordinate system, and a post-implant CT scan converted into an approximate ACPC coordinate system. The two scans need to be precisely coregistered, such that we can link the electrode locations in the post-implant anatomical image to their corresponding locations in the pre-implant anatomical image. This coregistration step is automatic and consists of determining a rigid body transformation that provides the best fit of the CT and MRI scan based on their common denominator, the skull (recall that CT scans show very little brain anatomy). After coregistration, the CT scan's coordinate system is updated to have spatial coordinates consistent with those of the MRI scan, as well as with pertinent files from the FreeSurfer operation performed on that scan. Overall, the coregistration procedure, also implemented in FieldTrip's *ft_volumerealign*, lays the groundwork for obtaining precise knowledge of the neuroanatomical locations of the electrodes, thereby providing a mechanism to isolate the local electrophysiological signal and dictating how subsequent data analysis is executed and interpreted. Now that the anatomical images are precisely coregistered, we can precisely identify and label the implanted electrodes in the electrode placement step we describe next.

27.3 Electrodes

At this point in the workflow, we have a CT image showing the implanted electrodes, but not the precise coordinates of the electrodes themselves. The next step is to localize, label, and sort the electrodes to match the channel names assigned in the raw electrophysiological recording file. Furthermore, we determine whether it will be necessary to compensate for electrode displacement due to "brain shift", the inward sinking of the brain following ECoG implantation. Electrode localization benefits from knowledge of the implantation procedure, including details of the number of electrodes along each sEEG shaft, the size of ECoG grids and strips and their possible overlap. There has been significant progress in automating the localization of electrodes in the post-implant CT or MRI scan, but none currently achieve 100% accuracy, rendering manual review necessary regardless [5, 21]. Furthermore, it is important that all electrodes are assigned labels that unambiguously link them to the channel names configured in the acquisition software and used during recording, which can vary across medical facilities as well as from clinician to clinician. Generally, a prefix is used that bears some anatomical relationship with the target site, such as RAM for an sEEG probe in the right amygdala, or LPG for a left parietal ECoG grid. In the numbering scheme of sEEG electrodes, electrode 1 is the electrode farthest away from the insertion site; this also applies to single row ECoG strip electrodes. To determine the numbering of multi-row ECoG grid and strip electrodes, careful notes must be taken during recording.

Brain shift refers to the phenomenon of tissue displacement during surgery and is contingent on the type of electrodes that are used for epilepsy monitoring. To implant multi-row ECoG grid or strip electrodes, clinicians must perform a craniotomy in which a C-shaped bone flap is surgically removed from the skull to access the cortical surface [38]. This procedure may cause the brain to sink inwards from its pre-surgical position due to subdural fluid loss and pressure exerted by the electrodes themselves. As a result, electrodes that are known to have been placed on the cortical surface occasionally appear buried within the cortical tissue from the pre-implant MRI after coregistration, sometimes more than a centimeter deep [30, 24]. The displacement is most pronounced directly below a craniotomy and is usually minimal for implants solely involving burr holes, including depth and single row strip electrodes [34]. We discuss strategies for brain shift compensation as well as subsequent anatomical registration in the following steps.

27.3.1 Localizing Electrodes in the Anatomical Image

FieldTrip offers an interactive electrode-placement GUI tool, *ft_electrodeplacement*, that displays anatomical images together with an electrode list. Clicking an electrode label in the list will directly assign that label to the current crosshair position in the CT scan, see Fig. 27.2. Several features facilitate precise navigation of the anatomical image including (1) zooming in and out of the crosshair section, (2) an intensity slider that allows thresholding the image's low and high values, (3) a toggle function that allows switching between active views of the CT and MRI scan, and (4) a magnet feature that transports the selected crosshair position to the nearest weighted maximum or minimum, within a certain voxel radius of the selected location. This enables users to efficiently pinpoint high-intensity electrodes or dark voids marking their locations in the post-implant CT and MRI scans, respectively. An interactive tool like this has several advantages for precisely identifying and labeling the electrodes. First, it allows loading the recording file with the list of electrode labels, obviating the need to manually sort and rename the electrodes to match the iEEG recordings. Second, the ability to toggle between the anatomical scans is helpful to precisely assign the crosshair position in cases where the quality of the image is less than ideal and to manually determine electrodes' anatomical labels.

27.3.2 Compensating for Electrode Displacement due to Brain Shift

After localizing the electrodes, re-alignment of electrode grids to the pre-implant cortical surface may be necessary to compensate for brain shift. To assess whether there is brain shift, we visualize ECoG electrodes together with a smooth hull around the cortical mesh generated by FreeSurfer, see leftmost panel of Fig. 27.3. This hull tracks the (exposed) outer surface on which the electrode grid rested. Several research labs have developed re-alignment techniques to compensate for electrode displacement, reducing localization error to under 3 mm based on surgical photographs [2, 7, 29, 41, 31, 4, 11, 17]. FieldTrip's *ft_electroderealign* currently supports two of

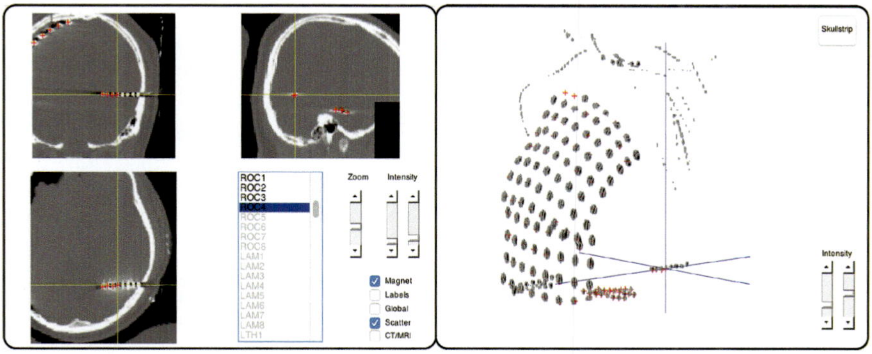

Fig. 27.2 Electrode localization. Clicking an electrode label in the main panel on the left will directly assign that label to the current crosshair position in the CT scan. Several features facilitate precise navigation of the anatomical CT, such as a zoom mode, a magnet option that transports the crosshair to the nearest weighted maximum (or minimum in case of a post-implant MRI), and the interactive 3D scatter figure shown on the right. Adapted from [33]

these techniques to project electrode grids back to the cortical surface. One method, illustrated in Fig. 27.3, uses an optimization algorithm that minimizes an energy function defined by inter-electrode distances and global deformation of the electrode configuration [11]. The other method projects electrodes to the cortical surface in the direction of the local norm vector of the electrode grid [17]. To prevent inwardly displaced electrodes from being incorrectly placed in nearby cortical sulci during back-projection, we re-align electrode grids to the smooth hull used above rather than to the highly folded cortical mesh directly. Furthermore, given that different grids can move independently from one another, we re-align electrode grids individually by running separate re-alignment procedures for each grid.

27.3.3 Registering Electrodes to an Anatomical Template and Atlas

Thus far in the workflow, we have obtained electrode locations aligned to the patient's neuroanatomy. To systematically combine and examine iEEG data from different patients, we need to generalize, or spatially normalize, the electrode coordinates by registering the patient's brain to a standardized model of brain anatomy. FieldTrip supports two methods to achieve this, embedded within *ft_electroderealign*. The first registers the patient's brain to a template brain by deforming the entire brain in three-dimensional space [1]. This volume-based registration technique considers the overall geometry of the brain and can be used for spatial normalization of all types of electrodes, whether at depth or on the surface. The second method, which relies on the FreeSurfer operation described above, registers the patient's brain to the template by considering features on the cortical surface [8]. This surface-based registration

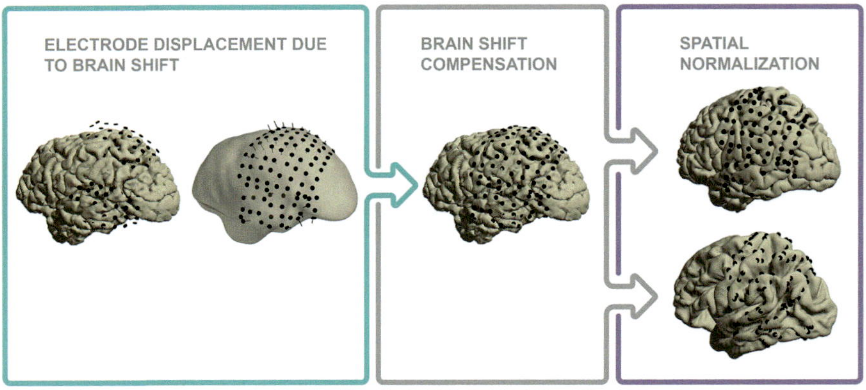

Fig. 27.3 Brain shift compensation and spatial normalization. In cases in which there is electrode displacement due to brain shift (left-most panel), compensation by re-aligning the electrode grids to the pre-implant cortical surface (middle panel) may be necessary. The re-aligned grids to the cortical surface can then be spatially normalized using volume-based or surface-based registration to a template brain (top and bottom in right-most panel, respectively). This allows data originating from multiple patients to be readily aggregated or overlaid with standardized anatomical atlases. Adapted from [33]

technique considers only the curvature patterns of the cortex and thus can be used for the spatial normalization of electrodes located on or near the cortical surface, most appropriately ECoG grid and strip electrodes. As seen in the right-most panel of Fig. 27.3, compared with volume-based registration, with surface-based registration, the original grid geometry is no longer preserved as electrodes are moved from one brain to another according to the curvature pattern of the cortex.

Another advantage of spatially normalizing the electrode locations is that they can be overlaid with standardized anatomical atlases. This allows looking up anatomical or functional labels corresponding to the electrodes and aggregate iEEG data from different patients on the basis of those labels. FieldTrip's *ft_volumelookup* supports looking up labels in a number of atlases, including the AFNI Talairach-Tournoux atlas [20], the AAL atlas [37], the BrainWeb dataset [6], the JuBrain Cytoarchitectonic Atlas [12], the VTPM atlas [39], the Brainnetome Atlas [13], the Yeo Atlases [42], and the Melbourne Subcortical Atlas [36], in addition to FreeSurfer's Desikan-Killiany and Destrieux atlases [9, 10]. Together, these atlases provide a variety of ways for combining electrodes from different patients, providing a data-driven means to overcome the challenge of spatial inconsistency across patients in iEEG analysis.

27.4 Electrophysiological Recordings

Electrophysiological data are acquired continuously post-implantation as part of ongoing epilepsy monitoring. Each data channel represents, as a function of time, the electric potential difference between an electrode and its reference. Most commonly, the electrodes are referenced to a single, common electrode during acquisition. This referential montage has the benefit that the recordings can be readily remontaged to a more preferred scheme during offline analysis [3]. Markers or triggers for stimulus-onset times and responses are typically recorded in a dedicated channel, allowing for precise synchronization of experimental scenarios with the electrophysiological recordings. The electrophysiological data, together with the electrode labels and other pertinent recording information, are written to a recording file in a data format dictated by the data-acquisition system. The final step in the workflow is to import and preprocess segments of interest from the raw recording file, followed by their integration with the electrode information we obtained in relation to the patient's neuroanatomy. When spatially normalized, this electrode information bridges the gap between the electrophysiological data and patient or template brain anatomy. We describe this final integrative step and how it benefits subsequent analysis next.

27.4.1 Preprocessing the Electrophysiological Recordings

The electrophysiological recordings consist of a mixture of signal-of-interest and noise, both neural and non-neural. The main objective of preprocessing is to improve the overall signal-to-noise ratio of the data, while preparing it in a format that is suitable for follow-up analysis. The specifics of preprocessing will be contingent on the experimental design and research question at hand, but the general order remains the same. First, we define the segments of data that will be used for further analysis, either time intervals of experimental interest or epileptogenic activity in the case of clinical studies. Then, we filter the data for high-frequency and power line-noise, followed by the removal of channels and/or segments of the data that are poor quality for experimental purposes, such as those that contain signal artifacts or epileptiform activity (see Chap. 4). Using FieldTrip, this is readily achieved with *ft_preprocessing*, a versatile function that combines a slew of preprocessing options with the capability to read data formats used by the most popular electrophysiological data-acquisition systems. Finally, we add the electrode information obtained during the previous steps to the preprocessed electrophysiological data. The advantage of adding the electrode information at this stage is that FieldTrip will keep it consistent with the electrophysiological data going forward, as during the re-referencing process we describe next.

27.4.2 Re-referencing and Subsequent Analysis

So far, we have linked the electrophysiological recordings with electrode locations corresponding to the patient's neuroanatomy. The initial foundation for this integration was laid during electrode localization, which offered the opportunity to directly link anatomical locations to electrode labels corresponding to the electrophysiological data as stored in the recording file. The integration was completed when we combined the electrode information with the preprocessed electrophysiological data. This benefits subsequent processing and analysis in at least two ways. First, it affords the flexibility to consider different referencing schemes, and create new channels depending on the type of derivation we choose and thereby modify all parts of this integration. For example, in the case of re-referencing to a bipolar montage, the same montage that is used for the electrophysiological data can also be automatically applied to the channel positions to reflect the update in neuroanatomical context, e.g., between the two electrodes that constitute a newly formed bipolar channel (for the pros and cons of different referent schemes, see Chap. 28). Second, the integration affords the flexibility to browse through various anatomically informed representations of the electrophysiological data during subsequent analysis. For instance, FieldTrip's plotting suite allows interactively launching new figures showing the recalculated distribution of activity for selected channels, frequency, or time intervals, thus without having to concern oneself with repeatedly matching electrophysiological signals with electrodes' labels and anatomical locations. In addition, by using the spatially normalized electrode coordinates in the analysis, electrophysiological data originating from multiple patients can be readily aggregated or overlaid on standardized models of brain anatomy. Together these represent just a few examples of how integrated representations of electrophysiological activity and brain anatomy can both simplify existing and open up new possibilities in iEEG analysis.

27.5 Discussion

We have described a workflow for obtaining precise knowledge of the neuroanatomical locations of surgically implanted electrodes, and efficiently integrating that knowledge with the electrophysiological data and its subsequent analysis. Human intracranial datasets are collected under unique circumstances in clinical settings. For these reasons, they come in different shapes and sizes, and can be processed for different purposes. These considerations suggest it is important for a workflow to strike the right balance between flexibility and efficiency. The need for flexibility is further amplified by intracranial data typically imposing greater demands on alternative options and strategies in the analysis than non-invasive data recorded in more standardized laboratory settings. As shown in Fig. 27.1, the present workflow takes as input various data combinations and offers several routes during preprocessing to accommodate data from different patients and situations, including non-ideal cases.

Efficiency pertains to the optimization of processing within and across patients. For within patient processing, the workflow reduces redundancy and facilitates integration across steps, as seen in how electrode locations can be directly assigned to labels collected from the electrophysiological recording file during localization. This design obviates the need to sort and rename electrodes to match the electrophysiological data during later stages of analysis. To optimize processing across patients, each step has an implementation counterpart in an analysis protocol embedded within the open-source FieldTrip toolbox [33]. The implementation as a toolbox enables the use of batch scripting to deal with repeated analyses within and across patients. In addition, it allows researchers to readily build on a continuously growing body of analysis methods that, over the past decade, have been developed and used by a large research community [26]. Several suitable alternative toolboxes exist that can implement aspects of the workflow we have described here, e.g., [2, 16, 19, 22].

Most notably, integrated iEEG analysis opens up the possibility to ask a different set of research questions relating to anatomically precise and consistent representations of electrophysiological activity. Recent work performing such analyses have begun to open up new dimensions in the study of neural processing and information flow, including the propagation of wave-like activity across cortical and hippocampal tissue [43, 32, 18]. Using a workflow which readily integrates electrophysiological recordings with the neuroanatomical context facilitates discoveries like these and more generally allows researchers to advance the field of human iEEG.

References

1. Ashburner J, Friston KJ (1999) Nonlinear spatial normalization using basis functions. Hum Brain Mapp 7(4):254–266
2. Blenkmann AO, Phillips HN, Princich JP, Rowe JB, Bekinschtein TA, Muravchik CH, Kochen S (2017) iElectrodes: a comprehensive open-source toolbox for depth and subdural grid electrode localization. Front Neuroinform. https://doi.org/10.3389/fninf.2017.00014
3. Boatman-Reich D, Franaszczuk PJ, Korzeniewska A, Caffo B, Ritzl EK, Colwell S, Crone NE (2010) Quantifying auditory event-related responses in multichannel human intracranial recordings. Front Comput Neurosci 4:4
4. Brang D, Dai Z, Zheng W, Towle VL (2016) Registering imaged ECoG electrodes to human cortex: a geometry-based technique. J Neurosci Methods. https://doi.org/10.1016/j.jneumeth.2016.08.007
5. Centracchio J, Sarno A, Esposito D, Andreozzi E, Pavone L, Di Gennaro G, Bartolo M et al (2021) Efficient automated localization of ECoG electrodes in CT images via shape analysis. Int J Comput Assist Radiol Surg 16(4):543–554
6. Cocosco CA, Kollokian V, Kwan RK, Pike GB, Evans AC (1997) BrainWeb: online interface to a 3D MRI simulated brain database. Proc 3rd Int Conf Funct Mapp Hum Brain 5(S425)
7. Dalal SS, Edwards E, Kirsch HE, Barbaro NM, Knight RT, Nagarajan SS (2008) Localization of neurosurgically implanted electrodes via photograph-mri-radiograph coregistration. J Neurosci Methods 174(1):106–115
8. Dale AM, Fischl B, Sereno MI (1999) Cortical surface-based analysis. I. Segmentation and surface reconstruction. Neuroimage 9(2):179–194
9. Desikan RS, Ségonne F, Fischl B, Quinn BT, Dickerson BC, Blacker D, Buckner RL et al (2006) An automated labeling system for subdividing the human cerebral cortex on MRI scans into gyral based regions of interest. Neuroimage 31(3):968–980
10. Destrieux C, Fischl B, Dale A, Halgren E (2010) Automatic parcellation of human cortical gyri and sulci using standard anatomical nomenclature. Neuroimage. https://doi.org/10.1016/j.neuroimage.2010.06.010
11. Dykstra AR, Chan AM, Quinn BT, Zepeda R, Keller CJ, Cormier J, Madsen JR, Eskandar EN, Cash SS (2012) Individualized localization and cortical surface-based registration of intracranial electrodes. Neuroimage 59(4):3563–3570

12. Eickhoff SB, Stephan KE, Mohlberg H, Grefkes C, Fink GR, Amunts K, Zilles K (2005) A new SPM toolbox for combining probabilistic cytoarchitectonic maps and functional imaging data. Neuroimage 25(4):1325–1335
13. Fan L, Li H, Zhuo J, Zhang Y, Wang J, Chen L, Yang Z, et al (2016) The human brainnetome atlas: a new brain atlas based on connectional architecture Cereb Cortex.https://doi.org/10.1093/cercor/bhw157
14. Flinker A, Korzeniewska A, Shestyuk AY, Franaszczuk PJ, Dronkers NF, Knight RT, Crone NE (2015) Redefining the Role of Broca's Area in Speech. Proc Natl Acad Sci USA 112(9):2871–2875
15. Glen DR, Taylor PA, Buchsbaum BR, Cox RW, Reynolds RC (2020) Beware (surprisingly common) left-right flips in your MRI data: an efficient and robust method to check MRI dataset consistency using AFNI. Front Neuroinform 14:18
16. Groppe DM, Bickel S, Dykstra AR, Wang X, Mégevand P, Mercier MR, Lado FA, Mehta AD, Honey CJ (2017) iELVis: an open source MATLAB Toolbox for localizing and visualizing human intracranial electrode data. J Neurosci Methods 281:40–48
17. Hermes D, Miller KJ, Noordmans HJ, Vansteensel MJ, Ramsey NF (2010) Automated electrocorticographic electrode localization on individually rendered brain surfaces. J Neurosci Methods 185(2):293–298
18. Kleen JK, Chung JE, Sellers KK, Zhou J, Triplett M, Lee K, Tooker A, Haque R, Chang EF (2021) Bidirectional propagation of low frequency oscillations over the human hippocampal surface. Nat Commun. https://doi.org/10.1038/s41467-021-22850-5
19. Kubanek J, Schalk G (2015) NeuralAct: a tool to visualize electrocortical (ECoG) activity on a three-dimensional model of the cortex. Neuroinformatics 13(2):167–174
20. Lancaster JL, Rainey LH, Summerlin JL, Freitas CS, Fox PT, Evans AC, Toga AW, Mazziotta JC (1997) Automated labeling of the human brain: a preliminary report on the development and evaluation of a forward-transform method. Hum Brain Mapp. https://doi.org/10.1002/(sici)1097-0193(1997)5:4%3c238::aid-hbm6%3e3.0.co;2-4
21. LaPlante RA, Tang W, Peled N, Vallejo DI, Borzello M, Dougherty DD, Eskandar EN, Widge AS, Cash SS, Stufflebeam SM (2017) The Interactive electrode localization utility: software for automatic sorting and labeling of intracranial subdural electrodes. Int J Comput Assist Radiol Surg. https://doi.org/10.1007/s11548-016-1504-2
22. Li G, Jiang S, Chen C, Brunner P, Wu Z, Schalk G, Chen L, Zhang D (2019). iEEGview: an open-source multifunction GUI-based matlab toolbox for localization and visualization of human intracranial electrodes J Neural Eng 17(1):016016
23. Li X, Morgan PS, Ashburner J, Smith J, Rorden C (2016) The first step for neuroimaging data analysis: DICOM to NIfTI conversion. J Neurosci Methods 264:47–56
24. Maurer CR, Hill DLG, Maciunas RJ, Barwise JA, Michael Fitzpatrick J, Wang MY (1998) Measurement of intraoperative brain surface deformation under a craniotomy. Med Image Comput Comput-Assist Interv—MICCAI'98. https://doi.org/10.1007/bfb0056187
25. Metzger BA, Magnotti JF, Wang Z, Nesbitt E, Karas PJ, Yoshor D, Beauchamp MS (2020) Responses to visual speech in human posterior superior temporal gyrus examined with iEEG deconvolution. J Neurosci: Off J Soc Neurosci 40(36):6938–6948
26. Oostenveld R, Fries P, Maris E, Schoffelen J-M (2011) FieldTrip: open source software for advanced analysis of MEG, EEG, and invasive electrophysiological data. Comput Intell Neurosci 2011:156869
27. Pacheco Estefan D, Zucca R, Arsiwalla X, Principe A, Zhang H, Rocamora R, Axmacher N, Verschure PFMJ (2021) Volitional learning promotes theta phase coding in the human hippocampus. Proc Natl Acad Sci U S A 118(10). https://doi.org/10.1073/pnas.2021238118
28. Parvizi J, Kastner S (2018) Promises and limitations of human intracranial electroencephalography. Nat Neurosci 21(4):474–483
29. Pieters TA, Conner CR, Tandon N (2013) Recursive grid partitioning on a cortical surface model: an optimized technique for the localization of implanted subdural electrodes. J Neurosurg 118(5):1086–1097

30. Roberts DW, Hartov A, Kennedy FE, Miga MI, Paulsen KD (1998) Intraoperative brain shift and deformation: a quantitative analysis of cortical displacement in 28 cases. Neurosurgery 43(4):749–758; discussion 758–760
31. Stieglitz LH, Ayer C, Schindler K, Oertel MF, Wiest R, Pollo C (2014) Improved localization of implanted subdural electrode contacts on magnetic resonance imaging with an elastic image fusion algorithm in an invasive electroencephalography recording. Neurosurgery 10(Suppl 4): 506–512; discussion 512–513
32. Stolk A, Brinkman L, Vansteensel MJ, Aarnoutse E, Leijten FS, Dijkerman CH, Knight RT, de Lange FP, Toni I (2019) Electrocorticographic dissociation of alpha and beta rhythmic activity in the human sensorimotor system. eLife 8. https://doi.org/10.7554/eLife.48065
33. Stolk A, Griffin S, van der Meij R, Dewar C, Saez I, Lin JJ, Piantoni G, Schoffelen J-M, Knight RT, Oostenveld R (2018) Integrated analysis of anatomical and electrophysiological human intracranial data. Nat Protoc 13(7):1699–1723
34. Sweet JA, Hdeib AM, Sloan A, Miller JP (2013) Depths and grids in brain tumors: implantation strategies, techniques, and complications. Epilepsia 54(Suppl 9):66–71
35. Talairach J, Tournoux P (1988) Co-planar stereotaxic atlas of the human brain: 3-dimensional proportional system: an approach to cerebral imaging. George Thieme Verlag
36. Tian Y, Margulies DS, Breakspear M, Zalesky A (2020) Topographic organization of the human subcortex unveiled with functional connectivity gradients. Nat Neurosci 23(11):1421–1432
37. Tzourio-Mazoyer N, Landeau B, Papathanassiou D, Crivello F, Etard O, Delcroix N, Mazoyer B, Joliot M (2002) Automated anatomical labeling of activations in SPM using a macroscopic anatomical parcellation of the MNI MRI single-subject brain. Neuroimage 15(1):273–289
38. Voorhies JM, Cohen-Gadol A (2013) Techniques for placement of grid and strip electrodes for intracranial epilepsy surgery monitoring: pearls and pitfalls. Surg Neurol Int 4:98
39. Wang L, Mruczek REB, Arcaro MJ, Kastner S (2015) Probabilistic maps of visual topography in human cortex. Cereb Cortex 25(10):3911–3931
40. Wellmer J, von der Groeben F, Klarmann U, Weber C, Elger CE, Urbach H, Clusmann H, von Lehe M (2012) Risks and benefits of invasive epilepsy surgery workup with implanted subdural and depth electrodes. Epilepsia. https://doi.org/10.1111/j.1528-1167.2012.03545.x
41. Yang AI, Wang X, Doyle WK, Halgren E, Carlson C, Belcher TL, Cash SS, Devinsky O, Thesen T (2012) Localization of dense intracranial electrode arrays using magnetic resonance imaging. Neuroimage 63(1):157–165
42. Yeo BT, Thomas FM, Krienen JS, Sabuncu MR, Lashkari D, Hollinshead M, Roffman JL et al (2011) The organization of the human cerebral cortex estimated by intrinsic functional connectivity. J Neurophysiol 106(3):1125–1165
43. Zhang H, Watrous AJ, Patel A, Jacobs J (2018) Theta and alpha oscillations are traveling waves in the human neocortex. Neuron 98(6):1269–81.e4

Chapter 28
How Should I Re-reference My Intracranial EEG Data?

George M. Parish, Sebastian Michelmann, and Simon Hanslmayr

Abstract The analysis of intracranial electrophysiological recordings requires processing choices. Electrical signals are recorded relative to a reference and the choice of that online reference may be sub-optimal depending on the goal of the subsequent analysis. Therefore, a secondary re-referencing operation is often undertaken aiming to increase the signal-to-noise ratio, which can entail transforming the signal in relation to a specific hypothesis. However, comparative studies on this much understudied issue of re-referencing are sparse, which can lead to habitual and ill-informed decision making. This chapter starts with giving a non-exhaustive overview over common re-referencing schemes before presenting three studies that explore what re-referencing means for cortical alpha and gamma power during a motor task as well as lower frequency power in the medial temporal lobe during a memory task. By revealing how different strategies lead to different observations in the iEEG signal and their modulation by task or behaviour, we demonstrate how significant this early transformative decision is for further analyses.

28.1 Why Do We Need a Reference?

Intracranial EEG offers the exciting prospect of measuring brain activity at high spatial *and* high temporal resolution, which is not possible with non-invasive methods

G. M. Parish · S. Hanslmayr (✉)
School of Psychology, Centre for Human Brain Health, University of Birmingham, Birmingham, UK
e-mail: Simon.Hanslmayr@glasgow.ac.uk

S. Michelmann
Department of Psychology, Princeton University, Princeton, NJ, USA

Princeton Neuroscience Institute, Princeton University, Princeton, NJ, USA

S. Hanslmayr
School of Neuroscience and Psychology, Centre for Cognitive Neuroimaging, University of Glasgow, Glasgow, UK

N. Axmacher (ed.), *Intracranial EEG*, Studies in Neuroscience, Psychology and Behavioral Economics, https://doi.org/10.1007/978-3-031-20910-9_28

where a researcher needs to compromise on either of the two dimensions. The motivation of an iEEG study therefore often is to utilize this unique strength in temporal and spatial resolution in an experiment. However, there is a fundamental biophysical reality that any researcher wishing to use iEEG faces, which naturally comes with the type of signal we record. This signal represents fluctuations in voltages over time which are caused by weak electrical fields generated by the summated activity of neurons [1]. An iEEG recording at a given channel therefore refers to the dynamic measurement of potential differences, or differences in "electrical pressure", between two points. This can be written as a simple subtraction formula:

$$V_{(t)} = V1_{(t)} - V2_{(t)}$$

where V(t) is the signal at time (t), and V1(t) is the electrical potential at one recording site at time (t) and V2(t) is the electrical potential at another recording site at time (t). This raises a fundamental question, which is "Is a change in V caused by a change in V1, or V2, or by a change in both?". Referencing describes the process by which we initially chose the recording points V1 and V2 to record a signal of interest V, which is in most cases dictated by clinical needs. Re-referencing refers to carrying out this subtraction offline, i.e. after the initial recordings have been made, to isolate a signal of interest based on a particular research question. Re-referencing is therefore one of the first steps in the preprocessing "pipeline" of an iEEG dataset. Because there is no "one-size-fits-all" approach (as will become clear in the remainder of this chapter), this initial step needs to be the result of an informed decision based upon careful considerations. An in-optimal reference may even lead to erroneous conclusions.

At this point you may be asking yourself, why don't we simply choose an electrically inactive site as a reference? Indeed, the very term "reference" implies a source that is non-active against which the activity of an "active" source can be measured. Unfortunately, this is a myth often encountered in EEG research. In such a hypothetical scenario any change in the signal can be interpreted unequivocally to the change in the "active" electrode. The biophysical reality of the brain, however is that there are no electrically inactive sources (see [2] for an excellent in-depth discussion on this issue[1]). While there are areas in the brain that are electrically more or less active, like grey matter (electrically more active) and white matter (electrically less active), there will be no area in the brain, or on the scalp that is electrically inactive [3]. Even electrodes placed on the bone or on the skin outside the brain will pick up electrical activity that is volume conducted to that site and therefore introduce this activity into recordings if being used as reference.

Let us consider a specific example of a recording where electrodes in the hippocampus are referenced against a scalp electrode placed on the mastoid (i.e., a point at the back of the head behind the ear, which is a popular choice for referencing [4]). After analysing the data, the researcher may find alpha power decreases that are modulated by the task and concludes that alpha power modulations in the

[1] This book chapter is mostly concerned with scalp EEG but many of the fundamental biophysical properties also apply to intracranial EEG.

hippocampus reflect a specific cognitive process. However, it may well be that the hippocampus itself does actually not show any modulations in alpha power at all, and that instead these alpha modulations are solely introduced by the mastoid reference which happens to pick up alpha signals volume conducted to the scalp. Such an example is demonstrated in [5], Fig. 28.2 and illustrates one possibility where a wrongful conclusion is made due to an in-optimal referencing choice.[2]

So far, we have only considered electrical fluctuations that are generated by the brain. However, several other sources also give rise to changes in electrical potentials that may be picked up by intracranial electrodes, or introduced by reference electrodes. Muscle activity induced by head movements, speech, and chewing for instance; another example is power line noise introduced by electrical sources near the patient (i.e., power sockets, patient bed, medical devices near the patient, etc.). These signals can be amplified by sub-optimal reference choices and, in extreme cases, may render a whole recording unfit for analysis. On the other hand, re-referencing can be a powerful tool to separate out such artefacts and may even salvage a "lost" dataset.

To summarise, we have clarified that any intracranial EEG researcher is forced to make a decision on how to re-reference the data. This decision has the potential to improve the data quality or to make it worse, and in extreme examples lead to wrongful conclusions. Because no electrically inactive source exists there is currently no gold standard in the field as to how to re-reference an iEEG dataset. Hence the answer to the title question of this chapter "*How should I re-reference my iEEG data*" will be different on a case-by-case basis, depending on the research questions and what the signal of interest is.

The aim of this book chapter is to introduce the reader into the complex issue of re-referencing by providing a (non-exhaustive) overview over the different most used referencing schemes. A secondary aim is to give the reader an intuition about the advantages/disadvantages of different referencing schemes by reviewing the results of two recent empirical studies exploring the effects of re-referencing on a signal of interest. Inevitably, this chapter will raise more questions than it will answer because our current knowledge on this issue is far from being complete. Our hope is to stimulate further research, and to give the reader a few tools to make a better-informed decision on which re-referencing scheme to choose for their dataset.

28.2 How Can We Re-reference iEEG Data?

The following (non-exhaustive) list of commonly used re-referencing approaches may guide the reader through the literature on this topic and help them to make an informed decision about the optimal choice of reference for the analyses they pursue (see Fig. 28.1). Before describing these different referencing schemes it is important

[2] This is not to say that the hippocampus does not show alpha oscillations that are modulated by a task. Indeed several studies suggest that the hippocampus expresses genuine alpha oscillations that are modulated by memory processes (e.g. [6]).

to define the terminology to avoid confusion. The term ***electrode*** is used to refer to the device that is inserted into the brain, or placed onto the brain which can be either a depth, strip or grid electrode.[3] The term ***contact*** refers to a singular location on that device where electrical contact with the brain tissue is made. The term ***channel*** will be used to refer to the recorded signal which is impacted by given referencing or re-referencing procedures.

28.2.1 Monopolar Reference

In intracranial and classical EEG studies, monopolar reference describes the referencing of all contacts against a single contact [1]. This is the typical scenario that is encountered in raw iEEG data, where data may be referenced to a subdural contact, or to a contact that is placed in the bone or on the mastoid [6, 7]. Notably, a researcher's preferred choice of such a reference may not correspond to the optimal choice in a clinical setting.

In the re-referencing step, the researcher can easily change the monopolar reference, by subtracting the newly selected reference channel from all other channels, i.e., at a given channel k that currently reflects the voltage difference between k and the online reference o, activity is recomputed such that

$$V_k' = (V_k - V_r)$$

where r is the index of the new reference channel. Because both V_k and V_r are capturing the voltage difference to the online reference o, the activity from o will cancel out during subtraction. Note that the removal of shared activity that stems from the online reference is a common goal that is shared between many re-referencing schemes.

28.2.2 Bipolar Reference

Bipolar re-referencing is one of the most applied re-referencing schemes in intracranial EEG. It describes the subtraction of each channel from its neighbour. Specifically, a new channel k is computed as

$$V_k' = (V_k - V_{k-1})$$

A key goal of bipolar referencing is to highlight activity that is local: by subtracting a channel from its neighbour (typically on the same electrode shaft), activity that is shared between the two channels will cancel out. Bipolar re-referencing can also

[3] In some occasions we also use electrode to refer to recording sites used in scalp EEG.

a $V'_k = V_k -$
$V'_{i,j} = V_{i,j} -$

A non-exhaustive list of re-referencing methods

Monopolar	Bipolar	Average	Gram-Schmidt (GS)	White matter
V_r	$V_{k-1};\ V_{i-1,j}$	$\bar{x}$	$\frac{\langle \vec{x}, \vec{v_k} \rangle}{\langle \vec{x}, \vec{x} \rangle}\vec{x};\ \frac{\langle \vec{x}, \vec{v_{i,j}} \rangle}{\langle \vec{x}, \vec{x} \rangle}\vec{x}$	$V_w;\ \bar{w}$

Laplacian	Independent component analyses (ICA)
$0.5(V_{k-1} + V_{k+1});$ $0.25(V_{i-1,j} + V_{i+1,j} + V_{i,j-1} + V_{i,j+1})$	$[V] = [A] \cdot [S]$ $[A^{-1}] \cdot [V] = [S]$

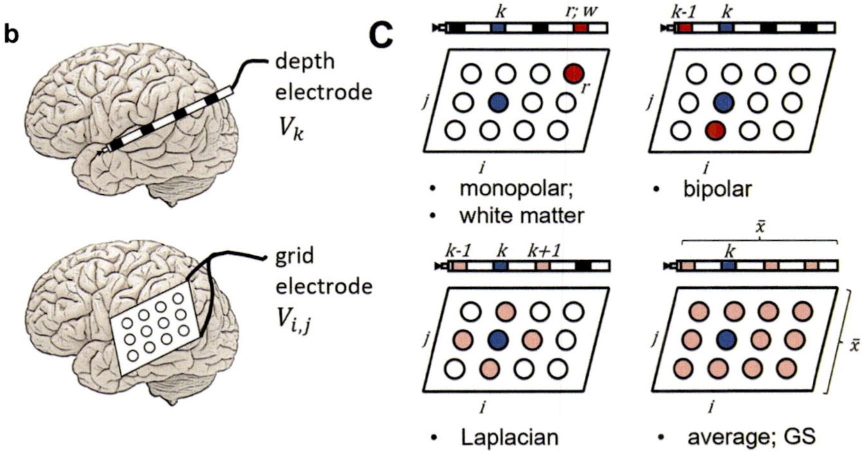

Fig. 28.1 A non-exhaustive list of commonly used re-referencing methods. **a** Equations for several re-referencing methods that can be applied to intracranial EEG data, for both 1-dimensional depth electrodes (Vk), 1-dimensional strip electrodes and 2-dimensional grid electrodes (Vij). **b** Visualisation of a depth electrode (top panel) and grid electrodes (bottom panel). **c** Visualisations of the application of the equations in (**a**) to the actual data. Blue nodes indicate the contact in question (Vk or Vij), red nodes indicate single target re-reference contacts and light red indicate a set of re-reference contacts that will be averaged over. Note that these visualisations do not depict possible re-reference choices that are exterior to the brain, for example in the case of the monopolar method, nor do they depict the possibility to average over all depth or grid electrodes within the brain, for example in the case of the average or Gram-Schmidt methods

be described as the spatial derivative; after re-referencing a channel captures the change in activity from one contact to the next [3, 8]. Because activity recorded from the reference contact should affect neighbouring channels to a similar extent, noise and activity from the online reference will cancel out with the bipolar referencing operation. Likewise, however, bipolar re-referencing removes any signal that is shared between two neighbouring channels, which disproportionally affects the low-frequency range. Depending on how much signal and noise is shared on two neighbouring electrodes, bipolar re-referencing can therefore either increase or decrease the signal to noise ratio on a channel (see: [5]). Another key disadvantage of bipolar re-referencing is the inevitable loss of information due to the reduced dimensionality of the re-referenced data: after re-referencing on, for instance, a depth electrode with 8 contacts, activity will be pulled together into 7 channels that reflect the difference between neighbours. The measured activity on the original 8 channels, however, cannot be reconstructed anymore by the linear combination of the new 7 channels, unless there were already linear dependencies beforehand (the new data has a reduced rank [5]).

28.2.3 Laplacian Reference

Laplacian reference pursues a similar goal to bipolar re-referencing, however, instead of considering a single neighbour, each channel is re-referenced to the average of both neighbouring channels for shaft and strip electrodes, or to the average of its 4 nearest neighbours for grid electrodes [3]. Specifically, for depth and strip electrodes a new channel k is computed as

$$V_k' = V_k - 0.5(V_{k-1} + V_{k+1})$$

On grid electrodes, the new channel at the 2-D position (i, j), can be computed as

$$V_{i,j}' = V_{i,j} - 0.25(V_{i-1,j} + V_{i+1,j} + V_{i,j-1} + V_{i,j+1})$$

The advantages and disadvantages of Laplacian re-referencing are comparable to bipolar re-referencing, however, because all neighbours are considered the researcher does not have to decide on a direction for the operation. A key disadvantage is again the inevitable loss of information due to the reduced dimensionality of the data (see above). This problem is even exacerbated with Laplacian re-referencing, the resulting data will be reduced in rank by 2 for each electrode [5].

28.2.4 Common Average Reference (CAR) and Median Reference

Average reference is a popular reference for scalp EEG. In scalp EEG a coverage of the head can be approximated as a sphere if sufficient (and approximately equidistant) electrode coverage is given. The sum of electrical potentials that are measured on opposite sides of the head should therefore—at least in theory—be zero. Therefore, in scalp EEG, a reasonable approximation of removing activity from the online reference (often mastoid, or Cz) can be achieved by subtracting the average [2, 9].

The broad electrode coverage to approximate the head as a sphere is almost certainly never achieved with intracranial EEG, where coverage is entirely determined by clinical considerations. Furthermore, intracranial contacts measure signal that is more local than the summed potential that is picked up by scalp EEG contacts [10] and depending on the location of a contact, large differences in electrical potential can be observed; the reader may consider, for instance, the difference in amplitude between contacts located in the Hippocampus and in nearby white matter.

Nonetheless, the average across all channels may entail an approximation of activity at the reference, especially if the online reference is very noisy and therefore accounts for a large proportion of the variance.

With average reference, a new channel k is computed as

$$V_k' = (V_k - \overline{x})$$

where $\overline{x}$ represents the average across all channels. Average re-referencing has the advantage of preserving information, it only reduces the total rank of the data by one (because the sum of all channels is 0, each channel is a linear combination of all others; it can be re-written as their negative sum). This advantage, however, may (but doesn't have to) come at the cost of a lower signal to noise ratio in the re-referenced channels and of a potential mislocalization of effects. Specifically, the average across all channels is sensitive to extreme values, e.g., sharp high-amplitude noise on single channels; subtracting the average from otherwise clean channels, may therefore reduce the signal to noise ratio on that channel. Furthermore, the average may be sensitive to high amplitude oscillations from sub-cortical structures. CAR can therefore lead the researcher to attribute effects to structures that are not involved in the measured neural process. On channels that strongly contribute to the average (e.g., high amplitude channels in the hippocampus), the signal will also appear attenuated (by a factor of 1/N, where N refers to the number of channels) after re-referencing.

The use of a common median reference is an attempt to alleviate the sensitivity of CAR to extreme values. Again, a new channel k is computed as ($V'_k = V_k - \overline{x}$), where $\overline{x}$ now represents the median across all channels, e.g., [11].

28.2.5 Gram-Schmidt Orthogonalization

A potential issue with re-referencing is that the subtraction of the new reference can introduce artifacts that were not present on a channel before the re-referencing step (see above). A recent approach to address this issue is the use of orthonormalization between a channel and the average across all other channels via the Gram-Schmidt process [12]. Specifically, the new channel k is computed as:

$$V_k' = V_k - \frac{\langle \overrightarrow{x}, \overrightarrow{v_k} \rangle}{\langle \overrightarrow{x}, \overrightarrow{x} \rangle} \overrightarrow{x}$$

where $\overrightarrow{v_k}$ represents the time-series of channel k, and $\overrightarrow{x}$ represents the time series of the average across all channels except k; the brackets denote the dot product. This procedure effectively removes the part of the signal at V_k that is shared with the average across all other channels, reducing the risk of introducing artefacts from other channels. As the average is scaled by the inner product of the channel and channel average, this operation can boost smaller amplitude local signal whilst bringing down higher amplitude global signal.

28.2.6 White Matter Reference

The idea behind the use of a white matter reference is to select a single channel or the average of a group of channels that pick up little to no signal. Because electric potentials that are associated with neural activity are generated in gray matter, contacts located in white matter are assumed to not pick up signal and only reflect shared noise, e.g., from the online reference.

The new re-referenced channel is then computed as

$$V_k' = (V_k - V_w)$$

where w is the index of the selected white matter channel or

$$V_k' = (V_k - \overline{w})$$

where $\overline{w}$ is the average activity across a group of the white matter channels [13, 14]. Importantly, the assumption of white matter contacts being "silent" is overly simplified; a recent study by [3] demonstrated that contacts in the white matter record a mixture of (zero lag) volume conducted signal from nearby gray matter and surprisingly from distant gray matter, i.e., signal that was carried by the white matter fibers themselves [3].

Identification of White Matter Contacts in Intracranial EEG

One of the challenges for the use of white matter reference is to determine which contacts are suitable. Contacts are often located via MRI scans from before and after surgical implantation of electrodes, that are then normalized against one another—a transformation that does not always give perfect anatomical information. To quantify the amount of white/gray matter surrounding a contact, Mercier et al. [3] propose a Proximal Tissue Density index (PTD) based on white/gray matter estimates from a Freesurfer parcellation. PTD is computed within a 3 × 3 × 3 mm cube around the centroid of an electrode contact as PTD = (VoxGray − VoxWhite)/(VoxGray + VoxWhite), where VoxGray and

VoxWhite describe the number of gray and white matter voxels respectively. This measure results in a value between (−1) for contacts that are surrounded entirely by white matter and (+1) for contacts that are surrounded entirely by gray matter.

When selecting white matter contacts for re-referencing, it is also useful to assess properties of the recorded signal. Parish et al. [15] propose an approach that is informed by electrophysiological signal properties. They reason that, as a 'silent' channel, a white matter contact should have a lower signal to noise ratio than other channels. They demonstrate that a white matter contact is characterized by less variability in the local field potential (LFP), and by less low frequency (2–12 Hz) power and gamma frequency (60–90 Hz) power than contacts that are situated in gray matter (e.g., in MTL gray matter). The authors propose to perform white matter channel-selection based on a threshold, where, for instance, WM theta power should not exceed 20% of that for MTL theta power (compare Fig. 28.2; note that contacts in the hippocampus typically express stronger signal than other channels [16]).

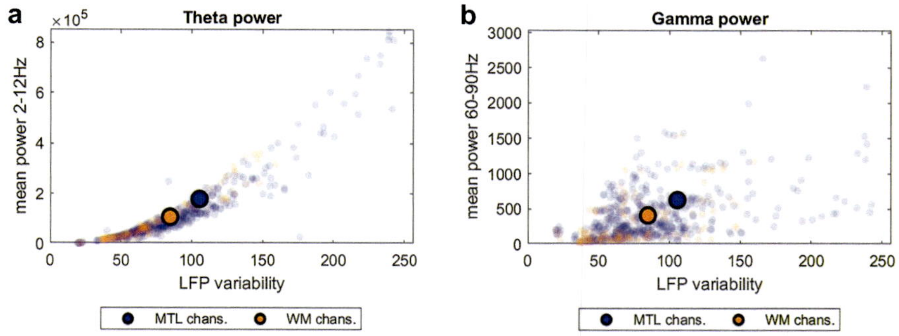

Fig. 28.2 Analyses from an episodic memory paradigm, using iEEG data recorded with depth electrodes, originally referenced using either a bipolar method or a reference contact close to the recording source. **a, b** Mean power in the theta (**a** 2–12 Hz) and gamma (**b** 60–90 Hz) frequency bands for contacts situated in the medial temporal lobe (MTL; blue) and white matter (WM; orange), against variability of the local field potential (LFP; calculated as 2 × standard deviation). Blue and orange dots indicate individual MTL or WM channels, respectively, whilst blue and orange circles indicate the centroid of all MTL or WM channels, respectively

28.2.7 Independent Component Analysis

Independent Component Analysis is a method that "unmixes" data into its underlying latent components [17]. It is widely used in EEG to subtract out activity that stems from ocular movements and blinks, but also electrical noise [18]. The reasoning to apply ICA as a re-referencing method, is that neural activity that is measured by a given channel reflects a mixture of local activity, online reference activity, electrical noise and volume conducted activity from nearby areas, but the researcher is blind to the nature of that superposition of activity. ICA can unmix the recorded data into its underlying independent sources (note that noise should be independent from neurally generated activity). The decomposition of the data into its underlying independent components can then be used to systematically eliminate certain components from the data. Because a linear combination of independent components leads back to the original data, it is possible to inspect how much each component affects each channel. Crucially, activity and noise from the online reference mixes into all channels to a similar extent because recorded channels reflect the voltage difference between their respective contact and the online reference contact. It follows that a component that captures activity from the online reference is very global. Discarding this global component should therefore eliminate undesired parts of the data, while leaving the rest of the data intact [5, 19].

This reference can be thought of as a data-driven re-referencing scheme because the coefficients of the computation are learned from the statistical properties of the data at hand. While average reference, for instance, assumes that the average is a good approximation of broad noise and noise at the online reference (using the same coefficient 1/N for every channel), ICA learns coefficients that optimally isolate components that are independent from the rest of the data. A spatially broad component can then be identified by the researcher and be removed from the data.

28.2.8 Spatio-spectral Decomposition and Tailored Spatial Filtering Approaches

The example of ICA illustrates that it is useful to think of the re-referencing step as spatial filtering operation that highlights certain properties of the data. The bipolar re-referencing of a channel, for instance, is a spatial filter with the fixed coefficients $[1, -1]$ on neighbouring contacts (and zero otherwise), whereas ICA learns coefficients to extract each component from the data. Consequently, the researcher might wonder whether tailored spatial filters should be applied directly for the purpose of highlighting desired properties of the signal. One recent such application is the use of spatio-spectral decomposition (SSD) [20] for intracranial EEG data [21].

SSD is a spatial filter that operates on the spectrum of the time series, to maximize power in a selected frequency band over its flanking frequencies. If the goal of the analysis is to extract specific oscillations with high signal to noise ratio, the researcher may therefore opt to use this method directly in lieu of re-referencing. Similarly, it is noteworthy that other spatial filtering methods (e.g., logistic regression, or linear discriminant they are often used for classification [11, 22]) can learn coefficients including those afforded by common re-referencing schemes. It may therefore be preferable to learn spatial filters on the data directly for analysis purposes, without an intermediate re-referencing step that could potentially result in a loss of information (see above).

28.3 What Does Re-referencing Do to My Data? An Empirical Comparison Between Different Re-referencing Schemes

In this section, we collect results from the relatively sparse number of studies that compare how your re-referencing choice reflects on your data. These studies work with data collected from two very different experiments: one an associative memory paradigm that is interested in low frequency theta oscillations in the hippocampus, the other a gesture decoding paradigm that is interested in mid to high frequency alpha and broadband gamma oscillations in the motor cortex. The relative differences in outcome indicate that your choice of re-referencing method should be part of a wider strategy that is driven by your hypothesis, as each method will transform your data relative to its own assumptions.

The data compared here is obtained via intracranial stereotaxic EEG (sEEG) obtained by recordings from depth electrodes which are implanted into the brain to target potential focal points of epileptic seizures for monitoring and assessment of surgical feasibility. As these electrodes penetrate the deeper structures of the limbic system, contacts along the length of the electrode offer recordings from lateral contacts which pick up signals from many neuronal groups—near and far. First, we will look at how different re-referencing methods perform as a means of data cleaning, as well as whether re-referencing affects the proportion of channels that are responsive to the task at hand. From there, it is important to ask what is left in your electrical signals after all this pre-processing, and more importantly—what is it telling you? This is described in Sect. 28.2.2 which illustrates the effect of re-referencing on basic signal metrics such as power in specific frequency bands, and task-related differences between these signals as observed in a memory task and a motor task.

28.3.1 What Is Left in Your Data After Re-referencing?

A recent study [13] compared how several re-referencing methods transformed the raw signal, with a specific interest in obtaining local signals for analyses in the broadband gamma and alpha frequency bands. During data acquisition, the data was originally referenced using the mean of two white matter contacts. This study found that the signal was quite often contaminated when applying a monopolar re-reference from white matter areas, whereby task-related activity that was previously localised to the monopolar reference channel was introduced to all other channels to become more global in nature (see Fig. 28.3a, b). This led to more channels being responsive to the task at hand (Fig. 28.3c, d), however, this did not translate into an increase in overall power in these frequency bands (Fig. 28.3e, f). Another important observation made in this study is that neighbour driven methods such as bipolar or Laplacian seem to be good at eliminating global activity from the combined set of macro contacts, leading to a relatively small amount of correlation between channels (see Fig. 28.3a, b). The ability of neighbour driven methods to remove global components and reduce the contamination of task-related activity enables further analyses to focus on the most responsive channels (Fig. 28.3c, d) that contain the most local amount of useful signal (Fig. 28.3e, f).

Neighbour driven re-referencing methods are therefore commonly applied to maximise local signals. However, [5] demonstrate that ICA may be better suited for this goal (see Fig. 28.4). The authors compare the performance of ICA in extracting local signal, to the performance of bipolar referencing. Simulations suggest that ICA outperforms bipolar referencing in sensitivity (i.e., in isolating signal from local sources) and specificity (discarding activity from distant sources). Indeed, bipolar reference only performed reasonably well, when signal was very local and noise levels were low, however, ICA still performed better under these conditions.

When choosing a re-referencing method, one thing to consider is how it will interact with the signal in relation to artefacts. Here, we consider an artefact to be an undesirable sharp transient in the grounded electrical signal, the causes of which can be varied. They might be introduced by electrical interference near to the recording or grounding sites, such as 50 Hz line noise or similar electrical bursts. Since sEEG depth electrodes are implanted to identify the origin of epileptic seizures, interictal epileptiform discharges (or IEDs) produced by a pathological area are a common occurrence. Whilst these are a desirable observation in a clinical setting, they can distort offline analyses of the electrical signal and understandably lower the patient's performance on memory or attentional tasks (see also Chap. 3). When pre-processing sEEG data, it is often considered important to identify and remove trials with IEDs and other artefacts. Artefact identification and removal can be done either manually or algorithmically depending on time and the volume of data.

Parish et al. [15] employed an associative memory paradigm and compared data retention after applying several commonly used re-referencing methods (described in Fig. 28.1) on sEEG data recorded from the medial temporal lobe (Fig. 28.5). Individual trials were removed after applying an automatic artefact rejection algorithm which was sensitive to sharp transient offsets in the signal. The retention rate of individual channels was then recorded, as well as the proportion of channels that were responsive to the task after an inter-trial phase consistency (ITPC) check between 1 and 20 Hz. To determine the rate by which re-referencing transforms artefacts that are identified in this way (Fig. 28.5c), automatic artefact rejection was applied to identify

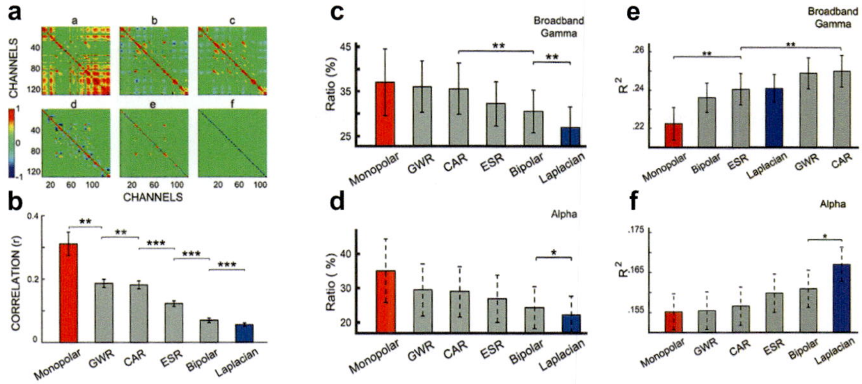

Fig. 28.3 Analyses from a motor task paradigm, using sEEG data originally referenced to white matter contacts. Time–frequency analyses here do not perform a 1/F correction. **a, b** Signal correlation for different referencing methods. **a** Correlation matrix from Subject 12 for the six referencing methods: (a) monopolar; (b) grey/white matter (GWR); (c) common average (CAR); (d) electrode shaft resonance (ESR; where channels are referenced to the average across the entire grid or depth electrode); (e) bipolar; and (f) Laplacian. Colors correspond to the correlation between two specific channels. The correlation between channels varies across the methods. **b** Average Pearson's correlation and standard error for the six referencing methods. Asterisks denote the significance of the difference between correlations established using paired t-tests: *** ($p < 0.001$), ** ($p < 0.01$). These statistical results are shown only for the nearest pairs that show a significant difference. **c, d** Fraction of all channels that are related to the task for different referencing methods. For each subject, we calculated the ratio of task-related channels by dividing the number of task-related channels by the number of all channels. **c** Mean (averaged across subjects) and standard error of the ratio of task-related channels for broadband gamma power. **d** Mean (averaged across subjects) and standard error of the ratio of task-related channels for alpha power. Asterisks denote the significance of the difference between the ratio of task-related channels for adjacent referencing methods, established using paired t-tests: ** ($p < 0.01$), * ($p < 0.05$). These statistical results are shown only for the nearest pairs that show a significant difference. **e, f**: Coefficient of determination (R2) for different referencing methods, which determines how strongly alpha or broadband gamma power was modulated by the task [23–25]. **e** Mean and standard error of R2 for broadband gamma power, calculated across all channels from all subjects. **f** Mean and standard error of R2 for alpha power. Asterisks denote significance of the difference (paired t-test) between R2 values for referencing methods: ** ($p < 0.01$), * ($p < 0{:}05$). These statistical results are shown only for the nearest pairs that show a significant difference

undesirable trials in both the raw and the re-referenced data. The relative difference in identified artefacts is then described as the number of newly introduced artefacts (i.e., trials that did not have an artefact in the raw signal but now do in the re-referenced signal) and the number of no longer existing artefacts (i.e., trials that did have artefacts in the raw data that are no longer present after re-referencing). In general, artefacts were both removed and introduced at a similar rate, indicating that re-referencing is not an optimal strategy for the removal of sharp transient spikes from the signal and should be used in tandem with more targeted methods. The data presented here is ordered by the amount of power in the theta frequency band (1–12 Hz), and as

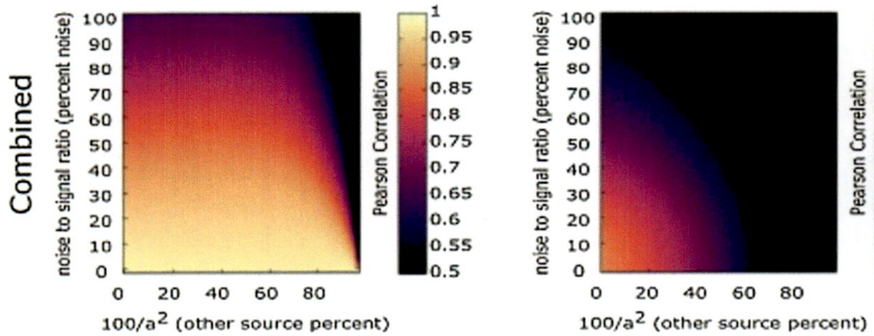

Fig. 28.4 Combined correlation (sensitivity * specificity) of extracted ICA components (left) and bipolar channel (right) with simulated sources. After re-referencing, the local component and the bipolar channel were correlated with an underlying simulated source (sensitivity) and with an interfering distant source (1-correlation, i.e., specificity) under various levels of noise (y-axis) and spatial mixing (x-axis; mixing from a distant source before re-referencing/unmixing, where $1/a^2$ is the scaling factor of the source at distance d that decays with $1/a^d$). Bipolar referencing only performed well when signal was very local and noise levels were low, yet ICA outperformed bipolar re-referencing under all conditions

such we can see a slight positive correlation between power and the percentage of responsive channels following an ITPC check—though a 1-way ANOVA suggests that one's choice of re-referencing scheme does not have a significant effect on channel responsiveness.

Whilst this comparative analysis does not equate to any tangible data quality metric, it can still be informative to see how re-referencing methods transform the data. In terms of overall data retention, there is little difference between most re-referencing methods except for the median method (MD), which resulted in the removal of far more trials per channel than any other method. Figure 28.5c shows that the median method is an outlier in how it transforms artefacts within the data: it fails to sufficiently reduce the amplitude of pre-existing artefacts due to its nature of being less sensitive to outliers. Furthermore, it appears to introduce artefacts to more channels than any other re-referencing scheme, which is surprising given the typical reason one might employ its usage, leaving this method with significantly less data than other methods after automatic artefact rejection. In comparison to the use of the mean average (AV), it appears that the high amplitude outliers of a localised artefact are sufficiently offset within the mean, such that the subtraction of the mean strikes the right balance between reducing the localised artefact without introducing it to the other contacts on the electrode. In contrast, the Gram-Schmidt method seems to preserve potentially useful signals by removing more artefacts than it introduces, resulting in a slightly higher trial retention than most of the other methods. This is likely due to the nature of the Gram-Schmidt method, which subtracts variance that is shared across the signal, making it less likely that more localised artefacts will be introduced to other channels. Most notable however, is the way in which ICA works to clean the data of artefacts. It seemingly transforms the data less than other methods (introducing and removing fewer artefacts), though it equally manages to

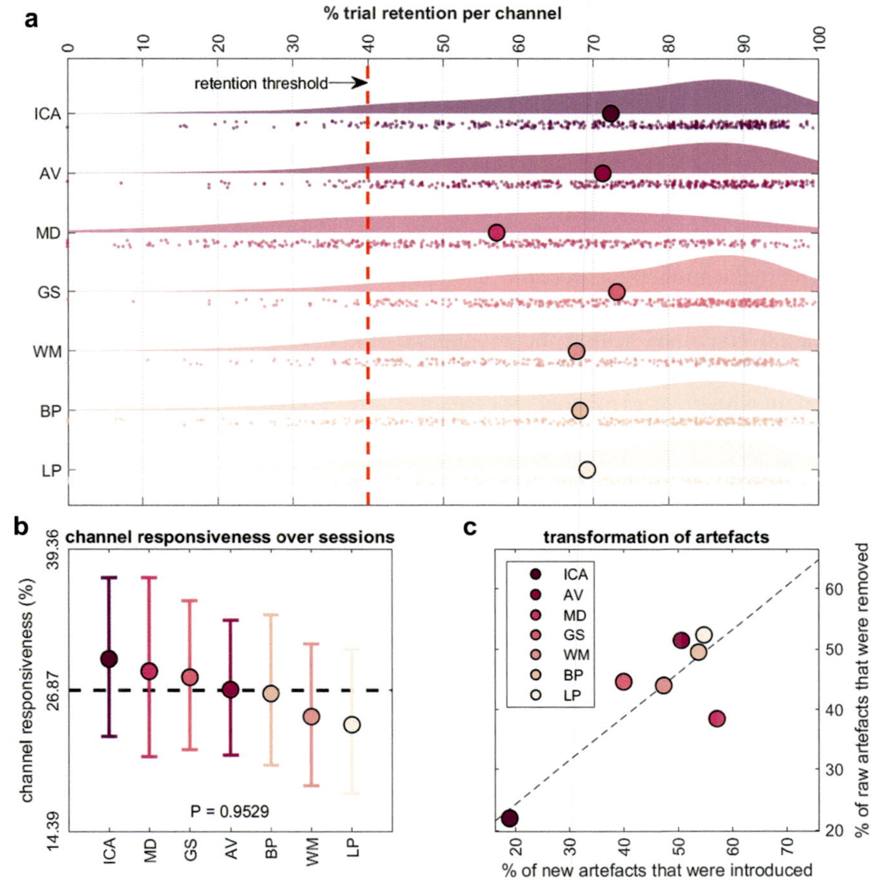

Fig. 28.5 Automatic artefact rejection on a large dataset of 16 subjects, looking at macro contact data from sEEG depth electrodes. If the variance of a trial, calculated as the root mean square (RMS) or Z-score normalisation of the LFP, was above a threshold of 4, then the trial was indexed as artefactual. **a** Visualisation of the retention percentage across channels for 7 commonly used re-referencing methods, independent component analyses (ICA), common average (AV), median (MD), Gram-Schmidt (GS), white matter (WM), bi-polar (BP), and Laplacian (LP). Re-referencing was applied before artefact rejection. The red dotted line indicates the threshold for retaining a channel in the analysis (i.e., 40%). The order of methods is determined by overall theta power (1–12 Hz; see Fig. 28.6). **b** The percentage of channels that showed significant inter-trial phase consistency between 1 and 20 Hz. A 1-way ANOVA indicates that the re-referencing method does not have a significant effect on channel responsiveness ($p = 0.9529$). **c** Proportion of artefacts from the raw data that were no longer present after re-referencing (y-axis), alongside the proportion of artefacts in the re-referenced data that were not present in the raw data (x-axis), essentially: artefacts removed and artefacts introduced by re-referencing method. A line of best fit for both was calculated from information taken from underlying channel datapoints, where scatter points represent averages across re-referencing methods. Figure taken from [15]

retain a high proportion of trials in comparison to other methods (Fig. 28.5a) and also the highest number of responsive channels (Fig. 28.5b). This suggests the power of using a data driven method to efficiently identify and remove common noise that exists across a set of independent channels.

In line with [13] (see Fig. 28.3) the Laplacian method here returns a lower proportion of channels that are responsive to the task—perhaps indicating the effect that a neighbour driven re-referencing method can have for producing a more localised signal with a lower proportion of task-related channels and a higher signal to noise ratio. Here, the monopolar method also returns a relatively lower proportion of responsive channels relative to other methods, in contrast to [13], where task-related information contained within the monopolar reference was thought to contaminate other channels. This might indicate the strength of using the data to identify appropriate white matter monopolar contacts for re-referencing (see Fig. 28.2).

28.3.2 What Is Your Data Telling You After Re-referencing?

We next consider what the data is telling you after re-referencing, looking at two studies. We first continue with the associative memory task [15], looking at power, peak frequencies and memory effects in lower frequency band, followed by another motor task that compared the effect of re-referencing methods on decoding accuracy in higher frequency bands [26].

First, we consider the power of specific frequencies from a signal obtained during an associative memory paradigm [15]. A Wavelet time–frequency decomposition with 1/F correction was applied to the re-referenced signal, where 1/F pink noise was reduced (estimated separately per channel) to enhance the analyses of low frequency oscillations. Low frequency oscillations tend to be more ubiquitous in the brain, as they are important for the long-range synchronisation of neuronal processes, whilst high frequency oscillations tend to be more derived from local neuronal circuits [27]. This means that lower frequencies will return higher amplitudes in the power spectrum, as more neurons are entrained together in phase by more global and low frequency rhythms. In this study the medial temporal lobe is under observation, where low frequencies are thought to be mostly driven by the theta frequency (typically 2–8 Hz) which dominates there, rather than the alpha rhythm (typically 8–15 Hz) that is dominant in cortical areas [14]. By plotting the average 1/F corrected power across all re-referencing methods (Fig. 28.6a), we can see that macro contacts seem to have two components in the theta frequency range and one component in the alpha range (slow theta at ~2–4 Hz and fast theta at 4–8 Hz and possibly alpha at 8–12 Hz).

Different re-referencing methods have varied effects on each of these low frequency components, which roughly fit in line with their respective assumptions that were described earlier. For example, bi-polar (BP) and Laplacian (LP) both vastly reduce the 4–12 Hz components (Fig. 28.6a, b: comparative power across all channels; and Fig. 28.6cii–iii: statistical comparison of binned mean power for common channels). As low frequency oscillations tend to be a more global phenomenon that

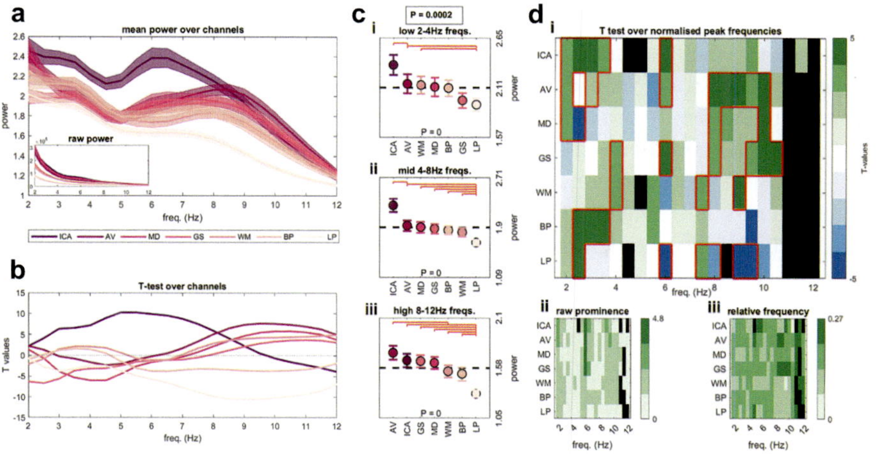

Fig. 28.6 Comparing power and peak frequencies in macro contact data for 7 commonly used re-referencing methods, independent component analyses (ICA), common average (AV), median (MD), Gram-Schmidt (GS), white matter (WM), bi-polar (BP), and Laplacian (LP). The order of methods is rank ordered by overall theta power (1–12 Hz). For this analysis we used all available channels regardless of location or responsiveness to the task, except those that were rejected after automatic artefact removal. **a** Comparative power differences across re-referencing schemes using all available channels per re-referencing method in an independent manner. Power has been 1/F corrected throughout this figure, though the raw power is shown in an inset of (**a**). **b** Paired t-test over channels, where each re-referencing scheme was compared to the average of all other re-referencing schemes at every frequency bin. **c** The mean power across re-referencing schemes was split into three windows between 2–4 Hz (**ci**), 4–8 Hz (**cii**), and 8–12 Hz (**ciii**), rank ordered by magnitude where paired t-tests indicate the nearest rank-ordered neighbour with significantly less power (red lines, $p <= 0.05$). A 2-way ANOVA (top p-value) indicates interaction of re-referencing scheme by frequency and 1-way ANOVAs indicate a main effect per frequency bin (p-values within subplots). **d** Peak frequency differences. Peak frequencies were detected on a channel-by-channel basis, where the prominence of a peak equates to the difference in power between a local maximum and neighbouring minima (**dii**). The frequency of detected peak frequencies was also recorded (i.e., the number of times a peak was found at any given frequency bin) (**diii**), which was normalised across re-referencing methods (such that each frequency bin summed to 1). After applying a 2-sample t-test on the multiplication of prominence and relative frequency, T-values (**di**) indicate where any given re-referencing scheme has a higher number of larger peaks than the average of all other re-referencing methods (or vice versa), where significance is indicated by red boxes ($p <= 0.05$, FDR corrected). Black squares in (**di**) indicate insufficient common channels for statistical comparison, black squares in (**dii–diii**) indicate that no peaks were found

is often sampled at many neighbouring recording sites, these signals are therefore reduced here by neighbour driven re-referencing methods, which will result in an emphasis being placed on the more localised faster frequencies as was seen in the previous study, where LP resulted in higher gamma power [13]. Another potential issue with such local BP and LP referencing methods is that they may enhance low frequency activity at tissue borders between areas where oscillations are homogenous within an area but different between areas which would appear as a ring-like shape of enhanced power. This is also reflected in the peak frequency comparison, where

LP in particular has significantly fewer and smaller peaks in the 6–10 Hz range when compared to other re-referencing methods. In line with previous studies [13], the BP operation does not reduce global components as much as the Laplacian, though interestingly it does return a higher slow theta component than most other methods.

Another method that works in line with its assumptions is the Gram-Schmidt method, which works to remove only the part of the signal that is shared with the average across all channels. Due to the nature of the signal, this typically entails reducing higher amplitude low frequencies whilst enhancing lower amplitude high frequencies (as shown by the upwards trending curve in Fig. 28.6b), which in effect produces a more localised signal. Similarly, other methods that also enhance the putative alpha frequencies (8–12 Hz) are the common average (AV), median (MD) and white matter (WM) methods (Fig. 28.6a–c), where AV in particular produces a larger number of prominent peaks in the 8–10 Hz range (Fig. 28.6di). This might suggest that these methods have a tendency to introduce signals from cortical areas where alpha rhythms (8–12 Hz) are dominant. By far the method that maintains the most amount of potentially useful theta is ICA. It produces significantly more power in the 2–8 Hz frequency bins, indicating that it does not introduce cortical alpha into the hippocampal signal. Similar to the neighbour driven BP and LP methods, ICA also produces a large number of peaks in 2–4 Hz frequency range, though unlike those methods ICA also enhances a 6 Hz component which is seemingly lost when applying most other methods.

From this comparative analysis, we might be able to ascertain that the 2–4 Hz component is thus the most localised rhythm within the MTL, as it is not significantly reduced by neighbour driven methods such as BP and LP. The 6 Hz theta component can therefore be considered a more global theta rhythm as it is almost entirely reduced by these neighbour-driven methods whilst also maximised by the data-driven nature of ICA. Additionally, it might be prudent to caution against the over interpretation of faster theta or alpha frequencies in the MTL if one has applied a more global re-referencing akin to the common average or white matter monopolar, as it might simply have been introduced from cortical areas (this same pattern is also implied by findings from [13]; see Fig. 28.3).

Next, [15] looked at the effect of re-referencing on memory effects (Fig. 28.7). Data was obtained by way of a cued recall memory paradigm, where subjects were presented with a cue followed by a pair of images (termed the encoding phase). After several trials, and a distractor task, cues were subsequently presented again and the subject indicated how many images they could remember (termed the retrieval phase), before identifying the paired images from a selection on screen. Memory effects refer to the differences between successful memory trials (i.e., hits) where the participant could correctly retrieve the memory, and unsuccessful memory trials (i.e., misses) where the participant could not correctly retrieve the memory. These memory effects were analysed during the memory formation (i.e., encoding) phase of the experiment, and during the retrieval phase of the experiment. Theta oscillations have long been thought to play a formative role for such memory operations [28], where both positive and negative effects (i.e., where hits have more theta power than misses, or vice versa) have been reported. It might be the case that this effect is made up of both a narrowband and broadband component, where contrasts in paradigm, recording technique and referencing scheme might emphasise one of these components over the other [29]. Therefore, it is important to consider whether or not re-referencing has any interaction with such memory effects.

Overall, the data from the study produced large negative fast-theta and alpha (6–12 Hz) effects—indicating that trials where subjects performed worse had a larger amount of this frequency band, as backed up by the literature [29]. As well as this, there might be a small positive slow-theta (2–4 Hz) retrieval memory effect—indicating that trials that were successfully remembered might have higher theta power (see GS data shown in Fig. 28.7c). However, there was no significant interaction between re-referencing method and frequency which suggests that re-referencing did not strongly affect memory effects for the macro contact data. Nevertheless, there were some subtle yet insignificant differences. In general, re-referencing with

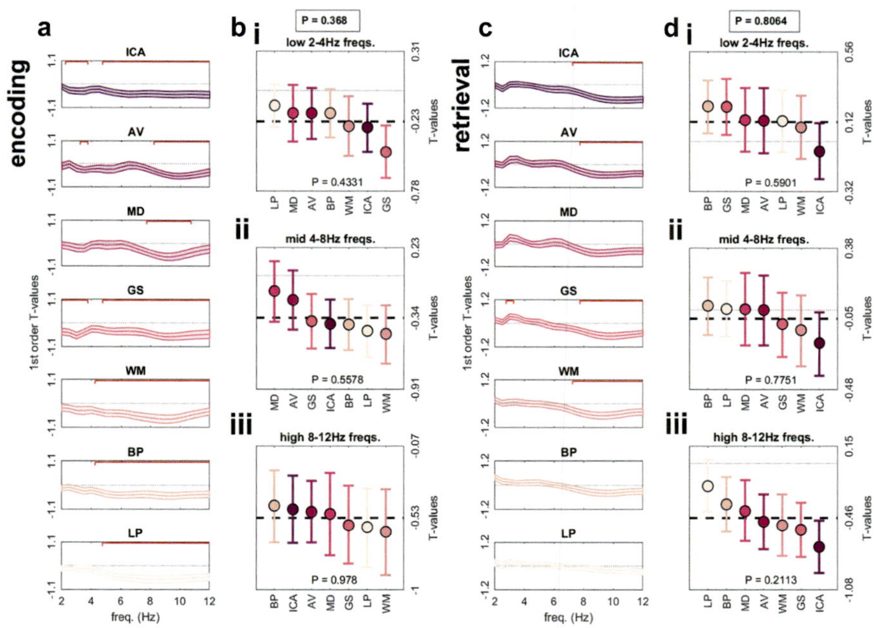

Fig. 28.7 Memory effects for contacts situated in the hippocampus, that indicated significant inter-trial phase consistency (1–20 Hz). Comparing between 7 commonly used re-referencing methods, independent component analyses (ICA), common average (AV), median (MD), Gram-Schmidt (GS), white matter (WM), bi-polar (BP), and Laplacian (LP). The order of methods is rank ordered by overall theta power (1–12 Hz; see Fig. 28.6). Independent t-tests were applied on a channel-by-channel basis for power across hit trials minus miss trials, both for encoding (**a, b**) and retrieval (**c, d**) trials. 1-sample t-test applied to the resultant T-values (**a/c**) to indicate significant differences to zero (red lines, $p <= 0.05$, FDR corrected). T-values were further split into 3 frequency bins 2–4; 4–8 and 8–12 Hz (**b/d**), where both a 2-way ANOVA (top p-value) indicates interaction of re-referencing scheme by frequency bin and 1-way ANOVAs indicate the main effect per frequency bin (p-values within subplots)

ICA, WM, AV, MD and GS seemed to produce more negative memory effects in the fast theta and alpha frequency range, particularly for retrieval effects, whereas local referencing schemes such as BP and LP did not show these effects. This may indicate that the memory related alpha power decreases [6] in the MTL originate from a spatially broadly distributed source as opposed to very local sources. Importantly, these negative fast theta and alpha memory effects are unlikely to be introduced by cortical signals as demonstrated by the ICA results. It is interesting to note that the only method to obtain significant theta effects in both memory contrasts was the Gram-Schmidt method, which produced a negative theta effect (hits < misses) at encoding and a positive theta effect at retrieval (hits > misses). It is also worth noting that the Laplacian method completely abolishes any observable effects at retrieval (see Fig. 28.7c), perhaps indicating that a neighbour-driven method reduces too much of the global signal in lower frequencies across both hit and miss data (see Fig. 28.6).

Another study [26] complements these low frequency findings related to memory, by looking at gesture decoding accuracy for a motor task across several frequency bands (see Fig. 28.8). This entailed usage of a classification algorithm that first built a feature vector based upon spectral power (normalised to a baseline period) obtained over overlapping segmentations of the signal, where features constituted the mean power across frequency bins. Next, a set of channels were selected that maximised these features using a search optimisation algorithm, where power was further normalised across trials to eliminate inter-channel differences. A support vector machine was then applied for the classification of multiple hand gestures. This process was repeated in a tenfold cross-validation process to obtain a comparable gesture decoding accuracy measure for each re-referencing method.

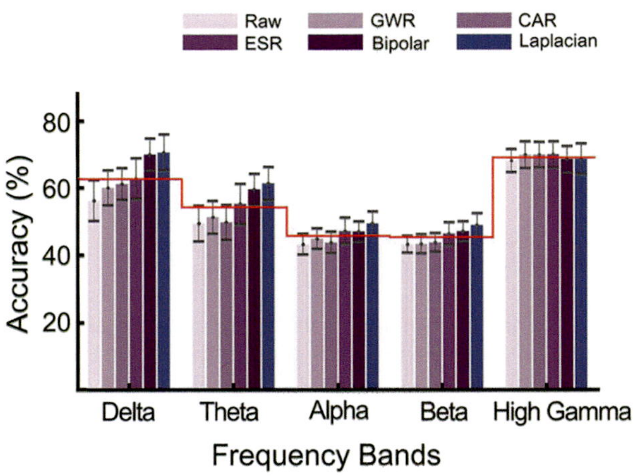

Fig. 28.8 Single-frequency band-based decoding accuracy (SDA) of data cleaning methods for multiple sub-bands. Comparison of SDAs of different frequency bands. Bars and error bars represented mean accuracy and standard error calculated across all subjects, respectively. Redline represented average SDAs across all methods for different bands, respectively. Figure taken from [26]

Importantly, it was found that the adoption of any re-referencing method made a significant improvement on decoding accuracy, implying that a sufficient amount of non-task related noise was removed by re-referencing. In general, this study found that the more localised the re-referencing method the higher the decoding accuracy across most frequency bands, likely due to the ability of localised re-referencing methods (such as the neighbour-driven Laplacian and bi-polar) to retain more task-related information by reducing globally shared signals. However, as the spatial topography of implanted electrodes was an essential feature of the decoding algorithm, then it would be expected that re-referencing methods that improve spatial resolution would also improve decoding performance.

28.4 Discussion of Results

Overall, the way in which you re-reference your intracranial sEEG data will have a transformative effect on your signal. We gave an overview of three comparative studies that looked at sEEG data collected from epilepsy patients who were implanted with depth electrodes for the purpose of monitoring and observation. Each of these studies compared a range of re-referencing methods and their effect on the research objective in question. Two of these studies [13, 26] focused on motor task paradigms, with a focus on analysing data collected from cortical areas where such processes are typically observable. The dominant oscillatory rhythms in these regions are faster alpha and gamma frequencies [27], which might be locally produced by individual cortical regions as they respond to the environment. The third study focused on an associative memory paradigm, with a focus on analysing data collected from the medial temporal lobe, which is thought to be essential for memory processes [30]. This region is thought to be dominated by theta oscillations [14], a global rhythm that enables long range communication and memory processes [28, 31].

This overview therefore did not consider any similar comparative study over other domains or regions, such as the visual domain or prefrontal cortex. The explicit focus of this chapter is time–frequency analyses and oscillations, so other important and understudied issues such as event-related potentials and related phasic phenomena were not considered. The predominant focus of the above analyses was on macro contact data for sEEG depth electrodes. A better understanding of how re-referencing effects sEEG data on grid and strip electrodes is still required. Another not well understood mode of data is that which is collected from micro-wires, an increasingly popular technique which can be used in tandem with the macro contacts and are capable of recording highly localised single unit and population activity (see also Chaps. 42–46). In studying this issue, it might prove useful to understand how re-referencing might introduce phase-reversals into highly localised data and the likely ensuing effect that this has on the identified waveforms of neuronal spikes. Indeed, the issue of phase-reversals might be a prescient one for any analyses of oscillatory phase in general (which neighbour driven re-referencing methods might be more susceptible to)—which also requires further study.

In general, the re-referencing methods described here each work according to their own assumptions to emphasise either local or global effects, dependent on the type of data you are working with and the paradigm you are interested in. In general, re-referencing can be a good way to increase the signal to noise ratio of your data [26]—though it might prove insufficient to reduce sharp transient noise such as IEDs [15]—and might even contaminate previously unaffected signals in this respect. More localised methods (such as the neighbour-driven Laplacian or bipolar) might best enhance local effects, especially if spatial resolution is an important element in the design of the analyses [13]. However, simulations have shown (see Fig. 28.4; [5]) that a data-driven method (such as independent component analyses or ICA), might be better suited to both isolate signal from local sources and discard activity from distant sources—especially in noisier conditions. Equally, more global methods (such as methods that make use of the average) might best enhance more global effects in the lower frequency range [15]. The data-driven ICA method seems to perform equally well at both isolating local signals [5], and enhancing global signals [15], depending on the hypothesis and parameters utilised—highlighting the versatility of data-driven methods to extract desirable components for further analyses.

In sum, more comparative work is required to build a full image of the transformative effect that re-referencing has on intracranial electrophysiological data that is collected from various neuronal sources and for a variety of hypothesis. Hopefully this chapter will encourage researchers to consider re-referencing as an active pre-processing decision that needs careful consideration.

References

1. Schomer DL, Lopes da Silva FH (2018) Niedermeyer's electroencephalography: basic principles, clinical applications, and related fields, 7th edn. Oxford University Press, New York, NY, xviii, 1239 pp
2. Nunez PL, Srinivasan R, Oxford University Press (2006) Electric fields of the brain the neurophysics of EEG. Oxford University Press, Oxford, p 1 (online resource) (xvi, 611 pp)
3. Mercier MR et al (2017) Evaluation of cortical local field potential diffusion in stereotactic electro-encephalography recordings: a glimpse on white matter signal. Neuroimage 147:219–232
4. Lei X, Liao K (2017) Understanding the influences of EEG reference: a large-scale brain network perspective. Front Neurosci 11:205
5. Michelmann S et al (2018) Data-driven re-referencing of intracranial EEG based on independent component analysis (ICA). J Neurosci Methods 307:125–137
6. Staresina BP et al (2016) Hippocampal pattern completion is linked to gamma power increases and alpha power decreases during recollection. Elife 5
7. Fell J et al (2001) Human memory formation is accompanied by rhinal-hippocampal coupling and decoupling. Nat Neurosci 4(12):1259–1264
8. Lachaux JP, Rudrauf D, Kahane P (2003) Intracranial EEG and human brain mapping. J Physiol Paris 97(4–6):613–628
9. Nunez PL (2010) REST: a good idea but not the gold standard. Clin Neurophysiol 121(12):2177–2180
10. Varela F et al (2001) The brainweb: phase synchronization and large-scale integration. Nat Rev Neurosci 2(4):229–239

11. Treder MS et al (2021) The hippocampus as the switchboard between perception and memory. Proc Natl Acad Sci U S A 118(50)
12. Roux F et al (2022) Oscillations support short latency co-firing of neurons during human episodic memory formation. biorXiv.org
13. Li G et al (2018) Optimal referencing for stereo-electroencephalographic (SEEG) recordings. Neuroimage 183:327–335
14. Griffiths BJ et al (2019) Directional coupling of slow and fast hippocampal gamma with neocortical alpha/beta oscillations in human episodic memory. Proc Natl Acad Sci U S A 116(43):21834–21842
15. Parish G et al. The effect of re-referencing on low frequency power and spike-field coherence in intracranial EEG data (in prep)
16. Hasselmo ME (2012) How we remember: brain mechanisms of episodic memory. MIT Press, Cambridge, Mass.; London, xii, 366 pp
17. Comon P (1994) Independent component analysis, a new concept. Signal Process 36(3):287–314
18. Delorme A, Makeig S (2004) EEGLAB: an open source toolbox for analysis of single-trial EEG dynamics including independent component analysis. J Neurosci Methods 134(1):9–21
19. Whitmore NW, Lin SC (2016) Unmasking local activity within local field potentials (LFPs) by removing distal electrical signals using independent component analysis. Neuroimage 132:79–92
20. Nikulin VV, Nolte G, Curio G (2011) A novel method for reliable and fast extraction of neuronal EEG/MEG oscillations on the basis of spatio-spectral decomposition. Neuroimage 55(4):1528–1535
21. Schaworonkow N, Voytek B (2021) Enhancing oscillations in intracranial electrophysiological recordings with data-driven spatial filters. PLoS Comput Biol 17(8):e1009298
22. Liu Y et al (2019) Human replay spontaneously reorganizes experience. Cell 178(3):640–652 e14
23. Kubanek J et al (2009) Decoding flexion of individual fingers using electrocorticographic signals in humans. J Neural Eng 6(6):066001
24. McFarland DJ et al (1997) Spatial filter selection for EEG-based communication. Electroencephalogr Clin Neurophysiol 103(3):386–394
25. Pfurtscheller G et al (2006) Mu rhythm (de)synchronization and EEG single-trial classification of different motor imagery tasks. Neuroimage 31(1):153–159
26. Liu S et al (2021) Investigating data cleaning methods to improve performance of brain-computer interfaces based on stereo-electroencephalography. Front Neurosci 15:725384
27. von Stein A, Sarnthein J (2000) Different frequencies for different scales of cortical integration: from local gamma to long range alpha/theta synchronization. Int J Psychophysiol 38(3):301–313
28. Buzsaki G (2002) Theta oscillations in the hippocampus. Neuron 33(3):325–340
29. Herweg NA, Solomon EA, Kahana MJ (2020) Theta oscillations in human memory. Trends Cogn Sci 24(3):208–227
30. Squire LR, Stark CE, Clark RE (2004) The medial temporal lobe. Annu Rev Neurosci 27:279–306
31. Solomon EA et al (2019) Dynamic theta networks in the human medial temporal lobe support episodic memory. Curr Biol 29(7):1100–1111 e4

Chapter 29
What Are the Pros and Cons of ROI Versus Whole-Brain Analysis of iEEG Data?

Carina Oehrn

Abstract Human invasive neural recordings stem from neurological and psychiatric patients with depth and subdural electrodes implanted for diagnostic or therapeutic purposes. Electrode placement is therefore dictated by clinical needs, and the sparseness in brain coverage and variability in electrode location across patients impedes group-level analyses using random-effects models. This chapter discusses the process of electrode selection for subsequent analyses and in particular the advantages and disadvantages of two approaches. First, one can select regions-of-interest (ROIs) guided by a priori hypotheses emerging from previous findings, e.g., non-invasive lines of research. This procedure allows for random-effects analyses and inferences about the population but can lead to a loss of spatial precision and an increase of type-II error. Alternatively, one can conduct whole-brain analyses and study all available electrodes, either for each individual or across patients by means of functional pre-selection of contacts or a fixed-effects analysis with pooled data across patients. This approach preserves anatomical precision but reduces the generalizability of the findings and requires rigorous correction for type-I error. Finally, I describe one procedure that combines the strengths of ROI selection and the analysis of all electrodes and thereby represents a good compromise between the two methods.

29.1 Introduction

Intracranial EEG (iEEG) offers the opportunity to record electrophysiological activity with exceptional anatomical precision. Nevertheless, combining data across participants without losing this spatial resolution is challenging. IEEG recordings in humans stem from neurological and psychiatric patients. In epilepsy, patients are usually implanted with depth or subdural electrodes for diagnostic purposes to determine the localization of seizure onset and to assess eligibility for epilepsy surgery (see Chap. 1). In indications for deep brain stimulation (DBS), such as Parkinson's disease

C. Oehrn (✉)
Department of Neurological Surgery, University of California, San Francisco, USA
e-mail: Carina.Oehrn@ucsf.edu

N. Axmacher (ed.), *Intracranial EEG*, Studies in Neuroscience, Psychology and Behavioral Economics, https://doi.org/10.1007/978-3-031-20910-9_29

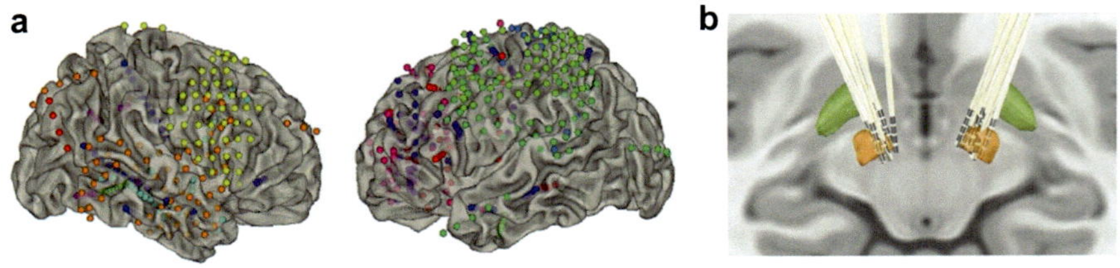

Fig. 29.1 Variability of electrode location across patients. **a** Example of variance in cortical electrode positions in 13 epilepsy patients [1]. Each color represents data from one patient. **b** Example of variability in subcortical electrode location. The illustration shows bilateral electrode placement in the subthalamic nucleus (STN, orange) of 14 patients with obsessive–compulsive disorder [2]

and obsessive–compulsive disorder, depth electrodes are chronically implanted for therapeutic purposes. Implantations in one patient can provide up to a hundred or even more distinct recording sites across the brain, allowing for simultaneous recordings within and across various brain structures. Electrode location is, however, dictated by clinical needs and varies across patients (Fig. 29.1). Further, coverage of brain areas is limited to few brain regions. For research purposes, however, it is necessary to compare task-related activity in a circumscribed anatomical brain area across patients in order to draw inferences about its function beyond the single-subject level. How to cluster electrodes for group-level analyses is therefore a major challenge in human iEEG research. Two general approaches are commonly applied for electrode selection. First, one can focus the analysis on electrodes within one or several regions of interest (ROIs). Alternatively, one can use all electrodes and conduct a whole-brain analysis. Type I and type II errors in the following sections refer to false positive and false negative findings, respectively, related to multiple comparisons and data selection in the spatial (not in the temporal or frequency) domain.

29.2 Steps Before Electrode Selection

In epilepsy patients, the first step of iEEG data analysis is commonly the rejection of data with epileptic artifacts. At a minimum, electrodes covering the seizure onset zone are excluded from the analysis [3]. More conservatively, the analysis may even be restricted to recordings from the hemisphere contralateral to the epileptogenic focus [4]. Next, the position of electrodes is determined manually or semi-automatically (see Chap. 27). Nowadays, many software platforms offer tools for determining electrode placement [5–7]. This is usually done by co-registration of pre-implant structural magnetic resonance imaging (MRI) to the post-implant computed tomography (CT) or MRI scan [4]. Subsequently, a common procedure is to project electrode location into Montreal Neurological Institute (MNI) or Talairach space (www.talairach.org). Each contact can thereafter be assigned to a corresponding anatomical label using atlases (e.g., Freesurfer's automated segmentation, https://surfer.nmr.mgh.harvard.edu/). For all selection procedures performed in normalized space, it is crucial to confirm electrode location by visual inspection in native space.

29.3 Regions of Interest Analysis

One common approach for electrode selection in iEEG research is the a priori definition of ROIs. The selection of ROIs is hypothesis-driven and based on findings from previous invasive or noninvasive studies in humans or animals demonstrating robust task-related activity in the respective brain area. The ROI can therefore represent a circumscribed anatomical region, such as the hippocampus, or the spatial location of a functional effect reported in a previous publication, for instance, MNI coordinates from a functional MRI study [8]. For each patient, all electrodes inside the ROI are chosen. This will lead to the exclusion of patients without electrodes inside the defined brain area from subsequent analyses. Due to the sparse sampling of iEEG data, many studies focus on regions most commonly covered by electrodes. The most frequent forms of focal epilepsy comprise temporal and frontal lobe epilepsy. Many studies in epilepsy patients therefore focus on the role of the frontal cortex and medial temporal lobe for cognitive functions. In DBS patients, electrodes are usually implanted subcortically, for instance in the subthalamic nucleus for Parkinson's disease or the ventral striatum for obsessive–compulsive disorder.

29.3.1 Anatomical ROI Definition

One option of ROI definition is the selection of anatomically defined regions that have previously been associated with the cognitive function of interest. If the ROI is macroscopically identifiable, electrodes can be selected in native space. This is usually the case for the hippocampus, the amygdala, the basal ganglia and other subcortical brain areas (Fig. 29.2a, e.g., [9, 10]). In this case, one can choose all electrodes in the respective brain area or focus on the analysis of sub-regions. Macroscopic ROI selection is more challenging when it comes to cortical regions. One approach is thereby the identification of electrodes in specific gyri in native or normalized space (Fig. 29.2b, e.g., [11, 12]). In addition, or alternatively, one can use Brodmann areas to guide electrode selection [11, 13]. To this end, contact positions are usually transferred into normalized space and the nearest grey matter is determined by means of an atlas. Subsequently, all electrodes inside the relevant brain area are selected.

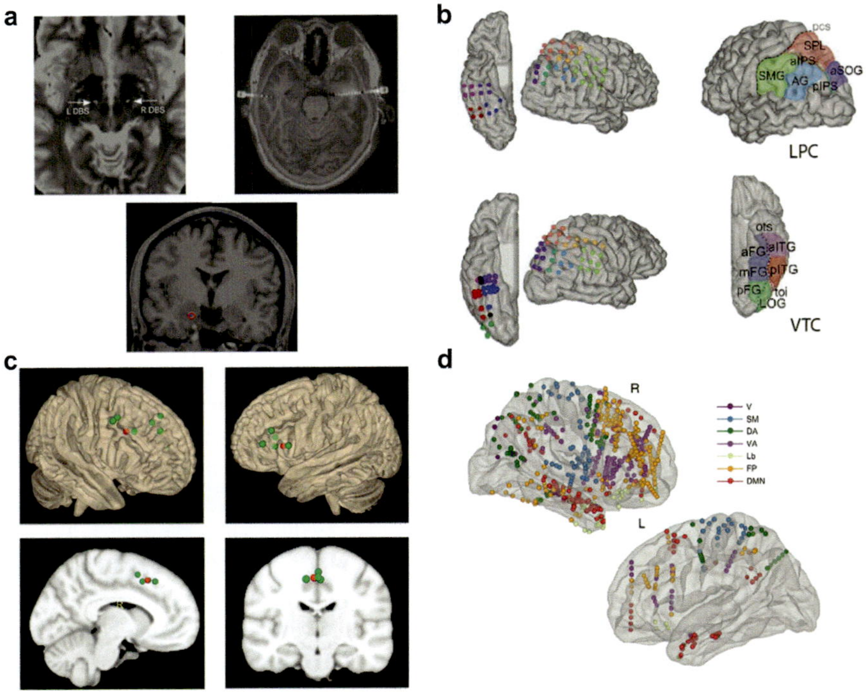

Fig. 29.2 Examples of methods for ROI selection. **a** Examples of macroscopically identifiable ROI on a co-registered CT/MRI of a single patient. Left: Location of the DBS electrodes in the subthalamic nucleus shown as a hypointense region with lead artifacts (white arrows) [14]. Right: illustration of one electrode laterally inserted in the hippocampus [10]. Below: red circle illustrates a selected electrode in the amygdala [15]. **b** Example of selecting contacts within the bounds of gyri in native space, here in the lateral parietal cortex (LPC) and ventral temporal cortex (VTC) of two epilepsy patients color-coded by anatomical subregion [12]. **c** Selection of one electrode per patient (green) within the dorsolateral and dorsomedial prefrontal cortex based on their proximity to MNI coordinates of activations observed in a previous fMRI study (red) [8]. **d** Grouping of electrodes based on functional connectivity networks: visual (V), sensori-motor (SM), limbic (Lb), default mode network (DMN), ventral attention (VA), fronto-parietal (FP) and dorsal attention (DA) [16]

29.3.2 Functional ROI Definition

ROIs can further be defined based on the location of functional effects, for instance the MNI coordinates of a task-related blood oxygenation level dependent (BOLD) effect from a previous functional MRI (fMRI) study. To this end, the Euclidian distance between the MNI coordinates of each contact and the respective target coordinate is calculated. Thereafter, one can choose one contact per patient based on the minimum distance to the target coordinate or select all electrodes within a certain radius (Fig. 29.2c, [8]). This procedure can also be applied to functional networks. For instance, one can assign electrodes to networks based on coordinates of functional resting-state or task-related connectivity obtained from an independent set of subjects (Fig. 29.2d, [16–19]). The sensitivity and specificity of the statistical analysis depend on the size of the ROI relative to spatial extent of the task-related effect. The choice of the radius around the target coordinates is therefore a crucial step in ROI definition. If the ROI is chosen too large, local effects can be diluted due to an inclusion of task-irrelevant electrodes. If a ROI is chosen too small, effects can be missed, as less electrodes and patients are entered into the analysis reducing statistical power. Both can lead to false negative findings. The effect of ROI size on statistical results has not yet been systematically assessed and there is no easy answer to this question. The spatial spread of task-related effects, i.e., the distance from the peak effect in which the effect is still measurable, depends on many inter-related factors, including the brain region and its microarchitecture, properties of the recording electrodes, the task, and the characteristics of the signal to be analyzed (compare Chap. 17). Further, the location of effects can vary across patients, e.g., up to 24 mm within the dorsolateral prefrontal cortex (DLPFC) [8]. Thus, it is difficult to provide universal recommendations for optimal ROI size. It is therefore important to assess the spatial variability of functional effects that have been described in the literature for a given cognitive process or task. Previous studies in humans and non-human primates suggest that electrocorticography (ECoG) and macro depth electrodes record activity from a radius of several milli- to centimeters ([20, 21]; see Chap. 17). Further, the normalization procedure and brain shift resulting from surgery can compromise the precision of electrode localization (see Chap. 27). Previous studies found task-related effects using a radius of 3–12.5 mm from ROI voxels and vertices (see Sect. 29.4.2.3), target coordinates, or functional networks [8, 17, 19]. A radius of 10 mm might therefore be used as a rule of thumb with larger ROI size in regions with higher inter-individual spatial variability of effects such as the DLPFC.

29.3.3 How Can I Handle a Different Number of Electrodes in the ROI for Each Patient?

Unless a single electrode per patient is selected, patients usually have different numbers of electrodes inside the ROI. One statistical approach that is still commonly used in clinically oriented studies is a fixed-effects analysis. In this type of analysis, one assumes that the underlying effect is the same in all participants and that any variance is due to sampling errors. Therefore, data from all electrodes across patients are entered into the analysis as if they originated from one patient. While fixed-effects analyses are suitable for some specific research questions, they have several limitations (see Chap. 36). First, they can inflate type I error, as they ignore dependency between within-patient measurements [22]. Second, this statistical approach does not support inferences about the general population, as it does not account for the heterogeneity between patients [23]. Conclusions drawn from the obtained results are only valid for the included group of subjects. Further, the effects in few patients with many samples contributing to the analysis can drive the results. To test whether the obtained results generalize to the population level, the statistical approach should consider that participants represent a randomly drawn sample from a larger population. For research questions in cognitive neuroscience, random-effects analyses are therefore preferable [22]. Thus, the same number of samples should be entered for each patient. To this end, the measure of interest can be averaged across electrodes within each patient before statistical analysis. An alternative approach using all data while considering participants as random factor is a mixed-effects model (e.g., [24, 25], Chap. 36).

29.3.4 What Are the Advantages and Disadvantages of ROI Selection?

Electrode selection based on ROI has several advantages. First, this approach allows for a rigorous group-level analysis. As every included patient has one or more electrodes in the ROI, one can perform a random-effects analysis and thus draw inferences about the respective population. In epilepsy patients, electrodes are implanted for a short diagnostic time period and frequently cover a broad range of brain areas. When analyses are restricted to electrodes covering non-epileptic tissue, studies often make inferences about the general population. In experiments with DBS patients, recordings are obtained from chronically implanted electrodes that are placed in brain areas inside a pathophysiological network. Inferences from effects found in these patients to cognitive functioning in healthy participants may not be impossible but need to be considered with caution. Second, ROI definition reduces the dimensionality of the data and thereby the multiple comparison problem that results from the analysis of a large number of electrodes. However, one main disadvantage of ROI selection is that statistical comparisons depend on the size of the defined ROI in relation to the extent of the effect, which is somewhat arbitrary (see Sect. 29.3.2). ROIs that are either too small or too large can lead to false negative findings. Further, summarizing all electrodes within a ROI reduces the spatial resolution of iEEG as a function of ROI size. As such, the original strength of iEEG in disentangling selective engagement of neuronal populations at the millimeter scale is compromised.

29.4 Whole-Brain Analyses

As an alternative to the a priori definition of ROI, one can enter all available electrodes per patient into the analysis to fully leverage the anatomical precision of iEEG recordings. Such whole-brain analyses can comprise the analysis of single electrodes (per patient or across patients) and group-level statistics. The latter can be based on fixed-effects analyses, a functional selection of electrodes in each patient, or mapping of all electrodes into normalized space.

29.4.1 Electrode-Level Whole-Brain Analyses

For some research questions, it can be appropriate to refrain from group analyses and study effects in each single participant. This can be particularly relevant when interested in inter-individual variability (see Chap. 10). It can further be suitable for data sets with low sample sizes, for instance when studying activity in rarely implanted brain areas. Some studies conducted statistical analyses per electrode for all patients, i.e. treating them as if they came from one participant, in order to cover a larger cortical surface (Fig. 29.3a, e.g., [1]). This approach avoids the problem of electrode selection as all contacts per patient are entered into the analysis. Thus, the high spatial resolution of iEEG recordings is preserved and the risk of type II error is reduced. However, one major drawback of these analyses is that they do not allow for conclusions about the general population, unless subsequent group-level statistics are performed (see Sect. 29.3.3). In case that electrodes for all patients are entered into the analysis, results can be driven by a single patient. Further, this approach bears an increased risk of spurious results, i.e., type I error, and requires rigorous correction for multiple comparisons.

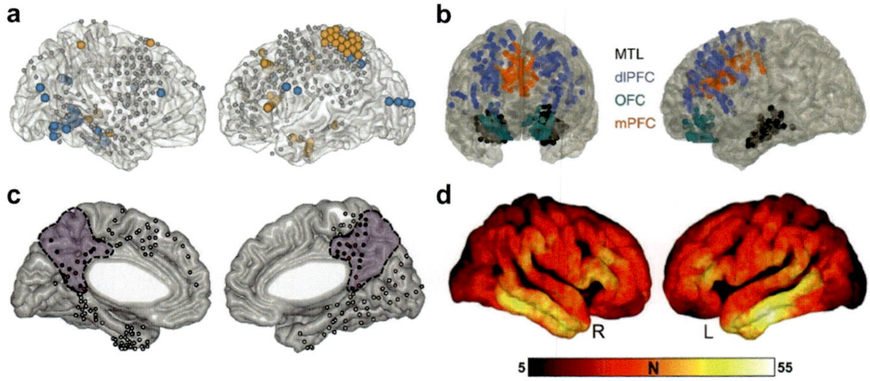

Fig. 29.3 Different approaches to whole-brain analyses. **a** Illustration of results from an analysis on single electrode level pooling contacts across 13 patients. Colors represent positive (orange) and negative (blue) task-related effects [1]. **b** Combination of ROI selection and fixed-effects analysis. Illustration shows electrodes across 15 subjects in the medial temporal lobe and three prefrontal ROIs, which were used for the fixed-effects analysis [26]. **c** A combination between ROI definition (posteromedial cortex, marked in purple) and the selection of responsive electrodes [3]. Electrode locations of eight epilepsy patients are illustrated and ROI electrodes responsive to the task are marked in red, opposed to unresponsive electrodes (black). Electrodes falling outside the anatomical boundaries of the ROI are filled in white. **d** Example of functional mapping into normalized space. Plots illustrate the number of patients contributing to cortical vertices and the functional maps in this study with 55 epilepsy patients. Contacts contributed to a vertex when the center of the vertex was located within a 10 mm vicinity around a given contact [29]. All graphs illustrate electrode location on template images

29.4.2 Group-Level Whole-Brain Analyses

29.4.2.1 Fixed Effects Analysis

A second approach for a whole-brain analysis of all electrodes per patient comprises fixed-effects group-level analyses by concatenating electrodes from all patients (Fig. 29.3b, [1, 26]). The advantage of this method is that it maintains the anatomical precision of iEEG recordings, as it does not require averaging across electrodes. As mentioned in Sect. 29.3.3, however, this statistical approach does not account for the heterogeneity between patients and the conclusions drawn from the obtained results cannot be extrapolated beyond the group included in the study. Further, results can be driven by few patients. If the fixed-effect analysis is not restricted to a ROI, no conclusion about a specific anatomical region can be made. Thus, it is recommended to combine this approach with other statistical methods [1].

29.4.2.2 Selecting Responsive Electrodes

One way of combining the analysis of all available electrodes and performing group-level analysis is by investigating activity in all electrodes of each patient and choosing contacts based on functional criteria, e.g. the response to a cue or highest sleep spindle power (Fig. 29.3c, e.g., [27, 28]). Using this approach, one needs to be particularly careful about avoiding circular analysis. Thus, electrode selection should not be based on comparisons subsequently performed on the group level. Common pitfalls of functional electrode selection are discussed in Chap. 37. One disadvantage of this approach is that it can be difficult to draw conclusions about circumscribed brain areas, as electrode locations can vary between patients. It is thus recommended to combine this approach with a ROI selection in order to constrain electrodes to a specific anatomical region [3, 19, 28]. This also reduces the number of statistical comparisons performed, for which correction is required.

29.4.2.3 Projecting Electrodes into Normalized Space

An alternative approach that combines a fine-grained spatial resolution with random-effects group-level analyses consists of creating whole-brain maps of functional effects in normalized space. To this end, standard brain templates are parcellated into a number of standardized sub-areas to which contacts are allocated. The outcome measure of interest is thereafter averaged across all contacts assigned to each sub-area within each patient. As the resulting averaged outcome measures in each sub-area contain one value per patient, this approach allows for random-effects group analyses in order to identify regions with significant task-related activity. The statistical analysis is usually limited to fields with a minimum number of patients (often $n = 5$). Several studies have used this approach. In one study, the authors assigned cortical contacts to the closest Brodmann area, which results in a spatial sampling of 52 sub-regions [30]. Alternatively, cortical electrodes can be projected to a standardized surface-based grid of ROIs with 2,400 or 600 points spaced at 5 or 10 mm distance, respectively [31]. In another study, the authors grouped normalized electrodes by segregating Talairach space into >53,000 overlapping spheres with 12.5 mm radius [13]. Contacts were thereby grouped based on their location inside a sphere. Using a similar approach, the authors of subsequent studies assigned contacts to vertices and voxels in MNI space (Fig. 29.3d, [29, 32]). In line with findings that electrodes record neural activity from a radius of several millimeters to centimeters [20, 21], they projected the measures of interest to all cortical and hippocampal parcellations within a 10–12.5 mm and 3 mm radius around each contact, respectively. The mentioned approaches result in one functional map per patient, and subsequent group-level analyses compare functional effects across patients. These procedures have several advantages. On one hand, they allow for a random-effects analysis and the extrapolation of results to the general population. Secondly, the analysis is not restricted to a priori selected brain areas, as effects from all electrodes are considered in this analysis (apart from electrodes covering rarely implanted brain areas). These approaches therefore reduce type II error compared to ROI selection. Further, some spatial resolution is sacrificed compared to single electrode analysis, but a high anatomical precision can be maintained. Moreover, these methods reduce the risk of type I error compared to the analysis of all available electrodes, as effects are averaged within brain areas, voxels or vertices leading to a reduced number of comparisons. One drawback of this method is that it is not suitable for small groups of patients, as sub-regions have to be covered by a minimum number of participants to be included into the analysis.

29.4.3 What Are the Advantages and Disadvantages of Whole-Brain Analyses?

The advantages and disadvantages of whole-brain analyses depend on the method used. The analysis of effects in every electrode of individual patients maintains the highest spatial resolution and the lowest probability of type II error, but also the highest chance of type I error. Fixed-effects analyses can provide broad spatial sampling of brain areas, but results can be driven by single patients. Both procedures do not allow for inferences beyond the studied group of subjects. The selection of electrodes based on functional criteria is associated with decreased type II error compared to a priori ROI selection and allows for random-effects group-level analyses, but inferences about the functional role of a particular brain area can be difficult to interpret, as the spatial location of effects can vary largely across patients. Further, attention needs to be paid to avoid circular analysis. One good compromise between maintaining anatomical precision and preparing data for random-effects group-level analysis is the projection and subsequent grouping of electrodes into normalized space. This approach sacrifices less spatial resolution than ROI selection, is associated with less probability for type II error and simultaneously reduces type I error compared to statistical contrasts for every single electrode. This procedure therefore offers a good compromise between ROI selection and statistics on single-electrode level. Further, the standardized mapping of electrode locations facilitates the comparability across studies. Nevertheless, correction for multiple comparisons is required.

29.5 Summary

In summary, both ROI and whole-brain analyses come with several advantages and disadvantages. Focusing the analysis on one or several ROIs, in which activity is averaged across electrodes, allows for random-effects group-level analyses and is associated with a lower risk of type I error compared to whole-brain analyses. However, ROI analysis sacrifices spatial resolution and bears a relatively high risk of type II error. In contrast, whole-brain analyses are associated with a lower chance of type II error compared to the selection of a few ROIs, as all covered brain regions are considered. Further, these methods can provide insights into human brain function beyond previous knowledge, as analysis is not restricted to a priori defined ROIs. Moreover, whole-brain analyses maintain a higher anatomical precision, which is, however, associated with a larger number of comparisons and thus increased type I error. Thus, these data-driven approaches require rigorous group-level analyses and corrections for multiple comparisons. One good compromise between the a priori ROI selection and analysis of all electrodes separately is the creation of functional maps in standardized space followed by group analyses. However, this method requires a larger number of patients, as a minimum number of patients per voxel or vertex must be reached. For data sets with few patients, a combination of electrode selection methods is therefore recommended in order to avoid spurious results, substantiate findings, and minimize type II error.

References

1. Fellner M-C, Gollwitzer S, Rampp S et al (2019) Spectral fingerprints or spectral tilt? Evidence for distinct oscillatory signatures of memory formation. PLoS Biol 17:e3000403. https://doi.org/10.1371/journal.pbio.3000403
2. Li N, Baldermann JC, Kibleur A et al (2020) A unified connectomic target for deep brain stimulation in obsessive-compulsive disorder. Nat Commun 11:3364. https://doi.org/10.1038/s41467-020-16734-3
3. Foster BL, Dastjerdi M, Parvizi J (2012) Neural populations in human posteromedial cortex display opposing responses during memory and numerical processing. Proc Natl Acad Sci U S A 109:15514–15519. https://doi.org/10.1073/pnas.1206580109
4. Watrous AJ, Deuker L, Fell J et al (2015) Phase-amplitude coupling supports phase coding in human ECoG. Elife 4. https://doi.org/10.7554/eLife.07886

5. Oostenveld R, Fries P, Maris E et al (2011) FieldTrip: open source software for advanced analysis of MEG, EEG, and invasive electrophysiological data. Comput Intell Neurosci 2011:156869. https://doi.org/10.1155/2011/156869
6. Horn A, Li N, Dembek TA et al (2019) Lead-DBS v2: towards a comprehensive pipeline for deep brain stimulation imaging. Neuroimage 184:293–316. https://doi.org/10.1016/j.neuroimage.2018.08.068
7. Groppe DM, Bickel S, Dykstra AR et al (2017) iELVis: an open source MATLAB toolbox for localizing and visualizing human intracranial electrode data. J Neurosci Methods 281:40–48. https://doi.org/10.1016/j.jneumeth.2017.01.022
8. Oehrn CR, Hanslmayr S, Fell J et al (2014) Neural communication patterns underlying conflict detection, resolution, and adaptation. J Neurosci 34:10438–10452. https://doi.org/10.1523/JNEUROSCI.3099-13.2014
9. Liu J, Zhang H, Yu T et al (2020) Stable maintenance of multiple representational formats in human visual short-term memory. Proc Natl Acad Sci U S A 117:32329–32339. https://doi.org/10.1073/pnas.2006752117
10. Tan RJ, Rugg MD, Lega BC (2020) Direct brain recordings identify hippocampal and cortical networks that distinguish successful versus failed episodic memory retrieval. Neuropsychologia 147:107595. https://doi.org/10.1016/j.neuropsychologia.2020.107595
11. Oehrn CR, Fell J, Baumann C et al (2018) Direct electrophysiological evidence for prefrontal control of hippocampal processing during voluntary forgetting. Curr Biol 28:3016–3022.e4. https://doi.org/10.1016/j.cub.2018.07.042
12. Daitch AL, Foster BL, Schrouff J et al (2016) Mapping human temporal and parietal neuronal population activity and functional coupling during mathematical cognition. Proc Natl Acad Sci U S A 113:E7277–E7286. https://doi.org/10.1073/pnas.1608434113
13. Burke JF, Long NM, Zaghloul KA et al (2014) Human intracranial high-frequency activity maps episodic memory formation in space and time. Neuroimage 85(Pt 2):834–843. https://doi.org/10.1016/j.neuroimage.2013.06.067
14. Shimamoto SA, Ryapolova-Webb ES, Ostrem JL et al (2013) Subthalamic nucleus neurons are synchronized to primary motor cortex local field potentials in Parkinson's disease. J Neurosci 33:7220–7233. https://doi.org/10.1523/JNEUROSCI.4676-12.2013
15. Chen S, Tan Z, Xia W et al. (2021) Theta oscillations synchronize human medial prefrontal cortex and amygdala during fear learning. Sci Adv 7. https://doi.org/10.1126/sciadv.abf4198
16. Arnal LH, Kleinschmidt A, Spinelli L et al (2019) The rough sound of salience enhances aversion through neural synchronisation. Nat Commun 10:3671. https://doi.org/10.1038/s41467-019-11626-7
17. Kragel JE, Ezzyat Y, Lega BC et al (2021) Distinct cortical systems reinstate the content and context of episodic memories. Nat Commun 12:4444. https://doi.org/10.1038/s41467-021-24393-1
18. Norman Y, Raccah O, Liu S et al (2021) Hippocampal ripples and their coordinated dialogue with the default mode network during recent and remote recollection. Neuron 109:2767-2780.e5. https://doi.org/10.1016/j.neuron.2021.06.020
19. Kucyi A, Daitch A, Raccah O et al (2020) Electrophysiological dynamics of antagonistic brain networks reflect attentional fluctuations. Nat Commun 11:325. https://doi.org/10.1038/s41467-019-14166-2
20. Kajikawa Y, Schroeder CE (2011) How local is the local field potential? Neuron 72:847–858. https://doi.org/10.1016/j.neuron.2011.09.029
21. Dubey A, Ray S (2019) Cortical electrocorticogram (ECoG) is a local signal. J Neurosci 39:4299–4311. https://doi.org/10.1523/JNEUROSCI.2917-18.2019
22. Aarts E, Verhage M, Veenvliet JV et al (2014) A solution to dependency: using multilevel analysis to accommodate nested data. Nat Neurosci 17:491–496. https://doi.org/10.1038/nn.3648
23. Penny WD, Holmes AJ (2007) Random effects analysis. In: Statistical parametric mapping. Elsevier, pp 156–165

24. Yin Q, Johnson EL, Tang L et al (2020) Direct brain recordings reveal occipital cortex involvement in memory development. Neuropsychologia 148:107625. https://doi.org/10.1016/j.neuropsychologia.2020.107625
25. Kam JWY, Helfrich RF, Solbakk A-K et al (2021) Top-down attentional modulation in human frontal cortex: differential engagement during external and internal attention. Cereb Cortex 31:873–883. https://doi.org/10.1093/cercor/bhaa262
26. Helfrich RF, Lendner JD, Mander BA et al (2019) Bidirectional prefrontal-hippocampal dynamics organize information transfer during sleep in humans. Nat Commun 10:3572. https://doi.org/10.1038/s41467-019-11444-x
27. Staresina BP, Bergmann TO, Bonnefond M et al (2015) Hierarchical nesting of slow oscillations, spindles and ripples in the human hippocampus during sleep. Nat Neurosci 18:1679–1686. https://doi.org/10.1038/nn.4119
28. Helfrich RF, Fiebelkorn IC, Szczepanski SM et al (2018) Neural mechanisms of sustained attention are rhythmic. Neuron 99:854–865.e5. https://doi.org/10.1016/j.neuron.2018.07.032
29. Kragel JE, Ezzyat Y, Sperling MR et al (2017) Similar patterns of neural activity predict memory function during encoding and retrieval. Neuroimage 155:60–71. https://doi.org/10.1016/j.neuroimage.2017.03.042
30. Sederberg PB, Schulze-Bonhage A, Madsen JR et al (2007) Hippocampal and neocortical gamma oscillations predict memory formation in humans. Cereb Cortex 17:1190–1196. https://doi.org/10.1093/cercor/bhl030
31. Trotta MS, Cocjin J, Whitehead E et al (2018) Surface based electrode localization and standardized regions of interest for intracranial EEG. Hum Brain Mapp 39:709–721. https://doi.org/10.1002/hbm.23876
32. Miller J, Watrous AJ, Tsitsiklis M et al (2018) Lateralized hippocampal oscillations underlie distinct aspects of human spatial memory and navigation. Nat Commun 9:2423. https://doi.org/10.1038/s41467-018-04847-9

Chapter 30
How to Detect and Analyze Traveling Waves in Human Intracranial EEG Oscillations?

Anup Das, Erfan Zabeh, and Joshua Jacobs

Abstract The brain is a complex, interconnected network and the large-scale spatiotemporal coordination of neuronal activity is vital for cognition and behavior. Prior studies have proposed that traveling waves of brain oscillations are one mechanism that helps coordinate complex neuronal processes and are crucial for cognition. Traveling waves consist of oscillations that propagate progressively across the cortex and previous studies have shown that these waves play a foundational role for learning, memory processing, and memory consolidation and a range of other behaviors across multiple species. The prevalence of traveling waves in cognition thus indicates that spatiotemporal patterns of neuronal oscillations may coordinate multiple neuronal brain networks and impact behavior. Even though there are several different approaches for analyzing traveling waves using electrophysiological recordings, computational tools targeting the analysis and visualization and understanding of traveling waves are still rare. We briefly review the literature on human intracranial electroencephalography (iEEG), which has shown that traveling waves play an important role in cognition. We then describe a statistical methodology based on circular–linear regression for the detection and analysis of traveling waves from human electrophysiological oscillations. We hope that this approach will provide a more mechanistic understanding of the coordination of neurons across space and time.

A. Das (✉) · E. Zabeh · J. Jacobs
Department of Biomedical Engineering, Columbia University, New York, NY 10027, USA
e-mail: ad3772@columbia.edu

J. Jacobs
e-mail: joshua.jacobs@columbia.edu

E. Zabeh
Mortimer B. Zuckerman Mind Brain Behavior Institute, Columbia University, New York, NY 10027, USA

J. Jacobs
Department of Neurological Surgery, Columbia University, New York, NY 10027, USA

N. Axmacher (ed.), *Intracranial EEG*, Studies in Neuroscience, Psychology and Behavioral Economics, https://doi.org/10.1007/978-3-031-20910-9_30

30.1 Introduction

Prior research has shown that neuronal oscillations play a fundamental role for learning, memory processing, and consciousness in the brain across species [1, 2] (see also Chap. 19). Starting from the discovery of alpha frequency band (8–12 Hz) oscillations by Hans Berger in 1929 [3] in scalp electroencephalograhy (EEG) recordings in humans to the advent of invasive electrocorticogram recordings in 1949 by Jasper and Penfield in humans [4], oscillations are now widely believed to play an important role for spatiotemporal coordination among multiple brain networks [2]. Growing evidence suggests that oscillations do not occur at individual neurons in an isolated way, rather they occur simultaneously across multiple neurons in a given small brain area or on a larger scale across multiple brain areas [5].

Oscillation-based temporal synchronization among neurons is a foundational mechanism for information transfer and coordination among neurons [2]. These synchronization mechanisms are usually thought to involve zero-phase-lag synchronization among neurons, where phases of the recordings from multiple electrodes are temporally aligned [6]. However, recent studies have found waves of electrical activity propagating across the cortex in the human brain [7–15]. This was possible because of improved methods in analyzing simultaneous intracranial EEG (iEEG) recordings from many brain areas [5], which have shown systematic spatial variation of instantaneous phases of the electrodes across the cortex. These systematic phase delays reveal the progressive propagation of neuronal activity across the cortex, known as *traveling waves*, which have been shown to be closely related to behavior [5].

Studies in animals showed that the frequency, strength, direction, and speed of traveling waves correlate with a broad range of behaviors in animals. These include visual perception [16–23], movement initiation [24–28], and memory processing [29, 30] in non-human primates as well as visual processing [31–33] and spatial navigation [34–36] in rodents. However, currently, there are no well-established methods for analyzing traveling waves in the human brain and rigorous computational tools specifically targeting the visualization and understanding of traveling waves are rare. Therefore, even though oscillations are seemingly ubiquitous in the human brain [2], many of the potential traveling waves associated with those oscillations have been missed. Although some iEEG studies in humans have demonstrated a role for traveling waves in cognition [5], systematic studies with rigorous analysis of traveling waves are still lacking.

In this chapter, our focus is on the development of novel approaches for the detection and analysis of traveling waves from human iEEG recordings. Most prior studies in the last two decades have detected traveling waves predominantly in scalp EEG and magnetoencephalography (MEG) recordings [37–42]. However, several recent iEEG studies in humans have also demonstrated the existence of traveling waves [7–15]. These iEEG recordings are from pharmaco-resistant epileptic patients who underwent neurological surgery for removal of their seizure onset zones. Traveling waves in these studies were usually detected and analyzed using the spatial gradient of the phases of the recordings from the iEEG electrodes [9, 12]. Traveling waves are present spontaneously during resting-state conditions such as eye-closure [9] (Fig. 30.1a) and passive fixation [43] (Fig. 30.1b). Traveling waves are also present during sleep spindles suggesting their putative role in memory consolidation and plasticity [8, 12] (Fig. 30.1c). More recently, traveling waves were also detected during movement imagery suggesting their putative role in coordinating complex movements [13] (Fig. 30.1d). Furthermore, traveling waves play a prominent role during speech processing [11], and also, during working memory, suggesting their putative role for memory processing as well [15] (Fig. 30.1e).

To provide a broader introduction for how to measure traveling waves in the human brain, we describe an approach based on *circular statistics* to capture and analyze traveling waves amidst neuronal oscillations in iEEG recordings. Our approach overcomes several challenges for analyzing these waves in the human brain. First, we describe novel approaches for detecting oscillations in the iEEG electrodes and introduce methods to detect groups of nearby electrodes each having an oscillation at nearly identical frequencies. We then describe methods to detect and analyze the features of waves of electrical activity propagating across the cortex corresponding to these detected oscillations. We also introduce a novel approach based on circular statistics to track the progressive variation of the phases across multiple electrodes and subsequently detect traveling waves based on the fitted parameters of a circular-linear regression model. And, finally, we describe several features of the detected traveling waves and how they are intimately related with human behavior. Our approach is rigorous and can be used to identify multiple foundational mechanisms underlying the propagation of traveling waves and to reveal their link with behavior.

30.2 Approach to Measure Traveling Waves of Neuronal Oscillations

Many previous studies of traveling waves in animal models used relatively simple analytical approaches based on the spatial gradient of phases. These approaches made sense because neural recording electrodes in animals are usually implanted in a relatively small area of the brain that was consistently placed across animals. However, there are several inherent aspects of human iEEG datasets that make it challenging to detect and analyze traveling waves. In particular, placements of electrodes across patients can be highly variable due to the complicated clinical protocols involved and can consist of multiple types of electrodes such as grid, strip, and depth electrodes [44, 45]. Furthermore, the frequencies of neuronal oscillations can vary substantially across human patients, even after controlling for task behavior and electrode placement [15]. Due to these challenges, an improved method for measuring traveling waves would be preferable if it were able to accommodate the specific features of the signals in a given patient's iEEG recordings.

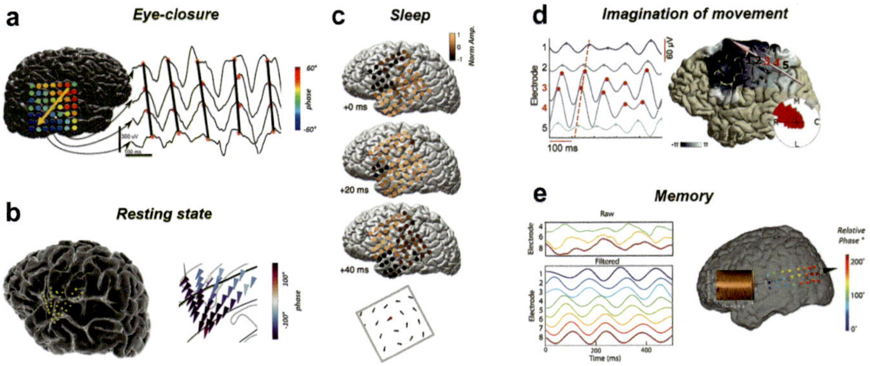

Fig. 30.1 Traveling waves in the human brain across multiple domains, detected using iEEG. **a** Traveling waves during eye-closure resting-state condition. Alpha oscillations in patients propagated as traveling waves in the cortex during eye-closure resting-state condition. Colors of electrodes represent the phases of their oscillations (left panel) and traces represent the raw voltages of the marked electrodes (right panel). Adapted with permission from Halgren et al. [9]. **b** Traveling waves during passive fixation resting-state condition in the insula. Beta oscillations in the human insula propagated as traveling waves during passive fixation resting-state condition. Shown is an example of implanted stereo EEG electrodes in a patient (left panel), with arrows denoting the direction of the traveling wave associated with each electrode (right panel). Adapted with permission from Das et al. [43]. **c** Traveling waves during sleep spindles. Sleep spindles are traveling waves in the human brain. When visualized on the cortex, individual spindle cycles are often organized as rotating waves traveling from temporal (+0 ms, top) to parietal (+20 ms, middle) to frontal (+40 ms, bottom) lobes. Adapted with permission from Muller et al. [12]. **d** Traveling waves during movement imagery in the sensorimotor cortex. Alpha rhythmic activity during imagined movement in a representative individual. Shown are filtered signals for the five electrodes numbered on the brain plot (left panel). Cortical phase maps indicate the average phase at each cortical site relative to a central sensorimotor reference electrode. Local arrows indicate the propagation direction at each electrode, with arrow size weighted by the local phase gradient magnitude (right panel). Large global arrow indicates the mean propagation direction across the sensorimotor cortex, with arrow size weighted by the alignment of sensorimotor gradients. Alpha rhythm propagation is maximal in a caudo-rostral direction (red distribution, denoted by a polar plot). Adapted with permission from Stolk et al. [13]. **e** Traveling waves during memory processing. Alpha-theta oscillations are traveling waves in humans while performing a working memory task. Example shows data from a patient with an 8.3-Hz traveling wave (right panel). Shown are raw signals from three selected electrodes (left panel), the selected electrodes are ordered from anterior (top) to posterior (bottom). Also shown are the filtered signals (filtered at 6–10 Hz) for the eight electrodes numbered on the brain plot. Adapted with permission from Zhang et al. [15]

Our method overcomes these challenges by customizing the analysis pipeline according to the iEEG recordings from each individual patient [15, 43]. This approach consists of two primary steps: (i) The first step consists of identification of spatially contiguous clusters of electrodes with narrowband oscillations at similar frequencies. Identifying a group of nearby electrodes with a single oscillation frequency is crucial since, by definition, a traveling wave involves a single frequency and whose phase progressively propagates through these electrodes, thus making it possible to detect the traveling wave when it passes by these electrodes. (ii) The second step consists of identification of systematic spatial variation of the instantaneous phases of the electrodes for each cluster, defined to be a traveling phase wave. This step is important since we want to capture the systematic phase delays of the wave across the group of electrodes in the oscillation cluster identified in step (i) above, thus enabling us to detect the presence or absence of a traveling wave. Once these systematic spatial variations in phases have been detected, we can then analyze the features of this spatial phase propagation and examine its relationship with human behavior. These procedures are detailed below. Our new approaches are flexible in the sense that they can be easily applied to other domains of brain imaging such as scalp EEG and MEG recordings as well as recordings from animal models.

30.2.1 Identification of Oscillations and Clustering Algorithm

By definition, a traveling wave involves a neuronal oscillation that appears with a time delay across multiple regions of the cortex. Therefore, our first step in identifying these patterns is to detect oscillations that appear at a single frequency at multiple nearby electrodes. To detect such patterns, we first identify spatially contiguous clusters of electrodes with oscillations at the same or similar frequencies [15, 43]. Critically, we perform this procedure in an adaptive fashion that is well suited for human iEEG data by accommodating differences in electrode positions and oscillation frequencies across individuals. This flexibility is especially important since iEEG electrodes in humans can be in the form of grid, strip, or depth electrodes and can also span multiple brain areas. Our approach can overcome this challenge by detecting waves which can travel through multiple types of electrodes and spanning many different brain areas, including both gray matter and white matter volumes.

The first step to detect oscillations in neuronal signals is estimating their power distribution in the frequency domain and distinguishing true narrowband rhythmic oscillations from background fluctuations such as the 1/f signal [15]. There are several methods that can been used for such steps [15, 46, 47] (see also Chap. 23). In our work, we have used Morlet wavelets to compute the power spectra of the neuronal oscillations. Morlet wavelets are useful particularly for analyzing intracranial recordings because of their superior ability to detect transient, possibly non-stationary, oscillatory dynamics [15, 48]. After using Morlet wavelets to measure each electrode's power spectrum, we then distinguish true narrowband oscillations as those that have peaks that are significantly greater than the background 1/f spectrum. We use a thresholding procedure to ensure that we specifically focus on significant narrowband oscillations that are reliably different from the background 1/f signal at an electrode [46].

We use this approach to identify narrowband oscillations at each recording site and eventually, to find multiple nearby electrodes oscillating at nearly the same frequency. To distinguish narrowband peaks in an electrode's power spectrum from the background signal, we fit a line to each patient's mean power spectrum in log–log coordinates using robust linear regression [15] (Fig. 30.2a). We then subtract the actual power spectrum from the regression line. This normalized power spectrum removes the 1/f background signal and emphasizes narrowband oscillations as positive deflections (Fig. 30.2a.1). We identify narrowband peaks in the normalized power spectrum as any local maximum greater than some predefined threshold. In our work [43], we used a threshold of one standard deviation above the mean, but other thresholds could be used depending on the experimenter's needs. This method reliably identifies the frequencies where individual electrodes show strong oscillations, as can be seen in Fig. 30.2a.1 which shows our approach from two example electrodes. One of these electrodes has a narrowband oscillation that we successfully detected at the theta frequency band and the other has one at alpha band, thus demonstrating the efficacy of this approach.

Next, to identify traveling waves (see below), we focus on the groups of contiguous electrodes that show oscillations at the same frequency. We focus on the contiguous electrode groups because our focus is to characterize the oscillations that are traveling waves by having each cycle propagating across contiguous regions of cortex. To identify these groups, or *oscillation clusters*, we implement a spatial clustering algorithm that we designed to find the contiguous groups of electrodes that exhibit narrowband oscillations at a given frequency (Fig. 30.2a.1). To identify the specific electrodes that comprise a spatially contiguous group, we first create a pairwise-adjacency matrix that indicates whether each pair of electrodes is contiguous. This matrix indicates whether each electrode pair is separated by less than some predefined threshold (such as <= 20 mm [15]). Finally, we use this adjacency matrix to identify mutually connected spatial clusters of electrodes by computing the connected components of this graph [49]. In our work, we only include clusters with at least four connected electrodes in our analysis. Further, in our work, we have allowed for some electrodes to show oscillations at nearby but nonidentical frequencies (such as within 10%; see [15]), which allows our procedure to accommodate oscillations that can slightly vary across frequencies. However, note that some parameters of this method could be tweaked according to the experimenter's needs. Once we identify a group of electrodes that oscillate at same or similar frequencies, we can then design methods to detect the presence or absence of a traveling wave. This is described below.

30.2.2 Identification of Traveling Waves

After identifying oscillation clusters in each person, which will distinguish the contiguous regions of cortex that oscillate at a single frequency, the next step in our framework is to identify traveling waves that propagate across that cluster [12, 15]. Quantitatively, we can define a traveling phase wave as a set of simultaneously recorded neuronal oscillations at the same frequency whose instantaneous phases vary systematically with the location of the electrodes, such that individual cycles of

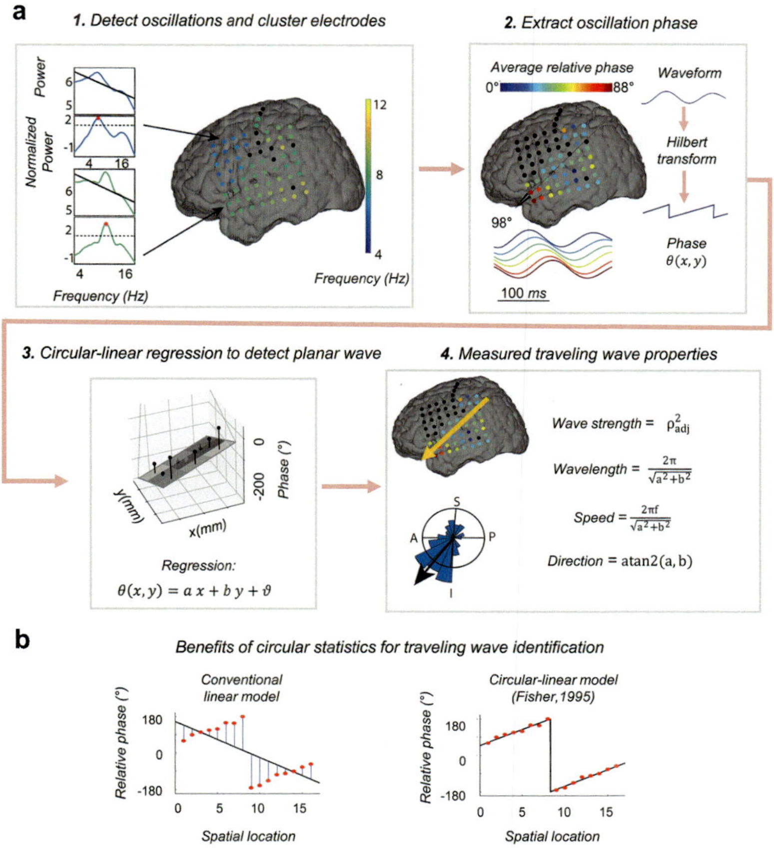

Fig. 30.2 Detection and analysis of traveling waves. **a** Flowchart illustrating the pipeline for detecting and analyzing traveling waves. First column: Identification of spatially contiguous clusters of electrodes with narrowband oscillations at similar frequencies. Narrowband oscillation at each electrode was identified by fitting a line to each electrode's mean power spectrum in log–log coordinates using robust linear regression and then subtracting the actual power spectrum from the regression line. This normalized power spectrum removes the 1/f background signal and emphasizes narrowband oscillations as positive deflections. Shown are two oscillation clusters, one oscillating at theta frequency and the other at alpha frequency. Second column: Calculation of phases of the electrodes for each oscillation cluster. Phases of the oscillations were calculated using Hilbert transform on the filtered signals. Colors in this figure show the phases of the electrodes corresponding to the alpha oscillation cluster in the first column. Third column: Circular-linear regression was used to identify traveling waves of phase progression for each oscillation cluster at each time point. For each spatial phase distribution, two-dimensional (2-D) circular–linear regression was used to assess whether the observed phase pattern varies linearly with the electrode's coordinates in 2-D. Shown are solid circles representing the actual phases, the fitted plane, which is the predicted phase, and bars denoting the residuals between the actual and predicted phases. Fourth column: Features of traveling waves were calculated based on the fitted parameters of the circular-linear regression model. Shown are electrodes with colors denoting the phases and arrow denoting the direction of the traveling wave, with distribution of directions in blue, denoted by a polar plot. **b** Advantages of the circular-linear regression model for the identification of traveling waves. Due to the circular nature of the phase, any potential linear association between phase and electrode coordinates can be better estimated by the circular-linear model (right) compared to the linear regression model (left)

the oscillation move across the cortex. A challenge for quantifying and tracking the traveling spatial patterns in human intracranial recordings is tracking the systematic presence of multiple cycles of oscillations across multiple electrodes. To perform this task, we use a circular-linear regression which models the relation between oscillation phase and electrode position (Fig. 30.2b). Since the phase wraps around every 360°, a circular-linear regression model, leveraging circular statistics [50], is important to use rather than a conventional linear model (Fig. 30.2b).

To identify traveling waves from the phases of electrodes in each oscillation cluster (Fig. 30.2a.2), we first measure the instantaneous phases of the signals from each electrode of a given cluster by applying a zero phase-lag filter at the peak frequency of the detected oscillation. In our analysis, we have used a Butterworth filter at the cluster's narrowband peak frequency (bandwidth [$f_p \times 0.85$, $f_p/0.85$] where f_p is the peak frequency). We then use the Hilbert transform on each electrode's filtered signal to extract the instantaneous phase at each time-point of the iEEG recordings [15]. However, other transforms such as the Fourier and wavelet transforms can also be used to extract the instantaneous phases of the electrodes as well. These instantaneous phase values are then fed-in to the circular-linear regression model described below.

In a traveling wave, the phases of an ongoing oscillation are spatially organized, with a systematic phase shift across space in the cortex. Accordingly, to measure such patterns, we use a 2D circular-linear regression to quantitatively measure the relation between oscillation phase and electrode position. This regression lets us assess whether the observed phase pattern varies linearly with the electrode's coordinates (Fig. 30.2a.3).

The structure for our circular-linear model is as follows. x_i and y_i represent the 2-D coordinates and θ_i the instantaneous phase of the ith electrode. Whereas the original electrode positions are of course in 3D in the brain's volumetric coordinates, we reduced the data to 2-D coordinates x_i and y_i corresponding to the cortical surface by projecting the 3-D Talairach coordinates of electrodes into the best-fitting 2-D plane using principal component analysis. Even though this procedure is most applicable to subdural grid electrodes, it can be applied to stereo EEG depth electrodes as well [43]. This procedure can also be carried out in the 3-D volumetric space in the brain, however, projecting the 3-D coordinates to 2-D helps in better visualizing and interpreting the traveling wave [43]. Based on the 2D electrode coordinates, to measure the phase propagation, we then fit a 2-D circular-linear model to the phase distribution at each timepoint. This model has the following structure,

$$\hat{\theta}_i = (ax_i + by_i + \vartheta) \bmod 360^\circ,$$

where $\hat{\theta}_i$ is the predicted phase, a and b are the phase slopes corresponding to the rate of phase change (or spatial frequencies) projected into each of the orthogonal dimensions, and ϑ is the phase offset.

Circular–linear models do not have an analytical solution and, hence, we fit them iteratively using numerical methods [50], which makes this procedure computationally complex. To simplify model fitting, we first convert the parameters of the model from cartesian coordinates to polar coordinates. We define $\alpha = a\tan 2(b, a)$ which denotes the angle of wave propagation and $\xi = \sqrt{a^2 + b^2}$ which denotes the spatial frequency (Fig. 30.2a.4). We fit α and ξ to the distribution of oscillation phases at each time point by conducting a grid search over $\alpha \in [0°, 360°]$ and $\xi \in [0, 180/\delta]$ in sufficiently small increments of phase and phase/space steps respectively (for example, the step sizes used by [43] are 5° and 0.5 °/mm for phase and phase/space respectively). Note that $\xi = 180/\delta$ corresponds to the spatial Nyquist frequency of $180/\delta$ °/mm, corresponding to the highest spacing δ mm between neighboring electrodes. We fit the model parameters ($a = \xi\cos(\alpha)$ and $b = \xi\sin(\alpha)$) for each time point to most closely match the phase observed at each electrode in the cluster. We compute the goodness of fit as the mean vector length $\overline{r}$ of the residuals between the predicted ($\hat{\theta}_i$) and actual (θ_i) phases [50],

$$\overline{r} = \sqrt{\left[\frac{1}{n}\sum_{i=1}^{n}\cos\left(\theta_i - \hat{\theta}_i\right)\right]^2 + \left[\frac{1}{n}\sum_{i=1}^{n}\sin\left(\theta_i - \hat{\theta}_i\right)\right]^2},$$

where n is the number of electrodes. We choose the selected values of α and ξ to maximize $\overline{r}$. We repeat this procedure for each oscillation cluster. To measure the statistical reliability of each fitted traveling wave, we examined the phase variance that was explained by the best fitting model. To do this, we compute the circular correlation ρ_{cc} between the predicted ($\hat{\theta}_i$) and actual (θ_i) phases at each electrode [43]:

$$\rho_{cc} = \frac{\sum_{i=1}^{n}\sin\left(\theta_i - \overline{\theta}\right)\sin\left(\hat{\theta}_i - \overline{\hat{\theta}}\right)}{\sqrt{\sum_{i=1}^{n}\sin^2\left(\theta_i - \overline{\theta}\right)\sum_{i=1}^{n}\sin^2\left(\hat{\theta}_i - \overline{\hat{\theta}}\right)}},$$

where bar denotes averaging across electrodes. Finally, to account for the variation in the number of electrodes across clusters, we apply an adjustment to control for number of fitted model parameters [43]:

$$\rho_{adj}^2 = 1 - \frac{\left(1 - \rho_{cc}^2\right)(n-1)}{n-k-1},$$

where k is the number of independent regressors ($k = 3$ in this case). We refer to ρ_{adj}^2 as the wave-strength of the traveling wave [15] as it quantifies the strength of the traveling wave (note that ρ_{adj}^2 has been referred to as phase gradient directionality (PGD) in some prior studies [12, 15, 26]). We note that ρ_{adj}^2 can now be compared across different clusters and subjects with varying number of electrodes. To test for the statistical significance of a traveling wave, we shuffle the coordinates of the electrodes and re-estimate the strength of the wave for each shuffling. In this way, we construct a histogram of surrogate wave-strength values against which we then compare the empirical wave-strength to test for the presence or absence of a traveling wave [43].

A step-by-step visual demonstration of the circular-linear regression approach to detect traveling waves has been illustrated in Fig. 30.3 for the iEEG recordings measured on two trials of a task, one with a traveling wave and the other without a wave. As a result of this fitting procedure, the direction α represents the spatial orientation at which the traveling wave propagates with a continuously increasing phase gradient through the cortex (Fig. 30.3). When this direction α is visualized as an arrow on a brain plot, the traveling wave's propagation can be seen visually. In these plots, individual oscillation cycles appear at relatively early timepoints on the electrodes near the tail of the directional arrow and at later latencies on electrodes near the head of the arrow (Fig. 30.1a).

30.3 Features of Traveling Waves

The 2D circular-linear model that we described above is a very useful quantitative tool for measuring the instantaneous properties of the traveling waves at each moment. The fitted coefficients a and b from the model can be used to calculate all the key features of the current traveling wave on that electrode cluster (Fig. 30.2a.4). For example, the parameters α and ξ in the polar coordinates denote the angle of wave propagation and the spatial frequency respectively. Other features of the traveling wave such as the wavelength (2π/spatial frequency) and the speed (wavelength $\times$ frequency) can be readily derived from these parameters as well.

Another defining advantage of our proposed approach is that traveling waves can be reliably detected on a single-trial level using our methods. In a working memory task [15], we were able to detect traveling waves in ~81% of clusters at the single-trial level and ~67% of clusters had consistent traveling waves at the single-trial level which also had a consistent propagation direction. In another study using a verbal working memory task [51], we found that frontal theta and temporal alpha traveling waves are more reliably detected during the earlier periods of a trial compared to late detection of reliable temporal theta traveling waves in a trial, during the memory encoding periods.

Current studies have indicated that multiple features of traveling waves are related to human behavior [11, 51] (Fig. 30.4). Traveling waves exist across a broad range of frequencies, starting from low frequency delta to higher frequency beta bands. Prior iEEG studies have detected traveling waves at alpha frequency during eye-closure resting-state in subdural grid electrodes [9], similar to the alpha waves observed in scalp EEG recordings [37]. Recently, we have found traveling waves of theta and beta oscillations during passive fixation resting-state in the insula using stereo-EEG depth electrodes [43], indicating that traveling waves are present in not only the surface of the cortex, but also deep brain structures such as the insula. This finding also suggested that whereas lower frequency alpha oscillations are a defining feature of the resting brain at the surface of the cortex, higher beta frequencies may play

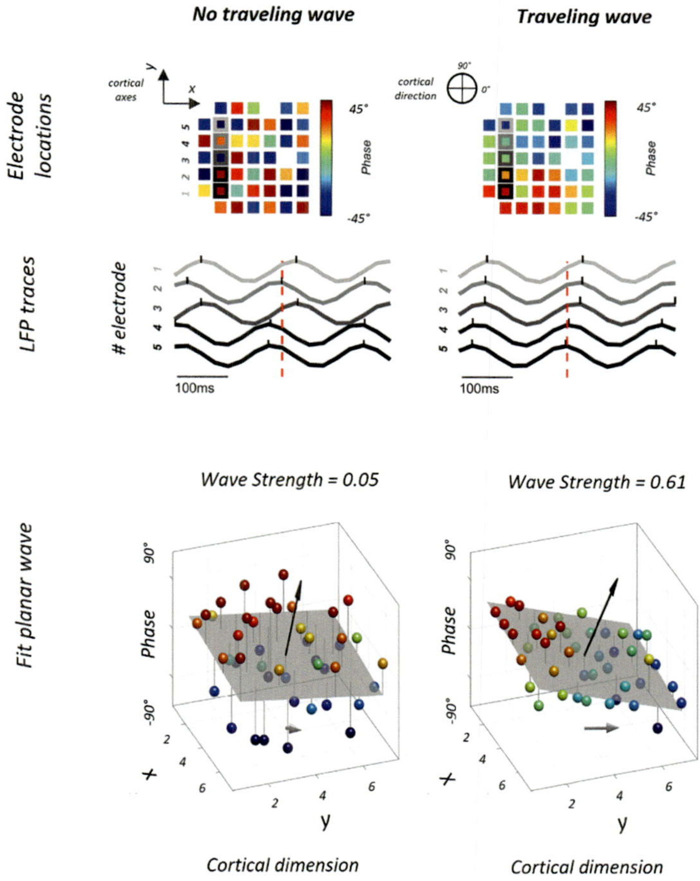

Fig. 30.3 Examples of detection of traveling waves for two representative trials. Left and right columns correspond to two example trials demonstrating the absence and presence of a traveling wave respectively. Top row: phase organization of electrodes in two representative trials of a cortical recording array with 49 electrodes. Colors represent the instantaneous phase of each electrode. Observe the systematic spatial variation of the phases of the electrodes in the right column indicating the presence of a traveling wave. Contrast this to the left column where there is no systematic spatial variation of the phases indicating the absence of a traveling wave. Middle row: filtered LFP traces of five adjacent electrodes as labelled in the top row. Observe that the amplitude peaks occur at successively later times for electrodes 1–5 in the right column in contrast to the left column. Red-colored, vertical dashed lines represent the time-instants at which the phase distributions were plotted in the top row. Bottom row: Fitted planes estimated from the circular-linear regression analysis showing the best fit between the phases and the locations of the electrodes. Solid circles denote the actual phases of the electrodes with the colors as in the top-row and the vertical bars denote the residuals between the actual phases and the predicted phases using the circular-linear regression. The thick black arrows indicate the orthogonal vectors of the fitted planes and gray arrows represent the projection of these orthogonal vectors on the X–Y plane (cortical plane). Observe the smaller vertical bars in the right column indicating the presence of a traveling wave in contrast to the larger vertical bars in the left column indicating the absence of a traveling wave. Features of the traveling wave can be estimated by the parameters of the fitted plane (Fig. 30.2a)

an enhanced role in deeper structures of the cortex such as the insula. Other studies have also shown that these traveling waves are also highly relevant for memory processing in the human brain, and not just in resting-state. Specifically these studies have shown a crucial role for theta and alpha oscillatory traveling waves during a working memory task [15] and also verbal episodic memory task [51], suggesting that lower frequency oscillations might be more relevant for memory processing in the human brain. Some other studies have also shown a role of low frequency delta, theta, and alpha oscillatory traveling waves during speech processing as well [11]. On the other hand, higher frequency beta oscillatory traveling waves play an important role during movement imagery [13] and sleep spindles [8, 12], in line with the role of beta frequency for sensorimotor neuronal activity processing and propagation of sleep spindles in the human brain, respectively.

The frequencies of these oscillations that we just described usually define the speed of propagation of traveling waves. Indeed, lower frequency traveling waves such as theta and alpha usually propagate in the range ~0.25–1 m/s [9, 15, 51], whereas higher frequency beta traveling waves usually propagate in the range ~0.5–5 m/s [12, 13]. In our previous studies, we have also found that the speed of a traveling wave increases with an increase in its oscillation frequency [15]. Moreover, prior work on computational modeling of weakly coupled oscillators has also shown that traveling waves can naturally emerge from spatially varying gradients of oscillations across different frequencies [52] and suggests that the propagation speed of these waves depends on the associated oscillation frequency. However, it is important to observe that these waves were detected using subdural grid electrodes on the surface of the cortex. In our more recent work, using stereo EEG depth electrodes, we have found that waves travel at ~0.7 m/s in the insula during passive fixation resting-state condition for both low frequency theta and higher frequency beta oscillations [43]. This indicates that the speed-frequency relation of the traveling waves that we observe on the cortical surface may not be relevant for deep brain structures such as the insula and suggests that putatively different mechanisms might be involved for the origin of these traveling waves for the cortical surface and deeper cortical regions. Further studies are needed to definitively examine the speed-frequency relationship of these traveling waves. Rigorous computational models of these waves can go a long way in providing important insights into the characteristics of these traveling waves and the link between the different features of these waves. It is worth noting that, all these types of traveling waves features, and more, can be detected at each moment using the methods we described here.

Another important feature of a traveling wave is its propagation direction. The direction of a traveling wave informs us about the spatiotemporal coordination of different brain regions and its relation to behavior. In our previous studies, we have found that waves usually propagate from higher frequency regions to lower frequency regions [15], suggesting that the different features of traveling waves are inter-linked. More importantly, several studies have found that the propagation direction of these waves is linked to human behavior. The directions of traveling waves can distinguish speech compared to non-speech trials [11], successful memory encoding compared to unsuccessful memory encoding and memory recall in a verbal episodic memory

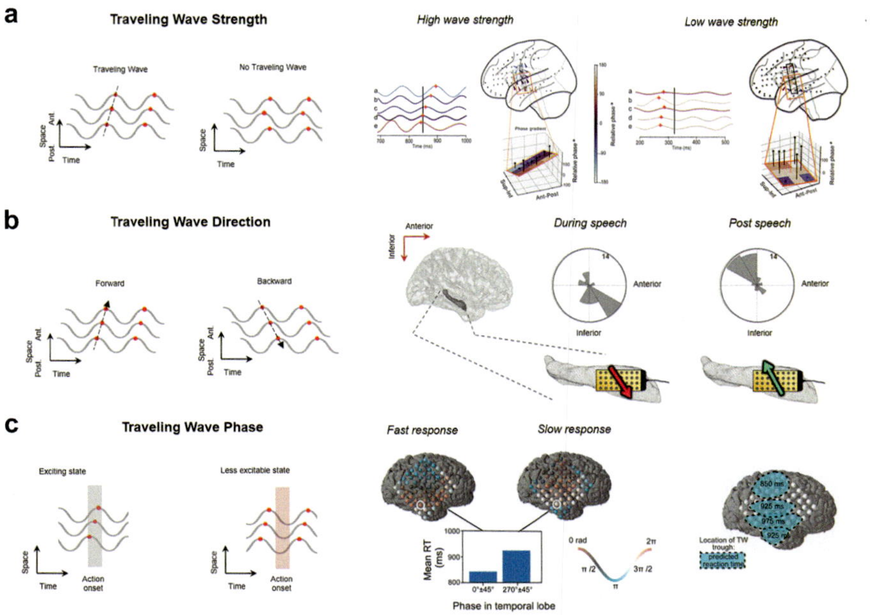

Fig. 30.4 Features of traveling waves and potential behavioral relevance. **a** Wave-strength. (Left): Cartoon demonstration showing that systematic variation of the phase in space indicates high wave-strength, while non-systematic variation of phase indicates low wave-strength. (Right): Filtered signals of electrodes from two different trials (corresponding to the two different brain plots) from a representative patient performing a verbal episodic memory task, demonstrating high and low strength of traveling waves. Arrows denote the direction of the waves with colors denoting the phases. Note the lower residuals, indicating higher wave-strength in the first column, with higher residuals indicating lower wave-strength in the second column. In humans, no association has been found between wave-strength and behavior [51]. Adapted with permission from Mohan et al. [51]. **b** Direction. (Left): Cartoon demonstration of forward and backward direction of traveling waves. (Right): iEEG recordings of the hippocampus from a patient demonstrating that the direction of traveling waves is dependent on the timing of a speech task, with waves traveling opposite to each other for speech compared to non-speech periods. Adapted with permission from Kleen et al. [11]. **c** Phase. (Left): Cartoon demonstration linking behavior to the phase of the traveling waves (excitable versus non-excitable). (Right): Recordings from electrocorticographic electrodes in a representative patient performing the Sternberg working memory task shows that the reaction time of the patient is correlated with the phase of the traveling wave in each electrode, shaping a spatial map for the preferred phase of the traveling wave for the optimal performance. Adapted with permission from Mohan et al. [51]

task [51], and fast response times compared to slow response times in a working memory task [15], demonstrating its behavior relevance. Moreover, the directions of traveling waves also shift across brain regions. During eye-closure resting-state [9], waves travel from anterosuperior to posteroinferior direction broadly across the cortex. In a working memory task [15], the waves travelled from posterior to anterior direction in the frontal and temporal lobes, however no definite wave direction was found in the occipital and parietal lobes.

Furthermore, the timing (or phase) of traveling waves also plays a critical role in human cognition. In our recent study, we have shown that the timing of a wave precisely defines fast and slow response times, in a working memory task [51]. These results may be similar to a set of findings in animals, where the phase of traveling beta oscillations predicted stimulus detection in visual perception [17].

Together, across this broad literature, these findings suggest that multiple features of traveling waves simultaneously define different behavioral states in humans. Since our proposed methodology can directly extract all these features of the traveling waves, it provides a useful tool for probing the direction of information flow for the precise spatiotemporal coordination of neuronal activity underlying different behaviors in humans.

30.4 Discussion

Oscillations play a prominent role in the brain and studies across multiple species have shown that they are correlated with learning, memory processing, and consciousness [1, 2]. Even though oscillations are seemingly ubiquitous in the human brain [2], how these oscillations spatiotemporally coordinate neuronal activity across multiple brain regions, has remained elusive, due to lack of well-established methods for rigorously analyzing these oscillations. Recent advances [5] in obtaining highly precise, simultaneous intracranial EEG recordings from many brain areas have shown systematic spatial variation of instantaneous phases of the electrodes in an oscillation cluster which lays the foundation for possible existence of traveling waves. Intracranial recordings from subdural grid, strip, or depth electrodes often contain dynamics which are complex and it's difficult to visualize time-periods of systematic spatiotemporal patterns across broad regions in the brain and often, can be missed by separately analyzing individual traces of electrodes as is often done by neurologists [45]. Using these types of traditional analysis, we can visualize the waves only when the recordings align with the direction of wave propagation and this may be the reason why many of the previous intracranial EEG studies might have missed these traveling waves [45], which are now known to be ubiquitous across the human brain [5]. Traveling waves exist across multiple cognitive domains such as resting-state, speech, memory processing, and sleep, in the human brain, using iEEG recordings. Previous studies detected and analyzed traveling waves using the spatial gradient of the phases of the iEEG recordings [9, 12].

Here we described a new approach based on circular statistics to capture and analyze traveling waves in iEEG recordings. Our approach is general and can quantitatively measure all key features of these traveling waves. This approach consists of two primary steps, (i) identification of spatially contiguous clusters of electrodes, and (ii) identification of systematic spatial variation of instantaneous phases of the electrodes for each cluster, defined to be a traveling phase wave. Even though we described our approach in an iEEG setting, our methods are also applicable to other modalities such as scalp EEG, MEG, and optical recordings as well as field potential and depth electrode recordings in animals. These may be promising areas of future work because there is evidence for traveling waves in these settings as well [31, 34, 37, 53–55]. Moreover, several features of traveling waves can be directly extracted from the parameters of our circular-linear regression model. We can then analyze the relationship of these features to different human behaviors as we have done in our previous studies [14, 15, 43, 51].

It is important to note that even with our analysis method, a number of features of the data must be satisfied in order to measure traveling waves. In particular, measuring traveling waves accurately requires adequate sampling of electrodes across the region that exhibits each oscillation. The detection of traveling waves is also constrained by the size of an oscillation cluster, and a sufficient number of electrodes, all oscillating at nearly similar oscillation frequencies, is necessary to capture a wave traveling across the cortex [15]. To find the features of traveling waves that reliably correlate with behavior, owing to inter-individual differences in oscillations [15], a large sample size of patients may be important to reveal the key features of traveling waves. In one of our previous studies, we detected traveling waves across 77 patients in a working memory task [15], and we had found substantial heterogeneity in oscillation frequency and direction of these waves across patients. To this end, open-source data sharing efforts (see Chap. 38) will be crucial to analyze inter-individual and gender-related differences of traveling waves across large cohorts of patients.

Given that we found evidence for the existence of traveling waves across several frequency bands such as the delta, theta, alpha, and beta ranges [14, 15, 43, 51], it raises an important question of how the waves in these bands relate to each other. Previous studies have shown that theta and beta traveling waves in the human insula travel independently of each other during resting-state [43]. In another study, gamma power was phase-locked to alpha traveling waves in the human neocortex [7], similar to the more traditional phase-amplitude coupling mechanism found in the human cortex [56]. However, how the interactions of these waves in different frequency bands relate to human behavior remains unknown and future studies specifically focusing on developing novel methods for analyzing the interactions between these waves at different frequencies and their links to human cognition are needed to fill this important gap.

The new methods that we have developed related to traveling waves could potentially be informative about information coding in local neuronal activity and how it is coordinated across larger brain networks. Many brain areas that show traveling waves, including the hippocampus [14] and the neocortex [15], are also regions that show gradients in neural coding. Because the timing of local neuronal activity is phase-locked to specific phases of traveling waves [7, 47], it suggests that traveling waves may underlie neuronal processing by supporting a type of temporal multiplexing, in which only certain subregions in particular cortical areas are active at a given moment [57]. We previously noted that traveling waves allow particular cortical regions to be consistently indexed by the phase delay of the overlying traveling waves, because human traveling waves maintain a consistent spatial frequency across trials [14]. Combined with the findings that various cortical regions such as the hippocampus and the frontal lobe show gradients in neural representations that match the direction of traveling waves propagation [58, 59], this suggests that traveling waves could be important for large-scale information coding by allowing different cortical representations to be indexed at specific phase delays. Our rigorous approaches to precisely estimate the different features of traveling waves would thus be informative of a new type of cortical communication involving the role of traveling waves to coordinate cortico–hippocampal interactions.

It is also important to note that traveling waves also exist during interictal spiking activity [60] as well as seizures [61]. It thus becomes critical to distinguish traveling waves arising from pathological activity from those arising from putative normal brain function, and further research is needed to develop more advanced methods for classification of normal and non-normal traveling waves [62]. Finding strong relations between the different features of these traveling waves and human behavior may help to avoid interpretational difficulty between putative normal and pathological traveling waves.

Even though we focused on methods to detect and analyze planar traveling waves here due to their behavioral relevance [11, 15, 51], more complex patterns of traveling waves such as radial and spiral waves have also been detected in the human brain, especially during sleep spindles [8, 12], and also recently, in monkey [29, 63] and rodent [64, 65] brains. It will be interesting to show whether and how these more complex radial and spiral traveling waves are relevant for other types of behaviors such as learning, and verbal episodic and spatial memory tasks in humans. These complex patterns of traveling waves might also indicate excitation/inhibition of neural ensembles in the brain. For example, the center of an outward spiral or a source traveling wave might putatively have elevated neuronal excitation compared to the rest of the cortex and an inward spiral might have comparatively decreased neuronal excitation, in light of computational models that showed that traveling waves propagate from areas with faster intrinsic rhythmicity to the slower ones [52]. This may help us identify brain regions with relatively distinctive levels of excitation/inhibition. Therefore, it is crucial to develop rigorous signal processing methods to carry out a comprehensive analysis of these complex patterns of traveling waves. Relatedly, some previous studies have suggested the use of curl and divergence analysis of spatial phase gradients to detect radial and spiral traveling waves [12, 31, 66, 67]. Building on this work, an interesting future research direction would be to develop new methods by extending the circular-linear regression model-based methods presented here to account for more complex patterns such as radial or spiral waves or any combinations of these wave patterns. In our recent work, we have adopted the circular-linear regression approach described here to detect localized traveling waves by fitting the phase-plane in a localized sub-cluster of electrodes (Figs. 30.1b, 30.4a) and shown that it is possible to estimate features of traveling waves for individual electrodes rather than the entire cluster of electrodes, and found more complex patterns of traveling waves beyond planar waves [43]. Additional work is necessary to fully characterize the spatiotemporal features of these complex patterns of traveling waves, and this could provide a key step towards more fully distinguishing the functional role of the spatiotemporal dynamics of brain oscillations in various types of cognition.

Acknowledgements We thank Drs. Uma Mohan and Honghui Zhang for technical assistance. This research was supported by a RISE award and an NSF CAREER award to J.J.

Conflict of Interest Statement The authors declare no competing financial interests.

References

1. Buzsáki G (2000) Rhythms of the brain. Oxford University Press, New York
2. Buzsaki G, Draguhn A (2004) Neuronal oscillations in cortical networks. Science 304(5679):1926–1929
3. Berger H (1929) Über das Elektrenkephalogramm des Menschen. Archiv f Psychiatrie 87:527–570
4. Jasper H, Penfield W (1949) Electrocorticograms in man: effect of voluntary movement upon the electrical activity of the precentral gyrus. Arch Psychiatr Nervenkr 183:163–174
5. Muller L, Chavane F, Reynolds J, Sejnowski TJ (2018) Cortical travelling waves: mechanisms and computational principles. Nat Rev Neurosci 19(5):255–268
6. Uhlhaas PJ, Pipa G, Lima B, Melloni L, Neuenschwander S, Nikolić D et al (2009) Neural synchrony in cortical networks: history, concept and current status. Front Integr Neurosci 3:17
7. Bahramisharif A, van Gerven MA, Aarnoutse EJ, Mercier MR, Schwartz TH, Foxe JJ et al (2013) Propagating neocortical gamma bursts are coordinated by traveling alpha waves. J Neurosci 33(48):18849–18854
8. Dickey CW, Sargsyan A, Madsen JR, Eskandar EN, Cash SS, Halgren E (2021) Travelling spindles create necessary conditions for spike-timing-dependent plasticity in humans. Nat Commun 12(1):1027
9. Halgren M, Ulbert I, Bastuji H, Fabó D, Erőss L, Rey M et al (2019) The generation and propagation of the human alpha rhythm. Proc Natl Acad Sci U S A 116(47):23772–23782
10. Hangya B, Tihanyi BT, Entz L, Fabó D, Erőss L, Wittner L et al (2011) Complex propagation patterns characterize human cortical activity during slow-wave sleep. J Neurosci 31(24):8770–8779
11. Kleen JK, Chung JE, Sellers KK, Zhou J, Triplett M, Lee K et al (2021) Bidirectional propagation of low frequency oscillations over the human hippocampal surface. Nat Commun 12(1):2764
12. Muller L, Piantoni G, Koller D, Cash SS, Halgren E, Sejnowski TJ (2016) Rotating waves during human sleep spindles organize global patterns of activity that repeat precisely through the night. Elife 5
13. Stolk A, Brinkman L, Vansteensel MJ, Aarnoutse E, Leijten FS, Dijkerman CH et al (2019) Electrocorticographic dissociation of alpha and beta rhythmic activity in the human sensorimotor system. Elife 8
14. Zhang H, Jacobs J (2015) Traveling theta waves in the human hippocampus. J Neurosci 35(36):12477–12487
15. Zhang H, Watrous AJ, Patel A, Jacobs J (2018) Theta and alpha oscillations are traveling waves in the human neocortex. Neuron 98(6):1269–1281.e4
16. Besserve M, Lowe SC, Logothetis NK, Schölkopf B, Panzeri S (2015) Shifts of gamma phase across primary visual cortical sites reflect dynamic stimulus-modulated information transfer. PLoS Biol 13(9):e1002257
17. Davis ZW, Muller L, Martinez-Trujillo J, Sejnowski T, Reynolds JH (2020) Spontaneous travelling cortical waves gate perception in behaving primates. Nature 587(7834):432–436
18. Gabriel A, Eckhorn R (2003) A multi-channel correlation method detects traveling gamma-waves in monkey visual cortex. J Neurosci Methods 131(1–2):171–184
19. Muller L, Reynaud A, Chavane F, Destexhe A (2014) The stimulus-evoked population response in visual cortex of awake monkey is a propagating wave. Nat Commun 5:3675

20. Nauhaus I, Busse L, Carandini M, Ringach DL (2009) Stimulus contrast modulates functional connectivity in visual cortex. Nat Neurosci 12(1):70–76
21. Townsend RG, Solomon SS, Chen SC, Pietersen AN, Martin PR, Solomon SG et al (2015) Emergence of complex wave patterns in primate cerebral cortex. J Neurosci 35(11):4657–4662
22. Vinck M, Lima B, Womelsdorf T, Oostenveld R, Singer W, Neuenschwander S et al (2010) Gamma-phase shifting in awake monkey visual cortex. J Neurosci 30(4):1250–1257
23. Zanos TP, Mineault PJ, Nasiotis KT, Guitton D, Pack CC (2015) A sensorimotor role for traveling waves in primate visual cortex. Neuron 85(3):615–627
24. Balasubramanian K, Papadourakis V, Liang W, Takahashi K, Best MD, Suminski AJ et al (2020) Propagating motor cortical dynamics facilitate movement initiation. Neuron 106(3):526–536.e4
25. Denker M, Zehl L, Kilavik BE, Diesmann M, Brochier T, Riehle A et al (2018) LFP beta amplitude is linked to mesoscopic spatio-temporal phase patterns. Sci Rep 8(1):5200
26. Rubino D, Robbins KA, Hatsopoulos NG (2006) Propagating waves mediate information transfer in the motor cortex. Nat Neurosci 9(12):1549–1557
27. Rule ME, Vargas-Irwin C, Donoghue JP, Truccolo W (2018) Phase reorganization leads to transient β-LFP spatial wave patterns in motor cortex during steady-state movement preparation. J Neurophysiol 119(6):2212–2228
28. Takahashi K, Kim S, Coleman TP, Brown KA, Suminski AJ, Best MD et al (2015) Large-scale spatiotemporal spike patterning consistent with wave propagation in motor cortex. Nat Commun 6:7169
29. Bhattacharya S, Brincat SL, Lundqvist M, Miller EK (2022) Traveling waves in the prefrontal cortex during working memory. PLoS Comput Biol 18(1):e1009827
30. Zabeh E, Foley NC, Jacobs J, Gottlieb JP (2022) Traveling waves in the monkey frontoparietal network predict recent reward memory. bioRxiv
31. Liang Y, Song C, Liu M, Gong P, Zhou C, Knöpfel T (2021) Cortex-wide dynamics of intrinsic electrical activities: propagating waves and their interactions. J Neurosci 41(16):3665–3678
32. Stroh A, Adelsberger H, Groh A, Rühlmann C, Fischer S, Schierloh A et al (2013) Making waves: initiation and propagation of corticothalamic Ca2+ waves in vivo. Neuron 77(6):1136–1150
33. Xu W, Huang X, Takagaki K, Wu JY (2007) Compression and reflection of visually evoked cortical waves. Neuron 55(1):119–129
34. Agarwal G, Stevenson IH, Berényi A, Mizuseki K, Buzsáki G, Sommer FT (2014) Spatially distributed local fields in the hippocampus encode rat position. Science 344(6184):626–630
35. Patel J, Fujisawa S, Berényi A, Royer S, Buzsáki G (2012) Traveling theta waves along the entire septotemporal axis of the hippocampus. Neuron 75(3):410–417
36. Patel J, Schomburg EW, Berényi A, Fujisawa S, Buzsáki G (2013) Local generation and propagation of ripples along the septotemporal axis of the hippocampus. J Neurosci 33(43):17029–17041
37. Alamia A, VanRullen R (2019) Alpha oscillations and traveling waves: signatures of predictive coding? PLoS Biol 17(10):e3000487
38. Alexander DM, Jurica P, Trengove C, Nikolaev AR, Gepshtein S, Zvyagintsev M et al (2013) Traveling waves and trial averaging: the nature of single-trial and averaged brain responses in large-scale cortical signals. Neuroimage 73:95–112
39. Lozano-Soldevilla D, VanRullen R (2019) The Hidden spatial dimension of alpha: 10-Hz perceptual echoes propagate as periodic traveling waves in the human brain. Cell Rep 26(2):374–380.e4
40. Mahjoory K, Schoffelen JM, Keitel A, Gross J (2020) The frequency gradient of human resting-state brain oscillations follows cortical hierarchies. Elife **9**
41. Massimini M, Huber R, Ferrarelli F, Hill S, Tononi G (2004) The sleep slow oscillation as a traveling wave. J Neurosci 24(31):6862–6870
42. Zich C, Quinn AJ, Bonaiuto JJ, O'Neill G, Mardell LC, Ward NS et al (2022) Spatiotemporal organization of human sensorimotor beta burst activity. bioRxiv

43. Das A, Myers J, Mathura R, Shofty B, Metzger BA, Bijanki K et al (2022) Spontaneous neuronal oscillations in the human insula are hierarchically organized traveling waves. Elife 11
44. Mercier MR, Dubarry AS, Tadel F, Avanzini P, Axmacher N, Cellier D et al (2022) Advances in human intracranial electroencephalography research, guidelines and good practices. Neuroimage 119438
45. Parvizi J, Kastner S (2018) Promises and limitations of human intracranial electroencephalography. Nat Neurosci 21(4):474
46. Donoghue T, Haller M, Peterson EJ, Varma P, Sebastian P, Gao R et al (2020) Parameterizing neural power spectra into periodic and aperiodic components. Nat Neurosci 23(12):1655–1665
47. Watrous AJ, Miller J, Qasim SE, Fried I, Jacobs J (2018) Phase-tuned neuronal firing encodes human contextual representations for navigational goals. Elife 7
48. Solomon EA, Stein JM, Das S, Gorniak R, Sperling MR, Worrell G et al (2019) Dynamic theta networks in the human medial temporal lobe support episodic memory. Curr Biol 29(7):1100–1111.e4
49. Tarjan R (1972) Depth-first search and linear graph algorithms. Siam J Comput 1(2):146–160
50. Fisher NI (1993) Statistical analysis of circular data. Cambridge University Press
51. Mohan UR, Zhang H, Jacobs J (2022) The direction and timing of theta and alpha traveling waves modulate human memory processing. bioRxiv
52. Ermentrout GB, Kopell N (1984) Frequency plateaus in a chain of weakly coupled oscillators I. SIAM J Math Anal 15(2):215–237
53. Alexander DM, Nikolaev AR, Jurica P, Zvyagintsev M, Mathiak K, van Leeuwen C (2016) Global neuromagnetic cortical fields have non-zero velocity. PLoS ONE 11(3):e0148413
54. Hindriks R, van Putten M, Deco G (2014) Intra-cortical propagation of EEG alpha oscillations. Neuroimage 103:444–453
55. Lubenov EV, Siapas AG (2009) Hippocampal theta oscillations are travelling waves. Nature 459(7246):534–539
56. Canolty RT, Edwards E, Dalal SS, Soltani M, Nagarajan SS, Kirsch HE et al (2006) High gamma power is phase-locked to theta oscillations in human neocortex. Science 313(5793):1626–1628
57. Friston KJ (2019) Waves of prediction. PLoS Biol 17(10):e3000426
58. Badre D, D'Esposito M (2009) Is the rostro-caudal axis of the frontal lobe hierarchical? Nat Rev Neurosci 10(9):659–669
59. Strange BA, Witter MP, Lein ES, Moser EI (2014) Functional organization of the hippocampal longitudinal axis. Nat Rev Neurosci 15(10):655–669
60. Smith EH, Liou JY, Merricks EM, Davis T, Thomson K, Greger B et al (2022) Human interictal epileptiform discharges are bidirectional traveling waves echoing ictal discharges. Elife 11
61. Smith EH, Liou JY, Davis TS, Merricks EM, Kellis SS, Weiss SA et al (2016) The ictal wavefront is the spatiotemporal source of discharges during spontaneous human seizures. Nat Commun 7:11098
62. Frauscher B, von Ellenrieder N, Zelmann R, Doležalová I, Minotti L, Olivier A et al (2018) Atlas of the normal intracranial electroencephalogram: neurophysiological awake activity in different cortical areas. Brain 141(4):1130–1144
63. Bhattacharya S, Donoghue JA, Mahnke M, Brincat SL, Brown EN, Miller EK (2022) Propofol anesthesia alters cortical traveling waves. J Cogn Neurosci 1–13
64. Hamid AA, Frank MJ, Moore CI (2021) Wave-like dopamine dynamics as a mechanism for spatiotemporal credit assignment. Cell 184(10):2733–2749.e16
65. Matityahu L, Gilin N, Atmna Y, Tiroshi L, Wickens JR, Goldberg JA (2022) Mechanism of dopamine traveling waves in the striatum: theory and experiment. bioRxiv
66. Ermentrout GB, Kleinfeld D (2001) Traveling electrical waves in cortex: insights from phase dynamics and speculation on a computational role. Neuron 29(1):33–44
67. Townsend RG, Gong P (2018) Detection and analysis of spatiotemporal patterns in brain activity. PLoS Comput Biol 14(12):e1006643

Chapter 31
How Can I Investigate Perceptual and Cognitive Function Using Neural Frequency Tagging?

Simon Henin, Caspar M. Schwiedrzik, Nai Ding, and Lucia Melloni

Abstract In this chapter we will introduce the concept of neural frequency tagging (NFT), a versatile tool that can be used to explore how the brain processes, segments, and tracks specific cognitive processes via rhythmic neural responses. First, we explain how NFT can be used to investigate perceptual and cognitive processes. We explore critical experimental design considerations and how they can be exploited by NFT to answer specific questions in cognition using intracranial electroencephalography (iEEG). Next, we describe how NFT is calculated and, crucially, we explain how results can be interpreted in the context of human cognition and possible limitations of its use to explore certain brain processes. We end the chapter by addressing specific signal processing issues and potential pitfalls in its implementation and

S. Henin (✉) · L. Melloni (✉)
Department of Neurology, NYU Langone Health, 222 East 41st Street, 9th Floor, New York, NY 10017, USA
e-mail: simon.henin@nyulangone.org

L. Melloni
e-mail: lucia.melloni@ae.mpg.de

Comprehensive Epilepsy Center, NYU Langone Health, 223 East 34th Street, New York, NY 10016, USA

C. M. Schwiedrzik
Neural Circuits and Cognition Lab, European Neuroscience Institute Göttingen – A Joint Initiative of the University Medical Center Göttingen and the Max Planck Society, Grisebachstrasse 5, 37077 Göttingen, Germany

Perception and Plasticity Group, German Primate Center – Leibniz Institute for Primate Research, Kellnerweg 4, 37077 Göttingen, Germany

N. Ding
Key Laboratory for Biomedical Engineering of Ministry of Education, College of Biomedical Engineering and Instrument Sciences, Zhejiang University, Hangzhou 310027, China

Research Center for Advanced Artificial Intelligence Theory, Zhejiang Lab, Hangzhou 311121, China

L. Melloni
Neural Circuits, Consciousness and Cognition Research Group, Max Planck for Empirical Aesthetics, Grüneburgweg 14, 60322 Frankfurt am Main, Germany

N. Axmacher (ed.), *Intracranial EEG*, Studies in Neuroscience, Psychology and Behavioral Economics, https://doi.org/10.1007/978-3-031-20910-9_31

explore promising new avenues of cognitive research using NFT, such as development across the lifespan, and discuss possible future directions for which NFT would be an ideal tool for tackling difficult neuroscientific topics.

31.1 Introduction

As we interact and apprehend the sensory environment, perceptual information unfolds in a more-or-less continuous manner over time. Yet, our experience of the world, and our memories, consists of a series of coherent and punctuated sequences that have demarcated beginnings, middles and ends. A core problem in cognitive neuroscience has been to understand how and why the continuous flow of experience is partitioned in meaningful events for the mind; and how those units are acquired as a function of experience and learning. Speech perception offers a paradigmatic example: language comprehension depends on the identification and successful integration of information delivered at different rates, i.e., phonemes are combined into syllables, syllables into words, and words into phrases and sentences; all of which unfold in time. More generally, perception, cognition, and action all occur not in isolation, but in the form of sequences. Whether we walk or talk, the brain needs to be able to extract and to construct coherent sequences. Yet, due to the inherent complexity of analyzing time-varying signals, research has mostly focused on investigating neuronal responses time-locked to the presentation of a given stimulus in isolation. It is known, however, that rhythmic and periodic stimulation can elicit rhythmic neural responses (e.g., [1, 2]): a phenomenon known as neural entrainment, steady-state responses, or neural frequency tagging (NFT). Here, neural responses, measured non-invasively through electro- or magneto-encephalography (EEG or MEG), or invasively through intracranial encephalography (iEEG), are selectively enhanced as a function of the periodicity of the stimulation. This neural entrainment can then be quantified in the frequency domain as a peak in the spectrum, whose frequency responses enable inferences as to the events that are being tracked in a stimulus stream. This method can be used to track low-level properties in a sensory stream, e.g., rhythmic changes in intensity of a sound—a pure tone, referred to as the auditory steady state response (aSSR) [3–5]; or the contrast of a visual stimulus—a Gabor patch, referred to as the steady state visual evoked response (SSVEP) [6]. Recently, neural entrainment has been extended to investigate more abstract stimulus properties. Low-frequency neural entrainment has been observed for rhythms of musical beats [7], speech segmentation in adults [8–11], infants [12–14] and in developmental dyslexia [15]; and also for tracking of linguistic structures in adults [16]. As such, the use of neural frequency tagging has received much attention in recent years, both due to its theoretical implications for parsing and integration of information [17, 18] but also since it can be used as tool for online tracking of high-level cognition in non-verbalizing populations such as infants [19, 20], patients with disorders of consciousness [21, 22], and more generally as a tool to investigate perception and cognition across species [23, 24].

31.2 Neural Frequency Tagging in Intracranial Electrocorticography (iEEG) Experiments

The general principle of applying NFT to neural signals is that a rhythmic stimulus will produce endogenous neural activity that is synchronized to the external input. However, the power of NFT, as an analysis technique, only becomes fully realizable when the rhythmic stimulus is manipulated in a manner that contains categorical elements (e.g., elements distinguishable from the base stimulus presentation rate). When hypothesized categorical elements are presented at a specified frequency (or rate) within the contiguous presentation stream, this allows for the meaningful separation of category-specific responses to be separated from stimulus-specific responses (see Fig. 31.1a). By exploiting this neural entrainment to different categorical elements of an input stream, one can deduce the possible mechanistic processes in the brain that subserve them. Moreover, by exploiting the superior localizability of iEEG signals to specific cortical and subcortical brain regions, we can deduce where (and sometimes when) in the brain these processes exist.

In this way, NFT is ideal for identifying and localizing task responsive electrodes in the brain (see Fig. 31.1b). For example, Jonas et al. [25] embedded images of faces into a visual stream of images containing multiple object categories. Images were presented at fixed rate of 6 Hz, with faces presented into the stream on every 5th presentation (e.g. $6/5 = 1.2$ Hz). Measuring iEEG from a large cohort of patients, they used NFT to dissociate between electrodes responsive to the visual stream and those responsive at the frequency of the face presentation (e.g. face-responsive electrodes). In such a way, they were able to localize face-selective regions of the ventral occipito-temporal cortex. Further, the strength of NFT was used to lateralize the face-selective responses, showing dominant responses to faces in right fusiform gyrus. Using similar techniques, NFT has been used to localize and dissociate the roles of specific brain regions in various types of visual categorical processing (e.g. faces vs. words; [26, 27]). Similarly, NFT can be used to scout for (or identify) responsive electrodes and carry them forward to further analyses (e.g. [11]; Fig. 31.1c).

An even more powerful use of NFT analysis can be realized when combined with stimulus paradigms aimed at uncovering mental representations/processes. Given a properly thought-out stimulus paradigm (e.g., a continuous stimulus stream with a hidden/covert structure at defined frequencies), NFT can be applied to neural signals (iEEG) to explore the brain's ability to uncover these hidden structures. For example, Ding et al. [16] devised an experiment in which the different parts of grammatical speech (e.g., words, phrases, whole sentences) could be decomposed via NFT. In that study, each word lasted 250 ms, while a phrase was composed of 2 words and therefore lasted 500 ms, and each sentence was composed of 4 words lasting 1 s. As such, the rate of words, phrases and sentences could be tagged at 4, 2, and 1 Hz, respectively. Using this technique, they were able to provide evidence for speech processing at multiple timescales corresponding to these linguistic features. Further, using iEEG they were able to localize these responses to distributed regions, most of which (e.g., Superior temporal gyrus, STG; Inferior frontal gyrus, IFG) have been

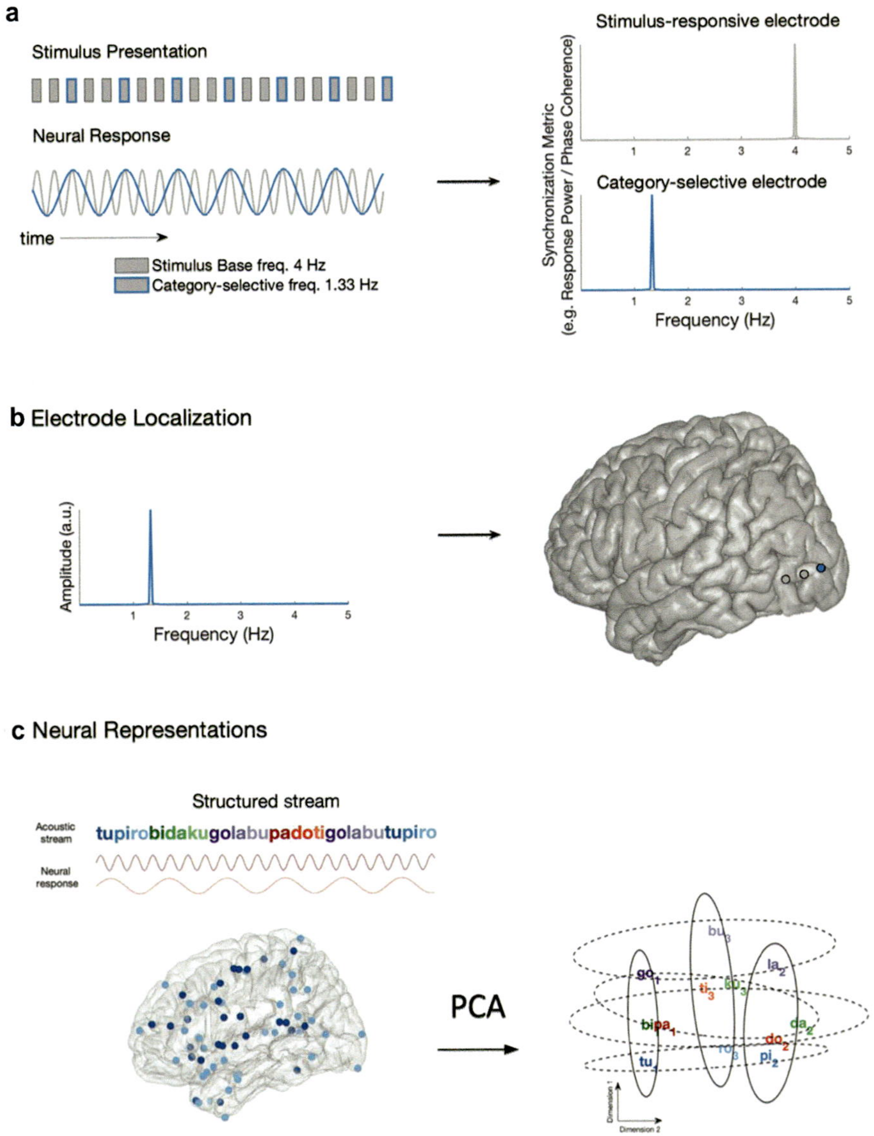

Fig. 31.1 Design and analysis of NFT paradigms using iEEG. **a** Abstracted schematic design of an NFT experimental paradigm (left) and subsequent NFT measures (right), **b** Example localization of "category-selective" electrodes, **c** Using electrodes identified via NFT to perform subsequent analyses to explore neural representations, adapted from Henin et al. [11]

shown to be critical for speech processing. In another experiment, Henin et al. [11] used NFT to identify electrodes that tracked different low- and high-level features of sequences using a statistical learning paradigm in both the auditory and visual domains. Specifically, in the auditory condition, they presented sequences of syllables to patients, which either followed a random sequence or a structured sequence containing four hidden words composed of 3 syllables that subjects were expected to implicitly learn after a brief exposure. As such, syllables (250 ms) and words (3 * 250 ms) could be tagged at different frequencies, i.e., 4 and 1.33 Hz respectively (see Fig. 31.1c). Taking advantage of the sensitivity of NFT, they showed that electrodes exhibiting different NFT response profiles (e.g., electrodes responding to words vs. words and syllables) mapped to different hierarchical processing strategies thought to be involved in statistical learning of novel sequence patterns [28].

The use of NFT in iEEG experiments enables one to uncover the mechanisms supporting the processing of complex stimuli and where these computations take place in the brain. Of particular note, the use of NFT paradigms often requires only short presentations of stimuli. For example, localization of "responsive" electrodes using NFT can be achieved in only minutes [11, 25] as compared with more traditional event-related experimental designs, which typically require many repetitions over longer experimental durations. The speed with which NFT paradigms can be run is advantageous, especially when measuring iEEG from patients in clinical settings, whose ability and attention during longer tasks is often limited [29]. Critically, NFT is exquisitely sensitive to track online learning across sensory modalities as demonstrated in Henin et al. (2021). The latter offers an added advantage over many other analysis techniques, especially when used in iEEG studies, as it enables tracking the temporal tuning of learning across different cortical areas.

31.3 How to Compute NFT

Broadly speaking, NFT analysis is the technique of uncovering rhythmic components of a neural signal that map to stimulus-related features. In its most simplistic form, this can mean analyzing the Fourier spectrum of a signal, with the resultant peaks in the spectrum analyzed considering the underlying frequencies of interest (e.g., frequencies representing different stimulus-related activations/representations).

In principle, windowing the signal for analyses is not necessary, and large analysis windows improve the frequency resolution. However, in practice, including replication (e.g., multiple trials) can help increase the signal-to-noise ratio of the technique. Using multiple trials, NFT can then be calculated using any measure that captures synchronized neural activity at different rates (i.e., in the frequency domain). However, this implies that the stimuli (and relevant rhythmic representations) should be resolvable in the frequency domain (e.g., representations should be at integer multiples of the frequency resolution of the fast Fourier transform). Then, assuming the neural signal is represented in the frequency domain by $Y(f) = \sum_f A_f e^{j\theta_f}$, NFT at frequency f can be computed across all trials, i = 1, 2 … K, using one of several different measures, for example,

$$\text{Response Power} \quad P_r(f) = \frac{1}{K}\sum_{i=1}^{K} \left|A_{f,i}\right|^2 \tag{31.1}$$

$$\text{Phase Coherence} \quad R^2(f) = \left(\frac{1}{K}\sum_{i=1}^{K} \cos\theta_{f,i}\right)^2 + \left(\frac{1}{K}\sum_{i=1}^{K} \sin\theta_{f,i}\right)^2 \tag{31.2}$$

While Eqs. (31.1) and (31.2) represent examples of power and phase-locked synchronization measures, several other measures exist for computing rhythmic neural interactions (cf. [30–32]) (see also Chap. 32), and are not reviewed here, however, their implementation in terms of NFT analysis is much the same. Some additional considerations when computing NFT are discussed at the end of this chapter.

31.4 How to Interpret NFT

There are two main hypotheses regarding the neural mechanisms underlying the neural response tracking stimulus rhythms. The first hypothesis is that spontaneous neural oscillations in the brain are synchronized/entrained to external rhythms through phase resetting [33, 34]. This hypothesis is appealing since it links the neural response to external rhythms with spontaneous neural activity. However, the properties of various spontaneous oscillations are not well understood and it is possible that their properties change when they entrain to an external stimulus. The second hypothesis is that the rhythmic response is a superposition of event-related potentials (ERPs) evoked by the external rhythmic stimulus [35]. This hypothesis links the neural response to external rhythms with ERPs. However, since the brain is a nonlinear and adaptive system, the ERP often depends on the stimulus and the brain's internal state. Furthermore, since neither hypothesis can be easily characterized by a computational model, it is challenging to quantify which hypothesis can better describe the neural response to external rhythms.

The ongoing debate between two hypotheses, however, mostly concerns how the neural response characterized by the NFT is linked to other neural phenomena. They do not cast concern on the interpretation that the NFT can characterize neural encoding of the frequency tagged stimulus feature. The debate, however, has motivated discussions about NFT analysis methods. For example, if the neural response is generated by resetting the phase of ongoing neural oscillations (without an increase in power of the oscillations), this will lead to an increase in the inter-trial response phase coherence (Eq. 31.2) but not the response power in single trials (Eq. 31.1). Single trial power is often characterized by averaging the frequency-domain response power across trials, which is distinct from the evoked power (i.e., power of the time-domain response averaged across trials). By contrast, if the stimulus evokes additional neural responses that are independent of ongoing background oscillations, the response power in single trials will increase. Based on these intuitions, early studies have attempted to distinguish the two hypotheses by analyzing the response power and indeed demonstrated that phase coherence is much more sensitive to the stimulus than the response power [36]. Nevertheless, simulation results show that regardless of the generation mechanism (e.g., phase-resetting or independent responses), NFT is much more reliably characterized by response phase coherence than single-trial power [37], as single-trial power is much more sensitive to the level of background activity (e.g. SNR of the response to background activity).

31.4.1 Advantage of Frequency-Domain Analyses

Neural signals are signals in time. Why bother to analyze them in the frequency domain? This question is especially obvious under the hypothesis that NFT is only analyzing superimposed ERPs. When the rhythmic stimulus consists of a sequence of discrete events, e.g., tone pips or pictures, the ERP to each event can be easily extracted. When comparing the ERP in two conditions, how to appropriately choose the time window is a question. If the comparison is done for each time point, this will involve a large number of statistical comparisons and how to correct the effect of multiple comparisons becomes an issue. If a window is chosen based on prior knowledge, it leaves possible that the effect may fall out of the window or the effect may have both negative and positive polarities and cancel out in the average. Frequency-domain analysis does not suffer from these problems: It only requires testing whether the neural response at the target frequency is significantly stronger than a baseline or appropriately chosen comparison condition.

In addition to analyzing the response at the target frequency, i.e., f_0, some studies have used NFT to investigate the responses at frequencies that are harmonically related to the target frequency, e.g., $2f_0$, $3f_0$, and $4f_0$ [35, 38]. However, typically only a very small number of harmonic frequencies are observed and are not well understood from a phenomenological perspective. Harmonics in NFT spectrum may arise from nonlinearities inherent in the brain responses (e.g. non-sinusoidal responses) or from the stimulus presentation frequencies themselves (e.g. if the neural response to the stimulus decays before the next presentation cycle, this can produce NFT with broad spectral components, see [35]), and exactly how to interpret these harmonics is still an ongoing debate [38]. In addition, some studies frequency tag multiple features/stimuli with multiple frequencies and these studies sometimes also concern the intermodulation responses, which often reflect the interactions between the neural encoding of different features/stimuli. For example, if two features are separately tagged at f_1 and f_2, intermodulation responses include responses at, e.g., $nf_1 \pm mf_2$, where n and m are integers. In these conditions, n and m are typically small numbers and the interactions are often not straightforward to characterize in the time domain [39–41].

31.5 Challenges and Pitfalls

When computing NFT, several issues should be considered. When segmenting trials into epochs, care should be taken regarding how data epochs overlap. In principle, data epochs should contain independent data segments. However, several studies have used overlapping data epochs to increase the temporal resolution of their analysis. When overlapping epochs are used, this will lead to artifactual estimates of entrainment at the frequency of the overlap (e.g., if 10-element segments of data are used with 9-element overlap, ITC estimates will show entrainment at the 1-element overlap frequency, Fig. 31.2a). If the increased temporal resolution afforded by overlapping window analyses is nevertheless essential, it is important to consider the effect of the overlap on the NFT spectrum and include well-designed control conditions (e.g., streams with no hidden structure) to ensure that observed NFT effects are not driven by artifacts in the analysis design [12].

Furthermore, the length of the analysis interval, whether windowed or not, should be chosen such that it exactly comprises an integer number of cycles. Otherwise, Fourier analysis will lead to additional peaks in the spectrum known as "overspill" artifacts (Fig. 31.2b). However, as neural oscillations are rarely pure sinusoids, overspill artifacts often occur, and again, careful controls should be considered. Similarly, care should be taken to filter out trends in the data, since their frequency decomposition can lead to a "sawtooth" pattern of peaks in the spectrum that is superimposed onto the NFT frequency peak (Fig. 31.2c) (see [42]). These potential pitfalls do not preclude the use of NFT as a valid analysis technique, but as in any experimental design, point to the need for careful consideration of potential artifacts in the interpretation of the results.

To assess whether a peak in the NFT spectrum is statistically significant, several statistical approaches have been proposed [37, 43]: the first is a comparison between the experimental and a control condition at the frequency of interest. For this, the appropriate parametric or non-parametric tests can be used. This approach has the advantage that the above-mentioned artificial peaks in the power spectrum that can arise from the use of overlapping analysis windows are automatically taken care of. However, this approach does not assess the frequency specificity of the NFT peak and often cannot be performed if no control condition was recorded. Another possibility is to compare the power of the peak frequency with the average of two (or more) neighboring frequencies. This determines the "peakiness" in the spectrum and simultaneously takes care of the above-mentioned sawtooth pattern in the data that might obscure true peaks. When using phase coherence, statistical assessment is even more straightforward, with the principled assumption that non-significant responses are uniformly distributed in the phase (independent of frequency and/or experimental conditions). Therefore, a simple phase shuffling test (permutation test)

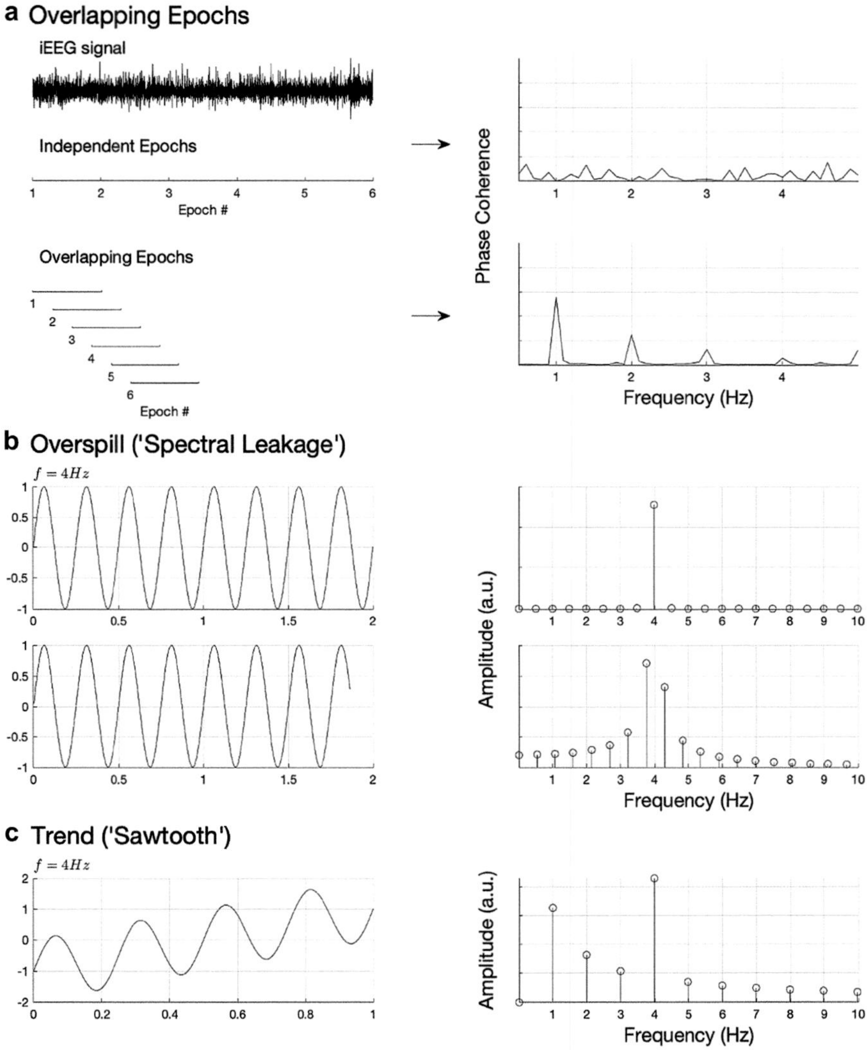

Fig. 31.2 Example of different types of artifacts to consider when using NFT. **a** Effect of using overlapping epochs. In this case, the iEEG signal considered has no inherent coherence. When analyzed using independent epochs, no significant phase coherence emerges (top). However, when using overlapping epochs (T = 1 Hz), artifacts are observed in the NFT spectrum at the frequency of overlap (bottom). **b** Overspill or "spectral Leakage" is caused when non-integer multiples of the target frequency are used in the FFT analysis window. **c** Low-frequency "Trend" artifacts are observed when slow EEG drifts are not removed from the data

or Rayleigh test [44] may be used to assess significance. Of course, all approaches can also be combined, e.g., through an appropriate conjunction test (e.g., [45]) or with a frequency × condition interaction in a repeated measures ANOVA.

31.6 Promises of NFT and Future Directions

Since its popularization in the 1970's in the domains of noninvasive electroencephalography and electroretinography, frequency tagging has seen continuous methodological improvement and topical extension. The popularity of frequency tagging stems, on the one hand, from the above-mentioned advantages in terms of signal to noise ratio and the high efficiency of the technique in terms of stimuli per time unit that vastly improves upon event-related potentials and similar measures that require sizeable baselines between trials. On the other hand, it has become clear that frequency tagging lends itself particularly well to the study of cognitive and perceptual phenomena that unfold in time. Here, the stimuli per time unit advantage of frequency tagging becomes even more clear: for example, in the language domain, neural data can be simultaneously acquired at the level of syllables, words, and sentences. Furthermore, frequency tagging has higher ecological validity than trial-based measures if temporally unfolding phenomena are studied.

While frequency tagging has traditionally been employed mostly in the domain of non-invasive electrophysiology, it is now also adopted for non-electrophysiological measures such as video-based pupil diameter tracking (e.g., [46]), and, as we laid out above, has seen tremendous success in iEEG research.

In particular, over the last years, iEEG-based NFT has proven to be suitable to study complex cognitive phenomena such as speech perception [16] and sequence learning [11] with exquisite spatio-temporal resolution. Beyond that, the power of the NFT approach, combined with appropriate paradigms, now makes it possible to investigate cognition even in populations that are otherwise hard to study, e.g., because they lack the ability to understand instructions or to give verbal reports, such as infants, certain patients, and other species. We thus anticipate that the near exponential growth of publication using NFT will continue, leading to exciting new insights into the brain and cognition.

References

1. Picton TW, John MS, Dimitrijevic A, Purcell D (2003) Human auditory steady-state responses. Int J Audiol 42(4):177–219. https://doi.org/10.3109/14992020309101316
2. Regan D (1977) Steady-state evoked potentials. J Opt Soc Am 67(11):1475–1489. https://doi.org/10.1364/josa.67.001475
3. Galambos R, Makeig S, Talmachoff PJ (1981) A 40-Hz auditory potential recorded from the human scalp. Proc Natl Acad Sci USA 78(4):2643–2647. https://doi.org/10.1073/pnas.78.4.2643

4. Ross B, Borgmann C, Draganova R, Roberts LE, Pantev C (2000) A high-precision magnetoencephalographic study of human auditory steady-state responses to amplitude-modulated tones. J Acoust Soc Am 108(2):679–691. https://doi.org/10.1121/1.429600
5. Wang Y, Ding N, Ahmar N, Xiang J, Poeppel D, Simon JZ (2012) Sensitivity to temporal modulation rate and spectral bandwidth in the human auditory system: MEG evidence. J Neurophysiol 107(8):2033–2041. https://doi.org/10.1152/jn.00310.2011
6. Norcia AM, Appelbaum LG, Ales JM, Cottereau BR, Rossion B (2015) The steady-state visual evoked potential in vision research: a review. J Vis 15(6):4. https://doi.org/10.1167/15.6.4
7. Nozaradan S, Peretz I, Missal M, Mouraux A (2011) Tagging the neuronal entrainment to beat and meter. J Neurosci 31(28):10234–10240. https://doi.org/10.1523/JNEUROSCI.0411-11.2011
8. Batterink LJ, Paller KA (2017) Online neural monitoring of statistical learning. Cortex 90:31–45. https://doi.org/10.1016/j.cortex.2017.02.004
9. Batterink LJ, Paller KA (2019) Statistical learning of speech regularities can occur outside the focus of attention. Cortex 115:56–71. https://doi.org/10.1016/j.cortex.2019.01.013
10. Buiatti M, Pena M, Dehaene-Lambertz G (2009) Investigating the neural correlates of continuous speech computation with frequency-tagged neuroelectric responses. Neuroimage 44(2):509–519. https://doi.org/10.1016/j.neuroimage.2008.09.015
11. Henin S, Turk-Browne NB, Friedman D, Liu A, Dugan P, Flinker A et al (2021) Learning hierarchical sequence representations across human cortex and hippocampus. Sci Adv 7(8). https://doi.org/10.1126/sciadv.abc4530
12. Benjamin L, Dehaene-Lambertz G, Fló A (2021) Remarks on the analysis of steady-state responses: spurious artifacts introduced by overlapping epochs. Cortex 142:370–378. https://doi.org/10.1016/j.cortex.2021.05.023PMID-34311971
13. Choi D, Batterink LJ, Black AK, Paller KA, Werker JF (2020) Preverbal infants discover statistical word patterns at similar rates as adults: evidence from neural entrainment. Psychol Sci 31(9):1161–1173. https://doi.org/10.1177/0956797620933237
14. Kabdebon C, Pena M, Buiatti M, Dehaene-Lambertz G (2015) Electrophysiological evidence of statistical learning of long-distance dependencies in 8-month-old preterm and full-term infants. Brain Lang 148:25–36. https://doi.org/10.1016/j.bandl.2015.03.005
15. Zhang M, Riecke L, Bonte M (2021) Neurophysiological tracking of speech-structure learning in typical and dyslexic readers. Neuropsychologia 158:107889. https://doi.org/10.1016/j.neuropsychologia.2021.107889
16. Ding N, Melloni L, Zhang H, Tian X, Poeppel D (2016) Cortical tracking of hierarchical linguistic structures in connected speech. Nat Neurosci 19(1):158–164. https://doi.org/10.1038/nn.4186
17. Buzsáki G (2006) Rhythms of the brain. Oxford University Press, Oxford, New York
18. Giraud AL, Poeppel D (2012) Cortical oscillations and speech processing: emerging computational principles and operations. Nat Neurosci 15(4):511–517. https://doi.org/10.1038/nn.3063
19. de Heering A, Rossion B (2015) Rapid categorization of natural face images in the infant right hemisphere. Elife 4:e06564. https://doi.org/10.7554/eLife.06564
20. Peykarjou S, Hoehl S, Pauen S, Rossion B (2017) Rapid categorization of human and ape faces in 9-month-old infants revealed by fast periodic visual stimulation. Sci Rep 7(1):12526. https://doi.org/10.1038/s41598-017-12760-2
21. Sokoliuk R, Degano G, Banellis L, Melloni L, Hayton T, Sturman S et al (2021) Covert speech comprehension predicts recovery from acute unresponsive states. Ann Neurol 89(4):646–656. https://doi.org/10.1002/ana.25995
22. Xu C, Gao J, Gao J, Li L, He F, Yu J et al (2022) Statistical learning in patients in the minimally conscious state. Cereb Cortex https://doi.org/10.1093/cercor/bhac222; Xu C, Gao J, Gao J, Li L, He F, Yu J et al (2022) Statistical learning in patients in the minimally conscious state. medRxiv 2022:2022.01.04.22268656. https://doi.org/10.1101/2022.01.04.22268656
23. Das A, Ray S (2021) Effect of cross-orientation normalization on different neural measures in macaque primary visual cortex. Cereb Cortex Commun 2(1):tgab009. https://doi.org/10.1093/texcom/tgab009

24. Rager G, Singer W (1998) The response of cat visual cortex to flicker stimuli of variable frequency. Eur J Neurosci 10(5):1856–1877. https://doi.org/10.1046/j.1460-9568.1998.00197.x
25. Jonas J, Jacques C, Liu-Shuang J, Brissart H, Colnat-Coulbois S, Maillard L et al (2016) A face-selective ventral occipito-temporal map of the human brain with intracerebral potentials. Proc Natl Acad Sci 113(28):E4088–E4097. https://doi.org/10.1073/pnas.1522033113PMID-27354526
26. Hagen S, Jacques C, Maillard L, Colnat-Coulbois S, Rossion B, Jonas J (2020) Spatially dissociated intracerebral maps for face- and house-selective activity in the human ventral occipito-temporal cortex. Cereb Cortex 30(7):4026–4043. https://doi.org/10.1093/cercor/bhaa022
27. Hagen S, Lochy A, Jacques C, Maillard L, Colnat-Coulbois S, Jonas J et al (2021) Dissociated face- and word-selective intracerebral responses in the human ventral occipito-temporal cortex. Brain Struct Funct 226(9):3031–3049. https://doi.org/10.1007/s00429-021-02350-4
28. Dehaene S, Meyniel F, Wacongne C, Wang L, Pallier C (2015) The neural representation of sequences: from transition probabilities to algebraic patterns and linguistic trees. Neuron 88(1):2–19. https://doi.org/10.1016/j.neuron.2015.09.019
29. Parvizi J, Kastner S (2018) Promises and limitations of human intracranial electroencephalography. Nat Neurosci 21(4):474–483. https://doi.org/10.1038/s41593-018-0108-2
30. Lachaux JP, Rodriguez E, Martinerie J, Varela FJ (1999) Measuring phase synchrony in brain signals. Hum Brain Mapp 8(4):194–208. https://doi.org/10.1002/(sici)1097-0193(1999)8:4%3c194::aid-hbm4%3e3.0.co;2-c
31. Mitra PP, Pesaran B (1999) Analysis of dynamic brain imaging data. Biophys J 76(2):691–708. https://doi.org/10.1016/S0006-3495(99)77236-X
32. Vinck M, van Wingerden M, Womelsdorf T, Fries P, Pennartz CM (2010) The pairwise phase consistency: a bias-free measure of rhythmic neuronal synchronization. Neuroimage 51(1):112–122. https://doi.org/10.1016/j.neuroimage.2010.01.073
33. Schroeder CE, Lakatos P (2009) Low-frequency neuronal oscillations as instruments of sensory selection. Trends Neurosci 32(1):9–18. https://doi.org/10.1016/j.tins.2008.09.012
34. Doelling KB, Assaneo MF, Bevilacqua D, Pesaran B, Poeppel D (2019) An oscillator model better predicts cortical entrainment to music. Proc Natl Acad Sci USA 116(20):10113–10121. https://doi.org/10.1073/pnas.1816414116
35. Zhou H, Melloni L, Poeppel D, Ding N (2016) Interpretations of frequency domain analyses of neural entrainment: periodicity, fundamental frequency, and harmonics. Front Hum Neurosci 10:274. https://doi.org/10.3389/fnhum.2016.00274
36. Luo H, Poeppel D (2007) Phase patterns of neuronal responses reliably discriminate speech in human auditory cortex. Neuron 54(6):1001–1010. https://doi.org/10.1016/j.neuron.2007.06.004
37. Ding N, Simon JZ (2013) Power and phase properties of oscillatory neural responses in the presence of background activity. J Comput Neurosci 34(2):337–343. https://doi.org/10.1007/s10827-012-0424-6
38. Retter TL, Rossion B, Schiltz C (2021) Harmonic amplitude summation for frequency-tagging analysis. J Cogn Neurosci 33(11):2372–2393. https://doi.org/10.1162/jocn_a_01763
39. Alp N, Kogo N, Van Belle G, Wagemans J, Rossion B (2016) Frequency tagging yields an objective neural signature of Gestalt formation. Brain Cogn 104:15–24. https://doi.org/10.1016/j.bandc.2016.01.008
40. Drijvers L, Jensen O, Spaak E (2021) Rapid invisible frequency tagging reveals nonlinear integration of auditory and visual information. Hum Brain Mapp 42(4):1138–1152. https://doi.org/10.1002/hbm.25282
41. Vergeer M, Kogo N, Nikolaev AR, Alp N, Loozen V, Schraepen B et al (2018) EEG frequency tagging reveals higher order intermodulation components as neural markers of learned holistic shape representations. Vis Res 152:91–100. https://doi.org/10.1016/j.visres.2018.01.007
42. Bach M, Meigen T (1999) Do's and don'ts in Fourier analysis of steady-state potentials. Doc Ophthalmol 99(1):69–82. https://doi.org/10.1023/a:1002648202420

43. Meigen T, Bach M (1999) On the statistical significance of electrophysiological steady-state responses. Doc Ophthalmol 98(3):207–232. https://doi.org/10.1023/a:1002097208337
44. Fisher NI (1993) Statistical analysis of circular data. Cambridge University Press, Cambridge
45. Nichols T, Brett M, Andersson J, Wager T, Poline JB (2005) Valid conjunction inference with the minimum statistic. Neuroimage 25(3):653–660. https://doi.org/10.1016/j.neuroimage.2004.12.005
46. Schwiedrzik CM, Sudmann SS (2020) Pupil diameter tracks statistical structure in the environment to increase visual sensitivity. J Neurosci 40(23):4565–4575. https://doi.org/10.1523/JNEUROSCI.0216-20.2020

Chapter 32
How Can I Analyze Connectivity in iEEG Data?

Ethan A. Solomon

Abstract Understanding how brain regions influence one another is foundational to modern neuroscience, which places a heavy emphasis on how behavior emerges from networks of inter-regional activity. Extracting, analyzing, and interpreting these brain networks from intracranial EEG remains an ill-defined process, with little by way of standardized protocols or guidance on navigating a vast suite of analytic tools. This chapter begins by reviewing methods for computing iEEG-based connectivity, including common spectral approaches such as phase locking, phase consistency, and coherence—as well as popular adaptations to account for volume conduction and several alternative metrics. Statistical frameworks for analyzing these data are presented in a cognitive neuroscience context, with a focus on how variability in connectivity relates to human cognition. This chapter also discusses challenges in the interpretation of iEEG connectivity, particularly in light of current debates over the causal-versus-correlative nature of functional connectivity. As intracranial data is becoming increasingly widespread and accessible, the chapter closes with a note on the need to standardize analytic approaches to iEEG-based connectivity.

32.1 What is Connectivity, and Why Study It?

Since the structure of neurons was first described, it has never been controversial that nervous systems are communication networks—signals are propagated from sense organs to the brain, from region-to-region within the brain, and finally to muscles that generate behavior. It is natural that cognitive neuroscientists, particularly those who rely on iEEG, want to understand how signal propagation within the brain relates to thought and behavior. In this chapter, we broadly define "functional connectivity" as a measure of correlated activity between two areas of the brain. This definition is deliberately vague and avoids invoking concepts of communication, causality, dependence, or "flow" [1].

E. A. Solomon (✉)
Stanford University, Stanford, CA, USA
e-mail: esolom@stanford.edu

N. Axmacher (ed.), *Intracranial EEG*, Studies in Neuroscience, Psychology and Behavioral Economics, https://doi.org/10.1007/978-3-031-20910-9_32

To be sure, it is plausible that correlated neural activity reflects any (or all) of these concepts, but the methods discussed here are insufficient to make strong claims about anything besides correlation. For instance, if region A and region B exhibit correlated activity during a successful task state, this could reflect the influence of axonal projections from region A to region B, projections via a relay and transformation in region C, independent drive from region C, or independent generators of similar activity patterns in both regions in response to external inputs (Fig. 32.1) [2]. Statistical methods to tease out some of these possibilities exist, but rely on strong assumptions about the brain and the meaning of our electrical readouts of brain function. Alternative approaches, discussed elsewhere in this text, would involve applying exogenous, targeted perturbations and observing causal responses throughout the brain (see Chap. 40).

But correlations are not useless. Inter-regional correlations in brain activity can be useful guides towards developing hypothesis about causality in brain function, and emerging data indeed confirms that measures of functional connectivity—that is, correlations between brain signals—are predictive of how exogenous stimulation propagates through the brain. Moreover, correlations provide valuable insight into the groupings of brain regions that may share common activity patterns, representations, or functional roles in the generation of complex behaviors—even if the mechanism

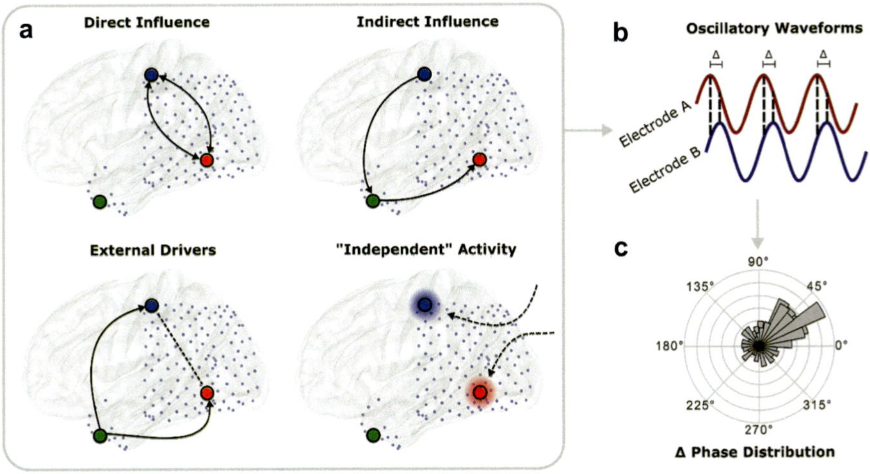

Fig. 32.1 Different underlying dynamics can generate similar patterns of functional connectivity. **a** "Functional connectivity" is deliberately defined in this chapter as an indicator of correlated iEEG activity between brain regions, not as an indicator of causality or information transfer. This is because most methods for calculating functional connectivity cannot distinguish between (1) direct influence between two regions via axonal projections, (2) indirect influence through a third, potentially unrecorded region, (3) an unrecorded region independently driving activity at two sensors, or (4) external inputs which happen to yield similar activity patterns despite no direct influence between the two. **b** In each of these cases, phase-lagged oscillations could emerge in two recorded regions, yielding **c** a tight distribution of phase lags across time (or trials), which is interpreted as positive "functional connectivity"

by which activity becomes synchronized is unknown. It is on this basis that broad networks with distinct roles in cognition have been identified, and there is growing evidence that understanding these networks has diagnostic or therapeutic utility.

Intracranial EEG presents unique challenges in the study of functional connectivity. Measures of functional connectivity are arguably more (though incompletely) protocolized and validated in the functional MRI community, which has long leveraged the relative consistency of scans across subjects to identify functional connections which relate to behavior. No such luck in the iEEG world—every subject has a different sampling of brain regions, often with different recording hardware, collected on different systems. The interpretation of frontal–temporal connectivity, for instance, could differ dramatically between a subject with an electrode in the anterior aspect of the middle temporal gyrus versus another with an implant 1 cm posterior. The high temporal resolution and concomitant frequency decomposition of iEEG signals is a powerful advantage over fMRI, but also opens a complicated world of variable interpretation: which set of frequencies "carry" functional connectivity, and could those frequencies change from moment-to-moment? Do signals at one frequency interact with neural responses in another? These questions—and many others—are too expansive and unexplored to answer in the course of this chapter, but are an exciting focus of much ongoing work [3].

32.2 IEEG-Based Connectivity Metrics

There is no single "right way" to measure functional connectivity, especially using iEEG. However, there are a number of metrics which are used more commonly in the literature. These metrics can be broken down into two broad categories, based on whether or not a correlative measure is directly dependent on the phase consistency between two iEEG signals. Because of their widespread use, we will focus primarily on phase-based measures of iEEG connectivity, but will further consider alternative metrics which do not directly rely on the computation of phase (see *Non-phase metrics of iEEG connectivity*).

32.2.1 Phase-Based Measures of Functional Connectivity

As described intensively in this text, iEEG signals can be easily broken down into component parts at distinct frequencies, enabling cognitive scientists to study the relationship between the frequency of neural activity, cognition, and downstream behavior (see also Chap. 49). This spectral decomposition of a signal is fundamental to EEG/iEEG studies and is most commonly employed to examine how spectral power—reflecting the amplitude of a signal—relates to a behavioral variable of interest. Analogously, functional connectivity can be computed for specific frequencies or frequency ranges ("bands"), reflecting how the *frequency-specific component* of iEEG signals correlates between two electrodes. Phase—or the measure of position through the course of a waveform—is useful for measuring inter-regional connectivity because it essentially indexes the configuration of a neural oscillation over time, in units that are standardized across electrodes (see *Interpreting iEEG connectivity* for further discussion).

Power and phase are interrelated components of EEG/iEEG signals. Power is a complex summation of neural activity that likely reflects different physiological processes depending on the frequency band of interest, but broadly derives from the spatially-summed postsynaptic potentials generated by dendritic inputs. High temporal correlation of such inputs across a wider spatial extent of neurons drives higher-amplitude oscillations and therefore higher power. Coarsely, greater power indicates greater synaptic input activity in a region, but note that other types of cellular activity—including action potentials—can also contribute to iEEG power, especially at higher frequencies [3]. The phase of these oscillations is an indication of whether an oscillation is at a trough, peak, or somewhere in-between, reflecting periods of greater or lesser synaptic and cellular activity. Phase is thought to play several roles, including the coordination of spiking activity [4], the modulation of power in other frequency bands [5], and the synchronization of oscillations between different brain areas [6, 7]. Phase can be measured independent of power, but as the power of an oscillation approaches zero, its measured phase will be increasingly unreliable.

Connectivity metrics will often be derived from the difference in phase between neural oscillations in two different electrodes. If the phase difference between two electrodes is highly consistent over trials, or time, those electrodes have high functional connectivity (Fig. 32.2). Conversely, if the phase difference between two oscillations is highly variable or inconsistent, those electrodes are said to have low functional connectivity in that frequency band. Critically, functional connectivity should not exist in a vacuum—*any* consistency in phase differences can be interpreted as positive connectivity, so should always be tested parametrically, compared against a control condition, or compared against a null distribution (see *Statistical frameworks for analyzing iEEG connectivity*).

How do we quantify the consistency of phase differences between two electrodes? In this section, we will review some common metrics for summarizing the consistency of distributions of phase differences (across time points or trials) between two iEEG sensors. These metrics have been developed to account for several factors that may alter the interpretation of "connectivity." The most prominent of these factors are volume conduction, baseline power of neural oscillations, and statistical bias, each reviewed here.

32.2.2 Phase-Locking Value

The phase-locking value (PLV) is one of the most straightforward and commonly-used metrics for computing inter-areal functional connectivity in iEEG data [8]. The metric reflects the consistency of phase differences between two oscillations, at a

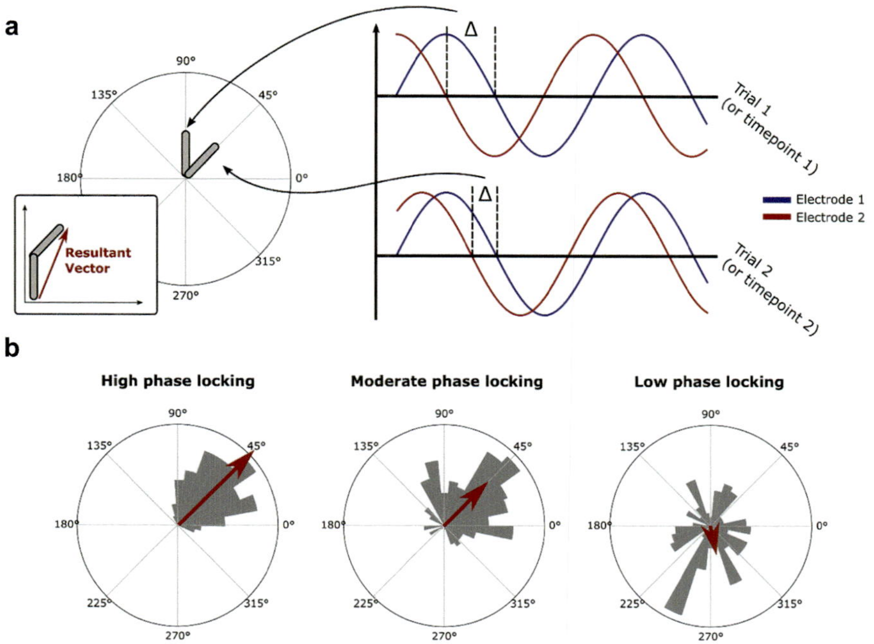

Fig. 32.2 Computation and interpretation of the phase-locking value (PLV). **a** PLV is estimated by first taking the phase difference between two oscillations from two separate spatial sources, and aggregating phase differences across time or trials. The length of the mean resultant vector—the vector sum of all of the polar data—is the phase locking value. **b** If all phase differences are closely aligned, vectors will sum additively, producing long resultant vectors and high PLV. If phase differences are more diffusely distributed around the unit circle, some vectors will cancel out, producing shorter resultant vectors and correspondingly lower PLV. Phase differences in a random uniform distribution around the unit circle—as would be the case in completely uncorrelated oscillations—will, on average, completely cancel out and produce very short resultant vectors (in the limit case, a PLV of zero)

given frequency. "Consistency" is taken to mean the constancy of values over either time or experimental trials. For example, two oscillations recorded from two different electrodes which maintain a constant phase difference for several cycles could be said to have a high PLV—and therefore a high functional connectivity—in that interval. Alternatively, if the difference between two oscillations is constant at a specific point in time following the onset of an experimental trial, across all trials, this would also indicate high FC in an across-trials sense. Neither time nor trials is "better" than the others—the choice over which dimension to integrate depends on the experimenter's hypothesis (but this choice should always be made clear to the reader!) [9]. Note that the physiologic interpretation between time vs. trial-based connectivity differs. Connectivity over time suggests that, through the course of several oscillatory cycles, two brain regions maintained a consistent phase lag which—in theory—facilitated communication or neuroplastic mechanisms during that interval. Connectivity over trials indicates that, relative to a point in time defined by experimental variables, two regions tended to have a similar phase lag on all trials. This could indicate that experimental conditions induced stereotyped iEEG responses among the recorded regions, which may also be a mechanism that supports inter-areal communication and plasticity.

The PLV can most intuitively be thought of by considering the distribution phase differences around the unit circle (Fig. 32.2). If phase differences are highly consistent, they will bunch together towards a specific direction (in circular terminology, a *von Mises* distribution with a low circular variance might be expected, as in the left panel of Fig. 32.2b). If phase differences are inconsistent, they will be distributed around the unit circle randomly, with no clear directionality to the distribution. In practice, this property of phase distribution is captured by the mean resultant vector length; this metric is calculated by treating each data point as a circular unit vector, summing the vectors, and normalizing by the total number of vectors (i.e. averaging the vectors; see Eq. 32.1). The resultant vector length (which will fall between 0 and 1 because of the normalization step) is the phase-locking value. A PLV of 1 would mean that all vectors—in other words, all phase differences—were exactly the same across all trials/timepoints. Conversely, a PLV of 0 means that phase differences were distributed exactly uniformly around the unit circle.

$$PLV_{tf} = |N^{-1} \sum_{n=1}^{N} e^{i(\varphi_{xtf} - \varphi_{ytf})}| \qquad (32.1)$$

Equation 32.1 Calculation of the mean resultant vector length, or the PLV. N is the total number of trials across which PLV is calculated, while φ refers to the phase angle of electrode x or y at timepoint t and frequency f. The phase differences are summed and then normalized by the total number of trials (or timepoints, if computing PLV over time). Finally, only the magnitude of the result is taken, since PLV itself is independent of the mean phase direction.

Several important points arise from this formulation. First, note that this measure of phase consistency between two electrodes is entirely independent of the average phase direction—in other words, so long as the phase distribution is highly clustered, PLV will be high, whether that cluster points to 0°, 90°, 180°, and so on. In practice, this means that the PLV (and its cousins, described below) measures functional connectivity regardless of the degree of lag between two signals, so long as that lag remains constant over time/trials. Later, we discuss modifications to the PLV which account for phase directions (see "PLI, wPLI, and the question of volume conduction").

Note also that the PLV is a measure of connectivity which derives solely from phase information extracted from EEG activity (in other words, PLV analysis utilizes only the imaginary component of a complex number). This offers some benefits, as it keeps connectivity analyses simple and easier to interpret, without the need to account for other variables that describe the amplitude, duration, other qualities of neural oscillations. Conversely, these qualities may be relevant to a neuroscientific question about connectivity, and will be missed entirely by a pure PLV analysis. For example, it may be relevant to know that the amplitude of an ongoing theta oscillation at electrode A decreases during times at which phase consistency with electrode B also appears to fall, suggesting that the phase measure itself might be subject to more statistical noise. Fortunately, there exist measures of functional connectivity which account for spectral power as well as phase.

32.2.3 Coherence

A relatively simple modification to the PLV formula can be introduced to account for changes in spectral power at the contributing electrodes, called spectral coherence [10]. For each unit vector of phase difference (as in Fig. 32.2a), the vector length is weighted by the multiplied signal magnitudes (the square root of signal power) at that specific time point or trial. In other words, if signal magnitudes happen to be low, the corresponding phase difference vector will contribute less to the overall measure of functional connectivity that we call coherence (Fig. 32.3). Note also that because baseline spectral power changes over the course of time or trials, the weighted vector sum is divided by the average power at both electrodes over time/trials, thereby rendering an output of coherence in a range from 0 to 1, just like the PLV (Eq. 32.2). As this normalization step uses average power, it does not "undo" the weighting by power at specific trials or instantaneous timepoints inherent to this metric. Additionally, it is important to observe that the phase differences which make the greatest contribution to coherence will occur alongside *correlated* changes in power at both electrodes; simultaneous increases in power at two electrodes, associated with consistent phase differences, will tend to drive high coherence values.

$$Coh_{xy} = \left| \frac{C_{xy}}{C_{xx}C_{yy}} \right| = \frac{\left| \frac{1}{N} \sum_{n=1}^{N} |m_{nx}| \left| m_{ny} \right| e^{i\varphi_{nxy}} \right|^2}{(\frac{1}{N} \sum^{N} |m_{nx}|^2)(\frac{1}{N} \sum^{N} |m_{ny}|^2)} \tag{32.2}$$

Equation 32.2 Calculation of the coherence between electrodes *x* and *y*. C_{xy}, C_{xx}, and C_{yy} refer to the cross-spectral density and auto-spectral densities of electrodes *x* and *y*, respectively. *m* refers to the signal magnitude at each electrode, trial *n*, and timepoint. φ refers to phase differences between electrodes. *N* is the total number of trials (or time interval) across which coherence is computed. Note that the numerator (C_{xy}) is simply a magnitude-weighted (and squared) version of the PLV, while the denominator serves to normalize the result by the average power at both electrodes, rendering a result between 0 and 1.

Coherence has been a popular choice in many recent iEEG studies, because investigators may see it as an easy-to-interpret measure of FC that incorporates two features of iEEG signals—power and phase—which are both understood to be important in neural dynamics. Several efficient implementations of coherence are available

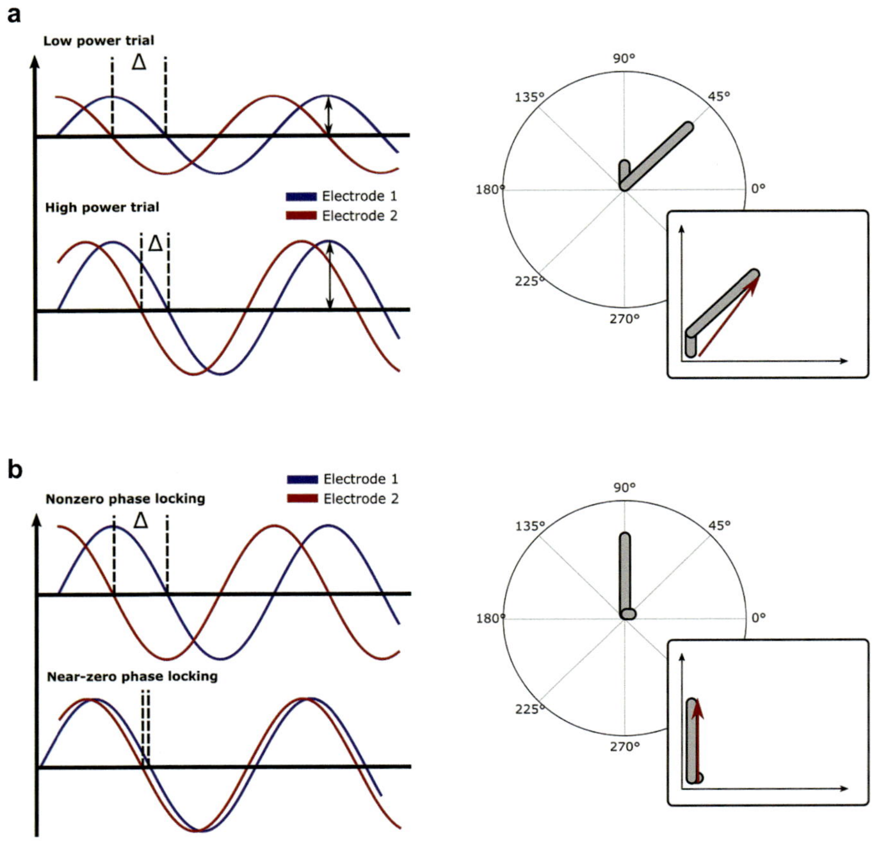

Fig. 32.3 Spectral coherence and weighted phase-lag index. **a** Spectral coherence reflects the power-weighted phase consistency between two signals. Phase vectors which are measured during periods of low power contribute less to the overall measure of phase consistency. **b** The weighted phase-lag index (wPLI) accounts for volume conduction by down-weighting the contribution of phase differences that are close to the zero axis. Phase differences of exactly zero would make no contribution to the measure, while phase differences of 90° or 270° make maximum contribution

in popular open-source scientific computing packages; we recommend the MNE Python implementation [11] (see https://mne.tools/0.13/generated/mne.connectivity.spectral_connectivity.html).

32.2.4 PLI, wPLI and the Question of Volume Conduction

The brain is not nearly as discrete as the sensors we use to record it. Indeed, the brain is a heterogenous mixture of fat, neurons, glia, vasculature, ionic fluids, intercellular proteins, and more. While some of these substances are more electrically conductive than others, the brain's general makeup of ions dissolved in water tends to propagate electrical field potentials well beyond their source, i.e. ionic gradients across neuronal cell membranes. Because of this, electrical field potentials—the substrate of iEEG recordings—may be detectable far from their point of neuronal origin. As such, human electrophysiologists often ask whether iEEG recorded by an electrode is a representation of local activity versus distant sources. This is important when measuring functional connectivity, because observed phase differences between two regions may not reflect inter-regional activity propagated through axonal projections, but instead conduction of electrical activity through aqueous solution—or "volume conduction" [12, 13].

Significant effort has been devoted to characterizing and modeling volume conduction, in an effort to free our functional measures of this potentially-contaminating influence. While these efforts are important, far less often is the question asked, "what does it matter?" Indeed, if volume conduction is a natural phenomenon of the brain, might we be losing important information by ignoring it? If our own crude sensors can be affected by volume-conducted activity, it is possible that neurons are too! Might electrical fields propagated through inter-neuronal space be a meaningful substrate of inter-areal communication in the brain? Perhaps, but perhaps not.

At the very least, it is worth considering how volume-conduction could affect iEEG data. Operating under the assumption that volume conducted effects are an undesirable confound, methods exist for mitigating its effect on iEEG data. These methods are predicated on the idea that volume-conducted signals, or contaminants introduced by referencing schemes, result in signal components with 0 degrees of phase lag between two electrodes [14]. (Similarly, volume conduction can also result in a measured phase difference of 180 degrees, depending on an electrode's position relative to a source dipole, i.e. whether it detects negative vs. positive voltage fluctuations.) This is because volume-conducted electrical activity is more-or-less simultaneously propagated to all electrodes, so it will be seen with essentially no time delay regardless of where an electrode is positioned relative to another (the same would apply to spurious phase correlations introduced via a common referencing scheme).

An early attempt to address such zero-degree lags came in the form of the phase lag index (PLI) [15]. The intuition behind the PLI is to measure the degree of asymmetry in phase differences across the 0–180 axis (also called the real axis in complex-number terminology); if phase differences across trials/time tend to consistently point above or below this axis (i.e. towards 90° or 270°), one can be confident that there is little zero-lag component, else there would have been vectors pointing towards 0 or 180°. Conversely, if phase differences are distributed symmetrically across the real axis, such as towards 0 or 180°, there are substantial zero-lag components that might derive from spurious sources. Note that this measure of asymmetry (Eq. 32.3) is irrespective of the numerical angle of phase lag—so long as lags are asymmetrically distributed around the real axis, PLI will be high. This relationship comes are the expense of the metric's stability, especially if phase lags are relatively close to the real axis.

$$PLI_{xy} = \left| \frac{1}{N} \sum_{n=1}^{N} sign\left(\varphi_{nxy}\right) \right| \tag{32.3}$$

Equation 32.3 Calculation of the phase-lag index (PLI). Notation as in prior equations, where "sign" is the operator that assigns 1 or -1 depending on whether the phase difference φ is above or below the zero axis (i.e. the 0–180 degree axis on a polar plot). Note again how this formula is a slightly modified version of the original PLV formula in Eq. 32.1.

To address this instability, Vinck et al. developed the "weighted PLI" [16]. The wPLI operates similarly to the PLI in that is discounts the influence of phase lags near the zero axis (i.e. lags close to 0° or 180°). But unlike the PLI, it is not agnostic to the value of phase lag; instead, phase lags are down-weighted the closer they are to the zero (or real) axis (stated differently, phase lags are weighted according to the magnitude of the imaginary component of the complex vector). In this way, phase lags of zero would not make any contribution to the wPLI, while lags of 90° or 270 would make maximum contributions (Fig. 32.3b). This type of vector-weighting is similar in principle to the way in which phase lags are weighted by power to derive spectral coherence. The wPLI generally provides a good estimate of functional connectivity independent of spurious zero-lag correlations. However, this metric has low sensitivity to high correlations with low phase lags, as very concentrated phase clustering around a value close to zero, like 10°, will still yield a "low" wPLI. Moreover, care should be taken when using the wPLI to compare relative synchronization between two conditions, as decreases would be observed if (1) phase lag distributions become less concentrated with a constant mean direction, or (2) the mean phase lag rotates towards zero while the distribution maintains constant concentration [17, 18]. Note that in the latter case, relative wPLI will still decrease even if phase lags rotate within a range unlikely to reflect volume conduction; e.g. 90° to 60°. (For further statistical considerations when using the wPLI and other phase-based metrics, see *Statistical frameworks for analyzing iEEG connectivity*).

Similar to volume conduction effects, an experimenter's choice of referencing scheme (see Chap. 28) can also introduce spurious correlations between signals in distant electrodes, if not properly accounted for [12, 13]. For example, the common-average reference—in which the average iEEG signal across all electrodes is subtracted from each individual trace—will also add common signal elements to all electrodes. This procedure could manifest in connectivity metrics as high correlations with zero phase lag, spuriously increasing the measured connectivity between a given electrode pair. The bipolar reference avoids such brain-wide contamination at the expense of reduced sensitivity to dynamics with relatively broad topography, as the reference essentially acts as a high-pass filter. In particular, this reference may disrupt the detection of low-frequency neural oscillations like the theta oscillation [19]. In both of these cases, careful statistical controls—such as a contrast between cognitive states or against a baseline period, can mitigate but not necessarily eliminate the effect of referencing.

32.3 Alternative Metrics of iEEG Functional Connectivity

While phase is one of the most common ways to probe intracranial functional connectivity, phase-free measures have been developed that offer unique windows into intracranial connectivity. An experimenter's choice to use these metrics as opposed to phase-based values depends on the underlying hypothesis being tested. However, the two metrics we will discuss here—Granger causality and broadband power correlations—have unique properties which prompt their usage in certain scenarios. Granger causality, as the name implies, can identify putatively causal relationships between brain regions (and identify directed interactions), whereas broadband power correlations are thought to be particularly reflective of general neural activation and align with findings from the fMRI literature.

32.3.1 Granger Causality

The metrics discussed so far have ignored a key question about iEEG functional connectivity: Is there a *direction* to the communication between brain regions? The PLV and its related metrics are all undirected interactions, meaning that the metric says nothing about whether the activity at one electrode precedes the other. "Granger causality" has been increasingly used in the neurosciences to identify directed interactions between timeseries that originate from different parts of the brain, and moreover to *statistically* suggest a level of causal interaction between such timeseries [20–22]. The logic is to use linear autoregression to measure how much electrode A's own timeseries predicts future values, and compare to the combined autoregressive prediction of electrode A and electrode B. In other words, does knowledge of electrode B's timeseries add useful information for predicting the future of electrode A? If so, electrode B exerts a positive Granger "causal" influence on electrode A. Note that the word "causality" is used loosely in this context, as the confounds presented in Fig. 32.1 are not addressed by the computation of Granger causality (in its simplest form). Formally, the autoregressive equation for electrode A is:

$$A_t = \sum_{n=1}^{N} x A_{t-n} + \varepsilon_{At}$$

where A is the iEEG value at time t, x is the autoregression coefficient, and ε_{At} is the error between the predicted and actual value at time t. N is the total length of time over which to estimate the autoregression. This equation essentially uses prior values of electrode A's timeseries to predict future values, weighted by a linear coefficient x. When information from two electrodes is available, the equation becomes:

$$A_t = \sum_{n=1}^{N} x A_{t-n} + \sum_{n=1}^{N} y B_{t-n} + \varepsilon_{ABt}$$

Electrode A's timeseries can now be predicted using information from electrode B, weighted by an additional coefficient y. The error term ε_{ABt} reflects the error between prediction versus actual after incorporating the new information from electrode B. If the error from the bivariate regression is less than the error from the univariate autoregression, we can infer that B exerts a Granger-causal influence on electrode A. To quantify this, we take the natural logarithm of the ratio of the error variances:

$$G = \ln\left(\frac{var(\varepsilon_A)}{var(\varepsilon_{AB})}\right)$$

Note that if electrode B adds no predictive value beyond knowledge of electrode A's timeseries, the ratio of variances will be 1 and the Granger value will be zero. If the error decreases with consideration of electrode B, then the ratio will be greater than 1 and G will take on a positive value.

32.3.2 Power Correlations

One of the simplest ways to conceptualize functional connectivity is as the correlated neural activity between different regions of the brain. iEEG spectral power is generally a reflection of such activity, be it synaptic potentials (more likely to drive oscillatory power at lower frequencies) or the direct consequence of population-level spiking activity (more likely to be observed at high frequencies) [23, 24]. In a similar manner to the way in which the fMRI BOLD signal is correlated between regions, the timecourse of spectral power in particular frequencies can be correlated as an alternative measure of functional connectivity that is not directly rooted in the phase alignment of oscillations. The general method is straightforward: for a pair of electrodes, compute spectral power over a suitably long interval (a minimum three cycles of the frequency of interest is a common rule-of-thumb), and next compute the correlation coefficient between the two timecourses. (Note that power values are often non-normal, necessitating use of the Spearman correlation to avoid biased results [9]). High correlation coefficients indicate co-activation of the two electrodes at a given frequency.

This method tends to work best at lower frequencies, which naturally evolve at slower timescales and which are often subject to substantial temporal smoothing in spectral analysis (i.e. Morlet wavelets). However, the correlation of high-frequency power timecourses poses a challenge (especially ~50 Hz and up), since these values will tend to fluctuate widely over time intervals that are most task-relevant (typically at least several hundred milliseconds to seconds). To observe the slower-timescale fluctuations *of* a high-frequency signal, the amplitude *envelope* is measured and correlated, essentially a form of temporal smoothing. This is best achieved by low-pass filtering the power timeseries before running a correlation [25, 26].

32.4 Statistical Frameworks for Analyzing iEEG Connectivity

Due to their somewhat abstract nature, it is exceptionally difficult to interpret phase-based measures of iEEG connectivity in a vacuum—it's clear that a PLV of 0.4 means less synchronization than a PLV of 0.8, but what does a PLV of 0.4 or 0.8 really *mean*? Here, proper statistical frameworks are essential for drawing meaningful conclusions from these kinds of data, but as with many aspects of iEEG connectivity, there is no definitively prescribed or "correct" way to do this. As such, this section will present important considerations and outline existing frameworks for analyzing iEEG connectivity data, couched among the broader message that investigators should carefully consider the origin, distribution, and specific characteristics of their own data when designing statistical tests.

Investigators should first take care to note that the PLV and its related metrics suffer from a bias problem—lower sample sizes tend to inflate a PLV-like measure of synchronization [27, 28]. This means that a direct comparison between PLVs measured between two conditions is inadvisable if those conditions consistently produce differing numbers of trials/timepoints (as is common in many neuroscience paradigms, such as comparisons between "correct" and "incorrect" judgments). A simple t-test between distributions of PLVs could therefore erroneously suggest a significant difference between conditions exists when, in reality, one condition simply produces fewer observations than the other. Fortunately, there are several methods by which one can account for this bias.

Perhaps the most straightforward approach to mitigating this bias is to design experiments or analyses to directly match the number of observations/trials/timepoints, but this is not always feasible. Nonparametric tests offer a compelling option to account for inherent bias in phase-based metrics without altering experimental design or sacrificing data (see also Chap. 35). For instance, bootstrapping methods could be used to subsample from the condition with more observations, compute and compare the PLV between count-matched data, and then continue resampling to construct a bootstrap estimate of the true difference between conditions. Similarly, labels between the two conditions could be repeatedly shuffled and the difference in PLV between conditions computed on each shuffle, generating a null distribution of PLV-differences to be expected *given* the inherent bias. The true PLV difference can then be compared to this null distribution to derive a p-value for the significance of the difference [29] (Fig. 32.4). Note that, in this method, the null distribution will likely not be centered around zero, as it will inherently identify the degree of bias due to sampling differences.

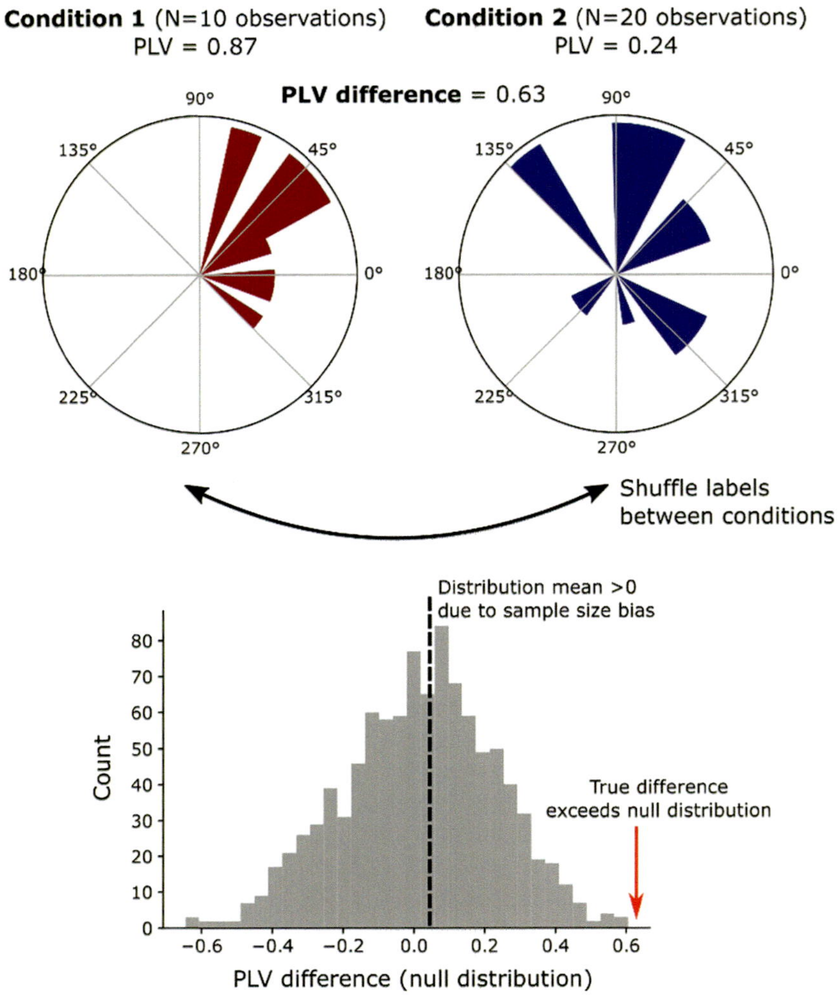

Fig. 32.4 Example nonparametric test for PLV difference. Trials are randomly shuffled between experimental conditions to give rise to a distribution of PLV differences that would be expected by chance, against which the true PLV difference can be compared

However, nonparametric methods are computationally expensive, and yet other methods also exist for mitigating the bias problem. Several modifications to PLV-based metrics have been introduced to produce new, bias-free measures of phase synchronization that can be safely used even in the case of differential observations between conditions. In 2010, Vinck, et al. published a description of the pairwise phase consistency (PPC), which is essentially the average dot product between possible pairs of phase vectors in a given distribution. Pairs of vectors with small angular distance between them—as is the case in a highly concentrated distribution, will have larger dot products and yield a larger PPC. The authors show that the PPC is an unbiased estimate of the squared PLV. An unbiased version of the PLI and a debiased version (i.e. may still reflect some bias at small sample sizes) of the wPLI were similarly developed [16].

32.5 Interpreting iEEG Connectivity (and Next Steps for the Field)

Statistical tests are necessary to establish a sense of how reliable patterns of connectivity might be, but they are insufficient to explain the neurobiological basis of functional connectivity, or how connectivity might support cognitive function and behavior. As noted previously, the effect of "functional connectivity"—that is, correlations between the electrical signals in different parts of the brain—can manifest similarly from very different underlying dynamics (see Fig. 32.1). Essentially, the underspecification of purely correlative measures will be an enduring problem in human neuroscience, which no amount of assumption-laden analytical complexity will truly solve. Ultimately, causal interventions in the brain—like stimulation—will be key to building a full understanding of how brain regions influence one another (see Chaps. 5, 39, 41 and 52 on stimulation approaches).

Until that time, what can we learn from studies of correlative intracranial functional connectivity? A large body of work has used the computational techniques reviewed in this chapter to generate compelling hypotheses about how functional connectivity gives rise to human behavior. One of the most intuitive and popular hypotheses is that synchronized oscillations between brain regions are a substrate of inter-areal communication; coordinated oscillations of the electrical field potential ensure that disparate neuronal assemblies co-activate and promote selective synaptic strengthening. For example, the gamma rhythm in particular serves to align the spiking output of one area with the optimally-receptive phase of another, facilitating transmission of neural representations [6, 30]. Related hypotheses posit that the theta rhythm synchronizes brain regions with a particularly important role in facilitating neural plasticity that gives rise to episodic memory [7, 19, 31, 32].

In line with these ideas, many recent studies have used the analytic techniques discussed in this chapter to better understand how functional connectivity relates to human cognition. A key insight common to all such studies is that moment-to-moment variability in the inter-areal synchronization correlates with periods of time when cognitive processes are operating smoothly. For example, transient increases in iEEG theta-band synchronization between the lateral temporal, medial temporal, and prefrontal cortex are associated with successful memory encoding, as measured using the PLV [17, 29], spectral coherence [33], and PPC [34]. Theta coherence between the prefrontal cortex and entorhinal cortex was also found to be correlated with navigational performance in human subjects [35]. When considered alongside the powerful notion that neuronal spiking is organized by theta oscillations [36], these results suggest theta oscillations align neural representations across brain regions in service of effective cognition (Fig. 32.5). More generally, several lines of evidence suggest that neural oscillations facilitate communication between brain regions when the phase lag matches the axonal conduction delay, thereby allowing the output of one region to arrive at a downstream area during a time of maximum neuronal excitability [7, 37].

An exhaustive review of this literature is outside the scope of this chapter and is addressed in more detail elsewhere in this text (see Chap. 25). Study of intracranially-based functional connectivity has given rise to a host of compelling frameworks for how the human brain works—our next steps are to use the new tools of human neuroscience to support or refute these frameworks with causal perturbations of human brain networks.

This does not mean that the age of correlative functional connectivity is over. Far from it—these metrics can offer powerful support for existing frameworks, or inspire entirely new conceptualizations of brain function. But to make the best use of iEEG-based functional connectivity in human neuroscience, investigators must agree on a common language. Too often, the overwhelming space of frequency bands, connectivity metrics, and analytical flexibility causes well-intentioned investigators to engage in forking-path analyses that can make results difficult to interpret or compare across laboratories. There is no clear "right answer," but the field must engage in a collective effort to standardize our approaches—what range defines a "frequency band"? When should we pick spectral coherence, as opposed to the PLV? Do we conceptualize volume conduction as a pure confound, or a feature of natural brain activity? This effort will surely be a perpetual work-in-progress, but one that will be ever more important as iEEG proliferates in human neuroscience.

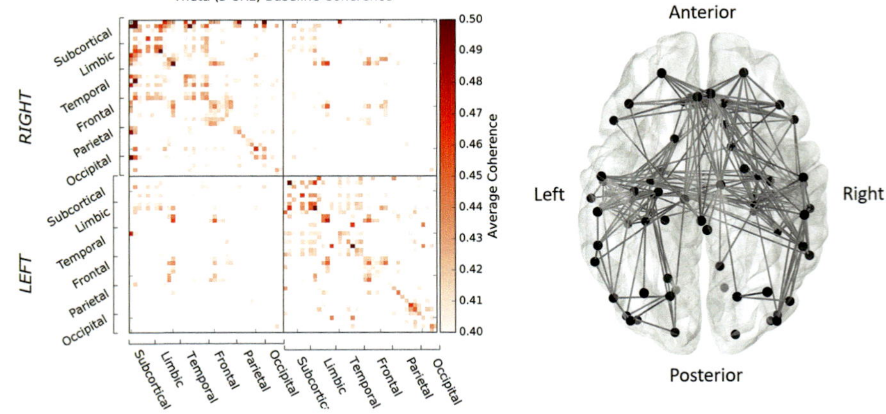

Fig. 32.5 Estimation of whole-brain patterns of iEEG theta coherence at rest. *Left*: Adjacency matrix representation of theta-band (3–8 Hz) coherence between Talairach brain regions, with each pairing represented by a given row/column combination. Data were averaged across electrode pairs and 287 neurosurgical patients with indwelling electrodes. Low coherence values were thresholded to zero for visualization purposes. *Right*: Visual representation of the suprathreshold theta-band coherences depicted on left. (Unpublished data, courtesy of E. Solomon and Michael Kahana.)

References

1. Marc D, Mehler A, Kording KP (2018) The lure of misleading causal statements in functional connectivity research
2. Stevenson IH, Rebesco JM, Miller LE, Körding KP (2008) Inferring functional connections between neurons. Curr Opin Neurobiol 18:582–588
3. Pesaran B et al (2018) Investigating large-scale brain dynamics using field potential recordings: analysis and interpretation. Nat Neurosci 21:903–919
4. Buzsáki G (2002) Theta oscillations in the hippocampus. Neuron 33:325–340
5. Onslow ACE, Bogacz R, Jones MW (2011) Quantifying phase-amplitude coupling in neuronal network oscillations. Prog Biophys Mol Biol 105:49–57
6. Fries P (2015) Rhythms for cognition: communication through coherence. Neuron 88:220–235
7. Fell J, Axmacher N (2011) The role of phase synchronization in memory processes. Nat Rev Neurosci 12:105–118
8. Lachaux JP, Rodriguez E, Martinerie J, Varela FJ (1999) Measuring phase synchrony in brain signals. Hum Brain Mapp 8:194–208
9. Cohen MX (2014) Analyzing neural time series data. Analyzing neural time series data. The MIT Press. https://doi.org/10.7551/mitpress/9609.001.0001
10. Walter DO (1963) Spectral analysis for electroencephalograms: Mathematical determination of neurophysiological relationships from records of limited duration. Exp Neurol 8:155–181
11. Gramfort A et al (2014) MNE software for processing MEG and EEG data. Neuroimage 86:446–460
12. Fein G, Raz J, Brown FF, Merrin EL (1988) Common reference coherence data are confounded by power and phase effects. Electroencephalogr Clin Neurophysiol 69:581–584
13. Schiff SJ (2005) Dangerous phase. Neuroinformatics 3:315–318
14. Nolte G et al (2004) Identifying true brain interaction from EEG data using the imaginary part of coherency. Clin Neurophysiol 115:2292–2307
15. Stam CJ, Nolte G, Daffertshofer A (2007) Phase lag index: assessment of functional connectivity from multi channel EEG and MEG with diminished bias from common sources. Hum Brain Mapp 28:1178–1193
16. Vinck M, Oostenveld R, van Wingerden M, Battaglia F, Pennartz CMA (2011) An improved index of phase-synchronization for electrophysiological data in the presence of volume-conduction, noise and sample-size bias. Neuroimage 55:1548–1565
17. Solomon EA et al (2019) Dynamic theta networks in the human medial temporal lobe support episodic memory. Curr Biol 29
18. Cohen MX (2015) Effects of time lag and frequency matching on phase-based connectivity. J Neurosci Methods 250:137–146
19. Herweg NA, Solomon EA, Kahana MJ, Kahana MJ (2019) Theta oscillations in human memory. Trends Cogn Sci 20
20. Seth AK, Barrett AB, Barnett L (2015) Granger causality analysis in neuroscience and neuroimaging. J Neurosci 35:3293–3297
21. Granger CWJ (1969) Investigating causal relations by econometric models and cross-spectral methods. Econometrica 37:424
22. Seth AK, Edelman GM (2007) Distinguishing causal interactions in neural populations. Neural Comput 19:910–933
23. Manning JR, Jacobs J, Fried I, Kahana MJ (2009) Broadband shifts in local field potential power spectra are correlated with single-neuron spiking in humans. J Neurosci Off J Soc Neurosci 29:13613–13620
24. Lachaux J-P, Axmacher N, Mormann F, Halgren E, Crone NE (2012) High-frequency neural activity and human cognition: past, present and possible future of intracranial EEG research. Prog Neurobiol 98:279–301
25. Keller CJ et al (2013) Neurophysiological investigation of spontaneous correlated and anticorrelated fluctuations of the BOLD signal. J Neurosci Off J Soc Neurosci 33:6333–6342

26. Foster BL, Rangarajan V, Shirer WR, Parvizi J (2015) Intrinsic and task-dependent coupling of neuronal population activity in human parietal cortex. Neuron 86:578–590
27. Aydore S, Pantazis D, Leahy RM (2013) A note on the phase locking value and its properties. Neuroimage 74:231–244
28. Vinck M, Van Wingerden M, Womelsdorf T, Fries P, Pennartz CMA (2010) The pairwise phase consistency: a bias-free measure of rhythmic neuronal synchronization. Neuroimage 51:112–122
29. Solomon EA et al (2017) Widespread theta synchrony and high-frequency desynchronization underlies enhanced cognition. Nat Commun 8:1704
30. Buzsáki G, Schomburg EW (2015) What does gamma coherence tell us about inter-regional neural communication? Nat Neurosci 18:484–489
31. Buzsáki G, Moser EI (2013) Memory, navigation and theta rhythm in the hippocampal-entorhinal system. Nat Neurosci 16:130–138
32. Colgin LL (2013) Mechanisms and functions of theta rhythms. Annu Rev Neurosci 36:295–312
33. Fell J et al (2003) Rhinal-hippocampal theta coherence during declarative memory formation: interaction with gamma synchronization? Eur J Neurosci 17:1082–1088
34. Watrous AJ, Tandon N, Conner CR, Pieters T, Ekstrom AD (2013) Frequency-specific network connectivity increases underlie accurate spatiotemporal memory retrieval. Nat Neurosci 16:349–356
35. Chen D et al (2021) Theta oscillations coordinate grid-like representations between ventromedial prefrontal and entorhinal cortex. Sci Adv **7**
36. O'Keefe J, Recce ML (1993) Phase relationship between hippocampal place units and the EEG theta rhythm. Hippocampus 3:317–330
37. Chapeton JI, Haque R, Wittig JH, Inati SK, Zaghloul KA (2019) Large-scale communication in the human brain is rhythmically modulated through alpha coherence. Curr Biol 29:2801-2811.e5

Chapter 33
How Can I Analyze Large-Scale Intrinsic Functional Networks with iEEG?

Aaron Kucyi and Sepideh Sadaghiani

Abstract An intrinsic functional brain network is a set of discrete, spatial elements that exhibit statistically dependent activity ("functional connectivity") with each other in a largely state-invariant manner (e.g. across wakeful rest, task performance, and sleep). *Large-scale* intrinsic networks—involving coupling between distant brain regions—were initially discovered with human functional neuroimaging (fMRI) based on hemodynamic signals. Though fMRI studies suggest critical relevance of these networks to brain function, findings remain challenging to interpret given the low temporal resolution and indirect nature of fMRI. Human iEEG is poised as a unique method that can deliver fundamental insights into the neurophysiological connectivity processes in intrinsic networks. In this chapter, we review iEEG analysis methods that have been used to identify electrophysiological networks closely resembling those found using classical fMRI functional connectivity. We focus on amplitude and phase coupling within multiple frequency bands as measures of iEEG intrinsic connectivity. We review evidence that iEEG connectivity shows state-invariant patterns of inter-regional coupling across multiple contexts. Moreover, we review applications of intrinsic iEEG connectivity patterns in predicting the roles of discrete neuronal populations in cognitive function. Finally, we explore how iEEG sheds light on the cognitive relevance of temporal dynamics within and between intrinsic networks.

A. Kucyi (✉)
Department of Psychological and Brain Sciences, Drexel University, Philadelphia, PA, US
e-mail: ak4379@drexel.edu

S. Sadaghiani
Psychology Department, University of Illinois at Urbana-Champaign, Urbana, IL, US

N. Axmacher (ed.), *Intracranial EEG*, Studies in Neuroscience, Psychology and Behavioral Economics, https://doi.org/10.1007/978-3-031-20910-9_33

33.1 Introduction

Cognitive neuroscience has classically emphasized the use of experimenter-administered task paradigms that actively engage targeted mental processes such as memory, attention, and decision-making. However, the brain exhibits highly coordinated patterns of spontaneous activity, even in the absence of explicit engagement in cognitive tasks or changes in external input. In a landmark finding with functional Magnetic Resonance Imaging (fMRI), researchers discovered that infraslow (<0.1 Hz) blood oxygenation level-dependent (BOLD) signals are highly correlated between distant regions in the brain when people are wakeful "at rest" (i.e., letting their minds wander freely) [1]. The spatial patterns of statistically dependent activity ("functional connectivity") in resting state fMRI reflect functional network integration at a *large-scale* level (i.e., coupling between distinct regions that are each typically 1mm^3 or larger). These large-scale fMRI networks partly (but not exclusively) reflect underlying anatomical connectivity [2] such as the coupling between homotopic regions of the two hemispheres that are structurally connected via commissural fibers [3]. The networks identified at rest also strongly resemble the most common patterns of brain-wide co-activation that are evoked during performance of cognitive tasks [4]. Strikingly, the same spatial architecture of functional connectivity observed at rest is also largely preserved during a wide variety of conditions, including sleep [5], anesthesia-induced unconsciousness [6, 7], and engagement in a diverse variety of cognitive tasks [8, 9]. Thus, the state-invariant nature of these networks suggest a persistent role in "intrinsic" brain functions that cannot be attributed to immediate inputs or outputs [10, 11].

Since their discovery, there has been enormous interest within neuroscience in investigating the significance of large-scale intrinsic networks to healthy cognitive function, clinical dysfunction, neurodevelopment, and aging [12]. Moreover, there is increasing interest in the temporal dynamics of these networks in relation to ongoing cognition [13, 14]. However, findings remain challenging to interpret given the low temporal resolution, low signal-to-noise ratio, and indirect nature of fMRI, which relies on hemodynamic changes that reflect complex interactions of underlying neural and metabolic activity [15]. Non-invasive techniques, including scalp EEG and magnetoencephalography (MEG), have offered some insight into the electrophysiological basis of large-scale intrinsic networks [16]. Yet those techniques also suffer from low signal-to-noise as well as limited ability to precisely localize the anatomical sources of activity.

Thus, within human neuroscience, iEEG is poised as a unique method that can deliver fundamental insights into the neurophysiology of large-scale intrinsic networks and their dynamics. Directly implanted electrodes offer a combination of high signal-to-noise (higher than fMRI and EEG/MEG), high temporal resolution (higher than fMRI and similar to EEG/MEG), and spatial resolution (similar to fMRI and higher than EEG/MEG). Although iEEG cannot offer whole-brain coverage due to clinical considerations, activity is typically sampled from multiple locations distributed throughout the brain, often including simultaneous recording from distinct nodes within one or multiple intrinsic networks. Another unique aspect of iEEG, relative to most non-invasive methods, is that recordings are semi-chronic. This permits detailed investigations of intrinsic networks across various conditions and states (e.g. experimental tasks, naturalistic behavior, sleep) at multiple time scales.

In this chapter, we review iEEG analysis methods that have been used to identify and interrogate large-scale intrinsic networks. Given the rich information available in iEEG signals, multiple approaches can be used to investigate these networks. We focus primarily on coupling of iEEG signals within multiple frequency bands in terms of both amplitude and phase, two analysis methods that capture distinct neurophysiological processes (see also Chap. 32) [17, 18]. After a brief overview of methodological considerations, we review evidence that iEEG connectivity shows state-invariant patterns of inter-regional coupling across multiple mental contexts. Moreover, we describe how iEEG intrinsic connectivity patterns predict task-evoked electrophysiological responses as well as the consequences of intracranial stimulation. Finally, we explore how iEEG sheds light on the functional significance of temporal dynamics within and between intrinsic networks.

33.2 Methodological Considerations

How are large-scale intrinsic networks typically analyzed in iEEG data? In contrast to fMRI, it is usually not appropriate to examine iEEG intrinsic networks based on simply the raw signals, as these signals contain a broad range of frequency information with dissociable components. The spectral content of iEEG, when characterized as power as a function of frequency f, is non-uniform in that it shows a $1/f$ pattern where low frequencies contribute disproportionately to the raw signal. As such, time–frequency decomposition is typically applied to isolate specific frequency bands prior to analyses of intrinsic network coupling [19]. There are various methodological options for performing time–frequency decomposition (e.g. Hilbert transform, wavelet-based methods), yet results have been largely concordant across studies of iEEG intrinsic networks regardless of the method choice [20–22]. Importantly, although time–frequency analysis methods decompose the data into distinct frequency bands, they do not guarantee that the derived time series contain oscillations (i.e., periodic/cyclical properties). Indeed, engagement of neuronal populations as measured with iEEG, either when evoked [23, 24] or spontaneous [25], is often characterized by a shift in *broadband*, non-oscillatory components of power amplitude. Recently, it has been suggested that band-limited oscillatory signals can be further disentangled from broadband $1/f$-like fluctuations using dedicated tools (see also Chaps. 22 and 23) [26, 27], but the potential value of such approaches to intrinsic network analyses has not yet been investigated.

Intrinsic networks are typically defined based on coupling of amplitude or phase of frequency band-limited activity across a time scale of seconds, minutes or longer (see Fig. 33.1 for overview). Amplitude coupling considers region pairs for which the strength (spectral power) of regional frequency band-limited activity waxes and wanes in an associated manner as being functionally connected. To estimate amplitude coupling, following re-referencing and minimal preprocessing of the raw iEEG data (see Chap. 28), time–frequency decomposition is first applied to each electrode signal (Fig. 33.1a, b). The decomposition can either be applied to an entire, continuous recording session of minutes or hours (a more computationally intensive approach) or to shorter time windows (e.g. sliding windows of seconds). Connectivity estimates from such short time windows can later be averaged to construct a 'static' connectivity organization, or alternatively can be employed to study time-varying dynamics in connectivity (cf. dedicated section below). After applying time–frequency analysis, estimates of the amplitude of band-limited power are obtained at each sample (termed "power amplitude" or "envelope") (Fig. 33.1c). At this stage, the continuous power amplitudes (or multiple estimates within shorter windows) from two different electrode locations can be correlated with one another to estimate functional connectivity in terms of amplitude coupling.

The major alternative family of within-frequency functional connectivity measures, phase coupling, is conceptualized as consistent phase relationships across brain regions. During periods of consistent phase difference (lag) of a given frequency across a pair of regions, the exchange of neural information is thought to be temporally well-aligned and efficient [28–30]. Many measures of phase coupling first estimate the time course of the phase of a particular band-limited oscillation. To this end, the Hilbert transform can be applied to the band-pass filtered signal time course at each electrode (the Wavelet transform offers an alternative). One common measure

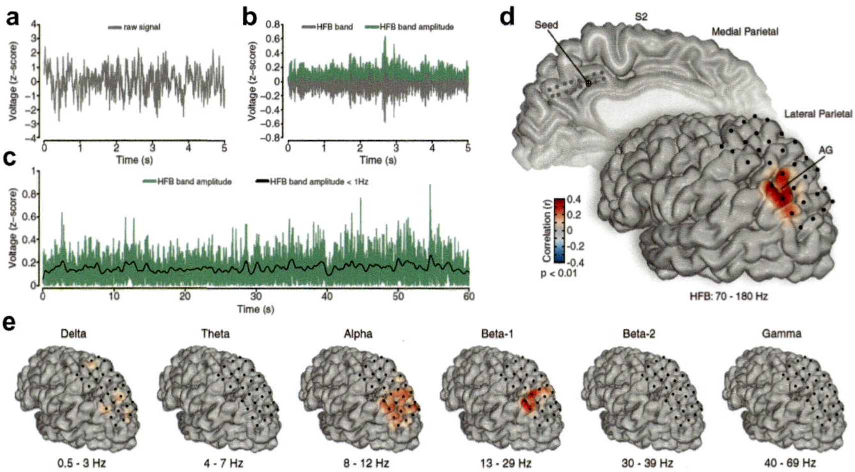

Fig. 33.1 Illustration of how iEEG functional connectivity can be computed based on frequency band-limited power amplitude coupling. (**a**) Raw time series from a single electrode. (**b**) Filtered electrode time series in the high-frequency broadband (HFB; 70–180 Hz; gray) and HFB amplitude (green). (**c**) Unfiltered (green) and < 1 Hz filtered (black) HFB amplitude. (**d**) Seed-based correlation between a region in medial parietal cortex and other electrode sites (based on < 1 Hz filtered HFB amplitude). (**e**) Same as (D) but for a range of lower frequency bands. Adapted with permission from [44]

of phase lag consistency, the Phase Locking Value (PLV), is based on the stability (i.e., inverse of the variance) of a set of instantaneous phase differences [31]. In the original PLV formulation for task-based and trial-locked connectivity, the set of phase pairs either spans across trials at one given time point or across time points in individual trials. Adopted to studies of intrinsic connectivity, this set comes from consecutive phase cycles within a given time window over several seconds (to permit sufficient oscillation cycles [18, 32]). Conversely, the Phase Lag Index (PLI) calculates the average of the *sign* of the phase difference over a given set of instantaneous phases [33] (see [34] for a more stable weighted version). An alternative and widely used approach to phase coupling is coherence (and variations thereof, e.g. [35]), which calculates cross-correlation in the frequency domain for a pair of electrodes [36]. Note however, that this measure does not only reflect interregional coupling of phases, but also amplitudes. Thus, coherence may be an appropriate choice if one conceptualizes the functional importance of phase synchrony to depend on the strength of the carrier oscillation. The choice of the particular measure of phase coupling therefore depends on the mechanistic view under consideration. Further, the different measures have different levels of susceptibility to volume conduction (imaginary coherence [35], imaginary PLV [32], and PLI/weighted PLI [33, 34] are thought to be less affected by this artifact; see below).

For completeness, we also note that beyond the within-frequency, band-limited measures most commonly employed and covered above, other approaches have also been explored to study intrinsic networks in iEEG. For example, instead of relying on spectral power or phase, intrinsic networks have been reported in cross-region correlations of the *direct* ECoG signal time courses in the infraslow range (<0.5 Hz) [37] and in canonical frequency bands [38]. Further, large-scale intrinsic network organization has been observed for *cross-frequency* coupling, where high-frequency broadband (HFB; ~ 70–180 Hz) in specific distributed regions is coupled to the phase of a slower oscillation in a seed region [39]. In light of the breadth of possible procedures to calculate electrophysiological long-range connectivity, we recommend careful deliberation of biological mechanisms of connectivity for a well-informed approach [16].

Important to the study of intrinsic networks, band-limited power amplitude time series and phase time courses contain frequencies that are much faster than BOLD fMRI signals. As such, when the goal is to investigate the electrophysiological connectivity processes that correspond to those known from fMRI, then temporal filtering is sometimes applied to the region-wise power amplitude estimates so that low frequencies can be isolated prior to computing inter-region correlations [40, 41]. Alternatively, one might be interested in the slow fluctuation range of the cross-region connectivity time course (when connectivity is calculated in a windowed, time-resolved manner). In this case, the time course of amplitude coupling or phase coupling can be temporally low-pass filtered (e.g. [42]). The temporal filtering method must be carefully considered [43]. When applied, distinct temporal filters are often explored and are commonly split into > 1 Hz, 0.1–1 Hz, and < 0.1 ranges [21, 22, 41, 44]. The < 0.1 Hz range closely matches the infraslow component at which intrinsic BOLD networks are typically investigated, but this approach involves severe loss of higher frequency information in the estimates of power amplitude or cross-region phase- and amplitude coupling.

Using either filtered or unfiltered measures, a "seed-based" connectivity map can be generated for the coupling of one electrode to all other electrodes (Fig. 33.1d), or more globally, an 'all-to-all' connectivity matrix can be computed between all implanted electrode sites (Fig. 33.3a) [18]. Connectivity maps may be generated for multiple carrier frequencies, which can yield different patterns of results (Fig. 33.1e).

We close our methodological considerations with a note on accounting for physical distance. In amplitude-, phase-, and phase-amplitude coupling connectivity analyses, iEEG correlations between electrode sites are systematically greater at shorter, relative to longer, distances. There are multiple possible reasons for this relationship. On one hand, electrodes that are spatially proximal to one another may sample from distinct neuronal populations that are within the same network, thereby leading to genuine estimates of strong coupling. On the other hand, adjacent electrodes may partially sample from the same neuronal populations due to volume conduction or instantaneous field spread (see Chap. 17). Such factors has been well-described within the context of non-invasive EEG/MEG and can create artifactual connectivity estimates [45]. The problem is less severe in iEEG due to much greater proximity between each electrode and the neuronal population that it is sampling from [46]. However, for improved interpretability, studies of intrinsic network coupling have typically accounted for physical distance in their analyses. For example, when comparing the similarity of spatial maps of BOLD and iEEG functional connectivity, or between iEEG maps obtained across different states, physical distance (typically Euclidean distance) has been regressed out via partial correlation or cubic spline regression [20–22, 37, 47]. When spatial similarities remain consistent even after accounting for distance, the identified intrinsic network patterns are likely to be genuine rather than a product of volume conduction.

33.3 Intrinsic Networks from Amplitude Coupling

To date, most iEEG studies on large-scale intrinsic networks have focused on coupling of *power amplitude* within frequency band-limited ranges, typically involving within-frequency band coupling between distinct electrode sites. This focus builds on a rich literature of cognitive task iEEG studies, where stimulus-/task-evoked changes (relative to a baseline prestimulus state) in regional frequency-specific power amplitude have been found across a broad variety of conditions [48, 49]. More recently, the self-organizing properties of task-independent, ongoing co-fluctuations in band-limited power amplitude have been investigated between regions across large-scale intrinsic networks [18, 20–22, 37, 41, 42]. Though a large range of carrier frequencies have been explored, much of this work has emphasized HFB or "high gamma" power amplitude, given that this component of the iEEG signal is known to correlate positively with the BOLD signal [50, 51] and with local neuronal population spiking rate near an iEEG electrode [52–54].

Using the above-described power amplitude connectivity analyses, researchers performed iEEG studies that confirmed an electrophysiological basis for large-scale intrinsic networks. Landmark studies by Nir et al. [41] and He et al. [37] provided initial evidence for neuroanatomically selective iEEG coupling patterns that resemble those originally identified with BOLD fMRI. A hallmark feature of intrinsic BOLD networks is inter-hemispheric coupling between homotopic regions [1, 3]. Analyzing bihemispheric iEEG recordings from right and left auditory cortex, Nir et al. [41] investigated this inter-hemispheric phenomenon across resting state, task, and sleep states via correlations of band-limited power amplitudes across a broad range of frequency bands within the 1–100 Hz range. They found that across all states, gamma band-limited power (40–100 Hz) showed the strongest inter-hemispheric correlation, although significant coupling was also found at lower frequencies. When the infraslow (<0.1 Hz) component of gamma power amplitude was examined via temporal filtering, strong inter-hemispheric correlations ($r = 0.5$ to 0.8) were found that approached the magnitude of correlations typically identified in fMRI data. This inter-hemispheric coupling was highly anatomically selective as it did not generalize to electrodes outside of the bilateral auditory regions. However, fMRI data were not available for direct comparison with iEEG. In complementary work, He et al. [37] more directly compared BOLD (obtained via pre- or post-surgical scans) versus ECoG functional connectivity of intrinsic networks in somatosensory, motor and surrounding regions. When investigating gamma (50–100 Hz) band-limited power, they found significant positive spatial correlations between BOLD and ECoG connectivity estimates, with some dependence on the state in which ECoG data were acquired (correlations were significant for waking data and REM sleep but not for slow-wave sleep). Together, these two studies [37, 41] provided the foundations for further work investigating intrinsic networks via iEEG power amplitude coupling across diverse behavioral states and broader brain systems.

Using similar power amplitude coupling analyses, iEEG intrinsic networks have now been identified within various systems extending well beyond auditory and sensorimotor regions, suggesting a general electrophysiological basis of intrinsic BOLD functional connectivity across individuals, brain networks, and behavioral states [21]. For example, iEEG has revealed intrinsic electrophysiological coupling across rest, task and sleep states within the default mode network (DMN) [55], one of the most frequently investigated intrinsic networks in neuroimaging research (Fig. 33.2). Anatomically-precise intrinsic coupling between nodes of the DMN has consistently been found based on fluctuations in HFB/high gamma power amplitude [20–22, 44]. Lower frequency ranges also show DMN organization that may be less anatomically specific [20, 21, 44], though future work is needed to quantify the potential inter-frequency differences in spatial patterns within networks. Additionally, electrodes implanted within other well-described cortical association networks, such as the "dorsal attention" (DAN) and "frontoparietal control" networks, have consistently revealed intrinsic connectivity within the HFB power range [20, 21]. It was further suggested that these networks can be distinguished from one another based on preferential coupling within low-frequency bands (e.g. 4–8 Hz theta in DMN vs. 8–12 alpha in DAN) [20], a phenomenon that requires more in-depth future study.

The state-invariant nature of intrinsic iEEG networks across wider systems has also been confirmed; similarities in spatial organization have been repeatedly observed across rest, task and sleep states [21, 44, 47].

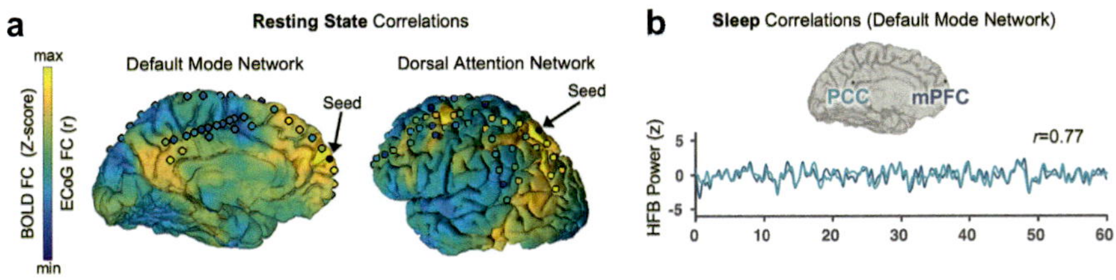

Fig. 33.2 Power amplitude coupling between core regions within major intrinsic networks originally described with fMRI. **(a)** Correspondence between presurgical resting state BOLD functional connectivity (shown on cortical surface) and ECoG high-frequency broadband (70–170 Hz; 0.1–1 Hz filtered envelope) functional connectivity within the same patient's brain (fMRI and ECoG data acquired separately). The medial prefrontal (mPFC) and posterior cingulate (PCC) cortices, within the default mode network (left), show strong, anatomically selective coupling in both modalities. The superior parietal lobule and frontal eyes fields, within the dorsal attention network (right), similarly show strong selective coupling. **(b)** Highly correlated fluctuations (high-frequency broadband power amplitude; 0.1–1 Hz filtered) between mPFC and PCC persists during sleep (in same patient as shown in (A)). Adapted with permission from [21]

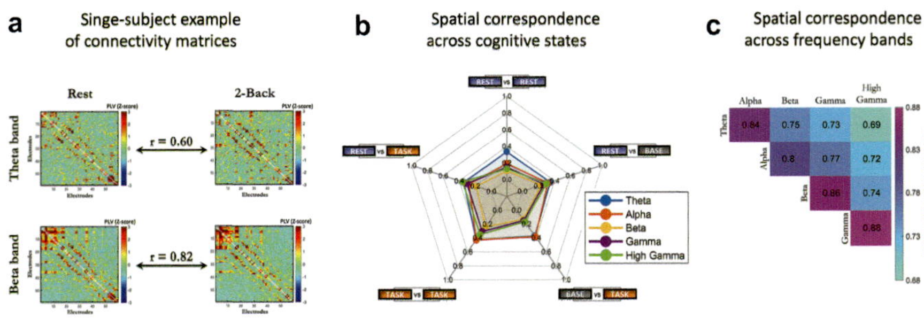

Fig. 33.3 Invariance of the intrinsic spatial organization of electrophysiological connectivity over cognitive states and oscillatory frequency bands. **(a)** All-to-all electrode phase coupling matrices for an example subject. For both the theta (top row) and beta band (bottom row), the connectivity pattern showed strong spatial correlation between task-free wakefulness (left) and a cognitively demanding working memory 2-back task (right). A (weaker) similarity can also be visually appreciated between the frequency bands (top vs. bottom). **(b)** Spatial correlation was assessed for a group of subjects between the connectivity matrices from resting wakefulness ("rest"), pre-stimulus baseline ("base"), and active trial-related processing during the post-stimulus period of various different tasks ("task"). Group-average values are shown for the cross-state dissimilarity (1 minus spatial correlation). In all frequency bands, this dissimilarity was no greater across mental states than between two resting state periods. **(c)** The group-average spatial correlations across frequency-specific connectivity patterns (consensus over mental states) showed strong spatial correlation between frequency bands. All visualized data are derived using Phase locking Value (PLV) on ECoG data. Equivalent results were observed for band-specific amplitude correlations. Adapted with permission from [47]

In addition to within-network coupling, the power amplitude approach has also been used to examine the electrophysiology of between-network phenomena. A major finding that emerged from intrinsic connectivity fMRI is a negative correlation (anticorrelation) between signals in the DMN and other networks, especially DAN and salience networks [56–58]. During active task performance, the DMN typically exhibits stimulus- or task-evoked BOLD deactivation when the DAN and salience network show BOLD activation, an inter-network pattern often referred to as 'functional antagonism' [59]. This phenomenon has been extensively validated with iEEG based on task-evoked HFB power amplitude [60–62]. However, whether or not *intrinsic* (state-invariant) anticorrelation, including presence during a resting state, is genuinely grounded in neurophysiological dynamics has been a matter of debate due to technical limitations of fMRI [63]. Within this context, iEEG has been used to investigate the neurophysiology of intrinsic DMN anticorrelation. In short duration (~3–6 min) resting state iEEG recordings, Keller et al. [22] reported DMN anticorrelations within the low-frequency (<1 Hz) component of high-frequency broadband (50–150 Hz) power amplitude fluctuations. However, when comparing resting state iEEG to BOLD correlations (using presurgical fMRI), only a subset of negative BOLD correlations corresponded to negative iEEG correlations. Moreover, iEEG anticorrelations were generally weaker than those found in BOLD data [22].

Importantly, however, power amplitude connectivity analysis methods typically assume zero-lag relationships between regions. Task-evoked iEEG responses have revealed that the onset of activations within the DAN and salience network precedes the onset of DMN deactivation by hundreds of milliseconds [61, 62], a phenomenon that fMRI has been blind to. Thus, Kucyi et al. [62] investigated *lagged* iEEG interactions between neuronal populations that were functionally localized within the DMN, DAN or salience network based on task-evoked response profiles. Using a lagged cross-correlation of HFB power amplitude, they found DMN anticorrelations in both rest and task conditions. However, relative to rest, the magnitude of these lagged anticorrelations was stronger during a continuous performance task that required sustained attention [62]. Future iEEG studies may shed light on the rapid, temporally dynamic processes that may underlie DMN anticorrelations as well as broader, intrinsic coordination between networks. Methodologically, this work highlights limitations of power amplitude coupling analyses that purely rely on zero-lag relationships, suggesting a need for future studies to account for lagged inter-regional relationships. Important contributions of lagged relationships to intrinsic network organization have been recently demonstrated based on slow fluctuations in fMRI and iEEG signals [64], but iEEG can also be leveraged to generate much more precise, millisecond-level insights.

33.4 Intrinsic Networks from Phase Coupling

While the majority of iEEG investigations of large-scale intrinsic networks have focused on amplitude coupling as discussed above, phase coupling approaches have delivered a network organization that is spatially convergent with that of amplitude coupling. Given this alignment, the current section provides only a brief discussion of some of the major phase coupling-based findings.

The stark mechanistic difference between amplitude coupling and phase coupling (see methodological considerations above; [17]) invites the question of how their intrinsic spatial organization relates to each other. Mostame and Sadaghiani [18] quantified the spatial similarity of ECoG-derived connectivity matrices across the two connectivity modes during short periods (1.5 or 2.5 s) of pre-stimulus baseline and post-stimulus task processing (finger flexing and verb generation). The study investigated *stimulus-related* changes from pre-stimulus to post-stimulus periods, as well as *intrinsic* connectivity present during both periods. The study confirmed that phase- and amplitude coupling constitute distinct connectivity processes, since *stimulus-related* changes in the two connectivity modes occurred in different connections and were temporally independent. Importantly however, when assessing the distributed pattern of *intrinsic* connectivity, amplitude correlations showed a spatial pattern substantially similar to that of phase coupling during both post- and pre-stimulus periods. This observation held true irrespective of the choice of phase coupling measure (PLV, weighted PLI and imaginary coherence). The stability of the shared, distributed intrinsic network organization across baseline and active (post-stimulus) periods further suggests that this organization may be largely invariant to cognitive contexts.

Indeed, direct investigations of connectivity patterns over various cognitive contexts confirms such state invariance. Kramer et al. [65] measured frequency-specific coherence in day-long ECoG recordings comprising various cognitive states. They estimated the core intrinsic organization from static connectivity over the full ~ 24-h recording. They then demonstrated that this core organization is observable in shorter recordings (> = 100 s) during a broad range of mental states consisting of wakefulness, drowsiness, as well as stage II and III sleep. In experimentally controlled task states and resting wakefulness, Mostame and Sadaghiani [47] directly compared spatial connectivity patterns between cognitive contexts. They found strong spatial resemblance of phase coupling during the active trial (post-stimulus) period of various tasks, task-free, pre-stimulus baseline, and task-free resting state in all frequency bands (Fig. 33.3b). Of note, a substantial part of the intrinsic connectivity pattern of individual frequency bands was spatially shared across frequencies (Fig. 33.3c). Equivalent results were observed for amplitude coupling. Collectively, these findings show that long-range electrophysiological connectivity is governed by a spatial organization that is relatively stable over mental and cognitive states, largely shared across the breadth of canonical oscillatory bands, and comparable across phase coupling and amplitude coupling.

33.5 Applications to the Study of Cognition

Once intrinsic networks are identified, how can they inform or predict the role of neuronal populations in cognitive functions? In this section, we cover evidence from two key types of applications that highlight the value of iEEG intrinsic connectivity to the study of cognition: (A) Predicting task-evoked electrophysiological responses; and (B) Predicting experiential effects elicited by intracranial stimulation.

First, iEEG studies have demonstrated that intrinsic connectivity is tightly related to the tendency of two electrode sites to co-activate with each other in response to a cognitive task or external stimulus [44, 66]. Predictions of task-evoked iEEG responses can be generated from intrinsic connectivity as estimated from the same individual's presurgical resting state fMRI [67] or from a population-level atlas of intrinsic networks (e.g. [68]) when fMRI data are not available (though the latter approach may be less accurate). Additionally, intrinsic connectivity (e.g. during a resting state) can be estimated directly from iEEG to predict task-evoked responses. For example, Foster et al. [44] showed that two key nodes of the DMN (posteromedial cortex and angular gyrus) responded with increased HFB power to a task condition involving autobiographical memory recall. The same sites showed strong, anatomically-selective coupling of slow (<1 Hz filtered) fluctuations in HFB power during resting and sleep states. In another example, Kucyi et al. [62] showed that task-evoked HFB response profiles could be predicted based on the location of a given electrode within the DMN, DAN or salience network (as predefined based on fMRI-based intrinsic network atlas) (Fig. 33.4). Such evidence aligns more generally with prior fMRI work, where it has been shown that at the global (whole-brain) level, regions exhibiting functional connectivity also tend to co-activate with one another during cognitive task performance [4, 69]. The reason that intrinsic networks tend to resemble task-evoked responses remains under debate, and future iEEG studies may shed light on the neurophysiological mechanisms that underlie such relationships. For example, it is possible that spontaneous activity is constrained by statistical histories of co-activation [70], provides predictive signals for perception [11, 71], and partly reflects ongoing experiences and cognitive processes [14].

Second, intrinsic connectivity can predict how intracranial stimulation will cause a change in a patient's subjective experience or behavior [72, 73]. High-frequency (typically 50–100 Hz) stimulation is often applied in the iEEG setting for the clinical purpose of functional mapping (see also Chap. 39). Importantly, the intrinsic network identity that an electrode is located within can determine whether or not an effect of stimulation will be found and what the specific experiential or behavioral qualities of an effect will be. Indeed, the classical behavioral effects that are evoked by intracranial stimulation of sensorimotor regions have been used to validate fMRI intrinsic connectivity parcellation methods [74]. In an application to more complex cognitive function, Parvizi et al. [72] showed that stimulation of electrodes implanted within the mid-cingulate cortex–when specifically localized to the salience network–evoked the subjective, narrative experience within patients that was termed by the experimenters a "will to persevere." Furthermore, in a larger whole-cortex analysis

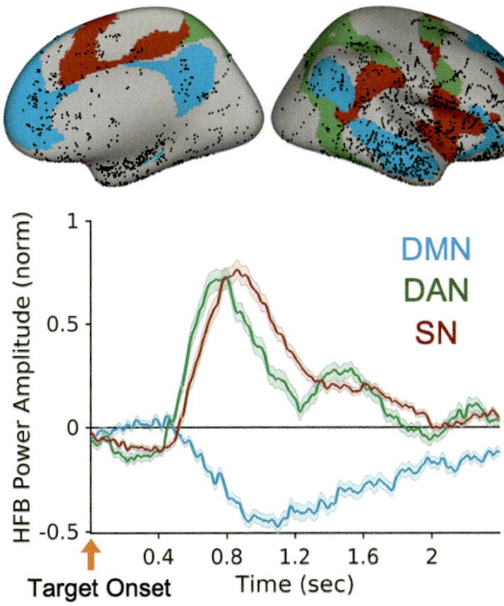

Fig. 33.4 Intrinsic network organization predicts task-evoked electrophysiological HFB response profile. (Top) Locations of electrodes implanted within cortical nodes of the default mode network (DMN; blue), dorsal attention network (DAN; green), and salience network (SN; red) in a cohort of 31 patients. (Bottom) Average task-evoked HFB response profiles of electrode sites that showed a significant response to a visual target stimulus that appeared in ~ 10% trials in a continuous performance task. Following target onset, increased HFB activation appears earlier in DAN than in SN sites, and DMN deactivations are later and more sustained than the activations in both other networks. Adapted with permission from [62] (Creative Commons license: https://creativecommons.org/licenses/by/4.0/)

across 67 patients, Fox et al. [75] showed that stimulation of a given electrode within the DMN, compared to stimulation of other networks, is less likely to result in any experiential or behavioral response. However, it was recently shown that intracranial stimulation of DMN regions (localized with presurgical fMRI intrinsic connectivity) affects performance in a creative thinking task, even though patients may not be aware of stimulation-evoked changes in subjective experience or behavior [76].

Taken together, these findings strongly highlight the relevance of intrinsic connectivity to the study of cognitive function with iEEG. Additionally, these findings may have important practical implications for research and clinical procedures conducted within the iEEG setting. For example, they raise the possibility that intrinsic functional connectivity, which can be estimated during relatively passive conditions (e.g. resting state, sleep), may serve as a substitute for, or initial screening prior to, cognitive task performance or electrical stimulation. Given that some cognitive tasks are highly demanding or difficult for patients to perform, and that electrical stimulation procedures can involve patient burden, intrinsic connectivity estimates can play an important role in determining which tasks should be administered or which electrodes should be stimulated in a given patient.

33.6 Temporal Dynamics of Intrinsic Networks

This chapter has so far considered intrinsic connectivity as 'static' in that the described analysis approaches and applications explicitly search for iEEG coupling that is stable across different states. However, though intrinsic network architecture is largely state-invariant, changes occur in the strength of coupling over time [65, 77]. Such connectivity changes can be externally elicited but can also occur spontaneously. These fluctuations are often subtle yet significant, suggesting that intrinsic activity 'explores' a dynamic functional repertoire of distinct brain states about a central core organization [78]. Time-varying functional connectivity analysis has recently risen in popularity in resting state fMRI [79], with some findings beginning to highlight how ongoing network fluctuations relate to spontaneous cognitive and affective processes, including arousal, attention, memory replay and mind wandering (see also Chap. 21) [14, 80]. Yet those findings are limited due to poor fMRI temporal resolution that cannot capture activity relevant to cognitive events that may arise on the time scale of milliseconds. It is thus within this area in the study of intrinsic networks–their temporal dynamics–where iEEG may have potential to offer its most unique scientific insights into the nature of cognition. The study of intrinsic network dynamics with iEEG is still in its infancy. We here briefly highlight select examples of preliminary insights, and point to future research directions, on the cognitive and behavioral relevance of iEEG network dynamics.

The spontaneous dynamics of iEEG networks at 'rest,' and their behavioral relevance, have been studied across multiple time scales [25, 65, 81, 82]. Importantly, ongoing changes in inter-electrode coupling, or regional activity at electrodes localized to intrinsic networks, have been related to changes in external variables that reflect ongoing cognition. For example, time-varying iEEG amygdalar and hippocampal network configurations across distinct 15–20 min resting state windows were predictive of self-reported mood fluctuations [83, 84]. This suggests that even at relatively slow time scales, intrinsic network connectivity is not fully stable and that its changes across time windows–even lasting tens of minutes–meaningfully relate to subjective experiences that underlie mood states. Future iEEG studies that employ experience sampling may shed more light on the relevant time scales of network dynamics that correspond to richer varieties of ongoing conscious phenomena.

At a faster time scale, Kucyi and Parvizi [25] showed that spontaneous, transient increases in HFB power amplitude ("activations") during a resting state were often of similar magnitude to task-evoked activations within a dorsal anterior insular cortex (daIC) region localized to the salience network. Moreover, both task-evoked and spontaneous daIC activations reliably preceded pupillary dilations by several hundred milliseconds, suggesting a role in autonomic arousal regardless of whether it is externally-elicited or self-generated [25]. Future work is needed to examine how these spontaneous, functionally significant, daIC events are dynamically coupled to wider brain networks to mediate the ongoing changes in arousal and cognition that naturally arise "at rest."

Another emerging area of importance in the study of temporal dynamics of intrinsic networks concerns the role of "sharp-wave ripple" (SWR) events that occur spontaneously in the hippocampus (see Chaps. 21 and 24). These robust neuronal synchronization events have been well characterized in animal models [85] and can be detected in the human brain uniquely using iEEG [86–89]. Hippocampal SWRs occur more frequently during resting states than during active task performance and are hypothesized to function in the consolidation, and spontaneous reactivation, of memory [85]. The dialogue between hippocampal SWRs and large-scale intrinsic network dynamics in the human brain, and its potential significance to ongoing cognition, remains poorly characterized. However, it was recently shown that during cued autobiographical memory recollection, hippocampal ripples occur in concert with DMN activation [90]. Further iEEG studies of intrinsic activity (e.g. during rest and 'naturalistic' states involving spontaneously generated behavior) may further reveal the potentially critical role of hippocampal SWRs in coordinating intrinsic brain network dynamics on multiple time scales.

33.7 Conclusion

In this chapter, we reviewed iEEG analysis methods for studying large-scale intrinsic brain networks and described how these approaches are playing an increasingly important role in the study of human cognition. To date, iEEG has been critical in validating evidence from other neuroimaging modalities for the existence of large-scale intrinsic networks. The convergence of evidence with that from other modalities and populations (e.g. neurotypical individuals) supports the idea that iEEG–despite its usual reliance on patients with epilepsy–provides meaningful insights into the nature of typical intrinsic brain organization. Critically, beyond simply validating other modalities, iEEG can deliver unique insights into the temporal dynamics of intrinsic networks that go well beyond what can be gleaned from other human neuroscience techniques. Future iEEG studies are expected to yield fundamental insights into how the brain dynamically self-organizes itself to support cognition.

References

1. Biswal B, Yetkin FZ, Haughton VM, Hyde JS (1995) Functional connectivity in the motor cortex of resting human brain using echo-planar MRI. Magn Reson Med 34:537–541
2. Honey CJ, Sporns O, Cammoun L et al (2009) Predicting human resting-state functional connectivity from structural connectivity. Proc Natl Acad Sci U S A 106:2035–2040
3. Lowe MJ, Mock BJ, Sorenson JA (1998) Functional connectivity in single and multislice echoplanar imaging using resting-state fluctuations. Neuroimage 7:119–132
4. Smith SM, Fox PT, Miller KL et al (2009) Correspondence of the brain's functional architecture during activation and rest. Proc Natl Acad Sci U S A 106:13040–13045

5. Fukunaga M, Horovitz SG, van Gelderen P et al (2006) Large-amplitude, spatially correlated fluctuations in BOLD fMRI signals during extended rest and early sleep stages. Magn Reson Imaging 24:979–992
6. Peltier SJ, Kerssens C, Hamann SB et al (2005) Functional connectivity changes with concentration of sevoflurane anesthesia. NeuroReport 16:285–288
7. Vincent JL, Patel GH, Fox MD et al (2007) Intrinsic functional architecture in the anaesthetized monkey brain. Nature 447:83–86
8. Gratton C, Laumann TO, Nielsen AN et al (2018) Functional brain networks are dominated by stable group and individual factors, not cognitive or daily variation. Neuron 98:439-452.e5
9. Cole MW, Bassett DS, Power JD et al (2014) Intrinsic and task-evoked network architectures of the human brain. Neuron 83:238–251
10. Fox MD, Raichle ME (2007) Spontaneous fluctuations in brain activity observed with functional magnetic resonance imaging. Nat Rev Neurosci 8:700–711
11. Sadaghiani S, Kleinschmidt A (2013) Functional interactions between intrinsic brain activity and behavior. Neuroimage 80:379–386
12. Zhang J, Kucyi A, Raya J et al (2021) What have we really learned from functional connectivity in clinical populations? Neuroimage 242:118466
13. Hutchison RM, Womelsdorf T, Allen EA et al (2013) Dynamic functional connectivity: promise, issues, and interpretations. Neuroimage 80:360–378
14. Kucyi A, Tambini A, Sadaghiani S et al (2018) Spontaneous cognitive processes and the behavioral validation of time-varying brain connectivity. Netw Neurosci 2:397–417
15. Logothetis NK (2008) What we can do and what we cannot do with fMRI. Nature 453:869–878
16. Sadaghiani S, Brookes MJ, Baillet S (2021) Connectomics of human electrophysiology. Neuroimage 247:118788
17. Engel AK, Gerloff C, Hilgetag CC, Nolte G (2013) Intrinsic coupling modes: multiscale interactions in ongoing brain activity. Neuron 80:867–886
18. Mostame P, Sadaghiani S (2020) Phase- and amplitude-coupling are tied by an intrinsic spatial organization but show divergent stimulus-related changes. Neuroimage 219:117051
19. Foster BL, He BJ, Honey CJ et al (2016) Spontaneous neural dynamics and multi-scale network organization. Front Syst Neurosci 10:7
20. Hacker CD, Snyder AZ, Pahwa M et al (2017) Frequency-specific electrophysiologic correlates of resting state fMRI networks. Neuroimage 149:446–457
21. Kucyi A, Schrouff J, Bickel S et al (2018) Intracranial electrophysiology reveals reproducible intrinsic functional connectivity within human brain networks. J Neurosci 38:4230–4242
22. Keller CJ, Bickel S, Honey CJ et al (2013) Neurophysiological investigation of spontaneous correlated and anticorrelated fluctuations of the BOLD signal. J Neurosci 33:6333–6342
23. Miller KJ, Leuthardt EC, Schalk G et al (2007) Spectral changes in cortical surface potentials during motor movement. J Neurosci 27:2424–2432
24. Ramot M, Fisch L, Harel M et al (2012) A widely distributed spectral signature of task-negative electrocorticography responses revealed during a visuomotor task in the human cortex. J Neurosci 32:10458–10469
25. Kucyi A, Parvizi J (2020) Pupillary dynamics link spontaneous and task-evoked activations recorded directly from human insula. J Neurosci 40:6207–6218
26. Wen H, Liu Z (2016) Separating fractal and oscillatory components in the power spectrum of neurophysiological signal. Brain Topogr 29:13–26
27. Donoghue T, Haller M, Peterson EJ et al (2020) Parameterizing neural power spectra into periodic and aperiodic components. Nat Neurosci 23:1655–1665
28. Varela F, Lachaux JP, Rodriguez E, Martinerie J (2001) The brainweb: phase synchronization and large-scale integration. Nat Rev Neurosci 2:229–239
29. Singer W (1999) Neuronal synchrony: a versatile code for the definition of relations? Neuron 24(49–65):111–125
30. Fries P (2005) A mechanism for cognitive dynamics: neuronal communication through neuronal coherence. Trends Cogn Sci 9:474–480

31. Lachaux JP, Rodriguez E, Martinerie J, Varela FJ (1999) Measuring phase synchrony in brain signals. Hum Brain Mapp 8:194–208
32. Sadaghiani S, Scheeringa R, Lehongre K et al (2012) α-band phase synchrony is related to activity in the fronto-parietal adaptive control network. J Neurosci 32:14305–14310
33. Stam CJ, Nolte G, Daffertshofer A (2007) Phase lag index: assessment of functional connectivity from multi channel EEG and MEG with diminished bias from common sources. Hum Brain Mapp 28:1178–1193
34. Vinck M, Oostenveld R, van Wingerden M et al (2011) An improved index of phase-synchronization for electrophysiological data in the presence of volume-conduction, noise and sample-size bias. Neuroimage 55:1548–1565
35. Nolte G, Bai O, Wheaton L et al (2004) Identifying true brain interaction from EEG data using the imaginary part of coherency. Clin Neurophysiol 115:2292–2307
36. Wang J, Tao A, Anderson WS et al (2021) Mesoscopic physiological interactions in the human brain reveal small-world properties. Cell Rep 36:109585
37. He BJ, Snyder AZ, Zempel JM et al (2008) Electrophysiological correlates of the brain's intrinsic large-scale functional architecture. Proc Natl Acad Sci U S A 105:16039–16044
38. Betzel RF, Medaglia JD, Kahn AE et al (2019) Structural, geometric and genetic factors predict interregional brain connectivity patterns probed by electrocorticography. Nat Biomed Eng 3:902–916
39. Weaver KE, Wander JD, Ko AL et al (2016) Directional patterns of cross frequency phase and amplitude coupling within the resting state mimic patterns of fMRI functional connectivity. Neuroimage 128:238–251
40. Leopold DA, Murayama Y, Logothetis NK (2003) Very slow activity fluctuations in monkey visual cortex: implications for functional brain imaging. Cereb Cortex 13:422–433
41. Nir Y, Mukamel R, Dinstein I et al (2008) Interhemispheric correlations of slow spontaneous neuronal fluctuations revealed in human sensory cortex. Nat Neurosci 11:1100–1108
42. Ko AL, Weaver KE, Hakimian S, Ojemann JG (2013) Identifying functional networks using endogenous connectivity in gamma band electrocorticography. Brain Connect 3:491–502
43. de Cheveigné A, Nelken I (2019) Filters: When, why, and how (Not) to use them. Neuron 102:280–293
44. Foster BL, Rangarajan V, Shirer WR, Parvizi J (2015) Intrinsic and task-dependent coupling of neuronal population activity in human parietal cortex. Neuron 86:578–590
45. Palva JM, Wang SH, Palva S et al (2018) Ghost interactions in MEG/EEG source space: A note of caution on inter-areal coupling measures. Neuroimage 173:632–643
46. Dubey A, Ray S (2019) Cortical electrocorticogram (ECoG) is a local signal. J Neurosci 39:4299–4311
47. Mostame P, Sadaghiani S (2021) Oscillation-based connectivity architecture is dominated by an intrinsic spatial organization, not cognitive state or frequency. J Neurosci 41:179–192
48. Crone NE, Miglioretti DL, Gordon B, et al (1998) Functional mapping of human sensorimotor cortex with electrocorticographic spectral analysis. I. Alpha and beta event-related desynchronization. Brain 121 (Pt 12):2271–2299
49. Lachaux J-P, Axmacher N, Mormann F et al (2012) High-frequency neural activity and human cognition: past, present and possible future of intracranial EEG research. Prog Neurobiol 98:279–301
50. Mukamel R, Gelbard H, Arieli A et al (2005) Coupling between neuronal firing, field potentials, and FMRI in human auditory cortex. Science 309:951–954
51. Hermes D, Miller KJ, Vansteensel MJ et al (2012) Neurophysiologic correlates of fMRI in human motor cortex. Hum Brain Mapp 33:1689–1699
52. Manning JR, Jacobs J, Fried I, Kahana MJ (2009) Broadband shifts in local field potential power spectra are correlated with single-neuron spiking in humans. J Neurosci 29:13613–13620
53. Ray S, Maunsell JHR (2011) Different origins of gamma rhythm and high-gamma activity in macaque visual cortex. PLoS Biol 9:e1000610
54. Nir Y, Fisch L, Mukamel R et al (2007) Coupling between neuronal firing rate, gamma LFP, and BOLD fMRI is related to interneuronal correlations. Curr Biol 17:1275–1285

55. Fox KCR, Foster BL, Kucyi A et al (2018) Intracranial electrophysiology of the human default network. Trends Cogn Sci 22:307–324
56. Fox MD, Snyder AZ, Vincent JL et al (2005) The human brain is intrinsically organized into dynamic, anticorrelated functional networks. Proc Natl Acad Sci U S A 102:9673–9678
57. Fransson P (2005) Spontaneous low-frequency BOLD signal fluctuations: an fMRI investigation of the resting-state default mode of brain function hypothesis. Hum Brain Mapp 26:15–29
58. Chai XJ, Castañón AN, Öngür D, Whitfield-Gabrieli S (2012) Anticorrelations in resting state networks without global signal regression. Neuroimage 59:1420–1428
59. Buckner RL, DiNicola LM (2019) The brain's default network: updated anatomy, physiology and evolving insights. Nat Rev Neurosci 20:593–608
60. Ossandón T, Jerbi K, Vidal JR et al (2011) Transient suppression of broadband gamma power in the default-mode network is correlated with task complexity and subject performance. J Neurosci 31:14521–14530
61. Raccah O, Daitch AL, Kucyi A, Parvizi J (2018) Direct cortical recordings suggest temporal order of task-evoked responses in human dorsal attention and default networks. J Neurosci 38:10305–10313
62. Kucyi A, Daitch A, Raccah O et al (2020) Electrophysiological dynamics of antagonistic brain networks reflect attentional fluctuations. Nat Commun 11:325
63. Murphy K, Fox MD (2017) Towards a consensus regarding global signal regression for resting state functional connectivity MRI. Neuroimage 154:169–173
64. Mitra A, Snyder AZ, Hacker CD et al (2016) Human cortical-hippocampal dialogue in wake and slow-wave sleep. Proc Natl Acad Sci U S A 113:E6868–E6876
65. Kramer MA, Eden UT, Lepage KQ et al (2011) Emergence of persistent networks in long-term intracranial EEG recordings. J Neurosci 31:15757–15767
66. Daitch AL, Foster BL, Schrouff J et al (2016) Mapping human temporal and parietal neuronal population activity and functional coupling during mathematical cognition. Proc Natl Acad Sci U S A 113:E7277–E7286
67. Dastjerdi M, Foster BL, Nasrullah S et al (2011) Differential electrophysiological response during rest, self-referential, and non-self-referential tasks in human posteromedial cortex. Proc Natl Acad Sci U S A 108:3023–3028
68. Yeo BTT, Krienen FM, Sepulcre J, et al (2011) The organization of the human cerebral cortex estimated by intrinsic functional connectivity. J Neurophysiol
69. Tavor I, Parker Jones O, Mars RB et al (2016) Task-free MRI predicts individual differences in brain activity during task performance. Science 352:216–220
70. Harmelech T, Malach R (2013) Neurocognitive biases and the patterns of spontaneous correlations in the human cortex. Trends Cogn Sci 17:606–615
71. Pezzulo G, Zorzi M, Corbetta M (2021) The secret life of predictive brains: what's spontaneous activity for? Trends Cogn Sci 25:730–743
72. Parvizi J, Rangarajan V, Shirer WR et al (2013) The will to persevere induced by electrical stimulation of the human cingulate gyrus. Neuron 80:1359–1367
73. Parvizi J, Braga RM, Kucyi A et al (2021) Altered sense of self during seizures in the posteromedial cortex. Proc Natl Acad Sci U S A 118:e2100522118
74. Wang D, Buckner RL, Fox MD et al (2015) Parcellating cortical functional networks in individuals. Nat Neurosci 18:1853–1860
75. Fox KCR, Shi L, Baek S et al (2020) Intrinsic network architecture predicts the effects elicited by intracranial electrical stimulation of the human brain. Nat Hum Behav 4:1039–1052
76. Shofty B, Gonen T, Bergmann E, et al (2022) The default network is causally linked to creative thinking. Mol Psychiatry 1–7
77. Chang C, Glover GH (2010) Time-frequency dynamics of resting-state brain connectivity measured with fMRI. Neuroimage 50:81–98
78. Kringelbach ML, Deco G (2020) Brain states and transitions: insights from computational neuroscience. Cell Rep 32:108128

79. Lurie DJ, Kessler D, Bassett DS et al (2020) Questions and controversies in the study of time-varying functional connectivity in resting fMRI. Netw Neurosci 4:30–69
80. Martin CG, He BJ, Chang C (2021) State-related neural influences on fMRI connectivity estimation. Neuroimage 244:118590
81. Casimo K, Madhyastha TM, Ko AL et al (2019) Spontaneous variation in electrocorticographic resting-state connectivity. Brain Connect 9:488–499
82. Breshears JD, Hamilton LS, Chang EF (2018) Spontaneous neural activity in the superior temporal gyrus recapitulates tuning for speech features. Front Hum Neurosci 12:360
83. Kirkby LA, Luongo FJ, Lee MB et al (2018) An Amygdala-hippocampus subnetwork that encodes variation in human mood. Cell 175:1688-1700.e14
84. Scangos KW, Khambhati AN, Daly PM, et al (2021) Closed-loop neuromodulation in an individual with treatment-resistant depression. Nat Med.
85. Buzsáki G (2015) Hippocampal sharp wave-ripple: A cognitive biomarker for episodic memory and planning. Hippocampus 25:1073–1188
86. Axmacher N, Elger CE, Fell J (2008) Ripples in the medial temporal lobe are relevant for human memory consolidation. Brain 131:1806–1817
87. Helfrich RF, Lendner JD, Mander BA et al (2019) Bidirectional prefrontal-hippocampal dynamics organize information transfer during sleep in humans. Nat Commun 10:3572
88. Norman Y, Yeagle EM, Khuvis S, et al (2019) Hippocampal sharp-wave ripples linked to visual episodic recollection in humans. Science 365:eaax1030
89. Vaz AP, Inati SK, Brunel N, Zaghloul KA (2019) Coupled ripple oscillations between the medial temporal lobe and neocortex retrieve human memory. Science 363:975–978
90. Norman Y, Raccah O, Liu S et al (2021) Hippocampal ripples and their coordinated dialogue with the default mode network during recent and remote recollection. Neuron 109:2767-2780.e5

Chapter 34
What Do I Need to Consider for Multivariate Analysis of iEEG Data?

Weizhen Xie, John H. Wittig Jr., and Kareem A. Zaghloul

Abstract Intracranial EEG (iEEG) offers an opportunity to directly record neural activity with a high signal-to-noise ratio from the human brain in order to study human cognition. Deciphering the rich information afforded by iEEG entails multivariate analyses of time-series data across recording sites, spectral frequencies, cognitive events, and participants. In this chapter, we discuss some approaches and issues to consider for multivariate analyses of iEEG data. We first overview the types and amount of data researchers may acquire through iEEG recording and discuss some common practices used for iEEG data extraction before multivariate analyses. We then review approaches to harness the multivariate structure of iEEG data, including similarity analyses of neural data and classification based on machine learning. We discuss how the research question can guide which type of analysis to pursue in order to advance our understanding of human cognition.

34.1 The Multivariate Nature of iEEG Data Analysis

Neural signals captured through intracranial electroencephalogram (iEEG) recordings reflect active neural processes superimposed at a recording location [1]. These signals can be recorded using either subdural electrodes placed on the brain surface or through penetrating depth electrodes placed in deep brain structures. In human studies, iEEG recordings are routinely acquired in epilepsy monitoring units from patients implanted with intracranial electrodes to localize their seizure onset zone for surgical intervention. As this clinical practice opens up important investigatory opportunities to study human brain functions [2], there is a strong scientific interest in understanding the rich information content captured by iEEG recordings.

One aspect of the richness of iEEG recordings is that the type of electrode used at each recording site has a substantial impact in the spatial extent of the neural activity being captured (Fig. 34.1a; see also Chap. 17). At the macroscale, subdural

W. Xie · J. H. Wittig Jr. · K. A. Zaghloul (✉)
Surgical Neurology Branch, National Institute of Neurological Disorders and Stroke, Bethesda, MD 20892, USA
e-mail: kareem.zaghloul@nih.gov

N. Axmacher (ed.), *Intracranial EEG*, Studies in Neuroscience, Psychology and Behavioral Economics, https://doi.org/10.1007/978-3-031-20910-9_34

and depth electrodes have a recording contact diameter of 1–2 mm. These macroelectrodes record population activity from local neuronal networks containing hundreds of thousands of neurons [3]. At the microscale, microelectrode probes (diameter <100 μm) organized as microwires or multielectrode arrays (MEAs) record local field potentials (LFPs) [1]. These LFPs contain both extracellular field activity of thousands of neurons as well as the spiking activity of individual neurons.

Another aspect of the richness of iEEG recordings is the diverse spectral information captured regardless of spatial scale. The raw neural signals often exhibit

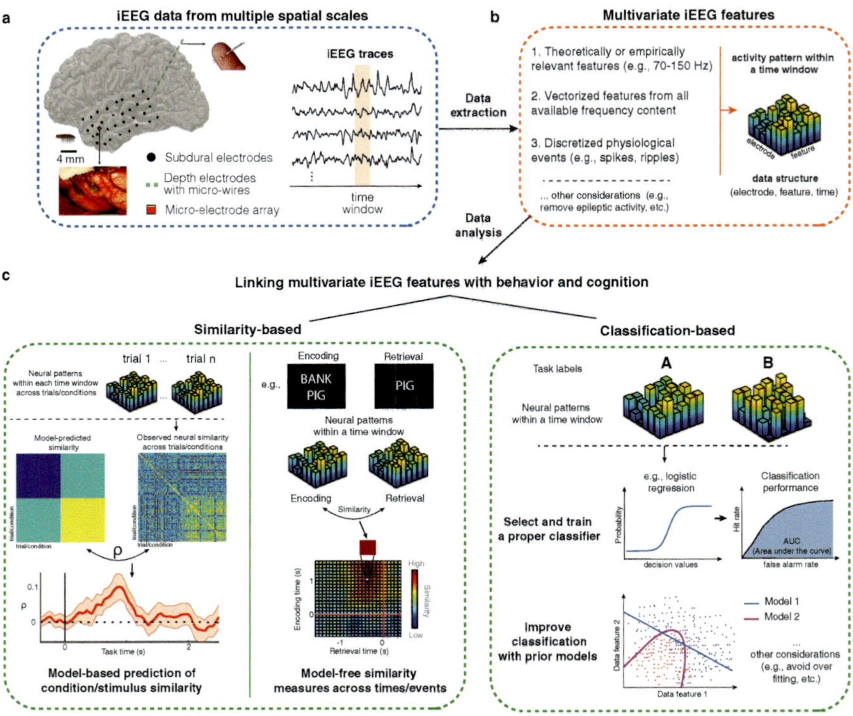

Fig. 34.1 Schematic overview of various considerations in multivariate analysis of iEEG data. **a** iEEG data capture neural activity from multiple spatial scales, with rich information from different spectral frequencies. **b** Several principled considerations in extracting multivariate features from iEEG data. Typically, a matrix of electrode by feature by time data structure is extracted for later analysis. **c** The multivariate features of iEEG data are commonly linked with behavior and cognition based on one of two principled approaches, namely similar-based analysis (left) or classification-based analysis based on machine learning (*right*). In similarity-based analysis, patterns of neural activity within a given time window can be associated with model-predicted similarity patterns across trials/conditions or with other neural patterns at another task event/time point (e.g., encoding vs. retrieval in a memory task). In classification-based analysis, a proper classifier is often chosen and trained to distinguish patterns of neural activity associated with different task labels. A priori models can be incorporated to improve classification performance; yet, overfitting should be avoided to ensure generalization

oscillations at different frequencies. The synchronization (or desynchronization) of iEEG activity at certain frequencies over time has been associated with several key cognitive functions. For example, similar to scalp EEG, iEEG data contain commonly identified lower frequency signals (e.g., $\leq$12 Hz). These low frequency rhythms may serve as carrier frequencies that communicate information content between distant nodes of a large-scale network [4]. Because of their high signal-to-noise ratio, iEEG data also contain rich broadband high-frequency signals (e.g., 70–150 Hz) that are traditionally thought to reflect neuronal spiking activity averaged across thousands of neurons immediately adjacent to a recording site [5] (see also Chaps. 16 and 44 on the relationship of spikes and local field potentials).

In addition to the spatial and spectral richness that are innate to iEEG recordings, experimental designs can introduce richness in the hierarchical structure of the data sets compiled across experimental sessions and participants, which can provide a scaffolding for statistical and multivariate analysis. For example, iEEG data at different recording sites are nested within individual recording sessions, which are then further nested within a participant across different experimental sessions. For single-unit spike data, this can mean that the recorded units may vary from day to day or session to session since the micro-electrode probes tend to shift over time. This hierarchical structure of iEEG data therefore gives rise to a complex and high-dimensional data matrix that has multiple dependent measurements to be explored across trial types, sessions, and participants. Thus, without some principled considerations, analyzing and interpreting the multivariate data can be computationally, analytically, and conceptually challenging. An approach to disentangle these hierarchies relies upon multi-level inference and mixed-effect modeling, which is discussed in a separate chapter (see Chap. 36). Here we primarily focus on several additional principled considerations to increase the interpretability of multivariate analysis for iEEG data.

34.2 Data Extraction

To begin with, iEEG data contain electrical field fluctuations that may be both neurophysiological and environmental (e.g., 60 Hz line noise). Data preprocessing is therefore necessary to further improve the signal-to-noise ratio. Moreover, since iEEG signals are routinely captured from patients who are being monitored for seizures, some amount of data preprocessing is required to identify and remove epileptic activity [6]. To reduce data complexity, one may also consider conventional data reduction procedures such as principal component analysis (PCA). These procedures enable investigators to focus on data components that carry the majority of the variance of the recorded iEEG signals [7]. In addition to these conventional preprocessing and data reduction considerations, there are several additional principled approaches that can be deployed to extract relevant information from iEEG time series for the study of the human brain and cognition (Fig. 34.1b).

First, one may consider extracting specific iEEG features that are theoretically or empirically relevant for specific information processing functions. For example, since broadband high-frequency activity reflects local neural activity around the recording site [5, 8], high-frequency power is often linked with the functional role of the recorded brain tissue. By tracking the average broadband high-frequency power (70–150 Hz) across different tasks in distributed brain areas, for example, Haller and colleagues [9] have identified patterns of persistent neural activation in the cortex that encode perceptual features of stimuli and predict motor outputs. Similarly, in the context of functional connectivity, consistent with the computational role of low-frequency activity in facilitating communication across large-scale brain networks [2, 4], Chapeton and colleagues [10, 11] have shown that iEEG electrode pairs in the temporal lobe with stable correlated broadband activity also reliably show coherent 8–12 Hz activity. In light of these observations, it is reasonable to focus on certain iEEG frequencies for theory-driven inquiries that aim to examine either functional localization (e.g., 70–150 Hz) or connectivity (e.g., 8–12 Hz).

A second approach that extends beyond a restricted range of the information space in iEEG data defined a priori is to incorporate available non-redundant information from iEEG data across the spatial and frequency domains for multivariate analyses. For example, when linking iEEG data with cognitive functions, one may extract the power values from pre-defined frequency bands (e.g., delta, theta, alpha, beta, gamma, and high-frequency broadband) across a set of electrodes for the purposes of neural pattern analyses [12] (also see later sections of this chapter). Variations of this approach include a more continuous extraction of spectral frequency contents across a wide frequency range (e.g., 1–150 Hz, 1-Hz steps from 1 to 29 Hz, 5-Hz steps from 30 to 150 Hz) [13] or within the broadband high-frequency range (e.g., 80–120 Hz in bands of 10 Hz) [14]. These variations take into account potential individual differences in frequency-specific contributions, which can be assessed by performing a leave-one-feature-out jackknife procedure [13, 15].

Third, in addition to continuous iEEG time series and their features (e.g., power or phase of certain spectral frequency), another approach is to characterize discrete physiological events that may serve a computational role for cognition. For example, discrete neuronal spiking information can be extracted from LFPs recorded by microscale electrodes via spike-sorting algorithms [16] (see also Chap. 43). The firing rate or spatial–temporal pattern of spiking activity (e.g., sequences) may then be further examined in relation to cognitive processes such as memory formation and retrieval [17, 18]. Likewise, continuous iEEG traces recorded by macroelectrodes can also be filtered to identify discrete synchronous high-frequency neural events (i.e., "ripples") [19], similar to those identified in animal models [20]. These discrete ripple events in the temporal lobe have been shown to play a critical role in supporting memory formation and retrieval [21, 22].

These data extraction considerations are beneficial for reducing the complexity in multivariate analyses and for increasing their interpretability. It is worth noting, however, that analytical considerations should remain secondary to the research questions and designs. Thus, while sophisticated models and algorithms become increasingly available for multivariate analyses, clarifying the physiological characteristics of recorded iEEG signals remains a key consideration in the research of brain and cognition.

34.3 Multivariate Analysis of iEEG Data via Similarity-Based Analyses and Classification

Once a set of iEEG features is identified, one can begin to analyze these multivariate iEEG features in a manner that extends beyond conventional univariate analyses that compare, for example, the mean power values of a certain frequency across different conditions. Two approaches for analyzing multivariate data have been identified. First, one can focus on the similarity or consistency of the patterns of iEEG activity across different experimental stimuli, trials, or conditions. This measure of consistency captures whether certain iEEG features reliably encode information embedded in a cognitive task, providing an estimate of task-related information in iEEG signals. A second approach that extends the investigation of pattern similarity is to apply machine-learning algorithms to decode information available in the extracted iEEG features. These machine learning algorithms provide a direct measure of whether the recorded iEEG signals contain information that can be predicted or read out by a computational model trained by an independent dataset. These two approaches are complementary to one another, and both provide information related to brain activity during a cognitive task. For both of these approaches, time-resolved analyses can be performed with the high temporal resolution afforded by iEEG. Here, we will briefly introduce these two approaches and discuss some principled considerations in interpreting the results from these analyses.

34.3.1 Similarity-Based Analysis

Representational similarity analysis (RSA) has been used as an analytical framework for linking brain-activity measurement, behavioral measurement, and computational modeling [23]. In this framework, "representation" is often operationalized as the evoked patterns of neural activity that contain information associated with a given stimulus or task event. The similarity between representations of any two events can be quantified using any one of several different measurements used for parametric or non-parametric correlation analysis between the vectors of iEEG features for each event. We favor the cosine similarity metric, as it preserves the sign of each feature vector, and thus penalizes feature vectors with similar shape but opposite signs [12]. Irrespective of the exact metric used, if the patterns of neural activity are consistent across a set of homogeneous stimuli or task events, this suggests that these patterns of neural activity retain stimulus- or task-specific information [24]. Notably, this similarity-based approach imposes minimal assumptions on how information is distributed over the neural signals and whether such measures of neural activity are independent. Hence, RSA is analytically appealing to investigate how iEEG measures, regardless of their dependence, are related to cognitive variables measured by a task. Building upon this premise, two subsequent approaches can be considered (Fig. 34.1c, *left*).

First, one can examine how the representational patterns captured by iEEG features are related to an a priori model that predicts similarity patterns across stimuli or task events [25]. For instance, the similarity between evoked patterns of iEEG activity across visual images can be directly correlated with the similarity predicted by deep neural network models between these same images [26] (see also Chap. 50). This approach allows us to identify the amount of shared variance between the neural data and the model. Time-resolved analyses across recording sites based on this approach can then provide further information about when and where the brain exhibits neural activity most similar to the proposed model.

Second, in the absence of a proposed underlying model that would make assumptions regarding the similarity between patterns of activity, one may also examine how a homogeneous set of stimuli or task events are coded across different repetitions during a recording session. For example, in visual perception, by examining how stimuli from the same category evoke similar iEEG patterns, Jacques and colleagues [27] identified iEEG category-selective signals in the human ventral temporal cortex that are consistent with findings revealed by functional magnetic resonance imaging (fMRI). In the context of memory, by examining the similarity of activity patterns at each encoding and retrieval timepoint of a verbal paired-associates recall task, Yaffe and colleagues [12, 28] demonstrated the reinstatement of iEEG activity in cortical macroelectrodes between memory encoding and successful retrieval. Notably, such neural reinstatement is also evident when examining the firing rates [18] and sequential patterns of spike activity [17] captured by microelectrodes placed in the temporal lobe. These findings suggest that similarity-based analyses can bridge multiple levels of neurophysiological phenomena [23] to link multivariate iEEG data with fundamental cognitive principles such as memory reinstatement [29].

34.3.2 Multivariate Classification

Recent advances in machine-learning techniques have provided diverse tools and platforms for decoding information present in iEEG data. This is particularly relevant for brain-machine interfaces, where information read-out is a fundamental requirement for decoding movement and for guiding neuromodulation [30, 31]. Classic approaches such as naive Bayes classification, linear discriminant analysis (LDA), and support vector machine learning (SVM) remain powerful tools to achieve these goals. More recently, the increased computational feasibility of implementing deep neural network modeling has also opened up new opportunities for examining the complex multivariate structure of iEEG data [32, 33]. The key to selecting and harnessing these powerful methods is understanding how each method's assumptions dictate data requirements (Fig. 34.1c, *right*).

First, as different classification algorithms are built upon different assumptions, it is important to consider whether a given iEEG dataset satisfies the assumptions of a chosen classifier. For example, some classifiers may assume independence of the data features (e.g., naive Bayes). iEEG data with dependent features, such as power values from adjacent spectral frequencies, would therefore not be suitable for these types of classifiers. Second, as classification algorithms rely upon training using a large amount of data with different features to test on an independent test set, the selection of relevant data features [34] and the separation of training and test sets remain an important issue to consider [35]. In principle, unbiased feature selection with independent training and test sets may help avoid overfitting. In addition, one may also evaluate model complexity, variance explained, and computational time, which are common considerations when applying machine-learning techniques to biomedical research [35, 36].

Besides these considerations above, the application of multivariate classification of iEEG data may also benefit from incorporating computational models that characterize the underlying cognitive processes to improve classification performance. For example, iEEG research has raised the possibility that speech may be decoded using neuroprosthetic devices [37, 38]. However, in practice, speech decoding performance has been unsatisfactory due to various experimental and modeling barriers [39]. These efforts can be substantially improved by incorporating natural language models that capture linguistic characteristics of words, such as word sequence probability, from our everyday language. For example, Moses and colleagues [40] recently demonstrated that speech decoding error rates can substantially drop from 60.5% to only 25.6% when including a natural language model in their speech classifier based on iEEG data. In light of these observations, incorporating models of how semantic information in our everyday language is represented and used to support diverse mental functions, such as memory search [41] and decision making [42], may offer a promising approach for exploring brain-machine interfaces in higher human cognition.

34.4 Summary

In this chapter, we have outlined the multivariate nature of iEEG data and discussed a few principled considerations for data extraction and the choice of analytical options. These include various ways to extract relevant data to reduce complexity and improve interpretability, similarity-based analyses with or without an a priori model, and approaches to ensure and improve the quality of multivariate classification. While these approaches and considerations may serve well as starting points, the quest for analyzing the rich information in multivariate iEEG data does not end here. As pointed out by John Tukey [43], data analysis in most cases is like detective work and requires one to embrace the flexibility of high-dimensional data with some guiding principles and converging evidence. In doing so, we may maximize the rich information provided by iEEG recordings in order to advance our understanding of the human brain and cognition.

References

1. Buzsáki G, Anastassiou CA, Koch C (2012) The origin of extracellular fields and currents-EEG, ECoG, LFP and spikes. Nat Rev Neurosci 13:407–420. https://doi.org/10.1038/nrn3241
2. Parvizi J, Kastner S (2018) Promises and limitations of human intracranial electroencephalography. Nat Neurosci 21:474–483
3. Miller KJ, Sorensen LB, Ojemann JG, den Nijs M (2009) Power-law scaling in the brain surface electric potential. PLoS Comput Biol 5:e1000609. https://doi.org/10.1371/journal.pcbi.1000609
4. Canolty RT, Edwards E, Dalal SS, Soltani M, Nagarajan SS, Kirsch HE, Berger MS, Barbare NM, Knight RT (2006) High gamma power is phase-locked to theta oscillations in human neocortex. Science 313(80):1626–1628. https://doi.org/10.1126/science.1128115
5. Manning JR, Jacobs J, Fried I, Kahana MJ (2009) Broadband shifts in local field potential power spectra are correlated with single-neuron spiking in humans. J Neurosci 29:13613–13620. https://doi.org/10.1523/JNEUROSCI.2041-09.2009
6. Bigdely-Shamlo N, Mullen T, Kothe C, Su K-M, Robbins KA (2015) The PREP pipeline: standardized preprocessing for large-scale EEG analysis. Front Neuroinform 9:1–20. https://doi.org/10.3389/fninf.2015.00016
7. Geyer K, Campbell F, Chang A, Magnotti J, Beauchamp M, Allen GI (2020) Interpretable visualization and higher-order dimension reduction for ECoG data. In: 2020 IEEE international conference on big data (big data). IEEE, pp 2664–2673
8. Wittig JH, Jang AI, Cocjin JB, Inati SK, Zaghloul KA (2018) Attention improves memory by suppressing spiking-neuron activity in the human anterior temporal lobe. Nat Neurosci 21:808–810
9. Haller M, Case J, Crone NE, Chang EF, King-Stephens D, Laxer KD, Weber PB, Parvizi J, Knight RT, Shestyuk AY (2018) Persistent neuronal activity in human prefrontal cortex links perception and action. Nat Hum Behav 2:80–91. https://doi.org/10.1038/s41562-017-0267-2
10. Chapeton JI, Haque R, Wittig JH, Inati SK, Zaghloul KA (2019) Large-scale communication in the human brain is rhythmically modulated through alpha coherence. Curr Biol 29:2801–2811. https://doi.org/10.1016/j.cub.2019.07.014
11. Chapeton JI, Inati SK, Zaghloul KA (2017) Stable functional networks exhibit consistent timing in the human brain. Brain 140:628–640. https://doi.org/10.1093/brain/aww337
12. Yaffe RB, Kerr MSD, Damera S, Sarma SV, Inati SK, Zaghloul KA (2014) Reinstatement of distributed cortical oscillations occurs with precise spatiotemporal dynamics during successful memory retrieval. Proc Natl Acad Sci United States Am 111:18727–18732
13. Estefan DP, Zucca R, Arsiwalla X, Principe A, Zhang H, Rocamora R, Axmacher N, Verschure PFMJ (2021) Volitional learning promotes theta phase coding in the human hippocampus. Proc Natl Acad Sci U S A 118:1–12. https://doi.org/10.1073/pnas.2021238118
14. Peters MAK, Thesen T, Ko YD, Maniscalco B, Carlson C, Davidson M, Doyle W, Kuzniecky R, Devinsky O, Halgren E, Lau H (2017) Perceptual confidence neglects decision-incongruent evidence in the brain. Nat Hum Behav 1:1–8. https://doi.org/10.1038/s41562-017-0139
15. Zhang H, Fell J, Staresina BP, Weber B, Elger CE, Axmacher N (2015) Gamma power reductions accompany stimulus-specific representations of dynamic events. Curr Biol 25:635–640. https://doi.org/10.1016/j.cub.2015.01.011
16. Harris KD, Henze DA, Csicsvari J, Hirase H, Buzsáki G (2000) Accuracy of tetrode spike separation as determined by simultaneous intracellular and extracellular measurements. J Neurophysiol 84:401–414. https://doi.org/10.1152/jn.2000.84.1.401

17. Vaz AP, Wittig JH, Inati SK, Zaghloul KA (2020) Replay of cortical spiking sequences during human memory retrieval. Science 367(80):1131–1134. https://doi.org/10.1126/science.aba0672
18. Jang AI, Wittig JH, Inati SK, Zaghloul KA (2017) Human cortical neurons in the anterior temporal lobe reinstate spiking activity during verbal memory retrieval. Curr Biol 27:1700–1705. https://doi.org/10.1016/j.cub.2017.05.014
19. Tong APS, Vaz AP, Wittig JH, Inati SK, Zaghloul KA (2021) Ripples reflect a spectrum of synchronous spiking activity in human anterior temporal lobe. Elife 10:1–25. https://doi.org/10.7554/eLife.68401
20. Buzsáki G (2015) Hippocampal sharp wave-ripple: a cognitive biomarker for episodic memory and planning. Hippocampus 25:1073–1188. https://doi.org/10.1002/hipo.22488
21. Vaz AP, Inati SK, Brunel N, Zaghloul KA (2019) Coupled ripple oscillations between the medial temporal lobe and neocortex retrieve human memory. Science 363(80):975–978. https://doi.org/10.1126/SCIENCE.AAU8956
22. Norman Y, Yeagle EM, Khuvis S, Harel M, Mehta AD, Malach R (2019) Hippocampal sharp-wave ripples linked to visual episodic recollection in humans. Science 365(80):eaax1030. https://doi.org/10.1126/science.aax1030
23. Kriegeskorte N, Mur M, Bandettini P (2008) Representational similarity analysis—connecting the branches of systems neuroscience. Front Syst Neurosci 2:1–28
24. Kriegeskorte N, Diedrichsen J (2019) Peeling the onion of brain representations. Annu Rev Neurosci 42:407–432. https://doi.org/10.1146/annurev-neuro-080317-061906
25. Nili H, Wingfield C, Walther A, Su L, Marslen-Wilson W, Kriegeskorte N (2014) A toolbox for representational similarity analysis. PLoS Comput Biol 10:e1003553
26. Liu J, Zhang H, Yu T, Ni D, Ren L, Yang Q, Lu B, Wang D, Heinen R, Axmacher N, Xue G (2020) Stable maintenance of multiple representational formats in human visual short-term memory. Proc Natl Acad Sci U S A 117:32329–32339. https://doi.org/10.1073/pnas.2006752117
27. Jacques C, Witthoft N, Weiner KS, Foster BL, Rangarajan V, Hermes D, Miller KJ, Parvizi J, Grill-Spector K (2016) Corresponding ECoG and fMRI category-selective signals in human ventral temporal cortex. Neuropsychologia 83:14–28. https://doi.org/10.1016/j.neuropsychologia.2015.07.024
28. Yaffe RB, Shaikhouni A, Arai J, Inati SK, Zaghloul KA (2017) Cued memory retrieval exhibits reinstatement of high Gamma power on a faster timescale in the left temporal lobe and prefrontal cortex. J Neurosci 37:4472–4480
29. Tulving E, Thomson DM (1973) Encoding specificity and retrieval processes in episodic memory. Psychol Rev 80:352–373
30. Shih JJ, Krusienski DJ, Wolpaw JR (2012) Brain-computer interfaces in medicine. Mayo Clin Proc 87:268–279. https://doi.org/10.1016/j.mayocp.2011.12.008
31. Volkova K, Lebedev MA, Kaplan A, Ossadtchi A (2019) Decoding movement from electrocorticographic activity: a review. Front Neuroinform 13:1–20. https://doi.org/10.3389/fninf.2019.00074
32. Antoniades A, Spyrou L, Martin-Lopez D, Valentin A, Alarcon G, Sanei S, Took CC (2018) Deep neural architectures for mapping scalp to intracranial EEG. Int J Neural Syst 28:1–15. https://doi.org/10.1142/S0129065718500090
33. Antoniades A, Spyrou L, Took CC, Sanei S (2016) Deep learning for epileptic intracranial EEG data. In: 2016 IEEE 26th international workshop on machine learning for signal processing (MLSP). IEEE, pp 1–6
34. Chandrashekar G, Sahin F (2014) A survey on feature selection methods. Comput Electr Eng 40:16–28. https://doi.org/10.1016/j.compeleceng.2013.11.024
35. Greener JG, Kandathil SM, Moffat L, Jones DT (2022) A guide to machine learning for biologists. Nat Rev Mol Cell Biol 23:40–55. https://doi.org/10.1038/s41580-021-00407-0
36. Zitnik M, Nguyen F, Wang B, Leskovec J, Goldenberg A, Hoffman MM (2019) Machine learning for integrating data in biology and medicine: Principles, practice, and opportunities. Inf Fusion 50:71–91. https://doi.org/10.1016/j.inffus.2018.09.012

37. Chang EF, Niziolek CA, Knight RT, Nagarajan SS, Houde JF (2013) Human cortical sensorimotor network underlying feedback control of vocal pitch. Proc Natl Acad Sci 110:2653–2658. https://doi.org/10.1073/pnas.1216827110
38. Chang EF, Rieger JW, Johnson K, Berger MS, Barbaro NM, Knight RT (2010) Categorical speech representation in human superior temporal gyrus. Nat Neurosci 13:1428–1432. https://doi.org/10.1038/nn.2641
39. Martin S, Millán J del R, Knight RT, Pasley BN (2019) The use of intracranial recordings to decode human language: challenges and opportunities. Brain Lang 193:73–83. https://doi.org/10.1016/j.bandl.2016.06.003
40. Moses DA, Metzger SL, Liu JR, Anumanchipalli GK, Makin JG, Sun PF, Chartier J, Dougherty ME, Liu PM, Abrams GM, Tu-Chan A, Ganguly K, Chang EF (2021) Neuroprosthesis for decoding speech in a paralyzed person with anarthria. N Engl J Med 385:217–227. https://doi.org/10.1056/nejmoa2027540
41. Xie W, Bainbridge WA, Inati SK, Baker CI, Zaghloul KA (2020) Memorability of words in arbitrary verbal associations modulates memory retrieval in the anterior temporal lobe. Nat Hum Behav 4:937–948. https://doi.org/10.1038/s41562-020-0901-2
42. Zhao WJ, Richie R, Bhatia S (2021) Process and content in decisions from memory. Psychol Rev. https://doi.org/10.1037/rev0000318
43. Tukey JW (1969) Analyzing data: sanctification or detective work? Am Psychol 24:83–91

Chapter 35
How Can I Conduct Surrogate Analyses, and How Should I Shuffle?

Hui Zhang

Abstract The surrogate analysis is a widely used assumption-free statistical method that can be applied to data sets which do not satisfy prerequisites for parametric statistical tests. Due to several practical reasons, the distribution or parameters of intracranial EEG (iEEG) data violates the assumption of parametric tests. In this case, the surrogate analysis provides a solution to accurately estimate the significance level by ranking the empirical statistical value within a data-specific null-hypothesis distribution of permuted statistical values. Different from parametric analyses, surrogate analyses are performed regardless of any assumptions about the distribution of the empirical data. As introduced in previous chapters, iEEG data is composed of multiple dimensions: It samples from different frequencies at different time points from different sensors. Applying statistical tests to multi-dimensional iEEG data is likely to introduce multiple comparison problems which can be corrected by the surrogate analysis. In this chapter, I will introduce in detail the surrogate analysis and its application for iEEG data analyses.

35.1 What is a Surrogate Analysis?

The surrogate analysis is an assumption-free statistical method. The essence of the surrogate analysis is to setup a data-specific null-hypothesis distribution of permuted statistical values that does not make any assumptions about the distribution and parameters of the empirical data. The significance level of a statistical test on the empirical data is computed by comparing the empirical statistical value to the null-hypothesis distribution of permuted statistical values which are calculated by using permutation methods discussed in Sect. 35.3.

H. Zhang (✉)
Department of Neuropsychology, Institute of Cognitive Neuroscience, Ruhr University Bochum, Bochum, Germany
e-mail: Hui.Zhang-c5u@rub.de

N. Axmacher (ed.), *Intracranial EEG*, Studies in Neuroscience, Psychology and Behavioral Economics, https://doi.org/10.1007/978-3-031-20910-9_35

35.2 When to Adopt a Surrogate Analysis?

A prerequisite for performing parametric statistical tests is that the data or transformed data (e.g., logarithmically transformed data) is normally distributed. Applying parametric tests to data that are not normally distributed induces inaccurate calculations of p values, confidence intervals, and effect sizes. In this case, the surrogate analysis provides a non-parametric approach which does not rely on the assumption of normally distributed data. In the following part, I will discuss situations in which the surrogate analysis is utilized in iEEG studies.

35.2.1 Not Normally Distributed Data

The normal distribution defines how a data set is distributed around the mean. It is a prerequisite for most parametric analyses. In order to obtain a normally distributed data set, researchers can either test a large number of subjects to decrease the variance across subjects or test a large number of trials within each subject to reduce the unexpected noise of each subject. Collecting a data set with a large number of subjects or a large number of trials for each subject is not easy in iEEG studies. Patient numbers in iEEG studies are usually small (Fig. 35.1) for several reasons. First, the number of implanted patients is small because not all epilepsy patients are treated with neurosurgery [1, 2]. Second, not all implanted patients are suitable for testing because of cognitive dysfunctions [3] (Chap. 2). Third, the subject number is even smaller when researchers are focusing on sensors located in one or several specific brain region(s). Even when patients are testable, they may not conduct as many testing trials as healthy participants due to several practical reasons (e.g., patients getting fatigued easily, uncomfortableness or other clinical reasons). As a result, experimental paradigms adapted for implanted epilepsy patients are usually shorter in duration and lower in difficulty level. Moreover, it is also hard to rigidly constrain patients' demographic information (e.g., age and gender) because the implantation procedure is purely serving the clinical purpose. The reduced control over participant recruitment, limited number of patients, and small number of trials per patient affect the skewness and kurtosis of the data. This likely results in non-normally distributed iEEG data. In this case, the significance level of a statistical test can be accurately estimated by setting up a distribution of permuted statistical values under the null-hypothesis which are calculated from shuffled data that have a similar skewness and kurtosis as the empirical data.

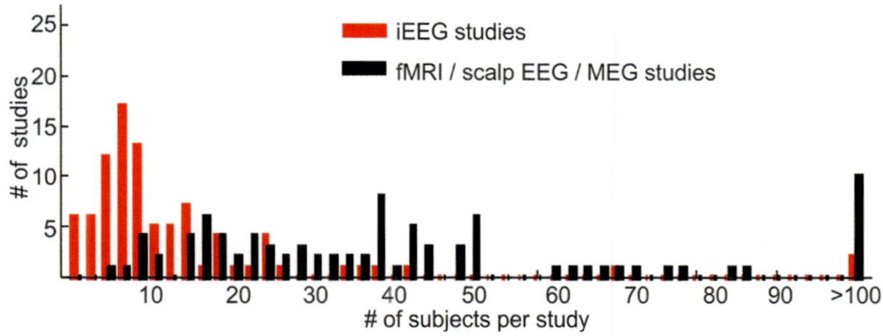

Fig. 35.1 Number of subjects per study in peer reviewed papers. Among all peer reviewed papers from 2016–2021 on pubmed (https://pubmed.ncbi.nlm.nih.gov/), 15 iEEG papers and 15 fMRI/scalp EEG/MEG papers were randomly selected from each year. The plot indicates that the number of subjects per iEEG study is in general smaller than the number of subjects per fMRI/scalp EEG/MEG study

35.2.2 Non-Matched Trial Numbers Between Conditions

When researchers design an experiment, they often plan to collect the same number of trials for matched experimental conditions. This rules out the possibility that differences between experimental conditions are caused by differences of trial numbers. IEEG studies may result in non-matched trial numbers for several reasons. When epilepsy patients are running tasks, it happens frequently that an experiment is terminated because patients are getting fatigued or because of clinical constraints. Early termination may result in different numbers of trials between experimental conditions. When researchers afterwards process the data, the numbers of artifact-contaminated trials may again differ between experimental conditions. After artifact rejection, the numbers of clean trials may thus differ between conditions. Finally, a difference in trial numbers may be induced by applying certain data analysis methods. For example, when researchers perform representational similarity analysis (RSA; [4]) between encoding and retrieval of n items, the generated similarity matrix is composed of n on-diagonal data points (i.e., similarities between same items) and $n \times (n\text{-}1)$ off-diagonal data points (i.e., similarities between different items). When conditions with different trial numbers are compared with each other (e.g., on-diagonal RSA data vs. off-diagonal RSA data), the significance level can be accurately estimated by ranking the empirical statistical value within the distribution of permuted statistical values which are computed based on shuffled conditions that have the same difference of trial numbers as the empirical data.

35.2.3 Correction for Multiple Comparisons

An iEEG data set usually has multiple dimensions (e.g., experimental condition X time X frequency X sensor). When different experimental conditions are compared in several of these dimensions, the same statistical test is applied to a large number of data points (e.g., time X frequency X sensor). Repeated application of the same statistical test causes multiple comparison problems that increase the type I error and generate false positive conclusions. It is thus imperative to correct for multiple comparisons. Importantly, different data points of iEEG data are not independent from each other. Successive time points, adjacent frequency bins, and nearby sensors are correlated with each other due to temporal autocorrelations, the nature of time–frequency decompositions, volume conduction and possibly other factors. The cluster-based surrogate analysis [5] is one of the methods to solve multiple comparison problems in iEEG data and will be described in detail in Sect. 35.4.

35.3 How to Perform Surrogate Analyses?

In surrogate analyses, it is important to obtain a null-hypothesis distribution of permuted statistical values using appropriate permutation approaches. In general, the application of a permutation approach depends on the research question, the characteristics of the data structure, and the statistical tests applied to the empirical data. Thus, the shuffling procedure can be applied to subject level, condition level, or trial level. Note that it is important to shuffle variables that one is interested in while maintaining other variables as similar to the empirical data as possible. For example, if one investigates the difference of gamma power between two brain regions within 200 ms after displaying pictures to participants, one could randomly switch labels of brain regions within each participant. On the other hand, if one investigates the difference of gamma power between time windows 0–200 ms and 200–400 ms after displaying pictures to participants, one could randomly switch time labels of each trial while maintaining labels of brain regions the same as in the empirical data. In this section, I will introduce different ways of permuting the data and thus different procedures of performing a surrogate analysis.

35.3.1 Switching Condition Labels

When target conditions are matched with each other, a surrogate analysis is performed based on shuffled data with shuffled conditions that are also matched with each other. For example, in order to maximize the statistical power/minimize the data collection, experimental conditions are designed as within-subject effects in many iEEG studies. The statistical test is performed by considering the matched feature of the empirical data (e.g., using a paired t-test). To accurately compute the p level of the empirical statistical value, the null-hypothesis distribution of the permuted statistical values is computed based on permuted conditions that are also matched with each other. Switching the condition label can generate permuted data by maintaining the matched feature (e.g., shuffled conditions are also from the same subject). It is easy to apply this method in practice. Within each permutation, a vector is randomly generated to indicate which condition pairs (e.g., subjects) should switch labels. After labels are switched, a set of permuted experimental conditions is generated. Then the same statistical test is applied to the permuted data. The obtained statistical value is saved as a permuted statistical value. If there are two experimental conditions from n subjects, the number of possible permutations is 2^n.

35.3.2 Shuffling Condition Labels

The shuffling label approach can be applied to different situations with slightly different ways of shuffling. In the first situation, experimental conditions are independent from each other. The selection of the statistical test (e.g., two sample T test) considers the fact that the empirical data are independent. The shuffling condition label approach also takes this into account. To this end, data sets from different conditions are first pooled together. Within each permutation, the same number of data sets as in the empirical data are then drawn from the pool for each condition. Afterwards, the same statistical test is applied to the permuted data. This generates a permuted statistical value. If there are two experimental conditions with one condition having m subjects and the other condition having n subjects, the number of all possible permutations is $C^n_{m+n} = \frac{(m+n)!}{n! \times m!}$.

As discussed in Sect. 35.2.2, there are situations in which trial numbers differ between matched experimental conditions. This possibly biases the estimation of the significance level of the statistical test. In order to control for this bias, the null-hypothesis distribution of permuted statistical values is calculated based on permuted conditions with the same bias in trial numbers. Within each permutation, one pools all trials together from matched conditions within each condition pair (e.g., within each subject). Afterwards, the same number of trials as in the empirical data is randomly drawn from the trial pool for each condition within each pair. At the end, the same statistical test is applied, generating a permuted statistical value. Let's assume that there are two experimental conditions with one condition having m trials and the other condition having n trials for one subject ($subj_j$). Using the shuffling condition label approach, the number of all possible permutations for that subject (p_{subj_j}) is C^n_{m+n}. The number of all possible permutations for all subjects is $p_{subj_1} \times p_{subj_2} \times \cdots \times p_{subj_{all}}$. Note that the shuffling approach applied to matched conditions shuffles at the trial level. By contrast, the shuffling approach applied to independent conditions in the above paragraph is shuffling at the subject level.

35.3.3 Shifting Permutation

In iEEG studies, several data analysis methods are based on continuous data series, e.g., calculating phase-locking values [6], phase-amplitude coupling [7], spike trains of single unit data [8, 9] etc. In these analyses, the continuous data structure is an important feature that affects the result. In order to maintain this continuous feature, a shifting permutation approach can be applied. Within each permutation, a data time series is randomly segmented into two epochs. These two epochs are reconnected to form a permuted series of data by switching the sequence of them (i.e., the end of the second epoch connects to the start of the first epoch). The same data analyses and statistical tests are then applied to the permuted data, resulting in a permuted statistical value. The continuity of data and the relations between adjacent data points (e.g., temporal autocorrelations) are largely preserved after shifting permutation. Thus, this approach can generate a null-hypothesis distribution of permuted statistical values that accurately estimates the significance level of the statistical test on the empirical data. If there are n data points in a series of data, the number of all possible permutations is n.

35.3.4 Procedure of the Surrogate Analysis

The surrogate analysis tests an assumption-free null-hypothesis that there is no statistical difference between experimental conditions (i.e., that data from different experimental conditions are drawn from the same pool). To test this null-hypothesis, the surrogate analysis is performed in three steps in practice. The first step is to perform the statistical test on the empirical data (e.g., a paired T test across subjects between two matched experimental conditions) and to extract the statistical value (e.g., the t value). The second step is to build a null-hypothesis distribution of permuted statistical values (e.g., a permuted t distribution) by repeatedly performing the permutation procedure (Sects. 35.3.1–35.3.3) multiple times. The last step is to identify the rank of the empirical statistical value within the null-hypothesis distribution of permuted statistical values to compute the p value. If an empirical statistical value is larger/smaller than 95% of all permuted values, it can be concluded that the null-hypothesis is rejected at an alpha level of 0.05 ($p = 0.05$).

35.4 Cluster-Based Permutation Analysis

As mentioned in Sect. 35.2.3, an iEEG data set is usually composed of multiple dimensions (e.g., experimental condition X time X frequency X sensor). Applying statistical tests between experimental conditions for these various dimensions (e.g., time X frequency X sensor) causes multiple comparison problems. It is thus imperative to correct for multiple comparisons to reduce the type I error. A conventional way is performing a Bonferroni correction. However, this method is too conservative because there are usually a large number of data points in iEEG studies. Let's assume that there are 200 data points. To achieve an alpha level at 0.05 after Bonferroni correction, each data point would need to correspond to an uncorrected p value lower than 0.05/200 = 0.00025. This would likely introduce large type II errors. In addition, the Bonferroni correction assumes that data points are independent from each other. This assumption is not realistic for iEEG data as described in Sect. 35.2.3. To solve this problem, previous researchers have proposed cluster-based permutation analysis [5] which has been applied to more than 2,000 studies up till now, including some from our own group [10–13].

One advantage of applying a cluster-based permutation analysis is that it does not rely on any assumption about the distribution of the empirical data. Another advantage is that it does not assume that data points are independent from each other. Instead, it naturally assumes that nearby data points are correlated with each other. This reflects important characteristics of iEEG data because temporally successive data points (e.g., data from 10 and 20 ms after stimulus onset), data points from adjacent frequencies (e.g., EEG power at 10 Hz and 11 Hz), and spatially nearby data points (e.g., two nearby sensors from the same brain region) are correlated.

One assumption underlying cluster-based permutation analysis is that all data points contribute equally to the to-be tested null-hypothesis. The homogenous distribution of iEEG data is sometimes difficult to be satisfied. In general, the iEEG data within each frequency band or brain region is homogenous. However, it is inhomogeneous if it is across a large frequency range or across different brain regions. To solve this problem, a region-of-interest (ROI) approach can be applied by focusing on a specific frequency band, time period, or brain region. In the frequency domain, this inhomogeneity problem can also be addressed either by applying different frequency resolutions in different frequency ranges (e.g., sampling at 1 Hz for low frequency bands and 5 Hz for high frequency bands [10, 11, 14]) or by selecting frequencies in a logarithmically transformed space (e.g., equally spaced samples of logarithmically transformed frequencies; [15]).

35.4.1 Criterions of Defining a Cluster

One important aspect when performing the cluster-based permutation analysis is defining a cluster. After performing statistical tests, each data point has a statistical value and a corresponding p value. A cluster is defined via those connected data points that each have a p value corresponding to a pre-defined alpha level (e.g., $p < 0.05$). There are different criteria for defining if two data points are connected. To better describe these criteria, let's assume that all data points are equally sized geometric figures (Fig. 35.2). For example, in one dimensional data all data points are line segments of equal length (e.g., when calculating ERP components; Fig. 35.2a), in two dimensional data they are equally sized squares (e.g., when computing time–frequency clusters; Fig. 35.2b, c), and in three dimensional data they are equally sized cubes (e.g., when identifying brain regions; Fig. 35.2d, f). In one dimensional data, two data points with a shared endpoint are defined as being connected (Fig. 35.2a). In two dimensional data, there are two ways of defining if two data points are connected: The first one is that two data points share an edge (Fig. 35.2b); the second one is

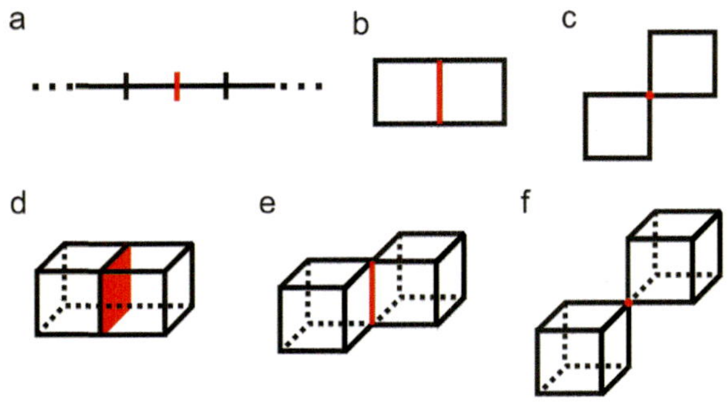

Fig. 35.2 Criteria for defining a cluster. (a) Two data points with a shared end are defined as connected in one dimensional data. (b, c) Two data points with a shared edge (b) or a shared vertex (c) can be defined as connected in two dimensional data. (d, f) Two data points with a shared surface (d), a shared edge (e) or a shared vertex (f) can be defined as connected in three dimensional data. The red dot/line/surface in each panel indicates the connected part of two data points

that two data points share a vertex (Fig. 35.2c). One can either define clusters by the first criterion or by both criteria. In three dimensional data, there are three ways of defining if two data points are connected: The first one is that two data points share the same surface (Fig. 35.2d); the second one is that two data points share the same edge (Fig. 35.2e); the third one is that two data points share the same vertex (Fig. 35.2f). One can choose either the first, the combination of the first and the second, or the combination of all three criteria. Note that data points that show effects in opposite directions (e.g., data points having positive statistical values vs. data points having negative statistical values) should not be clustered together even when these data points are connected.

35.4.2 Procedure of Performing the Cluster-Based Permutation Analysis

The overall procedure of performing the cluster-based permutation analysis does not differ much from other surrogate analyses (Sect. 35.3.4). To make the description easier to read, I will focus on clusters in which all data points have positive statistical values (i.e., positive clusters). The procedure is composed of three steps. The first step is to perform statistical tests on the empirical data and select all resulting clusters (Sect. 35.4.1). Within each cluster, the summed statistical value across all data points is extracted. The extracted value is assigned as the statistical value of that cluster. The second step is to perform an appropriate permutation analysis as discussed in Sect. 35.3. Within each permutation, one performs the same statistical test as in the empirical data and selects all clusters. Again, one assigns a statistical value to each cluster, using exactly the same criteria as for the empirical data. Afterwards, the cluster with the largest statistical value is saved for further analyses. If no cluster is found in one permutation, a statistical value of zero is saved for this permutation. The permutation procedure is repeated n times and generates n permuted statistical values. In the third step, the statistical value of each empirical cluster is ranked within the distribution of permuted statistical values to compute the alpha level of that cluster. A cluster having a statistical value that is larger than 95% of all permuted statistical values rejects the null-hypothesis at an alpha level of 0.05. This suggests a significant difference between experimental conditions within that cluster.

There are statistical tests which can generate negative clusters in which each data point has a negative statistical value (e.g., resulting from paired or two-sample T tests). The procedure to test the null-hypothesis of negative clusters is largely the same as for positive clusters except for two aspects: In the second step, the permuted statistical value is the *smallest* statistical value among all negative clusters in each permutation. In the third step, a cluster with a statistical value *smaller* than 95% of all permuted negative clusters rejects the null-hypothesis at the alpha level of 0.05. If both positive and negative clusters are obtained from the empirical data, the null-hypothesis distributions of both positive and negative permuted statistical values should be built separately. Usually, the null-hypothesis distributions of the positive and negative permuted statistical values are not symmetric around the y axis unless one uses the switching label approach (Sect. 35.3.1). The significance levels of positive and negative empirical clusters are estimated within the distribution of positive and negative permuted statistical values respectively.

35.5 How Many Permutations Are Appropriate?

In general, the more times one permutes, the more stable the result is (Fig. 35.3a, b). In practice, it is usually not possible to perform all possible permutations. For example, if one computes the significance level of a t value from a paired T test across 20 subjects using the switching condition label approach, the number of all possible permutations is 2^{20} which is more than one billion. The selection of the number of permutations is affected by several aspects. One is the computing load. Nowadays, it is typically no problem to perform a large number (e.g., 10,000) of iterations on a computer with high-end configuration. The second is if the obtained significance level is used for further computation. For example, the obtained significance level may be applied to a further Bonferroni correction together with 4 other p values. In order to survive the Bonferroni correction, each estimated p value should be smaller than 0.05/5 (0.01) after surrogate analysis. In this case, the iteration number of permutations should be not smaller than 100. The last but not the least consideration is that it depends on the requirement of the precision. In some cases, the empirical statistical value exceeds all permuted values. If the iteration number is 20, it can be reported that $p < 0.05$. The precision is obviously low. One can increase the iteration number to increase the precision. For example, 1,000 iterations can generate a smallest $p < 0.001$ which is precise enough for most studies.

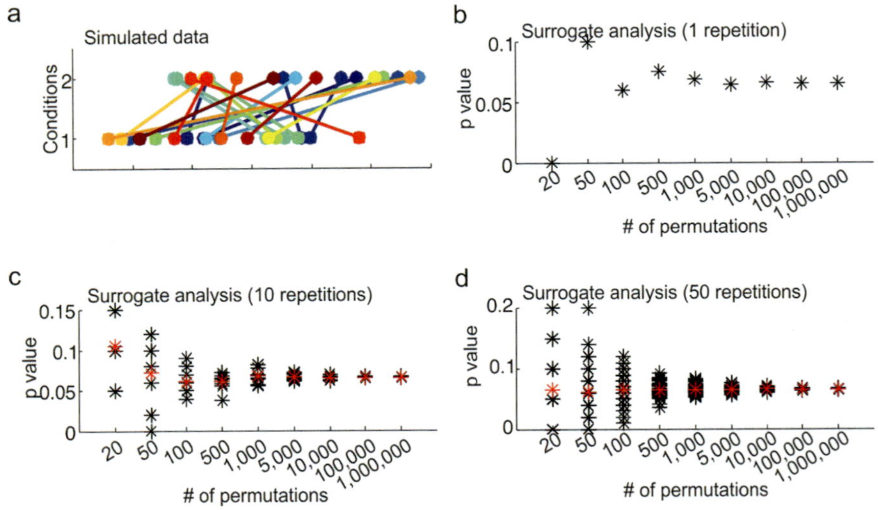

Fig. 35.3 Methods to accurately estimate the significance level. (a) Simulated data of two experimental conditions from 20 subjects. The null-hypothesis is that there is no difference between the two conditions. Each colourful dot indicates one simulated subject. The same colour indicates data from the same subject. (b) A paired T test is applied to the simulated data and the resulting *t* value is assigned as the empirical statistical value. The null-hypothesis distribution of permuted statistical values is obtained by randomly switching condition labels. With increasing numbers of permutations, the estimated *p* value becomes more and more stable. (c, d) To obtain a stable *p* value with smaller numbers of permutations, the surrogate analysis is repeated 10 (c) or 50 (d) times. This generates 10 or 50 estimated *p* values, respectively. Mean *p* values are computed by averaging across these 10 or 50 *p* values. As illustrated in the figure, the mean *p* value with many repetitions (50 times; d) of the surrogate analysis is more stable than with few repetitions (10 times; c) when the iteration number is small

When the number of all possible permutations and the actual iteration number of permutations differ much, this affects the accuracy of the significance level (Fig. 35.3a, b). To solve this problem, one can either increase the number of iterations or repeat the surrogate analysis multiple times. By repeating the surrogate analysis, one obtains a *p* distribution (Fig. 35.3c, d). The median or mean *p* value of the *p* distribution is more stable when the iteration number is small and also more similar to the *p* value computed with a large iteration number. By repeating a surrogate analysis that has a small number of iterations of the permutation procedure (e.g., 20 times as in Fig. 35.3d) for a small number of times (e.g., 50 times as in Fig. 35.3d), one could not only obtain a stable estimation of the significance level of the empirical data, but also dramatically decrease the overall computing load. It is especially helpful to reduce the influence of a biased extraction of data sets in a surrogate analysis.

References

1. Engel J, Wiebe S, French J, Sperling M, Williamson P, Spencer D, Gumnit R, Zahn C, Westbrook E, Enos B, Quality Standards Subcommittee of the American Academy of Neurology, American Epilepsy Society, American Association of Neurological Surgeons (2003) Practice parameter: temporal lobe and localized neocortical resections for epilepsy: report of the Quality Standards Subcommittee of the American Academy of Neurology, in association with the American Epilepsy Society and the American Association of Neurological Surgeons. Neurology 60:538–547. https://doi.org/10.1212/01.wnl.0000055086.35806.2d
2. Foldvary N, Bingaman WE, Wyllie E (2001) Surgical treatment of epilepsy. Neurol Clin 19:491–515. https://doi.org/10.1016/s0733-8619(05)70028-1
3. van Rijckevorsel K (2006) Cognitive problems related to epilepsy syndromes, especially malignant epilepsies. Seizure 15:227–234. https://doi.org/10.1016/j.seizure.2006.02.019
4. Kriegeskorte N, Mur M, Bandettini P (2008) Representational similarity analysis—connecting the branches of systems neuroscience. Front Syst Neurosci 2:4. https://doi.org/10.3389/neuro.06.004.2008
5. Maris E, Oostenveld R (2007) Nonparametric statistical testing of EEG- and MEG-data. J Neurosci Methods 164:177–190. https://doi.org/10.1016/j.jneumeth.2007.03.024
6. Lachaux JP, Rodriguez E, Martinerie J, Varela FJ (1999) Measuring phase synchrony in brain signals. Hum Brain Mapp 8:194–208. https://doi.org/10.1002/(sici)1097-0193(1999)8:4%3c194::aid-hbm4%3e3.0.co;2-c
7. Tort ABL, Komorowski R, Eichenbaum H, Kopell N (2010) Measuring phase-amplitude coupling between neuronal oscillations of different frequencies. J Neurophysiol 104:1195–1210. https://doi.org/10.1152/jn.00106.2010
8. Ekstrom AD, Kahana MJ, Caplan JB, Fields TA, Isham EA, Newman EL, Fried I (2003) Cellular networks underlying human spatial navigation. Nature 425:184. https://doi.org/10.1038/nature01964
9. Vaz AP, Wittig JH, Inati SK, Zaghloul KA (2020) Replay of cortical spiking sequences during human memory retrieval. Science 367:1131–1134. https://doi.org/10.1126/science.aba0672
10. Estefan DP, Sánchez-Fibla M, Duff A, Principe A, Rocamora R, Zhang H, Axmacher N, Verschure PFMJ (2019) Coordinated representational reinstatement in the human hippocampus and lateral temporal cortex during episodic memory retrieval. Nat Commun 10:2255. https://doi.org/10.1038/s41467-019-09569-0
11. Liu J, Zhang H, Yu T, Ni D, Ren L, Yang Q, Lu B, Wang D, Heinen R, Axmacher N, Xue G (2020) Stable maintenance of multiple representational formats in human visual short-term memory. PNAS 117:32329–32339. https://doi.org/10.1073/pnas.2006752117
12. Zhang H, Fell J, Staresina BP, Weber B, Elger CE, Axmacher N (2015) Gamma power reductions accompany stimulus-specific representations of dynamic events. Curr Biol 25:635–640. https://doi.org/10.1016/j.cub.2015.01.011
13. Zhang H, Fell J, Axmacher N (2018) Electrophysiological mechanisms of human memory consolidation. Nat Commun 9:4103. https://doi.org/10.1038/s41467-018-06553-y
14. Munia TTK, Aviyente S (2019) Time-frequency based phase-amplitude coupling measure for neuronal oscillations. Sci Rep 9:12441. https://doi.org/10.1038/s41598-019-48870-2
15. Thammasan N, Miyakoshi M (2020) Cross-frequency power-power coupling analysis: A useful cross-frequency measure to classify ICA-decomposed EEG. Sensors 20:7040. https://doi.org/10.3390/s20247040

Chapter 36
How Can iEEG Data Be Analyzed via Multi-Level Models?

Pengcheng Lv and Liang Wang

Abstract iEEG data are usually collected from a small number of patients with a number of electrode contacts who may perform different task conditions. For such nested data, traditional linear regression models are not ideally suited. Instead, we recommend using linear mixed effect (LME) models to analyze iEEG data. These models describe the relationship between a dependent variable and independent variables, with coefficients that can vary depending on group variables. A typical LME consists of two parts, fixed effects and random effects. Fixed-effects terms are the traditional linear regression part, while the random effects are associated with patients and the nested contacts. LME captures the hierarchical structure of the data, accounting for differential variability between patients and between and within contacts, and potentially allows for generalization to the population. In this chapter, we list three common models to illustrate how to apply LME to the iEEG data.

36.1 Introduction

IEEG data are local field potential signals measured directly from drug-resistant epilepsy patients who have electrodes implanted in their brain to localize the epileptic focus. This type of data has the advantages of high temporal and spatial resolution and high signal-to-noise ratio [2, 10]. It is a powerful tool to explore how the human brain works during cognitive processing [3, 7, 15]. Compared to non-invasive neuroimaging data, iEEG data are relatively difficult to collect (see also Chaps. 4 and 5). On the one hand, recruitment of patients is relatively slow and depending on clinical resources and considerations. On the other hand, given the fact that the locations of the implanted electrode are completely determined by clinical needs, whether the electrode contacts are located in the region of interest does not depend on basic research questions. Therefore, in most cases we may collect a relatively small number of patients (i.e., smaller than in fMRI studies) who are each implanted with variable numbers of electrodes and perform a cognitive task with different

P. Lv · L. Wang (✉)
CAS Key Laboratory of Mental Health, Institute of Psychology, Beijing 100101, China
e-mail: lwang@psych.ac.cn

N. Axmacher (ed.), *Intracranial EEG*, Studies in Neuroscience, Psychology and Behavioral Economics, https://doi.org/10.1007/978-3-031-20910-9_36

conditions. For such data, a traditional linear regression model is not ideally suited, because multiple contacts from the same patient cannot be considered as independent; both their signal properties and their noise level may differ systematically from electrodes in a different patient. Let us figure out how to solve this issue using LME methods.

We simulate one dataset in which five patients with 8, 7, 9, 8, and 11 hippocampal contacts, respectively, complete a spatial location memory task in a virtual environment. We want to address the question whether theta power changes along the anterior–posterior (A-P) axis of the hippocampus during translational movements. To answer this question, we first determine the relative position of each patient's hippocampal contacts along the anterior–posterior (A-P) hippocampal axis (as was done in [8]; see also Chaps. 27 and 28 for preprocessing) and Z-score the magnitude of theta power in each of these contacts across all translational movement periods. In a traditional linear regression analysis across all contacts, we would then build a linear regression model on all 43 pairs of values to analyze theta power as a function of location along the longitudinal axis of the hippocampus (Fig. 36.1). This analysis showed that theta power did not significantly change along the A-P axis ($t_{41} = 0.911$, $p = 0.368$; Fig. 36.1a; see the example script for more details). However, when analyzing the data of each patient separately, we found that theta power actually increased along the A-P axis for each individual patient (all $p < 0.05$; Fig. 36.1b).

Why are there such different results? When we look closely at Fig. 36.1b, we can understand the problem. The position of the hippocampal contacts is highly different for each patient, especially Patient 1 and Patient 5, whose contacts are located in the anterior and posterior hippocampus, respectively. However, the theta power in the anterior hippocampus of Patient 1 is significantly greater than the theta power in

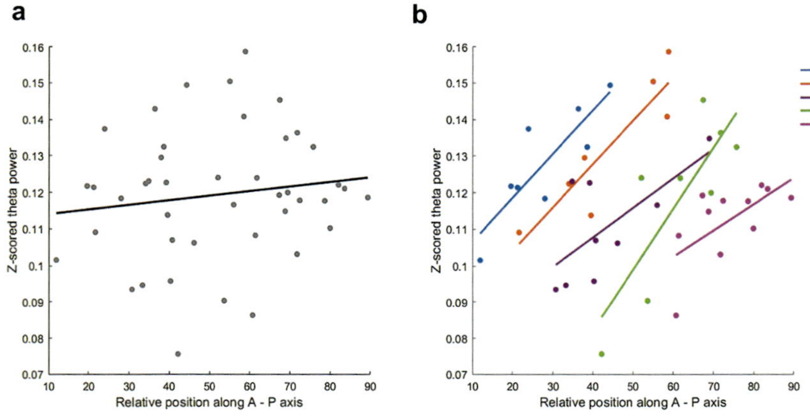

Fig. 36.1 Example data. (a) Simple linear regression of theta power as a function of longitudinal position across 43 hippocampal contacts in 5 patients. Each grey dot is a contact. The black line is the fitted line (not significant). (b) Simple linear regression of position and theta power separately in each patient. Each point indicates one contact. Each color-coded line is a fitted line for each patient (significant in every patient)

the posterior hippocampus of Patient 5 (two-sample t-test, $t_{17} = 2.606$, $p = 0.018$), although the theta power increases along the A-P axis for both patients (Patient 1: $t_6 = 3.967$, $p = 0.007$; Patient 5: $t_9 = 2.457$, $p = 0.036$). Looking back at the simple linear regression (Fig. 36.1a), it turns out that the data violates the assumption of independent observations because theta power is correlated acorss the contacts of each patient. Ignoring the correlation between samples and applying simple linear regression analysis across all channels biases statistical results [16]. What we learn from this example is that we should adopt appropriate statistical methods for the data we collect. For a multi-level data where contacts are nested in patients, we should use multi-level linear models [9, 11], also called linear mixed models.

36.2 Linear Mixed Effect Model

Before understanding LME, let's recall ordinary linear regression models. The linear regression model can be expressed in matrix form as:

$$y = X\boldsymbol{\beta} + \boldsymbol{\epsilon}$$

where $\boldsymbol{y}$ is the dependent variable expressed as an N × 1 column vector (N are the observed samples), $\boldsymbol{X}$ is an N × p matrix (p − 1 independent variables, plus an intercept term), $\boldsymbol{\beta}$ is the regression coefficient expressed as a p × 1 column vector, $\boldsymbol{\epsilon}$ is the random error expressed as an N × 1 column vector. As can be seen from the formula, the linear regression model only contains two parts, namely the fixed effect ($\boldsymbol{\beta}$) and the noise ($\boldsymbol{\epsilon}$).

The formula of LME can be expressed as:

$$y = X\boldsymbol{\beta} + Zu + \boldsymbol{\epsilon}$$

It can be seen that an LME is similar to the ordinary linear regression model, but includes the additional term $\boldsymbol{Zu}$. $\boldsymbol{Z}$ is an N × q matrix where each column is a random factor, $\boldsymbol{u}$ is the regression coefficient of the random factor, also known as the random effect, which is a column vector of q × 1. There are two main types of random factors: random intercept and random slope.

LME can also be understood from a different perspective, which may be helpful is is described in the next paragraph.

36.3 Understanding the Construction of LMEs from the Perspective of Multi-Level Linear Models

LMEs and multi-level linear models are essentially the same. They are both models that deal with nested data, but have different perspectives on the data. Briefly, a multi-level linear model first performs linear regression within each group using the lowest level data. Then, the obtained regression coefficients of each group are used as the dependent variable, and the linear regression is performed again with higher-level independent variables. Taking a simple two-level model as an example, we measure the independent variables X_1 (e.g., the relative position of contacts) and Z_1 (e.g., patient's age) at the first (i.e., individual) and second (i.e., group) level, respectively. Assuming that the second-level independent variable affects the first-level coefficients (i.e., intercept and slope), the first-level formula can be written as follows:

$$\mathrm{Y}_{ij} = \beta_{0j} + \beta_{1j} X_{1ij} + e_{ij}$$

where

Y_{ij} is the dependent variable measured at the ith individual of the jth group, such as the theta power of the ith contact of the jth patient,

X_{1ij} is the independent variable measured in the ith individual of the jth group, such as the relative position of the ith contact of the jth patient,

β_{0j} is the intercept of the jth group,

β_{1j} is the slope of the jth group,

e_{ij} is the random error of the ith individual in the jth group.

The second level formula is as follows:

$$\beta_{0j} = \gamma_{00} + \gamma_{01} Z_{1j} + u_{0j}$$

$$\beta_{1j} = \gamma_{10} + \gamma_{11} Z_{1j} + u_{1j}$$

where

Z_{1j} is the independent variable Z_1 measured in the jth group, such as the age of the jth patient,

γ_{00} is the intercept of each group of intercepts after controlling Z_1,

γ_{01} is the effect of each group of intercepts modulated by Z_1,

γ_{10} is the intercept of the slope of each group after controlling Z_1,

γ_{11} is the effect that the slopes of each group are modulated by Z_1,

u_{0j} is the random error of the intercept of the jth group,

u_{0j} is the random error of the slope of the jth group.

When we substitute the model of the second level into the first level to obtain the combined model, we can readily understand the LME.

$$\mathrm{Y}_{ij} = \gamma_{00} + \gamma_{01} Z_{1j} + u_{0j} + \left(\gamma_{10} + \gamma_{11} Z_{1j} + u_{1j}\right) * X_{1ij} + e_{ij}$$

$$Y_{ij} = (\gamma_{00} + \gamma_{10}X_{1ij} + \gamma_{01}Z_{1j} + \gamma_{11}Z_{1j}X_{1ij}) + (u_{0j} + u_{1j}X_{1ij}) + e_{ij}$$

From the combined model, we can see that $(\gamma_{00}+\gamma_{10}X_{1ij}+\gamma_{01}Z_{1j}+\gamma_{11}Z_{1j}X_{1ij})$ is the fixed effect part of the model, and the corresponding matrix form is $\boldsymbol{X\beta}$. $(u_{0j} + u_{1j}X_{1ij})$ is the random effect part, and the corresponding matrix form is $\boldsymbol{Zu}$. For the random effects, the intercept and slope of each group consist of random errors u_{0j} and u_{1j}, respectively. Parameter estimation, hypothesis testing, and extension to generalized linear mixed models are out of the scope of this chapter. If you want to learn more, you can refer to these books [9, 11, 14].

36.4 Examples of LME with MATLAB

Next, we will use three simple examples to illustrate how to perform LME analysis on iEEG data through the *fitlme* function in MATLAB. In this part, we only show the construction of the basic formula and the results (please see the example scripts). As mentioned above, iEEG data are inherently nested. Usually, the contact level and the patient level are included in LME, and in most cases the repeated measurements are included as the first level. Here we present the three most common types of models: A two-level model for a continuous independent variable, a three-level model for repeated measurements, and a model involving a cross-level interaction of continuous and categorical independent variables.

36.4.1 Two-Level Model: Contacts that Are Nested in Patients

Returning to the example data, we determined relative positions (i.e., percentile) along the A-P axis of the hippocampus of 5 patients and extracted their Z-scored theta power during periods of translational movement. The dependent variable was theta power (ThetaPower), the first level independent variable was the relative position (Position), and there was no independent variable at the second level. Given that theta power may be systematically different between patients, the patient number (PatientID) was used as a random effect. In LMEs, when we focus on the independent variables of the first level and their interactions, we need to decentralize them within groups [6]. Here, we subtract the average position of the patient's contacts from the position of each contact. Meanwhile, for model parsimony, we assumed that the effect of contact position on theta power was consistent across patients. That is, we used an LME with random intercept and fixed slope with the following formula:

ThetaPower ~1 + Position + (1 | PatientID).

where 1 represents the intercept term, and (1|PatientID) represents the random intercept for each patient. The results show that theta power during translational movements increases gradually along the A-P axis of the hippocampus ($t_{41} = 6.472$,

$p < 0.001$). Of course, if you find that the slopes vary significantly between patients, you can use a random slope model with the following formula:

ThetaPower ~1 + Position + (1 + Position | PatientID).

where (1 + Position|PatientID) indicates the random intercept and slope for each patient. The MATLAB example script shows that in our sample data, the results of the two models do not differ much. Therefore, in the following examples, we also use random intercept models as examples.

36.4.2 Three-Level Model: Repeated Measurements Are Nested in Contacts that Are Nested in Patients

In the previous example data, we also measured theta power at each hippocampal contact during stationary periods in all patients. Then, we wanted to analyze whether the theta power during stationary periods is lower than that during translational periods. At this point, the theta power was measured for each contact during both translation and stationary periods, and movement states (i.e., translation and stationary) are nested in the contacts that are nested in patients. The movement state (MoveState) is the first level independent variable. The second level is the contact, which has no independent variable that we are currently concerned about. We use the contact number (ContactID) as a random effect. The third level is the patient, and there is also no independent variable. The patient number (PatientID) is used as a random effect. The formula is as follows:

ThetaPower ~1 + MoveState + (1 | PatientID) + (1 | PatientID:ContactID).

where MoveState represents the categorical variable of movement state (stationary or translation) encoded with 1 for translational periods and 2 for stationary periods. (1 | PatientID:ContactID) indicates the random effect of the interaction between the contact and the patient, i.e. the random intercept of each contact within each patient. Other parts are the same as in the previous example. The results of the model showed that the theta power of patients during stationary periods was significantly lower than during translations ($t_{84} = -8.764$, $p < 0.001$).

36.4.3 Cross-Level Interactions Between Continuous and Categorical Independent Variables

In many studies, we are probably interested in the interactions between variables. Different from ordinary linear regression models, LMEs can deal with interactions between independent variables at different levels (or cross-level). Using the data from the two examples above, the question we may want to investigate is whether the hippocampal A-P axis-dependent pattern of theta power is affected by movement states. This results in a three-level structure of data. The dependent variable is still theta power (ThetaPower). The independent variable of the first level is the movement state (MoveState). The independent variable of the second level is the relative position (Position) of the contacts. The third level is the patient. No independent variable is measured, and the patient number is used as a random effect. The formula is as follows:

ThetaPower ~1 + Position*MoveState + (1 | PatientID) + (1 | PatientID:ContactID).

where Position*MoveState is equivalent to Position + MoveState + Position:MoveState, in which Position:MoveState represents the interaction between relative position and movement state. The other parts are the same as above. We can use the *anova* function in MATLAB to return the results of the F-tests for each fixed-effects term. For simple effects, we need to modify the reference condition when coding dummy variables and obtain simple effects under each condition separately (see the example script for details). Results showed that the interaction between relative position and movement state is significant ($F_{1,82} = 42.018, p < 0.001$). Simple effects analysis showed that patients' hippocampal theta power increase along the A-P direction during translation ($t_{82} = 6.533, p < 0.001$) and decreased during stationary periods ($t_{82} = -2.634, p = 0.01$).

36.5 Discussion

Given the natural nesting characteristics of iEEG data, we advocate using LMEs to perform statistical tests, rather than using traditional linear regression models. Many recent iEEG studies have used LMEs for data analysis [4, 5, 12, 13, 17]. In this chapter, we aimed to describe how to apply LMEs to iEEG data. Its stricter assumptions and parameter estimation are beyond the scope. If you are interested, you can refer to more detailed books on this topic [9, 11, 14]. Here, we provided three examples that reflect some cases which are most commonly found in iEEG research, and provide associated MATLAB example code. It is worth noting that other software packages can implement LME analyses, such as R or SPSS. Due to the different methods of parameter estimation used by each software, results may also be slightly different. However as far as we know, the results output from MATLAB and R are not fundamentally different for the simple LMEs that are commonly used for iEEG data. For more complex LMEs, such as those involving more than three levels and including more independent variables and interaction terms, we recommend using packages in R (such as *lme4* [1]) for analysis. In addition, R has many packages (such as *emmeans*) that can analyze interactions and simple effects, which MATLAB cannot handle very well. Recently, there has been a related literature on how to apply R to perform LME analysis on neuroscience data in details [18]. Here, our goal was to provide a basic introduction to LMEs. Do not stop there, you can explore more about LMEs based on the materials mentioned above.

References

1. Bates D, Maechler M, Bolker BM, Walker SC (2015) Fitting linear mixed-effects models using lme4. J Stat Softw 67(1):1–48. https://doi.org/10.18637/jss.v067.i01
2. Buzsaki G, Anastassiou CA, Koch C (2012) The origin of extracellular fields and currents - EEG, ECoG. LFP and spikes. Nat. Rev. Neurosci. 13(6):407–420. https://doi.org/10.1038/nrn3241
3. Buzsaki G, Draguhn A (2004) Neuronal oscillations in cortical networks. Science 304(5679):1926–1929. https://doi.org/10.1126/science.1099745
4. Chen, D., Kunz, L., Lv, P., Zhang, H., Zhou, W., Liang, S., et al. (2021). Theta oscillations coordinate grid-like representations between ventromedial prefrontal and entorhinal cortex. *Sci. Adv., 7*(44), eabj0200, https://doi.org/10.1126/sciadv.abj0200.
5. Chen, S., Tan, Z., Xia, W., Gomes, C. A., Zhang, X., Zhou, W., et al. (2021). Theta oscillations synchronize human medial prefrontal cortex and amygdala during fear learning. *Sci. Adv., 7*(34), eabf4198, https://doi.org/10.1126/sciadv.abf4198.
6. Enders C, Tofighi D (2007) Centering predictor variables in cross-sectional multilevel models: A new look at an old issue. Psychol Methods 12:121–138. https://doi.org/10.1037/1082-989X.12.2.121
7. Friston KJ, Bastos AM, Pinotsis DA, Litvak V (2015) LFP and oscillations—what do they tell us? Curr Opin Neurobiol 31:1–6. https://doi.org/10.1016/j.conb.2014.05.004
8. Goyal A, Miller J, Qasim SE, Watrous AJ, Zhang H, Stein JM et al (2020) Functionally distinct high and low theta oscillations in the human hippocampus. Nat Commun 11(1):2469. https://doi.org/10.1038/s41467-020-15670-6
9. Hox JJ, Moerbeek M, Van de Schoot R (2018) Multilevel analysis: Techniques and applications, 3rd edn. Routledge, New York
10. Jacobs J, Kahana MJ (2010) Direct brain recordings fuel advances in cognitive electrophysiology. Trends Cogn Sci 14(4):162–171. https://doi.org/10.1016/j.tics.2010.01.005
11. Jiang, J., & Nguyen, T. (2021). Linear and generalized linear mixed models and their applications (2nd ed., Vol. 1). New York: Springer.
12. Norman Y, Raccah O, Liu S, Parvizi J, Malach R (2021) Hippocampal ripples and their coordinated dialogue with the default mode network recent and remote recollection. Neuron 109(17):2767–2780. https://doi.org/10.1016/j.neuron.2021.06.020
13. Norman Y, Yeagle EM, Harel M, Mehta AD, Malach R (2017) Neuronal baseline shifts underlying boundary setting during free recall. Nat Commun 8:1301. https://doi.org/10.1038/s41467-017-01184-1
14. Raudenbush SW, Bryk AS (2002) Hierarchical linear models: Applications and data analysis methods (2nd ed., Vol. 1). London: Sage
15. Siegel M, Donner TH, Engel AK (2012) Spectral fingerprints of large-scale neuronal interactions. Nat Rev Neurosci 13(2):121–134. https://doi.org/10.1038/nrn3137
16. Stevens JP (2015) Applied multivariate statistics for the social sciences, 6th edn. Routledge, New York
17. Vikbladh OM, Meager MR, King JA, Blackmon K, Devinsky O, Shohamy D et al (2019) Hippocampal contributions to model-based planning and spatial memory. Neuron 102(3):683–693. https://doi.org/10.1016/j.neuron.2019.02.014
18. Yu Z, Guindani M, Grieco SF, Chen L, Holmes TC, Xu X (2022) Beyond t test and ANOVA: applications of mixed-effects models for more rigorous statistical analysis in neuroscience research. Neuron 110(1):21–35. https://doi.org/10.1016/j.neuron.2021.10.030

Chapter 37
How Can I Avoid Circular Analysis ("Double Dipping")?

Nora Alicia Herweg

Abstract Intracranial montages can include up to a few hundred electrodes per patient and data is sampled at least every few milliseconds. Like fMRI, scalp EEG and other neuroimaging methods, intracranial recordings therefore produce data sets that allow for numerous comparisons, raising the need for adequate multiple comparisons correction. To reduce the number of comparisons and to answer questions related to a specific subset of the data, analysis pipelines contain a selection process. Here, researchers select specific features of the recorded data, such as particular electrodes, frequencies, or time points, for further analyses. The selection of electrodes, in particular, poses a unique challenge in intracranial research, given that each patient has a unique electrode montage that is determined based on their clinical needs. If the criteria guiding the selection process are not independent of the subsequent statistical analyses, they can render the analyses circular and their results invalid. This chapter discusses strategies to avoid circularity and to derive valid statistical inferences when working with intracranial recordings, considering both univariate and multivariate analysis techniques.

37.1 What Is Circular Analysis and Why Is It a Problem?

When we conduct an intracranial EEG study, we typically acquire a large dataset, reflecting electrical potentials across many electrodes and time points. The number of electrodes typically ranges from several tens to few hundreds of electrodes per patient. The duration of a typical testing session can range from anywhere between a few minutes to a couple of hours, with sampling rates commonly ranging between 500 and 2000 Hz. A 1 h session recorded at 1000 Hz, for instance, will result in 3.6 Million data points for each electrode contact. When analyzing the data, we select subsets of it to conduct one or more statistical test that address our research questions. For instance, we may select electrodes localized to a particular region of the brain, or

N. A. Herweg (✉)
Department of Neuropsychology, Institute of Cognitive Neuroscience, Faculty of Psychology, Ruhr University, Bochum, Germany
e-mail: nherweg@sas.upenn.edu

N. Axmacher (ed.), *Intracranial EEG*, Studies in Neuroscience, Psychology and Behavioral Economics, https://doi.org/10.1007/978-3-031-20910-9_37

time points within a particular window relative to stimulus onset. We may further sort the data points, for instance by assigning them to relevant conditions using criteria such as task performance. Or we may explicitly (e.g. during multivariate regression or classification) or implicitly (e.g. when averaging data within conditions with unequal N) assign weights to each data point, that determine how much it contributes to our statistical result. *Selection*, of which *sorting* and *weighting* can both be understood as special cases [1], is a necessary step of the research process but it comes with the risk of *circularity*.

Circularity arises when the criteria that guide selection are not independent from subsequent statistical analysis. Let us assume, for example, that we conduct a statistical test to compare the neural activity elicited by two conditions A and B in each of 20 individual time-windows relative to stimulus onset. Under the null hypothesis of no difference between these conditions, an alpha threshold of 0.05 will on average lead us to observe an effect in 1 of the 20 time-windows. In an extreme case, researchers may choose to select the time-window for which they report the statistical result based on the statistical result itself (having observed an effect), rendering the statistical inference invalid.

Instead, it is necessary to report the results of all statistical tests and to appropriately control for the number of tests. Common approaches to address multiple comparison correction with intracranial EEG data differ from those used for other neuroimaging methods like fMRI. Intracranial recording contacts are implanted as grid, strip or depth electrodes. These electrodes provide coverage of portions of the brain surface (grid and strip) or deep brain structures (depth), while their positioning is determined based on the clinical needs of the patient (see Chap. 1). Therefore, the electrode montage as a whole does not sample neural activity evenly across space and subjects. As a result, cluster-based methods (discussed in Chap. 35) are mostly used to detect effects across contiguous samples of time, frequency or time–frequency dimensions, after selecting electrodes based on a priori criteria. Alternatively, researchers can choose to correct for the number of statistical comparisons at the level of individual samples (i.e. time points, frequencies, electrodes). Bonferroni correction is a simple but sometimes overly conservative method to control the family-wise error rate, in which the alpha level is divided by the number of statistical tests. Other methods, such as the Benjamini–Hochberg procedure instead control the false-discovery rate and are less conservative [2].

Adjusting the alpha threshold to correct for the increased probability of falsely rejecting the null hypothesis due to multiple comparisons is necessary but not sufficient to prevent circularity. Let us assume that we appropriately controlled the type-one error rate and observed an effect in 5 out of 20 time-windows. Whereas we can safely conclude that there is a statistically significant difference between conditions A and B in some time-windows, we do not know whether the 5 time-windows in which we observed an effect exactly match the true effect in the data. This is because the data we collected is a combination of signal and noise. In any given time-window, noise may either increase or decrease the difference between our conditions of interest. Therefore, we may fail to detect a time-window with a true effect if noise in that time window happened to attenuate the relevant effect; or we may observe an effect in a time-window in which there is no true effect if noise in that time-window happened to increase the relevant condition difference. The selection of time-windows that passed the significance threshold will thus be biased to include time-windows in which noise increased rather than decreased the difference between conditions A and B. Any effect size estimates derived from this specific selection of time windows will be inflated due to circularity [1, 3].

Selecting data points for the very effect in question is an extreme example of non-independence. Less extreme examples are less widely recognized but are likewise problematic. We may for instance want to identify electrodes that exhibit a difference between two conditions A and B, and then ask in a second step whether those same electrodes also exhibit a difference between conditions A and C. Since our selection favors electrodes that show extreme activation profiles in condition A (thus rendering them different from B), these electrodes will be biased to also exhibit a difference between A and C, resulting in invalid statistical inference [1]. These considerations of course also apply to the selection of frequency bands, and are not unique to intracranial EEG but affect most neuroimaging methods.

37.2 Can Multivariate Analyses Be Circular?

So far, we have considered the case of mass-univariate analyses, in which we relate the activity at multiple sampling locations (i.e. time points, frequencies, electrodes) to a variable of interest (e.g. task condition) with a statistical model for each sampling location (e.g. regression, t-test etc.). In recent years, multivariate analyses have gained popularity in M/EEG research [4], including in the realm of intracranial EEG (e.g. [5–7], see also Chap. 34). In multivariate analyses, we use a single statistical model to predict a variable of interest from the activity at multiple sampling locations (i.e. the feature set). If the to-be-predicted variable is categorical, this is called classification; in the continuous case, it is called regression. In either case, we estimate parameters that describe the relation between the feature set (x) and the variable of interest (y). We can think of these parameters as weights, or selection criteria, to be applied to each feature to arrive at the predicted value (e.g. in the linear model $y = \beta_1 x_1 + \ldots + \beta_n x_n + e$, the feature set x is weighted by β; [1]). Having estimated the parameters in the model, we next want to assess the model's performance in predicting y from x. If the model's predicted values are closer to the true y values than would be expected by chance, we conclude that the feature set carries information about y.

Circularity in multivariate analyses arises if parameter estimation (i.e. selection) is not independent from evaluation of model performance. Testing a model on the exact same data on which the model was trained constitutes an extreme example of non-independence that can lead to strongly inflated performance measures. With a large number of features, and hence parameters, multivariate models tend to *overfit* noise in the data, resulting in a high performance on the training data that does not generalize to new data points [8]. To avoid this problem, researchers usually split their data sets into a training set and a test set. The parameters are estimated based on the training set and the estimated model's performance is tested in the test set. However, splitting the data set in half means that we are losing power to detect an effect. To attenuate the reduction in power, researchers often use cross-validation, an iterative procedure in which the data set is split into several subsets [9]. On each iteration, one subset serves as the test set and the remaining subsets are used to train the model. In the end, performance is averaged across test sets/iterations.

Deciding how to split the data into training and test sets is a crucial decision that can impact the validity of the findings. This is because different ways of splitting the data can result in different levels of statistical dependency between training and test set. Intracranial EEG is temporally autocorrelated and splitting the data into small subsets such that neighboring subsets are strongly correlated can result in circularity. Dependence between training and test can also be introduced during an initial feature selection step. For instance, selecting electrodes, time window or frequencies as the feature set based on a univariate effect across the entire data set introduces statistical dependence between training and test sets and can generate spurious results [1]. Similarly, using information from the test set to scale or otherwise preprocess features in the training set can lead to invalid results [10, 11].

37.3 How Should I Select Data Points to Avoid Circularity?

If your research question and hypothesis allow to select data points (frequencies, time points, contacts) based on a priori criteria, you should prefer this type of selection over one that is driven by your current data set. You may want to consult prior studies that show effects relevant to your research question in a particular frequency band, time window or region of the brain (see also Chap. 29 for the definition of regions of interest). Whereas time windows and frequency bands can be exactly replicated in each participant, electrode selection is trickier given the fact that each subject has an individual electrode montage. Electrode selection is usually accomplished via a mask. This mask will ideally be derived from an anatomical atlas or from an independent data set (e.g. using a repository for fMRI results like https://neurosynth.org/; [12]). If you are working with a large sample size it may be possible to select specific small subregions and exclude subjects with missing electrode coverage in a given region. If your sample size is small, you may opt for larger regions of interest instead. In either case, the number of electrode contacts in each region will likely differ across subjects. To proceed, you can average each subject's data across electrodes in a given region, use a statistical model that can handle hierarchical random effects (see Chap. 36), or select a single electrode based on independent criteria (e.g. the most anterior contact in each patient). Instead of using a mask, it is also possible to select a single electrode directly, for instance by minimizing the distance to a particular MNI coordinate and setting a cut-off for the maximally allowed distance. Finally, some research questions may require a subject-specific selection of data points. In this case, you should plan for an independent contrast (e.g. using a separate localizer task) prior to data acquisition.

Whatever your selection criterion, you must make this decision prior to analyzing your data. Choosing a selection criterion among a set of criteria because it maximizes a certain effect, again renders the analysis circular and its result invalid [12]. The only way to demonstrate that hypotheses and data selection were independent a priori decisions is to preregister your study, for instance in a public repository such as the Open Science Framework (OSF, https://osf.io/). Doing so will not prevent you from performing additional exploratory analyses in your data set but it will ensure that the results of such analyses are not interpreted as confirmation of an a priori hypothesis but rather as the basis for a new hypothesis to be tested in an independent data set.

37.4 What if I Don't Have a Specific Hypothesis?

Some research questions may require a more exploratory rather than confirmatory approach. In this case, it may not be possible to pinpoint exactly which frequencies, time windows or contacts should exhibit the effect in question. It is important to interpret the results of these analyses appropriately—exploratory analyses generate hypotheses; they do not confirm them. If you do not have a specific hypothesis going into the analyses, you should be prepared to confirm your exploratory results in a separate data set. To do this you can split your data set in half prior to the analysis (e.g. at the level of subjects or experimental sessions), perform all exploratory analyses on the first half and generate specific hypothesis to be tested in the second half of the data. This approach reduces your power to detect an effect in either of the two halves, and hence should be used only if a priori selection is not possible, and if your dataset is sufficiently large.

37.5 How Do I Ensure Independence Between Training and Test Set in Multivariate Analyses?

Studies have demonstrated spurious effects in multivariate fMRI analyses driven by temporal autocorrelation. These effects are stronger within than across scanner runs (a run is a continuous period of image acquisition with fMRI that usually lasts a few minutes; an experiment usually consists of multiple such runs) and with shorter inter stimulus intervals. Therefore, it has ben suggested that data be split at the level of runs rather than within a run [13, 14]. Like fMRI data, (intracranial) EEG data is temporally autocorrelated but it is not acquired in runs. Due to the autocorrelation, a split that results in subsets that are large (e.g. experimental blocks or sessions; order of several minutes) rather than small (e.g. trials; order of several seconds) should generally be preferred because it reduces the correlation between training and test set. This is of particular relevance if the to-be-predicted variable is temporally structured as well [14, 15]. Furthermore, information from the test set should not be used during feature selection or preprocessing. Feature selection should either be based on a priori criteria, or on the training set only. Similarly, preprocessing of training data should not rely on information from the test set. Instead, the most commonly used approach is to calculate relevant parameters (e.g. mean and standard deviation when using z-scoring) in the training data and apply those parameters to both training and test set.

Finally, a permutation test can provide a measure of model performance that is robust to circularity if used appropriately (see Chap. 35). In a permutation test, the cross-validated performance is assessed repeatedly on a resampled data set [16]. On each permutation iteration, class labels are shuffled relative to the feature set and the full cross-validation procedure (including all relevant selection steps) is repeated on each reshuffled version. The result is a permutation distribution of model performance under the null hypothesis, which exhibits the same biases as the original performance metric. If the permutation distribution deviates from the theoretically expected chance distribution (e.g. exhibiting a mean of above 50% in balanced binary classification), this is suggestive of circularity in the analysis pipeline [8]. Comparing our model's performance to the permutation distribution instead of the theoretically expected chance level, can thus result in a more realistic performance estimate. However, the permutation distribution can only be interpreted in this way if the resampled labels could have arisen from the same process that generated the original label sequence [17]. In many situations, the original class labels are not fully randomized but instead exhibit temporal structure, for instance when the randomization of condition labels in the experiment was subject to constraints or when the labels reflect a behavioral variable with temporal autocorrelation [14]. In these situations, the resampling procedure has to be adapted to reproduce the statistical properties of the original label sequence [18].

37.6 Summary

Circularity is a major concern in intracranial research because statistical inference and effect size estimation are usually preceded by several selection steps. Researchers may analyze neural activity in a particular region of the brain or in a particular portion of time–frequency space. Whenever possible, the selection of data points should be driven by a priori criteria, which are independent from subsequent statistical analysis. Ideally these decisions would be documented prior to data analysis. In multivariate analyses, selection is part of model training and circularity arises if parameter estimation is not independent from subsequent evaluation of model performance. To avoid inflated performance measures, researchers need to ensure independence of training and test data.

References

1. Kriegeskorte N, Simmons WK, Bellgowan PS, Baker CI (2009) Circular analysis in systems neuroscience: the dangers of double dipping. Nat Neurosci 12:535–540. https://doi.org/10.1038/nn.2303
2. Benjamini Y, Hochberg Y (1995) Controlling the false discovery rate: a practical and powerful approach to multiple testing. J R Stat Soc Ser B 57:289–300
3. Vul E, Harris C, Winkielman P, Pashler H (2009) Puzzlingly high correlations in fMRI studies of emotion, personality, and social cognition. Perspect Psychol Sci 4:274–290. https://doi.org/10.1111/j.1745-6924.2009.01132.x
4. Grootswagers T, Wardle SG, Carlson TA (2017) Decoding dynamic brain patterns from evoked responses: a tutorial on multivariate pattern analysis applied to time series neuroimaging data. J Cogn Neurosci 29:677–697. https://doi.org/10.1162/jocn_a_01068
5. Ezzyat Y, Wanda PA, Levy DF, et al (2018) Closed-loop stimulation of temporal cortex rescues functional networks and improves memory. Nat Commun 9. https://doi.org/10.1038/s41467-017-02753-0
6. Anumanchipalli GK, Chartier J, Chang EF (2019) Speech synthesis from neural decoding of spoken sentences. Nature 568:493–498. https://doi.org/10.1038/s41586-019-1119-1
7. Leuthardt EC, Schalk G, Wolpaw JR et al (2004) A brain-computer interface using electrocorticographic signals in humans. J Neural Eng 1:63–71. https://doi.org/10.1088/1741-2560/1/2/001
8. Ball TM, Squeglia LM, Tapert SF, Paulus MP (2020) Double dipping in machine learning: problems and solutions. Biol Psychiatry Cogn Neurosci Neuroimaging 5:261–263. https://doi.org/10.1016/j.bpsc.2019.09.003
9. Raamana P, Engemann D, Schwartz Y et al (2016) Assessing and tuning brain decoders: cross-validation, caveats, and guidelines. Neuroimage 166–179
10. Hebart M, Görgen K, Haynes J-D (2015) The Decoding Toolbox (TDT): a versatile software package for multivariate analyses of functional imaging data. Front Neuroinform 8:1–18. https://doi.org/10.3389/fninf.2014.00088
11. Treder MS (2020) MVPA-Light: a classification and regression toolbox for multi-dimensional data. Front Neurosci 14:1–19. https://doi.org/10.3389/fnins.2020.00289
12. Button KS (2019) Double-dipping revisited. Nat Neurosci 22:681–690. https://doi.org/10.1038/s41593-019-0386-3
13. Mur M, Bandettini PA, Kriegeskorte N (2009) Revealing representational content with pattern-information fMRI—an introductory guide. Soc Cogn Affect Neurosci. https://doi.org/10.1093/scan/nsn044
14. Mumford JA, Davis T, Poldrack RA (2014) The impact of study design on pattern estimation for single-trial multivariate pattern analysis. Neuroimage 103:130–138. https://doi.org/10.1016/j.neuroimage.2014.09.026
15. Li R, Johansen JS, Ahmed H et al (2021) The perils and pitfalls of block design for EEG classification experiments. IEEE Trans Pattern Anal Mach Intell 43:316–333. https://doi.org/10.1109/TPAMI.2020.2973153
16. Valente G, Castellanos AL, Hausfeld L et al (2021) Cross-validation and permutations in MVPA: validity of permutation strategies and power of cross-validation schemes. Neuroimage 238:118145. https://doi.org/10.1016/j.neuroimage.2021.118145
17. Nichols TE, Holmes AP (2001) Nonparametric permutation tests for functional neuroimaging: a primer with examples. Hum Brain Mapp 15:1–25
18. Kriegeskorte N, Goebel R, Bandettini P (2006) Information-based functional brain mapping. Proc Natl Acad Sci 103:38633868. https://doi.org/10.1073/pnas.0600244103

Chapter 38
How Can Intracranial EEG Data Be Published in a Standardized Format?

Dora Hermes and Jan Cimbalnek

Abstract Sharing data or code with publications is not something new and licenses for public sharing have existed since the late 20s century. More recent worldwide efforts have led to an increase in the amount of data shared: funding agencies require that data are shared, journals request that data are made available, and some journals publish papers describing data resources. For intracranial EEG (iEEG) data, considering how and when to share data does not happen only at the stage of publication. Human subjects' rights demand that data sharing is something that should be considered when writing an ethical protocol and designing a study before data are collected. At that moment, it should already be considered what levels of data will be collected and potentially shared. This includes levels of data directly from the amplifier, reformatted or processed data, clinical information and imaging data. In this chapter we will describe considerations and scholarship behind sharing iEEG data, to make it easier for the iEEG community to share data for reproducibility, teaching, advancing computational efforts, integrating iEEG data with other modalities and allow others to build on previous work.

38.1 Introduction

The sharing of data is becoming a core aspect of scientific discourse both inside and outside of laboratories. Part of scientific advancement is to be able to test and compare different theories of brain function against different datasets. Conflicts between theories can stem from many sources, including differences between data supporting the theory or differences in implementation, visualization and code. Students often learn about these issues when they run into problems reproducing previous results from within or outside of a lab. Buckheit and Donoho lay out a scientific decision

D. Hermes (✉)
Department of Physiology and Biomedical Engineering, Mayo Clinic, Rochester, MN, USA
e-mail: hermes.dora@mayo.edu

J. Cimbalnek
International Clinical Research Center, St. Anne's University Hospital, Brno, Czech Republic

N. Axmacher (ed.), *Intracranial EEG*, Studies in Neuroscience, Psychology and Behavioral Economics, https://doi.org/10.1007/978-3-031-20910-9_38

tree that helps determine the source of the differences [1]. In case of disparities it should be questioned whether an idea is correct, whether an implementation is correct, whether measurements differ, whether parameters in analyses are correct and whether the results are displayed correctly. To be able to walk this decision tree and compare theories between and within labs, they argue that the actual scholarship of a manuscript is the complete code, environment and data to regenerate the figures, and that the manuscript by itself is a mere advertisement for this scholarship.

Intracranial EEG (iEEG) data are only collected in select centers with the essential neurosurgical and neurological expertise and provide a unique perspective on human brain function. Curating these data for publication, such that others can reproduce or test new theories against these data, is not a trivial process and goes well beyond a data format for the iEEG time series data. Data files often do not contain important information, such as experimental designs with complex sequences of event onsets, stimuli that were displayed, digitized electrode positions or subject level variables like age or the type of disease. Labs store this information in many different ways and tracking down this information can be extremely challenging even for subsequent lab members within the same lab. Maintaining a systematic way of data organization is part of the ongoing expertise of an iEEG lab.

Maintaining and sharing these valuable data is easier if there is a well documented framework in place. These frameworks are being developed by large community efforts that lay out standard structures for storing both data and metadata in a manner that is consistent between different types of neuroimaging, behavioral and electrophysiological data. Community driven standardization of data structures are believed to be essential to facilitate data sharing and ensure data accessibility and interoperability [2]. One such community effort is the Brain Imaging Data Structure (BIDS) [3], which is supported by the International Neuroinformatics Coordinating Facility (INCF, https://www.incf.org/). BIDS was initially developed for Magnetic Resonance Imaging (MRI) data to minimize data curation time, reduce errors, allow for the implementation of automated analysis workflows and allow validating datasets for the presence of essential metadata. BIDS is based on common lab practices and analyses and labs do not need extensive informatics expertise to use BIDS. With many scientists from the iEEG community, we therefore extended BIDS to iEEG data [4] and ensured that iEEG-BIDS is applicable to most common use cases. This effort was well aligned with the MEG and EEG BIDS extensions to ensure compatibility between electrophysiology data types [5, 6]. In addition to the relative ease of implementation and human readability, the machine readability in BIDS allows automated validation and processing at large scales.

Community driven standards for publishing electrophysiological and neuroimaging data do not only make data easier to parse for other users and machines, but can, hopefully, also relieve some of the burden related to data curation. Even if data are only meant to be shared within a group or a small consortium between a few labs, a shared structure helps walking the scientific decision tree. The same code can run on multiple datasets if they are in the same structure, which facilitates the implementation of quality checks at many different levels as well as analyzing and visualizing data, which are all part of the scientific iEEG scholarship described in this book.

38.2 Study Design and Human Subjects

Before digging into the details of iEEG data components, it is important to briefly consider the title of this chapter again, which asks how data can be *published* in a standardized format. If we want to publish iEEG data, the question of how this should be done should not be asked at the stage of publication. Human subjects' rights demand that data sharing is considered when designing a study and writing an ethical protocol. At that moment, it should be considered what types of data will be collected and potentially shared and informed consent forms need to match these plans. The IRB protocol should also detail what type of anatomical images, clinical information or behavioral data will be collected for the subjects. Also, it should be considered with whom data will be shared, only with other specific researchers or such that they can be used by anyone for teaching purposes or perhaps even commercially. Different countries have varying rules for human subjects protection. The Open Brain Consent provides guidelines that make navigating this process easier [7].

38.3 From Source Data to Raw Data

Curating data for sharing means structuring the iEEG data and metadata into a usable, preferably standardized, format that tools can automatically process. The iEEG source data directly from the recording device do typically not conform to the standardized BIDS file naming and format. There is thus a conversion step between the *'source'* data from the recording device in the amplifier format, to the *'raw'* data that are unprocessed or minimally processed due to conversion to the proper file format and in BIDS (Fig. 38.1). This is comparable to MRI research, where DICOM data are considered *source* data and NIFTI files with proper naming are *raw* data in BIDS. Subsequently, *raw* data that have undergone any level of processing are considered *derived* data. In contrast with iEEG source data, raw data and derived data can follow standardized structures for publication. An online BIDS validator is available to check whether data conform to the BIDS specification [3] and contain all the required data and metadata.

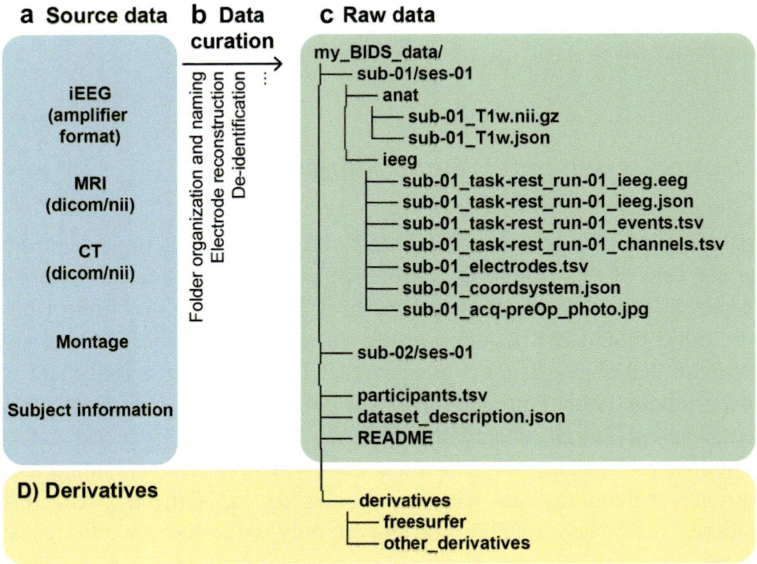

Fig. 38.1 Levels of iEEG data organization. **a** Source data refer to the data directly from the recording system, in the format that the data are recorded in. **b** Data curation from source data to raw data involves steps such as folder organization and BIDS compatible naming, electrode reconstruction, de-identification of data etc. **c** Raw data refer to the data that are used in the processing pipelines. Raw data at the general level within a lab may be organized in BIDS. Data can be organized across projects, where all subject specific information is contained. **d** Derivatives are also starting to be described within BIDS, and may include preprocessing steps or Freesurfer extractions

38.3.1 Creating the Overall Folder Structure

Creating the standardized BIDS folder structure for an experiment may be similar to marking the rooms in an empty new house for the movers: living room, dining room, bedroom, etc. When rooms are unlabeled, boxes with toys may end up in the dining room etc. Defining the basic folder structure makes sure that new, incoming data find their place. The folder structure identifies a subject and session level, followed by the acquisition type (e.g. ieeg, T1w, bold, Fig. 38.1). The filename can then also contain task, run and other optional labels. Note that the example in Fig. 38.1 is far from exhaustive, and the online evolving BIDS specification provides all necessary, current and up to date details: https://bids.neuroimaging.io/specification.html. Defining these subject, session and task labels allows writing code to create lists of the data in an automatic fashion. All iEEG data from subjects who performed a particular task can be listed [8], and the structure can be validated with the online BIDS validator. In addition, with this folder structure specified, data from a new subject can relatively easily flow into the correct folder in the data structure.

Metadata are present at several different positions within the BIDS structure to add information about the study, the participants, the sessions, the events etc. (Fig. 38.1).

All metadata files within BIDS are in field-value (.json) or tab separated file (.tsv) formats, since these are human and machine readable (examples in Fig. 38.2). At the project level of a BIDS structure (Fig. 38.1c), a dataset description file forces a scientist to think about sources of funding, a license under which data may be shared and contributors. A participants file further provides the option to add metadata like the age of each subject. These metadata are not unique to iEEG, but general to all types of BIDS data.

For each subject, iEEG study data can then be stored in the BIDS structure. This does not only concern the raw iEEG time series file. Before electrodes are implanted, imaging data and clinical information may be collected. After electrodes are implanted, operative photographs, post-implant imaging, the montage with electrode numbers and behavioral data may be collected. When electrodes are explanted, resected tissue may be labeled, imaging data and post surgical outcomes like Engel scores [9] may be collected. De-identifying and curating these data into a comprehensive BIDS format requires expertise with the clinical workflow and the different systems where the imaging, electrophysiology and behavioral data are collected (Fig. 38.2).

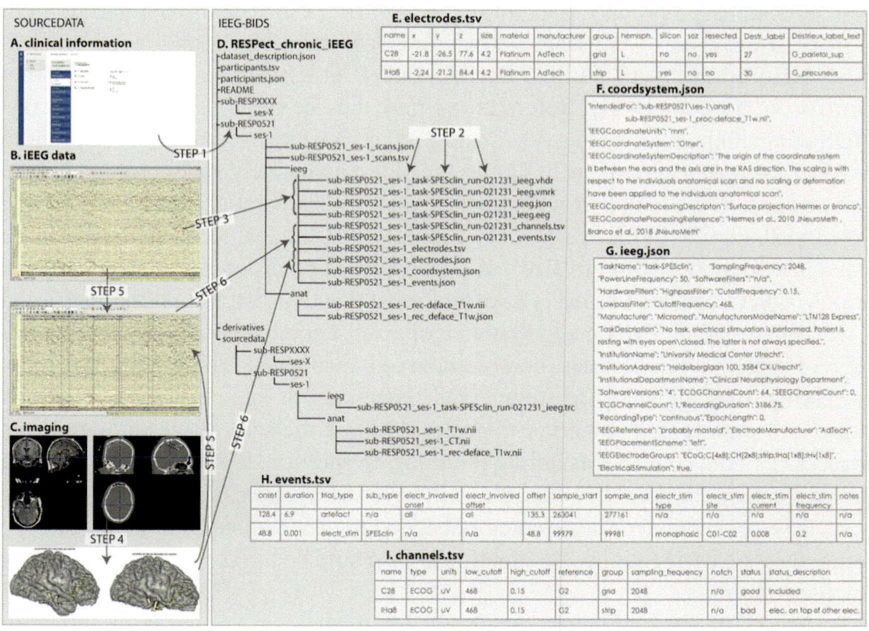

Fig. 38.2 One study identified 6 steps in the curation of clinical iEEG data in BIDS. Step 1: clinical information is extracted for a subject. Step 2: session, task and run labels are assigned. Step 3: source time series are converted to BIDS time series data format and corresponding _ieeg.json file. Step 4: Electrode positions are extracted. Step 5: Electrode positions are matched with corresponding electrode names. Step 6: Channel, event and electrode positions are saved in BIDS.tsv and.json file formats. Reproduced from [14]

Since BIDS has such large, expanding, community support, there are many tools available online to help curate BIDS datasets. Imaging data, for example, have many converters available from DICOM to BIDS [10, 11]. To curate iEEG data there is a BIDS starter kit that outlines the process (https://bids-standard.github.io/bids-starter-kit/), as well as tools within larger software packages such as Fieldtrip and MNE Python (https://bids.neuroimaging.io/benefits.html#converters). These community driven online resources provide the backbone of BIDS and the online content invites new users to help contribute to the standard by clarifying and extending the content and tools.

38.3.2 Curating the iEEG Time Series Data and Channel Information

For iEEG data, formatting the time series data in BIDS almost always requires a file conversion step, since source data are typically not in a BIDS appropriate format. There are many different types of amplifiers and companies that each have their own, potentially proprietary, data format. BIDS for iEEG accepts data in EDF (.edf [12]) or BrainVision Core Data Format format (.vhdr, .vmrk, and .eeg) and will allow MEF3, Neurodata Without Borders (NWB [13]) or eeglab, but many other formats do not comply [4]. Reformatting may thus be required to produce a raw BIDS compatible dataset (Fig. 38.2, step 3). For choosing a format it is relevant to consider the advantages of each. EDF is the oldest and widely spread format but the storage capacity is limited as well as its capabilities for parallel processing (https://en.wikipedia.org/wiki/European_Data_Format). It is suitable for datasets where the individual files are smaller in size and the numerical precision is 16 bits and lower. BrainVision Core Data Format is a newer format that does not have these issues and is well supported by Python and Matlab toolboxes. On the other hand, a less known MEF3 format provides lossless compression of the data and is designed for parallel processing, thus allowing for storage and processing of large volumes of data. A proper raw data format should be considered before the start of the study to avoid unnecessary data conversions during the study. In addition, if data are recorded on a clinical system, the amplifier data may still have video, audio, time-stamps of when the data were recorded or other notes embedded in the data that need to be removed before data are de-identified. De-identification should be a core part of the source to raw file conversion.

Source iEEG data may not contain all the information necessary to automatically populate all necessary BIDS metadata and some information may need to be added in a less automated fashion. One such example is the reference that was used during the recordings, which is not an easy to look up field in the source data file. Where this reference electrode was placed can help understand important differences between studies, especially when considering oscillations or traveling waves (Chap. 28). Figure 38.2 shows several steps in the workflow of curating an iEEG data collected for clinical epilepsy research purposes in BIDS (from [14]). It can be seen in this example that the _ieeg.json file describes the reference. The _ieeg.json sidecar also includes the hardware filters that were used, information that is sometimes only available in an amplifier manual, but is also essential for publication and data reuse.

38.3.3 Curating Electrode Positions

For iEEG data, source data also contain files necessary to reconstruct the positions of the electrode contacts (Chap. 27). A CT or MRI scan with electrodes in place, and sketches and text files with the electrode montage are necessary to reconstruct these positions. In the workflow of an iEEG lab these files are part of the source data and reconstructing the positions in an interpretable manner is a core part that is necessary before data can be analyzed. In addition to the positions of the electrode contacts, the type, size and manufacturer can be added. In terms of the scientific decision tree, electrode positions are necessary to answer important questions, such as the spatial location and spread of signals and the fall off of signals as a function of Euclidean distance between electrodes. Electrode positions can be in the original single subject space, where Euclidean distances are maintained or in a standard MNI or Talairach space where nonlinear transformations may have been applied. These differences directly affect the type of analyses that can be done with the data. The electrode positions and spaces are well described in BIDS during data curation.

38.3.4 Curating Task Information

Events in the data can be stored in BIDS as well. Many iEEG studies include some stimulus or task presentation (e.g. see Chap. 48), or electrical stimulation (see Chaps. 5, 39, 41, 52, 53). The timing of the experiments can be linked to the data through a photodiode or marker channel or sometimes in one integrated system. Data curation involves synchronizing and storing these events with the data. BIDS allows storing discrete and continuous events and also has options to link to images, audio files or videos. Further information about events can also be stored in a BIDS compatible manner using Hierarchical Event Descriptor (HED) tags [15]. Standardized task information allows rapid quality checks such as the presence of basic sensory and motor responses that are expected during a particular task (see also Chaps. 4 and 5 for practical guidelines on online and offline quality checks).

38.3.5 Derivatives

Derived data have undergone any level of processing, such as filtering, denoising, re-referencing or statistical processing. When raw data are stored in a standard structure, it becomes easier to develop automated analyses that produce these derivatives in a reproducible manner. BIDS apps provide one way to design automated processing pipelines, in which a container interfaces with the BIDS structure and requires little user input to perform an automated analysis [16]. Reproducible workflows have also been developed for EEG data [17]. In such workflows, different tools can be combined into one package, and include quality checks in addition to processing steps. The exact same pipeline can then be run on different subjects or datasets, assisting in the overall reproducibility.

38.3.6 Choosing a Publication Platform

A number of options exist for publication of the curated dataset. One option is to publish the dataset on a general data sharing platform for research such as Open Science Foundation platform (https://osf.io/), figshare (https://figshare.com/) or Dryad Digital Repository (https://datadryad.org/). The uploaded datasets receive a DOI and can be cited, however, the platforms are not dedicated to brain research and are often limited in provided space which can be a problem for iEEG data with high sampling frequencies. Another option is to share the dataset through one of the databases solely dedicated to storage of iEEG data, for example the US based ieeg.org (https://www.ieeg.org/), but this can provide problems for multimodal data. Another option is to publish the dataset with platforms dedicated to brain research developed as part of the BRAIN initiative in the US. OpenNeuro, for example, is specifically designed for and validates data in BIDS (https://www.openneuro.org/ [18]). Another option is DABI, which places less restrictions and validations on the data format (https://dabi.loni.usc.edu). The Human Brain Project in Europe also has platforms for sharing data (https://ebrains.eu/), and includes an example of published iEEG data during memory processing in BIDS [19]. Important to consider is the data sharing license that is used on a certain platform and that a permanent DOI is provided to track the data. Processing platforms are arising that can link to these data sharing platforms to facilitate cloud computing and large scale automated processing [20, 21].

38.4 Discussion

Sharing a well curated dataset with publication is different than just dropping a file with time series data online. The scholarship and knowledge of the clinical workflow behind iEEG data curation takes time to acquire, and involves not only running a task, but also writing the protocols that allow one to do the research. The cost of collecting and sharing iEEG data in a standardized format is not negligible, and has to be budgeted in proposals. It may be an easy question to ask which iEEG electrodes are recording from a seizure onset zone, but it takes time from a highly trained clinician to answer this question. It is argued that this cost behind sharing data is outweighed by the potential that is provided by sharing the data [22]. In addition to basic manuscripts, the scientific knowledge and scholarship of experimentalists can indeed be shared through well curated data.

Working with datasets in standardized formats can reduce processing time and effort and can help walk the scientific decision tree. A standardized format creates a checklist to help curate data with necessary aspects for publication. In addition to living, dynamic best practice documents, such as the COBIDAS guidelines for MEEG [23], or published guidelines with best practices [24], a standard format with validation tools helps ensure that important aspects are not missed. Differences between experiments can be due to many different recording settings, such as different lowpass filters, reference schemes or electrode types. These standardized formats and guidelines help curate data in a manner that such parameters are coded in a standard format. When these parameters are findable in a human and machine readable manner, it becomes easier to understand differences between datasets.

Standards for sharing and reporting data, such as BIDS and COBIDAS, are all living documents that are evolving around large communities and their needs. In the past, small groups standardized local data formats, and shared these with the community. While many of these efforts did not lack usefulness, they often lacked large-scale support across many labs located in different continents. With the growing Open Science movement, including Social Media, Github, Google docs [25], large communities can more easily work together to standardize data structures. The BIDS standard described in this chapter is driven by the communities' members, not by a few labs. This facilitates designing automated analyses, and comparing datasets and experiments, and advance reproducible iEEG research.

References

1. Buckheit JB, Donoho DL (1995) WaveLab and reproducible research. In: Antoniadis A, Oppenheim G (eds) Wavelets and statistics. Springer, New York, New York, NY, pp 55–81
2. Poline J-B, Kennedy DN, Sommer FT et al (2022) Is neuroscience FAIR? a call for collaborative standardisation of neuroscience data. Neuroinformatics. https://doi.org/10.1007/s12021-021-09557-0
3. Gorgolewski KJ, Auer T, Calhoun VD et al (2016) The brain imaging data structure, a format for organizing and describing outputs of neuroimaging experiments. Sci Data 3:1–9
4. Holdgraf C, Appelhoff S, Bickel S et al (2019) iEEG-BIDS, extending the Brain Imaging Data Structure specification to human intracranial electrophysiology. Sci Data 6:1–6
5. Pernet CR, Appelhoff S, Gorgolewski KJ et al (2019) EEG-BIDS, an extension to the brain imaging data structure for electroencephalography. Sci Data 6:103
6. Niso G, Gorgolewski KJ, Bock E et al (2018) MEG-BIDS, the brain imaging data structure extended to magnetoencephalography. Sci Data 5:180110

7. Bannier E, Barker G, Borghesani V et al (2021) The Open Brain Consent: Informing research participants and obtaining consent to share brain imaging data. Hum Brain Mapp 42:1945–1951
8. Yarkoni T, Markiewicz CJ, de la Vega A et al (2019) PyBIDS: Python tools for BIDS datasets. J Open Source Softw 4. https://doi.org/10.21105/joss.01294
9. Engel J Jr (2006) Report of the ILAE classification core group. Epilepsia 47:1558–1568
10. Halchenko Y, Goncalves M, di Oleggio Castello MV, et al (2021) nipy/heudiconv
11. Zwiers MP, Moia S, Oostenveld R (2021) BIDScoin: a user-friendly application to convert source data to brain imaging data structure. Front Neuroinform 15:770608
12. Stead M, Halford JJ (2016) Proposal for a standard format for neurophysiology data recording and exchange. J Clin Neurophysiol 33:403–413
13. Teeters JL, Godfrey K, Young R et al (2015) Neurodata without borders: creating a common data format for neurophysiology. Neuron 88:629–634
14. Demuru M, van Blooijs D, Zweiphenning W et al (2022) A practical workflow for organizing clinical intraoperative and long-term iEEG data in BIDS. Neuroinformatics. https://doi.org/10.1007/s12021-022-09567-6
15. Robbins K, Truong D, Appelhoff S et al (2021) Capturing the nature of events and event context using hierarchical event descriptors (HED). Neuroimage 245:118766
16. Gorgolewski KJ, Alfaro-Almagro F, Auer T et al (2017) BIDS apps: improving ease of use, accessibility, and reproducibility of neuroimaging data analysis methods. PLoS Comput Biol 13:e1005209
17. Pernet CR, Martinez-Cancino R, Truong D et al (2020) From BIDS-formatted EEG data to sensor-space group results: a fully reproducible workflow with EEGLAB and LIMO EEG. Front Neurosci 14:610388
18. Markiewicz CJ, Gorgolewski KJ, Feingold F et al (2021) The OpenNeuro resource for sharing of neuroscience data. Elife 10. https://doi.org/10.7554/eLife.71774
19. Cimbalnik J, Dolezal J, Topçu Ç et al (2022) Intracranial electrophysiological recordings from the human brain during memory tasks with pupillometry. Sci Data 9:6
20. Delorme A, Truong D, Youn C et al (2022) NEMAR: an open access data, tools, and compute resource operating on NeuroElectroMagnetic data. arXiv preprint arXiv:220302568
21. Avesani P, McPherson B, Hayashi S et al (2019) The open diffusion data derivatives, brain data upcycling via integrated publishing of derivatives and reproducible open cloud services. Sci Data 6:69
22. Voytek B (2016) The virtuous cycle of a data ecosystem. PLoS Comput Biol 12:e1005037
23. Pernet C, Garrido MI, Gramfort A et al (2020) Issues and recommendations from the OHBM COBIDAS MEEG committee for reproducible EEG and MEG research. Nat Neurosci 23:1473–1483
24. Mercier MR, Dubarry A-S, Tadel F et al (2022) Advances in human intracranial electroencephalography research, guidelines and good practices. Neuroimage 119438
25. Voytek B (2017) Social media, open science, and data science are inextricably linked. Neuron 96:1219–1222

Part IV
Advanced Topics

Chapter 39
What Are the Contributions and Challenges of Direct Intracranial Electrical Stimulation in Human Cognitive Neuroscience?

Jacques Jonas and Bruno Rossion

Abstract Direct electrical stimulation (DES) is an old and powerful technique to causally inform about the localization of human brain function for clinical and research purposes. However, DES faces important challenges particularly in research: poorly known mechanisms and localization of the effects, methodological limitations due to clinical settings, etc. Through contributions of DES studies performed in the ventral occipito-temporal cortex, in particular to understand human face recognition, this chapter illustrates how future DES studies can overcome these challenges. At the methodological level, increasing the value of DES in cognitive neuroscience will depend on the use of well-controlled and diverse experimental paradigms across enough trials and stimulations to objectively evaluate DES effects. The combination of DES with independent or simultaneous measurements with functional magnetic resonance imaging and intracranial electroencephalography, particularly with frequency-tagging, offers new promises for causal objective mapping of brain function. Single or multiple subjects' studies are both well suited to this purpose, depending on the evaluated function and the frequency of observed effects. At a theoretical level, since it is now well established that DES affects remote brain regions, future DES studies should focus on assessing the connectivity of the critical sites to identify the network affected by the stimulation.

J. Jonas (✉) · B. Rossion
Université de Lorraine, CNRS, CRAN, 54000 Nancy, France
e-mail: j.jonas@chru-nancy.fr

Université de Lorraine, CHRU-Nancy, Service de Neurologie, 54000 Nancy, France

N. Axmacher (ed.), *Intracranial EEG*, Studies in Neuroscience, Psychology and Behavioral Economics, https://doi.org/10.1007/978-3-031-20910-9_39

39.1 Introduction

Localizing and understanding sensorimotor and cognitive functions of the human brain are major goals in cognitive neuroscience. Among functional mapping techniques, the application of electrical current with an electrode to brain nuclei, to the cortex or white matter tracts, i.e., direct electrical stimulation (DES), holds a special place. Together with lesion studies, DES was one of the first methods used to investigate brain function, well before the advent of neuroimaging. Historically, DES provided to establish some of the foundations of modern neuroscience, such as the electrical nature of the brain, the localization of brain functions, and the functional anatomy of the sensorimotor cortex [1–3].

DES has also provided some of the most fascinating observations in human neuroscience (e.g., out-of-body experience [4]; self-face hallucination [5]). Most importantly for the purpose of this chapter, since it offers to draw a causal link between a specific brain region and a given function [6, 7], DES is often considered as the gold standard for functional localization, in particular for clinical purposes. That is, the rationale of DES is that applying an electrical current to the brain allows to temporarily disrupt the specific function of the stimulated region and therefore simulates what would be the behavioral effect if this region was removed or lesioned ("virtual lesion").

In practice, DES can be delivered extraoperatively via two types of intracranial electrodes in epileptic patients refractory to medication (Box 39.1): either intracerebral "depth" electrodes inserted inside the brain (stereo-electroencephalography, SEEG) or subdural electrodes applied onto the cortex after removing a part of the skull (electro-corticography, ECOG). DES can also be performed intraoperatively, mostly during brain tumor resection, by applying electrical currents over the cortex or the white matter via handheld probe electrodes [8]. In this chapter, we will focus on DES applied extraoperatively in epileptic subjects, i.e., through electrodes implanted intracranially for several days or weeks to define the localization and extent of epileptic seizures, allowing more carefully controlled stimulation procedures and concomitant electrophysiological recordings (intracranial EEG or iEEG).

Box 39.1: Two Types of Intracranial Approaches for Extraoperative DES in Epileptic Patients

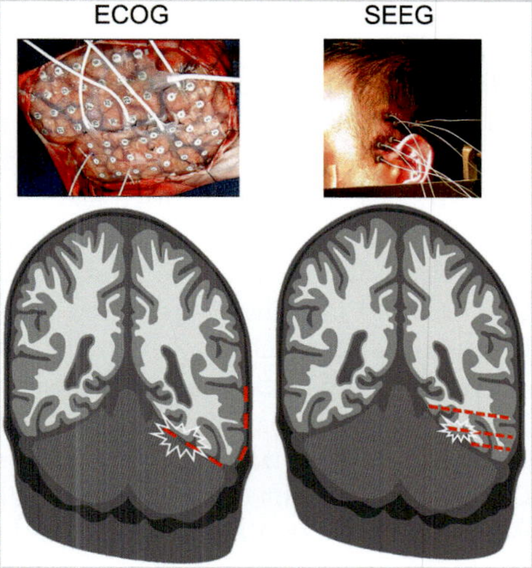

ECOG consists in applying electrodes onto the cortical surface after craniectomy (i.e., subdural electrodes). Subdural electrodes have a circular shape and are spatially arranged as grids or strips with typically 5–10 mm inter-electrode spacing (center-to-center). SEEG consists in inserting needle electrodes inside the brain through small holes in the skull (i.e., depth electrodes). The current intracerebral electrodes are thin cylinders (e.g., 0.8 mm diameter) typically containing 8–15 contiguous individual recordings sites (or contacts) separated by an insulating material (3.5 mm spacing, center-to-center). From the point of view of fundamental research, each technique has its own advantages: while ECOG offers a more extensive spatial coverage, SEEG provides recordings and stimulations directly inside the grey matter and allows the specific exploration of cortical sulci, white matter, and deep structures (e.g., amygdala, hippocampus). The usual variables of electrical stimulation parameters are intensity, waveform, duration and frequency of the pulses, total duration and montage (bipolar or monopolar). The commonly used parameters of these techniques have been reported in previous reviews [9–14]. While the stimulation settings are similar between ECOG and SEEG, current intensities are lower in SEEG (0.5–5 mA compared to 1–20 mA in ECOG) in order to deliver similar charge density, and pulse duration is usually longer in SEEG. DES parameters in clinical practice remain insufficiently standardized with significant variations across epilepsy centers [15], which is an issue for DES functional mapping (risk of false negatives and false positives, lack of reproducibility across centers and studies, etc.). In order to streamline DES procedure, French epilepsy centers published recommended stimulation parameters for SEEG [11]: bipolar and biphasic square wave current delivered either in low-frequency DES (frequency: 1 Hz, pulse duration: 0.5–3 ms, intensity: 0.5–4 mA, total duration: 20–60 s) or high-frequency DES (frequency: 50 Hz, pulse duration: 0.5–1 ms, intensity: 0.5–5 mA, total duration: 3–8 s). These 2 types of DES are recommended for functional localization, with low-frequency DES particularly well-suited for the primary cortices (especially the primary motor cortex) and to study functional connectivity, and high-frequency DES for the associative cortex. However, varying these parameters, beyond the usual recommendations but within the safety limits, produces differential neural effects that help to understand the physiology of the DES (e.g., [16]) and highlight some behavioural effects which may have been missed with standard settings; [17–20]; see also [21]). For example, a promising avenue for DES in the future would be to adapt stimulation parameters on the known physiology of the stimulated region or on individual electrophysiological analyses performed before DES (e.g., stimulation frequency matching the main endogenous oscillations of the stimulated site, [18]; theta burst stimulation parameters in the medial temporal lobe, [20], see also [22] for microstimulations).

Since cognitive functions are localized (to some extent) in specific brain regions and networks, stimulating discrete brain regions to observe behavioral consequences in real time represents a unique way to understand human brain function. Over several decades of investigation, DES with SEEG or ECOG has provided unique information about the anatomico-functional organization of the human brain [10, 13, 14, 23–26]. However, as early as the first DES studies on the motor cortex of dogs and non-human primates [1, 2], the relevance of DES effects to map and understand brain function has been debated, mainly because of the unknown effect of DES at local and distant sites. While it is sometimes claimed that "*if its rules of use are rigorously applied, the sensitivity of DES for detection of cortical and axonal eloquent structures is 100%*" [6], these claims have been criticized [27–29]. These criticisms are based on the fact that the physiology of DES is far from being fully understood, that DES effects could be due to the involvement of a large brain surface (both local and distant) and that these effects may be unpredictable, depending on local and remote factors. In sum, the danger of ignoring the complexity of DES may lead to oversimplistic and equivocal conclusions about the role of the stimulated region [29]. Moreover, since DES investigations in humans are performed in clinical settings with inherent limitations (e.g., limited testing time, limited number of experiments and trials, electrode positions based on clinical factors, patients with epilepsy or brain tumor, etc.), they are sometimes considered as anecdotical compared to other neuroscientific approaches, in particular recordings and stimulations performed in non-human primates [30, 31]. Therefore, despite the "gold standard" label for mapping brain function that is still attributed to human DES nowadays in clinical settings, the technique faces a substantial number of challenges that need to be addressed.

In the present chapter, we first illustrate these challenges with DES studies applied to the human ventral occipito-temporal cortex (VOTC), one of the most explored regions in temporal epilepsy patients. Then, we illustrate how DES could overcome these challenges and provide a unique body of knowledge about the functions of the VOTC and the human brain in general. To do so, we build upon a specific research program with SEEG DES that aims at understanding arguably the most complex recognition function for the human brain, *face identity recognition* (FIR). In a last section, we survey the methodological and theoretical issues necessary for DES to overcome these challenges and to be considered as a key technique to understand the neural basis of human brain function.

39.2 Challenges of DES Studies in the VOTC

In humans, the VOTC supports many cognitive functions including visual recognition (through the ventral visual pathway), semantic memory (through the anterior temporal lobe), episodic and spatial memory (through the medial temporal lobe), and some language-related functions (mainly reading and naming through the left lateralized ventral visual pathway and the basal temporal language area, i.e., BTLA, respectively).

Recognition based on vision (considered as the dominant sensory modality in primates, humans in particular; see [32]) is of one most complex human brain functions, enabling us to quickly and automatically behave adaptively in a rich, dynamic and fundamentally ambiguous sensory environment. This function is supported by a network of brain regions in the VOTC, forming the ventral visual stream [33]. The ventral visual stream emerges from low-level (i.e., retinotopic) visual areas in the occipital lobe (e.g., V1, V2, V4, etc.) and continues through the ventral temporal lobe where higher-level category-selective areas (e.g., for faces, scenes, letters/words or other categories such as body parts) have been identified in neuroimaging studies over the past three decades [33, 34].

DES performed in the VOTC evokes either visual hallucinations (simple such as phosphenes or complex such as faces or scenes), illusions (e.g., distortions of the face being perceived during stimulation) or recognition impairments (for reviews, see [25, 35, 36]). These DES studies have supported the hierarchical organization of the ventral visual pathway [37–39] and showed the causal role of some category-selective areas in the recognition of their preferred category (faces: [40, 41]; written words: [42–44]; scenes: [45]). However, these latter studies are mostly considered as confirming the utmost importance of these regions in the dedicated functions that were already defined by neuroimaging or lesion studies. For example, recent ECOG DES studies have shown a causal link between the most studied and well-known face-selective area, i.e., the Fusiform Face Area (FFA, [46]; Fig. 39.1) in the lateral section of the middle fusiform gyrus (LatMidFG), and face perception

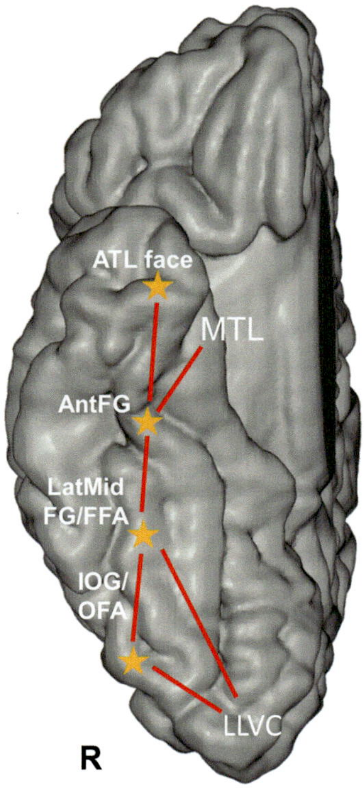

Fig. 39.1 Schematic illustration of the main VOTC face-selective regions (IOG/OFA, LatMidFG/FFA, AntFG and ATL face area) in the right hemisphere and of their hypothetical patterns of reentrant connectivity. FMRI and iEEG studies have recorded robust face-selective neural activity in these regions, except in the AntFG, in which face-selectivity has been disclosed only in iEEG recordings because of a strong BOLD signal drop-out affecting this region, making this region almost invisible in fMRI (see [97]). SEEG DES of the right IOG/OFA and AntFG have induced transient FIR impairment while SEEG DES to the LatMidFG has led to perceptual face distortion or palinopsia. According to a recent hypothesis [25], transient changes of the currently experienced face stimulus during SEEG DES of the right IOG and LatMidFG would be due to reentrant direct (i.e., monosynaptic) connections between these face-selective posterior regions and low-level (i.e., retinotopic) visual cortex (LLVC). In contrast, the face-selective AntFG is not directly connected to the low-level visual cortex (LLVC), but has direct connections with the medial temporal lobe (MTL), mainly the hippocampus, such that stimulation of the AntFG leads to transient failures to encode the visual stimulation experience in memory [98, 99]. Face-selective regions are shown at their approximate locations—considering a wide interindividual variability—as schematic yellow stars on a reconstructed cortical surface of the Colin27 brain

[30, 31, 40, 41]. By electrically stimulating the FFA in the right hemisphere, these studies reported either face-related perceptual changes (i.e., subjective change in the visual appearance of a face, usually a distortion of the experimenter's face in front of the subject) and face or face-part hallucinations. Although these studies showed that the FFA is causally involved in visually processing the category of faces, they provided no information about the specific role of the FFA in face recognition. Most strikingly, while lesion studies, neuroimaging and intracranial recordings point to an important role of the (right) FFA and its neighboring VOTC face-selective areas in face *identity* recognition (e.g., [47–51]), impairment at this function during DES to the FFA has never been demonstrated (see [25]). Moreover, these ECOG DES studies were performed without well-controlled experimental set-up, essentially requiring subjects to passively look at real faces or objects in the room and describe their subjective experience, thus reinforcing the anecdotal reputation of DES studies in neuroscientific research. For this reason, these studies have offered no or very little contribution to neurofunctional models of human face recognition, which are somewhat paradoxically more influenced by findings in non-human primate recording studies [52]. In the same vein, observations of transient deficits in written word reading (alexia) during DES to a region of the left VOTC (potentially the so-called Visual Word Form Area, VWFA; [53]) have been rather anecdotic, i.e., reported without quantitative analyses of the behavioral effects and independent functional mapping [42, 44], or only briefly described in the context of extensive iEEG recording investigations [43, 54].

Another example of the limited impact of DES in the definition of VOTC functions is the study of the so-called BTLA, a functionally defined region based on DES effects observed at the interface between the ventral visual stream, the semantic system in the anterior temporal lobe (ATL) and the perisylvian language system [55]. DES performed on the left VOTC (inferior temporal gyrus, fusiform gyrus and parahippocampal gyrus) has consistently elicited deficits in naming (in both visual and authority modalities) but also in reading and comprehension [56–59] (for review see [10]). However, beyond the clinical community, the concept of BTLA and the information derived from DES of this region have so far generated little interest in fundamental neuroscience research, and the outcome of these studies is not included in theoretical frameworks of the anatomo-functional organization of language (e.g., [60, 61]). Despite relatively frequent effects of DES observed in this region, two factors at least have been brought forward to reduce interest in the BTLA. First, post-operative language deficits are either weak, inconsistent, or transient following resection of this region [57, 62, 63]. Second, DES studies have recorded distant post-discharges or cortico-cortical evoked potentials (CCEP), suggesting strong anatomico-functional connections between the BTLA and perisylvian language regions [64–66]. These two observations suggest that language-related deficits observed upon BTLA stimulation are due to remote rather than local effects of DES [29]. In short, studies on the BTLA provide a good example of the type of arguments raised against the validity of the information derived from DES and its ability to precisely map brain regions that are critical for cognitive functions.

39.3 DES to Understand Human Face Identity Recognition

39.3.1 Why Studying Face Identity Recognition with DES?

In its interaction with the environment, the brain is primarily a biological recognition system: it needs to provide selective/specific responses to sensory stimuli, these responses being generalizable across a wide variety of viewing conditions and temporal contexts. In primates and particularly in humans, the recognition[1] of faces is particularly rich and socially relevant, allowing to rapidly e.g., tell apart males and females, decode others' emotional states from their expression, estimate their age, infer their ethnical origin, their attention from head and gaze orientation, attractiveness and even make social judgments of dominance or trustworthiness [68, 69].

For several reasons, recognizing someone's *identity* from their faces, i.e., face identity recognition (FIR), is arguably the most challenging human recognition function, across the board. First, while individual human faces, even in a genetically homogenous population, appear to differ more than in other animal species [70], they nevertheless look similar in their basic features and configuration, requiring relatively fine-grained visual discrimination processes. Second, a given face identity can vary substantially in appearance across viewing conditions [71], requiring high level generalization abilities. Third, in most modern societies, the number of facial identities to recognize is very large, usually from several hundreds to thousands of individual faces [72]. Fourth, the number of identities to recognize is often undetermined, i.e., changes across different contexts and over time, with familiar faces mixed up among unfamiliar faces in various contexts.

Considering the challenge at stake, human adults' FIR is impressive, i.e., with up to thousands of face identities recognized accurately [72], automatically and at a glance (e.g., [73]). Yet, this challenge also explains why there is so much natural interindividual variability in this ability in the normal population [74], an ability that is easily disrupted in many neurological, neurodevelopmental and psychiatric disorders [75].

For long, knowledge of the neural basis of FIR relied on the localization of lesions in patients with prosopagnosia (a category-selective impairment in FIR; [76, 77]), these lesions being consistently found in the VOTC, with a right hemispheric dominance [49, 78–81]. However, cases of prosopagnosia are rare (at least when properly diagnosed to exclude general visual object agnosia or semantic memory disorders; see [77, 82]) and have usually large lesions that limit the spatial resolution of these investigations. Functional magnetic resonance imaging (fMRI) studies have shown reduced

[1] The term 'Recognition' is often used in psychological research to refer to the judgment of previous occurrence [67]. However, the term is used here in a general biological meaning to refer to the production of a selective (i.e., discriminant) response to a given sensory input, a response that can be reproduced (i.e., generalized) across variable viewing conditions. Defined as such, even the decision or response signaling that a visual stimulus is a face is a recognition function (i.e., generic face recognition).

neural activity when repeating (unfamiliar or familiar) pictures of the same facial identities as compared to different identities in several face-selective (i.e., responding more to faces than non-face objects) VOTC regions, most notably the FFA in the LatMidFG and the occipital face area (OFA, Fig. 39.1) in the inferior occipital gyrus (IOG) [50, 83–86]. However, fMRI studies analyzing pattern variations of neural activity across voxels for different face identities have only reported small and/or inconsistent effects across experiments, sometimes outside face-selective regions, casting doubt on the contribution of the FFA and other posterior face-selective regions of the VOTC to FIR and pointing instead to the ATL [87–91]. Moreover, directly contrasting pictures of unfamiliar and familiar faces rarely leads to consistent differences in face-selective VOTC regions, and may instead recruit widespread unspecific regions in the brain, probably because of the richness of semantic information linked to familiar faces [50, 89, 92].

In this context, DES of face-selective VOTC regions, which are largely specific to humans or hominoids [93, 94] and largely inaccessible to transcranial magnetic stimulation (TMS), may provide invaluable information regarding the neural basis of this complex FIR function. Moreover, the human face recognition function is thought to be highly specialized and localized but also supported by a vast network of bilateral VOTC regions (Fig. 39.1; [95, 96]), making it particularly amenable to DES.

Below we briefly describe the contribution of DES studies performed in the last decade in the context of SEEG as an example case of promises and challenges of human DES in cognitive neuroscience.

39.3.2 DES of the Right Face-Selective IOG Impairs FIR

The first case of transient FIR impairment with DES (i.e., "transient prosopagnosia") was reported ten years ago [100]. During stimulation of a right IOG region (Fig. 39.2a), subject KV reported various subjective experiences (e.g., *"the face does not appear to me as a single entity"*, *"the facial elements were mixed"*) although some stimulation trials on the same contacts did not lead to perceptual changes. Regardless of her perceptual experience, DES to the right IOG caused a sudden inability for KV to recognize the face identity in front of her (i.e., the neurologist, or photographs of famous faces). This effect was transient, in fact ending as soon as the stimulation stopped, and highly consistent across trials [100] (see also [25], with all videos of the stimulation trials).

At the time of this original observation, only relatively limited behavioral observations could be performed on the case of KV, all with familiar faces. However, about a year later, KV underwent a second SEEG to perform radiofrequency-thermolesions of the epileptic focus. An intracerebral electrode was again inserted in KV's right IOG, near the location of the previous critical stimulation site (Fig. 39.2b; [101]). A behavioral paradigm with simultaneously presented pictures of unfamiliar face identities appearing on a computer screen next to each other at each trial was designed,

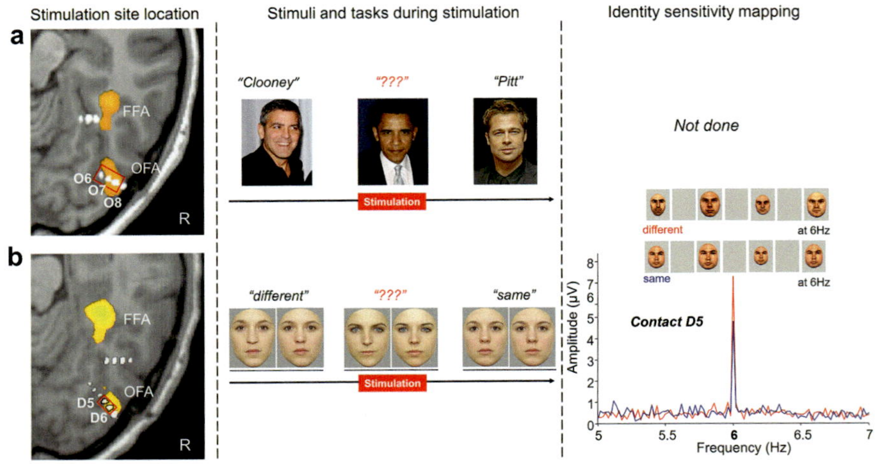

Fig. 39.2 Stimulating the face-selective right IOG induces transient prosopagnosia (subject KV). **a** Reprinted with permission from [100]. **b** [101]. In both studies, the left panel shows the fMRI face-selective activations in the right VOTC (axial slices) with the SEEG electrodes superimposed (white dots); the middle panel shows the stimuli presented during the stimulation procedure; the right panel shows SEEG recordings during a FPVS paradigm measuring sensitivity to face identity. In [100], the eloquent contacts O6, O7 and O8 (in the red rectangle) were located in the right face-selective IOG ("OFA") as shown by fMRI (shown here) and face-selective ERPs recorded on these contacts. Stimulation of these contacts induced a transient inability to recognize famous faces while object recognition was preserved. In [101], stimulating two contacts located within the right face selective IOG (D5/D6) evoked a transient inability to discriminate unfamiliar face identities. During SEEG, KV was tested with a FPVS adaptation paradigm measuring sensitivity to face identity at a fast rate of 6 Hz, with either identical faces or different faces [102]. The significantly largest difference for different versus same faces for upright faces was found on the critical contact D5 (right panel shows responses to different and same faces at 6 Hz in the frequency domain)

asking KV to determine whether they were of the same identity or not (Fig. 39.2b). To adjust the task to KV's excellent FIR ability (as pre-assessed with neuropsychological tests), faces that differed only by 40% along a morphed continuum were selected. While KV performed this task extremely well outside of stimulation, DES inside the face-selective right IOG led to systematic errors (i.e., answering "same" when different unfamiliar face identities were presented). She stated: *"I had a feeling of a strong resemblance*", "*there were two identical faces*", as if DES inside the right IOG interrupted her ability to grasp the physical differences between the two unfamiliar face identities. There was no visual distortion or rearrangement of facial elements reported ([101]; with videos of the stimulation trials). Sensitivity to unfamiliar face identity of each intracerebral contact was quantified independently in SEEG with a frequency-tagging approach (or Fast Periodic Visual Stimulation, FPVS) adaptation paradigm (fast periodic presentation of either different or same face identities; [102]) (see also Chap. 31 on frequency tagging in iEEG studies). Strikingly, among all electrode contacts implanted in KV's brain (N = 27), the largest face identity adaptation effect was recorded on the critical stimulation site (Fig. 39.2b).

These two successive reports of DES performed in the same brain region of the same patient [100, 101] at a one-year interval are not only unique to our knowledge, but they serve well to illustrate the progress that can be made in methodological control and systematicity as well as the refinement of hypotheses and correlations with independent electrophysiological measures to enrich the contribution of DES to our understanding of the neural basis of cognitive functions.

39.3.3 DES of the Right Face-Selective Anterior Fusiform Gyrus Impairs FIR

The anterior fusiform gyrus (AntFG) is located anteriorly to the LatMidFG/FFA but posteriorly to the ATL face area usually found in fMRI close to the temporal pole (Fig. 39.1). Unfortunately, little is known about its role in FIR mainly because this region is affected by a large BOLD signal drop-out arising from magnetic susceptibility artifact [97, 103, 104]. Consequently, only a handful of fMRI studies identified this region as face-selective (e.g., [105–107]). In contrast, large face-selective activity has consistently been reported with SEEG in this region [108, 109].

Transient impairment to recognize facial identity during right AntFG stimulation was reported initially a few years ago in a single case, CD (Fig. 39.3a; [98]). Upon DES to the right AntFG, CD was transiently unable to recognize any famous face picture presented. Visual object recognition was intact upon stimulation. As for subject KV, her behavioral impairment was clear, massive, and highly reproducible. After stimulation, CD said that she did not recognize the face identity, as if the face was shown for the first time. She did not report any perceptual change in the structure of the face. Subsequently, CD was unable to remember which were the face pictures presented during the stimulation, as if they did not enter her memory [98] (see also [25], with all videos of the stimulation trials).

As for the two successive observations on subject KV described above, this unique observation on CD was subsequently—recently—reproduced and extended by a behavioral investigation during right AntFG stimulation in another subject, ND (Fig. 39.3b; [99]). Based on the initial observation, multiple tasks that did not require a verbal output were designed, both with familiar and unfamiliar faces, quantifying performance in terms of both accuracy rates and response times. Upon stimulation of the right AntFG, DN was impaired at pointing out a familiar face among unfamiliar faces and at matching different pictures of the same identity, either familiar or unfamiliar. However, he had no difficulty at pointing famous names, and naming common objects and famous buildings. As for subject CD, DN never reported visual face-related changes, stating for example: "*I don't' know who he is*"; "*I don't know the 3 faces you showed me*", "*I had difficulties recognizing her*", "*I didn't recognize the face immediately*". DN also failed to remember all the visual items presented (face and non-face-items) during the stimulation. Sensitivity to unfamiliar face identity of each contact were measured independently with a FPVS paradigm [110, 111]. Again,

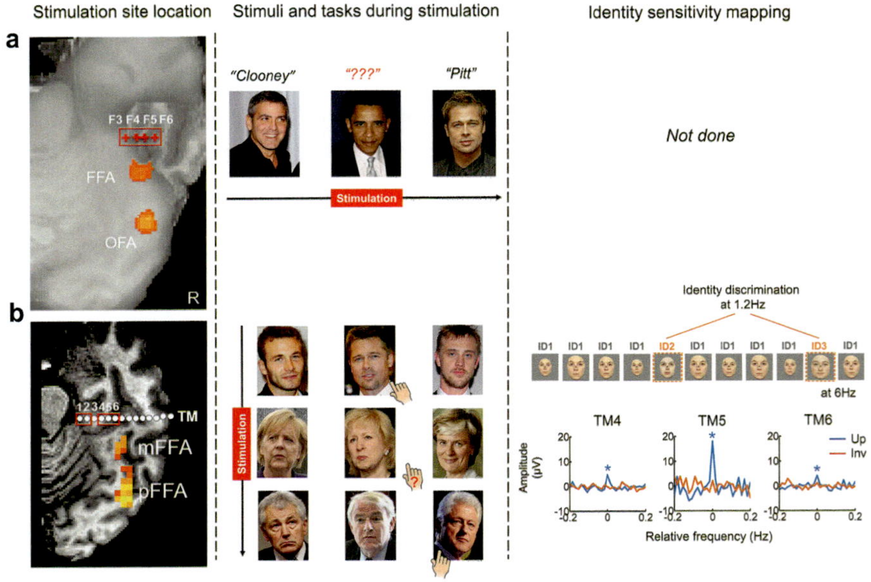

Fig. 39.3 Stimulating the right AntFG induces transient prosopagnosia. **a** Subject CD [98]. **b** Subject DN [99]. In both studies, the left panel shows fMRI face-selective activations in the right VOTC (axial slices) with SEEG electrodes superimposed (red crosses or white dots); the middle panel shows the stimuli presented during DES; the right panel shows SEEG recordings during a FPVS paradigm measuring sensitivity to face identity (for subject DN only, [110, 111]). **a** In subject CD, DES of contacts F3 to F6, located in the right AntFG, anteriorly to the FFA, induced a transient inability to recognize famous faces. Despite large face-selective responses in SEEG on these contacts, fMRI face-selective activations were not found because of a severe signal drop-out affecting the right AntFG (the left panel displays the raw functional images in light grey, showing the critical contacts being located in a region with very low fMRI signal). **b** In subject DN, DES of the same region (contacts TM1 to TM2 and TM4 to TM6) induced a transient inability to point out the familiar faces among unfamiliar faces (along with the inability to matching the identity of either familiar or unfamiliar faces). Again, no fMRI face-selective activations were found in the vicinity of the stimulation sites despite large SEEG face-selective responses recorded on these contacts. Of all the 141 recorded contacts in DN's brain, one of the critical contacts (TM5) recorded the largest face identity discrimination response amplitude in the upright condition (as well as the largest face inversion effect, i.e., upright-inverted) as measured by a FPVS paradigm (right panel shows the sum of identity discrimination responses at 1.2 Hz and harmonics centred on 0 Hz, in both upright and inverted conditions)

one of the few critical contacts recorded the largest neural face identity discrimination response (Fig. 39.3b; [99]).

39.3.4 DES to the FFA: Subjective and Objective Effects

The case studies presented above concern face-selective VOTC regions that are located posteriorly (IOG) and anteriorly (AntFG) to the FFA, which is not only, by far, the most studied and well-known face-selective human brain region [112], but also the right hemisphere region showing the most consistent and largest face-selective activity [108, 113]. However, to date, there has been no clear case of transient prosopagnosia following DES applied to this region. That is, DES with ECOG electrodes positioned over the right (but not the left) FFA have reported face-related perceptual changes and face or face-part hallucinations [30, 31, 40, 41, 114] (see also [115]) but without an objective FIR impairment (as only behaviorally tested by [40]; without any effect). With ECOG, these perceptual phenomena have been observed consistently enough to report studies in (small) groups of subjects (N = 8, [30]; N = 5, [114]), allowing to statistically demonstrate the prevalence of perceptual effects when stimulating on face-selective as compared to non-face-selective sites in the LatMidFG.

Importantly, two group studies have related DES to this region of the FFA (bilaterally) with objective face-related behavioral effects. Chong et al. [116] showed their participants (N = 8) ambiguous visual displays of a face and a house image superimposed in different transparency levels (Fig. 39.4a), asking them to recognize the images as face or nonface stimuli. This generic face recognition performance significantly decreased with brief (550 ms) electrical stimulation of the bilateral LatMidFG relative to no stimulation. The DES effect was small but significant thanks to the inclusion of several subjects and electrode contacts in the study. Moreover, subsequent analyses showed that this impairment by DES was confined to face-selective electrodes, with the amount of interference being positively correlated with the degree of face-selectivity of the electrodes as determined with iEEG (event-related potentials, ERPs).

Extending these findings, [117] applied DES to electrophysiologically face-selective sites of the bilateral LatMidFG in six subjects, and showed significant increases in response time at detecting whether a face stimulus was (slightly) distorted or not (Fig. 39.4b) (see also Chap. 40). The effect was also confined to face-selective sites defined independently, with no effect found for DES to neighboring non-face-selective (i.e., place-selective) regions. Importantly, response times were increased only when DES was applied 100 ms after visual onset (compared to 200 ms before or 500 after this onset), a latency that corresponds remarkably well to the onset time of face-selective activity in this region [117] (see also [118, 119]).

Unlike most of the above-cited studies reporting face-related perceptual changes (e.g., [31, 40]), these latter two studies did not localize the FFA in fMRI, measuring instead face-selectivity with independent electrophysiological recordings. Moreover, they used only brief stimulation durations. However, both of these studies provide a more systematic methodologically controlled approach to the effect of DES in the VOTC on human face recognition, here limited to the recognition of a visual stimulus as a (normal) face.

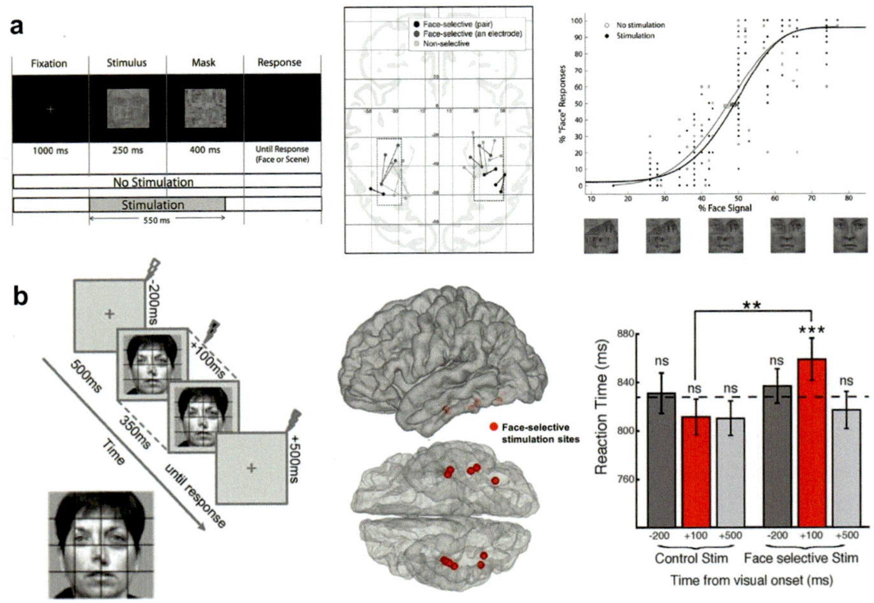

Fig. 39.4 Objective generic face recognition impairment during DES of the LatMidFG as assessed by quantitative measurements. For the 2 studies illustrated here, the left panel shows the experimental computerized paradigm during DES, the middle panel the location of the stimulated electrodes and the right panel the quantitative behavioral measurements across subjects. **a** Adapted from [116] with permission. Subjects (N = 8) had to decide whether compound stimuli mixing a certain percentage face and scene images were a face or a house while brief DES (550 ms) was applied to LatMidFG sites. Compared to no stimulation or stimulation of non-face-selective sites, DES of face-selective sites of the LatMidFG electrodes impaired categorization of the stimulus as a face, as shown by a small but significant increase of the point of subjective equality. **b** Adapted from [117] with permission. Subjects (N = 6) had to decide whether a face stimulus was distorted or not with DES applied to either face-selective and control (non-face selective) regions at −200, 100, and 500 ms with respect to the face presentation (face-selective sites are shown in red in a brain template). Only DES to face-selective sites and applied 100 ms after the face presentation induced a response time increase at the judgment of face distortion

How about face *identity* recognition linked to the FFA? To date, only one case of DES to the FFA apparently affecting FIR has been reported, with SEEG: subject MB who, during stimulation in this region of the right hemisphere experienced facial *palinopsia*, i.e., hallucinations of facial elements appropriately incorporated in the face identity in front of her [120]. MB experienced this phenomenon for the faces of a clinician or of an experimenter in front of her ("*I saw you with eyes and ears which were not yours*"). Although she stated that the superimposed features were those of a familiar face, she was unable to determine the identity of that face. She was also tested more systematically with photographs, stating, for example, that "*the photograph of Sarkozy* [former French president, first presented face picture during stimulation] *was transposed onto the other face identity* [second presented face picture during stimulation]". MB never reported aberrant facial configurations, and reproducibly stated that the facial structure was preserved ("*it was a normal face*"). Although MB was not tested behaviourally with unfamiliar faces during DES, the sensitivity to unfamiliar face identity of each contact was also measured independently with a frequency-tagging paradigm [110, 111]. Again, strikingly, of all the 137 recorded contacts, the critical contact in the right LatMidFG recorded, by far, the largest neural face individuation response [120].

39.3.5 What Can Be Learned from DES in Face-Selective VOTC Regions?

In summary, following anecdotal observations of the effect of DES on face-selective VOTC regions in the context of extensive and detailed reports of iEEG recordings [121, 122], a number of studies performed over the last decade (since [100]) have focused on the effects of DES on perceptual experience, behavior, and even electrophysiological activity evoked by concurrent face stimulation.

Collectively, these studies show first that category-selectivity, i.e., face-selectivity, is a key predictor of the effect of DES. Indeed, the effective cortical sites are almost systematically located in highly face-selective regions, either shown by independent iEEG recordings (using standard ERPs or a frequency-tagging approach) or a face-localizer in fMRI. Moreover, across all recorded contacts, the contacts leading to DES effects are generally the most face-selective [40, 41, 99, 100, 116, 120].

Second, while DES to bilateral face-selective cortical sites in the LatMidFG can affect even the simple classification of a visual stimulus as a (normal) face, spectacular phenomena such as a change of percept or a complete interruption of face identity recognition have been observed following stimulation in the right hemisphere only (with the exception of one left handed subject in [123] and one in [30]; see [124] for the role of handedness in face recognition lateralization). Altogether with the lack of FIR impairment when stimulating corresponding regions of the left hemisphere, as regularly tested in our clinical center, this suggests that the right hemisphere is both necessary and sufficient for FIR. This conclusion is an agreement with the long-standing view of the right hemispheric predominance of human face (identity) recognition, as supported by a wealth of evidence in cognitive neuroscience [125].

Third, while (small) group studies have been performed with DES in ECOG, they have restricted their investigation to stimulation over face-selective contacts of the LatMidFG (i.e., the FFA, including sometimes two functional regions, pFus-faces and mFus-faces; e.g., [30]), focusing either on spontaneous perceptual reports or on a single behavioral task. In contrast, DES effects with SEEG have been rarer, but concerned more extensively studied single cases, with stimulation effects observed beyond the (right) FFA. Specifically, the 4 SEEG cases (5 explorations) summarized above (i.e., KV 2 times; CD; DN; MB) show that intracerebral stimulation of spatially different regions, i.e., the right IOG, LatMidFG and AntFG can evoke highly reproducible transient impairments of FIR, both in terms of subjective reports and quantified behavioral measures, while the recognition of non-face images (common objects, famous buildings, famous names) is preserved [98–101, 120]. While the two posterior face-selective regions (IOG/OFA and LatMidFG/FFA) were already well identified (which does not mean that their critical role was established before DES), SEEG DES therefore extended the neural basis of FIR to the face-selective right AntFG (Fig. 39.1), a region with very little evidence of relationship to FIR (although see [126]), mainly because it is almost invisible in fMRI due to a large magnetic susceptibility artefact [77, 103]. Importantly, in all cases in which these tests were performed (KV second SEEG, MB, ND) the largest neural measures of sensitivity to the visual individuality of unfamiliar faces (i.e., independently from long-term familiarity) were also recorded on the very same contacts eliciting the transient FIR impairment.

Finally, while slight alterations of performance can be due to DES to several spatially dissociated contacts in the same patient, a FIR impairment appears to be restricted to the stimulation of one to three adjacent pairs of contacts, i.e., these sites being spatially confined. This does not imply that the effect of DES is limited to the stimulated contact, as demonstrated in some of these studies [117] and discussed below.

39.4 Interpretations, Practical and Theoretical Considerations for Future DES Studies

Based on the specific DES observations described above, in this last section of the chapter, we discuss practical and theoretical issues that are important to take into consideration to increase the value of DES in cognitive neuroscience.

39.4.1 Bringing the Lab into the Clinical Room

As mentioned in the introduction, the impact of DES studies in cognitive neuroscience has been limited in part because the experimental set-up is often less controlled than the experimental standards usually accepted in research laboratories worldwide. This gap is of course due to the fact that DES data is acquired in a clinical context, with limited testing time and experimental resources (for a description of the practical challenges of iEEG research and how they may be addressed, see also Chaps. 4 and 5). DES effects are sometimes unexpected and the time to adapt the experimental set-up is also limited in this context. As a consequence, DES studies usually include a restricted number of stimulations and trials, rely on retrospective clinical data, or are based only on the observation of the subjects' spontaneous behavior or subjective reports. Yet, in recent years, there has been an effort to bring the lab into the clinical room to match as much as possible the lab experimental standards, and this effort should be pursued in the future. As illustrated in the present chapter for FIR, some DES studies have used well-controlled experimental paradigms specifically designed to test specific hypothesis regarding the potential function of the stimulated brain region (e.g., [22, 99, 101, 116, 117, 127–130]; see Figs. 39.2b, 39.4). Ideally, critical sites should be tested with a sufficient number of different paradigms, including control tasks, in order to isolate as much as possible the nature of the function of the stimulated region. For these tasks, objective measurements in terms of accuracy rates and response times should be recorded. Moreover, for all of these tasks, a sufficient number of stimulations, i.e., trials, should be performed in order to objectively (i.e., statistically) define the DES effects relative to trials without stimulation or stimulation to non-critical sites.

The experimental set-up during DES can be computerized to better control the delay between the stimulation and the trial onsets [101], to facilitate the recording of accuracy and response times and to promote multicenter DES studies thanks to easily sharable testing software (e.g., NeuroMapper; [26]). These computer-based paradigms promote innovative testing methods (e.g., video, virtual reality), providing a deeper understanding of usually targeted functions but also expand the range of tested functions (episodic memory, socio-emotional cognitive function, spatial memory, e.g., [26, 131, 132]). At the same time, computer-based approaches are not always possible in the clinical context and may lack flexibility (e.g., quickly removing problematic trials with low accuracy outside DES, or quickly showing impaired trials again after the stimulation procedure to test the subject once more or to let the subject comment about what happened during DES). Moreover, using only computer-based approaches may prevent subjects from reporting phenomenological experiences that are not available in non-human animal research and may be particularly useful in understanding the nature of the affected brain function and/or guiding further experiments.

39.4.2 Group or Single Case Studies?

The general principle of DES group studies is to perform stimulations across several subjects with the same test(s). Such studies are particularly suitable to map critical sites for a given brain function across a large area of cortex, as long as this function can easily be disrupted by DES to reach a sufficient number of subjects with critical sites (e.g., naming across the left ATL). They enable for example to report the percentage of positive and negative sites across regions and to compute proportion maps of positive sites (e.g., [58, 59, 101]). When DES results are projected into a standard brain space, the individual anatomy and functional organization is blurred, a caveat that can be avoided by grouping DES results according to their location based on the individual anatomy (e.g., [58]). By focusing on a specific brain region, these studies can also reveal subtle DES effects, such as increases of average error rates or response times (e.g., [116, 117, 132]; Fig. 39.4). Multiple critical sites across multiple subjects allow such studies to correlate DES results with independent variables to better understand why some sites elicit positive effects while others do not (e.g., [116, 117]). Although they have the advantage to group subjects stimulated with similar cognitive tests and stimulation procedures, these tests are usually limited in number, usually one or two, restraining the full comprehension of the disturbed function. This limitation is usually due to practical reasons, either because a streamlined procedure is required to test all subjects the same way or because some studies rely on retrospective data acquired in a clinical context.

Other studies report in-depth explorations of a small number of subjects in the same article (e.g., [117, 129]). This type of study provides limited spatial sampling but deeper investigation of each subject and of the studied function. While our research group benefits from a large number of SEEG investigations per year at the University Hospital in Nancy (France) and has been able to map the neural basis of face, object and visual word recognition with large samples for SEEG recordings [108, 109, 133], the rarity of FIR impairments observed during DES (i.e., 5 cases in 10 years) has constrained us to report single case studies only. Despite this obvious limitation in number of cases and reproducibility of effects, this approach offered us the opportunity to report in-depth DES investigations of each subject, along with extensive multimodal explorations beyond DES (behavior, fMRI, SEEG recordings). First, well-controlled behavioral paradigms were specifically designed to test a specific function (e.g., [99, 101]). DES during these paradigms was repeated as often as possible to objectively measure DES effects with quantitative variables (error rates and/or response times increase during versus outside DES). Control tasks were also designed to isolate as much as possible the nature of the disturbed cognitive function. Second, beyond these objective measures, subjective reports and the semiology of these visual hallucinations or illusions were also investigated and clearly reported (e.g., facial palinopsia; [120]). Third, subjects were all tested with behavioral tasks before the DES procedure to ensure integrity of the tested function. In fact, subjects were even tested with the exact same tasks used subsequently during the DES procedure to ensure that their accuracy at these tasks was very high or even at ceiling, such that every failed trial during DES could be unambiguously classified as a stimulation-induced impairment (e.g., [99]). Finally, DES effects were interpreted in light of independent measures in every single case, mainly fMRI and SEEG recordings, either to map the whole network (e.g., the face network with face-selectivity measurements) or a specific function (e.g., sensitivity to face identity) (see also e.g., [41, 117, 129]).

As described above, all FIR impairments during DES were evoked by stimulating face-selective sites (as shown by both fMRI and SEEG recordings) and, when tested, by stimulating the regions with the highest sensitivity to unfamiliar face identity (as shown by SEEG recordings, [99, 101, 120]). The support of these independent measures is essential for DES studies for several reasons. They allow additional evidence of the critical role of the simulated region when DES and independent measurements evaluate the same function (e.g., face identity; [99, 101, 120]). When DES investigations remain ambiguous about the nature of the disrupted function because of a limited testing time or number of trials, these independent measures can tilt the balance in favor of one hypothesis according to the type of responses recorded on the critical contacts. For instance, when the largest iEEG amplitude is recorded specifically on a critical site, it helps interpreting this stimulated site as a key node in the cortical network (e.g., entry point, highly connected node, etc.). In our DES studies specifically, we greatly benefited from the frequency-tagging approach for SEEG measurements, not only for its high sensitivity but also for enabling us to objectively identify, quantify and classify response amplitudes on each contact in the frequency domain rather than in an inherently ambiguous and subjective time-domain representation [99, 101, 120].

In summary, group and multiple case studies both have their own advantages and drawbacks, and the choice of one rather than the other approach should be based on the aim of the study, the frequency of observations and the nature of the tested cognitive function. A single case approach is inevitable if the observed phenomena and effects are extremely rare or if the patient's characteristics are unique. When such DES single case reports are well conducted (sufficient number of DES with adapted paradigms, multimodal investigations with independent measurements), they should not be considered as anecdotal or as "case reports" with their associated pejorative connotation, but as opportunities to perform in-depth investigations of a cognitive function, generate and evaluate new hypotheses [134–137]. Hence, as illustrated in the present chapter, we firmly believe that these types of studies have a key role to play in the future of DES.

39.4.3 A SEEG Advantage Over ECOG for DES?

As described above, transient impairment of FIR has been reported so far in 4 cases with DES using depth electrodes (i.e., SEEG). In contrast, DES with subdural electrodes, i.e., ECOG, has been much less successful and limited to anecdotal reports without quantification in early studies [121, 122]. Subsequent ECOG studies applying DES to face-selective regions of the VOTC (mainly the FFA) evoked changes of the face percept (i.e., face distortions) in several cases [30, 31, 40, 41, 114]; see also [115], but never reported a FIR impairment. One potential reason is that ECOG DES studies rarely tested for such impairments with behavioral tasks (with the exception of [40], in which no effect of DES on naming of celebrity pictures was found). Moreover, while complete category-selective impairment of FIR following brain damage, i.e., prosopagnosia, remains extremely rare, the prevalence of significant drops in recognition accuracy rates or increases in response times during FIR tasks in right posterior brain-damaged patients is high [138, 139]. Hence, in addition to subjective perceptual reports, the effect of DES to face-selective regions of the VOTC with subdural electrodes may be objectively identified in future studies with behavioral FIR tasks measuring accuracy rates and response times (as performed during generic face recognition tasks; [116, 117]; Fig. 39.4).

Alternatively, it could be that depth electrodes as used in SEEG, thanks to the position of the stimulated contacts within the brain tissue and often within the cortex as well as to the small intercontact spacing (see Box 39.1), allow for more focal stimulation effects compared to subdural electrodes. In contrast, DES with the latter approach probably involves a relatively large cortical zone beyond regions specifically involved in FIR, explaining why visual distortions sometimes extend to non-face objects [40, 41] and can be experienced when stimulating non-face-selective sites [30]; see also [122].

Compared to the relatively low risk profile associated with the small burr holes of SEEG, ECOG requires a partial craniotomy. SEEG is therefore a safer surgical procedure with fewer post-operative complications such as hemorrhages and infections [140]. Specifically for DES to understand the neural basis of cognitive function, compared to ECOG, SEEG is also likely to leave the subject more alert cognitively to accurately report his/her perceptual experiences and to perform explicit behavioral tasks.

Despite a lower spatial coverage compared to ECOG, SEEG also has the advantage to access deep and medial structures that are inaccessible to ECOG (e.g., the medial temporal lobe, [141]; the precuneus, [5]; the cingulate gyrus, [142]; the transverse temporal gyrus, [143]; the insula, [144]). SEEG also allows one to specifically stimulate sulci, where critical effects are often found, sometimes more frequently than in the adjacent gyri [5, 42, 58, 99]. For example, the second highest rate of induced anomia in the left VOTC was found in a sulcus, the occipito-temporal sulcus [58], and the VWFA is also located primarily within this sulcus [145], making it particularly amenable to DES with SEEG [42]. SEEG DES enables also to independently stimulate grey and white matter [130].

While SEEG is already predominant in European clinical centers, these advantages, along with other clinical advantages not developed here, explain the recent rise of SEEG in the US compared to ECOG [13, 146]. For example, SEEG became the most frequently performed iEEG procedure in the US Medicare population in 2016 [13, 146]. While the selection of an iEEG procedure does not depend on advantages to answer fundamental research questions, we think that the increasing use of SEEG internationally will allow for more focal DES, targeting cognitive functions more specifically in the years to come.

39.4.4 Functional Specificity of Local and Remote DES Effects

As illustrated in the present chapter, several pieces of evidence show that DES effects are specific to the functional role of the stimulated region. Indeed, strikingly, almost all DES effects interfering with face perception (i.e., the subjective experience of seeing faces) were reported after stimulating cortical sites defined independently as face-selective [30, 31, 40, 41, 98, 100, 101, 114, 115, 120–122, 147]. Moreover, there is generally a highly positive correlation between the frequency or magnitude of DES effects and the amplitude of the face-selective [41, 99, 101, 116, 120] or face identity-sensitive [99, 101, 120] electrophysiological activity. Along these observations, numerous stimulations of (sometimes weak) face-selective sites do not lead to face recognition/perception effects [30, 38, 114, 117].

The SEEG DES studies described in the chapter allowed for a clarification of the functional characteristics of the critical sites, beyond their high-level of face-selectivity. Specifically, only the stimulation of cortical sites that share common characteristics could elicit behavioral effects (in the case of FIR: right-lateralized, anatomically restricted, located from the IOG to the AntFG, highly face-selective, highly sensitive to unfamiliar individual faces). In our experience, DES of face-selective sites located in the left hemisphere or anteriorly to the AntFG located in classical semantic ATL regions [51, 148] never produced FIR impairments. This shows that the stimulation current does not spread indifferently across the whole bilateral cortical face-selective network and that the function—FIR—is impaired only if, within the specialized network, a key node is *directly* stimulated.

At first glance, these observations could be interpreted as evidence for a strictly localized function, suggesting that DES effects on face recognition/perception are due exclusively to interference with the function of a highly functionally specific local region (e.g., [31]). However, given the high density of connectivity between brain regions [149] and our current knowledge of DES effects, viewing these effects as being strictly focal seems outdated [6, 29, 150–152]. Instead, DES effects may arise when directly stimulating a key node of a functionally specialized network, i.e., a dense population of highly specialized neurons with increased connectivity to other sites of the network ([117, 153]; Fig. 39.5a). Nevertheless, a current crucial issue for DES studies is to assess the functional *specificity* of the current propagation and remote effects (i.e., that DES remote effects specifically concern the functional network to which the stimulation site belongs).

Regarding human face recognition, several pieces of evidence show the functional specificity of the remote effects. SEEG DES showed that a FIR impairment could be evoked by stimulating several distant sites (right IOG, LatMidFG and antFG), all part of the network supporting (unfamiliar) FIR as evidenced in electrophysiology [51]. There were also subtle differences in DES effects across regions probably due to these regions' own connectivity pattern with other nearby networks/regions (Fig. 39.1). Yet, that did not prevent us from highlighting the main functional role of these regions. For instance, face-perceptual changes (i.e., changes in the phenomenological experience of faces) have been found when stimulating posterior face-selective regions (OFA, FFA) [100, 101, 120]; see also [40, 41], i.e., regions that are known to be directly connected with low-level visual areas [154], but never when stimulating the more anteriorly located AntFG [98, 99]. In contrast, direct stimulation of the right AntFG only led to a generic deficit in encoding the stimulation episode in memory, interpreted as reflecting a remote effect of DES to the connected medial temporal lobe episodic memory system [25].

The cortical face network view of DES effects, which is supported by independent evidence of tight anatomo-functional connectivity within this network [155], is also well illustrated by the study of [117]; (Fig. 39.5b), in which the effect of DES to the LatMidFG (affecting behavioral performance) spreads anteriorly and posteriorly to a

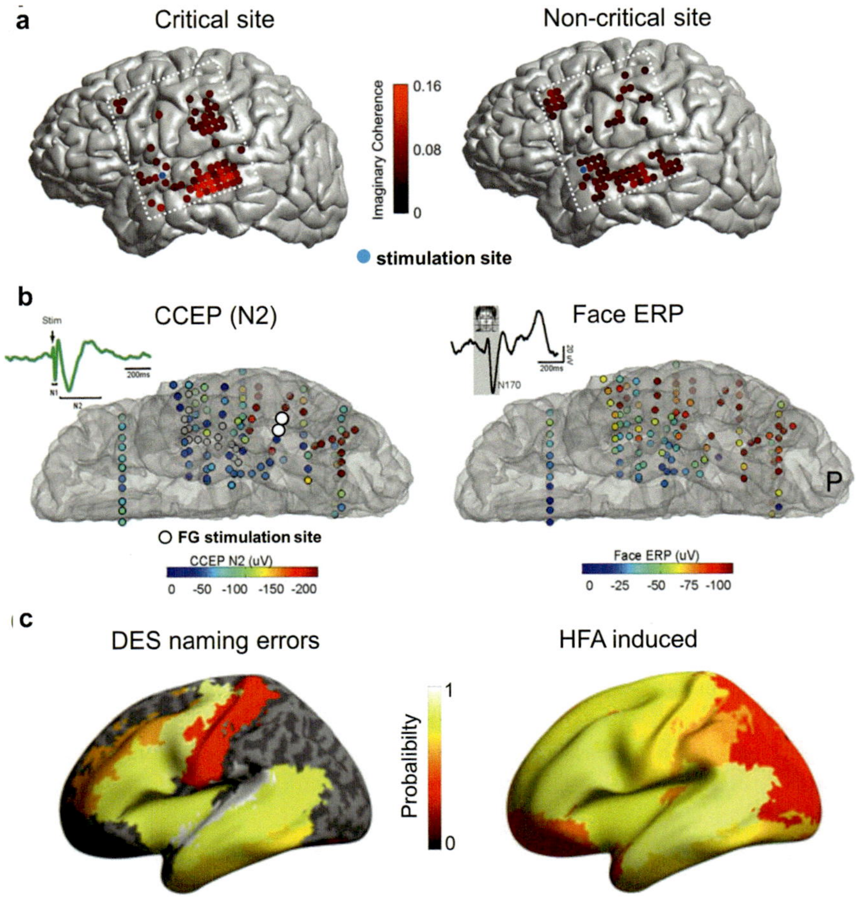

Fig. 39.5 Assessing the connectivity of critical DES sites. **a** Adapted from [153] with permission. A critical DES language site showed increased functional connectivity (left panel) compared to a non-critical DES site (right panel), as measured by imaginary coherence in the alpha band (example in one subject). **b** Adapted from [117] with permission. Single-pulse DES over the LatMidFG induced increase reaction time at detecting face distortions as well as distant CCEP along the VOTC (left panel, example in one subject). Interestingly, there was an amplitude correlation across electrodes between the N2 CCEP (left panel) and the face N170 recorded independently (right panel), showing that the stimulation current propagates preferentially within the face-selective network. **c** Adapted from [152] with permission. Critical DES naming sites (left panel shows the probability of evoking a naming impairment in the left hemisphere across 29 subjects) induced high-frequency activity (HFA) at distant sites, whose probability of occurrence is shown on right panel

number of recorded contacts along the VOTC but also to more distant regions of the temporal lobe and prefrontal cortex in some patients, as shown by distant recorded CCEP (see Chap. 40). Importantly, in that study, there was a strong amplitude correlation between the CCEP and the face-selective N170 potential in each patient and in a group average, showing that the spread of activity reflects the intrinsic organization of a functional network. Even if cortical face networks of humans and monkeys differ substantially in anatomical substrates, region size, lateralization and cortical distance between the nodes of this network [94, 156], such findings are in line with the observation that focal electrical microstimulation of a face-selective region in the monkey brain also spreads selectively within the cortical face-selective network [150].

39.4.5 Assessing the Connectivity of the Critical Sites

Once it is admitted that DES is likely to affect a network linked to the stimulated site rather than only an isolated functional brain region, the next step for future DES studies is to identify and characterize this network. A first approach consists of identifying the critical sites using DES and subsequently assessing the functional connectivity of these sites using independent connectivity measures. In particular, high-frequency DES has been combined with CCEP, especially in studies stimulating language regions. Critical language sites identified with high-frequency DES either in Broca's area, Wernicke's area or the BTLA showed strong connectivity within the language network, i.e., elicited CCEP in the 2 remaining regions [65, 66, 157], see also [158]. DES could also be combined with various measures of functional connectivity using iEEG recordings acquired independently (e.g., coherence, phase-locking values, debiased phase lag index, etc.; see Chaps. 32 and 33). Several studies have shown increased functional connectivity of critical DES sites with distant regions [17, 153, 159, 160]; Fig. 39.5a). Finally, DES could be also combined with structural connectivity methods [158, 161]. For example, with diffusion tensor imaging, it has been shown that critical BTLA sites were structurally connected to the temporal pole, medial temporal lobe, lateral temporal, and occipital cortex through the inferior longitudinal fasciculus [161].

A second approach consists of using the DES effect itself to study the connectivity of the critical sites by *simultaneously* combining DES and iEEG recordings. Using this approach, a number of studies have identified remote power spectral changes in various frequency bands induced by DES evoking behavioral effects [152, 160]; see also [162, 163]). For instance, Perrone-Bertolotti and colleagues [152] found induced high frequency activity by critical language DES in remote sites belonging to the language system (Fig. 39.5c). Interestingly, they found similar results using remote after-discharges, indicating that these discharges could also be used to index the connectivity of DES sites. When single-pulse DES was sufficient to evoke a behavioral effect, it could be used at the same time to study the effective connectivity of the stimulated region using a CCEP approach (e.g., [19] for low-frequency DES in the fornix inducing memory improvement along with CCEP in hippocampus and posterior cingulate).

Other approaches have proposed to identify the distant spectral lines corresponding to the stimulation waveform of high-frequency DES evoking clinical effects

[164] or to measure the functional connectivity change just after DES [165]. Finally, a promising approach has recently been introduced in human research: DES while simultaneously recording fMRI to map evoke distant BOLD activation [166, 167].

Overall, connectivity analyses should be highly promoted in future DES studies not only to avoid the simplistic interpretation of strictly focal DES effects but to better characterize the neurofunctional organization of the critical network. However, there are at least two avenues for improvement. First, studies will need to show the specificity of the connectivity pattern of critical sites (both qualitatively and quantitatively), for example by comparing the connectivity of critical and non-critical DES sites ([117, 153, 163, 165]; Fig. 39.5a). Second, all connectivity methods listed above are designed to identify the overall connectivity of a region, independently of the various functional sub-networks linked to the stimulated region. Our investigations of single cases showing FIR impairments during DES strongly suggest that the stimulation could be specific enough to only affect a sub-network (see also [6]). Future studies will need to identity the particular sub-network affected by DES, using more specific connectivity measures. For example, functional responses (e.g., low- and high frequency neural activity to specific stimuli) could be recorded with and without concomitant DES, with the local and distant modulation of these responses reflecting the specific functional network affected by DES [117, 168].

References

1. Fritsch G, Hitzig E (1870) Uber die elektrische Erregbarkeit des Grosshirns. Arch Anat Physiol Wiss Med 37:300–332
2. Ferrier D (1876) The functions of the brain. G P Putnam's Sons, New York
3. Penfield W, Boldrey E (1937) Somatic motor and sensory representation in the cerebral cortex of man as studied by electrical stimulation. Brain 60:389–443. https://doi.org/10.1093/brain/60.4.389
4. Blanke O, Ortigue S, Landis T, Seeck M (2002) Stimulating illusory own-body perceptions. Nature 419:269–270. https://doi.org/10.1038/419269a
5. Jonas J, Maillard L, Frismand S et al (2014) Self-face hallucination evoked by electrical stimulation of the human brain. Neurology 83:336–338. https://doi.org/10.1212/WNL.0000000000000628
6. Mandonnet E, Winkler PA, Duffau H (2010) Direct electrical stimulation as an input gate into brain functional networks: principles, advantages and limitations. Acta Neurochir (Wien) 152:185–193. https://doi.org/10.1007/s00701-009-0469-0
7. Mahon BZ, Miozzo M, Pilcher WH (2019) Direct electrical stimulation mapping of cognitive functions in the human brain. Cogn Neuropsychol 36:97–102. https://doi.org/10.1080/02643294.2019.1630375
8. Papagno C (2017) Studying cognitive functions by means of direct electrical stimulation: a review. Neurol Sci 38:2079–2087. https://doi.org/10.1007/s10072-017-3095-2
9. Ritaccio AL, Brunner P, Schalk G (2018) Electrical stimulation mapping of the brain: basic principles and emerging alternatives. J Clin Neurophysiol 35:86–97. https://doi.org/10.1097/WNP.0000000000000440
10. Aron O, Jonas J, Colnat-Coulbois S, Maillard L (2021) Language mapping using stereo electroencephalography: a review and expert opinion. Front Hum Neurosci 15:619521. https://doi.org/10.3389/fnhum.2021.619521

11. Isnard J, Taussig D, Bartolomei F et al (2018) French guidelines on stereoelectroencephalography (SEEG). Neurophysiol Clin 48:5–13. https://doi.org/10.1016/j.neucli.2017.11.005
12. Britton JW (2018) Electrical stimulation mapping with stereo-EEG electrodes. J Clin Neurophysiol 35:110–114. https://doi.org/10.1097/WNP.0000000000000443
13. Grande KM, Ihnen SKZ, Arya R (2020) Electrical stimulation mapping of brain function: a comparison of subdural electrodes and stereo-EEG. Front Hum Neurosci 14:611291. https://doi.org/10.3389/fnhum.2020.611291
14. Trébuchon A, Chauvel P (2016) Electrical stimulation for seizure induction and functional mapping in stereoelectroencephalography. J Clin Neurophysiol 33:511–521. https://doi.org/10.1097/WNP.0000000000000313
15. Hamberger MJ, Williams AC, Schevon CA (2014) Extraoperative neurostimulation mapping: results from an international survey of epilepsy surgery programs. Epilepsia 55:933–939. https://doi.org/10.1111/epi.12644
16. Mohan UR, Watrous AJ, Miller JF et al (2020) The effects of direct brain stimulation in humans depend on frequency, amplitude, and white-matter proximity. Brain Stimulat 13:1183–1195. https://doi.org/10.1016/j.brs.2020.05.009
17. Alagapan S, Riddle J, Huang WA et al (2019) Network-targeted, multi-site direct cortical stimulation enhances working memory by modulating phase lag of low-frequency oscillations. Cell Rep 29:2590-2598.e4. https://doi.org/10.1016/j.celrep.2019.10.072
18. Alagapan S, Lustenberger C, Hadar E et al (2019) Low-frequency direct cortical stimulation of left superior frontal gyrus enhances working memory performance. Neuroimage 184:697–706. https://doi.org/10.1016/j.neuroimage.2018.09.064
19. Koubeissi MZ, Kahriman E, Syed TU et al (2013) Low-frequency electrical stimulation of a fiber tract in temporal lobe epilepsy: Koubeissi et al: electrical stimulation in TLE. Ann Neurol n/a-n/a. https://doi.org/10.1002/ana.23915
20. Miller JP, Sweet JA, Bailey CM et al (2015) Visual-spatial memory may be enhanced with theta burst deep brain stimulation of the fornix: a preliminary investigation with four cases. Brain 138:1833–1842. https://doi.org/10.1093/brain/awv095
21. Zangaladze A, Sharan A, Evans J et al (2008) The effectiveness of low-frequency stimulation for mapping cortical function. Epilepsia 49:481–487. https://doi.org/10.1111/j.1528-1167.2007.01307.x
22. Titiz AS, Hill MRH, Mankin EA et al (2017) Theta-burst microstimulation in the human entorhinal area improves memory specificity. eLife 6:e29515. https://doi.org/10.7554/eLife.29515
23. Selimbeyoglu A, Parvizi J (2010) Electrical stimulation of the human brain: perceptual and behavioral phenomena reported in the old and new literature. Front Hum Neurosci 4:46. https://doi.org/10.3389/fnhum.2010.00046
24. Desmurget M, Song Z, Mottolese C, Sirigu A (2013) Re-establishing the merits of electrical brain stimulation. Trends Cogn Sci 17:442–449. https://doi.org/10.1016/j.tics.2013.07.002
25. Jonas J, Rossion B (2021) Intracerebral electrical stimulation to understand the neural basis of human face identity recognition. Eur J Neurosci 54:4197–4211. https://doi.org/10.1111/ejn.15235
26. Drane DL, Pedersen NP, Sabsevitz DS et al (2021) Cognitive and emotional mapping with SEEG. Front Neurol 12:627981. https://doi.org/10.3389/fneur.2021.627981
27. Strick PL (2002) Stimulating research on motor cortex. Nat Neurosci 5:714–715. https://doi.org/10.1038/nn0802-714
28. Karnath H-O, Borchers S, Himmelbach M (2010) Comment on "movement intention after parietal cortex stimulation in humans." Science 327:1200. https://doi.org/10.1126/science.1183735
29. Borchers S, Himmelbach M, Logothetis N, Karnath H-O (2012) Direct electrical stimulation of human cortex—the gold standard for mapping brain functions? Nat Rev Neurosci 13:63–70. https://doi.org/10.1038/nrn3140

30. Schrouff J, Raccah O, Baek S et al (2020) Fast temporal dynamics and causal relevance of face processing in the human temporal cortex. Nat Commun 11:656. https://doi.org/10.1038/s41467-020-14432-8
31. Schalk G, Kapeller C, Guger C et al (2017) Facephenes and rainbows: causal evidence for functional and anatomical specificity of face and color processing in the human brain. Proc Natl Acad Sci 114:12285–12290. https://doi.org/10.1073/pnas.1713447114
32. Hutmacher F (2019) Why is there so much more research on vision than on any other sensory modality? Front Psychol 10:2246. https://doi.org/10.3389/fpsyg.2019.02246
33. Grill-Spector K, Weiner KS (2014) The functional architecture of the ventral temporal cortex and its role in categorization. Nat Rev Neurosci 15:536–548. https://doi.org/10.1038/nrn3747
34. Kravitz DJ, Saleem KS, Baker CI et al (2013) The ventral visual pathway: an expanded neural framework for the processing of object quality. Trends Cogn Sci 17:26–49. https://doi.org/10.1016/j.tics.2012.10.011
35. Cicmil N, Krug K (2015) Playing the electric light orchestra—how electrical stimulation of visual cortex elucidates the neural basis of perception. Philos Trans R Soc B Biol Sci 370:20140206. https://doi.org/10.1098/rstb.2014.0206
36. Bosking WH, Beauchamp MS, Yoshor D (2017) Electrical stimulation of visual cortex: relevance for the development of visual cortical prosthetics. Annu Rev Vis Sci 3:141–166. https://doi.org/10.1146/annurev-vision-111815-114525
37. Lee HW, Hong SB, Seo DW et al (2000) Mapping of functional organization in human visual cortex: electrical cortical stimulation. Neurology 54:849–854. https://doi.org/10.1212/WNL.54.4.849
38. Murphey DK, Maunsell JHR, Beauchamp MS, Yoshor D (2009) Perceiving electrical stimulation of identified human visual areas. Proc Natl Acad Sci 106:5389–5393. https://doi.org/10.1073/pnas.0804998106
39. Jonas J, Frismand S, Vignal J-P et al (2014) Right hemispheric dominance of visual phenomena evoked by intracerebral stimulation of the human visual cortex: right hemispheric dominance of visual phenomena. Hum Brain Mapp 35:3360–3371. https://doi.org/10.1002/hbm.22407
40. Parvizi J, Jacques C, Foster BL et al (2012) Electrical stimulation of human fusiform face-selective regions distorts face perception. J Neurosci 32:14915–14920. https://doi.org/10.1523/JNEUROSCI.2609-12.2012
41. Rangarajan V, Hermes D, Foster BL et al (2014) Electrical stimulation of the left and right human fusiform gyrus causes different effects in conscious face perception. J Neurosci 34:12828–12836. https://doi.org/10.1523/JNEUROSCI.0527-14.2014
42. Sabsevitz DS, Middlebrooks EH, Tatum W et al (2020) Examining the function of the visual word form area with stereo EEG electrical stimulation: a case report of pure alexia. Cortex 129:112–118. https://doi.org/10.1016/j.cortex.2020.04.012
43. Hirshorn EA, Li Y, Ward MJ et al (2016) Decoding and disrupting left midfusiform gyrus activity during word reading. Proc Natl Acad Sci 113:8162–8167. https://doi.org/10.1073/pnas.1604126113
44. Mani J, Diehl B, Piao Z et al (2008) Evidence for a basal temporal visual language center: cortical stimulation producing pure alexia. Neurology 71:1621–1627. https://doi.org/10.1212/01.wnl.0000334755.32850.f0
45. Megevand P, Groppe DM, Goldfinger MS et al (2014) Seeing scenes: topographic visual hallucinations evoked by direct electrical stimulation of the parahippocampal place area. J Neurosci 34:5399–5405. https://doi.org/10.1523/JNEUROSCI.5202-13.2014
46. Kanwisher N, McDermott J, Chun MM (1997) The fusiform face area: a module in human extrastriate cortex specialized for face perception. J Neurosci 17:4302–4311. https://doi.org/10.1523/JNEUROSCI.17-11-04302.1997
47. Barton JJS, Press DZ, Keenan JP, O'Connor M (2002) Lesions of the fusiform face area impair perception of facial configuration in prosopagnosia. Neurology 58:71–78. https://doi.org/10.1212/WNL.58.1.71
48. Rossion B (2014) Understanding face perception by means of prosopagnosia and neuroimaging. Front Biosci-Elite 6:258–307. https://doi.org/10.2741/E706

49. Cohen AL, Soussand L, Corrow SL et al (2019) Looking beyond the face area: lesion network mapping of prosopagnosia. Brain 142:3975–3990. https://doi.org/10.1093/brain/awz332
50. Kovács G (2020) Getting to know someone: familiarity, person recognition, and identification in the human brain. J Cogn Neurosci 32:2205–2225. https://doi.org/10.1162/jocn_a_01627
51. Jacques C, Rossion B, Volfart A et al (2020) The neural basis of rapid unfamiliar face individuation with human intracerebral recordings. Neuroimage 221:117174. https://doi.org/10.1016/j.neuroimage.2020.117174
52. Duchaine B, Yovel G (2015) A revised neural framework for face processing. Annu Rev Vis Sci 1:393–416. https://doi.org/10.1146/annurev-vision-082114-035518
53. Cohen L, Lehéricy S, Chochon F et al (2002) Language-specific tuning of visual cortex? Functional properties of the Visual Word Form Area. Brain 125:1054–1069. https://doi.org/10.1093/brain/awf094
54. Woolnough O, Donos C, Rollo PS et al (2021) Spatiotemporal dynamics of orthographic and lexical processing in the ventral visual pathway. Nat Hum Behav 5:389–398. https://doi.org/10.1038/s41562-020-00982-w
55. Lüders H, Lesser RP, Hahn J et al (1986) Basal temporal language area demonstrated by electrical stimulation. Neurology 36:505–510. https://doi.org/10.1212/wnl.36.4.505
56. Lüders H, Lesser RP, Hahn J et al (1991) BASAL temporal language area. Brain 114:743–754. https://doi.org/10.1093/brain/114.2.743
57. Krauss GL, Fisher R, Plate C et al (1996) Cognitive effects of resecting basal temporal language areas. Epilepsia 37:476–483. https://doi.org/10.1111/j.1528-1157.1996.tb00594.x
58. Bédos Ulvin L, Jonas J, Brissart H et al (2017) Intracerebral stimulation of left and right ventral temporal cortex during object naming. Brain Lang 175:71–76. https://doi.org/10.1016/j.bandl.2017.09.003
59. Forseth KJ, Kadipasaoglu CM, Conner CR et al (2018) A lexical semantic hub for heteromodal naming in middle fusiform gyrus. Brain 141:2112–2126. https://doi.org/10.1093/brain/awy120
60. Poeppel D, Hickok G (2004) Towards a new functional anatomy of language. Cognition 92:1–12. https://doi.org/10.1016/j.cognition.2003.11.001
61. Poeppel D, Emmorey K, Hickok G, Pylkkanen L (2012) Towards a new neurobiology of language. J Neurosci 32:14125–14131. https://doi.org/10.1523/JNEUROSCI.3244-12.2012
62. Hermann BP, Perrine K, Chelune GJ et al (1999) Visual confrontation naming following left anterior temporal lobectomy: a comparison of surgical approaches. Neuropsychology 13:3–9. https://doi.org/10.1037//0894-4105.13.1.3
63. Abdallah C, Brissart H, Colnat-Coulbois S et al (2021) Stereoelectroencephalographic language mapping of the basal temporal cortex predicts postoperative naming outcome. J Neurosurg 1–11. https://doi.org/10.3171/2020.8.JNS202431
64. Ishitobi M, Nakasato N, Suzuki K et al (2000) Remote discharges in the posterior language area during basal temporal stimulation. NeuroReport 11:2997–3000. https://doi.org/10.1097/00001756-200009110-00034
65. Koubeissi MZ, Lesser RP, Sinai A et al (2012) Connectivity between perisylvian and bilateral basal temporal cortices. Cereb Cortex 22:918–925. https://doi.org/10.1093/cercor/bhr163
66. Araki K, Terada K, Usui K et al (2015) Bidirectional neural connectivity between basal temporal and posterior language areas in humans. Clin Neurophysiol 126:682–688. https://doi.org/10.1016/j.clinph.2014.07.020
67. Mandler G (1980) Recognizing: the judgment of previous occurrence. Psychol Rev 87:252–271. https://doi.org/10.1037/0033-295X.87.3.252
68. Bruce V, Young A (1998) In the eye of the beholder: the science of face perception. Oxford University Press, New York, NY, US
69. Todorov A (2017) Face value: the irresistible influence of first impressions. Princeton University Press
70. Sheehan MJ, Nachman MW (2014) Morphological and population genomic evidence that human faces have evolved to signal individual identity. Nat Commun 5:4800. https://doi.org/10.1038/ncomms5800

71. Burton AM, Kramer RSS, Ritchie KL, Jenkins R (2016) Identity from variation: representations of faces derived from multiple instances. Cogn Sci 40:202–223. https://doi.org/10.1111/cogs.12231
72. Jenkins R, Dowsett AJ, Burton AM (2018) How many faces do people know? Proc R Soc B Biol Sci 285:20181319. https://doi.org/10.1098/rspb.2018.1319
73. Yan X, Rossion B (2020) A robust neural familiar face recognition response in a dynamic (periodic) stream of unfamiliar faces. Cortex 132:281–295. https://doi.org/10.1016/j.cortex.2020.08.016
74. Wilmer JB (2017) Individual differences in face recognition: a decade of discovery. Curr Dir Psychol Sci 26:225–230. https://doi.org/10.1177/0963721417710693
75. Barton JJS, Davies-Thompson J, Corrow SL (2021) Prosopagnosia and disorders of face processing. In: Handbook of clinical neurology. Elsevier, pp 175–193
76. Bodamer J (1947) Prosop's agnosia; the agnosia of cognition. Arch Psychiatr Nervenkrankh Ver Mit Z Gesamte Neurol Psychiatr 118:6–53. https://doi.org/10.1007/BF00352849
77. Rossion B (2018) Damasio's error—prosopagnosia with intact within-category object recognition. J Neuropsychol 12:357–388. https://doi.org/10.1111/jnp.12162
78. Meadows JC (1974) The anatomical basis of prosopagnosia. J Neurol Neurosurg Psychiatry 37:489–501. https://doi.org/10.1136/jnnp.37.5.489
79. Sergent J, Signoret J-L (1992) Varieties of functional deficits in prosopagnosia. Cereb Cortex 2:375–388. https://doi.org/10.1093/cercor/2.5.375
80. Bouvier SE, Engel SA (2006) Behavioral deficits and cortical damage loci in cerebral achromatopsia. Cereb Cortex 16:183–191. https://doi.org/10.1093/cercor/bhi096
81. Barton JJS (2008) Structure and function in acquired prosopagnosia: lessons from a series of 10 patients with brain damage. J Neuropsychol 2:197–225. https://doi.org/10.1348/174866407X214172
82. Gainotti G (2010) Not all patients labeled as "prosopagnosia" have a real prosopagnosia. J Clin Exp Neuropsychol 32:763–766. https://doi.org/10.1080/13803390903512686
83. Gauthier I, Tarr MJ, Moylan J et al (2000) The fusiform "face area" is part of a network that processes faces at the individual level. J Cogn Neurosci 12:495–504. https://doi.org/10.1162/089892900562165
84. Gilaie-Dotan S, Gelbard-Sagiv H, Malach R (2010) Perceptual shape sensitivity to upright and inverted faces is reflected in neuronal adaptation. Neuroimage 50:383–395. https://doi.org/10.1016/j.neuroimage.2009.12.077
85. MichaelP E, Henson RN, Rowe JB et al (2013) Different neural mechanisms within occipitotemporal cortex underlie repetition suppression across same and different-size faces. Cereb Cortex 23:1073–1084. https://doi.org/10.1093/cercor/bhs070
86. Hermann P, Grotheer M, Kovács G, Vidnyánszky Z (2017) The relationship between repetition suppression and face perception. Brain Imaging Behav 11:1018–1028. https://doi.org/10.1007/s11682-016-9575-9
87. Kriegeskorte N, Formisano E, Sorger B, Goebel R (2007) Individual faces elicit distinct response patterns in human anterior temporal cortex. Proc Natl Acad Sci 104:20600–20605. https://doi.org/10.1073/pnas.0705654104
88. Nestor A, Plaut DC, Behrmann M (2011) Unraveling the distributed neural code of facial identity through spatiotemporal pattern analysis. Proc Natl Acad Sci U S A 108:9998–10003. https://doi.org/10.1073/pnas.1102433108
89. Natu V, O'Toole AJ (2011) The neural processing of familiar and unfamiliar faces: a review and synopsis. Br J Psychol Lond Engl 1953 102:726–747. https://doi.org/10.1111/j.2044-8295.2011.02053.x
90. Goesaert E, Op de Beeck HP (2013) Representations of facial identity information in the ventral visual stream investigated with multivoxel pattern analyses. J Neurosci 33:8549–8558. https://doi.org/10.1523/JNEUROSCI.1829-12.2013
91. Anzellotti S, Fairhall SL, Caramazza A (2014) Decoding representations of face identity that are tolerant to rotation. Cereb Cortex N Y N 1991 24:1988–1995. https://doi.org/10.1093/cercor/bht046

92. Gobbini MI, Haxby JV (2007) Neural systems for recognition of familiar faces. Neuropsychologia 45:32–41. https://doi.org/10.1016/j.neuropsychologia.2006.04.015
93. Bryant KL, Preuss TM (2018) A comparative perspective on the human temporal lobe. In: Bruner E, Ogihara N, Tanabe HC (eds) Digital endocasts. Springer, Japan, Tokyo, pp 239–258
94. Rossion B, Taubert J (2019) What can we learn about human individual face recognition from experimental studies in monkeys? Vision Res 157:142–158. https://doi.org/10.1016/j.visres.2018.03.012
95. Sergent J, Ohta S, Macdonald B (1992) Functional neuroanatomy of face and object processing: a positron emission tomography study. Brain 115:15–36. https://doi.org/10.1093/brain/115.1.15
96. Grill-Spector K, Weiner KS, Kay K, Gomez J (2017) The functional neuroanatomy of human face perception. Annu Rev Vis Sci 3:167–196. https://doi.org/10.1146/annurev-vision-102016-061214
97. Rossion B, Jacques C, Jonas J (2018) Mapping face categorization in the human ventral occipitotemporal cortex with direct neural intracranial recordings: intracranial mapping of face categorization. Ann N Y Acad Sci 1426:5–24. https://doi.org/10.1111/nyas.13596
98. Jonas J, Rossion B, Brissart H et al (2015) Beyond the core face-processing network: intracerebral stimulation of a face-selective area in the right anterior fusiform gyrus elicits transient prosopagnosia. Cortex 72:140–155. https://doi.org/10.1016/j.cortex.2015.05.026
99. Volfart A, Yan X, Maillard L et al (2022) Intracerebral electrical stimulation of the right anterior fusiform gyrus impairs human face identity recognition. Neuroimage 250:118932. https://doi.org/10.1016/j.neuroimage.2022.118932
100. Jonas J, Descoins M, Koessler L et al (2012) Focal electrical intracerebral stimulation of a face-sensitive area causes transient prosopagnosia. Neuroscience 222:281–288. https://doi.org/10.1016/j.neuroscience.2012.07.021
101. Jonas J, Rossion B, Krieg J et al (2014) Intracerebral electrical stimulation of a face-selective area in the right inferior occipital cortex impairs individual face discrimination. Neuroimage 99:487–497. https://doi.org/10.1016/j.neuroimage.2014.06.017
102. Rossion B, Boremanse A (2011) Robust sensitivity to facial identity in the right human occipito-temporal cortex as revealed by steady-state visual-evoked potentials. J Vis 11:16. https://doi.org/10.1167/11.2.16
103. Wandell BA (2011) The neurobiological basis of seeing words: seeing words. Ann N Y Acad Sci 1224:63–80. https://doi.org/10.1111/j.1749-6632.2010.05954.x
104. Axelrod V, Yovel G (2013) The challenge of localizing the anterior temporal face area: a possible solution. Neuroimage 81:371–380. https://doi.org/10.1016/j.neuroimage.2013.05.015
105. Nasr S, Tootell RB (2012) Role of fusiform and anterior temporal cortical areas in facial recognition. Neuroimage 63:1743–1753
106. Rossion B, Hanseeuw B, Dricot L (2012) Defining face perception areas in the human brain: a large-scale factorial fMRI face localizer analysis. Brain Cogn 79:138–157. https://doi.org/10.1016/j.bandc.2012.01.001
107. Pyles JA, Verstynen TD, Schneider W, Tarr MJ (2013) Explicating the face perception network with white matter connectivity. PLoS ONE 8:e61611. https://doi.org/10.1371/journal.pone.0061611
108. Jonas J, Jacques C, Liu-Shuang J et al (2016) A face-selective ventral occipito-temporal map of the human brain with intracerebral potentials. Proc Natl Acad Sci 113. https://doi.org/10.1073/pnas.1522033113
109. Hagen S, Jacques C, Maillard L et al (2020) Spatially dissociated intracerebral maps for face- and house-selective activity in the human ventral occipito-temporal cortex. Cereb Cortex 30:4026–4043. https://doi.org/10.1093/cercor/bhaa022
110. Liu-Shuang J, Norcia AM, Rossion B (2014) An objective index of individual face discrimination in the right occipito-temporal cortex by means of fast periodic oddball stimulation. Neuropsychologia 52:57–72. https://doi.org/10.1016/j.neuropsychologia.2013.10.022

111. Rossion B, Retter TL, Liu-Shuang J (2020) Understanding human individuation of unfamiliar faces with oddball fast periodic visual stimulation and electroencephalography. Eur J Neurosci 52:4283–4344. https://doi.org/10.1111/ejn.14865
112. Kanwisher N (2017) The quest for the FFA and where it led. J Neurosci Off J Soc Neurosci 37:1056–1061. https://doi.org/10.1523/JNEUROSCI.1706-16.2016
113. Gao X, Gentile F, Rossion B (2018) Fast periodic stimulation (FPS): a highly effective approach in fMRI brain mapping. Brain Struct Funct 223:2433–2454. https://doi.org/10.1007/s00429-018-1630-4
114. Sanada T, Kapeller C, Jordan M et al (2021) Multi-modal mapping of the face selective ventral temporal cortex–a group study with clinical implications for ECS, ECoG, and fMRI. Front Hum Neurosci 15:616591. https://doi.org/10.3389/fnhum.2021.616591
115. Mundel T, Milton JG, Dimitrov A et al (2003) Transient inability to distinguish between faces: electrophysiologic studies. J Clin Neurophysiol 20:102–110. https://doi.org/10.1097/00004691-200304000-00003
116. Chong SC, Jo S, Park KM et al (2013) Interaction between the electrical stimulation of a face-selective area and the perception of face stimuli. Neuroimage 77:70–76. https://doi.org/10.1016/j.neuroimage.2013.01.074
117. Keller CJ, Davidesco I, Megevand P et al (2017) Tuning face perception with electrical stimulation of the fusiform gyrus: tuning Face Perception with ES of the FG. Hum Brain Mapp 38:2830–2842. https://doi.org/10.1002/hbm.23543
118. Jacques C, Witthoft N, Weiner KS et al (2016) Corresponding ECoG and fMRI category-selective signals in human ventral temporal cortex. Neuropsychologia 83:14–28. https://doi.org/10.1016/j.neuropsychologia.2015.07.024
119. Kadipasaoglu CM, Conner CR, Baboyan VG et al (2017) Network dynamics of human face perception. PLoS ONE 12:e0188834. https://doi.org/10.1371/journal.pone.0188834
120. Jonas J, Brissart H, Hossu G et al (2018) A face identity hallucination (palinopsia) generated by intracerebral stimulation of the face-selective right lateral fusiform cortex. Cortex 99:296–310. https://doi.org/10.1016/j.cortex.2017.11.022
121. Allison T, Ginter H, McCarthy G et al (1994) Face recognition in human extrastriate cortex. J Neurophysiol 71:821–825. https://doi.org/10.1152/jn.1994.71.2.821
122. Puce A (1999) Electrophysiological studies of human face perception. III: effects of top-down processing on face-specific potentials. Cereb Cortex 9:445–458. https://doi.org/10.1093/cercor/9.5.445
123. Rangarajan V, Parvizi J (2016) Functional asymmetry between the left and right human fusiform gyrus explored through electrical brain stimulation. Neuropsychologia 83:29–36. https://doi.org/10.1016/j.neuropsychologia.2015.08.003
124. Bukowski H, Dricot L, Hanseeuw B, Rossion B (2013) Cerebral lateralization of face-sensitive areas in left-handers: only the FFA does not get it right. Cortex 49:2583–2589. https://doi.org/10.1016/j.cortex.2013.05.002
125. Rossion B, Lochy A (2022) Is human face recognition lateralized to the right hemisphere due to neural competition with left-lateralized visual word recognition? A critical review. Brain Struct Funct 227:599–629. https://doi.org/10.1007/s00429-021-02370-0
126. Behrmann M, Avidan G, Gao F, Black S (2007) Structural imaging reveals anatomical alterations in inferotemporal cortex in congenital prosopagnosia. Cereb Cortex 17:2354–2363. https://doi.org/10.1093/cercor/bhl144
127. Coleshill SG (2004) Material-specific recognition memory deficits elicited by unilateral hippocampal electrical stimulation. J Neurosci 24:1612–1616. https://doi.org/10.1523/JNEUROSCI.4352-03.2004
128. Lacruz ME, Valentín A, Seoane JJG et al (2010) Single pulse electrical stimulation of the hippocampus is sufficient to impair human episodic memory. Neuroscience 170:623–632. https://doi.org/10.1016/j.neuroscience.2010.06.042
129. Shimotake A, Matsumoto R, Ueno T et al (2015) Direct exploration of the role of the ventral anterior temporal lobe in semantic memory: cortical stimulation and local field potential evidence from subdural grid electrodes. Cereb Cortex 25:3802–3817. https://doi.org/10.1093/cercor/bhu262

130. Mankin EA, Aghajan ZM, Schuette P et al (2021) Stimulation of the right entorhinal white matter enhances visual memory encoding in humans. Brain Stimulat 14:131–140. https://doi.org/10.1016/j.brs.2020.11.015
131. Suthana N, Haneef Z, Stern J et al (2012) Memory enhancement and deep-brain stimulation of the entorhinal area. N Engl J Med 366:502–510. https://doi.org/10.1056/NEJMoa1107212
132. Jacobs J, Miller J, Lee SA et al (2016) Direct electrical stimulation of the human entorhinal region and hippocampus impairs memory. Neuron 92:983–990. https://doi.org/10.1016/j.neuron.2016.10.062
133. Lochy A, Jacques C, Maillard L et al (2018) Selective visual representation of letters and words in the left ventral occipito-temporal cortex with intracerebral recordings. Proc Natl Acad Sci 115. https://doi.org/10.1073/pnas.1718987115
134. Shallice T (1979) Case study approach in neuropsychological research. J Clin Neuropsychol 1:183–211. https://doi.org/10.1080/01688637908414450
135. Caramazza A (1986) On drawing inferences about the structure of normal cognitive systems from the analysis of patterns of impaired performance: the case for single-patient studies. Brain Cogn 5:41–66. https://doi.org/10.1016/0278-2626(86)90061-8
136. Bao Y, Pöppel E, Zaytseva Y (2017) Single case studies as a prime example for exploratory research: single case studies. PsyCh J 6:107–109. https://doi.org/10.1002/pchj.176
137. Rossion B (in press) Twenty years of investigation with the case of prosopagnosia PS to understand human face identity recognition. Part I: function. Neuropsychologia. https://doi.org/10.1016/j.neuropsychologia.2022.108278
138. Benton AL, Van Allen MW (1972) Prosopagnosia and facial discrimination. J Neurol Sci 15:167–172. https://doi.org/10.1016/0022-510X(72)90004-4
139. Valentine T, Powell J, Davidoff J et al (2006) Prevalence and correlates of face recognition impairments after acquired brain injury. Neuropsychol Rehabil 16:272–297. https://doi.org/10.1080/09602010500176443
140. Yan H, Katz JS, Anderson M et al (2019) Method of invasive monitoring in epilepsy surgery and seizure freedom and morbidity: a systematic review. Epilepsia 60:1960–1972. https://doi.org/10.1111/epi.16315
141. Vignal J-P, Maillard L, McGonigal A, Chauvel P (2007) The dreamy state: hallucinations of autobiographic memory evoked by temporal lobe stimulations and seizures. Brain J Neurol 130:88–99. https://doi.org/10.1093/brain/awl329
142. Chassagnon S, Minotti L, Kremer S et al (2008) Somatosensory, motor, and reaching/grasping responses to direct electrical stimulation of the human cingulate motor areas: clinical article. J Neurosurg 109:593–604. https://doi.org/10.3171/JNS/2008/109/10/0593
143. Trébuchon A, Alario F-X, Liégeois-Chauvel C (2021) Functional topography of auditory areas derived from the combination of electrophysiological recordings and cortical electrical stimulation. Front Hum Neurosci 15:702773. https://doi.org/10.3389/fnhum.2021.702773
144. Mazzola L, Royet J-P, Catenoix H et al (2017) Gustatory and olfactory responses to stimulation of the human insula: taste-odor and insula. Ann Neurol 82:360–370. https://doi.org/10.1002/ana.25010
145. Cachia A, Roell M, Mangin J-F et al (2018) How interindividual differences in brain anatomy shape reading accuracy. Brain Struct Funct 223:701–712. https://doi.org/10.1007/s00429-017-1516-x
146. Abou-Al-Shaar H, Brock AA, Kundu B et al (2018) Increased nationwide use of stereoencephalography for intracranial epilepsy electroencephalography recordings. J Clin Neurosci Off J Neurosurg Soc Australas 53:132–134. https://doi.org/10.1016/j.jocn.2018.04.064
147. Vignal JP, Chauvel P, Halgren E (2000) Localised face processing by the human prefrontal cortex: stimulation-evoked hallucinations of faces. Cogn Neuropsychol 17:281–291. https://doi.org/10.1080/026432900380616
148. Ralph MAL, Jefferies E, Patterson K, Rogers TT (2017) The neural and computational bases of semantic cognition. Nat Rev Neurosci 18:42–55. https://doi.org/10.1038/nrn.2016.150
149. Sporns O (2015) Cerebral cartography and connectomics. Philos Trans R Soc B Biol Sci 370:20140173. https://doi.org/10.1098/rstb.2014.0173

150. Moeller S, Freiwald WA, Tsao DY (2008) Patches with Links: a unified system for processing faces in the macaque temporal lobe. Science 320:1355–1359. https://doi.org/10.1126/science.1157436
151. Tolias AS, Sultan F, Augath M et al (2005) Mapping cortical activity elicited with electrical microstimulation using fMRI in the macaque. Neuron 48:901–911. https://doi.org/10.1016/j.neuron.2005.11.034
152. Perrone-Bertolotti M, Alexandre S, Jobb A et al (2020) Probabilistic mapping of language networks from high frequency activity induced by direct electrical stimulation. Hum Brain Mapp 41:4113–4126. https://doi.org/10.1002/hbm.25112
153. Rolston JD, Chang EF (2018) Critical language areas show increased functional connectivity in human cortex. Cereb Cortex 28:4161–4168. https://doi.org/10.1093/cercor/bhx271
154. Finzi D, Gomez J, Nordt M et al (2021) Differential spatial computations in ventral and lateral face-selective regions are scaffolded by structural connections. Nat Commun 12:2278. https://doi.org/10.1038/s41467-021-22524-2
155. Wang C, Pan R, Wan X et al (2020) A longitudinal study on the mental health of general population during the COVID-19 epidemic in China. Brain Behav Immun 87:40–48. https://doi.org/10.1016/j.bbi.2020.04.028
156. Rossion B (in press) What makes us human? Face identity recognition. In: Garcia, Ibanez (ed) The Routledge handbook of neurosemiotics. Routledge, NY
157. Matsumoto R, Nair DR, LaPresto E et al (2004) Functional connectivity in the human language system: a cortico-cortical evoked potential study. Brain 127:2316–2330. https://doi.org/10.1093/brain/awh246
158. Bratu F-I, Oane I, Barborica A et al (2021) Network of autoscopic hallucinations elicited by intracerebral stimulations of periventricular nodular heterotopia: an SEEG study. Cortex 145:285–294. https://doi.org/10.1016/j.cortex.2021.08.018
159. Kim K, Schedlbauer A, Rollo M et al (2018) Network-based brain stimulation selectively impairs spatial retrieval. Brain Stimulat 11:213–221. https://doi.org/10.1016/j.brs.2017.09.016
160. Natu VS, Lin J-J, Burks A et al (2019) Stimulation of the posterior cingulate cortex impairs episodic memory encoding. J Neurosci 39:7173–7182. https://doi.org/10.1523/JNEUROSCI.0698-19.2019
161. Enatsu R, Kanno A, Ookawa S et al (2017) Distribution and network of basal temporal language areas: a study of the combination of electric cortical stimulation and diffusion tensor imaging. World Neurosurg 106:1–8. https://doi.org/10.1016/j.wneu.2017.06.116
162. Solomon EA, Kragel JE, Gross R et al (2018) Medial temporal lobe functional connectivity predicts stimulation-induced theta power. Nat Commun 9:4437. https://doi.org/10.1038/s41467-018-06876-w
163. Beauchamp MS, Sun P, Baum SH et al (2012) Electrocorticography links human temporoparietal junction to visual perception. Nat Neurosci 15:957–959. https://doi.org/10.1038/nn.3131
164. Barborica A, Oane I, Donos C et al (2022) Imaging the effective networks associated with cortical function through intracranial high-frequency stimulation. Hum Brain Mapp 43:1657–1675. https://doi.org/10.1002/hbm.25749
165. Popa I, Barborica A, Scholly J et al (2019) Illusory own body perceptions mapped in the cingulate cortex—an intracranial stimulation study. Hum Brain Mapp 40:2813–2826. https://doi.org/10.1002/hbm.24563
166. Oya H, Howard MA, Magnotta VA et al (2017) Mapping effective connectivity in the human brain with concurrent intracranial electrical stimulation and BOLD-fMRI. J Neurosci Methods 277:101–112
167. Thompson WH, Nair R, Oya H et al (2020) A data resource from concurrent intracranial stimulation and functional MRI of the human brain. Sci Data 7:258. https://doi.org/10.1038/s41597-020-00595-y
168. Hansen N, Chaieb L, Derner M et al (2018) Memory encoding-related anterior hippocampal potentials are modulated by deep brain stimulation of the entorhinal area. Hippocampus 28:12–17. https://doi.org/10.1002/hipo.22808

Chapter 40
How Can I Investigate Causal Brain Networks with iEEG?

Yuhao Huang and Corey Keller

Abstract While many human imaging methodologies probe the structural and functional connectivity of the brain, techniques to investigate cortical networks in a causal and directional manner are critical but limited. The use of iEEG enables several approaches to directly characterize brain regions that are functionally connected and in some cases also establish directionality of these connections. In this chapter we focus on the basis, method and application of the cortico-cortical evoked potential (CCEP), whereby electrical pulses applied to one set of intracranial electrodes yields an electrically-induced brain response at local and remote regions. In this chapter, CCEPs are first contextualized within common brain connectivity methods used to define cortical networks and how CCEP adds unique information. Second, the practical and analytical considerations when using CCEP are discussed. Third, we review the neurophysiology underlying CCEPs and the applications of CCEPs including exploring functional and pathological brain networks and probing brain plasticity. Finally, we end with a discussion of limitations, caveats, and directions to improve CCEP utilization in the future.

40.1 Introduction

The brain connectome, a representation of the functional and structural connection amongst neural elements, has been indispensable for understanding normal and pathological brain activity. Tight interconnections between cortical regions are

Y. Huang
Department of Neurosurgery, Stanford University Medical Center, Stanford, CA 94305, USA

C. Keller (✉)
Department of Psychiatry and Behavioral Sciences, Stanford University Medical Center, Stanford, CA 94305, USA
e-mail: ckeller1@stanford.edu

Veterans Affairs Palo Alto Healthcare System, and the Sierra Pacific Mental Illness, Research, Education, and Clinical Center (MIRECC), Palo Alto, CA 94394, USA

N. Axmacher (ed.), *Intracranial EEG*, Studies in Neuroscience, Psychology and Behavioral Economics, https://doi.org/10.1007/978-3-031-20910-9_40

known to underlie important motor, perceptual, and cognitive processes. Characterizing the brain connectome involves mapping the neural elements and the interregional pathways that connect them, whether directly or indirectly. Several experimental approaches exist to probe the human brain connectome ranging from non-invasive imaging modalities to invasive electrophysiology. In this chapter, we present the notion of cortico-cortical evoked potentials (CCEP) and place it within the context of other brain connectivity measurements.

40.2 Types of Brain Connectivity

The anatomical connections between neural elements define the structural connectivity. At the microscale this constitutes neurons and the synaptic connections between neurons. Attempting to map the complete structural connectivity at this scale might be accomplished in animal models, but is not immediately feasible in humans. Instead, on a macroscale, structural connectivity can be characterized by a set of inter-areal pathways connecting regionally distinct brain regions. Although there is not a unified division of the human brain, numerous parcellation schemes are widely available and define anatomical brain areas at the macroscale [1, 2]. Commonly, diffusion tensor imaging (DTI) and computational tractography map white-matter tracts non-invasively in the human brain. The quantitative outputs from DTI approaches can be tracked longitudinally to assess for changes over time and can be compared between healthy and pathological states [3, 4]. Although the white matter structural connections amongst brain regions both enable and constrain information flow, they do not imply functionality nor directionality. For instance, fiber tracts between two regions may or may not be utilized in the context of certain cognitive states, and the flow of information cannot be established at a particular time based on the sole presence of anatomical links.

In contrast, the correlated neural activity amongst different brain regions defines functional connectivity. Two regions are functionally connected if the neurophysiological activities in those two regions are statistically dependent. This approach enables delineating brain networks in the context of different brain states, and enables characterization of inter-regional communication in a dynamic fashion. Commonly, brain activity on a macroscale is indexed by electrophysiological recordings (electroencephalography) or functional neuroimaging (functional MRI). One emerging method to determine functional connectivity is called cortico-cortical evoked potential (CCEP) mapping and is the focus of this chapter. Unlike other methods of mapping functional connectivity, CCEP is a measure of causal influence between brain regions studied. CCEP mapping can not only provide strong evidence for functional connectivity, but also provide evidence for flow of information, making it a versatile approach to studying brain network dynamics.

40.3 History of CCEP

CCEPs were first introduced by Matsumoto et al. in 2004 when they recruited subjects with intractable epilepsy implanted with invasive subdural electrodes and applied single pulse electrical stimulation [5]. These response curves or CCEPs have a characteristic waveform that is dependent on the area being recorded, distance from stimulation site, cortical orientation with respect to gyri/sulci, proximity to white matter tracts, degree of pathology, and brain state. Given that CCEPs have excellent spatiotemporal resolution, do not require participant performance on cognitive tasks, and provide measures of causality and directionality, CCEPs have been used extensively in recent years [5–16].

40.4 Methods and Quantification of CCEP

40.4.1 Eliciting and Recording CCEPs

CCEP mapping is an invasive electrophysiological approach relying on electrical measurements from implanted subdural electrodes (electrocorticography or ECoG) or depth electrodes (stereoelectroencephalography or sEEG). In both cases, electrodes are implanted in various brain regions clinically for seizure mapping in patients with epilepsy; however, clinical populations with implanted electrodes are now expanding to those with intractable pain, depression and other neuropsychiatric disorders [17]. To obtain CCEPs, electrical current varying between 1 and 10 mA is delivered to these intracranial electrodes. The current can be delivered in a monopolar or bipolar manner. In monopolar stimulation, the ground electrode can be chosen far away from the stimulation electrode, usually in distant white matter or extracranial space. In bipolar stimulation, the current is delivered between a pair of adjacent electrodes. The current amplitude chosen is dependent on the patient. Typically, less current is needed to activate cortical regions using bipolar compared to monopolar stimulation [18]. Ideally the current amplitude is maximized to improve signal to noise ratio (SNR), but not so high that it would be perceived by the participant (thus direct cortical evoked potentials would be confounded with non-specific sensory evoked potentials) or would induce undesired clinical side effects such as epileptiform discharges. Electrical stimulation pulses result in a local neural response as well as in a distal response. The magnitude of the distal responses is related to the absolute distance and the functional connectivity between the recording and stimulating locations [15]. Stimulation pulses are typically applied between 20 and 150 times at 0.2 to 1 Hz frequency [7, 15]. For studies with shorter inter-stimulus interval (ISI) time period, it is important (1) to add a small amount of jitter between pulses to prevent neural entrainment and (2) to ensure that there are no lasting effects such as long-term depression, which often occurs at 1 Hz ISI [15, 19]. The number of pulses required for sufficient signal-to-noise (SNR) vary, and in part depends on the quality of the recording amplifier, electrode resistance, and the strength of connection between the stimulating and recording sites.

40.4.2 Design of CCEP Experiments

There are several experimental considerations when designing a study with CCEP mapping. One common objective is to index the underlying effective connectivity between the site of stimulation and other recording sites [20]. In this case, repetitive pulses of electrical stimulation are delivered at several stimulation sites of interest and CCEPs at all other sites are obtained. CCEP mapping can also be used to index changes in brain connectivity after an experimental intervention. These interventions can be in the form of a behavioral task [13] or a stimulation paradigm designed to alter connectivity on a short timescale [7, 15]. For instance, to determine if effective connectivity is altered after a session of high-frequency stimulation, CCEP pulses can be delivered before and after the intervention to determine the impact on effective connectivity [7].

40.4.3 Analysis of CCEPs

Different institutions implement different analytical pipelines to analyze CCEPs. Here we outline a common approach to constructing a CCEP analysis pipeline. First, the local field potential (LFP) signal comprising CCEPs is sampled at ≥500 Hz to allow sufficient sampling of high gamma activity (70–200 Hz). Three main steps in CCEP analysis pipelines include pre-processing (removal of line noise, detrending, and demeaning), artifact removal, and re-referencing. First, line noise can be removed using a bandstop (Notch) filter or a discrete Fourier transform (DFT) filter and data should be detrended and demeaned. Second, electrical artifacts due to stimulation should be removed, especially if power analysis is to be performed. These artifacts are typically present for the first 5 or 10 ms after stimulation, depending on amplifier type, sampling rate, and stimulation amplitude. There are several methods to remove these artifacts including simple interpolation [21], template rejection [22] and replacement with underlying neural data [23]. Third, a re-referencing scheme is selected to remove common noise from the recording channels. A variety of re-referencing schemes exist with advantages and disadvantages associated with each one [24] (see also Chap. 28). For SEEG electrodes, bipolar or Laplacian re-referencing is commonly preferred [7]. For subdural electrodes, data are often re-referenced using common average or Laplacian methods [7]. After these steps are performed, data are epoched and CCEPs can be quantified using peak analysis in the time domain (N1, N2) or power analysis in the frequency domain (delta 1–4 Hz, theta 4–8 Hz, alpha 8–12 Hz, beta 12–25 Hz, gamma 25–70 Hz and high-gamma 70–200 Hz). In the time domain, the morphology of CCEPs varies significantly and depends on both the stimulation and

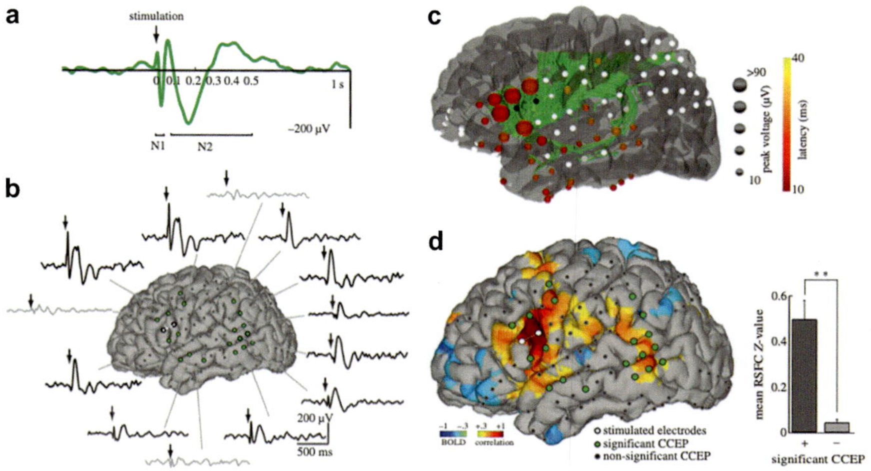

Fig. 40.1 CCEPs and their relationship to anatomical and functional connectivity. **a** CCEPs typically consist of an early N1 (10–50 ms) and a later N2 (50–250 ms) response. **b** Example of CCEP maps (electrical stimulation to white electrodes). **c** Comparison of CCEP (effective connectivity) and DTI (structural connectivity). Here, the number of white matter tracts measured with DTI are positively correlated with the strength of the CCEP N1 and negatively correlated with its latency. Adapted with permission from [8]. **d** Comparison of effective and functional connectivity. Here regions exhibiting strong CCEP N1 and N2 also exhibit strong fMRI connectivity. Adapted with permission from [16]

recording sites. Typically, CCEPs consist of an early potential (termed N1, 10–50 ms) and a later potential (termed N2, 50–300 ms, [5]) (Fig. 40.1a, b). The amplitudes of the N1 and the N2 potentials can be commonly quantified by averaging the signal, by calculating the peak to trough amplitude or by determining the area under the curve (AUC) within a specified timeframe. Notably, filtered CCEPs, in particular in the high-gamma range, have also been used to quantify effective connectivity [23].

40.5 Applications of CCEPs

40.5.1 Investigate Inter- and Intra-Regional Connectivity of Functional Brain Networks

CCEP mapping has been used extensively to map the causal, inter-regional connectivity in the human brain. CCEP mapping offers several advantages over other non-invasive mapping tools as it produces *directional* and *causal* measures of connectivity. Luders et al. first mapped the language network with CCEPs, reporting reciprocal connections between Broca's and Wernicke's regions [5] as well as within motor cortex [9]. In later work, CCEPs were utilized to demonstrate strong frontal and parietal intralobar connections with associated unidirectional frontal-to-temporal connections with rare temporal-to-frontal connections [10]. CCEPs have also been used to examine connections to and from the hippocampus [6, 14], as well as sensorimotor [16], visual [13], and recently default mode [25] networks. In summary, CCEP mapping has begun to reveal causal connectivity within and between well-known human functional networks. We predict this work will continue to expand as CCEP mapping becomes commonplace in the epilepsy monitoring unit as a part of routine brain mapping.

40.5.2 Comparing CCEP Mapping to Other Non-invasive Connectivity Methods

A comparison of CCEP mapping to other known non-invasive methodologies that probe anatomical connectivity (diffusion tensor imaging; DTI) and functional connectivity (resting state functional MRI; rs-fMRI) is necessary to (1) identify components of CCEPs that track standard connectivity measures and (2) provide direct electrophysiology grounding to non-invasive connectivity measures that indirectly map neural activity. For example, if the N1 (10–50 ms) reflects direct cortico-cortical connectivity, then DTI structural measures should correspond to the N1. Indeed, the number of tracts between two regions positively correlates with the strength of the N1 response and negatively with the latency of the N1 response in the CCEP [8] (Fig. 40.1c). CCEPs have also been compared to rs-fMRI measures, where co-variations of ultraslow (<0.1 Hz) fluctuations of the BOLD signal map functional brain networks. However, the neural underpinnings of rs-fMRI are not clear. We hypothesized that fast electrically-propagated potentials observed with CCEP mapping would travel in a similar trajectory as BOLD co-variations. We showed that regions with higher BOLD correlations demonstrated stronger CCEP N1 and N2s. These findings were replicated across patients and functional networks and demonstrated that temporal correlations of slow, spontaneous hemodynamics reflect similar functional interactions to those arising from fast electrically propagated activity [16] (Fig. 40.1d). Specifically, a recent study has shown that the CCEP derived network resembles functional connectivity (as measured by resting state correlation) in channels local to the stimulation site, whereas remote CCEP network correlates best with structural connectivity as measured by DTI [49]. In summary, CCEP mapping provides a direct measure of effective brain connectivity compared to non-invasive brain connectivity measures. The work summarized here demonstrates that the N1 of the CCEP partially reflects structural [8] and functional [16] connectivity while the N2 at least partially reflects functional connectivity [16].

40.5.3 CCEP Mapping to Measure Pathophysiological Networks

Just as CCEP mapping can probe functional networks, this methodology can also investigate the pathophysiology associated with neurological and psychiatric disorders. As CCEPs are typically performed during intracranial monitoring for seizure localization, epilepsy is by far the most common pathology explored. As seizures arise infrequently, clinicians typically welcome quantitative methods such as CCEP mapping that help explore regions that initiate and propagate seizures. Furthermore, as mounting evidence indicates that seizures arise from a set of interconnected brain regions, tools such as CCEP mapping have become increasingly helpful to delineate these interconnected brain regions that may require surgical resection. Between its potential aid in determining seizure networks and ease of implementation—CCEP mapping typically takes <1 h and does not require patient participation—CCEP mapping has become increasingly standard in the workup of seizure localization.

Because an abnormal excitation/inhibition balance may indicate regions that can generate seizures ('seizure onset zone'), the CCEP may reflect this imbalance when stimulating (or recording) in the seizure onset zone. Indeed, compared to control regions, larger amplitude N1 was observed following single pulse electrical stimulation to the seizure onset zone [12]. When stimulating the seizure onset zone, CCEP amplitude can also predict the onset of ictal events [26]. In addition to changes in the N1 or N2, 'afterdischarges' have been observed 200–1000 ms after electrical stimulation (after the N1 and N2). These later voltage deflections consisting of 'spikes' or 'sharp waves' due to enhanced excitation appear to localize to seizure onset zones [27] and a poorer surgical outcome has been observed when tissue eliciting these afterdischarges were not resected [27]. In summary, both standard CCEP N1 amplitude and afterdischarges are powerful complements to standard methods to localize pathological tissue (Fig. 40.2).

40.5.4 CCEP Mapping to Probe Brain Plasticity

CCEPs probe effective brain connectivity in a cross-sectional manner. However, CCEPs can also measure changes in cognitive state [13] and probe brain plasticity [7, 15]. To characterize how repetitive stimulation can induce brain plasticity, we applied electrical stimulation in a bipolar fashion patterned to mimic non-invasive transcranial magnetic stimulation (TMS) treatments. Electrical stimulation patterned at 10 Hz modulated a subset of brain regions measured by changes in CCEPs assessed pre/post intervention (Fig. 40.3). Modulated regions exhibited stronger baseline CCEPs and could be predicted based on a combination of anatomical (distance from stimulation site) and effective (CCEPs) connectivity. Furthermore, we were able to assess changes occurring *during* 10 Hz electrical stimulation using CCEPs elicited from the last pulse in each stimulation train [7]. 'Intratrain' CCEPs—those derived from

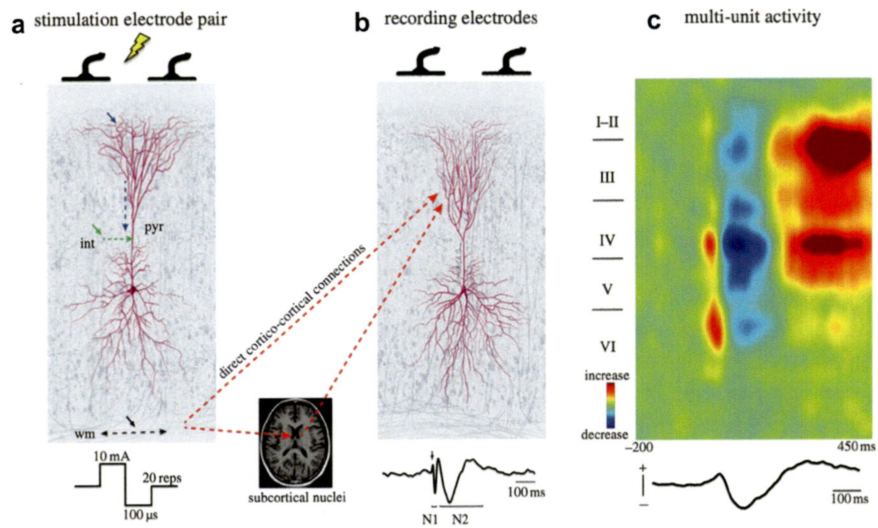

Fig. 40.2 Proposed mechanism of CCEP generation. **a** Electrical stimulation propagates through the cortex to local pyramidal cells via direct dendritic activity (blue arrows), adjacent interneurons (green arrows), or white matter traversing the stimulated region (black arrows). **b** Electrical activity from stimulation is propagated to downstream regions via direct cortico-cortical connections and indirect cortico-subcortico-cortical projections. **c** Multiunit response to electrical stimulation. CCEP below was derived from a recording from deep layers. The initial N1 response is accompanied by an increase in mid-deep increase in multiunit activity, while the later N2 is accompanied by a suppression/activation pattern. Adapted with permission from [33]

the last pulse in a stimulation train—changed with subsequent stimulation trains. These intratrain CCEPs also predicted CCEP changes that outlasted the stimulation intervention. Together, this work demonstrated that CCEPs can be used to probe brain plasticity and that baseline connectivity profiles can be used to predict regions susceptible to stimulation-induced brain changes. Future human plasticity studies will focus on how CCEPs are modulated as a function of stimulation frequency, duration, amplitude, and brain state.

40.6 Mechanistic Basis of CCEPs

The neurophysiological mechanisms underlying CCEPs are only partially known and future work to further elucidate these mechanisms will improve insights from CCEP studies. Here, we provide an overview of what is known about the mechanistic basis of CCEPs. Although CCEPs are applied using different stimulation parameters, the most common form is with *bipolar, biphasic stimulation*. Biphasic stimulation allows for a balanced charge to be delivered and bipolar stimulation provides a more local

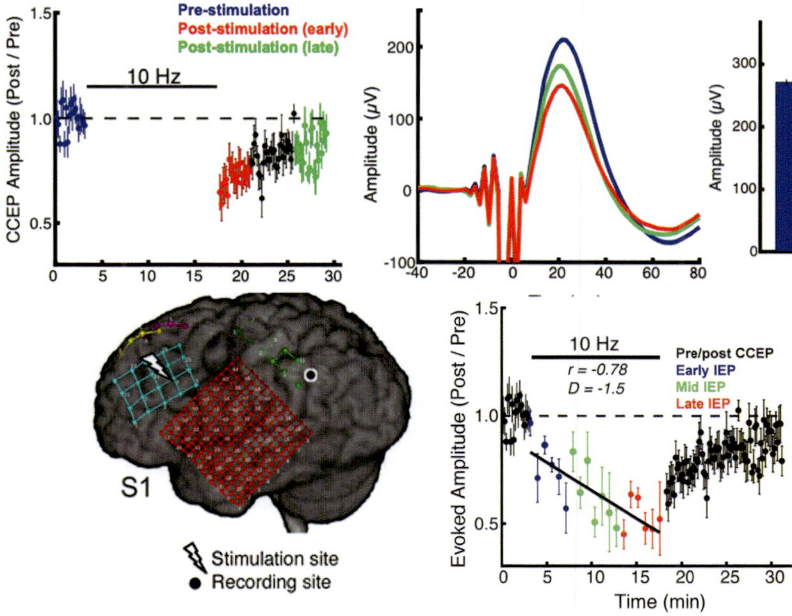

Fig. 40.3 Repetitive electrical stimulation modulates the CCEP. In this representative stimulation-recording pair in one subject, electrical stimulation was applied to the prefrontal cortex (lightning bolt) and recordings were measured in the parietal cortex. Repetitive 10 Hz electrical stimulation (5 s on, 10 s off, 3000 total pulses) modulated the CCEP for ~20 min. Top: quantification of CCEP amplitude over time. Bottom: by evaluating the CCEP after the last pulse of each 10 Hz stimulation, one can evaluate the evolution of plasticity effects induced by repetitive stimulation or other interventions. Adapted with permission from [7, 15]

stimulation of cortex compared to monophasic stimulation [28], thus minimizing the spatial spread of electrical stimulation.

40.6.1 Neurophysiology at Site of Stimulation

Electrical stimulation triggers multiple local cortical events that determine if pyramidal neurons fire and action potentials propagate to remote regions. Initial electrical stimulation depolarizes superficial dendrites of layer V pyramidal neurons and layer II/III inhibitory neurons that synapse on layer V neurons. The balance of these two events determine if layer V pyramidal neurons will fire; that is, the firing of many GABAergic neurons will hyperpolarize layer V neurons and inhibit their firing and orthodromic propagation of action potentials, whereas sufficient pyramidal dendritic depolarization without GABAergic activation will lead to pyramidal neuron firing and orthodromic propagation. In addition to this neuronal interplay, electrical stimulation will also directly depolarize long-range axons traversing the stimulated region and generate action potentials propagating orthodromically to distant synapses as well as antidromically backpropagating to pyramidal cell soma [29, 30]. In summary, electrical stimulation activates white matter tracts both physiologically through pyramidal cell firing from dendritic depolarization and non-physiologically through direct depolarization of traversing white matter tracts.

40.6.2 Potential Propagation Pathways

Evidence from animal studies have shown that (1) direct cortical stimulation elicits superficial-to-deep propagation of cortical layers locally at stimulated cortex [31] and (2) orthodromic axonal activation occurs far more than antidromic activation [32]. Data from human studies support the fact that electrical stimulation primarily activates neurons in middle layers [33] and that middle layer pyramidal neurons typically propagate to mono- and poly-synaptically connected regions via cortico-cortical and cortico-subcortical-cortical projections [31] (Fig. 40.2). Together this work suggests that CCEPs propagate via both a major pyramidal cell contribution via orthodromic cortico-cortical and cortico-subcortical-cortical projections as well as a minor antidromic contribution [9, 33].

40.6.3 Electrophysiology Underlying the N1 and N2 of the CCEP

Typical CCEP patterns consist of an early N1 peak within the first 50 ms and a later N2 slow wave lasting up to 500 ms [9, 20, 33]. Direct electrical stimulation has been shown to activate pyramidal neurons mono-synaptically connected to the site of stimulation within 4–8 ms of stimulation [34, 35]. Unfortunately, clinical grade amplifiers saturate for up to 10 ms, masking this initial response. Instead, the N1 observed after 10 ms in clinical amplifiers likely reflect oligo- and poly-synaptic pyramidal responses local or in close proximity to the applied stimulation. Several lines of evidence are consistent with this notion, including (1) excitatory neuronal responses within 50 ms (N1) after direct electrical stimulation of cat cortex [36]; (2) neuronal spiking within 50 ms (N1) after direct electrical stimulation in human cortex [37]; (3) increases in multi-unit activity in deep (layer IV–VI) cortical layers after single pulse stimulation in humans [33]. In contrast, the N2 slow wave in humans is time-locked to a reduction in both spiking [37] and middle-to-deep layer multi-unit activity [33], likely reflecting a prolonged inhibitory time period. In summary, single and multi-unit recordings of the responses to cortical stimulation suggest that the N1 evoked response reflects early pyramidal neuron activation while the N2 response represents long-lasting inhibition.

It is worth noting that this pattern of brief excitation followed by long-lasting inhibition is not specific to CCEPs and found in other physiological and pathophysiological neuronal events. These include the neural response to epileptic discharges [37, 38] and single pulses of TMS [39]. As such, studying these events will yield further insight into the neurophysiological basis of this excitation-inhibition wave and these processes can be evoked and perturbed.

40.7 Advanced Considerations

40.7.1 Limitations and Caveats

Although CCEP mapping represents an important method to causally and directionally probe the human brain, several limitations are worth discussing. First, as mentioned above, because common clinical amplifiers saturate during the first 10 ms after electrical stimulation, important mono- and di-synaptic connections may be masked by simulation artifacts. As such, novel methods to remove electrical stimulation artifacts analytically [23, 40] as well as the transition to research grade stop-and-hold amplifiers or those with high dynamic range are currently underway. These modifications will be critical to elucidate cortical responses within the first 10 ms. Second, without a clear understanding of the neural basis of the strength, latency, and polarity of the N1 and N2, as well as the degree that CCEPs reflect orthodromic versus antidromic propagation patterns, it is difficult to accurately interpret results from CCEP studies. As underlying neurobiological mechanisms are elucidated, more accurate interpretation of the CCEP will be possible. Third, the variability of experimental design across institutions and even across patients within an institution limits the ability to compare across studies. These parameters include electrode type (stereo-EEG vs. subdural implantation), stimulation configuration (monopolar, bipolar), current amplitude (from 1 to 10 mA), pulse duration (100–500 μs), inter-stimulation interval (0.5–10 s), and number of stimulation repetitions (10–500). Other parameters that are more difficult to control and often not reported include the patient's cognitive state, type and dosage of anti-epileptic medication, time of day, and level of sleep deprivation. A thorough reporting of each of these parameters in a reproducible manner is critical to assess differences across centers and studies. Fourth, a common criticism in CCEP studies (and all intracranial studies in epilepsy patients) is the lack of generalizability of results. However, several lines of evidence can minimize this concern: (1) as these patients vary with respect to seizure semiology and localization, consistent results across patients are not likely due to the patient's pathophysiology; (2) electrodes involved in the seizure onset zone and early seizure spread are often removed prior to analysis so that results are from non-pathological regions.

40.8 Future CCEP Mapping Approaches

In the future, three key improvements will be critical for the CCEP field. Over the past 5 years, an increasing number of hospitals have incorporated CCEP mapping as a standard approach for patients with implanted electrodes. As the community increases, more sophisticated quantitative methods will be applied to analyze CCEP data. The first key advance, as discussed before, is the transition to research grade amplifiers to observe neural information within 10 ms of the electrical pulse and potentially study mono- and di-synaptic connections. A second key advance comes from the improvement in computational power over the past decade. A typical CCEP study involves analysis of the N1 and N2 individually at each recording electrode for each stimulation site. Recent increases in computational power will enable data driven multivariate approaches by considering the N1 and N2 timing and amplitude *between* recording electrodes and stimulation sites. Examples include graph theory frequency decomposition and multivariate machine learning approaches. A typical CCEP study uses *divergent* analysis to examine CCEPs by comparing CCEP waveforms at multiple electrodes after stimulation of one region. In contrast, a recent study has explored the potential use of *convergent* CCEP analysis, where CCEP responses at the same electrode are compared after the sequential stimulation of multiple other regions, with sufficient time between stimulation of each region (Fig. 40.4). Convergent analysis controls for the variability of electrode orientation, gray/white matter, etc. in interpreting the response [41]. Another computationally-intensive method explores the use of *basis curves* to more accurately define the CCEP. Although the N1 and N2 are useful ways to describe time windows around the CCEP, in reality there is significant variability with respect to timing and polarity of each potential, even within a given patient. As such, Miller et al. recently used computationally-driven tools such as basis curves to extract the principal components of CCEP waveforms [50] (Fig. 40.4). Using basis curves to explore how these more fundamental components contribute to physiology and pathology across brain regions will open up new avenues of research. A third critical improvement comes from open science and big data initiatives [42]. Here, CCEP data format standardization will enable big data computational approaches not possible before (see also Chap. 38). For example the Functional Brain Tractography database (f-tract.eu) is a dataset with CCEP recordings from >1500 patients with standardized data formatting [43]. Together, these improvements will pay the way for the next generation of CCEP mapping: standardized, data-driven, and reproducible.

Several other research avenues for CCEPs should be explored. First, CCEPs are applied using different stimulation parameters across groups and it will be important to explore how the stimulation parameter space differentially affects brain dynamics. Second, recent studies have explored how brain state influences CCEPs, either using cognitive tasks [13] or awake/sleep patterns [44]. Future work will further explore how cognitive states differentially influence CCEP amplitude, latency, and propagation patterns. Third, the combination of intracranial and non-invasive stimulation and recording methods will be critical to bridge the knowledge gap that exists between

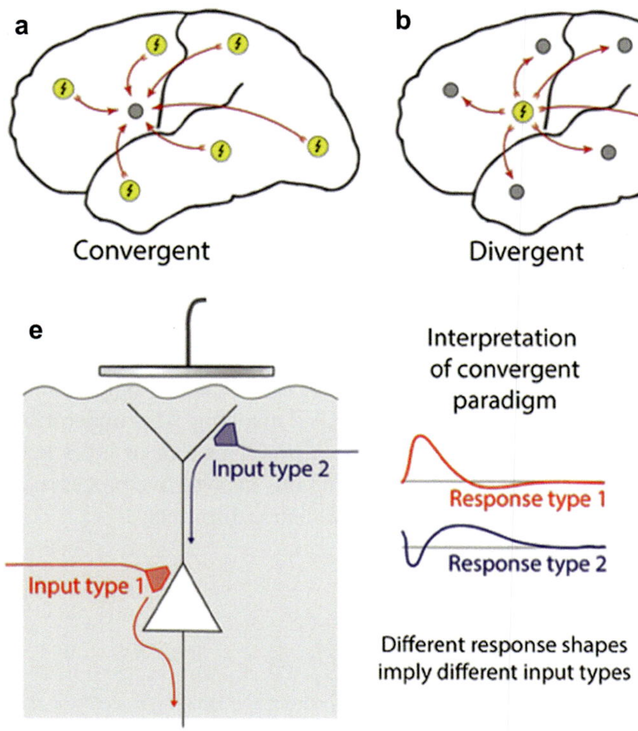

Fig. 40.4 Convergent and divergent CCEP analysis paradigms. **a** *Convergent*—CCEPs at one region (gray circle) are compared with the effect of stimulating all other regions (yellow circle). **b** Divergent—The CCEP at all regions are examined and compared in response to stimulation of a chosen region. Adapted with permission from [50]

intracranial and non-invasive brain recordings and move towards more generalizable tools that can be utilized in outpatient clinics. Recent combined recordings have mapped the relationship between CCEPs measured both intracranially and non-invasively [45]. Fourth, implantation of deep electrodes for the monitoring, understanding, and treatment of neuropsychiatric illness is increasingly being performed [17, 46–48]. For instance, depth electrode implantations are being done for patients with obsessive compulsive disorder (OCD) [51] or major depressive disorder (MDD) [52]. In both cases, depth electrodes implanted in multiple regions were used to assay network responses to tasks and to mood variations. Hence, one can expect that as electrodes are implanted for MDD, OCD, obesity, pain, and many other disorders, CCEP mapping will become a mainstream method of evaluation of cortical excitability and plasticity.

40.9 Conclusion

CCEP mapping represents an important methodological tool to quantify brain connectivity in humans. CCEP mapping has excellent spatial and temporal resolution with *electrophysiological grounding* as with any intracranial approach with the additional characteristics of *causality* and *directionality*. The N1 of the CCEP likely reflects early excitation of pyramidal neurons from feedforward connections while the N2 likely represents a long-lasting inhibitory period from feedback connections. The N1 reflects at least in part structural connectivity between two regions while the N2 can be influenced by brain state and cognitive demands. Future improvements in amplifier design, computational power, and big data sharing initiatives, as well as the expansion of implanted electrode approaches to indications other than epilepsy will pave the way for the next generation of CCEP mapping. This underutilized tool in basic and clinical neuroscience represents a powerful tool to investigate brain connectivity, the causal involvement of brain regions in cognitive processes, and the pathological networks in neurological and psychiatric disorders.

References

1. Desikan RS, Ségonne F, Fischl B et al (2006) An automated labeling system for subdividing the human cerebral cortex on MRI scans into gyral based regions of interest. Neuroimage 31:968–980. https://doi.org/10.1016/j.neuroimage.2006.01.021
2. Fischl B, van der Kouwe A, Destrieux C et al (2004) Automatically parcellating the human cerebral cortex. Cereb Cortex N Y N 1991 14:11–22. https://doi.org/10.1093/cercor/bhg087
3. Maffei C, Lee C, Planich M et al (2021) Using diffusion MRI data acquired with ultra-high gradient strength to improve tractography in routine-quality data. Neuroimage 245:118706. https://doi.org/10.1016/j.neuroimage.2021.118706
4. Yendiki A, Panneck P, Srinivasan P et al (2011) Automated probabilistic reconstruction of white-matter pathways in health and disease using an atlas of the underlying anatomy. Front Neuroinformatics 5:23. https://doi.org/10.3389/fninf.2011.00023
5. Matsumoto R, Nair DR, LaPresto E et al (2004) Functional connectivity in the human language system: a cortico-cortical evoked potential study. Brain J Neurol 127:2316–2330. https://doi.org/10.1093/brain/awh246
6. Kubota Y, Enatsu R, Gonzalez-Martinez J et al (2013) In vivo human hippocampal cingulate connectivity: a corticocortical evoked potentials (CCEPs) study. Clin Neurophysiol Off J Int Fed Clin Neurophysiol 124:1547–1556. https://doi.org/10.1016/j.clinph.2013.01.024
7. Huang Y, Hajnal B, Entz L et al (2019) Intracortical dynamics underlying repetitive stimulation predicts changes in network connectivity. J Neurosci Off J Soc Neurosci 39:6122–6135. https://doi.org/10.1523/JNEUROSCI.0535-19.2019
8. Conner CR, Ellmore TM, DiSano MA et al (2011) Anatomic and electro-physiologic connectivity of the language system: a combined DTI-CCEP study. Comput Biol Med 41:1100–1109. https://doi.org/10.1016/j.compbiomed.2011.07.008
9. Matsumoto R, Nair DR, LaPresto E et al (2007) Functional connectivity in human cortical motor system: a cortico-cortical evoked potential study. Brain J Neurol 130:181–197. https://doi.org/10.1093/brain/awl257
10. Matsumoto R, Nair DR, Ikeda A et al (2012) Parieto-frontal network in humans studied by cortico-cortical evoked potential. Hum Brain Mapp 33:2856–2872. https://doi.org/10.1002/hbm.21407

11. Enatsu R, Gonzalez-Martinez J, Bulacio J et al (2015) Connections of the limbic network: a corticocortical evoked potentials study. Cortex J Devoted Study Nerv Syst Behav 62:20–33. https://doi.org/10.1016/j.cortex.2014.06.018
12. Enatsu R, Piao Z, O'Connor T et al (2012) Cortical excitability varies upon ictal onset patterns in neocortical epilepsy: a cortico-cortical evoked potential study. Clin Neurophysiol Off J Int Fed Clin Neurophysiol 123:252–260. https://doi.org/10.1016/j.clinph.2011.06.030
13. Keller CJ, Davidesco I, Megevand P et al (2017) Tuning face perception with electrical stimulation of the fusiform gyrus. Hum Brain Mapp 38:2830–2842. https://doi.org/10.1002/hbm.23543
14. Mégevand P, Groppe DM, Bickel S et al (2017) The hippocampus and amygdala are integrators of neocortical influence: a corticocortical evoked potential study. Brain Connect 7:648–660. https://doi.org/10.1089/brain.2017.0527
15. Keller CJ, Huang Y, Herrero JL et al (2018) Induction and quantification of excitability changes in human cortical networks. J Neurosci Off J Soc Neurosci 38:5384–5398. https://doi.org/10.1523/JNEUROSCI.1088-17.2018
16. Keller CJ, Bickel S, Entz L et al (2011) Intrinsic functional architecture predicts electrically evoked responses in the human brain. Proc Natl Acad Sci USA 108:10308–10313. https://doi.org/10.1073/pnas.1019750108
17. Scangos KW, Khambhati AN, Daly PM et al (2021) Closed-loop neuromodulation in an individual with treatment-resistant depression. Nat Med 27:1696–1700. https://doi.org/10.1038/s41591-021-01480-w
18. Kundu B, Davis TS, Philip B et al (2020) A systematic exploration of parameters affecting evoked intracranial potentials in patients with epilepsy. Brain Stimulat 13:1232–1244. https://doi.org/10.1016/j.brs.2020.06.002
19. David O, Woźniak A, Minotti L, Kahane P (2008) Preictal short-term plasticity induced by intracerebral 1 Hz stimulation. Neuroimage 39:1633–1646. https://doi.org/10.1016/j.neuroimage.2007.11.005
20. Keller CJ, Honey CJ, Entz L et al (2014) Corticocortical evoked potentials reveal projectors and integrators in human brain networks. J Neurosci Off J Soc Neurosci 34:9152–9163. https://doi.org/10.1523/JNEUROSCI.4289-13.2014
21. Amengual JL, Vernet M, Adam C, Valero-Cabré A (2017) Local entrainment of oscillatory activity induced by direct brain stimulation in humans. Sci Rep 7:41908. https://doi.org/10.1038/srep41908
22. Caldwell DJ, Cronin JA, Rao RPN et al (2020) Signal recovery from stimulation artifacts in intracranial recordings with dictionary learning. J Neural Eng 17:026023. https://doi.org/10.1088/1741-2552/ab7a4f
23. Crowther LJ, Brunner P, Kapeller C et al (2019) A quantitative method for evaluating cortical responses to electrical stimulation. J Neurosci Methods 311:67–75. https://doi.org/10.1016/j.jneumeth.2018.09.034
24. Li G, Jiang S, Paraskevopoulou SE et al (2018) Optimal referencing for stereo-electroencephalographic (SEEG) recordings. Neuroimage 183:327–335. https://doi.org/10.1016/j.neuroimage.2018.08.020
25. Veit MJ, Kucyi A, Hu W et al (2021) Temporal order of signal propagation within and across intrinsic brain networks. Proc Natl Acad Sci USA 118:e2105031118. https://doi.org/10.1073/pnas.2105031118
26. David O, Blauwblomme T, Job A-S et al (2011) Imaging the seizure onset zone with stereo-electroencephalography. Brain J Neurol 134:2898–2911. https://doi.org/10.1093/brain/awr238
27. Valentín A, Alarcón G, Honavar M et al (2005) Single pulse electrical stimulation for identification of structural abnormalities and prediction of seizure outcome after epilepsy surgery: a prospective study. Lancet Neurol 4:718–726. https://doi.org/10.1016/S1474-4422(05)70200-3
28. Nathan SS, Sinha SR, Gordon B et al (1993) Determination of current density distributions generated by electrical stimulation of the human cerebral cortex. Electroencephalogr Clin Neurophysiol 86:183–192. https://doi.org/10.1016/0013-4694(93)90006-h

29. Stuart G, Schiller J, Sakmann B (1997) Action potential initiation and propagation in rat neocortical pyramidal neurons. J Physiol 505(Pt 3):617–632. https://doi.org/10.1111/j.1469-7793.1997.617ba.x
30. Kaiser KM, Zilberter Y, Sakmann B (2001) Back-propagating action potentials mediate calcium signalling in dendrites of bitufted interneurons in layer 2/3 of rat somatosensory cortex. J Physiol 535:17–31. https://doi.org/10.1111/j.1469-7793.2001.t01-1-00017.x
31. Steriade M, Amzica F (1996) Intracortical and corticothalamic coherency of fast spontaneous oscillations. Proc Natl Acad Sci USA 93:2533–2538. https://doi.org/10.1073/pnas.93.6.2533
32. Ezure K, Oshima T (1985) Lateral spread of neuronal activity within the motor cortex investigated with intracellular responses to distant epicortical stimulation. Jpn J Physiol 35:223–249. https://doi.org/10.2170/jjphysiol.35.223
33. Keller CJ, Honey CJ, Mégevand P et al (2014) Mapping human brain networks with cortico-cortical evoked potentials. Philos Trans R Soc Lond B Biol Sci 369. https://doi.org/10.1098/rstb.2013.0528
34. Finlay BL, Schiller PH, Volman SF (1976) Quantitative studies of single-cell properties in monkey striate cortex. IV. Corticotectal cells. J Neurophysiol 39:1352–1361. https://doi.org/10.1152/jn.1976.39.6.1352
35. Calvin WH, Sypert GW (1976) Fast and slow pyramidal tract neurons: an intracellular analysis of their contrasting repetitive firing properties in the cat. J Neurophysiol 39:420–434. https://doi.org/10.1152/jn.1976.39.2.420
36. Creutzfeldt OD, Watanabe S, Lux HD (1966) Relations between EEG phenomena and potentials of single cortical cells. I. Evoked responses after thalamic and erpicortical stimulation. Electroencephalogr Clin Neurophysiol 20:1–18. https://doi.org/10.1016/0013-4694(66)90136-2
37. Alarcón G, Martinez J, Kerai SV et al (2012) In vivo neuronal firing patterns during human epileptiform discharges replicated by electrical stimulation. Clin Neurophysiol Off J Int Fed Clin Neurophysiol 123:1736–1744. https://doi.org/10.1016/j.clinph.2012.02.062
38. Keller CJ, Truccolo W, Gale JT et al (2010) Heterogeneous neuronal firing patterns during interictal epileptiform discharges in the human cortex. Brain J Neurol 133:1668–1681. https://doi.org/10.1093/brain/awq112
39. Mueller JK, Grigsby EM, Prevosto V et al (2014) Simultaneous transcranial magnetic stimulation and single-neuron recording in alert non-human primates. Nat Neurosci 17:1130–1136. https://doi.org/10.1038/nn.3751
40. O'Shea DJ, Shenoy KV (2018) ERAASR: an algorithm for removing electrical stimulation artifacts from multielectrode array recordings. J Neural Eng 15:026020. https://doi.org/10.1088/1741-2552/aaa365
41. Miller KJ, Müller K-R, Hermes D (2021) Basis profile curve identification to understand electrical stimulation effects in human brain networks. PLoS Comput Biol 17:e1008710. https://doi.org/10.1371/journal.pcbi.1008710
42. Milham MP (2012) Open neuroscience solutions for the connectome-wide association era. Neuron 73:214–218. https://doi.org/10.1016/j.neuron.2011.11.004
43. Holdgraf C, Appelhoff S, Bickel S et al (2019) iEEG-BIDS, extending the brain imaging data structure specification to human intracranial electrophysiology. Sci Data 6:102. https://doi.org/10.1038/s41597-019-0105-7
44. Pigorini A, Sarasso S, Proserpio P et al (2015) Bistability breaks-off deterministic responses to intracortical stimulation during non-REM sleep. Neuroimage 112:105–113. https://doi.org/10.1016/j.neuroimage.2015.02.056
45. Mikulan E, Russo S, Parmigiani S et al (2020) Simultaneous human intracerebral stimulation and HD-EEG, ground-truth for source localization methods. Sci Data 7:127. https://doi.org/10.1038/s41597-020-0467-x
46. Wu H, Miller KJ, Blumenfeld Z et al (2018) Closing the loop on impulsivity via nucleus accumbens delta-band activity in mice and man. Proc Natl Acad Sci USA 115:192–197. https://doi.org/10.1073/pnas.1712214114

47. Scangos KW, Khambhati AN, Daly PM et al (2021) Distributed subnetworks of depression defined by direct intracranial neurophysiology. Front Hum Neurosci 15:746499. https://doi.org/10.3389/fnhum.2021.746499
48. Provenza NR, Sheth SA, Dastin-van Rijn EM et al (2021) Long-term ecological assessment of intracranial electrophysiology synchronized to behavioral markers in obsessive-compulsive disorder. Nat Med. https://doi.org/10.1038/s41591-021-01550-z
49. Crocker B, Ostrowski L, Williams ZM, Dougherty DD, Eskandar EN, Widge AS, Chu CJ, Cash SS, Paulk AC (2021) Local and distant responses to single pulse electrical stimulation reflect different forms of connectivity. Neuroimage 237:118094. https://doi.org/10.1016/j.neuroimage.2021.118094
50. Miller KJ, Müller KR, Hermes D (2021) Basis profile curve identification to understand electrical stimulation effects in human brain networks. PLoS Comput Biol 17(9):e1008710. https://doi.org/10.1371/journal.pcbi.1008710
51. Provenza NR, Sheth SA, Dastin-van Rijn EM, Mathura RK, Ding Y, Vogt GS, Avendano-Ortega M, Ramakrishnan N, Peled N, Gelin LFF, Xing D, Jeni LA, Ertugrul IO, Barrios-Anderson A, Matteson E, Wiese AD, Xu J, Viswanathan A, Harrison MT, Bijanki KR, Storch EA, Cohn JF, Goodman WK, Borton DA (2021) Long-term ecological assessment of intracranial electrophysiology synchronized to behavioral markers in obsessive-compulsive disorder. Nat Med 27(12):2154–2164. https://doi.org/10.1038/s41591-021-01550-z
52. Scangos KW, Khambhati AN, Daly PM, Makhoul GS, Sugrue LP, Zamanian H, Liu TX, Rao VR, Sellers KK, Dawes HE, Starr PA, Krystal AD, Chang EF (2021) Closed-loop neuromodulation in an individual with treatment-resistant depression. Nat Med 27(10):1696–1700. https://doi.org/10.1038/s41591-021-01480-w

Chapter 41
What Are the Promises and Challenges of Closed-Loop Stimulation?

Youssef Ezzyat

Abstract Intracranial electroencephalography (iEEG) offers a unique basis for recording and understanding the physiological activity of the human brain. Since its earliest applications researchers have also used intracranially-placed electrodes to electrically stimulate the brain to probe the neural mechanisms of cognition. A typical approach, known as *open-loop*, is to select stimulation parameters (e.g. timing, frequency, brain location) a priori and then apply stimulation either acutely or chronically, in order to observe its effects on cognitive function and neural activity. In recent years there has been substantial development in methods for *closed-loop* stimulation in which at least one of the parameters for each stimulation event is determined by the neural response to preceding stimulation events. This chapter discusses some promises and challenges of closed-loop stimulation through evaluation of recent studies that have used closed-loop stimulation to modulate neural function and behavior. Closed-loop frameworks are well-suited to test theories that relate ongoing neural dynamics to cognitive function, and further development of such approaches may pave the way for chronic closed-loop neuromodulation that is independent of experimental tasks. Several challenges remain to be addressed, including optimal modeling of stimulation's effects on physiology, cognition, and behavior, which will determine the eventual impact of closed-loop approaches.

41.1 Introduction

Intracranial EEG has many strengths as a method for studying brain function. One of the most important and earliest to be recognized is the ability to apply direct electrical stimulation to affect neural activity and cognitive function. In its earliest and most widespread application, direct brain stimulation was used to map the function of the cortex, revealing the organization of motor and language abilities in the brain, and providing patient-specific roadmaps for surgical resection [1, 2]. Alongside the

Y. Ezzyat (✉)
Department of Psychology and Program in Neuroscience and Behavior, Wesleyan University, Middletown, CT 06459, United States
e-mail: yezzyat@wesleyan.edu

N. Axmacher (ed.), *Intracranial EEG*, Studies in Neuroscience, Psychology and Behavioral Economics, https://doi.org/10.1007/978-3-031-20910-9_41

continued use of stimulation for surgical mapping [3], clinicians and researchers have made major advances in using stimulation to directly modulate brain function. This work has led to therapies for treating neurological dysfunction as well as causal methods for relating neural function to behavior.

Here, we review recent advances in applying intracranial direct brain stimulation in *closed-loop*. We first discuss the development of its precursor, open-loop stimulation, and selectively highlight clinical and research applications of open-loop stimulation approaches. We then describe how closed-loop approaches developed out of the clinical research setting, and conclude with a discussion of some promises and challenges that arise in the context of closed-loop stimulation.

41.2 Open-Loop Versus Closed-Loop

Before focusing on issues that are specific to closed-loop stimulation, we offer a brief example to draw the distinction between open-loop and closed-loop systems. The notions of *open-loop* and *closed-loop* come from engineering, where a standard problem is the design of systems that can reliably apply transformations to input(s) in order to produce output(s) that satisfy or reflect a desired end state. In an open-loop system, a predetermined input is applied to the system in order to produce an output. Using knowledge of the input itself and the transformation that the system performs, an appropriate input can be applied in order to reach the desired end state. A conventional microwave oven is an example of an open-loop system in which the user sets the microwave to operate at a given intensity and for a given duration in order to warm food. Open-loop systems are relatively simple to understand and design, making them an attractive choice for many problems. However, their simple design means that they have limitations, one of which being that they are not robust to disturbances within the system. For example, if the user does not know that the food is partially frozen, they may not set the microwave to operate at an appropriate intensity or duration.

Like open-loop systems, closed-loop systems also perform input–output transformations, however they use feedback from the output to modulate the input in order to achieve the desired end state. In the example of a microwave, sensors which detect the moisture level inside the microwave can be used as closed-loop feedback to determine when the food has been warmed to a desired state. The microwave uses the signals from these sensors as feedback to adjust the intensity and duration of cooking. For the situation in which the user does not realize the food is partially frozen, the feedback from the internal sensor will provide the information that is necessary to prolong or intensify the cooking. Information from the output therefore affects the input, allowing the closed-loop microwave to achieve the desired end state in the face of disturbance.

Although warming food in a microwave is somewhat different from direct brain stimulation, the example highlights issues relating to open and closed-loop systems that are common to both scenarios. In particular, closed-loop systems are most effective when (1) there is a well-defined end state; and (2) there is a model that relates changes to the system's input to changes in the output. For a microwave, the desired end state is food that is warmed to an appropriate temperature, and the input/output model relates the electromagnetic energy produced by the microwave to the amount of thermal energy generated in the food.

For the iEEG researcher interested in using closed-loop stimulation to study neural activity and behavior, the desired end state is usually a neurophysiological state (biomarker) that is associated with a behavioral outcome. Such closed-loop biomarkers can be defined on the basis of previous research or developed through novel paradigms. The model relating the system's input and output corresponds to a model of stimulation's effects on neurophysiology. As will be described below, this is a challenging problem due to the many ways in which stimulation delivery can be manipulated (e.g. timing, amplitude, frequency, pulse width), and our limited understanding of the effects of each of these parameters on local and network physiology and whether those effects vary based on the anatomical target. A further complication arises from differences in how stimulation parameters are described in the literature. A move toward standardization in reporting is an important step toward a unified understanding across studies [4].

There are several parameters available to the researcher that can be used to tailor the pattern of stimulation that is delivered through iEEG electrodes (for detailed discussions of general issues relating to direct brain stimulation with iEEG, see Chaps. 5, and 39). Broadly, these include: stimulation timing (when to stimulate); and stimulation pattern (how to stimulate). In *open-loop* stimulation designs, researchers select these parameters in advance (typically based on some combination of: the parameters used in the existing literature; modeling of the expected current flow; and the experimental hypotheses and task structure), and then apply stimulation in order to observe its effects on cognition, physiology, or both. The researcher may conduct repeated trials in order to gather sufficient data at the specified stimulation parameters before changing one or more of the parameters in order to collect comparison data. Such approaches have been highly valuable in providing causal evidence for the role of particular brain structures in aspects of high-level cognition [5–10].

For the remainder of this chapter we discuss the development of closed-loop approaches for iEEG stimulation. We briefly highlight first how clinical research has influenced this development, and how questions about the closed loop biomarker (the model relating stimulation to physiology) and the selection of stimulation parameters have been answered in the clinical domain. We then describe the expansion of closed-loop research to other domains and some promises and challenges for this work.

41.3 Clinical Development of Closed-Loop Stimulation

Because iEEG research is conducted in a clinical setting, it is perhaps not surprising that most studies that have evaluated closed-loop stimulation have done so with clinical outcomes in mind. Treating neurological dysfunction is an urgent societal problem and there is increasing evidence for, and interest in, the effectiveness of direct brain stimulation for a variety of such disorders [11, 12]. In a broader sense, however, the dual imperatives to achieve clinically meaningful symptom reduction while also limiting side effects have provided clinical research with a relatively clear set of objectives that have helped in focusing development of closed-loop stimulation systems that address these two constraints.

Parkinson's disease (PD) is a domain that has produced significant development of approaches for closed-loop stimulation. An established treatment for motor symptoms of PD is high-frequency open-loop stimulation to the globus pallidus or subthalamic nucleus. While such open-loop stimulation can be quite effective, one challenge to this approach is that precise stimulation parameters must be tuned by a clinician in order to achieve symptom relief that is well-tolerated. This tuning process can take weeks or months, exposing the patient to unnecessary stimulation and more time living with symptoms that reduce quality of life. Once effective individualized stimulation parameters are identified, they are often then further adjusted over time in response to changes in efficacy [13]. This situation, in which the inputs must be iteratively updated and tuned in order to maintain a desired outcome state, is precisely the sort that can benefit from a closed-loop approach.

Because of this, there has been significant interest in closed-loop approaches for treating PD with stimulation [14]. One goal of such research has been to tune stimulation more rapidly in response to changes in activity in the dysfunctional PD circuit. Work in non-human primate models showed the potential effectiveness of closed-loop stimulation for PD [15], and subsequent studies have suggested that stimulation amplitude and frequency can be adjusted via closed-loop feedback to reduce PD symptoms as well as biomarkers of PD, such as beta band power in the basal ganglia [16–18]. Closed-loop brain stimulation could also be used to broaden the space parameters that are manipulated, in order to identify those that provide the most effective treatment, which may vary from patient to patient. For example, while the standard PD stimulation treatment involves periodic bursts of high-frequency stimulation, computational modeling has suggested that closed-loop optimization algorithms could be used to reveal novel therapeutic waveforms [19].

Epilepsy is another disorder that has drawn intense development of methods for stimulation in closed-loop [20, 21]. For patients with medically-refractory epilepsy, seizure onset can be disrupted with stimulation that is applied to the seizure focus. However, because seizures are stochastic and temporally circumscribed, a key challenge is to selectively apply stimulation only when it is needed and not continuously. Research into stimulation for epilepsy has therefore sought to address the key question of when to stimulate. Using patient-specific parameters set by a clinician, responsive neurostimulators monitor neural activity in order to detect the onset of seizures, and deliver stimulation to a predetermined target in order to inhibit seizure onset [22]. This closed-loop system has been safe and effective in reducing seizure frequency [23–25].

The preceding examples from investigations of PD and epilepsy show how closed-loop designs can aid in identifying effective stimulation waveform parameters and timing. Another potential benefit of closed-loop stimulation systems compared to open-loop systems is that through the use of feedback, closed-loop systems can theoretically achieve the desired end state by delivering less stimulation overall. From an engineering perspective, if less stimulation needs to be delivered to reach or maintain desired end states, then it is possible to power implantable stimulation devices for longer.

For example, although continuous high-frequency stimulation is the standard stimulation protocol for direct brain stimulation treatment of tremor in Parkinson's, it is possible to alleviate motor symptoms while delivering less overall stimulation by triggering stimulation in response to an ongoing measurement of Parkinson's related pathological brain activity. One study did this by measuring power in the beta band (13–35 Hz) from electrodes implanted in the subthalamic nucleus, and triggering stimulation when beta band power exceeded a predetermined threshold [26]. This responsive stimulation approach improved a clinical assessment of motor function compared to continuous stimulation while reducing stimulation time by 56%, thereby reducing the overall device energy requirements. The study further showed that responsive stimulation improved motor symptoms compared to randomly delivered stimulation as well as an unstimulated control condition. An additional benefit of this approach was a reduction in non-motor (speech) side effects that are associated with continuous stimulation for Parkinson's [27]. A similar closed-loop approach also showed reduced stimulation time and motor improvement when measuring another clinical symptom of Parkinson's, essential tremor [28]. Another study implemented a closed-loop approach, triggering stimulation in response to an ongoing measurement of a *motor* biomarker, using the phase of a patient's tremor [29]. The study stimulated the ventrolateral thalamus and identified, for each individual patient, the most effective oscillatory tremor phase at which to deliver stimulation.

The preceding selective review highlights some of the ways in which clinical treatment has driven advancement in the development of closed-loop stimulation systems. In particular, closed-loop systems have aided in identifying effective stimulation parameters, and in deploying clinically-effective stimulation only when it is needed. The next section expands on some particular promises and challenges of closed-loop stimulation as they relate to understanding mechanistic links between brain and behavior.

41.4 Promises and Challenges

In this section we discuss some promises and challenges associated with closed-loop iEEG stimulation. Two fundamental questions must be addressed when developing closed-loop stimulation systems. The answer to the first, *When to stimulate?*, depends largely on the development of stable biomarkers, for example iEEG-derived physiological correlates of the behavioral or cognitive process under study. The answer to the second question, *How to stimulate?*, depends on understanding of the effects of various stimulation parameters on physiology and behavior. We then discuss potential future applications of closed-loop stimulation for causal investigation of the neural mechanisms of cognition.

Closed-loop stimulation for epilepsy provides an example case in which stimulation must be delivered at specific moments in response to a particular neural biomarker. In epilepsy this corresponds to abnormal ictal or interictal physiology [30]. In PD, biomarkers of motor symptoms such as beta band power in the affected cortico-basal ganglia-thalamic circuit [31] have been investigated for triggering stimulation [26]. Thus, an effective closed-loop stimulation system depends on a stable biomarker that can be used to trigger stimulation when it is needed.

In the clinical setting, a biomarker must bear some relation to measured symptoms and ideally reflect the underlying pathological mechanism. This link between measured outcome and biomarker remains vital when the outcome measure is instead a not-necessarily-pathological behavioral measurement of a cognitive process. Thus, a critical aspect of defining a stable biomarker is to develop experimental tasks and/or cognitive batteries that measure both the behavior and the physiological mechanism with sufficient specificity. Using experimental tasks that have been validated across species and are known to depend on specific brain structures [32] can aid in identifying biomarkers that can be used to trigger stimulation.

In order for the closed-loop biomarker to provide feedback that can guide stimulation to achieve the desired end state, the biomarker must either be stable with respect to the neural input (i.e. the measured physiology) or must adapt over time as the brain's internal dynamics shift. Several recent studies have used multi-session designs in an attempt to develop stable biomarkers that are not affected by variability in brain states that may be correlated or mediate the relation between the biomarker and the measured behavioral outcome. One study evaluated closed-loop brain stimulation in the context of treatment-resistant depression, capitalizing on and extending several advances in closed-loop methodology. A patient underwent multi-day intracranial monitoring, stimulation mapping, and biomarker identification, in order to develop a personalized stimulation approach that would reduce depression symptoms [33]. By using a data-intensive personalized approach, this study was able to address the patient-to-patient heterogeneity that has led to inconsistent outcomes in previous work that used non-personalized approaches. Another study also used a multi-day recording approach to develop stable patient-specific biomarkers of episodic memory based on broadband spectral power, in order to deliver closed-loop cortical stimulation to enhance memory function [34]. Physiological recordings across days were used to develop a patient-specific neural decoder that successfully generalized over test sessions across days to predict memory outcomes.

One of the most difficult and important questions to be answered about closed-loop brain stimulation is how to select stimulation waveform parameters. This is true even in domains in which there is relatively clear evidence for stimulation's behavioral effects. For example, although biomarkers for stimulation in Parkinson's Disease have been identified and are effective in closed-loop [14], there remains debate about stimulation's underlying mechanisms of action [35]. While some models emphasized stimulation's role in creating an informational lesion at the stimulation site [36, see 37 for review], other models have sought to understand how stimulation affects the broader pathological network [38, 39]. Thus, an important area for future research in other domains will be to develop predictive models that reveal how the stimulated region and its network are likely to be affected by a particular set of stimulation parameters.

Several studies investigating stimulation using macroelectrodes have focused on the role of white matter pathways in determining how stimulation of a particular brain target affects local and distributed physiology. For example, stimulation near white matter vs. gray matter has distinct excitatory and inhibitory effects near the electrode site [40]. Other studies have suggested that targeting white matter pathways with stimulation can transition broader brain networks between states that reflect distinct cognitive modes [41, 42]. Consistent with the idea that stimulation can be used to affect a broader network via white matter pathways, single pulse stimulation of white matter, compared to gray matter, leads to larger effects at more spatially distant recording sites [43]. Similar distinctions between the effects of white and gray matter stimulation have also been shown using functional connectivity to identify the broader network [44].

Using microelectrodes to deliver more targeted stimulation, in combination with predictive modeling to determine stimulation's effects, is another promising approach. Widespread use of this approach will depend on recent advances in modeling and stimulation technology developed in non-human models [45, 46] being translated to humans. For example, microwire recordings in the human hippocampus during a memory task (delayed match to sample) provided the data for a model, based on earlier rodent and non-human primate work [47, 48], that linked CA3 and CA1 unit activity. This model was used in later sessions to decode CA3 activity and to trigger spatiotemporally-patterned microstimulation to replicate the predicted CA1 firing pattern. Stimulation improved memory during the task and on delayed recognition, relative to random and sham stimulation control conditions [49].

41.4.1 Causal Tests of the Neural Basis of Cognition

As mentioned earlier, closed-loop designs are particularly useful for investigating the causal mechanisms of behavior. This can be especially true for processes that involve preparation or planning, which may differ in their timing from trial to trial and individual to individual. For example, the supplementary motor area (SMA) is active prior to the execution of motor movements, which is thought to reflect queuing of motor commands before their execution [50–52]. A recent iEEG study described a system that detected the onset of activity in SMA in order to trigger stimulation in closed-loop [53]. The researchers performed an initial biomarker identification stage to select an SMA electrode with enhanced high-frequency (>60 Hz) power prior to movement onset. Using a patient-specific threshold on high-frequency power, the patient then performed a separate session of the motor task in which closed-loop stimulation was delivered in response to the detected biomarker. Stimulation of SMA prior to movement onset (but not other non-SMA motor areas) delayed movement timing [53]. Stimulation's effects on movement timing were consistent across patients, in spite of the variability in the biomarker from patient to patient, in terms of electrode location, the frequency range of the power increase, and its timing. This suggests that using a closed-loop approach helped account for individual differences, although a specific open-loop stimulation comparison condition was not reported.

Closed-loop stimulation has also been used to test predictions of cognitive models of behavior. For example, theories of memory function suggest that shifts in a cognitive representation of context that is present during encoding, which can be measured with iEEG [54], can lead to divergent effects on associative memory. Accelerating the drift rate of temporal context should facilitate memory in a paired-associates task, perhaps by reducing interference [55] while it should reduce memory measured using free recall, where temporal associations across items support performance [56]. In one study, closed-loop stimulation during the encoding phase was found to affect the drift rate of a neural representation of context in the temporal lobe [57]. This change in the drift rate of neural context led to effects on memory that validated the predictions of the cognitive theoretical model, and more broadly showed how stimulation can be used to causally manipulate neural function to test theoretical models.

Many closed-loop stimulation studies have focused on building models that relate neural activity to behavior. However, it is also possible to take advantage of intracranially-implanted electrodes to trigger stimulation in closed loop while decoding *behavioral* measures. One approach used a latent variable model of participant response time to predict trial-to-trial fluctuations in cognitive control [58]. The latent variable model, a generalization of the Kalman filter approach that is common in brain-machine motor prosthetics, predicted participant response times in the Multi-Source Interference Task (MSIT). The authors used the trial-level model predictions to trigger stimulation to the internal capsule, part of the striatal-frontal circuit that underlies inhibition of prepotent responses at moments requiring cognitive control [59, 60]. When the model predicted lapses in performance, stimulation delivered in closed-loop led to faster response times (i.e. improved performance). A key aspect of this study was the inclusion of an open-loop comparison group in which stimulation was delivered randomly on half of the trials. The authors found that open-loop internal capsule stimulation also sped response times, although to a lesser extent than closed-loop [58]. By including an open-loop comparison condition, the data support the interpretation that behavioral modeling combined with intracranial stimulation can facilitate cognitive control. Given that the open and closed-loop comparison was a between-participants manipulation with small cohorts, future work should extend these findings by implementing within-participants designs or increasing the comparison group sizes.

41.4.2 Naturalistic Closed Loop

Many closed-loop studies remain linked to laboratory-style experimental paradigms, yet there is increasing recognition that neural dynamics must be understood outside of well-defined experimental structure [61]. Thus, an important question for future research is how to deploy closed-loop stimulation in more naturalistic environments in a way that can illuminate the neural basis of behavior. Development of systems for iEEG recording and stimulation in ambulatory human participants will be critical to this work. Research in this area that builds upon existing chronically implanted stimulators has led to tools for synchronizing physiological recordings with measurements of task performance and behavior [62, 63] (see also Chaps. 52 and 53). For example, synchronizing wireless iEEG with motion capture to measure walking trajectory and speed outside of the hospital setting allowed for detailed characterization of theta oscillations in the human medial temporal lobes [64]. Further developments have shown the ability to trigger direct brain stimulation and to synchronize invasive stimulation and physiological recordings with additional wearable sensors [65]. These systems will provide rich data for developing models for deploying closed-loop stimulation during naturalistic behavior.

With the technical development of tools for recording and stimulation during a broader range of behaviors, researchers will have the opportunity to stimulate the brain in response to a much broader range of brain states. However, this opportunity will also create challenges for closed-loop algorithms in their ability to differentiate brain states that should and should not trigger stimulation. Neural variability may be a feature that enables intelligent behavior, rather than a source of unwanted noise [66–68], and, if it is informative, is likely to exist at multiple timescales [69]. Technologies to collect and synchronize greater amounts of neural data with behavior will provide opportunities for detailed predictions for triggering closed-loop stimulation, however this will require more complex dynamic models relating neural activity to cognition and behavior. Such issues have already confronted clinical researchers who have access to years of closed-loop recording and stimulation data [70].

41.5 Conclusion

Here, we have discussed some promises and challenges of closed-loop stimulation, highlighting recent studies that have used such approaches to modulate neural function and behavior. Emerging from development for clinical applications, closed-loop research frameworks are poised to continue to benefit basic science research. Software tools are being developed and shared for decoding iEEG activity and triggering stimulation in closed loop [71], which can aid labs in the adoption of closed-loop approaches. Although in this chapter we have focused on issues relating to modeling of stimulation's effects on physiology, cognition, and behavior, additional challenges remain, such as accounting for stimulation-induced recording artifacts [72]. Nonetheless, iEEG stimulation in closed-loop has unique strengths as a research tool and will continue to provide insights into brain-behavior relations as the field moves forward.

References

1. Ojemann GA (1983) Brain organization for language from the perspective of electrical stimulation mapping. Behav Brain Sci 6:189–206
2. Penfield W, Perot P (1963) The brain's record of auditory and visual experience: a final summary and discussion. Brain 86:595–696
3. George DD, Ojemann SG, Drees C, Thompson JA (2020) Stimulation mapping using stereo-electroencephalography: current and future directions. Front Neurol. https://doi.org/10.3389/fneur.2020.00320
4. Suthana N, Aghajan ZM, Mankin EA, Lin A (2018) Reporting guidelines and issues to consider for using intracranial brain stimulation in studies of human declarative memory. Front Neurosci. https://doi.org/10.3389/fnins.2018.00905
5. Alagapan S, Lustenberger C, Hadar E et al (2019) Low-frequency direct cortical stimulation of left superior frontal gyrus enhances working memory performance. Neuroimage 184:697–706
6. Halgren E, Wilson CL, Stapleton JM (1985) Human medial temporal-lobe stimulation disrupts both formation and retrieval of recent memories. Brain Cogn 4:287–295
7. Jonas J, Rossion B (2021) Intracerebral electrical stimulation to understand the neural basis of human face identity recognition. Eur J Neuroci 54:4197–4211
8. Kim K, Ekstrom AD, Tandon N (2016) A network approach for modulating memory processes via direct and indirect brain stimulation: toward a causal approach for the neural basis of memory. Neurobiol Learn Mem 134:162–177
9. Mankin EA, Fried I (2020) Modulation of human memory by deep brain stimulation of the entorhinal-hippocampal circuitry. Neuron 106:218–235
10. Suthana N, Haneef Z, Stern J et al (2012) Memory enhancement and deep-brain stimulation of the entorhinal area. N Engl J Med 366:502–510
11. Cagnan H, Denison T, McIntyre C, Brown P (2019) Emerging technologies for improved deep brain stimulation. Nat Biotechnol 37:1024–1033
12. Krauss JK, Lipsman N, Aziz T et al (2021) Technology of deep brain stimulation: current status and future directions. Nat Rev Neurol 17:75–87
13. Moro E, Poon YYW, Lozano AM et al (2006) Subthalamic nucleus stimulation: improvements in outcome with reprogramming. Arch Neurol 63:1266–1272
14. Bouthour W, Mégevand P, Donoghue J et al (2019) Biomarkers for closed-loop deep brain stimulation in Parkinson disease and beyond. Nat Rev Neurol 15:343–352
15. Rosin B, Slovik M, Mitelman R et al (2011) Closed-loop deep brain stimulation is superior in ameliorating parkinsonism. Neuron 72:370–384
16. Daneshzand M, Faezipour M, Barkana BD (2018) Robust desynchronization of Parkinson's disease pathological oscillations by frequency modulation of delayed feedback deep brain stimulation. PLoS ONE 13:e0207761
17. Rosa M, Arlotti M, Ardolino G et al (2015) Adaptive deep brain stimulation in a freely moving Parkinsonian patient. Mov Disord 30:1003
18. Swann NC, de Hemptinne C, Thompson MC et al (2018) Adaptive deep brain stimulation for Parkinson's disease using motor cortex sensing. J Neural Eng 15:046006
19. Feng XJ, Greenwald B, Rabitz H et al (2007) Toward closed-loop optimization of deep brain stimulation for Parkinson's disease: concepts and lessons from a computational model. J Neural Eng 4:L14
20. Morrell MJ (2011) Responsive cortical stimulation for the treatment of medically intractable partial epilepsy. Neurology 77:1295–1304

21. Nair DR, Laxer KD, Weber PB et al (2020) Nine-year prospective efficacy and safety of brain-responsive neurostimulation for focal epilepsy. Neurology 95:e1244–e1256
22. Bergey GK, Morrell MJ, Mizrahi EM et al (2015) Long-term treatment with responsive brain stimulation in adults with refractory partial seizures. Neurology 84:810–817
23. Geller EB, Skarpaas TL, Gross RE et al (2017) Brain-responsive neurostimulation in patients with medically intractable mesial temporal lobe epilepsy. Epilepsia 58:994–1004
24. Heck CN, King-Stephens D, Massey AD et al (2014) Two-year seizure reduction in adults with medically intractable partial onset epilepsy treated with responsive neurostimulation: final results of the RNS System Pivotal trial. Epilepsia 55:432–441
25. Jobst BC, Kapur R, Barkley GL et al (2017) Brain-responsive neurostimulation in patients with medically intractable seizures arising from eloquent and other neocortical areas. Epilepsia 58:1005–1014
26. Little S, Pogosyan A, Neal S et al (2013) Adaptive deep brain stimulation in advanced Parkinson disease. Ann Neurol 74:449–457
27. Little S, Tripoliti E, Beudel M et al (2016) Adaptive deep brain stimulation for Parkinson's disease demonstrates reduced speech side effects compared to conventional stimulation in the acute setting. J Neurol Neurosurg Psychiatry 87:1388–1389
28. He S, Baig F, Mostofi A et al (2021) Closed-loop deep brain stimulation for essential tremor based on thalamic local field potentials. Mov Disord 36:863–873
29. Cagnan H, Pedrosa D, Little S et al (2017) Stimulating at the right time: phase-specific deep brain stimulation. Brain 140:132–145
30. Kossoff EH, Ritzl EK, Politsky JM et al (2004) Effect of an external responsive neurostimulator on seizures and electrographic discharges during subdural electrode monitoring. Epilepsia 45:1560–1567
31. Hammond C, Bergman H, Brown P (2007) Pathological synchronization in Parkinson's disease: networks, models and treatments. Trends Neurosci 30:357–364
32. Deadwyler SA, Hampson RE, Song D et al (2017) A cognitive prosthesis for memory facilitation by closed-loop functional ensemble stimulation of hippocampal neurons in primate brain. Exp Neurol 287:452–460
33. Scangos KW, Khambhati AN, Daly PM et al (2021) Closed-loop neuromodulation in an individual with treatment-resistant depression. Nat Med 27:1696–1700
34. Ezzyat Y, Wanda PA, Levy DF et al (2018) Closed-loop stimulation of temporal cortex rescues functional networks and improves memory. Nat Commun 9:1–8
35. Lozano AM, Lipsman N, Bergman H et al (2019) Deep brain stimulation: current challenges and future directions. Nat Rev Neurol 15:148–160
36. Grill WM, Snyder AN, Miocinovic S (2004) Deep brain stimulation creates an informational lesion of the stimulated nucleus. NeuroReport 15:1137–1140
37. Lozano AM, Lipsman N (2013) Probing and regulating dysfunctional circuits using deep brain stimulation. Neuron 77:406–424
38. Ashkan K, Rogers P, Bergman H, Ughratdar I (2017) Insights into the mechanisms of deep brain stimulation. Nat Rev Neurol 13:548–554
39. McIntyre CC, Anderson RW (2016) Deep brain stimulation mechanisms: the control of network activity via neurochemistry modulation. J Neurochem 139:338–345
40. Mohan UR, Watrous AJ, Miller JF et al (2020) The effects of direct brain stimulation in humans depend on frequency, amplitude, and white-matter proximity. Brain Stim 13:1183–1195
41. Khambhati AN, Kahn AE, Costantini J et al (2019) Functional control of electrophysiological network architecture using direct neurostimulation in humans. Netw Neurosci 3:848–877
42. Stiso J, Khambhati AN, Menara T et al (2019) White matter network architecture guides direct electrical stimulation through optimal state transitions. Cell Rep 28:2554–2566
43. Paulk AC, Zelmann R, Crocker B et al (2022) Local and distant cortical responses to single pulse intracranial stimulation in the human brain are differentially modulated by specific stimulation parameters. Brain Stim 15:491–508
44. Solomon EA, Kragel JE, Gross R et al (2018) Medial temporal lobe functional connectivity predicts stimulation-induced theta power. Nat Commun 9:1–13

45. Qiao S, Sedillo JI, Brown KA et al (2020) A causal network analysis of neuromodulation in the mood processing network. Neuron 107:972–985
46. Yang Y, Qiao S, Sani OG et al (2021) Modelling and prediction of the dynamic responses of large-scale brain networks during direct electrical stimulation. Nat Biomed Eng 5:324–345
47. Berger TW, Hampson RE, Song D et al (2011) A cortical neural prosthesis for restoring and enhancing memory. J Neural Eng 8:046017
48. Hampson RE, Song D, Opris I et al (2013) Facilitation of memory encoding in primate hippocampus by a neuroprosthesis that promotes task-specific neural firing. J Neural Eng 10:066013
49. Hampson RE, Song D, Robinson BS et al (2018) Developing a hippocampal neural prosthetic to facilitate human memory encoding and recall. J Neural Eng 15:036014
50. Orgogozo JM, Larsen B (1979) Activation of the supplementary motor area during voluntary movement in man suggests it works as a supramotor area. Science 206:847–850
51. Roland PE, Larsen B, Lassen NA, Skinhoj E (1980) Supplementary motor area and other cortical areas in organization of voluntary movements in man. J Neurophys 43:118–136
52. Zimnik AJ, Lara AH, Churchland MM (2019) Perturbation of macaque supplementary motor area produces context-independent changes in the probability of movement initiation. J Neurosci 39:3217–3233
53. Moore BD IV, Aron AR, Tandon N (2018) Closed-loop intracranial stimulation alters movement timing in humans. Brain Stim 11:886–895
54. Manning JR, Polyn SM, Baltuch GH et al (2011) Oscillatory patterns in temporal lobe reveal context reinstatement during memory search. Proc Natl Acad Sci USA 108:12893–12897
55. Davis OC, Geller AS, Rizzuto DS, Kahana MJ (2008) Temporal associative processes revealed by intrusions in paired-associate recall. Psychon Bull Rev 15:64–69
56. Sederberg PB, Miller JF, Howard MW, Kahana MJ (2010) The temporal contiguity effect predicts episodic memory performance. Mem Cognit 38:689–699. https://doi.org/10.3758/MC.38.6.689
57. El-Kalliny MM, Wittig JH, Sheehan TC et al (2019) Changing temporal context in human temporal lobe promotes memory of distinct episodes. Nat Commun 10:1–10
58. Basu I, Yousefi A, Crocker B et al (2021) Closed-loop enhancement and neural decoding of cognitive control in humans. Nat Biomed Eng 1–13
59. O'Reilly RC, Frank MJ (2006) Making working memory work: a computational model of learning in the prefrontal cortex and basal ganglia. Neural Comput 18:283–328
60. Sharpe MJ, Stalnaker T, Schuck NW et al (2019) An integrated model of action selection: distinct modes of cortical control of striatal decision making. Annu Rev Psychol 70:53–76
61. Huk A, Bonnen K, He BJ (2018) Beyond trial-based paradigms: Continuous behavior, ongoing neural activity, and natural stimuli. J Neurosci 38:7551–7558
62. Meisenhelter S, Testorf ME, Gorenstein MA et al (2019) Cognitive tasks and human ambulatory electrocorticography using the RNS System. J Neurosci Methods 311:408–417
63. Rao VR, Leonard MK, Kleen JK et al (2017) Chronic ambulatory electrocorticography from human speech cortex. Neuroimage 153:273–282
64. Aghajan ZM, Schuette P, Fields TA et al (2017) Theta oscillations in the human medial temporal lobe during real-world ambulatory movement. Curr Biol 27:3743–3751
65. Topalovic U, Aghajan ZM, Villaroman D et al (2020) Wireless programmable recording and stimulation of deep brain activity in freely moving humans. Neuron 108:322–334
66. Sheehan TC, Sreekumar V, Inati SK, Zaghloul KA (2018) Signal complexity of human intracranial EEG tracks successful associative-memory formation across individuals. J Neurosci 38:1744–1755
67. Uddin LQ (2020) Bring the noise: reconceptualizing spontaneous neural activity. Trends Cogn Sci 24:734–746
68. Waschke L, Kloosterman NA, Obleser J et al (2021) Behavior needs neural variability. Neuron 109:751–766
69. He BJ (2014) Scale-free brain activity: past, present, and future. Trends Cogn Sci 18:480–487

70. Sisterson ND, Wozny TA, Kokkinos V et al (2019) Closed-loop brain stimulation for drug-resistant epilepsy: towards an evidence-based approach to personalized medicine. Neurotherapeutics 16:119–127
71. Zelmann R, Paulk AC, Basu I et al (2020) CLoSES: A platform for closed-loop intracranial stimulation in humans. Neuroimage 223:117314
72. Iturrate I, Pereira M, Millán JDR (2018) Closed-loop electrical neurostimulation: challenges and opportunities. Curr Opin Biom Eng 8:28–37

Chapter 42
Which Are the Most Important Aspects of Microelectrode Implantation?

Angelique Sao-Mai S. Tay, Bassir Caravan, and Adam N. Mamelak

Abstract Chronically implanted intracranial electrodes containing microwire bundles capable of recording extracellular single unit action potentials provide significant insights into the ways neurons and neural networks regulate cognition, movement, and many aspects of human behavior, as well as insights into the neuronal basis of seizures. The most used electrode for these recordings is the commercially available Behnke-Fried hybrid electrode. In this design, a bundle of microwires protrude from the distal end of a macroelectrode. Adhering to a few simple techniques during insertion of these electrodes can maximize the number of neurons recorded and also achieve stable, high quality, and long-lasting recordings. Techniques include: (1) preloading the microwires into the macroelectrode bundle and cutting them to a 4–5 mm length with sharp scissors prior to use; (2) targeting the distal end of the macroelectrode to ensure that the microwires protrude into grey matter; (3) using orthogonal insertion trajectories whenever possible; (4) avoid kinking of the microwires; and (5) ensuring that the distal macroelectrode does not penetrate the grey matter where the microwires will be placed. This chapter will review the technical nuances for electrode insertion that lead to high quality recordings, a key first step to successful recording, sorting, and analysis of human single units.

42.1 Introduction

For patients with drug resistant epilepsy, the implantation of intracranial depth electrodes is often used to help identify the seizure focus, and plan subsequent therapies aimed at stopping or reducing the number of seizures [1, 2]. During this process, a

A. S.-M. S. Tay · B. Caravan · A. N. Mamelak (✉)
Department of Neurosurgery, Cedars-Sinai Medical Center, Los Angeles, CA, USA
e-mail: adam.mamelak@cshs.org

A. S.-M. S. Tay
Department of Neurosurgery, Cedars-Sinai Medical Center, 127 S. San Vicente Blvd, A6600, 90048 Los Angeles, United States

N. Axmacher (ed.), *Intracranial EEG*, Studies in Neuroscience, Psychology and Behavioral Economics, https://doi.org/10.1007/978-3-031-20910-9_42

study commonly referred to as Stereotactic Electroencephalography (SEEG), penetrating electrodes are placed in the brain and patients undergo continuous monitoring of EEG for several days to weeks in an epilepsy monitoring unit. During this monitoring process patients are awake, generally comfortable, and able to fully interact, thus providing a unique opportunity to perform behavioral testing and correlate behavioral responses with brain activity. Research in this arena has led to seminal discoveries about how the human brain functions in both health and disease. The simultaneous insertion of research-quality microelectrodes during the implantation of clinical macroelectrodes enables neuroscientists to record in vivo single unit extracellular action potentials, multi-units, and local field potentials. These microelectrode recording have proven instrumental in furthering our understanding of the correlation between single neuron firing patterns, inter-areal neuronal network activity, and human behavior. Given the size and fragility of these electrical contacts, as well as the relative rarity of being able to perform these studies, it is essential that proper technique be followed to maximize the quality and reliability of data obtained from these experiments.

Although microelectrode recordings are not currently used for clinical diagnosis, the use of hybrid depth electrodes containing microwires carried within a macroelectrode array has been shown to be safe and well-tolerated [3], thus justifying their use. The most used electrode, and the one we have exclusively used over the past 17 years, is the Behnke-Fried electrode (Adtech, Racine, WI), initially developed at UCLA in the 1960s through 1990s with several modifications [4, 5]. This electrode consists of two parts: an outer macroelectrode sheath and an inner microelectrode bundle (Fig. 42.1). The macroelectrode contains platinum iridium macroelectrode contacts spaced along the distal end of a hollow plastic sheath. The microelectrode is a bundle of eight 40um diameter platinum iridium wires held together in a resin, with the distal 15 mm of electrode wire protruding in a flower spray configuration. During the sterilization process these wires can get coated with ethylene oxide that can inhibit recording ability. Therefore, the distal ends must be cut to the desired length at the time of surgery, thus exposing the distal tips for recording. The microelectrode also contains a green sheath that slides over the distal end of the protruding microelectrodes to protect them and guide them into the center hollow core of the macroelectrode. This sheath is then fully retracted towards the base of the electrode at final insertion to make sure that it does not cover the protruding microelectrodes. Many publications describing human single unit activity have been reported using this electrode [2–4]. However, the number of single units recorded, quality, and stability of these recording appears to be highly variable between centers (multiple personal communications with ANM), suggesting that techniques for electrode insertion may plan a critical role in obtaining the highest quality data. The goal of this chapter is to summarize our experience inserting Behnke-Fried electrodes, with the aim of pointing out the technical nuances we have identified as most important to achieving high yield, high quality single unit recording in humans.

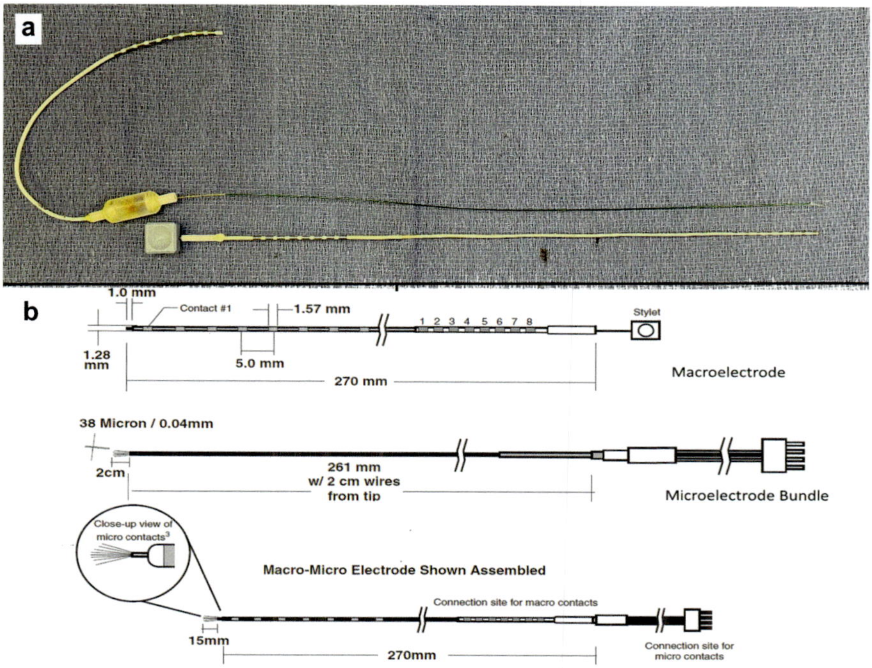

Fig. 42.1 **a** Behnke-Fried (Adtech) electrode. Pictured above is the green guide sheath around the microelectrode bundle. The sheath is used to cover the microwires during insertion into the macroelectrode and is retracted back towards the base of the microelectrode bundle at final insertion. The microwires extend beyond the green sheath when full inserted and the sheath is retracted. Below it is the hollow macroelectrode in which the microelectrode bundle is inserted. **b** Schematic drawings of macroelectrode (top), microelectrode (middle) and assembled macro–micro array (bottom). Reproduced with permission from Adtech, Racine WI

42.2 Target Selection

Patients with drug resistant epilepsy (DRE) are referred for Phase II stereotactic electroencephalography (SEEG) studies based on multidisciplinary evaluation of all pre-surgical data. Such pre-surgical data typically includes clinical history, analysis of seizure semiology, neuropsychological testing, scalp electrode-based video-EEG monitoring, magnetic resonance image (MRI), single photon emission computed tomography (SPECT), fluorodeoxyglucose-positron emission tomography (FDG-PET) and magnetoencephalography (MEG) [6]. The depth electrode targets and trajectories are then selected to maximize the chances of identifying the hypothetical seizure onset zone.

Decisions such as the number and location of electrodes can vary significantly between centers and individual situations. In our center we tend to routinely perform bilateral implantations, often targeting classic limbic regions such as amygdala, hippocampus, entorhinal cortex, anterior cingulate cortex, ventromedial prefrontal cortex, and pre-supplementary motor area. Whenever possible we aim for largely orthogonal trajectories for these structures to maximize sampling and facilitate easier data interpretation. Other regions include insula, visual cortices, regions of localized MEG spike, cortical dysplasia, or other regions of interest. For regions such as insula, visual cortices, or focal cortical dysplasia, we typically use non-orthogonal trajectories. Due to the possible risk associated with placement of depth electrodes, such as bleeding in the brain, stroke, infection, and death, it is unethical to insert electrodes in non-clinically relevant areas that are pre-determined by the pre-surgical area [7].

Prior to surgery, we acquire multiple sequences of Magnetic Resonance Imaging (MRI) of the brain, along with Computed Tomography Angiogram (CTA) and CT Venogram (CTV) studies. MRI sequences should be of high quality, devoid of movement artifact, and contain at least one sequence aimed to optimized visualization of grey matter and grey-white junction (e.g., 2 mm thick whole brain Spoiled Gradient Recalled Echo, SPGRE or Magnetization Prepared—Rapid Gradient Echo, MPRAGE). Use of a 3 T MRI scan is also preferred if possible. There are several pulse sequences optimized for patients with epilepsy [8, 9]. High quality pre-operative imaging without movement artifact is a critical first step to ensure accurate electrode targeting. At our center these images sets are loaded on an S8 Stealth Station (Medtronic, Louisville, Colorado) or ROSA robotic (Zimmer-Biomet) computer with stereotactic targeting planning software suite and co-registered. A plan is created for each electrode target and trajectory (Fig. 42.2). A variety of other planning systems are available, and the choice of planning and insertion method depends largely on the surgeon and center.

After selecting the target and entry points, the MRI, CTA and CTV, sequences are followed on the "trajectory" and "probe's eye" views from the surface to target. Plans are adjusted to avoid vessels and to minimize traversing of sulci. The thickness of the skull at the trajectory entry site is also measured and noted.

42.2.1 Technical "Pearls" for Electrode Planning

1. **Plan target depth of macroelectrode tip at grey-white junction, rather than at medial edge of cortex**. To record single unit action potentials, the microwires must sit in cortical grey matter. These wires typically extend from the distal tip of the macroelectrode by 4–5 mm, and the thickness of cortex is usually 3–6 mm. Therefore, the distal tip of the macroelectrode must be positioned to allow the microwires to protrude into the cortical grey matter (Fig. 42.3). When standard depth electrodes are placed, the distal tip usually targets the deepest or most medial aspect of the planned trajectory. However, in targeting the deepest macroelectrode contact, we aim for 4–5 mm shallower. We have not noted any

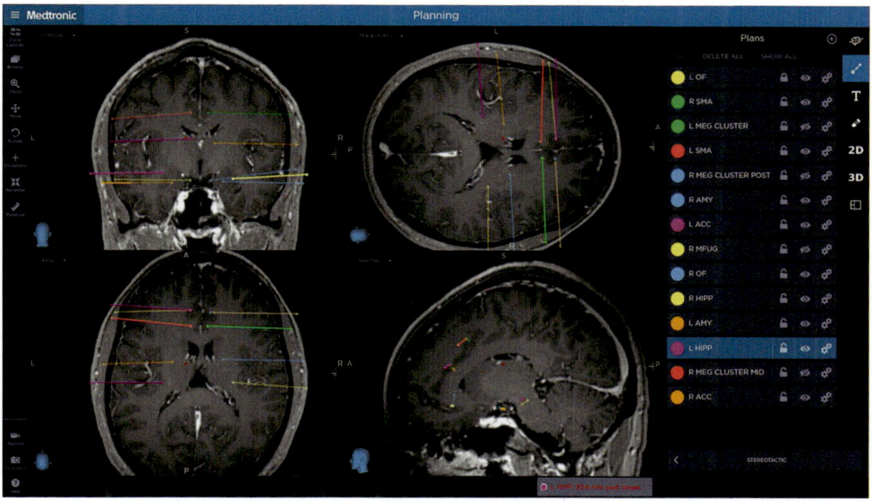

Fig. 42.2 Screenshot from a planning software workstation showing examples of electrode target and trajectory plans for a patient undergoing SEEG placement. Note that the distal end of the targets aim for the grey-white junction in cortical regions and for grey matter structures such as amygdala so that the 4nmm protrusion of the microwires will remain in grey matter. Orthogonal trajectories are preferred whenever feasible

difference in our ability to localize seizures with this modification. However, we have rouitnely noted that if the electrodes are placed too deep in the grey matter, then far less optimal neuronal recordings can be obtained. Placing the distal macroelectrode contact at the grey matter-white matter border substantially improves neuronal yield and quality.

2. **Use orthogonal insertion trajectories whenever possible**. We find this allows for maximal recordings from cortical surface and deep targets. Orthogonal trajectories tend to be more accurate than non-orthogonal ones, and are often the shortest path to target and not typically deflected by oblique bone and dura penetrations [10].

42.3 Device and Methodology for Electrode Insertion

Depth electrodes (either those with microelectrode bundles or without) are inserted according to well established neurosurgical methods and vary depending on the surgeon training and preferences. In the past we have used a rigid stereotactic frame with an orthogonal guide post positioned on the side rail of the frame base for orthogonal insertions, and a stereotactic arc for non-orthogonal ones [11]. More recently we have converted to a robotic manipulator method [12]. The use of stand-alone robotic systems such as the ROSA ONE® (Medtech, Montpellier, France) or Neuromate® (Renishaw, Dublin, Ireland) and methods based on frameless stereotactic guidance

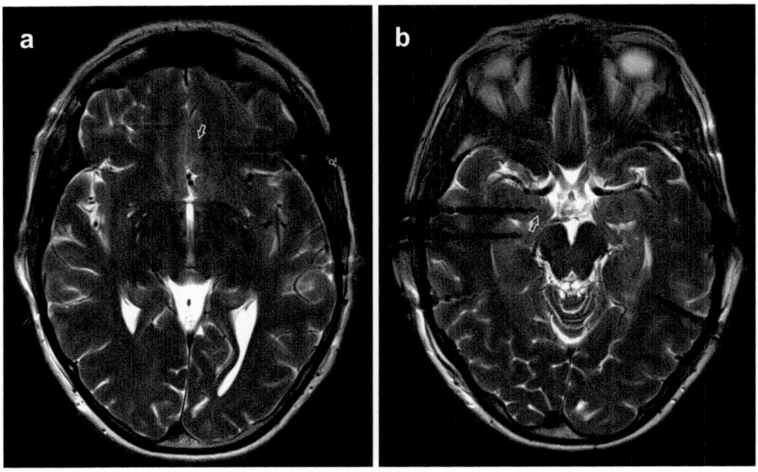

Fig. 42.3 Representative T2 weighted MRI images demonstrating correct depth placement targeting the grey -white junction so that the microwires protrude into grey matter. **a** Targeting of ventro-medial prefrontal cortex. **b** Targeting of amygdala and hippocampus. Arrows demonstrate protruding microwires at distal tip

[13–15] are also very popular. While all of these methods can work, for microelectrode recording purposes we strongly favor a more rigid insertion method, based on the improved accuracy of these systems when compared to frameless ones [16]. We suggest the following practices to help improve surgical flow, reduce risk of infection, and increase targeting accuracy:

1. **Patients should undergo a full head shave**. Fully shaving the head reduces any limitation to preferred trajectories, avoids the possibility that hair can distort the trajectory and reduces risk of infection. To reduce infection, the head shave should be done just before the surgery and should not be done by the patient at home.
2. **Rigid fixation of the head using a stereotactic frame rather than a three-point head holder**. This provides increased access to all orthogonal targets with minimal interference from the head holder. We utilize ear bars inserted into the auditory canals of the patient at the time of frame placement to ensure a perfectly centered and orthogonal frame.
3. **If non-robotic stereotactic targeting methods are used, we recommend the patient be brought to the CT scanner with the frame in place to acquire a registration image with the localizer in place**. We find this method has the highest registration accuracy. This image is merged to the previously acquired MRI to obtain coordinates for electrode placement. Intraoperative CT or intra-operative 3D fluoroscopy (e.g., O-arm®, Medtronic, Louisville, CO) can also be used, although CT is slightly more accurate than the O-arm®.
4. **Substantial effort should be made to ensure the highest degree of registration accuracy possible**. For the Autoguide® (Medtronic, Louisville, CO) system that

we previously used we obtained a head CT scan with skin placed fiducial markers the night before the day of surgery. The CT is co-registered to other image sets, and stereotactic space established using a standard registration algorithm. This marker-based registration is subsequently improved upon using an additional "skin tracer" registration. We require a registration accuracy of 2.5 mm or less to proceed (ideally <1.5 mm). Several registration methods have been described for other robotic systems, and care should be taken to ensure the highest degree of registration accuracy possible as this will improve targeting accuracy. The ROSA® and Neuromate® systems have their own registration methods, with registration acccuracies less than 1 mm being standard. Sterotactic arc-based systems also have a high degree of registration accuracy but are less commonly used now due to the slowness of this method and the need to manually set target coordinates for each insertion.

5. **The patient is positioned in a "lounge chair" position or flat**. We perform all electrode placements with the patient under general anesthesia. The patient is positioned in a semi-sitting "lounge chair" position with the head elevated about 30degrees above the chest, or perfectly flat with the frame attached to OR bed. We find this position provides maximal access for the surgeon for drilling and placing anchor bolts, while minimizing risk of brain shift due to CSF leak. For ROSA® and Neuromate® systems it is easiest to place the patient flat as this simplifies intraoperative confirmation imaging with CT. The entire head is prepped with an iodine based antiseptic solution, ensuring that the prep extends below the zygoma (cheek bone) on both sides and up to the frame rods. A sterile "U" drape is draped around the base of the frame but below the base ring and extended around the front of the patient with a full or three-quarter sheet placed to cover the patient body. This creates a sterile working area but does not limit access to any region of the head. In general the method of patient positioning is left to the discretion of the surgeon, based on experience and the exact insertion method being used.

42.4 Insertion of Electrodes

Insertion of depth electrodes with microwires is largely identical to standard insertion method and follows standard neurosurgical methods. These technical details are beyond the scope or goal of this chapter. Rather, we will focus on the specific technical nuances for insertion of hybrid Behnke-Fried electrodes.

For each electrode insertion the following general steps are carried out:

1. Select the correct targeting plan of the guidance system computer
2. Align the robotic arm guide tube, frameless stereotaxic arm, or stereotactic frame guide tube (depending on surgical method used) directly over the planned insertion site from the pre-measured target. Alignment is carried out based on pre-planned targets as discussed earlier, and the specifics of each device.

3. Make a stab incision in the skin and drill the pilot hole for the anchor bolt that will hold the electrode in place, with drill depth based on pre-measured bone thickness. Coagulate and then puncture the dura below the pilot hole.
4. Screw the anchor bolt into the skull using the guide tube to make sure it follows the exact trajectory as the drill.
5. Measure the length of the electrode to be inserted to planned target.
6. Insert the electrode to target.
7. Tighten the anchor bolt to lock the electrode in place.

Each electrode is placed using these steps. Then a full head wrap is applied completely covering up the electrodes.

42.4.1 Technical "Pearls" for Microelectrode Insertion

When utilizing Behnke-Fried microelectrodes the following technical tips will ensure the best chance for high quality, high yield, stable recordings. We consider each of these steps to be critical to achieving good results.

1. **Pre-load the microwires into the macroelectrode bundle**. On a back table in the operating room, the surgeon should open and insert the microelectrode bundle and inspect it. Cover the distal end of the microwire bundle with the green protective guide sheath and completely insert into the macroelectrode such that it protrudes through the distal end (Fig. 42.4a). While inserting, the green sheath must be pulled back so that it is flush with the electrode connector at the proximal end of the electrode to ensure that the green sheath does not protrude from the distal tip of the macroelectrode. This replicates how the electrode is situated in the brain. The distal exposed microwires protruding from the end of the macroelectrode are substantially longer (approximately 15 mm) than the length they will be in the brain (Fig. 42.4b) Once fully "loaded", the distal microwires should be cut with sharp tonotomy scissors so that they protrude 4–5 mm in length. We measure the length of the microwire bundle with a ruler to ensure accurate length (see Fig. 42.4c, d). Once cut, carefully remove the microbundle and place it on the back table. Proceed with insertion of the macroelectrode into the brain. When doing these steps care is taken to not kink the microbundle and ensure the green protective sheath overlying it is fully in its retracted position near the microbundle base.
2. **Cut the microwire bundle so that it will extend from the distal end of the macroelectrode by a length of 4–5 mm**. We have experimented with lengths ranging from 3 to 8 mm and found that 4 to 5 mm gives the best recordings.
3. **When cutting the microwires, make sure that the cut is quick and sharp**. A cut that is more of a gnawing motion could crush the distal ends of the microwires. Use sharp tonotomy scissors to cut the wires. DO NOT use a scalpel as this will crush the ends of the microwires.

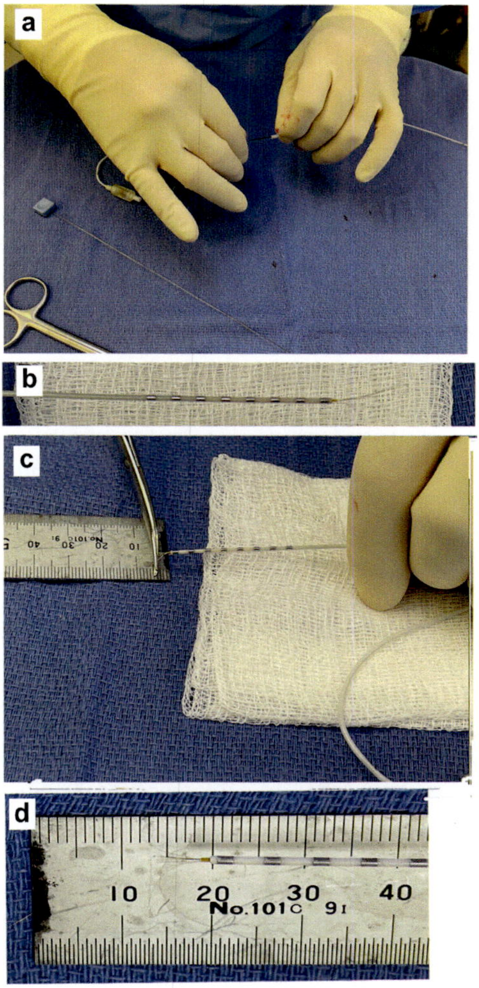

Fig. 42.4 Preparing the electrodes for insertion. **a** The microelectrode is preloaded into the macroelectrode sheath on a back table in the OR. Note the green sheath should be fully retracted at final insertion. **b** On full insertion the microelectrode bundle extends approximately 15 mm from the distal sheath. This must be cut smaller. **c** The microelectrode bundle is cut to protrude a length of 4–5 mm from the macroelectrode tip. **d** Final appearance of macro-micro electrode distal tip after cutting. During actual insertion the macro sheath is inserted to target and then the microbundle inserted through it, ensuring it will be 4–5 mm from the distal tip of the macro electrode target

4. **After inserting the macroelectrode through the anchor bolt remove the stylet but do not tighten the anchor bolt even temporarily before inserting the microbundle**. Tightening the bolt can crush the outer sheath of the macroelectrode and make insertion of the microwires more difficult. We find that the electrode will typically stay in place quite easily during this process, but if needed, an assistant can hold the macroelectrode in place.
5. **Avoid kinking the microwires during placement**. The microwire bundle is fragile and will kink easily if under any pressure. We typically have the surgical assistant hold the proximal end of the microwire and connector while the surgeon is inserting the distal end into the macroelectrode. The assistant ensures that the microwire bundle is not kinking as it is being inserted. The green sheath that covers the microwire bundle must be passed beyond the distal end of the

microelectrode bundle prior to inserting it into the macroelectrode and while passing to target. However, the green sheath must be fully retracted so that its proximal end sits on the base of the microelectrode connector (at the opposite end of the electrode from the tip) before fully inserting the electrode into the base of the macroelectrode. This ensures that the distal microwires protrude into the brain at the pre-measured length and are not covered by the plastic sheath, which will prevent any recording. Electrode kinking is most likely to occur during this step of insertion and the surgeon should take extra care to guide the microbundle slowly and without tension all the way into position while retracting the green outer sheath fully.

6. **Only tighten the anchor bolt screw after the entire microbundle is inserted**. Once the microelectrode is placed the anchor bolt is tightened. In general, the degree to which it is tightened is not critical, but we recommend not tightening to its full maximum to avoid crushing of the microwires. A snug closure will reliably hold the electrodes without risk of slipping. Confirmation of accurate insertion of microelectrode bundles can often be seen on skull X-rays (Fig. 42.5), although CT and MRI are more useful in providing specific localization information.
7. **Subgaleal reference electrodes improve stable single-neuron recordings and signal-to-noise ratio**. Strip electrodes are typically implanted below the scalp with contacts pointing away from the brain. We place a 1×4 contact subdural strip in the midline parietal areas (approximately Pz) and advance this in the subgaleal space. The electrode contacts are positioned facing out towards the scalp rather than towards the bone. A second 2 contact strip is placed more anteriorly (approximately Cz) with the contacts facing towards the skull. This electrode serves as a reference for the clinical EEG reccordings.
8. **Avoid kinking of electrodes when applying or changing the head wrap**. Unlike standard depth electrodes, the microwire bundles are more prone to break with bending of the outer sheath of the electrode. Therefore, care should be taken to minimize the amount of bend and flex in the electrodes when applying a head wrap or changing the wrap. We typically collect all the electrodes from a single side of the head and hold them together with a Steri-strip (3M) for strain relief. When the head wrap is applied, care is taken to make sure the electrode sites are padded and that all the electrodes exit the dressing with as little bending as possible. Minimizing the number of dressing changes and hookups in the epilepsy monitoring unit will improve the stability and quality of recording over time.

42.5 Summary

The primary goal of this chapter is to provide the researcher or clinician interested in using Behnke-Fried macro-micro electrode to record extracellular single unit action potential with several technical "pearls" to improve the reliability and yield of these electrodes when obtaining single unit recordings in humans. This chapter focused

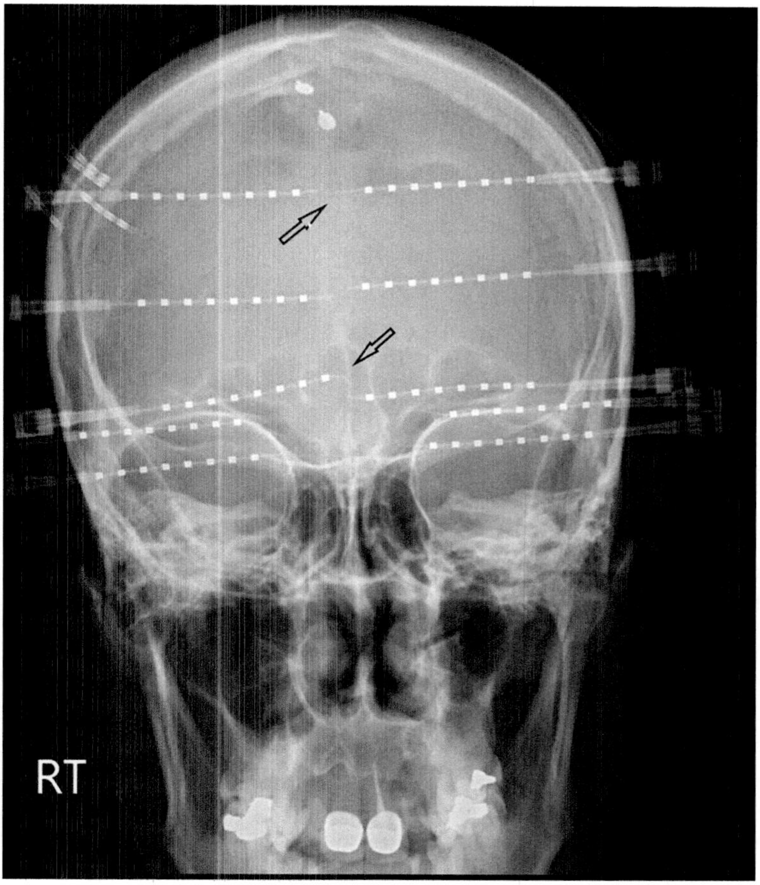

Fig. 42.5 Postoperative Skull X-ray showing a typical appearance of multiple hybrid depth electrodes in place. Protrusion of microelectrode bundles can be seen in several electrodes (see arrows for examples)

exclusively on surgical details, as other chapter will focus on experimental set-up and similar aspect of the electrophysiology recording process. This chapter is by no means an authoritative overview of the methods used to insert depth electrodes but is aimed specifically at aspect of microelectrode insertion for research application. Based on our extensive experience with the Behnke-Fried electrode, as well as feedback from several of our colleagues who also use these wires, we feel the pearls provided here will lead to the best possible results. Undoubtedly as other microelectrodes come into greater use similar technical pearls will be useful for those systems.

References

1. Obien MEJ, Deligkaris K, Bullmann T, Bakkum DJ, Frey U (2015) Revealing neuronal function through microelectrode array recordings. Review. Front Neurosci 8(423). https://doi.org/10.3389/fnins.2014.00423
2. Hefft S, Brandt A, Zwick S et al (2013) Safety of hybrid electrodes for single-neuron recordings in humans. Neurosurgery 73(1):78–85. https://doi.org/10.1227/01.neu.0000429840.76460.8c
3. Carlson AA, Rutishauser U, Mamelak AN (2018) Safety and utility of hybrid depth electrodes for seizure localization and single-unit neuronal recording. Stereotact Funct Neurosurg 96(5):311–319. https://doi.org/10.1159/000493548
4. Fried I, Wilson CL, Maidment NT et al (1999) Cerebral microdialysis combined with single-neuron and electroencephalographic recording in neurosurgical patients. Technical note. J Neurosurg 91(4):697–705. https://doi.org/10.3171/jns.1999.91.4.0697
5. Rutishauser U, Schuman EM, Mamelak AN (2006) Online detection and sorting of extracellularly recorded action potentials in human medial temporal lobe recordings, in vivo. J Neurosci Methods 154(1–2):204–224. https://doi.org/10.1016/j.jneumeth.2005.12.033
6. Iida K, Otsubo H (2017) Stereoelectroencephalography: indication and efficacy. Neurol Med Chir (Tokyo) 57(8):375–385. https://doi.org/10.2176/nmc.ra.2017-0008
7. Mamelak A (2014) Ethical and practical considerations for human microelectrode recording studies. Single neuro studies of the human brain. MIT Press
8. Mikulis DJ, Roberts TP (2007) Neuro MR: protocols. J Magn Reson Imaging 26(4):838–847. https://doi.org/10.1002/jmri.21041
9. Verma G, Delman BN, Balchandani P (2021) Ultra high field MR imaging in epilepsy. Magn Reson Imaging Clin N Am 29(1):41–52. https://doi.org/10.1016/j.mric.2020.09.006
10. Faraji AH, Remick M, Abel TJ (2020) Contributions of robotics to the safety and efficacy of invasive monitoring with stereoelectroencephalography. Front Neurol 11:570010–570010. https://doi.org/10.3389/fneur.2020.570010
11. Minxha J, Mamelak AN, Rutishauser U (2018) Surgical and electrophysiological techniques for single-neuron recordings in human epilepsy patients. In: Sillitoe RV (ed) Extracellular recording approaches. New York, Springer, pp 267–293
12. Tay AS-MS, Menaker SA, Chan JL, Mamelak AN (2022) Placement of stereotactic electroencephalography depth electrodes using the autoguide robotic system: technical methods and initial results. Oper Neurosurg
13. Sharma JD, Seunarine KK, Tahir MZ, Tisdall MM (2019) Accuracy of robot-assisted versus optical frameless navigated stereoelectroencephalography electrode placement in children. J Neurosurg Pediatr 23(3):297–302. https://doi.org/10.3171/2018.10.Peds18227
14. Cardinale F, Rizzi M, d'Orio P et al (2017) A new tool for touch-free patient registration for robot-assisted intracranial surgery: application accuracy from a phantom study and a retrospective surgical series. Neurosurg Focus 42(5):E8. https://doi.org/10.3171/2017.2.Focus16539
15. Vakharia VN, Sparks R, O'Keeffe AG et al (2017) Accuracy of intracranial electrode placement for stereoencephalography: a systematic review and meta-analysis. Epilepsia 58(6):921–932. https://doi.org/10.1111/epi.13713
16. Girgis F, Ovruchesky E, Kennedy J, Seyal M, Shahlaie K, Saez I (2020) Superior accuracy and precision of SEEG electrode insertion with frame-based vs. frameless stereotaxy methods. Acta Neurochir (Wien) 162(10):2527–2532. https://doi.org/10.1007/s00701-020-04427-1

Chapter 43
How Can We Process Microelectrode Data to Isolate Single Neurons in Humans?

Mar Yebra and Ueli Rutishauser

Abstract Extracellular recordings of single neurons are a commonly used method to study the neural mechanisms of cognition and disease. While extensively used in animal models, rare clinical cases also allow such recordings from the human brain using high-impedance microwires. These recordings allow the study of the activity of individual human neurons during cognitive tasks at single-spike resolution. Here, we discuss one such clinical scenario: Microwires embedded in depth electrodes implanted in epilepsy patients. We outline the three main processing steps to derive well isolated putative single neurons from such recordings: Signal processing, spike detection, and spike sorting. We provide an overview of the state of the art in the acquisition and processing of extracellular recordings with microwires, review a typical experimental setup, spike sorting and detection algorithms. We conclude by providing a step-by-step example, visualizing each intermediate processing step. Together, this chapter provides a practical guide on how to utilize signal processing, spike detection, and spike sorting to derive high-quality single-neuron recordings.

43.1 Introduction

Listening to the activity of individual neurons with extracellular recordings is one of the principal techniques to study the nervous system. Except for rare clinical opportunities, it is not possible to perform such recordings in humans due to their invasive manner. In the rare situations where such recordings are possible, however, they offer the exceptionally valuable opportunity to study the human nervous system at the level of single neurons in awake human beings. These recordings are performed in neurosurgical patients who are undergoing electrophysiological recordings as part of their treatment for a neurological disease or as part of participation in a clinical trial

M. Yebra · U. Rutishauser (✉)
Department of Neurosurgery, Cedars-Sinai Medical Center, Los Angeles, CA 90048, USA
e-mail: Ueli.Rutishauser@cshs.org

U. Rutishauser
Computation and Neural Systems, California Institute of Technology, Pasadena, CA 91125, USA

N. Axmacher (ed.), *Intracranial EEG*, Studies in Neuroscience, Psychology and Behavioral Economics, https://doi.org/10.1007/978-3-031-20910-9_43

for a brain machine interface. In this chapter, we describe the processing steps to go from raw continuous recordings to a series of timestamps of spikes associated with a putative single neuron during an experiment in an awake human being performing a task. Our focus in this chapter is on recordings from hybrid depth electrodes with embedded microwires implanted in patients with suspected focal epilepsy [5, 8, 17, 33]. In such recordings, each neuron is typically only visible on a single recording channel. We therefore focus here on algorithms that sort single channel data and that have been utilized to process such data from humans. The processing pipeline we describe here is applicable also to other recording setups used with human patients with few modifications, including intra-operative microelectrode recordings during DBS implantation [19] and recordings from chronically implanted Utah arrays [30].

43.1.1 Epilepsy Patients

Pharmaco-resistant epileptic patients with suspected focal seizures who are candidates for surgical resection or implantation of a responsive neuromodulation device are sometimes implanted with depth electrodes as part of their surgical workup. These electrodes are then used to monitor their brain activity before and during seizures to help localize the epileptic seizure onset zone (SOZ). Hybrid versions of the depth electrodes contain 30-40um diameter microwires that allow the recording of an extracellular signal from a high-impedance insulated contact that is only exposed at the tip (Fig. 43.1a). Implantation targets vary depending on clinical need but typically include areas within the Medial Temporal Lobe (MTL) such as the hippocampus and amygdala and various areas along the midline of the frontal lobe (Fig. 43.1b).

The experiments take place in the patient's room and require, in addition to the standard clinical recording system, a research electrophysiology data acquisition system. We briefly describe our typical recording setup to provide context for how the data is acquired whose processing is described below (Fig. 43.1c). We use the ATLAS electrophysiology system (Neuralynx Inc.), a laptop to present stimuli to the patient, and a response pad (Cedrus Inc.) to acquire button presses. The laptop used to display the experiment and save the behavioral results is synchronized with the acquisition system via transistor-to-transistor logic (TTL) parallel port connection to the data acquisition system. Microwires are first connected to head stages located on the head (see Fig. 43.1c). Microwire recordings are broadband (0.1 Hz-9 kHz) with a 32,000 Hz sampling rate. This high sampling rate is essential to resolve single neurons with enough datapoints (a typical waveform is ~1 ms long, i.e., 32 datapoints in this case).

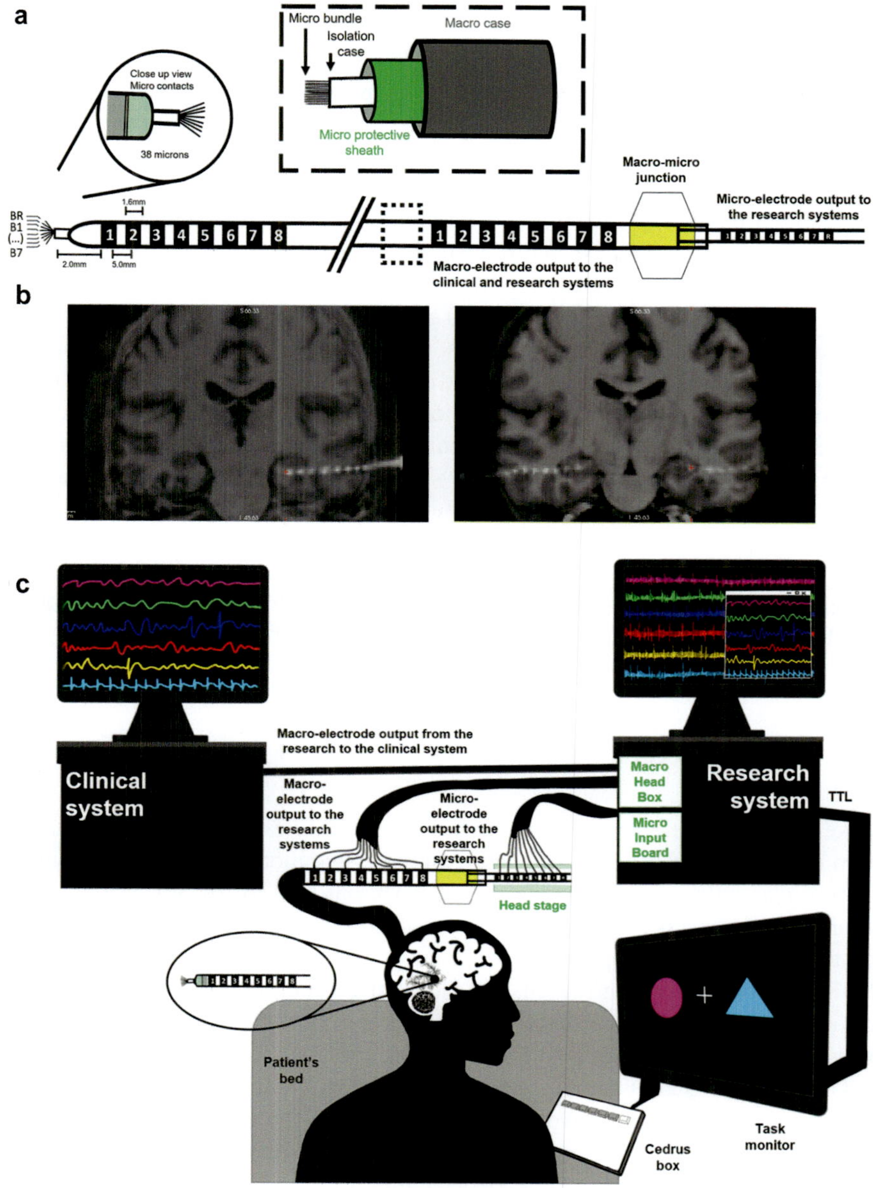

Fig. 43.1 Single unit recordings in human epilepsy patients. **a** Schematic of a typical Behnke-Fried Hybrid Depth (BFD) electrode, containing both macro (iEEG/ECOG) contacts and a bundle of microwires exiting at the tip. The version shown has two 'pig tail' style connectors (right side). (Redrawn similarly to Fig. 1B in [18]) **b** Two examples of patient's co-registered pre-operative MRI and post-operative CT. Note that the distorted electrode trajectories are due to the merge to the pre-operative MRI. Red dot indicates location of microwires. **c** Schematic of a typical experimental setup in the patient's room, consisting of three separate systems

43.1.2 Properties of Microwire Recordings

Extracellular spikes can be recorded from micro-wires that terminate within ~< 100 μm of a neuron [2]. If the group of isolated spikes satisfy certain criteria (see below), this series of spikes is assumed to be the generated by an individual neuron, i.e., Single Unit Activity (SUA). If an isolated group of spikes appears to be originating from multiple neurons, such activity is referred to as Multiunit Activity (MUA). The data we describe in this chapter was obtained using the hybrid macro–micro Behnke-Fried depth electrode (BFD). This electrode is hollow inside, a space which contains a bundle of microwires that are each insulated except for the tip. While there are different possible configurations, we typically use the version that consists of 8 recording wires of 38 μm diameter and one uninsulated reference wire. The wires terminate in a pigtail-type plug that connects to the preamplifiers on the head (see Fig. 43.1a). Typical impedances of intact microelectrodes measured ex-vivo in sterile saline ranged between 50 and 500 kΩ, 20–30 KΩ lower than post-implant measurements in-vivo [18]. This is in line with the in-vivo measurements provided in [32], which showed a range of 38–245 KΩ measured in two patients across a total of 56 electrodes. Several case series have been published that evaluate the safety profile of utilizing hybrid depth electrodes instead of standard depth electrodes. These studies conclude that microwires do not modify the risk profile of depth electrodes [3, 18].

43.2 Spike Sorting

Spike sorting is the computational process that identifies single putative neurons from raw extracellular recordings based on the assumption that the morphology of extracellular spikes differs between different neurons [15, 26, 28, 36]. Spike sorting algorithms use unsupervised clustering to group different spike shapes into different clusters, each of which represents a different putative neuron. If the signal-to-noise (SNR) ratio is not sufficiently high, or if there are too many similar waveform shapes in each recording, it is possible that a given cluster is the result of a merge of neurons from multiple neurons. Here, our focus is on identifying clusters representing single units (see 43.2.6 Quality metrics. for criteria to select the subset of such clusters).

There are several steps involved in spike sorting. The first step is filtering the signal to isolate frequencies of interest. The second step is to detect the points of time at which a spike occurred (spike detection), followed by waveform alignment and feature extraction. Third, the final step is to cluster the detected waveforms into a number of different clusters and to assign each detected spike to one of these clusters (see Fig. 43.2). Most commonly used spike sorting algorithms use all of these steps (see Table 43.1).

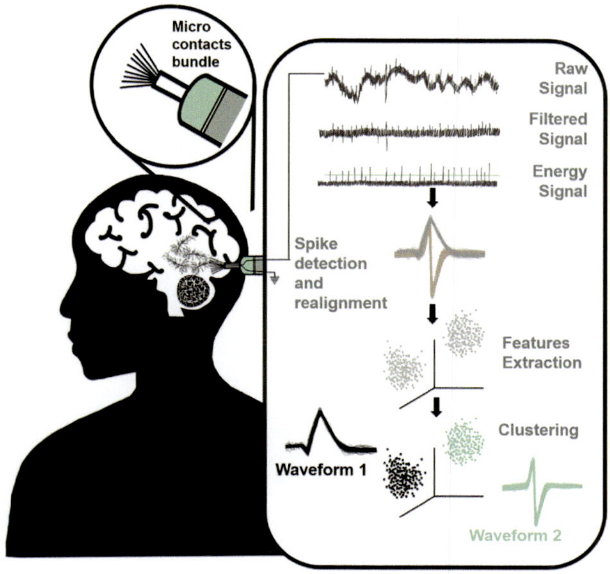

Fig. 43.2 Overview of processing steps involved in extracting single neurons from extracellular recordings in humans. (The data in this figure is for illustration only and is not real data)

43.2.1 Preprocessing

43.2.1.1 Filtering

The first step in processing the microelectrode data is to band-pass filter the raw recorded signal. Typically, the frequencies of interest are 300-3000 Hz (Fig. 43.2, see column 3 in Table 43.1), a range that is determined by the spectral content of the extracellular waveforms generated by the neurons in a given area [2] (note that, in some cases, this range has to be adjusted, for example for the broad waveforms of dopamine neurons [13]. Since the goal is to extract the waveform with as little distortions as possible, it is important that the filter applied does not introduce distortions [27]. For this reason, it is important that a non-causal zero-phase digital filter is used [17].

43.2.1.2 Common Average Reference

A common pre-processing step (before filtering) is to remove the common average of all wires recorded in a given brain area. This step, if applied, is used before spike detection in order to reduce artifacts shared across channels [23]. Also, if recordings are done with a local bipolar reference (as is often done in humans), it is possible that all channels contain the spikes of a single neuron that was near the reference wire.

Table 43.1 Spike sorting algorithms that have commonly been used to process human single-wire recordings

Name	Online	Preprocessing	Automatic	Detection	Waveform features	Cluster	Notes
Osort [36]	Yes	Filtering: 300–3000 Hz (Zero phase filtering when offline)	Yes, but manual curation needed	Thresholding using a multiple of std of the energy signal computed with 1 ms squared kernel	Raw decorrelated waveforms (256 datapoints upsampled at 100 kHz)	Distance between each detected raw signal and the cluster average waveform < T	Can be used online Realignment step before feature extraction with the maximum realigned at position 95 of the 256 points
KlustaKwik [12]	No	High pass filter 800 Hz. Hamming window-based finite impulse response filter, filter order 50	Manual	Thresholding using a multiple of std of the root mean-square power of the filtered signal Resample and realignment	PCA. 12 dimensions feature vector created for each spike by projecting each resampled waveform onto the 3 first components of each channel	Manual cluster cutting using gClust. Draw polygons in a two-dimensional space (2 features) scatter plot where the spikes were projected. By iteratively viewing different projections cluster boundaries can be refined	Initially designed for single wires or tetrodes. Lately adapted for high density (see next row Klusta)

(continued)

Table 43.1 (continued)

Name	Online	Preprocessing	Automatic	Detection	Waveform features	Cluster	Notes
Klusta [29]	No	Filtering: 3rd order Butterworth forward–backward mode 500–95% Nyquist frequency	Manual	Double-threshold flood fill algorithm (Spatio-temporal detection)	3 principal components of spatio-temporal signatures of the detected waveforms. Uses a mask vector to indicate if electrodes are spatially connected and allow temporal overlap in case they are not	GMM with Masked expectation maximization (EM) algorithm (High dimensional classification)	Suited for high density recordings
Wave_klus_v1 [28]	No	Filtering: 4 pole Butterworth 300-6000 Hz	Automatic offline sorting, manual cluster selection	Thresholding using a multiple of an estimate of noise of the amplitude of the filtered signal	(n × k)-array of wavelet coefficients, where n is the number of spikes and k the number of sampling points. 10 out of k dimensions in which the distribution of wavelet coefficients differs most from normality	Superparamagnetic clustering (SPC), single temperature is chosen for clustering	Can deal with polytrode data Wavelet does not require signal stationarity

(continued)

Table 43.1 (continued)

Name	Online	Preprocessing	Automatic	Detection	Waveform features	Cluster	Notes
Wave_klus_v2 [6]	No	Filtering: Zero phase filtering 300-3000 Hz	Yes. Proposed as a fully automated algorithm showing an improvement from previous version. However, manual curation can be done to assess the sorting performance	Thresholding using a multiple of an estimate of noise of the amplitude of the filtered signal	Optimized data-driven selection of the relevant features (i.e., wavelet coefficients). Variable number of wavelet coefficient are chosen with the highest deviation from normality	Superparamagnetic clustering (SPC), more than one temperature is chosen for clustering given that clusters appear at different temperatures. Inclusion criterion since same cluster could appear at different temperatures. Regime border detection to avoid overclustering at higher temperatures	Can deal with polytrode data

(continued)

Table 43.1 (continued)

Name	Online	Preprocessing	Automatic	Detection	Waveform features	Cluster	Notes
ClusterLizard [14]	Yes	First-order elliptic band-pass filter with 0.8 dB ripple 0.3–3.0 kHz	Can be. Manual curation recommended	Thresholding the median of the absolute values of the filtered data	Haar Wavelet decomposition. Number of features equal to the number of data points per whitened spike shape plus the number of wavelet components per non-whitened spike shape. Finally, those that most differ from normality are the selected features	Euclidean distance between each spike and the weighted and normalized features of every cluster. In each iteration clusters are tested for fusion	

In this case, this particular neuron can be isolated and removed from the remaining channels by subtracting the common average and using this common average signal as a virtual channel for spike detection and sorting [36].

43.2.2 Spike Detection

The first step in spike sorting is detecting the points of time at which a spike is present in the recorded signal. Commonly used approaches for single-wire recordings in humans include thresholding of the raw signal, thresholding an energy signal, and wavelet methods. These different methods have different advantages and disadvantages, which we next describe briefly.

To detect spikes using thresholding-based methods, the amplitude of action potentials needs to be larger than the background and the spike waveform needs to be relatively stable over time (which might not be the case in case of bursting or electrode movement). Also, levels of background noise may not be stationary over time. Therefore, the threshold(s) are typically defined as a function of the underlying noise properties in the window of time in which spikes are detected. Also, some methods directly threshold the raw signal, whereas others first compute a surrogate signal that amplifies spikes and suppresses non-spike background before thresholding. In human recordings, the following methods have often been used.

Amplitude of raw signal The simplest method is to set the threshold to a multiple (generally between 4–5) of the estimated standard deviation of the filtered raw signal [28]. Every time this threshold is crossed, a spike is extracted. A problem with this method of defining the threshold is that the threshold will become large for channels with high firing rates (due to the fact that there are many spikes). To avoid this problem, a common method is to define the threshold based on the median amplitude rather than the standard deviation:

$$T = n\ median\left(\frac{|x|}{0.6748}\right) \tag{43.1}$$

where n is a constant (typically n = 4) and the second term is an estimate of the standard deviation of the noise in the voltage x [28, 29, 31] in a relatively large window (several seconds) around the time of spike detection.

Energy signal An alternative method to thresholding the raw signal is to first compute a surrogate signal in which SNR is maximized for the detection of spikes relative to other signals present in extracellular recordings. This is achieved by first computing a surrogate 'energy' signal of the raw signal before applying a threshold [1]. This is commonly done for chronically implanted electrodes that cannot be moved to optimize SNR [12, 17, 34–36]. The overarching idea of the energy signal is to convolve the raw signal with a kernel that is approximately equivalent to a spike waveform, thereby amplifying aspects of the signal that look like spikes and suppressing other parts. Practically, this is implemented as a convolution of the filtered signal with a rectangular kernel with a width of approximately a single action potential (i.e. ~1 ms; this rectangular kernel is a matched filter) [1]. This energy signal (Fig. 43.2 and Fig. 43.3a) is then used to detect spikes using a threshold that is a multiple (normally between 3.5–6) of the standard deviation of the energy signal [36]. Note that in this approach, the energy signal is only used to define the times at which the threshold is crossed. The actual waveform is always extracted from the raw signal and not the energy signal.

Wavelets A more comprehensive (but computationally expensive) method is based on wavelets. Instead of convolving the filtered signal with a rectangular kernel as it is done when computing the energy signal (see above), the filtered signal is

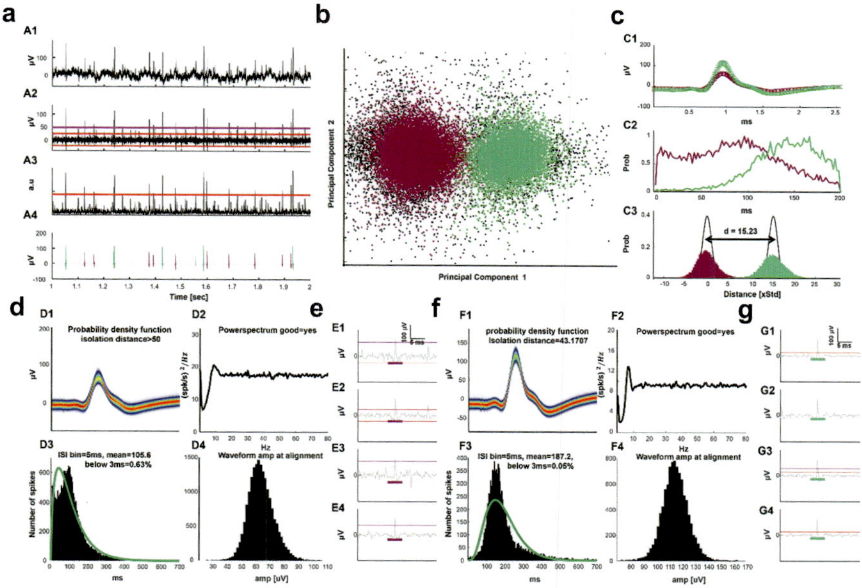

Fig. 43.3 Spike sorting example using OSort on a recording obtained from a hybrid depth electrode implanted in the right ACC. **a** A1 Raw signal. A2 Filtered signal (300–3000 Hz bandpass). A3 Energy signal. The threshold (red) corresponds to 5 std, and the magenta line is the median as specified in Eq. (1). A4 Detected spike times in different colors for different clusters. **b** Plot of the 2 first PCA components of the raw decorrelated waveforms explaining most of the variance. Axes are first and second components in the feature space, every point represents a detected neuron and color represents different clusters as a result of the spike sorting algorithm. Black points are detected waveforms which are not assigned to any of the five clusters (noise or unsortable). **c** The projection test result for the 2 neurons identified in this channel [24]. C1 The mean waveforms for each cluster, error bars pertain to the standard deviation of the waveforms. C2 Superimposed ISI for both neurons. C3 Projection test as the probability distribution of distances between spikes. The distance d is the distance between the two clusters (here, d = 15.23). **d** Neuron 1 summary output of Osort: D1 Spike waveforms. Colormap of realigned signals and isolation distance [39]. D2 Power spectrum of the spike trains. D3 Inter-spike interval (ISI) up to 700 ms. D4 Waveform amplitudes at alignment point probability distributions. **e** Example traces showing four instances of neuron 1 in the filtered signal. **f** and **g** same as panel d and, but for Neuron 2

convolved with multiple wavelet functions that resemble spike shapes at different scales. Wavelet family functions that are biphasic (and thus resemble spikes) and which are often used for this purpose are either biorthogonal wavelets (Bior) [7] or Haar wavelets [6, 22, 26, 28]. After applying the wavelet transform, the resulting coefficients represent a measure of similarity between the recorded signal and the basis function at different scales and translations (which represent different spike amplitudes and durations). If the recorded data contains signal (spikes), the statistical properties of the wavelet coefficients will depend on the structure of the noise. Assuming that the signal is sparse (as is the case for spikes) and assuming white gaussian noise, only a few coefficients will be different from zero (those corresponding to spikes). These relevant coefficients are again detected using thresholding, with the threshold being a function of the noise variance. The detection process is carried out by defining an acceptance threshold for every scale. The time associated with the maximal coefficient is an approximation of the time at which the spike occurred [21].

43.2.3 Spike Alignment

After a spike is detected, the next critical step is to determine the point of time at which to align the detected spike with all the other detected spikes. To do so, a snippet of the recorded signal with enough samples to contain the entire spike (~2.5 ms) is extracted from the filtered raw signal. In this snippet, the alignment point is then defined based on an algorithm. This algorithm, in many instances, simply defines the alignment point as the point of time at which the voltage is maximal (or minimal). However, note that that in human recordings with bipolar reference, the polarity of spikes is arbitrary, and it is therefore a-priori not known whether a spike should be aligned at the maximum or minimum. Due to this, some alignment algorithms first determine whether a given waveform has significant negative or positive peaks and based on this use the minimum or maximum. It is worth nothing thought that this process can fail in the case of spikes which have approximately equal amplitude in the negative and positive direction. In these cases, this algorithm would split the cluster in two due to alignment mistakes. Such cases will require either manually enforcing a single alignment point (negative or positive) or merging of the two clusters. In practice, alignment issues are a reason for a significant proportion of spike sorting issues that require manual intervention. A second difficulty is that the exact time at which the peak of the spike occurred is unknown due to the low sampling rate relative to the speed by which a spike is produced. Therefore, to improve peak finding, it is common to upsample the sampling rate before performing alignment. In OSort, this is achieved by upsampling waveforms to 100 kHz sampling rate after extraction [36].

43.2.4 Feature Extraction

The next step in spike sorting is deciding which features of the detected spikes to use for spike sorting. Methods to do so include selecting a small subset of waveform features directly from the raw waveforms, data-driven computing of the most distinguishing aspects between all the waveforms in a given recording, transformation into a different feature set followed by selection (wavelets), and the raw waveform itself. We summarize each approach briefly next.

Raw waveform or subsets thereof Some algorithms, including OSort, cluster directly in the raw waveform space. Others first extract a small subset of commonly used features to differentiate between different detected extracellular waveforms. These typically include the maximal amplitude, width at half-max, trough-to-peak width, and repolarization time [20]. Clustering is then performed in this relatively low dimensional feature space. Note that because these features are correlated, an additional orthogonalization step is often used before classifying.

Data-driven feature selection using dimensionality reduction An alternative to a priori feature selection is to select features in a data-driven way. Most commonly, a small (<10) number of features are selected by using principal component analysis (PCA). To achieve this, all spike waveforms detected on a channel are pooled, with each waveform a single datapoint in a space that has the same dimensionality as the original waveforms (i.e., 256 datapoints for a 2.56 ms long segment sampled at 100 kHz). PCA is then performed to extract a small number of dimensions that together explain a large proportion of the variance between spikes. Typically, < 10 dimensions will explain > 90% of the variance [12, 23, 29, 36, 40]. A downside of this approach, however, is that it cannot be used for online sorting as defining the feature space requires access to all to-be clustered spikes.

Wavelets Spike sorting can also be performed in the wavelet coefficient feature space rather than the raw space of the waveforms. To do so, after spikes are detected, the wavelet transform is applied to derive wavelet coefficients for each spike at different scales and times. Out of this large number of coefficients, a subset is then selected that best separate the spike waveforms. One approach to do so is to choose a fixed number of coefficients (i.e., the top-most X features that show the largest deviation from normality). This serves as a dimensionality reduction step for clustering [28]. Recently, new spike sorting algorithms have been introduced that automatically select a variable number of coefficients before sorting by identifying all features that have multi-model distributions for a given dataset [6].

43.2.5 Clustering

The final step in spike sorting is to perform unsupervised clustering within the feature space of the detected waveforms to decide how many clusters are present and to assign each detected spike to one of these clusters. While there are many algorithms that have been used for this purpose [43], we here focus on the subset that have been used extensively to automatically or semi-automatically process human microwire data (see Table 43.1).

Pairwise distance-based clustering One class of approaches is to compute the pairwise distance in feature space (e.g., Euclidian Distance, Mahalanobis distance) between all pairs of spikes and to assign spikes to the same cluster that are close to each other based on this metric. Such clustering can either be obtained through batch algorithms such as k-means [4] that require all spikes to be present at the time of clustering, or online iterative algorithms that compute distances only to all spikes detected in the past. An example of the latter is the OSort algorithm [36], which, for every new spike detected, computes the distance to all existing clusters and then either assigns the new spike to one of the existing clusters or creates a new cluster based on a threshold on the maximal distance that is considered to be part of an existing cluster. This threshold can be computed based on the standard deviation of the background noise. After a spike is assigned to an existing cluster, the mean waveform of that cluster is updated. Also, if the distance is less than the threshold for two or more clusters those are merged.

Gaussian Mixture Models (GMM): A second class of clustering algorithms uses maximum-likelihood fitting of mixtures of Gaussians (the most prominent example being KlustaKwik and never versions thereof). GMMs are probabilistic models that assume that the recorded extracellular signal is as a linear sum of the events due to foreground cells and the background noise modeled as Gaussian distributions with independent means, variances, and amplitudes [38]. The first step is selecting the model which implies selecting the number of cells (or Gaussians), followed by the inference process that determines which cells fired at which times and with which waveform shapes. These mixture models can be fit using Expectation Maximization or Variational Bayesian algorithms [42], including automatic determining of the best number of clusters [41].

Superparamagnetic clustering (SPC): A third method that has found extensive use for human single neuron processing is SPC as part of the wave_clus (and derivatives thereof) sorting package [28]. This method is based on simulated interactions of each waveform (in feature space) with its K-nearest neighbors. The interaction strength between a given waveform and one of its K-nearest neighbors is equal to exponentially decaying function of the Euclidian distance between the two datapoints. Clustering is then performed by first initializing each datapoint with a random state 1…q (with q the maximal number clusters), followed by a Monte Carlo algorithm that picks random points and changes their state and that of their k-nearest neighbors with a probability that is a function of the interaction strength and a free parameter referred to as the temperature (a procedure referred to as the Wolf algorithm). The temperature T determines the probability of points changing their cluster identities together, with intermediate values corresponding to the 'superparamagnetic' regime in which only closely interacting points will change identity together. This is the regime used for spike sorting [28]. More recently, an improved version of SPC has been introduced, in which different temperatures can be used for different clusters [6].

43.2.6 Quality Metrics

After clusters have been identified, the last step is to evaluate each cluster to decide whether it represents a putative single neuron or not. Also, at this stage, some clusters will also turn out to be over clustered and thus must be merged. We next describe the key criteria used to evaluate a given cluster.

Stable waveform The detected waveform should remain stable throughout the experiment. If not, this is indicative of electrode movement/drift or spike sorting errors. Stability is assessed by inspecting whether the waveform peak amplitude and width are similar for all the waveforms belonging to the same cluster (see Fig. 43.3d1 and f1).

Stable firing rate Firing rates should remain stable over the course of the experiment (except for transient modulation on relatively short time scales). If firing rates change abruptly or drift slowly, this could be indicative of a sorting problem.

Amplitude of waveform First, the amplitudes at the alignment point of the waveform should follow a unimodal distribution (see Fig. 43.3d4 and f4). If this distribution is multimodal, several clusters were merged into the cluster being inspected (unless the cell is bursty). Second, the standard deviation of the voltage distribution across all waveforms at all points of time should be approximately the same.

Inter-spike histogram Spike trains are often modelled as a Poisson process with spikes generated randomly, independent of each other and with a uniform probability of occurrence in time. Thus, the probability distribution of the Inter Spikes Interval (ISI) can be approximated by an exponential function. However, for real neurons the ISI should be larger than the refractory period (~3 ms, no less than 1 ms) so the ISI probability distribution can be better accounted by a gamma process (see Fig. 3 in [10] for more detail and gamma fits shown in light green in Fig. 43.3d3 and f3 for illustration purposes). Typically, the proportion of ISIs that violate this criterion should be less than 3% for a well isolated unit (see Fig. 43.3d3 and f3).

Autocorrelation of spike trains The autocorrelation captures second order changes in the dynamics of neuronal firing (such as bursts or modulation of firing probability by oscillations). It is defined as the average joint probability density of a spike at time t and $t + \tau$, minus their mean values. Therefore, if there is no violation of the refractory period a "dip" should be visible in the autocorrelation function of the spike train [10]. Thus, an approximately zero (or negative dip) autocorrelation for small (<3 ms) time lags is expected. Also, for long time periods (>100 ms, but this depends on the type of cell) the autocorrelation is expected to return to zero.

Power spectrum of spike trains The power spectrum reflects the frequency content of the spike train and intrinsic properties of the neurons such as the refractory period. It is computed as the Fourier transform of the autocorrelation function (see above). For a Poisson process, the Fourier transform of the autocorrelation yields a positive constant and will usually contain additional terms deviating from a flat spectrum when the spike train differs from Poisson (i.e., gamma). Thus, it normally shows an initial dip and a no return to zero for higher frequencies (for more details see Fig. 14b and Sect. 6.2 in [10] and [9]). Note that depending on the shape of the ISI probability distribution this initial dip can be less evident (i.e., shorter refractory period and broader range of output firing rates, see Figs. 1 and 2 in [9]). Moreover, the power spectrum of the spike train should contain no peaks at line frequency (60 Hz/50 Hz (US/ Europe) and/or harmonics thereof. Also, it should not contain sharp peaks at other frequencies, which are indicative of other types of non-physiological processes such as a periodic TTL artifact or a refresh-triggered noise from the screen (up to 240 Hz) (see Fig. 43.3d2).

Isolation Distance This metric was first introduced by [12] and measures how well separated the spikes in a given cluster are from other spikes recorded on the same multichannel electrode. One way of computing this metric is to calculate the smallest ellipsoid from the cluster center containing all of the cluster spikes and an equal number of noise spikes [39]. Thus, for cluster *Clus* containing n spikes the isolation distance can be computed as the Mahalanobis distance of the nth closest non cluster (*Clus*) spike to the center of the cluster *Clus*:

$$D^2_{i,clus} = (x_i - \eta_{clus})^T \sum_c^{-1} (x_i - \eta_{clus}) \tag{43.2}$$

where x_i is the feature vector for spike *i*, and η_{clus} is the mean feature vector for cluster *Clus*. $\sum_c^{-1}$ is the covariance matrix of spikes in cluster *Clus*. Note that this metric cannot be calculated (it is undefined) for cases in which the target cluster contains more spikes than there are other (noise) spikes. A higher value indicates that non-cluster spikes are located farther away. Note this value can be normalized by cluster size, so that clusters with a higher number of spikes do not necessarily have a higher isolation distance [16, 39].

Projection test to evaluate similarity between two clusters It is important to quantify how similar a pair of two clusters are to assess whether they are sufficiently different or should perhaps be merged. The metric that we have found to be useful for this purpose is the projection test, which was first suggested in [24, 25] and which is implemented as part of OSort [36] (Table 43.1). The overall idea is to, after appropriate normalization, connect the two centers of the clusters with a line in the original feature space and to project all datapoints onto this one-dimensional line. The distributions of the data points associated with these two clusters along this line are then evaluated using a goodness of fit test as well as a distance metric (in units of standard deviations) to assess how far apart the two distributions are. If the distance is too small one or both clusters must be rejected or merged. A distance of > 5 guarantees an overlap of $< 1\%$, a distance > 3.2 an overlap $< 5\%$ and a distance of > 2.8 an overlap of $< 7.5\%$ (see Fig. 43.3 c3).

As part of a well-documented human single neuron paper, it is expected that the quality of the data and the spike sorting is documented by providing histograms of above quality metrics across all clusters included in the data analysis. See Fig. S1 in [37] for an example.

43.2.7 A Practical Example

In this section we provide an example of how to sort and evaluate single neurons using the output figures and result metrics from OSort [36] with data recorded in an epileptic patient while performing a cognitive experiment.

Figure 43.3 depicts subsets of the output figures produced by OSort for a single channel recording from a microwire in the right Anterior Cingulate Cortex (ACC). In Fig. 43.3a1 the raw trace of the recorded signal is shown. Figure 43.3a2 shows the corresponding bandpass filtered trace (300–3000 Hz). Figure 43.3a3 shows the energy signal, which in this instance was used as the detection signal (red line shows the threshold used; see 43.2.2 Spike detection). Note that the energy signal provides better SNR. Figure 43.3a4 shows the detected spikes, colored according to the cluster they were assigned to. Figure 43.3b shows the waveforms in the feature space. The spikes are projected in the two first principal components of the raw decorrelated waveforms, every point represent a detected spike and the color represent a different cluster assignation. As can be seen in the figure clusters corresponding to the two neurons, depicted in green and purple respectively, are very well separable. In Fig. 43.3c1 the two waveforms are plotted together in order to check the actual waveforms assigned to different clusters are indeed different. Figure 43.3c2 shows the ISIs in ms for both identified clusters for easier comparison. In Fig. 43.3c3 the projection test probability density functions are shown [24]. In this plot the pairwise distance between spike counts is plotted for both neurons together with the distance (d) between the two distributions. Distances of > 5 guarantees an overlap of $< 1\%$, a distance > 3.2 an overlap $< 5\%$ and a distance of > 2.8 an overlap of $< 7.5\%$ as long as their distributions, ISIs and waveforms look different enough [36]. Figure 43.3d and f show a subset of the criteria that were used to decide whether the two detected clusters correspond to neurons. First, in Fig. 43.3d1 the waveform traces realigned to the peak (see 43.2.3 Spike alignment) are shown as a color map depicting the probability density function isolation distance. The isolation distance is a spike quality metric described in [39], representing the quality of a cluster of extracellularly recorded spikes by calculating how well separated the spikes in the cluster are from other spikes recorded on the same multichannel electrode (see 43.2.6 Quality metrics.). This distance should be high enough, values greater than $10^{1.5}$ are recommended [39]. (Note that slightly lower thresholds of 20 have been also used in [11]). This plot is useful to check whether the realigned traces are similar, and to verify that the peak amplitudes are similar and have reasonable values (20–200 uV). The power spectrum of the spike train shown in Fig. 43.3d2 can be used to check there are no peaks in the 50–60 Hz and/or harmonics due to electrical noise, other artifactual frequencies such as a TTL artifact or screen refresh artifact. Otherwise, that neuron should be rejected. Fig. 43.3d3 shows the Inter-Spike Interval (ISI) which represents a very important metric for the correct sorting of a neuron. There should be few ISIs lower than 3 ms in order not to violate the refractory period of a neuron. (Note that with this criteria bursty cells might be discarded, also because normally if a neuron fires bursts of spikes they would in addition violate the similar amplitude at the alignment point criterion). Interestingly enough, besides the initial dip in the ISI distribution in both depicted neurons in this example (purple and green) Fig. 43.3d3 and f3, there is also a peak in the distribution between 100-200 ms, possibly reflecting that these neurons fire at a certain theta rhythm. That would also be consistent with the bump observed at theta frequencies in the power spectrum plots. Lastly, in Fig. 43.3d4 and f4, the waveform amplitude at alignment point probability distribution is shown. This should be relatively narrow and unimodal distribution reflecting the waveforms have similar amplitude at alignment point. Figure 43.3e and g are four traces of the filtered signal showing instances of the sorted neuron for each example.

Acknowledgements This work was supported by the National Institutes of Health (U01NS117839 to U.R.).

References

1. Bankman IN, Johnson KO, Schneider W (1993) Optimal detection, classification, and superposition resolution in neural waveform recordings. IEEE Transanctions Biomed Eng 40:836–841
2. Buzsáki G, Anastassiou CA, Koch C (2012) The origin of extracellular fields and currents—EEG, ECoG, LFP and spikes. Nat Rev Neurosci 13:407–420
3. Carlson AA, Rutishauser U, Mamelak AN (2018) Safety and utility of hybrid depth electrodes for seizure localization and single-unit neuronal recording. J Stereotactic functional neurosurgery 96:311–319
4. Caro-Martín CR, Delgado-García JM, Gruart A et al (2018) Spike sorting based on shape, phase, and distribution features, and K-TOPS clustering with validity and error indices. Sci Rep 8:1–28
5. Cash SS, Hochberg LR (2015) The emergence of single neurons in clinical neurology. Neuron 86:79–91
6. Chaure FJ, Rey HG, Quian Quiroga R (2018) A novel and fully automatic spike-sorting implementation with variable number of features. J Neurophysiol 120:1859–1871
7. Daubechies I (1995) Ten Lectures on Wavelets. SIAM Appl Comput Harmon Anal, Philadelphia
8. Engel AK, Moll CK, Fried I et al (2005) Invasive recordings from the human brain: clinical insights and beyond. Nat Rev Neurosci 6:35–47
9. Franklin J, Bair W (1995) The effect of a refractory period on the power spectrum of neuronal discharge. SIAM J Appl Mathmatics 55:1074–1093
10. Gabbiani F, Koch C (1998) Principles of spike train analysis. Methods Neuronal Model 12:313–360
11. Harris KD, Csicsvari J, Hirase H et al (2003) Organization of cell assemblies in the hippocampus. Nature 424:552–556
12. Harris KD, Henze DA, Csicsvari J et al (2000) Accuracy of tetrode spike separation as determined by simultaneous intracellular and extracellular measurements. J Neurophysiol 84:401–414
13. Kamiński J, Mamelak AN, Birch K et al (2018) Novelty-sensitive dopaminergic neurons in the human substantia nigra predict success of declarative memory formation. J Current Biology 28(1333–1343):e1334
14. Knieling S, Sridharan KS, Belardinelli P et al (2016) An unsupervised online spike-sorting framework. Int J Neural Syst 26:1550042
15. Lewicki MS (1998) A review of methods for spike sorting: the detection and classification of neural action potentials. Netw: Comput Neural Syst 9:R53

16. Liu X, Wan H, Shi L (2014) Quality metrics of spike sorting using neighborhood components analysis. Open Biomed Eng J 8:60
17. Minxha J, Mamelak AN, Rutishauser U (2018) Surgical and electrophysiological techniques for single-neuron recordings in human epilepsy patients. In: Extracellular recording approaches. Springer, p 267–293
18. Misra A, Burke J, Ramayya A et al (2014) Methods for implantation of micro-wire bundles and optimization of single/multi-unit recordings from human mesial temporal lobe. J Neural Eng 11:026013
19. Mosher CP, Mamelak AN, Malekmohammadi M et al (2021) Distinct roles of dorsal and ventral subthalamic neurons in action selection and cancellation. Neuron 109(869–881):e866
20. Mosher CP, Wei Y, Kamiński J et al (2020) Cellular classes in the human brain revealed in vivo by heartbeat-related modulation of the extracellular action potential waveform. Cell Rep 30(3536–3551):e3536
21. Nenadic Z, Burdick JW (2004) Spike detection using the continuous wavelet transform. IEEE Trans Biomed Eng 52:74–87
22. Niediek J, Boström J, Elger CE et al (2016) Reliable analysis of single-unit recordings from the human brain under noisy conditions: tracking neurons over hours. PLoS ONE 11:e0166598
23. Pachitariu M, Steinmetz NA, Kadir SN et al (2016) Fast and accurate spike sorting of high-channel count probes with KiloSort. Adv Neural Inf Process Syst 29:4448–4456
24. Pouzat C, Delescluse M, Viot P et al (2004) Improved spike-sorting by modeling firing statistics and burst-dependent spike amplitude attenuation: a Markov chain Monte Carlo approach. J Neurophysiol 91:2910–2928
25. Pouzat C, Mazor O, Laurent G (2002) Using noise signature to optimize spike-sorting and to assess neuronal classification quality. J Neurosci Methods 122:43–57
26. Quiroga (2012) Spike sorting. 22:R45-R46
27. Quiroga (2009) What is the real shape of extracellular spikes? 177:194–198
28. Quiroga NZ, Ben-Shaul Y (2004) Unsupervised spike detection and sorting with wavelets and superparamagnetic clustering. Neural Comput 16:1661–1687
29. Rossant C, Kadir SN, Goodman DF et al (2016) Spike sorting for large, dense electrode arrays. Nat Neurosci 19:634–641
30. Rutishauser U, Aflalo T, Rosario ER et al (2018) Single-neuron representation of memory strength and recognition confidence in left human posterior parietal cortex. Neuron 97(209–220):e203
31. Rutishauser U, Cerf M, Kreiman G (2014) 6 Data analysis techniques for human microwire recordings: spike detection and sorting, decoding, relation between neurons and local field potentials. J Single Neuron Studies of the Human Brain: Probing Cognition:59
32. Rutishauser U, Mamelak AN, Schuman EM (2006) Single-trial learning of novel stimuli by individual neurons of the human hippocampus-amygdala complex. J Neuron 49:805–813
33. Rutishauser U, Reddy L, Mormann F et al (2021) The architecture of human memory: insights from human single-neuron recordings. J Neurosci 41:883–890
34. Rutishauser U, Ross IB, Mamelak AN et al (2010) Human memory strength is predicted by theta-frequency phase-locking of single neurons. Nature 464:903–907
35. Rutishauser U, Schuman EM, Mamelak AN (2008) Activity of human hippocampal and amygdala neurons during retrieval of declarative memories. Proc Natl Acad Sci 105:329–334
36. Rutishauser U, Schuman EM, Mamelak AN (2006) Online detection and sorting of extracellularly recorded action potentials in human medial temporal lobe recordings, in vivo. J Neurosci Methods 154:204–224
37. Rutishauser U, Ye S, Koroma M et al (2015) Representation of retrieval confidence by single neurons in the human medial temporal lobe. Nat Neurosci 18:1041–1050
38. Sahani M, Pezaris JS, Andersen RA (1998) On the separation of signals from neighboring cells in tetrode recordings. Advances in neural information processing systems:222–228
39. Schmitzer-Torbert N, Jackson J, Henze D et al (2005) Quantitative measures of cluster quality for use in extracellular recordings. Neuroscience 131:1–11

40. Shoham S, Fellows MR, Normann RA (2003) Robust, automatic spike sorting using mixtures of multivariate t-distributions. J Neurosci Methods 127:111–122
41. Souza BC, Lopes-Dos-Santos V, Bacelo J et al (2019) Spike sorting with Gaussian mixture models. Sci Rep 9:1–14
42. Takekawa T, Fukai T (2009) A novel view of the variational Bayesian clustering. Neurocomputing 72:3366–3369
43. Veerabhadrappa R, Ul Hassan M, Zhang J et al (2020) Compatibility evaluation of clustering algorithms for contemporary extracellular neural spike sorting. Front Syst Neurosci 14:34

Chapter 44
How Is Single-Neuron Activity Related to LFP Oscillations?

Salman E. Qasim and Lukas Kunz

Abstract Local field potentials (LFPs) can provide biomarkers of cognitive function at the meso-scale level of brain organization. LFPs represent the aggregation of thousands of transmembrane currents from local and non-local brain regions and it is thus still not understood how LFP fluctuations relate to action potentials of single neurons. Because these action potentials represent key units of computation in the human brain it is important to understand how they relate to and interact with LFPs during human cognitive functioning. Using intracranial probes which include both macro- and micro-electrodes, researchers can simultaneously measure synchronous changes in single-neuron action potentials and LFPs with respect to human cognition and behavior. In this chapter, we describe recent advances in recording and analyzing simultaneous single-neuron spiking and LFP oscillations. We provide a practical guide to estimating the relationship between neural spiking, LFP power, and LFP phase—with a specific focus on recent approaches to investigating how different LFP oscillations might modulate the timing of single-neuron spiking. We describe how the relationship between the LFP and single-neuron spiking is thought to be functionally important for representing information in the brain, and suggest that studying this relationship has broad relevance for understanding the neurophysiological mechanisms underlying human cognition.

44.1 Introduction

Action potentials are presumably the most prevalent and fundamental means for neurons to communicate with each other [1]. Action potentials "transfer information" between connected neurons by triggering the release of chemical neurotransmitters from the presynaptic neuron into the synaptic cleft. These neurotransmitters, in turn, may trigger action potentials in the connected, postsynaptic neurons. Recording and

S. E. Qasim
Department of Psychiatry, Icahn School of Medicine at Mt., Sinai, New York, NY, USA

L. Kunz (✉)
Department of Biomedical Engineering, Columbia University, New York, NY, USA
e-mail: drlukaskunz@gmail.com

N. Axmacher (ed.), *Intracranial EEG*, Studies in Neuroscience, Psychology and Behavioral Economics, https://doi.org/10.1007/978-3-031-20910-9_44

quantifying the occurrence of action potentials relative to cognitive stimuli that the subject experiences during an experimental paradigm (such as a memory task) may thus provide a direct measure of neuronal communication and computation with respect to behavior. Using this approach, multiple studies have provided insight into how the human brain processes perceptions, memories, and actions [2].

Single action potentials can be recorded from the human brain by tracking the current flow in the extracellular space using different types of invasive electrodes placed in the immediate vicinity of the neurons [3]. For example, "Behnke-Fried electrodes" have been developed, which feature microelectrodes with a diameter of 40 μm that are inserted through an inner hole of standard intracranial depth electrodes [4] (see Chap. 42 for details on the procedure of microelectrode implantation). These microelectrodes record extracellular potentials with a particularly high spatial and temporal resolution (often more than 30 kHz) and thus allow the identification of action potentials (which are very fast, transient events) in the high-frequency component of the recorded voltage difference. Because the shape of action potentials differs between different neurons depending on the spatial relationship between the microelectrode and the neurons, and because the refractory period of neurons is generally larger than one millisecond, it is possible to assign action potentials recorded on the same microelectrode to different neurons [5–9] (for further information on this topic, see Chaps. 43 and 45).

In addition to the identified action potentials, the slower components of the recorded potential are referred to as the local field potential (LFP), whose physiological basis is complex and not fully understood [10]. Because both signals are embedded in the extracellular potential (Fig. 44.1), many studies have aimed at elucidating the relationship between single-neuron action potentials and the LFP. Major topics of interest concern the question of how the LFP influences the precise occurrence of action potentials and how, in turn, populations of action potentials contribute to the LFP. Another important topic of interest relates to the idea that the brain represents information in the relationship between action potentials and the LFP, above and beyond the information that is represented by the rate of the neuron's action potentials [11–15].

In this chapter, we thus discuss several ways in which the activity of single neurons is associated with the LFP signal and the oscillations that are present in the LFP. Specifically, we will consider the relationship between neuronal action potentials and LFP power, which is the squared amplitude of the LFP, as well as the relationship between neuronal action potentials and LFP phase, which is the momentary deflection of the LFP [15, 16]. We suggest that understanding the exact relationships between single-neuron activity and LFP signals will help us understand the physiological determinants of single-neuron activity in the human brain, and will help us discover the ways in which the human brain represents information about various cognitive stimuli.

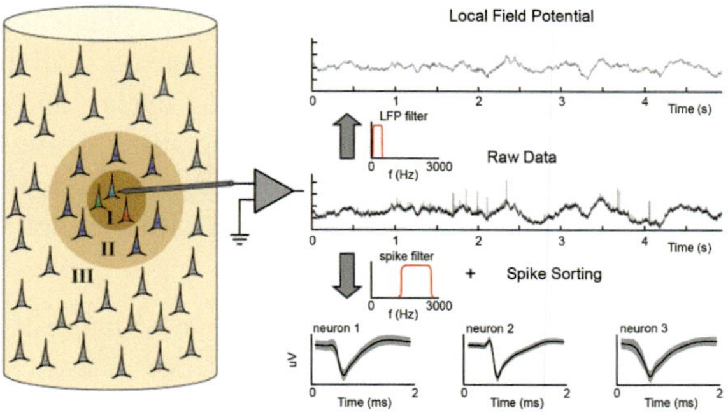

Fig. 44.1 Schematic of extracellular recording and signal separation. Left: Illustration of an extracellular electrode placed in the vicinity of single neurons in the brain. The action potentials of nearby neurons are detectable and separable (I), while the action potentials of neurons a little further away are detectable but not necessarily separable (II), and neurons even further away are not even detectable (III). Right: Schematic of recording and preprocessing of microelectrode data. The raw data are filtered in low (<300 Hz) and high (e.g., 300–3000 Hz) frequencies to yield the signals for LFPs and single-neuron spike sorting, respectively. The right bottom panel shows the action potentials from three different putative neurons recorded on the same microelectrode. Note the distinct waveform shape of each putative neuron. Figure modified with permission from Rey et al. [17]

44.2 Analyzing the Relationship Between Spikes and LFPs

In order to analyze the correlation between action potentials and LFPs, researchers must first perform several preprocessing steps, including (a) filtering the raw data to separate the LFP and the high-frequency signal containing possible action potentials, (b) detecting individual action potentials in the high-frequency signal and assigning these action potentials to distinct single neurons (or multi-units), and (c) characterizing the time-varying spectral power and phase of the LFP signal at different frequencies of interest. These individual steps are covered in practical detail in Chap. 43 of this book, have been extensively discussed in previous publications (for example, Refs. [5, 18, 19]), and various toolboxes have been developed to aid researchers to perform these steps [5, 18–22]. Here, we will pick up after these preprocessing steps to describe recent approaches to analyzing the relationship between spike times and neural oscillations. To begin, all of the following analyses will require two initial steps:

1. Estimate the time-varying spectral power or phase in a frequency band of choice (extensively described elsewhere [23]). Briefly, the most common methods for doing so utilize:

   - A bandpass filter followed by a Hilbert transform. Power can then be extracted as the squared amplitude and phase as the angle of the analytic signal.

- A continuous wavelet transform, which provides instantaneous power and phase by convolving the signal with a brief template oscillation (often a Morlet wavelet). By adjusting the number of wavelet cycles, an optimal balance between time and frequency resolution can be obtained for the data at hand.
- A short-time Fourier transform, which produces instantaneous measures of power and phase by computing a Fourier transform on short data segments using a sliding-window approach.

2. Isolate the spike train for the putative single neuron of interest. This results in a sequence of spike times, which is typically recorded as a vector of zeros (no action potentials within a given time bin) and ones (presence of action potentials within a given time bin) the same length as the epoch of interest.

The different methods for estimating power and phase each have their own pitfalls in terms of computational intensity, time-frequency resolution, and flexibility in regards to waveform shape and signal stationarity. In general, researchers should not adopt a one-size-fits-all approach to power and phase estimation and instead endeavor to identify the method most appropriate for their dataset, as improper power and phase estimates may lead to artifactual spike-LFP correlations. When filters are applied to the data, filter characteristics should be reported in detail and a plot of the impulse response should be included in the publication [24].

44.2.1 Computing the Relationship Between Spiking and Spectral Power

Single-neuron spiking can be examined with regard to how it covaries with features of the LFP power spectrum, such as the power within specific frequency bands or the overall shape of the power spectra. Time series such as the LFP are often formalized as the superposition of component signals of different frequencies. Measuring power at specific frequencies provides a measure of "how much" signal there is at a particular frequency. Meanwhile, single-neuron spiking is often described by the "firing rate" of a neuron, or how many action potentials are recorded over a period of time. As such, the most straightforward relationship between spiking rate and the LFP is to compute the correlation between power and firing rate as follows:

1. Bin the epoch of interest into equal-sized temporal bins, typically in the range of 100–500 ms, compute the firing-rate in each bin (n spikes/bin duration), and z-score the firing rates relative to a baseline epoch or relative to the entire experimental session.
2. Compute the average spectral power (at a particular frequency) in these temporal bins.
3. Compute the correlation between firing rate in each bin with spectral power in each bin.

This approach has been useful in demonstrating both positive and negative correlations between neuronal spiking rates and delta/theta power in the human auditory cortex (negative association) [25], alpha power in the monkey sensorimotor regions (negative) [26], beta power in monkey motor regions (either positive or negative) [27], and gamma power in the monkey somatosensory cortex and in various regions of the human brain (positive) [28, 29]. However, researchers should note that no singular relationship between LFP power and firing rates has been determined [15]. Instead, the magnitude and polarity of spiking-power correlations can vary even within individual neurons [27] or across different cortical depths [30]. Furthermore, recent work has shown that power within narrow, predefined frequency bands does not necessarily capture the most informative dimensions of the LFP signal. For example, human studies have shown that the overall power estimated across the broadband power spectrum (2–150 Hz) exhibits the strongest correlation with local neuronal firing rates [29]. By employing new methods that systematically parameterize the relative contribution of narrowband peaks and aperiodic spectral shifts to the overall power spectrum [31], future studies may help us to further clarify the exact relationship between spiking activity and LFP power and to characterize how neuronal spiking correlates with the full suite of features characterizing the LFP power spectrum.

44.2.2 Computing the Relationship Between Spiking and Oscillatory Phase

Neural spiking can also be examined with regard to how it covaries with the phase of the LFP at different frequencies. Phase refers to the momentary deflection of the LFP and can vary between 0° and 360°. Accordingly, the phase of a given action potential describes its within-cycle "position" on the LFP waveform. By decomposing the LFP into its different frequency components, we can measure the phase of the signal at different frequencies as a function of time. Upon identifying the phase, researchers can then compute whether the timing of action potentials is systematically aligned with particular phases of the LFP or whether action potentials occur more frequently at some phases than others. Two prominent phenomena are phase locking and phase precession, which are described below.

It is worth noting a few caveats that one should take into account when estimating phase. First, phase estimates are difficult to interpret in the absence of a meaningful oscillation. Thus, it is often worth computing the power in a given frequency before computing the phase estimate. Accordingly, certain methods only utilize phase estimates from data where power exceeds a certain baseline threshold [32, 33], and several toolboxes have been developed that estimate phase only during periods when power exceeds the 1/f background spectrum [13, 34]. Furthermore, Hilbert methods for phase estimation are most useful for stationary, sinusoidal signals, which is not always true of LFP signals. If analyzing non-sinusoidal LFP data, we recommend estimating phase using other approaches such as linear interpolation methods which make no assumptions about waveform shape or instantaneous frequency [35]. Finally, researchers should be careful to implement rigorous permutation testing protocols (described in this chapter) to ensure that erroneous spike-phase correlations do not emerge from phase slips or edge effects common to instantaneous phase estimation.

Phase locking Phase locking describes the phenomenon that a neuron activates at the same phase of an oscillation across cycles, a phenomenon prevalent in both animals [36–38] and humans [13, 16, 18]. Phase locking is presumably of fundamental relevance for neural computation by determining whether a stimulus will be remembered or forgotten [18, 39], by gating the transmission of action potentials (and, thus, the flow of information) between connected brain regions [36–38], and by coordinating the simultaneous arrival of multiple different inputs to a target region. Phase-locking is distinct from the spike-power correlation (as described above) because it measures the timing of spike occurrence relative to the LFP phase, not the overall increases in firing rate as a function of oscillatory power. Spike phase-locking can be computed as follows:

1. Align the spike train and LFP phase estimate for the frequency of interest in the time-domain. This should allow you to assign each spike a concurrent LFP phase estimate. It is often worth plotting segments of the phase estimate and the spike train together to verify that each spike is assigned a valid phase estimate.
2. Using these vectors of spike phases, compute the degree to which spiking tends to occur at specific phases of the LFP. The most common methods for doing so are:
    - Compute the Rayleigh statistic [16, 18].
    - Compute the spike-field coherence [40].
    - Compute the phase-locking value [41].
    - Compute the pairwise-phase consistency (PPC) [42].
3. To assess statistical significance, compare the computed value against an empirically derived null distribution drawn from surrogate data with matched spike counts:
    - Circularly rotate one signal with respect to the other to generate each of many surrogate datasets (e.g., >1000 surrogate datasets).
    - Compute the strength of phase-locking for each surrogate dataset.
    - Z-score the phase-locking metric for the original data against the distribution of surrogate metrics.
4. Identify the specific phase that the neuronal spiking is locked to by plotting a histogram of spike phases and by calculating the circular mean of the phases at which the spikes occurred [43].

These steps form the basis for a majority of studies measuring spike-phase coupling. It is critical to emphasize the importance of performing permutation statistics to determine significance—many, if not all, of these methods are biased by the spike rate or low sample size. Other unbiased methods have emerged to deal with this issue, relying on spike-triggered average LFPs [44] or machine-learning methods [45], which may be compared to the approaches described above.

Phase precession Phase precession describes the intriguing phenomenon that a neuron activates at successively earlier phases of consecutive cycles of an ongoing oscillation. Phase precession has been extensively described in place cells relative to the local theta rhythm [46]. Place cells are neurons that encode spatial information [47] by increasing their firing rates whenever the subject navigates across a particular location in the spatial environment (the so-called "place field" of a given place cell). Phase precession has also been observed in grid cells [38], which are another type of neurons that is important for spatial navigation and spatial memory [48]. Theta-phase precession may provide a means to encode spatial information above and beyond the information contained in the cells' firing rate [12, 49] and presumably plays an important role in the formation of memories for sequences of events [33, 50, 51].

Phase precession has typically been identified by computing the circular-linear relationship between the spike phase (which is the circular variable) and a linear variable of interest such as location in space as follows:

- Circular-linear method from Kempter et al. [52]:

 1. Align the spike train and phase estimate for the frequency of interest, and assign a phase to each spike.
 2. Align the spike phases to the linear variable of choice on each trial/epoch (e.g., position).
 3. Fit a circular-linear model to the data to estimate the slope and angular offset.
 4. Scale the linear variable with respect to the absolute value of the slope and take the result modulo 2π to transform the linear variable into a circular variable.
 5. Compute the circular-circular correlation between the spike phase and the transformed linear variable [43].

However, this approach depends on a linear metric of spike progression (such as spike position within the place field of a place cell). A different method was thus developed, which estimates phase precession without depending on a linear behavioral variable or assumptions about spike timing:

- Spike-phase autocorrelation method from Mizuseki et al. [53]:

 1. Unwrap the spike-phase vector to make it a linear variable.
 2. Compute the autocorrelation for the unwrapped spike-phase vector. The phase-bin size and window length can be manually determined as a function of the available number of spikes.
 3. Estimate the power spectrum of the frequency of spiking relative to the LFP by computing the Fourier transform of the spike-phase autocorrelation histogram.
 4. Spectral peaks at relative frequencies greater than 1.0 indicate that the cell is oscillating at a faster frequency than the reference LFP, which is the key property of phase precession.

5. To assess statistical significance of this peak, measure its height against the background spectrum relative to an empirically derived null distribution drawn from surrogate data with matched spike counts.
   - Circularly shuffle the spike-phases within each cycle of the reference LFP signal to disrupt cross-cycle dynamics while maintaining other spiking dynamics.
   - Compute the strength of phase precession for the surrogate dataset.
   - Z-score the height of the spectral peak of the original data against the distribution of surrogate metrics.

It is important to note that this method requires enough spike phases to compute a reasonable autocorrelation (on the order of >100 spike phases). Otherwise, it can be difficult to properly detect rhythmicity in the autocorrelation unless it is highly stereotyped. Note that this method can also be used to identify spike-phase locking, if the power spectra estimated from the spike-phase autocorrelograms exhibit peaks at relative frequencies equal to 1.0. However, given that this method relies on a relatively high number of spikes, it is not our recommended technique for computing spike-phase locking. Also, while all methods described here are largely meant to assess phase precession relative to oscillations at theta frequency, other studies have assessed phase precession relative to higher-frequency signals [54, 55].

44.3 Relevance for Human Behavior and Cognition

44.3.1 Spike-Power Associations During Human Cognition

Several studies examined the relationship between neuronal spiking and LFP power while human participants performed cognitive tasks. For example, while subjects passively viewed popular English-speaking videos, neuronal spiking in the auditory cortex was found to correlate positively with LFP power in frequencies above 30 Hz, but negatively with LFP power in frequencies below 30 Hz [25] (see also Chap. 21). This generally positive relationship between spiking activity and gamma LFP power was supported by follow-up analyses examining both periods of audio-visual sensory stimulation and spontaneous activity [56]. This study also pointed out, however, that the relationship between single-neuron firing and LFP gamma power is considerably mediated by the strength of firing-rate correlations between neighboring neurons, indicating that there can be situations in which the activity of single neurons is actually uncoupled from population activity reflected in the LFP [56]. Other research examined the relationship between neural spiking and LFP power during a virtual navigation task, showing that LFP power in the delta-, theta-, and alpha-frequency ranges are generally associated with decreased firing rates, whereas LFP power in the beta- and gamma-frequency ranges are associated with increased firing rates [29] (Fig. 44.2a). In addition, as described above, this study also demonstrated that neuronal spiking is in fact best predicted by broadband shifts in the LFP power spectrum and not by power changes in narrow frequency bands [29].

Future studies are needed to further clarify to what extent the changes in spike-power associations are mediated by particular cognitive processes or stimuli that the subjects experience during the tasks, particularly since the association between spiking activity and gamma power has been shown to vary considerably over time [56]. For example, a recent study showed that the positive correlation between neuronal spiking and LFP gamma power is more strongly pronounced when human subjects experience aversive emotional stimuli as compared to neutral emotional stimuli [57]. Although a large majority of the temporal variability may be traced back to between-neuron firing-rate correlations—meaning that higher spike-gamma correlations occur when the contributing neurons are more synchronized [56]—additional variance may be explained by the experimental conditions. Examining the exact relationship between spike-power associations and cognitive processes may provide critical information as to whether the correlation between spiking and LFP power simply represents a useful proxy for local population spiking activity, or whether these two signals can encode distinct information. A previous study, which examined single-neuron and LFP responses to specific items (i.e., virtual characters and landmarks) during a virtual navigation task, already pointed in this direction: item-specific responses in single-neuron activity appeared to be independent from item-specific responses in LFP theta and gamma power, recorded from the human hippocampus and entorhinal cortex [58]. Recent work utilizing laminar probes that span different cortical depths suggest that such dissociation between spiking and LFP power may rely on the depth and timing of the LFP [30], illustrating the need for future studies to carefully characterize the spatiotemporal profile of the LFP signal.

Additionally, spike-power associations may be important for human cognition by promoting spike-phase associations (Fig. 44.2b) (whose importance is described below). In particular, bursts of LFP oscillations (i.e., periods in which LFP power is particularly high and thus leads to easily identifiable oscillations in particular LFP frequencies) can induce transient spike-phase locking, which in turn is thought to be important for behavior and cognition [16], but may also reflect pathological processes: for example, bursts of beta oscillations in Parkinson's disease temporarily increase spike-phase locking in the subthalamic nucleus and are correlated with hypokinetic behavior [59]. Furthermore, elevated power in the gamma frequencies can reflect transient bursts of high-frequency LFPs such as ripples [60] (Fig. 44.2c). Ripples induce spike-phase locking and spike-phase precession and are thus thought to be highly important for cognitive functions including memory, learning, and planning [55, 60–62].

In addition, examining spike-power associations is not only relevant from the basic physiological perspective but also with respect to disease processes, because pronounced spike-power associations may constitute a cause or consequence of neurological disorders. In addition to the findings in Parkinson's patients described above, in epileptic patients, high-frequency oscillations are prevalent in seizure-generating brain regions and recruit increased neuronal spiking [63–67]. These increases in spiking-power associations in the high-frequency range are thus indicative of patho-

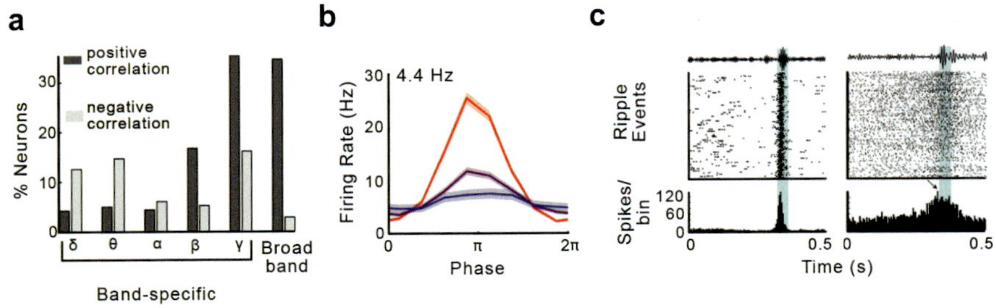

Fig. 44.2 Examples of spike-power correlations during human cognition. **a** Percentage of neurons showing significant correlation with power in different frequency bands of the LFP, as well as broadband (2–150 Hz) LFP power, while subjects performed a virtual navigation task. **b** Example of a neuron exhibiting phase-locking at 4.4 Hz during a virtual navigation task. Note that the neuron shows strongest phase locking during periods of high oscillatory power (red) as compared to periods of medium (purple) and low (blue) oscillatory power. **c** Examples of ripple-aligned spike histograms showing an increase in neuronal firing during high-frequency LFP bursts (80–200 Hz). Figures modified with permission from Manning et al. [29], Jacobs et al. [16] [Copyright 2007 Society for Neuroscience], and Le Van Quyen et al. [63]

logical neural activity and may explain some of the cognitive deficits observed in these patients [68, 69]. Future studies may examine whether high-frequency oscillations are particularly pathological when they lead to stronger firing-rate increases as compared to when the firing-rate increases are only weakly expressed.

44.3.2 Spike-Phase Associations During Human Cognition

While human participants performed virtual navigation tasks, strong phase locking of neuronal spiking to local oscillations (i.e., oscillations recorded on the same microelectrodes as the neurons) occurred in various regions of the human brain including the medial temporal lobe as well as frontal and parietal cortices [16] (Fig. 44.3a). Spike-phase locking appeared to be particularly pronounced with respect to theta and gamma frequencies. Interestingly, spikes locked to various oscillatory phases in the theta-frequency range, whereas they preferentially locked to the oscillatory troughs in the gamma-frequency range [16]. This observation is in line with theoretical models on theta-phase coding [70, 71], which suggest that information about cognitive stimuli or cognitive states is not only encoded in the firing rate of the neurons but also in the phase of the theta oscillations at which the neurons' action potentials occur [13, 72, 73]. Furthermore, spike-phase locking in the human hippocampus and amygdala has been described during successful memory formation [18] and when human subjects hold multiple working memories in mind [73]. By means of intraoperative recordings, another study found that single neurons lock to local theta and beta rhythms in the subthalamic nucleus while human subjects processed sensorimotor conflicts [74].

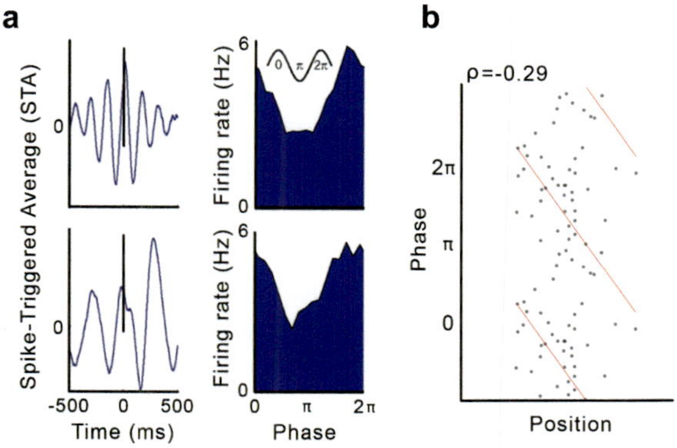

Fig. 44.3 Examples of phase locking and phase precession in humans performing memory and spatial navigation tasks. **a** Left: spike-triggered average LFPs in the human brain demonstrating that action potentials (black line) occur at the same phase of an ongoing oscillation over time. Time 0 denotes the timing of the spikes. Note the clear rhythmicity of the averaged LFP in the theta-frequency range. Right: histograms of firing-rate as a function of oscillatory phase also exhibit clear preferences for spiking at specific phases of the ongoing oscillations. **b** Recent evidence from humans demonstrates that phase precession is also present in the human hippocampal formation during virtual navigation. Figures modified with permission from Jacobs et al. [16] [Copyright 2007 Society for Neuroscience] and Qasim et al. [33]

In addition to local spike-phase locking, researchers also examined distant spike-phase locking, in which single neurons lock to neural oscillations recorded from electrodes implanted in distant brain regions. For example, neurons in various temporal lobe regions have been shown to phase-lock to neural oscillations in the hippocampus while subjects performed a virtual navigation task [75]. Furthermore, single neurons in the subthalamic nucleus lock to distant cortical gamma during movement preparation [76] as well as to both low- and high-beta-frequency oscillations in the motor cortex [77]. Together, these studies show that both local and distant spike-phase locking is a widespread phenomenon throughout the human brain—presumably playing essential roles in neural processing and information transfer during various types of human cognition. In addition, spike-phase locking has been described to vary as a function of disease symptoms in Parkinson's [78], suggesting that understanding spike-phase locking in the human brain may also help us understand some of the consequences neurological diseases have on fine-tuned neuronal mechanisms. It should be noted that spike-phase locking, as discussed here, is not thought to arise solely from independent, self-sufficient rhythmicity in the neuronal spiking, but instead due to a modulatory relationship between neuronal spiking and the LFP.

Moreover, based on influential findings in animal studies [32, 38, 46, 79], recent human single-neuron studies provided evidence that not only phase locking but also phase precession exists in humans [33, 80, 81]. Phase precession occurs preferentially relative to theta oscillations and because theta oscillations in the human brain are less prominent and less stable than in the rodent hippocampus [82], theta-phase precession is less evident in humans than in animals but may nevertheless exert important functions in human cognition. For example, phase precession was demonstrated to occur in spatially-modulated neurons of the human hippocampus and entorhinal cortex during virtual navigation, as well as in (non-spatially modulated) neurons of the human anterior cingulate cortex, orbitofrontal cortex, amygdala, and hippocampus when subjects searched for specific goal locations [33] (Fig. 44.3b).

Furthermore, phase precession was identified in human time cells, which represent specific time points during time intervals [80]. These time cells presumably help to temporally structure different events in a larger episode and phase precession in these cells may help to interconnect temporally neighboring events. Another human single-neuron study showed that phase precession occurs in hippocampal and entorhinal neurons that code for specific visual stimuli when subjects learned a fixed sequence of pictures [81]. When analyzing the neurons' preferred visual stimuli as well as the pictures that preceded and followed the preferred visual stimuli in the fixed sequence, the authors observed that action potentials of these neurons occurred at different theta phases for the preferred, preceding, and following pictures. Consistent with phase precession, action potentials occurred at progressively earlier phases as the sequence advanced from the preceding over the preferred to the following stimulus [81]. Together, these studies thus provide first empirical evidence that phase precession occurs in the human brain and that it may specifically support human sequence learning, which has long been suggested as an important cognitive function of phase precession [46, 50, 51].

44.4 Conclusion

In this chapter, we described techniques and analysis strategies for assessing the relationship between two primary types of signals acquired by direct neural recordings in humans: neuronal spiking and LFPs. Despite the important advances that have been made on this topic in recent years, much is still unknown about the different phenomena that can be observed when considering the relationship between neuronal spiking and LFPs. Based on studies in animals, we can make specific predictions what kind of spike-field phenomena may be associated with particular cognitive functions in humans (e.g., phase precession during sequence learning), but future studies are needed to systematically investigate the generalizability of observations from rodents to humans. Furthermore, adapting technological advances from animal models for human use, such as high-density microelectrode arrays [83] or laminar recordings [30], might provide the improved spatiotemporal resolution and signal-to-noise ratio required to rigorously determine how different neuronal dynamics correlate with or contribute to LFP fluctuations (see also Chap. 46). In this way, nascent methods or electrode technologies may provide new breakthroughs in understanding the relationship between spikes and LFPs in the human brain.

Acknowledgements L.K. received funding from the German Research Foundation (DFG; KU 4060/1-1; Projektnummer 447634521).

References

1. Kandel E, Schwartz J, Jessell T. Principles of neural science
2. Fried I, Rutishauser U, Cerf M, et al. Single neuron studies of the human brain: probing cognition
3. Renshaw B, Forbes A, Morison BR (1940) Activity of isocortex and hippocampus: electrical studies with micro-electrodes. J Neurophysiol 3(1):74–105
4. Fried I, Wilson C, Maidment N et al (1999) Cerebral microdialysis combined with single-neuron and electroencephalographic recording in neurosurgical patients. J Neurosurg 91:697–705
5. Quiroga RQ, Nadasdy Z, Ben-Shaul Y (2004) Unsupervised spike detection and sorting with wavelets and superparamagnetic clustering. Neural Comput 16:1661–1687
6. Rutishauser U, Mamelak A, Schuman E (2006) Single-trial learning of novel stimuli by individual neurons of the human hippocampus-amygdala complex. Neuron 49(6):805–813
7. Knieling S, Niediek J, Kutter E et al (2017) An online adaptive screening procedure for selective neuronal responses. J Neurosci Methods 291:36–42. https://doi.org/10.1016/j.jneumeth.2017.08.002
8. Chaure FJ, Rey HG, Quian Quiroga R (2018) A novel and fully automatic spike-sorting implementation with variable number of features. J Neurophysiol 120(4):1859–1871. https://doi.org/10.1152/jn.00339.2018
9. Buccino AP, Hurwitz CL, Garcia S et al (2020) Spikeinterface, a unified framework for spike sorting. Elife 9. https://doi.org/10.7554/eLife.61834
10. Buzsaki G, Anastassiou C, Koch C (2012) The origin of extracellular fields and currents—eeg, ecog, lfp and spikes. Nat Rev Neurosci 13:407–419
11. O'Keefe J, Recce ML (1993) Phase relationship between hippocampal place units and the EEG theta rhythm. Hippocampus 3:317–30
12. Huxter J, Burgess N, O'Keefe J (2003) Independent rate and temporal coding in hippocampal pyramidal cells. Nature 425:828–832
13. Watrous AJ, Miller J, Qasim SE, et al (2018) Phase-tuned neuronal firing encodes human contextual representations for navigational goals. eLife 7:e32554
14. Tingley D, Alexander AS, Quinn LK, et al (2018) Multiplexed oscillations and phase rate coding in the basal forebrain. Sci Adv 4(8):eaar3230. https://doi.org/10.1126/sciadv.aar3230
15. Kunz L, Maidenbaum S, Chen D et al (2019) Mesoscopic neural representations in spatial navigation. Trends Cogn Sci 23(7):615–630. https://doi.org/10.1016/j.tics.2019.04.011
16. Jacobs J, Kahana MJ, Ekstrom AD et al (2007) Brain oscillations control timing of single-neuron activity in humans. J Neurosci 27(14):3839–3844
17. Rey HG, Pedreira C, Quian Quiroga R (2015) Past, present and future of spike sorting techniques. Brain Res Bull 119(Pt B):106–17. https://doi.org/10.1016/j.brainresbull.2015.04.007
18. Rutishauser U, Ross I, Mamelak A et al (2010) Human memory strength is predicted by theta-frequency phase-locking of single neurons. Nature 464(7290):903–907
19. Niediek J, Boström J, Elger CE et al (2016) Reliable analysis of single-unit recordings from the human brain under noisy conditions: tracking neurons over hours. PLoS One 11(12):e0166598. https://doi.org/10.1371/journal.pone.0166598
20. Delorme A, Makeig S (2004) EEGLAB: an open source toolbox for analysis of single-trial EEG dynamics. J Neurosci Methods 134:9–21
21. Tadel F, Baillet S, Mosher JC et al (2011) Brainstorm: a user-friendly application for meg/eeg analysis. Comput Intell Neurosci 2011:879716. https://doi.org/10.1155/2011/879716
22. Oostenveld R, Fries P, Maris E et al (2011) Fieldtrip: open source software for advanced analysis of meg, eeg, and invasive electrophysiological data. Comput Intell Neurosci 2011:156869. https://doi.org/10.1155/2011/156869
23. Cohen MX. Analyzing neural time series data: theory and practice issues in clinical and cognitive neuropsychology
24. de Cheveigné A, Nelken I (2019) Filters: When, why, and how (not) to use them. Neuron 102(2):280–293. https://doi.org/10.1016/j.neuron.2019.02.039

25. Mukamel R, Gelbard H, Arieli A et al (2005) Coupling between neuronal firing, field potentials, and fMRI in human auditory cortex. Science 309(5736):951–954
26. Haegens S, Nácher V, Luna R et al (2011) Alpha-oscillations in the monkey sensorimotor network influence discrimination performance by rhythmical inhibition of neuronal spiking. Proc Natl Acad Sci U S A 108(48):19377–82. https://doi.org/10.1073/pnas.1117190108
27. Canolty RT, Ganguly K, Carmena JM (2012) Task-dependent changes in cross-level coupling between single neurons and oscillatory activity in multiscale networks. PLoS Comput Biol 8(12):e1002809. https://doi.org/10.1371/journal.pcbi.1002809
28. Ray S, Crone N, Niebur E et al (2008) Neural correlates of high-gamma oscillations (60–200 Hz) in Macaque local field potentials and their potential implications in electrocorticography. J Neurosci 28(45):11526
29. Manning JR, Jacobs J, Fried I et al (2009) Broadband shifts in local field potential power spectra are correlated with single-neuron spiking in humans. J Neurosci 29(43):13613–13620
30. Leszczyński M, Barczak A, Kajikawa Y, et al (2020) Dissociation of broadband high-frequency activity and neuronal firing in the neocortex. Sci Adv 6(33):eabb0977. https://doi.org/10.1126/sciadv.abb0977
31. Donoghue T, Haller M, Peterson EJ et al (2020) Parameterizing neural power spectra into periodic and aperiodic components. Nat Neurosci 23(12):1655–1665. https://doi.org/10.1038/s41593-020-00744-x
32. Eliav T, Geva-Sagiv M, Yartsev MM et al (2018) Nonoscillatory phase coding and synchronization in the bat hippocampal formation. Cell 175(4):1119-1130.e15. https://doi.org/10.1016/j.cell.2018.09.017
33. Qasim SE, Fried I, Jacobs J (2021) Phase precession in the human hippocampus and entorhinal cortex. Cell 184(12):3242-3255.e10. https://doi.org/10.1016/j.cell.2021.04.017
34. Watrous AJ, Buchanan RJ (2020) The oscillatory reconstruction algorithm adaptively identifies frequency bands to improve spectral decomposition in human and rodent neural recordings. J Neurophysiol 124(6):1914–1922. https://doi.org/10.1152/jn.00292.2020
35. Cole S, Voytek B (2019) Cycle-by-cycle analysis of neural oscillations. J Neurophysiol 122(2):849–861. https://doi.org/10.1152/jn.00273.2019
36. Siapas A, Lubenov E, Wilson M (2005) Prefrontal phase locking to hippocampal theta oscillations. Neuron 46:141–151
37. Sirota A, Montgomery S, Fujisawa S et al (2008) Entrainment of neocortical neurons and gamma oscillations by the hippocampal theta rhythm. Neuron 60(4):683–697
38. Hafting T, Fyhn M, Bonnevie T et al (2008) Hippocampus-independent phase precession in entorhinal grid cells. Nature 453(7199):1248–52. https://doi.org/10.1038/nature06957
39. Pavlides C, Greenstein YJ, Grudman M et al (1988) Long-term potentiation in the dentate gyrus is induced preferentially on the positive phase of theta-rhythm. Brain Res 439:383–387
40. Jarvis MR, Mitra PP (2001) Sampling properties of the spectrum and coherency of sequences of action potentials. Neural Comput 13(4):717–49. https://doi.org/10.1162/089976601300014312
41. Lachaux JP, Rodriguez E, Martinerie J et al (1999) Measuring phase synchrony in brain signals. Hum Brain Mapp 8(4):194–208
42. Vinck M, Lima B, Womelsdorf T et al (2010) Gamma-phase shifting in awake monkey visual cortex. J Neurosci 30(4):1250
43. Berens P (2009) Circstat: a matlab toolbox for circular statistics. J Stat Softw 31(10)
44. Li Z, Cui D, Li X (2016) Unbiased and robust quantification of synchronization between spikes and local field potential. J Neurosci Methods 269:33–8. https://doi.org/10.1016/j.jneumeth.2016.05.004
45. Zarei M, Jahed M, Daliri MR (2018) Introducing a comprehensive framework to measure spike-lfp coupling. Front Comput Neurosci 12:78. https://doi.org/10.3389/fncom.2018.00078
46. O'Keefe J (1993) Hippocampus, theta, and spatial memory 3:917–924
47. O'Keefe J, Dostrovsky J (1971) The hippocampus as a spatial map: preliminary evidence from unit activity in the freely-moving rat. Brain Res 34:171–175

48. Hafting T, Fyhn M, Molden S et al (2005) Microstructure of a spatial map in the entorhinal cortex. Nature 436:801–806. https://doi.org/10.1038/nature03721
49. Jeewajee A, Barry C, Douchamps V et al (2014) Theta phase precession of grid and place cell firing in open environments. Philos Trans R Soc Lond B Biol Sci 369(1635):20120532. https://doi.org/10.1098/rstb.2012.0532
50. Jaramillo J, Kempter R (2017) Phase precession: a neural code underlying episodic memory? Curr Opin Neurobiol 43:130–138. https://doi.org/10.1016/j.conb.2017.02.006
51. Reifenstein ET, Bin Khalid I, Kempter R (2021) Synaptic learning rules for sequence learning. eLife 10:e67171
52. Kempter R, Leibold C, Buzsáki G et al (2012) Quantifying circular-linear associations: hippocampal phase precession. J Neurosci Methods 207(1):113–124
53. Mizuseki K, Sirota A, Buzsáki G (2009) Theta oscillations provide temporal windows for local circuit computation in the entorhinal-hippocampal loop. Neuron 64:267–280
54. Lansink CS, Meijer GT, Lankelma JV et al (2016) Reward expectancy strengthens ca1 theta and beta band synchronization and hippocampal-ventral striatal coupling. J Neurosci 36(41):10598–10610. https://doi.org/10.1523/JNEUROSCI.0682-16.2016
55. Bush D, Ólafsdóttir HF, Barry C et al (2022) Ripple band phase precession of place cell firing during replay. Curr Biol 32(1):64-73.e5. https://doi.org/10.1016/j.cub.2021.10.033
56. Nir Y, Fisch L, Mukamel R et al (2007) Coupling between neuronal firing rate, gamma LFP, and BOLD fMRI is related to interneuronal correlations. Curr Biol 17(15):1275–1285
57. Fedele T, Tzovara A, Steiger B et al (2020) The relation between neuronal firing, local field potentials and hemodynamic activity in the human amygdala in response to aversive dynamic visual stimuli. Neuroimage 213:116705. https://doi.org/10.1016/j.neuroimage.2020.116705
58. Ekstrom AD, Viskontas I, Kahana MJ et al (2007) Contrasting roles of neural firing rate and local field potentials in human memory. Hippocampus 17(8):606–17
59. Baaske MK, Kormann E, Holt AB et al (2020) Parkinson's disease uncovers an underlying sensitivity of subthalamic nucleus neurons to beta-frequency cortical input in vivo. Neurobiol Dis 146:105119. https://doi.org/10.1016/j.nbd.2020.105119
60. Buzsáki G (2015) Hippocampal sharp wave-ripple: a cognitive biomarker for episodic memory and planning. Hippocampus 25(10):1073–188. https://doi.org/10.1002/hipo.22488
61. Joo HR, Frank LM (2018) The hippocampal sharp wave-ripple in memory retrieval for immediate use and consolidation. Nat Rev Neurosci 19(12):744–757. https://doi.org/10.1038/s41583-018-0077-1
62. Vaz AP, Inati SK, Brunel N et al (2019) Coupled ripple oscillations between the medial temporal lobe and neocortex retrieve human memory. Science 363(6430):975–978. https://doi.org/10.1126/science.aau8956
63. Le Van Quyen M, Bragin A, Staba R et al (2008) Cell type-specific firing during ripple oscillations in the hippocampal formation of humans. J Neurosci 28(24):6104–10. https://doi.org/10.1523/JNEUROSCI.0437-08.2008
64. Jacobs J, Staba R, Asano E et al (2012) High-frequency oscillations (hfos) in clinical epilepsy. Prog Neurobiol 98(3):302–15. https://doi.org/10.1016/j.pneurobio.2012.03.001
65. Truccolo W, Ahmed OJ, Harrison MT et al (2014) Neuronal ensemble synchrony during human focal seizures. J Neurosci 34(30):9927–44. https://doi.org/10.1523/JNEUROSCI.4567-13.2014
66. Smith EH, Merricks EM, Liou JY et al (2020) Dual mechanisms of ictal high frequency oscillations in human rhythmic onset seizures. Sci Rep 10(1):19166. https://doi.org/10.1038/s41598-020-76138-7
67. Guth TA, Kunz L, Brandt A et al (2021) Interictal spikes with and without high-frequency oscillation have different single-neuron correlates. Brain 144(10):3078–3088. https://doi.org/10.1093/brain/awab288
68. Holmes GL (2015) Cognitive impairment in epilepsy: the role of network abnormalities. Epileptic Disord 17(2):101–16. https://doi.org/10.1684/epd.2015.0739
69. Matsumoto JY, Stead M, Kucewicz MT, et al (2013) Network oscillations modulate interictal epileptiform spike rate during human memory. Brain 136(8)

70. Lisman J, Idiart MA (1995) Storage of 7 ± 2 short-term memories in oscillatory subcycles. Science 267:1512–1515
71. Lisman JE, Jensen O (2013) The theta-gamma neural code. Neuron 77(6):1002–1016
72. Watrous AJ, Fell J, Ekstrom AD et al (2015) More than spikes: common oscillatory mechanisms for content specific neural representations during perception and memory. Curr Opin Neurobiol 31:33–39
73. Kamiński J, Brzezicka A, Mamelak AN et al (2020) Combined phase-rate coding by persistently active neurons as a mechanism for maintaining multiple items in working memory in humans. Neuron 106(2):256-264.e3. https://doi.org/10.1016/j.neuron.2020.01.032
74. Zavala B, Damera S, Dong JW et al (2017) Human subthalamic nucleus theta and beta oscillations entrain neuronal firing during sensorimotor conflict. Cereb Cortex 27(1):496–508. https://doi.org/10.1093/cercor/bhv244
75. Schonhaut DR, Ramayya AG, Solomon EA, et al (2020) Single neurons throughout human memory regions phase-lock to hippocampal theta. bioRxiv. https://www.biorxiv.org/content/early/2020/07/01/2020.06.30.180174
76. Fischer P, Lipski WJ, Neumann WJ et al (2020) Movement-related coupling of human subthalamic nucleus spikes to cortical gamma. Elife 9. https://doi.org/10.7554/eLife.51956
77. Lipski WJ, Wozny TA, Alhourani A et al (2017) Dynamics of human subthalamic neuron phase-locking to motor and sensory cortical oscillations during movement. J Neurophysiol 118(3):1472–1487. https://doi.org/10.1152/jn.00964.2016
78. Shimamoto SA, Ryapolova-Webb ES, Ostrem JL et al (2013) Subthalamic nucleus neurons are synchronized to primary motor cortex local field potentials in parkinson's disease. J Neurosci 33(17):7220–33. https://doi.org/10.1523/JNEUROSCI.4676-12.2013
79. Skaggs WE, McNaughton BL, Wilson MA et al (1996) Theta phase precession in hippocampal neuronal populations and the compression of temporal sequences. Hippocampus 6:149–172
80. Umbach G, Kantak P, Jacobs J et al (2020) Time cells in the human hippocampus and entorhinal cortex support episodic memory. Proc Natl Acad Sci U S A 117(45):28463–28474. https://doi.org/10.1073/pnas.2013250117
81. Reddy L, Self MW, Zoefel B et al (2021) Theta-phase dependent neuronal coding during sequence learning in human single neurons. Nat Commun 12(1):4839. https://doi.org/10.1038/s41467-021-25150-0
82. Jacobs J (2014) Hippocampal theta oscillations are slower in humans than in rodents: implications for models of spatial navigation and memory. Philos Trans R Soc B: Biol Sci 369(1635):20130304
83. Paulk AC, Yang JC, Cleary DR et al (2021) Microscale physiological events on the human cortical surface. Cereb Cortex 31(8):3678–3700. https://doi.org/10.1093/cercor/bhab040

Chapter 45
How Can We Use Simultaneous Microwire Recordings from Multiple Areas to Investigate Inter-Areal Interactions?

Juri Minxha and Jonathan Daume

Abstract In the past few decades, significant progress has been made in understanding human cognition using intracranial electrophysiological recordings. Studies in this body of literature have focused on task-aligned tuning of spiking activity and features extracted from the local field potential (LFP). More recently, leveraging simultaneous and multi-site microelectrode recordings, it has become feasible to study interactions and information flow between multiple brain regions. Electrophysiological measures of coordination between brain areas are typically based on correlations of LFP-derived features, spike-timing, or some combination of the two. Any fluctuation in these measures has been interpreted to reflect changes in the unidirectional or bidirectional transmission of information between the two brain areas. Our goal in this chapter is to review: (1) the experimental evidence for inter-areal communication in microelectrode recordings, focusing on two test-cases of relevance, and (2) current methodological approaches and their possible (technical) limitations. The objective is to underscore the utility and importance of making such measurements in neural data, while cautioning against some common interpretational pitfalls and recommending validation approaches to avoid them.

45.1 Introduction

If there is one consensus emerging about brain function it would probably have to be that *everything is everywhere.* By that, we mean that most cognitive function is supported by the coordination of many distributed brain areas. This fact underscores

---

J. Minxha (✉) · J. Daume
Department of Neurosurgery, Cedars-Sinai Medical Center, Los Angeles, CA, USA
e-mail: jminxha@caltech.edu

J. Minxha
Division of Biology and Biological Engineering, California Institute of Technology, Los Angeles, CA, USA

Center for Theoretical Neuroscience, College of Physicians and Surgeons, Columbia University, New York, NY, USA

N. Axmacher (ed.), *Intracranial EEG*, Studies in Neuroscience, Psychology and Behavioral Economics, https://doi.org/10.1007/978-3-031-20910-9_45

the importance of neural recordings which simultaneously sample many different anatomical regions. In the animal literature, the number of neurons and regions sampled simultaneously has increased dramatically in the past decade (see [1, 2] as examples of state of the art). In human intracranial electrophysiology however, coverage of the brain is typically sparse and determined entirely by clinical need. And yet, the opportunity occasionally arises to record simultaneously from multiple brain regions. In such situations, it is possible to identify neural correlates of behavior that are not readily apparent with intra-regional analysis. For example, recent work has shown that the coupling of ripple oscillations between the medial temporal lobe (MTL) and neocortex predicts the successful retrieval of episodic memories [3, 4]. Our own work has shown that context-dependent decision-making and memory retrieval depends on the functional coupling between the medial frontal cortex (MFC) and the MTL [5].

Inter-areal communication can be discussed in the context of many kinds of recording technologies (ex. ECoG, sEEG), experimental setups (ex. chronic epilepsy recordings, intraoperative recordings), methodological approaches (ex. spike-field coherence, granger causality), and scientific questions (ex. memory, decision making, navigation). Our goal in this chapter is not to provide a comprehensive list of the literature on inter-areal interaction, but rather to demonstrate its feasibility, advantages, and potential pitfalls through case studies based on our own work. While we will discuss inter-areal interactions within this limited scope, the methodological approaches outlined generalize to any experimental configuration. We begin by discussing two recently published case studies where inter-areal communication played a central role, followed by an introduction to the theory and practical considerations of some of the popular metrics used in measuring inter-areal communication.

45.2 Case Studies

Ad-Tech Behnke-Fried depth electrodes [6] are commonly used for monitoring neural activity from epileptic patients. These electrodes allow for simultaneous low-impedance recordings from contacts on the shank of the depth electrode as well as high-impedance recordings from a bundle of microwires present at the tip of the electrode. Figure 45.1a shows an image of the most medial macro contact of the depth electrode, the fiber bundle of microwires protruding at the tip, and a sample recording of the local field potential from the latter. Any oscillations in the LFP can be picked up on the microwires, as shown in Fig. 45.1b. Event-related responses vary quite broadly across these two different types of signals (i.e., macro vs. micro, Fig. 45.1c) and although both have unique properties, in this chapter we will focus on the signals that are recorded from the microelectrodes.

Across the whole brain, there can be more than a hundred individual microwire contacts, with 8 contacts per brain area. Typical anatomical targets include amygdala and hippocampus in the MTL, dorsal anterior cingulate cortex (dACC) and pre-supplementary motor area (preSMA) in the MFC, and ventromedial prefrontal cortex

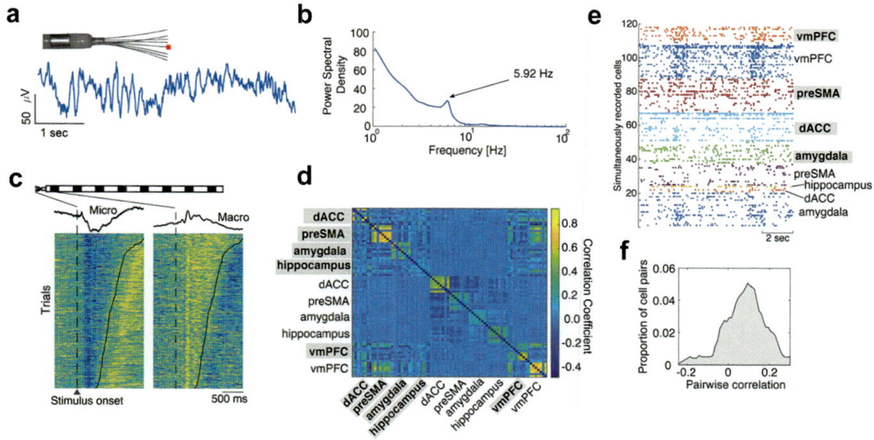

Fig. 45.1 Simultaneous recording of LFP and spiking activity across many brain regions. **a** Example hippocampal LFP recording from an Ad-Tech Behnke-Fried microelectrode. **b** Power spectrum of the recording in (A) showing a peak in the theta (θ) band. **c** Example trial-averaged response of a recording on a micro- and the most medial macro-electrode. **d** Correlation matrix across all areas (8 electrodes per area, n = 80 electrodes). Bold lettering with gray background indicates that the recording was from the right brain hemisphere. **e** A 10 s snapshot of simultaneously recorded cells across all brain areas, during a memory retrieval task. **f** Distribution of pairwise neuronal correlations between vmPFC and amygdala in the session shown in (E)

(vmPFC). While electrode placement depends exclusively on clinical relevance, recordings are *typically* done in both left and right hemispheres.

The recordings are locally referenced (bipolar), allowing us to explore the correlational structure across all brain areas without worrying about things like volume conduction (Fig. 45.1d). In addition to LFP, microwires also record spiking activity. With an average of 1 cell/electrode, we can record the activity of tens of cells simultaneously across all brain areas. Figure 45.1e shows a 10 s snippet of 118 simultaneously recorded cells. Notice that there is a lot of structure in the spike timing of the cells across all areas of the brain; the volleys of activity correspond to instances when the subject is presented with a visual stimulus. Given the simultaneity of the recordings, it is possible to compute neuronal correlations across brain areas (Fig. 45.1f).

These types of recordings allow us to use a variety of methods for measuring inter-areal interactions, including *field-field*, *spike-field*, and *spike-spike* metrics. For a more extensive treatment of the intracranial technique discussed here, see [7] and Chaps. 42–44. In what follows, we will show how this experimental configuration has enabled us to study inter-areal interactions to address specific scientific questions.

a. **Task-dependent modulation of functional connectivity between medial temporal lobe and medial frontal cortex**

A defining characteristic of intelligence is the ability to switch between different tasks with relative ease. Two key aspects of such cognitive flexibility are: (1) the retrieval of contextually relevant information from memory, and (2) the ability to selectively utilize relevant information depending on task demands. Both aspects depend on the MFC, the MTL, and functional interactions between them. We used the experimental setup described in the previous section to study the interaction between these brain areas as subjects alternate between two different tasks, an object recognition (i.e., categorization) task, and a memory retrieval task.

The subjects performed these tasks in blocks of trials and were given instructions only once at the beginning of the block; therefore, they had to *remember* what task they were performing going into each trial. Figure 45.2a shows the general structure of the task. We measured spiking activity and LFP simultaneously in the MFC and MTL (Fig. 45.2b). While there are many measures of inter-areal neural interactions, we focused here on just two, spike-field coherence (SFC), and field-field cross-correlations. A more detailed description of these metrics is available in the next section (see also Chap. 44). To avoid any issues related to event-related potentials (ERPs) induced by the presentation of a visual stimulus, we analyzed the neural activity in the baseline period of the trial (i.e., before stimulus presentation).

Since we recorded spiking activity and LFP in both the MFC and MTL, there are number of possible configurations for computing SFC. For example, we can

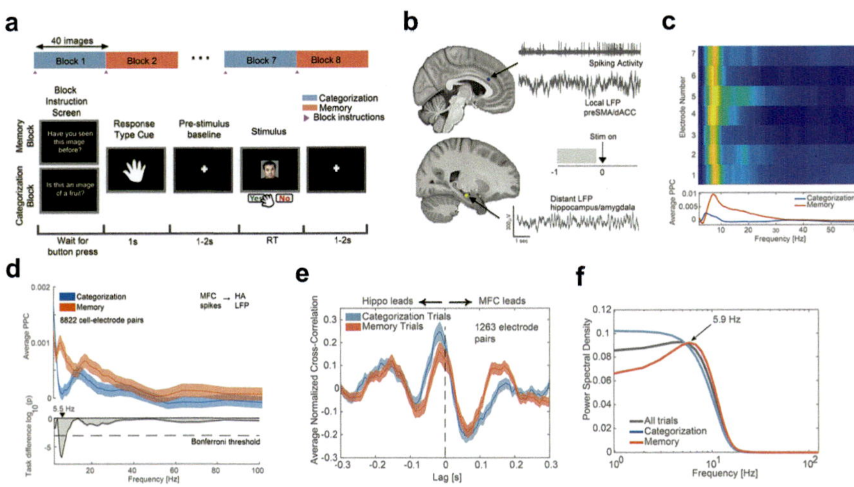

Fig. 45.2 A task-switching paradigm to highlight inter-regional brain synchrony. **a** Task switching paradigm in which subjects alternated between two tasks, (1) object recognition/categorization, and (2) memory retrieval. **b** As the subjects perform the task, we measure field-field and spike-field interactions between two brain regions believed to support memory retrieval, medial frontal cortex, and hippocampus/amygdala in the medial temporal lobe. **c** Example cell in the dACC showing strong SFC with hippocampal theta oscillations. Since there are many microwires in ipsilateral hippocampus, we compute SFC with each one separately. The top panel shows the average SFC across all microwires, separated as a function of task. **d** Average SFC across all cell-electrode pairs recorded in the task. **e** Average cross-correlation across 1263 hippocampus-MFC electrode pairs, split by task type. **f** Cross-correlation spectra, showing a peak in the theta band for the memory task. Panels A-D reproduced with permission from [5]

compute SFC using spiking activity in MFC and *distal* LFP recorded in MTL. Each configuration allows us to observe a different phenomenon and has its own scientific interpretation. In our study, we found that MFC cells were strongly phase-locked to hippocampal LFP, especially in the theta (3–8 Hz) band. Note that since there are many remote electrodes, we get one estimate of the SFC for all possible cell-electrode pairs. Figure 45.2c shows the SFC of a single MFC cell with 7 *distal* electrodes in the ipsilateral hippocampus. Recall that while there are 8 contacts in a bundle, here we are using one of them as the reference.

In the example shown in Fig. 45.2c, the SFC looks very similar across all cell-electrode pairs. This is not always the case however, and the results depend on the degree to which the neural recordings across all microwires in a bundle are correlated with each other. Once the procedure for measuring SFC was established, we investigated the degree to which it was modulated by the task that the subject was performing. Task-based differences can be seen in individual cells (Fig. 45.2c, bottom panel) and across the entire population, as shown in Fig. 45.2d.

We also corroborated our SFC results using a field-field approach, by computing the cross-correlation across all electrode pairs in the MFC and MTL. Much like the SFC, the average cross-correlation shows strong modulation by task (Fig. 45.2e). There are two features of the average cross-correlation that stand out. First, unlike the SFC, the cross-correlation allows for some measure of directionality of the interaction between MFC and MTL[1]; the peak of the average shown in Fig. 45.2e is at approximately −7 ms, which indicates that activity in the hippocampus precedes that in the MFC. Second, the average cross-correlation reveals a 3–8 Hz oscillation, suggesting that the coherence between MFC and MTL is predominantly driven by theta oscillations (Fig. 45.2f). Both the spike-field coherence and the cross-correlation provide converging evidence for the role of theta in mediating task-dependent communication between MFC and MTL. It is clear from these results that this channel of communication is enhanced when the task at hand requires memory retrieval.

There are several considerations in performing this kind of field-field and spike-field analysis which we have not mentioned so far. First and foremost, the way the microwires are referenced can have profound effects on the results. Local (i.e., bipolar) referencing is perhaps the most conservative when computing inter-area interactions, because it eliminates artifacts that might be introduced by a common reference as well as any concerns related to volume conduction. There is however a penalty for using this referencing scheme. Depending on the degree to which the microwire bundle splays out upon surgical insertion, the neural signals picked up by microwires may be highly correlated (i.e., the wires did not separate sufficiently). As a result, referencing the activity $V(t)$ on a microwire with what is approximately a copy, $V'(t)$, can diminish the underlying signal. The second consideration in doing this kind of analysis is electrode selection. Artifact removal and rejection of electrodes with interictal epileptiform discharges (IEDs) is of paramount importance for

[1] SFC can also be used to infer directionality, but for this, we must account for the contribution of local (i.e., spikes and LFP are recorded within the same region) coherence to the overall measured values.

interpretable results. The third consideration has to do with comparing inter-areal interactions across conditions. This requires proper balancing of trials, spikes, and the firing rates of the cells in the two conditions.[2] Further considerations which are outside the scope of this chapter include (a) computing inter-areal interaction metrics with the contralateral brain area as a validation method, (b) generating proper null distributions for statistical tests, and (c) estimating *within-condition* variability for a given metric, by resampling trials.

b. **Saccade-related neural communication between the hippocampus and amygdala in the human brain**

In the previous study, we measured inter-areal interactions between MFC and MTL as a function of task. In that case, changes in the underlying functional connectivity between brain areas occurred on the timescale of a few minutes, which is roughly the duration of a single block of trials. In this section, we will see an example where we measured changes in inter-areal interactions on the timescale of hundreds of milliseconds. The goal of the study described here, was to understand how unconstrained visual sampling (i.e., eye movements) coordinate neural communication between brain areas.

To this end, we asked patients to scan arrays of images containing objects sampled from a small set of visual categories: cars, fruits, fractals, flowers, human faces, and monkey faces. Figure 45.3a (left) shows an example of such an array, containing eight images placed at equal distance from the center of the screen. Subjects saw approximately fifty such arrays, each for four seconds. They were instructed to study the images within so that they could perform a memory recognition task at the very end (Fig. 45.3a, right). Subjects sampled the images around the array freely, averaging about 10 fixations/array (Fig. 45.3c, d). As they studied the images, we recorded activity in the amygdala and hippocampus simultaneously (see Fig. 45.3b for electrode locations). We expected the interaction between these areas to be coordinated by the eye movements, and the strength to be modulated by the visual category of the fixated item.

Figure 45.3e shows LFP snippets recorded in the amygdala (green) and hippocampus (yellow), aligned with the initiation of a saccade. The period of analysis is marked in gray. Aligning the analysis window 200 ms after the onset of the saccade avoids oculomotor artifacts which are frequently present in the LFP shortly after the initiation of a saccade. In our recordings, these artifacts are *strongly* attenuated because of our bipolar referencing scheme; given the proximity of the microwires, the artifacts look very similar across all of them, making them easy to remove by using one of the electrodes as the reference.

We used two methods to measure neural interactions, (1) frequency resolved imaginary coherence [10], which removes potential volume conduction effects by eliminating zero-phase lag coupling, and (2) granger causality [11, 12]. After grouping

[2] Note that balancing firing rates and number of spikes across conditions is not the same thing. Spike balancing simply requires equalizing the number of spikes (by downsampling for example) across conditions, whereas balancing firing rates means the exclusion of cells that have a statistically significant firing rate modulation across conditions.

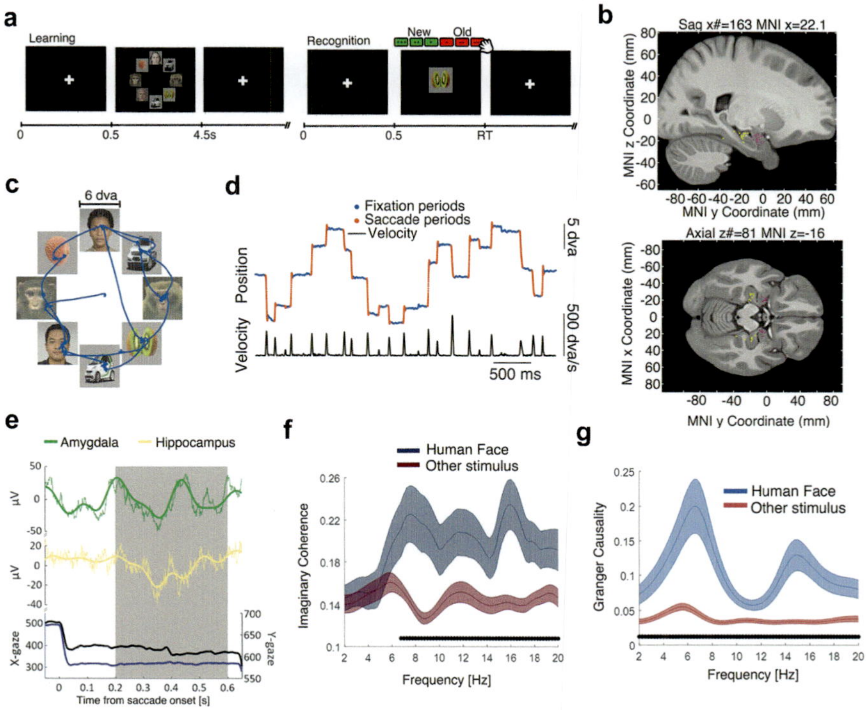

Fig. 45.3 Saccade-aligned synchrony between hippocampus and amygdala depends on the social relevance of fixated item. **a** Example of a learning through free-viewing (left) and new/old recognition (right) trial. **b** Recording locations in the amygdala (pink) and hippocampus (yellow). Each dot indicates the location of a wire bundle in a single patient, projected onto the CIT168 brain atlas in MNI coordinates. **c** Example scan path of a subject's eye movements during free viewing. **d** The horizontal position and velocity of the subject's eye movements. Periods in red indicate saccades and periods in blue indicate a fixation period. **e** Example, saccade aligned LFP traces from the amygdala and hippocampus. Shown at the bottom is the horizontal (x) and vertical (y) position of the subject's eye as a function of time. **f** Increased amygdala-hippocampus iCoh for human faces vs. all other stimuli. **g** Granger causality when fixating on human faces vs. other stimuli. Shown in this plot is amygdala driving hippocampus (i.e., A → H, peaks: 6.5 and 15 Hz). See [8, 9] for the complete set of results from this project. Reproduced with permission from [9]

fixations by the visual category of the underlying image, both the imaginary coherence (Fig. 45.3f) and the granger causality analysis (Fig. 45.3g) revealed preferential interaction between amygdala and hippocampus for human faces. Furthermore, the results from the granger causality suggest that the direction of interaction is from amygdala to hippocampus and not the other way around. Note that each method has its respective advantages; the imaginary coherence analysis reduces the possibility of effects due to volume conduction, whereas the granger analysis provides a measure of the directionality of the effect.

As in the previous section, we will end with a few additional considerations and notes of caution. First and foremost, we reiterate the important role of the referencing scheme and the removal of artifacts (i.e., IEDs) during the preprocessing of the LFP traces. Secondly, we were careful in filtering which saccades were included in the analysis, using only those trials that were free of subsequent saccades and blinks in the post-saccadic time window (200–600 ms) where the analysis was done. Lastly, we want to stress that each method for measuring inter-areal connectivity has its advantages and disadvantages, and to use converging evidence from multiple metrics to draw robust and reproducible conclusions.

45.3 Current Methodological Approaches and Their Limitations

Inter-areal interactions can be measured at various scales, from neural populations to the level of individual neurons. Most commonly, human intracranial studies of interactions between brain areas have focused on measuring the synchronization between LFPs (see also Chaps. 32 and 33). The LFP recorded on the microwires is thought to represent the summed activity of nearby neurons [13, 14]. Known as *field-field* interactions, these kinds of measurements are thought to indirectly reflect interactions between populations of neurons. With the ability to record LFPs as well as spiking activity from individual cells, microwire recordings enable us to study more detailed levels of interaction (see also Chap. 44). Of particular importance, mixed-modality *spike-field* metrics capture the level of coordination between spiking activity and LFP, recorded locally or distally (i.e., inter-areal). Lastly, *spike-spike* methods measure the degree of coordination in the spike timing of two simultaneously recorded cells. In this section, we review common methodological approaches used to assess degree of connectivity in microwire recordings.

a. **Field-field interactions**

The LFP, as measured by microwires, represents the summed synaptic activity of hundreds to thousands of neurons adjacent to the recording electrode [13, 14]. The broadband LFP signal may consist of the summed activity from multiple neural populations, each producing a different oscillatory signature [15]. A common practice in characterizing inter-areal interactions is to first filter the LFP within a certain frequency range in each area, thereby extracting population activity with specific spectral properties, and then measure synchrony between brain areas within the filtered traces. Inter-area field-field metrics are diverse and can be broadly grouped into a few categories depending on what features of the LFP are used. Many studies focus on inter-areal *phase synchronization* among neural populations, while others compute *amplitude correlations, cross-correlations,* or *cross-frequency coupling.* Insights gained from these methods reveal the different forms of communication that exist among neural populations [16–18].

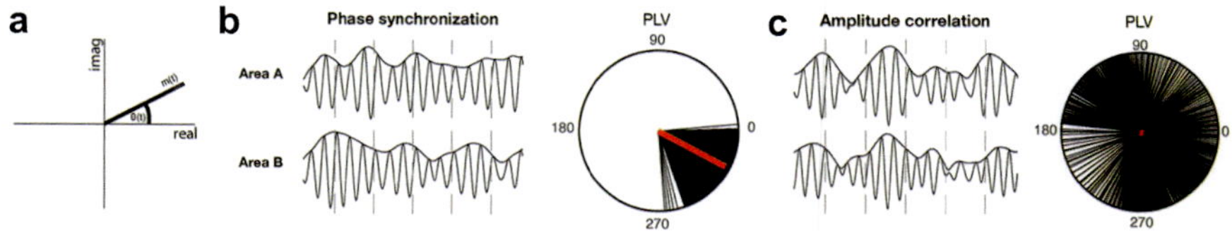

Fig. 45.4 Within-frequency field-field interactions. **a** Phase and amplitude of a signal **x** at time **t** can be represented by a vector in complex space, where the amplitude refers to the length m(t) and the phase to the angle θ(t) of the vector. **b** Schematic illustration of phase synchronization: the phases of the LFP signals from area A and B are highly synchronized over time as indicated by the dashed lines. Due to the high consistency of the phase differences between the signals across all time samples, the average vector (red) of the phase differences has a length close to 1. The amplitudes (i.e., envelope) of the signals, however, are only weakly correlated. **c** In this example, the phases of the signals are not synchronized, resulting in a length of the average vector close to 0. The signals, however, strongly correlate in their amplitude fluctuations

i. *Phase synchronization*

Neural interaction between two LFPs is often assessed by how tightly the phases of their rhythmic, frequency-specific activity are synchronized over time. This form of interaction is usually referred to as *phase synchronization* (Fig. 45.4b) [19, 20–23]. Phase synchronization can be quantified using a metric called the phase-locking value (PLV) [24], which estimates how consistent the phase differences between two signals are over time. Since phase is a circular variable, we cannot use standard statistical tests to measure deviation of a distribution of phase samples from a null model. The PLV is therefore based on computing the mean vector length (MVL) after averaging the phase differences across all time samples in complex space. Briefly, at each time point, the signal *x(t)* can be represented as a vector in complex space (Fig. 45.4a). The length of the vector *m* corresponds to the amplitude of the signal and the angle θ represents the phase. Note that the signal *x(t)* in this example, is a processed version of the broadband LFP recorded on a microwire, filtered in a narrow frequency band of interest. In Euler's notation, *x(t)* can be written as:

$$x(t) = m(t)e^{i\theta(t)} \tag{45.1}$$

where *m(t)* is the amplitude and θ*(t)* is the phase of the signal at time t. Using this representation, the time-dependent phase difference between two LFP traces, A and B, is simply the difference between the respective phases, θ_A and θ_B. Since the PLV metric in its simplest form ignores the amplitudes of the two signals, the phase difference can be represented as follows:

$$\Delta_{AB}(t) = 1e^{i(\theta_{A(t)} - \theta_{B(t)})} \tag{45.2}$$

To estimate how consistent the phase differences are over time, the length of the mean vector, i.e., the PLV, can be computed as:

$$PLV = \left| \frac{1}{n} \sum_{t=1}^{n} e^{i(\theta_{A(t)} - \theta_{B(t)})} \right| \tag{45.3}$$

where n is the number of time samples. When the phase differences across all time samples between the two signals are uniformly distributed, averaging across the vectors results in a mean vector that has a length close to zero (see Fig. 45.4c). However, if the phase differences are highly consistent across time, the resulting length of the mean vector will be close to 1 (Fig. 45.4b). Thus, the PLV results in values between 0 and 1, with high values representing strong phase synchronization and low values representing weak to no synchronization. Note that the absolute phase difference between the two signals, and the strength of the amplitude correlation over time do not affect the PLV. It is therefore a measure that is purely based on the phase difference between two signals. Also note that this metric only makes sense if the spectra of the two LFP signals is bandlimited to a narrow range of frequencies.

An alternative measure that not only determines the consistency of phase differences of two signals across time, but also takes their amplitudes into account, is called *coherence* [25, 26]. Coherence is computed by measuring the magnitude-squared cross-spectral density C_{AB} of two signals in areas A and B at a given frequency and normalizing it with the product of the power of each of the signals. Mathematically this is expressed as:

$$C_{AB} = \left| \frac{1}{n} \sum_{t=1}^{n} \left| m_{A(t)} \right| \left| m_{B(t)} \right| e^{i(\theta_{A(t)} - \theta_{B(t)})} \right|^2 \tag{45.4}$$

$$Coh_{AB} = \frac{C_{AB}}{\left(\frac{1}{n} \sum_{t=1}^{n} \left| m_{A(t)} \right|^2 \right) \left(\frac{1}{n} \sum_{t=1}^{n} \left| m_{B(t)} \right|^2 \right)} \tag{45.5}$$

where C_{AB} is the cross-spectral density between signal A and B at a given frequency, $m_{A(t)}$ is the amplitude of signal A at time t, and $m_{B(t)}$ is the amplitude of signal B at time t. Like the PLV, coherence also varies between 0 and 1, with higher values representing stronger coupling between signals.

ii. *Amplitude correlation*

Besides phase synchronization, electrophysiological recordings can also yield important insights into interactions among brain areas that are reflected by co-fluctuations of amplitudes [18]. In contrast to phase synchronization methods, amplitude correlation methods usually measure the degree to which two distant signals correlate in their amplitude envelopes for a given frequency range but ignore how strongly their phases synchronize (see Fig. 45.4c) [22]. The two different approaches could therefore highlight different underlying neural communication strategies among brain areas [17]. To assess interregional amplitude correlations, a common approach is to compute a linear correlation between the amplitude envelopes of two signals across time. Thus, amplitude correlations can also reveal *anti*-correlations between amplitude signals, which is an important advantage over the phase synchronization methods described above. The temporal lag between the signals thus matters.

iii. *Cross-frequency coupling*

Phase synchronization and amplitude correlations both measure inter-areal interactions *within* the same frequency range. Neural populations however can also interact *across* frequency bands, which is usually referred to as cross-frequency coupling (CFC) [27, 28]. CFC is thought to play important roles in integrating cognitive functions taking place in different frequency bands [16, 20, 29, 30]. Most studies of CFC so far have focused on an intra-area analysis [31–34]. In studies that sample activity across many anatomical regions, CFC can also be assessed across brain areas [35–37].

Different forms of CFC exist in the brain. Phase-amplitude coupling (PAC) [32, 34], where the phase of a lower frequency band modulates the amplitude of a higher frequency band, and phase-phase coupling [38, 36, 37], where the phase of a higher frequency band is nested into a lower frequency rhythm (see Fig. 45.5), are among the most studied forms of CFC. There are however other forms, such as amplitude-amplitude coupling [39] or phase-frequency coupling [40] which have been documented in the literature. Despite the variety of approaches to computing CFC, the initial steps in these different pipelines are very similar. First, the signals are filtered within each frequency band of interest. In an inter-areal setting for example, the signal from area A is filtered in a lower frequency range, while the signal from area B is filtered in a higher frequency range. Using the Hilbert transform, we compute the analytic representation of each signal, to extract the instantaneous phase and/or amplitude at each time point. From here the methods diverge, depending on which form of cross-frequency coupling is relevant. In the remainder of this subsection, we focus on the most studied forms of CFC, PAC and xPLV.

PAC is by far the most studied form of CFC. Several methods to estimate PAC have been suggested but two are among the most utilized. One method, introduced by Canolty and colleagues [32], has strong commonalities with the PLV introduced

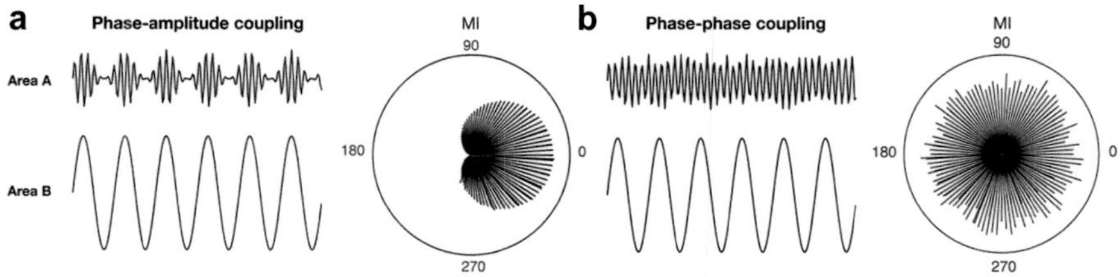

Fig. 45.5 Cross-frequency field-field interactions. **a** Schematic illustration of inter-regional phase-amplitude coupling, where the amplitude of a higher-frequency signal from area A correlates with the phase of a lower-frequency signal from area B. The average vector (red), here determining the strength of the phase-amplitude interaction, has a length different from 0 and thus indicates the existence of significant PAC. **b** Illustration of phase-phase coupling, where the phase of a slower rhythm is coupled to the phase of a faster rhythm. Since no PAC exists in this example, the length of the average vector determining PAC is close to 0

in the section above. At each time point, the frequency-specific phase of the signal in area A and the amplitude in area B is extracted. The combination of these two values can be represented as a vector in complex space that has a length equal to the amplitude of signal B and an angle equal to the phase in signal A (see Fig. 45.5a). In Euler's format, this can be written as:

$$m_{B(t)} e^{i\theta_{A(t)}} \tag{45.6}$$

where $m_{B(t)}$ is the amplitude of signal B and $\theta_{A(t)}$ the phase of signal A at time t. The modulation index (MI), which measures how strongly the amplitude of signal B is modulated by the phase of signal A, is again measured using the MVL after averaging across all time samples, and is computed by:

$$MI = \left| \frac{1}{n} \sum_{t=1}^{n} m_{B(t)} e^{i\theta_{A(t)}} \right| \tag{45.7}$$

If there is no systematic relationship between the phase in signal A and the amplitude in signal B, the vectors from all time samples of the two signals will be uniformly distributed around the circle. Their average thus results in a MVL that has a length of zero (Fig. 45.5b). If, however, there is a systematic relationship between phase and amplitude, that is, certain phases in signal A always correspond to a narrow distribution of amplitudes in signal B, then the mean vector will have a length that is significantly different from zero (Fig. 45.5a). Note that strong modulation will usually not result in values close to 1, as is the case for the PLV, but its value is dependent on the absolute amplitude of signal B. Normalizing the amplitude before computing the MI is therefore recommended, especially when the MI is compared between frequencies or conditions.

A significant drawback of the method introduced by Canolty becomes apparent when the amplitude in signal B is multi-modally distributed across the phase of signal A. In a symmetric bimodal scenario for example, the two amplitude peaks in signal B could cancel each other out and lead to an MI with length zero. A method introduced by Tort et al. [41, 42] circumvents this problem. In their method, after extracting the phase and amplitude in each of the signals using the Hilbert transform, the phase in signal A is discretized into bins and the corresponding amplitude of signal B is averaged within each bin, producing a distribution of amplitude values as a function of phase. Using the Kullback–Leibler distance metric, any significant divergence of this distribution from uniformity is indicative of significant CFC between areas A and B.

Cross-frequency *phase-phase* coupling (Fig. 45.5b), starts with the same initial steps of filtering and Hilbert transformation. The difference here is that we measure phase-phase instead of phase-amplitude coupling, and therefore extract the phase of the higher frequency signal instead of the amplitude. The most common method to estimate phase-phase coupling strongly resembles the PLV method introduced above (thus termed xPLV) [36]. In a cross-frequency setting, the phase of the lower frequency signal gets "accelerated" to match the frequency of the higher frequency signal before assessing the consistency of the phase differences between the two signals, which can be formulated as:

$$xPLV = \left| \frac{1}{n} \sum_{t=1}^{n} e^{i(\frac{f_B}{f_A}\theta_{A(t)} - \theta_{B(t)})} \right| \tag{45.8}$$

where f_A and f_B are the center frequencies of the signals in area A and B, respectively.

A range of limitations should be considered when assessing CFC metrics, both in an intra- as well as inter-areal setting. Discussing these here, however, would be outside the scope of this chapter. An useful overview of the different caveats inherent to CFC analyses and their assessments can be found in Aru et al. [43].

b. Single neuron interactions

Assessing field-field interactions is an indispensable tool for studying neural communication among populations of neurons and can reveal important insights into network-level processes of cognition. Field-field interactions are however a proxy for what we would *really* like to measure, communication between cells. This is where multisite microwire recordings reveal their full potential. Having access to multiple electrodes implanted in different sites of the brain enables us to study long-range neuronal communication at the level of individual cells [5]. Here we discuss two classes of single-neuron metrics: (a) interaction of single neurons with the LFP, and (b) spike-spike interactions between pairs of cells.

i. Spike-field interactions

Methods that assess spike-field interactions can provide valuable insights into how single neurons interact with a *local* (i.e., same area) or *distal* (i.e., different area) population of neurons. They are thought to represent a rather indirect measure of the relationship between neural spiking and synaptic population activity [44]. A common approach to measuring spike-field interactions is the *spike-field coherence* (SFC), which is based on the PLV introduced above [45, 46]. The SFC measures the consistency of spike times with respect to the phase of an underlying oscillation and can be stated mathematically as:

$$SFC = \left| \frac{1}{n} \sum_{s=1}^{n} e^{i\theta_{x(s)}} \right| \tag{45.9}$$

where $\theta_{x(s)}$ is the phase of signal x at each spike time *s*, and *n* is the number of spikes. The SFC is a normalized value which ranges between 0 (low consistency) and 1 (high consistency) (Fig. 45.6). An alternative way to compute the SFC, which in some situations can be more robust, relies on the spike-triggered average of a short piece of LFP centered on each spike instead [45, 47]. This method has the advantage that it normalizes for power differences, which might yield a more reliable estimate of the phase.

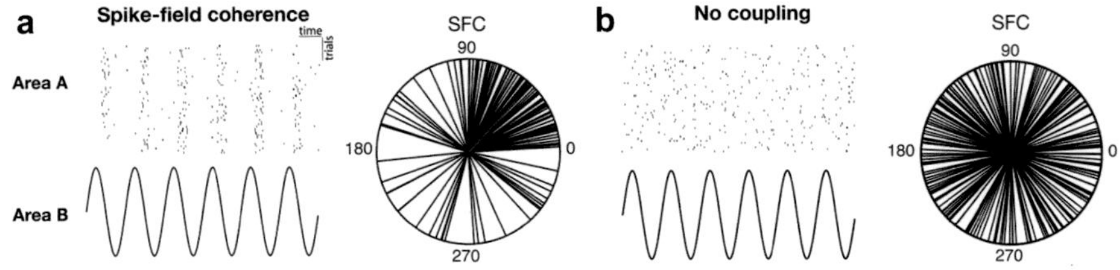

Fig. 45.6 Spike-field interactions. **a** Spikes of a neuron recorded in area A occur consistently at a certain phase of an LFP signal recorded in area B, a phenomenon called spike-field coherence (SFC). The length of the average vector (red) determining SFC is close to 1 and indicates strong coupling. **b** In this example, the neuron fires randomly, thus no systematic relation to the LFP in area B exists. SFC, i.e., the length of the average vector, is therefore close to 0

ii. Spike-spike interactions

Spike-spike interactions reflect the most resolved form of neural communication that can be studied with human microwire recordings. Using spike correlations (in its various forms), we can resolve strength of inter-areal interaction, directionality, whether the interaction is excitatory or inhibitory [48, 49], as well as modulation by state (or task condition) [50]. Spike-spike interactions can be measured by computing cross-correlations between two spike trains. Cross-correlations measure the similarity between the spike train of neuron A and the (lagged) spike train of neuron B. For example, if neuron B often fires 2-3 ms after a spike from neuron A, the cross-correlation will have a peak between 2-3 ms (see Fig. 45.7). An increase in the probability of neuron B spiking after each spike from neuron A, suggests an excitatory influence of neuron A over neuron B (Fig. 45.7a). A reduction on the other hand, indicates that neuron A has an inhibitory effect on neuron B (Fig. 45.7b). Computing spike correlations is the gold standard in measuring inter-areal interactions, but often it is not feasible to do so. Estimates from individual cell pairs are noisy and therefore to get a reliable measure of inter-areal coordination, we must average across many cell pairs. This is difficult to do given that we can only record from a handful of cells from each brain area at a time.

iii. Limitations

Inter-areal interactions suffer from several limitations and their results should always be interpreted with caution. One of the most important considerations when determining field-field interactions between electrodes is volume conduction, which refers to the fact that two or more electrodes can pick up a signal from the same underlying source [51]. When not properly accounted for, volume conduction can lead to spurious interactions among recording electrodes and to a misinterpretation of results [10, 52]. Volume conduction is exacerbated when the two areas are anatomically close or when using a global reference electrode. As pointed out in the case studies, an effective way to account for volume conduction is using local bipolar referencing schemes, where each electrode is referenced against a neighboring channel. Bipolar referencing ensures that only local sources are picked up by each electrode, making it

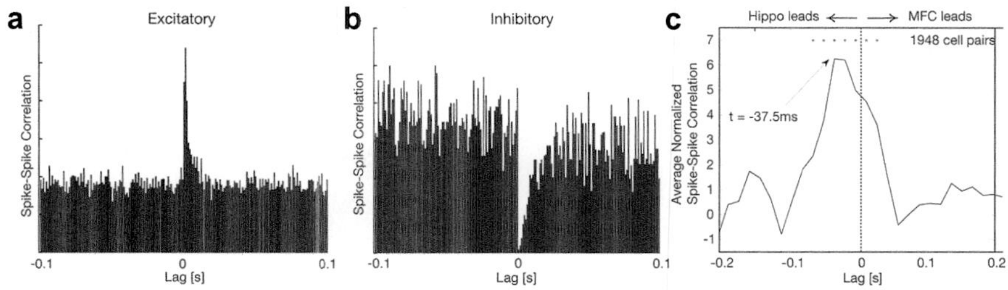

Fig. 45.7 Spike-spike interactions. **a** Cross-correlations of the spike trains from two simulated neurons. In this example, spikes from neuron B consistently occur shortly after spikes from neuron A. This indicates that neuron A and B are monosynaptically connected to each other and that neuron A has an excitatory effect on neuron B. **b** Here, neuron B often fails to spike after spikes from neuron A. Neuron A thus has an inhibitory effect on neuron B. **c** Human spike-spike interactions on the population level. Cross-correlations were computed for all simultaneously recorded cell pairs from the Hippocampus and the medial frontal cortex (MFC) and then averaged. A peak left to the dashed midline indicates that spikes of cells in the MFC on average followed spikes of cells recorded in the Hippocampus, which suggests that the communication between the two areas was led by the Hippocampus. Data in (C) from [5]

unlikely that distant recordings sites are picking up the same source [12]. In addition, methods such as *imaginary coherence* [10] and the *weighted phase-lag index* [52], which are slightly modified variants of the phase synchronization methods discussed above, as well as *orthogonalized amplitude correlations* [53] and *partial correlations* [54] account for zero-(phase)lag interactions and therefore minimize the effect of volume conduction (see Chap. 32). These methods can be helpful tools to enhance the interpretability of inter-areal interaction results.

One notable limitation of the methods discussed here is the lack of directionality information. That is, whether area A sends information to area B or vice-versa cannot be inferred from metrics such as the PLV or PAC. There are however methods, such as Granger causality [11] or cross-frequency directionality [55], which can provide evidence for the directionality of an interaction and can be used in concert with the synchronization metrics discussed above.

Moreover, interaction metrics can tell us if area A is interacting with area B, but they cannot resolve why this might be the case. For example, it may be that two brain areas receive concurrent inputs from a third source and might in fact *not* communicate with one another directly. For a more concrete example, imagine that the onset of a visual stimulus causes phase resetting[3] in theta oscillations both in area A as well as in area B. These phase resets will lead to an increase in theta phase synchronization between area A and B shortly after each stimulus onset. You can imagine other variations of this scenario. For example, neuronal populations in areas A and B might both respond with an increase in firing rates to the onset of the stimulus, which in turn, will inflate cross-correlations between those regions. In

[3] A phenomenon where the phase of an ongoing oscillations (ex., theta) resets to a particular value. This has been suggested as a mechanism for coordinating large-scale neural networks during goal-directed behavior [56].

both scenarios, it is unclear if areas A and B directly communicate with one-another or if they just receive concurrent input from other sources. There are a few, method-specific practices that can mitigate these factors. For example, it is often advisable to remove ERPs prior to the analysis of inter-areal interactions or to focus the analysis on time periods that are free of abrupt signal changes. For spike-spike correlations, computing noise-correlations in addition to signal-correlations can prove insightful. In general, it is advisable to interpret inter-areal interaction results with a healthy amount of scientific caution.

The waveform of neuronal spikes can further strongly influence phase estimations of the LFP, especially when considering oscillations in higher frequency bands [57]. An important step when assessing SFC is therefore to clean the LFP from spike waveforms to avoid observing spurious spike-field interactions. Although less severe in an inter-areal setting, where the spike waveform itself is not present in the LFP recording, phase estimations of the LFP might still be systematically distorted by spiking activity of other neurons in that region, especially in the presence of inter-areal spike-spike correlations. Cleaning the signal from all spike waveforms prior to computing SFC can therefore minimize this problem [57].

In addition, the SFC (like the other PLV-based measures described here) is a measure that is biased by the number of samples (i.e., spikes) taken into account [58]. Comparing the SFC between different conditions, each having a different number of spikes, can therefore be misleading (see Fig. S1 in [47] for an illustration of this effect). A possible way to avoid this problem is by adjusting the number of spikes across all conditions through random subsampling [59]. Repeating this step several hundred times and averaging the SFC across all iterations thereby ensures robustness of the results. A common alternative measure to the SFC is *pairwise phase consistency* (PPC) [58], which has the specific advantage that it not biased by the number of spikes (or samples) considered and is therefore well-suited in situations where the spike counts are low.

Lastly, phase estimation quality is heavily dependent on the signal-to-noise ratio of the data. Phase-based metrics such as the PLV, SFC, or CFC are therefore strongly influenced by the amount of spectral power available in the considered frequency bands (i.e., it is easier to estimate the phase if the oscillation is strong). Observed differences in phase-based metrics between conditions or areas can therefore easily be confounded with differences in spectral power. It is therefore advisable to balance these confounding factors (through resampling of spikes/trials) across conditions before computing synchrony between areas.

45.4 Conclusions

Measurements of inter-areal connectivity, when possible, can offer invaluable insight into brain function. In this chapter we have described two studies where making such measurements has played a big role in shaping our understanding of context-dependent memory retrieval and how visual exploration coordinates communication between brain areas. In addition, we have also discussed some of the most popular measures of inter-areal connectivity as well as their respective advantages and drawbacks. It is important to note that all the methods discussed in this chapter are correlational in nature and cannot be used to draw any conclusions regarding causality (see Chaps. 39 and 40 for stimulation approaches to investigate connectivity). Furthermore, none of the approaches discussed can discern *real* connectivity between two areas from instances where this might be an epiphenomenon, a byproduct of common input for example. There are however mitigating factors, and in general we recommend using several converging analysis pipelines in drawing conclusions about inter-areal interactions. We expect that these kinds of measurements will become increasingly more common as we continue to recognize the highly distributed nature of all cognitive function and as the recording technologies in human intracranial electrophysiology improve.

Acknowledgements This work was supported by the National Institutes of Health (U01NS117839) and the German National Academy of Sciences Leopoldina.

References

1. Steinmetz NA, Zatka-Haas P, Carandini M et al (2019) Distributed coding of choice, action and engagement across the mouse brain. Nature 576:266–273
2. Stringer C, Pachitariu M, Steinmetz N et al. (2019) Spontaneous behaviors drive multidimensional, brainwide activity. Science 364
3. Vaz AP, Inati SK, Brunel N et al (2019) Coupled ripple oscillations between the medial temporal lobe and neocortex retrieve human memory. Science 363:975–978
4. Vaz AP, Wittig JH, Inati SK et al (2020) Replay of cortical spiking sequences during human memory retrieval. Science 367:1131–1134
5. Minxha J, Adolphs R, Fusi S et al (2020) Flexible recruitment of memory-based choice representations by the human medial frontal cortex. Science 368:eaba3313
6. Fried I, Wilson CL, Maidment NT et al (1999) Cerebral microdialysis combined with single-neuron and electroencephalographic recording in neurosurgical patients. J Neurosurg 91:697–705
7. Minxha J, Mamelak AN, Rutishauser U (2018) Surgical and electrophysiological techniques for single-neuron recordings in human epilepsy patients. In: Extracellular recording approaches. Springer, pp 267–293
8. Minxha J, Mosher C, Morrow JK et al (2017) Fixations gate species-specific responses to free viewing of faces in the human and macaque amygdala. Cell Rep 18:878–891
9. Staudigl T, Minxha J, Mamelak AN et al (2022) Saccade-related neural communication in the human medial temporal lobe is modulated by the social relevance of stimuli. Sci Adv
10. Nolte G, Bai O, Wheaton L et al (2004) Identifying true brain interaction from EEG data using the imaginary part of coherency. Clin Neurophysiol 115:2292–2307
11. Granger CW (1969) Investigating causal relations by econometric models and cross-spectral methods. Econometrica: J Econ Soc 424–438
12. Trongnetrpunya A, Nandi B, Kang D et al (2016) Assessing granger causality in electrophysiological data: removing the adverse effects of common signals via bipolar derivations. Frontiers Syst Neurosci 9:189

13. Buzsáki G, Anastassiou CA, Koch C (2012) The origin of extracellular fields and currents—EEG, ECoG, LFP and spikes. Nat Rev Neurosci 13:407–420
14. Parvizi J, Kastner S (2018) Promises and limitations of human intracranial electroencephalography. Nat Neurosci 21:474–483
15. Buzsáki G (2006) Rhythms of the brain. Oxford University Press, New York
16. Canolty RT, Knight RT (2010) The functional role of cross-frequency coupling. Trends Cogn Sci 14:506–515
17. Engel Andreas K, Gerloff C, Hilgetag Claus C et al. (2013) Intrinsic coupling modes: multiscale interactions in ongoing brain activity. Neuron 80:867–886
18. Siegel M, Donner TH, Engel AK (2012) Spectral fingerprints of large-scale neuronal interactions. Nat Rev Neurosci 13:121–134
19. Engel AK, Fries P, Singer W (2001) Dynamic predictions: oscillations and synchrony in top–down processing. Nat Rev Neurosci 2:704–716
20. Fell J, Axmacher N (2011) The role of phase synchronization in memory processes. Nat Rev Neurosci 12:105–118
21. Fries P (2005) A mechanism for cognitive dynamics: neuronal communication through neuronal coherence. Trends Cogn Sci 9:474–480
22. Nolte G, Galindo-Leon E, Li Z et al (2020) Mathematical relations between measures of brain connectivity estimated from electrophysiological recordings for gaussian distributed data. Front Neurosci-switz 14:577574
23. Palmigiano A, Geisel T, Wolf F et al (2017) Flexible information routing by transient synchrony. Nat Neurosci 20:1014–1022
24. Lachaux JP, Rodriguez E, Martinerie J et al (1999) Measuring phase synchrony in brain signals. Hum Brain Mapp 8:194–208
25. Nunez PL, Silberstein RB, Shi Z et al (1999) EEG coherency II: experimental comparisons of multiple measures. Clin Neurophysiol 110:469–486
26. Nunez PL, Srinivasan R, Westdorp AF et al (1997) EEG coherency I: statistics, reference electrode, volume conduction, Laplacians, cortical imaging, and interpretation at multiple scales. Electroen Clin Neuro 103:499–515
27. Hyafil A, Giraud A-L, Fontolan L et al (2015) Neural cross-frequency coupling: connecting architectures, mechanisms, and functions. Trends Neurosci 38:725–740
28. Jensen O, Colgin LL (2007) Cross-frequency coupling between neuronal oscillations. Trends Cogn Sci 11:267–269
29. Lisman John E, Jensen O (2013) The theta-gamma neural code. Neuron 77:1002–1016
30. Palva JM, Palva S (2018) Functional integration across oscillation frequencies by cross-frequency phase synchronization. Eur J Neurosci 48:2399–2406
31. Axmacher N, Henseler MM, Jensen O et al (2010) Cross-frequency coupling supports multi-item working memory in the human hippocampus. Proc National Acad Sci 107:3228–3233
32. Canolty RT, Edwards E, Dalal SS et al (2006) High gamma power is phase-locked to theta oscillations in human neocortex. Science 313:1626–1628
33. Daume J, Graetz S, Gruber T et al (2017) Cognitive control during audiovisual working memory engages frontotemporal theta-band interactions. Sci Rep-uk 7:12585
34. Daume J, Gruber T, Engel AK et al (2017) Phase-amplitude coupling and long-range phase synchronization reveal frontotemporal interactions during visual working memory. J Neurosci 37:313–322
35. Friese U, Köster M, Hassler U et al (2013) Successful memory encoding is associated with increased cross-frequency coupling between frontal theta and posterior gamma oscillations in human scalp-recorded EEG. Neuroimage 66:642–647
36. Siebenhühner F, Wang SH, Arnulfo G et al (2020) Genuine cross-frequency coupling networks in human resting-state electrophysiological recordings. Plos Biol 18:e3000685
37. Siebenhühner F, Wang SH, Palva JM et al (2016) Cross-frequency synchronization connects networks of fast and slow oscillations during visual working memory maintenance. Elife 5:e13451

38. Sauseng P, Klimesch W, Gruber WR et al (2008) Cross-frequency phase synchronization: a brain mechanism of memory matching and attention. Neuroimage 40:308–317
39. Helfrich RF, Herrmann CS, Engel AK et al (2016) Different coupling modes mediate cortical cross-frequency interactions. Neuroimage 140:76–82
40. Lowet E, Roberts MJ, Bosman CA et al (2016) Areas V1 and V2 show microsaccade-related 3–4-Hz covariation in gamma power and frequency. Eur J Neurosci 43:1286–1296
41. Tort ABL, Komorowski R, Eichenbaum H et al (2010) Measuring phase-amplitude coupling between neuronal oscillations of different frequencies. J Neurophysiol 104:1195–1210
42. Tort ABL, Kramer MA, Thorn C et al (2008) Dynamic cross-frequency couplings of local field potential oscillations in rat striatum and hippocampus during performance of a T-maze task. Proc National Acad Sci 105:20517–20522
43. Aru J, Aru J, Priesemann V et al (2015) Untangling cross-frequency coupling in neuroscience. Curr Opin Neurobiol 31:51–61
44. Okun M, Naim A, Lampl I (2010) The subthreshold relation between cortical local field potential and neuronal firing unveiled by intracellular recordings in awake rats. J Neurosci 30:4440–4448
45. Fries P, Reynolds JH, Rorie AE et al (2001) Modulation of oscillatory neuronal synchronization by selective visual attention. Science 291:1560–1563
46. Vinck M, Battaglia FP, Womelsdorf T et al (2012) Improved measures of phase-coupling between spikes and the Local Field Potential. J Comput Neurosci 33:53–75
47. Rutishauser U, Ross IB, Mamelak AN et al (2010) Human memory strength is predicted by theta-frequency phase-locking of single neurons. Nature 464:903–907
48. Barthó P, Hirase H, Monconduit LC et al (2004) Characterization of neocortical principal cells and interneurons by network interactions and extracellular features. J Neurophysiol 92:600–608
49. Csicsvari J, Hirase H, Czurko A et al (1998) Reliability and state dependence of pyramidal cell-interneuron synapses in the hippocampus an ensemble approach in the behaving rat. Neuron 21:179–189
50. Doiron B, Litwin-Kumar A, Rosenbaum R et al (2016) The mechanics of state-dependent neural correlations. Nat Neurosci 19:383–393
51. Nunez PL, Srinivasan R (2006) Electric fields of the brain. Oxford University Press. Inc., New York, NY, USA
52. Vinck M, Oostenveld R, Wingerden MV et al (2011) An improved index of phase-synchronization for electrophysiological data in the presence of volume-conduction, noise and sample-size bias. Neuroimage 55:1548–1565
53. Hipp JF, Hawellek DJ, Corbetta M et al (2012) Large-scale cortical correlation structure of spontaneous oscillatory activity. Nat Neurosci 15:884–890
54. Cohen M (2014) Analyzing neural time series data: theory and practice. MIT Press, Cambridge
55. Jiang H, Bahramisharif A, Van Gerven MA et al (2015) Measuring directionality between neuronal oscillations of different frequencies. Neuroimage 118:359–367
56. Voloh B, Womelsdorf T (2016) A role of phase-resetting in coordinating large scale neural networks during attention and goal-directed behavior. Frontiers Syst Neurosci 10:18
57. Zanos TP, Mineault PJ, Pack CC (2011) Removal of spurious correlations between spikes and local field potentials. J Neurophysiol 105:474–486
58. Vinck M, Wingerden MV, Womelsdorf T et al (2010) The pairwise phase consistency: a bias-free measure of rhythmic neuronal synchronization. Neuroimage 51:112–122
59. Kamiński J, Brzezicka A, Mamelak AN et al (2020) Combined phase-rate coding by persistently active neurons as a mechanism for maintaining multiple items in working memory in humans. Neuron 106:256-264.e253

Chapter 46
How Can Laminar Microelectrodes Contribute to Human Neurophysiology?

Mila Halgren

Abstract Human laminar microelectrodes (linear arrays implanted acutely or semichronically in surgical patients) present an exciting new frontier of intracranial electrophysiology. Though most iEEG is limited to imaging networks, laminars can resolve the cortical microcircuits underlying cognition. Normally implanted in animal models, laminar probes can record the current-source-density, which reflects transmembrane currents, as well as single and multi-unit activity (MUA) throughout the cortical depth. These measures of neural activity allow the mapping of laminar physiology underlying diverse neural phenomena in humans. For instance, several studies have shown laminar activity sensitive to language and perception. They've also discovered motifs of different rhythms during sleep (slow waves, spindles) and wakefulness (delta/theta, alpha). Intriguingly, these studies suggest an outsize role for superficial layers in cortical oscillations which may be human specific. Human laminar recordings have also shed light on cortical physiology in general, such as the spatiotemporal dissociation of high-gamma-power (HGP) and MUA. In disease, laminars have elucidated the structure of epileptiform discharges. However, key conceptual and methodological issues like proper referencing, clinical constraints and comparisons with animal models remain. These difficulties notwithstanding, new innovations in recording density, simultaneous surface-laminar recordings and extracranial source-inference will enable laminars to answer fundamental questions in human neurophysiology.

46.1 Introduction

As reviewed in other chapters, intracranial human electrophysiology (iEEG) has provided vital insights into human cognition. The exquisite spatiotemporal precision of these recordings allows for the direct study of human cortex. However, a limitation of typical iEEG is the lack of layer specific recordings. Neocortex is characterized by

M. Halgren (✉)
Department of Brain & Cognitive Sciences, McGovern Institute for Brain Research, Massachusetts Institute of Technology, Cambridge, MA, USA
e-mail: mhalgren@mit.edu

N. Axmacher (ed.), *Intracranial EEG*, Studies in Neuroscience, Psychology and Behavioral Economics, https://doi.org/10.1007/978-3-031-20910-9_46

its layered structure, composed of six laminae with distinct cell types, connectivity and physiology. This architecture is highly consistent across cortical areas, leading to the hypothesis that there is a canonical laminar microcircuit [1, 2].

Though laminar physiology has been fruitfully studied in rodents and non-human-primates, humans have considerably different cortical architecture. For instance, human cortex has novel neurons and glia not found in rodents [3–5]. Even homologous cells have strikingly different electrophysiological and transcriptomic properties across species [4, 6–11]. For example, layer V pyramidal cells are 7× more likely to be bursty in mice than humans in vitro [6]. On the network level, human cortex is much more recurrent than rodent cortex [12–15]. These interspecies differences are particularly pronounced in supragranular layers, which disproportionately expanded in humans [16, 17], are more cellularly heterogeneous than in rodents [5], and may be critical for human specific cognitive abilities [18–20]. Therefore, the use of animal models for laminar physiology must be validated directly in humans. This is made possible by human laminar microelectrodes, or linear arrays which span the cortical depth perpendicularly.

The only laminar arrays used in humans (save Neuropixels) were designed and developed by Mr. László Papp, Prof. György Karmos, Prof. Eric Halgren, Prof. István Ulbert and Neuronelektród Ltd. in the late 1990s [21], and are manufactured with platinum-iridium for increased biocompatibility (especially in potentially irritable epileptogenic cortex). Harboring 24 contacts with 150 μm spacing, they are long enough to span the cortical depth while sampling each cortical layer with multiple channels (Fig. 46.1a–c). This fine spatial sampling enables the collection of action potentials (AP) as well as their proxies, multi-unit-activity (MUA) and high-gamma-power (HGP). Additionally, transmembrane currents due to synaptic activity and active channels can be estimated by the current-source-density (CSD) within individual laminae (Fig. 46.1d). The CSD measures current sinks and sources, reflecting net transmembrane current flow in and out of neurons (respectively). Pairs of sinks and sources, or current dipoles, generate the electrical and magnetic signals measured with iEEG, EEG and MEG (see Sect. 46.3.1). Though diverse cells such as glia and interneurons contribute to transmembrane current flow, the cortical CSD is likely dominated by pyramidal cells due to their numerosity, laminar orientation and large dipole moments [22]. Being able to derive the CSD is a critical advantage of laminar recordings, as the closely related monopolar (distantly referenced) local field potential (LFP) and its first derivative, the local field potential gradient (LFPg) are contaminated by distant sources via volume-conduction (see Sect. 46.3.1).

These platinum-iridium laminars come in surface and depth varieties. The former anchor to the cortical surface via a thin silicone sheet which sits on the pia (Fig. 46.1a). This holds the probe in place and keeps it normal to the cortical surface/perpendicular to cortical layers. Conversely, depth laminars protrude from the end of macro-electrode stereo-EEG probes and allow recordings from medial areas such as the hippocampus and cingulate. Recordings can be made acutely (within the operating room, ~1 h) and/or semi-chronically (within the epilepsy-monitoring-unit, ~1 week). In the semi-chronic case, investigators can record multiple tasks and behavioral states (wakefulness, sleep, anesthesia) from the same patient, yielding large amounts

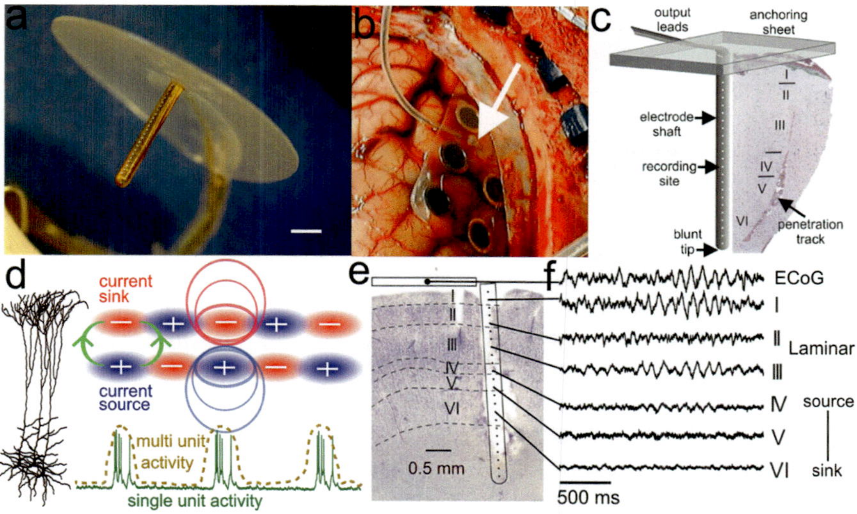

Fig. 46.1 **a** Photomicrograph of a laminar array with anchoring silicone sheet, provided by László Papp of Neuronelektrod Ltd., Budapest, Hungary (scale bar: 1 mm). **b** Picture of ECoG with simultaneously implanted laminar array (arrow). **c** Schematic of implanted laminar array, with co-histology from the implantation in **b**. **b**, **c** Reproduced with permission from [97]. **d** Schematic of the two principal forms of data measured with a laminar probe: the CSD (reflecting transmembrane currents) and single/multi-unit activity. **e** Co-histology of laminar track. **f** Representative traces of CSD recorded from the probe, in addition to overlying ECoG. **e**, **f** Reproduced with permission from [61]

of data from a single implantation. Though previous experiments have used only platinum-iridium laminars, two groups recently recorded with Neuropixels [23, 24]. Neuropixels can record from 384 channels with 20 μm spacing simultaneously, a drastic advance on platinum-iridium probes [25]. Though human laminar recordings are in their infancy, with only 27 papers which analyze this data, numerous insights into cortical physiology have been made. Previous findings from human laminar recordings are briefly reviewed, before discussing methodological challenges and future directions.

46.2 Insights

46.2.1 Oscillations

Oscillations, or rhythmic fluctuations in neural activity, dominate human cortex and are critical for cognition and behavior [26, 27]. Despite a long tradition of study with extracranial EEG and MEG, little is known of the basic physiology underlying these oscillations. Though many seminal studies have been made in animals or in vitro, it is unknown how these translate to in vivo human cortex.

The key questions human laminar recordings can resolve, over and above macro-electrode iEEG, are: 1. Which laminae do the transmembrane currents which generate these rhythms stem from? 2. Which layers have cell bodies which fire phasically with these rhythms? And 3, how might this laminar profile in humans differ from animal models? Question 1 can be answered by measuring the power of a given oscillation's CSD across the cortical depth; 2 by measuring the synchrony of this oscillation with a simultaneous metric of unit activity; and 3, by recording in areas and behavioral states comparable to an animal model. Most studies have found that low frequency rhythms in humans are dominated by *currents and firing in superficial laminae*; whether this generalizes to other animals is unclear.

46.2.2 Sleep Rhythms

Slow waves and K-complexes (0.5–2 Hz) are the largest graphoelements in the human EEG, and coordinate memory consolidation [28] and metabolic maintenance [29]. In rodents, the slow oscillation primarily reflects sinks and sources in deep cortical layers driven by layer V pyramidal cells [30–34]. Surprisingly, Cash et al. found that human K-complexes reflect layer I/II sinks and sources which primarily modulate supragranular firing (Fig. 46.2a) [35]. Human slow waves have a highly similar laminar profile to K-complexes, comprised mostly by layer I/II currents and supragranular firing (Fig. 46.2c) [36].

Spindles, like slow waves, are oscillations critical for learning and memory [37–40]. Two studies on their laminar profile in humans [41, 42] found that spindles were dominated by currents and firing in superficial and middle layers, the latter reporting that superficial currents were stronger (Fig. 46.2b). Controversy remains over whether these spindles can be separated into hypothesized "matrix" and "core" types, respectively driven by focal or non-specific thalamic input [43].

46.2.3 Wake Rhythms

Delta and theta oscillations are found throughout human cortex [44], and play a crucial role in memory and cognition [45]. Two reports found that delta/theta rhythms were particularly pronounced in superficial cortex throughout a wide variety (19 in the former) of cortical association areas [46, 47]. These rhythms were phase-reset by oddball stimuli, and likely contribute to the novelty-induced P300 and N400 (see Sect. 46.2.4). An important caveat is that slow delta is sometimes caused by drowsiness or epileptic pathology [48], and does not necessarily reflect healthy or alert cortical activity.

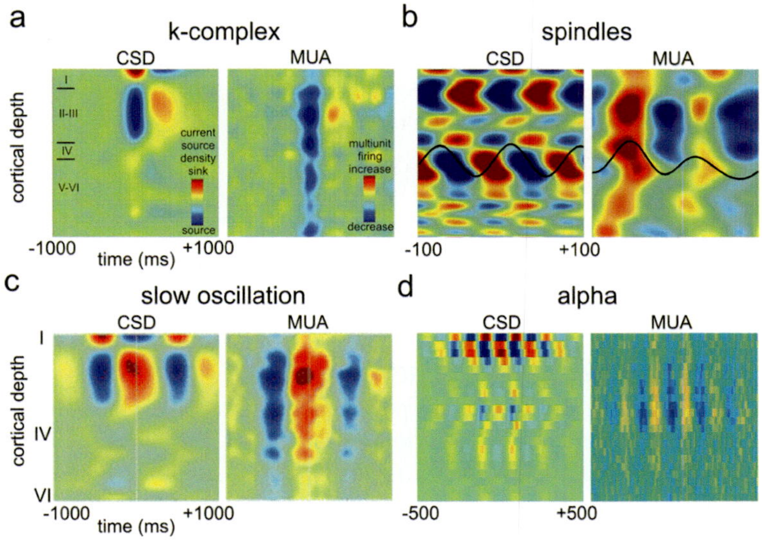

Fig. 46.2 **a** The laminar profile of k-complexes, with the averaged CSD on left and averaged MUA right. **b–d** Same as **a**, but for spindles, slow waves and alpha. **a–d** are reproduced with permission from [35, 36, 42, 61], respectively

Alpha oscillations (7–13 Hz) [26] dominate the waking EEG, and are linked to attention [49], perception [50, 51] and functional inhibition [52]. Within cortex, in vitro recordings and animal models suggest that alpha originates from infragranular layers driven by layer V pyramidal cells [53–58] (but see [59, 60]). In contrast, human alpha-band currents are strongest in very superficial (~I/II) laminae and primarily modulate layer III firing (Fig. 46.2d) [61]. Therefore, the human alpha rhythm likely reflects currents on the apical dendrites of layer III pyramidal cells (see Sect. 46.3.1). Further experiments should determine if this is a general laminar motif, or if alpha modulates supragranular cells in some cortical regions and infragranular cells in others [60].

In sum, human laminar studies have converged on superficial layers as a primary locus for cortical oscillations. In contrast, animal literature on low-frequency oscillations has emphasized the role of deep layers [30, 53, 57, 58, 62]. A possible explanation is that oscillatory networks shifted from infragranular to supragranular cortex between rodents and primates. This may be due to the enlargement of supragranular layers in humans, occupying ~60% of the cortical depth versus ~40% in rodents [16]. Human supragranular cells are also more recurrently connected [15] and have stronger h-currents than rodents [11], both of which are critical for rhythmicity. Interestingly, human supragranular pyramidal cells are more transcriptomally and electrophysiologically diverse than in mice, suggesting that laminar changes in gene expression and physiology may account for supragranular rhythms in humans [5]. A critical caveat: even if the currents and firing related to low-frequency oscillations are supragranular, a sparse infragranular population may be causally necessary to generate the rhythm [30]; therefore, stating a rhythm is "generated" in superficial layers simply means that current dipole measured extracranially and the cells which fire phasically with the rhythm are supragranular, not that superficial layers alone are causally sufficient to drive an oscillation.

46.2.4 Cortical Physiology

Besides oscillatory dynamics, human laminar recordings have yielded insights into cortical physiology. These include the laminar structure of event-related potentials (ERPs) and the differential origins of HGP and MUA.

The N400 (scalp-negative, ~400 ms post stimulus) is an ERP elicited by semantically meaningful (particularly anomalous) stimuli [63]. Hypothesized to underly lexico-semantic integration [64] and/or predictive processing [65], it has long been used to probe language and semantic cognition. Despite this, little is known of its underlying physiology as it cannot be studied in (non-linguistic) animal models. Therefore, Halgren et al. made laminar recordings in the anteroventral temporal lobe, a known N400 generator [66], during a semantic priming task [67]. Word presentation induced a prominent excitatory current sink in layer IV and passive return source in layers II/III, with a prominent peak ~400 ms post stimulus (Fig. 46.3a). This sink was longest in response to semantically anomalous words, much like the scalp N400. Because layer IV receives feedforward input, this suggests that the N400 reflects first-pass lexico-semantic encoding of words within the cortical language network. A further study found that, rather than reflecting a novel potential per-se, the laminar N400 arises from the phase-reset of ongoing theta rhythms [47]. A study in cingulate and temporal areas of the closely related P300 (evoked by anomalous stimuli regardless of semantic content) found it was dominated by superficial currents due to a phase-reset of ongoing slow rhythms (in this case delta) [46] (Fig. 46.3b), suggesting a similar circuit for processing stimulus deviance via the phase-reset of pre-stimulus oscillations. These results suggest that human theta might have a somewhat deeper (involving layer IV) and distinct laminar distribution than cortical delta; this should be addressed in future experiments.

A further contribution of human laminars is differentiating the origins of MUA (~300–3000 Hz) and HGP (~70–190 Hz). Though HGP is lower frequency than traditional MUA, it's strongly correlated with single-unit firing [68–70] and is therefore the standard proxy for unit activity in macroelectrode iEEG [71]. To determine possible physiological differences in unit firing and HGP, Leszczyński et al. measured HGP and MUA simultaneously in macaque and human laminar recordings [72]. Surprisingly, they found that HGP and MUA had distinct laminar origins; while MUA was driven by middle/deep layers, HGP was found within both middle/deep and superficial layers (Fig. 46.3c). Intriguingly, this superficial HGP could occur without concurrent deep MUA. A parallel finding was observed in a human laminar study of HGP and MUA modulation by the alpha rhythm [61]. Though HGP was modulated by alpha throughout superficial layers (~layers I-III), MUA was only

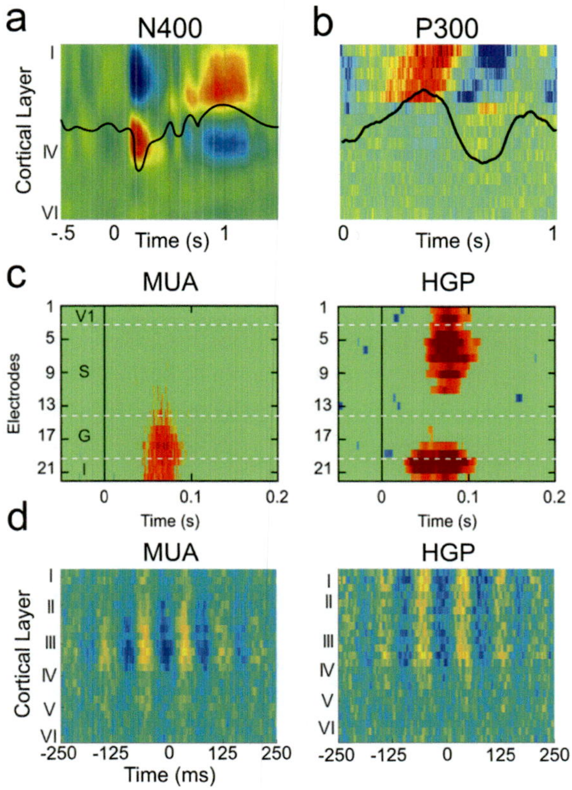

Fig. 46.3 **a** CSD profile of the N400, reproduced with permission from [67]. **b** LFPg profile of the P300, reproduced with permission from [46]. **c** MUA and HGP in macaque V1 evoked by flashes, reproduced with permission from [72]. Note that MUA is strictly in deep cortex, whereas HGP is evoked in superficial and deep laminae. **d** MUA and HGP locked to ongoing alpha in human parietal cortex, reproduced with permission from [61]; MUA is modulated within deep layer III, whereas HGP extends throughout supragranular cortex

modulated within layer III (Fig. 46.3d). MUA stemming from deeper layers than HGP, as well as being dissociable from it, is consistent with HGP reflecting (superficial) dendritic processes such as calcium spikes dissociable from (deeper) somatic firing. Further evidence to this effect is that the NMDA antagonist phencyclidine attenuated HGP without affecting MUA [72]. Relating discrete calcium spikes (in principle recordable extracellularly [73]) to ongoing HGP, and quenching both via pharmacological manipulation, would provide further evidence that HGP reflects dendritic calcium spikes rather than unit firing per se.

46.3 Challenges

46.3.1 Referencing

How do we interpret the field potentials recorded by laminar arrays? Electric potentials in cortex are primarily generated by transmembrane currents. These currents arise from synaptic activity and voltage-dependent active channels such as calcium spikes, h and t currents, etc. [22]. Maxwell's equations necessitate that the sum of potential differences in a circuit is zero [74]; therefore, these currents form balanced current dipoles of sinks (into cells) and sources (out of cells). Due to cortex's laminar architecture, these dipoles are arranged vertically (Fig. 46.1d). Therefore, cortical field potentials are due to vertically arranged transmembrane current dipoles generated by synaptic and active channels. It can be shown that given some assumptions such as radially symmetric currents and isotropic and homogeneous tissue conductance, the second spatial derivative of the field potential vertically (i.e. along a laminar probe) yields the net transmembrane current at a channel [75, 76], also called the current-source-density (CSD).

Consequently, monopolar (distantly referenced) LFPs do not themselves reflect underlying transmembrane currents, but instead the summed fields generated by dipoles volume conducted throughout surrounding tissue. This means that laminar LFPs are not a faithful proxy for local currents. Indeed, the strong influence of volume conduction make the distantly-referenced field potential misleading to analyze. Firstly, this is because potentials evoked by distant cortical or subcortical areas can lead to large field potentials in the absence of local activity [77–79]. Secondly, even within an area, potentials will volume conduct across cortical layers. Alarmingly, infragranular LFPs are often susceptible to volume conduction from supragranular sources [80] (Fig. 46.4a, b). A common objection to the use of local (bipolar, CSD) referencing is that they can artificially silence synchronous, shared activity between electrodes (see Chap. 28). However, this is emphatically not the case for laminar recordings. This is because these electrodes lie within, not outside of, the cortex (unlike ECoG or scalp EEG). If a current dipole lies between a series of contacts, field potentials will exhibit changes in amplitude which (when the second derivative is taken) yield a dipole. If the monopolar field potential amplitude does not change, or changes linearly between contacts, the signal must be due to volume-conduction from a dipole in a different cortical layer and/or area. Concerningly, these volume-conducted potentials are often orders of magnitude larger than those produced by local currents [77]. As such, volume conduction can lead to high amplitude, coherent potentials at locations without any local synaptic activity. Therefore, the finding of a perfectly synchronous monopolar potential across some contacts implies that its generator does not lie between these contacts (as the second derivative, or CSD, of a constant is zero). This makes the analysis of monopolar field potentials very difficult to interpret.

The importance of proper referencing is illustrated in the controversy surrounding the origin of the cortical alpha rhythm. Several primate laminar studies have emphasized infragranular layers, primarily due to high alpha power in the monopolar field potential in deep cortex [53, 56, 58]. However, laminar studies in humans and macaques which used local referencing found greater alpha power in superficial layers [59, 61]. This is most consistent with a supragranular alpha source which volume conducts to deep layers; because contacts in superficial layers lie within the generating dipole, the monopolar field potential is silent.

Volume conduction makes the use of local referencing, such as CSD, necessary to localize transmembrane currents to specific cortical layers. However, there are

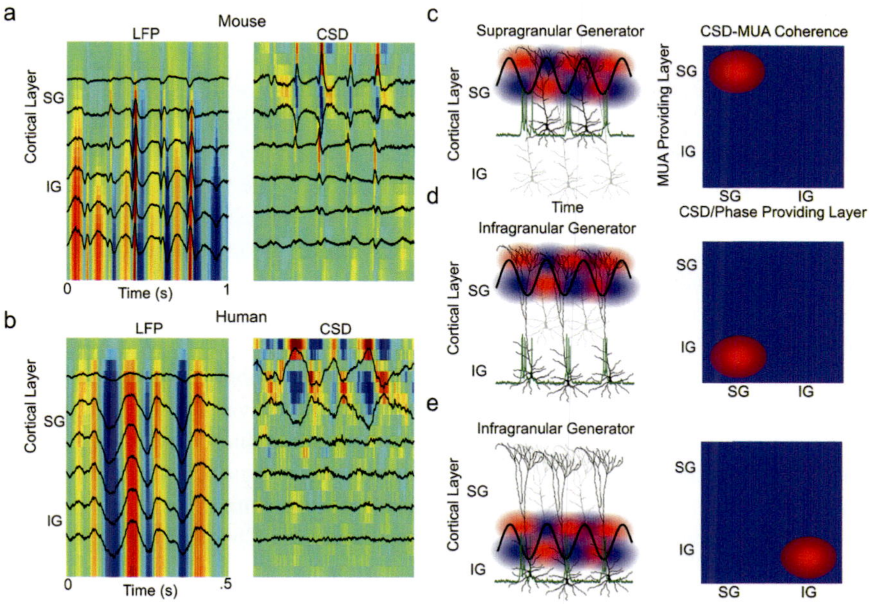

Fig. 46.4 **a** LFP and CSD derivations of the same activity from a mouse laminar recording [114]. **b** LFP and CSD from a human laminar recording. In **a-b**, supragranular currents volume conduct to deeper layers. **c** Example of a rhythm generated by supragranular pyramidal cells. This can be represented in the CSD-MUA coherence plot on the right, which shows that supragranular currents are synchronous with supragranular firing. **d** Same as c, but for a rhythm generated by supragranular currents onto the apical dendrites of infragranular pyramidal cells. Supragranular currents are now synchronous with infragranular spiking. **e** as **c** and **d**, but for currents and firing in infragranular layers

several challenges to accurate CSD estimation; firstly, because the second spatial derivative is ill-defined at the edge of an array, the top and bottom channels must be discarded. This can be ameliorated by the Vaknin Correction, which takes the edge monopolar channels (i.e. most superficial and deep), and then pads the referential recording with these signals before computing the CSD [81]. This is justified by the minimal decay of the monopolar field potential above and below the array. A more serious concern is the magnification of spatial noise by the second derivative; small deviations in amplitude across channels, due to differing impedances, spacing, noise, etc. will be exaggerated by CSD. Solutions for this include spatial (Gaussian) smoothing prior to CSD calculation, or kernel-CSD methods [82]. Alternatively, the local-field-potential-gradient (LFPg), or first derivative of the monopolar LFP, can be used instead. This measure has similar spatial localization to CSD by attenuating volume conduction [59], but is less interpretable than the CSD (as it represents the integral of local sinks and sources, not transmembrane currents per se). Lastly, a critical assumption of CSD analysis is that electrodes are situated normal to the cortical surface [75]. This can be difficult to ensure, particularly with poor control of implantation in a clinical environment, and with a lack of co-histology. One solution is to anchor the probe to a silicone sheet which adheres to the cortical surface (Fig. 46.1a, b).

How do we interpret the relationship of cortical firing (as measured by MUA or single units) and the local transmembrane currents yielded by the CSD? Critically, the CSD indicates which layer a given current lies in, but not the layer of the somas being excited or inhibited by these currents. Because of the rapid spatial decay of the fields generated by action potentials, unit activity is usually only observable near the soma [22]. Therefore, by measuring the synchrony or coherence of currents and unit firing, one can infer the likely layers of both the dendrites and the somas involved in a graphoelement (Fig. 46.4c–e). For example, the fact that human alpha-band currents are in layers I/II does not (in itself) indicate that alpha-modulated cells reside in superficial layers. These currents could be onto the apical dendrites of either layer II/III or layer V/VI pyramidal cells. The only way of distinguishing these possibilities is to record unit activity simultaneously, which presumably reflects somatic spiking, and measure the coherence between this somatic firing and (usually dendritic) currents (Fig. 46.4c–e). For alpha oscillations in human association cortex, this CSD-MUA coherence was highest between layer I/II CSD and layer III MUA, suggesting a supragranular pyramidal source for alpha.

How do we determine if a CSD sink/source is excitatory or inhibitory? Though it may seem, trivially, that inward currents are excitatory and outward currents reflect inhibition, this is not necessarily the case. Because net current is conserved, every excitatory ("active") current sink must be balanced by a return ("passive") current source, and vice-versa. Therefore, current sinks may reflect active excitation or passive return from inhibitory current, and current sources may reflect active inhibition or the passive return from an excitatory sink. We can determine if a current sink/source is excitatory/inhibitory by seeing if it coincides with increased or decreased firing. If a current sink co-occurs with increased firing, it likely represents an active excitatory current, with its paired source reflecting passive return currents to extracellular space. Conversely, a current source accompanied by decreased firing indicates an active inhibitory current, with a paired passive current sink. Because currents can always be active or passive, unit activity must be recorded simultaneously with CSD to infer whether a sink or source is active (excitatory or inhibitory), or merely a passive return current. For example, in recordings of human alpha oscillations, current sinks/sources in layer I were matched to increased/decreased firing (respectively), with matched return currents in layer II/III (Fig. 46.2d). This indicates that layer I sinks/sources are "active" (reflecting excitation/inhibition) and layer II/III sinks/sources are passive return currents. However, because all extracellular CSD signals represent an unknown mixture of cells, definitive evidence for these excitatory and inhibitory circuits must come from single-cell electrophysiology.

46.3.2 Recording Conditions

As discussed (Sect. 46.2.3), human laminar recordings have strong currents in layers I-III not found in animal studies [35, 36, 46, 61]. This might indicate a critical interspecies distinction in LFP generation due to differences in cortical physiology. Human pyramidal cells have extremely elaborate layer I/II arborization not paralleled in other species [83, 84], as well as other anatomical specializations not seen in rodents (see Sects. 46.1 and 46.2.3). However, the risk remains that these currents are illusory due to poor CSD estimation at the edge of the laminar array. Fine sampling of the gray matter/CSF boundary, possible with human Neuropixels [23, 24], could resolve whether this is artifactual or a real interspecies difference.

Relatedly, it's usually unclear which cortical layers are recorded by each contact on a laminar probe. In rodents, this is resolved by explanting the tissue surrounding the probe and finding the track caused by the electrode (often with the assistance of electrolytic lesions and/or dye). However, neither electrolytic lesions nor probe dyeing are approved for use in human surgical cases. Some reports have performed histology on the explanted tissue, and find the probe track from tissue damage [23, 36, 42, 85]. When co-histology is not available (as is often the case), there are several strategies for assigning contacts to layers. First, layers may be estimated by current-source-density analysis of stimulus-evoked activity. Within sensory and neighboring cortex, stimuli should evoke a feedforward sink in layer IV, allowing the identification of supragranular and infragranular layers. Unfortunately, it's not clear whether this generalizes to human association areas. Alternatively, laminae might be estimated from anatomical measurements of laminar depth in human staining studies (Hutsler et al. 2005). On average, layers I–III occupy the first 60% of the cortical depth, layer IV 6%, and layers V/VI the bottom 33%. In this case, it is important to determine if the laminar probe spans the cortical depth. This can be done by examining MUA and CSD, both of which should sharply attenuate in white matter. In cases where laminar depth cannot be confidently estimated, it's best to restrict conclusions about cortical dynamics to superficial versus deep layers. Though which layer channels in the middle third of the array lie in will be ambiguous, it's highly likely that the top and bottom third of channels will correspond to supragranular and infragranular laminae, respectively.

A last challenge of human laminar recordings is the lower unit yield than comparable rodent studies. This is due to a few factors; first, clinical equipment in the OR and EMU create significant amounts of electrical noise. Though animal studies can reduce noise by changing grounds/references, turning off other electronic devices and using Faraday cages, this troubleshooting is difficult in a clinical environment. A related issue is that (unlike in animals) one cannot make multiple penetrations of cortex until an area with high spike rates is found. A further difficulty is the trauma suffered by cortex upon the insertion of a microelectrode array. Though animal experimentalists can wait for a long period of time post electrode insertion to allow cortex to adapt, within acute OR experiments, time is of the essence and recordings must usually be started immediately. Lastly, as most platinum-iridium human laminars have intercontact separations on the order of ~150 um, units cannot be detected across multiple contacts, which makes spike sorting more challenging than for dense microelectrode arrays. Human Neuropixel recordings don't have this drawback due to their high spatial sampling (20 μm), and have isolated up to 202 single units in an acute recording [23, 24]; future human laminars might also improve their yield via similarly dense electrodes. A useful alternative to single-units is MUA, or the envelope of the signal filtered from ~300–3000 Hz. Analyzing MUA sidesteps the difficulties of single-unit isolation to make general conclusions about firing rates in different laminae, instead of the properties of individual cells. HGP may also provide complementary information about firing to MUA; however, as discussed in Sect. 46.2.4, HGP does not reflect spiking as tightly as MUA (particularly spatially) [61, 72, 86].

46.4 Promises

46.4.1 Macroelectrode-Laminar Correspondence

Because of the technical and clinical challenges involved in laminar recordings, most iEEG research will continue to rely on macroelectrode recordings such as ECoG. These recordings do not, by themselves, allow for conclusions to be drawn about individual laminae. However, laminar-specific patterns can be inferred from ECoG alone if the correspondence of laminar to ECoG activity is known. This is done by recording events with ECoG and laminar electrodes simultaneously, and then determining how different layers contribute to the ECoG signal. The results of this analysis for a given oscillation or potential can then be used to interpret previous and future ECoG experiments. For instance, Fabo et al. used laminars in temporal cortex and subiculum to study the physiology of inter-ictal discharges (IIDs) and found distinct IIDs corresponding to apical or somatic depolarization [87]. These IIDs were simultaneously recorded with ECoG, allowing future research conducted solely with macroelectrodes to infer the laminar structure of subicular IIDs without a laminar probe.

In general, recent work suggests ECoG primarily reflects supragranular activity. For instance, Ujma et al. recorded sleep spindles simultaneously with ECoG and laminar electrodes. Contrary to the hypothesized split of spindles into "matrix" and "core" types with different laminar profiles, they found that ECoG-recorded spindles consistently co-occurred with laminar spindles with maximal power in superficial layers [42]. Likewise, Halgren et al. found that ECoG recordings were most synchronous with superficial laminae during spontaneous wakefulness and sleep (Fig. 46.5a). Similar analysis of simultaneous ECoG and laminar recordings of slow-waves and alpha found high correlations between ECoG and superficial (but not deep) contacts [36, 61] (Fig. 46.5b); the former study found a Pearson correlation of $r > 0.9$ between LFPg recorded in layer II and neighboring ECoG in 5 different patients

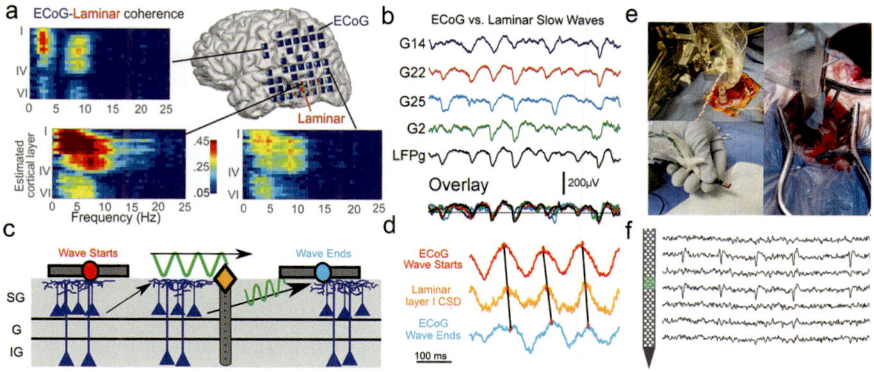

Fig. 46.5 **a** Coherence between individual ECoG contacts and a laminar array during spontaneous wakefulness, as a function of frequency and depth. Reproduced with permission from [46]. **b** LFPg from layer II and simultaneous ECoG during SWS in three patients. Reproduced with permission from [36]. **c** Schema of recording travelling waves between two ECoG contacts and a laminar array simultaneously. **d** Simultaneous ECoG and CSD of the alpha rhythm during spontaneous wakefulness. **c** and **d** reproduced with permission from [61]. **e** Implantation of a Neuropixel in a human surgical patient. **f** Action potential traces from a human Neuropixel recording. **e** and **f** from Dr. Angelique Paulk, with permission

[36]. HGP recorded by ECoG also reflects superficial layers, as the primary HGP generators are in supragranular cortex [72].

An important factor in the correlation between ECoG and superficial layers is the proximity of superficial laminae to ECoG contacts; a layer I dipole will a field significantly stronger than a layer VI dipole of equal current strength at the cortical surface [88]. However, even ECoG contacts several centimeters away from a laminar probe (for which the distance between superficial and deep layers becomes irrelevant), are coherent with superficial LFPg [46] (Fig. 46.5a). This may be due to the further finding that, between two laminar arrays, activity is most synchronous between the superficial contacts of the two probes [46]. The higher amplitude within and synchrony between superficial layers, in addition to their proximity to nearby ECoG contacts, likely explain the correspondence between superficial activity and simultaneous ECoG.

46.4.2 Laminar Structure of Travelling Waves and Propagating IIDs

Oscillations often propagate coherently across the cortical surface. Rhythms which exhibit this wave-like spread are called travelling waves (see Chap. 30). The speed, direction and frequency of these waves are linked to sleep, memory and perception [89–94], and could serve many computational roles [95]. Though their underlying physiology is unknown, propagation speeds and computational modeling point to horizontal fibers within superficial layers as the substrate for these rhythms [96]. ECoG-laminar recordings could allow for the imaging of travelling waves as they "pass through" the laminar array, and reveal which laminae travelling waves propagate "into" and "out of". For instance, if travelling waves generally propagate via intracortical short-range superficial fibers, we should see superficial current sinks coinciding with travelling waves as they propagate through the cortex in which a laminar is implanted (Fig. 46.5c). This was seen in an ECoG-laminar study of alpha travelling waves, which found that currents in superficial layers had a phase in between that of neighboring ECoG electrodes, suggesting a propagation within superficial laminae [61] (Fig. 46.5d).

Like cortical rhythms, IIDs also exhibit wave-like propagation through cortex. Ulbert et al. used simultaneous laminar-ECoG to study IIDs which were generated local to the laminar probe, versus IIDs which propagated towards the laminar array from a distant location. When IIDs were generated locally, they initiated within layer V. Conversely, IIDs propagating from a distal location were first seen in granular or supragranular layers [97]. The laminar structure and speed of this propagation ($\geq$1 m/s) is consistent with polysynaptic intracortical propagation. Fine-grained analysis of these events should yield further insights into the laminar physiology of propagating rhythms and IIDs.

46.4.3 Exotic (Non Somatic Action Potential) Waveform Physiology

Though most studies of extracellular activity focus on stereotyped action potential waveforms, other microscale events elude identification. These include W-shaped deflections, large positive spikes and waveforms too slow to reflect somatic spikes [98]. A tantalizing prospect is mapping these events to cellular phenomena studied in vitro, such as dendritic (NMDA & AMPA) spikes, plateau potentials, postsynaptic potentials, axonal spikes, etc. This would allow for the study of these sub-cellular events in behaving humans. A compelling example is the measurement of backpropagating action potentials with laminar recordings in mice, rats and rabbits [99–102]. Identifying backpropagating action potentials in human laminar recordings would allow experimentalists to see how these events are related to complex cognitive tasks involving feedback or top-down input.

An important step in this direction was performed by Paulk et al., who searched laminar recordings for these non-action potential exotic waveforms [103]. They found two distinct waveforms, dubbed Type 2 and Type 3 events. Type 2 events had a timescale of ~15 ms, and might correspond to backpropagating action potentials [100]. Type 3 events were significantly slower (~200 ms) and concentrated within superficial layers. This suggests that they may reflect dendritic calcium spikes, which were found to be supragranular and have a similar timescale in extracellular rodent recordings [73]. Critically, Paulk et al. also discovered these events in other recording modalities (PEDOT:PSS surface grids, Utah Arrays, Neuropixels) and species (macaques and mice), and could modulate them with behavioral tasks and pharmacological manipulation. In a further study using Neuropixels in humans, Paulk et al. found many positive-going waveforms, which likely correspond to axonal spikes [24]. Future human Neuropixel recordings have the fine spatial sampling to characterize these and other exotic events (Fig. 46.5e, f) [23, 24]. Fortuitously, this may be easier in humans than animal models due to the large size of human neurons. This approach could be validated by simultaneous patch-clamp and extracellular recordings performed in animal models and in vitro [98, 102, 104].

46.4.4 Validation of Extracranial Laminar Inference

Human laminar recordings can also validate and calibrate non-invasive alternatives to imaging laminar-specific activity in humans, such as high resolution fMRI and MEG [105–109]. Being able to make laminar recordings via fMRI and MEG, without the constraints of iEEG, would allow for ambitious future experiments. However, both methods make critical assumptions regarding how MEG/BOLD signals are related to neural activity which drastically impact estimated laminar sources. Though laminar fMRI has exquisite spatial resolution, BOLD activity does not reflect neural firing per se, but instead the underlying vascular architecture and cerebral blood oxygenation and volume, as well as MRI imaging parameters [110]. For instance, differences in baseline cerebral blood volume across the cortical depth can lead to differences in the BOLD response across laminae which do not reflect differences in neural activity. Because the vasculature structure and CBV varies across cortical regions, these confounds must be addressed on an area-specific basis [110]. This might be done with human laminar arrays that simultaneously record electrophysiological and hemodynamic data [111]. High-precision MEG has also made inroads into laminar imaging, largely via individualized headcasts and complex source-reconstruction techniques [107–109, 112]. Experiments using these advances have localized low and high frequency oscillations to different layers [108] and furthered a biophysical model of beta generation [112]. Despite this technique's promises, laminar source-reconstruction depends on assumed parameters such as source sparseness and SNR. Different assumptions concerning relative supragranular/infragranular source strength and spatial spread profoundly impact MEG estimates of laminar activity [107]. These assumptions may explain why MEG localizes alpha to deep layers, while human laminar recordings find supragranular alpha [61]. Measuring these parameters with ground truth laminar recordings can calibrate and refine the assumptions made by laminar MEG and fMRI [113].

46.5 Conclusion

Laminar recordings promise deep insights into human cortical physiology. Understanding how different layers interact to produce cognition and behavior allows iEEG to move beyond network-level characterizations to mapping cortical microcircuitry. By comparison with analogous animal experiments, we can also gain insights into how human neocortex diverges from other mammals. Despite being only two decades old, human laminar recordings have already yielded important insights into cortical physiology. Striking findings include the dominance of currents and firing within superficial layers during low-frequency oscillations, the laminar origins of ERPs and IIDs, and the dissociation of HGP and MUA. Further innovations in high-density probes and simultaneous macro-microelectrode recordings will allow for a deeper understanding of how different sub-laminae and cell types contribute to high-level cognition.

Acknowledgements I thank Sydney S. Cash, Vincent D Tang, Eric Halgren and Harnett Lab members for useful feedback.

References

1. Douglas RJ, Martin KAC (2004) Neuronal circuits of the neocortex. Annu Rev Neurosci 27:419–451. https://doi.org/10.1146/annurev.neuro.27.070203.144152
2. Adesnik H, Naka A (2018) Cracking the function of layers in the sensory cortex. Neuron
3. Boldog E, Bakken TE, Hodge RD et al (2018) Transcriptomic and morphophysiological evidence for a specialized human cortical GABAergic cell type. Nat Neurosci. https://doi.org/10.1038/s41593-018-0205-2
4. Bakken TE, Jorstad NL, Hu Q et al (2021) Comparative cellular analysis of motor cortex in human, marmoset and mouse. Nature 598:111–119. https://doi.org/10.1038/s41586-021-03465-8
5. Berg J, Sorensen SA, Ting JT et al (2021) Human neocortical expansion involves glutamatergic neuron diversification. Nature 598:151–158. https://doi.org/10.1038/s41586-021-03813-8
6. Beaulieu-Laroche L, Brown NJ, Hansen M et al (2021) Allometric rules for mammalian cortical layer 5 neuron biophysics. Nature 600:274–278. https://doi.org/10.1038/s41586-021-04072-3
7. Eyal G, Verhoog MB, Testa-Silva G et al (2016) Unique membrane properties and enhanced signal processing in human neocortical neurons. Elife 5:e16553. https://doi.org/10.7554/eLife.16553
8. Fang R, Xia C, Zhang M et al (2021) Conservation and divergence in cortical cellular organization between human and mouse revealed by single-cell transcriptome imaging. https://doi.org/10.1101/2021.11.01.466826
9. Hodge RD, Bakken TE, Miller JA et al (2019) Conserved cell types with divergent features in human versus mouse cortex. Nature 573:61–68. https://doi.org/10.1038/s41586-019-1506-7
10. Beaulieu-Laroche L, Toloza EHS, van der Goes MS et al (2018) Enhanced dendritic compartmentalization in human cortical neurons. Cell. https://doi.org/10.1016/j.cell.2018.08.045
11. Kalmbach BE, Buchin A, Long B et al (2018) h-channels contribute to divergent intrinsic membrane properties of supragranular pyramidal neurons in human versus mouse cerebral cortex. Neuron. https://doi.org/10.1016/j.neuron.2018.10.012

12. Campagnola L, Seeman SC, Chartrand T et al (2021) Local connectivity and synaptic dynamics in mouse and human neocortex. https://doi.org/10.1101/2021.03.31.437553
13. Molnár G, Rózsa M, Baka J et al (2016) Human pyramidal to interneuron synapses are mediated by multi-vesicular release and multiple docked vesicles. Elife 5. https://doi.org/10.7554/eLife.18167
14. Molnár G, Oláh S, Komlósi G et al (2008) Complex events initiated by individual spikes in the human cerebral cortex. PLoS Biol. https://doi.org/10.1371/journal.pbio.0060222
15. Seeman SC, Campagnola L, Davoudian PA et al (2018) Sparse recurrent excitatory connectivity in the microcircuit of the adult mouse and human cortex. Elife. https://doi.org/10.7554/eLife.37349
16. Hutsler JJ, Lee D-G, Porter KK (2005) Comparative analysis of cortical layering and supragranular layer enlargement in rodent carnivore and primate species. Brain Res 1052:71–81. https://doi.org/10.1016/j.brainres.2005.06.015
17. Hofman MA (1988) Size and shape of the cerebral cortex in mammals. II. The cortical volume. Brain Behav Evol 32:17–26. https://doi.org/10.1159/000116529
18. Heyer DB, Wilbers R, Galakhova AA et al (2021) Expansion of cortical layers 2 and 3 in human left temporal cortex associates with verbal intelligence. https://doi.org/10.1101/2021.02.07.430103
19. Krienen FM, Yeo BTT, Ge T et al (2016) Transcriptional profiles of supragranular-enriched genes associate with corticocortical network architecture in the human brain. Proc Natl Acad Sci 113:E469. https://doi.org/10.1073/pnas.1510903113
20. Schmidt ERE, Zhao HT, Park JM et al (2021) A human-specific modifier of cortical connectivity and circuit function. Nature 599:640–644. https://doi.org/10.1038/s41586-021-04039-4
21. Ulbert I, Halgren E, Heit G, Karmos G (2001) Multiple microelectrode-recording system for human intracortical applications. J Neurosci Methods 106:69–79. https://doi.org/10.1016/S0165-0270(01)00330-2
22. Buzsáki G, Anastassiou C, a., Koch (2012) The origin of extracellular fields and currents—EEG, ECoG, LFP and spikes. Nat Rev Neurosci 13:407–420. https://doi.org/10.1038/nrn3241
23. Chung JE, Sellers KK, Leonard MK et al (2021) High density single-unit human cortical recordings using the Neuropixels probe. https://doi.org/10.1101/2021.12.29.474489
24. Paulk AC, Kfir Y, Khanna AR et al (2022) Large-scale neural recordings with single neuron resolution using Neuropixels probes in human cortex. Nat Neurosci. https://doi.org/10.1038/s41593-021-00997-0
25. Jun JJ, Steinmetz NA, Siegle JH et al (2017) Fully integrated silicon probes for high-density recording of neural activity. Nature. https://doi.org/10.1038/nature24636
26. Berger H (1935) Das Elektrenkephalogramm des Menschen. Naturwissenschaften 23:121–124. https://doi.org/10.1007/BF01496966
27. Buzsáki G, Draguhn A (2004) Neuronal oscillations in cortical networks. Science 304:1926–1929. https://doi.org/10.1126/science.1099745
28. Born J (2010) Slow-wave sleep and the consolidation of long-term memory. World J Biol Psychiatry 11:16–21. https://doi.org/10.3109/15622971003637637
29. Fultz NE, Bonmassar G, Setsompop K et al (1979) (2019) Coupled electrophysiological, hemodynamic, and cerebrospinal fluid oscillations in human sleep. Science. https://doi.org/10.1126/science.aax5440
30. Beltramo R, D'Urso G, Dal Maschio M et al (2013) Layer-specific excitatory circuits differentially control recurrent network dynamics in the neocortex. Nat Neurosci. https://doi.org/10.1038/nn.3306
31. Chauvette S, Volgushev M, Timofeev I (2010) Origin of active states in local neocortical networks during slow sleep oscillation. Cereb Cortex 20:2660–2674. https://doi.org/10.1093/cercor/bhq009
32. Sakata S, Harris KD (2009) Laminar structure of spontaneous and sensory-evoked population activity in auditory cortex. Neuron. https://doi.org/10.1016/j.neuron.2009.09.020

33. Senzai Y, Fernandez-Ruiz A, Buzsáki G (2019) Layer-specific physiological features and interlaminar interactions in the primary visual cortex of the mouse. Neuron. https://doi.org/10.1016/j.neuron.2018.12.009
34. Sirota A, Csicsvari J, Buhl D, Buzsaki G (2003) Communication between neocortex and hippocampus. Proc Natl Acad Sci 100:2065–2069
35. Cash SS, Halgren E, Dehghani N et al (1979) (2009) The human k-complex represents an isolated cortical down-state. Science 324:1084–1087. https://doi.org/10.1126/science.1169626
36. Csercsa R, Dombovári B, Fabó D et al (2010) Laminar analysis of slow wave activity in humans. Brain 133:2814–2829
37. Latchoumane CF, v., Ngo HV v., Born J, Shin (2017) Thalamic spindles promote memory formation during sleep through triple phase-locking of cortical, thalamic, and hippocampal rhythms. Neuron 95:424-435.e6. https://doi.org/10.1016/j.neuron.2017.06.025
38. Klinzing JG, Niethard N, Born J (2019) Mechanisms of systems memory consolidation during sleep. Nat Neurosci. https://doi.org/10.1038/s41593-019-0467-3
39. Maingret N, Girardeau G, Todorova R et al (2016) Hippocampo-cortical coupling mediates memory consolidation during sleep. Nat Neurosci. https://doi.org/10.1038/nn.4304
40. Durkin J, Suresh AK, Colbath J et al (2017) Cortically coordinated NREM thalamocortical oscillations play an essential, instructive role in visual system plasticity. Proc Natl Acad Sci U S A. https://doi.org/10.1073/pnas.1710613114
41. Hagler DJ, Ulbert I, Wittner L et al (2018) heterogeneous origins of human sleep spindles in different cortical layers. J Neurosci. https://doi.org/10.1523/jneurosci.2241-17.2018
42. Ujma PP, Hajnal B, Bódizs R et al (2021) The laminar profile of sleep spindles in humans. Neuroimage. https://doi.org/10.1016/j.neuroimage.2020.117587
43. Piantoni G, Halgren E, Cash SS (2016) The contribution of thalamocortical core and matrix pathways to sleep spindles. Neural Plasticity
44. Sachdev RNS, Gaspard N, Gerrard JL et al (2015) Delta rhythm in wakefulness: evidence from intracranial recordings in human beings. J Neurophysiol 114:1248–1254. https://doi.org/10.1152/jn.00249.2015
45. Herweg NA, Solomon EA, Kahana MJ (2020) Theta oscillations in human memory. trends in cognitive sciences
46. Halgren M, Fabó D, Ulbert I et al (2018) Superficial slow rhythms integrate cortical processing in humans. Sci Rep 8:2055. https://doi.org/10.1038/s41598-018-20662-0
47. Halgren E, Kaestner E, Marinkovic K et al (2015) Laminar profile of spontaneous and evoked theta: rhythmic modulation of cortical processing during word integration. Neuropsychologia 76:108–124. https://doi.org/10.1016/j.neuropsychologia.2015.03.021
48. Gaspard N, Manganas L, Rampal N et al (2013) Similarity of lateralized rhythmic delta activity to periodic lateralized epileptiform discharges in critically Ill patients. JAMA Neurol 70:1288–1295. https://doi.org/10.1001/jamaneurol.2013.3475
49. Saalmann YB, Pinsk MA, Wang L et al (1979) (2012) The pulvinar regulates information transmission between cortical areas based on attention demands. Science 337:753–756. https://doi.org/10.1126/science.1223082
50. Samaha J, Bauer P, Cimaroli S, Postle BR (2015) Top-down control of the phase of alpha-band oscillations as a mechanism for temporal prediction. Proc Natl Acad Sci 112:201503686. https://doi.org/10.1073/pnas.1503686112
51. Busch NA, Dubois J, VanRullen R (2009) The phase of ongoing EEG oscillations predicts visual perception. J Neurosci 29:7869–7876. https://doi.org/10.1523/JNEUROSCI.0113-09.2009
52. Jensen O, Mazaheri A (2010) Shaping functional architecture by oscillatory alpha activity: gating by inhibition. Front Hum Neurosci 4:186. https://doi.org/10.3389/fnhum.2010.00186
53. van Kerkoerle T, Self MW, Dagnino B et al (2014) Alpha and gamma oscillations characterize feedback and feedforward processing in monkey visual cortex. Proc Natl Acad Sci 111:14332–14341. https://doi.org/10.1073/pnas.1402773111

54. Silva L, Amitai Y (1979) Connors B (1991) Intrinsic oscillations of neocortex generated by layer 5 pyramidal neurons. Science 251:432–435. https://doi.org/10.1126/science.1824881
55. Buffalo EA, Fries P, Landman R et al (2011) Laminar differences in gamma and alpha coherence in the ventral stream. Proc Natl Acad Sci U S A 108:11262–11267. https://doi.org/10.1073/pnas.1011284108
56. Bollimunta A, Mo J, Schroeder CE, Ding M (2011) Neuronal mechanisms and attentional modulation of corticothalamic alpha oscillations. J Neurosci 31:4935–4943. https://doi.org/10.1523/JNEUROSCI.5580-10.2011
57. Womelsdorf T, Valiante TA, Sahin NT et al (2014) Dynamic circuit motifs underlying rhythmic gain control, gating and integration. Nat Neurosci 17:1031–1039
58. Bastos AM, Loonis R, Kornblith S et al (2018) Laminar recordings in frontal cortex suggest distinct layers for maintenance and control of working memory. Proc Natl Acad Sci 115:1117–1122. https://doi.org/10.1073/pnas.1710323115
59. Haegens S, Barczak A, Musacchia G et al (2015) Laminar profile and physiology of the α rhythm in primary visual, auditory, and somatosensory regions of neocortex. J Neurosci 35:14341–14352. https://doi.org/10.1523/JNEUROSCI.0600-15.2015
60. Bollimunta A, Chen Y, Schroeder CE, Ding M (2008) Neuronal mechanisms of cortical alpha oscillations in awake-behaving macaques. J Neurosci 28:9976–9988. https://doi.org/10.1523/JNEUROSCI.2699-08.2008
61. Halgren M, Ulbert I, Bastuji H et al (2019) The generation and propagation of the human alpha rhythm. Proc Natl Acad Sci 13092. https://doi.org/10.1073/pnas.1913092116
62. Neske GT (2016) The slow oscillation in cortical and thalamic networks: mechanisms and functions. Front Neural Circ
63. Kutas M, Federmeier KD (2011) Thirty years and counting: finding meaning in the N400 component of the event-related brain potential (ERP). Annu Rev Psychol 62:621–647. https://doi.org/10.1146/annurev.psych.093008.131123
64. Brown C, Hagoort P (1993) The processing nature of the N400: evidence from masked priming. J Cogn Neurosci 5:34–44. https://doi.org/10.1162/jocn.1993.5.1.34
65. Bornkessel-Schlesewsky I, Schlesewsky M (2019) Toward a neurobiologically plausible model of language-related, negative event-related potentials. Front Psychol 10
66. Halgren E, Dhond RP, Christensen N et al (2002) N400-like magnetoencephalography responses modulated by semantic context, word frequency, and lexical class in sentences. Neuroimage 17:1101–1116. https://doi.org/10.1006/nimg.2002.1268
67. Halgren E, Wang C, Schomer DL et al (2006) Processing stages underlying word recognition in the anteroventral temporal lobe. Neuroimage 30:1401. https://doi.org/10.1016/J.NEUROIMAGE.2005.10.053
68. Mukamel R, Gelbard H, Arieli A et al (1979) (2005) Neuroscience: Coupling between neuronal firing, field potentials, and fMRI in human auditory cortex. Science. https://doi.org/10.1126/science.1110913
69. Manning JR, Jacobs J, Fried I, Kahana MJ (2009) Broadband shifts in local field potential power spectra are correlated with single-neuron spiking in humans. J Neurosci 29:13613–13620. https://doi.org/10.1523/JNEUROSCI.2041-09.2009
70. Ray S, Crone NE, Niebur E et al (2008) Neural correlates of high-gamma oscillations (60–200 Hz) in macaque local field potentials and their potential implications in electrocorticography. J Neurosci 28:11526–11536. https://doi.org/10.1523/JNEUROSCI.2848-08.2008
71. Lachaux J-P, Axmacher N, Mormann F et al (2012) High-frequency neural activity and human cognition: past, present and possible future of intracranial EEG research. Prog Neurobiol 98:279–301. https://doi.org/10.1016/j.pneurobio.2012.06.008
72. Leszczyński M, Barczak A, Kajikawa Y et al (2020) Dissociation of broadband high-frequency activity and neuronal firing in the neocortex. Sci Adv 6:977–989. https://doi.org/10.1126/SCIADV.ABB0977/SUPPL_FILE/ABB0977_SM.PDF
73. Suzuki M, Larkum ME (2017) Dendritic calcium spikes are clearly detectable at the cortical surface. Nat Commun 8:276. https://doi.org/10.1038/s41467-017-00282-4

74. Gratiy SL, Halnes G, Denman D et al (2017) From Maxwell's equations to the theory of current-source density analysis. Eur J Neurosci 45:1013–1023. https://doi.org/10.1111/ejn.13534
75. Nicholson C, Freeman JA (1975) Theory of current source-density analysis and determination of conductivity tensor for anuran cerebellum. J Neurophysiol 38:356–368. https://doi.org/10.1121/1.3569737
76. Pesaran B, Vinck M, Einevoll GT et al (2018) Investigating large-scale brain dynamics using field potential recordings: analysis and interpretation. Nat Neurosci 21:903–919. https://doi.org/10.1038/s41593-018-0171-8
77. Herreras O (2016) Local field potentials: myths and misunderstandings. Front Neural Circuits. https://doi.org/10.3389/fncir.2016.00101
78. Kajikawa Y, Schoeder E (2012) How local is the local field potential? Neuron 72:847–858. https://doi.org/10.1016/j.neuron.2011.09.029.How
79. Kajikawa Y, Smiley JF, Schroeder CE (2017) Primary generators of visually evoked field potentials recorded in the macaque auditory cortex. J Neurosci. https://doi.org/10.1523/JNEUROSCI.3800-16.2017
80. Kajikawa Y, Schroeder CE (2015) Generation of field potentials and modulation of their dynamics through volume integration of cortical activity. J Neurophysiol 113:339–351. https://doi.org/10.1152/jn.00914.2013
81. Vaknin G, DiScenna PG, Teyler TJ (1988) A method for calculating current source density (CSD) analysis without resorting to recording sites outside the sampling volume. J Neurosci Methods 24:131–135. https://doi.org/10.1016/0165-0270(88)90056-8
82. Potworowski J, Jakuczun W, Łęski S, Wójcik D (2012) Kernel current source density method. Neural Comput 24:541–575. https://doi.org/10.1162/NECO_a_00236
83. Elston GN, Benavides-Piccione R, DeFelipe J (2001) The pyramidal cell in cognition: a comparative study in human and monkey. J Neurosci 21:RC163 LP-RC163. https://doi.org/10.1523/JNEUROSCI.21-17-j0002.2001
84. Mohan H, Verhoog MB, Doreswamy KK et al (2015) Dendritic and axonal architecture of individual pyramidal neurons across layers of adult human neocortex. Cereb Cortex 25. https://doi.org/10.1093/cercor/bhv188
85. Ulbert I, Karmos G, Heit G, Halgren E (2001) Early discrimination of coherent versus incoherent motion by multiunit and synaptic activity in human putative MT+. Hum Brain Mapp 13:226–238. https://doi.org/10.1002/HBM.1035
86. Rich EL, Wallis JD (2017) Spatiotemporal dynamics of information encoding revealed in orbitofrontal high-gamma. Nat Commun 8:1–13. https://doi.org/10.1038/s41467-017-01253-5
87. Fabó D, Maglóczky Z, Wittner L et al (2008) Properties of in vivo interictal spike generation in the human subiculum. Brain 131:485–499. https://doi.org/10.1093/BRAIN/AWM297
88. Griffiths 1942-DJ (2013) David J introduction to electrodynamics, 4th edn. Pearson, Boston [2013] ©2013
89. Zhang H, Watrous AJ, Patel A, Jacobs J (2018) Theta and alpha oscillations are traveling waves in the human neocortex. Neuron
90. Muller L, Piantoni G, Koller D et al (2016) Rotating waves during human sleep spindles organize global patterns of activity that repeat precisely through the night. Elife 5. https://doi.org/10.7554/eLife.17267.001
91. Massimini M (2004) The sleep slow oscillation as a traveling wave. J Neurosci. https://doi.org/10.1523/jneurosci.1318-04.2004
92. Pang 庞兆阳 Z, Alamia A, VanRullen R (2020) Turning the stimulus on and off changes the direction of α traveling waves. eNeuro 7:ENEURO.0218–20.2020. https://doi.org/10.1523/ENEURO.0218-20.2020
93. Davis ZW, Muller L, Martinez-Trujillo J et al (2020) Spontaneous travelling cortical waves gate perception in behaving primates. Nature 587:432–436. https://doi.org/10.1038/s41586-020-2802-y

94. Dickey CW, Sargsyan A, Madsen JR et al (2021) Travelling spindles create necessary conditions for spike-timing-dependent plasticity in humans. Nat Commun 12:1027. https://doi.org/10.1038/s41467-021-21298-x
95. Muller L, Chavane F, Reynolds J, Sejnowski TJ (2018) Cortical travelling waves: mechanisms and computational principles. Nat Rev Neurosci 19:255–268. https://doi.org/10.1038/nrn.2018.20
96. Davis ZW, Benigno GB, Fletterman C et al (2021) Spontaneous traveling waves naturally emerge from horizontal fiber time delays and travel through locally asynchronous-irregular states. Nat Commun 12:6057. https://doi.org/10.1038/s41467-021-26175-1
97. Ulbert I, Heit G, Madsen J et al (2004) Laminar analysis of human neocortical interictal spike generation and propagation: current source density and multiunit analysis in vivo. Epilepsia 45:48–56. https://doi.org/10.1111/J.0013-9580.2004.04011.X
98. Gold C, Henze DA, Koch C (2007) Using extracellular action potential recordings to constrain compartmental models. J Comput Neurosci 23:39–58. https://doi.org/10.1007/s10827-006-0018-2
99. Bereshpolova Y, Amitai Y, Gusev AG et al (2007) Dendritic backpropagation and the state of the awake neocortex. J Neurosci 27:9392–9399. https://doi.org/10.1523/JNEUROSCI.2218-07.2007
100. Buzsáki G, Kandel A (1998) Somadendritic Backpropagation of Action Potentials in Cortical Pyramidal Cells of the Awake Rat. J Neurophysiol 79:1587–1591. https://doi.org/10.1152/jn.1998.79.3.1587
101. Jia X, Siegle JH, Bennett C et al (2019) High-density extracellular probes reveal dendritic backpropagation and facilitate neuron classification. J Neurophysiol 121:1831–1847. https://doi.org/10.1152/jn.00680.2018
102. Marques-Smith A, Neto JP, Lopes G et al (2020) Recording from the same neuron with high-density CMOS probes and patch-clamp: a ground-truth dataset and an experiment in collaboration. bioRxiv 370080. https://doi.org/10.1101/370080
103. Paulk AC, Yang JC, Cleary DR et al (2021) Microscale physiological events on the human cortical surface. Cereb Cortex. https://doi.org/10.1093/cercor/bhab040
104. Gold C, Henze DA, Koch C, Buzsáki G (2006) On the origin of the extracellular action potential waveform: a modeling study. J Neurophysiol 95:3113–3128. https://doi.org/10.1152/jn.00979.2005
105. Lawrence SJD, Formisano E, Muckli L, de Lange FP (2019) Laminar fMRI: applications for cognitive neuroscience. Neuroimage 197:785–791. https://doi.org/10.1016/j.neuroimage.2017.07.004
106. Chen G, Wang F, Gore JC, Roe AW (2013) Layer-specific BOLD activation in awake monkey V1 revealed by ultra-high spatial resolution functional magnetic resonance imaging. Neuroimage 64:147–155. https://doi.org/10.1016/j.neuroimage.2012.08.060
107. Bonaiuto JJ, Rossiter HE, Meyer SS, et al (2018) Non-invasive laminar inference with MEG: comparison of methods and source inversion algorithms. Neuroimage 167:372–383. https://doi.org/10.1016/j.neuroimage.2017.11.068
108. Bonaiuto JJ, Meyer SS, Little S et al (2018) Lamina-specific cortical dynamics in human visual and sensorimotor cortices. Elife 7:e33977. https://doi.org/10.7554/eLife.33977
109. Troebinger L, López JD, Lutti A et al (2014) Discrimination of cortical laminae using MEG. Neuroimage 102(Pt 2):885–893. https://doi.org/10.1016/j.neuroimage.2014.07.015
110. Uludağ K, Blinder P (2018) Linking brain vascular physiology to hemodynamic response in ultra-high field MRI. Neuroimage 168:279–295. https://doi.org/10.1016/j.neuroimage.2017.02.063
111. Keller CJ, Cash SS, Narayanan S et al (2009) Intracranial microprobe for evaluating neuro-hemodynamic coupling in unanesthetized human neocortex. J Neurosci Methods 179:208–218. https://doi.org/10.1016/J.JNEUMETH.2009.01.036
112. Bonaiuto JJ, Little S, Neymotin SA et al (2021) Laminar dynamics of high amplitude beta bursts in human motor cortex. Neuroimage 242:118479. https://doi.org/10.1016/j.neuroimage.2021.118479

113. Rosen BQ, Krishnan GP, Sanda P et al (2019) Simulating human sleep spindle MEG and EEG from ion channel and circuit level dynamics. J Neurosci Methods 316:46–57. https://doi.org/10.1016/J.JNEUMETH.2018.10.002
114. Stringer C, Pachitariu M, Steinmetz N et al (2019) Spontaneous behaviors drive multidimensional, brainwide activity. Science. https://doi.org/10.1126/science.aav7893

Chapter 47
How Does Artificial Intelligence Contribute to iEEG Research?

Julia Berezutskaya, Anne-Lise Saive, Karim Jerbi, and Marcel van Gerven

Abstract Artificial intelligence (AI) is a fast-growing field focused on modeling and machine implementation of various cognitive functions with an increasing number of applications in computer vision, text processing, robotics, neurotechnology, bio-inspired computing and others. In this chapter, we describe how AI methods can be applied in the context of intracranial electroencephalography (iEEG) research. IEEG data is unique as it provides extremely high-quality signals recorded directly from brain tissue. Applying advanced AI models to this data carries the potential to further our understanding of many fundamental questions in neuroscience. At the same time, as an invasive technique, iEEG lends itself well to long-term, mobile brain-computer interface applications, particularly for communication in severely paralyzed individuals. We provide a detailed overview of these two research directions in the application of AI techniques to iEEG. That is, (1) the development of computational models that target fundamental questions about the neurobiological nature of cognition (AI-iEEG for neuroscience) and (2) applied research on monitoring and identification of event-driven brain states for the development of clinical brain-computer interface systems (AI-iEEG for neurotechnology). We explain key machine learning concepts, specifics of processing and modeling iEEG data and details of state-of-the-art iEEG-based neurotechnology and brain-computer interfaces.

J. Berezutskaya (✉) · M. van Gerven
Donders Institute for Brain, Cognition & Behaviour, Radboud University, Thomas van Aquinostraat 4, 6525 GD Nijmegen, The Netherlands
e-mail: yuliya.berezutskaya@donders.ru.nl

M. van Gerven
e-mail: marcel.vangerven@donders.ru.nl

A.-L. Saive
Institut Paul Bocuse Research Center, Lyon, France
e-mail: anne-lise.saive@institutpaulbocuse.com

K. Jerbi
Cognitive & Computational Neuroscience Lab, University of Montreal, Pavillon Marie-Victorin 90, Avenue Vincent d'Indy, Quebec, Canada
e-mail: karim.jerbi@umontreal.ca

N. Axmacher (ed.), *Intracranial EEG*, Studies in Neuroscience, Psychology and Behavioral Economics, https://doi.org/10.1007/978-3-031-20910-9_47

47.1 AI-iEEG for Neuroscience

The field of computational cognitive neuroscience uses mathematical models to describe neural processes underlying cognition and behaviour [1–3]. Advances in machine learning and the rapid increase in computational power have made it possible to apply sophisticated analysis methods to large amounts of brain data collected via increasingly sophisticated recording techniques [4].

Inspired by the computational metaphor of the brain as an information processing device, this has led to the emergence of so-called *encoding models* and *decoding models* in cognition and perception [5–8]. Neural encoding models capture how information is represented and processed in the brain. An approach related to encoding models, that has become increasingly popular in cognitive neuroscience, is *representational similarity analysis* (RSA) [9]. RSA more directly compares stimulus features as encoded in computational models and encoded in patterns of brain activity. Decoding models also aim to understand the representation of information in the brain but constitute the reverse approach of inferring information features from observed brain activity. Apart from the practical value of decoding approaches for neurotechnology (see Sect. 47.2), decoding models can also be informative in relating stimulus features to neural signals.

Given that many cognitive functions, including perception, memory, language and complex sensorimotor behaviour, are supported by large-scale distributed processes in the brain, data from neuroimaging (fMRI) and whole-brain electrophysiological (EEG/MEG) experiments in healthy individuals has been the primary source for construction and validation of computational models. The fMRI, MEG and EEG communities have provided large amounts of experimental data used for modeling cognitive processes in the brain [10–15].

However, limitations of non-invasive brain recording modalities have created a need for the use of intracranial data, such as intracranial electroencephalography (iEEG), which is a general term that refers to both surface electrocorticography grids (ECoG) and depth stereo-encephalography electrodes (sEEG). This is because iEEG offers a number of benefits compared to non-invasive modalities, such as sampling directly from the cortical tissue, high signal-to-noise ratio and high temporal and spatial resolution. Recent work has seen successful efforts in creation and validation of encoding and decoding models, RSA and models of temporal neural dynamics using iEEG.

In this section, we will describe encoding and decoding models, and include discussion of RSA and complex temporal dynamics in iEEG data. First, we will discuss linear encoding models for iEEG signals, including hand-engineered and deep-learning-based feature extraction for these models, and emergence of advanced non-linear neural encoding models. Then, we will examine RSA applications to iEEG. We will finish the section with a description of decoding models applied to iEEG data, including multivariate pattern analysis, and a discussion of associated temporal generalization concerns.

47.1.1 Encoding Models of Perception and Cognition

47.1.1.1 Linear Encoding Models

A linear neural encoding model represents a linear mapping of a set of *features* (model input) to the observed brain activity (model output, see also Chap. 48). Fitting such a model is based on the assumption that the brain activity is a weighted linear combination of features and normally distributed random noise:

$$\mathbf{Y} = \mathbf{X}\mathbf{B} + \boldsymbol{\epsilon}, \tag{47.1}$$

where $\mathbf{Y} \in \mathbb{R}^{n \times m}$ is a matrix of observed neural responses with n corresponding to the number of observations over time or trials, m corresponding, for example, to the number of iEEG electrodes or fMRI voxels, $\mathbf{X} \in \mathbb{R}^{n \times p}$ is a matrix of input features over n time points and p feature dimensions, $\mathbf{B} \in \mathbb{R}^{p \times m}$ is a matrix of regression coefficients and $\boldsymbol{\epsilon} \sim \mathcal{N}(\mathbf{0}, \boldsymbol{\sigma}^2\mathbf{I})$ is a noise term with $\mathbf{I}$ being the identity matrix.

In this formulation, the model input—the feature matrix $\mathbf{X}$—represents properties of the stimulus that may have triggered the observed brain responses. Linear encoding models are widely used in fitting brain activity evoked by external stimuli (Fig. 47.1). IEEG examples that use this simple form of the linear model include neural encoding of visual [16, 17] and auditory [18–21] perceptual features, motor information [22–24], memory, language [25–27] and decision making concepts [28].

Typically, models in these studies are fitted to predict the temporal envelope of the high-frequency broadband component (>60 Hz) in iEEG data (obtained with time-frequency analysis), as it is known to reflect local neural firing rates evoked by the stimulus [29, 30]. Alternative encoding approaches have been employed to model frequency-domain representations of iEEG signals [31–34]. Time-domain iEEG responses on average show less straightforward modulation by a cognitive task and are more varied compared to frequency-domain responses [35], and are therefore less typically used in linear fitting with stimulus features.

When working with a time-frequency representation of iEEG signals, kernel functions $g(\cdot)$ can be used to transform input stimulus features into a new representation that could better reflect temporal iEEG structure, yielding an encoding of the form

$$\mathbf{Y} = g(\mathbf{X})\mathbf{B} + \boldsymbol{\epsilon}. \tag{47.2}$$

This is similar to fMRI research, where an input feature matrix is convolved with the hemodynamic response function for the general linear model fit. For iEEG, Gaussian kernels can be applied to stimulus features to better approximate smoother iEEG time courses. Examples of studies that use such feature transformations include [36, 37].

Another property of the spectrotemporal iEEG signal is its high temporal resolution. It is usually addressed in models of continuous iEEG dynamics (as opposed to fitting trial-based data) by inclusion of lags and temporal integration over input features, such that

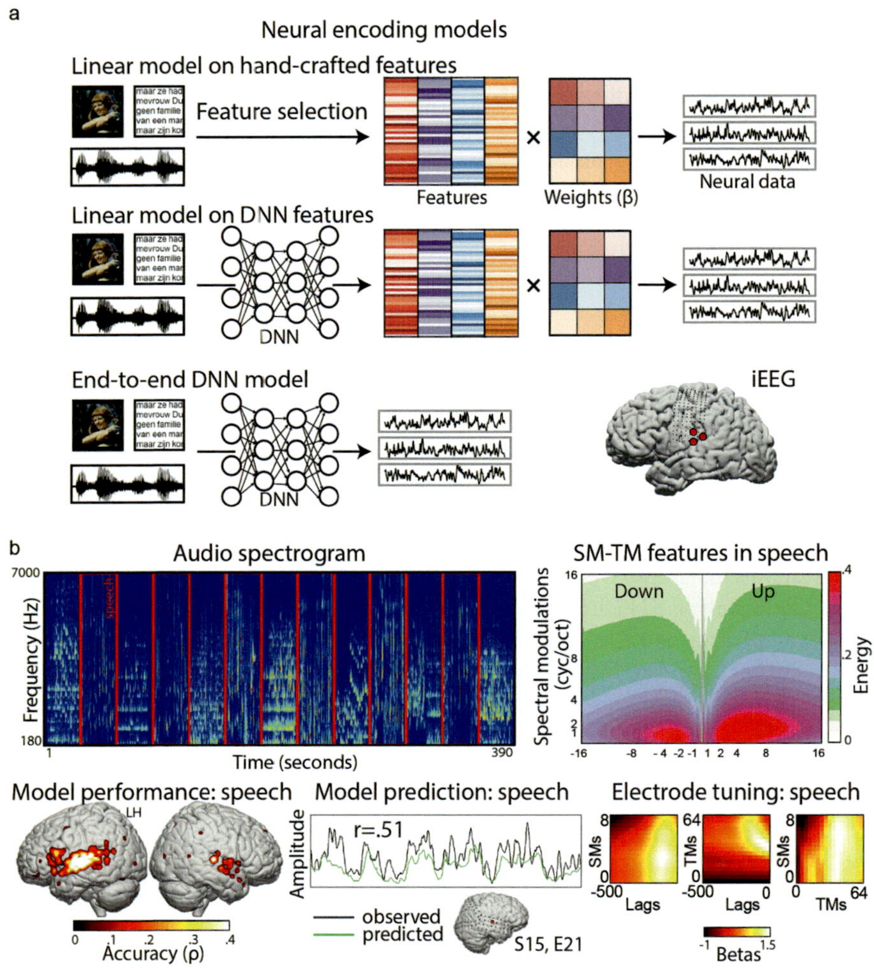

Fig. 47.1 **a** Overview of neural encoding models: linear encoding model on hand-crafted stimulus features, linear encoding model on features automatically extracted from the stimulus using deep learning models and end-to-end non-linear encoding model based on a deep learning approach. Each model in this example aims to predict high frequency band iEEG responses of three electrodes. **b** A linear encoding model that predicts ECoG high frequency band activity during speech dialogues in a short audiovisual film [20]. A set of spectrotemporal modulations (SM-TM) was extracted from the sound spectrogram and used to model associated iEEG responses throughout the perisylvian cortex. Each iEEG electrode exhibited tuning to specific features at different time latencies. Adapted from [20]. Copyright 2017 Berezutskaya et al.

$$\mathbf{y}_t = \bar{\mathbf{B}}\bar{\mathbf{x}}_t + \boldsymbol{\epsilon}, \tag{47.3}$$

where $\mathbf{y_t} \in \mathbb{R}^m$ is a vector of iEEG responses per time point t, $\bar{\mathbf{B}} \in \mathbb{R}^{m\times(p\times\tau)}$ a matrix of regression coefficients, τ is the time lag relative to t and $\bar{\mathbf{x}}_t$ a stack of vectors $\mathbf{x}_{t-s}$

with $s \in [0, \tau]$. Features at multiple lags have been used in models of encoding auditory, speech and movement information in iEEG [20, 23].

A standard linear regression problem is typically solved using ordinary least squares estimation of regression coefficients **B**, which can be achieved analytically:

$$\mathbf{B} = (\mathbf{X}^\top \mathbf{X})^{-1} \mathbf{X}^\top \mathbf{Y} \,. \tag{47.4}$$

To assess model performance, various metrics are used for comparing predicted and observed brain responses. A popular choice is a Pearson correlation coefficient, which measures the normalized covariance of predictions and their targets. To assess generalization potential of the modeling results, encoding models are trained on a subset of all data (typically 80–90%), and tested on a held-out test set (remaining data). To account for potential autocorrelation in iEEG signals, data from separate recording sessions and different experiments can be used as an independent test set [38]. Model performance is often *cross-validated* to obtain a more unbiased estimate of encoding accuracy.

To prevent *overfitting* to training data (and subsequent lack of generalization on test data), *regularization* methods are often used. They provide constraints, or priors, on model coefficients [9], resulting in a less flexible model fit and therefore are less prone to overfitting on the training data. Regularization is effective in models that use large numbers of correlated features. A popular method is L2 regularization that constrains **B** to small values. A linear model that uses L2 regularization is called a ridge regression and has the following analytical solution:

$$\mathbf{B} = (\mathbf{X}^\top \mathbf{X} + \lambda \mathbf{I})^{-1} \mathbf{X}^\top \mathbf{Y}, \tag{47.5}$$

where $\lambda \geq 0$ is a regularization parameter, typically chosen via nested cross-validation.

Neural information processing includes high non-linear transformations of its input. Therefore, the linear models we often use have severe limitations. They can be mitigated by fitting linear encoding models on non-linear feature representations. The latter can be either manually defined or extracted from computational models such as deep and recurrent neural networks.

47.1.1.2 Feature Engineering for Linear Encoding Models

To fit a linear model of brain activity as a response to external stimuli, one requires a set of features that represent stimulus information. Traditionally and similar to other recording modalities, such as fMRI, MEG and EEG, *hand-engineered features* have been used for this. For example, spectrotemporal features were used in encoding of auditory and speech information in iEEG [18–20], two-dimensional Gaussian features were used in population receptive field work in vision [16, 32] and kinematic trajectories and grasp features were used for modeling sensorimotor iEEG signals [23, 37].

The use of hand-engineered features to represent information means that the choice of features is at the discretion of the researcher. Features could be manually assigned data labels (for example, language labels, such as words or phonemes, classes of visually presented images, broadly defined experimental conditions) or a representation of the stimulus extracted via a preprocessing step or a mathematical model (for example, spectrotemporal audio features or edge-filtered image features). In the case of a sufficiently good fit compared against a model with random predictors, researchers analyse neural tuning profiles, or stimulus receptive fields. Many studies compare the fit of models that use different feature sets in an attempt to identify features that best explain observed brain data [39–43].

Despite its overall success, previous work (also with other neural recording techniques) highlighted some problems with using hand-engineered features, specifically, the inability to determine which feature sets should be deemed the optimal representation of information in a specific brain region; and how existing features need to be modified to explain brain responses better. Moreover, the use of hand-engineered features lacked the ability to explain how lower-level feature sets (such as image edges or orientations and spectrogram sound features) transformed to higher-level features (such as object categories and different words) as a result of information processing throughout the brain. Recent progress in *deep learning*, has opened up new possibilities in addressing the problem of feature engineering by minimizing manual feature selection and delegating feature extraction to powerful non-linear models.

In a vanilla deep artificial neural network (DNN), input features are passed to the first layer, where their weighted linear combination goes through a non-linear transformation, called an activation function, to obtain layer activations **a**. This computation is repeated at the next layers. The final layer activations are used to predict the targets. This yields the following sequence of non-linear transformations:

$$\mathbf{a}^{(1)} = g\left(\mathbf{B}^{(1)}\mathbf{x}_t + \mathbf{b}^{(1)}\right) \tag{47.6}$$

$$\mathbf{a}^{(2)} = g\left(\mathbf{B}^{(2)}\mathbf{a}^{(1)} + \mathbf{b}^{(2)}\right) \tag{47.7}$$

$$\hat{\mathbf{y}} = g\left(\mathbf{B}^{(k)}\mathbf{a}^{(k-1)} + \mathbf{b}^{(k)}\right) \tag{47.8}$$

where $k \in \mathbb{N}$ is the number of layers and **b** is a layer-specific bias vector.

DNNs are trained to optimize an *objective function* associated with the task they need to solve, for example labeling visual objects in the presented image, or predicting next words in a sentence. Minimizing mean squared error is a popular objective function for regression tasks (such as time series prediction), and cross-entropy loss is used in most classification tasks (such as object recognition). To optimize the objective function a *learning algorithm*, typically, gradient descent, is applied iteratively. This yields incremental updates of learnable parameters of the DNN, such as weights of its artificial neurons. In most cases, backpropagation is used for efficient gradient calculation. As a result of this learning, DNN layer activations can capture increasingly complex non-linear transformations of simple inputs (such as images, texts and sounds) necessary for solving the task. For a more detailed introduction to DNNs see [44–46].

Depending on the type of data and problem, various DNN architectures that outline network structure and connections can be used. *Multi-layer perceptrons* (MLPs) are feed-forward fully-connected DNNs as described by Eq. 47.8. *Convolutional neural networks* are feed-forward neural networks that use convolutions instead of standard matrix multiplication. They have had tremendous success in computer vision tasks, such as visual object recognition, image colorization and super-resolution. *Recurrent neural networks* [47] are cyclic models best suited for learning complex temporal dynamics in the data and have therefore been popular in language processing tasks, such as language modeling, machine translation and speech recognition. Recurrent architectures have recently been outperformed by *transformer* architectures [48]—feed-forward neural networks with an encoder and decoder component and attention modules for efficient input processing. Deep belief networks, autoencoders [49], graph DNNs and other DNN variants also have their practical applications. For a detailed overview of DNN architectures we refer readers to dedicated reviews [50–53].

Convolutional and recurrent DNNs have been successfully used for automated complex feature extraction. High-level learning tasks, such as object recognition or language modeling, and hierarchical structure of deep neural networks allow extraction of increasingly more complex representations from basic naturalistic inputs, such as images, sounds and texts. The resulting features across neural network layers have been used for linear mapping onto brain activity, first with non-invasive techniques [42, 54] and more recently using iEEG [55, 56] (Fig. 47.2). Many studies have demonstrated that DNN features explain and correlate with brain activity better than alternative, typically hand-engineered features in language [26, 27], audio perception [54, 57] and vision [42, 58, 59] (Fig. 47.2). Some iEEG work explored gradients of feature complexity throughout the model, similar to analogous work in neuroimaging [55, 56].

47.1.1.3 End-to-End Encoding Models Using Deep Learning

Success of deep learning models in extracting features that accurately predict observed activity throughout the brain, has led to the "deep learning revolution" in cognitive neuroscience. DNN feature models and activations became widely used for exploration and interpretation of the neural mechanisms underlying cognition and perception. Studies identified DNN properties crucial for accurate modeling of observed neural responses, such as hierarchical structure with increasingly complex features [54, 60], local recurrence [61, 62], feedback connections, non-linear mixed selectivity of single neurons [63] with ensemble-based feature separability [64]. See [44, 65–75] for perspectives and reviews and [76–80] for critical reports.

Following this line of thinking, several researchers in the field have argued that the next logical step in this direction is development of *end-to-end* deep learning models that map stimuli and behavior to brain activity directly [44, 81]. Attempts

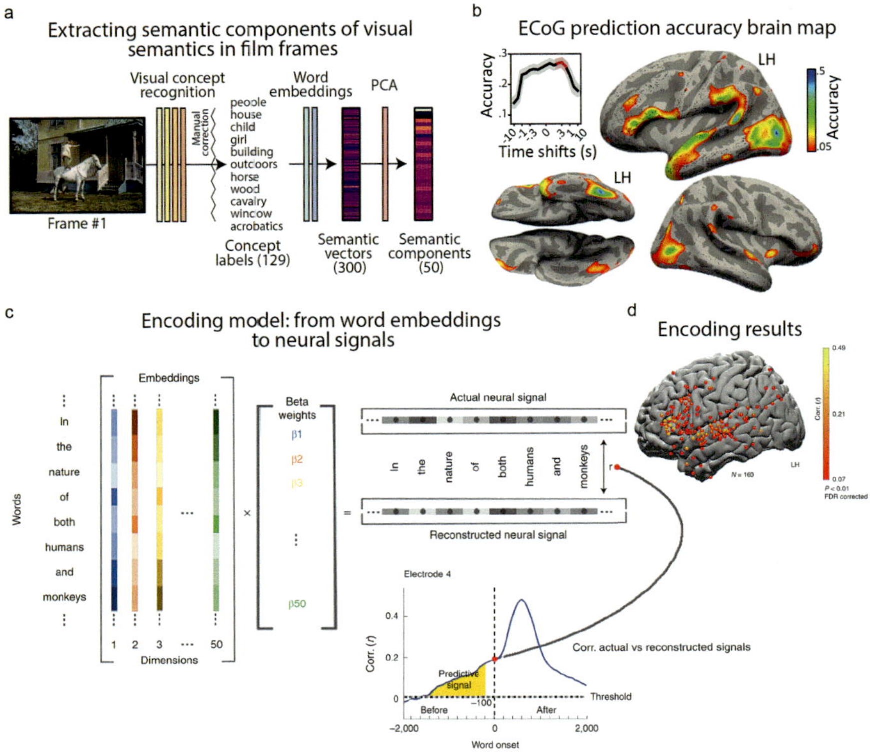

Fig. 47.2 **a**, **b** A linear encoding model that predicts ECoG high frequency band activity based on high-level visual features of the movie stimulus [55]. Prediction accuracy, measured as correlation between predicted and observed brain activity, is projected from individual electrode center coordinates to a regular grid in a common space and interpolated on the brain surface. Authors showed that components of visual semantics identified in a data-driven way map onto distinct cortical activation networks. Copyright 2020 Berezutskaya et al. **c**, **d** A linear encoding model that predicts ECoG high frequency band activity during a story listening task [27]. Using contextualized word embeddings in their encoding model, the authors demonstrated that neural activity prior to word onset contained information about upcoming words and that this information is used to calculate post-word-onset surprise. Copyright 2022 Goldstein et al.

for training and validating end-to-end DNNs that perform non-linear processing of simple input features (such as image pixels or sound waveforms) to directly predict observed brain dynamics have already been made in animal [82–86] and human [87–89] research. Recent work applied this approach to iEEG during audio and speech processing [38, 90] and extracted data-driven features from input sound data in a way that maximized the model fit to observed neural responses (Fig. 47.3). Such end-to-end DNN encoding models explained more variance in neural signals compared to linear encoding models and generalized well to novel experimental data.

The ultimate goal of such models is not limited to novel strategies to extract input features that best explain brain data. Such models can aim to provide an accurate

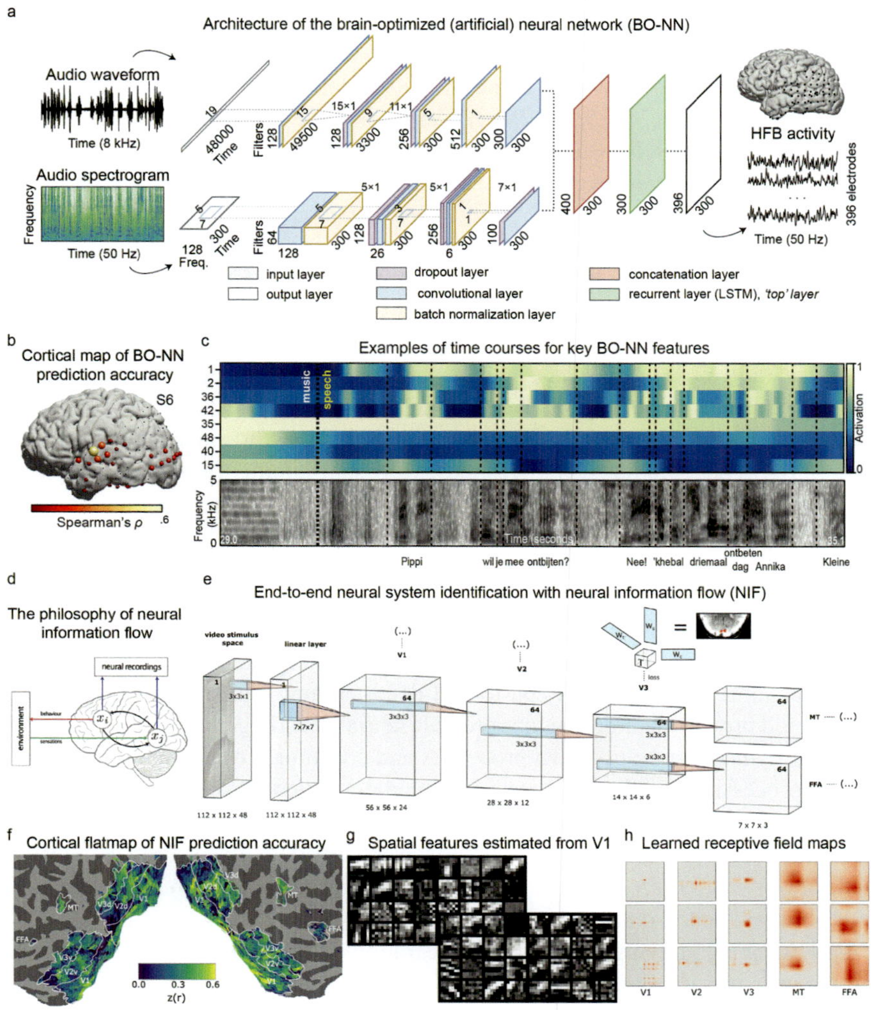

Fig. 47.3 **a–c** End-to-end brain-optimized encoding model based on a recurrent convolutional neural network (BO-NN) [38]. The model was trained to predict ECoG responses to a movie soundtrack. Model predictions generalized to a novel movie stimulus watched by a separate group of subjects. Extracted features encoded speech acoustic and temporal information and revealed a gradient of information propagation in the brain during speech perception. Copyright 2020 Berezutskaya et al. **d–h** End-to-end encoding model based on the neural information flow (NIF) [95]. The model was trained to fit fMRI responses to a movie stimulus throughout the ventral visual pathway (V1 to MT and FFA). This approach allowed for bottom-up estimation of input visual features that drove responses throughout the cortical hierarchy of visual processing and area-specific receptive fields. Copyright 2021 Seeliger et al.

computational account of neural processing mechanisms underlying perception and cognition. This, however, remains a challenging goal to achieve. For example, there has been a discussion of the biological feasibility of current DNN architectures and their learning mechanisms [65, 68, 69, 74, 91]. Furthermore, DNNs are often referred to as "black-box models", excellent for learning a complex mapping between inputs and outputs, but difficult when it comes to interpretation of their intermediate computations [67, 70–72, 74, 75, 78, 79]. It is especially the case for neural modeling as opposed to use of DNNs as feature models, where several visualization and computational tricks have been developed to help interpret what the model has learned [68, 74, 75, 92–94].

Extending upon the previous ideas of end-to-end encoding and incorporating some of the mentioned concerns, the neural information flow (NIF) framework was recently developed [95]. The framework is based on an end-to-end deep learning model of brain activity, constructed in a modular way, such that each artificial neuronal population (i.e. individual DNN layer), predicts, or projects to, a corresponding biological neuronal population (i.e. region of interest in the brain). In the case of modeling the ventral pathway of visual processing, illustrated by the authors, each convolutional layer (I, II, III, etc.) predicted brain responses to visual stimuli in a single corresponding brain area along the visual hierarchy (V1, V2, V3, etc.). Moreover, modifications in the model architecture allowed NIFs to model hemodynamic response function of fMRI (also see previous work on this [88]) explicitly. The model was validated with fMRI responses to naturalistic visual stimuli (Fig. 47.3). It could accurately predict observed brain responses through learning of gradually increasing visual receptive fields, plausible hemodynamic response rates and emerging high-level features of visual semantics. Overall, this end-to-end deep learning framework demonstrates that it could provide data-driven interpretable neural encoding models based on deep learning by incorporating known properties and constraints of neural systems, perceptual processing and the brain recording modality. Such models could be used for validation of existing models of neural processing (such as in the ventral visual pathway for object processing), model comparison and creation of new, fully data-driven models, whose structure, connectivity and learned perceptual features are optimized to fit observed neural data.

Following the growing tendency for integration of deep learning in neuroscience and the ever increasing amounts of neural data recorded from parallel neural sites over extended periods of naturalistic stimulation, it is likely that this approach will become dominant in the field. With the help of Bayesian approaches, effective model comparison and hyperparameter optimization can be performed for improving the explanatory power of end-to-end models.

Guided interpretable designs, including model architecture, objective function and learning algorithm, in the spirit of explainable AI will be key in modeling brain data using end-to-end deep learning models of information encoding. Key principles of neural computation known from previous work may need to be incorporated in the architectures of these models, including hierarchical processing with integrated local and feedback recurrence loops [96, 97], neural adaptation [98–100], sparse coding principles [101], temporal stability for noise robustness and code invariance [102–104], stochasticity in neural signals [105, 106] and oscillatory dynamics [107, 108]. However, given that detailed workings of many of these principles remain debated in neuroscience, end-to-end DNN encoding models could also provide an excellent framework for testing associated theory-driven hypotheses in silico [72].

Regarding the objective function, end-to-end deep learning models aim to mimic neural processes. Elements of this have been used in representational distance learning—an approach in which DNN feature models were trained with auxiliary objective functions [109]. The latter constrained DNN layer activations forcing them to approximate stimulus similarity structure observed in neural responses across brain areas. Similarly, end-to-end models can implement composite objective functions that represent constraints on DNN dynamics and representations.

Gradient descent via backpropagation has long been considered a biologically implausible learning algorithm. However, several studies have shown biologically grounded approximations thereof [110–113]. Moreover, a lot of promising work is done towards the development of more biologically plausible learning mechanisms similar to backpropagation [114–117]. At the same time, alternatives to error-guided supervised learning, such as reward-based learning or unsupervised (Hebbian) learning, are also an active area of research [112, 118–120].

The outlined end-to-end encoding approach using deep learning models is particularly promising in the case of iEEG, as more and more human iEEG data is becoming publicly available [121–126]. The long-duration and continuous nature of iEEG recordings makes them an excellent candidate for training and validation of flexible, interpretable end-to-end models of information encoding in the brain, albeit the application of this approach to iEEG is associated with additional iEEG-specific challenges. Such challenges relate to generalization across subjects due to individual coverage and electrode differences, the scale of neural signal representation and use of rate-based approaches as opposed to integration of single neuron activity in spiking neural networks for modeling local field potentials; and overall incomplete understanding of the nature of iEEG signals, its broadband spectrotemporal and oscillatory features. Some of these challenges can be addressed by incorporation of iEEG forward models [127, 128] and attempts at transfer learning in the search for common space across individual subject datasets [129], yet new creative approaches may be required to tackle data complexity.

Taken together, the application of deep learning models to neuroscience data, and iEEG in particular, remains an exciting and promising direction of research in creating a detailed computational account of neural mechanisms underlying perceptual and cognitive processes.

47.1.1.4 Representational Similarity Analysis (RSA) in iEEG

Representational Similarity Analysis (RSA) is a multivariate approach, initially proposed by Kriegeskorte and colleagues [9, 130, 131], which focuses on assessing distances between patterns rather than providing decoding accuracy. The principle of RSA is to estimate how the similarity of neural responses to a set of stimuli matches the similarity of perceptual or cognitive evaluations. Distances can also be computed based on predicted responses according to a specific computational model in order to test its fit in distributed regions across the brain or in different behavioral tasks. One of the major advantages of RSA distance matrices is to provide a robust solution to compare results across brain regions, subjects, neural recording modalities (e.g., fMRI, EEG, MEG), and even species (i.e., humans, monkeys). As such, the RSA method could greatly help integrating experimental research across laboratories and connecting methodological branches of systems neuroscience [9, 130, 132].

A handful of studies have combined RSA with the high temporal resolution of iEEG to assess the stability and changes of neural representations during cognitive tasks (see also Chaps. 34 and 50). Chang et al. [133] first used RSA to highlight the categorical neural representations of speech sounds, mapping their acoustic properties, in the posterior superior temporal gyrus. Zhang et al. [134] have characterized time-resolved gamma-band activity patterns in a navigational task, demonstrating the dynamic changes of path representations during encoding and retrieval. More recently, RSA has been used to characterize the dynamics of semantic coding combining current models of semantic memory and neural representations inside the ventral and anterior temporal cortex [135, 136].

It is also noteworthy that RSA is increasingly used to estimate similarity across biological and artificial neural networks [42, 77, 137, 138], enabling testing for nonlinear contributions of features in models [139]; see [75] for a review. The combination of AI algorithms and RSA has only been applied to human iEEG data in very few recent vision-related studies. Kuzovkin et al. [56] demonstrated that visual complexity along the ventral visual pathway, also visible in layers of deep convolutional neural networks (DCNN), was best predicted by gamma activity. Grossman et al. [140] revealed face-selective responses in the brain to match the structure of intermediate layers of the DCNN. This research direction is in its early stages but already demonstrates its high potential to evidence key functional principles of the human brain.

47.1.2 Decoding Models of Perception and Cognition

47.1.2.1 Supervised Machine Learning

Supervised decoding approaches have become common practice in cognitive neuroscience [141]. They generally involve training a model to classify brain signals into target categories. The latter are assigned "labels" and may reflect distinct groups (e.g. controls and patients), experimental conditions (e.g. familiar and unfamiliar face stimuli) or brain states (e.g. wakefulness, non-REM sleep and REM sleep). The basic principle is straightforward: The data is first split into "train" and "test" sets (or into train/validation/test sets). The training data and associated labels are used to train a model for class prediction. The training can in principle be formalized as a data-driven learning process by which a model learns to tune the parameters of a decision function to maximize the correct predictions. Model generalization is then explored by evaluating its predictions on the test data (i.e. samples unseen during training).

One way to ensure that the performance of the classifier is not biased by a lucky (or unlucky) choice of train-test split, the entire train-test procedure is generally repeated multiple times (i.e. cross-validation). In brain decoding work, the input to the classifier consists of brain variables which are, broadly speaking, of two types: (1) Hand-crafted features derived from raw brain activity (e.g. spectral power values of iEEG signals). This feature extraction process is generally guided by domain-specific knowledge. This process of selecting, manipulating, and transforming raw data into features that can be used in supervised learning is often referred to as feature engineering. (2) Original brain signals (such as raw iEEG) can also be used as direct input to the classifier. While shallow learning approaches often involve the use of hand-crafted features, techniques such as deep learning generally take the raw data as input, with little or no preprocessing at all (e.g. continuous iEEG raw time series).

A noteworthy issue that often arises when applying supervised machine learning in brain decoding, is the question of how it relates to using classical statistical approaches, such as comparison of means in inferential statistics. An important distinction between the two approaches is that while classical statistics are generally conducted on all the available data, machine learning focuses on out-of-sample generalization (see [142] for a more detailed discussion). Furthermore, the use of machine learning cannot be seen as a way to avoid testing the statistical significance of the reported results. In fact, in most cases, an assessment of the reliability of the decoding accuracy of a classifier requires statistical evaluation, for instance using permutation tests [143] (see also Chap. 35).

Conventional Single and Multi-feature Classification of iEEG Data

Shallow ML is based on an a priori selection of features and aims at identifying the best predictors of distinct cognitive states. In electrophysiology, brain activity has been primarily examined by quantifying increases and decreases in evoked-response potentials and spectral power in distinct frequency bands (very low frequency component [0.1–1.5 Hz], delta (δ) [2–4 Hz], theta (θ) [5–7 Hz], alpha (α) [8–13 Hz], beta (β) [13–30 Hz], low-gamma (low γ) [30–60] and broadband gamma (high γ) [60–200 Hz]). More recently, other spectral properties such as phase, phase-amplitude coupling, inter-trial coherence, and phase-locking value depicting both local and large-scale neural mechanisms have complemented the features palette used to advance our understanding of the human brain (Fig. 47.4a).

In this context, shallow ML models have proven to be helpful in addressing cognitive neuroscience questions by unraveling the contribution of various intracranial EEG-based features in cognitive processes. More specifically, by computing the decoding accuracy achieved by a specific feature—for example, how well gamma power discriminates between two conditions—we can make interpretations on the importance of this feature in the task. The application of ML as a brain decoding technique is increasingly common in different subfields of systems and cognitive neu-

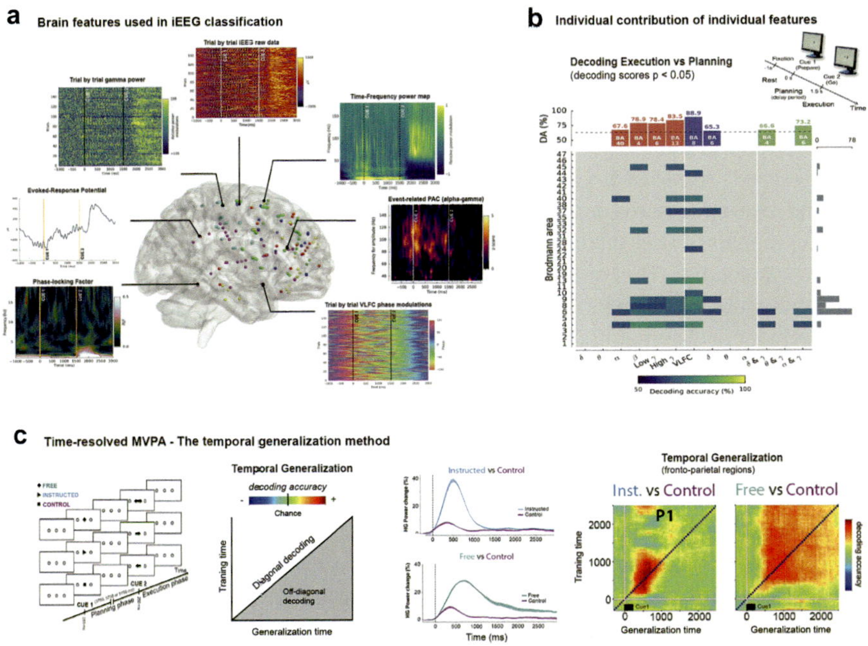

Fig. 47.4 **a–c** Decoding iEEG data using supervised ML models. **a** Examples of brain features computed from iEEG signal and used for classification. **b** Decoding accuracies of single-trial classifications of motor-states using power, phase and PAC features in significant iEEG sites. **c** Example of temporal generalization of choice-type decoding (instructed vs. control and free vs. control) using high-gamma (HG) activity across significant fronto-parietal iEEG sites. Generalization matrices show decoding performance plotted as a function of training time (vertical axis) and generalization time (horizontal axis). Reprinted with permission from [144, 145]

roscience including motor control (e.g., [144, 146]), sensory perception (e.g., [147–149]), and high-order cognitive processes such as cognitive interference (e.g., [150]), memory (e.g., [151, 152]) and decision making (e.g., [145, 153]).

ML Models

Various methods have been used to build ML models using intracranial EEG. These include support vector machines (e.g., [154–156]), linear discriminant analyses (e.g., [145, 153, 157]), logistic regressions (e.g., [158–160]), and decision trees such as random forests (e.g., [156]). The choice of the decoding model impacts the performance and the interpretation of the results (e.g., [151]). As all models do not rely on linear algorithms, it is critical to compare their predictions and explore nonlinear or unexpected interactions between variables.

Overview of Decoding Studies in iEEG

Using single-feature approach, some researchers have compared how predictive single-trial phase values in rhinal and hippocampal cortices are about encoding memory success, in a word recognition paradigm [155]). As another example, decoding results have been compared across frontal and parietal cortices during a delayed oculomotor decision task, demonstrating how free-choice in the brain is supported by sustained high-gamma activity during the delay period [145]. A similar data-driven approach, has allowed to disentangle the distinct contributions, on a trial-by-trial basis of power, phase, and phase-amplitude coupling to the planning and execution of a goal-directed motor task [144] (Fig. 47.4b). These studies illustrate how single-feature supervised classification can be employed with iEEG data to advance our understanding of how distinct electrophysiological patterns in individual brain regions are involved in goal directed behavior. In the next section, we will address some promising avenues using multi-features ML models.

One important aspect to watch out for when developing ML pipelines is the risk of data overfitting, where the model performs well on training data but does not provide accurate predictions on the previously unseen test data. An overfitted model learns patterns that accurately describe the training data, but do not generalize to the test data. This is a prominent issue in data with a small sample size (n) but large number of features, or predictors (p), a situation often referred to as "wide data", or the "p»n problem". This can become an important limitation in intracranial EEG studies where the number of patients is often quite limited whereas the feature space can be very important, which is in principle a perfect recipe for a model to drastically overfit.

A general recommendation to deal with overfitting is to simplify the model, for instance by decreasing the number of features through feature selection or dimensionality reduction techniques. In intracranial EEG, rather than examining generalization across subjects, intra-subject analysis is often preferred partly because of the heterogeneity of the electrode implantation across individuals. In this case, rather than reflecting the number of participants, the sample size becomes the number of trials, and a new model is trained on each subject using single-trial features with the aim of generalizing across trials, not subjects. This generally reduces the risk of overfitting, assuming there are more trials than features. This also explains why multi-feature classification is not very common in intracranial EEG data sets with modest trial numbers. As a general rule, to reduce the risk of overfitting, it is important to make sure one is not using an overly complex model with too many parameters.

The occurrence of overfitting when applying ML to intracranial EEG data can be detected by cross-validation (i.e. measuring performance on left-out data the algorithm has never seen during training). It is worth noting that sample size directly impacts the standard error of cross-validation, with small sample size studies yielding higher classification performance [161]. In addition, with small sample sizes it is particularly important to verify the statistical significance of observed decoding accuracies, as probabilistic chance levels only (e.g. 50% for binary classification) hold for infinite amounts of data. Large deviations from these thresholds can easily occur by chance in small samples. A recommended approach to handling this issue is to use permutations (i.e. randomly shuffling class labels and recomputing decoding accuracies) to derive a null distribution of classifier performance, and hence thresholds for statistical significance assessments of decoding accuracies [143] (see Chap. 35 for details). Again, this issue may be particularly important when analyzing small data sets of intracranial EEG.

Multivariate Analyses for iEEG (MVPA, Temporal Generalization)

Among common supervised decoding frameworks, Multivariate Pattern Analysis methods (MVPA) have been attracting growing attention in cognitive neuroscience (for reviews see, [162, 163]). MVPA assesses distributed neural representations—also called patterns—across multiple voxels (fMRI) or channels (EEG and MEG) simultaneously. Classifiers are trained to capture the spatial relationships, invisible with traditional univariate tests, between different brain locations across different experimental conditions. Their main advantage is to allow for content-specific decoding in brain activity, and thus to disentangle it from general cognitive functions. In addition, by combining spatial information, MVPA increases signal-to-noise ratio and facilitates single-trial prediction especially relevant in iEEG. Their usefulness has been demonstrated quite extensively in vision, including evidence for sensitivity to orientation in V1 [164, 165], and for distinct pattern of responses for visual categories (e.g., faces, houses) in the ventral temporal cortex [166]. A time-resolved application of MVPA called the temporal generalization method has also been introduced (see for a review [167]). This method allows for tracking how neural representations unfold in time by training and testing classifiers to discriminate at least two experimental conditions at all points in time [168, 169]. This approach allows for detecting and comparing periods of optimal decodability across time and cognitive operations.

Overview of iEEG Decoding Studies Using MVPA and Temporal Generalization

The application of multivariate analysis techniques to iEEG is still in its early days, though the high spatial and temporal resolutions of these signals would constitute a true advantage to discriminate neural patterns. Tsuchiya and colleagues [170] have applied multivariate models to ECoG power signals, finding reliable distinct representations for happy and fearful faces in the ventral temporal cortex. Spatial pattern analyses have also been adopted to identify successful word encoding in ECoG signals [152, 171] and iEEG [160], and to show the reinstatement of similar processes during successful encoding and retrieval of words in iEEG [172]. Using the temporal decoding method, Thiery et al. [145] have recently provided the first evidence for delayed high-gamma activity in the fronto-parietal cortex in mediating free choices compared to instructed choices (Fig. 47.4c).

Despite the significant advances that multivariate ML methods allow, certain limitations should be considered when interpreting classification results. Neural patterns obtained from combining depth electrodes can strongly vary between patients and often cover distant brain regions that may not be anatomically connected. This raises the question about the actual computational usage and biological relevance of the information included in the analysis. In fact, it is possible that decoding results are driven by a small number of electrodes rather than widely distributed patterns. Finally, when running MVPA analyses, one should keep in mind that decoding results strongly rely on feature selection, e.g. when using statistical feature ranking or dimensionality reduction techniques.

47.1.2.2 Deep Learning for Decoding

Deep learning is increasingly used in neuroscience research [173]. Although still in its early days, the application of deep learning to iEEG is very promising and is starting to gain momentum (see also Chap. 50). Using deep learning in the context of iEEG decoding work is in part appealing because it allows us to go beyond investigation based on handcrafted features (see Sect. 47.1.1.2). Indeed, representation learning [174] is a major motivation for using deep neural networks. In principle, representation learning (or feature learning) is the process by which a system automatically uncovers the representations needed for classification directly from the raw data. Although the features learned through deep learning are often described as being abstract, interpretation can be facilitated using an array of feature visualization techniques [175] including deconvolutional methods such as guided backpropagation [176]. The identification of the most discriminative samples through these approaches can be combined with signal processing tools commonly used in iEEG work, such as spectral analyses, to further enhance interpretability.

An interesting illustration of how deep learning can be applied to iEEG was provided by a study that used deep convolutional neural networks (CNNs) to probe which frequency bands of the iEEG data are correlated with feature transformations of ascending complexity along the ventral visual pathway during object recognition [56]. The study revealed that gamma activity (30–70 Hz) reflects the increasing complexity of visual feature representations in the deep CNN (see also [55]). These results illustrate how CNN activity may capture essential electrophysiological features of biological object recognition not only in space and time, but also in the frequency domain.

Recent work on decoding in cognitive neuroscience has shown the potential of other deep learning architectures, such as recurrent neural networks [177], generative adversarial neural networks [178–180] and transformers [181] to reconstruct stimulus information based on non-linear transformations of brain inputs. In decoding from iEEG data, deep learning models have been used as powerful feature extractors from raw iEEG signals and as a tool for combining data across subjects with varying implantation coverage [129]. Deep learning has also been applied to iEEG in the context of epilepsy, including seizures detection, e.g. [182–190] (see Sect. 47.2.3) as well as for brain-computer interfaces [191–194] (see Sect. 47.2).

47.2 AI-iEEG for Neurotechnology

The ever growing amount of neuroscientific knowledge has gradually made it possible for a novel field of neurotechnology to emerge. This field is focused on development of brain-computer interface (BCI) devices, and one of its fundamental goals is to offer technological solutions for real-world clinical problems. This includes technology for (1) restoration of lost cognitive and motor functions, such as visual prosthesis for the blind, BCIs for severely paralyzed individuals or cochlear implants for people with impaired hearing (see also Chap. 51); and (2) therapy and treatment of various chronic neurological conditions, such as deep brain stimulation devices for the Parkinson's disease, dystonia, epilepsy and severe mental disorders (see also Chaps. 52 and 53). Despite the availability and lower safety concerns of non-invasive brain recording techniques, intracranial technology has proven to be best suited for effective management of severe neurological conditions and restoration of severed cognitive or motor functions. This is because it provides superior neural signal quality [195], allows access to highly localized surface and deep brain structures, and overall has a better potential for long-term, continuous brain recordings and autonomous home use of the device. In that regard, ECoG and sEEG recordings have been instrumental in development, testing and deployment of various examples of emerging clinical neurotechnology. See [196–207] for reviews.

Neurotechnology is a highly multidisciplinary field with contributions from neuroscience, signal and data processing and electrical engineering disciplines. Fundamental neuroscience provides the theory of neural processes that support cognition and behavior, and the understanding of neural signals in health and disease. Electrical engineering and material science develop cutting-edge neural recording and stimulation technology and necessary hardware to enable operation and power supply of the device. Statistics, computational modeling and machine learning provide the methodology and algorithmic base for processing neural signals and addressing the clinical problem at hand, via control of external devices (a wheelchair, exoskeleton or computer) or brain stimulation to alleviate symptoms of a chronic neural condition. More recently, BCI methodology began to incorporate advanced computational models that make use of machine learning and deep learning approaches. These AI-powered algorithms are being used increasingly more often for (1) neural signal preprocessing and extraction of informative temporal and spectral features in brain data; (2) decoding models of target neural events, such as intended movements in paralyzed individuals or biomarkers of pathological activity in individuals with a neurological condition; (3) optimization of BCI algorithms and development of energy-efficient computing systems; (4) integration with external domain-specific tools and applications, such as virtual reality, natural language processing models and robotic components, for boosting the development of cutting-edge BCI solutions.

In this section, we will describe neurotechnology applications based on the combination of AI methodology and iEEG neural data. The main examples will include BCIs for speech and communication in severely paralyzed individuals, brain implants for motor control of a wheelchair, exoskeleton or robotic arm, and deep brain stimulation devices for neurological and psychiatric disorders.

47.2.1 IEEG BCI for Speech and Communication

Individuals who suffer from a motor neuron disease, such as amyotrophic lateral sclerosis, (ALS) or a brainstem stroke can develop severe motor paralysis. In extreme cases, it can lead to a "locked-in syndrome" (LIS) [208] and result in a complete loss of muscle control. Such individuals may lose the ability to perform voluntary body movements, such as walking, object interaction, speech, facial expressions, blinking, swallowing and, in extreme cases, breathing. Their means of communication are often reduced to limited eye movements or residual control of a few facial muscles. People with a complete LIS may lose any ability to communicate. Given modern standards of healthcare "locked-in" individuals can survive in their severely paralyzed state for many years and even decades [209, 210]. Interestingly, studies report that they can experience good quality of life [211–215], and that one of the major predictors of good quality of life is retaining the ability to communicate with the outside world [216, 217].

In many cases, motor paralysis in people with LIS is caused by damage to connections between the brain motor cortex and the spinal cord, or the spinal cord and nerves that lead to the muscle tissue. Unlike patients in a vegetative state, "locked-in" individuals remain conscious. Cortical activity in individuals who have suffered a brainstem stroke is largely unaffected and exhibits patterns similar to that of the healthy able-bodied individuals. Effects of motor neuron degeneration on cortical activity in individuals with a motor neuron disease, such as ALS, remain less well-understood. Several studies report a decrease in alpha [218] and theta [219] power compared to healthy controls, while other studies report an increase in alpha power [220] or no difference [221]. A recent study indicates that population-level sensorimotor dynamics in ALS subjects may be comparable to sensorimotor responses in non-human primates [222]. Consistent with this finding, several studies reported successful decoding of motor information from sensorimotor cortex of individuals with ALS [223, 224]. This work demonstrates the potential of BCI technology to detect intended communication signals from brain activity of "locked-in" individuals and translate them to computer commands, thereby unlocking a means of communication with the world.

Preparatory research towards BCI technology based on an iEEG implant relies on pre-clinical studies in able-bodied patients with medication-resistant epilepsy. These patients are temporarily implanted with iEEG electrodes (typically, for 7–10 days) for clinical monitoring of their condition with a goal to identify and subsequently remove neural sources of epilepsy. While implanted with iEEG, such patients can participate in various cognitive and motor tasks, and the collected iEEG signals are subsequently analyzed with a goal to identify, or decode, neural events relevant for BCI research. Most of this work rests on the assumption that speech production and other motor activity in able-bodied subjects engage brain mechanisms and cortical areas similar to those of attempted speech and imagined actions in paralyzed individuals. This assumption is supported by the recent work on attempted movements in amputees [225, 226] and paralyzed individuals [227, 228]. Given the reports on potentially shared neural basis of performed and imagined movements, including speech [229–233], it has been proposed to use imagined movement paradigms in able-bodies participants as a proxy for development of BCI technology for individuals with LIS. Other reports, however, provide conflicting evidence regarding the shared neural mechanisms of imagined and performed motor activity [234, 235], and in general, a distinction between attempted (no actual movement possible) and imagined (actual movement possible and likely inhibited) movement should be made. Further research comparing neural activity during performed, imagined and attempted actions is needed to gain a better understanding of these processes and inform BCI development.

The pre-clinical work on iEEG BCIs has been focused on identifying three key components of the emerging technology for communication: (1) optimal location for implanting iEEG electrodes, (2) optimal targets or features for decoding, and (3) optimal neural decoding model. State-of-the-art BCI technology for long-term autonomous use is limited to implants with a small number of electrodes (four to sixteen) [223, 224]. Larger numbers will require higher power consumption for continuous signal recording and analysis and may lead to overheating of the implanted components beyond reasonable temperatures. Moreover, implantation of electrodes over large parts of the cortex may lead to higher risks during and following the implantation surgery and may result in longer recovery times. Until these concerns are mitigated, targeting a smaller brain region for BCI implantation is preferred. Currently, sensorimotor cortex is the primary candidate for such BCIs as its involvement in muscle control during movement and communication has been studied extensively with iEEG in non-human primates [236–238] and human participants [23, 239–241]. Other brain areas have also been considered in decoding of language and speech, such as inferior frontal gyrus, temporal and parietal regions [230–232, 242–247]. These regions are part of the language processing network in the brain [248, 249]. However, language-related neural signals tend to be highly spatially distributed and varied across subjects, which results in less overall consensus about their potential for BCI use. It remains to be seen how the development of BCIs for communication can incorporate current neurolinguistic theory and benefit from signals recorded in the language processing network of the brain.

Another key component of a BCI for communication is the target of decoding, and given the focus on sensorimotor cortex, one of the prominent lines of research is decoding of motor information. This work includes decoding of performed movements, imagined or attempted movements of, for example, the upper limb, and relies on the previously mentioned assumption about the shared neural basis of performed and attempted movements. This research includes decoding of various types of movement, such as simple hand and finger movements [193, 250], gestures [241, 251–254] and elements of sign language [255]. Accurate decoding of several classes of facial movements, such as basic mouth movements [256] and facial expressions [257] has also been demonstrated.

The long-term goal of approaches based on accurate decoding of several discrete motor commands, is to enable communication via development of a BCI with several degrees of freedom for flexible cursor control [229, 258] or use of a computer-based language speller [259]. Decoded motor commands could then be used to control such interface to move the cursor (in a discrete: up, down, left, right, or continuous fashion) and perform item selection. Despite the potential of this approach, most recent work has been focused on making BCI communication faster and more efficient, and rely on a more natural way of communication, such as attempted speech. For this, BCI researchers have turned to decoding of discrete speech-related features, such as individual phonemes [230, 260–263], syllables [264], words [242, 247] and sentences [265]. Researchers have also used a closed set of decoded elements, such as phonemes, as building bricks for potential open-vocabulary decoding of full words and sentences [246, 266]. Some work explored decoding of acoustic features of speech [231, 267] and used external AI models to synthesize speech from them [233, 268, 269]. Another recent study used microelectrode arrays over the hand knob in sensorimotor cortex to decode handwriting patterns during word spelling [270]. Decoding of speech and speech-related features has proven to be more difficult compared to other motor movements. On the other hand, it can offer the potential to provide a faster, more natural and convenient way to communicate. Therefore, speech and speech-related features remain a more attractive and overall preferred target of decoding in BCI research.

The third component—optimal models for decoding—also remains a topic of debate. From the beginning, BCI development has relied on various statistical and machine learning algorithms to perform signal processing, dimensionality reduction, time series analysis and decoding [271, 272]. Various classification methods, such as logistic regression, support vector machines, k-nearest neighbours, random forests, artificial neural networks and others have been used and compared in terms of their decoding accuracy [245, 263, 264, 273, 274]. There have been attempts to boost the performance by aggregating results from an ensemble of classifiers, each generating its own decoding output. Latest advances in AI have led to the focus on the so-called *neuroengineering* approach that relies on deep learning for decoding. This includes the use of deep learning models for extracting features that could be used as targets of decoding [275], as well as end-to-end deep learning models for learning a complex mapping from neural data to speech and language features [233, 264, 268, 269, 274–277]. Deep learning has been used for transfer learning in iEEG to account for the variability in electrode placement across subjects and pre-train decoding models for new subjects on previously collected datasets [129, 276]. Another recent trend is the use of external AI tools for boosting the decoding results, such as language models [278, 279], audio synthesis models [233, 268, 269, 275], speech-to-text models [276] and models of articulatory-to-acoustic inversion [274]. Overall, it appears that AI-powered tools for decoding are beginning to dominate the BCI scene due to their superior accuracy and potential for sophisticated data-driven solutions.

Several human clinical trials of iEEG-based BCIs for communication aiming to validate the envisioned technology in the daily life of end users have already started. A recent study by Vansteensel et al. validated a fully-implantable BCI device for communication in an individual with late-stage ALS [223]. Using a four-channel ECoG implant over the subject's sensorimotor cortex and a simple linear decoding model the authors decoded binary "click" events based on attempted hand movements with 90% accuracy. The decoded clicks were used to control the on-screen computer menu and language speller. For the first time, an implantable BCI technology for communication offered its user the possibility of autonomous 24/7 home use. More recently, another study reported preliminary data from a fully-implantable BCI system in two ALS individuals [224]. Minimally invasive stent-electrodes [280] were implanted within participants' vein adjacent to their motor cortex. The study reported high-fidelity decoding of two and three classes of motor events (on average, 93% accuracy of binary decoding in the typing task) and thereby demonstrated high potential of their BCI system for autonomous use in daily life. Another recent study reported the proof-of-concept of decoding attempted speech using advanced AI models in a patient with anarthria (inability to speak) following a brainstem stroke [279]. The authors implanted a 128-channel ECoG grid (not suitable for 24/7 home use due to power limits) over the sensorimotor cortex, and built a deep recurrent neural network to detect attempted speech in neural signals in real time. After detection, another set of deep neural networks decoded individual words out of a closed set of 50 words. In addition, the system used a hidden Markov model approach and an external language model to decode short full sentences made of the set of 50 words. The study reported highly accurate (word error rate of 26%) decoding results, highlighted the importance of using external language models in achieving this result, and concluded that the developed BCI was an unprecedentedly fast (15 words a minute), accurate and naturalistic means of communication available for their participant.

Currently this work remains at the stage of clinical trials and research, but given its promising results and continuous progress in iEEG technology, BCI hardware, signal processing and decoding models, it may achieve commercialization in the foreseeable future. This will make BCI-based communication technology available for "locked-in" individuals. Once this technology becomes part of the healthcare system, it will allow its users to communicate more efficiently, thereby helping them re-integrate in the society and further improving the quality of their lives.

47.2.2 IEEG BCI for Motor Control

Up to 50% of individuals who undergo a spinal cord injury can develop a condition called tetraplegia, which refers to a complete paralysis of four limbs [281]. BCI for restoration of motor control aims to improve the lives of these individuals by decoding intended movements from neural signals in their sensorimotor cortex, and use them to control an external device. The latter can be a cursor on a screen, a virtual reality avatar or a mobility device, such as a robotic arm, exoskeleton or wheelchair. As this neurotechnology is focused on restoration of a lost function, its target population, goals and approach overlap with those of the BCIs for communication. Yet despite the severe paralysis, individuals with tetraplegia retain their ability to communicate, and this BCI technology aims to further decrease the reliance of its users on patient care by allowing them to move and interact with objects independently. Since these individuals suffer from paralysis of all four limbs, and because they are not likely to recover from this condition, long-term invasive BCI solutions may be best for restoring these persons' motor control.

The field of iEEG-based BCI for motor control has been based on work on single-cell recordings and motor-related action potentials, and has therefore been dominated by microelectrode Utah arrays. Successful preliminary work on movement decoding in animals [282–287] has laid down the pathway for clinical trials in humans (see Chap. 51). In 2006, first decoding results using a microelectrode Utah array implanted in a tetraplegic individual were published [288]. The study participant learned to control a cursor using attempted hand movements. In each trial, the participant was cued with one of four target positions (up, down, left and right) and was asked to attempt to move the cursor from the center of the screen to that position. The associated single-cell activity on the motor cortex showed modulation that was picked up by the decoding algorithm and used to update the cursor position. The participant reached accuracy of 73–95% across six sessions. This study was part of the first clinical trial BrainGate (Cyberkinetics, Inc.) aimed at development of microelectrode-based BCIs for motor control in paralysed human subjects (BrainGate).

Later studies, including work from other groups (University of Pittsburgh and Ohio/Neurolife), expanded upon these results and demonstrated microelectrode BCIs in tetraplegic subjects for a real-time control of various external devices: computer cursor [289, 290], robotic arm [291, 292], driving and flight simulators [293, 294]. These studies rely on decoding of a few discrete motor commands, for example, up, down, left, right [288]; complex multi-joint movements, such as reach and grasp [291, 295]; various movement parameters, such as velocity, translation, orientation, torque, etc. [292] and continuous muscle activity [296]. Several studies explored the possibility of restoring proprioception via electrical stimulation of the implant [297, 298] and showed that implementing sensory feedback from interaction of the BCI-controlled robotic arm with objects results in faster, more accurate and naturalistic motor control [299]. Several studies combined microelectrode BCIs with functional electrical stimulation [300] of peripheral nerves to reanimate participant's paralysed limb [296, 301, 302]. See [303] for an overview of the state of the art in upper limb decoding for BCI motor control.

Despite promising results achieved with microelectrodes, this invasive technology suffers from signal decay over time [304]. It has also been reported that for chronic long-term use of the implant, repeated calibration sessions are required [305, 306]. Moreover, microelectrode devices have not been thoroughly tested for autonomous home use (although see some preliminary recent work [307]) and are not currently certified for indefinite implantation, which largely affects and restricts the user experience. As a less invasive technique, iEEG does not suffer from signal decay over long periods of time as strongly as microelectrode arrays do. Interestingly, some work in non-human primates indicates that decoding from iEEG may be more accurate compared to microelectrode arrays [308]. Moreover, iEEG shows less signal instability over time and therefore does not require frequent re-calibration [223, 258, 305, 309–311]. Autonomous home use of iEEG implants in human subjects has been successfully demonstrated [223], and iEEG electrode technology and hardware for power supply continue to evolve towards providing even better long-term recording solutions [312, 313]. Altogether, these factors contribute to the emergence of iEEG-based BCI technology for restoration of motor control.

Unlike speech, motor control via implanted neurotechnology can be assessed in animals, and several previous studies have showed promising results of decoding performed and attempted movements in non-human primates implanted with ECoG grids [309, 314–316]. At the same time, similar to BCIs for communication, much of initial research on decoding motor information from neural activity has been done pre-clinically, in able-bodied patients temporarily implanted with iEEG for epilepsy monitoring. This work reported on successful decoding of discrete movements and postures of hand [317–321], fingers [322–324], foot [317, 325] and other body parts; as well as decoding of continuous movement trajectories [326–330] and movement properties, such as force, velocity, direction and speed [22], decoding of complex multi-joint movements and gestures [241, 251–254]. See [331–333] for reviews. Most of iEEG studies on motor decoding to date have focused on ECoG. However, some recent work in able-bodied subjects demonstrates the potential of sEEG in development of BCIs for motor restoration. These studies report successful cursor control [334, 335] and control of a robotic arm [336] using motor signals decoded from sEEG. See [203, 333] for reviews of pre-clinical work with sEEG.

Clinical trials in individuals with tetraplegia tested the possibility of real-time clinical applications of ECoG-based BCIs for motor control. Work by Wang and colleagues was among the earlier studies to validate the ECoG-based decoding of intended motor commands in a human subject with tetraplegia [337] (but see also [338, 339]). Following a spinal cord injury, the study participant lost volitional control of arm and hand. The participant was subsequently implanted with a high-density ECoG grid over the sensorimotor cortex and trained to perform 2D control (x and y coordinates) of a computer cursor using attempted hand and elbow movements (thumb for left, elbow for right, both thumb and elbow for up and neither for down). After 11 days of training, in their final session the participant reached an accuracy of 87% (chance was 8%) over 176 cursor control trials. Broadband neural activity in high frequency band (gamma and high gamma) showed largest modulation by the task. In another study, Silversmith et al. [305] tested a cursor control BCI system in a patient with severe tetraparesis, or partial tetraplegia. A closed-loop BCI system with a high-density ECoG implant was built to process motor neural signals (imagined arm and head movements) and use them for control of a cursor on a screen. The system achieved high performance and stability over time without the need of re-calibration.

A series of recent studies demonstrated the use of an ECoG-based BCI for control of an exoskeleton [311, 340, 341]. Benabid and colleagues for the first time showed a wireless ECoG device WIMAGINE [312] implanted epidurally and bilaterally over the sensorimotor cortex of a patient with tetraplegia (Fig. 47.5). In two years of training, the implanted individual learned to walk and perform complex upper limb movements by controlling a four-limb exoskeleton or a virtual avatar (virtual reality) with the ECoG implant. Motor control was achieved with an adaptive linear model that decoded parameters of movement to the target location (distance to its 3D position and angles between current and target location). After training the decoder, the participant could perform voluntary movement control with a success rate above 70% for walking and 71% for 3D bimanual reach and touch tasks. This performance was stable without the need for frequent re-calibration after 7 weeks [340] and later than 6 months [341] after implantation.

Another recent study used an ECoG implant in combination with a functional electrical stimulation device to enable cortical control of the paralyzed hand [342]. Through training, the study participant learned to trigger electrical stimulation of his hand with 89% accuracy and successfully perform hand grasping tasks. After the initial period of in-lab testing, the participant was able to use the developed BCI system autonomously at home.

Similar to BCI for communication, restoration of motor control relies on various decoding models from statistics and machine learning (see [331, 343] for reviews of models for decoding movement). Simpler approaches based on decoding of discrete classes use various classification algorithms, such as linear discriminant analysis, k-nearest neighbors, support vector machines and others. Models that decode continuous kinematic trajectories can use linear methods, such as the linear regression. For decoding of position and velocity of movement, models of linear dynamical systems, such as the Kalman filter, are typically employed [327, 329]. They compute hidden states of the system, such as intended movement, based on movement from the previous time stamp, and model observed states, such as neural activity, as linearly dependent on the computed hidden state data. Noise associated with observed and hidden states is modeled separately. Such formulation allows for better estimation of time series data and helps boost performance on noisy non-stationary iEEG data. Next to the Kalman filter, hidden Markov models are also often used to handle more complex temporal dynamics in a model with hidden and observable states [340]. For better temporal stability, adaptive and switching decoding models [340] have been employed as well.

Recently, computational approaches based on artificial neural networks [193, 253, 302, 344–347] have become increasingly popular. These models consistently appear to provide more accurate and robust decoding compared to simpler baselines [193, 302, 344]. Moreover, these approaches have the potential to handle concerns associated with real-time decoding for practical BCI applications. Namely, they can provide superior speed of the BCI response and stable performance over sessions and days [302], while offering relatively computationally inexpensive, and even power-efficient solutions. The latter is further explored by development of spiking neural networks in combination with power-efficient memristive hardware as BCI

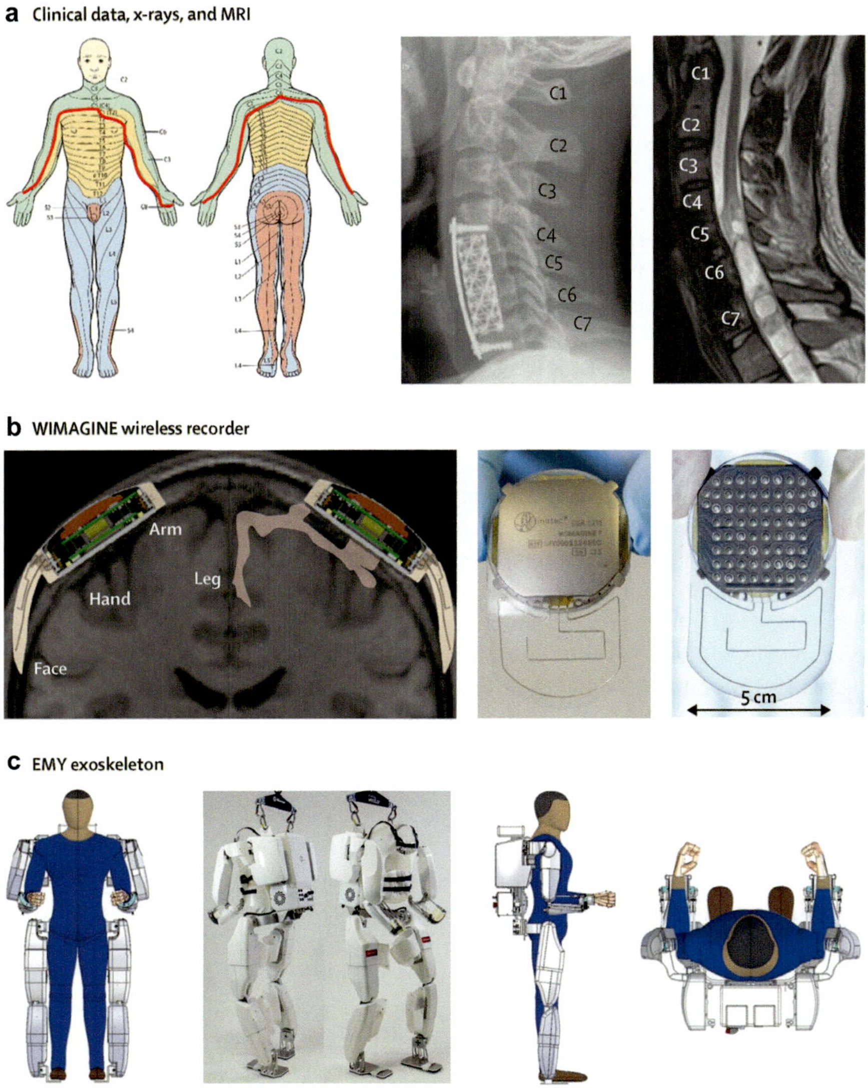

Fig. 47.5 ECoG-controlled exoskeleton: a proof of concept [340], pp. 1112–1122, reprinted with permission from Elsevier

solutions [348]. Another interesting AI application in motor decoding is analysis of video recordings collected simultaneously with ECoG and data-driven extraction of movement feature from video data [345] for decoding from the brain. Thus, similar to communication, external AI tools for extracting relevant features and building physical models can guide motor decoding and further improve the state-of-the-art performance. Altogether, as in the case of BCIs for communication, it appears that state-of-the-art BCIs for motor control, implemented either on microelectrode or iEEG implant, will continue to be powered by AI decoding algorithms and external tools for achieving highly accurate, sophisticated, reliable control of external devices by paralyzed individuals.

47.2.3 iEEG for Deep Brain Stimulation

Invasive neurotechnology for healthcare does not only focus on restoration of a lost function, such as speech, hearing, vision or motor control. Another branch of invasive neurotechnology targets chronic neurological and psychiatric conditions that do not respond to conventional treatment. Such conditions include movement disorders, such as Parkinson's disease, essential tremor, dystonia (involuntary muscle contractions); epilepsy; mental disorders, such as major depression, schizophrenia, obsessive-compulsive disorder, chronic pain and others. These conditions affect large numbers of people. For example, it is estimated that over 160 million people worldwide suffer from major depression [349]. About 50–80 million worldwide suffer from epilepsy [350, 351], of which 30–40% do not respond to medication [352]). Parkinson's disease affects over 6 million of the general population [353].

Many of these conditions, as well as the affected behavioral and cognitive function, have been associated with neural activity in deeper brain structures: basal ganglia in Parkinson's disease [354, 355], nucleus accumbens and median forebrain bundle in major depression [356], amygdala and hippocampus in temporal lobe epilepsy [357]. Severe cases that are resistant to medication can be treated with an invasive form of therapy called *deep brain stimulation* (DBS). This neurotechnology assumes implantation of a small electrode wire connected to a pulse generation device, typically placed under the skin around the chest [358], to modulate function of deeper brain structures. The device attempts to inhibit pathological neural activity via electrical stimulation of the neural tissue surrounding the implant and thereby improve the patient's condition.

Among the earliest applications of an implantable DBS technology in patients with Parkinson's disease was work by Benabid and colleagues [359]. The authors showed that high-frequency stimulation of ventral intermediate nucleus resulted in up to 88% improvement of tremor symptoms. The effect persisted up to 29 months after implantation. Subsequently, DBS in subthalamic nuclei was developed and successfully tested in patients with Parkinson's disease [360, 361]. It has since been shown to provide more relief and emotional well-being to patients compared to conventional medical treatment [362–364], and is now considered one of the state-of-the-art solutions for treating Parkinson's disease. Analogous applications were developed and successfully demonstrated in patients with medication-resistant epilepsy. Earlier work reported that modulation of anterior thalamic nucleus could lead to reduction in seizure frequency [365, 366]. Following successful large-scale trials by Fisher and colleagues [367], most modern DBS devices target this region for reduction of epileptic seizures. At the same time, stimulation of other deep brain structures, such as hippocampal formation [368], subthalamic nucleus and substantia nigra [369], centromedian nucleus [370] and cerebellar regions [371], has also been explored. See [357, 372, 373] for reviews. Apart from Parkinson's disease and epilepsy, DBS applications for tackling symptoms of chronic mental disorders are also being developed. These include DBS treatment of major depression [374], obsessive-compulsive disorder [375] and other conditions [376, 377].

DBS systems have already been able to help many people worldwide. Yet despite the accumulated success, commercially available DBS implementations have several limitations. Most systems are based on a single wire with a few electrodes (typically four) [358] and therefore provide only limited brain coverage that may not be suitable for comprehensive monitoring of the pathology, as brain signals in health and disease are believed to reflect large-scale states of dynamic neural systems [378, 379]. Moreover, commercially available DBS systems are not suitable for using recorded signals (with the exception of the Percept PC [380] and the Activa PC+S (Medtronic) neurostimulators), in guiding the stimulation strategy (closed-loop solutions). Instead, most systems rely on open-loop solutions, which means that they deliver a constant stimulation pulse to the implanted tissue based on a simple trial and error approach [377, 381] and do not process observed brain signals to adapt their stimulation behavior [381, 382]. Such DBS systems are less effective and could drain the battery during continuous stimulation even when it may not be necessary. As a consequence, DBS shows high variability of success and can lead to DBS-induced adverse effects [383–386]. Altogether, there is a need for better understanding of the neural signal, larger-scale recordings and active, adaptive monitoring of the system. Next-generation devices could represent a closed-loop system based on neuromodulation, which means that they could continuously monitor brain activity in compromised regions, identify neural patterns associated with healthy and pathological states, and devise a stimulation plan for treatment (see also Chaps. 41, 52 and 53).

A number of prototypes for adaptive closed-loop solutions have already been demonstrated [387–390]. Among those are studies that employ iEEG recordings for continuous recording and processing of neural signals to observe pathological activity and effects of stimulation, and adapt stimulation parameters accordingly. For example, several recent studies used ECoG grids implanted in addition to DBS wires for monitoring movement-related activity on the motor cortex [389, 390]. Herron et al. used ECoG electrodes to detect intended hand movements in a patient with essential tremor [389]. In this disorder, tremor occurs primarily during intentional movement, and applying stimulation after movement intention has been identified has led to a more energy-efficient DBS system without losing much of the therapeutic value. Similar results were obtained in patients with Parkinson's disease by Swann et al., who monitored high gamma ECoG activity on the motor cortex for markers of involuntary movements called dyskinesia, and used them to update DBS parameters. Another ECoG-based monitoring system for adaptive stimulation was tested in epilepsy and showed an up to 70% reduction of seizure frequency and persisting long-term positive effects [391].

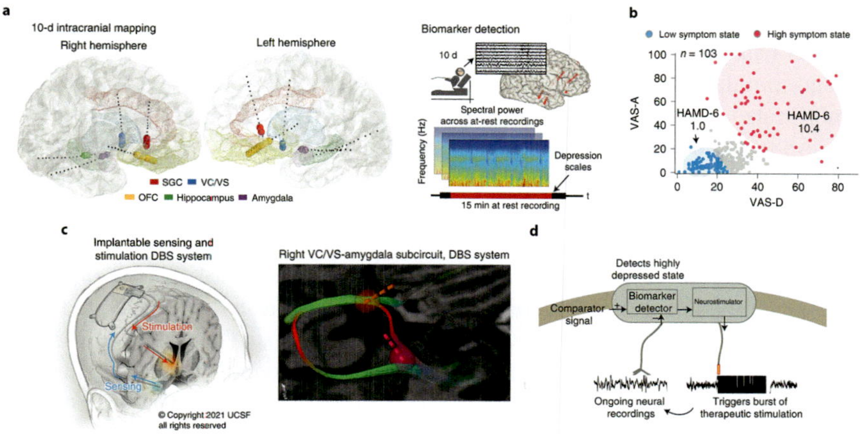

Fig. 47.6 SEEG-guided deep brain stimulation system for treatment of major depression [392]. Reprinted with permission from Springer Nature

In the context of DBS, another type of iEEG recordings is starting be be particularly attractive. As a less invasive technology compared to ECoG, designed for monitoring neural activity in deeper brain regions, sEEG has a lot of potential for successful guidance of adaptive DBS systems. A combination of continuous registration of activity throughout the brain and advanced computational algorithms for processing large-scale multi-channel data with complex temporal dynamics and connectivity structure, may provide the necessary understanding of the nature of neural signals in health and disease and ultimately provide appropriate DBS treatment. For example, in a recent work, Scangos et al. implanted sEEG electrodes in an individual with major depression to monitor neural signals in their hippocampus, amygdala, orbitofrontal, cingulate and striatum cortex [392]. The researchers were able to train online classifiers to continuously monitor sEEG activity in these regions and identify patient-specific biomarkers of pathological states. This allowed them to link neural activity in the patient's amygdala to their most severe depression symptoms. This knowledge was used to guide the subsequently implanted DBS system by detecting previously identified biomarkers of depression and applying stimulation to the affected sites (Fig. 47.6). The study showed that this approach resulted in a significant improvement of the patient's symptoms. Similarly, in another study, the authors demonstrated a successful mapping of neural states monitored with sEEG to individual symptoms of the underlying treatment-resistant depression [393]. Moreover, the authors used sEEG data to optimize parameters of stimulation applied to the affected brain regions via a standard DBS implant. SEEG-informed stimulation led to a large improvement of the patient's symptoms and their overall quality of life. Altogether, this work demonstrates the potential of sEEG implants for development of the next-generation closed-loop DBS systems with data-driven neuromodulation and personalized approaches to real-world clinical applications.

A line of research that takes this work one step further is based on the development of iEEG-based DBS systems, an integrated bidirectional approach to brain monitoring and stimulation through the same set of iEEG electrodes. Two recent studies have demonstrated a proof of concept for such a device. Basu et al. developed a system for monitoring cognitive control in individuals with mental disorders based on decoding of neural events associated with behavioral task performance [394]. Saucedo et al. reported on a similar approach applied to epilepsy and hippocampal sclerosis [395].

An important component of these adaptive DBS systems is machine learning and AI models. It has been shown that advanced algorithms that can handle high dimensionality of large-scale recordings and temporal connectivity structure may be better suited for signal processing and identification of brain states in health and disease. These algorithms are being increasingly used for identification of biomarkers of healthy and pathological states in real time and subsequent adaptation of stimulation parameters depending on how they affect behavior [396, 397]. In several applications, the potential of AI models for making complex predictions can be particularly valuable, for example, for prediction of seizure onset in epilepsy [398] or symptom severity in major depression [392]. Recently, progress in this area has been accelerated by an increase in availability of large public datasets during clinical recordings of pathological and healthy brain activity. The latter has stimulated the emergence of data science competitions, where data analysts compete in developing the algorithm that best solves the task, for example, seizure prediction. Next to that, the theory and computational models of large-scale network activity in health and disease are growing ever more comprehensive and complex [379, 399, 400]. These sophisticated models of dynamical systems and neural connectomics have the potential to drive forward our understanding of complex neural processes monitored with continuous multi-channel recordings, such as iEEG. In DBS neurotechnology, these contributions can lead to novel closed-loop real-world solutions that provide more effective personalized treatments for severe chronic neurological conditions and thereby substantially improve, quality of life of the affected individuals.

Acknowledgements This work was supported by the Netherlands Organisation for Scientific Research (NWO) and is part of the Language in Interaction consortium (NWO Gravitation Grant No. 024.001.006) and INTENSE consortium (NWO grant No. 17619). We thank Mariska Vansteensel, Jordy Thielen, Linda Geerligs and Pieter Kubben for their helpful comments on the initial version of the manuscript.

References

1. Ashby FG, Helie S (2011) J Math Psychol 55(4):273
2. O'Reilly RC, Munakata Y (2019) Computational explorations in cognitive neuroscience. MIT Press, Cambridge, MA. https://doi.org/10.7551/mitpress/2014.001.0001
3. van Gerven MAJ (2017) Front Comput Neurosci 11(December):1. https://doi.org/10.3389/fncom.2017.00112
4. Churchland PS, Sejnowski TJ (2016) Nat Rev Neurosci 17(11):667. https://doi.org/10.1038/nrn.2016.114

5. Holdgraf CR, Rieger JW, Micheli C, Martin S, Knight RT, Theunissen FE (2017) Front Syst Neurosci 11:61
6. van Gerven MA (2017) J Math Psychol 76:172
7. Kriegeskorte N, Douglas PK (2019) Curr Opin Neurobiol 55:167
8. King JR, Gwilliams L, Holdgraf C, Sassenhagen J, Barachant A, Engemann D, Larson E, Gramfort A (2020) Cogn Neurosci 6:691
9. Diedrichsen J, Kriegeskorte N (2017) PLoS Comput Biol 13(4):e1005508
10. Hanke M, Baumgartner FJ, Ibe P, Kaule FR, Pollmann S, Speck O, Zinke W, Stadler J (2014) Sci Data 1(1):1
11. Wakeman DG, Henson RN (2015) Sci Data 2(1):1
12. Seeliger K, Sommers R, Güçlü U, Bosch SE, Van Gerven M (2019) bioRxiv, p 687681
13. Schoffelen JM, Oostenveld R, Lam NH, Uddén J, Hultén A, Hagoort P (2019) Sci Data 6(1):1
14. Nastase SA, Liu YF, Hillman H, Zadbood A, Hasenfratz L, Keshavarzian N, Chen J, Honey CJ, Yeshurun Y, Regev M et al (2021) Sci Data 8(1):1
15. Armeni K, Güçlü U, van Gerven M, Schoffelen J (2022) Sci Data 9(1):1. https://doi.org/10.1038/s41597-022-01382-7
16. Harvey BM, Vansteensel MJ, Ferrier CH, Petridou N, Zuiderbaan W, Aarnoutse EJ, Bleichner MG, Dijkerman H, van Zandvoort MJ, Leijten FS et al (2013) Neuroimage 65:424
17. de Jong MC, Vansteensel MJ, van Ee R, Leijten FS, Ramsey NF, Dijkerman HC, Dumoulin SO, Knapen T (2020) Curr Biol 30(16):3089
18. Mesgarani N, Cheung C, Johnson K, Chang EF (2014) Science 343(6174):1006
19. Hullett PW, Hamilton LS, Mesgarani N, Schreiner CE, Chang EF (2016) J Neurosci 36(6):2014
20. Berezutskaya J, Freudenburg ZV, Güçlü U, van Gerven MA, Ramsey NF (2017) J Neurosci 37(33):7906
21. Oganian Y, Chang EF (2019) Sci Adv 5(11):eaay6279
22. Anderson NR, Blakely T, Schalk G, Leuthardt EC, Moran DW (2012) Exp Brain Res 223(1):1
23. Chartier J, Anumanchipalli GK, Johnson K, Chang EF (2018) Neuron 98(5):1042
24. Merrick CM, Dixon TC, Breska A, Lin J, Chang EF, King-Stephens D, Laxer KD, Weber PB, Carmena J, Knight RT et al (2022) Elife 11
25. Berezutskaya J, Baratin C, Freudenburg ZV, Ramsey NF (2020) Hum Brain Mapp 41(16):4587
26. Schrimpf M, Blank IA, Tuckute G, Kauf C, Hosseini EA, Kanwisher N, Tenenbaum JB, FedorenkoE (2021) Proc Natl Acad Sci 118(45)
27. Goldstein A, Zada Z, Buchnik E, Schain M, Price A, Aubrey B, Nastase SA, Feder A, Emanuel D, Cohen A et al (2022) Nat Neurosci 25(3):369
28. Saez I, Lin J, Stolk A, Chang E, Parvizi J, Schalk G, Knight RT, Hsu M (2018) Curr Biol 28(18):2889
29. Crone NE, Boatman D, Gordon B, Hao L (2001) Clin Neurophysiol 112(4):565
30. Ray S, Crone NE, Niebur E, Franaszczuk PJ, Hsiao SS (2008) J Neurosci 28(45):11526
31. Miller KJ, Sorensen LB, Ojemann JG, Den Nijs M (2009) PLoS Comput Biol 5(12):e1000609
32. Winawer J, Kay KN, Foster BL, Rauschecker AM, Parvizi J, Wandell BA (2013) Curr Biol 23(13):1145
33. Hermes D, Miller K, Wandell B, Winawer J (2015) Cereb Cortex 25(9):2951
34. Gaglianese A, Harvey BM, Vansteensel MJ, Dumoulin SO, Ramsey NF, Petridou N (2017) Hum Brain Mapp 38(1):293
35. Miller KJ, Schalk G, Hermes D, Ojemann JG, Rao RP (2016) PLoS Comput Biol 12(1):e1004660
36. Salari E, Freudenburg Z, Vansteensel M, Ramsey N (2018) IEEE Trans Neural Syst Rehabil Eng 26(5):1084
37. Branco MP, Geukes SH, Aarnoutse EJ, Vansteensel MJ, Freudenburg ZV, Ramsey NF (2019) J Neural Eng 16(5):056009
38. Berezutskaya J, Freudenburg ZV, Güçlü U, van Gerven MA, Ramsey NF (2020) PLoS Comput Biol 16(7):e1007992
39. Leeds DD, Seibert DA, Pyles JA, Tarr MJ (2013) J Vis 13(13):25

40. Khaligh-Razavi SM, Henriksson L, Kay K, Kriegeskorte N et al (2014) bioRxiv, p 009936
41. Wehbe L, Murphy B, Talukdar P, Fyshe A, Ramdas A, Mitchell T (2014) PLoS ONE 9(11):e112575
42. Yamins DL, Hong H, Cadieu CF, Solomon EA, Seibert D, DiCarlo JJ (2014) Proc Natl Acad Sci 111(23):8619
43. Jozwik KM, Kriegeskorte N, Mur M (2016) Neuropsychologia 83:201
44. Kriegeskorte N (2015) Annu Rev Vis Sci 1:417
45. LeCun Y, Bengio Y, Hinton G (2015) Nature 521(7553):436
46. Kriegeskorte N, Golan T (2019) Curr Biol 29(7):R231
47. Hochreiter S, Schmidhuber J (1997) Neural Comput 9(8):1735
48. Vaswani A, Shazeer N, Parmar N, Uszkoreit J, Jones L, Gomez AN, Kaiser Ł, Polosukhin I (2017) Adv Neural Inf Process Syst 30
49. Hinton GE, Salakhutdinov RR (2006) Sci 313(5786):504
50. Shrestha A, Mahmood A (2019) IEEE Access 7:53040
51. Khamparia A, Singh KM (2019) Expert Syst 36(3):e12400
52. Livezey JA, Glaser JI (2021) Brief Bioinform 22(2):1577
53. Abdel-Jaber H, Devassy D, Al Salam A, Hidaytallah L, EL-AmirM (2022) Algorithms 15(2):71
54. Kell AJ, Yamins DL, Shook EN, Norman-Haignere SV, McDermott JH (2018) Neuron 98(3):630
55. Berezutskaya J, Freudenburg ZV, Ambrogioni L, Güçlü U, van Gerven MA, Ramsey NF (2020) Sci Rep 10(1):1
56. Kuzovkin I, Vicente R, Petton M, Lachaux JP, Baciu M, Kahane P, Rheims S, Vidal JR, Aru J (2018) Commun Biol 1(1):1
57. Güçlü U, Thielen J, Hanke M, Van Gerven M (2016) Adv Neural Inf Process Syst 29
58. Cadieu CF, Hong H, Yamins DL, Pinto N, Ardila D, Solomon EA, Majaj NJ, DiCarlo JJ (2014) PLoS Comput Biol 10(12):e1003963
59. Cichy RM, Khosla A, Pantazis D, Oliva A (2017) Neuroimage 153:346
60. Güçlü U, van Gerven MA (2015) J Neurosci 35(27):10005
61. Kietzmann TC, Spoerer CJ, Sörensen LK, Cichy RM, Hauk O, Kriegeskorte N (2019) Proc Natl Acad Sci 116(43):21854
62. Kar K, Kubilius J, Schmidt K, Issa EB, DiCarlo JJ (2019) Nat Neurosci 22(6):974
63. Chaisangmongkon W, Swaminathan SK, Freedman DJ, Wang XJ (2017) Neuron 93(6):1504
64. Parde CJ, Colón YI, Hill MQ, Castillo CD, Dhar P, O'Toole AJ (2021) J Vis 21(8):15
65. Marblestone AH, Wayne G, Kording KP (2016) Front Comput Neurosci 94
66. Yamins DL, DiCarlo JJ (2016) Nat Neurosci 19(3):356
67. Grill-Spector K, Weiner KS, Gomez J, Stigliani A, Natu VS (2018) Interface Focus 8(4):20180013
68. Kietzmann TC, McClure P, Kriegeskorte N (2018) BioRxiv, p 133504
69. Richards BA, Lillicrap TP, Beaudoin P, Bengio Y, Bogacz R, Christensen A, Clopath C, Costa RP, de Berker A, Ganguli S et al (2019) Nat Neurosci 22(11):1761
70. Sinz FH, Pitkow X, Reimer J, Bethge M, Tolias AS (2019) Neuron 103(6):967
71. Kell AJ, McDermott JH (2019) Curr Opin Neurobiol 55:121
72. Cichy RM, Kaiser D (2019) Trends Cogn Sci 23(4):305
73. Hasson U, Nastase SA, Goldstein A (2020) Neuron 105(3):416
74. Saxe A, Nelli S, Summerfield C (2021) Nat Rev Neurosci 22(1):55
75. Lindsay GW (2021) J Cogn Neurosci 33(10):2017
76. Ullman S, Assif L, Fetaya E, Harari D (2016) Proc Natl Acad Sci 113(10):2744
77. Rajalingham R, Issa EB, Bashivan P, Kar K, Schmidt K, DiCarlo JJ (2018) J Neurosci 38(33):7255
78. Kay KN (2018) Neuroimage 180:101
79. Zador AM (2019) Nat Commun 10(1):1
80. Jacob G, Pramod R, Katti H, Arun S (2021) Nat Commun 12(1):1
81. Saxena S, Cunningham JP (2019) Curr Opin Neurobiol 55:103

82. Joukes J, Hartmann TS, Krekelberg B (2014) Front Syst Neurosci 8:239
83. Batty E, Merel J, Brackbill N, Heitman A, Sher A, Litke A, Chichilnisky E, Paninski L (2016)
84. Klindt D, Ecker AS, Euler T, Bethge M (2017) Adv Neural Inf Process Syst 30
85. Ecker AS, Sinz FH, Froudarakis E, Fahey PG, Cadena SA, Walker EY, Cobos E, Reimer J, Tolias AS, Bethge M (2018) arXiv:1809.10504
86. Cadena SA, Denfield GH, Walker EY, Gatys LA, Tolias AS, Bethge M, Ecker AS (2019) PLoS Comput Biol 15(4):e1006897
87. Berezutskaya J, Freudenburg Z, Ramsey N, Güçlü U, van Gerven M, Duivesteijn W, Pechenizkiy M, Fletcher G, Menkovski V, Postma E et al (2017) Benelearn, pp 149–153
88. Güçlü U, Van Gerven MA (2017) Front Comput Neurosci 11:7
89. Wen H, Shi J, Zhang Y, Lu KH, Cao J, Liu Z (2018) Cereb Cortex 28(12):4136
90. Keshishian M, Akbari H, Khalighinejad B, Herrero JL, Mehta AD, Mesgarani N (2020) Elife 9:e53445
91. Lake BM, Ullman TD, Tenenbaum JB, Gershman SJ (2017) Behav Brain Sci 40
92. Zeiler MD, Fergus R (2014) European conference on computer vision. Springer, pp 818–833
93. Zhou B, Bau D, Oliva A, Torralba A (2018) IEEE Trans Pattern Anal Mach Intell 41(9):2131
94. Barrett DG, Morcos AS, Macke JH (2019) Curr Opin Neurobiol 55:55
95. Seeliger K, Ambrogioni L, Güçlütürk Y, van den Bulk LM, Güçlü U, van Gerven M (2021) PLoS Comput Biol 17(2):e1008558
96. Douglas RJ, Martin KA (2007) Neuron 56(2):226
97. Bastos AM, Usrey WM, Adams RA, Mangun GR, Fries P, Friston KJ (2012) Neuron 76(4):695
98. Noguchi Y, Inui K, Kakigi R (2004) J Neurosci 24(28):6283
99. Krekelberg B, Boynton GM, van Wezel RJ (2006) Trends Neurosci 29(5):250
100. Benda J (2021) Curr Biol 31(3):R110
101. Olshausen BA, Field DJ (2004) Curr Opin Neurobiol 14(4):481
102. Földiák P (1991) Neural Comput 3(2):194
103. Kording KP, Kayser C, Einhauser W, Konig P (2004) J Neurophysiol 91(1):206
104. Clopath C, Bonhoeffer T, Hübener M, Rose T (2017) Philos Trans R Soc B: Biol Sci 372(1715):20160161
105. Fiser J, Berkes P, Orbán G, Lengyel M (2010) Trends Cogn Sci 14(3):119
106. Orbán G, Berkes P, Fiser J, Lengyel M (2016) Neuron 92(2):530
107. Ward LM (2003) Trends Cogn Sci 7(12):553
108. Buzsaki G, Draguhn A (2004) Science 304(5679):1926
109. McClure P, Kriegeskorte N (2016) Front Comput Neurosci 10:131
110. Guerguiev J, Lillicrap TP, Richards BA (2017) ELife 6:e22901
111. Scellier B, Bengio Y (2017) Front Comput Neurosci 11:24
112. Whittington JC, Bogacz R (2017) Neural Comput 29(5):1229
113. Sacramento J, Ponte Costa R, Bengio Y, Senn W (2018) Adv Neural Inf Process Syst 31
114. Balduzzi D, Vanchinathan H, Buhmann J (2015) Proceedings of the AAAI conference on artificial intelligence, vol 29
115. Lillicrap TP, Cownden D, Tweed DB, Akerman CJ (2016) Nat Commun 7(1):1
116. Pozzi I, Bohté S, Roelfsema P (2018) arXiv:1811.01768
117. Ahmad N, van Gerven MA, Ambrogioni L (2020) Adv Neural Inf Process Syst 33:10913
118. Movellan JR (1991) Connectionist models. Elsevier, pp 10–17
119. Hinton GE (2002) Neural Comput 14(8):1771
120. Detorakis G, Bartley T, Neftci E (2019) Neural Netw 114:1
121. Miller KJ (2019) Nat Hum Behav 3(11):1225
122. Nejedly P, Kremen V, Sladky V, Cimbalnik J, Klimes P, Plesinger F, Mivalt F, Travnicek V, Viscor I, Pail M et al (2020) Sci Data 7(1):1
123. Berezutskaya J, Vansteensel MJ, Aarnoutse EJ, Freudenburg ZV, Piantoni G, Branco MP, Ramsey NF (2022) Sci Data 9(1):1
124. Woolnough O, Kadipasaoglu CM, Conner CR, Forseth KJ, Rollo PS, Rollo MJ, Baboyan VG, Tandon N (2022) Sci Data 9(1):1
125. Peterson SM, Singh SH, Dichter B, Scheid M, Rao RP, Brunton BW (2022) Sci Data 9(1):1

126. Nieto N, Peterson V, Rufiner HL, Kamienkowski JE, Spies R (2022) Sci Data 9(1):1
127. Vermaas M, Piastra M, Oostendorp T, Ramsey N, Tiesinga P (2020) J Neural Eng 17(5):056031
128. Vermaas M, Piastra MC, Oostendorp TF, Ramsey NF, Tiesinga PH (2020) Neuroinformatics 18(4):569
129. Peterson SM, Steine-Hanson Z, Davis N, Rao RP, Brunton BW (2021) J Neural Eng 18(2):026014
130. Kriegeskorte N, Mur M, Bandettini P (2008) Front Syst Neurosci 2
131. Kriegeskorte N (2009) Front Neurosci 3(3):363. https://doi.org/10.3389/neuro.01.035.2009
132. Schrimpf M, Kubilius J, Lee MJ, Ratan Murty NA, Ajemian R, DiCarlo JJ (2020) Neuron 108(3):413. https://doi.org/10.1016/j.neuron.2020.07.040
133. Chang EF, Rieger JW, Johnson K, Berger MS, Barbaro NM, Knight RT (2010) Nat Neurosci 13(11):1428. https://doi.org/10.1038/nn.2641
134. Zhang H, Fell J, Staresina BP, Weber B, Elger CE, Axmacher N (2015) Curr Biol: CB 25(5):635. https://doi.org/10.1016/j.cub.2015.01.011
135. Chen Y, Shimotake A, Matsumoto R, Kunieda T, Kikuchi T, Miyamoto S, Fukuyama H, Takahashi R, Ikeda A, Lambon Ralph MA (2016) Cortex J Devot Study Nerv Syst Behav 79:1. https://doi.org/10.1016/j.cortex.2016.02.015
136. Rogers TT, Cox CR, Lu Q, Shimotake A, Kikuchi T, Kunieda T, Miyamoto S, Takahashi R, Ikeda A, Matsumoto R, Lambon Ralph MA (2021) eLife 10:e66276. https://doi.org/10.7554/eLife.66276
137. Reddy L, Cichy RM, VanRullen R (2021) eNeuro 8(3):ENEURO.0362. https://doi.org/10.1523/ENEURO.0362-20.2021
138. Bashivan P, Kar K, DiCarlo JJ (2019) Science 364(6439):eaav9436. https://doi.org/10.1126/science.aav9436
139. Khaligh-Razavi SM, Henriksson L, Kay K, Kriegeskorte N (2017) J Math Psychol 76:184. https://doi.org/10.1016/j.jmp.2016.10.007
140. Grossman S, Gaziv G, Yeagle EM, Harel M, Mégevand P, Groppe DM, Khuvis S, Herrero JL, Irani M, Mehta AD, Malach R (2019) Nat Commun 10(1):4934. https://doi.org/10.1038/s41467-019-12623-6
141. Glaser JI, Benjamin AS, Farhoodi R, Kording KP (2019) Prog Neurobiol 175:126. https://doi.org/10.1016/j.pneurobio.2019.01.008
142. Bzdok D (2017) Front Neurosci 11
143. Combrisson E, Jerbi K (2015) J Neurosci Methods 250:126. https://doi.org/10.1016/j.jneumeth.2015.01.010
144. Combrisson E, Perrone-Bertolotti M, Soto JL, Alamian G, Kahane P, Lachaux JP, Guillot A, Jerbi K (2017) Neuroimage 147:473. https://doi.org/10.1016/j.neuroimage.2016.11.042
145. Thiery T, Saive AL, Combrisson E, Dehgan A, Bastin J, Kahane P, Berthoz A, Lachaux JP, Jerbi K (2020) PLoS Biol 18(12):e3000864. https://doi.org/10.1371/journal.pbio.3000864
146. Wang M, Li G, Jiang S, Wei Z, Hu J, Chen L, Zhang D (2020) J Neural Eng 17(4):046043. https://doi.org/10.1088/1741-2552/ab9987
147. Jiang H, Schuele S, Rosenow J, Zelano C, Parvizi J, Tao JX, Wu S, Gottfried JA (2017) Neuron 94(1):207. https://doi.org/10.1016/j.neuron.2017.03.021
148. Liu H, Agam Y, Madsen JR, Kreiman G (2009) Neuron 62(2):281. https://doi.org/10.1016/j.neuron.2009.02.025
149. Isik L, Singer J, Madsen JR, Kanwisher N, Kreiman G (2018) Neuroimage 180(Pt A):147. https://doi.org/10.1016/j.neuroimage.2017.08.027
150. van de Nieuwenhuijzen ME, Axmacher N, Fell J, Oehrn CR, Jensen O, van GervenMJ (2016) NeuroImage 137:132. https://doi.org/10.1016/j.neuroimage.2016.05.008
151. Famili A, Krishnan G, Davenport E, Germi J, Wagner B, Lega B, Montillo A (2017) IEEE International symposium on biomedical imaging proceedings, vol 2017, p 587. https://doi.org/10.1109/ISBI.2017.7950589
152. Jun S, Kim JS, Chung CK (2021) Front Neurosci 15
153. Ter Wal M, Platonov A, Cardellicchio P, Pelliccia V, LoRusso G, Sartori I, Avanzini P, Orban GA, Tiesinga PHE (2020) Nat Commun 11(1):3075. https://doi.org/10.1038/s41467-020-16854-w

154. Watrous AJ, Deuker L, Fell J, Axmacher N (2015) eLife 4. https://doi.org/10.7554/eLife.07886
155. Höhne M, Jahanbekam A, Bauckhage C, Axmacher N, Fell J (2016) Neuroimage 139:127. https://doi.org/10.1016/j.neuroimage.2016.06.021
156. Kuzovkin I, Vidal JR, Perrone-Bertolotti M, Kahane P, Rheims S, Aru J, Lachaux JP, Vicente R (2020) Sci Rep 10(1):1
157. Davidesco I, Zion-Golumbic E, Bickel S, Harel M, Groppe DM, Keller CJ, Schevon CA, McKhann GM, Goodman RR, Goelman G, Schroeder CE, Mehta AD, Malach R (2014) Cereb Cortex (New York, N.Y.) 24(7):1879. https://doi.org/10.1093/cercor/bht038
158. Whaley ML, Kadipasaoglu CM, Cox SJ, Tandon N (2016) J Neurosci: Off J Soc Neurosci 36(4):1173. https://doi.org/10.1523/JNEUROSCI.2985-15.2016
159. Nourski KV, Steinschneider M, Oya H, Kawasaki H, Jones RD, Howard MA (1991) Cereb Cortex (New York, N.Y.) 24(2):340. https://doi.org/10.1093/cercor/bhs314
160. Ezzyat Y, Kragel JE, Burke JF, Levy DF, Lyalenko A, Wanda P, O'Sullivan L, Hurley KB, Busygin S, Pedisich I, Sperling MR, Worrell GA, Kucewicz MT, Davis KA, Lucas TH, Inman CS, Lega BC, Jobst BC, Sheth SA, Zaghloul K, Jutras MJ, Stein JM, Das SR, Gorniak R, Rizzuto DS, Kahana MJ (2017) Curr Biol: CB 27(9):1251. https://doi.org/10.1016/j.cub.2017.03.028
161. Varoquaux G (2018) Neuroimage 180(Pt A):68. https://doi.org/10.1016/j.neuroimage.2017.06.061
162. Norman KA, Polyn SM, Detre GJ, Haxby JV (2006) Trends Cogn Sci 10(9):424. https://doi.org/10.1016/j.tics.2006.07.005
163. Haynes JD (2015) Neuron 87(2):257. https://doi.org/10.1016/j.neuron.2015.05.025
164. Haynes JD, Rees G (2006) Nat Rev Neurosci 7(7):523. https://doi.org/10.1038/nrn1931
165. Kamitani Y, Tong F (2005) Nat Neurosci 8(5):679. https://doi.org/10.1038/nn1444
166. Haxby JV, Gobbini MI, Furey ML, Ishai A, Schouten JL, Pietrini P (2001) Science 293(5539):2425. https://doi.org/10.1126/science.1063736
167. King JR, Dehaene S (2014) Trends Cogn Sci 18(4):203. https://doi.org/10.1016/j.tics.2014.01.002
168. King JR, Gramfort A, Schurger A, Naccache L, Dehaene S (2014) PLoS ONE 9(1):e85791. https://doi.org/10.1371/journal.pone.0085791
169. Stokes MG, Kusunoki M, Sigala N, Nili H, Gaffan D, Duncan J (2013) Neuron 78(2):364. https://doi.org/10.1016/j.neuron.2013.01.039
170. Tsuchiya N, Kawasaki H, Oya H, Iii MAH, Adolphs R (2008) PLoS ONE 3(12):e3892. https://doi.org/10.1371/journal.pone.0003892
171. Gonzalez A, Hutchinson JB, Uncapher MR, Chen J, LaRocque KF, Foster BL, Rangarajan V, Parvizi J, Wagner AD (2015) Proc Natl Acad Sci 112(35):11066. https://doi.org/10.1073/pnas.1510749112
172. Kragel JE, Ezzyat Y, Sperling MR, Gorniak R, Worrell GA, Berry BM, Inman C, Lin JJ, Davis KA, Das SR, Stein JM, Jobst BC, Zaghloul KA, Sheth SA, Rizzuto DS, Kahana MJ (2017) Neuroimage 155:60. https://doi.org/10.1016/j.neuroimage.2017.03.042
173. Richards BA, Lillicrap TP, Beaudoin P, Bengio Y, Bogacz R, Christensen AJ, Clopath C, Costa RP, de Berker AO, Ganguli S, Gillon CJ, Hafner D, Kepecs A, Kriegeskorte N, Latham PE, Lindsay GW, Miller KD, Naud R, Pack CC, Poirazi P, Roelfsema PR, Sacramento J, Saxe AM, Scellier B, Schapiro AC, Senn W, Wayne G, Yamins DLK, Zenke F, Zylberberg J, Thérien D, Kording KP (2019) Nat Neurosci 22:1761
174. Bengio Y, Courville A, Vincent P (2013) IEEE Trans Pattern Anal Mach Intell 35(8):1798. https://doi.org/10.1109/TPAMI.2013.50
175. Grün F, Rupprecht C, Navab N, Tombari F (2016) ArXiv: abs/1606.07757
176. Springenberg JT, Dosovitskiy A, Brox T, Riedmiller MA (2015) CoRR abs/1412.6806
177. Li H, Fan Y (2019) Neuroimage 202:116059
178. Seeliger K, Güçlü U, Ambrogioni L, Güçlütürk Y, van Gerven MA (2018) Neuroimage 181:775
179. Le L, Ambrogioni L, Seeliger K, Güçlütürk Y, van Gerven M, Güçlü U (2021) BioRxiv

180. Dado T, Güçlütürk Y, Ambrogioni L, Ras G, Bosch S, van Gerven M, Güçlü U (2022) Sci Rep 12(1):1
181. Malkiel I, Rosenman G, Wolf L, Hendler T (2021) arXiv:2112.05761
182. Abou Jaoude M, Jing J, Sun H, Jacobs CS, Pellerin KR, Westover MB, Cash SS, Lam AD (2020) Clin Neurophysiol: Off J Int Fed Clin Neurophysiol 131(1):133. https://doi.org/10.1016/j.clinph.2019.09.031
183. Abou Jaoude M, Jacobs CS, Sarkis RA, Jing J, Pellerin KR, Cole AJ, Cash SS, Westover MB, Lam AD (2022) JAMA Neurol 79(6):614. https://doi.org/10.1001/jamaneurol.2022.0888
184. Ahmedt-Aristizabal D, Fookes C, Dionisio S, Nguyen K, Cunha JPS, Sridharan S (2017) Epilepsia 58(11):1817. https://doi.org/10.1111/epi.13907
185. Antoniades A, Spyrou L, Martin-Lopez D, Valentin A, Alarcon G, Sanei S, Cheong Took C (2017) IEEE Trans Neural Syst Rehabil Eng (IEEE Engineering in Medicine and Biology Society) 25(12):2285. https://doi.org/10.1109/TNSRE.2017.2755770
186. Antoniades A, Spyrou L, Martin-Lopez D, Valentin A, Alarcon G, Sanei S, Took CC (2018) Int J Neural Syst 28(8):1850009. https://doi.org/10.1142/S0129065718500090
187. Burrello A, Schindler K, Benini L, Rahimi A (2020) IEEE Trans Biomed Eng 67(2):601. https://doi.org/10.1109/TBME.2019.2919137
188. Caffarini J, Gjini K, Sevak B, Waleffe R, Kalkach-Aparicio M, Boly M, Struck AF (2022) Sci Rep 12(1):5397. https://doi.org/10.1038/s41598-022-09429-w
189. Chung YG, Jeon Y, Choi SA, Cho A, Kim H, Hwang H, Kim KJ (2020) Front Neurol 11:594679. https://doi.org/10.3389/fneur.2020.594679
190. Yamamoto S, Yanagisawa T, Fukuma R, Oshino S, Tani N, Khoo HM, Edakawa K, Kobayashi M, Tanaka M, Fujita Y, Kishima H (2021) J Neural Eng 18(5). https://doi.org/10.1088/1741-2552/ac23bf
191. Hashimoto H, Kameda S, Maezawa H, Oshino S, Tani N, Khoo HM, Yanagisawa T, Yoshimine T, Kishima H, Hirata M (2021) Int J Neural Syst 31(11):2050056. https://doi.org/10.1142/S0129065720500562
192. Pradeepkumar J, Anandakumar M, Kugathasan V, Lalitharatne TD, De Silva AC, Kappel SL (2021) Annual international conference of the IEEE engineering in medicine and biology society, vol 2021, p 420. https://doi.org/10.1109/EMBC46164.2021.9630958
193. Śliwowski M, Martin M, Souloumiac A, Blanchart P, Aksenova T (2022) J Neural Eng 19(2):026023
194. Xie Z, Schwartz O, Prasad A (2018) J Neural Eng 15(3):036009. https://doi.org/10.1088/1741-2552/aa9dbe
195. Ball T, Kern M, Mutschler I, Aertsen A, Schulze-Bonhage A (2009) Neuroimage 46(3):708
196. Wolpaw JR, Birbaumer N, McFarland DJ, Pfurtscheller G, Vaughan TM (2002) Clin Neurophysiol 113(6):767
197. Birbaumer N, Cohen LG (2007) J Physiol 579(3):621
198. Leuthardt EC, Schalk G, Roland J, Rouse A, Moran DW (2009) Neurosurg Focus 27(1):E4
199. Moran D (2010) Curr Opin Neurobiol 20(6):741
200. Shih JJ, Krusienski DJ, Wolpaw JR (2012) Mayo clinic proceedings, vol 87. Elsevier, pp 268–279
201. Chakrabarti S, Sandberg HM, Brumberg JS, Krusienski DJ (2015) Biomed Eng Lett 5(1):10
202. Martin S, Millán JR, Knight RT, Pasley BN (2019) Brain Lang 193:73
203. Herff C, Krusienski DJ, Kubben P (2020) Front Neurosci 14:123
204. Miller KJ, Hermes D, Staff NP (2020) Neurosurg Focus 49(1):E2
205. Müller-Putz G (2021) Neuroprosthetics and brain-computer interfaces in spinal cord injury. Springer, pp 217–232
206. Saha S, Mamun KA, Ahmed K, Mostafa R, Naik GR, Darvishi S, Khandoker AH, Baumert M (2021) Front Syst Neurosci 4
207. Rapeaux AB, Constandinou TG (2021) Curr Opin Biotechnol 72:102
208. Workgroup A (1995) Arch Phys Med Rehabil 76:205
209. Doble JE, Haig AJ, Anderson C, Katz R (2003) J Head Trauma Rehabil 18(5):435. https://doi.org/10.1097/00001199-200309000-00005

210. Laureys S, Pellas F, Van Eeckhout P, Ghorbel S, Schnakers C, Perrin F, Berre J, Faymonville ME, Pantke KH, Damas F et al (2005) Prog Brain Res 150:495
211. Robbins R, Simmons Z, Bremer B, Walsh S, Fischer S (2001) Neurology 56(4):442
212. Neudert C, Wasner M, Borasio GD (2004) J Palliat Med 7(4):551
213. Lulé D, Häcker S, Ludolph A, Birbaumer N, Kübler A (2008) Dtsch Arztebl Int 105(23):397
214. Lulé D, Zickler C, Häcker S, Bruno MA, Demertzi A, Pellas F, Laureys S, Kübler A (2009) Prog Brain Res 177:339
215. Linse K, Rüger W, Joos M, Schmitz-Peiffer H, Storch A, Hermann A (2017) Ann Neurol 81(2):310
216. Bruno MA, Bernheim JL, Ledoux D, Pellas F, Demertzi A, Laureys S (2011) BMJ Open 1(1):e000039
217. Rousseau MC, Baumstarck K, Alessandrini M, Blandin V, Billette de Villemeur T, Auquier P (2015) Orphanet J Rare Dis 10(1):1
218. Santhosh J, Bhatia M, Sahu S, Anand S et al (2005) Neurol India 53(1):99
219. Babiloni C, Pistoia F, Sarà M, Vecchio F, Buffo P, Conson M, Onorati P, Albertini G, Rossini PM (2010) Clin Neurophysiol 121(11):1816
220. Iyer PM, Egan C, Pinto-Grau M, Burke T, Elamin M, Nasseroleslami B, Pender N, Lalor EC, Hardiman O (2015) PLoS ONE 10(6):e0128682
221. Geronimo A, Simmons Z, Schiff S (2016) J Neural Eng 13(2):026002
222. Pandarinath C, Gilja V, Blabe CH, Nuyujukian P, Sarma AA, Sorice BL, Eskandar EN, Hochberg LR, Henderson JM, Shenoy KV (2015) Elife 4:e07436
223. Vansteensel MJ, Pels EG, Bleichner MG, Branco MP, Denison T, Freudenburg ZV, Gosselaar P, Leinders S, Ottens TH, Van Den Boom MA et al (2016) N Engl J Med 375(21):2060
224. Oxley TJ, Yoo PE, Rind GS, Ronayne SM, Lee CS, Bird C, Hampshire V, Sharma RP, Morokoff A, Williams DL et al (2021) J Neurointerv Surg 13(2):102
225. Kikkert S, Kolasinski J, Jbabdi S, Tracey I, Beckmann CF, Johansen-Berg H, Makin TR (2016) Elife 5:e15292
226. Bruurmijn ML, Pereboom IP, Vansteensel MJ, Raemaekers MA, Ramsey NF (2017) Brain 140(12):3166
227. Guenther FH, Brumberg JS, Wright EJ, Nieto-Castanon A, Tourville JA, Panko M, Law R, Siebert SA, Bartels JL, Andreasen DS et al (2009) PLoS ONE 4(12):e8218
228. Freudenburg ZV, Branco MP, Leinders S, Vijgh BH, Pels EG, Denison T, Berg LH, Miller KJ, Aarnoutse EJ, Ramsey NF et al (2019) Front Neurosci 1058
229. Leuthardt EC, Gaona C, Sharma M, Szrama N, Roland J, Freudenberg Z, Solis J, Breshears J, Schalk G (2011) J Neural Eng 8(3):036004
230. Pei X, Barbour DL, Leuthardt EC, Schalk G (2011) J Neural Eng 8(4):046028
231. Martin S, Brunner P, Holdgraf C, Heinze HJ, Crone NE, Rieger J, Schalk G, Knight RT, Pasley BN (2014) Front Neuroeng 7:14
232. Brumberg JS, Krusienski DJ, Chakrabarti S, Gunduz A, Brunner P, Ritaccio AL, Schalk G (2016) PLoS ONE 11(11):e0166872
233. Angrick M, Ottenhoff MC, Diener L, Ivucic D, Ivucic G, Goulis S, Saal J, Colon AJ, Wagner L, Krusienski DJ et al (2021) Commun Biol 4(1):1
234. Hermes D, Vansteensel MJ, Albers AM, Bleichner MG, Benedictus MR, Orellana CM, Aarnoutse E, Ramsey N (2011) J Neural Eng 8(2):025007
235. Maegherman G, Nuttall HE, Devlin JT, Adank P (2019) Front Human Neurosci 195
236. Shin D, Watanabe H, Kambara H, Nambu A, Isa T, Nishimura Y, Koike Y (2012) PLoS ONE 7(10):e47992
237. Chen C, Shin D, Watanabe H, Nakanishi Y, Kambara H, Yoshimura N, Nambu A, Isa T, Nishimura Y, Koike Y (2013) PLoS ONE 8(12):e83534
238. Umeda T, Koizumi M, Katakai Y, Saito R, Seki K (2019) Neuroimage 197:512
239. Pulvermüller F, Huss M, Kherif F, del Prado Martin FM, Hauk O, Shtyrov Y (2006) Proc Natl Acad Sci 103(20):7865
240. Bouchard KE, Mesgarani N, Johnson K, Chang EF (2013) Nature 495(7441):327

241. Branco MP, Freudenburg ZV, Aarnoutse EJ, Bleichner MG, Vansteensel MJ, Ramsey NF (2017) Neuroimage 147:130
242. Kellis S, Miller K, Thomson K, Brown R, House P, Greger B (2010) J Neural Eng 7(5):056007
243. Pei X, Leuthardt EC, Gaona CM, Brunner P, Wolpaw JR, Schalk G (2011) Neuroimage 54(4):2960
244. Ikeda S, Shibata T, Nakano N, Okada R, Tsuyuguchi N, Ikeda K, Kato A (2014) Front Hum Neurosci 8:125
245. Kanas VG, Mporas I, Benz HL, Sgarbas KN, Bezerianos A, Crone NE (2014) IEEE Trans Biomed Eng 61(4):1241
246. Herff C, Heger D, De Pesters A, Telaar D, Brunner P, Schalk G, Schultz T (2015) Front Neurosci 9:217
247. Martin S, Brunner P, Iturrate I, Millán JR, Schalk G, Knight RT, Pasley BN (2016) Sci Rep 6(1):1
248. Hickok G, Poeppel D (2007) Nat Rev Neurosci 8(5):393
249. Friederici AD (2011) Physiol Rev 91(4):1357
250. Bundy DT, Pahwa M, Szrama N, Leuthardt EC (2016) J Neural Eng 13(2):026021
251. Bleichner MG, Freudenburg ZV, Jansma JM, Aarnoutse EJ, Vansteensel MJ, Ramsey NF (2016) Brain Struct Funct 221(1):203
252. Li Y, Zhang S, Jin Y, Cai B, Controzzi M, Zhu J, Zhang J, Zheng X (2017) Behav Neurol 2017
253. Pan G, Li JJ, Qi Y, Yu H, Zhu JM, Zheng XX, Wang YM, Zhang SM (2018) Front Neurosci 555
254. Verwoert M, Vansteensel MJ, Freudenburg ZV, Aarnoutse EJ, Leijten FS, Ramsey NF, Branco MP (2021) J Neural Eng 18(5):056065
255. Leonard MK, Lucas B, Blau S, Corina DP, Chang EF (2020) Curr Biol 30(22):4342
256. Salari E, Freudenburg Z, Branco M, Aarnoutse E, Vansteensel M, Ramsey N (2019) Sci Rep 9(1):1
257. Salari E, Freudenburg ZV, Vansteensel MJ, Ramsey NF (2020) Ann Neurol 88(3):631
258. Blakely T, Miller KJ, Zanos SP, Rao RP, Ojemann JG (2009) Neurosurg Focus 27(1):E13
259. Cecotti H (2011) J Physiol-Paris 105(1–3):106
260. Blakely T, Miller KJ, Rao RP, Holmes MD, Ojemann JG (2008) 2008 30th Annual international conference of the IEEE engineering in medicine and biology society. IEEE, pp 4964–4967
261. Mugler EM, Patton JL, Flint RD, Wright ZA, Schuele SU, Rosenow J, Shih JJ, Krusienski DJ, Slutzky MW (2014) J Neural Eng 11(3):035015
262. Bouchard KE, Chang EF (2014) 2014 36th Annual international conference of the IEEE engineering in medicine and biology society. IEEE, pp 6782–6785
263. Ramsey NF, Salari E, Aarnoutse EJ, Vansteensel MJ, Bleichner MG, Freudenburg Z (2018) Neuroimage 180:301
264. Livezey JA, Bouchard KE, Chang EF (2019) PLoS Comput Biol 15(9):e1007091
265. Zhang D, Gong E, Wu W, Lin J, Zhou W, Hong B (2012) 2012 Annual international conference of the IEEE engineering in medicine and biology society. IEEE, pp 3292–3295
266. Herff C, Diener L, Angrick M, Mugler E, Tate MC, Goldrick MA, Krusienski DJ, Slutzky MW, Schultz T (2019) Front Neurosci 13:1267
267. Pasley BN, David SV, Mesgarani N, Flinker A, Shamma SA, Crone NE, Knight RT, Chang EF (2012) PLoS Biol 10(1):e1001251
268. Kohler J, Ottenhoff MC, Goulis S, Angrick M, Colon AJ, Wagner L, Tousseyn S, Kubben PL, Herff C (2021) arXiv:2111.01457
269. Berezutskaya J, Freudenburg ZV, Vansteensel MJ, Aarnoutse EJ, Ramsey NF, van Gerven MA (2022) bioRxiv
270. Willett FR, Avansino DT, Hochberg LR, Henderson JM, Shenoy KV (2021) Nature 593(7858):249
271. Van Gerven M, Farquhar J, Schaefer R, Vlek R, Geuze J, Nijholt A, Ramsey N, Haselager P, Vuurpijl L, Gielen S et al (2009) J Neural Eng 6(4):041001

272. Nicolas-Alonso LF, Gomez-Gil J (2012) Sensors 12(2):1211
273. Brumberg JS, Wright EJ, Andreasen DS, Guenther FH, Kennedy PR (2011) Front Neurosci 5:65
274. Anumanchipalli GK, Chartier J, Chang EF (2019) Nature 568(7753):493
275. Akbari H, Khalighinejad B, Herrero JL, Mehta AD, Mesgarani N (2019) Sci Rep 9(1):1
276. Makin JG, Moses DA, Chang EF (2020) Nat Neurosci 23(4):575
277. Sun P, Anumanchipalli GK, Chang EF (2020) J Neural Eng 17(6):066015
278. Moses DA, Mesgarani N, Leonard MK, Chang EF (2016) J Neural Eng 13(5):056004
279. Moses DA, Metzger SL, Liu JR, Anumanchipalli GK, Makin JG, Sun PF, Chartier J, Dougherty ME, Liu PM, Abrams GM et al (2021) N Engl J Med 385(3):217
280. Oxley TJ, Opie NL, John SE, Rind GS, Ronayne SM, Wheeler TL, Judy JW, McDonald AJ, Dornom A, Lovell TJ et al (2016) Nat Biotechnol 34(3):320
281. Anderson KD (2004) J Neurotrauma 21(10):1371
282. Serruya MD, Hatsopoulos NG, Paninski L, Fellows MR, Donoghue JP (2002) Nature 416(6877):141
283. Taylor DM, Tillery SIH, Schwartz AB (2002) Science 296(5574):1829
284. Carmena JM, Lebedev MA, Crist RE, O'Doherty JE, Santucci DM, Dimitrov DF, Patil PG, Henriquez CS, Nicolelis MAL (2003) PLoS Biol 1(2):e42. https://doi.org/10.1371/journal.pbio.0000042
285. Paninski L, Fellows MR, Hatsopoulos NG, Donoghue JP (2004) J Neurophysiol 91(1):515
286. Velliste M, Perel S, Spalding MC, Whitford AS, Schwartz AB (2008) Nature 453(7198):1098
287. Moritz CT, Perlmutter SI, Fetz EE (2008) Nature 456(7222):639
288. Hochberg LR, Serruya MD, Friehs GM, Mukand JA, Saleh M, Caplan AH, Branner A, Chen D, Penn RD, Donoghue JP (2006) Nature 442(7099):164
289. Simeral J, Kim SP, Black M, Donoghue J, Hochberg L (2011) J Neural Eng 8(2):025027
290. Gilja V, Pandarinath C, Blabe CH, Nuyujukian P, Simeral JD, Sarma AA, Sorice BL, Perge JA, Jarosiewicz B, Hochberg LR et al (2015) Nat Med 21(10):1142
291. Hochberg LR, Bacher D, Jarosiewicz B, Masse NY, Simeral JD, Vogel J, Haddadin S, Liu J, Cash SS, Van Der Smagt P et al (2012) Nature 485(7398):372
292. Collinger JL, Wodlinger B, Downey JE, Wang W, Tyler-Kabara EC, Weber DJ, McMorland AJ, Velliste M, Boninger ML, Schwartz AB (2013) The Lancet 381(9866):557
293. Kryger M, Wester B, Pohlmeyer EA, Rich M, John B, Beaty J, McLoughlin M, Boninger M, Tyler-Kabara EC (2017) Exp Neurol 287:473
294. Dunlap C, Bird L, Burkhart I, Eipel K, Colachis S, Annetta N, Ganzer P, Sharma G, Friedenberg D, Franklin R et al (2019) 2019 IEEE International conference on systems, man and cybernetics (SMC). IEEE, pp 277–284
295. Downey JE, Weiss JM, Flesher SN, Thumser ZC, Marasco PD, Boninger ML, Gaunt RA, Collinger JL (2018) Front Neurosci 801
296. Bouton CE, Shaikhouni A, Annetta NV, Bockbrader MA, Friedenberg DA, Nielson DM, Sharma G, Sederberg PB, Glenn BC, Mysiw WJ et al (2016) Nature 533(7602):247
297. Flesher SN, Collinger JL, Foldes ST, Weiss JM, Downey JE, Tyler-Kabara EC, Bensmaia SJ, Schwartz AB, Boninger ML, Gaunt RA (2016) Sci Transl Med 8(361):361ra141
298. Ganzer PD, Colachis SC 4th, Schwemmer MA, Friedenberg DA, Dunlap CF, Swiftney CE, Jacobowitz AF, Weber DJ, Bockbrader MA, Sharma G (2020) Cell 181(4):763
299. Flesher SN, Downey JE, Weiss JM, Hughes CL, Herrera AJ, Tyler-Kabara EC, Boninger ML, Collinger JL, Gaunt RA (2021) Science 372(6544):831
300. Ethier C, Miller LE (2015) Neurobiol Dis 83:180
301. Ajiboye AB, Willett FR, Young DR, Memberg WD, Murphy BA, Miller JP, Walter BL, Sweet JA, Hoyen HA, Keith MW et al (2017) The Lancet 389(10081):1821
302. Schwemmer MA, Skomrock ND, Sederberg PB, Ting JE, Sharma G, Bockbrader MA, Friedenberg DA (2018) Nat Med 24(11):1669
303. Bockbrader M (2019) Curr Opin Biomed Eng 11:85
304. Colachis SC, Dunlap CF, Annetta NV, Tamrakar SM, Bockbrader MA, Friedenberg DA (2021) J Neural Eng 18(4):0460d7

305. Silversmith DB, Abiri R, Hardy NF, Natraj N, Tu-Chan A, Chang EF, Ganguly K (2021) Nat Biotechnol 39(3):326
306. Pandarinath C, Nuyujukian P, Blabe CH, Sorice BL, Saab J, Willett FR, Hochberg LR, Shenoy KV, Henderson JM (2017) Elife 6:e18554
307. Simeral JD, Hosman T, Saab J, Flesher SN, Vilela M, Franco B, Kelemen JN, Brandman DM, Ciancibello JG, Rezaii PG et al (2021) IEEE Trans Biomed Eng 68(7):2313
308. Kanth ST, Ray S (2020) J Neurosci 40(12):2430
309. Chao ZC, Nagasaka Y, Fujii N (2010) Front Neuroeng 3:3
310. Degenhart AD, Eles J, Dum R, Mischel JL, Smalianchuk I, Endler B, Ashmore RC, Tyler-Kabara EC, Hatsopoulos NG, Wang W et al (2016) J Neural Eng 13(4):046019
311. Larzabal C, Bonnet S, Costecalde T, Auboiroux V, Charvet G, Chabardes S, Aksenova T, Sauter-Starace F (2021) J Neural Eng 18(5):056026
312. Mestais CS, Charvet G, Sauter-Starace F, Foerster M, Ratel D, Benabid AL (2014) IEEE Trans Neural Syst Rehabil Eng 23(1):10
313. Alahi MEE, Liu Y, Xu Z, Wang H, Wu T, Mukhopadhyay SC (2021) Mater Today Commun 29:102853
314. Watanabe H, Sato M, Suzuki T, Nambu A, Nishimura Y, Kawato M, Isa T (2012) J Neural Eng 9(3):036006
315. Shimoda K, Nagasaka Y, Chao ZC, Fujii N (2012) J Neural Eng 9(3):036015
316. Hu K, Jamali M, Moses ZB, Ortega CA, Friedman GN, Xu W, Williams ZM (2018) Sci Rep 8(1):1
317. Toro C, Deuschl G, Thatcher R, Sato S, Kufta C, Hallett M (1994) Electroencephalogr Clin Neurophysiol/Evoked Potentials Sect 93(5):380
318. Levine SP, Huggins JE, BeMent SL, Kushwaha RK, Schuh LA, Rohde MM, Passaro EA, Ross DA, Elisevich KV, Smith BJ (2000) IEEE Trans Rehabil Eng 8(2):180
319. Mehring C, Nawrot MP, de Oliveira SC, Vaadia E, Schulze-Bonhage A, Aertsen A, Ball T (2004) Decoding and interfacing the brain: from neuronal assemblies to cyborgs. J Physiol-Paris 98(4):498. https://doi.org/10.1016/j.jphysparis.2005.09.016. https://www.sciencedirect.com/science/article/pii/S0928425705000173
320. Leuthardt EC, Schalk G, Wolpaw JR, Ojemann JG, Moran DW (2004) J Neural Eng 1(2):63
321. Pistohl T, Schulze-Bonhage A, Aertsen A, Mehring C, Ball T (2012) Neuroimage 59(1):248
322. Kubanek J, Miller KJ, Ojemann JG, Wolpaw JR, Schalk G (2009) J Neural Eng 6(6):066001
323. Miller KJ, Zanos S, Fetz E, Den Nijs M, Ojemann J (2009) J Neurosci 29(10):3132
324. Hotson G, McMullen DP, Fifer MS, Johannes MS, Katyal KD, Para MP, Armiger R, Anderson WS, Thakor NV, Wester BA et al (2016) J Neural Eng 13(2):026017
325. Satow T, Matsuhashi M, Ikeda A, Yamamoto J, Takayama M, Begum T, Mima T, Nagamine T, Mikuni N, Miyamoto S et al (2003) Clin Neurophysiol 114(7):1259
326. Schalk G, Kubanek J, Miller KJ, Anderson N, Leuthardt EC, Ojemann JG, Limbrick D, Moran D, Gerhardt LA, Wolpaw JR (2007) J Neural Eng 4(3):264
327. Pistohl T, Ball T, Schulze-Bonhage A, Aertsen A, Mehring C (2008) J Neurosci Methods 167(1):105
328. Gunduz A, Sanchez JC, Carney PR, Principe JC (2009) Neural Netw 22(9):1257
329. Kellis S, Hanrahan S, Davis T, House PA, Brown R, Greger B (2012) 2012 Annual international conference of the IEEE engineering in medicine and biology society. IEEE, pp 4091–4094
330. Nakanishi Y, Yanagisawa T, Shin D, Fukuma R, Chen C, Kambara H, Yoshimura N, Hirata M, Yoshimine T, Koike Y (2013) PLoS ONE 8(8):e72085
331. Volkova K, Lebedev MA, Kaplan A, Ossadtchi A (2019) Front Neuroinform 13:74
332. Branco MP, de Boer LM, Ramsey NF, Vansteensel MJ (2019) Eur J Neurosci 50(5):2755
333. Chandrasekaran S, Fifer M, Bickel S, Osborn L, Herrero J, Christie B, Xu J, Murphy RK, Singh S, Glasser MF et al (2021) Bioelectron Med 7(1):1
334. Vadera S, Marathe AR, Gonzalez-Martinez J, Taylor DM (2013) Neurosurg Focus 34(6):E3
335. Murphy BA, Miller JP, Gunalan K, Ajiboye AB (2016) PLoS ONE 11(3):e0150359
336. Li G, Jiang S, Xu Y, Wu Z, Chen L, Zhang D (2017) 2017 8th International IEEE/EMBS conference on neural engineering (NER). IEEE, pp 375–378

337. Wang W, Collinger JL, Degenhart AD, Tyler-Kabara EC, Schwartz AB, Moran DW, Weber DJ, Wodlinger B, Vinjamuri RK, Ashmore RC et al (2013) PLoS ONE 8(2):e55344
338. Márquez-Chin C, Popovic M, Cameron T, Lozano A, Chen R (2009) Spinal Cord 47(11):802
339. Yanagisawa T, Hirata M, Saitoh Y, Goto T, Kishima H, Fukuma R, Yokoi H, Kamitani Y, Yoshimine T (2011) J Neurosurg 114(6):1715
340. Benabid AL, Costecalde T, Eliseyev A, Charvet G, Verney A, Karakas S, Foerster M, Lambert A, Morinière B, Abroug N et al (2019) Lancet Neurol 18(12):1112
341. Moly A, Costecalde T, Martel F, Martin M, Larzabal C, Karakas S, Verney A, Charvet G, Chabardes S, Benabid AL et al (2022) J Neural Eng 19(2):026021
342. Cajigas I, Davis K, Meschede-Krasa B, Prins N, Gallo S, Naeem J, Palermo A, Wilson A, Guerra S, Parks B et al (2021)
343. Tam W, Wu T, Zhao Q, Keefer E, Yang Z (2018) BMC Biomed Eng 1(1):1
344. Xie Z, Schwartz O, Prasad A (2018) J Neural Eng 15(3):036009
345. Wang N, Farhadi A, Rao R, Brunton B (2018) Proceedings of the AAAI conference on artificial intelligence, vol 32
346. Skomrock ND, Schwemmer MA, Ting JE, Trivedi HR, Sharma G, Bockbrader MA, Friedenberg DA (2018) Front Neurosci 763
347. Choi H, Lim S, Min K, Ahn K, Lee KM, Jang DP (2021) J Neural Eng 18(6):066022
348. Liu Z, Tang J, Gao B, Yao P, Li X, Liu D, Zhou Y, Qian H, Hong B, Wu H (2020) Nat Commun 11(1):1
349. James SL, Abate D, Abate KH, Abay SM, Abbafati C, Abbasi N, Abbastabar H, Abd-Allah F, Abdela J, Abdelalim A et al (2018) The Lancet 392(10159):1789
350. Brodie M, Shorvon S, Canger R, Halasz P, Johannessen S, Thompson P, Wieser H, Wolf P (1997) Epilepsia 38(11):1245
351. Kwan P, Brodie MJ (2000) N Engl J Med 342(5):314
352. Kobau R, Zahran H, Thurman DJ, Zack MM, Henry TR, Schachter SC, Price PH (2008)
353. Vos T, Allen C, Arora M, Barber RM, Bhutta ZA, Brown A, Carter A, Casey DC, Charlson FJ, Chen AZ et al (2016) The Lancet 388(10053):1545
354. Graybiel AM (2000) Curr Biol 10(14):R509
355. Obeso JA, Rodríguez-Oroz MC, Benitez-Temino B, Blesa FJ, Guridi J, Marin C, Rodriguez M (2008) Mov Disord: Off J Mov Disord Soc 23(S3):S548
356. Schlaepfer TE, Bewernick BH, Kayser S, Hurlemann R, Coenen VA (2014) Neuropsychopharmacology 39(6):1303
357. Laxpati NG, Kasoff WS, Gross RE (2014) Neurotherapeutics 11(3):508
358. Coffey RJ (2009) Artif Organs 33(3):208
359. Benabid AL, Pollak P, Louveau A, Henry S, De Rougemont J (1987) Stereotact Funct Neurosurg 50(1–6):344
360. Pollak P, Benabid A, Gross C, Gao D, Laurent A, Benazzouz A, Hoffmann D, Gentil M, Perret J (1993) Rev Neurol 149(3):175
361. Limousin P, Pollak P, Benazzouz A, Hoffmann D, Le Bas JF, Perret J, Benabid A, Broussolle E (1995) The Lancet 345(8942):91
362. Deuschl G, Schade-Brittinger C, Krack P, Volkmann J, Schäfer H, Bötzel K, Daniels C, Deutschländer A, Dillmann U, Eisner W et al (2006) N Engl J Med 355(9):896
363. Williams A, Gill S, Varma T, Jenkinson C, Quinn N, Mitchell R, Scott R, Ives N, Rick C, Daniels J et al (2010) Lancet Neurol 9(6):581
364. Odekerken VJ, van Laar T, Staal MJ, Mosch A, Hoffmann CF, Nijssen PC, Beute GN, van Vugt JP, Lenders MW, Contarino MF et al (2013) Lancet Neurol 12(1):37
365. Cooper I, Upton A, Amin I (1980) Stereotact Funct Neurosurg 43(3–5):244
366. Andrade D, Zumsteg D, Hamani C, Hodaie M, Sarkissian S, Lozano A, Wennberg R (2006) Neurology 66(10):1571
367. Fisher R, Salanova V, Witt T, Worth R, Henry T, Gross R, Oommen K, Osorio I, Nazzaro J, Labar D et al (2010) Epilepsia 51(5):899
368. Boëx C, Seeck M, Vulliemoz S, Rossetti AO, Staedler C, Spinelli L, Pegna AJ, Pralong E, Villemure JG, Foletti G et al (2011) Seizure 20(6):485

369. Handforth A, DeSalles AA, Krahl SE (2006) Epilepsia 47(7):1239
370. Valentín A, García Navarrete E, Chelvarajah R, Torres C, Navas M, Vico L, Torres N, Pastor J, Selway R, Sola RG et al (2013) Epilepsia 54(10):1823
371. Velasco F, Carrillo-Ruiz JD, Brito F, Velasco M, Velasco AL, Marquez I, Davis R (2005) Epilepsia 46(7):1071
372. Loddenkemper T, Pan A, Neme S, Baker KB, Rezai AR, Dinner DS, Montgomery EB Jr, Lüders HO (2001) J Clin Neurophysiol 18(6):514
373. Theodore WH, Fisher RS (2004) Lancet Neurol 3(2):111
374. Mayberg HS, Lozano AM, Voon V, McNeely HE, Seminowicz D, Hamani C, Schwalb JM, Kennedy SH (2005) Neuron 45(5):651
375. Blomstedt P, Sjöberg RL, Hansson M, Bodlund O, Hariz MI (2013) World Neurosurg 80(6):e245
376. Perlmutter JS, Mink JW (2006) Annu Rev Neurosci 29:229
377. Lozano AM, Lipsman N, Bergman H, Brown P, Chabardes S, Chang JW, Matthews K, McIntyre CC, Schlaepfer TE, Schulder M et al (2019) Nat Rev Neurol 15(3):148
378. Richardson MP (2012) J Neurol Neurosurg Psychiatry 83(12):1238
379. Breakspear M (2017) Nat Neurosci 20(3):340
380. Goyal A, Goetz S, Stanslaski S, Oh Y, Rusheen AE, Klassen B, Miller K, Blaha CD, Bennet KE, Lee K (2021) Biosens Bioelectron 176:112888
381. Kringelbach ML, Jenkinson N, Owen SL, Aziz TZ (2007) Nat Rev Neurosci 8(8):623
382. Sun FT, Morrell MJ (2014) Neurotherapeutics 11(3):553
383. Kupsch A, Benecke R, Müller J, Trottenberg T, Schneider GH, Poewe W, Eisner W, Wolters A, Müller JU, Deuschl G et al (2006) N Engl J Med 355(19):1978
384. Zauber SE, Watson N, Comella CL, Bakay RA, Metman LV (2009) J Neurosurg 110(2):229
385. Follett KA, Weaver FM, Stern M, Hur K, Harris CL, Luo P, Marks WJ Jr, Rothlind J, Sagher O, Moy C et al (2010) N Engl J Med 362(22):2077
386. Schrader C, Capelle HH, Kinfe T, Blahak C, Bäzner H, Lütjens G, Dressler D, Krauss J (2011) Neurology 77(5):483
387. Yamamoto T, Katayama Y, Ushiba J, Yoshino H, Obuchi T, Kobayashi K, Oshima H, Fukaya C (2013) Neuromodul: Technol Neural Interface 16(3):230
388. Malekmohammadi M, Herron J, Velisar A, Blumenfeld Z, Trager MH, Chizeck HJ, Brontë-Stewart H (2016) Mov Disord 31(3):426
389. Herron JA, Thompson MC, Brown T, Chizeck HJ, Ojemann JG, Ko AL (2016) J Neurosurg 127(3):580
390. Swann NC, de Hemptinne C, Thompson MC, Miocinovic S, Miller AM, Ostrem JL, Chizeck HJ, Starr PA et al (2018) J Neural Eng 15(4):046006
391. Jobst BC, Kapur R, Barkley GL, Bazil CW, Berg MJ, Bergey GK, Boggs JG, Cash SS, Cole AJ, Duchowny MS et al (2017) Epilepsia 58(6):1005
392. Scangos KW, Khambhati AN, Daly PM, Makhoul GS, Sugrue LP, Zamanian H, Liu TX, Rao VR, Sellers KK, Dawes HE et al (2021) Nat Med 27(10):1696
393. Sheth SA, Bijanki KR, Metzger B, Allawala A, Pirtle V, Adkinson JA, Myers J, Mathura RK, Oswalt D, Tsolaki E et al (2021) Biol Psychiatry
394. Basu I, Yousefi A, Crocker B, Zelmann R, Paulk AC, Peled N, Ellard KK, Weisholtz DS, Cosgrove GR, Deckersbach T et al (2021) Nat Biomed Eng 1–130
395. Saucedo-Alvarado PE, Velasco AL, Aguado-Carrillo G, Cuellar-Herrera M, Trejo-Martínez D, Márquez-Franco R, Velasco-Campos F (2022) J Neurosurg 1(aop):1
396. Neumann WJ, Rodriguez-Oroz MC (2021) Machine learning will extend the clinical utility of adaptive deep brain stimulation
397. Merk T, Peterson V, Lipski WJ, Blankertz B, Turner RS, Li N, Horn A, Richardson RM, Neumann WJ (2022) eLife 11:e75126
398. Kuhlmann L, Lehnertz K, Richardson MP, Schelter B, Zaveri HP (2018) Nat Rev Neurol 14(10):618
399. Sussillo D, Jozefowicz R, Abbott L, Pandarinath C (2016) arXiv:1608.06315
400. Meier JM, Perdikis D, Blickensdörfer A, Stefanovski L, Liu Q, Maith O, Dinkelbach HÜ, Baladron J, Hamker FH, Ritter P (2022) Exp Neurol 114111

Chapter 48
How Can I Identify Stimulus-Driven Neural Activity Patterns in Multi-Patient ECoG Data?

Jeremy R. Manning

Abstract Identifying stimulus-driven neural activity patterns is critical for studying the neural basis of cognition. This can be particularly challenging in intracranial datasets, where electrode locations typically vary across patients. This chapter first presents an overview of the major challenges to identifying stimulus-driven neural activity patterns in the general case. Next, we will review several modality-specific considerations and approaches, along with a discussion of several issues that are particular to intracranial recordings. Against this backdrop, we will consider a variety of within-subject and across-subject approaches to identifying and modeling stimulus-driven neural activity patterns in multi-patient intracranial recordings. These approaches include generalized linear models, multivariate pattern analysis, representational similarity analysis, joint stimulus-activity models, hierarchical matrix factorization models, Gaussian process models, geometric alignment models, inter-subject correlations, and inter-subject functional correlations. Examples from the recent literature serve to illustrate the major concepts and provide the conceptual intuitions for each approach.

Keywords Stimulus-driven · Multi-subject · Signal processing · Computational models · Dynamics

48.1 Overview

Studying brain function often requires identifying brain responses to a given stimulus or set of stimuli. For some stimuli, and for some systems, this identification problem is relatively straightforward. For example, when a photopigment in a retinal photoreceptor absorbs light, this triggers a cascade of responses that is ultimately sent from the retina to other brain areas via the optic nerve [71]. In the general sense, however (i.e., for arbitrarily complex stimuli and arbitrary brain areas), the problem

J. R. Manning (✉)
Dartmouth College, Hanover, NH, USA
e-mail: jeremy.r.manning@dartmouth.edu

N. Axmacher (ed.), *Intracranial EEG*, Studies in Neuroscience, Psychology and Behavioral Economics, https://doi.org/10.1007/978-3-031-20910-9_48

of identifying neural responses to known (or unknown) stimuli can be incredibly challenging [73].

48.1.1 Why Is It Challenging to Identify Stimulus-Driven Brain Activity?

To illustrate the enormity of the challenge of identifying stimulus-driven brain responses in the general sense, it can be useful to start by considering what form a complete solution might take. First, we need some means of defining (and measuring) what brain activity *is*. For example, should we concern ourselves with measuring membrane potentials or firing rates of individual neurons? Or population activity in a given brain structure or network? And is it more appropriate to analyze or interpret activity patterns in the **time domain** (e.g., firing rate or voltage as a function of time), or in the **frequency domain** (e.g., characterizing the signal through the relative contributions of its constituent sinusoidal components at different frequencies)? Should we consider neurons and/or brain structures in isolation, or should we instead interpret each "unit" of activity within the context of the network(s) it participates in or contributes to? We discuss several different approaches to these questions (and their relative trade-offs) in Sect. 48.1.2.

Second, we need some means of characterizing (and ideally, quantifying) the stimulus itself. For a simple stimulus, such as a single photon of light, emitted from a known location in an otherwise completely dark room, constructing a sufficiently comprehensive model of the stimulus might be straightforward—and perhaps even trivial. For other stimuli, such as real-world experiences, constructing a comprehensive model of "what is happening" can be highly complex (at best). Essentially, building a stimulus model entails quantifying how different **features**, or properties, change over time. In our single-photon example, we might represent the stimulus as a timeseries of zeros (no photon present) and ones (photon present). For more complex stimuli, however, it may not even be clear what the features *are*. We discuss considerations and approaches to building explicit stimulus models in Sect. 48.1.3.

Characterizing brain activity and building stimulus models are each complex challenges in their own right. Linking the two provides its own set of additional challenges. We discuss these issues in Sect. 48.2.

48.1.2 How Can We Measure Neural "Activity" in the Human Brain?

The brain is a complex organ comprising myriad cell types that interconnect to form a vast network. When neuroscientists use the term *brain activity*, this can refer to a variety of physiological phenomena. To contextualize what brain activity means, it can be helpful to first consider the brain's structure and function.

Neurons are the dominant cell type in the brain; the adult human brain contains roughly 100 billion neurons. The **cerebral cortex**, commonly associated with high-level brain function, comprises roughly 80% of the adult human brain's mass, but only roughly 20% of its neurons [59]. In general, **brain activity** refers to changes in cellular processes that neurons undergo. These processes can take many forms, including [77]: rapid changes in membrane voltage that result in neurotransmitter release, called **action potentials** [63]; sub-threshold changes in the membrane voltages of individual neurons or populations of neurons; neurotransmitter release; changes in the composition or distribution of cell surface proteins (including ion channels); metabolic changes (such as increased or decreased blood flow); changes in ion concentrations; structural or anatomical changes, and so on. Because the term brain activity often refers to the activities of neurons specifically, the term **neural activity** can be a more precise way of referring to these phenomena. However, it is worth noting that neurons are not the *only* cell type in the brain. For example, the brain also contains roughly 100 billion **glial cells** [4, 103], which also play an important role in supporting neuronal function, synapse formation, and brain metabolism [7].

After considering the various forms that brain (neural) activity can take, it can also be useful to define a relevant spatiotemporal scale (Fig. 48.1). For example, a biologist concerned with the structure and function of individual ion channels embedded in the neuronal cell membrane may be most interested in processes that happen over a span of picoseconds or nanoseconds (e.g., the amount of time it takes for an ion to pass through an ion channel, or the amount of time it takes for an ion channel to change its conformation). They may also be most interested in spatial resolutions on the orders of angstroms (e.g., the approximate scale of an individual ion channel). At another extreme, a neuroanatomist studying the comparative anatomy of different species of primates might be most interested in timescales on the order of decades (e.g., an entire lifetime) and spatial scales on the order of decimeters (e.g., the size of an adult human brain). As summarized in Fig. 48.1, different neuroimaging approaches are each associated with a range of temporal and spatial scales that they are best suited to measure.

If we are specifically interested in stimulus-driven neural activity, this implies focusing in on a limited range of spatiotemporal resolutions (gray shading in Fig. 48.1). Neuroimaging approaches that enable insights at those resolutions may provide particularly useful measures of stimulus-driven neural activity.

Non-invasive approaches. Neuroimaging approaches that rely on measurements taken using sensors that are placed without requiring surgery are referred to as **non-invasive**. In general, non-invasive neuroimaging entails placing one or more sensors on or near the subject's head. Examples include **scalp electroencephalography** (EEG; i.e., recording voltages from small electrodes placed on the scalp); **magnetoencephalography** (MEG; i.e., measuring tiny changes in the magnetic field outside of the head driven by local field potentials); and **functional magnetic resonance imaging** (fMRI; i.e., inferring changes in blood flow associated with neural activity using a powerful magnet placed around the head). A related approach, **magnetic**

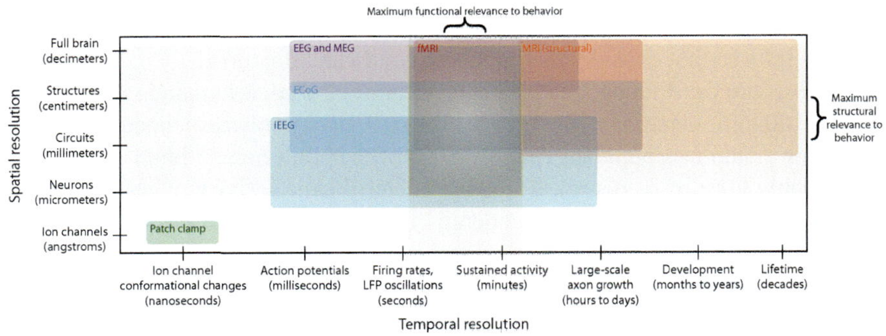

Fig. 48.1 Spatial versus temporal resolution. Each colored region represents the temporal (x-axis) and spatial (y-axis) limits of a recording method or neuroimaging modality [figure inspired by 31, 134]. Green shading denotes in vitro methods. Blue shading denotes invasive in vivo methods. Purple, red, and orange denote non-invasive in vivo methods. The gray shading denotes suggested ranges of maximum structural and functional relevance to experimental-scale behaviors. The shaded region outlined in yellow denotes overlap between the horizontal and vertical gray regions. *Note* Axes are not drawn to scale

resonance imaging (MRI) uses strong magnetic fields, magnetic field gradients, and radio waves to produce a static anatomical image of the brain. Each of these approaches is widely used by neuroscientists interested in studying the neural basis of cognition and behavior. A benefit of relying on non-invasive neuroimaging is that these approaches are low-risk and may be safely used on healthy (non-patient) participants, and without the supervision of a physician. The main drawback of non-invasive neuroimaging is that, because these approaches all rely on sensors placed outside of the head, any relevant activity that is filtered out by the skull, or that is too weak to be measured from distant sensors, cannot be captured. This means that non-invasive neuroimaging approaches tend to have lower spatiotemporal resolution than invasive approaches.

Invasive approaches. Neuroimaging approaches that require surgery are referred to as **invasive**. Invasive in vivo techniques entail placing sensors directly on the surface of and/or in direct contact with deep structure inside of a living person's brain. Examples include **intracranial electroencephalography** (iEEG; i.e., recording voltages from tiny wires implanted in the brain) and **electrocorticography** (ECoG; i.e., recording voltages from small electrodes lying directly on the brain's cortical surface). Both iEEG and ECoG are similar to (non-invasive) EEG, in that all three approaches entail recording aggregate voltages from populations of many neurons. The key differences between these approaches are the locations and sizes of the electrodes. When sensors are larger and are placed far from the signal sources (i.e., neurons), as in EEG, the sensors pick up on relatively large populations of neurons that are spread over a large portion of the brain. When sensors are smaller and placed in direct contact with signal sources, as in ECoG and iEEG, the sensors pick up on smaller populations of neurons that are closer to the recording surface of the electrodes. When tiny **microwires** are used to generate iEEG recordings, it is even possible to record action potentials from individual neurons (see Chaps. 42, 43, 44 and 45).

Invasive in vitro neuroimaging approaches entail taking measurements from brain slices or other structures that have been excised from an intact brain and placed in an isolated environment such as a petri dish. For example, **patch clamp** recordings use tiny glass micropipettes placed directly on the cell membrane to capture changes in membrane potential associated with the opening and closing of individual ion channels. In vitro approaches are not generally used to study stimulus-driven neural activity in humans, since excising the to-be-recorded tissues typically isolates the corresponding neurons from their sensory inputs. In vitro approaches also require extracting the to-be-recorded tissues from the brain, which presents ethical and safety concerns.

Because invasive neuroimaging approaches require physically cutting into the brain, they are not appropriate for use in non-patient populations. Rather, invasive recordings in humans are typically taken from neurosurgical patients who have their electrodes implanted as part of a diagnostic or treatment protocol. For example, people suffering from drug-resistant epilepsy may elect to undergo invasive monitoring (from implanted electrodes) in order to help neurologists localize the most likely source(s) of their seizures. During an extended hospital stay, the patients may elect to participate in research studies that are not directly related to their treatment, in the interest of advancing scientific knowledge by providing access to high quality recordings from their brain (also see Chaps. 4 and 5). Stimulus-driven responses in small circuits that unfold over sub-millisecond timescales can only be measured using invasive approaches like iEEG and ECoG. By providing measurements at both high spatial resolution and high temporal resolution, intracranial recordings can be ideally suited to studying stimulus-driven neural activity patterns (Fig. 48.1).

Single-channel neural signals. A single electrode implanted in a patient's brain measures changes in membrane potential (voltage) in individual neurons, other cells and signal sources, and populations of cells. Because neurons can most effectively transmit signals to other cells via action potentials, the timings of individual action potentials from a given neuron, or the firing rate of a neuron, can provide putative insights into that cell's function (Fig. 48.2a). For example, if a neuron changes its firing rate when the patient is exposed to a particular stimulus or experience, this could suggest that the neuron plays some role in processing information pertaining to that stimulus or experience.

Local field potentials (LFPs) reflect the aggregate neuronal firing and subthreshold changes in membrane potential across thousands of neurons near the recording surface of an electrode (Fig. 48.2b). When LFPs change during exposure to a stimulus or experience, this can suggest that the underlying *population* of neurons plays some role in processing that stimulus or experience. These changes may be aperiodic, as in Fig. 48.2b, or sinusoidal, as in Fig. 48.2c. Rhythmic (sinusoidal or periodic) changes in the LFP tend to reflect coordinated firing patterns across neurons in the population, whereas uncoordinated changes are reflected as changes in the volatility (or variance) of the LFP [20, 35, 46, 94].

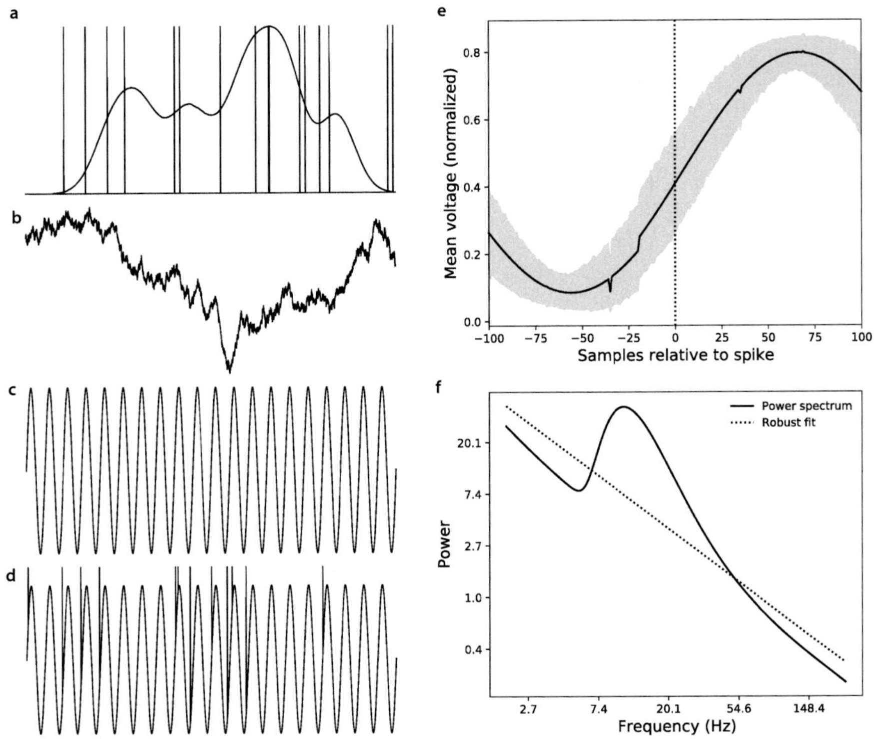

Fig. 48.2 Measuring and processing single-channel neural signal. All of the examples shown in this figure are constructed using simulated data. **a–d** Neural signals. The illustrated examples show voltage (y-axis, arbitrary units) as a function of time (x-axis, arbitrary units). **a** Neuronal firing. The vertical lines illustrate the times at which an artificial neuron fired action potentials (spikes). The smooth curve shows the timeseries of firing rates, computed by convolving the spike timeseries with a Gaussian kernel. **b** Local field potentials (LFPs). LFPs reflect the aggregate neuronal firing and sub-threshold changes in membrane potential across thousands of neurons near the recording surface of an electrode. **c** Oscillations. When the local field potential exhibits sinusoidal fluctuations, this can reflect coordinated changes in membrane potential across a population of neurons. **d** Phase coding. The timing of an individual neuron's action potentials with respect to oscillations in the local field potential can code information via the *phase* (angle) relative to a sine wave at the oscillation's frequency. **e** Spike-triggered average. Phase coding may be identified by sampling the LFP before and after each spike and then averaging across all spikes. The spike-triggered average in this panel is computed using the artificial LFP displayed in Panel **d** (also see Chap. 44). **f** Power spectrum and broadband power. Oscillatory contributions to the local field potential may be summarized as a *power spectrum* that shows the extent to which oscillations at each frequency contribute to the LFP. The power spectrum in this panel is computed by convolving second order Morlet wavelets with the artificial LFP displayed in Panel **c**. The underlying height of the power spectrum, called *broadband power* may be estimated by using robust regression to fit a line to the power spectrum in log-log space. The area under the robust fit line may be used to estimate the firing rates of neurons in the underlying population [94]

When both spike timing information and LFP recordings are available, it is possible to examine whether a given neuron's spikes are modulated according to the activity of the surrounding population (also see Chaps. 44 and 45). For example, **phase-locked neurons** [72] tend to fire action potentials during a particular phase (angle) of an oscillation that appears in an LFP (Fig. 48.2d). Other neurons exhibit **phase coding** by changing their preferred phase according to properties of the stimulus, ongoing experience, or behavior. For these neurons, the phases (of LFP oscillations) at which spikes occur can carry additional information beyond firing rate alone [70, 76, 90, 140]. One way of characterizing phase-depending firing is to compute a **spike-triggered average** of the LFP preceding and proceeding each spike (Fig. 48.2e). When a neuron exhibits a phase preference, its spike-triggered average will look like an oscillation centered at the neuron's preferred phase.

Although oscillations can sometimes be detected visually by examining a raw LFP recording, a more rigorous approach is to use signal processing methods to quantify the presence of oscillatory components of the LFP (also see Chaps. 22 and 23). A **power spectrum** (Fig. 48.2f) plots the **power** at each frequency—i.e., the extent to which oscillations at each frequency contribute to the LFP. When the LFP exhibits an oscillation, this appears as a peak (centered on the oscillation's frequency) in the LFP's power spectrum. An LFP may also exhibit *multiple* oscillations, which appear as multiple peaks in the power spectrum.

In addition to true (sinusoidal) oscillations in the LFP, the volatility of the LFP can also change its power spectrum. For example, an increase in the standard deviation of the LFP's changes in voltages across successive timepoints will result in an increase in power at *all* frequencies. So-called **broadband shifts** in power can occur when the neurons in the underlying population change their firing rates [94].

Given the many ways to measure and characterize neural responses, which approach is best? The answer depends in part on what we hope to learn. For example, if we are interested in processes that we expect to depend on very precise timing and relatively simple neural computations, then neuron-centric signals like spike timing and firing rate may be especially promising. If we are instead interested in processes that we expect to depend on large-scale computations carried out by populations of thousands of neurons, then we may instead benefit from focusing on periodic and aperiodic features of local field potentials recorded from relatively large electrodes. In general, lower-level processes (e.g., signal transduction) tend to rely on smaller numbers of neurons and occur over shorter timescales. Approaches that operate over few neurons and that support high temporal resolution are often best-suited to studying these low-level processes. By contrast, high-level processes (e.g., scene understanding, complex planning, emotional processing) tend to rely on large populations of neurons and occur over relatively long timescales [6, 29]. Approaches that record from larger populations of neurons and that measure processes or changes that unfold over longer timescales are often best-suited to studying these high-level processes. While these general principles have *tended* to hold across many studies and recording modalities, it is worth noting some exceptions. For example, single-neuron responses in humans [122] can sometimes exhibit selectivity for high-level stimuli and semantic concepts.

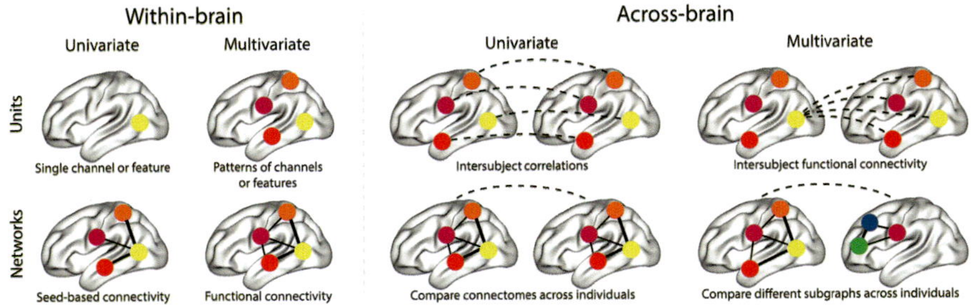

Fig. 48.3 Univariate and multivariate patterns. Individual electrodes (channels) or features (e.g., firing rate, phase, power at a given frequency, etc.) may be considered in isolation (univariate patterns) or in combination with other channels or features (multivariate patterns). Each feature or pattern may be considered as an "independent" functional unit, or within the context of its broader network. Finally, analyses may be carried out on data from a single patient's brain (within-brain) or in aggregate across multiple individuals (across-brain). *Note* Portions of this figure are adapted from [111]

Units and patterns versus networks. After identifying a set of signals that we think will be appropriate for studying our phenomenon or cognitive process of interest, the next key decisions regard whether we should treat those signals in isolation, or as part of a broader network. Essentially this comes down to a decision about how to combine signals and features within and across participants (Fig. 48.3).

Early single-neuron recordings (in cats and non-human primates) played a central role in Hubel and Wiesel's Nobel Prize-winning work on mapping out receptive fields of visual cortical neurons [66, 67]. They mapped out the **receptive fields** of neurons in the primary visual cortex by measuring their firing rates as a function of the visual stimulus shown on the retina. In general, a neuron's receptive field describes the stimulus to which it is maximally responsive. Hubel and Wiesel's work showed that the primary visual cortex is organized into **orientation columns** of neurons whose receptive fields are tilted dark or light bars at a particular orientation relative to horizontal. Several decades later, researchers used high-field fMRI to show analogous orientation columns in human primary visual cortex [157]. The receptive fields of individual neurons can be enormously complex. In contrast to the simple stimuli preferred by primary sensory neurons, neurons in other brain regions can have receptive fields that correspond to high-level concepts. For example, hippocampal **place cells** fire preferentially when an animal travels to a particular location in an environment [40, 82, 110]. Other work has shown that some medial temporal lobe neurons appear to increase their firing in response to photographs of specific faces, animals, objects, or scenes [122].

When recordings from several neurons are available, the set of firing rates across the population can provide additional information beyond that contained in the firing rates of individual neurons [3, 120]. For example, if a single place cell responds to one area of an environment, a population of many place cells that each respond to a different location in the environment can provide a rich cognitive map [148].

Although they do not provide information about spike timing, macro-scale LFP recordings also reflect population-level neuronal activity. Multi-electrode LFP recordings from one or more brain areas can be especially informative. For example, multi-channel LFPs may be used to decode visual stimuli [10, 11], auditory stimuli [11, 141], speech production [52, 113, 121], acute pain onset and intensity [162], and even semantic representations [39, 97].

Intracranial recordings (of individual neurons, populations of neurons, and LFPs) may also be considered within the context of the larger brain networks to which they belong or contribute [9]. One set of approaches to characterizing brain networks is informed by **graph theory**, a branch of mathematics concerned with characterizing network architectures, influence, and membership [8, 13, 19, 64, 111, 127, 139, 143, 144]. For example, a timeseries of recorded responses from multiple channels may be used to infer functional or causal interactions between the associated neural populations. After mapping out a network of pairwise connections between the responses, graph theoretic measures may be applied to estimate or compare the influence of a given channel or set of channels. Considering interactions can provide information beyond the responses of individual channels or patterns. For example, patterns of interactions between neurons or populations can show selective modulation in response to stimuli or features, even when the underlying individual neurons or populations do *not* appear responsive to the stimulus when considered in isolation [126].

Static versus dynamic measures of brain activity. When we attempt to discover the neural patterns associated with a particular stimulus or representation, we need to consider two fundamental questions about how the relevant patterns might change over time. The first question is whether brain representations are fundamentally stable. For example, each time you think of a concept, like the meaning of the word "automobile" do the brain areas relevant to representing that concept display the same basic activity patterns? Or do the neural representations of concepts change in meaningful ways over time, such that the representation of a concept looks fundamentally different each time we measure it? The second question is about whether representations themselves are static or dynamic. For example, when you think of the concept "cat," does the entire representation essentially become activated as a single unit? Or do different components of the representation (e.g., "fur," "mammal," "whiskers," "tail," etc.) come online in sequence, perhaps in a stereotyped way that adds additional nuance or meaning?

Some of our conceptual knowledge, and presumably the underlying neural representations of that knowledge, is acquired over timescales on the order of several years. For example, as they develop, children acquire new representations of concepts and how they are related or organized [14, 85]. Changes in neural representations that occur over the course of years are unlikely to be captured by intracranial recordings, which are typically made over timescales on the order of days or weeks. Nevertheless, *some* conceptual representations may change on faster timescales that can be captured by intracranial recordings, such as when concepts become emotionally charged [23, 156].

Another process that leads to changes in the neural representations is **pattern separation**. Pattern separation refers to the phenomenon of differentiating the neural representations of two or more related stimuli or concepts. For example, pattern separation can occur when we learn to identify and focus in on subtle differences between stimuli or concepts that initially seemed (nearly) identical. Pattern separation can occur over relatively short timescales, and can be identified using intracranial recordings [91].

Although research on the development of conceptual knowledge and pattern separation shows that neural representations *can* change over time, there is also substantial evidence that neural representations are at least somewhat stable over timescales of hours to days, and even across different individuals [57, 101, 136]. For the most part, this body of work treats the neural representations of concepts and stimulus responses as essentially static. By assuming that the neural patterns evoked by a fixed stimulus are stable (within and across presentations), one can use a machine learning approach called **pattern classification** to learn mappings between neural patterns and stimulus labels [109]. Once these mappings are learned, they may be applied to new neural patterns to estimate the stimulus or "thoughts" associated with those neural patterns. This allows researchers to estimate the cognitive dynamics that occur during neuroimaging [30, 51, 93, 118].

Some stimuli, such as words, images, pure tones, etc., are perceptually static. Most or all of the information in the stimulus is "made available" to our sensory systems at the same time. Certainly it may take some time for higher-order information to unfold in our minds, for example when we are presented with a complex or thought-provoking image. However, those dynamics are driven by internal processes rather than (directly) by the stimulus itself. Other stimuli, such as movies, motion sequences, and dynamic sounds like speech or music, are fundamentally dynamic. For example, if we were presented with only the average (across time) visual or auditory information in a popular feature-length film, we would be missing nearly all of the structure that made that film engaging or interesting. Dynamic stimuli, particularly **naturalistic stimuli** with rich spatiotemporal structure reminiscent of real-world experiences, can evoke dynamic neural responses that are often highly reliable across repeated presentations to a single participant participant, as well as across participants [24, 26, 43, 44, 53, 68, 69, 79, 80, 98, 105, 111]. While it is possible to temporally average across the timepoints of responses to some classes of dynamic stimuli while still achieving high reliability [105], this is not universally true. For example, in the domains of speech comprehension and speech production, temporal information is a primary indicator of meaning. Studying neural responses to speech therefore requires considering how neural correlates of speech unfold over time [52, 121].

Independent of whether a given stimulus is fundamentally static or dynamic, neural responses can also change according to which other stimuli were experienced nearby in time. For example, interpreting the neural response to 'B' in the sequence A B C D B might entail accounting for whether the given instance of 'B' is the one that follows 'A' or 'D'. Randomizing stimulus order and averaging over repeated trials effectively removes this sort of contextual information. However, in some cases, the

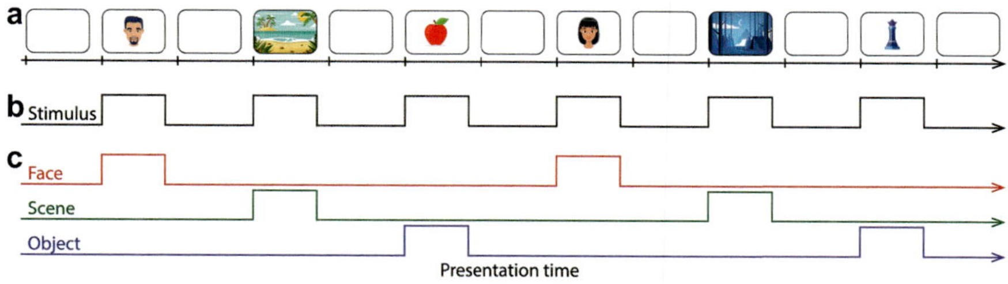

Fig. 48.4 Discrete stimulus timeseries. **a** Example stimulus sequence. A succession of images are presented to the participant on a computer screen, interspersed by intervals of blank screen. Images are drawn from three categories: faces, outdoor scenes, and concrete objects. **b** Onset and offset timing. Stimulus timing, but not stimulus category, may be conveyed using a single binary timeseries. **c** Stimulus identity. Stimulus identity (e.g., category) may be represented using a single binary timeseries for each stimulus category or feature

context in which a stimulus occurs—i.e., the set of other stimuli and thoughts that were experienced nearby in time—can play a critical role in how we process, interpret, and remember incoming information [92]. For example, priming participants using different cues can reliably bias them to interpet an ambiguous narrative in a particular way [159]. Accounting for these sorts of contextual effects often entails factoring in stimulus order or content to the corresponding analyses and models.

48.1.3 Building Explicit Stimulus Models

Identifying neural responses to a stimulus requires formalizing what the stimulus in question *is* and when the participant is exposed to it. Broadly, this entails building explicit or implicit models that describe the composition and/or dynamics of the stimulus. The features, outputs, or predictions of these models may then be related to neural patterns.

What is a stimulus "model"? From an analytic perspective, describing a stimulus typically entails characterizing how different aspects of the stimulus change over time. If we are solely interested in the presence or absence of a stimulus, or the timings of a discrete sequence of trials (Fig. 48.4a), then the "stimulus" might be describable as a simple binary sequence (Fig. 48.4b). One could then examine or compare neural activity patterns recorded during the "on" timepoints versus "off" timepoints, or estimate how neural responses change during the transitions between those binary states. If stimuli are drawn from a well-defined set of categories, then category information may be conveyed using one binary sequence per category (Fig. 48.4c).

In other instances, we may be interested in understanding how neural responses relate to specific stimulus values, or how those values change over time. For example, we might describe the brightness or salience of a visual stimulus (or the loudness of an auditory stimulus, etc.) as a sequence of real-valued numbers (Fig. 48.5a).

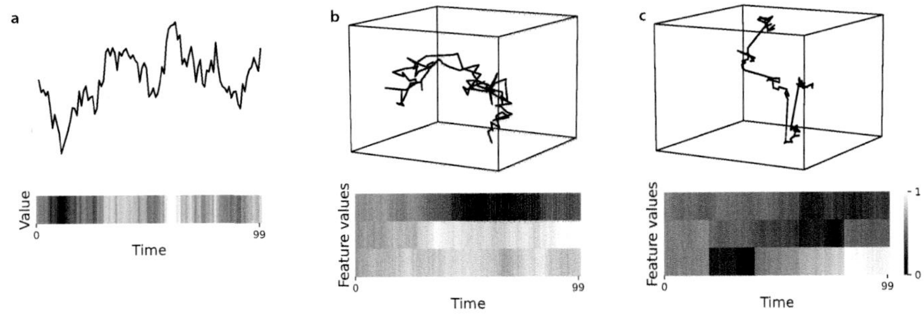

Fig. 48.5 Continuous stimulus timeseries. **a** Univariate real-valued stimulus. The stimulus takes on any value at any timepoint. In this example the stimulus values are autocorrelated. **b** Multivariate real-valued stimulus. The stimulus comprises multiple features, each of which can take on any value at any moment. When projected into 3D, the stimulus traces out a **trajectory** describing how the values of its features change over time. **c** Multivariate real-valued stimulus with event-level or trial-level dynamics. This stimulus is similar to the one displayed in Panel **b**, but here the stimulus values exhibit occasional event boundaries

This could enable us to understand graded neural responses to the stimulus, such as how neural responses change as a function of the stimulus values. **Event-triggered averages** (analogous to spike-triggered averages such as Fig. 48.2e) can also provide insights into how the stimulus tended to change during a time window centered on a particular neural event (e.g., an action potential or the appearance of a specific activity pattern).

Some stimuli are best described by multivariate real-valued feature vectors whose elements (i.e., "features") describe the absence, presence, or values of specific stimulus properties (Fig. 48.5b). Describing a stimulus as a timeseries of multivariate feature vectors can facilitate more nuanced mappings between those stimulus properties and different aspects of neural activity. For example, the firing rates of different neurons, or the patterns of power spectra across the electrodes in a given region of interest, might display different sensitivity to different stimulus features.

A fourth (general) way of describing how a stimulus changes over time is to use a multivariate timeseries with explicit event-level or trial-level boundaries (Fig. 48.5c). For example, in a movie, the scene cuts could constitute **event boundaries**—i.e., moments of transition where the stimulus features exhibit rapid "jumps" that are substantially larger than usual between-timepoint changes. Event boundaries can delineate changes in the focus of an ongoing conversation, scenes in a story or movie, environmental changes, or other transitions in the low-level or high-level content of the stimulus. Experimental trials can also be considered as a sort of event boundary. For example, a multivariate timeseries like that in Fig. 48.5c could also be used to describe the content of a sequence of short video clips presented in succession. Within each clip, the features might change comparatively less than across clips.

As described next, there are many ways to define what the stimulus features *are*. This requires making assumptions about which aspects of the stimulus "matter" (e.g., in terms of evoking neural responses, predicting behaviors, etc.), and about how the moments of the stimulus timecourse should be matched up with moments of a neural recording. While these assumptions can have a large influence on the outcome of an analysis, there is unfortunately no universal way of describing or modeling stimuli (or of relating stimulus dynamics to neural responses). Rather, one must make informed decisions about how to proceed based on the sorts of insights one hopes to gain from an analysis, or based on what approaches have performed well in prior related work.

Manual approaches. In trial-based experiments where stimuli (e.g., words, sounds, images, etc.) have well-defined onset and offset times, the stimulus onset and offset timeseries (e.g., Fig. 48.4b) can serve as a simple stimulus "model." A binary sequence that solely describes stimulus onset and offset times ignores **stimulus identity** (e.g., the category or label) and **stimulus features** (e.g., the values of the corresponding feature vectors). In this way, modeling the stimulus using a binary sequence makes the implicit assumptions that (a) stimulus timing is the main factor of interest with respect to the associated neural responses and (b) stimulus identity and stimulus features may be safely ignored.

To model stimulus identity *and* timing (when both identity and timing information are well-defined, e.g. as in trial-based experiments), the stimulus may also be modeled using a *set* of binary timeseries (Fig. 48.4c). For example, one might define a separate binary timeseries describing the onsets and offsets of only one stimulus category or trial type. This approach makes the implicit assumption that the specific identities of different exemplars (e.g., within a stimulus category—such as face images of different people's faces) may be safely ignored in favor of prioritizing coarser-scale information such as broad stimulus category or trial type labels.

When parameterized stimuli are constructed to vary along one or more explicit stimulus dimensions (e.g., visual stimuli that vary in brightness, contrast, spatial frequency, etc.), the values along each dimension and/or the parameters themselves can serve as a representation of the stimulus (e.g., Fig. 48.5). Each stimulus dimension (or parameter) may be represented by its own real-valued timeseries. This approach makes the implicit assumption that the only relevant sources of variation in the stimulus are those characterized by the specified stimulus dimensions or parameters. All other stimulus features or properties are effectively ignored.

Some stimuli cannot be adequately characterized or described using their associated parameters and/or features that can be directly mapped onto specific physical or perceptual properties. For example, such stimuli might be best described using high-level perceptual or conceptual properties of the stimulus such as the presence of specific objects or high-level content, emotional tone, etc. When these properties may be readily judged or rated by human observers, normed ratings or judgements (typically collected by an independent set of participants) may be used as another means of quantifying stimulus features. These judgements may take the form of integer-valued or real-valued responses (e.g., rating how "happy" an image or sound is, on a particular scale) or binary "yes/no" judgements (e.g., indicating whether or not a tree is present in an image). Stimulus features may also be derived directly from participants' own responses or behaviors (e.g., like versus dislike, remembered versus forgotten, familiarity ratings, etc.). In these cases the features would reflect not only properties of the stimuli themselves, but also potentially aspects of participants' idiosyncratic responses to those stimuli, cognitive operations, behaviors, and so on.

Automated approaches. Many types of stimuli, including natural images, text, complex sounds (e.g., speech, music, recordings of natural environments, etc.), movies, and others, cannot always be easily categorized or manually labeled or rated. In other cases, even if manually labeling stimuli might be possible in principle, in practice it may be too expensive in money or time to generate manual labels. Automated approaches to building stimulus models can scale to millions of stimuli and thousands of stimulus features.

One class of automated approaches to generating stimulus feature models entails applying probabilistic models and deep neural networks such as convolutional neural networks [84, 88], text embedding models [12, 15, 17, 22, 33, 37, 81, 87, 99, 100, 102, 107, 116, 117, 145, 154], transformers [18, 38, 124], and others [78] to the stimuli. The activations of the hidden or output layers of these networks may be used as feature vectors for the corresponding stimuli.

Another class of automated approaches includes visual entity tagging [27], image-to-text models [5, 36, 146], speech-to-text models [47, 74, 155, 163], and other algorithms for generating text data from non-text stimuli like still images, video, and sound. After training these models on large corpora of labeled examples, novel stimuli may be automatically tagged with text annotations. In turn, these annotations may be passed to text embedding models to construct feature vectors for each stimulus. Alternatively, the annotations may be treated directly as stimulus features, for example by treating each unique keyword as a binary feature that is either present or absent in each stimulus.

Human-in-the-loop techniques [160] provide a balance between purely manual and purely automated methods. These techniques entail combining human feedback with classic machine learning approaches. For example, **multidimensional scaling** [21, 41, 149] may be applied to pairwise similarity judgements from human participants to derive n-dimensional feature vectors whose pairwise correlations or (inverse) distances are consistent with those judgements.

To facilitate comparisons or other analyses, it can be useful to partition continuous stimuli into discrete "states" or "events". For example, **hidden Markov models** may be applied to a multidimensional timeseries of observations (e.g., feature vectors) to estimate the moments of transition that are interspersed between periods of relative stability [6, 61, 123]. Neural responses during different events, or at transition points, may then be examined further.

48.1.4 What Are Some Modality-Specific Challenges to Identifying Stimulus-Driven Brain Activity from Intracranial Recordings?

While intracranial recordings provide high spatiotemporal resolution data about neural activity (Fig. 48.1), the **coverage** afforded by intracranial recordings—i.e., the proportion of the brain volume captured across all electrodes—is relatively poor compared with popular non-invasive approaches like fMRI, MEG, and scalp EEG. The locations of electrodes implanted in a representative patient's brain are shown in Fig. 48.6a. Across the 169 electrodes, coverage is limited primarily to the left frontal and temporal lobes. The recordings from this example patient cannot provide direct information about activity outside of these regions.

The precise electrode numbers and implantation locations are determined by teams of clinicians whose primary goal is (typically) to locate the seizure focus for that patient. For some patients, the clinical team may have a relatively good sense of

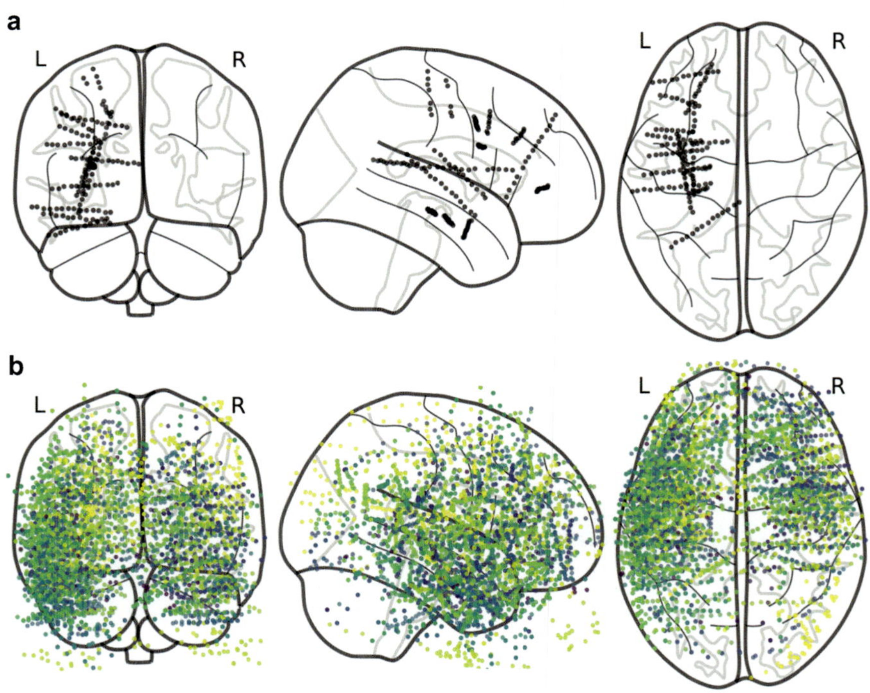

Fig. 48.6 Within-patient versus across-patient electrode coverage. Each dot denotes the location of the recording surface of one neurosurgically implanted electrode. **a** Example patient. The locations of the 169 electrodes implanted in one patient's brain are displayed. **b** Across-patient electrode locations. The locations of $n = 5023$ electrodes are displayed. Colors denote different patients (electrodes from $m = 53$ patients are displayed). *Both panels* Implantation locations are taken from [42] and filtered using a thresholding procedure to remove noisy signals reported by [112]

where the patient's seizures likely originate. These patients may be implanted with relatively few electrodes, spread over a relatively small area of the brain. For other patients where the seizure focus is less clear, they may be implanted with a larger number of electrodes spread throughout the brain. Taken together, these factors mean that each patient is implanted with a different number of electrodes, in different locations, according to a unique clinical plan. Whereas non-invasive recordings can often be easily aligned across people (since the sensor locations are typically held relatively constant across people), the alignment problem is highly non-trivial for intracranial recordings.

One benefit to having varied electrode locations across patients is that, for a sufficiently large number of patients, it becomes possible to obtain data from most of the brain (Fig. 48.6b). Full-brain maps and analyses may be obtained from intracranial recordings by constructing maps that are stitched together across patients. These maps obtained from intracranial recordings may then be compared to analogous maps obtained using other neuroimaging approaches to provide clues about the reliability of the across-patient findings. However, because full-brain maps derived from intracranial recordings typically require combining data across patients, it can be difficult to identify reliable within-patient effects, or to compare responses across patients (also see Chaps. 9 and 29).

48.2 Identifying Stimulus-Driven Neural Activity

Thus far, we have surveyed a variety of approaches for measuring or characterizing neural activity patterns and stimuli. When applied in conjunction, the result of these approaches is a set of two timeseries: one describing the patient's neural responses and the other describing the stimulus the patient experienced as they exhibited those neural responses. The final step is to combine these characterizations in order to relate changes in neural activity to changes in the stimulus. Broadly, this combination step may be carried out within participant or across participant.

48.2.1 Within-Participant Approaches

Within-participant analyses are carried out on data from a single person. Identifying stimulus-driven neural activity using within-participant analyses typically entails combining data over time (e.g., across runs, conditions, trials, etc.). The objective is to estimate maps, patterns, or response profiles that are unique to each individual. These within-participant estimates may then be combined across participants to examine general tendencies in the population and/or individual-specific markers.

Generalized linear models and multivariate pattern analysis. Given a neural recording (Sect. 48.1.2) and a stimulus model (Sect. 48.1.3), the two most widely used approaches to identifying stimulus-driven neural activity are generalized linear models and multivariate pattern analysis. Broadly, **generalized linear models** (GLMs) are an approach to predicting a set of labels, **Y**, typically represented by one or more feature vectors, from a set of an equal number of observations, also represented by feature vectors, **X** [106]. Formally, we say that

$$y_t = f(x_t, \beta),$$

where y_t is the N-dimensional vector of output labels for observation t, x_t is the M-dimensional tth observation, and β is an $M \times N$ matrix of weights. The **link function**, $f(\cdot)$ takes as inputs a set of observations and weights and produces as output a transformed version of the N-dimensional vector $\beta^\top x_t$. For example, if the elements $\beta^\top x_t$ are Real-valued and lie within the interval $(-\infty, +\infty)$, then a sigmoidal link function (e.g., the logistic or hyperbolic tangent functions) would transform the elements to lie within the interval $(-1, +1)$. The "power" of the generalized linear model framework comes from the flexibility in how the link function may be defined. By choosing an appropriate link function, it is possible to take arbitrary Real-valued inputs and transform them into outputs that match a wide variety of useful formats—e.g., (unbounded) Real-valued outputs; probability-like values bounded between 0 and 1; indicator vectors (i.e., vectors where all values are 0s except for one element whose value is 1); binary-like values (i.e., where extreme values are "pulled" towards one of two boundaries, as in sigmoid functions); and many more. When **X** reflects neural activity and **Y** reflects the stimulus features during the corresponding moments, the fitted GLM weights (i.e., β) describe how different aspects of neural activity relate to different stimulus features. These fitted weights may also be used to "decode" stimulus features from new, previously unobserved, neural data.

Multivariate pattern analysis (MVPA) describes a second class of approaches for connecting stimulus features and neural activity [56, 109]. Like GLMs, the goal of MVPA is to predict a set of labels from a set of observations. Typically the "labels" comprise stimulus features and the "observations" comprise neural responses. GLMs are a special case of MVPA for which the output features reflect a (potentially transformed) linear combination of the input features. However, MVPA also includes a variety of other approaches for which the relations between input and output features are non-linear, and potentially even non-monotonic. The umbrella term for such algorithms is **pattern classifiers**, and includes GLMs, support vector machines [16], boosting [130], naive Bayes classifiers [34, 89, 108], nearest neighbor-based classifiers [45], and (deep) neural network-based classifiers [151, 161], among others.

Representational similarity analysis. Comparing the **neural temporal correlation matrix** (i.e., the correlations between the neural patterns recorded at every pair of timepoints) to the **stimulus temporal correlation matrix** (i.e., the correlations between the stimulus feature vectors at every pair of timepoints) can reveal similarities and differences between how neural and stimulus feature change over time. **Representational similarity analysis** (RSA) entails computing the element-wise correlation between the upper triangles of the neural and stimulus correlation matrices [83]. Following the logic of [83], to the extent that neural patterns show a similar temporal correlation structure to stimulus features, we can interpret this as evidence that the neural and stimulus features are related.

RSA may also be carried out across a series of **searchlights**. Similar to how a searchlight illuminates a well-defined area of darkness, a searchlight analysis provides insights into the functionality or responses profiles of a focused region of interest. For each of a series of spherical volumes tiled throughout the brain, RSA may be performed for each volume by limiting the neural features under consideration to only those captured by electrodes within that sphere's radius. This yields, for each sphere (i.e., searchlight) a single correlation coefficient between that sphere's neural temporal correlation matrix and the stimulus temporal correlation matrix. Examining which searchlights displayed high versus low correlations can highlight which brain areas might represent the stimulus in a way consistent with a given stimulus model (i.e., the model used to construct or estimate the stimulus features).

A convenient property of RSA is that, unlike approaches like GLMs or MVPA, RSA does not require learning an explicit mapping between neural features and stimulus features. This is because RSA is driven solely by pattern similarities across timepoints, rather than by the specific properties of the patterns themselves. In this way, RSA can sometimes be a more sensitive way of identifying stimulus-driven neural patterns (e.g., compared with GLMs and MVPA). For example, high levels of noise during many of the exposures to a particular stimulus category will mean that neural decoding approaches will likely fail to effectively learn mappings between the neural and stimulus features for that category. However, RSA analyses effectively "average" over all timepoints, thereby highlighting aspects of neural activity with a similar temporal correlation structure to *any* stimulus features (even if the associations with a subset of the stimulus features are noisy).

Joint stimulus-activity models. Thus far, we have reviewed two approaches to identifying stimulus-driven neural activity. MVPA attempts to learn mappings between stimulus features and neural features, and RSA attempts to identify stimulus features and neural features that exhibit similar temporal correlation patterns. A third (related) approach entails building models that jointly consider the timecourses of neural and stimulus features. Whereas MVPA and RSA implicitly treat stimulus and neural features as a "ground truth," joint stimulus-activity models allow the stimulus features and neural features to mutually inform each other. These models assume that, while the *mappings* between stimulus and neural features may be relatively stable, there may be some times when the stimulus features provide a more reliable signal and other times when neural features provide a more reliable signal [150].

Figure 48.7 provides some geometric intuitions for the idea of joint stimulus-activity models. First, consider how we might go about estimating the reliability of some neural responses we measure as people experience a given stimulus, such as watching a movie. One approach might be to expose a single participant to the same stimulus multiple times. Or alternatively, we might expose several different people to the same stimulus. We could then examine the similarities and differences between the neural responses recorded across trials or individuals.

Geometrically, the neural features recorded from one individual, during one moment, can be conceptualized as a single point in a high dimensional feature space

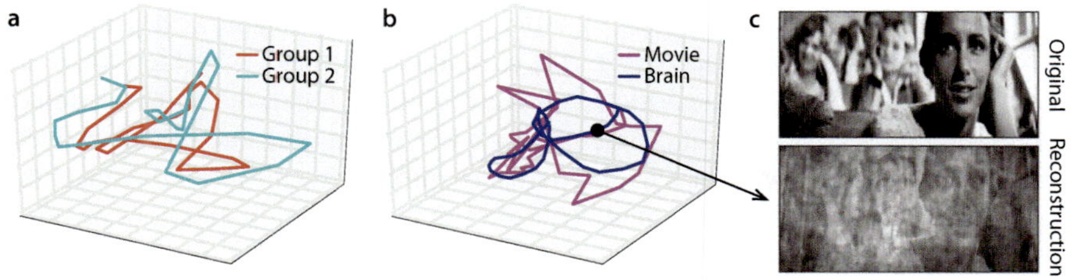

Fig. 48.7 Joint geometric models of stimulus and neural features. **a** Neural features from different individuals. Group-averaged trajectory of fMRI activity from ventral visual cortex split into two randomly-selected groups of subjects (group 1: $n = 6$, group 2: $n = 5$) watching the same movie [57]. **b** A common geometric space for stimulus and neural features. Group-averaged trajectory of fMRI activity from ventral visual cortex and trajectory of movie (pixel intensities over time) hyperaligned [57] to a common space. In Panels **a** and **b**, the high-dimensional stimulus and neural trajectories have been projected onto three dimensions to facilitate visualization [62]. **c** Interpreting coordinates in the common feature space. Stimuli (e.g., movie frames) may be reconstructed from ventral visual brain activity by mapping the coordinates of neural features in the common space onto coordinates in the stimulus space. *Note* This figure is adapted from [62]

(whose dimensions each correspond to a single neural feature). Over time, as the individual's neural activity changes, the successive activity patterns trace out a **trajectory** through this neural activity space. When the timecourses of neural activity patterns are similar across trials or individuals, this results in similar neural trajectory shapes (Fig. 48.7a).

The stimulus features from a single moment, as well as the timecourse describing how stimulus features change over time, may also be conceptualized as a trajectory through a (different) high dimensional feature space, whose dimensions each correspond to a single stimulus feature. It could be interesting to ask whether the neural and stimulus trajectories are similarly shaped, or whether there are particular moments or circumstances under which the shapes converge or diverge. However, because the dimensions of neural trajectories and stimulus trajectories do not (typically) match, an additional step is needed before such comparisons may be made. The **procrustean transformation** is a geometric transformation for bringing two sets of coordinates into an optimal point-by-point alignment. This entails computing the affine transformations (i.e., rotations, reflections, and scalings) that, when applied to coordinates in one set, minimize the average Euclidean distance between the corresponding coordinates in the second set. The resulting aligned coordinates may then be directly compared, since the transformations map the coordinates in the first set into the same coordinate system (with the same dimensions) as the second set [57]. Mapping a neural trajectory into the stimulus feature space (Fig. 48.7b, c) provides a common coordinate system for describing both stimulus features and neural features.

We can use this geometric framework to conceptualize what it means to jointly model the stimulus and neural responses. Consider, for example, the stimulus and neural trajectories displayed in Fig. 48.7b. While both trajectories look similar in some respects (e.g., they have roughly similar coarse-scale shapes), they also differ in potentially important ways (e.g., the stimulus trajectory is "spikier" than the neural trajectory). Which trajectory is "correct"? On one hand, the stimulus trajectory provides a relatively clean characterization of the stimulus that exactly reflects specific measurable aspects of what participants were exposed to. In this sense, the stimulus trajectory is not "corrupted" by measurement noise, inattention, or other factors unrelated to the stimulus itself. On the other hand, the stimulus trajectory is (by definition) a reflection only of the specific stimulus features that we, as the experimentalists, decided were likely to be important. Those features might be at best different and at worst unrelated to the stimulus properties that participants actually care about or respond to. In this sense, one could argue that neural responses reflect the most direct representation of aspects of the stimulus the brain is responding to, since those neural responses are uncorrupted by the experimentalists' assumptions about which stimulus features are important. Taken together, it is clear that neither the stimulus trajectory nor the neural response trajectory, in isolation, provide a complete reflection of an individual's internal mental representations of the stimulus. Instead, it might be most accurate to incorporate aspects of both the stimulus and neural trajectories. This joint stimulus-activity modeling approach acknowledges that the true representation(s) to which an individual's brain is responding may lie somewhere between the stimulus and neural trajectories. It is important to note that the specific mappings one learns through this approach will necessarily depend on both the neural responses (e.g., recording modality, recording location, neural features, etc.) and stimulus features (e.g., which features are included, how features are estimated or computed, etc.). In this way, the mappings between stimulus and neural trajectories should be interpreted as reflecting not only "pure" stimulus-driven responses or representations, but also the particular choices of neural and stimulus features.

48.2.2 Across-Participant Approaches

Across-participant analyses are carried out on data from multiple individuals. Identifying stimulus-driven neural activity using across-participant analyses typically entails building an across-participant model or developing analyses that characterize similarities or differences in responses across participants. The objective is to estimate a single map, pattern, or response profile that is common across individuals. Some approaches also attempt to estimate individual differences that characterize how each individual's responses differ from the group's (aggregated) responses.

Across-participant models. Building across-participant models for identifying stimulus-driven neural activity requires defining a common representation for describing neural activity (and, potentially, linking neural features with stimulus features). Each participant's data must first be mapped into the common representation space. This may be carried out using anatomical [2, 60, 147], functional [25, 57, 58, 156], or other [128, 135] alignment methods. Next, the **inference procedure** (i.e., the algorithm for estimating model parameters from the observed data) must learn both **local parameters** (i.e., parameters that are specific to each individual) and **global parameters** (i.e., parameters that are shared across individuals). The final step is often to learn a mapping or linking function for connecting local and/or global parameters to stimulus features. This last step can be carried out within-participant (by learning mappings between local parameters and stimulus features) or across-participant (by learning mappings between global parameters and stimulus features).

Hierarchical matrix factorization models. One reason that models (in the general sense) are "useful" is that they provide a means of re-representing complex measurements or phenomena using features or rules that compress, reduce, or otherwise simplify the original. When measurements or phenomena are represented as feature vectors, **dimensionality reduction algorithms** provide a general purpose framework for mapping those (typically high-dimensional) feature vectors onto a lower-dimensional space. These lower-dimensional representations of the original feature vectors can be simpler to visualize, analyze, conceptualize, and/or compute with than the original feature vectors. The most widely used approaches to dimensionality reduction utilize a family of mathematical approaches termed **matrix factorization**. We will next turn to a formal description of matrix factorization, including connections between matrix factorization and dimensionality reduction. In turn, this will enable us to define a formal framework for building multi-subject neural activity models.

Matrix factorization entails decomposing a matrix into the product of several other matrices. This family includes a large number of machine learning models, including Topographic Factor Analysis (TFA) [96, 98], Topographic Latent Source Analysis (TLSA) [50], Principal Components Analysis (PCA) [115], Exploratory Factor Analysis (EFA) [142], and Independent Components Analysis (ICA) [32, 75], among others. Within the domain of neuroimaging, the general formulation is to first organize the neural feature vectors (from a single subject) into a T by N data matrix, $\mathbf{Y}$ (where T is the number of observations and N is the number of neural features). We can then decompose $\mathbf{Y}$ as follows:

$$\mathbf{Y} \approx \mathbf{WF},$$

where $\mathbf{W}$ is a T by K **weight matrix** (which describes how each of K factors are activated for each observation) and $\mathbf{F}$ is a K by N matrix of factor images (which describes how each factor maps onto the neural features). In the general case, there are infinitely many solutions for this decomposition. Different matrix factorization approaches converge on specific choices for $\mathbf{W}$ and $\mathbf{F}$ by placing different constraints on the forms the matrices must take or by choosing optimization metrics that emphasize different aspects of $\mathbf{Y}$ to be preserved. For example, when $K \ll N$, the approximation of $\mathbf{Y}$ will be inexact.

Intuitively, matrix factorization means approximation of each row of $\mathbf{Y}$ with a weighted average of the rows of the **factor matrix**, $\mathbf{F}$ (where the weights are specified in the corresponding row of $\mathbf{W}$). When $K < N$, we can treat $\mathbf{W}$ as a low-dimensional (i.e., simpler, smaller, and more tractable) representation of $\mathbf{Y}$. Formally, "dimensionality reduction" means using **W** to approximate some aspects or properties of **Y**.

To illustrate how matrix factorization models may be constructed to capture multi-subject data, we can examine the details of two related models: hierarchical and non-hierarchical variants of the same matrix factorization model, topographic factor analysis. In its non-hierarchical framing, TFA specifies that each row of **F** is parameterized by the center parameter, μ, and the width parameter, λ, of a **radial basis function**. If a radial basis function has center μ and width λ, then its activity $\mathrm{RBF}(\mathbf{r}|\mu, \lambda)$ at location **r** is:

$$\mathrm{RBF}(\mathbf{r}|\mu, \lambda) = \exp\left\{-\frac{||\mathbf{r} - \mu||^2}{\lambda}\right\}.$$

The factor images are filled in by evaluating each radial basis function, defined by the corresponding parameters for each factor, at the location(s) of each electrode or brain region of interest. In contrast to the factors obtained using PCA or ICA, TFA's more constrained factors may be represented much more compactly; each factor corresponds to the structure or group of structures in the brain over which the factor spreads its mass (which is governed by μ and λ). TFA's factors may be conceptualized as nodes located in 3D space whose activity patterns influence the observed brain data (Fig. 48.8a, b).

Hierarchical Topographic Factor Analysis (HTFA) works similarly to TFA, but places an additional constraint over the factors to bias all of the subjects to exhibit similar factors. Whereas TFA attempts to find the factors that best explain an individual subject's data, HTFA also attempts to find the factors that are common across a group of subjects (Fig. 48.8c). This is important, because it allows the model to jointly consider data from multiple subjects.

HTFA handles multi-subject data by defining a *global template*, which describes in general where each radial basis function is placed, how wide it is, and how active its node tends to be. In addition to estimating how factors look and behave in general (across subjects), HTFA also estimates each individual's *subject-specific template*, which describes each subject's particular instantiations of each radial basis function (i.e. that subject's radial basis function locations and widths) and the factor weights (i.e. the activities of each of that subject's radial basis function factors in each of that subject's observed neural activity patterns). Because the subject-specific templates are related to each other (hierarchically [48], via the global template), a given factor's radial basis function will tend to be located in about the same location, and be about as large, across all of the subject-specific templates. Because each subject has the same set of factors (albeit in slightly different locations and with slightly different sizes) we can run analyses that relate the factors across subjects.

The general approach of learning global and subject-specific factors may be applied to many matrix factorization models. For example, TLSA [50], PCA [153], and ICA [152] each have hierarchical framings as well. The particular benefit of using (H)TFA to decompose and describe intracranial data is that the radial basis function factors may be evaluated at the unique locations of each individual patient's elec-

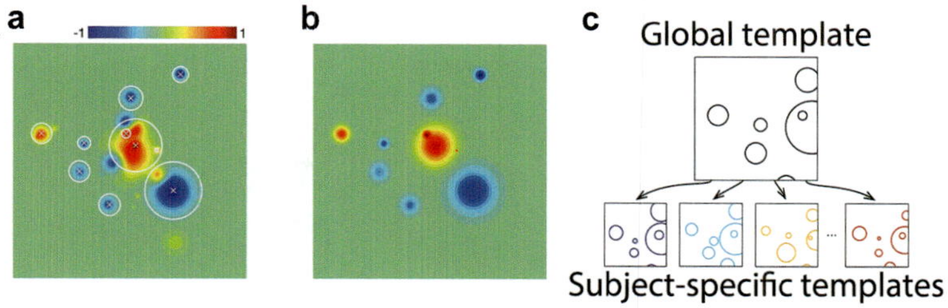

Fig. 48.8 Topographic factor analysis. **a** Spherical factors describe contiguous regions of similar activity. Each factor is represented as a radial basis function. A factor's image may be constructed by evaluating its radial basis function anywhere within the brain volume. Level curves for several example factors fit to a synthetic 3D image are outlined in white; ×s denote the factor centers projected onto the 2D slice displayed in the panel. **b** Brain images are described by weighted sums of the factors' images. After computing each factor's image (using its radial basis function), arbitrary brain images may be approximated using weighted combinations of the images for each factor. The per-image weights may be used as a low-dimensional embedding of the original data. A 2D slice of the reconstruction for the image displayed in panel **a** demonstrates how contiguous clusters of locations are approximated using weighted activations of spherical factors. **c** The global template serves as a prior for subject-specific parameters. The global template defines the numbers of factors, their locations, and their sizes, for the prototypical participant. Each individual participant's parameters (factor locations and sizes) are fit using the global template as a prior. This provides a linking function between different participants' factors, thereby enabling across-subject comparisons. A subset of the factors outlined in Panel **a** are displayed in the *global template* cartoon. The positions of these factors in each individual participant's *subject-specific template* are displayed in different colors. *Note* This figure is adapted from [86]

trodes, even though the electrode locations will differ across patients. And because the subject-specific templates are associated via the global template, aspects of the subject-specific templates may be compared across patients. For example, after using HTFA to learn the global and subject-specific templates, these templates may then be treated as neural features and examined in relation to stimulus features. Specifically, each patient's **W** matrix may be treated as neural features—but whereas the "raw" neural features in the original dataset will not be consistent across patients, the columns of each patient's **W** matrix may be directly combined or compared. In addition, the global and subject-specific factors (rows of **F**) may be examined to identify how the columns of **W** map onto different brain areas or structures.

Gaussian process models. **Gaussian process regression** [125] is an approach for estimating "missing" (unobserved) data by using related observed data. Gaussian process regression is particularly well-suited to applications where nearby datapoints are expected to take on similar values. For example, if we assume that nearby locations in the brain will exhibit similar neural activity patterns, we could use an approach like Gaussian process regression to estimate the most probable activity patterns from locations that were nearby (but not necessarily exactly overlapping with) the electrode implantation sites for that patient [112]. An overview of this technique is shown in Fig. 48.9.

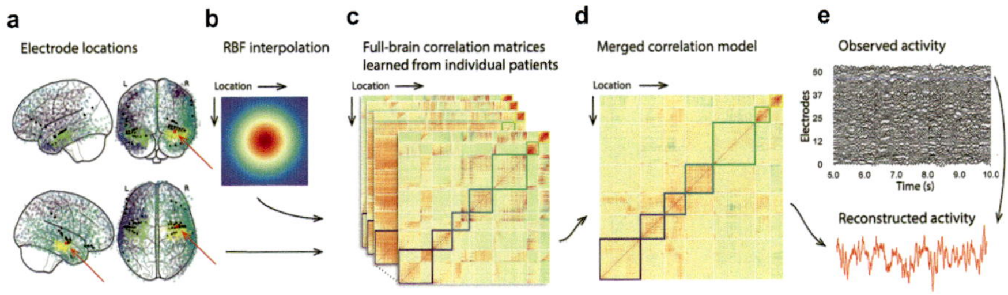

Fig. 48.9 Building across-patient models using Gaussian process regression. **a** Electrode locations. Each dot reflects the location of a single electrode implanted in the brain of one patient. A held-out recording location from one patient is indicated in red, and the patient's remaining electrodes are indicated in black. The electrodes from the remaining patients are colored by k-means cluster (computed using the full-brain correlation model shown in Panel **d**). **b** Radial basis function kernel. Each electrode contributed by the patient (black) weights on the full set of locations under consideration (all dots in Panel **a**). The weights fall off with positional distance (in MNI152 space) according to a radial basis function. **c** Per-patient correlation matrices. After computing the pairwise correlations between the recordings from each patient's electrodes, correlations between all locations may be estimated using radial basis function-weighted averages. This yields one estimated full-brain correlation matrix for each patient. **d** Merged correlation model. Combining the per-patient correlation matrices (Panel **c**) yields a single full-brain correlation model that captures information contributed by every patient. Here the rows and columns are sorted to reflect k-means clustering labels [using $k = 7$; 158], whereby locations are grouped according to their correlations with the rest of the brain (i.e., rows of the matrix displayed in the panel). The boundaries denote the cluster groups. The rows and columns of Panel **c** have been sorted using the Panel **d**-derived cluster labels. **e** Reconstructing activity throughout the brain. Given the observed recordings from the given patient (shown in black; held-out recording is shown in blue), along with a full-brain correlation model (Panel **d**), applying Gaussian process regression yields the most probable activity at the held-out location (red). *Note* This figure is adapted from [112], and data are from [95, 97, 131–133]

To build an across-subject model of neural activity patterns, we first need to define a set of locations in the brain to include in the combined model (Fig. 48.9a). These locations comprise the set of brain coordinates where we will want to estimate activity patterns for every patient. Next, we estimate a **correlation model** that describes how activity exhibited by each pair of locations is related. To estimate the correlation model, we first compute the pairwise correlations between activity recorded from each individual patients' electrodes, and then we use spatial blurring (Fig. 48.9b) to interpolate those correlations over the full set of target locations in the model. This yields a single estimated correlation matrix for each patient (Fig. 48.9c). We can then use weighted averaging to combine the patient-specific correlation matrices into a single correlation model (Fig. 48.9d). Essentially, an entry in a given patient's individual correlation matrix will be weighted more heavily in the combined model if that patient had electrodes nearby to the pair of locations that entry reflects. Conceptually, this combined correlation model reflects "global" information from multiple patients, whereas each individual patient's correlation matrix reflects "local" information from that patient alone. Given a correlation model (learned from multiple patients) and a set of recordings (observed from one patient), Gaussian process regression may be applied to reconstruct (i.e., estimate) neural activity at any location in the combined model—even if the given patient did not have any electrodes at that location (Fig. 48.9e).

Using Gaussian process regression to estimate full-brain activity patterns from a limited number of electrodes can be useful for identifying stimulus-driven neural activity. For example, whereas raw intracranial recordings are typically taken from different locations across patients (Fig. 48.6), the above approach may be used to estimate activity patterns at a common set of locations across people. Second, whereas raw intracranial recordings from a single patient typically lack full-brain coverage (Fig. 48.6a), the set of locations in the combined correlation model may be chosen to cover arbitrarily much of the brain, at arbitrarily high spatial resolution. In turn, this can enable researchers to train or apply other across-participant models, such as pattern classifiers, from different patients' intracranial recordings [129].

Hyperalignment and the shared response model. Even when we record across subjects from an (ostensibly) overlapping set of locations or neural features, individual differences in stimulus-driven neural responses, behavior, internal representations, and even neuroanatomy can lead to different observed responses. When working with intracranial recordings, where the recording locations rarely overlap across people, and where non-standard neuroanatomical traits are relatively common, these factors are even more prevalent. Hierarchical matrix factorization and Gaussian process models make the simplifying assumption that different individual's underlying neural representations are spatially similar. But what if the same *functional* representations are reflected by different *spatial* activity patterns across different people? Models that match up neural features primarily according to their spatial attributes will fail to capture or correctly identify (non-spatial) functional similarities across individuals. In contrast, **functional alignment models** attempt to discover functional overlap in neural activity patterns across individuals, even when the neural features across those individuals are incompatible or out of spatial alignment.

Hyperalignment [57] uses the procrustean transformation to align the neural trajectories of different individuals into a common feature space (Fig. 48.7). This entails computing the linear re-combination of neural features (for each individual) that brings the group's neural trajectories into the closest point-by-point alignment. Because the procrustean transformation is invertible, neural features may be mapped between different individuals, or between specific individuals and the common feature space. The **Shared Response Model** [SRM; 28] is similar to hyperalignment in that it provides a means of defining a common neural feature space that is shared across individuals. SRM extends hyperalignment by combining the alignment step with a dimensionality reduction step that attempts to specifically find a lower-dimensional common neural feature space.

Although hyperalignment and SRM are most often applied to fMRI data [58], in principle these models are modality-independent. For example, one recent study found that applying SRM to intracranial recordings, taken taken as patients watched a movie, revealed a set of shared components that co-varied with the affective content of the movie [156]. The study replicated (using intracranial recordings) several key findings from related fMRI work [23]. Another recent study, using intracranial recordings taken from the rodent hippocampus, demonstrated that hyperaligning neural features across animals enabled reliable across-individual decoding of the animals' spatial locations [25].

Inter-subject correlation and inter-subject functional correlation. **Inter-subject correlations** [ISC; 54] and **inter-subject functional correlations** [ISFC; 138] entail computing the correlations between time-aligned signals recorded from different individuals as they perform a common task (Fig. 48.3). ISC operates on the same (or equivalent) neural features across individuals, and ISFC operates on different *pairs* of neural features across individuals.

To compute ISC for a particular neural feature, we first isolate that feature's timeseries in each individual's brain. (If the same neural feature is not present across individuals, ISC may be performed after employing another approach to equating or mapping between neural features across individuals.) Next, for a single "reference" individual, we correlate their timeseries (for the given neural feature) with the timeseries for the same neural feature averaged across all *other* individuals. This yields a single correlation coefficient for that individual, for the given neural feature. Repeating this calculation using each individual in turn as the reference yields one correlation coefficient for each individual. Finally, we average the correlations across individuals to obtain a single ISC value for the given neural feature.

Computing ISFC is similar to computing ISC. However, whereas ISC correlates the *same* neural feature across individuals, ISFC correlates *different* neural features across individuals. The result is a symmetric matrix of correlation values that summarize how (on average, across individuals) each pair of neural features are correlated.

ISC and ISFC are particularly effective at capturing stimulus-driven activity patterns during naturalistic tasks (e.g., story listening, movie viewing, natural conversation, etc.) when constructing a reliable model of the stimulus timecourse can be challenging [137]. Effectively, ISC and ISFC treat the average signals recorded from *other* participants as a "model" of the stimulus dynamics. Since non-stimulus-driven activity patterns are not expected to be correlated across people, ISC and ISFC are designed to specifically identify timecourses of stimulus-driven neural patterns. Although these approaches are most commonly applied to non-invasive recordings, they have been successfully applied to intracranial recordings as well [55, 65, 104, 119].

48.3 Summary and Concluding Remarks

Identifying stimulus-driven neural activity requires selecting an appropriate recording modality and experimental paradigm, defining neural (Sect. 48.1.2) and stimulus (Sect. 48.1.3) features, and then building explicit or implicit linking functions between neural and stimulus features (Sect. 48.2). We reviewed two general strategies for building these links: within-participant approaches and across-participant approaches.

Within-participant approaches include generalized linear models, multivariate pattern analysis, representational similarity analysis, and joint stimulus-activity models. These approaches each attempt to identify individual-specific maps, patterns, and/or response profiles. Across-participant approaches include hierarchical matrix factorization models, Gaussian process models, geometric alignment models, inter-subject correlation, and inter-subject functional correlation. These approaches each attempt to identify stimulus-driven neural activity that is similar across individuals.

We also identified several challenges that are unique to intracranial recordings. These challenges primarily stem from two factors. First, building across-participant models requires accounting for differences in electrode placement, number of electrodes, and electrode type, across individuals (Fig. 48.6). Second, because intracranial electrodes must be implanted surgically, the subject population in human intracranial experiments is limited to neurosurgical patients with serious neurological symptoms such as drug-resistent epilepsy. These symptoms often result from brain abnormalities (e.g., trauma or other forms of physical damage, developmental abnormalities, and/or other structural or functional issues). These issues provide challenges both to comparing findings across individuals within an intracranial experiment, and also to generalizing any findings to the broader population.

As a field, cognitive neuroscience is still decades away from being able to link neural and stimulus features at high levels of detail. This is partly due to recording quality (even in high-fidelity modalities like intracranial recordings) and coverage, and partly due to insufficient quality or fidelity of stimulus models and decoding algorithms. Nevertheless, insights into the associations between stimuli and neural responses can help to elucidate the neural basis of cognition in ways that behavior alone cannot. For example, when behaviors are ambiguous (e.g., a response could convey several meanings, a response could arise from several equally reasonable or likely cognitive processes, etc.) or when there *are* no behaviors for a given cognitive phenomenon (e.g., forgetting, unshared internal thoughts, etc.), additional signal is needed to resolve those ambiguities. In addition, understanding the neural underpinnings of cognition requires measuring neural activity in some form. Recent developments in natural language processing and deep learning, along with advances in tools for more easily constructing neural, stimulus, and decoding models [e.g., 1, 49, 114] suggest a bright future for this important area of neuroscientific inquiry.

References

1. Abadi M, Barham P, Chen J, Chen Z, Davis A, Dean J, Devin M, Ghemawat S, Irving G, Isard M, Kudlur M, Levenberg J, Monga R, Moore S, Murray DG, Steiner B, Tucker P, Vasudevan V, Warden P, Wicke M, Yu Y, Zheng X (2016) TensorFlow: a system for large-scale machine learning. In: Proceedings of the USENIX symposium on operating systems design and implementation, pp 265–283
2. Archip N, Clatz O, Whalen S, Kacher D, Fedorov A, Kot A, Chrisochoides N, Jolesz F, Golby A, Black PM, Warfield SK (2007) Non-rigid alignment of pre-operative MRI, fMRI, and DT-

MRI with intra-operative MRI for enhanced visualization and navigation in image-guided neurosurgery. NeuroImage 35:609–624
3. Averbeck BB, Latham PE, Pouget A (2006) Neural correlations, population coding and computation. Nat Rev Neurosci 7:358–366
4. Azevedo FAC, Carvalho LRB, Grinberg LT, Farfel JM, Ferretti REL, Leite REP, Filho WJ, Lent R, Herculano-Houzel S (2009) Equal numbers of neuronal and nonneuronal cells make the human brain an isometrically scaled-up primate brain. J Comp Neurol 513:532–541
5. Baboud L, Čadik M, Eisemann E, Seidel HP (2011) Automatic photo-to-terrain alignment for the annotation of mountain pictures. IEEE Xplore. https://doi.org/10.1109/CVPR.2011.5995727
6. Baldassano C, Chen J, Zadbood A, Pillow JW, Hasson U, Norman KA (2017) Discovering event structure in continuous narrative perception and memory. Neuron 95:709–721
7. Barrs BA (2008) The mystery and magic of glia: a perspective on their roles in health and disease. Neuron 60:430–440
8. Bassett DE, Bullmore ED (2006) Small-world brain networks. Neuroscientist 12:512–523
9. Bassett DS, Sporns O (2017) Network neuroscience. Nat Neurosci 20:353–364
10. Belitski A, Gretton A, Magri C, Murayama Y, Montemurro MA, Logothetis NK (2008) Local field potentials and spiking activity in primary visual cortex convey independent information about natural visual stimuli. J Neurosci
11. Belitski A, Panzeri S, Magri C, Logothetis NK, Kayser C (2010) Sensory information in local field potentials and spikes from visual and auditory cortices: time scales and frequency bands. J Comput Neurosci 29:533–545
12. Bengio Y, Ducharme R, Vincent P (2003) A neural probabilistic language model. J Mach Learn Res 3:1137–1155
13. Betzel RF, Medaglia JD, Kahn AE, Soffer J, Schonhaut DR, Bassett DS (2017) Inter-regional ECoG correlations predicted by communication dynamics, geometry, and correlated gene expression. arXiv:1706.06088
14. Blaye A, Bernard-Peyron V, Paour JL, Bonthoux F (2006) Category flexibility in children: distinguishing response flexibility from conceptual flexibility; the protracted development of taxonomic representations. Eur J Dev Psychol 3:163–188
15. Blei DM, Ng AY, Jordan MI (2003) Latent Dirichlet allocation. J Mach Learn Res 3:993–1022
16. Boser BE, Guyon IM, Vapnick V (1992) A training algorithm for optimal margin classifiers. In: ACM, fifth annual workshop on computational learning theory
17. Brown PF, deSouza PV, Mercer RL (1992) Class-based n-gram models of natural language. Comput Linguist 18:467–480
18. Brown TB, Mann B, Ryder N, Subbiah M, Kaplan J, Dhariwal P, Neelakantan A, Shyam P, Sastry G, Askell A, Agarwal S, Herbert-Voss A, Krueger G, Henighan T, Child R, Ramesh A, Ziegler DM, Wu J, Winter C, Hesse C, Chen M, Sigler E, Litwin M, Gray S, Chess B, Clark J, Berner C, McCandlish S, Radford A, Sutskever I, Amodei D (2020) Language models are few-shot learners. arXiv:2005.14165
19. Bullmore E, Sporns O (2009) Complex brain networks: graph theoretical analysis of structural and functional systems. Nat Rev Neurosci 10:186–198
20. Buzsáki G (2006) Rhythms of the brain. Oxford University Press, New York, NY
21. Carroll JD, Chang J (1970) Analysis of individual differences in multidimensional scaling via an n-way generalization of "Eckart-Young" decomposition. Psychometrika 35:283–319
22. Cer D, Yang Y, Kong SY, Hua N, Limtiaco N, John RS, Constant N, Guajardo-Cespedes M, Yuan S, Tar C, Sung YH, Strope B, Kurzweil R (2018) Universal sentence encoder. arXiv:1803.11175
23. Chang LJ, Jolly E, Cheong JH, Rapuano K, Greenstein N, Chen PHA, Manning JR (2021) Endogenous variation in ventromedial prefrontal cortex state dynamics during naturalistic viewing reflects affective experience. Sci Adv 7. https://doi.org/10.1126/sciadv.abf7129
24. Chang LJ, Manning JR, Baldassano C, de la Vega A, Fleetwood G, Geerlings L, Haxby J, Lahnakoski J, Parkinson C, Shappell H, Shim WM, Wager T, Yarkoni T, Yeshurun Y, Finn E (2020) Naturalistic data analysis. Zenodo. https://doi.org/10.5281/zenodo.3937848

25. Chen HT, Manning JR, van der Meer MAA (2021) Between-subject prediction reveals a shared representational geometry in the rodent hippocampus. Curr Biol 31:4293–4304
26. Chen J, Leong YC, Honey CJ, Yong CH, Norman KA, Hasson U (2017) Shared memories reveal shared structure in neural activity across individuals. Nat Neurosci 20:115
27. Chen M, Zheng A, Weinberger K (2013) Fast image tagging. Proc Mach Learn Res 28:1274–1282
28. Chen PH, Chen J, Yeshurun Y, Hasson U, Haxby J, Ramadge PJ (2015) A reduced-dimension fMRI shared response model. In: Cortes C, Lawrence ND, Lee DD, Sugiyama M, Garnett R (eds) Advances in neural information processing systems. Curran Associates, Incorporated, pp 460–468
29. Chiang FK, Wallis JD, Rich EL (2022) Cognitive strategies shift information from single neurons to populations in prefrontal cortex. Neuron 110:709–721
30. Chiu YC, Wang TH, Beck DM, Lewis-Peacock JA, Sahakyan L (2021) Separation of item and context in item-method directed forgetting. NeuroImage 235:117983
31. Churchland PA, Sejnowski TJ (1988) Perspectives on cognitive neuroscience. Science 242:741–745
32. Comon P, Jutten C, Herault J (1991) Blind separation of sources, part II: problems statement. Signal Process 24:11–20
33. Conneau A, Kiela D, Schwenk H, Barrault L, Bordes A (2018) Supervised learning of universal sentence representations from natural language inference data. arXiv:1705.02364
34. Cortes C, Vapnick V (1995) Support vector networks. Mach Learn 20:273–297
35. Crone NE, Korzeniewska A, Franaszczuk PJ (2011) Cortical gamma responses: searching high and low. Int J Psychophysiol 79:9–15
36. de Andrade DOS, Maia LF, de Figueirêdo HF, Viana W, Trinta F, de Souza Baptista C (2018) Photo annotation: a survey. Multimedia Tools Appl 77:423–457
37. Deerwester S, Dumais ST, Furnas GW, Landauer TK, Harshman R (1990) Indexing by latent semantic analysis. J Am Soc Inf Sci 41:391–407
38. Devlin J, Chang MW, Lee K, Toutanova K (2018) BERT: pre-training of deep bidirectional transformers for language understanding. arXiv:1810.04805
39. Dezfouli MP, Daliri MR (2020) Single-trial decoding from local field potential using bag of word representation. Brain Topogr 33:10–21
40. Ekstrom AD, Kahana MJ, Caplan JB, Fields TA, Isham EA, Newman EL, Fried I (2003) Cellular networks underlying human spatial navigation. Nature 425:184–187
41. Ennis DM, Palen J, Mullen K (1988) A multidimensional stochastic theory of similarity. J Math Psychol 32:449–465
42. Ezzyat Y, Kragel JE, Burke JF, Levy DF, Lyalenko A, Wanda P, O'Sullivan L, Hurley KB, Busygin S, Pedisich I, Sperling MR, Worrell GA, Kucewicz MT, Davis KA, Lucas TH, Inman CS, Lega BC, Jobst BC, Sheth SA, Zaghloul K, Jutras MJ, Stein JM, Das SR, Gorniak R, Rizzuto DS, Kahana MJ (2017) Direct brain stimulation modulates encoding states and memory performance in humans. Curr Biol 27:1–8
43. Finn ES (2021) Is it time to put rest to rest? Trends Cogn Sci. https://doi.org/10.1016/j.tics.2021.09.005
44. Finn ES, Bandettini PA (2021) Movie-watching outperforms rest for functional connectivity-based prediction of behavior. NeuroImage 235:117963
45. Fix E, Hodges JL (1951) Discriminatory analysis. Nonparametric discrimination: consistency properties Technical report, USAF School of Aviation Medicine
46. Fries P, Nikolić D, Singer W (2007) The gamma cycle. Trends Neurosci 30:309–316
47. Furui S, Kikuchi T, Shinnaka Y, Hori C (2004) Speech-to-text and speech-to-speech summarization of spontaneous speech. IEEE Trans Speech Audio Process 12:401–408
48. Gelman A, Hill J (2007) Data analysis using regression and multilevel/hierarchical models. Cambridge University Press
49. Gelman A, Lee D, Guo J (2015) Stan: a probabilistic programming language for Bayesian inference and optimization. J Educ Behav Stat 20:530–543

50. Gershman SJ, Blei DM, Pereira F, Norman KA (2011) A topographic latent source model for fMRI data. NeuroImage 57:89–100
51. Gershman SJ, Schapiro AC, Hupbach A, Norman KA (2013) Neural context reinstatement predicts memory misattribution. J Neurosci 33:8590–8595
52. Hamilton LS, Oganian Y, Hall J, Chang EF (2021) Parallel and distributed encoding of speech across human auditory cortex. Cell 184:4626–4639
53. Hasson U, Malach R, Heeger DJ (2010) Reliability of cortical activity during natural stimulation. Trends Cogn Sci 14:40–48
54. Hasson U, Nir Y, Levy I, Fuhrmann G, Malach R (2004) Intersubject synchronization of cortical activity during natural vision. Science 303:1634–1640
55. Haufe S, DeGuzman P, Henin S, Arcaro M, Honey CJ, Hasson U, Parra LC (2018) Elucidating relations between fMRI, ECoG, and EEG through a common natural stimulus. NeuroImage 179:79–91
56. Haxby JV, Gobbini MI, Furey ML, Ishai A, Schouten JL, Pietrini P (2001) Distributed and overlapping representations of faces and objects in ventral temporal cortex. Science 293:2425–2430
57. Haxby JV, Guntupalli JS, Connolly AC, Halchenko YO, Conroy BR, Gobbini MI, Hanke M, Ramadge PJ (2011) A common, high-dimensional model of the representational space in human ventral temporal cortex. Neuron 72:404–416
58. Haxby JV, Guntupalli JS, Nastase SA, Ma F (2020) Hyperalignment: modeling shared information encoded in idiosyncratic cortical topographies. eLife 9:e56601
59. Herculano-Houzel S (2009) The human brain in numbers: a linearly scaled-up primate brain. Front Hum Neurosci 3. https://doi.org/10.3389/neuro.09.031.2009
60. Hermes D, Miller KJ, Noordmans HJ, Vansteensel MJ, Ramsey NF (2010) Automated electocorticographic electrode localization on individually rendered brain surfaces. J Neurosci Methods 185:293–298
61. Heusser AC, Fitzpatrick PC, Manning JR (2021) Geometric models reveal behavioral and neural signatures of transforming naturalistic experiences into episodic memories. Nat Hum Behav 5:905–919
62. Heusser AC, Ziman K, Owen LLW, Manning JR (2018) HyperTools: a Python toolbox for gaining geometric insights into high-dimensional data. J Mach Learn Res 18:1–6
63. Hodgkin AL, Huxley AF (1952) A quantitative description of membrane current and its application to conduction and excitation in nerve. J Physiol 117:500–544
64. Honey CJ, Kötter R, Breakspear M, Sporns O (2007) Network structure of cerebral cortex shapes functional connectivity on multiple time scales. Proc Nat Acad Sci, USA 104:10240–10245
65. Honey CJ, Thesen T, Donner TH, Silbert LJ, Carlson CE, Devinsky O, Doyle JC, Rubin N, Heeger DJ, Hasson U (2012) Slow cortical dynamics and the accumulation of information over long timescales. Neuron 76:423–434
66. Hubel DH, Wiesel TN (1962) Receptive fields, binocular interaction and functional architecture in the cat's visual cortex. J Physiol 160:106
67. Hubel DH, Wiesel TN (1968) Receptive fields and functional architecture of monkey striate cortex. J Physiol 195:215
68. Huth AG, de Heer WA, Griffiths TL, Theunissen FE, Gallant JL (2016) Natural speech reveals the semantic maps that tile human cerebral cortex. Nature 532:453–458
69. Huth AG, Nisimoto S, Vu AT, Gallant JL (2012) A continuous semantic space describes the representation of thousands of object and action categories across the human brain. Neuron 76:1210–1224
70. Huxter JR, Timothy TJ, Allen K, Csicsvari J (2008) Theta phase-specific codes for two-dimensional position, trajectory and heading in the hippocampus. Nat Neurosci 11:587–594
71. Jacobs GH (2021) Color vision. In: Encyclopedia of biological chemistry, IIIrd edn. Elsevier
72. Jacobs J, Kahana MJ, Ekstrom AD, Fried I (2007) Brain oscillations control timing of single-neuron activity in humans. J Neurosci 27:3839–3844

73. Jonas E, Kording KP (2017) Could a neuroscientist understand a microprocessor? PLoS Comput Biol 13:e1005268
74. Jones DA, Wolf F, Gibson E, Williams E, Fedorenko E, Reynolds DA, Zissman MA (2003) Measuring the readability of automatic speech-to-text transcripts. In: Proceedings of the Eurospeech conference, pp 1585–1588
75. Jutten C, Herault J (1991) Blind separation of sources, part I: an adaptive algorithm based on neuromimetic architecture. Signal Process 24:1–10
76. Kamondi A, Acsady L, Wang XJ, Buzsáki G (1998) Theta oscillations in somata and dendrites of hippocampal pyramidal cells in vivo: activity-dependent phase-precession of action potentials. Hippocampus 8:244–261
77. Kandel ER, Schwartz JH, Jessell TM (2000) Principles of neural science. Appleton and Lange
78. Kharitonov E, Lee A, Polyak A, Adi Y, Copet J, Lakhotia K, Nguyen TA, Rivière M, Mohamed A, Dupoux E, Hsu WN (2021) Text-free prosody-aware generative spoken language modeling. arXiv:2109.03264
79. Khosla M, Ngo GH, Jamison K, Kuceveski A, Sabuncu MR (2021) Cortical response to naturalistic stimuli is largely predictable with deep neural networks. Sci Adv 8. https://doi.org/10.1126/sciad.abe7547
80. Ki JJ, Kelly SP, Parra LC (2016) Attention strongly modulates reliability of neural responses to naturalistic narrative stimuli. J Neurosci 36:3092–3101
81. Kiros R, Zhu Y, Salakhutdinov R, Zemel R, Torralba A, Urtasun R, Fidler S (2015) Skip-thought vectors. In: Advances in neural information processing systems, pp 3294–3302
82. Knierim JJ, Kudrimoti HS, McNaughton BL (1995) Place cells, head direction cells, and the learning of landmark stability. J Neurosci 15:1648–1659
83. Kriegeskorte N, Mur M, Bandettini P (2008) Representational similarity analysis—connecting the branches of systems neuroscience. Front Syst Neurosci 2:1–28
84. Krizhevsky A, Sutskever I, Hinton GE (2012) ImageNet classification with deep convolutional neural networks. Adv Neural Inf Process Syst 25:1097–1105
85. Kroll JF, Sunderman G (2003) Cognitive processes in second language learners and bilinguals: the development of lexical and conceptual representations. In: Doughty CJ (ed) The handbook of second language acquisition. Wiley
86. Kumar M, Anderson MJ, Antony JW, Baldassano C, Brooks PP, Cai MB, Chen PH, Ellis CT, Henselman-Petrusek G, Huberdeau D, Hutchinson JB, Li YP, Lu Q, Manning JR, Mennen AC, Nastase SA, Richard H, Schapiro AC, Schuck NW, Shvartsman M, Sundaram N, Suo D, Turek JS, Turner D, Vo VA, Wallace G, Wang Y, Williams JA, Zhang H, Zhu X, Capota M, Cohen JD, Hasson U, Li K, Ramadge PJ, Turk-Browne NB, Willke TL, Norman KA (2021) BrainIAK: the brain image analysis kit. Aperture (in press)
87. Landauer TK, Dumais ST (1997) A solution to Plato's problem: the latent semantic analysis theory of acquisition, induction, and representation of knowledge. Psychol Rev 104:211–240
88. LeCun Y, Bengio Y (1995) Convolutional neural networks for images, speech, and time series. Handb Brain Theory Neural Netw 3361:1–14
89. Lee KH, Kung SY, Verma N (2012) Low-energy formulations of support vector machine kernel functions for biomedical sensor applications. J Signal Process Syst 69:339–349
90. Lisman J (2005) The theta/gamma discrete phase code occurring during the hippocampal phase precession may be a more general brain coding scheme. Hippocampus 15:913–922
91. Lohnas LJ, Duncan K, Doyle WK, Thesen T, Devinsky O, Davachi L (2018) Time-resolved neural reinstatement and pattern separation during memory decisions in human hippocampus. Proc Nat Acad Sci, USA. https://doi.org/10.1073/pnas.1717088115
92. Manning JR (2020) Context reinstatement. In: Kahana MJ, Wagner AD (eds) Handbook of human memory. Oxford University Press
93. Manning JR, Hulbert JC, Williams J, Piloto L, Sahakyan L, Norman KA (2016) A neural signature of contextually mediated intentional forgetting. Psychon Bull Rev 23:1534–1542
94. Manning JR, Jacobs J, Fried I, Kahana MJ (2009) Broadband shifts in LFP power spectra are correlated with single-neuron spiking in humans. J Neurosci 29:13613–13620

95. Manning JR, Polyn SM, Baltuch G, Litt B, Kahana MJ (2011) Oscillatory patterns in temporal lobe reveal context reinstatement during memory search. Proc Nat Acad Sci, USA 108:12893–12897
96. Manning JR, Ranganath R, Norman KA, Blei DM (2014) Topographic factor analysis: a Bayesian model for inferring brain networks from neural data. PLoS One 9:e94914
97. Manning JR, Sperling MR, Sharan A, Rosenberg EA, Kahana MJ (2012) Spontaneously reactivated patterns in frontal and temporal lobe predict semantic clustering during memory search. J Neurosci 32:8871–8878
98. Manning JR, Zhu X, Willke TL, Ranganath R, Stachenfeld K, Hasson U, Blei DM, Norman KA (2018) A probabilistic approach to discovering dynamic full-brain functional connectivity patterns. NeuroImage 180:243–252
99. Mikolov T, Chen K, Corrado G, Dean J (2013) Efficient estimation of word representations in vector space. arXiv:1301.3781
100. Mikolov T, Yih WT, Zweig G (2013) Linguistic regularities in continuous space word representations. In: Proceedings of the national association for computational linguistics, pp 746–751
101. Mitchell TM, Shinkareva SV, Carlson A, Chang KM, Malave VL, Mason RA, Just MA (2008) Predicting human brain activity associated with the meanings of nouns. Science 320:1191
102. Mnih A, Hinton G (2009) A scalable hierarchical distributed language model. Adv Neural Inf Process Syst
103. Mota B, Herculano-Houzel S (2014) All brains are made of this: a fundamental building block of brain matter with matching neuronal and glial masses. Front Neuroanat 8. https://doi.org/10.3389/fnana.2014.00127
104. Mukamel R, Gelbard H, Arieli A, Hasson U, Fried I, Malach R (2005) Coupling between neuronal firing, field potentials, and fMRI in human auditory cortex. Science 309:951–954
105. Nastase SA, Halchenko YO, Connolly AC, Gobbini MI, Haxby JV (2018) Neural responses to naturalistic clips of behaving animals in two different task contexts. Front Neurosci 12. https://doi.org/10.3389/fnins.2018.00316
106. Nelder JA, Wedderburn RWM (1972) Generalized linear models. J R Stat Soc 135:370–384
107. Nelson DL, McEvoy CL, Schreiber TA (2004) The University of South Florida free association, rhyme, and word fragment norms. Behav Res Methods Instrum Comput 36(3):402–407
108. Noble WS (2004) Support vector machine applications in computational biology. Kernel Methods Comput Biol 71–92
109. Norman KA, Polyn SM, Detre GJ, Haxby JV (2006) Beyond mind-reading: multi-voxel pattern analysis of fMRI data. Trends Cogn Sci 10:424–430
110. O'Keefe J (1979) A review of hippocampal place cells. Prog Neurobiol 13:419–439
111. Owen LLW, Chang TH, Manning JR (2021) High-level cognition during story listening is reflected in high-order dynamic correlations in neural activity patterns. Nat Commun 12. https://doi.org/10.1038/s41467-021-25876-x
112. Owen LLW, Muntianu TA, Heusser AC, Daly P, Scangos KW, Manning JR (2020) A Gaussian process model of human electrocorticographic data. Cereb Cortex 30:5333–5345
113. Pasley BN, David SV, Mesgarani N, Flinker A, Shamma SA, Crone NE, Knight RT, Chang EF (2012) Reconstructing speech from human auditory cortex. PLoS Biol 10:e1001251
114. Paszke A, Gross S, Massa F, Lerer A, Bradbury J, Chanan G, Killeen T, Lin Z, Gimelshein N, Antiga L, Desmaison A, Köpf A, Yang E, DeVito Z, Raison M, Tejani A, Chilamkurthy S, Steiner B, Fang L, Bai J, Chintala S (2019) PyTorch: an imperative style, high-performance deep learning library. Adv Neural Inf Process Syst 8026–8037
115. Pearson K (1901) On lines and planes of closest fit to systems of points in space. London Edinburgh Dublin Philos Mag J Sci 2:559–572
116. Pennington J, Socher R, Manning CD (2014) GloVe: global vectors for word representation. In: Proceedings of the conference on empirical methods in natural language processing
117. Peters ME, Neumann M, Iyyer M, Gardner M, Clark C, Lee K, Zettlemoyer L (2018) Deep contextualized word representations. arXiv:1802.05365

118. Polyn SM, Natu VS, Cohen JD, Norman KA (2005) Category-specific cortical activity precedes retrieval during memory search. Science 310:1963–1966
119. Potes C, Brunner P, Gunduz A, Knight RT, Schalk G (2014) Spatial and temporal relationships of electrocorticographic alpha and gamma activity during auditory processing. NeuroImage 97:188–195
120. Pouget A, Dayan P, Zemel R (2000) Information processing with population codes. Nat Rev Neurosci 1:125–132
121. Proix T, Saa JD, Christen A, Martin S, Pasley BN, Knight RT, Tian X, Poeppel D, Doyle WK, Devinsky O, Arnal LH, Mégevand P, Giraud AL (2022) Imagined speech can be decoded from low- and cross-frequency intracranial EEG features. Nat Commun 13. https://doi.org/10.1038/s41467-021-27725-3
122. Quiroga RQ, Reddy L, Kreiman G, Koch C, Fried I (2005) Invariant visual representation by single neurons in the human brain. Nature 435:1102–1107
123. Rabiner LR (1989) A tutorial on Hidden Markov Models and selected applications in speech recognition. Proc IEEE 77:257–286
124. Radford A, Wu J, Child R, Luan D, Amodei D, Sutskever I (2019) Language models are unsupervised multitask learners. OpenAI Blog 1
125. Rasmussen CE (2006) Gaussian processes for machine learning. MIT Press
126. Rigotti M, Barak O, Warden MR, Wang XJ, Daw ND, Miller EK, Fusi S (2013) The importance of mixed selectivity in complex cognitive tasks. Nature
127. Rubinov M, Sporns O (2010) Complex network measures of brain connectivity: uses and interpretations. NeuroImage 52:1059–1069
128. Sabuncu MR, Singer BD, Conroy B, Bryan RE, Ramadge PJ, Haxby JV (2010) Function-based intersubject alignment of human cortical anatomy. Cereb Cortex 20:130–140
129. Scangos KW, Khambhati AN, Daly PM, Owen LLW, Manning JR, Ambrose JB, Austin E, Dawes HE, Krystal AD, Chang EF (2021) Biomarkers of depression symptoms defined by direct intracranial neurophysiology. Front Hum Neurosci (in press)
130. Schapire RE (2003) The boosting approach to machine learning: an overview. In: Nonlinear Estimation Classif 149–171
131. Sederberg PB, Kahana MJ, Howard MW, Donner EJ, Madsen JR (2003) Theta and gamma oscillations during encoding predict subsequent recall. J Neurosci 23:10809–10814
132. Sederberg PB, Schulze-Bonhage A, Madsen JR, Bromfield EB, Litt B, Brandt A, Kahana MJ (2007) Gamma oscillations distinguish true from false memories. Psychol Sci 18:927–932
133. Sederberg PB, Schulze-Bonhage A, Madsen JR, Bromfield EB, McCarthy DC, Brandt A, Tully MS, Kahana MJ (2007) Hippocampal and neocortical gamma oscillations predict memory formation in humans. Cereb Cortex 17:1190–1196
134. Sejnowski TJ, Churchland PS, Movshon JA (2014) Putting big data to good use in neuroscience. Nat Neurosci 17:1440–1441
135. Sekihara K, Nagarajan SS, Poeppel D, Marantz A, Miyashita Y (2002) Application of an MEG eigenspace beamformer to reconstructing spatio-temporal activities of neural sources. Hum Brain Mapp 15:199–215
136. Shinkareva SV, Mason RA, Malave VL, Wang W, Mitchell TM, Just MA (2008) Using fMRI brain activation to identify cognitive states associated with perception of tools and dwellings. PLoS One e1394:1–9
137. Simony E, Chang C (2020) Analysis of stimulus-induced brain dynamics during naturalistic paradigms. NeuroImage 216:116461
138. Simony E, Honey CJ, Chen J, Hasson U (2016) Dynamic reconfiguration of the default mode network during narrative comprehension. Nat Commun 7:1–13
139. Sizemore AE, Giusti C, Kahn A, Vettel JM, Betzel RF, Bassett DS (2018) Cliques and cavities in the human connectome. J Comput Neurosci 44:115–145
140. Skaggs WE, McNaughton BL, Wilson MA, Barnes CA (1996) Theta phase precession in hippocampal neuronal populations and the compression of temporal sequences. Hippocampus 6:149–172

141. Smith E, Kellis S, House P, Greger B (2013) Decoding stimulus identity from multi-unit activity and local field potentials along the ventral auditory stream in the awake primate: implications for cortical neural prostheses. J Neural Eng 10:16010
142. Spearman C (1904) General intelligence, objectively determined and measured. Am J of Psychol 15:201–292
143. Sporns O, Betzel RF (2016) Modular brain networks. Ann Rev Psychol 67:613–640
144. Sporns O, Honey CJ (2006) Small worlds inside big brains. Proc Nat Acad Sci, USA 103:19219–19220
145. Steyvers M, Shiffrin RM, Nelson DL (2004) Word association spaces for predicting semantic similarity effects in episodic memory. In: Healy AF (ed) Cognitive psychology and its applications: Festschrift in Honor of Lyle Bourne, Walter Kintsch, and Thomas Landauer. American Psychological Association, Washington, DC
146. Stone Z, Zickler T, Darrell T (2018) Autotagging Facebook: social network context improves photo annotation. IEEE Xplore. https://doi.org/10.1109/CVPRW.2008.4562956
147. Studholme C, Constable RT, Duncan JS (2000) Accurate alignment of functional EPI data to anatomical MRI using a physics-based distortion model. IEEE Trans Med Imaging 19:1115–1127
148. Tolman EC (1948) Cognitive maps in rats and men. Psychol Rev 55:189–208
149. Torgerson WS (1958) Theory and methods of scaling. Wiley, New York, NY
150. Turner BM, Forstmann BU, Wagenmakers EJ, Brown SD, Sederberg PB, Steyvers M (2013) A Bayesian framework for simultaneously modeling neural and behavioral data. NeuroImage 72:193–206
151. Wan EA (1990) Neural network classification: a Bayesian interpretation. IEEE Trans Neural Netw 303–305
152. Wang Y, Guo Y (2019) A hierarchical independent component analysis model for longitudinal neuroimaging studies. NeuroImage 189:380–400
153. Westerhuis JA, Kourti T, MacGregor JF (1998) Analysis of multiblock and hierarchical PCA and PLS models. J Chemom 12:301–321
154. Wieting J, Kiela D (2019) No training required: exploring random encoders for sentence classification. arXiv:1901.10444
155. Wilpon JG, Rabiner LR, Lee CH, Goldman ER (1990) Automatic recognition of keywords in unconstrained speech using hidden Markov models. IEEE Trans Acoust Speech Signal Process 38:1870–1878
156. Xie T, Cheong JH, Manning JR, Brandt AM, Aronson JP, Jobst BC, Bujarski KA, Chang LJ (2021) Minimal functional alignment of ventromedial prefrontal cortex intracranial EEG signals during naturalistic viewing. bioRxiv. https://doi.org/10.1101/2021.05.10.443308
157. Yacoub E, Harel N, Uğurbil K (2008) High-field fMRI unveils orientation columns in humans. Proc Nat Acad Sci, USA 105:10607–10612
158. Yeo BTT, Krienen FM, Sepulcre J, Sabuncu MR, Lashkari D, Hollinshead M, Roffman JL, Smoller JW, Zollei L, Polimieni JR, Fischl B, Liu H, Buckner RL (2011) The organization of the human cerebral cortex estimated by intrinsic functional connectivity. J Neurophysiol 106:1125–1165
159. Yeshurun Y, Swanson S, Simony E, Chen J, Lazaridi C, Honey CJ, Hasson U (2017) Same story, different story: the neural representation of interpretive frameworks. Psychol Sci 28:307–319
160. Zanzotto FM (2019) Human-in-the-loop artificial intelligence. J Artif Intell Res 64:243–252
161. Zeng D, Liu K, Lai S, Zhou G, Zhao J (2014) Relation classification via convolutional deep neural network. In: Proceedings of the international conference on computational linguistics, pp 2335–2344
162. Zhang Q, Xiao Z, Huang C, Hu S, Kulkarni P, Martinez E, Tong AP, Garg A, Zhou H, Chen Z, Wang J (2018) Local field potential decoding of the onset and intensity of acute pain in rats. Sci Rep 8. https://doi.org/10.1038/s41598-018-26527-w
163. Ziman K, Heusser AC, Fitzpatrick PC, Field CE, Manning JR (2018) Is automatic speech-to-text transcription ready for use in psychological experiments? Behav Res Methods 50:2597–2605

Chapter 49
How Can We Identify Electrophysiological iEEG Activities Associated with Cognitive Functions?

Michal T. Kucewicz, Krishnakant Saboo, and Gregory A. Worrell

Abstract Electrophysiological activities of the brain are engaged in its various functions and give rise to a wide spectrum of low and high frequency oscillations in the intracranial EEG (iEEG) signals, commonly known as the brain waves. The iEEG spectral activities are distributed across networks of cortical and subcortical areas arranged into hierarchical processing streams. It remains a major challenge to identify these activities in the frequency spectrum, time, and anatomical space, especially during memory and higher-order cognitive functions. Traditionally, this has been done manually by visual inspection of the activities induced in iEEG signals, or semi-automatically by supervised signal detections of computer algorithms. Emerging machine-learning and artificial intelligence tools enable more automated, efficient, objective, and accurate solutions than the traditional expert review. In this chapter, we showcase example applications of features and methods to study cognitive functions and to identify brain areas for therapeutic interventions, including electrical stimulation. These resulted in unexpected findings about the spatiotemporal organization of memory processing and the effects of stimulating the brain. Thus identified iEEG activities offer electrophysiological biomarkers for mapping not only cognitive but also other sensorimotor functions, and are ideally suited for new brain-computer interface approaches to intelligently modulate specific neural processes.

M. T. Kucewicz (✉) · G. A. Worrell
BioTechMed Center, Brain & Mind Electrophysiology lab, Faculty of Electronics, Telecommunications and Informatics, Multimedia Systems Department, Gdansk University of Technology, Gdansk, Poland
e-mail: michal.kucewicz@pg.edu.pl

M. T. Kucewicz · K. Saboo · G. A. Worrell
Department of Physiology & Biomedical Engineering, Mayo Clinic, Rochester, MN, USA

M. T. Kucewicz · G. A. Worrell
Department of Neurology, Mayo Clinic, Rochester, MN, USA

K. Saboo
Department of Electrical and Computer Engineering, University of Illinois, Urbana-Champaign, IL, USA

N. Axmacher (ed.), *Intracranial EEG*, Studies in Neuroscience, Psychology and Behavioral Economics, https://doi.org/10.1007/978-3-031-20910-9_49

49.1 Challenges of Mining Large-Scale Electrophysiology

In the myriad of electrophysiological activities generated by the brain, one of the greatest challenges of neuroscience has been to identify the ones that are associated with particular brain functions. There is a wide spectrum of low and high frequency oscillations that can be detected in the intracranial EEG signals recorded from cortical or subcortical electrodes during performance of sensorimotor or cognitive tasks. Especially in the case of tasks involving cognitive functions like attention, memory and executive control, there is an entire spectrum of iEEG activities distributed across widespread networks of sensory, motor and association areas. Each of these areas generates signal oscillations in the classic EEG frequency bands and beyond, ranging from slow and delta-frequency waves through the theta, alpha, beta, and gamma rhythms, up to the ripple and high frequency activities of the spectrum. These are very dynamic in time and space, emerging on the timescale of tens to hundreds of milliseconds at various electrode locations together with the cognitive processes. Capturing this plethora of brain activities across the large spatial scale of neural architecture and the temporal space of cognitive processing is already a difficult task, which requires a large-scale electrophysiological approach, including special electrodes and signal processing methods [13]. Identification of selected activities that are actually associated with a given function is more challenging still.

Hence, before we can start identifying and associating specific iEEG activities with cognitive processes, we first need to consider the spatiotemporal limitations of our signal recordings. The temporal scale of activities that can be resolved in a given recording depends both on the sampling rate of signal acquisition and on the spectral frequency of the fastest activity. For instance a sampling rate of at least 2000 Hz would be required to resolve the waveforms of a fast ripple oscillation at 500 Hz [25] and dissociate it from other sources of broadband power increases [11, 23]. Likewise, a rate of 4000 Hz would be required to sample the waveform of an ultra-fast ripple oscillation at 1000 Hz in the iEEG signal [5], according to a general recommendation of at least two times greater sampling rate than the suggested limit of the Nyquist theorem [6]. The spatial scale, on the other hand, is determined by the electrode contact size. Micro-electrode contacts with diameters less than 50 um can capture a fast ripple generated on the scale of an individual cortical column [2, 3] or action potentials from specific neurons. The traditional macro-electrode contacts used in the iEEG sample from a larger electrical field potential that may be limited in resolving very local neural activities (see Chaps. 16 and 17). Density of sampling the electrical fields with multiple geometrically arranged electrode contacts is another factor to consider in capturing the spectral activities across the spatial scale. Hybrid electrode designs with arrays of multiple micro- and macro-electrode contacts offer new opportunities for overcoming the spatiotemporal limitations with so-called large-scale electrophysiology [13]. Signals from hundreds of electrode contacts can now be acquired with the emerging technologies for high-density recordings and tools for automated signal processing and 'big data' science leveraged to analyse those recordings.

Comprehensive and automated identification of electrophysiological activities in iEEG signals is the main topic of this chapter. Once our iEEG signals are recorded and the spatiotemporal limits of our technique are known, we face the challenge of finding the spectral activities of interest in the large volumes of acquired data. As the new technologies for high-density, large-scale electrophysiological recordings are developed and the amount of acquired data grows rapidly, there is a pressing need for computational tools and methods for efficient signal processing and data analysis. Traditional manual expert review of processed signals is no longer realistic for the daunting sizes of electrophysiological data collected nowadays in the terabyte scale. Reviewing signals from thousands of electrode channels from multiple subjects is neither efficient nor reliable, even if this is performed with pre-processed spectral and other features. Neuroscience studies are transformed from being primarily driven by the expert manual review that relies primarily on human intelligence to becoming catalyzed and complemented by machine-learning and artificial intelligence tools (see Chap. 47). These tools are ideally suited for the aims of large-scale electrophysiology to automatically and comprehensively probe all spectral, spatial and temporal dimensions of the recorded signals and identify activities associated with cognitive processes.

49.2 Manual and Automatic Detection of Signal Activities

There are three general approaches to identification of an activity in the signal. The first approach is fully manual review of the raw signals or processed signal features by an expert. In this approach, an expert is manually browsing through the recordings to visually detect responses of interest across time and electrode channels. A classic example would be an epileptologist screening raw iEEG recordings to mark seizures or interictal epileptiform activities. The same screening process could be enhanced by filtering the signal in a ripple frequency range to facilitate manual detection of pathological High Frequency Oscillations (HFOs) (see Chap. 24). Calculating the power of the filtered signal to obtain a spectral feature could further enhance spotting significant events across time on any one electrode contact. Given that the spectral power in the high frequency ranges of the iEEG spectrum correlates with the local spiking activity of neuronal populations (see Chaps. 16, 18, and 44), the same high frequency spectral feature could be used to detect activations associated with a cognitive function. Increases in the spectral power would signify enhanced neuronal activities at the times of task performance and lead to identification of the 'active' electrode channels implanted in particular brain areas. In contrast to seizures and interictal discharges, however, these physiological activities are more subtle and thus more difficult to see with a naked eye in this manual review process.

The second approach is to incorporate automated analysis into the manual review process. In the above-mentioned cases, detection of seizures, interictal pathological, or the physiological activities could be performed automatically and then reviewed visually by an expert. By applying a threshold on the spectral power, the raw, or the filtered signal feature, any significant event crossing the threshold in time can be rapidly and effortlessly detected and quantified by a simple computer algorithm across large volumes of data. Electrode channels showing more or less detections than the others provide candidates for identification of brain locations and neural activities of interest. This semi-automated approach is incomparably faster than the manual approach but is still primarily dependent on the expert.

The third approach, on the other hand, is fully automated and requires only to be supervised by an expert. Detection of the significant events and classification of the 'active' channels is performed completely by computer algorithms. The classification algorithms vary in their complexity from simple programs with hard-coded threshold values for assigning the active class label, through machine-learning classifiers that determine the threshold values based on pre-programmed data separation methods, all the way to artificial intelligence algorithms that separate the data based on features and classification thresholds learned from previous 'training' datasets. Expert supervision is still more or less needed to plan and program the analysis, choose or direct feature extraction, and assess the outcomes. There are now a number of automated methods for identifying electrophysiologically 'active' channels in a particular brain function.

Methods that use a "training" dataset to determine separation criteria for identification of active electrode contacts are referred to as wrapper techniques. These techniques employ the so-called "supervised learning" paradigm of machine learning to select the active contacts based on task-related classification [17, 24, 27]. Features computed from the iEEG signals at each electrode are used to predict cognitive task performance—contacts that are most predictive of the performance outcome are deemed as 'active'. The training data typically consists of signal features and task-performance information (e.g. correct or error). Methods that do not use a training dataset employ the so-called "unsupervised learning" paradigm of machine learning. Electrode contacts are classified based on the distribution of the iEEG features to separate the contacts that show features reflecting task-related physiological activities [16]. Other methods use filtering techniques that separate out the active contacts that satisfy a priori chosen criteria for any of the features. A general design for methodological steps in this identification process is summarized in Fig. 49.1. After all, every method type has specific advantages that need to be considered for particular applications as well as for the data available [1].

49.3 Electrophysiological Features of Neural Activities

Whatever is your general approach and particular detection or classification method, the identification of iEEG activities will only be as good as the features that are used. These can be categorized into univariate and bivariate, meaning derived from individual or paired electrode contacts, respectively. A classifier can use multiple univariate and/or bivariate features to enable multivariate classification (see Chap. 34). One of the most basic univariate features is amplitude of the signal in

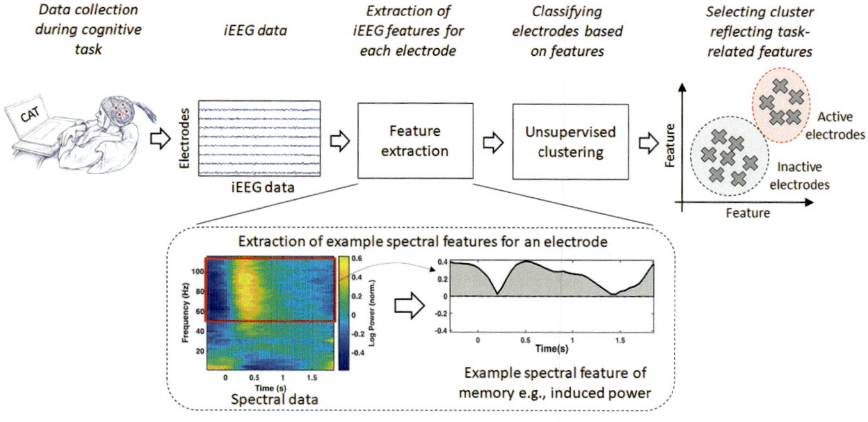

Fig. 49.1 Pipeline for automated identification of active electrodes. Features are extracted for each electrode from iEEG data collected during a cognitive task. Electrodes are classified as active or inactive using machine learning methods based on the features

response to a task event, e.g. the event related potential (ERP). Sensory and also association areas can be identified based on ERPs obtained from specific contacts at various latencies in response to stimulus presentation. Spectral power of the signal in a given frequency band (aka power-in-band) is one of most commonly used univariate features. Bivariate features can be derived from the raw or filtered signal or from the spectral components of the signal. Correlation of two signals in a given amount of time is one simple example of a bivariate feature. Spectral coherence is another common bivariate feature of the phase relationship in a given frequency component of the signals. It is usually not enough to take raw values of these features, since different electrode contacts will record iEEG signals at various amplitude scales due to differences in impedance both across anatomical sites and time of recording. Hence, most features have to be normalized.

Change from baseline is a classic example of feature normalization that can be applied both to the evoked (ERP) and induced responses (the former is time-locked to the stimulus onset). The obtained feature value during the studied brain process is subtracted from the value outside of the process (typically before the stimulus presentation) for every contact independently. This baseline normalization is very suitable for the evoked features, but presents challenges for the induced features. The evoked features are tightly time-locked to stimulus presentation in a given task with a clear-cut baseline period before the presentation and a response period following the stimulus. The induced features like the spectral power-in-band are not as tightly locked and can appear before the stimulus presentation driven by attentional or anticipatory processes. In case of the latter, it is less clear what period to use as the baseline. Alternative normalization techniques include z-score transformation, which scales all feature values to standard deviations above and below the mean of the analyzed signal. Other techniques include dividing each feature value in a given frequency band by the mean power across all bands, for instance, to scale the values from different contacts, sessions or subjects. What we want to obtain in the end are normalized features that extract the total amount of change or response in a given electrophysiological activity that can be compared across the anatomical sites of the electrode contacts, and ideally also across the spectrum of neural activities.

The Fig. 49.1 example shows a univariate feature of the spectral power in a high gamma frequency band around the time of stimulus presentation, normalized by z-score transformation. Total amount of induced power-in-band, which has been averaged across multiple trials of stimulus presentation, is integrated from absolute values of both positive and negative deviation from the mean. In this way, any increases or decreases in the high gamma power that were induced either before or during stimulus presentation will be captured in this one feature. The same feature can be obtained for the other frequency bands and compared in one normalized scale of standard deviations from the mean. In the example, two power-in-band features are then used to plot their distributions across all electrode contacts. The active contacts can be classified using a Gaussian mixture model with any one of the power-in-band or other features. This simple approach allows different features to be compared for how well each can classify the data relative to a gold-standard manual review by an expert [16]. More importantly, it can identify the anatomical sites and the neural activities underlying these features and address physiological questions about cognitive and other brain functions.

49.4 Applications for Investigating Memory and Cognition

Once electrode contacts that show electrophysiological responses in a given task are identified, we can look more closely at their features to understand the spectral, temporal and spatial dimensions of the underlying neural activities. One basic question that can be addressed is: what is the spatial scale at which a given spectral activity is generated? Let's say that features that separated the active and inactive contacts were derived from a specific frequency range of the EEG spectrum. We can then look at the spatial distribution of the active electrode contacts. High frequency oscillations (HFOs) are a good example of an activity that is detected only on individual micro-contacts and the corresponding closest macro-contact (if available) but not necessarily on the neighboring micro-contacts [3, 4, 26]. Based on these findings it has been estimated that HFOs are generated at a spatial scale of <1 mm^3, corresponding roughly to a single cortical column [25]. Likewise, a micro-seizure generation can be limited to a single micro-contact and be initially undetectable on the neighboring macro-contacts until further spreading of the seizure activity [20]. In cognitive processes, HFO activities in the high gamma/ripple frequency range were detected at the same time on both the macro- and the nearby micro-channels implanted on the cortical surface [21]. In all three cases above of studying spatial scale, the electrophysiological activities were relatively easy to detect—they were all spatially limited to very localized discharges and defined frequency range. These can be efficiently identified and studied with recordings from a small number of contacts in a single subject without a need for a fully automated method.

The situation is different, however, in case of large studies involving multiple subjects implanted with hundreds of electrode contacts and recorded at various centers. Such iEEG datasets are becoming more common now, and enable studying processes that are widely applicable across subjects. However, their large sizes present many challenges for efficient study of neural activities. Apart from the obvious technical issues relating to particular ways that data are recorded at various centers with different formats, sampling rates, electric noise levels, electrode coordinate systems etc., analyzing the neural activities across a wide frequency spectrum, anatomical space, and temporal span of the iEEG recordings necessitates employment of automated tools. Here, we provide three example studies from an iEEG project that involved over two hundred epilepsy patients, who performed the same free recall verbal memory tasks at multiple clinical centers. We asked different questions about the spectral, spatial and temporal dimensions of the neural activities.

In the first study, a multi-feature classification of spectral power from a wide range of frequency bands was used to predict successful and unsuccessful encoding of words for subsequent free recall [7]. Spectral features in the theta and gamma frequency bands in specific brain areas showed the greatest memory effect of predicting recall. Even though the classification itself did not reveal a high performance in terms of specificity and sensitivity, it was found helpful in determining target neural activities, brain areas, and cognitive states for electrical stimulation to improve memory performance. In this particular case, pre-selecting active electrode contacts yielded no further improvement for the classification but significantly shortened the computation time [17]. The same classifier was also used to select parameters of electrical current and trigger stimulation of the lateral temporal cortex in a fully automated closed-loop design [8]. Although the resultant improvement in memory performance was of the same magnitude as in the original discovery of open-loop stimulation in the lateral temporal cortex [12], it was still a milestone methodological achievement to develop patient-specific classification and modulation of brain states that was fully automated (see Chap. 41).

In the second study, electrophysiological activities induced in this free recall task were mapped across the brain areas and time of memory encoding [12, 15]. Univariate features of spectral power were calculated for a fully automated unsupervised classification of electrode contacts [16], which was performed separately in low, intermediate, and high frequency bands to identify contacts that recorded significant responses to stimulus presentation. This revealed unexpected findings about the spatiotemporal organization of the induced neural activities (Fig. 49.2). First of all, the vast majority of the identified macro-contacts were classified as active in only one or two frequency bands, meaning that any one recorded area of the cortex showed spectral activities at a distinct frequency range in this task. These findings are in agreement with the spectral fingerprint hypothesis [18], resulting in a mosaic-like pattern of theta, alpha, beta and gamma patches of cortical neural activities. Furthermore, the low, intermediate and high frequency activities were induced differently in the two hemispheres and at different times of memory encoding. The

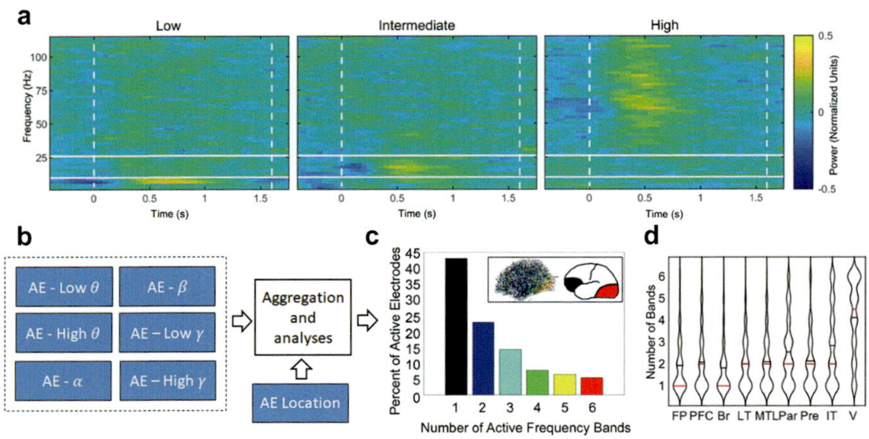

Fig. 49.2 Spatial organization of induced neural activities. **a** Representative examples of trial-averaged spectrograms during stimulus presentation from active electrodes at different anatomical locations showing induced power changes confined to one of the three frequency ranges. **b** Active electrodes were identified separately for each band (low theta: 2–4 Hz, high theta: 5–9 Hz, alpha: 10–15 Hz, beta: 16–25 Hz, low gamma: 25–55 Hz, high gamma: 65–115 Hz) and combined with the anatomical location of the electrode to study their spatiotemporal organization. **c** Percentage of electrodes that were found active in one or multiple different (up to 6) frequency range(s). Inset figure: Distribution of the electrodes active in one or more of the frequency bands reveals the highest overlap in the posterior areas of the occipital and parietal visual cortex, which was gradually decreasing in more anterior cortical areas as shown in the average brain surface plot and the cartoon summary of all active electrodes. **d** Violin plots (black and red lines indicate the mean and the median, respectively) of distribution of number of bands in which each electrode was active for nine cortical areas (V—visual; IT—inferior temporal; Pre—precuneus; Par—lateral parietal; MTL—mesial temporal lobe; LT—lateral temporal; Br—Broca's area; PFC—prefrontal cortex; FP—frontal pole)

induced power in each frequency band had a specific propagation pattern across the sensory and higher-order cortical areas of the processing stream (Fig. 49.3), reminiscent of travelling waves (see Chap. 30). In the end, the automated identification of active contacts in this memory task enabled an efficient and comprehensive mapping of human memory encoding and provided insights into the distributed independent dynamics and functions of low, intermediate, and high frequency neural activities [15].

The third study aimed to find the brain areas and neural activities that were most predictive of successful memory encoding [22]—the subsequent memory effect. For a number of reasons, including identification of specific contacts in particular regions, the initial studies failed to localize one confined area and activity that would have a greater magnitude of the effect than the others. A hotspot of this subsequent memory effect was found in the theta frequency activities recorded from electrode contacts in the left anterior prefrontal cortex (Fig. 49.3). The memory effect in this new hotspot location was found to be even greater than for the previously reported successful memory enhancement in the lateral temporal cortex, providing a new target for

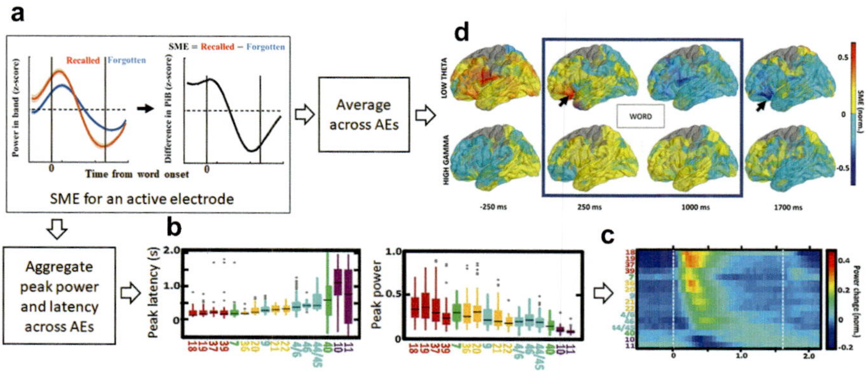

Fig. 49.3 Temporal organization of induced neural activities and a hotspot of memory encoding. **a** Subsequent Memory Effect (SME) for an active electrode is computed by taking the difference between the power in a specific frequency band for recalled and forgotten words. **b**, **c** Peak latency and peak power of the power-in-band signals aggregated across anatomical locations (Brodmann areas) of active electrodes reveal a hierarchical propagation of memory processing across the cortex, as visualized in the heatmap matrix. **d** SME values were averaged across active electrode localizations and interpolated on an average brain surface to reveal a hotspot for memory encoding in low theta neural activities of the anterior prefrontal cortex

electrical brain stimulation. Thus, the automated identification of the neural activities induced during memory encoding resulted in rapid mapping and localization of targets for modulating cognitive functions. This was accomplished using basic univariate features—other bivariate measures of connectivity, including correlation, spectral coherence or cross-frequency coupling, may prove even more successful for these mapping and heuristic purposes.

49.5 New Technologies and Future Directions

Rapid and efficient identification of neural activities recorded at particular electrode contacts becomes a necessity for the emerging neurotechnologies. Implantable devices for concurrent recording and stimulation of iEEG activities require automated signal processing and analysis. Even the very choice of target electrode contacts for stimulation therapies becomes more and more reliant on the employment of automated intelligent identification methods. In the classic example of the DBS (Deep Brain Stimulation) devices for treating movement disorders, target contacts for the therapeutic stimulation are selected manually based on the information acquired from the intra-operative recordings and anatomical localization of electrode implantation. Even these manual procedures would benefit from rapid localization and quantification of neural activities. New technologies for responsive and adaptive DBS already take full advantage of automated identification of neural activities to target the timing and localization of the stimulation. This automated approach is even more useful in similar devices implanted for management of seizures in epilepsy, where the target location and parameters of electrical stimulation are more difficult to determine both in the anatomical space of the brain and in time. In fact, the pathological HFOs observed in epilepsy are known to be a 'moving target' and change with time [9]. Likewise, the neural activities associated with cognitive processing are known to differ between individuals and even change within an individual across the time of brain development (see Chap. 10). Given that the new devices are designed to be implanted commonly for a life-time, manual selection and adjustment of the sensing and stimulation parameters is less and less efficient, feasible, and realistic.

The automated methods, on the other hand, are opening up to vast opportunities provided by the emerging neurotechnologies. Identification of the neural activities and optimization of their modulation can now be performed outside of the local signal acquisition and stimulation, which was traditionally all performed within a given iEEG system or an implantable device. Algorithms, which use computationally intensive machine-learning or artificial intelligence methods, are no longer limited to the local hardware capabilities but can use recordings streamed from local devices and be implemented across other devices and cloud computing facilities [10, 19]. Possibilities for implementing automated adaptive identification of neural activities and intelligent optimization of brain modulation are almost limitless once distributed to the virtual reality of internet space.

As a result, probing cognition is no longer bound to the laboratory space but can be performed remotely anywhere and at any time, as long as access to the virtual space of the internet is provided. Personal hand-held devices that stream iEEG activities between local and remote computing devices are ideally positioned to deliver cognitive tasks. The tasks themselves can be supervised manually or automatically triggered by unsupervised online analysis of the neural activities identified in the recordings. Additional features that are commonly utilized in the tasks, like accelerometry or eye-tracking, are easily implemented with wearable technologies on almost any smartphone or tablet computer. The eye-tracking measures like gaze position or pupil size are now commonly used to complement the iEEG features in studying cognitive functions (see Chap. 14) or in new Brain-Computer Interface applications [14], see Chap. 51). Empowered by all these technological capabilities, new paradigms for exploring human memory and cognition in virtual, augmented or natural environments are developed to take advantage of the emerging technological opportunities (see Chaps. 52 and 53). Cognitive functions of the human brain are now being probed and modulated in real time and home environments to eventually improve individual lives. Efficient identification of electrophysiological activities across the large scale of their organization is a pivotal step for these exciting new avenues.

Funding This work was supported from the IDUB grant Aurum 'BME iEEG network' at the Gdansk University of Technology and from the First Team grant of the Foundation for Polish Science co-financed by the European Union under the European Regional Development Fund (Grant No. POIR.04.04.00-00-4379/17).

Conflict of Interest Statement Authors declare no conflicts of interest.

Data availability statement Code used for identifying active electrodes is freely available at: https://github.com/kvsaboo/TaskActiveElectrodeIdentification.

References

1. Alotaiby T, Abd FE, El-Samie SA, Alshebeili (2015) A review of channel selection algorithms for EEG signal processing. EURASIP J Adv Sig Process 2015(1):1–21
2. Bragin A, Engel J Jr, Wilson CL, Fried I, Mathern GW (1999) Hippocampal and entorhinal cortex high-frequency oscillations (100–500 Hz) in human epileptic brain and in kainic acid-treated rats with chronic seizures. Epilepsia 40(2):127–137
3. Bragin A, Mody I, Wilson CL, Engel Jr, (2002) Local Generation of Fast Ripples in Epileptic Brain. The Journal of Neuroscience: The Official Journal of the Society for Neuroscience 22(5): 2012–21
4. Bragin A, Wilson CL, Staba RJ, Reddick M, Fried I, Engel JE Jr (2002) Interictal high-frequency oscillations (80–500 Hz) in the human epileptic brain: entorhinal cortex. Ann Neurol 52(4):407–415
5. Brázdil M, Pail M, Halámek J, Plešinger F, Cimbálník J, Roman R, Klimeš P et al (2017) Very high-frequency oscillations: novel biomarkers of the epileptogenic zone. Ann Neurol 82(2):299–310
6. Cimbalnik J, Kucewicz MT, Worrell G (2016) Interictal high-frequency oscillations in focal human epilepsy. Curr Opin Neurol 29(2):175–181
7. Ezzyat Y, Kragel JE, Burke JF, Levy DF, Lyalenko A, Wanda P, O'Sullivan L et al (2017) Direct brain stimulation modulates encoding states and memory performance in humans. Curr Biol CB 27(9):1251–1258
8. Ezzyat Y, Wanda PA, Levy DF, Kadel A, Aka A, Pedisich I, Sperling MR et al (2018) Closed-loop stimulation of temporal cortex rescues functional networks and improves memory. Nat Commun 9(1):365
9. Gliske SV, Irwin ZT, Chestek C, Hegeman GL, Brinkmann B, Sagher O, Garton HJL, Worrell GA, Stacey WC (2018) Variability in the location of high frequency oscillations during prolonged intracranial EEG recordings. Nat Commun 9(1):2155
10. Kremen V, Brinkmann BH, Kim I, Guragain H, Nasseri M, Magee AL, Attia TP et al (2018) Integrating brain implants with local and distributed computing devices: a next generation epilepsy management system. IEEE J Transl Eng Health Med 6(September):2500112
11. Kucewicz MT, Berry BM, Kremen V, Brinkmann BH, Sperling MR, Jobst BC, Gross RE et al (2017) Dissecting gamma frequency activity during human memory processing. Brain: A J Neurol 140 (5):1337–1350
12. Kucewicz MT, Berry BM, Miller LR, Khadjevand F, Ezzyat Y, Stein JM, Kremen V et al (2018) Evidence for verbal memory enhancement with electrical brain stimulation in the lateral temporal cortex. Brain: A J Neurol 141(4):971–978
13. Kucewicz MT, Michael Berry B, Worrell GA (2018) Simultaneous macro- and microrecordings. Invasive Stud Hum Epileptic Brain. https://doi.org/10.1093/med/9780198714668.003.0036
14. Lech M, Berry B, Topcu C, Kremen V, Nejedly P, Lega B, Gross RE et al (2021) Direct electrical stimulation of the human brain has inverse effects on the theta and gamma neural activities. IEEE Trans Biomed Eng 68(12):3701–3712
15. Marks VS, Saboo KV, Topçu Ç, Lech M, Thayib TP, Nejedly P, Kremen V, Worrell GA, Kucewicz MT (2021) Independent dynamics of low, intermediate, and high frequency spectral intracranial EEG activities during human memory formation. Neuroimage 245(October):118637

16. Saboo KV, Varatharajah Y, Berry BM, Kremen V, Sperling MR, Davis KA, Jobst BC et al (2019) Unsupervised machine-learning classification of electrophysiologically active electrodes during human cognitive task performance. Sci Rep 9(1):17390
17. Saboo KV, Varatharajah Y, Berry BM, Sperling MR, Gorniak R, Davis KA, Jobst BC et al (2019) A computationally efficient model for predicting successful memory encoding using machine-learning-based EEG channel selection. In: 2019 9th international IEEE/EMBS conference on neural engineering (NER), pp 323–27. ieeexplore.ieee.org
18. Siegel M, Donner TH, Engel AK (2012) Spectral fingerprints of large-scale neuronal interactions. Nat Rev Neurosci 13(2):121–134
19. Sladky V, Nejedly P, Mivalt F, Brinkmann BH, Kim I, St. Louis EK, Gregg NM et al (n.d.) Distributed brain co-processor for tracking electrophysiology and behavior during electrical brain stimulation. https://doi.org/10.1101/2021.03.08.434476.
20. Stead M, Bower M, Brinkmann BH, Lee K, Richard Marsh W, Meyer FB, Litt B, Van Gompel J, Worrell GA (2010) Microseizures and the spatiotemporal scales of human partial epilepsy. Brain: A J Neurol 133(9):2789–2797
21. Topçu Ç, Marks VS, Saboo KV, Lech M, Nejedly P, Kremen V, Worrell GA, Kucewicz, MT (2022) Hotspot of human verbal memory encoding in the left anterior prefrontal cortex. EBioMedicine, 82: p.104135. https://doi.org/10.1016/j.ebiom.2022.104135.
22. Vaz AP, Wittig Jr JH, Inati SK, Zaghloul KA (2020) Replay of cortical spiking sequences during human memory retrieval. Science 367(6482):1131–1134
23. Waldert S, Lemon RN, Kraskov A (2013) Influence of spiking activity on cortical local field potentials. J Physiol 591(21):5291–5303
24. Wei Q, Lu Z, Chen K, Ma Y (2010) Channel selection for optimizing feature extraction in an electrocorticogram-based brain-computer interface. J Clin Neurophysiol. https://doi.org/10.1097/wnp.0b013e3181f52f2d
25. Worrell GA, Jerbi K, Kobayashi K, Lina JM, Zelmann R, Le Van M (2012) Recording and analysis techniques for high-frequency oscillations. Prog Neurobiol 98(3):265–278
26. Worrell GA, Gardner AB, Matt Stead S, Hu S, Goerss S, Cascino GJ, Meyer FB, Marsh R, Litt B (2008) High-frequency oscillations in human temporal lobe: simultaneous microwire and clinical macroelectrode recordings. Brain: A J Neurol 131(Pt 4):928–937.
27. Yang J, Singh H, Hines EL, Schlaghecken F, Iliescu DD, Leeson MS, Stocks NG (2012) Channel selection and classification of electroencephalogram signals: an artificial neural network and genetic algorithm-based approach. Artif Intell Med 55(2):117–126

Chapter 50
How Can We Track Cognitive Representations with Deep Neural Networks and Intracranial EEG?

Daniel Pacheco Estefan

Abstract Studies harnessing the heuristic potential of Deep Neural Networks (DNNs) and the high spatiotemporal resolution of intracranial EEG (iEEG) data are providing fine-grained details about the nature of brain representations. Originally employed to simulate visual processing and object recognition in the ventral visual stream (VVS), DNNs have recently been applied as models of brain representation in higher-order cognitive domains. In addition to perception, this approach has recently yielded unprecedented insights in memory research, revealing the representational formats and temporal dynamics of working and episodic memory representations. In this chapter, the potential and the limitations of modeling cognitive representations with DNNs and iEEG data are reviewed.

50.1 Introduction

The goal of cognitive neuroscience is to develop a mechanistic understanding of how the brain represents information to implement cognitive functions. A fundamental tool to achieve this objective are computational models, i.e., formal descriptions of information processing in the brain that describe the operations applied to sensory inputs and their transformation to produce cognition and behavior. Computational models have been described at different levels of abstraction, ranging from biophysically detailed models of individual neurons to rate-based models of neural populations. In recent years, a family of architectures originally developed in the field of Machine Learning (ML) has been increasingly employed to model brain function in cognitive neuroscience: the class of Deep Neural Networks (DNNs). Despite being highly abstract versions of biological brains, DNNs have achieved human-level performance in a variety of 'cognitive' tasks such as visual object recognition, text generation or Atari video games [1–3]. To solve these tasks, DNNs employ multiple layers of internal representations of stimuli, which have been used to model

D. P. Estefan (✉)
Institute of Cognitive Neuroscience, Department of Neuropsychology, Ruhr University Bochum, 44801 Bochum, Germany
e-mail: Daniel.PachecoEstefan@ruhr-uni-bochum.de

N. Axmacher (ed.), *Intracranial EEG*, Studies in Neuroscience, Psychology and Behavioral Economics, https://doi.org/10.1007/978-3-031-20910-9_50

activity in brains of animals and humans [4–6]. Most human studies using DNNs have been conducted with neuroimaging and Magnetoencephalography (MEG) data (e.g., Refs. [2, 7–9]). Recently, however, investigations have started to apply them to other recording modalities, including electrocorticography (ECoG) and intracranial EEG (iEEG). In this chapter, we review these advancements, highlighting the adoption of Representational Similarity Analysis as a modelling framework to compare representations in DNNs and biological systems (RSA; Box 1; [10]). Studies employing RSA, DNNs and iEEG to characterize the transformative nature of memory representations and their relation to hippocampal theta oscillations are reviewed. Finally, the strengths, limitations and future opportunities of current modeling approaches are discussed.

50.2 Deep Neural Networks in Cognitive Neuroscience

The development of convolutional DNNs (cDNNs) in the domain of computer vision largely contributed to the virtuous interaction of Artificial Intelligence (AI) and neuroscience. CDNNs are a particular class of neural network architectures trained to classify images—i.e., to assign object labels to perceptually invariant visual inputs-, whose connectivity structure is loosely inspired by the visual system. After training, these networks develop layered representations of stimuli that hierarchically encode increasingly abstract visual properties: Early layers reflect low-level features of images such as edges or color patches, while deeper layers are sensitive to more complex visual information, such as the presence of objects or object parts [11]. Several studies demonstrated the validity of cDNNs as models of biological vision, showing that they capture relevant features of information processing in the ventral visual stream (VVS) of animals and humans [1, 12, 13]. This was initially shown in the macaque Inferotemporal (IT) cortex in two studies [7, 14]. The validation of cDNNs as models of visual processing promoted their application to other animal species (e.g., rodents [15]) and recording modalities. Several human studies have applied cDNNs to model visual perception with fMRI and RSA [14, 16–18]. These investigations revealed a match in the hierarchy of representations in the VVS and DNN layers, with superficial layers mapping to early visual areas and deep layers to higher-level regions [13, 14]. The anatomical correspondence of DNN layers, however, is currently a matter of debate. A recent study showed that all regions along the VVS match better with deep cDNN layers, suggesting that high-level categorical information is present at all levels of visual processing [19]. It has been acknowledged that the correspondence of layers of very deep feedforward cDNNs—currently the state of the art in ML—and particular brain regions is limited [12, 20–22]. Nevertheless, efforts to quantify the quality of models in predicting brain activity, such as the 'brain-score' metric [23], evaluate the predictive power of DNNs for each region of the VVS separately. Moreover, studies in visual perception have delineated a sharp line between VVS and medial temporal lobe (MTL) supported computations, using DNNs as models of VVS only [24].

Beyond the debate regarding the anatomical correspondence of DNN layers, recent MEG and iEEG studies have provided fine-grained details about the neurophysiological patterns that underlie DNN-like computations in the brain, characterizing their chronometry [9, 25], spectral properties [26], and architectural constraints [27, 28]). Among these advancements, architectural factors have been considered particularly important [13, 29]. For instance, it has been shown that recurrent connections play a crucial role in the predictive power of convolutional DNNs as models of visual processing. Indeed, shallow recurrent architectures are better predictors of VVS activity than deep feedforward networks [13, 23, 27–31]. Feedback connectivity is particularly important given the critical role of top-down connections, such as those originating in the prefrontal cortex (PFC; [27, 28]) on VVS activity.

Beyond the domain of vision, investigations are starting to use DNNs as models of higher-level cognitive representation (Fig. 50.1b). In general, two approaches can be distinguished: (1) studies that trained DNNs to perform complex cognitive tasks and compared their internal representations with those formed in the brains of animals and humans (i.e., a similar approach as in the visual domain; Fig. 50.1b, Left), and (2) studies that used DNNs optimized for lower-level tasks (e.g., visual perception) to model aspects of brain computation during higher-level cognitive functions (e.g., visual working memory; Fig. 50.1b, Right).

Studies following the first approach have so far mostly been conducted with fMRI. For instance, Cross et al. employed a convolutional DNN to map images extracted from Atari games to actions leading to reward in the games (Fig. 50.1b, Top left; [2]). Trained to maximize reward within a reinforcement learning (RL) setting, this network approximates the Q-value function in RL, reflecting the value of taking an action in a particular state (thus named Deep-Q-Network, DQN). Representations formed across layers of DQN were observed in a widespread network of sensorimotor regions in humans playing Atari games, demonstrating a convergent solution to the problem of state representation in biological and artificial brains [2]. Similarly, EmoNet, a convolutional DNN trained to classify images according to their emotional content, developed representations of high-level emotional categories (Fig. 50.1b, Bottom left; [8]). The structure of representations in deep layers of EmoNet was predictive of activity along the VVS of humans watching emotional movies. This result indicates that high-level emotional categories are represented in visual areas and may be taken to suggest that these categories can be extracted from visual representations alone. Interesting results have also been observed in auditory perception, where task-optimized convolutional DNNs are predictive of cortical responses to auditory inputs in macaques [32]. In the domain of human language, recent studies have shown that internal representations formed in transformer language models (e.g., GPT-2) map to distributed language networks in the brain [3, 33–35]. Transformers are a novel class of DNN architectures that rely on self-attention and contextual representations to capture long-range dependencies in their inputs [36]. Interestingly, the ability of the models to predict brain activity is correlated with their performance in the task of next word prediction, suggesting that the human brain is also engaged in this task during language processing [33, 34].

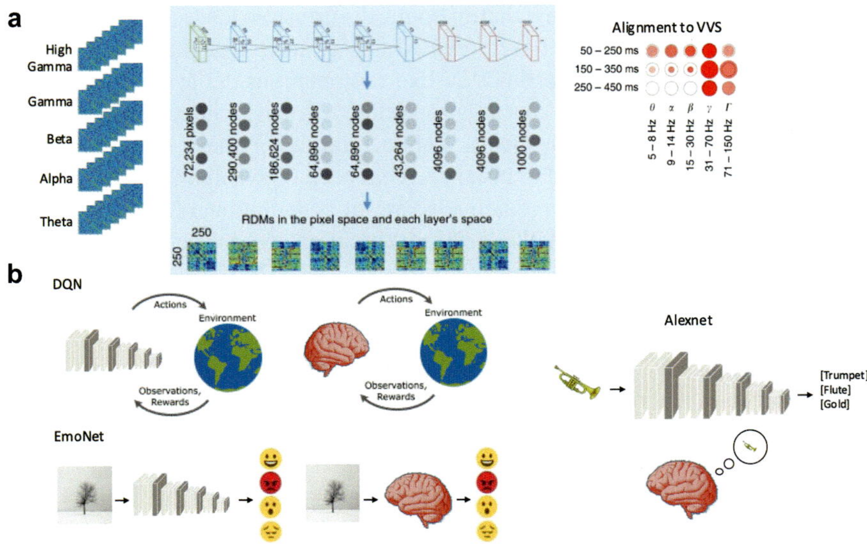

Fig. 50.1 Modeling cognitive representations with DNNs. **a** Convolutional DNNs have been employed to model visual object recognition in the VVS with iEEG. In a recent study [26], neural RDMs were created for each frequency band and compared to different layers of the DNN Alexnet (Left). This revealed a specific role of low gamma oscillations (31–70 Hz) in conveying feedforward information during visual processing. Consistent with neuroimaging research, this study showed that superficial layers of Alexnet fit well early visual regions in the VVS, while deep layers are better matched to high-level visual regions (Right). **b** DNNs have been employed to model higher-level cognitive functions beyond the domain of visual perception. Two approaches can be distinguished. Left: Representations in biological and artificial brains performing the same task are compared. Top: The Deep-Q-Network is trained to solve Atari Games and used as a model of state-space representations in the human brain [2]. Bottom: The EmoNet network is trained to classify emotions from images and used to track representations of emotional content [8]. Right: Alternatively, studies using iEEG in memory research have used pretrained architectures from computer vision and natural language processing to investigate mnemonic processing of visual and semantic information

Studies following the second approach—of using DNNs trained on lower-level tasks to investigate representations during higher-level cognitive function—, have mostly employed iEEG. Using DNNs that were pre-trained for visual and language processing, these studies are providing crucial details about the employment of different visual representational formats during working and episodic memory processes, the transformation of these representations, and their dependence on deep brain oscillations. These recent developments are reviewed in the next section.

50.3 DNNs and iEEG: Insights from Memory Research

Harnessing the representational power of DNNs and the high temporal resolution of iEEG, recent studies have described the formats and dynamics of memory representations with an unprecedented level of detail. For instance, several investigations have now successfully applied the neural network Alexnet [37] to investigate the moment-to-moment representational transformations underlying the mnemonic processing of images [38, 39]. Alexnet is a feedforward convolutional network composed of five convolutional layers and 3 fully connected layers [37]. Using the Alexnet network and RSA, a recent study showed that neural representations of items during visual working memory encoding (VSWM) are matched to deep but not early layers of the network (Fig. 50.2, [39]; see also Chap. 8). This occurs in a relatively early time-period reflecting fast representational transformations from low- to high-level visual information (250–770 ms after stimulus onset). Interestingly, in a later encoding time-period (1–2 s), representations were not captured by Alexnet layers but by a semantic network that was not trained with images but with text corpora [40], highlighting the relevance of abstract representations for VSWM (Fig. 50.2a, b, Left).

Consistently, a second study also using Alexnet revealed the transformation of representations from low- to high-level visual and semantic formats in a time frame of ~1.5 s at encoding [38]. Critically, representational dynamicity—the degree of representational transformation across time—was correlated with subsequent memory performance in the task [38]. Moreover, this study revealed that representations retrieved from long-term memory were more similar to those during short-term maintenance than during encoding, revealing the highly transformative nature of these neural patterns [38].

In addition to characterizing the dynamic nature of memory, studies using DNNs, RSA and iEEG have provided insights into the organizing role of brain oscillations during memory processes (see also Chap. 19). Theories have proposed that low frequency oscillations organize and structure representations that are encoded in high-frequency or single unit activity [41] (see also Chap. 44). Using a word2vec model [42] to describe the semantic relationships among visual images in an episodic memory task, a recent iEEG study confirmed that neural representations of items are organized along hippocampal theta oscillations according to their semantic content during episodic memory retrieval [43]. In this study, similarity between time-resolved distributed oscillatory patterns was calculated between encoding and retrieval using Spearman's rho. Distributions of encoding retrieval similarity (ERS) values over hippocampal theta phase were constructed by computing the specific retrieval phase of the ERS time-series (Fig. 50.2b, Middle). The pairwise similarity of the phase-ERS distributions was then compared with the distance extracted from the word2vec embeddings, revealing a segregation of semantically proximal items into more distant theta (3–8 Hz) phases (Fig. 50.2b, Right). This result is consistent with previous research showing that hippocampal theta oscillations reflect semantic distances between words during memory retrieval [44]. Using a similar approach of characterizing semantic proximities between stimuli using word2vec embeddings,

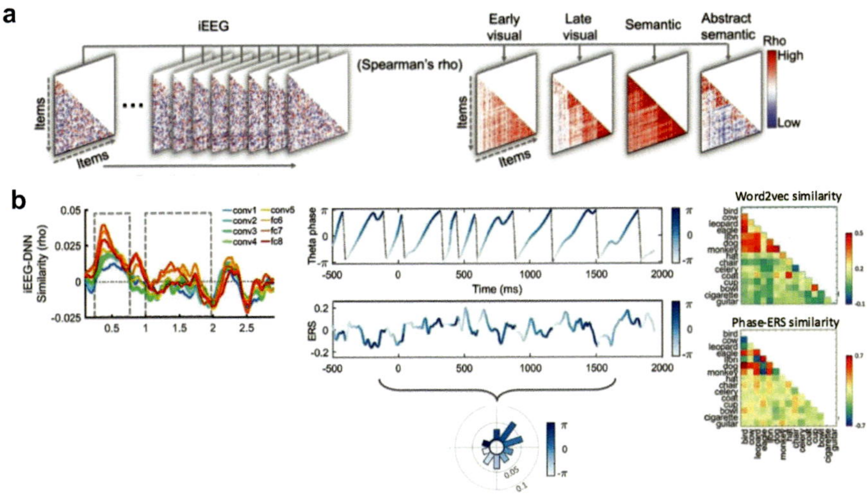

Fig. 50.2 Specific advantages of modelling cognitive representations with iEEG. **a** DNNs can be applied to assess the representational structure of stimuli in a time-resolved manner, by defining feature vectors for each item and computing pairwise similarities among them in sliding time windows. These RDM 'movies' can then be compared to different model RDMs (e.g., from early layers of a DNN). **b** Different uses of time-resolved signals: Left: The moment-to-moment representational transformations during working memory encoding are reflected in the traces representing the fit of a particular network layer to the neural RDMs [38]. High-level visual information is present during an early time-period at encoding, as observed in the significant fit of deep layers of the Alexnet network. Middle: Hippocampal theta phases (top row) were linked to time-resolved encoding-retrieval similarity (ERS) values in each trial (middle row), resulting in phase-ERS distributions reflecting the preferred phase of item reactivations [43] (bottom row). Distances between these distributions were subsequently compared to semantic distances among items as assessed by a word2vec DNN. Correspondent matrices of one participant in the experiment are shown (Right)

this study revealed that semantic distances correlate with theta power in a free recall task, establishing the role of hippocampal theta oscillations in organizing semantic information during episodic memory processes.

In addition to hippocampal theta oscillations, oscillatory activity in neocortical regions have been linked to relevant neural computations such as top-down inhibitory control (alpha and beta; [45, 46]) or bottom-up information flow (gamma; [47]). By tracking the relationship of representational signals and oscillatory power at different frequencies and electrodes, a recent study found that ERS is negatively correlated with beta and positively correlated with gamma power during episodic memory retrieval [43], consistent with results obtained in a combined fMRI-EEG experiment [48]. In these studies, DNNs were not employed but item-specific activity was extracted using RSA. Nevertheless, this research shows the potential of relating representational patterns extracted from DNNs to neurophysiological signals such as oscillatory power and/or phase.

Finally, iEEG studies have applied the DNN-RSA framework to characterize the specific frequencies in which information is encoded and dynamically represented during memory processes, as previously achieved in visual perception [26]. A recent study tracked frequency resolved, DNN-derived representations extracted from iEEG in a working memory paradigm involving a retro cue. In this study, participants were required to maintain three items in working memory and only one of them (indicated by the cue) was tested after the maintenance period. The geometry of activity patterns in the prefrontal cortex (PFC) after the presentation of the cue was captured by the deepest layers of a recurrent DNN [49] specifically in the beta (16–28 Hz) band, suggesting an inhibitory top-down representation [50].

50.4 Discussion

In this chapter, two approaches for modeling cognitive representations with DNNs were reviewed. The first approach used task-optimized DNNs as models of brain representations and computations in visual processing and other higher-level cognitive domains. The second approach employed DNN architectures from computer vision and natural language processing to track cognitive representations underlying memory. These two approaches have both strengths and limitations, and present opportunities for future research.

Modeling neural data with task-optimized DNNs has revealed critical insights about the nature of brain computations. In vision studies, DNNs have been used to model several different types of data, including rodent [15] and monkey [14, 16] electrophysiology but also human fMRI, MEG, and iEEG [9, 12, 13, 23, 28, 51]. The variety of methods and number of studies has allowed for a highly detailed characterization of the temporal and spectral dynamics of visual processing in the brain. In other cognitive domains, DNNs have only been used with neuroimaging, which represents an opportunity for upcoming iEEG research. For instance, while fMRI studies revealed the sensorimotor networks involved in the transformation of visual information into a representation of states of the world in the context of Atari games [2], the frequencies and temporal dynamics of the neurophysiological patterns underlying these representations are currently unknown. A second interesting direction to explore is to characterize the representation of sensory and cognitive variables in the hippocampus, a structure to which iEEG offers direct access. Theories have proposed that hippocampal place cells encode the representation of world states during sharp-wave ripples events [52] and theta oscillations [53]. Such dynamic state-space representation might serve adaptive roles such as learning of action-reward associations and plan future actions. It would be interesting to test these accounts by characterizing the representation of states in the human hippocampus with deep RL models (e.g., DQN) and iEEG data. Similarly, the study of emotional processing in the amygdala could capitalize on recently developed architectures for emotion classification [8].

The second approach reviewed in this chapter—using architectures trained in lower-level tasks to model higher-level cognitive functions—provides the practical advantage of avoiding computationally extensive network training. More importantly, this modeling framework has been supported by theoretical considerations

[54, 55]. Indeed, representational accounts of memory have argued that regions representing particular stimulus properties (e.g., low-level visual features in early visual regions) are involved in the representation of these features irrespective of the cognitive process in which they are engaged [55]. While these accounts make specific predictions about the anatomical brain regions that are activated for particular representations, iEEG studies have assessed neural representations more widely. For instance, convolutional DNNs have been used to model representations extracted from neural patterns covering several brain regions [38, 39, 43]. This contrasts with modeling approaches in visual neuroscience that typically target to map activity in particular layers of DNNs to specific anatomical areas [23].

While it is important to describe the computational role of particular brain regions, global representational patterns may provide functionally relevant information not observable at a local scale. Interestingly, studies have characterized the relationship of representations at these different levels of analysis. Typically, iEEG global representational patterns are computed by correlating neural activity from all available electrodes in a particular subject, while local patterns are calculated at individual electrodes using as features only frequencies and times [43]. Studies have shown that the contribution of each electrode to global representational patterns correlates with its local levels of pattern similarity during memory retrieval [56]. Electrode contribution to the global pattern is assessed through a 'jackknife' procedure, in which a baseline value of pattern similarity is computed when including all features, and then again after removing these features one by one [43, 56]. This electrode-wise metric of contribution has been linked to oscillatory power at different frequencies revealing the critical role of theta, beta and gamma oscillations in encoding and retrieval of item-specific representations [43]. While this analysis was not applied to DNN-derived representations, it could be easily adopted to RSA time-series extracted from DNN activations.

An additional relevant line of research that could benefit from the representational power of DNNs involves the study of the coordination of different brain regions representing different content during cognitive processes. The high spatiotemporal resolution of iEEG makes it an appropriate data type to address this issue. For instance, iEEG has been used to investigate the coordination of memory reinstatement during episodic retrieval between the hippocampus and the Lateral Temporal Cortex (LTC; [57]). This study found that representations of items in the LTC and of item-context associations in the hippocampus are coordinated at the trial level during memory retrieval, supporting a theory of hippocampal function according to which the hippocampus forms and retains an index of neocortical areas activated by experiential events [57, 58]. Similarly, a recent study using MEG [59] characterized how representations extracted from DNNs that were mapped to different brain regions were related during visual processing (e.g., how representations of low-level features of images in early visual regions relates to the representation of more abstract visual information in higher-level areas). Applying Granger Causality to RSA time courses extracted from DNN layers, this study found visual information travels through feedforward connections, while visual information is transformed into semantic representations through feedforward and feedback activity [59]. A similar approach could be applied to iEEG, which could also investigate the transformation of perceptual signals into semantics in deep MTL structures, including the hippocampus, where abstract conceptual representations have been observed [60].

It must be noticed that some of the studies following the first approach reviewed in this chapter have employed transfer learning techniques that facilitate network training on customized tasks. Transfer learning consists of partially retraining a network, typically by adjusting the architecture and weights in the final layers while keeping the rest unmodified. For instance, the EmoNet network [8] is fully based on the Alexnet architecture and trained parameters except from its final classification layer, which was adjusted to classify emotions instead of objects from images. Transfer learning has been used to evaluate the impact of particular architectural features, e.g., lateral connectivity, in the performance of DNN models [61]. For instance, a recent study demonstrated that adding recurrent connections to Alexnet's Fc7 layer significantly improved the network performance and fit to human behavior [61]. In addition to transfer learning, studies have fully retrained off-the-shelf architectures to investigate changes in the internal representations formed in the networks after changing their training objective. Following this approach, a recent study evaluated the performance of the VGG16 network [20], a deep convolutional architecture originally designed to classify images, after retraining it to recognize only faces or objects [62]. This study showed that training for these tasks separately led the network to develop very different representations of these two categories, while training on both tasks simultaneously made it segregate itself into two separate systems for faces and objects. Transfer learning and retraining off-the-shelf networks avoids unnecessary network design and training and allows to evaluate task-performing models customized to particular tasks.

In addition to re-training existing architectures, the development of novel deep learning models will be critical in future iEEG-DNN research. With the exception of visual processing, which has incorporated novel types of models, most of currently employed task-performing networks in other cognitive domains lack recurrent connectivity and therefore temporal dynamics [29]. To model higher-level cognitive functions, it will be crucial to assess how inputs of the system are dependent on contextual information and memory. Promising new results have been obtained in the domain of NLP, where networks trained in next-word prediction based on contextual embeddings are currently the best predictive models of brain activity during human language tasks [3, 33, 35, 36]. In reinforcement learning, models have also incorporated recurrent connectivity to deal with the issue of memory [63, 64]. Modeling cognitive representations extracted from iEEG data with these novel architectures constitutes a fascinating opportunity for future research.

While many of the modelling approaches discussed in this chapter assign a prominent role to prediction of neural activity patterns based on the responses observed in DNNs, this does not by itself provide a mechanistic account of cognitive processing. A match in the representational geometry of representations formed in biological and artificial brains does not imply that the same computations are being performed in the two systems, because similar representations could be achieved through different computations [4, 5]. In that sense, the ability of current DNN models to predict neural activity patterns is one relevant aspect that should be complemented with other methods in future research.

Box 1: Representational Similarity Analysis and Its Application to iEEG Data

Representational Similarity Analysis (RSA; [10]) is a widely used multivariate analysis framework for characterizing the representation of stimuli in the brain. RSA characterizes a system's representations by means of a representational dissimilarity matrix (RDM), traditionally defined via the pairwise distances (e.g., 1-correlation) among stimuli. In a cognitive neuroscience experiment, a neural RDM is typically built by computing these pairwise distances among activity patterns representing items and/or particular experimental conditions. These might comprise multivariate activity from multiple fMRI voxels or time–frequency-electrodes patterns in iEEG. To test specific hypotheses about the structure of this neural RDM, a model RDM is defined. Model RDMs can be hand-crafted or taken from other representational systems such as DNNs. Critically, comparisons between model and neural representational geometries and are performed at a second level of analysis and do not depend on the individual features that are used in each system to build the RDMs. This provides RSA with great flexibility to compare biological and artificial systems irrespective of their particular modality of representation (e.g., artificial units in a DNN or fMRI voxel activity humans). The fit of the model RDM to the neural RDM quantifies the evidence supporting the model hypothesis. The fit of the model can be assessed by correlating vectorized RDMs or using General Linear Models (GLMs). In GLMs, a set of RDM predictors representing different factors that could affect the neural representational geometry are defined. These predictors might not be orthogonal to each other, and studies have applied jackknife procedures to assess the specific contribution of each predictor [31]. When using DNNs as models of representation, and given that different layers of DNNs are correlated, studies have used partial regression or correlation techniques to assess representations in individual layers [38, 39].

In humans, RSA has been mostly applied to neuroimaging data. Recently, however, it has also been used with other recording modalities such as MEG, ECoG or iEEG. RSA can be flexibly adjusted to exploit the high temporal and spectral resolution of the latter recording modalities. In time-resolved analyses, representational feature vectors characterizing individual items are computed in sliding time-windows. For each time-window, the representational geometry of the stimuli is defined, resulting in an RDM time-series (or an RDM 'movie'). This can be contrasted to "static" models (e.g., a particular layer in a feedforward convolutional DNN, Fig. 50.2a) or "dynamic" models (e.g., RDM movies extracted from recurrent DNNs). Similarly, RDMs can be specified for particular frequency bands or individual frequencies, and specific anatomical locations (by selecting electrodes or voxels as features). RSA can also be applied to whole-brain patterns and/or broad frequency bands. While traditional approaches have assessed compartmentalized regions in the VVS, recent iEEG studies have started to use brain-wide patterns including electrodes located in distributed regions. To assess the contribution of individual brain regions, some studies have performed local RSA by extracting activity at one specific electrode [57]. Time and/or frequency resolved analyses, together with spatially localized RSA, provide a powerful way to link representations extracted with DNNs to the vast neurophysiological literature on iEEG oscillations and their role in cognitive function.

References

1. Kriegeskorte N (2015) Deep neural networks: a new framework for modeling biological vision and brain information processing. Annu Rev Vis Sci 1:417–446
2. Cross L, Cockburn J, Yue Y, O'Doherty JP (2021) Using deep reinforcement learning to reveal how the brain encodes abstract state-space representations in high-dimensional environments. Neuron 109:724–738
3. Goldstein A, Zada Z, Buchnik E et al (2022) Shared computational principles for language processing in humans and deep language models. Nat Neurosci 25. https://doi.org/10.1038/s41593-022-01026-4
4. Richards BA, Lillicrap TP, Beaudoin P et al (2019) A deep learning framework for neuroscience. Nat Neurosci 22:1761–1770
5. Saxe A, Nelli S, Summerfield C (2021) If deep learning is the answer, what is the question? Nat Rev Neurosci 22:55–67. https://doi.org/10.1038/s41583-020-00395-8
6. Kriegeskorte N, Douglas PK (2018) Cognitive computational neuroscience. Nat Neurosci 21:1148–1160
7. Khaligh-Razavi S-M, Kriegeskorte N (2014) Deep supervised, but not unsupervised, models may explain IT cortical representation. PLoS Comput Biol 10:e1003915
8. Kragel PA, Reddan MC, LaBar KS, Wager TD (2019) Emotion schemas are embedded in the human visual system. Sci Adv 5:eaaw4358
9. Cichy RM, Khosla A, Pantazis D et al (2016) Comparison of deep neural networks to spatio-temporal cortical dynamics of human visual object recognition reveals hierarchical correspondence. Sci Rep 6:1–13
10. Kriegeskorte N, Mur M, Bandettini PA (2008) Representational similarity analysis-connecting the branches of systems neuroscience. Front Syst Neurosci 2:4
11. Zeiler MD, Fergus R (2014) Visualizing and understanding convolutional networks. In: European conference on computer vision, pp 818–833
12. Lindsay GW (2021) Convolutional neural networks as a model of the visual system: past, present, and future. J Cogn Neurosci 33:2017–2031
13. Yamins DLK, DiCarlo JJ (2016) Using goal-driven deep learning models to understand sensory cortex. Nat Neurosci 19:356–365
14. Yamins DLK, Hong H, Cadieu CF et al (2014) Performance-optimized hierarchical models predict neural responses in higher visual cortex. Proc Natl Acad Sci 111:8619–8624
15. Vinken K, de Beeck H (2021) Using deep neural networks to evaluate object vision tasks in rats. PLoS Comput Biol 17:e1008714
16. Cadieu CF, Hong H, Yamins DLK et al (2014) Deep neural networks rival the representation of primate IT cortex for core visual object recognition. PLoS Comput Biol 10:e1003963

17. Eickenberg M, Gramfort A, Varoquaux G, Thirion B (2017) Seeing it all: convolutional network layers map the function of the human visual system. Neuroimage 152:184–194
18. Xu Y, Vaziri-Pashkam M (2021) Limits to visual representational correspondence between convolutional neural networks and the human brain. Nat Commun 12:1–16
19. J SN, C LB (2022) Reassessing hierarchical correspondences between brain and deep networks through direct interface. Sci Adv 8:eabm2219. https://doi.org/10.1126/sciadv.abm2219
20. Simonyan K, Zisserman A (2014) Very deep convolutional networks for large-scale image recognition. arXiv preprint arXiv:14091556
21. Szegedy C, Liu W, Jia Y et al (2015) Going deeper with convolutions. In: Proceedings of the IEEE conference on computer vision and pattern recognition, pp 1–9
22. Szegedy C, Ioffe S, Vanhoucke V, Alemi A (2016) Inception-v4, inception-resnet and the impact of residual connections on learning (2016). arXiv preprint arXiv:160207261
23. Kubilius J, Schrimpf M, Hong H et al (2019) Brain-like object recognition with high-performing shallow recurrent ANNs. In: Wallach H, Larochelle H, Beygelzimer A et al (eds) Neural information processing systems (NeurIPS). Curran Associates, Inc., pp 12785–12796
24. Bonnen T, Yamins DLK, Wagner AD (2021) When the ventral visual stream is not enough: a deep learning account of medial temporal lobe involvement in perception. Neuron 109:2755–2766
25. Cichy RM, Pantazis D, Oliva A (2014) Resolving human object recognition in space and time. Nat Neurosci 17:455–462
26. Kuzovkin I, Vicente R, Petton M et al (2018) Activations of deep convolutional neural networks are aligned with gamma band activity of human visual cortex. Commun Biol 1:1–12
27. Kar K, DiCarlo JJ (2021) Fast recurrent processing via ventrolateral prefrontal cortex is needed by the primate ventral stream for robust core visual object recognition. Neuron 109:164–176
28. Kar K, Kubilius J, Schmidt K et al (2019) Evidence that recurrent circuits are critical to the ventral stream's execution of core object recognition behavior. Nat Neurosci 22:974–983
29. van Bergen RS, Kriegeskorte N (2020) Going in circles is the way forward: the role of recurrence in visual inference. Curr Opin Neurobiol 65:176–193
30. Nayebi A, Sagastuy-Brena J, Bear DM et al (2021) Goal-driven recurrent neural network models of the ventral visual stream. bioRxiv
31. Kietzmann TC, Spoerer CJ, Sörensen LKA et al (2019) Recurrence is required to capture the representational dynamics of the human visual system. Proc Natl Acad Sci 116:21854–21863
32. Kell AJE, Yamins DLK, Shook EN et al (2018) A task-optimized neural network replicates human auditory behavior, predicts brain responses, and reveals a cortical processing hierarchy. Neuron 98:630–644
33. Caucheteux C, King J-R (2022) Brains and algorithms partially converge in natural language processing. Commun Biol 5:134. https://doi.org/10.1038/s42003-022-03036-1
34. Schrimpf M, Blank IA, Tuckute G et al (2021) The neural architecture of language: integrative modeling converges on predictive processing. Proc Natl Acad Sci 118:e2105646118
35. Caucheteux C, King J-R (2021) Language processing in brains and deep neural networks: computational convergence and its limits. bioRxiv 2020.07.03.186288. https://doi.org/10.1101/2020.07.03.186288
36. Vaswani A, Shazeer N, Parmar N et al (2017) Attention is all you need. Adv Neural Inf Process Syst 30
37. Krizhevsky A, Sutskever I, Hinton GE (2012) Imagenet classification with deep convolutional neural networks. Adv Neural Inf Process Syst 25
38. Liu J, Zhang H, Yu T et al (2021) Transformative neural representations support long-term episodic memory. Sci Adv 7:eabg9715
39. Liu J, Zhang H, Yu T et al (2020) Stable maintenance of multiple representational formats in human visual short-term memory. Proc Natl Acad Sci 117:32329–32339

40. Song Y, Shi S, Li J, Zhang H (2018) Directional skip-gram: explicitly distinguishing left and right context for word embeddings. In: Proceedings of the 2018 conference of the North American Chapter of the Association for Computational Linguistics: human language technologies, volume 2 (Short Papers), pp 175–180
41. Lisman JE, Jensen O (2013) The theta-gamma neural code. Neuron 77:1002–1016. https://doi.org/10.1016/j.neuron.2013.03.007
42. Mikolov T, Chen K, Corrado G, Dean J (2013) Efficient estimation of word representations in vector space. arXiv preprint arXiv:13013781
43. Pacheco Estefan D, Zucca R, Arsiwalla X et al (2021) Volitional learning promotes theta phase coding in the human hippocampus. Proc Natl Acad Sci 118
44. Solomon EA, Lega BC, Sperling MR, Kahana MJ (2019) Hippocampal theta codes for distances in semantic and temporal spaces. Proc Natl Acad Sci 116:24343–24352. https://doi.org/10.1073/pnas.1906729116
45. Fellner M-C, Waldhauser GT, Axmacher N (2020) Tracking selective rehearsal and active inhibition of memory traces in directed forgetting. Curr Biol 30:2638–2644
46. ten Oever S, Sack AT, Oehrn CR, Axmacher N (2021) An engram of intentionally forgotten information. Nat Commun 12:6443. https://doi.org/10.1038/s41467-021-26713-x
47. Richter CG, Thompson WH, Bosman CA, Fries P (2017) Top-down beta enhances bottom-up gamma. J Neurosci 37:6698–6711
48. Gb J, George P, Frederic R et al (2019) Directional coupling of slow and fast hippocampal gamma with neocortical alpha/beta oscillations in human episodic memory. Proc Natl Acad Sci 116:21834–21842. https://doi.org/10.1073/pnas.1914180116
49. Spoerer CJ, Kietzmann TC, Mehrer J et al (2020) Recurrent neural networks can explain flexible trading of speed and accuracy in biological vision. PLoS Comput Biol 16:e1008215
50. Pacheco D, Fellner MC, Kunz L, Zhang H et al (2022) Maintenance and transformation of representational formats during working memory prioritization. bioRxiv, 2023–02
51. Zhuang C, Yan S, Nayebi A et al (2021) Unsupervised neural network models of the ventral visual stream. Proc Natl Acad Sci 118
52. Mattar MG, Daw ND (2018) Prioritized memory access explains planning and hippocampal replay. Nat Neurosci 21:1609–1617. https://doi.org/10.1038/s41593-018-0232-z
53. McNamee DC, Stachenfeld KL, Botvinick MM, Gershman SJ (2021) Flexible modulation of sequence generation in the entorhinal–hippocampal system. Nat Neurosci 24:851–862. https://doi.org/10.1038/s41593-021-00831-7
54. Cowell RA, Barense MD, Sadil PS (2019) A roadmap for understanding memory: decomposing cognitive processes into operations and representations. Eneuro 6
55. Murray EA, Bussey TJ, Saksida LM (2007) Visual perception and memory: a new view of medial temporal lobe function in primates and rodents. Annu Rev Neurosci 30:99–122
56. Zhang H, Fell J, Staresina BP et al (2015) Gamma power reductions accompany stimulus-specific representations of dynamic events. Curr Biol 25:635–640. https://doi.org/10.1016/j.cub.2015.01.011
57. Pacheco Estefan D, Sánchez-Fibla M, Duff A et al (2019) Coordinated representational reinstatement in the human hippocampus and lateral temporal cortex during episodic memory retrieval. Nat Commun 10:1–13
58. Teyler TJ, DiScenna P (1986) The hippocampal memory indexing theory. Behav Neurosci 100:147
59. Clarke A, Devereux BJ, Tyler LK (2018) Oscillatory dynamics of perceptual to conceptual transformations in the ventral visual pathway. J Cogn Neurosci 30:1590–1605
60. Quiroga RQ (2012) Concept cells: the building blocks of declarative memory functions. Nat Rev Neurosci 13:587–597
61. Tang H, Schrimpf M, Lotter W et al (2018) Recurrent computations for visual pattern completion. Proc Natl Acad Sci 115:8835–8840
62. Dobs K, Martinez J, Kell AJE, Kanwisher N (2022) Brain-like functional specialization emerges spontaneously in deep neural networks. Sci Adv 8:eabl8913. https://doi.org/10.1126/sciadv.abl8913

63. Lin D, Richards BA (2021) Time cell encoding in deep reinforcement learning agents depends on mnemonic demands. bioRxiv 2021.07.15.452557. https://doi.org/10.1101/2021.07.15.452557
64. Banino A, Barry C, Uria B et al (2018) Vector-based navigation using grid-like representations in artificial agents. Nature 557:429–433. https://doi.org/10.1038/s41586-018-0102-6

Chapter 51
How Can I Use Utah Arrays for Brain-Computer Interfaces?

Christian Klaes

Abstract To record neuronal data from single neurons requires sophisticated hard- and software. Signals can be acquired acutely, for example using single Tungsten electrodes, or chronically, for example using an electrode array. Chronically implanted electrodes can provide spiking data for long periods of times so that long lasting experiments or clinical applications become viable. In case of a brain-computer interface (BCI) where information directly from the brain is used to drive an extracorporeal device, like a robotic prosthesis, it is desirable to use chronically implanted electrodes that can work for multiple years, ideally for a lifetime. One candidate device that has been used in countless animal experiments and several human clinical trials is the Utah electrode array (UEA) which was developed in the late 1980s at the University of Utah. At the time of writing the Utah electrode array is marketed by Blackrock Neurotech and 31 people have been implanted with one or multiple UEAs. In this chapter an overview over the UEA is given and its use for BCI applications will be described in detail.

51.1 What Is a Brain-Computer Interface?

A Brain-Computer Interface (BCI) provides a direct connection between the brain and a computer. Information of intended actions that arise, for example in the motor related areas of the brain, are read-out by specific sensors. The recorded neuronal information is then passed on to a so-called 'decoder' which interprets the neuronal signals and outputs according control signals. These signals are then used to drive an effector such as a neuroprosthetic device, e.g. a mouse cursor, a robotic arm or an exoskeleton. A bidirectional BCI can also write-in information, so that instead of receiving information via the periphery, e.g. the body surface, the areas of the brain that would normally receive this input are directly electrically stimulated. Primary targets for reading-out motor related information are the primary motor cortex (M1),

---

C. Klaes (✉)
Faculty of Medicine, Department of Neurotechnology, Ruhr University Bochum, Universitätsstraße 150, 44801 Bochum, Germany
e-mail: christian.klaes@rub.de

N. Axmacher (ed.), *Intracranial EEG*, Studies in Neuroscience, Psychology and Behavioral Economics, https://doi.org/10.1007/978-3-031-20910-9_51

premotor cortices or the posterior parietal cortex. For writing-in information the primary sensory cortex (S1) can be targeted. A technical distinction between various BCI types can be made by looking at the invasiveness of the device. The most common BCI are non-invasive and use electrodes placed on the scalp, i.e. electroencephalography (EEG). EEG based BCI have a high temporal resolution but lack spatial resolution which makes it difficult to control devices with them. Long training phases, few degrees of freedom to control and a lot of concentration required from the user make them less suitable for continuously operating a neuroprosthesis. Another candidate is functional magnetic resonance imaging (fMRI) which has a better spatial resolution and signals from within the brain can be as easily recorded as surface areas. Unfortunately, fMRI has a very low temporal resolution and is impractical for size and cost reasons. Functional near infrared spectroscopy (fNIRS) can be used in a similar setup as EEG or in conjunction with EEG. It offers some potential advantages for the signal to noise ratio but suffers from low temporal resolution since it only indirectly measures brain activity similar to fMRI. In addition to problems with reading out signals due to low spatial and temporal resolution there are also no viable options to write-in information as feedback for the brain. Currently this can be done non-invasively only using transcutaneous direct current stimulation (tDCS) or transcranial magnetic stimulation (TMS). Both of these methods have been used to stimulate the brain but only on a very broad level and by activating large parts of the brain which makes them less specific. For use in a neuroprosthetic to provide somesthetic feedback tDCS and TMS are not precise enough.

If invasive BCI are used the signal-to-noise ratio and the spatial resolution can get much better. Electrodes can be placed either on top of the dura mater which is considered to be less invasive or beneath it. For placement on top or under the dura mater, soft electrode grids are used which can stay in the body for several days to a few weeks to record electrocorticographic (ECoG) signals. ECoG signals are similar to EEG signals but since they are closer to the cortex and no skull is in the way the signal-to-noise ratio can be much higher. The local field potentials (LFP) that they record are less detailed than single neuron activity but they are a highly stable signal. There are also micro ECoG grids available which can detect signals close to single neuron level [1]. Generally, ECoG based BCI systems strike a good balance between invasiveness and signal quality and have been used in research settings extensively [2, 3].

The most invasive systems used today are electrodes implanted beneath the dura mater, often in the form of larger arrays containing hundreds of electrodes. These arrays are placed beneath the dura and are penetrating the outer layers of the cortex. With such arrays single neuron activity can be measured in real-time. The signal quality of these electrodes can be excellent although it is possible that only few electrodes are close enough to neurons to pick up single cell spiking activity depending on the location. Currently, the only available electrode array for implantation in humans with FDA approval is the Utah electrode array (UEA).

51.2 Anatomy of the Utah Array

The UEA is a silicon-based probe that consists of 100 electrodes which are arranged in a 10 by 10 grid. The four electrodes in the corners of the square-shaped array are used as reference electrodes, so that 96 electrodes can be used for recording spiking activity. The length of each electrode can vary between 1,000 and 1,500 μm. Usually, all electrodes are of the same length, but slanted designs are also possible in which the electrode length decreases from one side of the array to the other. At the base the electrodes have a thickness of 80 μm and taper to a sharp point at the tip. The original Utah array was developed in the late 1980s at the University of Utah [4]. The UEA can be used to record intracortical voltage changes caused by single spiking neurons. It also can be used to apply electrical stimulation to neural tissue. Different coating types are available if the array is primarily used for stimulation (iridium oxide tips) or recording (platinum tips). The UEA was originally designed for retinal or cortical stimulation, for example to be used in a visual neuroprosthesis. The array has been used in many primate and rodent research experiments. Recently it has also been implanted in a few dozen severely paralyzed humans. The impedance of each electrode in saline is in the range of 10–20 kΩ [5]. Figure 51.1 shows a UEA at different magnifications. A percutaneous connector is usually attached at the end of a wire bundle coming from the array (Fig. 51.1 right). The percutaneous connector is anchored to the skull by eight titanium screws and connects to a preamplifier if used for recording. The preamplifier then connects to a neural signal processor which filters and digitizes the signals. The digitized signals can then be further processed by a decoder.

51.3 Implantation

At the time of this writing the UEA is the only electrode array that has been implanted in humans. There are dozens of laboratories around the world that use UEAs for BCI purposes in monkeys [6, 7] or other small animals [8] and several clinical studies with humans [9–12]. Array implantations in humans to be used in a BCI have to be carefully planned and several key points have to be addressed:

(a) **Localizing the right position for implantation**. Depending on the purpose of the BCI several brain regions could be the target for an UEA. In the most common case arrays are implanted in the primary motor cortex (M1), dorsal premotor cortex (PMd) or more rarely posterior parietal cortex (PPC). Anatomical landmarks, like the hand knob, can be used for placement but as an alternative a preoperative fMRI scan can be performed. In a scanning session the patient needs to imagine different movements. The recorded activity in the targeted areas can pinpoint the best array placement for the planned BCI tasks.

(b) **Placement of the percutaneous connector**. Currently available UEAs use a percutaneous connector to connect the array to an amplifier. Computer hardware

Fig. 51.1 The Utah electrode array. Size comparison (top left). Close up of the array showing a full row of 10 electrodes (bottom left). Utah electrode assembly showing the array on the left connected to the percutaneous connector on the right (right)

for spike sorting and decoding remain outside of the body. A good placement of the bone anchored connector is very important for patient comfort and a secure connection. If multiple arrays are implanted connector location and orientation are important. Enough space between the connectors is necessary and also a medial positioning should be preferred so that the study participant is not bothered when sleeping.

(c) **Array insertion**. The UEA is usually inserted using a pneumatic insertion tool provided by Blackrock Neurotech. The force should be applied perpendicular to the cortex surface. Blood vessels need to be avoided even if the optimal placement cannot be achieved then anymore. It should also be avoided to place the array too close to sulci to avoid that some of the electrodes do not enter the cortex at all. A firm, flat placement needs to be verified before the dura is closed.

(d) **Backup array**. The electrodes of the UEA are very fragile and can easily be destroyed even by slight touch. Handling them very carefully is therefore paramount. The cables connecting them with the percutaneous connector can be slightly bent to stay in the right position. Also, care has to be taken for optimal cable management, especially if multiple arrays are implanted or if some of the arrays are used for stimulation. Even with all precautions it is recommendable to have a backup array ready in the operating room.

After implantation and recovery, which usually takes a few weeks, the arrays can be connected to a computer using the percutaneous connector. Since scar tissue formation and encapsulation of the electrodes are still unsolved problems neural recordings are best right after implantation and slowly decline over time. If possible, data quality can be checked already in the operating room. If that is not possible

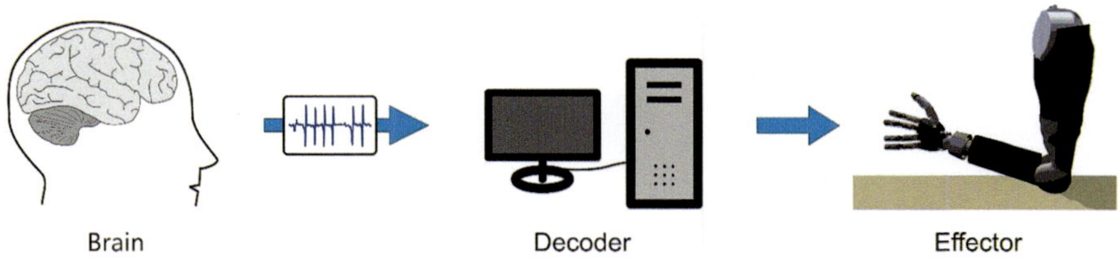

Fig. 51.2 Basic layout of a BCI. Neural data is recorded from the brain (left) and passed on to a decoder software. The decoder learns correlations between brain signals and intended actions. The decoder then sends control commands to an effector. This can for example be a cursor on a screen, a robotic arm or an exoskeleton

functioning of the implanted arrays should be verified as soon as possible after surgery (Fig. 51.2).

51.4 Data Acquisition

UEAs can record extracellular activity ('spiking') of nearby neurons. UEAs are typically using a sampling rate of 30 kHz which allows for a detailed waveform analysis and classification. The waveform shape varies depending on the geometric position of an electrode in regard to a nearby neuron. It is therefore possible to discriminate activity from multiple nearby neurons. Spike sorting is the process of identifying and classifying action potentials of neurons based on this waveform information. The sampled spike candidates are manually, semi-automatically or fully automatically sorted and assigned to either belong to a neuron or being noise. The spike sorting process is complex and often involves human experts [13] (see also Chap. 43). There have been some attempts to automatize spike sorting using statistical methods [14] or for example using deep learning [15, 16]. For decoding purposes, it is important to make sure that spiking activity is attributed to a specific neuron which is not always possible. Combining spikes from multiple neurons (multi-unit activity; MUA) should be avoided since neurons might have very different selectivities, i.e. one neuron could be selective for movements to the right while another could be selective for a movement to the left. The resulting mixed selectivity might compromise decoding.

Once a UEA has been implanted its position cannot be changed anymore. Still, it can happen that the array involuntarily shifts. Long term stability of the array is important especially if the array is used in a neuroprosthetic application. Some algorithms can track changing waveforms of neurons that shifted, but this is not a trivial task and not always possible. Another problem is scar tissue formation after array insertion and encapsulation of the electrodes as result of immune response. There are no optimal solutions yet for these problems, but spike classification and sorting need to take these long-term effects into account. After the spike sorting process information is passed to the decoder which works in a loop to analyze incoming spiking activity.

51.5 Decoder Loop

A BCI is a complex system that does not only consist of the recording hardware but also includes a decoder and an effector. The decoder is a software that "decodes" neural data, aligning actions with neural signals. It is a core component of a BCI system and a lot of development is focused on improving algorithms to achieve higher accuracy and faster processing. In many neuroprosthetic scenarios the decoder is trained on the movement intentions of the user. These intentions can be read-out for example from the primary motor cortex [17], premotor cortex or the parietal cortex. Machine learning is one of the main techniques to build a decoder. Usually two modes of operation exist: training and inference. In the training mode the decoder learns which neural signals correspond to which action. This is a supervised machine learning task since the correct responses are labeled. Due to the shifting of arrays, explained in the previous section, re-training a decoder can be necessary. In the inference mode, which is the normal operating mode of a decoder, novel neural signals are passed in and actions have to be inferred based on the previous training. One of the earliest algorithms for decoding is the so-called population vector method [18] in which the average selectiveness of a population of neurons is used to predict a variable like movement direction. Later Kalman filters [19] and support vector machines [20] became more prominent. The decoding problem is well suited for deep learning algorithms and newer methods use them as well [21]. The decoded neural signals are then used to create a control command for the effector. The effector can for example be a cursor on a computer screen, a robotic arm or an exoskeleton. The control signals can also be passed into an assistant system that uses additional information to plan the trajectory of the effector. An explanation for how a simple decoder could work is shown in Fig. 51.3.

51.6 Somesthetic Feedback

The outlined decoder loop is widely used and established. One problem with such a setup is that the BCI user does not get any feedback from the effector. It is known that effective feedback is very important to perform any action [22–24]. Without it the BCI user has to rely on visual observation alone which can be exhausting and requires constant attention. One solution to this problem is to sensorize the effector. The sensor information from the effector is translated using an encoder which is the counterpart to a decoder. It translates external sensor data into signals that the brain can understand. The electrical impulses are then applied directly to the cortex via electrodes using intracortical microstimulation [25–28]. The UEA can be used as a

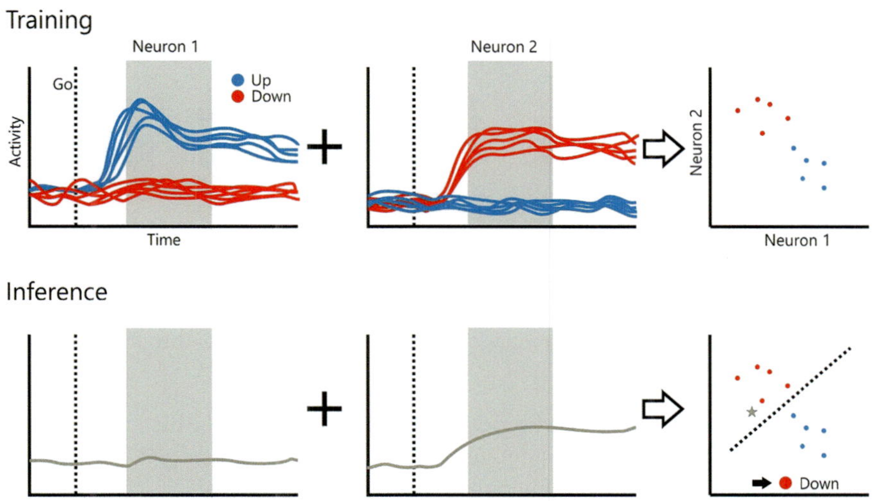

Fig. 51.3 Illustration of a simple decoder. The top row shows the training mode in which the decoder learns to associate particular brain states to intended actions. In this case the neural signal consists of the mean firing rates of two neurons (neuron 1 and neuron 2). Neuron 1 is selective for imagined hand movements up and neuron 2 is selective for imagined hand movements down. Since firing rates are variable, multiple sessions (5 in this example) in which the BCI user has to think of up or down movements are recorded. Mean activity is measured in the greyed-out time window. Single brain states in neural space which are labeled (red or blue) according to the direction imagined can be constructed (right). In inference mode (bottom) the decoder is confronted with an unlabeled brain state, represented by activity of the same two neurons and projected in neural space (bottom right). Since the unlabeled brain state is located above the decision boundary that has been learned in the previous training episode the decoder infers that the intended action is a movement down

stimulating array as well in which case it is usually coated with iridium oxide. Target regions are typically the primary sensory cortex (S1 or S2). Ideally the stimulated regions of the brain correspond to the areas that are typically receiving input from corresponding parts of the actual body. So, for example if a hand prosthesis is the effector, then the hand region within the primary sensory cortex can be the target of stimulation. The exact positioning for a stimulation array is not as easy to define as for a recording array. If a patient does not receive any somesthetic feedback then it is not possible to use a fMRI task to determine the region based on activity since it cannot be stimulated by touching the corresponding body part. Therefore, anatomical landmarks of the brain surface, like the "hand knob", need to be taken into account. Another possibility would be to touch adjacent regions of the body that the patient can still feel to determine the likely optimal placement for the stimulation array. A third alternative is to use non-invasive stimulation like transcranial magnetic stimulation (TMS) to directly stimulate the primary sensory areas and have the patient report what he or she feels to determine the array placement.

Both, recording and stimulating UEAs can be used in parallel to form a bidirectional BCI (BBCI). In a BBCI scenario care has to be taken that the stimulating pulses

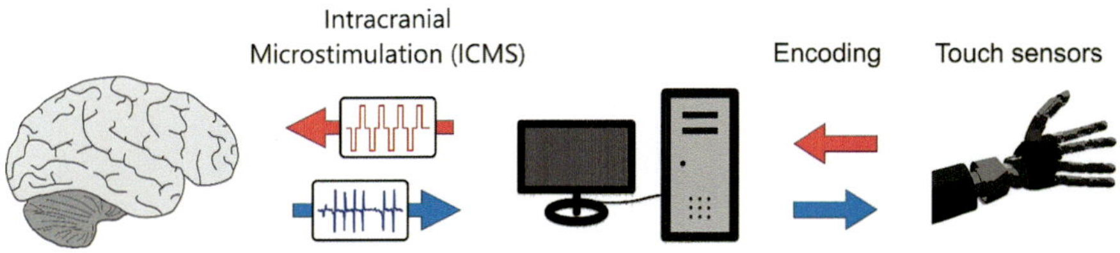

Fig. 51.4 Schematic overview of a bidirectional BCI. The information flow of the regular decoding mode is shown in blue (compare to Fig. 51.2). The new feedback flow of information is shown in red. Touch sensors register a touch on a robotic hand. This information is encoded into a signal that can be interpreted by the brain. A microstimulator applies intracortical microstimulation (ICMS) through an UEA or other type of stimulating electrode to corresponding brain areas

do not interfere with the recorded neural signals. Depending on the distance between the arrays and the stimulation amplitude this can easily happen and interference needs to be either prevented or filtered out form the neural signal (Fig. 51.4).

51.7 Outlook

Current developments in BCI research are likely to result in commercially available neuroprostheses in the near future. Several companies are trying to commercialize fully implantable systems for severely paralyzed individuals. Some of them use UEAs and some other electrode designs. Regardless of the system, it is conceivable that multiple diseases could be treated using BCI systems. Paralysis, certain types of memory loss or sensory deficiencies are among them (see also Chaps. 47, 52 and 53). One bottleneck for future BCIs is the bandwidth, i.e. the number of electrodes that can be recorded through and those that can be used for stimulation. A higher number of electrodes means that a much more fine-grained interface can be realized that is more precise and 'feels' more natural. Another issue that is not entirely solved is biocompatibility. Smaller electrodes tend to generate less immune response and scar tissue formation. Long term encapsulation and neuron death is a problem for UEAs and other electrode types which leads to degrading signals over time. These problems need to be addressed for implants that are meant to work for a lifetime. Nevertheless, current systems are a viable option for patients that otherwise have no option of treatment. The electronic components of a fully implantable BCI need to be powered which poses additional challenges for low power microchips and high-density batteries. In conclusion, commercially available BCI systems for severely impaired individuals are on the horizon. The challenges currently faced are not of fundamentally neuroscientific but engineering problems that need to be solved.

References

1. Kellis S, Miller K, Thomson K, Brown R, House P, Greger B (2010) Decoding spoken words using local field potentials recorded from the cortical surface. J Neural Eng 7(5):056007
2. Leuthardt EC, Schalk G, Wolpaw JR, Ojemann JG, Moran DW (2004) A brain–computer interface using electrocorticographic signals in humans. J Neural Eng 1(2):63
3. Schalk G, Kubánek J, Miller KJ, Anderson NR, Leuthardt EC, Ojemann JG et al (2007) Decoding two-dimensional movement trajectories using electrocorticographic signals in humans. J Neural Eng 4(3):264
4. Campbell PK, Jones KE, Huber RJ, Horch KW, Normann RA (1991) A silicon-based, three-dimensional neural interface: manufacturing processes for an intracortical electrode array. IEEE Trans Biomed Eng 38(8):758–768
5. Jones KE, Campbell PK, Normann RA (1992) A glass/silicon composite intracortical electrode array. Ann Biomed Eng 20(4):423–437
6. Wessberg J, Stambaugh CR, Kralik JD, Beck PD, Laubach M, Chapin JK et al (2000) Real-time prediction of hand trajectory by ensembles of cortical neurons in primates. Nature 408(6810):361–365
7. Serruya MD, Hatsopoulos NG, Paninski L, Fellows MR, Donoghue JP (2002) Brain-machine interface: instant neural control of a movement signal. Nature 416(6877):141–142
8. Chapin JK, Moxon KA, Markowitz RS, Nicolelis MA (1999) Real-time control of a robot arm using simultaneously recorded neurons in the motor cortex. Nat Neurosci 2(7):664–670
9. Kennedy PR, Bakay RAE, Moore MM, Adams K, Goldwaithe J (2000) Direct control of a computer from the human central nervous system. IEEE Trans Rehabil Eng 8(2):198–202
10. Collinger JL, Wodlinger B, Downey JE, Wang W, Tyler-Kabara EC, Weber DJ et al (2013) High-performance neuroprosthetic control by an individual with tetraplegia. The Lancet 381(9866):557–564
11. Aflalo T, Kellis S, Klaes C, Lee B, Shi Y, Pejsa K et al (2015) Decoding motor imagery from the posterior parietal cortex of a tetraplegic human. Science 348(6237):906–910
12. Hochberg LR, Serruya MD, Friehs GM, Mukand JA, Saleh M, Caplan AH et al (2006) Neuronal ensemble control of prosthetic devices by a human with tetraplegia. Nature 442(7099):164–171
13. Lewicki MS (1998) A review of methods for spike sorting: the detection and classification of neural action potentials. Netw Comput Neural Syst 9(4):R53–R78
14. Rey HG, Pedreira C, Quian QR (2015) Past, present and future of spike sorting techniques. Brain Res Bull 1(119):106–117
15. Saif-ur-Rehman M, Lienkämper R, Parpaley Y, Wellmer J, Liu C, Lee B et al (2019) SpikeDeep-tector: a deep-learning based method for detection of neural spiking activity. J Neural Eng 16(5):056003
16. Saif-ur-Rehman M, Ali O, Dyck S, Lienkämper R, Metzler M, Parpaley Y et al (2020) SpikeDeep-classifier: a deep-learning based fully automatic offline spike sorting algorithm. J Neural Eng [Internet] [cited 2022 Feb 23]. https://doi.org/10.1088/1741-2552/abc8d4
17. Kalaska JF (2009) From intention to action: motor cortex and the control of reaching movements. In: Sternad D (ed) Progress in motor control: a multidisciplinary perspective [Internet]. Springer, Boston, MA, pp 139–78 [cited 2022 Feb 23] (Advances in experimental medicine and biology). https://doi.org/10.1007/978-0-387-77064-2_8
18. Georgopoulos A, Schwartz A, Kettner R (1986) Neuronal population coding of movement direction. Science 233(4771):1416–1419
19. Wu W, Black MJ, Mumford D, Gao Y, Bienenstock E, Donoghue JP (2004) Modeling and decoding motor cortical activity using a switching Kalman filter. IEEE Trans Biomed Eng 51(6):933–942
20. Olson BP, Si J, Hu J, He J (2005) Closed-loop cortical control of direction using support vector machines. IEEE Trans Neural Syst Rehabil Eng 13(1):72–80
21. Livezey JA, Glaser JI (2021) Deep learning approaches for neural decoding across architectures and recording modalities. Brief Bioinform 22(2):1577–1591

22. Rothwell JC, Traub MM, Day BL, Obeso JA, Thomas PK, Marsden CD (1982) Manual motor performance in a deafferented man. Brain 105(3):515–542
23. Ghez C, Gordon J, Ghilardi MF, Christakos CN, Cooper SE (1990) Roles of proprioceptive input in the programming of arm trajectories. Cold Spring Harb Symp Quant Biol 1(55):837–847
24. Sainburg RL, Ghilardi MF, Poizner H, Ghez C (1995) Control of limb dynamics in normal subjects and patients without proprioception. J Neurophysiol 73(2):820–835
25. O'Doherty JE, Lebedev MA, Li Z, Nicolelis MAL (2012) Virtual active touch using randomly patterned intracortical microstimulation. IEEE Trans Neural Syst Rehabil Eng 20(1):85–93
26. Berg JA, Dammann JF, Tenore FV, Tabot GA, Boback JL, Manfredi LR et al (2013) Behavioral demonstration of a somatosensory neuroprosthesis. IEEE Trans Neural Syst Rehabil Eng 21(3):500–507
27. Klaes C, Shi Y, Kellis S, Minxha J, Revechkis B, Andersen RA (2014) A cognitive neuroprosthetic that uses cortical stimulation for somatosensory feedback. J Neural Eng 11(5):056024
28. Flesher SN, Collinger JL, Foldes ST, Weiss JM, Downey JE, Tyler-Kabara EC et al (2016) Intracortical microstimulation of human somatosensory cortex. Sci Transl Med 8(361):361ra141–361ra141

Chapter 52
Can Chronically Implanted iEEG Sense and Stimulation Devices Accelerate the Discovery of Neural Biomarkers?

Kristin K. Sellers and Edward F. Chang

Invasive electrophysiology in humans has historically been possible under very constrained conditions in a small number of individuals. People with epilepsy may undergo intracranial monitoring, during which iEEG electrodes are temporarily implanted to localize seizure foci. The intraoperative environment also provides an opportunity for acute recordings of human neurophysiology. The advent of chronically implanted neurostimulation devices capable of sense and stimulation ("bidirectional") has provided a window into human neurophysiology not previously attainable. Such devices are most commonly used to treat patients with epilepsy, Parkinson's Disease, and essential tremor, with many other indications under active study. However, the neural recordings during daily life afforded by these devices allows for the discovery of biomarkers, or neural activity correlates, of physiological and pathological functions. Furthermore, the longitudinal nature of these recordings can provide new insight into the stability or lack thereof of these biomarkers. Here, we describe the devices available for such biomarker discovery, how biomarkers recorded using these devices can be used for research of physiological and disease states, provide example experimental workflows, and finally discuss the technological and practical limitations of currently available devices, as well as our projection of what research will be enabled by the next generation of devices.

K. K. Sellers
Department of Neurological Surgery, University of California, San Francisco, CA, USA

E. F. Chang (✉)
Weill Institute for Neurosciences, University of California, San Francisco, CA, USA
e-mail: Edward.Chang@ucsf.edu

N. Axmacher (ed.), *Intracranial EEG*, Studies in Neuroscience, Psychology and Behavioral Economics, https://doi.org/10.1007/978-3-031-20910-9_52

52.1 Introduction

Our understanding of human neurophysiology increases as a function of the availability of technology to record neural signals. Invasive electrophysiology of local field potential (LFP), single-unit (SU), or multi-unit (MU) spiking activity is commonly conducted in animal models. However, the ethical acquisition of such signals from humans is much more limited due to the invasive nature of placing or implanting a recording device, and the high burden to demonstrate safety and clinical efficacy. Opportunities for such recordings have historically been limited to a small population of individuals undergoing neurosurgical procedures for clinical need, such as surgical resection of tumors or localization of epileptic foci. But the development of new device technology, and the validation of its efficacy in treating disease, yet again opens the door for more comprehensive neurophysiological studies in humans.

Intracranial electroencephalography (iEEG) provides a temporally precise readout of population electrical activity at the mesoscale. This modality includes electrocorticography (ECoG) recorded using subdural grid and strip electrodes and stereotactic EEG (sEEG) recorded using penetrating depth electrodes. These techniques have seen increasing adoption over the last 3 decades [1]. Numerous scientific breakthroughs in cognitive neuroscience have been enabled by these invasive recordings in humans; see [2–4] for reviews. In particular, studies in language, emotion, and volition greatly benefit from the ability to record iEEG in humans, as animal models are poorly suited to address these topics. However, these studies share constraints imposed by traditional acquisition of iEEG in humans—restricted timeframe during an inpatient hospitalization, experiments conducted shortly following a neurosurgical procedure, and limited mobility and range of behaviors imposed by participants being tethered to external racks of recording equipment [2].

A seemingly unrelated advance in clinical treatment and resulting technology provided an answer to multiple of these challenges. We have long sought methods by which to modulate brain activity in order to test or restore physiological function or alleviate disease. One such approach has been the application of electrical stimulation, dating back to ancient times and gradually being tested with increasing scientific rigor [5–9]. The mechanisms of electrical stimulation are still under active study, with evidence for suppression of aberrant signaling (a functional ablation), targeted activation, disruption of input and output signals, and a host of other possibilities [10–12]. During its modern-day revival, electrical stimulation was primarily investigated for its ability to reduce pain [13, 14] and control tremors in movement disorders. However, beneficial effects did not persist upon termination of stimulation [15]. Therefore, stimulation devices were developed which could be chronically implanted. The first generations of these devices delivered stimulation continuously. Treatment from such devices was revolutionary for movement disorders, as reversible and programmable stimulation was nearly as effective at reducing symptoms as surgical lesioning of thalamic areas [16, 17], which was permanent and could lead to irreversible side effects.

However, some patients experienced side-effects from this constant stimulation, including but not limited to worsened speech and gait, paresthesias, muscle contractions, visual flashes, worsening akinesia, and cognitive or psychiatric deterioration [18–23]. Therefore, a new generation of devices was developed and eventually deployed with the added capability of recording neural activity, conducting computations on that activity in real-time, and delivering stimulation based on pre-programmed criteria. We call these devices "bidirectional", as they are capable of both sensing neural activity (input) and delivering stimulation (output). This technology has been highly beneficial for reduction of seizures in individuals with medically refractory epilepsy [24, 25]. While the sense capability of bidirectional devices enables a new paradigm of closed-loop or responsive stimulation, as a fortuitous byproduct, we also now have both the technology and ethical use-cases to chronically record iEEG from humans during their daily lives.

In the remainder of this chapter, we focus on the research applications afforded by chronically-implanted bidirectional sense- and stimulation-enabled neurostimulation devices. While these implants are typically restricted to patient populations, concurrent research can be conducted on either healthy, physiological functioning or on changes to circuits related to disease. In particular, we focus on the ability to determine neural correlates, or biomarkers, for physiological and disease states and symptoms.

52.2 What Bidirectional, Chronically Implanted iEEG Devices Are Available?

There are only a handful of chronically implanted devices currently available with the capability to sense iEEG neural activity and deliver stimulation. While stimulation is not needed for studies strictly interested in biomarker discovery and validation, it is typically the motivating factor for implanting these devices; stimulation provides the therapy which justifies the neurosurgery. There are very specific cases of neural device implants for sense only, typically in the context of brain-computer interface (BCI) applications, such as for cursor or prosthetic limb control [26–28] to restore some function for individuals who have experienced spinal cord injury, stroke, are paralyzed, or have select neurodegenerative conditions. While many of these devices use single units or LFP signals recorded using Utah arrays [29, 30] (see Chap. 51), use of iEEG has also proven successful [31–34]. A notable difference with these devices is the percutaneous connector, to which a cable or wireless transmitter [35] is physically connected during use. A limited number of devices do not require a percutaneous connector, but have an external inductively coupled connector attached to the head using paste [36]. While these devices can record data over long periods of time, the range of behaviors and states which can be captured is inherently limited by the nature of this tether and the typical support needed for connection to enable data acquisition.

The NeuroPace RNS® System (NeuroPace, Mountain View, California) is an FDA-approved device for the personalized treatment of epilepsy [37]. Tailor-made to detect neural activity patterns which are indicative of seizures, the cranially contained RNS system is comprised of two 4-contact leads connected to an implantable neurostimulator (INS). The system can store modest periods of iEEG data from up to 4 channels simultaneously, which are selected as bipolar pairs from the electrode contacts. On-board filters are default at 5–70 Hz, while special settings allow for filters at 1–90 Hz. The INS contains memory for up to ~32 channel-minutes of data sampled at 250 Hz, so the user must regularly transfer data from the INS onto a computer, which uploads to the cloud. Recordings can be scheduled based on time of day, automatically triggered by a programmed neural activity detector, or manually triggered by the patient using an external magnet.

Medtronic's bidirectional neurostimulation devices consist of leads implanted in the brain and a non-cranially contained INS, which is typically implanted in the chest and connected to the leads via cables tunneled through the neck. The Activa PC+S (Medtronic, Minneapolis, Minnesota) model is no longer sold and the Summit RC+S (Medtronic, Minneapolis, Minnesota) [38] system was only available to investigators under an FDA Investigational Device Exemption (IDE) through 2022. Their commercially available Percept PC (Medtronic, Minneapolis, Minnesota) device includes two leads, each with 4 contacts. The device has a sampling rate of 250 Hz and imposes two low-pass filters at 100 Hz, one high-pass filter at 1 Hz, and a second high-pass filter that is programmable to 1 or 10 Hz. The patient can trigger 30 s snapshots of LFP recording at home, but only the on-device calculated power from this snapshot is saved [39]. In clinic, iEEG from one bipolar channel from each lead can be recorded; which contacts can be selected for bipolar recording is limited based on stimulation configuration. The overall duration of live streaming using clinician peripherals is unlimited, although in practice a new recording must be started every 10 min to prevent data export errors [40]. The device has a primary cell battery and 1 h of streaming shortens battery life by approximately one day. Research studies using these devices are discussed below in the section "How can we discover biomarkers using bidirectional iEEG devices?".

There are other device systems in development, including the Smart Neurostimulation System (SNS) by Nia Therapeutics (Philadelphia, PA) for the treatment of memory loss caused by traumatic brain injury. The SNS is planned to have four 16-channel depth electrodes, a cranial INS, and an external wearable earpiece to provide power and therapy control. CorTec (Freiburg, Germany) is currently validating its device system, which contains two electrode arrays, an implanted electronics unit, and an external unit which provides power and communication through inductive coupling. Current technical specifications indicate 32 recording channels, 1 kHz sampling rate, and on-board filters between 0.1 and 450 Hz. Additional devices have been developed and tested in animal models [41, 42].

52.3 What Can Chronically Implanted iEEG Devices Provide That (Sub)acute iEEG Cannot?

Most invasive iEEG in humans occurs during acute intraoperative placement of recording leads or subacute implant (<28 days) of electrode arrays with externalized tails. Externalized array tails are connected to cables and amplifiers, digitizers, and acquisition computers (see Chaps. 4 and 5 for the practical implications of this setup for researchers). Multiple brain regions can be implanted with leads which each contain multiple contacts, and signals can be simultaneously recorded from all these contacts. However, the externalized electrode leads, heavy and bulky connectorization, and external acquisition equipment necessitate that implanted individuals stay in specialized units in the hospital, typically restricted to being relatively immobile in bed during recording. Recording under these conditions has proved to be valuable for the localization of epileptic foci, although also note that some seizures are associated with movement [43, 44], and thus limited mobility can be problematic. However, one can readily see how the range of naturalistic behaviors is greatly limited by the inpatient environment, being tethered to equipment, and proximity to major neurosurgery. Furthermore, externalized tails may prevent concurrent recording with EEG, eye trackers, or other non-invasive metrics (some of these topics further discussed in Chaps. 12–15).

There is overwhelming evidence that cognitive or behavioral state is a critical context in which to understand neural correlates [45]. The same stimulus or task can elicit different neural responses dependent upon state. Therefore, many studies seek to hold state (e.g. arousal, attention, movement) constant across experiments and participants in order to map neural correlates for a given state. Or studies seek to purposely measure the same behavior across a sub-selection of defined states. Indeed, it is the variation of neural activity along one axis of state—symptom or specific behavior state—that can be used to uncover neural biomarkers; this will be discussed more in the next section. The historical use of iEEG in the operating room or inpatient unit has both benefited and suffered in this regard—the contexts in which experiments can be conducted has been relatively confined, and there is generally insufficient time to study a multitude of states. With the availability of iEEG from chronically implanted devices, studies can be conducted across a broader range of states to better understand how this context affects neural correlates. One study testing electrical stimulation during concurrent iEEG demonstrated differential effects of stimulation on modulating depression symptoms based on state [46], while another demonstrated differential effects of stimulation based on memory encoding state [47] (see Chap. 41). In addition, studies during ambulation are now feasible—see Chap. 53 for more on mobile iEEG.

(Sub)acute iEEG in the context of cognitive neuroscience research also suffers from the limited timeframe of recording. Electrodes are typically explanted after 7–10 days. Longer implant time with externalized lead tails is associated with increased infection risk [48] and degraded signal quality from repeated mechanical strain on the cables and connections. This severely limits the ability to conduct longitudinal recordings. With so many cognitive neuroscience studies conducted at just one time-point, is there even a need for access to longitudinal recordings? We argue yes for several reasons. In particular, the characteristic time scale of the cognitive process of interest should be considered [49]. If the behavior of interest has a fast characteristic time scale (e.g. perception of a visual stimulus), it is relatively easy to present the stimulus multiple times in order to determine the neural correlates of perception. High trial count facilitates a good signal-to-noise ratio (SNR) of neural activity underlying the behavior of interest relative to other neural processing (e.g. thinking about what's for lunch, ignoring an itchy arm). However, when studying something with a much slower characteristic time scale or something which must be generated internally rather than through experimental presentation (e.g. internally oriented thoughts), it is much harder to disentangle neural activity related to the metric of interest vs all other neural processing. During a typical inpatient monitoring visit, there may be opportunity for only one or two such samples of the behavior of interest. While data can be pooled across participants, this heterogeneity again complicates determining neural correlates of the behavior. Access to recordings over weeks and months facilitates increasing SNR through repeated measurements of the behavior of interest.

Regardless of the characteristic time scale of the biological process of interest, there are other factors which make longitudinal recordings important. There are many processes in the human brain that are subject to periodic biological clocks. We are perhaps most familiar with the circadian rhythm—physical, mental, and behavioral changes that follow a 24 h cycle. Substantial literature demonstrates that circadian rhythm profoundly affects cognitive functions [50–52]. Some studies on behavior as a function of circadian rhythm may be possible during in-patient hospitalizations, but the timeframe and environment are often too restrictive to provide sufficient data. There also appear to be cycles of longer timescales that affect human brain functioning. Studies using iEEG devices have revealed a multidien cycling component to seizures in men and women [53, 54]. The female menstrual cycle is another often overlooked rhythm with corresponding changes in hormones and neurotransmitter function [55, 56]. Historically, behavioral research with mouse models has primarily been conducted in male mice to avoid the contribution of these cycling hormones [57]. While limited work using EEG and other noninvasive measures have demonstrated changes in oscillations and functional connectivity across the monthly cycle [58–61], little to no work using higher spatial resolution invasive electrophysiology has looked at cognitive neuroscience behaviors as a function of menstrual cycle.

Lastly, longitudinal recordings allow insight into the reorganization of networks as a function of elapsed time, an intervention, or other processes studied in the same person. Of particular interest is remodeling of networks over the course of disease treatment or progression. Use of scalp EEG has shown that tracking of neural activity from pre- to post- cognitive behavioral therapy can be used to predict efficacy of the therapy in reducing symptoms for social anxiety disorder [62]. Similar use of EEG has also been used to investigate how neural networks underlying emotion regulation change during trauma treatment [63]. Use of iEEG in such studies would provide greater spatial resolution and access to deeper brain structures but is largely not possible with subacute implants. Furthermore, iEEG provides SNR up to 100× higher compared to EEG [64], so detection of more subtle biomarkers may be possible with invasive chronic recordings.

52.4 How Can We Discover Biomarkers Using Bidirectional iEEG Devices?

Biomarkers are quite literally "biological markers", or something measurable that provides an objective indication of the state inside the body. The Biomarkers Definitions Working Group, organized by the National Institutes of Health, defines a biomarker as "a characteristic that is objectively measured and evaluated as an indicator of normal biological processes, pathogenic processes, or pharmacological responses to a therapeutic intervention" [65]. Biomarkers characteristically belong to four categories: molecular, histologic, radiographic, and physiologic. Molecular biomarkers have biophysical properties, and this category includes measurements of biological samples such as blood and cerebrospinal fluid. Radiographic biomarkers are obtained from imaging studies, such as x-rays and MRI scans. Histologic biomarkers assess biochemical or molecular alterations in cells, as may be tested in suspected cancer tissues. Lastly, physiologic biomarkers measure processes of the body. We will focus on this fourth category of biomarkers. Oftentimes, biomarkers are used in the context of screening for disease, assessing susceptibility or risk, aiding in diagnosis, indicating disease prognosis, monitoring progression or response to an intervention, and assessing safety [66]. We adopt a broader conceptualization of the utility of biomarkers. Specifically, neural activity (objectively) measured during behaviors can serve as biomarkers indicative of internal (neural) biological processes. As neural activity patterns are present on a spectrum, and are rarely completely present or absent, these can also be referred to as neural activity correlates of behavior.

The idea of neural activity biomarkers is not new. Many studies of neurophysiology in the context of behavior seek to uncover neural activity biomarkers of a particular behavior or process. Subacute iEEG has even previously been used to determine biomarkers. In a study of 13 patients undergoing iEEG epilepsy monitoring, a neural biomarker correctly classified 78% of patients as belonging to the group with a high burden of depressive symptoms or a control group with minimal depressive symptoms [67]. Power spectrum slope from iEEG can serve as a biomarker for arousal level, differentiating wakefulness from propofol anesthesia, rapid-eye-movement (REM) sleep, and non-REM sleep [68]. Individuals with epilepsy are implanted such that iEEG activity can be used to localize tissue with specific interictal and ictal biomarkers to inform resection surgery of epileptogenic areas [69]. Unique biomarkers can likely identify related but distinct processes, such as acute pain, chronic pain, and pain relief [70]. Carefully disentangling these constructs will provide insight into both the mechanisms controlling these processes, as well as provide objective metrics for monitoring symptom status and even potentially therapeutic efficacy.

A limited number of studies have started employing chronically implanted bidirectional iEEG devices to investigate neural correlates [71]. In individuals with Parkinson's disease, motor cortex gamma activity was found to be a biomarker for dyskinesia [72] and prefrontal neural correlates were found for anxiety and depression symptoms [73]. Neurophysiological correlates of speech perception and production were found to be stable over 1.5 years [74], providing valuable information about the stability of some neural biomarkers. A biomarker identified during subacute iEEG recordings was confirmed following chronic implant, and used to successfully control closed-loop neurostimulation therapy for a person with severe depression [75]. And in a slightly different application, the Medtronic PC+S device was implanted in a person with amyotrophic lateral sclerosis, and activity recorded over the motor cortex was decoded to reliably control a typing program [76].

As gleaned from these studies, a general experimental paradigm can be followed for the discovery of biomarkers using bidirectional iEEG devices (Fig. 52.1). The integral data required include concurrent or temporally proximate measures of neural activity and the behavior or symptom of interest. For example, this might be the participant completing a survey on mood at the same time as triggering a neural recording. This should be done over the course of weeks or months to build a sufficient dataset. Critically, there must be variation in the response of interest (e.g. severity of symptom, performance on a task). In the most simplistic model, we are interested in a regression between features of the neural activity patterns (neural correlates) and the metric of interest. For example, if the participant only does recordings when in a poor mood, there is no variation by which to enable modeling of how neural activity is related to fluctuations in mood. It is oftentimes helpful for the participant to complete ratings or surveys assessing other variables in addition to the metric of interest. This can provide the researcher information about potential covariates that may be influencing relationships seemingly detected between the metric of interest and neural activity. For example, level of sleepiness (and resulting strongly influenced neural activity) may need to be accounted for during analysis.

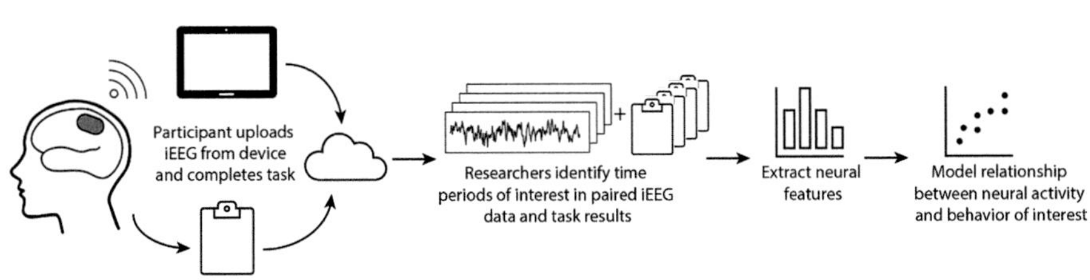

Fig. 52.1 General biomarker discovery pipeline: iEEG from the chronically implanted device is recorded concurrently with a behavior or measure of interest. The dataset of these paired samples requires variability in the behavior or measure of interest. Researchers identify time periods of interest based on the dynamics of the behavior or measure of interest, extract relevant neural features, and model the relationship between the neural activity feature(s) and the behavior

Next, we are faced with determining the relevant neural activity patterns, or neural features, of interest. Historically, power in canonical frequency bands has been widely studied [77]. Coherence, or the frequency-specific amplitude and/or phase correlated activity between regions, may also be of interest [78–80]. There may be more sophisticated or specific neural activity correlates of interest, depending on the behavior or symptom. For example, features of evoked potentials measured using EEG have been identified as biomarkers of cognitive decline in patients with Alzheimer's Disease and mild cognitive impairment [81, 82]. Calculation of neural features may be hypothesis driven, especially if there is prior literature from animal models or non-invasive human neurophysiology, or more exploratory. Either way, one must be careful to investigate alternative behaviors that may be co-occurring with the metric of interest.

Another important consideration is the timeframe over which to assess a relationship between neural activity and a metric of interest. As discussed above, the characteristic time scale of the biological process of interest should be considered. It likely does not make sense to look at five minutes of neural activity when looking for a biomarker of auditory perception. However, five minutes or even longer may be appropriate for neural activity snapshots related to emotion or other cognitive processes. When the characteristic time scale of the biological process is unknown, it can be helpful to look at multiple time periods to assess the most appropriate duration of neural activity for investigation [75, 78].

Lastly, a mathematical model must be used to relate the neural features with the metric of interest. These models can be linear or non-linear; the non-linear relationship between brain activity and behaviors can be represented using linear models by incorporating transforms or other representations of input and/or output features. Preprocessing or data transform steps may include ICA, PCA, thresholding, or K-means clustering. For the calculation of neural activity biomarkers, supervised models are used, with the behavioral or symptom output serving as the label. Commonly used models include logistic or linear regression, linear discriminant analysis, support vector machine, or random forest. Appropriate standards for regularization, train and test sets, or k-fold cross validation should be followed to avoid overfitting. Biomarkers can be either personalized, calculated for an individual [83], or generalized across a population. In either case, biomarkers are only meaningful if they generalize across some time period.

52.5 What Are the Challenges in Using Chronically Implanted iEEG Devices?

By definition, chronically implanted iEEG devices are invasive and require surgical implantation. Leads are fixed and later adjustment of position requires subsequent surgery. Because of the invasive nature, chronically implanted iEEG devices are currently limited to patients requiring stimulation for a clinical condition. This also imposes limitations on studies across the lifespan—young children are typically not implanted with these devices, and surgery may be contraindicated in elderly populations. Thus, as with (sub)acute iEEG recordings in humans, the participant population is restricted to individuals who have a diagnosed neurological or psychiatric condition. The majority of iEEG research to date has focused on brain regions commonly implicated in epilepsy or Parkinson's disease. Specifically, cortical areas in the temporal and parietal lobes and subcortical areas such as the hippocampus, amygdala, and insula are most commonly recorded in patients with epilepsy, and subthalamic nucleus is commonly sampled in Parkinson's disease. These are the same regions which are currently targeted for chronic iEEG implant. Research driven studies for neuropsychiatric disorders are expanding the regions that are implanted, including but not limited to subgenual cingulate, nucleus accumbens, and ventral capsule. In some cases, contact spacing on leads allows access to additional brain regions for recording, but this is limited. Thus, experimenters do not have ubiquitous access to all brain regions that may be of interest in studying a particular cognitive or behavior process. As the therapeutic indications for chronic stimulation with iEEG devices increase, there will likely be additional brain regions which are regularly implanted.

Despite technological advances having provided the ability to recording chronic iEEG, current technology also results in challenges for recording chronic iEEG. One of the primary limitations of the RNS device is the limited on-board memory capability. While the device is continually sensing neural activity and computing neural features to determine if stimulation should be delivered, the vast majority of those neural signals are not saved or available for offline analysis. Data can be continuously streamed from the device with the clinician tablet, but this requires an external wand be placed within a few centimeters of the patient's INS. Furthermore, the device has a primary cell battery, which requires surgical intervention for replacement. The Medtronic RC+S device had a rechargeable battery, removing some limitations of continuous streaming. However, substantial data post-processing is needed to transform data coming off the device to human-readable formats that are commonly used for neuroscience analyses [84]. Use of the Medtronic Percept device is hindered by extremely limited at-home recording capability and restricted configuration of bipolar recording channels flanking a stimulation contact. All currently available devices suffer from having only a couple of leads with a small number of contacts; simultaneous recording is possible only from a subset of these contacts. Sampling rates are low, and devices impose on-board filters which prevent ultra-low and high frequency activity from being effectively recorded.

Further, many cognitive neuroscience studies suffer from the Hawthorne effect. Participants know they are being observed, and that either consciously or subconsciously changes their behavior. The ability to record neural activity continuously, away from the laboratory environment, may decrease this and provide greater ecological validity in studies. Of course, participants should only have their neural recordings included in cognitive neuroscience studies with full consent and knowledge [85]. Experimenters must think carefully about how to mark, determine, and extract time periods that are relevant to the behavior being studied. This may require synchronization between data collected on the chronically implanted device and external equipment which measures a stimulus or other task feature.

52.6 What Will Be Possible for Biomarker Discovery with the Next Generation of Chronically Implanted Bidirectional iEEG Devices?

As the history of neuroscience has demonstrated, technological advances beget new discoveries, and need begets new technological advances. Academia and industry are actively working on the next generation of chronically implanted bidirectional iEEG devices. Currently available bidirectional iEEG devices are quite limited in their spatial coverage. Even compared to subacute iEEG implants, chronic devices have a small number of leads with fewer contacts. The spacing of these contacts is generally large (4.5–10 mm). The next generation of devices may incorporate ultra-high-density arrays [86] in order to better localize neural activity. Furthermore, a greater number of electrode contacts on the same lead is important to mitigate the current margin of error in surgery—use of stereotaxic frames allows for an average precision of 1–2 mm from the target area [87, 88], and the brain can shift by 2–4 mm during surgery [89–91] making targeting more difficult. In the future, chronically implanted iEEG devices will contain a combination of on-board memory and automatic data transfer to external peripheral equipment that will enable virtually continuous streaming with minimal burden on the user. The richness and quantity of these data will allow for new studies on single-trial investigations of complex naturalistic behavior. Custom-designed electronics, taking advantage of application-specific integrated circuits, will provide more advanced computation on-device with a smaller INS size. These electronics will bring recording capability—in terms of number of channels, sampling rate, onboard filtering, noise floor, etc.—more in line with capabilities of external acquisition rigs. Overall, this will enable the study of more sophisticated neural correlates of behavior. Advances in battery technology will expand the lifetime of devices. Even rechargeable batteries have a limited number of charge cycles, so replacement surgery is needed for all currently available devices. Continued development of battery-free devices, powered through internal power harvesting or far field transfer [92], may one day remove the need for battery replacement surgery altogether. Chronically implanted iEEG devices will also see integration with other physiological measures or wearables. For many applications, the biomarker discovery pipeline described previously still requires the implanted individual to perform tasks or respond to questions that are outside of their standard daily routine. Continued research will shed light into how we can use everyday behaviors to track symptoms or other states to then correlate with brain activity (e.g. digital biomarkers, speech rate, digital phenotyping) [93, 94].

Taken together, these ongoing technological advances promise an exciting future for cognitive neuroscience research using chronically implanted iEEG devices.

References

1. Parvizi J, Kastner S (2018) Promises and limitations of human intracranial electroencephalography. Nat Neurosci 21:474–483. https://doi.org/10.1038/s41593-018-0108-2
2. Mukamel R, Fried I (2012) Human intracranial recordings and cognitive neuroscience. Annu Rev Psychol 63:511–537. https://doi.org/10.1146/annurev-psych-120709-145401
3. Jacobs J, Kahana MJ (2010) Direct brain recordings fuel advances in cognitive electrophysiology. Trends Cogn Sci 14:162–171. https://doi.org/10.1016/j.tics.2010.01.005
4. Engel AK, Moll CKE, Fried I, Ojemann GA (2005) Invasive recordings from the human brain: clinical insights and beyond. Nat Rev Neurosci 6:35–47. https://doi.org/10.1038/nrn1585
5. Penfield W, Boldrey E (1937) Somatic motor and sensory representation in the cerebral cortex of man as studied by electrical stimulation. Brain 60:389–443. https://doi.org/10.1093/brain/60.4.389
6. Penfield W, Rasmussen T (1950) The cerebral cortex of man: a clinical study of localization of function. Macmillan, New York
7. Bartholow R (1874) Experimental investigations into the functions of the human brain. Am J Med Sci 305–313
8. Sironi VA (2011) Origin and evolution of deep brain stimulation. Front Integr Neurosci 5:42. https://doi.org/10.3389/fnint.2011.00042
9. Gildenberg PL (2005) Evolution of neuromodulation. Stereotact Funct Neurosurg 83:71–79. https://doi.org/10.1159/000086865
10. McIntyre CC, Savasta M, Kerkerian-Le Goff L, Vitek JL (2004) Uncovering the mechanism(s) of action of deep brain stimulation: activation, inhibition, or both. Clin Neurophysiol 115:1239–1248. https://doi.org/10.1016/j.clinph.2003.12.024
11. Chiken S, Nambu A (2016) Mechanism of deep brain stimulation. Neuroscientist 22:313–322. https://doi.org/10.1177/1073858415581986
12. Jakobs M, Fomenko A, Lozano AM, Kiening KL (2019) Cellular, molecular, and clinical mechanisms of action of deep brain stimulation-a systematic review on established indications and outlook on future developments. EMBO Mol Med 11:e9575. https://doi.org/10.15252/emmm.201809575
13. Hosobuchi Y, Adams JE, Rutkin B (1973) Chronic thalamic stimulation for the control of facial anesthesia dolorosa. Arch Neurol 29:158–161. https://doi.org/10.1001/archneur.1973.00490270040005
14. Heath RG (1963) Electrical self-stimulation of the brain in man. Am J Psychiatry 120:571–577. https://doi.org/10.1176/ajp.120.6.571
15. Benabid AL, Pollak P, Louveau A, Henry S, de Rougemont J (1987) Combined (thalamotomy and stimulation) stereotactic surgery of the vim thalamic nucleus for bilateral Parkinson disease. Stereotact Funct Neurosurg 50:344–346. https://doi.org/10.1159/000100803
16. Gross RE, Lozano AM (2000) Advances in neurostimulation for movement disorders. Neurol Res 22:247–258. https://doi.org/10.1080/01616412.2000.11740667
17. Benabid AL, Pollak P, Gervason C, Hoffmann D, Gao DM, Hommel M, Perret JE, de Rougemont J (1991) Long-term suppression of tremor by chronic stimulation of the ventral intermediate thalamic nucleus. Lancet Lond Engl 337:403–406. https://doi.org/10.1016/0140-6736(91)91175-t
18. Shipton EA (2012) Movement disorders and neuromodulation. Neurol Res Int 2012:309431. https://doi.org/10.1155/2012/309431
19. Chen CC, Brücke C, Kempf F, Kupsch A, Lu CS, Lee ST, Tisch S, Limousin P, Hariz M, Brown P (2006) Deep brain stimulation of the subthalamic nucleus: a two-edged sword. Curr Biol 16:R952–R953. https://doi.org/10.1016/j.cub.2006.10.013

20. Tripoliti E, Zrinzo L, Martinez-Torres I, Frost E, Pinto S, Foltynie T, Holl E, Petersen E, Roughton M, Hariz MI, Limousin P (2011) Effects of subthalamic stimulation on speech of consecutive patients with Parkinson disease. Neurology 76:80–86. https://doi.org/10.1212/WNL.0b013e318203e7d0
21. Huebl J, Brücke C, Schneider G-H, Blahak C, Krauss JK, Kühn AA (2015) Bradykinesia induced by frequency-specific pallidal stimulation in patients with cervical and segmental dystonia. Parkinsonism Relat Disord 21:800–803. https://doi.org/10.1016/j.parkreldis.2015.04.023
22. Castrioto A, Lhommée E, Moro E, Krack P (2014) Mood and behavioural effects of subthalamic stimulation in Parkinson's disease. Lancet Neurol 13:287–305. https://doi.org/10.1016/S1474-4422(13)70294-1
23. Volkmann J, Daniels C, Witt K (2010) Neuropsychiatric effects of subthalamic neurostimulation in Parkinson disease. Nat Rev Neurol 6:487–498. https://doi.org/10.1038/nrneurol.2010.111
24. Morrell MJ, RNS System in Epilepsy Study Group (2011) Responsive cortical stimulation for the treatment of medically intractable partial epilepsy. Neurology 77:1295–1304. https://doi.org/10.1212/WNL.0b013e3182302056
25. Heck CN, King-Stephens D, Massey AD, Nair DR, Jobst BC, Barkley GL, Salanova V, Cole AJ, Smith MC, Gwinn RP, Skidmore C, Van Ness PC, Bergey GK, Park YD, Miller I, Geller E, Rutecki PA, Zimmerman R, Spencer DC, Goldman A, Edwards JC, Leiphart JW, Wharen RE, Fessler J, Fountain NB, Worrell GA, Gross RE, Eisenschenk S, Duckrow RB, Hirsch LJ, Bazil C, O'Donovan CA, Sun FT, Courtney TA, Seale CG, Morrell MJ (2014) Two-year seizure reduction in adults with medically intractable partial onset epilepsy treated with responsive neurostimulation: final results of the RNS system pivotal trial. Epilepsia 55:432–441. https://doi.org/10.1111/epi.12534
26. Hochberg LR, Serruya MD, Friehs GM, Mukand JA, Saleh M, Caplan AH, Branner A, Chen D, Penn RD, Donoghue JP (2006) Neuronal ensemble control of prosthetic devices by a human with tetraplegia. Nature 442:164–171. https://doi.org/10.1038/nature04970
27. Collinger JL, Wodlinger B, Downey JE, Wang W, Tyler-Kabara EC, Weber DJ, McMorland AJ, Velliste M, Boninger ML, Schwartz AB (2013) High-performance neuroprosthetic control by an individual with tetraplegia. The Lancet 381:557–564. https://doi.org/10.1016/S0140-6736(12)61816-9
28. Gilja V, Pandarinath C, Blabe CH, Nuyujukian P, Simeral JD, Sarma AA, Sorice BL, Perge JA, Jarosiewicz B, Hochberg LR, Shenoy KV, Henderson JM (2015) Clinical translation of a high-performance neural prosthesis. Nat Med 21:1142–1145. https://doi.org/10.1038/nm.3953
29. Hatsopoulos NG, Donoghue JP (2009) The science of neural interface systems. Annu Rev Neurosci 32:249–266. https://doi.org/10.1146/annurev.neuro.051508.135241
30. Nordhausen CT, Maynard EM, Normann RA (1996) Single unit recording capabilities of a 100 microelectrode array. Brain Res 726:129–140
31. Moses DA, Metzger SL, Liu JR, Anumanchipalli GK, Makin JG, Sun PF, Chartier J, Dougherty ME, Liu PM, Abrams GM, Tu-Chan A, Ganguly K, Chang EF (2021) Neuroprosthesis for decoding speech in a paralyzed person with anarthria. N Engl J Med. https://doi.org/10.1056/NEJMoa2027540
32. Milekovic T, Sarma AA, Bacher D, Simeral JD, Saab J, Pandarinath C, Sorice BL, Blabe C, Oakley EM, Tringale KR, Eskandar E, Cash SS, Henderson JM, Shenoy KV, Donoghue JP, Hochberg LR (2018) Stable long-term BCI-enabled communication in ALS and locked-in syndrome using LFP signals. J Neurophysiol 120:343–360. https://doi.org/10.1152/jn.00493.2017
33. Leuthardt EC, Schalk G, Wolpaw JR, Ojemann JG, Moran DW (2004) A brain-computer interface using electrocorticographic signals in humans. J Neural Eng 1:63–71. https://doi.org/10.1088/1741-2560/1/2/001
34. Schalk G, Leuthardt EC (2011) Brain-computer interfaces using electrocorticographic signals. IEEE Rev Biomed Eng 4:140–154. https://doi.org/10.1109/RBME.2011.2172408

35. Simeral JD, Hosman T, Saab J, Flesher SN, Vilela M, Franco B, Kelemen JN, Brandman DM, Ciancibello JG, Rezaii PG, Eskandar EN, Rosler DM, Shenoy KV, Henderson JM, Nurmikko AV, Hochberg LR (2021) Home use of a percutaneous wireless intracortical brain-computer interface by individuals with tetraplegia. IEEE Trans Biomed Eng 68:2313–2325. https://doi.org/10.1109/TBME.2021.3069119
36. Guenther FH, Brumberg JS, Wright EJ, Nieto-Castanon A, Tourville JA, Panko M, Law R, Siebert SA, Bartels JL, Andreasen DS, Ehirim P, Mao H, Kennedy PR (2009) A wireless brain-machine interface for real-time speech synthesis. PLoS ONE 4:e8218. https://doi.org/10.1371/journal.pone.0008218
37. Sun FT, Morrell MJ (2014) The RNS System: responsive cortical stimulation for the treatment of refractory partial epilepsy. Expert Rev Med Devices 11:563–572. https://doi.org/10.1586/17434440.2014.947274
38. Stanslaski S, Herron J, Chouinard T, Bourget D, Isaacson B, Kremen V, Opri E, Drew W, Brinkmann BH, Gunduz A, Adamski T, Worrell GA, Denison T (2018) A chronically implantable neural coprocessor for investigating the treatment of neurological disorders. IEEE Trans Biomed Circuits Syst 12:1230–1245. https://doi.org/10.1109/TBCAS.2018.2880148
39. Jimenez-Shahed J (2021) Device profile of the percept PC deep brain stimulation system for the treatment of Parkinson's disease and related disorders. Expert Rev Med Dev 18:319–332. https://doi.org/10.1080/17434440.2021.1909471
40. Thenaisie Y, Palmisano C, Canessa A, Keulen BJ, Capetian P, Jiménez MC, Bally JF, Manferlotti E, Beccaria L, Zutt R, Courtine G, Bloch J, van der Gaag NA, Hoffmann CF, Moraud EM, Isaias IU, Contarino MF (2021) Towards adaptive deep brain stimulation: clinical and technical notes on a novel commercial device for chronic brain sensing. J Neural Eng 18. https://doi.org/10.1088/1741-2552/ac1d5b
41. Borton DA, Yin M, Aceros J, Nurmikko A (2013) An implantable wireless neural interface for recording cortical circuit dynamics in moving primates. J Neural Eng 10:026010. https://doi.org/10.1088/1741-2560/10/2/026010
42. Yin M, Borton DA, Aceros J, Patterson WR, Nurmikko AV (2013) A 100-channel hermetically sealed implantable device for chronic wireless neurosensing applications. IEEE Trans Biomed Circ Syst 7:115–128. https://doi.org/10.1109/TBCAS.2013.2255874
43. Lishman WA, Symonds CP, Whitty CWM, Willison RG (1962) Seizures induced by movement. Brain 85:93–108. https://doi.org/10.1093/brain/85.1.93
44. Iriarte J, Sánchez-Carpintero R, Schlumberger E, Narbona J, Viteri C, Artieda J (2001) Gait epilepsy. A case report of gait-induced seizures. Epilepsia 42:1087–1090. https://doi.org/10.1046/j.1528-1157.2001.0420081087.x
45. McCormick DA, Nestvogel DB, He BJ (2020) Neuromodulation of brain state and behavior. Annu Rev Neurosci 43:391–415. https://doi.org/10.1146/annurev-neuro-100219-105424
46. Scangos KW, Makhoul GS, Sugrue LP, Chang EF, Krystal AD (2021) State-dependent responses to intracranial brain stimulation in a patient with depression. Nat Med 27:229–231. https://doi.org/10.1038/s41591-020-01175-8
47. Ezzyat Y, Kragel JE, Burke JF, Levy DF, Lyalenko A, Wanda P, O'Sullivan L, Hurley KB, Busygin S, Pedisich I, Sperling MR, Worrell GA, Kucewicz MT, Davis KA, Lucas TH, Inman CS, Lega BC, Jobst BC, Sheth SA, Zaghloul K, Jutras MJ, Stein JM, Das SR, Gorniak R, Rizzuto DS, Kahana MJ (2017) Direct brain stimulation modulates encoding states and memory performance in humans. Curr Biol CB 27:1251–1258. https://doi.org/10.1016/j.cub.2017.03.028
48. Wiggins GC, Elisevich K, Smith BJ (1999) Morbidity and infection in combined subdural grid and strip electrode investigation for intractable epilepsy. Epilepsy Res 37:73–80. https://doi.org/10.1016/s0920-1211(99)00037-6
49. Papo D (2013) Time scales in cognitive neuroscience. Front Physiol 4:86. https://doi.org/10.3389/fphys.2013.00086
50. Xu S, Akioma M, Yuan Z (2021) Relationship between circadian rhythm and brain cognitive functions. Front Optoelectron 14:278–287. https://doi.org/10.1007/s12200-021-1090-y

51. Reid KJ, McGee-Koch LL, Zee PC (2011) Cognition in circadian rhythm sleep disorders. Prog Brain Res 190:3–20. https://doi.org/10.1016/B978-0-444-53817-8.00001-3
52. Schmidt C, Collette F, Cajochen C, Peigneux P (2007) A time to think: circadian rhythms in human cognition. Cogn Neuropsychol 24:755–789. https://doi.org/10.1080/02643290701754158
53. Baud MO, Kleen JK, Mirro EA, Andrechak JC, King-Stephens D, Chang EF, Rao VR (2018) Multi-day rhythms modulate seizure risk in epilepsy. Nat Commun 9:88. https://doi.org/10.1038/s41467-017-02577-y
54. Karoly PJ, Freestone DR, Boston R, Grayden DB, Himes D, Leyde K, Seneviratne U, Berkovic S, O'Brien T, Cook MJ (2016) Interictal spikes and epileptic seizures: their relationship and underlying rhythmicity. Brain 139:1066–1078. https://doi.org/10.1093/brain/aww019
55. Barth C, Villringer A, Sacher J (2015) Sex hormones affect neurotransmitters and shape the adult female brain during hormonal transition periods. Front Neurosci 9:37. https://doi.org/10.3389/fnins.2015.00037
56. Fink G, Sumner BEH, Rosie R, Grace O, Quinn JP (1996) Estrogen control of central neurotransmission: effect on mood, mental state, and memory. Cell Mol Neurobiol 16:325–344. https://doi.org/10.1007/BF02088099
57. Shansky RM (2019) Are hormones a "female problem" for animal research? Science. https://doi.org/10.1126/science.aaw7570
58. Solís-Ortiz S, Ramos J, Arce C, Guevara MA, Corsi-Cabrera M (1994) EEG oscillations during menstrual cycle. Int J Neurosci 76:279–292. https://doi.org/10.3109/00207459408986010
59. Haraguchi R, Hoshi H, Ichikawa S, Hanyu M, Nakamura K, Fukasawa K, Poza J, Rodríguez-González V, Gómez C, Shigihara Y (2021) the menstrual cycle alters resting-state cortical activity: a magnetoencephalography study. Front Hum Neurosci 15:411. https://doi.org/10.3389/fnhum.2021.652789
60. Hidalgo-Lopez E, Mueller K, Harris T, Aichhorn M, Sacher J, Pletzer B (2020) Human menstrual cycle variation in subcortical functional brain connectivity: a multimodal analysis approach. Brain Struct Funct 225:591–605. https://doi.org/10.1007/s00429-019-02019-z
61. Pletzer B, Harris T-A, Scheuringer A, Hidalgo-Lopez E (2019) The cycling brain: menstrual cycle related fluctuations in hippocampal and fronto-striatal activation and connectivity during cognitive tasks. Neuropsychopharmacology 44:1867–1875. https://doi.org/10.1038/s41386-019-0435-3
62. Moscovitch DA, Santesso DL, Miskovic V, McCabe RE, Antony MM, Schmidt LA (2011) Frontal EEG asymmetry and symptom response to cognitive behavioral therapy in patients with social anxiety disorder. Biol Psychol 87:379–385. https://doi.org/10.1016/j.biopsycho.2011.04.009
63. Schlumpf YR, Nijenhuis ERS, Klein C, Jäncke L, Bachmann S (2019) Functional reorganization of neural networks involved in emotion regulation following trauma therapy for complex trauma disorders. NeuroImage Clin 23:101807. https://doi.org/10.1016/j.nicl.2019.101807
64. Ball T, Kern M, Mutschler I, Aertsen A, Schulze-Bonhage A (2009) Signal quality of simultaneously recorded invasive and non-invasive EEG. Neuroimage 46:708–716. https://doi.org/10.1016/j.neuroimage.2009.02.028
65. Biomarkers Definitions Working Group (2001) Biomarkers and surrogate endpoints: preferred definitions and conceptual framework. Clin Pharmacol Ther 69:89–95. https://doi.org/10.1067/mcp.2001.113989
66. Amur S Biomarker terminology: speaking the same language
67. Scangos KW, Ahmad HS, Shafi A, Sellers KK, Dawes HE, Krystal A, Chang EF (2020) Pilot study of an intracranial electroencephalography biomarker of depressive symptoms in epilepsy. J Neuropsychiatry Clin Neurosci 32:185–190. https://doi.org/10.1176/appi.neuropsych.19030081
68. Lendner JD, Helfrich RF, Mander BA, Romundstad L, Lin JJ, Walker MP, Larsson PG, Knight RT (2020) An electrophysiological marker of arousal level in humans. eLife 9:e55092. https://doi.org/10.7554/eLife.55092

69. Kuroda N, Sonoda M, Miyakoshi M, Nariai H, Jeong J-W, Motoi H, Luat AF, Sood S, Asano E (2021) Objective interictal electrophysiology biomarkers optimize prediction of epilepsy surgery outcome. Brain Commun 3:fcab042. https://doi.org/10.1093/braincomms/fcab042
70. Shirvalkar P, Sellers KK, Schmitgen A, Prosky J, Joseph I, Starr PA, Chang EF (2020) A deep brain stimulation trial period for treating chronic pain. J Clin Med 9. https://doi.org/10.3390/jcm9103155
71. Meisenhelter S, Testorf ME, Gorenstein MA, Hasulak NR, Tcheng TK, Aronson JP, Jobst BC (2019) Cognitive tasks and human ambulatory electrocorticography using the RNS system. J Neurosci Methods 311:408–417. https://doi.org/10.1016/j.jneumeth.2018.09.026
72. Swann NC, de Hemptinne C, Miocinovic S, Qasim S, Wang SS, Ziman N, Ostrem JL, Luciano MS, Galifianakis NB, Starr PA (2016) Gamma oscillations in the hyperkinetic state detected with chronic human brain recordings in Parkinson's disease. J Neurosci 36:6445–6458. https://doi.org/10.1523/JNEUROSCI.1128-16.2016
73. de Hemptinne C, Chen W, Racine CA, Seritan AL, Miller AM, Yaroshinsky MS, Wang SS, Gilron R, Little S, Bledsoe I, San Luciano M, Katz M, Chang EF, Dawes HE, Ostrem JL, Starr PA (2021) Prefrontal physiomarkers of anxiety and depression in Parkinson's disease. Front Neurosci 15:748165. https://doi.org/10.3389/fnins.2021.748165
74. Rao VR, Leonard MK, Kleen JK, Lucas BA, Mirro EA, Chang EF (2017) Chronic ambulatory electrocorticography from human speech cortex. Neuroimage 153:273–282. https://doi.org/10.1016/j.neuroimage.2017.04.008
75. Scangos KW, Khambhati AN, Daly PM, Makhoul GS, Sugrue LP, Zamanian H, Liu TX, Rao VR, Sellers KK, Dawes HE, Starr PA, Krystal AD, Chang EF (2021) Closed-loop neuromodulation in an individual with treatment-resistant depression. Nat Med 27:1696–1700. https://doi.org/10.1038/s41591-021-01480-w
76. Vansteensel MJ, Pels EGM, Bleichner MG, Branco MP, Denison T, Freudenburg ZV, Gosselaar P, Leinders S, Ottens TH, Van Den Boom MA, Van Rijen PC, Aarnoutse EJ, Ramsey NF (2016) Fully implanted brain–computer interface in a locked-in patient with ALS. https://doi.org/10.1056/NEJMoa1608085. https://doi.org/10.1056/NEJMoa1608085. Accessed 22 Dec 2021
77. Buzsáki G, Draguhn A (2004) Neuronal oscillations in cortical networks. Science 304:1926–1929. https://doi.org/10.1126/science.1099745
78. Kirkby LA, Luongo FJ, Lee MB, Nahum M, Van Vleet TM, Rao VR, Dawes HE, Chang EF, Sohal VS (2018) An amygdala-hippocampus subnetwork that encodes variation in human mood. Cell 175:1688-1700.e14. https://doi.org/10.1016/j.cell.2018.10.005
79. Bowyer SM (2016) Coherence a measure of the brain networks: past and present. Neuropsychiatr Electrophysiol 2:1. https://doi.org/10.1186/s40810-015-0015-7
80. Chapeton JI, Haque R, Wittig JH, Inati SK, Zaghloul KA (2019) Large-scale communication in the human brain is rhythmically modulated through alpha coherence. Curr Biol 29:2801-2811.e5. https://doi.org/10.1016/j.cub.2019.07.014
81. Olichney JM, Yang J-C, Taylor J, Kutas M (2011) Cognitive event-related potentials: biomarkers of synaptic dysfunction across the stages of Alzheimer's disease. J Alzheimers Dis JAD 26:215–228. https://doi.org/10.3233/JAD-2011-0047
82. Jackson CE, Snyder PJ (2008) Electroencephalography and event-related potentials as biomarkers of mild cognitive impairment and mild Alzheimer's disease. Alzheimers Dement J Alzheimers Assoc 4:S137-143. https://doi.org/10.1016/j.jalz.2007.10.008
83. Sani OG, Yang Y, Lee MB, Dawes HE, Chang EF, Shanechi MM (2018) Mood variations decoded from multi-site intracranial human brain activity. Nat Biotechnol 36:954–961. https://doi.org/10.1038/nbt.4200
84. Sellers KK, Gilron R, Anso J, Louie KH, Shirvalkar PR, Chang EF, Little SJ, Starr PA (2021) Analysis-rcs-data: open-source toolbox for the ingestion, time-alignment, and visualization of sense and stimulation data from the Medtronic summit RC+S system. Front Hum Neurosci 15:398. https://doi.org/10.3389/fnhum.2021.714256
85. Chiong W, Leonard MK, Chang EF (2018) Neurosurgical patients as human research subjects: ethical considerations in intracranial electrophysiology research. Neurosurgery 83:29–37. https://doi.org/10.1093/neuros/nyx361

86. Chang EF (2015) Towards large-scale, human-based, mesoscopic neurotechnologies. Neuron 86:68–78. https://doi.org/10.1016/j.neuron.2015.03.037
87. Bjartmarz H, Rehncrona S (2007) Comparison of accuracy and precision between frame-based and frameless stereotactic navigation for deep brain stimulation electrode implantation. Stereotact Funct Neurosurg 85:235–242. https://doi.org/10.1159/000103262
88. Mobin F, Salles AAFD, Behnke EJ, Frysinger R (1999) Correlation between MRI-based stereotactic thalamic deep brain stimulation electrode placement, macroelectrode stimulation and clinical response to tremor control. Stereotact Funct Neurosurg 72:225–232. https://doi.org/10.1159/000029730
89. Winkler D, Tittgemeyer M, Schwarz J, Preul C, Strecker K, Meixensberger J (2005) The first evaluation of brain shift during functional neurosurgery by deformation field analysis. J Neurol Neurosurg Psychiatry 76:1161–1163. https://doi.org/10.1136/jnnp.2004.047373
90. Khan MF, Mewes K, Gross RE, Škrinjar O (2008) Assessment of brain shift related to deep brain stimulation surgery. Stereotact Funct Neurosurg 86:44–53. https://doi.org/10.1159/000108588
91. Hunsche S, Sauner D, Maarouf M, Poggenborg J, Lackner K, Sturm V, Treuer H (2009) Intra-operative X-Ray detection and MRI-based quantification of brain shift effects subsequent to implantation of the first electrode in bilateral implantation of deep brain stimulation electrodes. Stereotact Funct Neurosurg 87:322–329. https://doi.org/10.1159/000235804
92. Won SM, Cai L, Gutruf P, Rogers JA (2021) Wireless and battery-free technologies for neuroengineering. Nat Biomed Eng 1–19. https://doi.org/10.1038/s41551-021-00683-3
93. Dagum P (2018) Digital biomarkers of cognitive function. Npj Digit Med 1:1–3. https://doi.org/10.1038/s41746-018-0018-4
94. Robin J, Harrison JE, Kaufman LD, Rudzicz F, Simpson W, Yancheva M (2020) Evaluation of speech-based digital biomarkers: review and recommendations. Digit Biomark 4:99–108. https://doi.org/10.1159/000510820

Chapter 53
The Future of iEEG: What Are the Promises and Challenges of Mobile iEEG Recordings?

Sabrina L. Maoz, Matthias Stangl, Uros Topalovic, and Nanthia Suthana

Abstract Traditional approaches to recording deep brain activity in humans require participants to remain immobile, limiting the ecological validity and breadth of cognitive neuroscience questions that can be asked. Individuals with neurostimulator devices that are chronically implanted for clinical purposes present a rare opportunity to record intracranial electroencephalography (iEEG) from the human brain while participants are mobile and interacting with their environment in a natural way. Research-related benefits of such chronic neurostimulator devices include resistance to motion artifacts, access to deep brain structures, measurement of neural activity with high temporal resolution, as well as the possibility to perform closed-loop neuromodulation through stimulation that can be associated with specific behavioral or neurophysiological features. Furthermore, recent technical developments have streamlined the integration of numerous wearables with wireless iEEG recordings, including virtual and augmented reality headsets, which substantially broadens the scope of possible cognitive neuroscience experiments that can be implemented. Here, we provide an overview of the methodological and technical aspects of mobile iEEG recordings in human research participants and discuss associated promises and challenges. With this overview, we aim to inspire innovative future applications of mobile iEEG to advance our understanding of rich human behaviors in health and disease.

S. L. Maoz · N. Suthana (✉)
Department of Bioengineering, University of California, Los Angeles, California, USA
e-mail: nanthia@g.ucla.edu

M. Stangl · N. Suthana
Department of Psychiatry and Biobehavioral Sciences, Jane and Terry Semel Institute for Neuroscience and Human Behavior, University of California, Los Angeles, California, USA

U. Topalovic
Department of Electrical and Computer Engineering, University of California, Los Angeles, California, USA

N. Suthana
Department of Neurosurgery, David Geffen School of Medicine, University of California, Los Angeles, California, USA

N. Axmacher (ed.), *Intracranial EEG*, Studies in Neuroscience, Psychology and Behavioral Economics, https://doi.org/10.1007/978-3-031-20910-9_53

53.1 Introduction

The human brain undergoes complex cognitive processes throughout our daily lives by integrating proprioceptive and kinesthetic cues with rich sensory stimuli and information. Understanding how complex behaviors are encoded in the brain thus requires the ability to study human cognition and record brain activity during naturalistic paradigms and behaviors. Traditional methods of recording neural activity in humans are limited by large recording equipment and motion artifacts. Specifically, functional magnetic resonance imaging (fMRI), magnetoencephalography (MEG), and intracranial stereo-electroencephalographic (sEEG) recordings all require patients to be still, within or tethered to large recording equipment, or at the very least that their heads remain motionless.

Notable progress, however, has been made in mobile brain and body imaging technology using scalp electroencephalography (EEG) recordings, which has advanced our understanding of cognitive variables in conditions of naturalistic movement. Recent studies have used mobile scalp EEG to explore questions such as how walking modulates task switching [20], how movement speed modulates neural representations of focusing on a visual target [11], how attention varies across stationary vs walking conditions [10], how cognitive motor interference is modulated by movement complexity [19], and how the dynamics of cortical brain regions support specific features of active spatial navigation [5], among many others. Although these studies provide first-insight into how movement modulates human cognition, scalp EEG signals are also complicated by motion-induced artifacts and limited to recording broad neural population signals primarily from cortical structures and with limited spatial resolution. Thus, these constraints reduce the breadth and ecological validity of behavioral, neural, and cognitive processes and relationships that can be studied.

Recently approved implantable medical devices for treating neurological diseases (Table 53.1) have created a unique clinical opportunity to record more localized deep brain activity via iEEG recordings in humans who can be mobile and behave in naturalistic settings. These chronically implanted neural devices that enable wireless iEEG recording and neurostimulation are surgically implanted intracranially and can remain there permanently throughout a person's lifetime. Patients with such implants can enjoy day-to-day life as these devices are not externally visible nor do they interfere with any standard activities of daily living, including freely-moving activities of interest. Thus, the opportunity to record from such chronically implanted neurostimulation devices provides a unique window into cognition allowing for a broad—rather endless—range of ambulatory activities and cognitive tasks to be explored.

In the present chapter, we describe technical aspects as well as the promises and challenges related to performing mobile iEEG recordings in individuals who have permanent brain implants, which thereby enable scientists to advance the field of cognitive and clinical neuroscience. Specifically, we discuss challenges such as the limited number of brain regions that can be sampled in mobile iEEG studies, ethical considerations that should be considered, disease-related confounds, and difficulties related to equipment setup and synchronization of multiple data streams. We

Table 53.1 Shown are the general characteristics of three of the most commonly available human-use chronic implantable neurostimulator devices. MDD: Major Depressive Disorder, PTSD: Post-traumatic stress disorder, LOC: Loss of control (obesity), OCD: obsessive–compulsive disorder, SCI: Spinal cord injury

	Responsive neurostimulator (RNS)	Percept	RC+S
Company	Neuropace	Medtronic	Medtronic
Sampling frequency (Hz)	250	250	1000
Battery life	~8 years (320 model); ~4 years (300 model)	~5 years	rechargeable
Location of battery	Intracranial	Intrathoracic	Intrathoracic
Synchronization method	Use of the "Mark" or "Magnet" signal to inject an artifactual signal across devices	Stimulation to inject an artifactual signal across devices (e.g., Scalp EEG and iEEG)	Conversion to Unix time on each data stream
Example studies	Scangos [21, 22] Stangl et al. [25] Topalovic et al. [28] Henin et al. [9] Rao et al. [18] Meisenhelter et al. [14] Molina et al. [16] Aghajan et al. [1]	Shirvalkar et al. [24]	Gilron et al. [7] Sellers et al. [23]
Clinical trials	MDD, PTSD, LOC, OCD, epilepsy	MDD, OCD, SCI	MDD, epilepsy
Typical electrode regions	Hippocampus, amygdala, entorhinal cortex, parahippocampal cortex	Subthalamic nucleus, ventral intermediate nucleus of thalamus	Determined by investigational use criteria
# of implanted patients	Few thousand	Several hundreds	<30

also highlight how recording iEEG activity from chronic neurostimulators during complex cognitive tasks provides an exciting and rare opportunity to explore neural mechanisms of higher-order naturalistic human behavior. Key benefits include the ability to record iEEG activity from deep brain structures during ambulatory behaviors without motion-artifacts and with high spatial and temporal precision. Altogether, we argue that the promises outweigh the challenges associated with mobile iEEG studies, ultimately allowing for a rare window into discovery of deep brain mechanisms underlying naturalistic and ambulatory behaviors in humans.

53.2 Chronically Implanted Neural Sensing Devices

While deep brain stimulation (DBS) devices have been used in over a hundred thousand patients for over thirty years to treat neurological disorders such as Parkinson's disease, essential tremor, and dystonia, it is only within the last decade that there has been an increase in the availability of DBS devices that allow for sensing of neural activity. The first of these devices, the Responsive Neurostimulator System (RNS®, Neuropace, Inc.), was approved by the FDA in 2013 for the treatment of intractable epilepsy and has since been implanted in over 2500 patients and increasing by several hundred individuals each year. The FDA-approved clinical use of the RNS System is to detect seizure-related activity and subsequently deliver neurostimulation to reduce seizures frequency and pathological symptoms. The RNS System can support up to two intracranial depth or cortical strip electrode leads each of which have four contacts and are implanted in hypothesized seizure-onset zones. Exact locations of the RNS System electrodes vary by patient with both cortical (e.g., orbitofrontal, lateral temporal) and subcortical locations (e.g., amygdala, hippocampus, entorhinal cortex) as common targets. Contact-spacing on electrodes can also vary based on clinical needs, ranging from 3.5 to 10 mm (center-to-center). The RNS System also includes a hermetically-encased neurostimulator which is implanted in the skull and is thus shielded and resistant to movement artifacts and externalized electrical noise sources. This neurostimulator package contains the battery and stores iEEG data in small units (~240 s) with a sampling rate of 250 Hz on four bipolar channels until it is downloaded wirelessly (via an external wand device, Fig. 53.1a, b) to a server that can be accessed by researchers or clinicians. The RNS System can also be programmed to initiate stimulation on selected channels based on real-time analysis of incoming neural activity (e.g., seizure activity or other clinical/behavioral neurophysiological biomarkers).

The second FDA-approved DBS device that allows for recording of iEEG activity is the Percept PC Neurostimulator (Medtronic, Inc.) used to treat Parkinson's disease. Since its approval in 2020, it has been implanted in a few hundred patients and expected to increase rapidly as patients with older DBS non-sensing devices are upgraded during battery replacement procedures. The Percept can record iEEG activity with a sampling rate of 250 Hz and on up to 6 bipolar channels with the neurostimulator package and battery implanted in the chest near the clavicle in the thoracic cavity. FDA-approved electrode placement sites for the Percept include the subthalamic nucleus (STN) and ventral intermediate nucleus of the thalamus (VIM) [8], however, other sites can be targeted with FDA investigational device exemption (IDE) approval. Brain activity can be recorded in two formats: (1) 10 min windows, which are then averaged over time and reported, or (2) continuously while connected to the implantable pulse generator (IPG) which provides power.

The Activa PC+S and RC+S Neurostimulators (Medtronic, Inc.) are also available and allow for much more research flexibility (e.g., higher sampling rates, increased programmability of closed-loop capability and wider variety of stimulation parameters), however, are only available with FDA IDE approval and thus have been used in

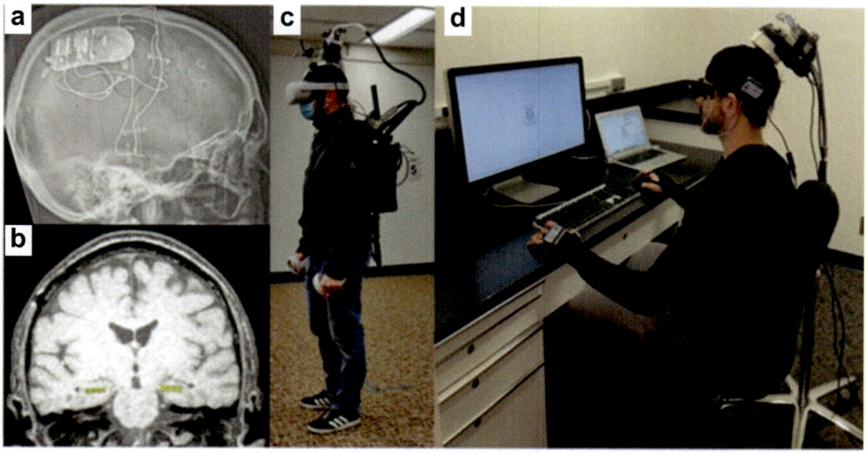

Fig. 53.1 An example CT (**a**) and MRI (**b**) showing the RNS System implanted intracranially with two leads implanted in the medial temporal lobe. **c** Mobile iEEG recording set-up. As part of the mobile deep brain recording and stimulation (Mo-DBRS) platform, a backpack can be worn, which includes a malleable metal arm holding the wand in place above the participant's RNS implant and is connected to a metal-framed backpack [28]. The participant shown is wearing a VR headset, carrying the associated hand-held controllers, and has a rigid body motion sensor antenna fixed to the top of their head for precise motion tracking. Full-body motion capture suits can also be worn. **d** Stationary iEEG recording set-up. Participant shown is wearing biometric sensors for recording heart rate and skin conductance (on hands), as well as an eye tracking headset that allows for pupillometry and eye-gaze tracking, all of which can also be worn during mobile studies. The wand is placed above the RNS implant using a portable wand holder

only a small number of patients (<30 total) across various clinical trials for epilepsy, depression, essential tremor, obsessive–compulsive disorder (OCD), dystonia, and Parkinson's disease.

53.3 Current Findings

Multiple studies have validated long-term recordings obtained from these chronically implanted devices by comparing findings of cognition to those identified using acute iEEG recordings [9, 18]. A rapid increase in the number of individuals chronically implanted with sensing DBS devices over the past decade has enabled scientists the ability to use mobile iEEG recordings in humans to carry-out, for the first time, cognitive and clinical neuroscience studies in freely-moving humans. One such area of study has been to determine whether findings from freely-moving rodent studies of spatial navigation translate to humans. The first research study of this kind investigated whether oscillatory activity in the medial temporal lobe (MTL) was modulated by walking speed [1] (Fig. 53.2a) given that decades of findings emphasized a critical role for rodent MTL theta oscillations (4–8 Hz) in spatial navigation and memory [3].

While this study was the first to discover the presence of speed-modulated theta oscillations in the human MTL during freely-moving walking behavior, it also highlighted fundamental species-specific differences between rodents and primates (including humans) worthy of additional investigation in future mobile iEEG studies: Theta oscillations were found to increase in prevalence during faster compared to slower walking speeds in ambulating humans, but their overall presence was less prominent (~10% of the time) compared to that found in freely-moving rodents, a result consistent with recent findings in freely-moving non-human primates [13] and replicated in an additional study [25]. Another more recent mobile iEEG study found similar levels of theta activity during walking (~10%), but further identified separate behavioral and environmental variables that also modulate MTL theta activity, such as one's own (or another person's) proximity to environmental boundaries (e.g. walls). Furthermore, this spatial modulation of theta power was dependent on cognitive state [25]. Future studies are needed to determine the complex relationship between MTL oscillatory activity and cognitive (e.g., task goal), behavioral (e.g., movement speed, direction, position) and environmental variables (e.g., boundaries, presence of others), which are now enabled with mobile iEEG recordings in humans. Another research group has explored patient-specific neurophysiological biomarkers relating to inadequate or excess movements in Parkinson's disease in five participants implanted with the Summit RC+S device with electrodes in the motor cortex and subthalamic nucleus. Streaming of neural activity in the home setting was collected and decoded in relation to movement state information obtained from wearable monitors. During data recordings, participants performed normal activities of daily living while wearing a watch that measured movement activity to distinguish bradykinesia and dyskinesia. The authors found that individual patients had unique neural biomarkers (e.g. frequency band associated with movement state) for changes between active and inactive movement states. Across participants, many exhibited coherence between the motor cortex and subthalamic nucleus that discriminated between mobile and immobile states (Fig. 53.2b) [7].

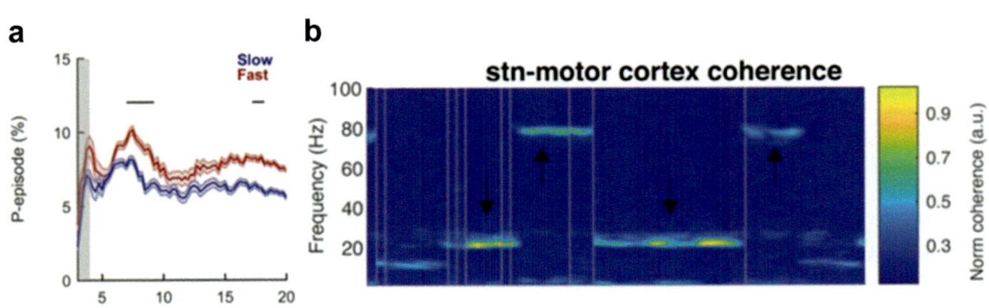

Fig. 53.2 Example findings from mobile iEEG studies. **a** Theta oscillations increase in prevalence during fast versus slow walking speeds in the medial temporal lobe (MTL, adapted from [1]). **b** Beta-gamma coherence between the subthalamic nucleus (STN) and motor cortex distinguishes mobile (low dyskinesia, upward arrows) from immobile states (high dyskinesia, downward arrows) in an example participant. Adapted from [7]

Moreover, there is an ever-increasing scope of research areas that can be investigated using mobile iEEG recordings to advance our understanding of other neurologic and psychiatric diseases. Research studies have begun exploring new clinical applications for chronically implanted sensing and stimulation devices that pave the way for future opportunities to record mobile iEEG from a broader range of brain regions. One group implanted the RNS System in the centromedian-parafascicular region of the thalamus in a case of medically-refractory Tourette syndrome [16]. A review proposed a possible approach to treat chronic pain by implanting a chronic sensing and stimulating device in a number of potential target thalamic, cingulate, and other regions [24]. Recent efforts have used responsive neurostimulation for the treatment of loss of control of eating in participants with treatment-refractory obesity by implanting the RNS System in the nucleus accumbens [29]. Another group implanted the RNS System in the amygdala and striatum to significantly improve depressive symptoms in a case of treatment resistant-depression [21, 22]. The RNS System has even been implanted in the occipital lobe of blind patients and used to investigate whether stimulation could be used to induce visual percepts [4].

Altogether, these results are an exciting foundation highlighting the utility and versatility of a new generation of neurostimulator devices for advancing cognitive neuroscience research. These studies have capitalized on mobile iEEG during a variety of naturalistic behaviors. Additionally, an increasing number of studies are exploring the use of chronic sensing and stimulating devices for the treatment of a broad range of other neurologic and psychiatric disorders. Looking ahead to the future, these studies foreshadow an increasing diversity of brain regions that can be wirelessly recorded from in naturalistic and chronic environments, expanding the possible scope of cognitive neuroscience research using chronic sensing and stimulating devices.

53.4 Technical Challenges

Designing and carrying out mobile iEEG studies in humans with chronically implanted neurostimulation devices comes with several technical challenges since current FDA-approved devices are designed primarily with clinical- and not research-needs in mind. For example, the RNS System and Percept devices themselves do not allow for real-time wireless access and/or control of the implanted neurostimulator, which is critical for designing and carrying out well-controlled mobile iEEG studies. There have, however, been externalized platforms recently developed that allow for external wireless control of and communication with chronic neurostimulation devices [7, 14, 28]. For example, one such platform is the mobile deep brain recording and stimulation (Mo-DBRS) platform [28], which provides researchers with open-source tools to enable real-time wireless control of the timing of stimulation, start/stop of iEEG recording, and accurate synchronization of iEEG data with wearable sensors and equipment (e.g., VR/AR headsets, eye trackers, motion sensors). While the Mo-DBRS platform has been primarily tested with the RNS

System (Fig. 53.1a, b), most features including integration with wearable sensors can be extended for use with other implantable neurostimulation devices. Another such system has been developed and shared with the scientific community to enable similar features using the RC+S System, including an open source toolbox that supports pre-processing of raw data, time-alignment across data streams, and basic power calculations [7, 23].

Here, we briefly discuss the technical features of one of these platforms, specifically, the Mo-DBRS platform, which includes a Wand (Fig. 53.1c, d) that continuously communicates with the intracranially implanted device via near-field telemetry. With this platform, the Wand is physically connected to a laptop or tablet programmer (carried by the participant in a backpack) to allow for wireless user-based control (e.g., to start and stop iEEG recordings). A custom-built programmer tool interfaces with the laptop or tablet programmer so that the user can programmatically control the Wand to stop/start iEEG data storage and deliver stimulation or synchronization pulses. The programmer tool and consequently the Wand can then be controlled using a wireless control device (e.g., Raspberry Pi) that can be wirelessly controlled by the researcher (manually) or researcher's program. During mobile iEEG experiments, the Wand can be affixed to the participant's head via mechanical solutions that relieve its weight and allow for several hours of comfortable ambulatory movement (Fig. 53.1c) or stationary experiments (Fig. 53.1d). All of the other Mo-DBRS accessories and tools (e.g., laptop or tablet programmer, programmer tool, Raspberry Pi) can fit comfortably within a backpack (Fig. 53.1c) along with any other wearable equipment (e.g., mobile scalp EEG amplifier) during ambulatory behavioral tasks. With these capabilities, the Mo-DBRS platform can thus be fluidly combined and synchronized with wearable sensors for physiological recordings (e.g., heart rate, respiration, skin conductance), eye tracking (for gaze and/or pupillometry), scalp EEG, precise kinematic tracking, and virtual reality (VR) or augmented reality (AR) headsets (see Table 53.2). Precise kinematic tracking, including head position and rotation, as well as information about arm, leg, and torso movements can be captured by wearing a full-body motion capture suit or placing motion capture markers at body joints or appendages of interest.

A critical challenge of experimental design in mobile iEEG studies using any platform is accurate and precise synchronization of neural recordings with behavioral task features and other data streams. For example, a single experiment may incorporate a task delivered on a VR headset with simultaneous full-body motion and eye tracking on independent devices. Such an experiment would result in several separate data streams (iEEG activity, VR task, positional and kinematic variables, and gaze positions), each of which would be recorded with unique time clocks and data acquisition start/stop times. Aligning these separate data streams into the same time scale is critical for relating the behavioral and neural variables that occur at any given point in time. To facilitate data alignment, it is therefore useful to deliver synchronization pulses simultaneously across all data streams. One approach is to use brief electrical deflections ("Mark" artifacts) or stimulation artifacts that are reflected in the iEEG that can be used as synchronization pulses to note the same timepoint as it occurs on all data streams. In the case of Mo-DBRS, the Mark command can be wirelessly

Table 53.2 Example wearable sensors and accessories that can be integrated with mobile iEEG recordings. From top to bottom: Eye-tracking headsets can record pupillometry, saccades and gaze positions. Other on-body sensors can record heart rate, respiration, skin conductance. Wall-mounted motion capture cameras can track position, speed and other movement variables. Mobile scalp EEG set up is shown with a lightweight backpack carrying the amplifier and connected to the scalp EEG cap. A wearable camera can be used to capture audiovisual data such as for comprehensive documentation of participants' field of view and visible or auditory events. Example virtual and augmented reality headsets are also shown

	Example product	
Eye tracking	Pupil Core (Pupil Labs GmbH) Eyetracking (Tobii AB)	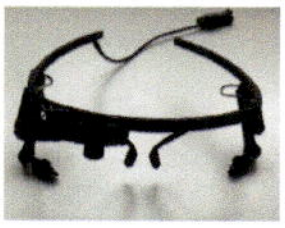
Heart rate	Smart center bionomadix (Biopac Systems Inc.)	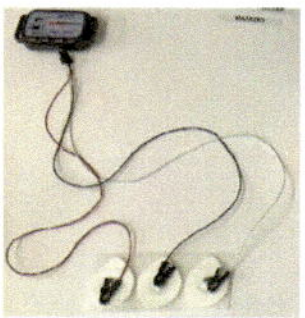
Respiratory rate	Smart center bionomadix (Biopac Systems Inc.)	
Skin conductance	Smart center bionomadix (Biopac Systems Inc.)	
Motion tracking	OptiTrack Cameras and Motive software (NaturalPoint, Inc.) (Xsens Inc.)	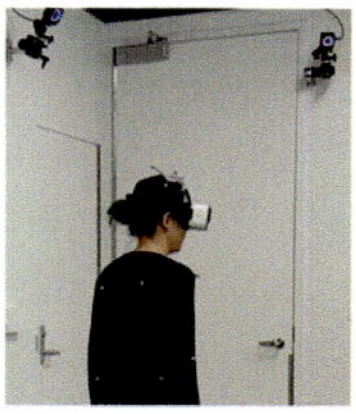
Scalp EEG	eego sports (ANT Neuro)	

(continued)

Table 53.2 (continued)

	Example product	
Audiovisual	GoPro 360 (GoPro Inc.)	
Virtual reality	Meta Quest 2 (Facebook Technologies, LLC) Pico neo 2 (Pico Interactive, Inc.) Vive Pro (HTC Corporation)	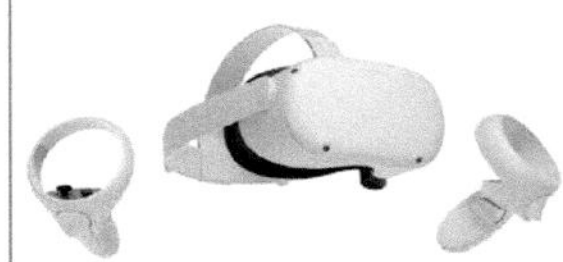
Augmented reality	MagicLeap (MagicLeap, Inc.) Hololens (Microsoft)	

initiated by an externalized wireless control device (e.g., Raspberry Pi) and delivered to the implanted device via the head-mounted Wand. In the above example, the same control device can send simultaneous signals to other data streams (e.g., VR task, motion tracking software) to synchronize the corresponding timepoints across all data streams. Previous studies have characterized sufficiently accurate and precise synchronization latencies using this method [28].

Lastly, the recording of numerous data streams through the use of wearables, audiovisual recordings, and mobile iEEG can result in enormous and complex data sets that are difficult to parse. The use of deep learning and artificial intelligence strategies to analyze these resulting large datasets during complex naturalistic behavioral studies will likely be beneficial for future mobile iEEG studies in humans [15] (see also Chap. 47). Furthermore, synchronization of multiple data streams can be challenging and result in unknown temporal latencies that make relating neural activity to precise behavioral events difficult. Future studies that add/adopt novel data types should characterize synchronization latencies prior to collecting data in participants with chronic sensing neurostimulators.

53.5 Clinical Confounds

While recording iEEG during ambulatory behaviors in humans presents a significant scientific opportunity, it is important to be cognizant of the limitations associated with doing such studies. Currently, individuals are implanted with chronic electrodes solely for clinical purposes and thus electrodes are often placed within disease-modified brain tissue. For example, in patients with epilepsy, electrodes are implanted in hypothesized seizure-onset zones. However, because electrode contacts can span up to 10 mm of tissue and often in bilateral brain regions, some contacts may be implanted in tissue that is less affected by disease. One way researchers attempt to characterize the presence of aberrant disease-related neural activity is by isolating any abnormal samples, for example interictal epileptiform discharges (IEDs) in the case of epilepsy. There are several methods currently used to detect IEDs and exclude the affected iEEG data [1, 6, 25]. Additionally, multiple studies have found that neural activity patterns outside the seizure onset zone are unaffected and that normal neural activity patterns occur between epileptic events [2, 6, 12, 17]. Another way to mitigate the potential contamination of neural recordings with disease-related episodes is by selecting participants that have a low frequency of these events, which can be characterized in advance given these patients have long-term at-home recordings available to the researcher and/or clinician. In the case of epilepsy, IED event frequency in the recent months preceding the study can thus be available to the researcher and used as a selection criterion for recruiting participants with less disease-related events.

It is important to recognize that mobile iEEG studies involve working with patient groups who may suffer from cognitive (e.g., memory impairments in epilepsy) and behavioral (e.g. difficulty walking in PD) deficits, and other comorbidities (e.g., depression in epilepsy or PD) (see also Chap. 52). This can introduce variability in a study and limit the generalizability of findings. However, as the number of approved indications for which the use of chronic sensing neurostimulation devices increases, the sample of individuals from which to select from will also increase. Thus, those individuals who are better suited for a given study can be preferentially selected in order to minimize variability across the study sample. Further, findings replicated across different clinical groups can lessen the probability that results are due to specific disease-related abnormalities and consequently increase confidence in the generalizability of findings to healthy individuals.

53.6 Limited Sampling of Brain Regions

Another constraint in mobile iEEG recordings is the limited number of electrode channels available per participant (up to 6 channels) and thus brain regions which can be sampled given that the placement of electrodes is driven solely by clinical criteria. Nonetheless, there exists a large pool of potential participants (thousands) from which to select from in order to acquire sufficient amounts of data from a given brain region while minimizing variability. Currently, studies with hypotheses related to medial temporal and striatal function are understandably the most common given electrodes are implanted most often in these brain regions. Furthermore, as the types of neurologic and psychiatric conditions treated by implanting chronic neurostimulators expands, a greater diversity of brain regions will likely be accessible, which in turn will open the door to a greater variety of cognitive and clinical neuroscience research questions that can be answered.

The spatial resolution of mobile iEEG data is another important variable to consider when designing a study using chronic neurostimulators, which is limited by the diameter of the implanted electrodes (1.5 mm), spacing of electrode contacts (up to 10 mm) if using bipolar recordings, and the spatial resolution of neuroimaging-based localization procedures (e.g., co-registration of MRI and CT images). However, in contrast to most intracranial neurophysiology studies performed in acute epilepsy monitoring units, mobile iEEG study participants are able to complete experiments of longer durations (up to 6–8 h per day, and over multiple days with flexible timelines) since they have not recently undergone neurosurgery or have any other competing medical/surgical procedures that co-occur. Finally, participants are often eager to contribute to research and are highly motivated, especially if they report positive clinical outcomes due to their chronic neurostimulator treatment and thus large sample studies are indeed possible to counteract the limited number of brain sites that can be sampled within a single participant.

53.7 Ethical Considerations

As with other human iEEG studies, research with patients who have chronic neural implants require a protocol approved by an IRB (internal review board) and should minimize risks associated with informed consent. Additional risks associated with mobile iEEG studies should be minimized including to prevent substantial battery drainage, frequency of seizures during the study, and discomfort during behavioral tasks, as well as to maintain confidentiality and data security during wireless iEEG data transfers. Participants should be made aware of the risks through a fully informed consent process and it is good practice to allow participants to choose whether stimulation therapy remains on during the iEEG recordings. While mobile iEEG studies are ideally carried out without the confound of artifacts due to responsive neurostimulation, the potential risk of increased symptoms (e.g. seizures) can be reduced by inviting participants who have an overall lower frequency of seizure-related activity in the months prior to the research study and thus lower the likelihood of stimulation events occurring during a given research study. Additionally, it is good practice to have a neurologist on the research team that can be available during the study to monitor iEEG activity for the presence of pathological events (e.g., IEDs) or behavioral symptoms. For neurostimulation studies, the level of stimulation administered should be controlled, ideally with a charge density lower than that used for treatment (e.g., $<10\ \mu C/cm^2$) [26, 27]. Finally, there is a risk of battery depletion associated with mobile iEEG studies that would lead to the need for more frequent battery-replacement surgeries, especially for research studies that use stimulation and/or real-time data viewing of iEEG data. To minimize this risk, research studies should limit the number of study visits per participant (especially if using stimulation and/or real-time recordings) and/or sample more widely from the population of patients who have chronic sensing neurostimulation devices implanted to prevent the same group of patients from participating repeatedly. In the case of RC+S and other existing and/or future rechargeable battery-powered devices, battery drainage associated with participation in research studies is not a concern.

53.8 Promises and Future Opportunities

Mobile iEEG studies provide an exciting and unique opportunity for first-in-human exploration of behavior and cognition in mobile and naturalistic settings. Such studies will help to bridge the gap between findings in freely-moving animal models to understand the human neural mechanisms that underlie everyday behaviors such as for example spatial navigation, physical movement, and social interaction. Importantly, mobile iEEG studies in humans provide an avenue for cognitive neuroscience studies to no longer be constrained by immobility and limited to artificial laboratory environments. Future research studies can thus use more realistic environments with higher ecological validity, either by constructing them using advanced 3D-modeling and VR/AR technologies to allow for experimental control or by simply having participants step into novel and complex real-world environments (e.g., outdoor or complex indoor spaces, social experiences) that may be difficult to reconstruct in lab settings. Mobile iEEG, especially combined with VR/AR and the opportunity to enter real-world situations opens the door to countless new questions across countless fields and topics in neuroscience: spatial navigation, episodic memory, emotion, social interaction, exercise, activities of daily living, psychiatric conditions, and sleep, to name a few.

Additionally, ongoing clinical trials using chronically implanted sensing neurostimulators for neuropsychiatric conditions foreshadow increased implantation of these devices in areas relating to altered emotional processing and pathological brain states (e.g., depression, PTSD, OCD, loss of control, panic disorder). This future opportunity opens the field up to many new questions that can be answered to better understand the pathophysiology underlying neuropsychiatric disease without relying only on animal models, as well as advance theories related to how emotional states shape cognition in everyday life experiences.

Finally, chronic neurostimulator devices also provide an opportunity for the development of novel closed-loop stimulation paradigms in relation to neural or behaviorally relevant cues (see also Chap. 41). A closed-loop approach allows the experiment to program a change in stimulation pattern in response to a condition that is met. This condition could be detection of a particular neural feature of interest, by online and real-time analysis of neural signals, or it could be more complex and relate a neural feature with a behavior that is detected. With the increasing data from mobile iEEG paired with multiple biometric data streams, there is a broad range of questions that can be investigated using closed-loop approaches that compare neural activity and behavioral states with and without the presence of stimulation of a particular brain region. This notable methodological advance provides researchers with the ability to test causal questions relating neural activity and behavior under naturalistic settings.

Glossary

Deep brain stimulator A medical device that has stimulation capabilities at one or more electrodes that are intracranially implanted in deep brain structures (i.e. below the cortex)

Electrode lead A probe inserted intracranially that can contain one or more contacts for recording and/or stimulation. Depending on the diameter, an electrode may be described as a micro-electrode or a macro-electrode. Micro-electrodes can record single-neuron and local field potential (LFP) activity and are currently only available in acute post-surgical hospital settings, while macro-electrodes record intracranial EEG and are currently used with chronic sensing and stimulation devices

Electrode contact A metal contact on an electrode that is typically constructed from silicon, platinum, tungsten, or other metallic substances. There may be one or more contacts on an electrode

Bipolar channel A recording channel that records neural activity by subtracting the difference in voltage across two (often adjacent) electrode contacts, where one is treated as the positive contact and the other is the negative contact

Seizure onset zone (SOZ) A region of the brain where clinical seizures originate from. Electrode placements are selected to target hypothesized seizure onset zones, in the case of epilepsy

Interictal epileptiform discharge (IED) A typical neural activity pattern that commonly occurs in the SOZ of epilepsy patients. These are not clinical-level seizures

Synchronization A strategy used to align the timestamps across multiple continuous and simultaneous data streams. One example strategy may involve a signal sent from one device simultaneously to all data streams to define an absolute timepoint that can be used to align timestamps across different data streams

Latency Temporal delays introduced when multiple devices send electronic messages serially. A good practice is to characterize the temporal latencies of a system to ensure that they remain lower than the temporal resolution of the neural signals of interest

Closed-loop system A system in which calculations are computed in real-time during ongoing continuous neural recordings. Typically, a real-time calculation is used to detect a neural biomarker or feature. Detection of this event of interest will then initiate some program that may involve a particular stimulation pattern or some other change of variables

Responsive neurostimulator A closed-loop device that continuously records neural activity and delivers a pulse of current stimulation when an imminent seizure or IED event is predicted, in order to prevent the manifestation of a clinical seizure

FDA Food and Drug Administration in the United States which reviews and approves pharmaceuticals as well as medical devices for clinical use

CT, MRI, fMRI Computed tomography (CT), magnetic resonance imaging (MRI), and functional magnetic resonance imaging (fMRI). CT and MRI brain scans provide anatomical information while fMRI illustrates changes in brain region activity, measured according to rapid changes in a particular brain region's hemodynamic response

References

1. Aghajan ZM, Schuette P, Fields TA, Tran ME, Siddiqui SM, Hasulak NR, Tcheng TK, Eliashiv D, Mankin EA, Stern J, Fried I, Suthana N (2017) Theta oscillations in the human medial temporal lobe during real-world ambulatory movement. Curr Biol 27:3743–3751
2. Akkol S, Kucyi A, Hu W, Zhao B, Zhang C, Sava-Segal C, Liu S, Razavi B, Zhang J, Zhang K, Parvizi J (2021) Intracranial electroencephalography reveals selective responses to cognitive stimuli in the periventricular heterotopias. J Neurosci 41:3870–3878
3. Bland BH (1986) The physiology and pharmacology of hippocampal formation theta rhythms. Prog Neurobiol 26:1–54
4. Caspi A, Barry MP, Patel UK, Salas MA, Dorn JD, Roy A, Niketeghad S, Greenberg RJ, Pouratian N (2021) Eye movements and the perceived location of phosphenes generated by intracranial primary visual cortex stimulation in the blind. Brain Stimul 14:851–860
5. Do TTN, Lin CT, Gramann K (2021) Human brain dynamics in active spatial navigation. Sci Rep 11:13036
6. Gelinas JN, Khodagholy D, Thesen T, Devinsky O, Buzsáki G (2016) Interictal epileptiform discharges induce hippocampal-cortical coupling in temporal lobe epilepsy. Nat Med 22:641–648
7. Gilron R, Little S, Perrone R, Wilt R, de Hemptinne C, Yaroshinsky MS, Racine CA, Wang SS, Ostrem JL, Larson PS, Wang DD, Galifianakis NB, Bledsoe IO, San Luciano M, Dawes HE, Worrell GA, Kremen V, Borton DA, Denison T, Starr PA (2021) Long-term wireless streaming of neural recordings for circuit discovery and adaptive stimulation in individuals with Parkinson's disease. Nat Biotechnol 39:1078–1085
8. Goyal A, Goetz S, Stanslaski S, Oh Y, Rusheen AE, Klassen B, Miller K, Blaha CD, Bennet KE, Lee K (2021) The development of an implantable deep brain stimulation device with simultaneous chronic electrophysiological recording and stimulation in humans. Biosens Bioelectron 176:112888
9. Henin S, Shankar A, Hasulak N, Friedman D, Dugan P, Melloni L, Flinker A, Sarac C, Fang M, Doyle W, Tcheng T, Devinsky O, Davachi L, Liu A (2019) Hippocampal gamma predicts associative memory performance as measured by acute and chronic intracranial EEG. Sci Rep 9:593
10. Ladouce S, Donaldson DI, Dudchenko PA, Ietswaart M (2019) Mobile EEG identifies the re-allocation of attention during real-world activity. Sci Rep 9:15851
11. Lee Y-E, Shin G-H, Lee M, Lee S-W (2021) Mobile BCI dataset of scalp- and ear-EEGs with ERP and SSVEP paradigms while standing, walking, and running. Sci Data 8:315
12. Liu S, Parvizi J (2019) Cognitive refractory state caused by spontaneous epileptic high-frequency oscillations in the human brain. Sci Transl Med 11:514
13. Mao D, Avila E, Caziot B, Laurens J, Dickman JD, Angelaki DE (2021) Spatial modulation of hippocampal activity in freely moving macaques. Neuron 109:3521–3534
14. Meisenhelter S, Testorf ME, Gorenstein MA, Hasulak NR, Tcheng TK, Aronson JP, Jobst BC (2019) Cognitive tasks and human ambulatory electrocorticography using the RNS System. J Neurosci Meth 311:408–417
15. Mobbs D, Wise T, Suthana N, Guzmán N, Kriegeskorte N, Leibo JZ (2021) Promises and challenges of human computational ethology. Neuron 109:2224–2238
16. Molina R, Okun MS, Shute JB, Opri E, Rossi PJ, Martinez-Ramirez D, Foote KD, Gunduz A (2018) Report of a patient undergoing chronic responsive deep brain stimulation for Tourette syndrome: proof of concept. J Neurosurg 129:308–314

17. Parvizi J, Kastner S (2018) Promises and limitations of human intracranial electroencephalography. Nat Neurosci 21:474–483
18. Rao VR, Leonard MK, Kleen JK, Lucas BA, Mirro EA, Chang EF (2017) Chronic ambulatory electrocorticography from human speech cortex. Neuroimage 153:273–282
19. Reiser JE, Wascher E, Arnau S (2019) Recording mobile EEG in an outdoor environment reveals cognitive-motor interference dependent on movement complexity. Sci Rep 9:13086
20. Richardson DP, Foxe JJ, Mazurek KA, Abraham N, Freedman EG (2021) Neural markers of proactive and reactive cognitive control are altered during walking: a mobile brain-body imaging (MoBI) study. NeuroImage 118853
21. Scangos KW, Khambhati AN, Daly PM, Makhoul GS, Sugrue LP, Zamanian H, Liu TX, Rao VR, Sellers KK, Dawes HE, Starr PA, Krystal AD, Chang EF (2021) Closed-loop neuromodulation in an individual with treatment-resistant depression. Nat Med 27:1696–1700
22. Scangos KW, Makhoul GS, Sugrue LP, Chang EF, Krystal AD (2021) State-dependent responses to intracranial brain stimulation in a patient with depression. Nat Med 27:229–231
23. Sellers KK, Gilron R, Anso J, Louie KH, Shirvalkar PR, Chang EF, Little SJ, Starr PA (2021) Analysis-rcs-data: open-source toolbox for the ingestion, time-alignment, and visualization of sense and stimulation data from the Medtronic summit RC+S system. Front Hum Neurosci 15:398
24. Shirvalkar P, Sellers KK, Schmitgen A, Prosky J, Joseph I, Starr PA, Chang EF (2020) A deep brain stimulation trial period for treating chronic pain. J Clin Med 9:3155
25. Stangl M, Topalovic U, Inman CS, Hiller S, Villaroman D, Aghajan ZM, Christov-Moore L, Hasulak NR, Rao VR, Halpern CH, Eliashiv D, Fried I, Suthana N (2021) Boundary-anchored neural mechanisms of location-encoding for self and others. Nature 589:420–425
26. Suthana N, Aghajan ZM, Mankin EA, Lin A (2018) Reporting guidelines and issues to consider for using intracranial brain stimulation in studies of human declarative memory. Front Neurosci 12:905
27. Suthana N, Haneef Z, Stern J, Mukamel R, Behnke E, Knowlton B, Fried I (2012) Memory enhancement and deep-brain stimulation of the entorhinal area. New Engl J Med 366:502–510
28. Topalovic U, Aghajan Z, Villaroman D, Hiller S, Christov-Moore L, Wishard T, Stangl M, Hasulak N, Inman C, Fields T, Eliashiv D, Fried I, Suthana N (2020) Wireless programmable recording and stimulation of deep brain activity in freely moving humans. Neuron 108:322–334
29. Wu H, Adler S, Azagury DE, Bohon C, Safer DL, Barbosa DAN, Bhati MT, Williams NR, Dunn LB, Tass PA, Knutson BD, Yutsis M, Fraser A, Cunningham T, Richardson K, Skarpaas TL, Tcheng TK, Morrell MJ, Roberts LW, Malenka RC, Lock JD, Halpern CH (2020) Brain-responsive neurostimulation for loss of control eating: early feasibility study. Neurosurgery 87:1277–1288

GPSR Compliance

The European Union's (EU) General Product Safety Regulation (GPSR) is a set of rules that requires consumer products to be safe and our obligations to ensure this.

If you have any concerns about our products, you can contact us on ProductSafety@springernature.com

In case Publisher is established outside the EU, the EU authorized representative is:

Springer Nature Customer Service Center GmbH
Europaplatz 3
69115 Heidelberg, Germany

Batch number: 10756371

Printed by Printforce, the Netherlands